AMINO ACID POOL

Distribution, Formation and Function of F

* Invited contribution not presented at the conference.

AMINO ACID POOLS

Distribution, Formation and Function of Free Amino Acids

PROCEEDINGS OF A SYMPOSIUM ON FREE AMINO ACIDS
HELD AT THE CITY OF HOPE MEDICAL CENTER,
DUARTE, CALIF. (U.S.A.), MAY 1961

Edited by

JOSEPH T. HOLDEN

Sponsored by
The Office of Naval Research, Washington 25, D.C.

Under the Auspices
of The Institute for Avanced Learning in the Medical Sciences

ELSEVIER PUBLISHING COMPANY
AMSTERDAM – LONDON – NEW YORK
1962

SOLE DISTRIBUTORS FOR THE UNITED STATES AND CANADA
AMERICAN ELSEVIER PUBLISHING COMPANY, INC.
52 VANDERBILT AVENUE, NEW YORK 17, N.Y.

LIBRARY OF CONGRESS CATALOG CARD NUMBER: 62-10360

WITH 1038 ILLUSTRATIONS AND 151 TABLES

PRINTED IN THE NETHERLANDS BY
N.V. BOEKDRUKKERIJ F. E. MACDONALD, NIJMEGEN

Organizing Committee for the Symposium:

JOSEPH T. HOLDEN
Chairman

EUGENE ROBERTS

GEORGE ROUSER

Round Table Discussions

edited by

MILTON WINITZ

and

ERICH HEINZ

PREFACE

The knowledge that living cells contain sizeable amounts of apparently uncombined, readily-extractable amino acids came to be widely appreciated at the same time that rapid chromatographic methods ideally suited to their study were developed. The prompt application of these procedures to the examination of protein-free tissue extracts has led to the accumulation of a huge volume of observations concerned with the so-called free amino acids. Despite apparent interrelations to metabolic processes, such as protein synthesis and active transport which also received intensive experimental study in these years, there has as yet been no comprehensive effort to collect, organize and evaluate this information. Some of us who have participated in various aspects of this work were acutely aware of the need for such a definitive summary and over the years have discussed numerous proposals to achieve this goal.

In discussions with representatives of the Office of Naval Research, Dr. EUGENE ROBERTS proposed the solution which is represented by this volume: a conference of leading investigators at which much of the available information would be summarized, followed by publication of the formal and informal discussions. A tentative program, developed during discussions between Dr. EUGENE ROBERTS, Dr. GEORGE ROUSER and the editor, was enthusiastically received by many of the prospective participants, and was approved for support by the Office of Naval Research. Encouraged by these favorable responses, the committee, under the editor's chairmanship, formally undertook the organization of the symposium which was held at the City of Hope Medical Center on May 19–22, 1961 under the title *Conference on Free Amino Acids*. The program consisted of three days of prepared reports interspersed with informal discussions followed by a final half-day session devoted to two spontaneous round-table discussions.

Participants were encouraged to submit extensive accounts of the work they wished to discuss in the hope that this volume would serve as a guide to a major portion of the literature on free amino acids. This goal was achieved in many, but certainly not all, cases. Authors were also encouraged to submit for publication as much original chromatographic evidence as they felt was required to document their conclusions. In the past, many journals have been reluctant to publish photographs of original chromatograms with the result that observations in this field frequently have not been adequately documented. While justified partly by the high costs involved, this editorial judgement can lead to the publication of conclusions derived from technically unsound chromatographic evidence. The publishers of this volume are to be commended for their understanding of the necessity to present this evidence in a form which permits its evaluation by the reader.

The function of the editor in this enterprise deserves comment. Participants were invited with the understanding that they would be given great latitude in the scope and method of presentation of their material. It seemed inappropriate, therefore, to embark on a vigorous program of revision in an effort to attain unity of format or a more comprehensive treatment of the subject. Only the most extreme points of view and methods of presentation were modified, generally in the interests of accuracy and clarity of expression. Although the discussions received a less permissive treatment, an effort was made in the portions which survived editorial scrutiny to retain as much as possible of the author's mode of expression. Transcripts of the round-table discussions were edited separately by elected participants. The conferees also recommended a less specific title for this volume than was used for the conference. It was clear that some participants retain a serious doubt that the freely-extractable amino acids are uncombined within cells. To accommodate all points of view the title *Amino Acid Pools* was adopted despite some reservations of an aesthetic nature.

Many will recognize the absence of contributions from some investigators who have made pre-eminent findings in this field. In most instances, prior commitments prevented their attendance and a serious effort was made in the interests of historical accuracy and comprehensiveness of treatment to have the respective subjects discussed by their associates or collaborators.

The success of this enterprise was due in large measure to the contributions of several dedicated associates. On behalf of the organizing committee I should like to identify these as follows: Mr. DANIEL SEEDMAN, who supervised travel, housing and other physical arrangements; Mrs. HELEN WENGER, Mrs. HELEN PINKERTON and Mrs. MARGERY SIDDONS from our secretarial staff; Mr. MAX LENZ who supervised the sound equipment; and Mr. FRANKLIN A. STEINKO who produced an accurate stenotyped record of the informal discussions.

Principal financial support for the conference came from a contract between this institution and the Office of Naval Research (Contract NONR-(G)0009-61). We are indebted to Mr. LEO SHINN, Chief of the Biochemistry Branch and Dr. JOHN LOEFER of the Pasadena Branch Office for their able and sympathetic management of the requisite negotiations. The conference was held at this institution under the auspices of the Institute for Advanced Learning in the Medical Sciences which also contributed generously to its support. Additional contributions were provided by the UPJOHN COMPANY, Kalamazoo, Mich. and by the DON BAXTER COMPANY, Glendale, Calif.

JOSEPH T. HOLDEN

CONTENTS

* A paper by DR. STANLEY L. MILLER entitled: "Synthesis of amino acids in the primitive earth" was read at the conference but is not represented by an article in this volume.

VII. Round Table Discussion

PART TWO — DYNAMIC ASPECTS OF CELLULAR FREE AMINO ACID POOLS

I. Permeability and amino acid transport

* Invited discussion not presented at the conference.

INTRODUCTION TO CONFERENCE

E. ROBERTS

Department of Biochemistry, City of Hope Medical Center, Duarte, Calif. (U.S.A.)

To date, most of the effort of the biochemist has been devoted to the description of the properties of the chemical substances present in living matter. The list is far from complete, as evidenced by the daily appearance of reports dealing with the characterization of new molecular entities of biological interest. Although it has been recognized that substances can pass in and out of living systems and can interact and be transformed in these systems in manners potentially explicable by principles of physics and chemistry, in no single instance have the *details* of these processes been described adequately for a *given* substance in a particular type of living cell. Although biochemistry had its beginnings as a handmaiden to medicine in attempting to correlate manifestations of disease processes with gross chemical changes in blood, urine and tissue, in recent years the biochemist has joined with the cytologist and microbiologist in combining knowledge and techniques in the study of the chemistry and function of parts of various cells, *viz.*, membrane, nucleus, mitochondria, microsomes, etc. However, these efforts have not yet answered the question: How do all of the verifiable observations add up to what one observes as life?

Some of the key questions still to be answered relate to the properties of the bricks and mortar of life, the lipids, polysaccharides, proteins, the desoxy- and ribonucleic acids, and various complexes and mixtures of these materials. These substances are formed largely from smaller molecules by hereditarily determined processes resembling directed polymerization rather than by random assembly of smaller units. The cellular chemistry of even the smaller units is incompletely known.

The point of departure chosen for this conference was the discussion of free or easily extractable amino acids and related substances in living cells, small molecules which are not only the building blocks of proteins, but also which have myriad other functions within cells. Not only is it of interest to know the quantity of an amino acid in a given amount of a particular tissue or cell type, but it is of ultimate interest to be able to determine the past history, present position, and future fate of a particular molecule in a specific cell. To our knowledge the information relevant to these problems which has been accumulated to date in various laboratories throughout the world has not been summarized in one place so that an adequate perspective could be attained. We are fortunate in having present a number of people who have contributed importantly to this field and look forward to an exciting exchange of information.

PART ONE

OCCURRENCE OF FREE AMINO ACIDS

I. ANALYTICAL METHODS

IDENTIFICATION OF THE ELUSIVE AMINO ACID

MILTON WINITZ

Laboratory of Physiology, National Cancer Institute, National Institutes of Health, Public Health Service, U.S. Department of Health, Education and Welfare, Bethesda, Md. (U.S.A.)

CRITERIA FOR ACCEPTANCE OF NATURAL OCCURRENCE OF AN α-AMINO ACID

In 1931, in a brilliant review entitled *The history of the discovery of the amino acids*, VICKERY AND SCHMIDT[1] proposed certain criteria that should be met before any report of the existence of a new amino acid as a constituent of proteins be generally accepted as valid. These criteria are given in what follows:

1. In order that an amino acid shall be accepted as a definite product of the hydrolysis of proteins, it must also have been isolated by some worker other than its discoverer.
2. Its constitution must have been established by synthesis and by demonstration of identity between the synthetic product and the natural racemized product, or by actual resolution of the synthetic product and preparation of the optically active natural isomer.
3. The substance must be liberated by hydrolysis from a preparation of a protein of demonstrated purity and must be adequately characterized by analysis of salts and of typical derivatives.

The protein-derived amino acids selected by VICKERY AND SCHMIDT, in 1931, which were acceptable according to these criteria, included all of those presently believed to occur in proteins with the exceptions of threonine, hydroxylysine, asparagine, glutamine and triiodothyronine, which had not yet been discovered, and included too β-hydroxyglutamic acid[2, 3]. Some 2 years later, SCHMIDT[4] recommended that norleucine be added to the list of accepted amino acids because of its apparent conformity with all of the required criteria. However, the seemingly irrefutable evidence upon which the acceptance of both β-hydroxyglutamic acid and norleucine rested was subsequently shown to be so tenuous that neither, at the present time, receives serious consideration as a constituent of proteins[5]. Such exceptions notwithstanding, the VICKERY-SCHMIDT criteria have served an invaluable role in that they prompted a generally more critical, as well as a more conservative attitude toward the acceptance of a new amino acid. Thus, amino acids such as diaminoglutaric acid[6, 7], hydroxyaspartic acid[6, 7], sarcosine[8], α-aminoadipic acid[9], norvaline[10, 11] and a host of others have, at one time or another, been reported to occur in proteins, but such claims either have subsequently been proven questionable or have not yet received the experimental support necessary for their general acceptance.

In addition to the amino acids which are known constituents of proteins, a number of amino acids have been found to occur in nature, either in the free state or in chemical combination in non-protein compounds of varying molecular size. Such amino acids have been detected in the biological fluids and tissues of various plants and animals, in the circulatory system, among the products of excretion, as inter-

References p. 22/24

mediates in the metabolic processes, as components of antibiotics and as bacterial decomposition products. Reports in the chemical literature allude to well over a hundred different amino acids of non-protein derivation that have been isolated from such sources[12]. That they may be possessed of a most diverse composition and structure is indicated by the few examples shown in Fig. 1. Although some of these amino acids have been unequivocally established as occurring in nature by a strong body of experimental data, somewhat more have been supported by data that would, at best,

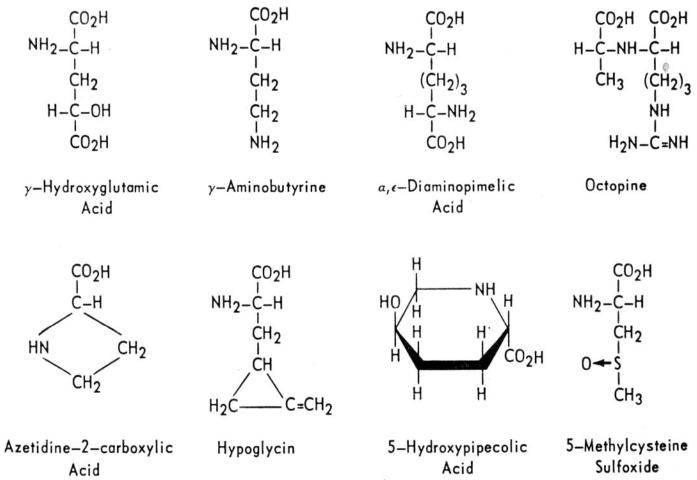

Fig. 1. Some naturally occurring amino acids of non-protein derivation.

provide only strong presumptive evidence for their natural occurrence. With most of the amino acids, however, sufficient evidence has not yet been amassed to demonstrate their existence in nature with even a modicum of assurance. These considerations, in addition to the appearance of a continually increasing number of reports of new amino acids of both the D- and L-varieties in nature, point out the urgent need for criteria which, if met, would reliably establish the natural occurrence of these amino acids. Proposals for such criteria are offered below:

1) The new amino acid should be physically isolated in sufficient quantity, and purified by several crystallizations from water, water–alcohol, or some other suitable solvent system, in order to obtain accurate and meaningful carbon, hydrogen, nitrogen and other possible elemental analyses.

2) Chromatographic procedures, whether two-dimensional techniques on paper employing different solvent systems, or elution techniques with columns employing various solvent mixtures, should show it to possess a single spot or peak, respectively, in order to assure the possibility of molecular homogeneity.

3) The molecule should be degraded by chemical organic procedures or unequivocal enzymatic procedures to identifiable products or fragments in order to obtain a knowledge of its structure.

4) Verification of the structure ultimately assigned to the molecule should be achieved through synthesis of the material employing unequivocal chemical procedures.

5) The synthetic material should be resolved into its optical antipodes and the physical, chemical and biological properties of the antipodes so secured compared with those of the natural material; in the case of amino acids with more than one asymmetric center, separation of the diastereomers should precede the resolution of each racemate into its optical antipodes.

6) Determination of the optical configuration of the molecule should be accomplished by any of a number of chemical, physical, optical or biological techniques.

7) The natural occurrence of the new amino acid should be independently confirmed by some investigator other than the discoverer.

IDENTIFICATION BY CHROMATOGRAPHIC AND COLORIMETRIC MEANS

At the present time, a facile and convenient means for establishing the identity of an amino acid from natural sources involves a comparison of its chromatographic behavior with reference amino acids in a variety of solvent systems. Ofttimes, no other criteria are employed. Now it is well known that certain structurally related amino acids, such as leucine and isoleucine, reveal a chromatographic behavior in numerous solvent systems wherein appreciable differences in mobility are the exception rather than the rule. It is therefore evident that through the use of chromatographic procedures alone, the danger is ever present that an amino acid of as yet unsuspected structure will remain hidden behind, or occupy the same position as a closely related amino acid of known structure, and hence will be overlooked. This probability was dramatically emphasized only recently with the announcement by OGLE, LOGAN AND ARLINGHAUS[13] that a new imino acid had been isolated from acid hydrolysates of collagen, a protein that had hitherto been the subject of chromatographic analysis by many investigators in many laboratories. In a private communication from Dr. OGLE, it was learned that during a study of the peptide sequence of collagen, a tripeptide was isolated which, after acid hydrolysis, displayed a previously unobserved peak running just before hydroxyproline upon passage of the hydrolysate through a Dowex-50 column employing 0.2 *N* sodium phosphate buffer as the eluant and a pH of 3.1. Analysis revealed that the new material was an imino acid and was of the same elemental composition as γ-hydroxyproline. That the compound was neither proline nor γ-hydroxyproline was indicated by a comparison of infrared spectra and the melting points of various derivatives, by the color reaction obtained with ninhydrin in glacial acetic acid and by the fact that it failed to react with dimethylaminobenzaldehyde following oxidation with chloramine-T. As degradation of the material with permanganate yielded β-alanine, the new material was assigned the structure of β-hydroxyproline. *It should be noted that the paper-chromatographic behavior of this new material was identical with that of γ-hydroxyproline in all solvent systems tested.*

That amino acids of varying structure may possess identical mobilities in a number of different solvent systems is an occurrence that is by no means rare. One such occurrence that was noted in our laboratory[14] some years ago involved a material which was obtained from the copper-catalyzed condensation of pyruvic acid and glycine in an alkaline medium (Fig. 2), and which possessed an empirical formula of $C_5H_9NO_5$. This formula could correspond to β-hydroxy-β-methylaspartic acid if it were assumed that the condensation involved the α-carbon atom of the pyruvic acid reactant. If, however, the condensation involved the β-carbon atom of the pyruvic acid, then

References p. 22/24

γ-hydroxyglutamic acid, which possesses the same empirical formula, could result. As authentic samples of γ-hydroxyglutamic acid were available, a chromatographic comparison of this material with the unknown synthetic material was made in three different solvent systems. The R_F values were identical in every instance. However, it was noted that in contradistinction to the usual rapid purple color given by γ-hydroxyglutamic acid with the ninhydrin reagent, the unknown synthetic product exhibited some unique color reactions. Thus, when a solution of this material was spotted on filter paper and the paper sprayed with a solution of ninhydrin in acetone and subsequently heated, a bright yellow spot appeared. The yellow color gradually changed to gray-brown after 4–8 h, and finally to purple after 18–24 h. If, however,

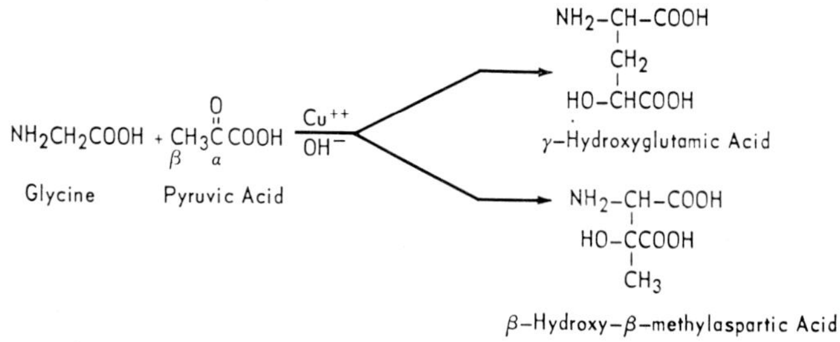

Fig. 2. Potential products arising from copper-catalyzed condensation of pyruvic acid and glycine.

the paper was sprayed with a basic ninhydrin solution and heated, the yellow spot appeared but almost immediately turned purple. Such behavior indicated that the new material was other than the suspected γ-hydroxyglutamic acid. Had it, however, responded to ninhydrin in the usual manner, it might have been all too tempting, on the basis of its mobility behavior alone, to assign it the erroneous structure.

From what has already been said, it becomes readily apparent that visual observation of the color-staining process on paper will sometimes permit distinctions between amino acids that cannot be made on the basis of movement analysis alone. A few instances of the specific color reactions revealed by the different amino acids after development with various color-inducing reagents are listed in Table I. Thus, it is commonly known that whereas most amino acids exhibit a blue or reddish-purple color after treatment with the ninhydrin reagent, proline displays a yellow color, tyrosine gives a dull greenish-purple, hydroxyproline yields a brown-yellow color (which serves to distinguish it from the nearby alanine spot[15]), glycine shows a rather grayish-purple color, asparagine reveals an orange-brown color (which serves to distinguish it from the nearby glycine spot[16]), and aspartic acid is characterized by a rather bright blue. Treatment of the paper chromatogram with weak alkali intensifies the blue color of the ninhydrin spot of phenylalanine[17], whereas treatment of the paper with cyclohexylamine prior to spraying with ninhydrin leads to a blue color for aspartic acid, orange for cystine, grayish-green for histidine, bluish-gray for phenylalanine, yellow for proline, carmine for hydroxyproline, grayish-purple for threonine, gray for tyrosine, and reddish-brown for glycine, with all of the other protein-derived amino acids yielding the more usual purple color[18]. In addition, a variety of other staining reagents are available which are more or less specific for certain amino

References p. 22/24

TABLE I

COLORIMETRIC DETECTION OF AMINO ACIDS ON PAPER

Amino acid	*Developing agent*	*Color*
Arginine	Sakaguchi reaction	Red
Asparagine	Ninhydrin	Orange-brown
Aspartic acid	Ninhydrin	Bright blue
Citrulline	Dimethylaminobenzaldehyde	Yellow
Cysteine	Nitroprusside	Red
Cystine	Ninhydrin–cyclohexylamine	Orange
	Nitroprusside–cyanide	Red
Glycine	Ninhydrin	Grayish-purple
	Ninhydrin–cyclohexylamine	Reddish-brow
	Naphthoquinone-4-sulfonate	Green
	o-Phthalaldehyde	Green
Histidine	Ninhydrin–cyclohexylamine	Grayish-green
	Pauly reaction	Red
	o-Phthalaldehyde	Bluish-green
	Sakaguchi reaction	Yellow
Hydroxyproline	Isatin	Blue
	Naphthoquinone-4-sulfonate	Red
	Ninhydrin	Brownish-yellow
	Ninhydrin–cyclohexylamine	Carmine
Phenylalanine	Ninhydrin–cyclohexylamine	Bluish-gray
	Ninhydrin–alkali	Intense blue
Proline	Isatin	Blue
	Naphthoquinone-4-sulfonate	Red
	Ninhydrin	Yellow
	Ninhydrin–cyclohexylamine	Yellow
Serine	Hypochlorite–dinitrobenzene	Violet
	Periodate–Nessler	Reddish-brown
Sulfur amino acids	Platinic iodide	Bleaching action
Threonine	Hypochlorite–dinitrobenzene	Violet
	Ninhydrin–cyclohexylamine	Grayish-purple
	Periodate–Nessler	Reddish-brown
Tryptophan	Dimethylaminobenzaldehyde	Yellow
	o-Phthalaldehyde	Gray
	Sakaguchi reaction	Brown
Tyrosine	Ninhydrin	Greenish-purple
	Ninhydrin–cyclohexylamine	Gray
	Pauly reaction	Reddish-brown

acids. Thus, histidine and tyrosine may be selectively detected by the Pauly reaction whereby the paper is treated with freshly diazotized sulfanilamide followed by sodium carbonate solution, with the histidine revealing a red color, tyrosine a reddish-brown, and other amino acids a yellow color[19, 20]. Proline and hydroxyproline, by reaction with isatin in butanol–acetic acid, followed by heating, reveal a blue color, with all other common amino acids yielding a weak rose[21–23]; subsequent treatment with normal hydrochloric acid causes the blue spot of proline to deepen in color whereas the spots due to the other amino acids fade away[24]. Arginine may be qualitatively identified as a red spot by the Sakaguchi reaction whereby the paper is sprayed with an alkaline α-naphthol solution, followed by a sodium hypochlorite or hypobromite solution; employing this process, the spot for histidine becomes yellow and that for tryptophan brown[25, 26]. Tryptophan and citrulline, by reaction with *p*-dimethylaminobenzaldehyde in hydrochloric acid, yield a yellow color[27], the reaction with citrulline serving

References p. 22/24

to distinguish it from the neighboring purple glutamine spot[15]. Serine and threonine may be identified either by the red-brown color developed upon treatment of the paper first with periodate and subsequently with the Nessler reagent [25–27], or by the violet color produced by reaction first with alkaline hypochlorite and then with 1, 2-dinitrobenzene[28]. Sulfur-containing amino acids may be detected by their bleaching action when the paper is sprayed with a solution of platinic iodide[25, 26, 29], and glycine by the green color it gives with *o*-phthalaldehyde[30]; with this latter reagent, histidine and tryptophan exhibit a blue-green color and a gray color, respectively. An interesting color reagent is 1, 2-naphthoquinone-4-sulfonate which in sodium carbonate solutions produces reddish spots with proline and hydroxyproline, a green spot with glycine, and blue-gray to violet colors with all other common amino acids[31, 32].

These represent only a few of the available, more or less specific color reactions exhibited by individual amino acids. They have been treated here at length to re-emphasize the fact that all too many amino acids are already known which display identical chromatographic behavior in a variety of solvent systems, and may even reveal identical elemental analyses. Hence, in the search for new amino acids, additional criteria may sometimes be necessary in order to expose those elusive amino acids that may otherwise evade detection by remaining concealed behind known amino acids of closely related structure.

STRUCTURAL DETERMINATION BY MEANS OF CHEMICAL DEGRADATION

If we now return to the condensation product of pyruvic acid and glycine considered earlier, it will be recalled that the unique color reactions exhibited by this material, by eliminating from consideration γ-hydroxyglutamic acid, suggested that it might be β-hydroxy-β-methylaspartic acid. In order to establish the structure of this material more firmly, degradative studies were undertaken. Now it is well known that most α-amino acids liberate one mole of carbon dioxide per mole of compound upon being subjected to manometric ninhydrin–CO_2 analysis[33]. Certain compounds such as aspartic acid and the diaminodicarboxylic acids, like α, ε-diaminopimelic acid and cystine, are exceptional in that two moles of carbon dioxide per mole of amino acid are liberated. Some years ago, we had occasion to synthesize and subsequently analyze various substituted aspartic acids, such as β-methylaspartic acid[34], β-hydroxyaspartic acid[35] and α-aminotricarballylic acid[36], and noted that these compounds also liberated two moles of carbon dioxide per mole of amino acid (Fig. 3). As the unknown synthetic condensation product behaved in a comparable manner upon manometric ninhydrin–CO_2 analysis, a substituted aspartic acid structure was suggested and this constituted further evidence in favor of a β-hydroxy-β-methylaspartic acid structure. γ-Hydroxyglutamic acid, on the other hand, behaved like the usual α-amino acid in that only the expected one mole of carbon dioxide was released. Further evidence favoring the β-hydroxy-β-methylaspartic acid structure was revealed by the fact that the material was converted to β-methylaspartic acid upon reduction with hydriodic acid and that it showed a positive periodate reaction, typical of an α-amino-β-hydroxy acid. The structural elucidation of this compound illustrates only a few of the degradative processes employed in the characterization of a new amino acid. As these processes are numerous in number and vary markedly with the nature and structure of the amino acid under consideration, their further elaboration is here precluded.

References p. 22/24

ESTABLISHMENT OF STRUCTURE VIA SYNTHESIS AND RESOLUTION

Although degradative methods may serve to identify the structure of a new natural amino acid with a fair degree of reliability, the structure can by no means be considered as having been unequivocally established until the material has been obtained by synthesis and subsequent resolution, and the physical, chemical and biological

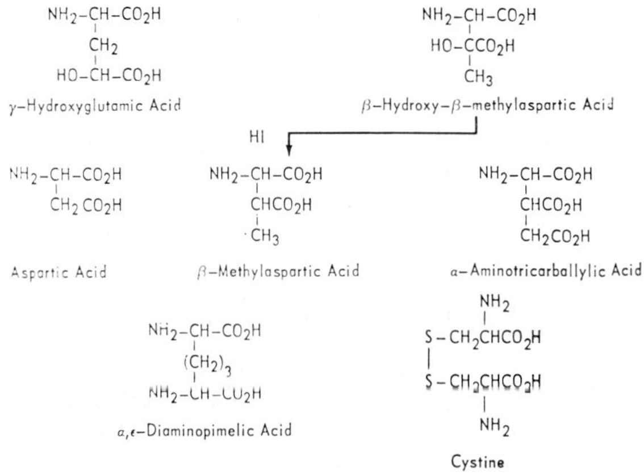

Fig. 3 Structures of mono- and diaminopolycarboxylic acids as related to CO_2 liberation with ninhydrin.

properties of the synthetic and natural products proven identical. Several of the more general methods employed for the synthesis of amino acids include: *(a)* the Strecker synthesis; *(b)* amination of α-halogen acids; *(c)* reductive amination; *(d)* amination via molecular rearrangement; *(e)* condensation of an aldehyde with an active methylene group; and *(f)* condensations with *N*-substituted aminomalonic esters (*cf.* ref. 37). Probably the most versatile of the presently available methods is the last-mentioned procedure, which involves the sodium ethoxide-mediated condensation either of acetylaminomalonic ester (I; R = CH_3) or ethyl acetamidocyanoacetate (V; R = CH_3), both of which are commercially available, with a suitable

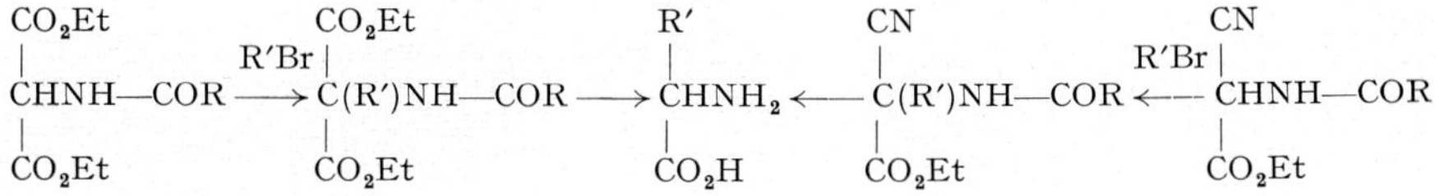

I II III IV V

alkyl or acyl halide (R'Br). Acid hydrolysis of the pertinent condensation product (II or IV) then yields the desired amino acid (III).

The synthetic procedures listed above may be considered as general only insofar as the assumption is valid that the requisite starting materials are available. Such is not invariably the case, however, for the side chains of amino acids are possessed of a most diverse structure and may require the expenditure of considerable effort for

References p. 22/24

their synthesis. Thus, the synthesis of a given amino acid by a general method may afford many problems unique unto itself.

If the synthesis leads to an amino acid which contains more than one asymmetric center, then a diastereomeric mixture of the racemates generally results. Separation of the racemates must precede their resolution into the optical antipodes, and for this purpose two general approaches have been used, namely, differential solubility and partition or ion-exchange chromatography. Although certain instances are known wherein it is possible to separate diastereomeric amino acids on the basis of differential solubility, such a process more frequently requires the prior conversion of the amino acid into a suitable derivative. It would, of course, prove most advantageous if partition or ion-exchange chromatography could be employed to separate the diastereomeric racemates, and although such procedures have been employed for the separation of threonine from allothreonine[38], hydroxylysine from allohydroxylysine[39, 40], and hydroxyproline from allohydroxyproline[40], among others, the procedures have been applicable only to the separation of micromolar amounts of material and hence have proven inadequate for preparative purposes. The problem of increasing the scale of such chromatographic separations to the preparative range was undertaken by Dr. MICHI in our laboratory[41] some 2 years ago. Attempts were first made to separate isoleucine from alloisoleucine, and after a long and tedious process of juggling conditions in many trials, conditions were finally found that would permit the separation of several of these diastereomers. The conditions evolved are shown in Fig. 4 A. Since the use of the usual citrate or phosphate buffers as eluants here would have

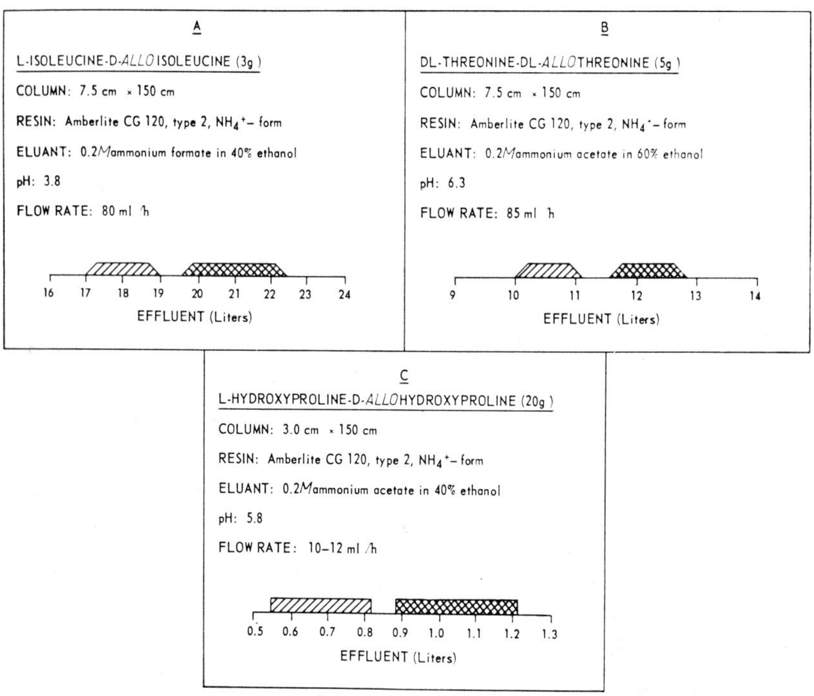

Fig. 4. Conditions for separation of diastereomers of various amino acids in macro quantities via ion-exchange chromatography.

References p. 22/24

resulted in the ultimate isolation of large amounts of buffer salts that would have interfered with the isolation of the desired components, volatile acetate and formate buffers patterned after those used by MOORE and STEIN, were employed. The buffer used here was 0.2 *M* with respect to ammonium ion and the resin was Amberlite CG 120 in the ammonium form. After equilibration of the column with buffer, 3 g of a diastereomeric mixture containing approximately equal parts of L-isoleucine and D-alloisoleucine, dissolved in the appropriate buffer, was poured onto the column. As is indicated by the graph (Fig. 4 A), the amino acid emerged in two distinctly separate fractions after the passage of 17 l. Evaporation of each separate fraction *in vacuo* resulted in a residual mass which was subjected to sublimation under high vacuum at 50° in order to remove contaminating buffer salts. The residual amino acid was recovered in approx. 95% over-all yield, was found to be analytically and chromatographically pure, and required no further purification or recrystallization.

Fig. 4 B illustrates the conditions employed in the separation of threonine from allothreonine. The starting material here was a 50 : 50 mixture of threonine and allothreonine, the buffer ammonium acetate, and the resin again Amberlite CG 120. It was observed that separation of the threonines appeared to be appreciably more sensitive to alcohol concentration than to pH. Thus, although successful separation of several hundred milligrams of the diastereomeric mixture could be obtained in 40% alcohol in the pH range of 5.8–6.6, it was only after the alcohol concentration was increased to 60% that the resolution became great enough to permit the separation of several grams of material. Recovery of the amino acid from the separate fractions was effected in over 90% yield and, as in the case of the isoleucines, the material was found to be analytically and chromatographically pure.

In Fig. 4 C, the conditions employed for the separation of L-hydroxyproline from D-allohydroxyproline are given. Extremely large quantities of a mixture of the two diastereomers could be separated under these conditions, as is evidenced by the fact that 20 g of a 50 : 50 mixture of hydroxyproline and allohydroxyproline could be separated on a column only 3 cm in dia. and 150 cm long, in contrast to the 7.5 × 150-cm column which was employed in the separation of the isoleucines and of the threonines. Recovery of analytically and chromatographically pure amino acids from the separate fractions proceeded in high yield as before. In any event, the data showed that the use of ion-exchange chromatography would permit the separation of diastereomeric amino acids on a scale sufficiently large for preparative purposes.

The next step in establishing the identity of a naturally occurring amino acid necessitates resolution of the synthetic racemic material into its optical antipodes by any one of a number of chemical or biological procedures. A distinct advantage which accrues from the use of biological procedures over purely chemical procedures is that they yield isomers of known optical configuration by virtue of the known optical specificity of the biological agent itself. In addition, biological procedures permit a more general approach and a more uniform resolution procedure, and do not require the tedious and time-consuming manipulations so generally encountered with chemical procedures. Probably the most satisfactory of all presently available biological resolution procedures is that developed by the late Dr. J. P. GREENSTEIN, who employed as the biological agents two separate enzyme systems isolated from hog kidney. One enzyme system, known as acylase I, is capable of acting asymmetrically only on the L-isomers of *N*-acylated DL-amino acids, while the other enzyme system, known as amidase,

References p. 22/24

is capable of acting asymmetrically only on the L-isomers of DL-amino acid amides[42, 43], as follows:

$$\left.\begin{array}{l}\text{L-RCONHCHR'CO}_2\text{H}\\ \\ \text{D-RCONHCHR'CO}_2\text{H}\end{array}\right] \xrightarrow[\text{carboxy-peptidase}]{\text{acylase}} \begin{array}{c}\text{L-NH}_2\text{CHR'CO}_2\text{H (organic solvent—insoluble)}\\ + \text{ RCO}_2\text{H } +\\ \text{D-RCONHCHR'CO}_2\text{H (organic solvent—soluble)}\end{array}$$

$$\left.\begin{array}{l}\text{L-NH}_2\text{CHR'CONH}_2\\ \\ \text{D-NH}_2\text{CHR'CONH}_2\end{array}\right] \xrightarrow{\text{amidase}} \begin{array}{c}\text{L-NH}_2\text{CHR'CO}_2\text{H (organic solvent—insoluble)}\\ + \text{ NH}_3 +\\ \text{D-NH}_2\text{CHR'CONH}_2\text{ (organic solvent—soluble)}\end{array}$$

The renal acylase and amidase systems have proven effective in the resolution of nearly all α-amino acids studied. Because of the optical specificity of the enzymes involved, the method of resolution is extremely simple. The racemic amino acids are *N*-acylated, generally by reaction with acetic anhydride or chloroacetyl chloride, the purified compound dissolved at pH 7.0 and at 0.1 *M* concentration in water, and the solution treated with the appropriate amount of purified enzyme. Hydrolysis at 37° proceeds to exactly 50% of the racemate, which may be readily checked by manometric ninhydrin measurements on aliquots, and goes no further irrespective of how long the digest stands. The enzyme is removed, the solution condensed, and the free L-amino acid separated from the *N*-acyl-D-amino acid either by precipitation with ethanol or passage over an ion-exchange resin. A similar procedure applies for the amino acid amides, which also reach a 50% hydrolysis end point.

Table II lists some of the amino acids resolved in our laboratory[42, 43] by the enzymatic procedures just described. The power of this general approach as a tool in the identification of naturally occurring amino acids is revealed by the fact that it permitted the configurational elucidation of natural α-aminoadipic acid, which was resolved[44] prior to its discovery in nature, that is has been employed in our laboratory by Dr. E. WORK[45, 46] for the purpose of establishing the configuration of α,ε-diaminopimelic acid and by Dr. FONES to determine the configuration both of natural hydroxylysine[47] and of α-methylserine (*cf.* ref. 48), that it has been recently employed[49] to establish the configurational identity of the γ-hydroxyglutamic acid of VIRTANEN AND HIETALA[50], and finally, that it will ultimately serve to establish the configuration of the α-aminopimelic acid of VIRTANEN AND BERG[51] after a sufficient amount of the natural material has been isolated to permit a comparison of its optical-rotation value with that of the synthetically obtained antipodes[52].

It is an unfortunate occurrence that all too many of the reports concerned with the isolation and structural elucidation of new naturally occurring amino acids neglect to include optical-rotation data, presumably because of a scarcity of material. Yet, from the standpoint of identification, the optical behavior of the new amino acid is probably its most important characteristic, not only because it may ultimately prove of utility in assigning an optical configuration to the material, but primarily because it represents a means for providing unequivocal evidence for or against the assigned structure after synthesis and subsequent resolution of the synthetic material.

References p. 22/24

TABLE II

ENZYMIC PREPARATION OF L- AND D-ISOMERS OF α-AMINO ACIDS

Amino acid	*Derivative employed*	*Enzyme employed*
Alanine	Acetyl	Acylase I
Butyrine	Acetyl	Acylase I
Valine	Acetyl	Acylase I
Norvaline	Acetyl	Acylase I
Isovaline	Chloroacetyl	Acylase I
Leucine	Acetyl	Acylase I
Norleucine	Acetyl	Acylase I
Isoleucine	Acetyl	Acylase I
Alloisoleucine	Acetyl	Acylase I
Heptyline	Chloroacetyl	Acylase I
Capryline	Chloroacetyl	Acylase I
Nonyline	Chloroacetyl	Acylase I
Decyline	Chloroacetyl	Acylase I
Undecyline	Chloroacetyl	Acylase I
Dodecyline	Amide	Amidase
Serine	Chloroacetyl	Acylase I
Homoserine	Chloroacetyl	Acylase I
δ-Hydroxynorvaline	Chloroacetyl	Acylase I
ε-Hydroxynorleucine	Chloroacetyl	Acylase I
Threonine	Chloroacetyl	Acylase I
allothreonine	Chloroacetyl	Acylase I
o-Methylthreonine	Chloroacetyl	Acylase I
o-Methylallothreonine	Formyl	Acylase I
Methionine	Acetyl	Acylase I
Ethionine	Acetyl	Acylase I
S-Benzylcysteine	Acetyl	Acylase I
S-Benzylhomocysteine	Acetyl	Acylase I
Aspartic acid	Acetyl	Acylase II
β-Methylaspartic acid	Acetyl	Acylase II
Glutamic acid	Carbobenzoxy	Acylase I
γ-Hydroxyglutamic acid	Chloroacetyl	Acylase I
allo-γ-Hydroxyglutamic acid	Chloroacetyl	Acylase I
α-Aminoadipic acid	Chloroacetyl	Acylase I
α-Aminopimelic acid	Chloroacetyl	Acylase I
β-Aminoalanine	α, β-Diacetyl	Acylase I
γ-Aminobutyrine	α, γ-Dichloroacetyl	Acylase I
Ornithine	α, δ-Dichloroacetyl	Acylase I
Lysine	α, ε-Dichloroacetyl	Acylase I
Histidine	α-Acetyl	Acylase I
Arginine	α-Acetyl	Acylase I
β-Phenylalanine	Chloroacetyl	Acylase I
Tyrosine	Chloroacetyl	Carboxypeptidase
Tryptophane	Chloroacetyl	Carboxypeptidase
Proline	Amide	Amidase
α-Phenylglycine	Chloroacetyl	Acylase I
α-Cyclohexylglycine	Chloroacetyl	Acylase I
β-Cyclohexylalanine	Chloroacetyl	Acylase I
β-Phenylserine	Trifluoroacetyl	Carboxypeptidase
allo-β-Phenylserine	Trifluoroacetyl	Carboxypeptidase
tert.-Leucine	Amide	Amidase
δ-Hydroxylysine	α-Chloroacetyl, ε-carbobenzoxy	Acylase I
allo-δ-Hydroxylysine	α-Chloroacetyl, ε-carbobenzoxy	Acylase I
α-Methylserine	Chloroacetyl	Acylase I
α, ε-Diaminopimelic acid	α, ε-Diamide	Amidase

References p. 22/24

DETERMINATION OF OPTICAL CONFIGURATION

Other methods are, of course, available for determining the optical configuration of a naturally occurring amino acid aside from synthesis, and subsequent resolution of the synthetic material with a biological agent of established optical specificity. For the present purposes, such methods may be roughly categorized as chemical, biological, physical and optical (*cf.* ref. 53).

Chemical methods are concerned primarily with the transformation of the natural amino acid either into another amino acid of established optical configuration or into some other compound of known optical configuration, using organic chemical procedures that do not induce rupture of any of the bonds surrounding the asymmetric carbon atom. These chemical methods cannot be effected by any "set" or standard procedure but, rather, vary one from the other because of differences both in the structure and in the nature of the functional groups of the side chains of the amino acids under consideration. They must therefore depend, for the most part, on the ingenuity of the chemist to devise a chemical route from one compound to another, on his ability and discretion to draw properly from the most pertinent of a vast wealth of available synthetic and degradative reactions, and on his skill as a technician to guide such reactions to their satisfactory culmination. When properly executed, this approach forms the most powerful, as well as the most unequivocal of all available methods for determining the optical configuration of an amino acid. The chemical method, however, generally necessitates the expenditure of a good deal more time and effort than most of us can reasonably afford to devote to it, so that recourse is more frequently made to biological and physical methods.

Biological methods for the determination of configuration depend upon the remarkable ability of living organisms, as well as certain tissues and enzyme fractions derived therefrom, to preferentially metabolize, incorporate, or chemically alter one antipode of a racemic amino acid, or a suitable derivative thereof, while leaving the other antipode essentially intact. An amino acid of unknown optical structure can therefore be assigned a D- or an L-designation with a high degree of reliability if one or the other of its optical isomers is susceptible to the action of a biological agent whose specificity toward amino acids of known configuration has been previously established. Toward this end, use has been made of the ability of certain molds, yeasts, bacteria, and even mammals, to more or less stereospecifically metabolize the L-isomer of a given amino acid at an appreciably faster rate than its corresponding D-antipode. Much greater use, however, has been made of the stereospecific oxidative and decarboxylative action of amino acid oxidases and decarboxylases, respectively, the hydrolytic action of acylases, amidases and carboxypeptidases, and the ability of certain proteases to catalyze peptide bond formation[53]. Probably the most convenient, and one of the most reliable of the enzymatic methods for the determination of optical configuration entails the use of amino acid oxidases. An amino acid oxidase generally exhibits an antipodal specificity which, depending on its origin, may be either L-directed or D-directed. In any case, the reaction, which may be conveniently followed in a Warburg respirometer, essentially involves an oxidative deamination of one molecule of the susceptible amino acid antipode, with the consumption of one molecule of oxygen, to yield one molecule each of the corresponding α-keto acid, ammonia and hydrogen peroxide, as follows:

References p. 22/24

$$RCH(NH_2)CO_2H + O_2 + H_2O \rightarrow RCOCO_2H + NH_3 + H_2O_2$$

A limitation of the method arises from the fact that not all α-amino acids which possess the required optical configuration are necessarily susceptible to the enzymatically-induced oxidation, so that the inability of a given amino acid to undergo oxidative deamination by an L- or D-amino acid oxidase does not imply that its configuration becomes known by default. Although amino acid decarboxylases have also been employed in configurational analysis, their utility is somewhat limited in that their action is generally confined either to a single substrate or a small number of very closely related substrates. It should further be noted that both L- and D-decarboxylases are known to occur in the plant kingdom[54], so that their optical specificity with respect to substrates of known configuration must be established with certainty before any measure of confidence can be reposed in their use.

The bland or bitter taste revealed by nearly all amino acids of the L-configuration as contrasted with the sweet taste revealed by D-amino acids[55, 56] may also be employed to indicate the configuration of an α-amino acid[57], but as the rule is not invariable and as the method is of a highly subjective nature, it cannot be employed with any degree of confidence. As the intestinal absorption of amino acids is presumably mediated by an active transport mechanism that is L-directed[57], and as toxicity studies with rats have indicated that the D-isomers of nearly all amino acids tested were less toxic than their corresponding L-forms[58, 59], these phenomena too may form a theoretical, although admittedly impractical basis for configurational analysis[57].

Although physical methods, such as X-ray analysis, kinetic reactions, and the melting-point curves of quasi-racemic compounds have also been utilized to determine the configuration of amino acids, the bulk of such physical methods, by far, have been concerned with the optical behavior of amino acids or their derivatives[53]. Thus, the L- or D-configuration of an amino acid can be reliably determined, in a single measurement, by the direction of rotation exhibited by the hydantoin[60, 61], N- or C-terminal glycine[62, 63], aminofluorene[64, 65], aminobiphenyl[64, 65] or benzidine[64, 65] derivative. As a marked relationship exists between the configuration of an amino acid and its rotatory dispersion curve, the measurement of the optical rotation of an amino acid at various wavelengths will also serve to indicate its configuration[66–68]. This method, in conjunction with kinetic studies, was successfully employed in our laboratory, in 1957, to establish the configuration of natural octopine, after attempts at several other methods had failed[69, 70]. Probably the most convenient, and one of the most reliable of the optical methods is that which involves the shift in optical rotation induced by ionization[71]. The principle upon which this method is based is contained in the CLOUGH-LUTZ-JIRGENSONS rule. This rule states that: *if the molecular rotation of an optically active amino acid is shifted toward a more positive direction upon the addition of acid to its aqueous solution, the amino acid is of the L-configuration; a negative direction of shift, however, is characteristic of a D-amino acid.* Its use is exemplified in the case of alanine as follows:

	5 N HCl		*Water*	*Difference*
L-Alanine	+13.0°	minus	+1.6°	+11.4°
D-Alanine	−13.0°	minus	−1.6°	−11.4°

Application of this rule in our laboratory to over 70 different L-amino acids revealed no exceptions as long as the measurements were confined only to amino acids which possessed a single asymmetric center. Certain exceptions to the rule were found, however, in the case of diasymmetric amino acids[71, 72].

In any event, from what has already been said, it is clear that a number of relatively simple biological and optical methods are available for determining the optical configuration of an amino acid which require, in many instances, only a few mg of material. It is therefore puzzling that many papers reporting the occurrence of a new amino acid still appear in which no attempt at all has been made to ascertain the configuration of the material.

IDENTIFICATION OF NATURAL γ-HYDROXYGLUTAMIC ACID

In conclusion, I should like to discuss the identification and characterization of natural γ-hydroxyglutamic acid, firstly, because its natural occurrence may be considered as established according to the criteria presented earlier; and secondly, because it covers the somewhat more complex case of an amino acid with two asymmetric centers.

The natural occurrence of γ-hydroxyglutamic acid was first observed by VIRTANEN AND HIETALA[50] in 1955 when two-dimensional paper chromatograms, with butanol–acetic acid and phenol–ammonia, of a 70% ethanolic extract of the green parts of the plant, *Phlox decussata*, revealed a hitherto unknown ninhydrin-reactive spot just above that of aspartic acid. Approx. 124 mg of the material was isolated from 3.5 kg of the fresh plant by column chromatography. Elemental analytical data, together with the fact that reduction of the material with hydriodic acid and red phosphorus at 140° led to the formation of glutamic acid, indicated that the new compound was a hydroxyglutamic acid. Oxidation of the material with potassium permanganate in sulfuric acid solution at 140° proceeded with the formation of aspartic acid, while comparison of the R_F value of the new compound in butanol–acetic acid with that of authentic β-hydroxyglutamic acid indicated that these compounds were not identical. It therefore appeared probable that the new acidic compound was a γ-hydroxyglutamic acid. No optical rotation data were presented for the new compound, nor was its optical configuration ascertained.

During the same year that γ-hydroxyglutamic acid was isolated from natural sources, its synthesis was described by BENOITON AND BOUTHILLIER in Canada[73] (*cf.* ref. 49). The reaction sequence employed is given in Fig. 5. Since this amino acid

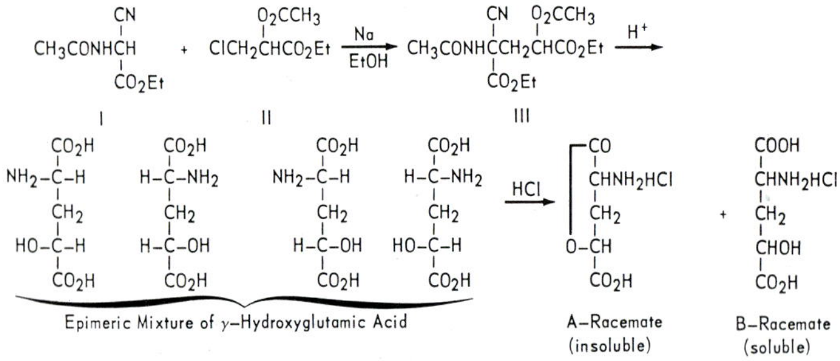

Fig. 5. Synthetic route to γ-hydroxyglutamic acid.

References p. 22/24

contains two asymmetric carbon atoms, the synthetic product was obtained as a mixture of two diastereometic racemates. Separation of the two racemic modifications, and subsequent resolution of each racemate into its optical antipodes was achieved some 2 years later in our laboratory[49] in Bethesda. Separation of the racemates was accomplished by exploiting the differential solubilities of the hydrochloride salts of the two racemates. Thus, after saturation of an aqueous solution of the diastereomeric mixture with hydrogen chloride gas, and subsequent chilling at −10°, some half of the starting material deposited as a crystalline substance which analyzed for the lactone hydrochloride and to which the designation of *A* was assigned. The soluble racemate, which was designated as the *B* form, could subsequently be isolated in good yield from the mother liquors.

The efficacy of this procedure for the separation of the two racemic diastereomers could be readily assessed with the aid of column chromatography, as shown in Fig. 6.

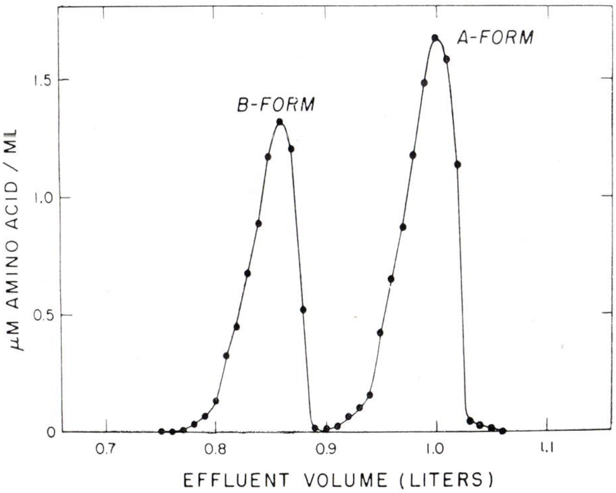

Fig. 6. Separation of racemic diastereomers of γ-hydroxyglutamic acid on Dowex-1-acetate, elution with 0.5 *N* acetic acid. *A* is allo-form, *B* is normal form.

Use was made of the fact that the original epimeric mixture undergoes complete separation into two peaks upon passage through the Dowex-1 acetate system. As chromatography of a 15-mg sample of once recrystallized lactone A, or the once recrystallized B form, revealed but a single peak in each instance, the optical integrity of each racemate was demonstrated. Resolution of the A and B forms into their respective optical antipodes was then achieved by the conversion of each racemate into its corresponding *N*-chloroacetyl derivative, succeeded by stereospecific enzymatic hydrolysis of the L-form of each of these derivatives with hog renal acylase I. As this enzyme is L-directed, the resolution procedure permitted both the eventual isolation of the four optically pure stereoisomers and the assignment of an L- or D-configuration to the α-carbon atom of each.

Now only the configuration of the α-asymmetric carbon atom of each of the four isomers of γ-hydroxyglutamic acid remained to be determined. This was accomplished by the conversion of each stereoisomer to the corresponding dihydroxyglutaric acid isomer through deamination with nitrous acid, as shown in Fig. 7. Now kinetic studies

References p. 22/24

have amply demonstrated that such conversion of an α-amino acid to the corresponding α-hydroxy acid proceeds in the absence of inversion of configuration. Hence, the α-configuration of each of the isomers of the dihydroxy acid could be considered as identical with that of the α-amino acid from which it was derived. The two dihydroxyglutaric acid isomers corresponding to the D- and L-antipodes of γ-hydroxyglutamic acid A revealed molecular rotation values of $+31°$ and $-31°$, respectively, as solutions of their barium salts in water, whereas the two isomers whose parent amino acids

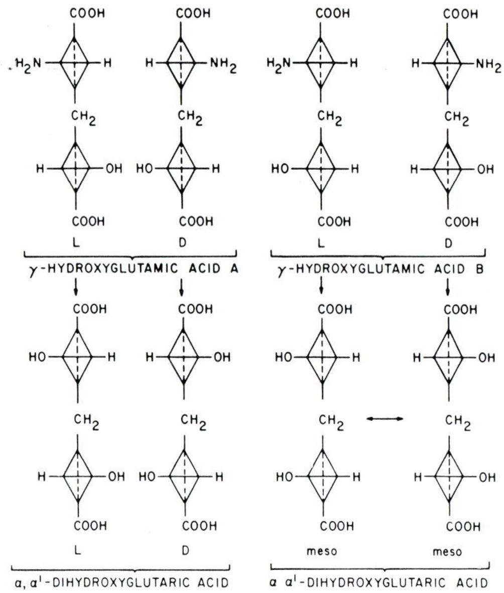

Fig. 7. Projections of the four stereomers of γ-hydroxyglutamic acid and the corresponding α, α'-dihydroxyglutaric acids derived therefrom.

represented the antipodes of the B form were completely devoid of optical activity. This data now sufficed to establish the γ-configuration of each of the four antipodes of the amino acid.

Assignments of configuration based on these results become readily explicable through examination of the Fischer projections of the four isomers of γ-hydroxyglutamic acid and the dihydroxy acids corresponding thereto. Thus, although the dihydroxy acid molecule embodies two centers of optical asymmetry, it also possesses symmetry such that, of the four theoretically possible isomers, only two constitute an externally compensated optically active pair, whereas the other two constitute the identical optically inactive *meso* form by virtue of internal compensation. Inasmuch as deamination of the antipodes of the A form of the amino acid results in two optically active enantiomorphic forms of the dihydroxy acid, it follows that the α-amino and γ-hydroxyl groups of the parent amino acid must assume a *trans* representation in the Fischer formulation. On the other hand, the functional groups of the enantiomorphic B forms must occupy a *cis* position by virtue of the fact that deamination of the amino acid leads to *meso*-dihydroxyglutaric acid.

References p. 22/24

Sufficient information is here provided to permit the stereochemical correlation of each of the four isomers of γ-hydroxyglutamic acid with those of the corresponding γ-hydroxyprolines, the configurations of which have been previously established. Such correlations are represented in Fig. 8. Thus, L-hydroxyproline (structure I) may be configurationally correlated with γ-hydroxy-L-glutamic acid B (structure II), whereas

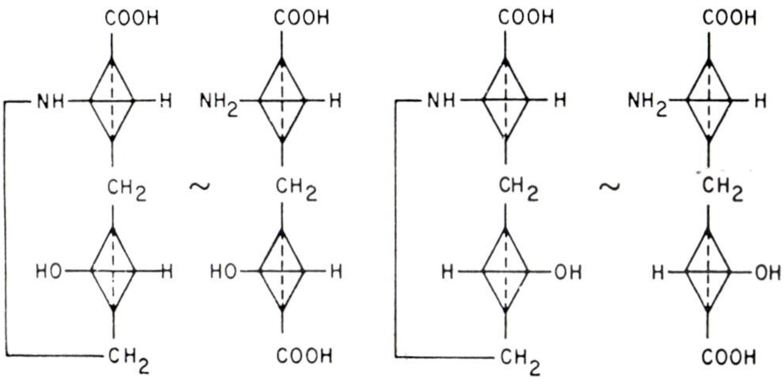

I II III IV

Fig. 8. Stereochemical relationship of the stereoisomeric forms of γ-hydroxyglutamic acid with the corresponding forms of γ-hydroxyproline.

L-allohydroxyproline (structure III) is the stereochemical equivalent of γ-hydroxyglutamic acid A (structure IV). The validity of these correlations was recently demonstrated by ADAMS AND GOLDSTONE[74], who showed that γ-hydroxy-L-glutamic acid B (structure II) is a metabolic product, in the mammal, of L-hydroxyproline (structure I). With this in mind, it now becomes possible to eliminate the A and B designations for the diastereomeric amino acids and adopt the prefix *allo* in their stead. The configurational designation and nomenclature of the γ-hydroxyglutamic acids could thereby be represented as shown in Table III.

As the configurations and optical rotations of each of the four antipodes of γ-hydroxyglutamic acid are now known, comparison of these with the optical rotation of the natural amino acid would suffice to establish the configuration of the latter. However,

TABLE III

CONFIGURATIONAL DESIGNATION AND NOMENCLATURE OF THE ISOMERIC γ-HYDROXYGLUTAMIC ACIDS

Isomer	*Configuration of γ-carbon atom*	*Recommended designation*
γ-Hydroxy-L-glutamic acid A	L_s or D_g	allo-γ-Hydroxy-L-glutamic acid
γ-Hydroxy-D-glutamic acid A	D_s or L_g	allo-γ-Hydroxy-D-glutamic acid
γ-Hydroxy-L-glutamic acid B	D_s or L_g	γ-Hydroxy-L-glutamic acid
γ-Hydroxy-D-glutamic acid B	L_s or D_g	γ-Hydroxy-D-glutamic acid

γ-Hydroxy-L-glutamic acid (mol. wt. 163.1):
$[M]_D = +31.8°$ (H_2O) and $+61.6°$ (5 N HCl); $c = 1\%$, $T = 25°$.
allo-γ-Hydroxy-L-glutamic acid (mol. wt. 163.1):
$[M]_D = -22.3°$ (H_2O) and $+5.0°$ (5 N HCl); $c = 1\%$, $T = 25°$.

References p. 22/24

it will be recalled that VIRTANEN AND HIETALA[50], in their initial isolation of γ-hydroxyglutamic acid, obtained the material in amounts insufficient for optical rotation values to be taken. The answer as to the configurational identity of natural γ-hydroxyglutamic acid, as well as independent experimental verification of its existence in nature, was subsequently provided by an optical rotation determination of the natural material[75], which was isolated by STEWARD AND POLLARD[76] in several g yield from *Phlox decussata* seed and which was shown by them to possess a chromatographic behavior identical with that of an authentic sample of the synthetic *allo* form[49] of the amino acid. An optical rotation determination of the natural material in water and 4 *N* HCl revealed it to possess the same values, within the limits of experimental error, as that of the L-isomer, and hence sufficed to unequivocally establish its identity as allo-γ-hydroxy-L-glutamic acid[75].

RECAPITULATION

Thus, from the data presented, it becomes evident that the natural occurrence of allo-γ-hydroxy-L-glutamic acid may be considered as established according to the criteria presented earlier. If these criteria seem unnecessarily rigid, it is well to remember that the analogous criteria, proposed some 30 years ago by VICKERY AND SCHMIDT[1], were accompanied by the following admonition: "Although these criteria may appear somewhat arbitrary, they are essential, unless one is prepared to accept a host of preparations that have been described, from time to time, but for the exact nature of which convincing proof has not been presented. Too many errors have been committed to warrant any but a thoroughly conservative attitude towards newly discovered amino acids." They went on further to state: "Even the greatest leaders have made mistakes. FISCHER himself described diaminotrioxydodecanoic acid in 1904, but withdrew it in 1917." The rapidly increasing numbers of naturally occurring amino acids that are currently being reported, all too often without adequate substantiation, make this admonition equally as cogent today.

REFERENCES

1 H. B. VICKERY AND C. L. A. SCHMIDT, *Chem. Revs.*, 9 (1931) 169.
2 H. D. DAKIN, *Biochem. J.*, 12 (1918) 290.
3 H. D. DAKIN, *Biochem. J.*, 13 (1919) 388.
4 C. L. A. SCHMIDT, *Ann. Rev. Biochem.*, 2 (1933) 72.
5 J. P. GREENSTEIN AND M. WINITZ, in *Chemistry of the Amino Acids*, J. Wiley and Sons, New York, 1961, p. 20.
6 Z. H. SKRAUP, *Ber.*, 37 (1904) 1596.
7 Z. H. SKRAUP, *Z. physiol. Chem., Hoppe-Seyler's*, 42 (1904) 274.
8 R. D. HAWORTH, R. MACGILLIVARY AND D. H. PEACOCK, *Nature*, 167 (1951) 1068.
9 E. WINDSOR, *J. Biol. Chem.*, 192 (1951) 595.
10 E. ABDERHALDEN AND A. BAHN, *Ber.*, 63 (1930) 914.
11 E. ABDERHALDEN AND F. REICH, *Z. physiol. Chem., Hoppe-Seyler's*, 193 (1930) 198.
12 J. P. GREENSTEIN AND M. WINITZ, in *Chemistry of the Amino Acids*, J. Wiley and Sons, New York, 1961, p. 25.
13 J. D. OGLE, M. A. LOGAN AND R. B. ARLINGHAUS, *Federation Proc.*, 20 (1961) 1.
14 L. BENOITON, M. WINITZ, R. F. COLMAN, S. M. BIRNBAUM AND J. P. GREENSTEIN, *J. Am. Chem. Soc.*, 81 (1959) 1726.
15 C. E. DENT, *Biochem. J.*, 43 (1948) 169.
16 W. R. MIDDLEBROOK, *Biochim. Biophys. Acta*, 7 (1951) 547.
17 A. E. PASIEKA AND J. F. MORGAN, *Biochim. Biophys. Acta*, 19 (1956) 366.
18 T. L. HARDY, D. O. HOLLAND AND J. H. C. NAYLOR, *Anal. Chem.*, 27 (1955) 971.
19 E. V. MCCOLLUM AND A. A. RIDER, *Arch. Biochem. Biophys.*, 40 (1952) 20.

[20] H. Frank and H. Petersen, *Z. physiol. Chem., Hoppe-Seyler's*, 299 (1955) 1.
[21] C. Fromageot, *Cold Spring Harbor Symposia Quant. Biol.*, 14 (1950) 49.
[22] A. Saifer and J. Oreskes, *Science*, 119 (1954) 124.
[23] J. Barrollier, J. Heilman and E. Watzke, *Z. physiol. Chem., Hoppe-Seyler's*, 304 (1956) 21.
[24] A. E. Pasieka and J. F. Morgan, *Proc. Soc. Exptl. Biol. Med.*, 93 (1956) 54.
[25] R. Consden, A. H. Gordon and A. J. P. Martin, *Biochem. J.*, 40 (1946) 33.
[26] R. Acker and C. Crocker, *Biochim. Biophys. Acta*, 9 (1952) 704.
[27] R. J. Block and D. Bolling, in *The Amino Acid Composition of Proteins and Foods*, Charles C. Thomas, Springfield, Ill., 1951.
[28] W. R. Fearon and W. A. Boggust, *Analyst*, 79 (1954) 101.
[29] H. M. Winegard, G. Toennies and R. J. Block, *Science*, 108 (1948) 506.
[30] A. K. Patton and E. M. Foreman, *Science*, 109 (1949) 339.
[31] D. Müting, *Naturwissenschaften*, 39 (1952) 303.
[32] K. V. Giri and A. Nagabhashanam, *Naturwissenschaften*, 39 (1952) 548.
[33] J. P. Greenstein and M. Winitz, in *Chemistry of the Amino Acids*, J. Wiley and Sons, New York, 1961, p. 1320.
[34] L. Benoiton, S. M. Birnbaum, M. Winitz and J. P. Greenstein, *Arch. Biochem. Biophys.*, 81 (1959) 434.
[35] T. Otani and M. Winitz, *Arch. Biochem. Biophys.*, in the press.
[36] J. P. Greenstein, N. Izumiya, M. Winitz and S. M. Birnbaum, *J. Am. Chem. Soc.*, 77 (1955) 707.
[37] J. P. Greenstein and M. Winitz, in *Chemistry of the Amino Acids*, J. Wiley and Sons, New York, 1961, p. 697.
[38] A. T. Shulgin, O. G. Lien, Jr., E. M. Gal and D. M. Greenberg, *J. Am. Chem. Soc.*, 74 (1952) 2427.
[39] P. B. Hamilton and R. A. Anderson, *J. Biol. Chem.*, 213 (1955) 249.
[40] K. A. Piez, *J. Biol. Chem.*, 207 (1954) 77.
[41] K. Michi, S. M. Birnbaum and M. Winitz, unpublished results.
[42] J. P. Greenstein, *Advances in Protein Chem.*, 9 (1954) 121.
[43] J. P. Greenstein and M. Winitz, in *Chemistry of the Amino Acids*, J. Wiley and Sons, New York, 1961, p. 715.
[44] J. P. Greenstein, S. M. Birnbaum and M. C. Otey, *J. Am. Chem. Soc.* 75 (1953) 1994.
[45] E. Work, S. M. Birnbaum, M. Winitz and J. P. Greenstein, *J. Am. Chem. Soc.*, 77 (1955) 1916.
[46] R. Wade, S. M. Birnbaum, M. Winitz, R. J. Koegel and J. P. Greenstein, *J. Am. Chem. Soc.*, 79 (1957) 648.
[47] W. S. Fones, *J. Am. Chem. Soc.*, 75 (1953) 4865.
[48] J. P. Greenstein and M. Winitz, in *Chemistry of the Amino Acids*, J. Wiley and Sons, New York, 1961, p. 2573.
[49] L. Benoiton, M. Winitz, S. M. Birnbaum and J. P. Greenstein, *J. Am. Chem. Soc.*, 79 (1957) 6192.
[50] A. I. Virtanen and P. K. Hietala, *Acta Chem. Scand.*, 9 (1955) 175.
[51] A. I. Virtanen and A. M. Berg, *Acta Chem. Scand.*, 8 (1954) 1085, 1725.
[52] R. Wade, S. M. Birnbaum, M. Winitz, R. J. Koegel and J. P. Greenstein, *J. Am. Chem. Soc.*, 79 (1957) 648.
[53] J. P. Greenstein and M. Winitz, in *Chemistry of the Amino Acids*, J. Wiley and Sons, New York, 1961, p. 46.
[54] S. Mizushima and K.-I. Sakaguchi, *Nippon Nôgei-kagaku Kaishi*, 20 (1956) 131.
[55] C. P. Berg, *Physiol. Revs.*, 33 (1953) 145.
[56] A. Meister, in *Biochemistry of the Amino Acids*, Academic Press, New York, 1957, p. 76.
[57] J. P. Greenstein and M. Winitz, in *Chemistry of the Amino Acids*, J. Wiley and Sons, New York, 1961, p. 150.
[58] P. Gullino, M. Winitz, S. M. Birnbaum, J. Cornfield, M. C. Otey and J. P. Greenstein, *Arch. Biochem. Biophys.*, 58 (1955) 253.
[59] P. Gullino, M. Winitz, S. M. Birnbaum, J. Cornfield, M. C. Otey and J. P. Greenstein, *Arch. Biochem. Biophys.*, 64 (1956) 319.
[60] J. P. Greenstein and M. Winitz, in *Chemistry of the Amino Acids*, J. Wiley and Sons, New York, 1961, p. 94.
[61] G. W. Clough, *J. Chem. Soc.*, 113 (1918) 526.
[62] J. P. Greenstein and M. Winitz, in *Chemistry of the Amino Acids*, J. Wiley and Sons, New York, 1961, p. 95.
[63] J. P. Greenstein, S. M. Birnbaum and M. C. Otey, *J. Biol. Chem.*, 204 (1953) 307.
[64] J. P. Greenstein and M. Winitz, in *Chemistry of the Amino Acids*, J. Wiley and Sons, New York, 1961, p. 97.
[65] J. P. Greenstein, M. Winitz and S. M. Birnbaum, *J. Am. Chem. Soc.*, 77 (1955) 5721.

[66] J. P. GREENSTEIN AND M. WINITZ, in *Chemistry of the Amino Acids*, J. Wiley and Sons, New York, 1961, p. 105.

[67] M. C. OTEY, J. P. GREENSTEIN, M. WINITZ AND S. M. BIRNBAUM, *J. Am. Chem. Soc.*, 77 (1955) 3112.

[68] J. W. PATTERSON AND W. R. BRODE, *Arch. Biochem.*, 2 (1943) 247.

[69] J. P. GREENSTEIN AND M. WINITZ, in *Chemistry of the Amino Acids*, J. Wiley and Sons, New York, 1961, p. 205.

[70] N. IZUMIYA, R. WADE, M. WINITZ, M. C. OTEY, S. M. BIRNBAUM, R. J. KOEGEL AND J. P. GREENSTEIN, *J. Am. Chem. Soc.*, 79 (1957) 652.

[71] J. P. GREENSTEIN AND M. WINITZ, in *Chemistry of the Amino Acids*, J. Wiley and Sons, New York, 1961, p. 83.

[72] M. WINITZ, S. M. BIRNBAUM AND J. P. GREENSTEIN, *J. Am. Chem. Soc.*, 77 (1955) 716.

[73] L. BENOITON AND L. P. BOUTHILLIER, *Can. J. Chem.*, 33 (1955) 1473.

[74] E. ADAMS AND A. GOLDSTONE, *J. Biol. Chem.*, 235 (1960) 3504.

[75] M. WINITZ, J. POLLARD AND F. C. STEWARD, unpublished results.

[76] F. C. STEWARD AND J. POLLARD, private communication.

DISCUSSION

Chairman: EUGENE ROBERTS

ROCKLAND: Are there any significant differences in the infrared spectra of the isomeric forms of γ-hydroxyglutamic acid?

WINITZ: There are quite significant differences. The infrared spectrum of the B-form of the amino acid is very sharp and shows quite definite peaks. The A-form of the amino acids shows a spectrum of indefinite character in which the peaks are not sharp but, rather, merge into each other. Of course, the D- and L-forms of the amino acid show the identical infrared pattern.

POLLARD: In Dr. WINITZ's criteria for purity of amino acids, he indicated that one must obtain a homogeneous peak on ion exchange chromatography or a single spot on a paper chromatogram. We have noticed in the past that hydroxyamino acids, particular γ-hydroxyamino acids, *e.g.*, γ-hydroxyvaline, will give two peaks during ion exchange chromatography using the method of PARTRIDGE AND BRIMLEY. Acid catalysis during the operation synthesizes some of the lactone which then runs as a basic rather than a neutral amino acid. The same is true for paper chromatography at certain pH's; where the lactone has formed, this will appear as a separate spot on the chromatogram even though the crystalline starting material was perfectly pure.

WINITZ: Lactone formation is a rather common phenomenon among the γ- and δ-hydroxyamino acids. Perhaps what should be done in an instance such as this is to make the observation, just as you have done, and then run a comparison with a highly pure synthetic preparation to ascertain whether or not it gives the same phenomenon. In this manner, you can unequivocally establish the purity of the natural material.

POLLARD: Actually, the phenomenon of lactone formation has its uses, since the formation of these lactones is frequently employed in proposing a tentative structural assignment, prior to eventual synthesis and resolution of the appropriate stereoisomer.

E. ROBERTS: I would like to compliment Dr. WINITZ on having done well an absolutely necessary job. One thing that many of us who were brought up in the old school have noticed is that the chemistry is getting more careless in many laboratories. While work has been accelerated greatly by paper-chromatographic methods, the precision of chemical characterization occasionally has become comparably more vague. This kind of discussion at the outset of a conference at which the occurrence of new substances may be discussed and spots on chromatograms given names is very appropriate.

II. PLANTS

THE SOLUBLE, NITROGENOUS CONSTITUENTS OF PLANTS

F. C. STEWARD AND J. K. POLLARD

Department of Botany, Cornell University, Ithaca, N.Y. (U.S.A.)

This account relates to developments, made since the first use of amino acid chromatography, largely from one laboratory; in this way, certain trends may be emphasized. However, reference may and should be made to other more complete reviews (*e.g.*, STEWARD AND THOMPSON[30]; STEWARD, ZACHARIUS AND POLLARD[33]; STEWARD AND POLLARD[29]; GREENSTEIN AND WINITZ[14]) for fuller citations and for more complete reference to the contributions from other laboratories than can here be given. Reference may also be made to the volume on nitrogen metabolism, subedited by MOTHES[20a].

In view of the importance of what plants do with nitrogen, the incomplete knowledge and often ingenuous interpretation of nitrogen metabolism, even in recent times, has been surprising. The discovery of asparagine (1806) and later of glutamine (1883) focused attention upon these amides, which with arginine (especially in conifers) accumulate as nitrogen-rich storage or mobile substances (see CHIBNALL[8] for historical citations). The views of PFEFFER, SCHULZE, and later PRIANISHNIKOW gave to ammonia and the amides an important place in sequences of protein synthesis and breakdown and recognized that these nitrogen-rich compounds, with organic acids derived from sugar as the carbon skeletons, represented the organic starting material for protein metabolism. It was, however, through the work and writings of VIRTANEN, CHIBNALL, and VICKERY that the keto acids, which are salient intermediates or the Krebs cycle, first appeared as the principal "ports of entry" for nitrogen into organic combinations; in this way the gap between carbohydrate and nitrogen metabolism was bridged. SCHULZE and PRIANISCHNIKOW spent much of their respective careers in the attempt to detect the free occurrence of the amino acids which are necessary to build protein, and they believed them to be formed from ammonia which PRIANISCHNIKOW termed the α and ω of nitrogen metabolism. However, the discovery of group transfer reactions upheld the amino acids alanine (from pyruvic acid), aspartic acid (from oxaloacetic acid) and glutamic acid (from α-ketoglutaric acid) as the main starting points, the "Grundaminosäuren" of VIRTANEN, for nitrogen metabolism.

Despite the combined activities of SCHULZE and PRIANISCHNIKOW, they listed in 1906 only ten amino acids as capable of occurring free in plants and even these had been demonstrated usually singly in selected situations. This list (arginine, histidine, isoleucine, leucine, lysine, phenylalanine, proline, tryptophane, tyrosine and valine), strangely, did not contain aspartic and glutamic acids, for they were then regarded when found free as apt to be decomposition products of the generally occurring amides. Thus early attention was focused on the detection in plants of those free amino acids which are protein building blocks and, apart from work directed to the recognition of many of the nitrogen bases and intermediates of alkaloid synthesis, surprisingly

References p. 42

little was known in the first decade of this century about many of the now familiar simple nitrogenous compounds that do occur free in plants.

The big change in this position came about through the applications of chromatography and of the use of isotopic carbon in the study of metabolism. The technique of two-directional chromatography on paper provided the means to separate even closely related compounds and such reagents as the ninhydrin and Ehrlich's reagent, or the technique of RYDON AND SMITH[25], provide sufficiently general means to detect on paper many even unsuspected compounds, and the use of ^{14}C-labeling techniques

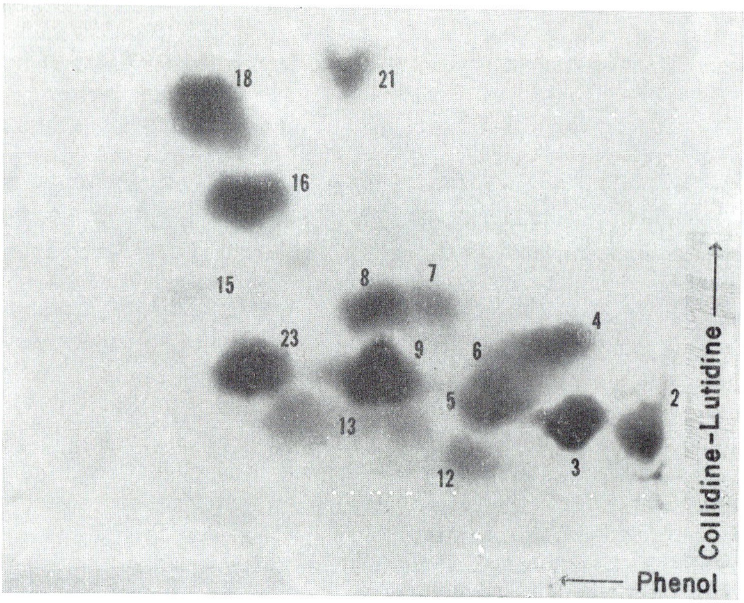

Fig. 1. The ninhydrin-reactive, alcohol-soluble nitrogen compounds of the potato tuber. Reproduction from an original Kodachrome picture (somewhat scarred) of a chromatogram dating back to the period of DENT, STEPKA AND STEWARD[10] (1947). The spots are designated by number according to the convention then adopted, namely: 2, aspartic acid; 3, glutamic acid; 4, serine; 5, glycine; 6, asparagine; 7, threonine; 8, alanine; 9, glutamine; 12, lysine; 13, arginine; 15, proline; 16, valine; 18, leucines; 21, tyrosine; 22, β-alanine; 23, γ-aminobutyric acid; 18, leucine and isoleucine, not here separated, but *cf.* Fig. 7.

also permit the separated compounds to be detected by their carbon skeletons instead of by their functional groups.

The rich harvest to follow was foreshadowed when phenol: collidine–lutidine chromatograms were first made on alcohol extracts of the potato tuber. Although a major investigation had been made of the amides of the potato tuber (STEWARD AND STREET[31]) to show that asparagine and glutamine accounted for all the amide nitrogen present, this information and much more could now readily be obtained by chromatography in a matter of hours. When all other expected compounds had been accounted for on the first potato chromatograms (see Fig. 1), these extracts disclosed other substances, the most prominent of which was quickly ascribed to γ-aminobutyric acid. In rapid succession, other compounds were seen and later identified in extracts of other plants and organs, *e.g.* pipecolic acid in the green bean and γ-methyleneglutamine in tulip bulbs (*cf.* refs. 29, 30).

From this beginning, large numbers of discrete compounds have been detected on paper chromatograms and their chromatographic characteristics have been recorded on "maps" (STEWARD, ZACHARIUS AND POLLARD[33]). Moreover, as methods of converting the free keto acids to their amino acid analogues were adopted, the number of these keto acids to be recognized increased greatly, far beyond those that were known from their presence as Krebs-cycle intermediates[29]. As more plants and families have been examined, the number of these unidentified substances has become very large indeed, but concomitantly an ever expanding number have been isolated and critically identified. The surprise is that so many of the recently discovered substances are of low molecular weight and that their presence was totally unexpected in plants, even in common food plants.

Systematic surveys of plant families by the newer methods are few. One such, on the the Liliaceae, was made with special reference to seven recently discovered nitrogenous compounds (pipecolic acid, hydroxypipecolic acid, γ-methyleneglutamine and its acid, γ-methylglutamic acid, γ-methyl-γ-hydroxyglutamic acid, γ-aminobutyric acid and azetidinecarboxylic acid), all of which were known to be present in some liliaceous plants. Although the distribution of these compounds in the family was very variable, some like γ-methyleneglutamine being confined to a few species, the family also yielded 45 unidentified substances (FOWDEN AND STEWARD[13]). Also, the members of the Leguminoseae, conspicuous because of their role in the discovery of the substituted piperidines and the γ-substituted glutamyl compounds are also found to contain many other compounds which have yet to be identified as shown on the "map" of compounds recognized in this laboratory by ZACHARIUS[36] (1952) and later by BARRALES[3] (1959, see Fig. 2), or the maps published from VIRTANEN's laboratory[20] (MIETTINEN, 1955).

Precision in the mapping of the positions of compounds as they occur on chromatograms is achieved by rigorous control of the variables that apply during the chromatography (especially temperature); the use of pure solvents and precise mixtures; and at least one of the now somewhat standard combinations such as phenol saturated at pH 5.5, followed by collidine–lutidine (1 : 3) saturated with water, or by butanol–acetic acid–water (9 : 1 : 2.9). Even so, it is better to refer the position of a substance to that of an internal standard like alanine (R. alanine) than to the solvent front (R_F). Position alone, even when recorded in several solvents, is only a guide to the identity of a component. The final identification requires supplementary techniques, and there is no general substitute for the isolation and characterization of a pure substance, followed by synthesis and critical matching with the unknown. Where two asymmetric centers exist in a given compound (for examples see GROBBELAAR, POLLARD AND STEWARD[16]), the task is not completed until the ultimate stereoisomeric configuration is known (*cf.* ref. 14 and the contribution of WINITZ to this Symposium).

A few selected sequences of compounds recognized or identified in this way will be mentioned in order to emphasize certain general conclusions which can now be drawn.

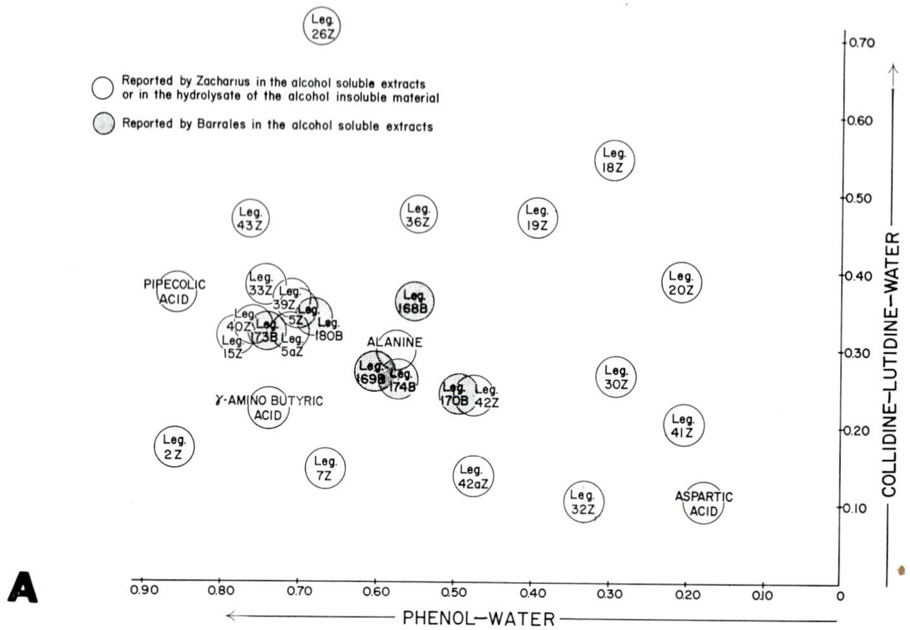

Fig. 2 A. Map showing the location on phenol: collidine–lutidine chromatograms of the unidentified ninhydrin-reactive compounds that occur in legumes.

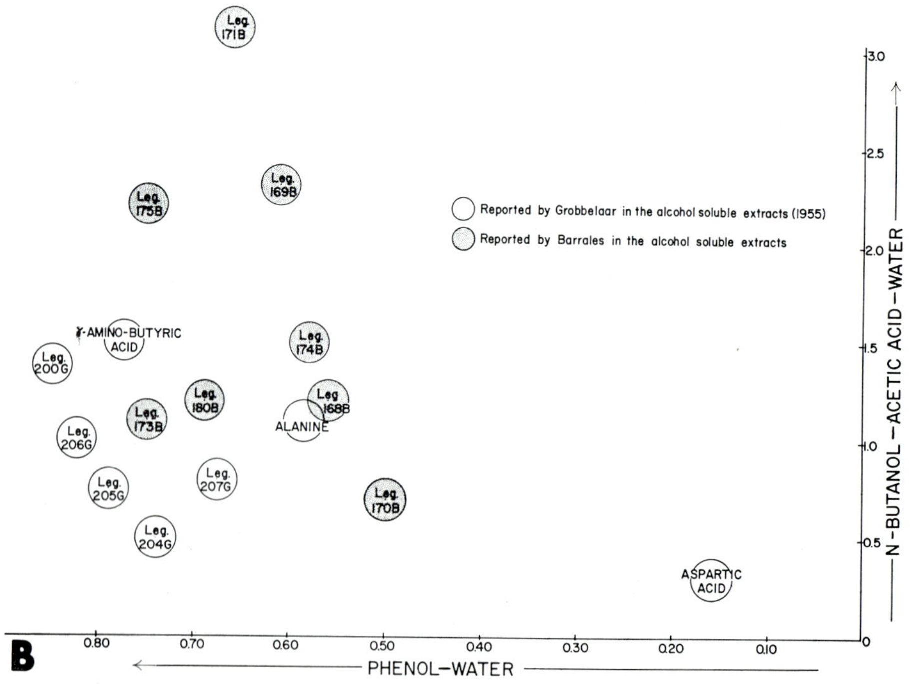

Fig. 2 B. Map showing the location on phenol–acetic acid–butanol chromatograms of the unidentified ninhydrin-reactive compounds that occur in legumes.

References p. 42

γ-Aminobutyric acid (γ-AB)

Long known as a bacterial byproduct, γ-aminobutyric acid was first detected (DENT, STEPKA AND STEWARD[10], *cf.* Fig. 1) and later proved (THOMPSON, POLLARD AND STEWARD[34]) to be prominent in higher plants from work done initially on the potato tuber, where it is one of the most prominent compounds in the soluble nitrogen fraction. The compound soon appeared to be ubiquitous in the plant kingdom and was one of the first non-α-amino, non-protein amino acids to be conspicuous. Although structurally related to glutamic acid, as if by decarboxylation of the α-COOH, it now appears that it is not formed in this way in carrot cultures, for it readily gives rise to glutamic acid, or rather to glutamine, whereas glutamine (or glutamic acid) supplied exogenously does not easily form γ-AB *in vivo* (STEWARD, BIDWELL AND YEMM[27]). It is now believed that γ-AB originates from a nitrogen-free "port of entry", probably succinic semi-aldehyde and that its nitrogen may be incorporated eventually in protein, though not of course as γ-AB which has never been demonstrated to be a protein constituent. Thus a compound first noticed in higher plants as a spot on a chromatogram in 1947 is now recognized as a well authenticated and widespread plant product with unexpected metabolic implications, and it has also received very active consideration in the metabolism of brain and other tissues (ROBERTS AND BREGOFF[24]; BAXTER AND ROBERTS[4]). The range of possible or demonstrable reactions involving γ-AB were summarized by the authors (1956), and since then the presumed reactions through succinic acid and its semi-aldehyde to γ-AB have been substantiated by enzymatic studies (BACHRACH[1]).

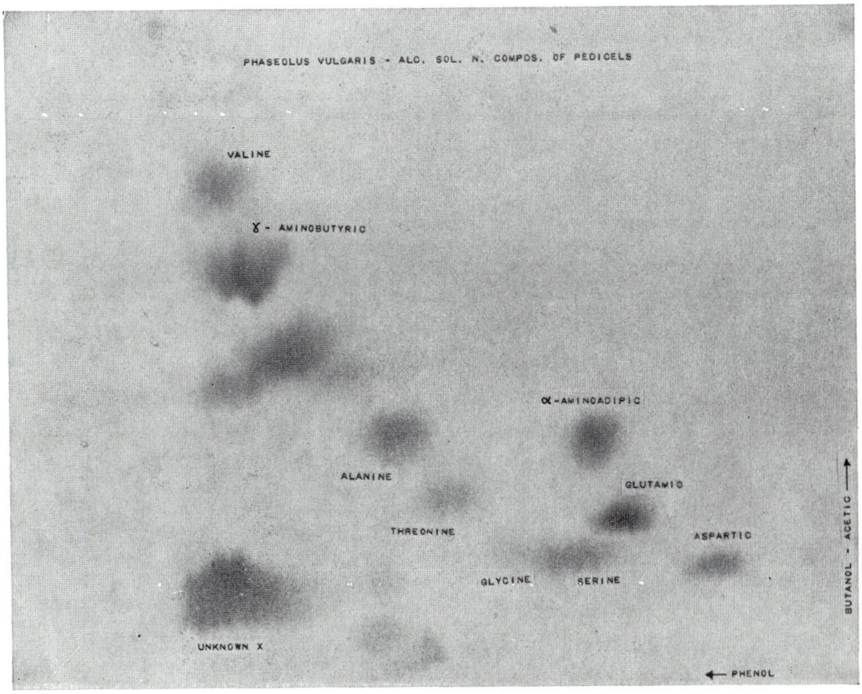

Fig. 3. Chromatogram to show the presence of α-aminoadipic acid, which became radioactive when ^{14}C-lysine was supplied.

References p. 42

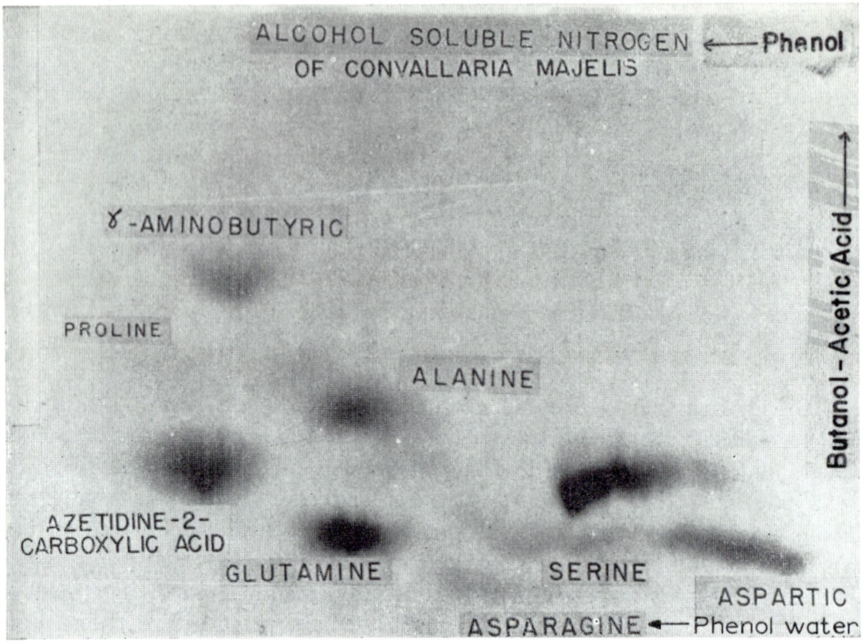

Fig. 4. The ninhydrin-reactive, alcohol-soluble compounds of the lily of the valley *(Convallaria majelis)*. Note: azetidine-2-carboxylic acid (brown).

Pipecolic acid; the substituted piperidines; cyclization from diamino acids

Piperidine-2-carboxylic acid was seen by ZACHARIUS (STEWARD AND THOMPSON[30]; ZACHARIUS[36]) in this laboratory in extracts of the ovary wall and seed of the green bean and later identified (ZACHARIUS, THOMPSON AND STEWARD[37]); it arises by elimination of ammonia and by ring formation from lysine (GROBBELAAR AND STEWARD[17]). From this point this acid has been reported from many angiosperms. Apparently lysine which is diverted to pipecolic acid is lost to protein synthesis, for any subsequent metabolism seems to proceed via α-aminoadipic acid (seen by GROBBELAAR[15], 1955, in the pedicels of *Phaseolus*, as shown in Fig. 3) and thence to glutaric acid

TABLE I

DISTRIBUTION OF SOME RECENTLY IDENTIFIED AMINO ACIDS THROUGHOUT THE LILIACEAE[13]

Amino acid	*Found in*	
	Species	*Genera*
γ-Methyleneglutamic acid	23	7
γ-Methyleneglutamine	17	2
γ-Methylglutamic acid	19	6
γ-Hydroxy-γ-methylglutamic acid	16	6
γ-Hydroxyglutamic acid	3	2
Azetidine-2-carboxylic acid	23	17
Pipecolic acid	9	9

References p. 42

which may be metabolized. The analogy thus disclosed between lysine and pipecolic acid, with glutamic acid and proline, now finds an unexpected but possible parallel in the compounds with four carbon atoms, as suggested by the unexpected new compound azetidine-2-carboxylic acid (FOWDEN[11, 12]; FOWDEN AND STEWARD[13]), which

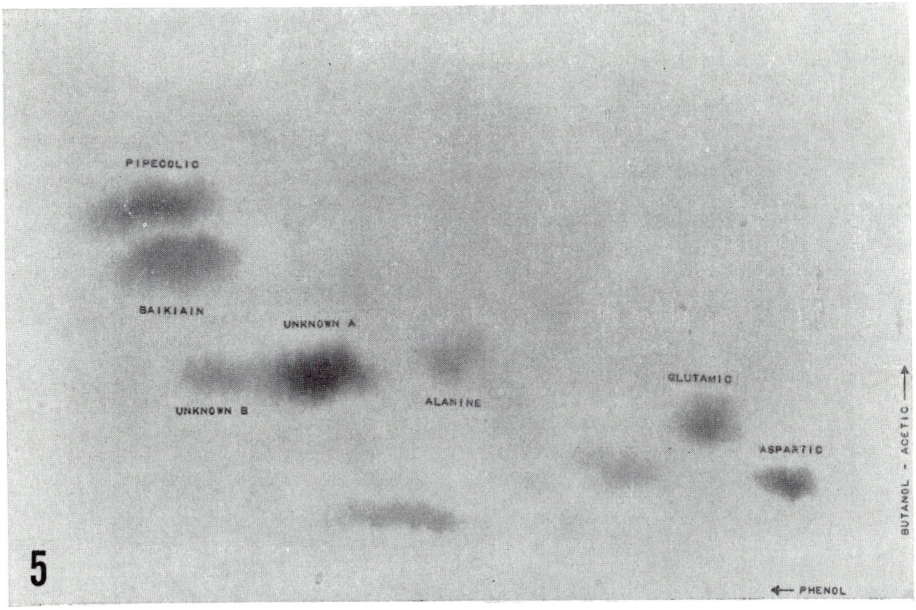

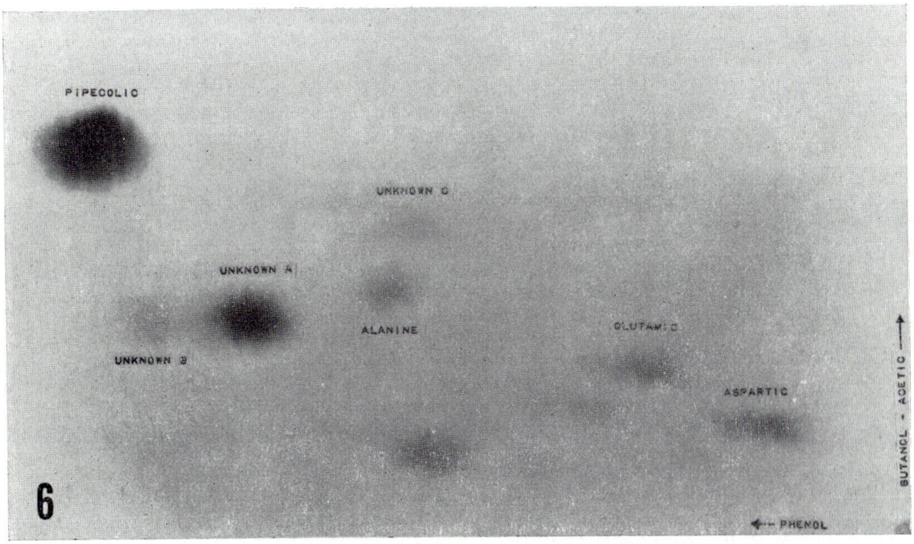

Figs. 5 and 6. The ninhydrin-reactive, alcohol-soluble compounds of the seeds of *Baikiaea plurijuga*. Fig. 5: Prior to reduction. Fig. 6: After catalytic hydrogenation; note the conversion of baikiain (brown) to pipecolic acid (purple). "Unknown A" (purple) is now known to be 5-hydroxypipecolic acid (*cf.* Fig. 7).

References p. 42

occurs in quantity in *Convallaria* (see Fig. 4) and in many other liliaceous plants (see Table I).

Substitution in the piperidine ring, foreshadowed by the discovery of baikiain (see Fig. 5), is seen by the presence of 5-hydroxypipecolic acid in legumes (Figs. 5 and 6) and in dates (Fig. 7) not to be a isolated occurrence. As yet, the analogous situation in the proline and azetidine rings are not known, for hydroxyproline seems to arise in other ways (POLLARD AND STEWARD[23]). The photographs of the typical chromatograms (which are here published as Figs. 4–8) not only show the location of these

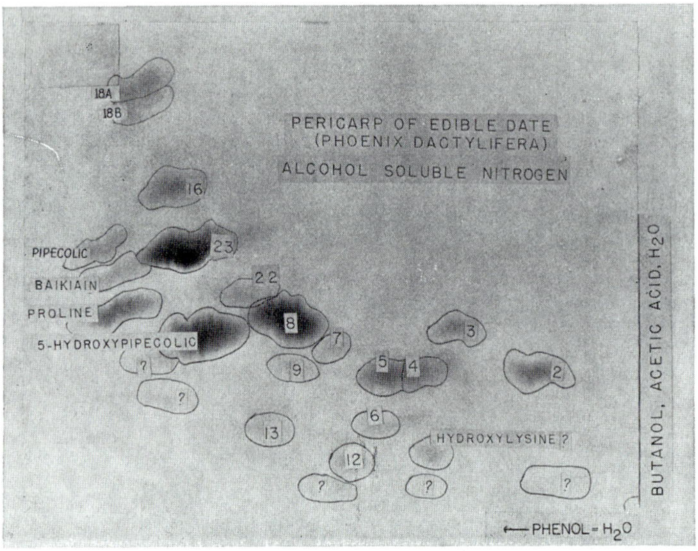

Fig. 7. The ninhydrin-reactive, alcohol-soluble compounds of the edible date. Spots designated by numbers 2–23 follow the same convention as in Fig. 1, but on this chromatogram leucine (18B) and isoleucine (18A) were separated. Note the occurrence of three piperidine derivatives.

compounds on papers prepared from extracts of the plants in question, but they emphasize how prominent these compounds are in the soluble nitrogen fraction of the organs in question.

The above shows that attention was early directed to the legumes because of the new compounds that were disclosed when extracts of these plants were chromatographed. Other examples of compounds disclosed in this way will be mentioned below. Maps, compiled by BARRALES[3] of unidentified ninhydrin-reactive compounds that had been revealed up to 1959 in this laboratory alone by the work of ZACHARIUS[36], GROBBELAAR[15] and BARRALES (see Fig. 2) had upon them some 39 distinct locations.

The rich variety of free nitrogen compounds to be detected in this way can be illustrated by observations made on phenol–H_2O: butanol–acetic acid chromatograms of *Saraca indica* seeds (Fig. 8) by BARRALES[3] with one of us (F.C.S.) This chromatogram showed three clearly defined but unidentified substances; some of the more usual constituents of such extracts; the now not unexpected and conspicuous amount

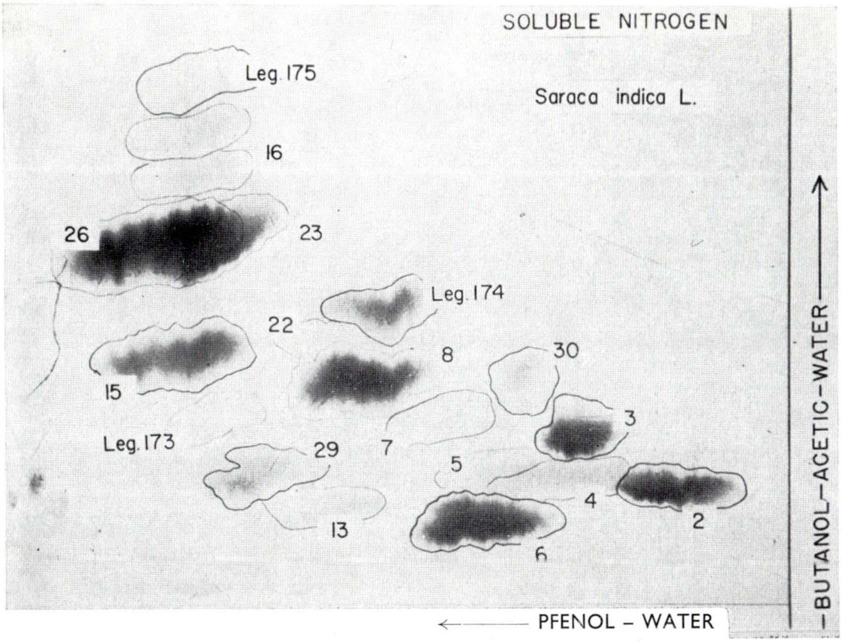

Fig. 8. The ninhydrin-reactive, alcohol-soluble compounds of the seeds of *Saraca indica L.* Spots numbered 2–23 follow the convention of Fig. 1; No. 26 represents pipecolic acid; No. 29 (brown) represents γ-methyleneglutamine; No. 30 (brown) represents γ-methyleneglutamic acid. Spots designated Leg. Nos. 173–175 represent unidentified substances encountered in this and other legumes (*cf.* Fig. 2).

of pipecolic acid and, in addition, two compounds that can be recognized from work which is now to be summarized, as γ-methyleneglutamine (No. 29 on the figure) and γ-methyleneglutamic acid (No. 30).

γ-Substituted glutamyl compounds

Discovered first in legumes and in tulip, γ-methyleneglutamic acid and its amide seemed as though they could be derived theoretically from the aldol condensation of two molecules of pyruvic acid. However, another compound now proves to be of more general significance. Although prepared from the keto acid (pyruvic aldol) before it was known to occur in nature, this amino acid was discovered first in *Adiantum* (Fig. 9, STEWARD, WETMORE AND POLLARD[32]). This compound, as synthesized from pyruvic aldol, is γ-hydroxy-γ-methylglutamic acid (GROBBELAAR, POLLARD AND STEWARD[16]). As in the case of the hydroxypipecolic acids, here also there are 4 diastereoisomeric possibilities, and it is not yet known which, or how many, of these are naturally occurring.

From the γ-hydroxy-γ-methylglutamic acid, however, many possible metabolic pathways may stem, leading to a wide variety of metabolites (STEWARD AND POLLARD[28]). On careful examination it transpires that this compound occurs in small amount on chromatograms of extracts of tulip (Fig. 10), although to show this the chromatograms needed to be somewhat overloaded with respect to the other con-

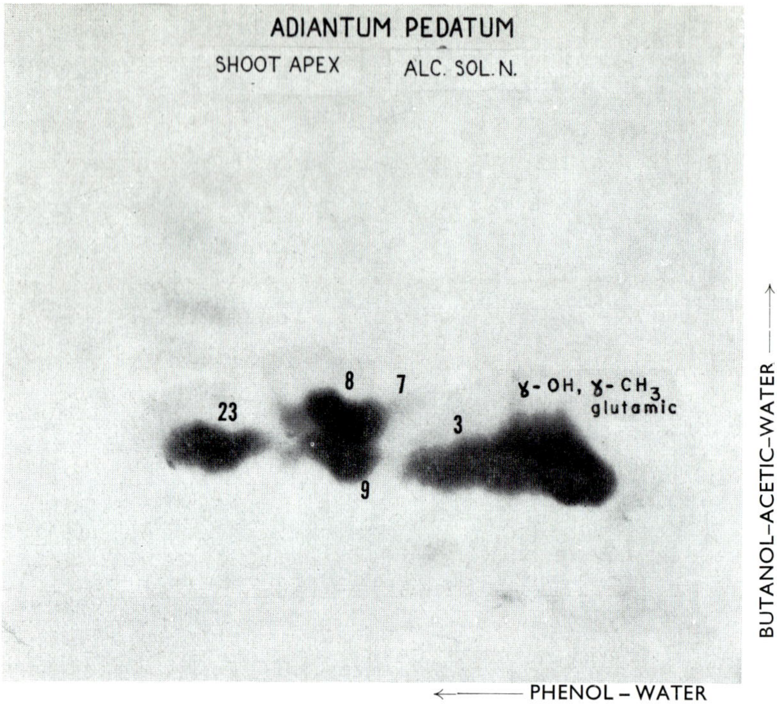

Fig. 9. The ninhydrin-reactive, alcohol-soluble compounds of the shoot apex of the maiden hair fern *(Adiantum pedatum)* as shown on a phenol: collidine–lutidine chromatogram. Compounds designated by Nos. 2–23 follow the convention of Fig. 1.

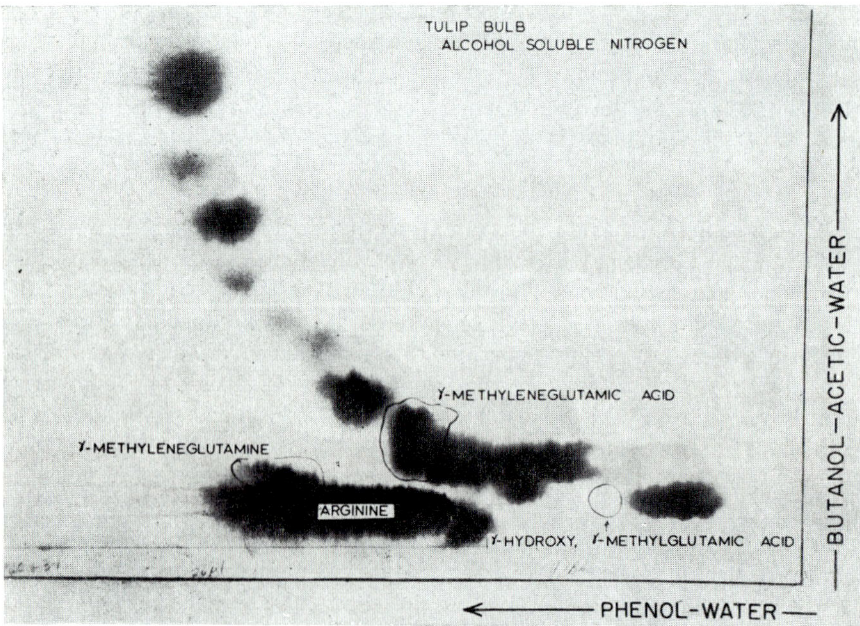

Fig. 10. The ninhydrin-reactive, alcohol-soluble compounds of the tulip bulb. Chromatogram arranged to show the designated γ-glutamyl compounds.

References p. 42

stituents which include the γ-methyleneglutamyl compounds; the latter are probably derived by loss of water from γ-hydroxy-γ-methylglutamic acid.

Obviously a metabolite which occurs in such diverse plants as a fern *(Adiantum)* and a flowering plant *(Tulipa)* must play a somewhat general role. In fact, it is surprising how widespread the distribution is of some compounds that owed their discovery to chromatography. A survey of the Liliaceae permitted one to show how frequently such compounds as those listed in Table I occurred in the species and genera of this family; and, as the figures show, these compounds are of such wide distribution that they may be regarded as general metabolites which happen to accumulate in detectable amounts in certain plants or organs or under certain genetic or nutritional conditions. Despite this, however, the Liliaceae revealed a baffling number of distinct ninhydrin-reacting compounds of which 45 were recognized in the 88 plants of this family that were examined[13].

The long list of new, or recently discovered, natural nitrogen compounds cannot be fully mentioned here. It has been reviewed on a number of occasions (*cf.* refs. 14, 29). It includes new hydroxyamino acids (γ-hydroxyvaline and homoserine), an *o*-acetyl derivative as of homoserine; new sulphur compounds (for summary see ref. 12) and several new derivatives of glutamine (*cf.* ref. 28), some of which are as yet only partially identified (*cf.* the contribution of THOMPSON to this symposium).

Some hydroxy amino acids and their derivatives

On a phenol–butanol–acetic acid chromatogram (Fig. 11 from a chromatogram made by GROBBELAAR[15]) of the shoot of *Pisum*, homoserine can be clearly seen and also

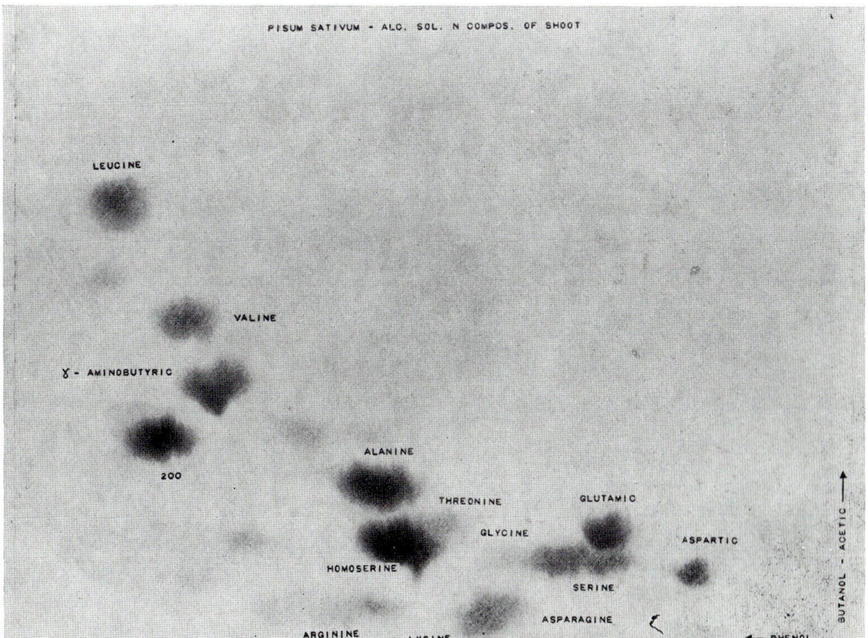

Fig. 11. The ninhydrin-reactive, alcohol-soluble compounds of the pea *(Pisum sativum L.)*. Note: The occurrence of homoserine and of another unidentified substance (No. 200) now known to be *o*-acetylhomoserine.

References p. 42

another compound (labeled 200 on the chromatogram). When isolated, the compound designated 200 proved to be *o*-acetylhomoserine, and its constitution was confirmed by GROBBELAAR AND STEWARD[18]. The relations of this substance to homoserine are shown in the formulae below:

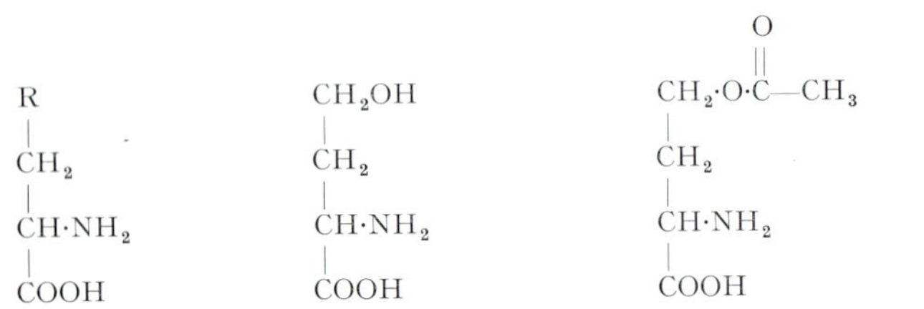

Homoserine   *o*-Acetylhomoserine

One may now recognize an array of compounds which substitute different groups "R" in what would otherwise be alanine. R is CH_3 in the familiar four-carbon acid α-aminobutyric acid and a variety of other groups give the known compounds, shown in Table II.

TABLE II

SOME NATURALLY OCCURRING DERIVATIVES OF HOMOSERINE AND ASPARTIC ACID

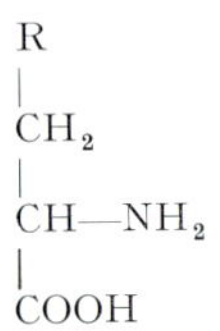

The formula on the left illustrates the basic four-carbon α-amino acid structure of this family of compounds. The indicated modifications at the γ-carbon produce the natural products listed.

"R" equals	*Derivative*
$—CH_3$	α-Aminobutyric acid
$—CH_2OH$	Homoserine
—CHO	Aspartic, β-Semialdehyde
—COOH	Aspartic acid
$—CH_2NH_2$	α, γ-Diaminobutyric acid
$—CH_2—O—Ac$	*o*-Acetylhomoserine
$—CH_2—O—NH_2$	Canaline
$—CH_2—O—NH—C(=NH)—NH_2$	Canavanine
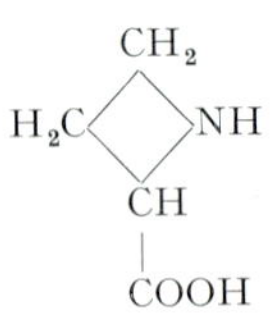	Azetidine-2-carboxylic Acid

In a species of Kalanchoe *(K. daigremontiana)* compounds were seen on chromatograms (Nos. 75 and 78 on Figs. 12 A and B) which proved to be two forms of the same substance because, when isolated from columns, the same substance was obtained in two fractions of the eluate. It subsequently transpired that these substances were

References p. 42

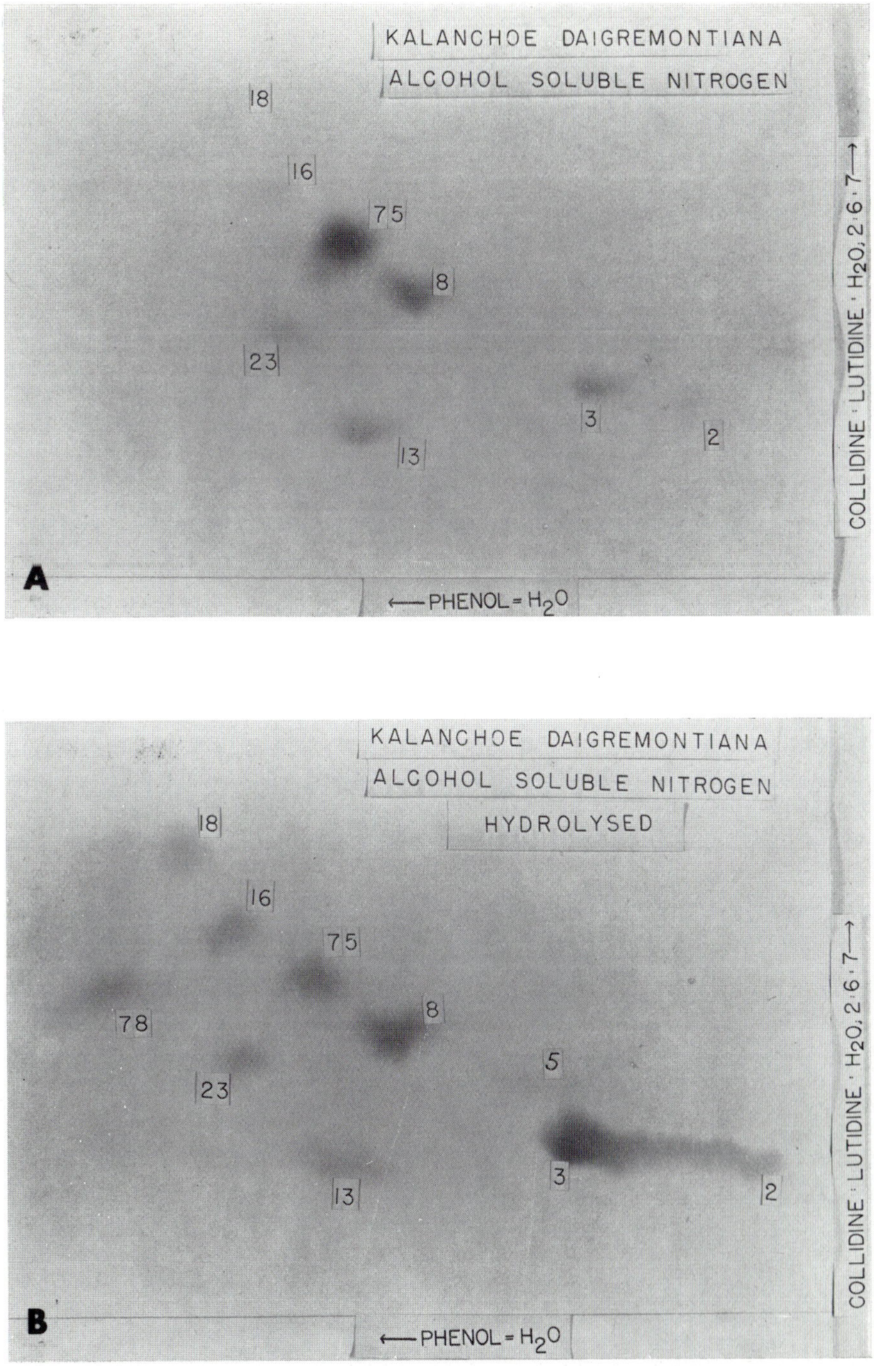

Fig. 12. The ninhydrin-reactive, alcohol-soluble compounds of *Kalanchoe daigremontiana*; the compounds designated by Nos. 2–23 follow the convention of Fig. 1. A, chromatogram showing γ-hydroxyvaline (purple) as No. 75. B, chromatogram showing both γ-hydroxyvaline (No. 75) and its lactone (purple) as No. 78 after acid hydrolysis.

References p. 42

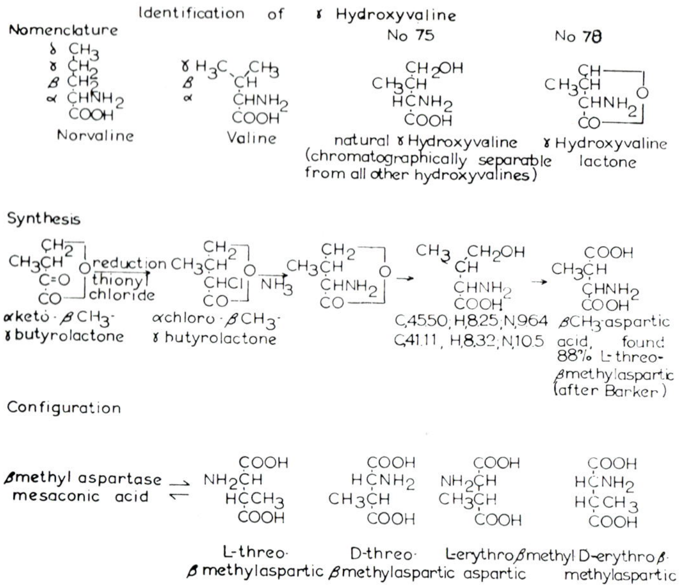

Fig. 13. The structure and synthesis of natural γ-hydroxyvaline and its lactone and certain criteria used to prove its identity.

related as a free hydroxyamino acid (hydroxyvaline) and its lactone. The synthesis of γ-hydroxyvaline by SONDHEIMER (working with the authors, see Fig. 13) established the identity of the natural compound as the γ-hydroxy compound by chromatographic comparison of the natural products, both free base and lactone, with the synthetic material (POLLARD, SONDHEIMER AND STEWARD[22]). When further evidence

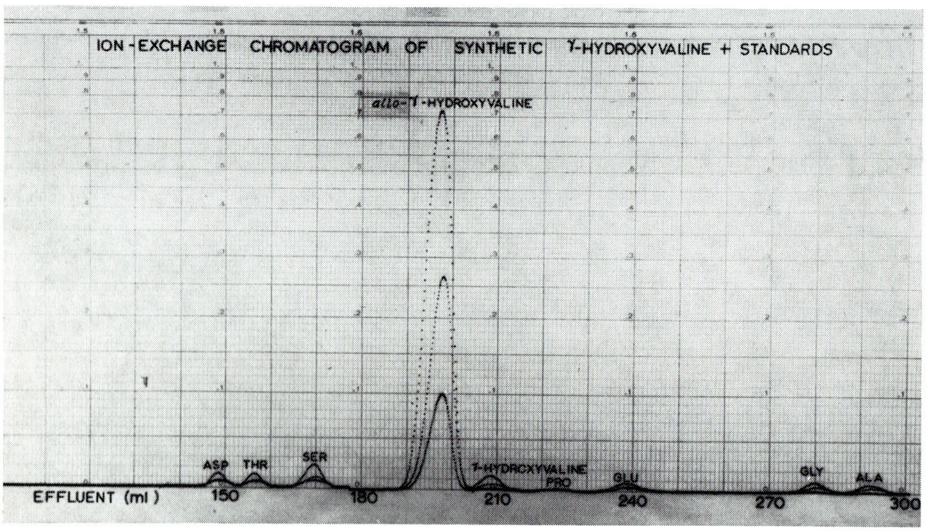

Fig. 14. Resolution of synthetic γ-hydroxyvaline into γ- and allo-γ-hydroxyvaline on ion-exchange resin. (Reference amino acids were inserted in the sample at low concentration.)

was sought by co-chromatography of natural and synthetic products on ion-exchange resins by the method of SPACKMAN, STEIN AND MOORE[26], the evidence in Fig. 14 was obtained. This shows that the synthetic product was 98% the unnatural diastereomer, namely allo-γ-hydroxyvaline, and it contained only 2% of the natural diastereomer. These two compounds had superimposed upon paper chromatograms. Final proof requires a synthesis of sufficient of the natural diastereomer to permit optical resolution and direct comparison of the optically active forms with the natural product.

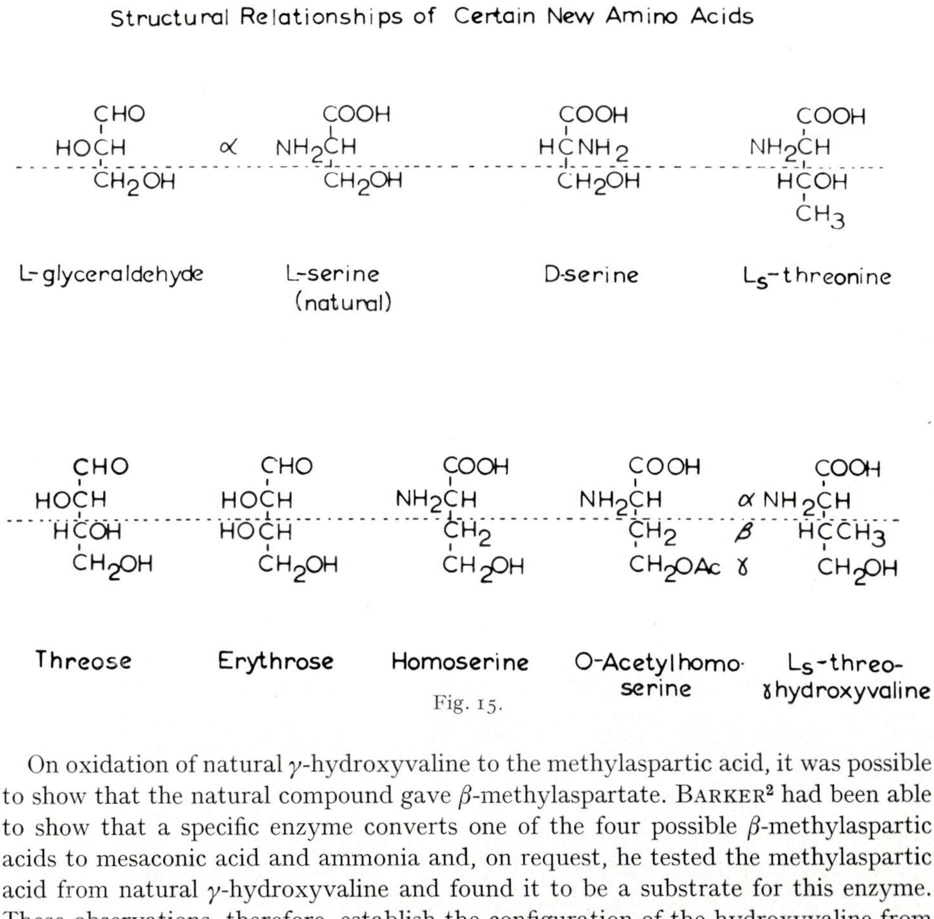

Fig. 15.

On oxidation of natural γ-hydroxyvaline to the methylaspartic acid, it was possible to show that the natural compound gave β-methylaspartate. BARKER[2] had been able to show that a specific enzyme converts one of the four possible β-methylaspartic acids to mesaconic acid and ammonia and, on request, he tested the methylaspartic acid from natural γ-hydroxyvaline and found it to be a substrate for this enzyme. These observations, therefore, establish the configuration of the hydroxyvaline from Kalanchoe, and its relations to serine and to threose are shown in the chart above, together with the structure of homoserine and its *o*-acetyl compound (Fig. 15).

It is now possible to state the structure of another hydroxy acid which occurs in plants. From apples, an amino acid was isolated by URBACH and by HULME. These two isolates were shown to be identical by their infrared-absorption spectra (HULME AND STEWARD[19]), but there was no conclusive distinction between the two alternative structures A and B (see p. 40).

By the application of mass spectroscopy to these compounds, BIEMANN has now shown that the masses corresponding to formula B as shown at C establish formula B as

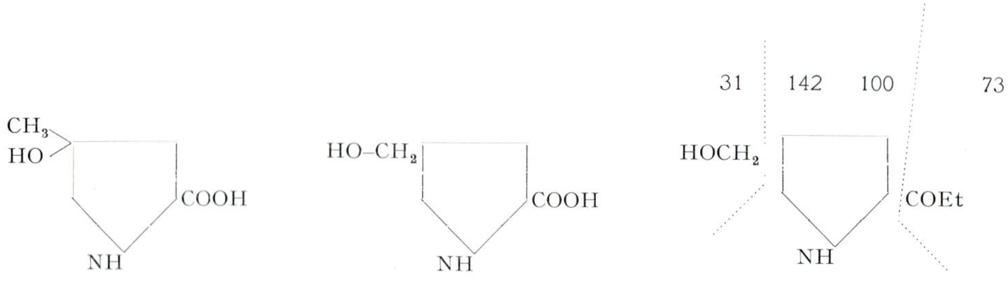

(A) (B) (C)

the structure of this compound (BIEMANN, DEFFNER AND STEWARD[6], 1961). Therefore, although the compound corresponding to structure A has not yet been found in plants, it may occur.

A compound present in *Phlox* (Fig. 16), which also occurs in *Hemerocallis* (Fig. 17), yields upon hydrolysis hydroxyglutamic acid, ammonia and a still unidentified carbon residue. The hydroxyglutamic acid also occurs free in *Phlox* (Fig. 16; *cf.* VIRTANEN AND HIETALA[35]). A recent isolation by one of us (J.K.P.) has permitted the unequivocal proof of the identity of this acid as γ-hydroxyglutamic acid. Since WINITZ has recently resolved the hydroxyglutamic acids into its four stereoisomers, it is now, therefore, possible to identify the natural product with its synthetic counterpart (*cf.* the contribution of WINITZ to this Symposium, and also BENOITON *et al.*[5]).

The list of new nitrogenous compounds, with unexpected structure, continues to mount. A new histidine isomer which contains the pyrazole ring, isolated and proven

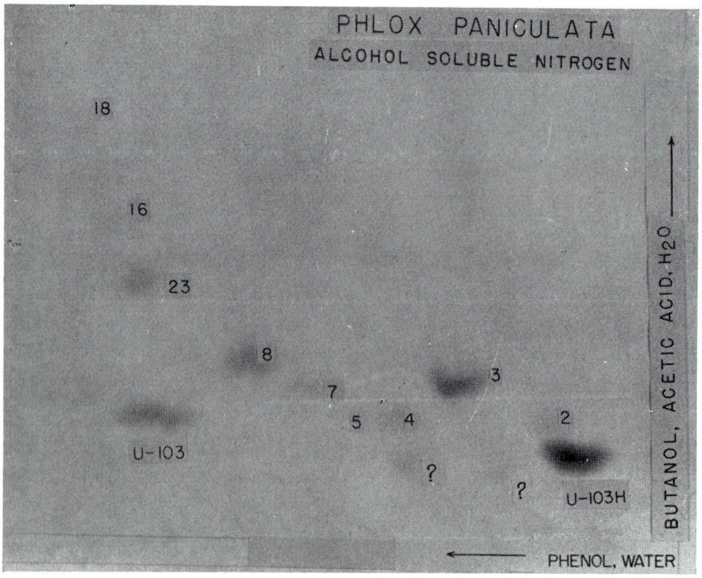

Fig. 16. The ninhydrin-reactive, alcohol-soluble compounds of the seeds of *Phlox paniculata*. Compounds designated by Nos. 2–23 follow the convention of Fig. 1. U-103H is now known to be γ-hydroxyglutamic acid. U-103 is now known to yield γ-hydroxyglutamic acid, ammonia and an unidentified residue on hydrolysis.

References p. 42

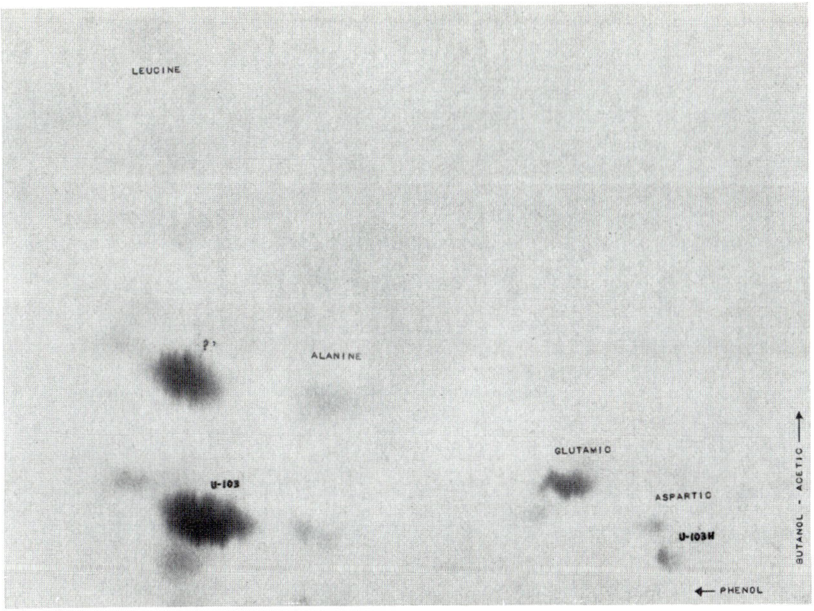

Fig. 17. The occurrence, in an extract of the fruits of *Hemerocallis* sp., of γ-hydroxyglutamic acid and the compound U-103 which also occurs in *Phlox* (*cf.* Fig. 16).

by NOE AND FOWDEN[21], and cases of new S-amino acids are examples. Also, other compounds which contain both the cyclopropane ring, the first example of which was disclosed by BURROUGHS[7], have come to light (see the communication to this Symposium by FOWDEN) and still other compounds with the hitherto rare terminal methylene group.

One should recognize, however, that what has been said above for the compounds detectable by ninhydrin can be extended to other classes of compounds. Ehrlich's reagent (dimethylaminobenzaldehyde) makes possible the detection of an array of ureides and of indole compounds, and both of these classes of compounds are more copiously represented in plants than at first was thought. Some nitrogen bases, detectable by ninhydrin (putrescine, tyramine, tryptamine, ethanolamine), occur as evident decarboxylation products, often under conditions of anomalous nutrition, but the simple nitrogen bases and betaines may well prove to be more widely occurring than is commonly supposed. Nitro compounds, of which β-nitropropionic acid in *Indigofera* (COOKE[9]) was an early example, may also be reduced and detected on paper as amino compounds.

The significance of the array of nitrogen compounds in plants

If ever the idea that the soluble nitrogen pool merely comprised the pre-fabricated compounds needed for condensation into protein was plausible, it cannot be so now. Many compounds not known to be used directly in protein synthesis exist, and they often occur in large amount. Their presence suggests that many normal metabolic routes have passed unsuspected. One compound or another may accumulate in a local situation by virtue of a metabolic block brought about by genetic, nutritional, or environmental means, or often it may be due to circumstances inherent in normal

References p. 42

development. In fact, some of the compounds in question may well function as inhibitors or antimetabolites which arrest metabolic sequences that would otherwise occur. This possibility exists in a number of storage organs in which growth and protein synthesis are arrested so that soluble nitrogenous compounds may accumulate.

ACKNOWLEDGEMENTS

In presenting this brief survey as the background for further discussion of free amino acids in plants, the authors speak for, and acknowledge their debt to, several collaborators who have worked in the senior author's laboratory, but who may not all be mentioned by name.

REFERENCES

1 U. BACHRACH, *Biochem. J.*, 77 (1960) 417.
2 H. A. BARKER, R. D. SMYTH, E. J. WAWSZKIEWICS, M. N. LEE AND R. M. WILSON, *Arch. Biochem. Biophys.*, 78 (1958) 468.
3 H. L. BARRALES, *Ph. D. Thesis*, Cornell University, Ithaca, N.Y., 1959.
4 C. F. BAXTER AND E. ROBERTS, *J. Biol. Chem.*, 233 (1958) 1135.
5 L. BENOITON, M. WINITZ, S. M. BIRNBAUM AND J. P. GREENSTEIN, *J. Am. Chem. Soc.*, 79 (1957) 6192.
6 K. BIEMANN, G. G. J. DEFFNER AND F. C. STEWARD, *Nature*, 191 (1961) 380.
7 L. F. BURROUGHS, *Nature*, 179 (1957) 360.
8 A. C. CHIBNALL, in *Protein Metabolism in the Plant*, Yale University Press, New Haven, 1939, p. 54.
9 A. R. COOKE, *Arch. Biochem. Biophys.*, 55 (1955) 114.
10 C. E. DENT, W. STEPKA AND F. C. STEWARD, *Nature*, 160 (1947) 682.
11 L. FOWDEN, *Nature*, 176 (1955) 347.
12 L. FOWDEN, *Ann. Repts. on Progr. Chem. (Chem. Soc. London)*, 56 (1959) 359.
13 L. FOWDEN AND F. C. STEWARD, *Ann. Botany (London)*, 21 (1957) 53.
14 J. P. GREENSTEIN AND M. WINITZ, in *Chemistry of the Amino Acids*, J. Wiley and Sons, New York, 1961, p. 25.
15 N. GROBBELAAR, *Ph. D. Thesis*, Cornell University, Ithaca, N.Y., 1955.
16 N. GROBBELAAR, J. K. POLLARD AND F. C. STEWARD, *Nature*, 175 (1955) 703.
17 N. GROBBELAAR AND F. C. STEWARD, *J. Am. Chem. Soc.*, 75 (1953) 4341.
18 N. GROBBELAAR AND F. C. STEWARD, *Nature*, 182 (1958) 1358.
19 A. C. HULME AND F. C. STEWARD, *Nature*, 175 (1955) 703.
20 J. K. MIETTINEN, *Ann. Acad. Sci. Fennicae, Ser. A. II.*, 60 (1955) 520.
20a K. MOTHES, (subed.), *Encyclopedia of Plant Physiology: Nitrogen Metabolism*, Vol VIII, (1958).
21 F. F. NOE AND L. FOWDEN, *Nature*, 184 (1959) B.A. 69.
22 J. K. POLLARD, E. SONDHEIMER AND F. C. STEWARD, *Nature*, 182 (1958) 1356.
23 J. K. POLLARD AND F. C. STEWARD, *J. Exptl. Botany*, 10 (1959) 17.
24 E. ROBERTS AND H. M. BREGOFF, *J. Biol. Chem.*, 201 (1953) 393.
25 H. N. RYDON AND P. W. G. SMITH, *Nature*, 169 (1952) 922.
26 D. H. SPACKMAN, W. H. STEIN AND S. MOORE, *Anal. Chem.*, 30 (1958) 1190.
27 F. C. STEWARD, R. G. S. BIDWELL AND E. W. YEMM, *Nature*, 178 (1956) 734, 789.
28 F. C. STEWARD AND J. K. POLLARD, in W. D. MCELROY AND B. GLASS (Eds.), *Inorganic Nitrogen Metabolism*, The Johns Hopkins Press, Baltimore, 1956, p. 377.
29 F. C. STEWARD AND J. K. POLLARD, *Ann. Rev. Plant Physiol.*, 8 (1957) 65.
30 F. C. STEWARD AND J. F. THOMPSON, *Ann. Rev. Plant Physiol.*, 1 (1950) 233.
31 F. C. STEWARD AND H. E. STREET, *Plant Physiol.*, 21 (1946) 155.
32 F. C. STEWARD, R. H. WETMORE AND J. K. POLLARD, *Am. J. Botany*, 42 (1955) 946.
33 F. C. STEWARD, R. M. ZACHARIUS AND J. K. POLLARD, *Ann. Acad. Sci. Fennicae, Ser. A. II.*, 60 (1955) 321.
34 J. F. THOMPSON, J. K. POLLARD AND F. C. STEWARD, *Plant Physiol.*, 28 (1953) 401.
35 A. I. VIRTANEN AND P. K. HIETALA, *Acta Chem. Scand.*, 9 (1955) 175.
36 R. M. ZACHARIUS, *Ph. D. Thesis*, University of Rochester, Rochester, N.Y. 1952.
37 R. M. ZACHARIUS, J. F. THOMPSON AND F. C. STEWARD, *J. Am. Chem. Soc.*, 74 (1952) 2949.

SOME RECENTLY-CHARACTERIZED AMINO ACIDS FROM PLANTS

L. FOWDEN AND D. O. GRAY

Department of Botany, University College, London (Great Britain)

During the last 15 years a gradual re-appraisal of the overall character of the free amino acid pools of plants has been necessary. In this time the number of amino and imino acids recognized as plant constituents has increased from some 25–30 (mainly those present also in proteins) to about 100. Each year several new acids are isolated and characterized and at present there seems to be no foreseeable limit to the number that will be recognized ultimately. Indeed, the situation regarding the amino acids of plants is approaching rapidly in complexity that encountered with the alkaloids.

Several reviews of this subject are available, the most recent being that of FOWDEN[1], which cites the earlier accounts. The present paper attempts to augment these with brief descriptions of acids either reported within the last year or at present under investigation in our laboratory.

Although the amount of available information on plant amino acids is increasing continually, it is still not possible to make many generalisations regarding their distribution. The protein amino acids are present normally in the free amino acid pools but in relative amounts that are very different for different species, or even for different organs of the same plant. For many of the other acids, the distribution within and between families is apparently haphazard. Some acids do seem to be especially characteristic of certain plant families, *e.g.* citrulline for the Cucurbitaceae, and azetidine-2-carboxylic acid for the Liliaceae (FOWDEN AND STEWARD[2]), whilst callus-type growth sometimes results in a characteristic production of particular acids, *e.g.* lysopine in crowngall tissue from salsify *(Scorzonera hispanica)*, tobacco *(Nicotiana tabacum)*, Virginia creeper *(Parthenocissus tricuspidata)*, and Jerusalem artichoke *(Helianthus tuberosus)* (BIEMANN *et al.*[3]), and hydroxyproline, present in the protein, of rapidly growing tissue explants of carrot root and potato tuber (STEWARD, THOMPSON AND POLLARD[4]). The largest proportion of the newer acids have been isolated from members of the families Liliaceae, Leguminosae and Cucurbitaceae.

The metabolic interrelationships of the majority of the more recently characterized acids are shrouded with uncertainty; few of them have been integrated into metabolic schemes that may be regarded as part of the basic pattern of nitrogen metabolism of plants.

ACIDS ISOLATED RECENTLY FROM MACROSCOPIC PLANTS

Isolations of three new sulphur-containing amino acids have been reported. *S*-(2-Carboxy-1-methylethyl)-L-cysteine (I) was obtained from seeds of *Acacia millefolia* and *Acacia willardiana* (GMELIN AND HIETALA[5]), whilst VIRTANEN[6] has reported that *S*-*n*-propyl-L-cysteine sulphoxide (II, dihydroalliin) occurs in onion bulbs *(Allium cepa)*. Chondrine (III, L-1,4-thiazan-3-carboxylic acid 1-oxide) has been

References p. 53

isolated from the red alga, *Chondria crassicaulis* (KURIYAMA, TAKAGI AND MURATA[7]). The latter acid is structurally similar to cycloalliin (IV, 5-methyl-L-1,4-thiazan-3-carboxylic acid 1-oxide) described earlier as a constituent of onion bulbs by VIRTANEN AND MATIKKALA[8].

$$\left.\begin{matrix}HOOC\cdot CH_2\\ CH_3\end{matrix}\right\rangle CH\cdot S\cdot CH_2\cdot CH(NH_2)\cdot COOH \qquad CH_3\cdot CH_2\cdot CH_2\cdot \underset{\underset{O}{\downarrow}}{S}\cdot CH_2\cdot CH(NH_2)\cdot COOH$$

I II

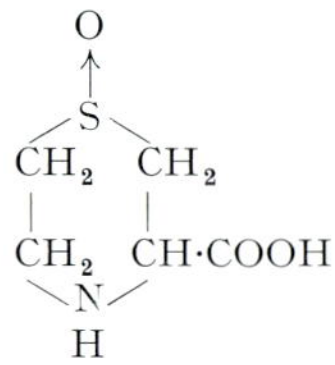

III

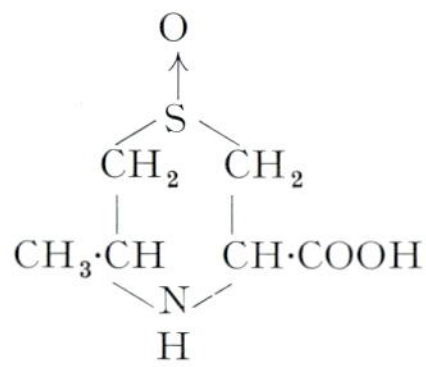

IV

Additions have been made to the group of substituted amides occurring in plants. JADOT, CASIMIR AND RENARD[9] have isolated N^5-*p*-hydroxyphenyl-L-glutamine (V) from the fungus, *Agaricus hortensis*. They have demonstrated also that N^5-ethyl-L-glutamine (VI, theanine), previously isolated from leaves of the tea plant (SAKATO[10]), occurs in another fungus, *Xerocomus badius* (CASIMIR, JADOT AND RENARD[11]). Another new γ-glutamyl derivative, β-*N*-(γ-L-glutamyl)-4-hydroxymethylphenylhydrazine (VII, trivial name agaritine) has been isolated from *Agaricus bisporus* by LEVENBERG[12] who has shown that soluble extracts of this mushroom contain a highly active enzyme that cleaves VII to yield L-glutamic acid and 4-hydroxymethylphenylhydrazine. N^4-Ethyl-L-asparagine (VIII) has been obtained from the squirting cucumber, *Ecballium elaterium* (GRAY AND FOWDEN[13]), whilst another member of the family Cucurbitaceae, *Bryonia dioica* (white bryony), contains both N^4-ethyl-L-asparagine and N^4-(2-hydroxyethyl)-L-asparagine (IX).

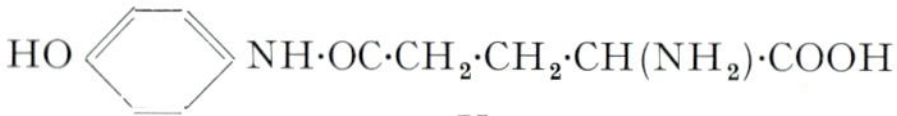

V

$$CH_3\cdot CH_2\cdot NH\cdot OC\cdot CH_2\cdot CH_2\cdot CH(NH_2)\cdot COOH$$

VI

VII

$$CH_3\cdot CH_2\cdot NH\cdot OC\cdot CH_2\cdot CH(NH_2)\cdot COOH \qquad HO\cdot CH_2\cdot CH_2\cdot NH\cdot OC\cdot CH_2\cdot CH(NH_2)\cdot COOH$$

VIII IX

Other new acids isolated include lysopine (X) (LIORET[14]; BIEMANN *et al.*[3]), an N^2-substituted derivative of lysine bearing a close structural analogy with octopine (derived from scallop and octopus muscle; KNOOP AND MARTIUS[16]; MORIZAWA[15]), which is the corresponding derivative of arginine. α-(Methylenecyclopropyl)glycine (XIV, see p. 48) has been isolated from the seeds of *Litchi chinensis*; lathyrine

References p. 53

($C_7H_{10}O_2N_4$), which is probably a heterocyclic amino acid whose ring opens upon hydrogenation to give an amino guanidino or a substituted amino guanidino acid, is present in seeds of *Lathyrus tingitanus* (BELL[17]), whilst 4-aminopipecolic acid (XI) has been isolated from *Strophanthus scandens* (SCHENK AND SCHÜTTE[18]).

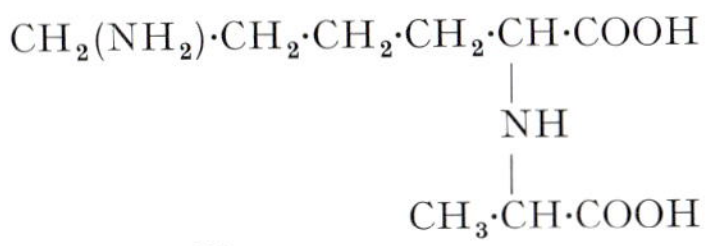

X

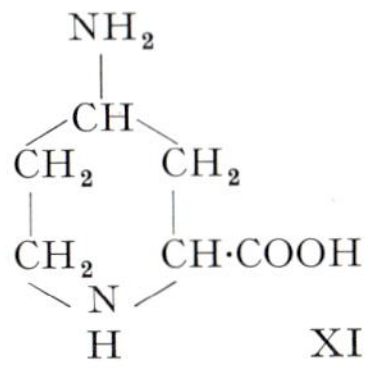

XI

Table I lists some of the properties of these newly isolated compounds.

Amino Acids of the Cucurbitaceae

The family of plants has been a source from which several non-protein amino acids have been isolated.

Citrulline was isolated many years ago from the juice of water melons, *Citrullus vulgaris* (WADA[19]). More recently it has been isolated from members of other plant families, *e.g.* nodules of alder, *Alnus* (MIETTINEN AND VIRTANEN[20]), and from red algae (KURIYAMA, TAKAGI AND MURATA[21]). During the course of our work on the substituted asparagines of *Bryonia*, we have isolated approx. 3 g of citrulline from 20 kg of fresh shoots.

β-Pyrazol-1-ylalanine (XII) was isolated more recently from seeds of water melon (NOE AND FOWDEN[22, 23]), and shown to be present in the seeds of the following additional members of the Cucurbitaceae: *Cucurbita pepo* (vegetable marrow), *Cucurbita ficifolia*, *Cucumis melo* (melon), *Cucumis sativa* (cucumber), *Ecballium elaterium*, *Bryonia dioica*, and *Echinocystis lobata*. This compound provided the first instance of a naturally occurring pyrazole derivative and its structure suggests that it may act as an antimetabolite of histidine. Perhaps this possibility stimulated attempts at chemical synthesis for several methods are now available (FINAR AND UTTING[24]; NOE AND FOWDEN[23]; SUGIMOTO, WATANABE AND IDE[25]; TAKESHITA, KIMURA, YABUUCHI, NISHIZUKA AND HAYAISHI[26]; also private communications from Professor A. E. BRAUNSTEIN, Moscow, and Dr. G. TALBOT, Saskatoon).

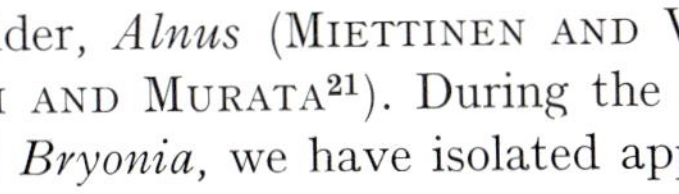

XII

No definite biosynthetic pathway is yet known but experiments in which pyrazole was supplied to germinating seeds of *Cucumis melo* suggested that the heterocyclic base may be utilized directly (NOE AND FOWDEN[23]). Biosynthesis may be found ultimately to resemble that of another heterocyclic derivative of alanine, namely tryptophan from indole. One pathway for the bio-degradation of β-pyrazol-1-ylalanine,

References p. 53

TABLE I

SOME PROPERTIES OF RECENTLY-ISOLATED AMINO AND IMINO ACIDS

Amino or imino acid	*m.p. (°C)**	*$[\alpha]_D$ in H_2O***	*$[\alpha]_D$ in HCl***	*R_F values Aqueous phenol****	*But–Acet§*	*Ninhydrin§§ colour*
S-(2-Carboxy-1-methylethyl)-L-cysteine	202	+6.6 (*c* = 2)	1 *N* +31 (*c* = 1.94)	0.48 (+)	0.35	BV
S-*n*-Propyl-L-cysteine sulphoxide	—	—	—	0.81 (+)	0.28	—
Chondrine	255–257 (d)	+20.9 (*c* = 2)	6 *N* +30.1 (*c* = 2)	0.70 (+)	0.09	B → Br
N^5-*p*-Hydroxyphenyl-L-glutamine	231.5	—	1 *N* +30 (*c* = 0.2)	0.75 (−)	—	V
β-*N*-(γ-L-glutamyl)-4-hydroxy-methylphenylhydrazine	(d)	+26.2 (*c* = 0.65)	—	—	—	V
N^4-Ethyl-L-asparagine	243 (d)	−5 (*c* = 2.4)	—	0.74 (+)	0.26	Br
N^4-(2-Hydroxyethyl)-L-asparagine	199–203 (d)	−2.9 (*c* = 5)	—	0.65 (+)	0.11	Br
N^4-Methyl-L-asparagine	241–244 (d)	−4.2 (*c* = 5.5)	—	0.66 (+)	0.16	Br
Lysopine	157–160	+18 (*c* = 14)	—	0.75 (+)	0.09	V
α-(Methylenecyclopropyl)glycine	(d)	+83.5 (*c* = 3)	5 *N* +110 (*c* = 2.5)	0.74 (+)	0.41	Br → V
4-Aminopipecolic acid	—	—	—	0.85 (+)	—	Y → Gr → GB
Lathyrine	215 (d)	−55.9 (*c* = 12)	—	0.57 (+)	0.13	OR

* (d) indicates melting with decomposition.
** *c*, concentration in %.
*** Normally water-saturated phenol was used. The (+) and (−) indicate presence and absence of NH_3 vapour, respectively.

§ The compositions of *n*-butanol–acetic acid–water mixtures used in various laboratories differ slightly. The values quoted are relative to alanine, R_F = 0.22.

§§ Key to colours: B, blue; V, violet; Br, brown; Y, yellow; Gr, green; G, grey; OR, orange-red.

demonstrated in a species of *Pseudomonas*, led to the production of pyrazole, ammonia and pyruvic acid (TAKESHITA, NISHIZUKA AND HAYAISHI[27]); here again a ready analogy with tryptophan metabolism may be drawn for tryptophanase action yields indole, ammonia and pyruvic acid (HOPKINS AND COLE[28]).

β-Pyrazol-1-ylalanine has a slight ability to undergo transamination with α-keto-glutarate or pyruvate when disrupted mitochondria obtained from etiolated seedlings of *Citrullus vulgaris* or *Cucumis melo* were used as a source of transaminases (Table II).

TABLE II

TRANSAMINATION BETWEEN AMINO ACIDS

Shows the amounts of transamination occurring in 1 h at 37° between various amino acids and α-ketoglutarate or pyruvate when reactions were catalysed by extracts of mitochondria obtained from *Citrullus vulgaris* and *Cucumis melo* seedlings. Transamination is expressed as µmoles glutamate or alanine formed per ml of reaction mixture. Reaction mixtures contained the following amounts of substances per ml: α-ketoglutarate or pyruvate (adjusted to pH 7.4), 50 µmoles; L-amino acids, 50 µmoles; potassium phosphates, pH 7.4, 50 µmoles.

Amino acid donor	*µmoles glutamate formed from α-ketoglutarate*		*µmoles alanine formed from pyruvate*
	Citrullus vulgaris extract	*Cucumis melo extract*	*Citrullus vulgaris extract*
β-Pyrazol-1-ylalanine	0.91	0.25	0.37
Aspartic acid	6.7	4.2	2.4
Alanine	8.6	7.1	—
Histidine	0.94	0.31	0.31
γ-Aminobutyric acid	0.41	0.32	0.39

N^4-Ethyl-L-asparagine and N^4-(2-hydroxyethyl)-L-asparagine. Extracts of certain species examined during the pyrazolylalanine investigations also contained compounds giving brown ninhydrin reacting spots on paper chromatograms. Two of these have now been characterized as N^4-ethyl-L-asparagine (VIII) and N^4-(2-hydroxyethyl)-L-asparagine (IX). The former was isolated first from the green parts of *Ecballium elaterium*; yield about 1 g from 4.5 kg fresh material (GRAY AND FOWDEN[13]). It has been obtained more recently in larger amount from shoots of *Bryonia dioica* (7.1 g from 20 kg plant material). During the latter isolation, 1.2 g of N^4-(2-hydroxyethyl)-L-asparagine was obtained. The compounds responsible for the other brown spots on chromatograms of bryony extracts remain to be identified. However, it is apparent already that plants may contain a series of aspartic acid or asparagine derivatives that may be shown ultimately to be as extensive as the class of compounds now known to be based on glutamic acid or glutamine.

Compounds VIII and IX were decomposed by acid hydrolysis; VIII gave an equimolar mixture of aspartic acid and ethylamine whilst IX yielded aspartic acid and ethanolamine. Both compounds were hydrolysed considerably more slowly than asparagine by 2 N HCl solutions at 100° and evidence obtained from paper chromatograms suggested that intermediate compounds, possibly cyclized forms of VIII and IX, were present during the earlier part of the hydrolysis period.

References p. 53

The metabolism of the substituted amides has not been investigated in plants as yet. It would seem reasonable to assume that deamidase-type enzymes would hydrolyse them to give products identical with those obtained after acid hydrolysis. Biosynthetic pathways leading to the amides may be dependent upon synthetase-type enzymes utilizing aspartic acid and ethylamine or ethanolamine as substrates (compare glutamine and asparagine synthetases), or upon transferase-type enzymes where asparagine and the appropriate amine would act as substrates (compare the action of glutaminotransferase upon mixtures of glutamine and hydroxylamine).

Extracts of fresh rat liver, which are known to possess strong asparaginase activity (GREENSTEIN AND CARTER[29]), hydrolysed compounds VIII and IX more slowly than asparagine; aspartic acid together with either ethylamine or ethanolamine were identified as reaction products. Although the degradation of asparagine is also catalysed by transaminases when keto acids are added to liver extracts (GREENSTEIN AND CARTER[29]; MEISTER, SOBER, TICE AND FRASER[30]), the addition of pyruvate did not cause a noticeable increase in the rate of breakdown of the substituted asparagines.

The two amides were synthesized chemically by a similar method in which an ethanolic solution of 4-ethyl hydrogen-L-aspartate, prepared by the method of CURTIS AND KOCH[31], was heated for several hours at 100° with either ethylamine or ethanolamine. Natural and synthetic products were shown to be identical by comparisons of m.p., R_F, and $[\alpha]_D$ and by infrared spectroscopy.

Our observations have indicated that several additional, unidentified amino acids are present in bryony and in other members of this family of plants.

Cyclopropylamino acids from the family Sapindaceae

Hypoglycin A (XIII), isolated from the fruits of the tropical plant, *Blighia sapida*, has been a centre of interest because it possesses both an intriguing chemical structure and a distinct physiological action in reducing the blood sugar levels of animals (HASSALL AND REYLE[32, 33]; PATRICK[34]). Its structure was elucidated by work in several different laboratories (ANDERSON *et al.*[35]; VON HOLT AND LEPPLA[36]; RENNER, JÖHL AND STOLL[37]; DE ROPP, VAN METER, DE RENZO, MCKERNS, PIDACKS, BELL, ULLMAN, SAFIR, FANSHAWE AND DAVIS[38]; WILKINSON[39]; ELLINGTON, HASSALL AND PLIMMER[40]) and has been confirmed by chemical synthesis. Now we have shown an analogous amino acid, α-(methylenecyclopropyl)glycine (XIV), to be a constituent of the fruit of the subtropical species, *Litchi chinensis*.

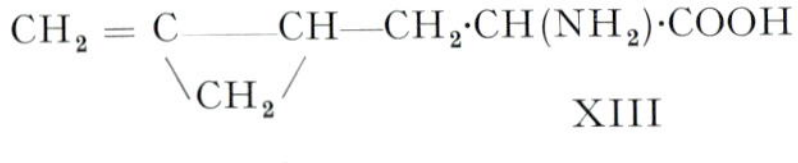
XIII

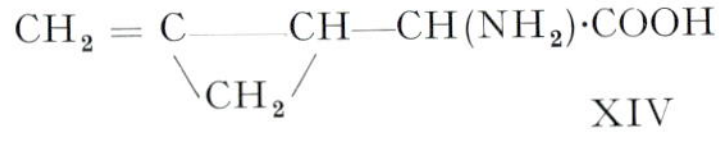
XIV

XIV was detected in extracts of *Litchi* seeds by paper chromatography; the extract contained an acid which moved to a position between γ-aminobutyric acid and valine on chromatograms developed with a *n*-butanol–acetic acid–water mixture (see Table I for R_F in phenol). The ninhydrin colour was brown initially but changed to violet over a period of 2 h. The crude acid (1.35 g) was isolated from 24 kg of fresh seeds (water content approx. 50%) by ion-exchange and paper-partition chromatographic methods. After recrystallization from 70% (v/v) ethanol, 0.68 g of pure compound was obtained; decomposition began at 202°, $[\alpha]_D^{21.5} + 83.5°$ ($c = 3.0\%$ in water), $[\alpha]_D + 110°$ ($c = 2.5\%$ in 5 N HCl).

References p. 53

Elementary analysis indicated the formula $C_6H_9NO_2$, which contains four hydrogen atoms less than a saturated, open-chain C_6 amino acid such as leucine. The presence of a terminal $=CH_2$ was inferred from its infrared spectrum which showed strong absorptions at 5.9 and 11.0 μ (see Fig. 1, curve B). ETTLINGER AND KENNEDY[41] have demonstrated that the exocyclic methylene group as it occurs in Feist's acid (1-methylenecyclopropane-*trans*-2,3-dicarboxylic acid) is characterized by strong infrared-absorption bands at 5.9, 7.1 and 11.0 μ. The 7.1 μ band cannot be used as a criterion for recognition of a $=CH_2$ group in an amino acid for the symmetrical stretching frequency of the $—COO^-$ group gives a superimposed peak at this wavelength. This

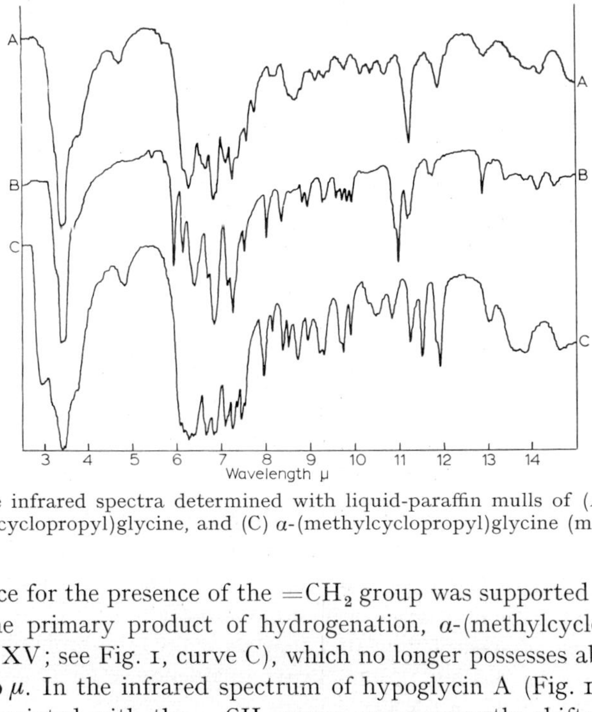

Fig. 1. Shows the infrared spectra determined with liquid-paraffin mulls of (A) hypoglycin A, (B) α-(methylenecyclopropyl)glycine, and (C) α-(methylcyclopropyl)glycine (major isomer, XV).

spectral evidence for the presence of the $=CH_2$ group was supported by the infrared spectrum of the primary product of hydrogenation, α-(methylcyclopropyl)glycine (major isomer, XV; see Fig. 1, curve C), which no longer possesses absorption peaks at 5.9 and 11.0 μ. In the infrared spectrum of hypoglycin A (Fig. 1, curve A), the absorptions associated with the $=CH_2$ group are apparently shifted slightly; the 11.0 μ band of XIV is seen at 11.2 μ for XIII (compare ANDERSON *et al.*[35]) and the absorption band at 5.9 μ presumably is shifted to slightly longer wavelength at 6.2–6.3 μ where it is overlapped by the band attributed to antisymmetrical stretching of the $—COO^-$ group.

The acid was decomposed completely when treated with 5 *N* HCl at 100° for 24 h. The main products included three ninhydrin-reacting compounds but these were not identical with any commonly-encountered amino acids.

The clue to the structure of the acid came from a study of the products obtained during catalytic hydrogenation in the presence of PtO_2 catalyst. The course of the reduction was followed by paper chromatography using a *tert.*-amyl alcohol–water–acetic acid mixture (10 : 10 : 1, v/v; upper phase used). A diagrammatic representation of the results is shown in Fig. 2. The final products included isoleucine (XVII), leucine (XVIII), and norleucine (XIX); isoleucine was formed in greater amount

References p. 53

than the other two isomers. The two earliest reduction products are presumably formed after uptake of only two hydrogen atoms, and represent geometrical isomers of α-(methylcyclopropyl)glycine (XV and XVI). The major isomer (XV, Fig. 2) was isolated by paper chromatographic methods from the mixture of amino acids resulting after hydrogenation of 75 mg of α-(methylenecyclopropyl)glycine. Elementary analysis gave: C, 49.2%; H, 8.7%; N, 9.0%. $C_6H_{11}NO_2{\cdot}H_2O$ requires C, 49.0%; H, 8.8%; N, 9.5%. The reduction sequence may then be written as:

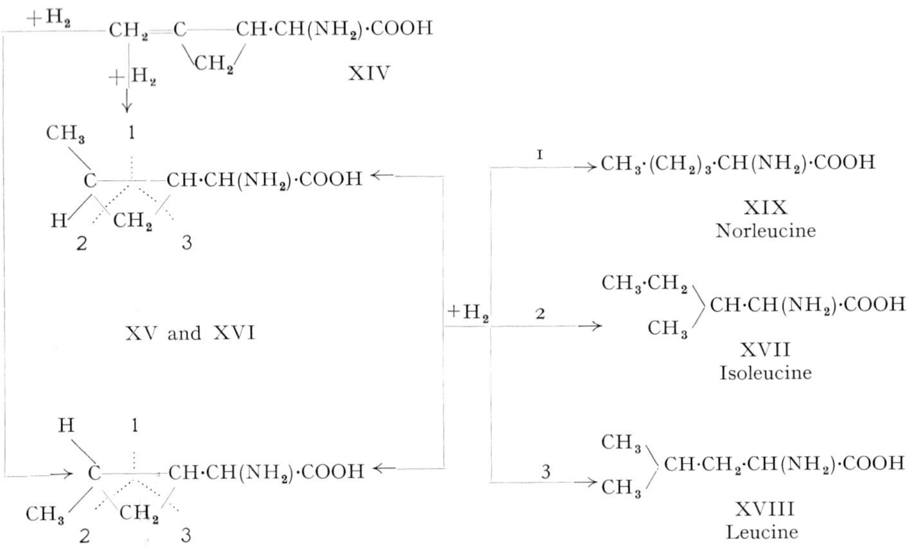

XIV

When hypoglycin A was hydrogenated under identical conditions, the time-course of the reaction showed the production of a similar series of amino acids, each moving slightly faster on paper chromatograms than the corresponding product from XIV, a behaviour characteristic of a series of homo-amino acids.

Whilst the infrared spectra indicated that the C=C was probably exocyclic, the hydrogenation results did not provide critical evidence to eliminate other possible structures, *e.g.* XX, XXI and XXII, below:

XX

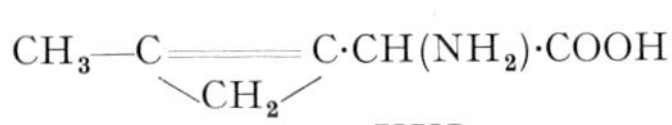

XXI

XXII

However, the nuclear magnetic-resonance spectrum of the compound dissolved in CF_3COOD (determined and interpreted by Dr. ABRAHAM, National Physical Laboratory, Teddington) eliminated these possibilities and supported structure XIV as the only one fitting all the experimental observations.

The stereochemical configuration of the two asymmetric carbon atoms was determined by isolation of the isoleucine produced by catalytic hydrogenation. We have

References p. 53

found that it is possible to separate isoleucine from alloisoleucine on paper chromatograms (Whatman No. 3MM paper) developed for 4 or 5 days in a *tert*.-amyl alcohol–water–acetic acid mixture (20 : 20 : 1, v/v; upper phase used; isoleucine runs slightly faster than alloisoleucine) and in this way the isolate was shown to be isoleucine. This isoleucine had $[\alpha]_D + 43 \pm 6°$ ($c = 0.5\%$ in 5 N HCl); authentic L-isoleucine has $[\alpha]_D + 39.5°$ ($c = 0.5$–2.0% in 5 N HCl). The four stereoisomers of α-(methyl-

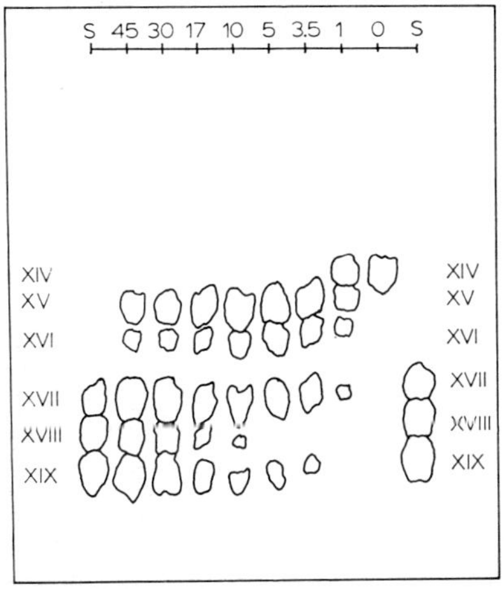

Fig. 2. A diagrammatic representation of a chromatogram, developed in t-amyl alcohol–water–acetic acid (20 : 20 : 1, v/v, upper phase), produced from samples of the reaction mixture obtained after the hydrogenation of α-(methylenecyclopropyl)glycine for various times (expressed in min at the origin of the chromatogram). The spots were identified as: XIV, α-(methylenecyclopropyl)-glycine; XV and XVI, geometrical isomers of α-(methylcyclopropyl)glycine; XVII, isoleucine; XVIII, leucine; and XIX, norleucine.

enecyclopropyl)glycine are represented using Fischer's convention by structures XXIII–XXVI; the appropriate isoleucine isomer that would be obtained after hydrogenation is listed below each representation. Structure XXIV represents the natural isomer isolated from *Litchi* seeds.

The L-configuration at the α-carbon atom is supported by the observation that the value of $[\alpha]_D$ measured for an α-(methylenecyclopropyl)glycine solution in 5 N HCl had a higher (+) value than when the determination was made in aqueous solution.

As yet we have not been able to assign exact stereochemical configurations to the

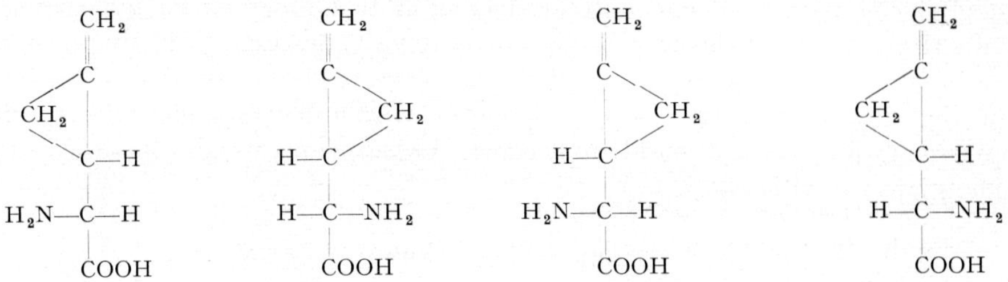

XXIII D-isoleucine | XXIV L-isoleucine | XXV D-alloisoleucine | XXVI L-alloisoleucine

References p. 53

two α-(methylcyclopropyl)glycine isomers (XV and XVI) produced during the hydrogenation.

Compound XIV was present only in the seeds of the fruits; no trace of XIV was found in the fleshy arils which form the edible portion of the fruit^. A small scale survey of the leaves of some members of the Sapindaceae has been made but neither XIV nor hypoglycin A has been detected in twelve genera examined. The survey will be extended to seeds of other members of the family as they become available.

The possibility that α-(methylenecyclopropyl)glycine exerts hypoglycaemic activity has been tested using mice. The limited amounts of material available prevented the

TABLE III

BLOOD-SUGAR LEVELS AND α-(METHYLENECYCLOPROPYL)GLYCINE

Reducing sugars of blood taken from starved mice 6.5 h after subcutaneous injection of varying doses of α-(methylenecyclopropyl)glycine.

α-(Methylenecyclopropyl)glycine injected (mg/kg body wt.)	*Blood-sugar levels. (mean of two determinations*, mg%)*
0	103**
0	71**
60	83
130	62
230	57
400	35

* Individual determinations show variations of ± 2 mg% about the mean values.
** Values in the range 55–112 mg% have been observed by us for control starved animals in other experiments.

use of larger animals such as rats, guinea-pigs and kittens used in the experiments with hypoglycin A (HASSALL AND REYLE[32, 33]; PATRICK[34]). Mice (approx. wt., 30 g) were starved for 22 h and then solutions of α-(methylenecyclopropyl)glycine were injected subcutaneously. The amounts supplied ranged up to 400 mg/kg body wt. The mice were killed 6.5 h after receiving the injections. Blood-sugar levels determined are shown in Table III. A fairly clear progressive reduction of blood-sugar levels occurred as increasing doses were supplied to the mice. The general activity of the mice was related to the dose; the animals receiving the higher doses were moribund at the end of the 6.5-h period. The hypoglycaemic activity of α-(methylenecyclopropyl)glycine is perhaps not quite as striking as that observed for hypoglycin A in rats (PATRICK[34]) but this may be due partly to varying species' sensitivity. Observations on the glycogen contents and the histology of the livers of mice receiving α-(methylenecyclopropyl)glycine are in progress. Higher doses produced livers which appeared much whiter and considerably enlarged when compared with those of control animals injected with water.

ACKNOWLEDGEMENTS

We would like to acknowledge grants-in-aid for this work from the Central Research Fund of the University of London. D. O. GRAY held an Agricultural Research Council

References p. 53

postgraduate studentship during the course of the work. We are grateful for the assistance given by Dr. R. D. HARKNESS in the animal experiments, and Miss J. CHESSUN as a technical assistant.

REFERENCES

1 L. FOWDEN, *Repts. Progr. Appl. Chem.*, 56 (1960) 359.
2 L. FOWDEN AND F. C. STEWARD, *Ann. Botany (London), N.S.*, 21 (1957) 53.
3 K. BIEMANN, C. LIORET, J. ASSELINEAU, E. LEDERER AND J. POLONSKY, *Biochim. Biophys. Acta*, 40 (1960) 369.
4 F. C. STEWARD, J. F. THOMPSON AND J. K. POLLARD, *J. Exptl. Botany*, 9 (1958) 1.
5 R. GMELIN AND P. K. HIETALA, *Z. physiol. Chem., Hoppe-Seyler's*, 322 (1960) 278.
6 A. I. VIRTANEN, *Schweiz. Z. allgem. Pathol. u. Bakteriol.*, 21 (1958) 970.
7 M. KURIYAMA, M. TAKAGI AND K. MURATA, *Bull. Fac. Fisheries, Hokkaido Univ.*, 11 (1960) 58.
8 A. I. VIRTANEN AND E. J. MATIKKALA, *Acta Chem. Scand.*, 13 (1959) 623.
9 J. JADOT, J. CASIMIR AND M. RENARD, *Biochim. Biophys. Acta*, 43 (1960) 322.
10 Y. SAKATO, *J. Agr. Chem. Soc. Japan*, 23 (1950) 262.
11 J. CASIMIR, J. JADOT AND M. RENARD, *Biochim. Biophys. Acta*, 39 (1960) 462.
12 B. LEVENBERG, *J. Am. Chem. Soc.*, 83 (1961) 503.
13 D. O. GRAY AND L. FOWDEN, *Nature*, 189 (1961) 401.
14 C. LIORET, *Compt. rend.*, 244 (1957) 2171.
15 K. MORIZAWA, *Acta Schol. Med. Univ. Kioto*, 9 (1927) 285.
16 F. KNOOP AND C. MARTIUS, *Z. physiol. Chem., Hoppe-Seyler's* 258 (1939) 238.
17 E. A. BELL, *Biochim. Biophys. Acta*, 47 (1961) 602.
18 W. SCHENK AND H. R. SCHÜTTE, *Naturwissenschaften*, 48 (1961) 223.
19 M. WADA, *Biochem. Z.*, 224 (1930) 420.
20 J. K. MIETTINEN AND A. I. VIRTANEN, *Physiol. Plantarum*, 5 (1952) 540.
21 M. KURIYAMA, M. TAKAGI AND K. MURATA, *Bull. Fac. Fisheries, Hokkaido Univ.*, 10 (1959) 277.
22 F. F. NOE AND L. FOWDEN, *Nature*, 184 (1959) 69.
23 F. F. NOE AND L. FOWDEN, *Biochem. J.*, 77 (1960) 543.
24 I. L. FINAR AND K. UTTING, *J. Chem. Soc.*, (1960) 5272.
25 N. SUGIMOTO, H. WATANABE AND A. IDE, *Tetrahedron Letters*, 11 (1960) 231.
26 M. TAKESHITA, R. KIMURA, T. YABUUCHI, Y. NISHIZUKA AND O. HAYAISHI, cited by M. TAKESHITA, Y. NISHIYUKA AND O. HAYAISHI[27].
27 M. TAKESHITA, Y. NISHIZUKA AND O. HAYAISHI, *Biochim. Biophys. Acta*, 48 (1961) 409.
28 F. G. HOPKINS AND S. W. COLE, *J. Physiol. (London)*, 29 (1903) 451.
29 J. P. GREENSTEIN AND C. E. CARTER, *J. Natl. Cancer Inst.*, 7 (1946) 57.
30 A. MEISTER, H. A. SOBER, S. U. TICE AND P. E. FRASER, *J. Biol. Chem.*, 197 (1952) 319.
31 T. CURTIS AND F. KOCH, *J. prakt. Chem.*, 38 (1888) 473.
32 C. H. HASSALL AND K. REYLE, *Nature*, 173 (1954) 356.
33 C. H. HASSALL AND K. REYLE, *Biochem. J.*, 60 (1955) 334.
34 J. PATRICK, *J. Appl. Physiol.*, 7 (1954) 140.
35 H. V. ANDERSON, J. L. JOHNSON, J. W. NELSON, E. C. OLSON, M. E. SPEETER AND J. J. VAVRA, *Chem. and Ind. (London)*, (1958) 330.
36 C. VON HOLT AND W. LEPPLA, *Angew. Chem.*, 70 (1958) 25.
37 U. RENNER, A. JÖHL AND W. G. STOLL, *Helv. Chim. Acta*, 41 (1958) 588.
38 R. S. DE ROPP, J. C. VAN METER, E. C. DE RENZO, K. W. MCKERNS, C. PIDACKS, P. H. BELL, E. F. ULLMAN, S. R. SAFIR, W. J. FANSHAWE AND S. B. DAVIS, *J. Am. Chem. Soc.*, 80 (1958) 1004.
39 S. WILKINSON, *Chem. and Ind. (London)*, (1958) 17.
40 E. V. ELLINGTON, C. H. HASSALL AND J. R. PLIMMER, *Chem. and Ind. (London)*, (1958) 329.
41 M. G. ETTLINGER AND F. KENNEDY, *Chem. and Ind. (London)*, (1956) 166.

INVITED DISCUSSION

γ-GLUTAMYL PEPTIDES IN PLANTS

JOHN F. THOMPSON, CLAYTON J. MORRIS,
WILFRED N. ARNOLD AND DONELLA H. TURNER

U.S. Plant, Soil and Nutrition Laboratory, Soil and Water Conservation Research Division, Agricultural Research Service, U.S.D.A., and the Botany Department, Cornell University, Ithaca, N.Y. (U.S.A.)

Since 1958, when the first γ-glutamyl dipeptide was found in legume seeds[1–3], nine additional γ-glutamyl dipeptides and a γ-glutamyl tripeptide have been isolated from plant tissues[4–8]. Also, there is chromatographic evidence for two more such dipeptides[8, 9], and chemical proof for other γ-glutamyl compounds in which the non-glutamyl portion is a nitrogen containing compound but not an amino acid[10–14].

This paper discusses the evidence for the identification of γ-glutamyl compounds from plants including the proof for two dipeptides which have not previously been reported. Determination of the content of γ-glutamyl peptides and hydrolytic products in various tissues has demonstrated that peptides may account for a large percentage of the non-protein nitrogen and that the peptide form of an amino acid may be many times higher than that of the free form. Evidence was found for the disappearance of one peptide during the germination of kidney bean seeds. Preliminary data for a γ-glutamyl transpeptidase indicates a possible pathway for the synthesis and breakdown of γ-glutamyl peptides.

EVIDENCE FOR THE OCCURRENCE OF γ-GLUTAMYL COMPOUNDS IN PLANTS

Although peptides are to be emphasized in this discussion, other substances in which a non-amino acid nitrogenous compound is attached to the γ-carboxyl of glutamic acid are included because of their obvious similarity to peptides. Except for glutathione and glutamine there is no evidence for the existence of γ-glutamyl compounds in animal tissues. With the exception of glutamine and glutathione, the natural occurrence of γ-glutamyl compounds has been known for less than a decade. It is therefore pertinent to examine the evidence for their identification.

γ-L-Glutamyl-S-methyl-L-cysteine. This dipeptide has been isolated from Lima bean seed *(Phaseolus limensis)* by buffer chromatography on a cation exchange resin[3] and from kidney bean seed *(Phaseolus vulgaris)* by dilute acetic acid chromatography on an anion exchange resin[1, 2]. In both cases the elemental analyses agreed closely with the theoretical values and the hydrolytic products were identical with L-glutamic acid and S-methyl-L-cysteine. γ-Glutamyl-S-methylcysteine was prepared by the methylation of reduced glutathione and action of carboxypeptidase on the resultant methylated glutathione. This preparation had an identical infrared spectrum to that of the material isolated from kidney bean seeds[2]. Chromatographic evidence for this peptide in other bean seeds has been reported[3].

References p. 64

γ-L-Glutamyl-S-methyl-L-cysteine sulfoxide. This was reported to be present in Lima bean seeds[3] on the basis of cochromatography. It is possible that this may have been formed from γ-glutamyl-S-methylcysteine during purification.

γ-Glutamylleucine. This substance was purified by chromatography on a cation exchange resin and on cellulose powder from an extract of Lima bean seeds[3]. The structure was established by the chromatographic identification of glutamic acid and leucine among the hydrolysis products. By the formation of a DNP derivative, glutamic acid was shown to be N-terminal. By cochromatography and lability to acid, this peptide was tentatively identified in seeds of California small white beans, pinto beans, Blackeye beans and kidney beans.

γ-L-Glutamyl-β-aminoisobutyric acid. This was isolated in crystalline form from a 50% alcohol extract of dormant bulbs of Wedgewood iris *(Iris tingitana)*[4] by chromatography on an anion exchange resin with dilute acetic acid. Hydrolytic products were identical with L-glutamic acid and β-aminoisobutyric acid by cochromatography and infrared spectra. The elemental analyses of peptide and hydrolytic products agreed with theory. Glutamic acid was shown to be N-terminal by DNP derivative and the γ-linkage was proven by the reaction with nitrous acid.

γ-L-Glutamyl-β-alanine. Pure crystalline material was obtained from the mother liquor left from the crystallization of γ-glutamyl-β-aminoisobutyric acid. Only L-glutamic acid and β-alanine were obtained from hydrolysis of the crystals. Synthetic γ-glutamyl-β-alanine gave an identical infrared spectrum to that of the isolated substance[7].

γ-L-Glutamyl-L-tyrosine. This was isolated in crystalline form from the non-protein fraction of soy bean seed *(Glycine max)* by similar techniques as those used in purification of γ-glutamyl-β-aminoisobutyric acid. The proof of structure was obtained by elemental analysis, degradation, synthesis and infrared spectra of peptide and degradation products as in the case of γ-glutamyl-β-alanine[8].

γ-L-Glutamyl-L-phenylalanine. An acid labile substance has been isolated in crystalline form from onion bulbs *(Allium cepa)*[5] and soy bean seeds[8]. The hydrolytic products proved to be L-glutamic acid and L-phenylalanine by chromatography and elemental analysis. The γ-glutamyl linkage was proven by decarboxylation with ninhydrin and reaction with fluorodinitrobenzene. Synthetic γ-L-glutamyl-L-phenylalanine was identical with isolated material[8].

γ-L-Glutamyl-S-(β-carboxy-N-propyl)cysteinylglycine. A tripeptide related to glutathione has been isolated from onion bulbs by VIRTANEN AND MATIKKALA[5]. Hydrolysis of the isolated material yielded L-glutamic acid, glycine and a S-substituted cysteine. The latter compound had a similar infrared spectrum to that of S-(β-carboxyl-*N*-propyl)cysteine. By the reaction of β-bromoisobutyric acid and glutathione in the presence of sodium hydroxide, a tripeptide which chromatographed like the substance from onions was obtained. The infrared spectra of the two compounds were similar.

References p. 64

γ-L-Glutamyl-hypoglycin (hypoglycin B). Two crystalline substances were isolated[15] from the fruit of the ackee *(Blighia sapida)* which causes Jamaican vomiting sickness and a reduction in blood-sugar level. The crystalline materials cause hypoglycemia. One of these (hypoglycin A) has been shown to be an α-amino acid with a cyclopropane ring (L-α-amino-β-(2-methylenecyclopropyl)propionic acid)[15]. The second substance (hypoglycin B) has less physiological action on a weight basis. It was found to be the γ-glutamyl peptide of hypoglycin A by degradation to glutamic acid and hypoglycin A (ref. 6). The γ-glutamyl linkage was demonstrated by reaction of fluorodinitrobenzene with the glutamic acid portion and formation of 2 moles of nitrogen per mole of peptide in the nitrous acid reaction. Final proof was obtained by synthesis of a γ-glutamyl peptide from isolated hypoglycin A which was the same as the isolated peptide.

γ-Glutamylvaline. This dipeptide was obtained from the non-protein fraction of the onion bulbs in crystalline form[5]. Hydrolytic degradation gave L-glutamic acid and valine.

γ-Glutamylisoleucine. Another dipeptide was isolated from onion bulbs by chromatography on Dowex-1 (ref. 5) with acetic acid. Acid hydrolysis gave two compounds which cochromatographed with isoleucine and glutamic acid.

γ-Glutamylalanine. VIRTANEN AND BERG[9] isolated material from a spot on paper chromatograms of 70% ethanol extracts of pea seedlings *(Pisum sativum)*. This material hydrolyzed to two compounds which cochromatographed with glutamic acid and alanine.

γ-Glutamylmethionine. An acid-labile substance has been purified from an alcohol extract of kidney bean seed by paper chromatography. The unhydrolyzed material gave a positive test for a sulfur amino acid with the iodoplatinate reagent[16] and a positive test for an alpha amino acid with pyridoxal[17]. The hydrolytic products cochromatographed with glutamic acid and methionine.

γ-Glutamylethylamine (Theanine). This ethyl amide was isolated from the non-protein fraction of tea leaves *(Camellia sinensis)* and mushrooms *(Xerocomus badius)*[12, 13]. The hydrolytic products were found to be L-glutamic acid and ethylamine and this was consistent with the elemental analysis. The structure was confirmed by synthesis[18]. This compound could be a decarboxylation product of either γ-glutamyl-β-alanine or γ-glutamylalanine.

γ-Glutamyl-β-aminopropionitrile (Lathyrus factor). A substance which is capable of producing skeletal abnormalities in rats has been isolated in pure form from Sweet pea seeds *(Lathyrus odoratus)*[19]. The compound was hydrolyzed to L-glutamic acid, ammonia and β-alanine. Elemental analysis indicated that the original compound was a nitrile. The γ-glutamyl-β-propionitrile was synthesized and found to be identical with the isolated substance by infrared and melting point[11]. This compound has an obvious chemical similarity to γ-glutamyl-β-alanine and it remains to be seen whether there is a biochemical relationship.

References p. 64

β-N-(γ-L-Glutamyl)-4-hydroxymethylphenylhydrazine (Agaritine). From the press juice of commercial mushrooms *(Agaricus bisporus)*, a ninhydrin-reactive material was isolated in crystalline form. Acid hydrolysis liberated L-glutamic acid and a compound which oxidized to an aryldiazonium salt. The latter could be hydrolyzed to *p*-hydroxybenzylalcohol. This evidence plus elemental analyses, and liberation of CO_2 with ninhydrin indicated the structure[10].

N-(γ-L-Glutamyl)-4-hydroxyaniline. An acid-labile substance was isolated from the non-protein fraction of a mushroom *(Agaricus hortensis)*, by chromatography on activated charcoal and Dowex-1 (ref. 14). The crystalline material hydrolyzed to L-glutamic acid and *p*-hydroxyaniline as shown by comparison of infrared spectra. Final proof was obtained by comparison of synthetic N-(γ-L-glutamyl)-4-hydroxyaniline with the isolated compound by infrared.

γ-Glutamylcysteine. γ-Glutamylcysteine has been known to be an intermediate in glutathione synthesis in plants and animals[20, 21]. This peptide must have widespread occurrence although it may never accumulate to a sufficient extent to be isolated in quantity.

TABLE I

OCCURRENCE OF γ-GLUTAMYL COMPOUNDS IN PLANTS

Moiety attached to γ-carboxyl of glutamic acid	*Source*	*Evidence*	*Reference*
L-Phenylalanine	Onion bulb	Isolation and degradation	5
	Soy bean seed	Isolation, degradation and synthesis	8
L-Tyrosine	Soy bean seed	Isolation, degradation and synthesis	8
L-Valine	Onion bulbs	Isolation and degradation	5
L-Isoleucine	Onion bulbs	Isolation and degradation	5
Leucine	Lima bean seeds	Isolation and chromatography	3
	Kidney bean seed	Chromatography and synthesis	8
Alanine	Pea seedlings	Chromatography	9
Methionine	Kidney bean seed	Chromatography	8
S-Methyl-L-cysteine	Kidney bean seed	Isolation, degradation and synthesis	2
	Lima bean seed	Isolation and degradation	3
β-Alanine	Iris bulb	Isolation, degradation and synthesis	7
β-Aminoisobutyric acid	Iris bulbs	Isolation and degradation	4
α-Amino-β-(2-methylene-cyclopropyl)propionic acid (hypoglycin A)	Ackee fruit	Isolation and degradation	6
L-Cysteinylglycine	Most plants	Complete	—
S-(β-Carboxy-*n*-propyl)-L-cysteinylglycine	Onion bulb	Isolation, degradation and synthesis	5
Ethylamine	Tea leaves	Isolation, degradation and synthesis	12, 18
	Mushrooms	Isolation, degradation and synthesis	13
β-Aminopropionitrile	Sweet pea seed	Isolation, degradation and synthesis	11, 19
p-Hydroxyphenylhydrazine	Mushrooms	Isolation and degradation	10
p-Hydroxyaniline	Mushrooms	Isolation, degradation and synthesis	14
Ammonia	Most plants	Complete	—

References p. 64

Concluding comments. The above discussion of γ-glutamyl compounds is summarized in Table I. Of the 19 compounds listed, adequate proof of structure is available for 16. The non-glutamyl portion is a protein amino acid in five cases and is a non-protein amino acid in three.

Dipeptides have been isolated mainly from storage tissues such as seeds and bulbs but also have been obtained from other tissues. Fungi, monocotyledons and dicotyledons were sources of isolated compounds. These facts indicate the possibility that glutamyl dipeptides may have widespread occurrence in various types of tissues throughout the plant kingdom.

QUANTITY OF PEPTIDES IN PLANTS

The content of several peptides and related compounds in certain plant tissues is presented in Table II. These data reveal several interesting points. The dipeptides make an appreciable contribution to the total non-protein amino nitrogen fraction. For example, γ-glutamyl-*S*-methylcysteine accounts for over 40% of the free amino nitrogen in kidney beans. At the same time the content of glutamic acid is less than 10% and that of methylcysteine is far less than this.

Even though glutamic acid is usually a prominent constituent of the non-protein fraction of plants, the γ-glutamyl peptides often are present in concentrations as high or higher than that of glutamic acid. The situation is even more striking with respect to the non-glutamic acid moiety of the peptides where the molar ratio of peptide-form to free form may be in the order of 100 or 1000 : 1 (Table II).

The data in Table II indicate amounts of peptides in tissues from which they were

TABLE II

THE QUANTITY OF γ-GLUTAMYL PEPTIDES AND RELATED COMPOUNDS IN PLANT TISSUE

Compound	Tissue	Content* $\mu g/g$ fresh wt.	Content* μmoles/100 μatoms non-protein nitrogen**	Mole ratio peptide/free form
γ-Glutamyl-S-methylcysteine	Kidney bean seed	4600	42.6	
Glutamic acid	Kidney bean seed	400	6.6	6.45
S-Methylcysteine	Kidney bean seed	221	3.2	13.3
γ-Glutamyl-β-amino-isobutyric acid	Iris bulb	632	15.8	
γ-Glutamyl-β-alanine	Iris bulb	97	2.6	
Glutamic acid	Iris bulb	452	17.8	1.03
β-Aminoisobutyric acid	Iris bulb	2.06	0.01	1580
β-Alanine	Iris bulb	0.90	0.005	520
Glutathione	Iris bulb	138	2.57	
γ-Glutamyltyrosine	Soy bean seed	384	4.94	
γ-Glutamylphenylalanine	Soy bean seed	393	4.66	
Glutamic acid	Soy bean seed	636	17.3	0.55
Tyrosine	Soy bean seed	15.7	0.18	27.5
Phenylalanine	Soy bean seed	57	0.70	6.7

* Data on kidney bean and soy bean seeds was obtained by quantitative paper chromatography[22], after purification on resins[2, 3]. Data from iris bulb was obtained by the method of MOORE *et al.*[24, 25].
** Nitrogen determined by reaction with ninhydrin after removal of ammonia[26].

isolated. The data in Tables III and IV show that these peptides occur in tissues throughout the plant. Although the iris bulb still contains appreciable quantities of the dipeptides during the flowering stage there are sizable amounts distributed throughout the plant. They are present in concentrations of the same order of magnitude as the acidic amino acids but considerably higher than glutathione. These data give no indication as to where the peptides are being formed and whether they are translocated.

VIRTANEN AND MATIKKALA found γ-glutamyl peptides only in the onion bulb and not in other parts of the plant. In this plant the peptides are not apparently translocated.

BREAKDOWN OF γ-GLUTAMYL PEPTIDES

The previous discussion has emphasized the existence of γ-glutamyl peptides in diverse tissues of various plants. The metabolism of these compounds should be even more interesting since it may give some clue as to their function.

Since the dry kidney bean seed contained appreciable amounts of γ-glutamyl-S-methylcysteine (Table II), the changes in the peptide during germination were deter-

TABLE III

PEPTIDE AND ACIDIC AMINO ACID CONTENTS OF VARIOUS PARTS OF WEDGEWOOD IRIS PLANTS AT THE FLOWERING STAGE

Tissue	*Quantity of compound* (μmoles/100 μatoms of non-protein nitrogen**)*				
	Aspartic acid	*Glutamic acid*	*γ-Glutamyl-β-amino-isobutyric acid*	*γ-Glutamyl-β-alanine*	*Glutathione*
Root	23.9	7.06	3.86	1.41	0.099
Bulb	1.23	5.70	6.87	1.33	0.00
Flower stalk	4.36	3.36	1.35	0.76	0.034
Leaves	8.84	8.96	2.65	0.83	0.017

* Determinations by quantitative paper chromatography[22] after purification[23].
** See footnote** in Table II.

TABLE IV

THE DISTRIBUTION AND TOTAL AMOUNT OF γ-L-GLUTAMYL-S-METHYL-L-CYSTEINE IN GERMINATING ETIOLATED SEEDLINGS OF KIDNEY BEAN *(Phaseolus vulgaris)*

Days	*γ-Glutamyl-S-methylcysteine* (μmoles per plant part)*				
	Cotyledons	*Hypocotyl*	*Roots*	*Primary leaves*	*Total*
0	9.62	—	—	—	9.62
4	7.65	0.16	0.01	0.02	7.84
7	2.68	0.84	0.02	0.09	3.81
9	1.22	0.74	0.19	0.06	2.21

* Measurements made as in Table III.

References p. 64

mined (Table IV). There was a considerable net loss of peptide during 9 days of germination even though there may have been a concomitant synthesis taking place. RINDERKNECHT *et al.*[3] have also reported a disappearance of this peptide during the germination of Lima bean seeds. Presumably hydrolysis had taken place. In the breakdown of glutathione in purified swine-kidney preparations[27], the first step is transpeptidation[27–32] followed by hydrolysis of the resultant cysteinylglycine as indicated below:

$$\text{glutathione} + \text{amino acid} \xrightleftharpoons{\text{transpeptidase}} \gamma\text{-glutamyl amino acid} + \text{cysteinylglycine}$$

$$\text{cysteinylglycine} + H_2O \xrightleftharpoons{\text{peptidase}} \text{cysteine} + \text{glycine}$$

Although transpeptidation may account for the disappearance of γ-glutamyl-S-methylcysteine during germination, this process appears unlikely because there was no concomitant increase in other acidic peptides. Transpeptidation alone would result in no net decrease in γ-glutamyl peptides. HIRD AND SPRINGELL[33] have shown that γ-glutamyl transpeptidases from sheep kidney transfer γ-glutamyl groups to water or amino compounds depending on the conditions. However, with a more purified enzyme preparation, BINKLEY[27] observed no hydrolysis under any conditions. Since there is some evidence for a transpeptidase and hydrolytic activity in bean extract (see below), a mechanism for peptide breakdown can be visualized. The ordinary proteinases and peptidases do not act on γ-glutamyl peptide bonds. A cell-free preparation from *Bacillus subtilis*[34] which hydrolyzes γ-glutamyl polypeptides has not been observed in higher plants. Hence it seems unlikely that the enzymes which are hydrolyzing the proteins of the cotyledons of the germinating bean are also hydrolyzing γ-glutamyl-S-methylcysteine.

Microorganisms appear to have an enzyme for the hydrolysis of the γ-glutamylcysteine bond but not other γ-glutamyl bonds. HAC, SNELL AND WILLIAMS[35] found that glutathione can substitute for glutamic acid in the growth of *L. arabinosus*. We have confirmed this observation and extended the test to the first two γ-glutamyl

TABLE V

THE UTILIZATION OF γ-GLUTAMYL PEPTIDES BY *Lactobacillus arabinosus**

Compound	*Optical density at 560 mμ (μmoles of compound per tube)*				
	0	*0.1*	*0.2*	*0.6*	*1.0*
L-Glutamic acid	0.008	0.012	0.106	0.272	0.377
Reduced glutathione	0.008	0.007	0.012	0.211	0.360
γ-Glutamyl-S-methylcysteine	—	—	—	—	0.028**
γ-Glutamyl-β-aminoisobutyric acid	—	—	—	—	0.008

* Each tube had 2.5 ml of the medium of HAC, SNELL AND WILLIAMS[35] without glutamic acid and glutamine. Samples were incubated 26 h at 37° and then diluted with 5 ml of water for absorbancy reading.

** 3 weeks incubation. 0.4 μmole of glutamic acid gave an absorbancy of 0.186.

peptides that were available to us (Table V). The latter peptides did not support growth of *L.arabinosus* and this agrees with the findings of WAELSCH *et al.*[36] with the same organism. It is desirable to extend these tests to the γ-glutamyl peptides of cysteine, β-alanine, tyrosine and phenylalanine. *L. arabinosus* apparently has the ability to split the peptide bond between the γ-carboxyl group of glutamic acid and cysteine but lacks this ability where the second amino acid is not cysteine. When this result is considered along with the evidence that in plant preparations[37] cysteine is the only protein amino acid which will form a γ-glutamyl dipeptide from glutamic acid, it appears that the enzymes involved in the formation and rupture of the γ-glutamylcysteine bond are specific for this bond. However, synthesis and breakage of this bond occur by different mechanisms.

FORMATION OF γ-GLUTAMYL PEPTIDES

Because of the widespread occurrence of glutathione, its metabolism has been extensively studied. It is obvious that the biosynthesis of the γ-glutamyl dipeptides may be analagous to the first step of glutathione synthesis wherein γ-glutamylcysteine is formed[20, 21]. The reaction for γ-glutamylcysteine synthesis may be formulated as follows for both plants and animals:

$$\text{glutamic acid} + \text{cysteine} + \text{ATP} \underset{Mg^{2+}}{\overset{\text{Synthetase}}{\rightleftarrows}} \gamma\text{-glutamylcysteine} + \text{ADP} + P_i$$

This reaction is formally similar to the formation of glutamine and carnosine (β-alanylhistidine)[39, 40] but no evidence has been forthcoming that γ-glutamyl peptides other than γ-glutamylcysteine[20, 21] are formed in an analogous fashion. We have made unsuccessful attempts to demonstrate the formation of γ-glutamylmethylcysteine in kidney bean seedling extracts in the same way that WEBSTER obtained the synthesis of γ-glutamylcysteine[41].

There is much evidence available[27–32] for the formation of γ-glutamyl peptides from glutathione and free amino acids in animal tissue preparations (see above). The intervention of ATP is not required because there is little change in free energy. WILLIAMS AND THORNE[42, 43] have observed an enzyme in the culture media of *Bacillus subtilis* which will form γ-glutamylglutamic acid and larger poly-γ-glutamyl peptides from glutamine. Other amino acids[42] not only apparently do not accept γ-glutamyl groups but some inhibit the transfer of glutamic acid. In plants, HANES *et al.*[29] have reported a glycyl-transferring enzyme in a press juice of cabbage, and VIRTANEN AND BERG[9] suggest that γ-glutamylalanine can be formed by transpeptidation in pea-seedling extracts.

Utilizing the method of GOLDBARG *et al.*[44], in which γ-glutamylaniline is the substrate and the release of aniline measures breaking of the γ-glutamyl bond, preliminary evidence has been obtained for both γ-glutamyl transpeptidase and peptidase in kidney bean seed extracts (Table VI). In the absence of added amino acids, the release of aniline is presumably due to peptidase action. Although this production of aniline could be the result of transpeptidase action, induced by endogenous amino acids, this appears unlikely because the ammonium sulfate precipitation and sub-

TABLE VI

EVIDENCE FOR TRANSPEPTIDASE AND A γ-GLUTAMYL BOND PEPTIDASE ACTIVITY IN BEAN SEED PREPARATIONS

*Fraction**	*Additions*	*mμmoles of aniline** liberated beyond zero-time control*
0–30% saturation Ammonium sulfate	None	10.7
	L-Methionine	16.1
	S-Methyl-L-cysteine	16.6
30–40% saturation Ammonium sulfate	None	13.0
	L-Methionine	16.5
	S-Methyl-L-cysteine	16.7

* Bean seed extract was prepared by grinding 25 g of dry kidney bean seeds with 50 ml of 0.040 *M* $NaHCO_3$.
** The reaction mixture contained 0.5 ml of bean extract with these additions (in μmoles): γ-glutamylaniline, 1.12; Tris buffer (pH 8.6), 50; amino acids, 5.62; to a total volume of 1.5 ml. Incubation was for 2 h at 25°. Aniline was determined by diazotization and coupling with β-naphthylethylenediamine[44].

sequent dialysis should have removed free amino acids. The additional release of aniline as a result of the addition of methionine or methylcysteine provides some evidence of the presence of a transpeptidase. Due to the considerable difference in the ratio of peptidase to transpeptidase activity in the two ammonium sulfate fractions (Table VI), it appears that these results are not explained by a hydrolysis or an aminolysis induced only by a transpeptidase[33]. The peptide-bond cleavage induced by bean extracts is consistent with the loss of γ-glutamylmethylcysteine in germinating kidney bean seeds (Table IV).

DISCUSSION: POSSIBLE PHYSIOLOGICAL ROLE OF γ-GLUTAMYL COMPOUNDS

Because γ-glutamyl compounds are numerous (Table I), occur in relatively high concentrations in some tissues (Table II) and exist in diverse plants and tissues (Tables I–III), it is not unreasonable to assume that they play a role in normal plant metabolism. All of these compounds are closely related chemically and may be formed in a similar manner.

Transpeptidases have been extensively studied in animal extracts but not in plant extracts although dipeptides have been found in plants but not in animals[36]. Although the role of transpeptidation in normal metabolism has not been elucidated, with the discovery of γ-glutamyl peptides in plants, a possible function for transpeptidases can be visualized. In plants they may be involved in formation of dipeptides while in animals they may act to help hydrolyze dipeptides that are in the diet. The glutaminotransferases[45] of animals and higher plants transfer γ-glutamyl groups to amines but not amino acids and hence are not involved in γ-glutamyl peptide metabolism. Certain bacterial enzyme preparations do transfer γ-glutamyl groups to amines and to amino acids but the only amino acid acceptor is glutamic acid[42, 43].

References p. 64

It has been suggested that transpeptidation might be involved in protein synthesis. Glutathione and other γ-glutamyl peptides are not apparently involved in protein synthesis[46]. Since several of the naturally occurring γ-glutamyl peptides contain amino acids not found in proteins (S-methylcysteine, β-alanine, β-aminoisobutyric acid and hypoglycin A), a specific role in protein synthesis seems unlikely.

The γ-glutamyl peptides have been found primarily in storage organs and constitute an appreciable proportion of the non-protein nitrogen. Since most of the nitrogen of bulbs and seeds occurs as proteins (80–90%), storage of nitrogen as peptides is not reasonable unless they are special storage forms for the non-glutamic acid portion of the peptide.

The data in Table III indicate that γ-glutamylmethylcysteine may be transported from the cotyledons to other parts of the plant. Dipeptides may act as transport forms of nitrogen, glutamic acid or the non-glutamic acid portion. Because of its central role in nitrogen metabolism, glutamic acid would be an excellent source of nitrogen for any tissue. It is equally possible that the non-glutamic moiety is transported in this way or that the peptide form protects the glutamic acid from being metabolized in the conductive tissue before it reaches active metabolic regions. In the onion plant peptides are not found in tissues other than the bulb suggesting that translocation of intact peptides does not occur.

Since the non-glutamic acid portions (β-aminopropionitrile, *p*-hydroxymethylphenylhydrazine, methylcysteine, *p*-hydroxyaniline) of several γ-glutamyl compounds are unusual groups, they may be toxic to the plant. Combination with glutamic acid might well reduce toxicity. However, this concept would not fit the case where normal protein amino acids are bound. It is, of course, equally puzzling as to why the plant synthesizes β-aminopropionitrile and *p*-hydroxymethylphenylhydrazine.

The discovery of γ-glutamyltyrosine and γ-glutamylphenylalanine in soy bean meal may have some bearing on the toxicity of soy bean meal. BORCHERS has found[47] that the toxicity of raw soy bean meal is reduced by tyrosine and not by phenylalanine. Possibly these peptides interfere with the hydroxylation of phenylalanine to form tyrosine.

It may well be that the various dipeptides have different functions in the diverse places where they occur.

Poly-γ-glutamyl peptides also occur in nature[48–50]. These appear to have a different metabolism from the aforementioned γ-glutamyl compounds.

ADDENDUM

Since the time that this manuscript was submitted, VIRTANEN AND MATILLA[51] reported the isolation of γ-glutamyl-S-allyl-L-cysteine from garlic. Also, VIRTANEN AND MATIKKALA[52] have crystallized from onion three more γ-glutamyl peptides; γ-glutamylmethionine, γ-glutamyl-S-methylcysteine and γ-glutamylleucine. Further proof of the identity of γ-glutamylleucine from kidney beans has been obtained by comparison of the infra red spectra of the isolated and synthetic peptides. The spectra were identical. The new peptides indicate that one can expect that many more γ-glutamyl peptides will be found.

References p. 64

REFERENCES

[1] R. M. Zacharius, C. J. Morris and J. F. Thompson, *Arch. Biochem. Biophys.*, 73 (1958) 281.
[2] R. M. Zacharius, C. J. Morris and J. F. Thompson, *Arch. Biochem. Biophys.*, 80 (1959) 199.
[3] H. Rinderknecht, D. Thomas and S. Aslin, *Helv. Chim. Acta*, 41 (1958) 1.
[4] C. J. Morris, J. F. Thompson, S. Asen and F. Irreverre, *J. Biol. Chem.*, 236 (1961) 1181.
[5] A. I. Virtanen and E. J. Matikkala, *Z. physiol. Chem., Hoppe-Seyler's*, 322 (1960) 8.
[6] C. H. Hassall and D. I. John, *J. Chem. Soc.*, (1960) 4112.
[7] C. J. Morris, J. F. Thompson, S. Asen and F. Irreverre, unpublished data, 1961.
[8] C. J. Morris and J. F. Thompson, unpublished data, 1961.
[9] A. I. Virtanen and A. M. Berg, *Acta chem. Scand.*, 8 (1954) 1089.
[10] B. Levenberg, *J. Am. Chem. Soc.*, 83 (1961) 503.
[11] E. D. Schilling and F. M. Strong, *J. Am. Chem. Soc.*, 77 (1955) 2843.
[12] Y. Sakato, *Nippon Nôgei-Kagaku Kaishi*, 23 (1950) 262.
[13] J. Casimir, J. Jadot and M. Renard, *Biochim. Biophys. Acta*, 39 (1960) 462.
[14] J. Jadot, J. Casimir and M. Renard, *Biochim. Biophys. Acta*, 43 (1960) 322.
[15] E. V. Ellington, C. H. Hassall, J. R. Plimmer and C. E. Seaforth, *J. Chem. Soc.*, (1959) 80.
[16] G. Toennies and J. J. Kolb, *Anal. Chem.*, 23 (1951) 823.
[17] G. D. Kalyankar and E. E. Snell, *Nature*, 180 (1957) 1069.
[18] Y. Sakato, T. Hashizume and Y. Kishimoto, *Nippon Nôgei-Kagaku Kaishi*, 23 (1950) 269.
[19] G. F. McKay, J. J. Lalich, E. D. Schilling and F. M. Strong, *Arch. Biochem. Biophys.*, 52 (1954) 313.
[20] G. C. Webster and J. E. Varner, *Arch. Biochem. Biophys.*, 52 (1954) 22.
[21] S. Mandeles and K. Bloch, *J. Biol. Chem.*, 214 (1955) 639.
[22] J. F. Thompson and C. J. Morris, *Anal. Chem.*, 31 (1959) 1031.
[23] J. F. Thompson, C. J. Morris and R. Gering, *Anal. Chem.*, 31 (1959) 1028.
[24] S. Moore, D. H. Spackman and W. H. Stein, *Anal. Chem.*, 30 (1958) 1185.
[25] D. H. Spackman, W. H. Stein and S. Moore, *Anal. Chem.*, 30 (1958) 1190.
[26] S. Moore and W. H. Stein, *J. Biol. Chem.*, 211 (1954) 907.
[27] F. Binkley, *J. Biol. Chem.*, 236 (1961) 1075.
[28] C. S. Hanes, F. J. R. Hird and F. A. Isherwood, *Nature*, 166 (1950) 288.
[29] C. S. Hanes, F. J. R. Hird and F. A. Isherwood, *Biochem. J.*, 51 (1952) 25.
[30] P. J. Fodor, A. Miller and H. Waelsch, *J. Biol. Chem.*, 202 (1953) 551.
[31] F. J. R. Hird and P. H. Springell, *Biochem. J.*, 56 (1954) 417.
[32] J. H. Kinoshita and E. G. Ball, *J. Biol. Chem.*, 200 (1953) 609.
[33] F. J. R. Hird and D. H. Springell, *Biochim. Biophys. Acta*, 15 (1954) 31.
[34] C. B. Thorne, C. G. Gomez, H. E. Noyes and R. D. Housewright, *J. Bacteriol.*, 68 (1954) 307.
[35] L. R. Hac, E. E. Snell and R. J. Williams, *J. Biol. Chem.*, 159, (1945) 273.
[36] H. Waelsch, in *Glutathione*, Academic Press Inc., New York, 1954, p. 151.
[37] G. C. Webster, personal communication.
[38] J. F. Speck, *J. Biol. Chem.*, 168 (1947) 403.
[39] G. D. Kalyankar and A. Meister, *J. Biol. Chem.*, 234 (1959) 3210.
[40] R. E. Winnick and T. Winnick, *Biochim. Biophys. Acta*, 31 (1959) 47.
[41] G. C. Webster, *Plant Physiol.*, 28 (1953) 728.
[42] W. J. Williams and C. B. Thorne, *J. Biol. Chem.*, 210 (1954) 203.
[43] W. J. Williams and C. B. Thorne, *J. Biol. Chem.*, 212 (1955) 427.
[44] J. A. Goldbarg, O. M. Friedman, E. P. Pineda, E. E. Smith, R. Chatterji, E. H. Stein and A. M. Rutenberg, *Arch. Biochem. Biophys.*, 91 (1960) 61.
[45] H. Waelsch, *Advances in Enzymol.*, 13 (1952) 237.
[46] R. W. Hendler and D. M. Greenburg, *Biochem. J.*, 57 (1954) 641.
[47] H. Borchers, *Federation Proc.*, 18 (1959) 517.
[48] G. Ivanovics and L. Erdos, *Z. Immunitätsforsch.*, 90 (1937) 5.
[49] B. L. Hutchings, E. L. R. Stokstad, J. H. Mowat, J. H. Boothe, C. W. Waller, R. B. Angier, J. Semb and Y. Subba Row, *J. Am. Chem. Soc.*, 70 (1948) 10.
[50] J. J. Pfiffner, D. G. Calkins, E. S. Bloom and B. L. O'Dell, *J. Am. Chem. Soc.*, 68 (1946) 1392.
[51] A. Virtanen and I. Matilla, *Suomen Kemistilehti B*, 34 (1961) 44.
[52] A. Virtanen and E. J. Matikkala, *Suomen Kemistilehti B*, 34 (1961) 53.

INVITED DISCUSSION

FREE AMINO ACIDS IN NORMAL AND MALIGNANT PLANT TISSUES GROWN IN TISSUE CULTURE

DAISY G. SIMONSEN AND EUGENE ROBERTS

Department of Biochemistry, Medical Research Center, City of Hope Medical Center, Duarte, Calif. (U.S.A.)

PHILIP R. WHITE

Roscoe B. Jackson Memorial Laboratory, Bar Harbor, Me. (U.S.A.)

Numerous aspects of the problems related to the free amino acids of plants have just been discussed. As described elsewhere in this Symposium (ROBERTS, SIMONSEN, KITTREDGE and ROBERTS), in both vertebrate and invertebrate animal species, it has been found that each tissue or cell type in a particular species has a characteristic distribution of free or easily extractable amino acids. However, quite similar patterns of free amino acids were found in many different types of tumors in rodents. It was of interest to examine in a similar manner the free amino acids in various normal and malignant plant tissues grown in synthetic medium[1–4].

Extracts from six to ten separate samples were chromatographed for each tissue. Aliquots were employed corresponding to fresh weights of each type of plant which

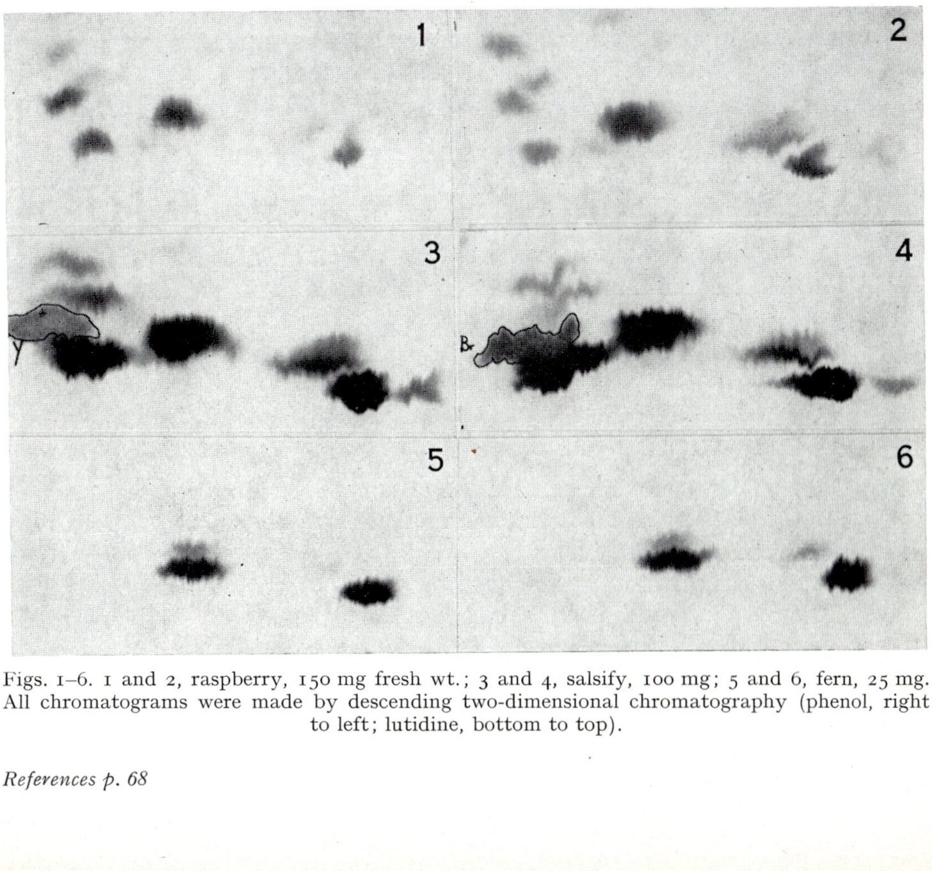

Figs. 1–6. 1 and 2, raspberry, 150 mg fresh wt.; 3 and 4, salsify, 100 mg; 5 and 6, fern, 25 mg. All chromatograms were made by descending two-dimensional chromatography (phenol, right to left; lutidine, bottom to top).

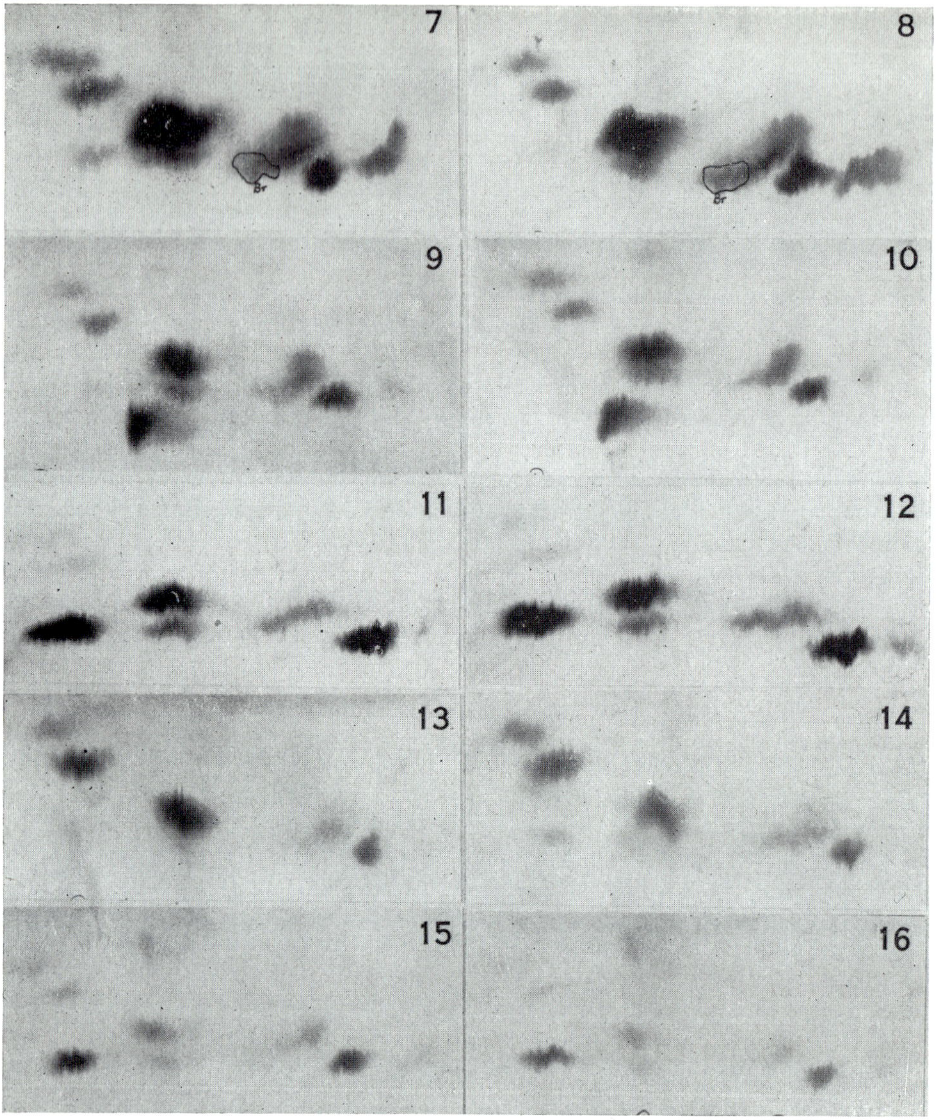

Figs. 7–16. 7 and 8, carrot, 200 mg; 9 and 10, pink periwinkle, 75 mg; 11 and 12, Boston ivy, 25 mg; 13 and 14, cactus, 200 mg; 15 and 16, Virginia creeper, 200 mg.

would give good chromatograms. In Figs. 1–6 are shown the chromatograms obtained from extracts of two separate cultures of three normal tissues which had been grown in tissue culture for 2 months prior to harvesting. There was a remarkable constancy of pattern of detectable constituents in a particular species: Figs. 1 and 2, *Rubus fruticosa* (raspberry)*, Figs. 3 and 4, *Scorzonera hispanica* (salsify)**,

The footnotes give the name of the investigator who first isolated the cultures and the place and approximate date on which the isolation was performed.

* Morel, France, 1945.

** Gautheret, France, 1945.

References p. 68

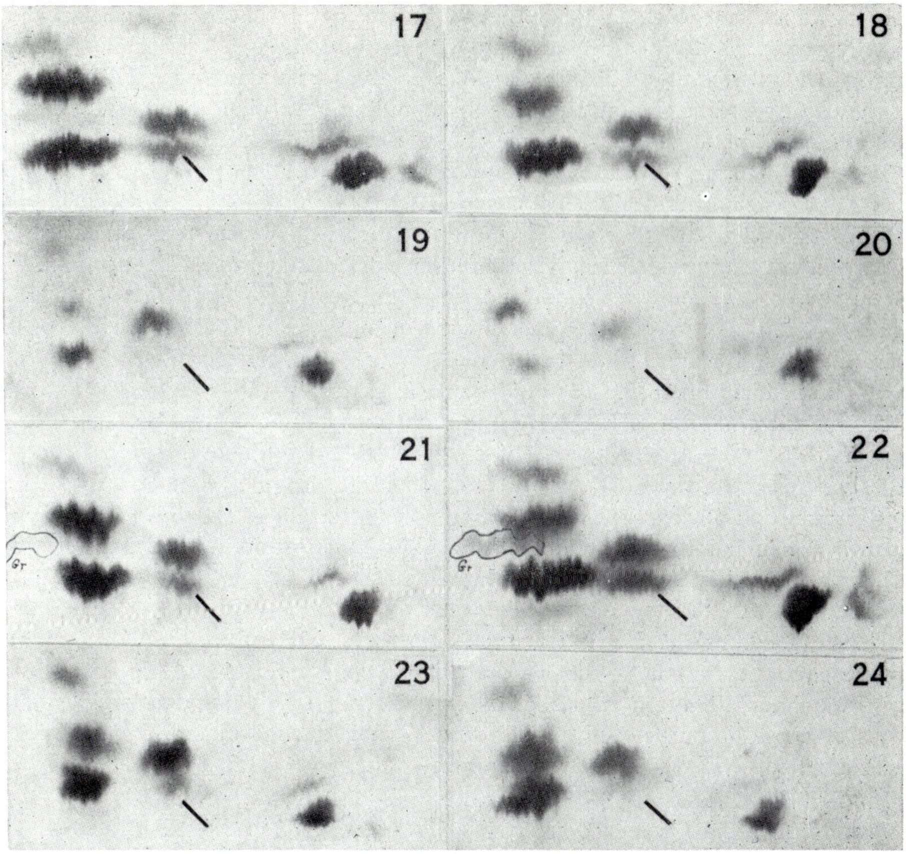

Figs. 17–24. Tobacco, all samples from 150 mg fresh wt.; 17 and 18, strain NA, 2 months; 19 and 20, strain N, 2 months; 21 and 22, strain NA, 4–5 months; 23 and 24, strain N, 4–5 months. The lines on the chromatograms point to glutamine.

and Figs. 5 and 6, *Pteridium aquifolium* (fern)*. The relative amounts of the various detectable constituents were distinctively different for each of the different species studied. These findings are analogous to those in animal tissues. As in normal plant tissues, the patterns were characteristic for each type of tumor, and the various tumors differed from each other. Results are shown in Figs. 7–16 for five malignant plant tissues: Figs. 7 and 8, *Daucus carota* (carrot)**, Figs. 9 and 10, *Vinca rosea* (pink periwinkle)***, Figs. 11 and 12, *Parthenocissus tricuspidata* (Boston ivy)§, Figs. 13 and 14, *Opuntia microacantha* (cactus)§§; and Figs. 15 and 16, *Parthenocissus quinquifolia* (Virginia creeper)§§§. In the animal

The footnotes give the name of the investigator who first isolated the cultures and the place and approximate date on which the isolation was performed.

* Morel, Boston, 1950.

** Gautheret, France, 1937.

*** White, Princeton, 1942

§ Morel, France, 1945.

§§ Morel, Boston, 1950.

§§§ Morel, Boston, 1945.

References p. 68

tumors which have been studied, the glutamine levels have been extremely low, usually below the level of detection by chromatographic methods, while relatively large amounts were found in normal tissues[5]. This amino acid was easily detectable in the cultured plant tumors and no distinction could be made between the normal and malignant plant tissues on the basis of glutamine content.

In Figs. 17–24 are shown the chromatograms of extracts of samples of two separately isolated hybrid strains of tobacco *(Nicotinia; N. glauca × N. Langsdorfii)*. In Figs. 17 and 18 are shown extracts of a strain (NA) which was isolated by Dr. WHITE in 1949 and which behaves more like normal tissue than the strain (N) of the same hybrid which was isolated in 1937 (Figs. 19 and 20) and which is tumorous. The NA strain showed a much higher concentration of ninhydrin-reactive constituents per unit fresh weight of tissue than strain N at 2 months of culture and also in 4–5 month cultures (Figs. 21 and 22, NA strain; Figs. 23 and 24, N strain). Although the patterns of the two strains obviously differ from each other, there appears to be more resemblance between them than noted between any of the previously discussed samples shown for either normal or tumor samples. Thus, there is a distribution of constituents in the samples of strain N shown in Figs. 23 and 24 which is similar to that found in some samples of strain NA (Figs. 17, 18, and 21). In all instances there was less glutamine in strain N.

The above results show clearly that each of the excised plant tissues grown on the same medium attains its own characteristic pattern of ninhydrin-reactive constituents and that no aspect of the pattern can be related to the "normality" or "malignancy" of these tissues. These results cannot be held to be strictly analogous to those obtained with animal tissues, since a wide range of plant species was examined, while the animal tumors were only from rats and mice.

ACKNOWLEDGEMENT

This work was supported in part by grants C-2568 (to E.R.) and C-2061 (to P.R.W.) from the National Cancer Institute, National Institutes of Health.

REFERENCES

1 J. REINERT AND P. R. WHITE, *Physiol. Plantarum*, 9 (1956) 177.
2 P. R. WHITE, *Plant Physiol.*, 12 (1937) 793.
3 P. R. WHITE, *Plant Physiol.*, 14 (1939) 527.
4 P. R. WHITE, *Ann. Rev. Plant Physiol.*, 2 (1950) 231.
5 E. ROBERTS AND D. G. SIMONSEN, in J. T. EDSALL, *Amino Acids, Proteins and Cancer Biochemistry*, Academic Press Inc., New York, 1960, p. 121.

INVITED DISCUSSION

THE USE OF ^{14}C-LABELED COMPOUNDS IN METABOLIC STUDIES: EFFECTS DUE TO AUTOCLAVING ^{14}C-UREA

J. K. POLLARD*

Department of Botany, Cornell University, Ithaca, N.Y. (U.S.A.)

In the course of certain studies on the metabolism of urea (LYNDON, POLLARD AND STEWARD[1]), urea seemed to form a compound toxic to carrot tissue cultures on autoclaving. To help characterize the change in urea on autoclaving, samples of ^{14}C-urea were autoclaved in the medium used to grow the carrot tissue cultures, and then analyzed by paper chromatography and radioautography with the results shown in Fig. 1.

Apparently, urea at least, forms a bewildering array of unidentified products when it is autoclaved in a somewhat complex medium. While none of the compounds so produced have been specifically identified, this note merely emphasizes the range of substances easily formed from a compound as simple as urea when it is autoclaved. There are two implications in these results. The first is that compounds produced by autoclaving could be mistakenly attributed to metabolic transformations of the

Fig. 1 A + B. Radioautographs of the chromatograms of (A) ^{14}C-urea and (B) of ^{14}C-urea autoclaved in modified White's medium containing 10% coconut milk.

* Present address: Calbiochem, 3625 Medford St., Los Angeles 63, Calif., U.S.A.

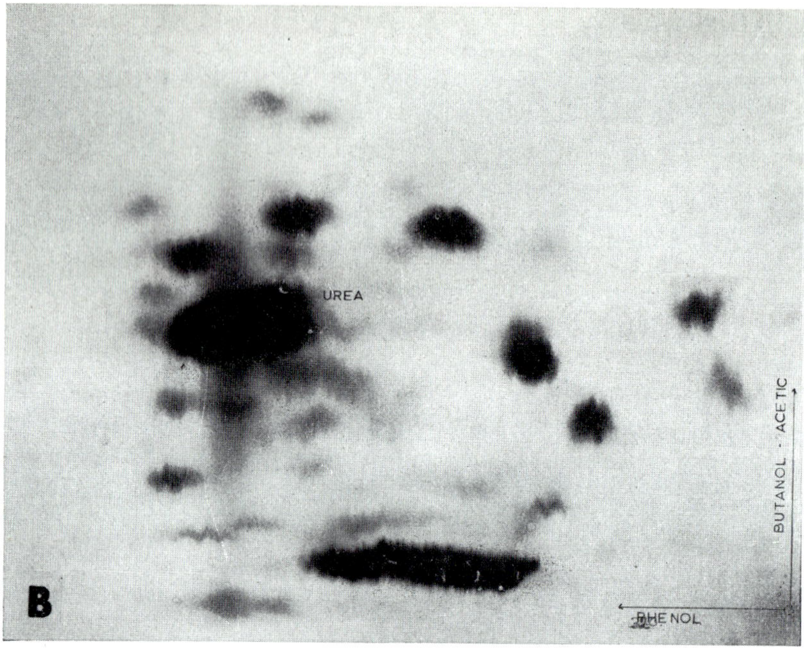

Fig. 1B. For legend see p. 69.

^{14}C-urea. The second is that if the substance in question had not been radioactive, the variety of compounds to which it may give rise in the medium would never have been detected.

Consequently, in studies of this sort, care must be taken to evaluate the effects of autoclaving on the actual composition of the final medium which is used in tracer experiments.

REFERENCE

[1] R. F. LYNDON, J. K. POLLARD AND F. C. STEWARD, *Plant Physiol.*, Suppl., 34 (1959) xx.

DISCUSSION

Chairman: EUGENE ROBERTS

POLLARD: Recently, we have repeated the isolation of an amino acid from *Phlox*, which was first described by VIRTANEN AND HIETALA[1]. Although the optical rotation was not determined by VIRTANEN, this substance was characterized chemically as γ-hydroxyglutamic acid. The original published analysis showed that this particular γ-hydroxyglutamic acid was not a hydrate and could be presumed to be identical with γ-hydroxyglutamic acid "A", later synthesized and resolved by BENOITON *et al.* (*cf.* WINITZ's contribution to this symposium). FOWDEN[2] has also isolated γ-hydroxyglutamic acid from *Hemerocalis* and has determined its optical rotation. With Dr. WINITZ's assistance, we have also determined the optical properties of the γ-hydroxyglutamic acid in *Phlox*. Consequently, by all the criteria now proposed, the γ-hydroxyglutamic acid from *Phlox* is found to be identical with the synthetic material called by BENOITON *et al.* "L-allo-γ-hydroxyglutamic acid".

The term *allo* was applied to this particular isomer because it was correlated stereochemically with L-allo-γ-hydroxyproline. This use of the term *allo* really implies more stereochemical significance than the term usually conveys. *Allo* simply means "the other one". Wherever possible, as in the case under discussion, the term should be used to designate that isomer, whatever its absolute configuration, which is different from the first discovered isomer, in this case, a natural product. Since no absolute rule governs the assignment of the term *allo*, some element of choice remains. However, the most important thing is the determination of absolute structure in the manner for which Dr. WINITZ stands. Therefore, so long as the absolute structure is unequivocally established, the choice of names, within the existing framework of rules, remains somewhat secondary so that the name as proposed by BENOITON *et al.* can certainly be retained. However, it is suggested that this use of the term *allo* should not become a precedent to cover generally a structural relationship with some other allo compound, for if this were adopted, the original usefulness of the term would be lost.

WINITZ: With regard to Dr. POLLARD's comments concerning the nomenclature of the γ-hydroxyglutamic acids, it will be recalled that γ-hydroxyglutamic acid was first discovered in nature by VIRTANEN in about 1955. As its configuration and optical rotation were not then ascertained, there was no way of knowing which of the four possible stereoisomers it actually represented. Such was the situation in 1957 when we prepared the material synthetically, separated the racemic diastereomers, labeled one racemate "A" and the other "B" according to the FISCHER convention, resolved each racemate, and then correlated the configurations of the antipodes of the A and B forms with those of the antipodes of hydroxyproline and allohydroxyproline. Some four years later, only a few short weeks ago, the material was isolated by STEWARD AND POLLARD from *Phlox decussata* seed, and it was found that the rotation of this natural material corresponded to that of the synthetic antipode to which we had previously assigned the designation of allo-γ-hydroxy-L-glutamic acid. Now, Dr. POLLARD presumes to question the use of the term *allo* in this situation because neither does its application to the naturally occurring form appeal to his esthetic senses nor were we prophetic enough to predict the optical identity of the natural material some four years prior to its actual characterization.

It should be emphasized that the term *allo* has no configurational meaning whatsoever. It just means the other form, and it makes no difference whether we designate the natural form as *allo* or *normal* as long as there is no confusion with regard to its optical configuration. In this connection, it should be recalled that both the *normal* and *allo* forms of several amino acids, such as the hydroxyprolines and the isoleucines, have been found in nature. It should further be recalled that ADAMS AND GOLDSTONE demonstrated that an enzyme from mammalian tissues converts hydroxyproline to the normal form of γ-hydroxyglutamic acid; allohydroxyproline, however, was not susceptible to the action of this enzyme and not converted to the allo-γ-hydroxyglutamic acid isolated from plants. It is obvious, therefore, that the use of the designations of *normal* and *allo* in connection with the natural occurrence of a given amino acid not only has no significance, but is a use for which these terms were never intended.

AXELROD: The isolation of β-aminoisobutyric has been mentioned. I should just like to add an

References p. 72

interesting side note here, that EVANS AND TSAI in my laboratory fed radioactive thymine to root sections of the Wedgewood strain of iris and isolated β-aminoisobutyric acid. Of course, it would have been quite surprising to have found it otherwise.

THOMPSON: Yes. Of course, we more or less assumed it came from thymine.

E. ROBERTS: HANES' early work with the γ-glutamyl peptides suggested the possibility that these compounds might be related to important metabolic sequences in living cells. Would somebody care to hazard a guess as to what role such compounds, besides glutathione and glutamine, might be playing in the metabolism of cells, in protein synthesis, or other processes.

THOMPSON: Well, of course, we have thought about this quite a bit. In our paper we suggested that they may have something to do with storage and with translocation, but we really do not have any evidence. The occurrence of such large quantities of these compounds certainly is very intriguing. We have seven or eight γ-glutamyl compounds, and it appears that many more may be found. At present, one can only surmise that their occurrence in these rather large quantities indicates that they play a role in normal metabolism. On the other hand, HIRD suggested that the only reason he could see for the occurrence of the γ-glutamyl transpeptidase, which is the enzyme that HANES originally reported on in 1950, was that it provides employment for biochemists.

HENDLER: I might mention that I spent a couple of years working on the significance of γ-glutamyl peptides in protein biosynthesis in Dr. GREENBERG's laboratory. We synthesized γ-glutamyl glycine labeled with radioactive glycine, and compared the rate of uptake of free glycine and glycine, so bound, into protein in about five or six different protein synthesizing systems. In every case free glycine was much more efficiently taken into the protein than γ-glutamyl peptide bound glycine. In no case did we ever find an uptake of radioactive amino acid into γ-glutamyl peptides when labeled amino acids and various unlabeled γ-glutamyl compounds supplied as donors of the γ-glutamyl group were incubated together. One other point that was advanced favoring the participation of γ-glutamyl peptides in protein synthesis was their resistance to hydrolysis by cellular peptidases. However, we found them to be quite readily hydrolyzed. We found no evidence whatsoever that γ-glutamyl peptides played a role in amino acid incorporation in the variety of systems that we studied.

E. ROBERTS: I wonder if there has been any evidence that the γ-glutamyl group can be activated by amino acid-activating systems.

LOFTFIELD: TUBOI AND HUZINO[3] in Japan found activation of a number of peptides by amino acid-activating enzymes. I do not know whether they included any glutamyl peptides. Other than that there is, of course, the synthesis of glutathione, but this is quite a different process from the ordinary amino acid-activating system.

E. ROBERTS: I still have a question about the problem that Dr. STEWARD brought up with regard to the synthesis of γ-aminobutyric acid from succinic semialdehyde as a natural pathway in plants. I am willing to be convinced, but it does seem to me that there is, at present, no known metabolic source of succinic semialdehyde that has been proven. Of course, succinic semialdehyde could come from α-ketoglutarate if the oxidase system were not working properly. In *in vitro* experiments using preparations from animal tissues, the formation of succinic semialdehyde from α-ketoglutarate has been shown to take place only when the α-ketoglutarate oxidase system has been partially inactivated.

REFERENCES

1 A. I. VIRTANEN AND P. K. HIETALA, *Acta Chem. Scand.*, 9 (1955) 175.
2 L. FOWDEN, *Ann. Rep. Progr. Chem.*, 56 (1959) 359.
3 S. TUBOI AND A. HUZINO, *Arch. Biochem. Biophys.*, 86 (1960) 309.

III. MICROORGANISMS

THE COMPOSITION OF MICROBIAL AMINO ACID POOLS

JOSEPH T. HOLDEN

Department of Biochemistry, Medical Research Institute, City of Hope Medical Center, Duarte, Calif. (U.S.A.)

This review summarizes a representative sample of the available information concerning the composition of amino acid pools in bacteria, yeasts, molds and other fungi. Most of the reports cited concern studies in which the primary intention was a description of the complete pool or a major part of it. Reports on individual components are less frequently included, usually only when a substance of uncommon structure or function is described. Studies on pool turnover, active transport, protein synthesis, etc., occasionally include descriptions of pool contents. Such information, when encountered, has been included unless extensive duplication in other sections of this volume could be anticipated.

From a historical viewpoint the first clear indication that microorganisms contain a diverse collection of apparently free amino acids must be credited to ERNEST GALE *et al.*[54, 56, 57, 178]. Using amino acid decarboxylases as an analytical tool, his group demonstrated that various bacteria and yeast contain large amounts of readily extractable amino acids. ROINE[151, 152, 153] simultaneously reported the occurrence in yeast of non-protein amino nitrogen compounds and while he identified alanine, glutamic acid and glutamine, the classical methods of isolation and identification used precluded a complete description of the pool. The work of GALE AND TAYLOR also provided an incomplete description of the pool since the amino acid decarboxylases used for quantitative analysis restricted their study to those amino acids which were substrates for these enzymes. The presence of a varied pool, however, was clearly apparent, and the development and widespread application of paper chromatographic methods which occurred at this time made possible the more complete studies which will be considered here.

Prior to this time, there had been little work done on microbial amino acid composition and such studies were concerned entirely with the composition of cell hydrolysates[27, 175]. The development of microbiological assay methods for the amino acids provided a convenient tool for such studies, but its extension to studies of the freely extractable pool did not occur until later and it was some time before the assay organisms themselves were subjected to analysis.

DESCRIPTION OF POOL CONTENTS

Examples of the available information concerning amino acid pool composition will be found in Tables I–XI. A number of general comments are required in view of the divergent techniques, organisms and objectives encountered in these studies.

A number of extraction procedures have been used to liberate pools from microbial cells. GALE[56] and GALE AND TAYLOR[57] originally found that boiling water, mechanical disruption and treatment with detergent substances liberated the pool completely. Most investigators since then have used boiling water or warm ethanol, although cold trichloroacetic and perchloric acids recently have found increasing use. The small number of reports encountered in which the comparative effectiveness of different extraction procedures was established, suggests that this factor is infrequently investigated. HANCOCK[71] recently compared nine extraction methods using *Staphylococcus aureus* and found them to liberate essentially equal amounts of amino acids. However, small differences in the amounts of individual components can be observed in his chromatograms. SPIEGELMAN *et al.*[172] have verified the efficacy of brief treatment with boiling water as a means of liberating unbound glutamic acid from yeast cells. However, occasional discrepancies are encountered as in the study of LINDENBERG AND MASSIN[111] who observed that different amounts of tyrosine are liberated when yeast cells are treated with cold trichloroacetic acid and boiling water. ALLEGRA *et al.*[2] also observed differences in ethanolic and water extracts of *Salmonella bareilly*. Such results suggest that completeness of extraction should not be taken for granted especially when a new organism or unusual cultural conditions are used. They also bear on the possibility that the pool is not a homogenous entity.

With few exceptions, microbial pools have been examined by two-dimensional paper chromatography. The recognized limitations of one-dimensional chromatography have led to its virtual abandonment for examination of the total pool, although recent work with the multiple redevelopment technique[140] suggests that it may find renewed use. Circular chromatography also has been used occasionally, but again the relatively uncertain resolution achieved when complex mixtures are examined jeopardizes the significance of the observations. Column chromatography, enzymatic and microbial methods have been employed relatively infrequently. It must be recognized that the pool may contain many components at levels below the detectability of the ninhydrin reagent and, therefore, may be much more complex than the reports summarized here indicate.

Quantitative estimation of microbial pool components, when attempted at all, most often has involved some form of photometric measurement of ninhydrin-spot color intensity. In most cases, however, amino acids are reported only as present or absent, or on three- or four-step scales of relative amounts arrived at by visual inspection of chromatograms. The data in Tables I–XI are presented in one of these three systems whichever corresponds most closely to the method used in the original report. Quantitative values have all been recalculated when necessary and are given as μmoles/100 mg dry cells; semiquantitative systems have been converted to a three-step scale (▲, ▲▲, ▲▲▲); present or absent (usually reported as + or −) are indicated by the presence or absence of a filled circle. Examination of a large number of tables suggests that the use of geometric symbols more successfully conveys the impression of a chromatogram and permits quicker identification of differences between organisms. It should be recognized that all such systems are crude approximations of the original chromatographic observations and that photographs of original chromatographs should be referred to whenever possible.

With few exceptions, spots on chromatograms have been named on the basis of relative position using a map of known amino acids for guidance. The potential hazards of this method seem to have been avoided only because the majority of microbial pools, at least when examined by the paper chromatographic method, appear to be composed largely of amino acids commonly found in protein. If one were to apply the criteria proposed by WINITZ (*cf.* this Symposium) for identification of amino acids as a qualification for consideration, this review would be confined to a discussion of no more than 10% of the reports which are cited.

Bacteria

The pool composition of a few gram-positive bacteria is presented in Table I. The lactic acid bacteria have been studied more intensively than any other bacterial group, largely through the efforts of CHEESEMAN, BERRIDGE, MATTICK and others at the University of Reading who have investigated the application of pool chromatography as an aid in taxonomic studies (*cf.* refs. 13, 30–34, 62, 63, 119, 161). BOTTAZZI[17–20], originally a collaborator of the Reading group, also has presented results of a similar study. In many instances, a large number of strains of a given organism have been compared. These results have been averaged for inclusion in Table I. The observations of this group are atypical to the extent that many of the commonly encountered amino acids are not found on their chromatograms whereas a large number of unknowns are reported. Cells were extracted with acetic acid, a method which appears to favor the demonstration of these components. This was useful for the purposes of their study since differences in distribution of unknowns has aided the separation of some species. However, results for the same organisms examined in different laboratories using different growth media and extractants show a markedly different pattern of amino acids (*e.g.*, *Lactobacillus arabinosus* and *S. aureus*). Such comparisons will be drawn repeatedly to illustrate a major conclusion of this survey, that microbial pools are subject to such wide variability depending on strain, cultural conditions, medium composition and age that comparisons appear to be valid only when all experimental conditions are standardized. Therefore, for bacteria, and it will be seen later that the same is true also of yeasts and molds, it is not possible to speak of characteristic patterns of pool amino acids without specific description of all cultural conditions.

Another example of the wide discrepancy in qualitative and quantitative pool composition is seen in the two studies on *Bacillus subtilis* (Table I). It should be noted, however, that under closely controlled conditions the pool can have a strikingly stable composition. Figs. 1–4 illustrate results presented by the Reading group showing that chromatograms even of different strains of an organism are virtually superimposable. The study of PFENNIG[142] also illustrates that major modifications in intracellular metabolism, as when a bacterial spore germinates to form a vegetative cell, are accompanied by profound, reproducible changes in pool composition, suggesting that the latter might be utilized to provide clues to the detailed metabolic changes.

With the exception of the results reported by the Reading group, the data shown in Table I demonstrate that most investigators using two-dimensional paper chromatography without recourse to adjuncts such as radioisotopes find that the pool consists largely of the amino acids present in protein. Pool constituents of unusual structure or interest such as peptides, D amino acids, etc., have been reported less frequently and will be considered separately below.

TABLE I

AMINO ACID POOLS IN GRAM-POSITIVE BACTERIA*

Organism	*Lactobacillus arabinosus*	*Lactobacillus arabinosus*	*Lactobacillus delbrueckii*	*Streptococcus lactis*	*Staphylococcus aureus*	*Staphylococcus aureus*	*Bacillus subtilis*	*Bacillus subtilis (cell)*	*Bacillus subtilis (spore)*
Ref. numbers	119	88	161	20	63	71	115	142	142
Growth medium**	CO	DO	CO	CO	CO	CO	CO	CO	CO
Age (h)	24	14	24	20	18	4	48	14	264
Growth phase***	ES	ES	ES	ES	ES	LE	LE		
Glutamic acid	▲▲▲	▲▲	▲▲▲	▲	▲▲	4.0	23.0	21.0	2.7
Aspartic acid	▲▲	▲	▲▲	▲▲	▲▲	3.8	2.5	1.1	0.3
Glutamine		▲▲	▲						
Asparagine	▲	▲	▲	▲				3.3	0.6
Alanine	▲▲	▲▲▲	▲▲	▲▲	▲▲	0.81	11.0	4.5	0.5
Glycine	▲	▲	▲▲	▲▲		0.28	1.1	3.3	1.6
Threonine	▲	▲	▲▲	▲		0.10	5.5	3.4	
Serine	▲		▲▲	▲▲	▲▲	0.34	3.8	0.8	0.4
Lysine		▲	▲▲▲		▲▲	0.22	6.2	0.2	0.3
Arginine			▲			0.22	1.4		
Histidine						0.17	1.0		
Leucine/Isoleucine		▲	▲			0.26/0.85	9.9	1.1	0.3
Valine		▲	▲				7.5	1.3	0.3
Methionine		▲				0.67	1.2		
Proline		▲			▲	1.7			
Hydroxyproline					▲				
Tyrosine			▲			0.24	1.7		
Phenylalanine						0.13		0.5	
Tryptophane						0.13			
γ-Aminobutyric acid		▲		▲▲					
Cyst(e)ine		▲				0.55			
Unknowns	4	2	4	6	6	1		●	●

* In this and the following tables, the presence of an amino acid is indicated in one of three systems of symbols. When only presence and absence was reported: ● indicates presence, a blank space indicates absence. All semi-quantitative systems have been converted to the following three-step scale: ▲, present in small amounts; ▲▲, moderate amounts; ▲▲▲, large amounts. Comparisons can be made only within a column and not between columns. Quantitative data have been converted when necessary and are shown as μmoles/100 mg dry weight of cells.

** The following abbreviations are used in this and subsequent tables. CO, complex organic media in which extracts of tissues such as yeast, liver, and beef are used to provide amino acids, vitamins and other growth factors. DO, defined organic media in which amino acids and vitamins are provided as known substances. This category includes media in which vitamin free acid hydrolysates of casein are used to supply amino acids. CS, carbohydrate–salts media in which carbon is provided in a single substance generally glucose, and nitrogen is supplied as a salt or, in a few instances, as a single amino acid. I, completely inorganic sources of carbon and nitrogen.

*** The apparent phase of growth is indicated as follows: EE, early exponential; ME, mid-exponential; LE, late exponential; ES, early stationary; S, stationary. Whenever this information was not provided directly by the author an estimate is presented if at all possible. The same comment applies to age which often was not exactly specified.

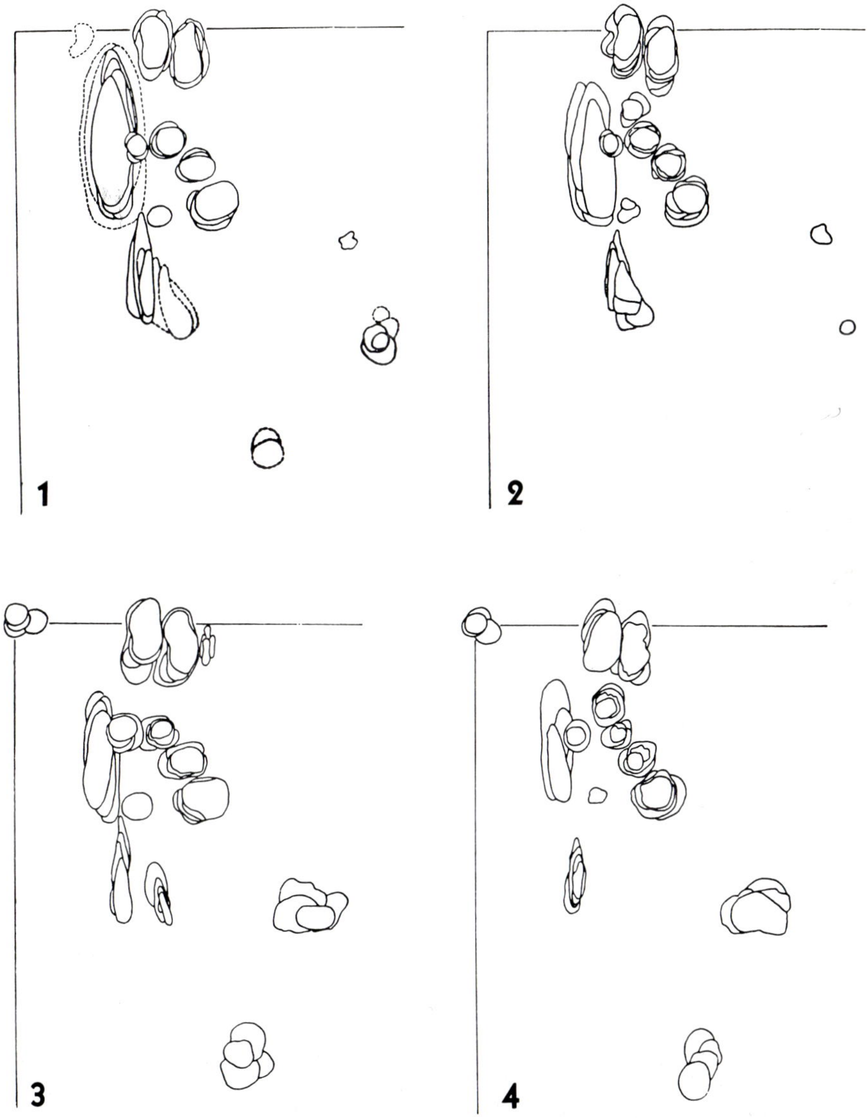

Figs. 1–4. From CHEESEMAN *et al.*[30]. Fig. 1. Superimposed tracings of chromatograms from six *L. casei* Group-1 strains. Fig. 2. Superimposed tracings of chromatograms from seven *L. casei* strains. Fig. 3. Superimposed tracings of chromatograms from four *L. plantarum* Group-1 strains. Fig. 4. Superimposed tracings from four *L. plantarum* Group-2 strains.

Descriptions of pools in the following gram-positive bacteria also have been reported: *L. acidophilus, L. helveticus, L. bulgaricus*[19, 119]; *L. brevis, L. fermenti*[119]; *Strep. cremoris, S. lactis*[20, 119]; *S. diacetilactis*[20]; *L. casei, L. plantarum*[18, 30]; *Staph. saprophyticus, S. lactis, S. roseus, S. afermentans*[63]; *L. lycopersici, L. brassicae fermentatae, L. pastorianus, L. caucasicus, L. buchneri, L. bifidus, L. viridescens, L. frigidus, L. cellobiosus, L. gayonii, L. mannitopoeus*[31]; *L. lactis*[19, 161], *L. leichmannii, L. delbrueckii, L. salivatius*[161]; *B. megaterium* (spore)[193]; *B. cereus*[127]; *Sarcina lutea*[45].

TABLE II

AMINO ACID POOLS IN GRAM-NEGATIVE BACTERIA*

Organism	*Escherichia coli*	*Pseudomonas saccharophila*	*Salmonella bareilly*	*Red Sulfur Bacterium*
Ref. numbers	116	118	2	129
Growth Medium**	CS	CS	DO	I
Age (h)		24	36	480
Growth phase***	LE	S	S	S
Glutamic acid	●	●	●	
Aspartic acid	●	●	●	
Glutamine		●		
Asparagine				
Alanine	●	●	●	
Glycine	●	●	●	●
Threonine	●	●	●	
Serine	●	●		●
Lysine	●	●	●	
Arginine	●		●	
Histidine		●	●	
Leucines	●	●	●	
Valine	●	●	●	
Methionine	●	●	●	●
Proline			●	
Hydroxyproline			●	
Tyrosine	●	●	●	
Phenylalanine	●		●	
Tryptophane				
γ-Aminobutyric acid				●
Cyst(e)ine				
Unknowns	●	3		

* See footnote * Table I.
** See footnote ** Table I.
*** See footnote *** Table I.

Following the report of TAYLOR[178], the gram-negative bacteria were thought to lack an amino acid pool. The data in Table II demonstrate that this impression was in error. A large number of investigators subsequently have found a diversified pool in these organisms, especially in *Escherichia coli*. This discrepancy appears partly to have risen from the osmotic lability of the pool in gram-negative bacteria[22, 98]. In contrast to gram-positive bacteria and yeast which retain their pools during repeated washing with water or solutions of low osmotic strength, organisms such as *E. coli* rapidly lose the pool under these conditions. In addition, the native pools in freshly harvested gram-negative bacteria appear to be smaller than those found in gram-positive organisms making their detection more difficult. Gram-negative in contrast to gram-positive bacteria also are frequently grown in media lacking preformed amino acids and, therefore, do not have an opportunity to augment the pool by accumulation of extracellular amino acids.

The assortment of amino acids found is usually as varied as that encountered in gram-positive bacteria. Generalizations regarding consistent differences between the two groups of bacteria are difficult to make. In contrast to the gram-positive bacteria, which frequently contain asparagine, pools in gram-negative bacteria appear not to contain this amino acid. Glutamine and tryptophane also are seldom reported. Despite their interesting nutritional requirements, little work on the autotrophic bacteria has been reported. Although the findings of MUKHERJEE[129] (Table II) indicate a very limited pool, judgement should be reserved until a more extensive study is presented including at least an examination of the effect of culture age and washing procedures. Interestingly, the algae, with a comparable ability to synthesize nutrients, contain a large and varied pool of amino acids[53, 137]. MUKHERJEE's report is of interest also because γ-aminobutyric acid and taurine appeared when the extracted cell residue was hydrolyzed, suggesting the occurrence of bound forms of these substances. On the other hand, such hydrolysates did not contain glutamic acid, threonine, the leucines, lysine, arginine or histidine indicating a unique protein composition.

SKARZYNSKI AND OSTROWSKI[168] have reported the occurrence of a variety of sulfur compounds (cysteine, cystine, methionine, cysteic acid, cysteine sulfinic acid and

TABLE III

COMPARISON OF TOTAL AMINO ACID POOL IN GRAM-POSITIVE AND -NEGATIVE BACTERIA

Organism	*Total Pool (µmoles/100 mg dry wt.)*	*Ref. numbers*
Staphylococcus aureus	28	71, 73
Bacillus subtilis	35	142
Bacillus subtilis	30–85	115
Escherichia coli	2–5	116
Staphylococcus aureus (+ chloramphenicol)	70	73
Escherichia coli (+ chloramphenicol)	11	116

α-thiooctanoic acid) in ethanolic extracts of *Thiobacillus thioparus*. In this case, whole-cell hydrolysates contained a typical mixture of protein amino acids. This group also has reported the interesting finding that this organism utilizes only the outer sulfur atom of thiosulfate leaving the inner sulfur atom in the medium as sulfate[169, 170].

Amino acid pools in the following gram-negative bacteria also have been described: *Aerobacter* sp.[166]; *Agrobacterium tumefaciens*[139]; *E. coli*[171]; *Pseud. hydrophila*[39]; *Pseud. savastanoi*[139]; *Pseud.* sp. [132]; *Rhizobium meliloti*[98]; *Vibrio cholera*[16].

Despite the uncertainty and controversy concerning the occurrence and size of amino acid pools in gram-negative bacteria, only a few reliable quantitative studies have been reported. Table III summarizes some values for gram-positive and -negative bacteria. The pool in freshly harvested cells of *E. coli* is about one tenth as large as the pools in gram-positive organisms. In both types, incubation in the presence of chloramphenicol causes a sizeable increase in pool size, but the *E. coli* pool is still relatively smaller. When incubated with a single amino acid many organisms have been observed to accumulate large amounts of this amino acid intracellularly. These intracellular levels also can be used as an index of the pool capacity. Results of representative studies are summarized in Table IV. Under these circumstances gram-positive bacteria accumulate single amino acids in amounts corresponding to total pool levels observed in freshly harvested cells (Table III). *E. coli* has been observed to accumulate moderate

References p. 105/108

amounts of proline, but when the extracellular osmotic strength is increased materially, proline or a mixture of amino acids can be accumulated to levels as high as those found in gram-positive bacteria. It is clear that while the native pool in *E. coli* is distinctly lower than the comparable pool in gram-positive organisms, this bacterium has the ability to retain very large pools under favorable osmotic conditions. The relation between extracellular osmotic strength and intracellular accumulation of single amino acids is discussed in detail elsewhere in this volume (HOLDEN, BRITTEN).

Table V summarizes results for the tubercle bacillus, Corynebacterium, and related organisms. *Mycobacterium tuberculosis* has been studied extensively and found to contain a diversified pool whose composition varies with age. In the two studies cited cells were cultured under closely similar conditions, but the qualitative com-

TABLE IV

POOL CAPACITY FOR EXOGENOUSLY SUPPLIED AMINO ACIDS

Organism	*Extracellular amino acid*	*Intracellular level (µmoles/100 mg dry wt.)*	*Ref. numbers*
Staphylococcus aureus	Glutamic acid	18–54	59
Streptococcus faecalis	Glutamic acid	45	87
Lactobacillus arabinosus	Glutamic acid	50–95	84
Escherichia coli	Proline	2–24	21, 23
Escherichia coli (+ sucrose)	Proline or amino acid mix	100	23

position found differs markedly. The results of another study[90, 91] disagree with the two examples shown. When cultured in an amino acid containing medium, *Corynebacterium diphtheriae* contains a remarkably varied pool including hydroxylysine which is an infrequently encountered constituent of microbial pools. WORK's investigation[188] was one of the first complete pool studies to follow GALE's reports and resulted in the detection of an unknown which was subsequently isolated and identified by WORK as diaminopimelic acid[189, 190]. This is one of the few examples in which chromatographic study of a microbial pool has revealed the existence of unusual components whose further investigation has provided insight to new metabolic pathways, in this case cell wall composition and synthesis (*cf.* WORK[191], SALTON[155]). BLASS AND MACHEBOEUF[15] independently encountered an unknown in *Vibrio chlolera* which they subsequently recognized to be diaminopimelic acid[16] following WORK's identification of this substance. The PW-8 strain of *C. diphtheriae* also has been studied by ISKIERKO[92] who found it to have a pool composition materially different from that reported by WORK. As indicated previously, this situation recurs with all classes of organisms and illustrates the sensitivity to change of microbial amino acid pools. The *Actinomyces* pool (Table V) also has a diversified composition even when the organism is cultured with an inorganic nitrogen source. On the other hand, another report on *A. violaceus*[11] shows a more limited pool composition. The taxonomically related organism *Nocardia rugosa* has a distinctly restricted pool.

TABLE V

AMINO ACID POOLS IN CORYNEBACTERIUM, MYCOBACTERIUM AND RELATED ORGANISMS*

Organism	*Mycobacterium tuberculosis var. hominis H37RV*	*Mycobacterium tuberculosis var. hominis H37RV*	*Corynebacterium diphtheriae PW-8*	*Actinomyces phaeochromogens*	*Nocardia rugosa*
Ref. numbers	141	51	188	14	10
Growth Medium**	CS	CS	SO	CS	CO
Age (days)	21	20	7		3
Glutamic acid	●	●	●	●	●
Aspartic acid	●	●	●	●	●
Glutamine		●			
Asparagine					
Alanine	●	●	●	●	●
Glycine	●	●	●	●	●
Threonine			●	●	●
Serine			●	●	
Lysine	●	●	●	●	
Hydroxylysine			●		
Arginine		●	●	●	
Histidine				●	
Leucines	●	●	●	●	●
Valine	●	●	●	●	●
Methionine			●	●	
Proline			●	●	
Tyrosine		●	●	●	
Phenylalanine			●	●	
Tryptophane					
γ-Aminobutyric acid	●		●		
β-Alanine			●		
α-Aminobutyric		●	●		
Cyst(e)ine			●	●	
Diaminopimelic acid			●		
Unknowns	●		●		

* See footnote * Table I.
** See footnote ** Table I.

The bacterial spore with its unique physiological properties has been examined by a number of investigators. The study of PFENNIG on *B. subtilis* spores already has been cited (Table I, ref. 142) to show that the pool is less varied and about one fifth the size of the vegetative cell pool. Dipicolinic acid (pyridine-2,6-dicarboxylic acid) has a unique occurrence in bacterial spores (*cf.* ref. 145) and considerable interest in its relation to the germination process and heat resistance of the spore[36] has developed. Dipicolinic acid appears on the basis of recent studies to be in the free state within the mature spore. A number of studies have shown, for example, that mechanical disruption as well as other procedures such as electrodialysis and treatment with detergents all release dipicolinic acid in a soluble, apparently uncombined form[144, 150, 193]. YOUNG[194] also observed release by mechanical disruption and hot water extraction, but her chromatographic findings suggest that this substance may occur in a combined form.

TABLE VI

AMINO ACID POOLS IN YEAST*

Organism	*Baker's yeast*	*Brewer's yeast*	*Saccharomyces cerevisiae*	*Saccharomyces cerevisiae*	*Torulopsis utilis*	*Candida utilis*
Ref. numbers	110	110	173	67	123	41
Growth Medium**	—	—	DO	CO	CS	CS
Age (h)	—	—	15	12		
Growth Phase***	—	—	LE	ME	LE	ME
Glutamic acid	6.2	4.8	6.5	10.0	8.1	29.0
Aspartic acid	0.9	2.9	2.2	3.6	0.4	0.9
Glutamine		1.0	9.5		1.8	
Asparagine	1.0	1.5	9.8		0.9	
Alanine	2.5	3.8	4.5		5.9	24.0
Glycine	0.1	0.5	2.9		1.0	11.0
Threonine	1.2		4.1	2.2	0.3	0.8
Serine	0.4	0.7	2.3	3.7	0.5	0.8
Homoserine					●	
Lysine	0.6	1.0	2.8	4.3	2.3	2.4
Ornithine		2.4	2.0		●	
Arginine	0.4	1.9	4.9	1.2	1.8	6.3
Citrulline			1.7		●	
Histidine	0.4	0.3	0.2	1.0	●	
Leucine/Isoleucine	0.5/0.5	0.5/0.5	1.1	0.5/0.6	0.1/0.1	12.0
Valine	1.2	1.9	1.9	1.6	0.2	6.5
Methionine			1.0	0.6	0.1	
Proline	1.2	1.9	●	0.9	●	0.7
Tyrosine	0.6	0.5	●	0.2	0.1	
Phenylalanine	1.2		●	0.3	0.1	
Tryptophane			●	0.03		
γ-Aminobutyric acid	0.3	0.5			●	
α-Aminobutyric		0.1	0.3			
Cyst(e)ine				0.1		
Cysteic acid			●			
Glutathione	1.9	1.4	1.7		●	
Ethanolamine					●	
Glycerylphosphoryl-ethanolamine					●	
Unknowns					●	

* See footnote * Table I.
** See footnote ** Table I.
*** See footnote *** Table I.

Yeast

Representative examples of yeast pool analyses are presented in Table VI. Following the early studies of TAYLOR[178, 179] and ROINE[151, 152, 153], the investigation of LINDAN AND WORK[110] was one of the first relatively complete descriptions of a yeast pool to appear. Unfortunately, commercial preparations of dried yeast were used so that the reliability of the values is open to question. The qualitative composition reported for brewer's yeast, however, compares closely with that found in another early study by LJUNGDAHL AND SANDEGREN[41] except for the detection by the latter workers of nine unidentified substances. The yeast pool is clearly very rich in amount and number of constituents. A number of compounds not usually found in bacteria have been reported including homoserine, ornithine, citrulline, ethanolamine, glycerylphosphorylethanolamine and α-aminobutyric acid. Glutathione is present in sizeable amounts in a number of yeast pools. It too is infrequently encountered in bacteria although ŠORM AND ČERNÁ[171] and the Carnegie Institution group[149] have reported it in *E. coli*. In addition, MIETTINEN[123] has reported an unknown in *Torulopsis utilis* which appears to contain citrulline and the constituents of glutathione. The values shown for *Candida utilis* were derived from the ^{14}C content of chromatographically separated amino acids extracted from cells grown on ^{14}C-labeled fructose. The amounts of glutamic acid, alanine, glycine, the leucines and valine are unusually high in comparison to the other organisms. HALVORSON AND SPIEGELMAN's data were obtained using microbiological assay. No value is reported for glutamine even though it does appear on their chromatograms[66] since this substance would be measured as glutamic acid. Alanine also is not reported. Its universal presence in other yeast strains suggests that this is a reflection of the difficulty of measuring this amino acid microbiologically at the time these analyses were performed.

Some of the most comprehensive investigations on the composition and turnover of amino acid pools have been carried out with yeast. In addition to the studies of COWIE *et al.* (*cf.* ref. 41 and COWIE, this Symposium) there have been the extensive studies of HALVORSON AND SPIEGELMAN *et al.*[66–70, 172] (see also HALVORSON, this Symposium) on various species and strains of *Saccharomyces* and of MIETTINEN on *Torulopsis utilis*[123]. As shown in Table VI, these studies have provided much information on the composition of the pool and have served also to demonstrate that the pool is utilized as a source of amino acids during protein synthesis. This subject is considered in detail in other sections of this volume.

In addition to the examples cited above, the amino acid pools of the following yeasts have been described: Baker's yeast[8]; *Candida pelliculosa*[180]; *Endomycopsis vernalis*[174]; *Saccharomyces ellipsoideus*[130]; *S. carlsbergensis*, *S. fragilis*, *S. chevalieri*, *S. ludwigii*, *S. cerevisiae*, and various other strains[70, 172].

Molds and other fungi

Representative values for pool amino acids in selected molds are shown in Table VII. A number of investigations on *Neurospora* have been reported, among them the elegant study by FUERST AND WAGNER[55] who examined a large number of biochemical and morphological mutants to determine whether consistent differences from the wild-type pool pattern could be demonstrated. As a precaution against pool variability not specifically associated with the mutation under study all strains which appeared to show differences were back-crossed to wild type and the mutant recovered. This process was repeated five times with the expectation that the background genetic material in wild type and mutant would thereby be substantially similar. Only a small number of mutants with distinctly altered pool compositions were encountered. Adenine-less mutants lacked leucine or phenylalanine and contained less glycine, γ-aminobutyric acid and lysine. Methionine mutants contained more threonine and glutamine. Although relations could be perceived between such changes and known metabolic pathways, pool studies did not seem to provide substantial advantages over the conventional method of examining culture filtrates of mutants as a means of detecting modifications in intracellular metabolism. A total of 35 ninhydrin-reactive substances

TABLE VII

AMINO ACID POOLS IN MOLDS*

Organism	*Neurospora crassa EM 5297a*	*Neurospora crassa 2a*	*Neurospora sitophila (conidia)*	*Penicillium chrysogenum*	*Penicillium chrysogenum*	*Aspergillus oryzae*	*Aspergillus flavus*
Ref. numbers	55	136	138	93	147	162	143
Growth Medium**	CS	DO	CS	—	CO	DO	CS
Age (days)	5	2	3	—	2	3	5
Glutamic acid	0.70	4.5	0.50	5.8	▲▲	●	3.7
Aspartic acid	0.09	1.9	0.05	1.9	▲▲	●	3.0
Glutamine	0.50		1.22		▲	●	
Asparagine			0.67				
Alanine	1.85	6.2	3.40	8.4	▲▲▲	●	9.9
Glycine	0.12	2.6	0.36	5.1	▲▲▲***	●	1.9
Threonine	0.29	2.6	0.38	3.2			1.5
Serine	1.25	2.6	0.37	14.3		●	1.4
Lysine	0.32	2.4	0.19	2.6		●	2.1
Ornithine					▲§	●	
Arginine			0.11	2.9	▲		2.0
Histidine			0.10	1.7	▲	●	
Leucine/Isoleucine	0.68§§	3.7	0.68	6.5/9.5	▲▲	●	1.8/1.3
Valine	0.48	3.2	0.45	12.8	▲	●§§§	2.9
Methionine			0.47	2.6			●
Proline			0.04	7.4	▲	●	1.7
Tyrosine	0.09	1.5	0.21	0.4			1.3
Phenylalanine			0.20	1.5		●	0.7
Tryptophane				7.4			0.5
γ-Aminobutyric acid	0.05		1.30			●	2.2
β-Alanine	0.18						
α-Aminobutyric acid	0.17						
Cyst(e)ine				●			2.9
Cysteic acid			0.17			●	
Glutathione	0.31						
Unknown	2		11				

* See footnote * Table I.
** See footnote ** Table I.
*** Reported as glycine and/or serine.
§ Reported as ornithine and/or lysine.
§§ Reported as leucine–phenylalanine.
§§§ Reported as valine–methionine.

were encountered in all the strains studied including taurine which is rarely found in microorganisms. As can be seen in Table VII, individual strains did not contain all these substances. In the other report on *Neurospora* shown in Table VII, the mold was grown in media containing one of several amino acids. Although fewer amino acids were reported (one-dimensional chromatography) all of them were present at levels at least twice as high as those reported by FUERST AND WAGNER. The latter authors did investigate the effect on the wild type pattern of growing the mold with single amino acids, purines and vitamins, and observed both increases and decreases in levels of individual amino acids. Conidia of *Neurospora sitophila* have been examined

TABLE VIII

AMINO ACID POOLS IN VARIOUS FUNGI*

Organism	*Fusarium javanicum*	*Fusarium javanicum*	*Fusarium decemcellulare (conidia)*	*Puccinia graminis (urediospores)*	*Tilletia caries (spores)*	*Phytophthora cactorum*	*Colletotrichum capsici*	*Fusarium vasinfectum*	*Claviceps litoralis*
Ref. numbers	37	185	182	26	131	37	160	135	12
Growth Medium**	CS	CS	CO	—	CS	CS	CS	CS	CO
Age (days)	24	14	15–20	—	14	29	3	16	20
Glutamic acid	▲▲▲	▲▲	●	●	▲▲▲	▲▲▲	●	1.4	0.20
Aspartic acid	▲▲▲	▲▲	●	●	▲▲▲	▲▲	●	1.4	0.12
Glutamine	▲▲▲	▲		●		▲▲▲	●		
Asparagine	▲▲▲			●		▲▲▲	●		
Alanine	▲▲▲	▲	●	●	▲▲▲	▲▲▲	●	1.2	0.28
Glycine	▲▲	▲		●	▲▲▲	▲▲	●	0.88	0.09
Threonine	▲▲	▲	●	●	▲	▲▲	●	0.54	●
Serine	▲▲	▲	●	●		▲▲	●	0.72	0.17
Lysine	▲▲	▲		●	▲▲▲	▲▲	●	0.65	0.04***
Ornithine						▲			
Arginine	▲▲	▲	●	●	▲▲▲	▲▲▲	●	0.41	
Histidine	▲▲			●	▲	▲▲			
Leucine/Isoleucine	▲▲		●	●	▲▲	▲▲	●	1.2	0.15
Valine	▲▲§	▲	●	●	▲▲	▲▲§	●	0.46	0.24
Methionine				●					0.20
Proline		▲			▲	▲▲▲	●		
Hydroxyproline							●		
Tyrosine	▲	▲		●	▲	▲	●	0.35	0.08
Phenylalanine		▲		●	▲		●	0.48	
Tryptophane									
γ-Aminobutyric acid	▲		●	●	▲▲▲	▲▲▲	●	0.48	
β-Alanine	▲					▲▲	●		
Cyst(e)ine		▲		●			●	0.59	0.08***
Cysteic Acid	▲					▲	●		
Djenkolic acid					▲▲				
Unknowns				11	●				

* See footnote * Table I.
** See footnote ** Table I.
*** Appears in pool at 35 days.
§ Reported as methionine and/or valine.

by OWENS *et al.*[138] for their content of free amino acids in the course of an extensive analysis which has accounted for 98% of the cell solids. A total of 30 ninhydrin-reactive compounds were reported including 11 unidentified substances. In contrast to bacterial spores, *Neurospora* conidia would seem to have a more varied pool than the vegetative mycelium. However, the discussion in preceding sections should suffice to establish the necessity of examining mycelial and conidial pools in a single investigation before such a conclusion can be expressed confidently.

There have been a number of attempts to correlate changes in medium and cellular free amino acids with the course of penicillin formation. For example, PYLE observed[147] that penicillin production follows the exhaustion of the amino acids in the medium. MELNIKOVA AND SURIKOVA[122] also observed that penicillin formation occurred when the intracellular pool amino acids were relatively high. However, identical changes were also observed in strains which do not produce the antibiotic. In any case, such studies have provided considerable analytical data for *Penicillium chrysogenum* some of which are shown in Table VII. The data of JANICKI AND SKUPIN[93] are particularly striking in view of the relatively high levels of serine, leucines, valine, proline and tryptophane reported.

SIMONART AND CHOW have studied the pool in *Aspergillus oryzae* and described the changes in its composition produced by variation in cultural and nutritional conditions[162–165]. The composition shown in Table VII includes all the compounds seen when the mold is grown at pH values between 3 and 7. Ornithine and γ-aminobutyric acid are found only at the lower pH's, whereas at higher values most of the amino acids occur in larger amounts. This study will be considered again later.

A large number of fungi have been examined, particularly organisms such as the smuts and rusts which parasitize commercial crops. Table VIII shows the pool composition of various other fungi selected to illustrate the variety of pool types in this group of organisms, the lack of obvious taxonomic correlation and again the lack of agreement between separate studies. The pools generally contain a wide assortment of amino acids but the few quantitative studies available indicate that the total amounts of amino acids contained therein are not large. γ-Aminobutyric acid and β-alanine are frequently found whereas hydroxyproline has been cited in only one report[160] and tryptophane appears uniformly not to be encountered although it was found in some of the organisms included in Table VII. CLOSE[37] has reported on the pool in eight fungi selected to represent various taxonomic and nutritional types. No obvious correlation between pool composition and these properties was encountered. CLOSE mentioned the possible occurrence of α-aminoadipic acid, 3,4-dihydroxyphenylalanine, ethanolamine and taurine in some of these organisms. Except for *Tilletia caries* mycelium[131] which may contain ethanolamine, and *Neurospora*[55] some strains of which contain ethanolamine and taurine, these substances have not been reported in the fungi. The two reports for *Fusarium javanicum* are in substantial disagreement, again most likely due to difference in strain and cultural conditions. The report by VENKATA RAM[185] includes data for 22 species of this genus. *Fusarium* conidia appear to contain a less varied pool than the vegetative mycelium. On the other hand, *Puccinia graminis* urediospores contain a more diversified pool, as do spores of *Tilletia caries*. In the latter, germination is accompanied by the appearance of glutamine, ethanolamine and large amounts of serine but without other substantial changes in the pool. This organism also contains djenkolic acid, an infrequently reported substance.

In addition to the fungi cited in Tables VII and VIII and the accompanying discussion the following organisms have been examined: *Aspergillus niger*[148]; *Chromocrea spinulosa*[37]; *Calviceps litoralis* and other ergots[100, 101]; *Craterellus cornucopeoides*[46]; *Dictyostelium discoideum*[105]; *Emericellopsis mirabilis, E. terricola*[113]; *Fusarium culmorum*[37, 185]; *Fusarium buharicum, F. bulbigenum* var. *lycopersici, F. caucasicum, F. chlamydosporum, F. conglutinans, F. dimerum, F. equiseti, F. lateritium, F. lini, F. moniliforme, F. orthoceras, F. oxysporum, F. poae, F. sambucinum, F. scirpi, F. semitectum, F. solani, F. sporotrichioides, F. udum, F. vasinfectum*[185]; *F. lycopersici*[52];

TABLE IX

QUANTITATIVE VALUES FOR MICROBIAL FREE AMINO ACID POOL FRACTIONS

Organism	Ref. numbers	Total amount (μmoles/100 mg dry wt.)	Amino acid fractions: Dicarboxylic								
			Acids	Amides	Basic	Alanine, glycine	Leucines, valine	Hydroxy	Aromatic and cyclic	Sulfur	γ-amino butyri acid β-alanine
			% of total pool								
Staphylococcus aureus	71	14.3	54.5		4.2	7.6	7.8	3.1	14.7	8.5	
Bacillus subtilis	115	75.8	33.2		11.0	15.6	22.4	12.0	3.7	1.5	
Bacillus subtilis	142	40.5	55.0	8.2	0.5	19.2	5.9	10.3	1.5		
*Escherichia coli**	116	3.88	17.2		12.4	55.3	12.1	**	2.8		
Baker's yeast	110	21.1	33.6	4.7	6.6	12.3	10.4	7.6	14.2	9.0	1.4
Brewer's yeast	110	28.1	26.0	8.0	20.0	15.3	9.3	2.5	7.7	4.5	2.1
Saccharomyces cerevisiae	173	59.4	14.7	32.5	19.5	12.4	5.1	10.8	***	4.6	
Saccharomyces cerevisiae	67	30.8	44.0	§	21.1		8.8	19.2	4.5	2.3	
Torulopsis utilis	123	23.7	35.9	11.4	17.3	29.0	1.7	3.4	0.8	0.4	
Candida utilis	41	96.4	31.0		9.1	36.3	19.2	1.7	0.7		
Neurospora crassa	55	7.1	11.1	7.0	4.5	27.7	16.3	21.7	1.3	4.4	5.7
Penicillium chrysogenum	93	94.0	8.2		7.7	14.4	30.6	18.6	17.8	2.8	
Aspergillus flavus	143	40.8	16.4		10.0	28.9	14.7	7.1	10.3	7.1	5.4
Fusarium vasinfectum	135	10.8	25.9		9.8	19.2	15.3	11.7	7.7	5.6	4.4
Claviceps litoralis	12	1.65	19.4		2.4	22.4	23.6	10.3	4.8	17.0	

* Incomplete analysis, often of fractions containing several amino acids.
** Serine included in alanine, glycine value.
*** Present in small amounts.
§ Value for dicarboxylic acids includes glutamine.

F. solani f. phaseoli[120]; *Hypomyces aurantius*[181]; *Mucor adventitius aurantiacus*[148]; *Phycomyces nitens, Pythium ultimum, Stereum purpureum, Thamnidium elegans*[37]; *Ustilago zeae*[48]; *Verticillium albo-atrum*[154]; *Zygorhynchus moelleri*[128].

Summary of quantitative studies

Some of the quantitative data included in Tables I–VIII are collected in modified form in Table IX to show the relative amounts of different amino acid types found in the total pool. It should be recognized that only a limited number of quantitative studies are available and that these pools are highly variable. Nonetheless, an appraisal of these data suggests that some of the apparent relations may survive more intensive investigation. The pool in gram-positive bacteria, for example, appears to be dominated by the dicarboxylic amino acids. The yeasts also contain large quantities of glutamic and aspartic acids, and an additional significant portion of the pool is made up by their amides. The basic amino acids generally do not form a major part of the pool except in the yeasts where values on the order of 20% are encountered. The fungi, with the possible exception of the yeasts, appear to be consistently high in the neutral aliphatic amino acids. γ-Aminobutyric acid and β-alanine form a significant part of the pool only in the molds and other fungi, although there are qualitative reports of their occurrence in bacteria (*e.g.*, Fig. 7; ref. 83). Suggestive trends in some of the other groups can be seen, but there are too many exceptions to permit a formal statement.

Occurrence of additional pool components

Most of the substances observed in the studies summarized to this point are the amino acids generally found in protein and in the pools of plant and animal organisms. When the commonly employed extraction and chromatographic procedures are used, these amino acids clearly form the major part of the pool in most microorganisms. It must be recognized, however, that the influence of the cell disruption and analytical procedures used on the composition of the extracted pool has not been examined exhaustively. The experiments cited by MITCHELL AND SIMMONS (p. 136) show clearly that the relatively commonplace pools observed by many workers in Drosophila in fact may be found to contain many more components when other analytical procedures are used.

It should be borne in mind also that no systematic effort has been made here to list all unusual amino acids or amino acid derivatives encountered in unautolyzed cell extracts. Rather, attention has been focused largely on investigations in which the total pool was examined. Many of these studies, however, have disclosed substances of unusual interest and some of these observations will be discussed below.

Peptides. Impressive evidence for the occurrence of sizeable peptide pools in microorganisms is found in the work of TURBA AND ESSER[183, 184] on *Torula utilis* and of CONNELL AND WATSON[38, 39] on the gram-negative bacterium *Pseudomonas hydrophila*. In both cases, cells were exposed to an isotopically labeled carbon source (acetate or glucose), harvested at intervals of a few second to a few hours, extracted with boiling ethanol and fractionated by paper electrophoresis into basic, neutral and acidic components each of which was fractionated further by paper chromatography. TURBA

AND ESSER observed 41 peptides most of which contained a large number of amino acid residues. Acidic peptides predominated and were more heavily labeled. McMANUS[121] also has observed the occurrence and formation of peptides in *T. utilis*. On the other hand, COWIE AND WALTON[40] failed to detect peptides in trichloroacetic acid extracts of *Torulopsis utilis* following exposure of cells to ^{14}C-fructose.

Dialyzed ethanolic extracts from *Ps. hydrophila* were found to contain 43 peptide fractions, 36 of which were examined for constituent amino acids. At least five amino acids were found in all fractions after hydrolysis and most contained eight or nine amino acids. Although basic peptides were usually detected, except for histidine, incorporation of isotope during 2 h incubation occurred only in the neutral and acidic fractions. These authors estimated that approx. 6 mg/g dry weight of free amino acids occur in this organism (approx. 4 μmoles/100 mg) and that an additional 3 mg/g of peptides occur. This is a significant amount of material, particularly if the generalization is substantiated by further study that gram-negative bacteria contain relatively small pools.

TURBA AND ESSER suggest a relation of these peptides to the synthesis of protein. CONNELL AND WATSON adopted a more cautious view towards the significance of their results, and point out that one third of the peptide fractions analyzed contain diaminopimelic acid, suggesting a relation of such peptides to cell wall biosynthesis.

It is clear that the detection of large and varied peptide pools so far has been the exceptional finding. The more common experience has been the detection of one or a few unknown spots on chromatograms which disappear when extracts are heated with strong acids. In a number of instances, some of them listed in Table X, more definitive evidence for the presence of peptides has been reported. ŠORM and his coworkers have isolated two peptides from *E. coli* and identified them as glutamylalanine and glutamyl-γ-aminobutyric acid. These substances are found only in small amounts in the normal strain used. They are present in much larger amounts in a chloramphenicol-resistant strain and in the normal strain grown with this antibiotic. More recently

TABLE X

REPORTS OF PEPTIDES IN MICROBIAL EXTRACTS

Organism	*Number of peptides*	*Composition*	*Ref. numbers*
Pseud. hydrophila	43	Most protein amino acids and DAPA	39
E. coli	2	Glutamylalanine Glutamyl-γ-aminobutyric acid	171
E. coli	30	10 amino acids; most often glu, cyst, gly	65
Pseud. saccharophila	1	Thre, gly, 2 unknowns	118
Staph. aureus	1	Glu, lys, ala, gly	71
Strep. faecalis	1	D-alanyl-D-alanine	89
Mycobact. tuberculosis	1	Asp, leu, met/val, ser, gly, cyst, glu, ala, unknown	141
Coryn. diphtheriae	1	Ala, gly, lys or arg, glu, DAPA	189
Torula utilis	41	Most protein amino acids	184
Torula utilis	2	Glu, gly, ala, leu, phe	121
Torulospis utilis	1	Citr, (orn?), glu, cyst, gly	123
Pen. chrysogenum	1	δ-(α-aminoadipyl)cysteinylvaline	6

References p. 105/108

this group has reported the occurrence in *E. coli* of 30 unknown substances which release a variety of amino acids on acid hydrolysis[65]. The unknown substance reported by HANCOCK to occur in extracts of *S. aureus* was shown by hydrolysis of isolated material to contain glutamic acid, lysine, alanine and glycine. This composition corresponds closely to the major amino acids found in cell walls of this organism suggesting a metabolic relation between the two. The material isolated by IKAWA AND SNELL from *S. faecalis*[89] also seems to be an intermediate in the synthesis of the mucopeptide portion of the cell wall.

In her early study on the amino acid pool of Corynebacterium[188] WORK detected two unknown ninhydrin-positive substances. One of these subsequently was shown to be diaminopimelic acid. The other appeared to be a relatively stable peptide which on hydrolysis with strong acid or base released varying amounts of alanine, glycine, lysine (or arginine), glutamic acid and diaminopimelic acid. In retrospect, it is very likely that this material also has a close relation to cell wall metabolism. An aminoadipyl peptide, reported by ARNSTEIN *et al.*[5, 6] to occur in *Pen. chrysogenum* along with free aminoadipic acid, may be involved in the biosynthesis of penicillin by this organism.

In addition to the substances cited here a large number of antibiotic peptides have been isolated from microbial cultures. The earlier literature has been summarized by SYNGE[177] and references to the more recent reports can be found in current volumes of the *Annual Review of Biochemistry* (*e.g.*, ref. 158). It should be noted that in most instances these substances are isolated from culture filtrates.

Nucleotide-peptides. The occurrence in bacteria of uridine-muramate-peptides and their involvement as intermediates in cell wall biosynthesis has been reviewed extensively[97, 155, 176] and, therefore, will not be considered further here. Recently, a considerable number of nucleotide-peptides have been isolated from yeast and Chlorella which, by virtue of their occurrence in these organisms, would appear to have no relation to cell wall biosynthesis. The most prevalent suggestion of a metabolic function for these substances is an involvement in protein synthesis. In most of these compounds the peptide is linked to uridine and usually yields an hydroxamate on treatment with hydroxylamine.

KONINGSBERGER *et al.*[104] originally detected such compounds in extracts of quick-frozen yeast. Subsequent reports from various laboratories substantiated the natural occurrence of these substances. GILBERT AND YEMM[60], for example, isolated a uridine-containing fraction from ethanol extracts of exponentially growing *Torulopsis utilis* which on hydrolysis yielded aspartic and glutamic acids, arginine and alanine. In an extensive study[43, 44, 75, 76, 77] HARRIS *et al.* have detected a variety of compounds in *Saccharomyces cerevisiae*. In one instance a compound was isolated which on alkaline hydrolysis yielded UMP and a tetrapeptide containing two units each of alanine and arginine [76]. A sizeable number of other amino acids have been shown to occur in such compounds, in many cases in peptidic form and linked to 5′-uridylic acid as well as to 5′-adenylic acid. This group has presented kinetic evidence supporting the proposal that the nucleotide-peptides are intermediates of protein synthesis[77].

HASE *et al.* also have reported extensively on nucleotide-peptides in yeast and Chlorella[78–81], and for the latter organism have suggested that they participate in reactions associated with nuclear and cell division. In these studies, ultraviolet-absorbing fractions isolated by column chromatography which yielded amino acids after acid hydrolysis also contained adenine and guanine in addition to uracil. Besides cyst(e)ine, a number of unidentified sulfur-containing compounds were observed in hydrolysates[80], as well as substances suspected to be amines[81]. More recently, JONES AND LEWIN[94] have demonstrated the occurrence of a large number of nucleotide-peptides in the green alga, *Chlamydomonas moewusii*. A variety of nucleotide bases were found and the peptide moiety most frequently contained glutamic acid, cystine, glycine and serine.

There also have been some reports that bacteria contain nucleotide-peptides other than the uridine and cytidine compounds implicated in cell wall biosynthesis. In contrast to the yeast compounds which usually contain uracil, adenine predominates in bacterial compounds. BROWN[24], for example, has detected the occurrence in *S. faecalis* of small amounts of adenyl peptides. JONSEN *et al.*[95] isolated a chromatographic fraction from *E. coli* which yielded adenine, alanine, glutamic acid and glycine on hydrolysis. Two nucleotide-containing fractions from *B. subtilis* also were investigated, one of which yielded ten ninhydrin-reactive spots after acid hydrolysis. In a separate report[96] this group described the isolation from bacterial cells and spores as well as yeast of an unknown which yields amino acids on hydrolysis, contains no ribose, but does contain a substance distinguishable from known bases which absorbs in the region of 260 mμ.

In summary, there have been a sufficient number of reports from different laboratories to justify the conclusion that a considerable number of nucleotide-peptides other than substances concerned in cell wall biosynthesis occur in various classes of microorganisms. The metabolic role of these substances remains in the realm of speculation although most investigators anticipate a relation to protein synthesis. The amounts of amino acid bound in these compounds are small in comparison to the apparently uncombined forms considered previously.

Miscellaneous compounds. AUBERT *et al.*[7] have reported that large quantities of *N*-succinyl-L-glutamic acid occur in *Bacillus megaterium* during spore formation. This substance which comprised up to 8.5% of the dry weight of the sporulating cell was not found in the vegetative cell prior to this stage and, except for traces, also not in the formed spore.

It is likely that careful study of organisms producing some of the newly discovered antibiotics having unusual amino acid constituents (see, for example, ref. 158) will be found to contain these subunits in an easily extractable form. It also should be borne in mind that many metabolically important substances may occur in the pool in amounts below the limit of detectability of the ninhydrin-chromatography technique. Such substances might be revealed by increasing the sensitivity of the method, for example, by using isotopes to label the pool. Examples of this approach can be found in the work of the group at the Carnegie Institution[149] or of DOWNEY AND BLACK[49]. The latter provided yeast with ^{35}S-labeled methylmercaptan. Cell extracts contained nine radioactive substances, none of which was a disulfide and only one a thiol. One of these fractions was identified as a stereoisomer of β-methyllanthionine, which on desulfuration yielded L-alanine and D-α-amino-n-butyric acid.

One of the greatest oversights in the study of microbial pools has been the lack of attention paid to the stereoisomeric form of the amino acids detected on chromatograms. In animal tissues the D-amino acids probably occur in small amounts or more likely not at all, and this has not been a serious defect of such studies. However, a number of D-amino acids are now known to occur in sizeable quantities in bacterial cells, particularly in the cell wall, and examination of cell extracts has revealed their presence in this fraction as well[83, 84, 89]. HOLDEN AND HOLMAN have shown that the freely-extractable pool in freshly harvested cells of *L. arabinosus* can contain as much as 5 μmoles of D-glutamic acid per 100 mg dry weight of cells and that this can increase to 37 μmoles/100 mg when resting cells accumulate L-glutamic acid from the external buffer. A systematic study of the occurrence of D-amino acids in microbial pools appears not to have been reported.

Extracellular amino acids. A detailed survey of this subject is beyond the scope of this review and only a few reports will be mentioned as a guide to the reader. These are largely empirical observations which so far have provided little insight to the factors which control the relative distribution of amino acids between intra- and extracellular fluid. Generally, there are two types of studies, those in which cells are grown in a complex medium and the removal of amino acids monitored at intervals throughout the growth phase, and those in which cells are grown in a simpler, carbohydrate–salts medium and the appearance in the extracellular fluid described of amino acids synthesized internally during cultivation. Many of the studies on the internal pool cited previously have included such chromatographic investigation of culture filtrates (see, for example, refs. 51, 52, 115, 120, 141, 143, 147, 149, 157, 188).

Among the earliest studies are those of PROOM AND WOIWOOD who examined the culture filtrates of 300 bacterial strains[146] for the most part describing the removal of amino acids from the medium. Subsequently[187], the release of newly-synthesized ninhydrin-reactive substances to the medium also was described. DAGLEY AND JOHNSON[42] grew *E. coli*, *Ps. aeruginosa* and a vibrio in mineral-salts medium, and followed the appearance of amino acids extracellularly. Of some interest was the finding that *E. coli* released amino acids even in the early exponential phase when protein degradation could be expected to be minimal, and that this release followed an ordered pattern in which an increasing number of amino acids appeared at progressively later stages of the exponential growth phase. The appearance of amino acids in the growth medium during cultivation of *E. coli* and *Ps. aeruginosa* also has been studied respectively by KAWANO[99] and CESAIRE *et al.*[28].

Recently, a large number of reports have appeared dealing with the production of amino acids by microbial fermentation (*cf.* refs. 29, 102, 103). In some instances, a 50% yield of glutamic acid based on the amount of glucose used has been obtained. It can be expected that future studies with these organisms may provide clues to the factors which control intracellular synthesis and retention of amino acids.

CONDITIONS WHICH MODIFY POOL SIZE AND COMPOSITION

Having described the composition of the free amino acid pools in various commonly studied groups of microorganisms, we shall now consider what is known about the conditions which control pool composition. As indicated previously, many investigators have understood that the pool can have a variable composition and have attempted to identify some of the factors which contribute to this variability. In the

References p. 105/108

following discussion, only illustrative examples of such observations will be presented and no attempt will be made to treat the subject exhaustively. This section will deal largely with cultural and metabolic factors which influence pool composition. Studies on the effects of disruptive chemical and physical forces which pertain more closely to the question of the intracellular state of the pool will be dealt with separately.

Effect of age

Growth of gram-positive and -negative bacteria is completed generally within 24 h of incubation. In most studies with gram-positive bacteria the total pool size is relatively constant during this period of cultivation, especially towards the end of the exponential growth phase[13, 62, 71], although in some cases, a slight drop in the early portion of the exponential growth phase has been reported[71]. Compensatory changes in the levels of individual amino acids may occur which in summation give the impression of a relatively constant pool[13, 73]. The difficulties in formulating generalizations can be seen by comparing Figs. 5–10, 11–16, and 17–22 which show the extractable pool of *L. arabinosus*, *Leuco. mesenteroides* and *S. faecalis* at various times of incubation in the same growth medium. In all cases growth was completed within 24 h of incubation. *L. arabinosus* had a large, varied pool very early in the exponential phase which remained relatively constant except for the appearance of γ-aminobutyric acid until the end of the active growth phase after which it declined sharply. With *Leuco.*

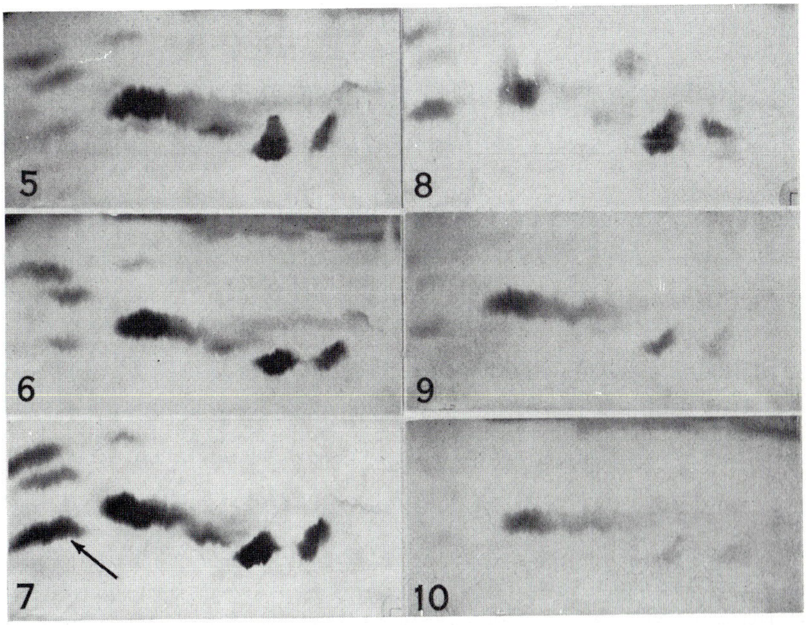

Figs. 5–10. Free amino acid pool of *L. arabinosus* 17-5, grown for varying times in a defined medium containing acid hydrolyzed casein. Incubation times were as follows: Fig. 5, 12 h; Fig. 6, 16 h; Fig. 7, 22 h; Fig. 8, 39 h; Fig. 9, 68 h; Fig. 10, 100 h. Maximum cell density was attained between 16 and 22 h. Chromatograms prepared using hot water extracts from 6.6 mg d.w. of cells. Chromatographic solvents: first direction, right to left, phenol, NH_3, H_2O; second direction, bottom to top, lutidine, H_2O. The substance indicated in Fig. 7 is γ-aminobutyric acid.

mesenteroides, on the other hand, the pool increased in size throughout the growth phase and was maintained at a high level for at least 15 h beyond the period of active growth after which it declined in size and variety. *S. faecalis* maintained a relatively constant pool throughout the period of active growth and while there were some small losses thereafter, a large and varied pool was retained by the cells even 3 days after active growth of the culture had ceased. In gram-negative bacteria equally inconstant observations have been encountered. MURUYAMA[132] observed a large increase with age in the pool of a pseudomonad, while MANDELSTAM[116] reported a marked and regular decline in *E. coli*. It is likely that the occurrence and retention of an amino acid in the

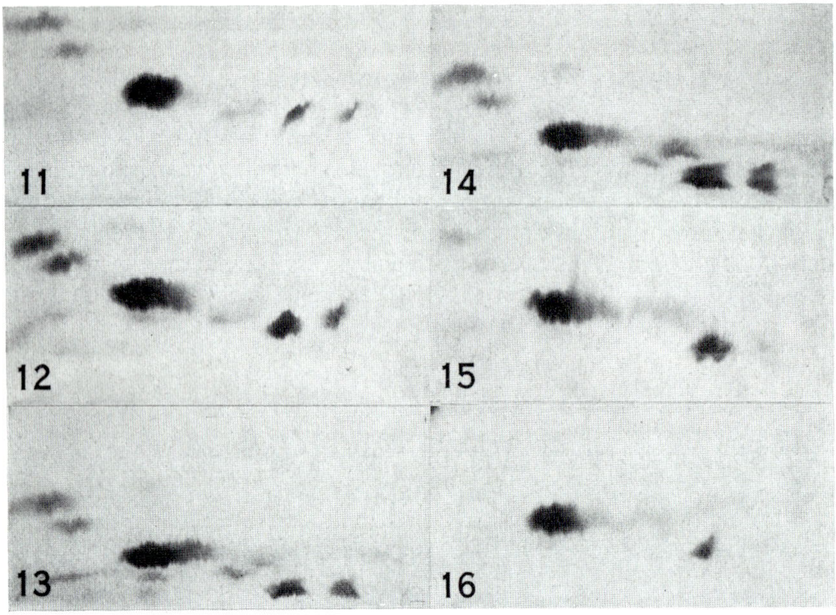

Figs. 11–16. Free amino acid pool of *Leuco. mesenteroides* P-60, grown for varying times in a defined medium containing acid hydrolyzed casein. Incubation times were as follows: Fig. 11, 12 h; Fig. 12, 16 h; Fig. 13, 22 h; Fig. 14, 39 h; Fig. 15, 68 h; Fig. 16, 100 h. Other details as in Figs. 5–10.

pool is the result of a balance in the rates of a considerable number of reactions which influence its concentration and that the component reactions differ among organisms and for the same organism at different phases of growth, thus accounting for these divergent patterns of response.

It should be noted that a number of investigations[87, 125, 159] have shown that cells from early exponential-phase cultures have less cell wall substance than cells harvested from stationary-phase cultures and, therefore, that they are more susceptible to damage by unfavorable physical and osmotic forces. This property was not widely appreciated at the time most of the studies cited here were performed. Thus, there is considerable possibility that some artifactual changes in pool composition mistakenly attributed to culture age have been described. Some of the changes, however, are unlikely to have originated in this way. It is of considerable interest that cells taken at different times from populations dividing exponentially in a medium containing all nutrients in excess should show changes in levels of individual amino acids. In this regard, an investigation of pool composition in synchronously dividing cells might be a worthwhile undertaking.

Among the fungi, growth periods are very much longer than those required to culture bacteria, and in many of these organisms sizeable pools can be retained for 20 and even 30 days[12, 52, 135, 143]. RITTER[148] has shown that pools are retained for periods even longer than this, although autolytic events contribute to the pool contents. The two studies on *Fusarium* again show a markedly different response to culture age and as with the bacteria, fluctuation in amount of amino acids predominate

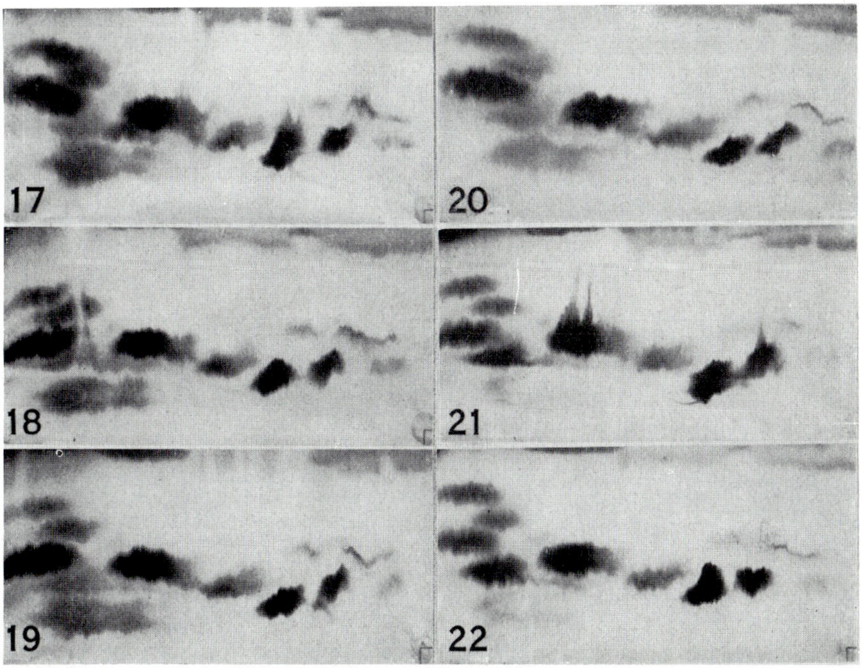

Figs. 17–22. Free amino acid pool of *Strep. faecalis* R, grown for varying times in a defined medium containing acid hydrolyzed casein. Incubation times were as follows: Fig. 17, 12 h; Fig. 18, 16 h; Fig. 19, 22 h; Fig. 20, 39 h; Fig. 21, 68 h; Fig. 22, 100 h. Other details as in Figs. 5–10.

over qualitative changes in pool composition. The effect of age on the pool in *Pen. chrysogenum*[122, 147] and *N. crassa*[55] has been studied. In the latter organism BARBESGAARD AND WAGNER[9] have shown that free phenylalanine and tyrosine decline at the onset of protoperithecia formation. In the slime mold, *Dictyostelium discoideum*, KRIVANEK AND KRIVANEK[105] showed that the qualitative pattern is a function of the stage of differentiation, and WRIGHT AND ANDERSON[192] observed that the size of the methionine pool was a function entirely of the development stage.

CHEESEMAN AND SILVA[31] have observed distinctly different chromatographic patterns in some heterofermentative lactobacilli grown at 30° and 37°. In general, the effect of varying incubation temperature which would change the division rate has seldom been studied.

References p. 105/108

Effect of growth-medium composition

Most investigations in which this question was considered have dealt with changes in the pool produced when the source of carbon and/or nitrogen is modified (*e.g.*, refs. 14, 55, 56, 62, 70, 82, 88, 90, 91, 139, 162, 163, 164, 178, 181, 195). Invariably, pools are found to be larger and more diversified when the growth medium is supplemented with a mixture of amino acids in place of an inorganic-nitrogen source. Generally, these larger pools are retained during thorough washing procedures and most likely are formed by the operation of metabolically dependent amino acid accumu-

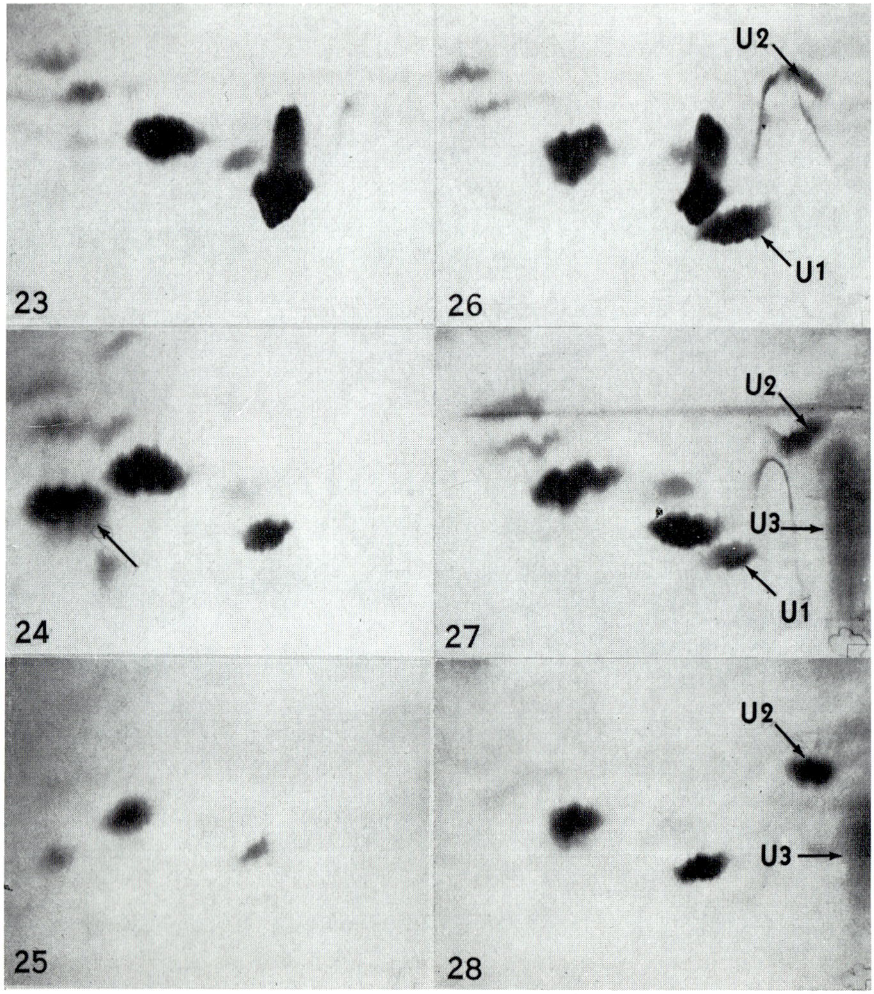

Figs. 23–28. Effect of a severe vitamin B_6 deficiency on the amino acid pool of *L. arabinosis* 17–5. Figs. 23, 24 and 25 show the pool in control cells grown in a synthetic complete medium for 16, 38 and 62 h, respectively. The compound indicated in Fig. 24 is γ-aminobutyric acid. Figs. 26, 27 and 28 show the pool in cells grown for 16, 38 and 62 h, respectively, in a medium from which vitamin B_6 was omitted. Maximum growth in the deficient culture was 20% of that achieved by the control culture and equal dry weights of cells were extracted for chromatography. U_1, U_2 and U_3 indicate unidentified substances. Chromatography as described in Figs. 5–10.

lation systems. However, the form in which inorganic nitrogen is provided also can influence the content of the pool[181, 195]. The effect of providing nitrogen as a single amino acid has been a popular experiment with obvious origins in earlier explorations of nitrogen assimilation in microorganisms. Numerous changes in qualitative pool composition are usually observed as shown for *Aspergillus oryzae* by SIMONART AND CHOW[163]. Table XI taken in modified form from this work shows the markedly different amino acid-pool composition that can be found depending on the amino acid used to support growth. Speculations concerning the reasons for such differences involve predictions from the known operation of the tricarboxylic acid cycle and the amino acid transaminases. Conversely, such observations show that when the potential enzymatic activity of an organism is known, observations of the fluctuating content of the pool can be applied to detect the shifting balance in reaction rates affecting different pool components.

Although less frequently studied, variations in the source of carbon also produce changes in the composition of the pool. For example, SIMONART AND CHOW[164] observed striking differences when a variety of dicarboxylic acids were used. Most investigators have encountered quantitative rather than qualitative changes in the pool[14, 70]. It should be noted that when an organic form of nitrogen is provided, the organism in effect is also supplied with considerable quantities of a new carbon source.

The effect of nutritional deficiencies on pool composition has received surprisingly little attention. GALE noted a reduction in pool level when *S. faecalis* was grown in media containing minimal amounts of required amino acids. However, the amount of amino acid provided sufficed to support protein synthesis and growth, and the changes noted probably reflected reduced accumulation of exogenous amino acids. Indeed, interference with protein synthesis appears from the work of MANDELSTAM[116] and HANCOCK[73] to increase the amounts of amino acids in the pool. The effects of vitamin deficiency on various properties of the lactic acid bacteria including pool composition and amino acid transport have been studied in this laboratory. Figs. 23–28 illustrate the effects of a vitamin B_6 deficiency on the pool of *L. arabinosus*. At the end of active growth nutritionally normal cells form γ-aminobutyric acid, whereas vitamin B_6-deficient cells do not, an observation which corresponds to the lack of glutamic acid decarboxylase activity in the latter cells. On the other hand, the pool in B_6-deficient cells contains three unknown substances not found in normal cells. Incubation of normal cells with isonicotinic acid hydrazide induces the appearance of these substances in smaller quantities. In addition, biotin-deficient cells accumulate an unidentified substance having the chromatographic properties of diaminopimelic acid. Observations such as these, of course, remain interesting curiosities unless the substances found are identified unequivocally and the metabolic basis for their appearance in the pool established by enzymatic studies.

Effect of incubation in nutritionally incomplete media

One of the most extensive investigations of this type has been carried out with *S. cerevisiae* by HALVORSON, SPIEGELMAN *et al.* (*cf.* refs. 70, 172). Incubation of the yeast in a nitrogen-free medium leads to depletion of the pool, but only when a metabolizable carbon source is provided. Inhibition of energy production prevents this loss, and provision of a poorly utilized carbon source can promote increases in pool levels. The depleted pool can be restored rapidly by exposure to inorganic nitrogen or a mixture of amino acids. Single amino acids also restore the pool, but affect individual components of the pool differently, an observation reminiscent of the results shown in Table XI. Depletion of the pool by nitrogen starvation can be reversed by a brief exposure to ultraviolet radiation. MIETTENEN[123] also has documented changes in the amino acid pool of *Torulopsis utilis* during nitrogen starvation.

In bacteria, somewhat different observations have been obtained. HOLDEN AND ROBERTS[83] demonstrated that incubation of *L. arabinosus* in phosphate buffers led,

TABLE XI

POOL COMPOSITION IN *Aspergillus oryzae* GROWN WITH SINGLE AMINO ACIDS*

(Adapted from SIMONART AND CHOW[163]).

Amino acid in growth medium	*Pool amino acids*															
	Alanine	*Glutamic acid*	*Aspartic acid*	*Glutamine*	*Glycine*	*Serine*	*Threonine*	*γ-Aminobutyric acid*	*Ornithine*	*Arginine*	*Citrulline*	*Valine*	*Leucine*	*Isoleucine*	*Proline*	*Hydroxyproline*
Aspartic acid	●	●	●	●												
Alanine	●	●	●	●												
Serine	●	●				●										
Threonine	●	●	●	●	●		●	●								
Glycine	●	●	●	●	●				●	●						
Valine	●	●	●	●				●				●				
Leucine	●	●	●	●	●			●				●	●			
Isoleucine	●	●	●	●	●			●	●					●		
Proline	●	●	●		●											
Hydroxy-proline	●	●	●		●											●
Arginine	●	●	●	●	●			●	●	●						
Citrulline	●	●	●	●	●			●	●	●	●	●	●			
Ornithine	●	●	●	●	●			●	●	●						

* See footnote * Table I.

with the exception of alanine and glutamic acid, to rapid loss of the pool unless glucose was present. Similarly, MANDELSTAM[116] observed that nitrogen starvation of *E. coli* carried out in the presence of glucose led to a rise in the pool. In a study on the release of amino acids from *Streptococcus faecium,* CHESBRO AND EVANS[35] observed only a small loss of amino acids from the pool during incubation in buffer at pH 8.5. Depletion of the pool, however, could be achieved by incubation at pH 10.5, and its replenishment observed during incubation at pH 9.5 in the presence of glucose.

In view of the possibility that there may be metabolically distinguishable pools[41], COWIE, BRITTEN, this Symposium) involving endogenously formed amino acids at sites or having exchange properties different from amino acids originating exogenously, the behavior of pools during such experiments may require re-examination with this possible distinction in mind.

Effects of metabolic inhibitors

Increases in pool size and accumulation of peptides during exposure to chloramphenicol or under conditions which interfere with protein synthesis have been discussed above. In contrast, ALLEGRA *et al.*[3] found that the addition of sub-inhibitory amounts of chloramphenicol to growing cultures of *Salmonella bareilly* had the opposite effect of reducing the number of amino acids in the pool. Penicillin causes a marked loss of amino acids from bacterial pools [25, 72, 114].This may occur indirectly by virtue of an inhibition of cell-wall formation and the appearance of osmotically labile cells from which the pool is readily lost. However, a more specific change in the permeability properties of the membrane has not been excluded. In the experiments with *Sarcina lutea*[25] proline accounted for all or most of the cellular free amino acid nitrogen lost during short exposure to penicillin suggesting a heterogenous loss of pool amino acids.

Exposure of *Verticillium albo-atrum* mycelium to 10 and 30 p.p.m. of the polyene antibiotic fungichromin resulted in complete loss of the free amino acid pool[154]. However, with 2 p.p.m. there was no inhibition of growth, and a 3-fold increase in cellular soluble N including marked increases in the leucines, arginine, lysine and proline. LAMPEN *et al.*[107, 117] have observed inhibitory effects of another polyene antifungal agent, nystatin, which may be related to changes in membrane permeability in yeast. Streptomycin caused a loss of amino acids from *Mycobacterium tuberculosis* during long exposure periods[50]. It is uncertain whether this is related to recent observations suggesting an effect of streptomycin on bacterial membranes[4] or on energy metabolism[74]. In view of the divergent behavior shown by the pool in cells provided with or deprived of an energy source one is not surprised to learn that exposure to DNP, azide and other agents of this type also produces different effects in different organisms. Thus in *E. coli*, azide and DNP lowered the pool levels of most amino acids and inhibited the increases produced by glucose although not all amino acids in the pool responded in the same direction[116]. On the other hand, in yeast glucose-dependent pool depletion was inhibited by DNP and azide[70].

Heavy metal salts appear to reduce pool sizes. For example, using conidia of *Fusarium decemcellulare*, TROGER[182] observed that copper and silver salts were especially effective in causing the release of amino acids to the suspending medium. Unfortunately, a corresponding decline of intraconidial amino acids was not demonstrated. In another fungus, *Colletotrichum capsici*, treatment with copper again reduced the level of most pool amino acids[160]. The direct cause of these changes has not been established.

In *N. crassa*, hydrocortisone appears to raise or lower pool amino acid levels to bring them closer to so-called base-line levels[136]. On the other hand, deoxycorticosterone has been shown by LESTER *et al.*[108,109] to have a marked inhibitory effect on amino acid, carbohydrate and inorganic ion uptake by this organism.

High salt concentrations have been shown by NANI AND GIOLITTI[134] to increase the pool amino acid levels in *B. subtilis*. On the other hand, the usually diversified yeast pool contained only glutamic and aspartic acids, alanine and histidine when cells were grown in the presence of high concentrations of salt[133].

Effects of radiation, polyploidy and antibiotic resistance

Treatment of *S. cerevisiae* with α-radiation, X-radiation and triethylenemelanine caused a decrease in most pool amino acid levels but a marked increase in the amounts of arginine, lysine, histidine, glutathione and two unidentified compounds[173]. Treatment of this yeast with ultraviolet radiation during nitrogen starvation causes a rapid and pronounced replenishment of the pool[70]. JORDAN has reported that leakage of ^{14}C-labeled pool components from *Rhizobium meliloti* is increased by X-radiation. Most of these observations are consistent with the general impression that radiation damages cellular permeability barriers, although other changes also must occur to account for the increases in pool contents observed by HALVORSON and by SPOERL AND CARLETON.

The effect of polyploidy has been commented on occasionally and a more detailed study of its effects should benefit our understanding of the manner in which intracellular pools are retained. The study of SARACHEK[156] using various *Saccharomyces* strains is notable in this regard. This study showed clearly that per unit dry weight of cell the glutamate and aspartate pools were constant regardless of the degree of ploidy. Since the dry weight per cell increased regularly with increased ploidy (haploid, diploid, triploid and tetraploid), the relative pool size per cell corresponded very closely to the degree of ploidy. Unfortunately, the relation between ploidy and cell volume and surface area were not described.

Considerable attention has been focused on changes in pool composition during acquisition of antibiotic resistance by microorganisms. GALE AND RODWELL[58] observed that penicillin resistance in *Staph. aureus* was accompanied by a decline in ability to form a large glutamic acid pool during incubation with buffers containing this amino acid. MABBITT AND GREGORY[114] likewise found a decreased pool size in a strain of *S. aureus* trained to grow in the presence of high concentrations of penicillin. On the other hand, naturally resistant strains contained large pools characteristic for this organism. GOLDFARB[61] also has described changes in the *S. aureus* pool as penicillin resistance was acquired, finding that while alanine, histidine and serine decreased, the pool contained increased amounts of valine and glycine. DESSI *et al.*[47] have reported that acquisition of terramycin resistance by *Brucella melitensis* is associated with the disappearance of valine, methionine, leucine and glutamic acid from the pool and a reduction in its content of histidine.

It should be noted that these studies, while very intriguing, must be appraised with caution since acquisition of these new metabolic characteristics is frequently accompanied by changes in cell properties which are normally used to establish its identity. Thus the danger of selecting contaminants is very high. This emphasizes again the value of performing such studies with induced mutants, suitably marked to ensure that a derivative of the original strain is being studied. It is possible furthermore, in these circumstances, to backcross the new strain as in the study of FUERST AND WAGNER[55] and thus to establish that the change in pool composition is in fact related to the acquisition of the new metabolic property.

INTRACELLULAR STATE OF THE AMINO ACID POOL

Despite the large number of reports which describe the composition of pools contained within diverse organisms, the question of the location in the cell and the intracellular condition of these pools remains a subject of dispute. Indeed in only a few of the studies dealing with the occurrence of pools has serious attention been directed to this problem. The most perceptive studies instead have come from those investigators con-

References p. 105/108

cerned with amino acid transport and pool turnover. Since these subjects are discussed thoroughly elsewhere in this Symposium, only brief mention will be made here of the conclusions reached in such investigations.

Conditions required to liberate the pool uniformly are consistent with the hypothesis that it is retained by a peripheral or intracellular membrane, although none of the studies taken alone would exclude the possibility of retention within an intracellular particle or attachment to intracellular polymers. Bacterial pools can be liberated by many relatively mild procedures including: (a) shaking with glass beads in the cold; (b) freezing and thawing; (c) grinding with abrasives; (d) sonication; (e) osmotic lysis of protoplasts; (f) treatment with detergents, cold trichloracetic or perchloric acid, boiling water, warm ethanol or cold butanol; (g) extremes in pH such as 1.5 or 10.5. It is obvious that in addition to destroying hypothetical permeability barriers such treatments expose intracellular structures to foreign and potentially disruptive osmotic and chemical environments, so that the possibility cannot be excluded confidently, that the pool is retained in association with intracellular polymers. It is of interest that a number of investigators have studied cells washed with acetone[82, 166] and lyophilyzed cells subsequently incubated in buffers[45] and found them to retain an amino acid pool.

CHESBRO AND EVANS[35] have found that the release of amino acids from *Streptococcus faecium* is dependent on hydroxyl-ion concentration and that the kinetics of amino acid appearance in the extracellular buffer and disappearance from the pool did not coincide. It is apparent that more than one process was involved since, in addition, much larger amounts of glutamic and aspartic acids disappeared from the pool at pH 10.5 than appeared in the extracellular phase. Furthermore, a number of amino acids such as glycine, serine, threonine, ornithine and γ-aminobutyric acid appeared in the extracellular pool even though their levels in the intracellular pool did not appear to change. Of course, such experiments are not comparable to the procedures normally used to liberate pools where metabolic change is avoided as much as possible, but there is a possibility that they might help to establish the mechanism of pool retention.

COWIE AND MCCLURE[41] have shown that exposure of yeast to high hydrostatic pressures increases the ease with which the pool contents can be removed. Since such leakage is dependent on a supply of exogenous carbohydrate, it appears that exit of the intracellular amino acids is not simply a diffusion process through a disrupted membrane. However, a direct demonstration of association between pool amino acids and intracellular polymers or particles has not been achieved, although LACHS AND GROS[106] have reported that a small part (possibly 5%) of the *E. coli* pool is combined with a soluble nucleic acid fraction.

In bacteria, MITCHELL[126] has attempted to demonstrate that the intracellular solutes express an osmotic activity which one would expect from their predicted concentration in the cell assuming them to be in free solution. This was done by equilibrating cell pastes with the atmospheres above sucrose solutions of various concentrations and measuring the weight of water taken up by the cells. However, the expected contribution to this activity of the amino acid pool would not exceed 20% of the total, and the possible error in the method arising from the long incubation required is too great to decide with certainty from these data whether or not the amino acids are free or bound. A cogent discussion of the evidence bearing on the intracellular state of low molecular-weight solutes in bacteria has been presented by MITCHELL[124].

References p. 105/108

Another approach to the demonstration of osmotic activity by intracellular amino acids is found in experiments showing that bacterial protoplasts swell during uptake of amino acids. ABRAMS[1] (also *cf.* discussions this Symposium) has shown that *S. faecalis* protoplasts swell when they glycolyze in the presence of high concentrations of some amino acids used as osmotic stabilizers. Experiments demonstrating the swelling of protoplasts during accumulation of carbohydrate derivatives[167] and aliphatic acids[186] have been presented as evidence of the intracellular osmotic activity of these substances. Comparable observations have been made in this laboratory using amino acids[87]. All such studies, however, suffer from the defect that the identity of the osmotically active molecule is not known. In no case has it been excluded that the entering solute does not displace intracellularly some substance which is then responsible for the osmotic activity.

A number of studies have shown that the size of the bacterial free amino acid pool can be controlled by extracellular osmotic activity (*cf.* Table IV; HOLDEN, BRITTEN, this Symposium). In gram-positive bacteria extremely large pools normally are retained even during water washes at which time concentration gradients between intra- and extracellular compartments of several thousand-fold would appear to exist. However, in gram-negative bacteria the pool is much more sensitive to loss under these conditions. Furthermore, the pool in gram-positive bacteria becomes osmotically labile when the amount of cell-wall substance is reduced slightly ([85, 86] HOLDEN, p. 582). While such observations demonstrate the osmotic sensitivity of a cell structure responsible for retention of the amino acid, they do not identify it as the cytoplasmic membrane and, therefore, do not justify a definitive comment on the intracellular state of the amino acid.

There is a strong possibility that microbial amino acid pools are heterogenous. In yeast, COWIE AND MCCLURE were able to distinguish a concentrating pool in which extracellular amino acids initially accumulate in a form very sensitive to loss by exposure to solutions of low osmotic strength, and a smaller internal pool which contains internally synthesized amino acids as well and which is less sensitive to osmotic shock. BRITTEN using *E. coli* also has suggested the occurrence of two proline pools which differ in specificity for this amino acid. The highly specific pool may be related to the similarly specific pool described by LACHS AND GROS to be associated with soluble RNA.

In conclusion, certain knowledge concerning the intracellular state of the free amino acid pool in microorganisms has not yet been achieved. It is reasonable to expect that some part of it will be entrained in the protein synthetic apparatus. There is serious question, however, whether more than a small fraction is bound in this way.

FUNCTION OF THE POOL AND APPLICATION OF POOL STUDIES

A large number of studies have shown that the cellular protein composition is unrelated to the composition of the free amino acid pool. However, it is now well established that the amino acid pool in microorganisms provides precursor material for the synthesis of protein. This and related aspects of pool studies form a separate part of this Symposium and will not be discussed further here. In some organisms the pool appears to have other uses as well. DAWES AND HOLMES[45] showed that in *Sarcina lutea* the pool provides the direct substrates for maintenance of endogenous respira-

References p. 105/108

tion. Oxygen uptake declined as the pool was depleted and there appeared to be no utilization of carbohydrate, protein or lipid stores. This, of course, is not a typical finding since, for example, yeast utilizes carbohydrate as a reservoir substrate for endogenous respiration. GROS AND GROS[64] have suggested that a complete amino acid pool is necessary for RNA synthesis. An effort was made to exclude the alternate interpretation that protein synthesis must occur simultaneously by measuring RNA synthesis in the presence of chloramphenicol. The validity of the conclusion, therefore, rests on the completeness of the inhibition in protein synthesis.

There is a sizeable literature recording efforts to correlate pool composition and other cell processes such as antibiotic synthesis[115, 122, 147], pathogenicity[48, 120], mating type[48], etc. With few exceptions, clear-cut correlations have not been encountered. Such negative findings possibly are not unexpected, but need not necessarily be taken as an indication that pool studies have little value in this regard. It seems likely in fact that the opposite will be found to be true if, in addition to its composition, the flow of substances through the pool is examined using tracers. Except in the investigations relating to protein synthesis this approach has not been employed extensively.

One of the most prevalent applications of microbial pool studies has been in the area of taxonomy. Among the fungi the findings appear uniformly to inspire pessimism, *i.e.*, neither separation of species nor genera seems possible on the basis of

TABLE XII

A SUMMARY OF THE DISTINGUISHING CHARACTERISTICS ON CHROMATOGRAMS OF 63 STRAINS OF *L. casei* AND *L. plantarum*

(Adapted from CHEESEMAN[33].)

Property	*Nos. of strains of*			
	L. casei		*L. plantarum*	
	Typical	*Atypical*	*Typical*	*Atypical*
Origin — in *casei* + in *plantarum*	29	0	30	4
α-Aminobutyric acid — in *casei* + in *plantarum*	29	0	30	4
Proline — in *casei* + in *plantarum*	27	2	34	0
Asparagine — or faint in *casei* strong + in *plantarum*	26	3	33	1
Ratio glycine/threonine ⩾ 0.8 in *casei* < 0.8 in *plantarum*	20*	8	24*	5
Ratio 50, 70/alanine > 1.0 in *casei* < 1.0 in *plantarum*	27*	1	27*	2

* Excluding five strains of *L. plantarum* and one strain of *L. casei* which were not examined quantitatively.

References p. 105/108

chromatographic patterns (*cf.* VENKATA RAM[185], CLOSE[37]). By far the most ambitious undertaking of this sort has been the examination of the lactobacilli and micrococci by the group at the University of Reading cited earlier[13, 30–34, 62, 63, 119, 161]. This work has included a study of 73 strains of the *L. casei-plantarum* group, 55 strains of micrococci and 91 strains of heterofermentative lactobacilli. Recognizing the pitfalls arising from the variability of microbial pools and the necessity of using carefully controlled cultural and analytical conditions, apparently reliable separations of species have been achieved in some cases. An example of the separation of *L. casei* and *L. plantarum* is shown in Table XII. A common experience in this work was the finding that strains within a species could show considerable pool differences but would retain certain characters in a constant relation thereby permitting their distinction from the strains of another species. The separation of the heterofermentative lactobacilli into discrete groups was not attained even after an analysis of the data by an electronic computer. On the whole, the authors retain a cautious optimism that the technique as an adjunct to conventional biochemical tests may simplify the classification of the lactobacilli.

CONCLUSION

One of the most obvious conclusions of this survey is the lack of constancy of the microbial pool. This is most likely a consequence of the dependence of the pool composition on the external environment and the ease with which the latter can be changed consciously or inadvertently. It is likely that this is not a property peculiar to this class of organisms and that pool variability will be generally encountered in cells cultured *in vitro*. Thus, whereas ROBERTS *et al.* (p. 284, this Symposium) has observed a remarkable constancy in the cellular pools in various mammalian tissues even in animals subjected to severe physiological stresses, EAGLE *et al.* (p. 694, this Symposium) find that mammalian cells grown in tissue culture can have a highly variable pool composition distinctly dependent on changes in composition of the extracellular fluid.

Accordingly, the value of purely descriptive studies of pool composition in microorganisms seems very much open to question. This would not extend, of course, to the examination of pools as an adjunct to studies of physiological and metabolic processes in a given microbial strain. However, even here, the high incidence of negative findings suggests that the examination of the pool should go beyond a description of the static state to include the turnover of its components. Indeed by combining isotopes and the chromatographic method to permit rapid monitoring of various components, examination of the multicomponent pool can provide a penetrating insight into the pathways of intracellular metabolism as they relate to physiological processes.

ACKNOWLEDGEMENTS

The preparation of this manuscript and the original observations of the author described here were made possible by Grant E-1487 from the U.S. Public Health Service, and by Contract No. Nonr-2702(00) between this institution and the Office of Naval Research.

REFERENCES

1 A. Abrams, G. Yamasaki, J. Redman and C. G. Mackenzie, *Federation Proc.*, 19 (1960) 129.
2 G. Allegra, G. Giuffrida and R. Niutta, *Giorn. microbiol.*, 4 (1957) 129.
3 G. Allegra, R. Niutta and G. Giuffrida, *Naples Inst. Sieroterap. Ital.*, 34 (1959) 47.
4 N. Anand and B. D. Davis, *Nature*, 185 (1960) 22.
5 H. R. V. Arnstein, M. Artman, D. Morris and E. J. Toms, *Biochem. J.*, 76 (1960) 353.
6 H. R. V. Arnstein and D. Morris, *Biochem. J.*, 76 (1960) 357.
7 J. P. Aubert, J. Millet, E. Pineau and G. Milhaud, *Compt. rend.*, 249 (1959) 1956.
8 W. Bair and G. Rouser, *Federation Proc.*, 11 (1952) 183.
9 P. O. Barbesgaard and S. Wagner, *Hereditas*, 45 (1959) 564.
10 U. Bardi and G. Boretti, *Giorn. microbiol.*, 6 (1958) 91.
11 M. N. Bekhtereva, *Mikrobiologiya*, 27 (1958) 560.
12 M. L. Berman, H. W. Youngken, Sr., *J. Am. Pharm. Assoc., Sci. Ed.*, 47 (1958) 888.
13 N. J. Berridge, G. C. Cheeseman, A. T. R. Mattick and V. Bottazzi, *J. Appl. Bacteriol.*, 20 (1957) 195.
14 A. M. Besborodov, *Mikrobiologiya*, 28 (1959) 690.
15 J. Blass and M. Macheboeuf, *Bull. soc. chim. biol.*, 29 (1947) 903.
16 J. Blass, O. Lecomte and M. Macheboeuf, *Bull. soc. chim. biol.*, 33 (1951) 1552.
17 V. Bottazzi, *Ann. microbiol.*, 7 (1957) 164.
18 V. Bottazzi, *Ann. microbiol.*, 8 (1958) 40.
19 V. Bottazzi, *Ann. microbiol.*, 8 (1958) 52.
20 V. Bottazzi, *Ann. microbiol.*, 9 (1959) 35.
21 R. J. Britten, R. B. Roberts and E. F. French, *Proc. Natl. Acad. Sci. U.S.*, 41 (1955) 863.
22 R. J. Britten, *Science*, 124 (1956) 935.
23 R. J. Britten and F. McClure, *Bacteriol Rev.*, in the press.
24 A. D. Brown, *Biochim. Biophys. Acta*, 30 (1958) 447.
25 J. W. Brown and S. B. Binkley, *J. Biol. Chem.*, 221 (1956) 579.
26 J. W. Broyles, *Phytopathology*, 42 (1952) 3.
27 M. N. Camien, A. J. Salle and M. S. Dunn, *Arch. Biochem.*, 8 (1945) 67.
28 O. G. Césaire, E. Neuzil and H. Boiron, *Bull. soc. chim. biol.*, 40 (1958) 1435.
29 K. Chao and J. W. Foster, *J. Bacteriol.*, 77 (1959) 715.
30 G. C. Cheeseman, N. J. Berridge, A. T. R. Mattick and V. Bottazzi, *J. Appl. Bacteriol.*, 20 (1957) 205.
31 G. C. Cheeseman and M. T. Silva, *J. Appl. Bacteriol.*, 22 (1959) 294.
32 G. C. Cheeseman and N. J. Berridge, *J. Appl. Bacteriol.*, 22 (1959) 307.
33 G. C. Cheeseman, *J. Appl. Bacteriol.*, 22 (1959) 341.
34 G. C. Cheeseman, *J. Dairy Research*, 27 (1960) 33.
35 W. R. Chesbro and J. B. Evans, *J. Bacteriol.*, 79 (1960) 682.
36 B. D. Church and H. Halvorson, *Nature*, 183 (1959) 124.
37 R. Close, *Nature*, 185 (1960) 609.
38 G. E. Connell and R. W. Watson, *Biochim. Biophys. Acta*, 24 (1957) 227.
39 G. E. Connell and R. W. Watson, *Can. J. Microbiol.*, 4 (1958) 633.
40 D. B. Cowie and B. P. Walton, *Biochim. Biophys. Acta*, 21 (1956) 211.
41 D. B. Cowie and F. T. McClure, *Biochim. Biophys. Acta*, 31 (1959) 236.
42 S. Dagley and A. R. Johnson, *Biochim. Biophys. Acta*, 21 (1956) 270.
43 J. W. Davies and G. Harris, *Biochim. Biophys. Acta*, 45 (1960) 28.
44 J. W. Davies and G. Harris, *Biochim. Biophys. Acta*, 45 (1960) 39.
45 E. A. Dawes and W. H. Holmes, *Biochim. Biophys. Acta*, 30 (1958) 278.
46 A. M. Denantes, C. Fabert and L. Kientzler, *Bull. soc. sci. Nancy*, 18 (1959) 343.
47 P. Dessi, A. M. Gianni, G. Pasini and M. Greco, *Giorn. microbiol.*, 8 (1960) 122.
48 J. E. DeVay, *Phytopathology*, 44 (1954) 583.
49 P. F. Downey and S. Black, *J. Biol. Chem.*, 228 (1959) 171.
50 G. Dragoni, *Patol. sper.*, 43 (1955) 436.
51 G. Dragoni, *Boll. ist. sieroterap. milan.*, 36 (1957) 266.
52 V. Fluck and R. H. Richle, *Phytopathol. Z.*, 24 (1955) 455.
53 L. Fowden, *Nature*, 167 (1951) 1030.
54 J. C. Freeland and E. F. Gale, *Biochem. J.*, 41 (1947) 135.
55 R. Fuerst and R. P. Wagner, *Arch. Biochem. Biophys.*, 70 (1957) 311.
56 E. F. Gale, *J. Gen. Microbiol.*, 1 (1947) 53.
57 E. F. Gale and E. S. Taylor, *J. Gen. Microbiol.*, 1 (1947) 77.
58 E. F. Gale and A. W. Rodwell, *J. Gen. Microbiol.*, 3 (1949) 127.
59 E. F. Gale, *Advances in Protein Chem.*, 8 (1953) 285.
60 D. A. Gilbert and E. W. Yemm, *Nature*, 182 (1958) 1745.
61 M. M. Goldfarb, *Antibiotiki*, 2 (1957) 35.

[62] M. Gregory and L. A. Mabbitt, *J. Appl. Bacteriol.*, 20 (1957) 218.
[63] M. Gregory and L. A. Mabbitt, *J. Appl. Bacteriol.*, 20 (1957) 226.
[64] F. Gros and F. Gros, *Exptl. Cell Research*, 14 (1958) 104.
[65] D. Grünberger, J. Černá and F. Šorm, *Experientia*, 16 (1960) 54.
[66] H. O. Halvorson and S. Spiegelman, *J. Bacteriol.*, 64 (1952) 207.
[67] H. O. Halvorson and S. Spiegelman, *J. Bacteriol.*, 65 (1953) 496.
[68] H. O. Halvorson and S. Spiegelman, *J. Bacteriol.*, 65 (1953) 601.
[69] H. O. Halvorson, S. Spiegelman and R. L. Hinman, *Arch. Biochem. Biophys.*, 55 (1955) 512
[70] H. Halvorson, W. Fry and D. Schwemmin, *J. Gen. Physiol.*, 38 (1955) 549.
[71] R. Hancock, *Biochim. Biophys. Acta*, 28 (1958) 402.
[72] R. Hancock, *Biochem. J.*, 70 (1958) 15P.
[73] R. Hancock, *Biochim. Biophys. Acta*, 37 (1960) 47.
[74] R. Hancock, *Biochem. J.*, 76 (1960) 69P.
[75] G. Harris, J. W. Davies and R. Parsons, *Nature*, 182 (1958) 1565.
[76] G. Harris and J. W. Davies, *Nature*, 184 (1959) 788.
[77] G. Harris and G. E. Neal, *Biochim. Biophys. Acta*, 47 (1961) 122.
[78] E. Hase, Y. Morimura and H. Tamiya, *Arch. Biochem. Biophys.*, 69 (1957) 149.
[79] E. Hase, H. Otsuka, S. Mihara and H. Tamiya, *Biochim. Biophys. Acta*, 35 (1959) 180.
[80] E. Hase, S. Mihara, H. Otsuka and H. Tamiya, *Biochim. Biophys. Acta*, 32 (1959) 298.
[81] E. Hase, S. Mihara, H. Otsuka and H. Tamiya, *Arch. Biochem. Biophys.*, 83 (1959) 170.
[82] D. S. Hoare, *J. Gen. Microbiol.*, 12 (1955) 534.
[83] J. T. Holden and E. Roberts, *Federation Proc.*, 15 (1956) 276.
[84] J. T. Holden and J. Holman, *J. Biol. Chem.*, 234 (1959) 865.
[85] J. T. Holden, *Biochim. Biophys. Acta*, 33 (1959) 581.
[86] J. T. Holden, in *Proc. 5th Internat. Congr. Biochem., Moscow, 1961.*
[87] J. T. Holden, in preparation.
[88] J. T. Holden, unpublished results.
[89] M. Ikawa and E. E. Snell, *Arch. Biochem. Biophys.*, 78 (1958) 338.
[90] J. Iskierko, *Med. Doświadczalna i Mikrobiol.*, 9 (1957) 179.
[91] J. Iskierko, *Med. Doświadczalna i Mikrobiol.*, 9 (1957) 369.
[92] J. Iskierko, *Med. Doświadczalna i Mikrobiol.*, 12 (1960) 43.
[93] J. Janicki and J. Skupin, *Acta Microbiol. Polon.*, 7 (1958) 139.
[94] R. F. Jones and R. A. Lewin, *Exptl. Cell Research*, 22 (1961) 86.
[95] J. Jonsen, S. Laland and L. Smith-Kielland, *Acta Chem. Scand.*, 13 (1959) 836.
[96] J. Jonsen, S. Laland, L. Smith-Kielland and R. Sŏmme, *Acta Chem. Scand.*, 13 (1959) 838.
[97] J. Jonsen and S. Laland, *Adv. in Carbohydrate Chem.*, 15 (1960) 201.
[98] D. C. Jordan, *Can. J. Microbiol.*, 5 (1959) 131.
[99] T. Kawano, *Yakugaku Zasshi*, 78 (1958) 126.
[100] T. Kawatani, M. Katayanagi and S. Kiyooka, *J. Pharm. Soc. Japan*, 71 (1951) 385.
[101] T. Kawatani, M. Katayanagi and S. Kiyooka, *J. Pharm. Soc. Japan*, 71 (1951) 389.
[102] S. Kinoshita, K. Tanaka, S. Udaka and S. Akita, *Proc. Intern. Symposium Enzyme Chem., Tokyo and Kyoto*, (1957) 464.
[103] S. Kinoshita, *Advances in Appl. Microbiol.*, 1 (1959) 201.
[104] V. V. Koningsberger, C. O. van der Grinten and J. T. G. Overbeek, *Biochim. Biophys. Acta*, 26 (1957) 483.
[105] J. O. Krivanek and R. C. Krivanek, *Biol. Bull.*, 116 (1956) 265.
[106] S. Lacks and F. Gros, *J. Mol. Biol.*, 1 (1959) 301.
[107] J. O. Lampen and P. Arnow, *Proc. Soc. Exptl. Biol. Med.*, 101 (1959) 792.
[108] G. Lester, D. Stone and O. Hechter, *Arch. Biochem. Biophys.*, 75 (1958) 196.
[109] G. Lester and O. Hechter, *Proc. Natl. Acad. Sci. U.S.*, 44 (1958) 1141.
[110] O. Lindan and E. Work, *Biochem. J.*, 48 (1951) 22.
[111] A. B. Lindenberg and M. Massin, *Compt. rend.*, 247 (1958) 2216.
[112] L. Ljungdahl and E. Sandegren, *Acta Chem. Scand.*, 4 (1950) 1150.
[113] G. W. Maag, L. W. Durrell and M. G. Payne, *Bull. Torrey Botan. Club.*, 86 (1959) 120.
[114] L. A. Mabbitt and M. Gregory, *Biochim. Biophys. Acta*, 37 (1960) 139.
[115] S. K. Majumdar and S. K. Bose. *Biochim. Biophys. Acta*, 29 (1958) 509.
[116] J. Mandelstam, *Biochem. J.*, 69 (1958) 103.
[117] F. Marini, P. Arnow and J. O. Lampen, *Bacteriol. Proc.*, (1960) 160.
[118] A. Markovitz and H. P. Klein, *J. Bacteriol.*, 70 (1955) 149.
[119] A. T. R. Mattick, G. S. Cheeseman, N. J. Berridge and V. Bottazzi, *J. Appl. Bacteriol.*, 19 (1956) 310.
[120] C. McAnnelly, *Phytopathology*, 49 (1959) 734.
[121] I. R. McManus, *J. Biol. Chem.*, 231 (1958) 777.
[122] A. A. Melnikova and E. I. Surikova, *Mikrobiologiya*, 26 (1957) 35.

123 J. K. MIETTINEN, *Ann. Acad. Sci. Fennicae, Ser. A. II.*, 58 (1954) 113 pp.
124 P. MITCHELL AND J. MOYLE, in E. T. C. SPOONER AND B. A. D. STOCKER, *Bacterial Anatomy*, Cambridge Univ. Press, 1956, p. 150.
125 P. MITCHELL AND J. MOYLE, *J. Gen. Microbiol.*, 16 (1957) 184.
126 P. MITCHELL AND J. MOYLE, *J. Gen. Microbiol.*, 20 (1959) 434.
127 T. MIWATANI, *Nippon Saikingaku Zasshi*, 12 (1957) 283; *Biol. Abstr.*, 32 (1958) 9325.
128 V. MOSES, *J. Gen. Microbiol.*, 20 (1958) 184.
129 H. MUKHERJEE, *Nature*, 184 (1959) 1742.
130 T. MURAYAMA, *Mem. Coll. Sci. Univ. Kyoto, Ser. B.*, 24 (1957) 225.
131 H. E. MURRAY AND F. P. ZSCHEILE, *Phytopathology*, 46 (1956) 363.
132 Y. MURUYAMA AND H. MITUI, *J. Biochem.*, 45 (1958) 169.
133 S. NAGAI, *J. Inst. Polytech. Osaka City Univ.*, 4 (1953) 35.
134 S. NANI AND G. GIOLITTI, *Boll. ist. sieroterap. milan*, 38 (1959) 124.
135 S. NATARAJAN, *J. Ind. Bot. Soc.*, 37 (1958) 233.
136 S. L. NEIDLEMAN AND A. R. KEMMERER, *Arch. Biochem. Biophys.*, 88 (1960) 172.
137 C. OGINO, *J. Tokyo Univ. Fisheries*, 41 (1955) 107.
138 R. G. OWENS, H. M. NOVOTNY AND M. MICHELS, *Contrib. Boyce Thompson Inst.*, 19 (1958) 355.
139 A. D. PANEK AND O. VERONA, *1st Botan. Univ. Lab. Crittogam., Pavia Atti*, 15 (1958) 224.
140 A. E. PASEICKA, *Can. J. Biochem. and Physiol.*, 38 (1960) 1137.
141 G. PAULETTA AND A. DEFRANCESCHI, *Biochim. Biophys. Acta*, 9 (1952) 271.
142 N. PFENNIG, *Arch. Mikrobiol.*, 26 (1957) 345.
143 N. C. PILLAI AND K. S. SRINIVASAN, *J. Gen. Microbiol.*, 14 (1956) 248.
144 J. F. POWELL AND R. E. STRANGE, *Biochem. J.*, 63 (1955) 663.
145 J. F. POWELL, in H. O. HALVORSON, *Spores*, American Institute of Biological Sciences Publication, No. 5 (1957) 72.
146 H. PROOM AND A. J. WOIWOOD, *J. Gen. Microbiol.*, 3 (1949) 319.
147 A. J. H. PYLE, *J. Gen. Microbiol.*, 11 (1954) 191.
148 R. RITTER, *Arch. Mikrobiol.*, 22 (1955) 248.
149 R. B. ROBERTS, P. H. ABELSON, D. B. COWIE, E. T. BOLTON AND R. J. BRITTEN, *Studies of Biosynthesis in Escherichia coli*, Carnegie Institution of Washington Publication No. 607, 1955.
150 L. J. RODE AND J. W. FOSTER, *J. Bacteriol.*, 79 (1960) 650.
151 P. ROINE, *Suomen Kemistilehti, B*, 19 (1946) 37; *Chem. Abstr.*, 41 (1947) 5573.
152 P. ROINE, *Suomen Kemistilehti, B*, 19B (1946) 40; *Chem. Abstr.*, 41 (1947) 5573.
153 P. ROINE, *Suomen Kemistilehti, B*, 19B (1946) 73; *Chem. Abstr.*, 41 (1947) 5574.
154 J. P. ROSS, *Phytopathology*, 49 (1959) 422.
155 M. R. J. SALTON, in *Microbial Cell Walls*, John Wiley and Sons, Inc., New York, 1960.
156 A. SARACHEK, *Experientia*, 14 (1958) 360.
157 M. SCHLÜTER AND F. BUKATSCH, *Zentr. Bakteriol. Parasitenk., Abt. II Ref.*, 112 (1959) 509.
158 R. SCHWYZER, *Ann. Rev. Biochem.*, 29 (1960) 183.
159 G. D. SHOCKMAN, J. J. KOLB AND G. TOENNIES, *J. Biol. Chem.*, 230 (1958) 961.
160 M. R. SIEGEL AND D. F. CROSSAN, *Phytopathology*, 50 (1960) 680.
161 M. T. SILVA AND G. S. CHEESEMAN, *J. Appl. Bacteriol.*, 22 (1959) 317.
162 P. SIMONART AND K. Y. CHOW, *Antonie van Leeuwenhoek, J. Microbiol. Serol.*, 19 (1953) 121.
163 P. SIMONART AND K. Y. CHOW, *Antonie van Leeuwenhoek, J. Microbiol. Serol.*, 19 (1953) 245.
164 P. SIMONART AND K. Y. CHOW, *Antonie van Leeuwenhoek, J. Microbiol. Serol.*, 20 (1954) 174.
165 P. SIMONART AND K. Y. CHOW, *Antonie van Leeuwenhoek, J. Microbiol. Serol.*, 20 (1954) 210.
166 B. N. SINGH, S. MATHEW AND N. ANAND, *J. Gen. Microbiol.*, 19 (1958) 104.
167 W. R. SISTROM, *Biochim. Biophys. Acta*, 29 (1958) 579.
168 B. SKARZYNSKI AND W. OSTROWSKI, *Nature*, 182 (1958) 933.
169 B. SKARZYNSKI, W. OSTROWSKI AND A. KRAWCYK, *Bull. acad. polon. sci.*, Classe II, 5 (1957) 159.
170 B. SKARZYNSKI, W. OSTROWSKI AND A. KRAWCYK, *Acta Biochim. Polon.*, 4 (1957) 249.
171 F. ŠORM AND J. ČERNA, *J. Coll. Czech. Chem. Communs.*, 25 (1960) 565.
172 S. SPIEGELMAN, H. O. HALVORSON AND R. BEN-ISHAI, in W. D. MCELROY AND B. GLASS, *Amino Acid Metabolism*, The Johns Hopkins Press, Baltimore, Maryland, 1955, p. 124.
173 E. SPOERL AND R. CARLETON, *J. Biol. Chem.*, 210 (1954) 521.
174 M. STEINER, *Symp. Soc. Exptl. Biol.*, 13 (1959) 177.
175 J. L. STOKES AND M. GUNNESS, *J. Bacteriol.*, 52 (1945) 195.
176 J. L. STROMINGER, *Physiol. Revs.*, 40 (1960) 55.
177 R. L. M. SYNGE, *Quart. Revs. (London)*, 3 (1949) 245.
178 E. S. TAYLOR, *J. Gen. Microbiol.*, 1 (1947) 86.
179 E. S. TAYLOR, *J. Gen. Microbiol.*, 3 (1949) 211.
180 M. A. TER-KARAPETYAN, *Doklady Akad. Nauk S.S.S.R.*, 122 (1958) 870; *Chem. Abstr.*, 53 (1959) 2354e.

181 J. M. Touzé-Soulet and C. Montant, *Compt. rend. soc. biol.*, 152 (1958) 874.
182 R. Tröger, *Arch. Mikrobiol.*, 29 (1958) 430.
183 F. Turba and H. Esser, *Angew. Chem.*, 65 (1953) 256.
184 F. Turba and H. Esser, *Biochem. Z.*, 327 (1955) 93.
185 C. S. Venkata Ram, *Proc. Natl. Inst. Sci. India*, Pt. B., 22 (1956) 227.
186 J. T. Wachsman and R. Storck, *J. Bacteriol.*, 80 (1960) 600.
187 A. J. Woiwood and H. Proom, *J. Gen. Microbiol.*, 4 (1950) 501.
188 E. Work, *Biochim. Biophys. Acta*, 3 (1949) 400.
189 E. Work, *Biochim. Biophys. Acta*, 5 (1950) 204.
190 E. Work, *Biochem. J.*, 49 (1951) 17.
191 E. Work, *Nature*, 179 (1957) 841.
192 B. E. Wright and M. L. Anderson, *Biochim. Biophys. Acta*, 43 (1960) 67.
193 M. Yoneda and M. Kondo, *Biken's J.*, 2 (1959) 365.
194 I. E. Young, *Can. J. Microbiol.*, 5 (1959) 197.
195 G. N. Zaizeva, A. N. Belozersky and V. J. Bykhovsky, *Mikrobiologiya*, 28 (1959) 675.

FREE AMINO ACIDS IN PROTOZOA

JOHN B. LOEFER AND OTTO H. SCHERBAUM

Office of Naval Research and California Institute of Technology, Pasadena, Calif. (U.S.A.); University of California, Los Angeles, Calif. (U.S.A.)

Reports on the occurrence of free amino acids (FAAs) in protozoa are available for only three genera, *Vorticella*, *Paramecium* and *Tetrahymena*. Although arrays of bound or protein amino acids (PAAs) have been determined in more cases, investigations of this type on protozoa have, indeed, been restricted by the lack of axenic culture methods. In a recent review (LOEFER AND SCHERBAUM[1]) covering both PAAs and FAAs, findings were recorded for ten species other than those belonging to *Tetrahymena* and for six strains of *T. pyriformis*, in addition to the results they obtained for *T. limacis* and seven strains of *T. pyriformis*. This report summarizes investigations on the FAAs reported for protozoa.

Even in cases where investigators have used axenic cultures, as has been true for the reports on FAAs in *Tetrahymena* (*cf.* Table I), a comparison of results and evaluation of data may be difficult for various reasons, such as variation in methods of culture or analysis of amino acids. In some cases, quantitative data were not available, or if available, were not readily susceptible to comparative interpretation. Hence, it will be noted in Table I that records of qualitative results only are reported for *Paramecium* and *Vorticella*.

Comparative results for various strains of *Tetrahymena* will be noted (Table I) after reviewing our own results on *T. limacis* and seven strains of *T. pyriformis*.

The methods employed had been worked out in connection with earlier studies (LOEFER AND SCHERBAUM[1]; LOEFER AND SCHERBAUM[2]; SCHERBAUM *et al.*[3]) and details of the procedures are described in these papers.

Essentially, the method involved use of concentrated washed organisms, axenically grown, from stationary-phase cultures. In most cases they were grown at 25°, although some analyses were also made on material from 10°, 29°, and 35° cultures. FAAs were extracted in the manner previously reported (SCHERBAUM *et al.*[3]). The method involves principally extraction with hot ethanol (80%). Lipids were removed with $CHCl_3$ and the aqueous phase evaporated. The dried residue was dissolved in 10% isopropanol for chromatography. In Figs. 2–6 each chromatogram was spotted with material extracted from 10 mg of dried organisms. The descending method of development was used with butanol–acetic acid–water for the first, and phenol–water for the second dimension. Exposure time for each solvent was approx. 15 h at 29°. Color spots following ninhydrin treatment were recorded photographically. R_F values obtained are indicated in Fig. 1, the standard against which the spots in Figs. 2–6 may be compared for identification.

Preliminary work on the quantitative estimation of amino acids had been made

* Grateful acknowledgement is made to the *J. Protozool.* for permission to use some previously published data and figures.

References p. 114

TABLE I

COMPARATIVE TABULATION FOR FREE AMINO ACIDS AND RELATED COMPOUNDS REPORTED IN PROTOZOA

/,	a combined determination;
P,	present (relative quantity not indicated);
O,	not mentioned;
—,	very weak or absent;
±,	low concentration;
+,	abundant;
++,	very abundant.
( )	these sulfur-containing substances and "unknowns" are the variables among the strains shown in the next table.

Free amino acids and related compounds	*Vorticella microstoma* (FINLEY, WILLIAMS[8])	*Paramecium aurelia* (LEE, RENE[9])	*Paramecium caudatum* (LEE, RENE[9])	*Paramecium multimicronucleatum* (LEE, RENE[9])	*Tetrahymena pyriformis* W (SCHLEICHER[6])	*Tetrahymena pyriformis* Var. 6, 3 strains (WELLS[7])	*Tetrahymena pyriformis* E (WU, HOGG[4])	*Tetrahymena pyriformis* GL (CHRISTENSSON[5])	*Tetrahymena pyriformis* GL (SCHERBAUM, JAMES and JAHN[3])	*Tetrahymena pyriformis* 7 strains, and *T. limacis* (LOEFER, SCHERBAUM[1])
Alanine	P	P	P	P	P	P	++	++	++	++
α-Aminobutyric acid	O	P	P	P	—	O	O	O	O	O
Arginine	P	P	P	P	P	P	++	++	+	+
Asparagine	O	P	P	P	O	O	+	O	±	±
Aspartic acid	P	P	P	P	P	P	+	+	+	+
Cysteine	O	—	—	—	—	P	O	±	±**	( )
Cystine	P	O	O	O	P*	/ P	±	O	/ ±	/ ( )
Cysteic acid	P	O	O	O	—	O	O	O	±	( )
Glutamic acid	P	P	P	P	P	P	++	++	++	++
Glutamine	O	O	O	O	O	O	±	O	+	+
Glycine	P	P	P	P	P	P	++	++	++	++
Hydroxyproline	P	—	—	—	O	—	O	O	O	O
Leucine	O	P	P	P	P	P	++	++	++	++
Isoleucine	P	/ P	/ P	/ P	P	P	/ ++	/ ++	/ ++	/ ++
Phenylalanine	P	P	—	—	P	P	++	+	/ ++	/ ++
Lysine	P	P	P	P	P	P	++	++	+	+
Histidine	O	—	—	—	P	—	±	±	/ +	/ +
Proline	P	P	P	P	P*	P	+	+	±	±
Serine	O	—	—	—	P	P	+	+	++	++
Threonine	P	—	—	—	P	P	+	+	++	++
Tryptophane	O	—	—	—	O	—	±	O	O	O
Taurine	O	O	O	O	—	O	±	O	±	( )
Tyrosine	P	P	P	P	P	P	+	+	+	+
Valine	P	—	—	—	P	P	++	+	++	++
Methionine	P	P	P	P	P	P	±	±	/ ++	/ ++
Methionine sulfoxide	O	O	O	O	P	O	O	O	O	O
Total number amino acids present	16	14	13	13	18	17	21	17	22	18–21
Number of unknowns	O	O	O	O	O	O	O	O	1	(x_1 x_2 x_3)

* Listed as absent in text, but shown as present in table 14.

** Not designated, but may be represented in cystine spot.

References p. 114

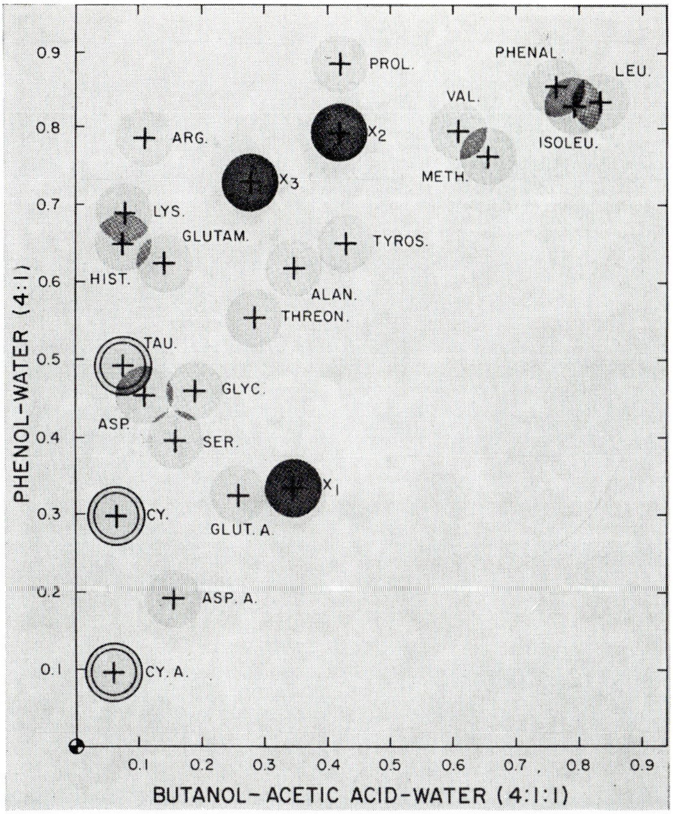

Fig. 1. Diagram showing R_F values of amino acids and related substances in phenol–water and butanol–acetic acid–water systems. This is the standard for identification of spots in Figs. 2–6. Abbreviations refer to amino acids and related compounds, including unknowns referred to in Table I.

on chromatograms of the FAAs of strain GL (SCHERBAUM *et al.*[3]). Correlations between spot size and actual amounts of eluted materials enabled us to devise a four-category semi-quantitative scheme for indicating relative amounts of FAAs observed on the original chromatograms. This scheme is utilized in the comparisons of FAAs shown in Tables I and II.

Except for the sulfur-containing amino acids cystine, cysteine and cysteic acid, taurine and the unknowns, there is little variation in the size and color intensity of spots between material obtained from the several strains of *Tetrahymena* tested. Neither are there striking differences in spot size evident between the several cultures incubated at 10° and 35°. In all cases 14 ninhydrin spots, probably representing 18 FAAs were observed, with additional S-containing amino acids and taurine in cases as noted below.

The FAA variations observed on chromatograms among the eight strains of *Tetrahymena* studied with respect to the S-containing amino acids, cystine, cysteine, cysteic acid, taurine and several unknowns are compared in Table II. When these three sets of spots for each strain were compared, no two were quite alike, except

TABLE II

AMINO ACIDS AND RELATED COMPOUNDS IN TETRAHYMENA

Modified from LOEFER AND SCHERBAUM[1].
Key to relative color intensity of spots:
++, very strong;
+, strong;
±, low concentration;
—, very weak or absent;
( ), these sulfur-containing substances and "unknowns" are the variables among the strains shown.

Substance	Regular patterns* (These 14 spots (19 AAs) in all strains) PAA	Regular patterns* (These 14 spots (18 AAs) in all strains) FAA	These strains had additional ninhydrin-positive spots as shown below: L1, Gf-J T. limacis	F	L3	PR	WH52 10°	WH52 25°	WH52 35°	HS 10°	HS 25°	HS 35°
10 Misc. AAs**	++	++										
4 Misc. AAs***	++	+										
Asparagine	—	±										
Cysteine/cystine	±	( )	—	±	+	+	+	+	—	+	—	—
Cysteic acid	±	( )	—	—	—	±	—	—	—	—	±	±
Glutamine	—	+										
Proline	±	±										
Taurine	—	( )	—	++	++	++	—	—	—	+	—	—
Tyrosine	+	+										
Unknowns: X_1	—	( )	—	++	++	+	—	—	+	++	++	—
X_2	—	( )	—	—	—	—	++	—	—	++	—	—
X_3	—	( )	—	—	—	—	+	—	—	—	—	—

* All data for 25° cultures, except where indicated for WH52 and HS.
** This group of ten included: alanine, glutamic acid, glycine, leucine/isoleucine/phenylalanine, serine, threonine, valine/methionine.
*** This group of four included: arginine, aspartic acid, lysine/histidine.

those for *T. limacis* and strains Gf-J and L1 of *T. pyriformis*, in which strains all of these spots were very weak or absent. In all cases cystine, cysteine and cysteic acid occurred in relatively low concentrations, as compared with most of the other substances. The failure to detect cystine in any of its forms in several strains may be an indication, not of its absence, but of the marginal quantity present in these strains.

In three strains at 35° (F, L3 and PR) taurine was quite abundant, the spot intensity was strong in strain HS at 10°, but very weak or absent in the others.

One unknown, X_3, was observed only in WH52-10° culture chromatograms (Fig. 5); another, X_1, in strains L3, PR, F and HS at 25° (Figs. 2 and 3) and in HS-10° and WH52-35° material (Table II).

The unknown X_2, was detected in 10° cultures of HS and WH52 (Fig. 5).

Fig. 6 is a photograph of a chromatogram of GL-29° 2-day stationary phase material. Inspection reveals the unknown X_2, seen in Fig. 5, as well as another, which has been designated X_4. SCHERBAUM *et al.*[3] observed an unknown corresponding to the latter in culture samples collected just prior to the beginning of synchronous division. When aliquots of the material used for spotting the chromatogram shown in Fig. 6

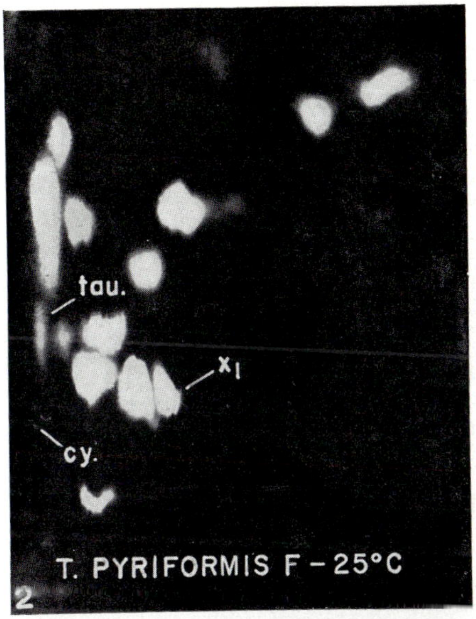

Fig. 2

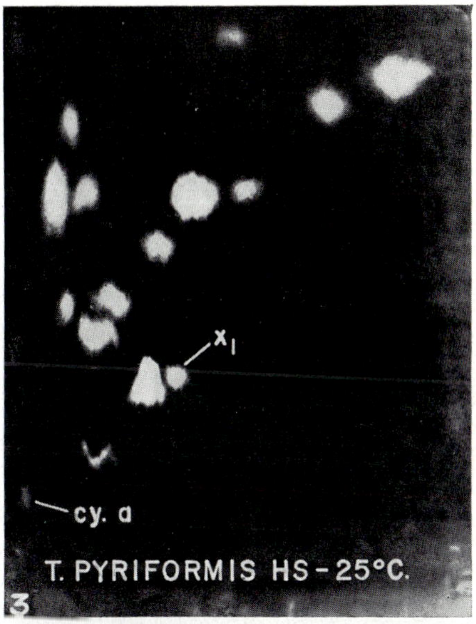

Fig. 3

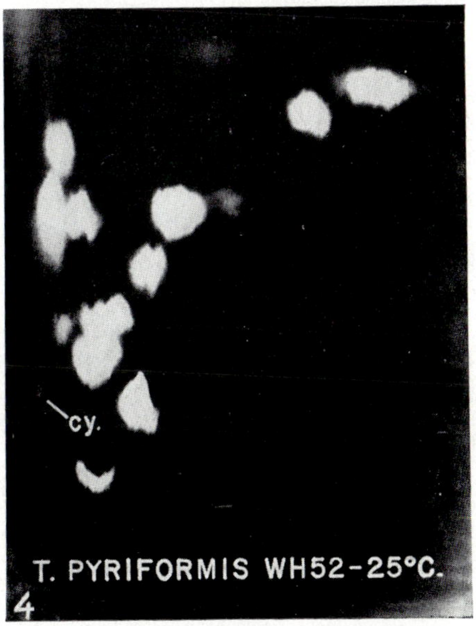

Fig. 4

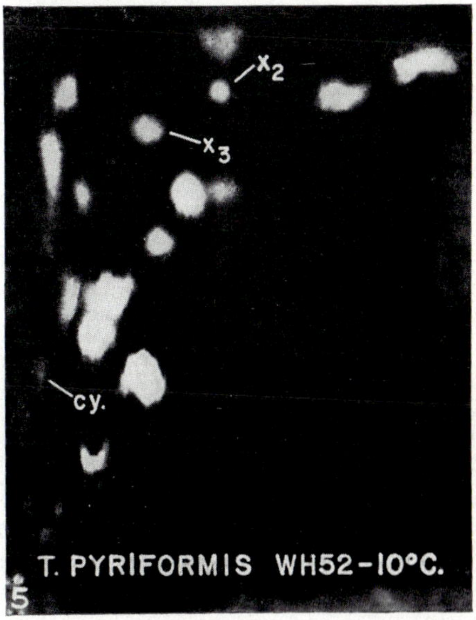

Fig. 5

were hydrolyzed prior to spotting, unknowns X_2 and X_4 were no longer detectable. It would appear, therefore, that unknowns X_2 and X_4 represent peptides. The precise nature of these four unknown substances requires further investigation.

The semi-quantitative scheme used above was extended to include comparisons with the findings of WU AND HOGG[4]. The latter employed both chromatography and microbiological methods. Their method revealed the presence of tryptophane, not shown by others. Their method also throws light on the relative amounts of amino

References p. 114

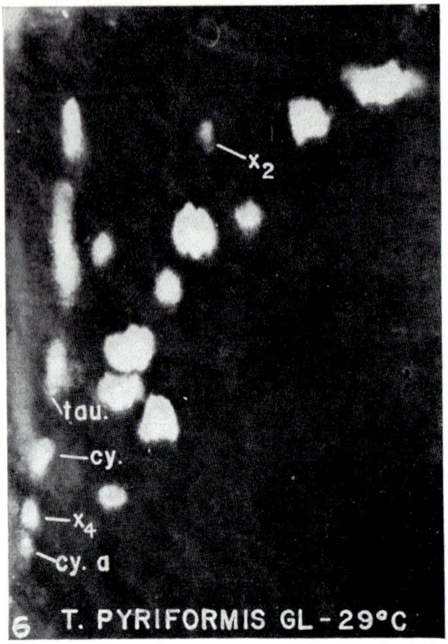

Fig. 6

acids that may be present in some of our combined determinations. For instance, valine is probably predominantly responsible for the valine–methionine spot; and lysine for the lysine–histidine spot. Except in Figs. 2 and 6 (strains F and GL), histidine appears to account for a relatively small portion of the lysine–histidine spot.

CHRISTENSSON[5], using synchronized cells and one-dimensional chromatography, obtained quantitative results of the same comparative order as ours. He did not report the occurrence of taurine, asparagine or glutamine, which were found in strain GL by SCHERBAUM *et al.*[3].

Other strains of *Tetrahymena* (Table I) analyzed by SCHLEICHER[6] and WELLS[7], although not comparable in our semi-quantitative scheme, nevertheless reveal a qualitative array of FAAs basically in agreement with those reported above.

Although factual differences noted in eight reports for protozoa on the occurrence of FAAs and related compounds appear to be evident from Tables I and II, particularly among strains of *Tetrahymena*, any conclusion as to the significance of these differences would at the present time appear to be premature.

REFERENCES

1 J. B. LOEFER AND O. H. SCHERBAUM, *J. Protozool.*, 8 (1961) 184.
2 J. B. LOEFER AND O. H. SCHERBAUM, *Anat. Record*, 131 (1958) 571.
3 O. H. SCHERBAUM, T. W. JAMES AND T. L. JAHN, *J. Cellular Comp. Physiol.*, 53 (1959) 119.
4 C. WU AND J. F. HOGG, *Arch. Biochem. Biophys.*, 62 (1956) 70.
5 E. G. CHRISTENSSON, *Acta Physiol. Scand.*, 45 (1959) 339.
6 J. D'A. SCHLEICHER, *Paperchromatography Analyses of Amino Acids in Protozoa: Some Aspects of the Metabolism of Aspartic Acid* (Cath. Univ. Amer. Biol. Studies 53), Catholic University of America Press, Washington, 1959, 74 pp.
7 C. WELLS, *J. Protozool.*, 7 (1960) 7.
8 H. E. FINLEY AND H. B. WILLIAMS, *J. Protozool.*, 1 (1954) 7.
9 J. W. LEE AND A. RENE, *Proc. La. Acad. Sci.*, 18 (1955) 13.

IV. INSECTS

FREE AMINO ACIDS IN INSECTS

P. S. CHEN

Institute of Zoology and Comparative Anatomy, University of Zurich, Zurich (Switzerland)

There is a large volume of information on the free amino acids in insects. Especially during the last decade, numerous studies have been carried out to analyze the patterns of free amino acids in different insects, and the changes of these patterns during morphogenetic development. Progress in this particular field of insect biochemistry has been facilitated by two important facts. First, insect blood is known to have an unusually high amino acid concentration which in some species may be more than 60 times higher than that in human blood. Second, since the advent of such new techniques as microbiological assays and paper partition chromatography, it has become possible to follow quantitative changes in those components which are present even in extremely low concentrations.

Recent work in various laboratories has demonstrated clearly that the amino acids, in addition to their function as protein constituents, enter into diverse metabolic pathways and participate in many activities of living organisms. There is also evidence which shows that certain amino acids play a specific role in reproduction and development. However, owing to the large variations found for different insects, evaluation of the observed results becomes difficult. The present paper is a review of some of the latest studies with an attempt to summarize the available data in an integrated manner, especially in relation to taxonomic, embryological, genetical and physiological problems.

OCCURRENCE OF FREE AMINO ACIDS IN DIFFERENT INSECTS

Patterns of free amino acids and related compounds in 20 insect species are summarized in Table I. Most data were obtained from analysis of hemolymph at either larval or imaginal stage. In those cases where the amount of hemolymph is limited, extracts of whole bodies were prepared. It is evident that no direct quantitative comparison between various studies can be made, as long as no reliable information concerning age and nutritional state of the specimens used is available. This may account for some of the contradictory findings of authors who have worked on the same species. In the following, a survey of the variations in the free amino acid composition of insects belonging to seven different orders (Lepidoptera, Diptera, Coleoptera, Orthoptera, Odonata, Hymenoptera and Hemiptera) will be considered.

Lepidoptera

Extensive data on Lepidoptera have been reported. By using paper partition chromatography, CHEN AND KÜHN[41] demonstrated that 24 free ninhydrin-reacting components, including γ-aminobutyric acid and three peptides, are present in the meal moth

References p. 132/135

Ephestia kühniella (Table I). One particular peptide (P_3), which is located very near glycine on the two-dimensional chromatogram, seems to be specific for this moth, because it was not found in any other insects investigated by them (see also ref. 39). Tryptophane was detected only in the *a*-mutant. Among these substances, the contents of proline and glutamine are especially high.

In a summarized paper DUCHÂTEAU AND FLORKIN[59], using microbiological assays, obtained data on the contents of 15 amino acids in the blood of a large variety of insects, including 29 different species of Lepidoptera. According to these authors the concentration of free amino acids in the Lepidoptera is high, particularly that of proline, lysine, histidine, glutamic acid and arginine. Since they used hydrolyzed dialysates for analyses, a large part of the glutamic acid may have been derived from glutamine. This is in agreement with the findings of CHEN AND KÜHN[41] who also reported a high content of glutamine in *Ephestia kühniella*. In the work of DUCHÂTEAU AND FLORKIN[59] it was noted that diapause pupae of different species (*Actias selene, Antheraea pernyi, Papilio machaon* and *Lasiocampa quercus*) showed a strikingly constant pattern of free amino acids. In *Smerinthus ocellatus* the pupae at diapause showed a higher value of total concentration than the developing larvae, mainly due to a higher content of arginine, lysine and proline. The same is true for *Sphinx ligustri*[56]. On the other hand, for *Euproctis chrysorrhoea*, no such difference was detected between larvae and pupae which were in active development.

Free amino acids in the hemolymph of the silkworm *Bombyx mori* have been investigated by several authors (DRILHON AND BUSNEL[54]; YOSHITAKE AND ARUGA[196]; FUKUDA[75]; ISHIMORI AND MUTO[104]). In a more recent paper WYATT, LOUGHHEED AND WYATT[195], also employing chromatography, reported the presence of 17 amino acids, two amides and several peptides in this insect (see Table I). But cystine (or cysteine) was detected in only one sample and tryptophane was not recorded by them. According to FUKUDA, KIRIMURA, MATUDA AND SUZUKI[77] tryptophane occurs in the silkworm blood, but its concentration is obviously very low. The quantitative analyses of WYATT *et al.*[195] revealed a relatively high concentration of glutamine, histidine, lysine, serine, glycine, and a large variation in the contents of histidine, tyrosine and proline. However, the general picture of free amino acids is in agreement with that provided by microbiological assays in this insect (SARLET, DUCHÂTEAU AND FLORKIN[161]; DUCHÂTEAU AND FLORKIN[57]; FUKUDA *et al.*[77]).

In the larval blood of the wax moth *Galleria mellonella* the presence of 21 free amino acids including α-amino-n-butyric acid has been demonstrated (PRATT[152]; AUCLAIR AND DUBREUIL[9]). However, tryptophane, cystine and α-amino-n-butyric acid were not recorded by WYATT *et al.*[195] who also worked on this species. These amino acids apparently are present in small quantities. According to AUCLAIR AND DUBREUIL[9] taurine and hydroxyproline appear only at metamorphosis. It should be mentioned that taurine, α-amino-n-butyric acid and β-alanine are normally not incorporated into proteins.

In the larvae of *Agrotis ypsilon, Heliothis virescens* and *Estigmene acraea*, CLARK AND BALL[43] reported the consistent presence of alanine, arginine, glutamic acid, glycine, histidine, leucine and/or isoleucine, proline, serine, threonine, tyrosine, valine and glutamine. Asparagine, α-amino-n-butyric acid and hydroxyproline were not identified in any one of these species, whereas aspartic acid, cysteic acid, lysine, methionine, phenylalanine, tryptophane and taurine were found only in one or two of them.

TABLE I

OCCURRENCE OF FREE AMINO ACIDS, AMIDES AND PEPTIDES IN DIFFERENT INSECTS

Amino acids, amides and peptides	*Lepidoptera*					*Diptera*				*Coleoptera*				*Orthoptera*				*Odonata*	*Hymenoptera*	*Hemiptera*
	Ephestia kühniella (Larva, pupa, adult)[41]	*Bombyx mori* (Larval blood)[195]	*Galleria mellonella* (Larval blood)[9,152]	*Euproctis chrysorrhoea* (Larval blood)[59,60]	*Sphinx ligustri* (Larval blood)[56,59]	*Drosophila melanogaster* (Larva)[39]	*Culex pipiens* (Larva, pupa, adult)[30,31]	*Musca domestica* (Larval and adult blood)[152]	*Gastrophilus intestinalis* (Larval blood)[121]	*Tenebrio molitor* (Larval blood)[9]	*Leptinotarsa decemlineata* (Adult blood)[59]	*Epilachna varivestis* (Larval blood)[9]	*Hydrophilus piceus* (Adult blood)[160,161]	*Blatta orientalis* (Adult blood)[43]	*Blatella germanica* (Adult blood)[152]	*Locusta migratoria* (Nymphal blood)[59,60]	*Schistocerca gregaria* (Adult blood)[175]	*Aschna cyanea* (Nymphal blood)[154]	*Apis mellifera* (Larval blood)[152]	*Oncopeltus fasciatus* (Adult blood)[152]
Alanine	+	+	+	+	+	+	+	+	+	+	+	+	+	+	+	+	+	+	+	+
Arginine	+	+	+	+	+	+	+	+	+	+	+	−	+	+	+	+	−	+	+	−
Aspartic acid	+	+	+	+	+	+	+	+	+	+	+	+	+	+	+	+	−	(+)	+	+
Cysteine	−	−	−	−	−	−	−	−	−	−	(+)*	−	−	+	−	−	−	−	−	−
Cystine	+	−	+	−	−	(+)	+	+	−	−	(+)	−	−	−	+	−	−	−	+	+
Glutamic acid	+	+	+	+	+	+	+	+	+	+	+	+	+	+	+	+	+	(+)	+	+
Glycine	+	+	+	+	+	+	+	+	+	+	+	+	+	+	+	+	+	+	+	+
Histidine	+	+	+	+	+	+	+	−	−	+	+	−	+	+	+	+	+	(+)	+	−
Hydroxyproline	−	−	−	−	−	(+)	−	−	−	−	−	+	−	−	−	−	−	−	+	−
Isoleucine	−	+	+	+	+	+	+	+	+	+	−	+	+	+	+	+	+	+	+	+
Leucine	+	+	+	+	+	+	+	+	+	+	+	+	+	+	+	+	+	+	+	+
Lysine	+	+	+	+	+	+	+	+	+	+	+	+	+	(+)	+	+	−	+	+	+
Methionine	−	+	+	+	+	(+)	+	+	−	+	(+)	+	−	+	+	+	−	(+)	+	+
Methionine sulfoxide	−	−	−	−	−	(+)	+	−	−	−	−	−	−	−	−	−	−	−	−	−
Phenylalanine	+	+	−	+	+	−	−	+	+	−	+	−	+	−	−	+	−	(+)	(+)	−
Proline	+	+	+	+	+	+	+	+	+	+	+	+	+	+	+	+	+	+	+	+
Serine	+	+	+	−	−	+	+	+	+	+	(+)	+	(+)	+	+	−	+	+	+	+
Threonine	+	+	+	+	+	+	+	+	−	+	+	+	+	+	+	+	+	(+)	+	−
Tryptophane	−	(+)	+	−	−	(+)	−	−	−	+	(+)	+	−	+	−	−	−	−	+	−
Tyrosine	+	+	+	+	+	+	+	+	+	+	(+)	+	+	+	+	+	+	+	+	+
Tyrosine-*O*-phosphate	−	−	−	−	−	(+)	−	−	−	−	−	−	−	−	−	−	−	−	−	−
Valine	+	+	+	+	+	+	+	+	+	+	+	+	+	+	+	+	+	+	+	+
Asparagine	−	+	+	(+)	(+)	(+)	−	−	−	−	(+)	+	(+)	−	+	(+)	−	−	+	−
Glutamine	+	+	+	(+)	(+)	+	+	+	+	+	(+)	+	(+)	+	−	(+)	+	−	+	+
α-Aminobutyric acid	−	−	+	−	−	−	−	−	−	−	−	−	−	−	−	−	−	−	+	−
γ-Aminobutyric acid	+	−	−	−	−	+	−	−	−	−	−	−	−	−	−	−	−	−	−	−
β-Alanine	+	+	+	−	−	+	+	+	−	+	−	+	−	+	+	−	−	−	+	+
Ornithine	+	−	−	−	−	(+)	−	−	−	−	−	−	−	−	−	−	−	−	−	−
Taurine	+	−	−	−	−	(+)	+	+	+	+	−	−	−	+	−	−	−	−	+	+
Peptides	+	+	−	−	−	+	+	−	+	−	−	−	−	−	−	−	−	−	−	−

* (+) indicates that the substance was not listed in the paper cited, but reported by other authors who worked on the same species.

With the exception of hydroxyproline, taurine, asparagine and tryptophane, AUCLAIR AND DUBREUIL[9] identified all the amino acids referred to above in three other species of Lepidoptera *(Phlegethontius quinquemaculatus, Malacosoma americana* and *Archips cerasivorana)* but α-amino-*n*-butyric acid was found only in *P. quinquemaculatus*.

Diptera

In connection with genetic and developmental problems, the free amino acids in larvae and adults of *Drosophila melanogaster* have been extensively studied by a number of investigators (HADORN AND MITCHELL[90]; AUCLAIR AND DUBREUIL[9]; CHEN AND HADORN[39]; BENZ[19]; FOX[70]; KAPLAN, HOLDEN AND HOCHMAN[107]). For larvae aged 72 and 96 h CHEN AND HADORN[39] noted the presence of 22 free ninhydrin-reacting substances, including four peptides (Table I). AUCLAIR AND DUBREUIL[9] found asparagine and hydroxyproline, but detected no β-alanine, histidine, and γ-amino-*n*-butyric acid. In addition to the above amino acids, BENZ[19] reported the presence of taurine and ornithine. Taurine and methionine sulfoxide appear in larger quantities only in later developmental stages and in the adults (CHEN AND DIEM[38]). FAULHABER[64] identified tryptophane, but this amino acid apparently accumulates in larger quantities only in the *v*-mutant (GREEN[82]). The occurrence of γ-amino-*n*-butyric acid in this insect is understandable, as this amino acid has been found in yeast extract (REED[155]). But the possibility that it is formed by decarboxylation of glutamate is not excluded. Very recently MITCHELL, CHEN AND HADORN[141] demonstrated the occurrence of tyrosine-*O*-phosphate in *Drosophila* larvae (see also paper by MITCHELL and SIMMONS in this conference). This compound has R_F values corresponding to peptides 1 plus 2 described by CHEN AND HADORN[39]. The last authors also observed a high concentration of tyrosine in the hydrolysates of these peptides.

Sexual differences in the pattern of free ninhydrin-reacting components of this insect have been noted by several workers. KAPLAN, HOLDEN AND HOCHMAN[107], by both chromatography and microbiological assays, showed that adult females contain almost twice as much methionine as males. FOX[69] as well as FOX, MEAD AND MUNYON[71] reported the presence of a "sex peptide" in male adults, but not in females. Very recently, CHEN AND DIEM[38] recorded a ninhydrin-positive substance which is specifically located in the accessory glands (paragonia) of the male flies and probably corresponds to the "sex peptide" described by FOX[69]. Its quantity increases in the course of imaginal development, apparently due to the gradual increase in size of these glands.

The mosquitoes are another group of dipterous insects the free amino acids of which have been studied in detail. In the body extract and hemolymph of *Culex pipiens* CHEN[30, 31] identified the presence of methionine sulfoxide and two peptides in addition to 19 other free ninhydrin-positive components (Table I). The same pattern was found for *Culex fatigans* (GEIGER[78]). Most of these amino acids were also reported for *Culex tarsalis, Culex stigmatosoma, Culex quinquefasciatus, Aedes varipalpus* and *Culiseta incidens* (CLARK AND BALL[42, 43]; BALL AND CLARK[15]). The main difference is that both tryptophane and α-amino-*n*-butyric acid were noted in these insects, but not in *C. pipiens* and *C. fatigans* (see ref. 33). CLARK AND BALL[43] identified cysteic acid which, as noticed by these authors, was probably derived from cystine. The patterns of free amino acids for *Culex salinarius, Aedes aegypti, Aedes sollicitans, Culiseta inornata* and *Anopheles quadrimaculatus* are very similar to that found for *C. pipiens*, except for

References p. 132/135

the absence of aspartic acid in *C. salinarius* (MICKS AND ELLIS[138, 139]). Analyses on protein hydrolyzates of the excreta of three mosquito species (*A. aegypti*, *C. pipiens*, *A. quadrimaculatus*) revealed the presence of galactosamine and glucosamine in addition to 16 amino acids (IRREVERRE AND TERZIAN[102]).

In the larval blood of *Calliphora erythrocephala* FINLAYSON AND HAMMER[65] detected alanine, glycine, histidine, aspartic acid, lysine, isoleucine, leucine, phenylalanine, proline, tyrosine, serine and valine. Working on the same species AGRELL[1] reported in addition, glutamic acid, glutamine, β-alanine, methionine, taurine, threonine, two peptides and one hydrolysable substance in the body extracts of pupae. The latter author, however, did not mention histidine and phenylalanine. For *Calliphora augur* HACKMAN[86] observed four more amino acids: arginine, asparagine, hydroxyproline and cystine/cysteine. The concentration of these substances is apparently very low.

In both larvae and adults of the house fly *Musca domestica* PRATT[152] identified at least 17 amino acids (Table I). Similar results have been reported recently by REIFF[156]. In the larval hemolymph of *Corethra plumicornis* CHEN AND HADORN[39] observed at least 16 free amino acids and a peptide. The concentration of glutamine is particularly high in this insect, and proline seems to be absent in its blood.

Coleoptera

In the larval hemolymph of the mealworm *Tenebrio molitor* DRILHON[50] estimated seven amino acids, the relative concentration of which, beginning from the highest one, is as follows: serine, glycine, alanine, tyrosine, valine, proline, histidine. AUCLAIR AND DUBREUIL[9], also by paper chromatography, included 11 more amino acids (Table I). The contents of lysine, alanine and tryptophane are especially low. In adult hemolymph of the Colorado beetle *Leptinotarsa decemlineata* SARLET *et al.*[161] reported 12 amino acids, among which glutamic acid and proline occur in strikingly high concentrations. The larval blood of this insect is especially rich in free amino acids. All together 21 ninhydrin-positive compounds including one unknown substance have been noted (AUCLAIR AND DUBREUIL[9]; DRILHON[50]).

USSING[176] reported the presence of lysine, arginine, histidine, tyrosine, leucine, valine, tryptophane and perhaps glutamine and hydroxyproline in the adult blood of the cockchafer *Melolontha vulgaris* and the wood-feeding beetle *Oryctes nasicornis*. Asparagine is found only in *M. vulgaris*. Hydroxyproline and four additional amino acids (glycine, serine, alanine, proline) were also detected in larvae of these two insects (DRILHON[50]).

Quantitative determinations of 14 amino acids in the adult hemolymph of the water beetle *Hydrophilus piceus* showed that glutamic acid and proline have by far the highest concentration (SARLET *et al.*[161]). As mentioned previously, glutamine was included in the glutamic acid data. While in the paper of SARLET *et al.*[161] serine was not mentioned, DRILHON[50] reported its presence in this insect. In the hemolymph of the swimming beetle *Dytiscus marginalis*, FLORKIN AND DUCHÂTEAU[68] observed the occurrence of histidine and tyrosine, but the absence of arginine, tryptophane, phenylalanine and cystine. In a later work of DRILHON[50] the presence of valine, serine, leucine, glutamic acid, aspartic acid, tyrosine and histidine was identified.

AUCLAIR AND DUBREUIL[9] established the amino acid composition of the larval blood of the Mexican bean beetle *Epilachna varivestis* (see Table I). For *Xylotrechus*

References p. 132/135

nauticus, CLARK AND BALL[43] also found these amino acids, except proline, β-alanine and asparagine. They included, however, arginine, cysteic acid, histidine, hydroxyproline and phenylalanine. More recently, PO-CHEDLEY[150] reported in the blood of the oriental beetle *Anomala orientalis* the presence of α-amino-*n*-butyric acid, arginine, cystine, histidine, ornithine, phenylalanine and taurine in addition to those found for *E. varivestis*.

Orthoptera

CLARK AND BALL[43] recorded in the adult blood of the oriental cockroach *Blatta orientalis* the occurrence of 19 amino acids including cysteic acid (Table I). AUCLAIR AND DUBREUIL[9] in addition found lysine, which occurs however at a low level.

Adults of the American roach *Periplaneta americana* contain the following free amino acids: alanine, cystine, glutamic acid, glycine, isoleucine/leucine, methionine, proline, serine, tyrosine, valine and glutamine (PRATT[152]). This insect apparently has fewer amino acids than the German cockroach *Blatella germanica* (see Table I).

In adult females of the stick insect *Dixippus morosus*, DRILHON[50] estimated the relative proportion of amino acid concentration as follows: valine, glycine, serine, tyrosine, leucine. DUCHÂTEAU, SARLET AND FLORKIN[61] detected in addition, alanine, lysine, methionine, phenylalanine, proline and threonine. Among these, histidine occurs at the highest level.

According to the recent work of TREHERNE[175] the concentration of glycine and serine in the hemolymph of adult females of the locust *Schistocerca gregaria* is particularly high. Certain amino acids such as leucine, alanine and valine show an especially high degree of variation.

Odonata

In the hemolymph of the dragon-fly nymph *Aeschna cyanea*, RAPER AND SHAW[154] estimated that glycine, alanine, valine and leucine have the greatest concentration, whereas serine and lysine are present in small quantities (Table I). DUCHÂTEAU AND FLORKIN[59] found, in addition, aspartic acid, glutamic acid, histidine, methionine, phenylalanine and threonine. The latter authors did not identify the species investigated.

Hymenoptera

Patterns of amino acids in the blood of worker, drone and queen larvae of the honey bee *Apis mellifera* have been analyzed by PRATT[152]. Histidine and α-amino-*n*-butyric acid were found only in the queen larvae. DUCHÂTEAU, SARLET AND FLORKIN[61] determined in the bee larvae the concentration of 15 amino acids among which glutamic acid, lysine and proline occur at a markedly high level. It is known that the worker and drone larvae are fed on pollen, whereas the queen larvae on royal jelly. AUCLAIR AND JAMIESON[11] reported 21 different free amino acids including histidine and α-amino-*n*-butyric acid in several pollen species. But PRATT AND HOUSE[153] included no cystine, histidine, hydroxyproline, phenylalanine, threonine and tryptophane in the list of free amino acids found by them in the royal jelly. Further studies are needed to clarify such discrepancies. KALLHAMMER[106] found in the hydrolyzates of young bees at hatching 15 amino acids including phenylalanine. Tyrosine showed a large variation.

References p. 13/135

Moreover, differences in the contents of histidine and cystine in the hydrolyzates of various organs were noticed.

Hemiptera

The adult blood of the milkweed bug *Oncopeltus fasciatus* contains 16 amino acids, including β-alanine and taurine (Table I). One interesting finding reported by AUCLAIR AND PATTON[14] is the occurrence of D-alanine in the hemolymph of this bug. According to the last two authors the unnatural isomer of this amino acid was not derived from the food. Furthermore, no D-amino acid oxidase was found in the fat body or Malpighian tubules of this insect (AUCLAIR[5]). The possibility that the D-alanine was formed by microorganisms could not be excluded. Studies on aseptic materials are needed to ascertain this point.

Some information on the amino acid composition in the blood-sucking bug *Rhodnius prolixus* is available. HARINGTON[93] reported the presence of leucine, histidine, histamine, taurine, glycine, valine, phenylalanine and alanine in the dry excreta of this insect. Histidine occurs also in a high concentration in the hemolymph, but histamine is absent. Since histidine was not found in serum-fed individuals, the author concluded that histidine was derived from the globin of hemoglobin. Obviously histidine gives rise to histamine by decarboxylation. But the occurrence of decarboxylase in this insect has not yet been established.

General remarks

The previous survey shows that, in general, the aliphatic amino acids play a dominant part in the chemical composition of the hemolymph. This is indicated by their large quantities and the constancy of their occurrence (see DRILHON[50]). Aromatic acids such as tyrosine, histidine and proline appear also in considerable amounts. Tyrosine is known to be involved in the tanning reaction of cuticles, while proline serves as an important constituent of the cuticular protein (for references see CHEN[30, 31]). The work of FUKUDA *et al.*[77] on the silkworm larvae demonstrated that the basic amino acids like histidine and lysine are present in a much higher quantity in the blood of this insect as in mammalian plasma. Free amino acids in insect hemolymph have been discussed in a recent review by WYATT[194].

Tyrosine-*O*-phosphate has been so far detected only in *Drosophila*. Hydroxyproline and cysteine are of relatively less common occurrence. The latter amino acid can be easily oxidized to cystine during the preparation of samples. Methionine sulfoxide has been found in the dipterous insects and in the egg of the silkworm and the grasshopper (see p. 118). It must be mentioned that many free amino acids identified in insects can be bound with other organic compounds. For instance, WESTLEY, WREN AND MITCHELL[186] found aspartic acid, cystine, glutamic acid and serine in the hydrolyzates of phospholipids from *Drosophila* larvae. They reported that in early developmental stages, essentially all of the non-protein amino acids were found in the lipid-soluble fraction. The technique used in preparing samples for amino acid analysis should be carefully considered.

Several amino acids like β-alanine, taurine, ornithine, α- and γ-aminobutyric acid, which are not found in protein molecules have been identified in the blood of various insects. This fact suggests that these compounds are particularly involved in intermediary metabolic pathways and fulfill special functions in insect metabolism.

The two amides, asparagine and glutamine, are of wide distribution in insects. Both of them are able to transfer amino groups in the transamination reactions (see p. 131). There is, however, very little information available on the peptides. In those studies, where no hydrolysis of samples has been carried out, some peptides have been possibly designated simply as unknown substances.

Other physiological factors may account for the high degree of variations in amino acid concentration found in insects. In higher organisms HATZ[94] observed that injection of amino acids such as tryptophane, tyrosine and histidine causes a temporary rise of alanine. AUCLAIR[6] also emphasized that in *Blatella germanica* certain amino acids, when fed alone, could affect the quality and concentration of free amino acids in the blood, but other amino acids did not show this effect. The concentration of different amino acids are obviously interrelated. For mammals LI, GESCHWIND AND EVANS[125] reported that growth hormone effects a distinct drop in amino acid contents of the blood. Although no such experiments have been yet carried out on insects, it is conceivable that the complicated hormone system in these organisms, especially during development, influences directly or indirectly the patterns of free amino acids.

The most obvious effect is of course the nutritional state. The study of CHEN AND HADORN[40] on *Drosophila* larvae showed that during starvation essential amino acids such as valine and leucine disappear completely from the chromatogram, whereas the contents of certain peptides increase rapidly. They observed that larvae fed on saccharose alone had a very high concentration of alanine, which was obviously synthesized from pyruvate through transamination. In this connection, one has to consider also the absorption of different amino acids contained in the diet. For the German cockroach *Blatella germanica*, AUCLAIR[6] reported that DL-homocystine could not be absorbed. Working on the locust *Schistocerca gregaria*, TREHERNE[175] reported that among the amino acids tested by him, glycine and serine were absorbed most rapidly from the mid-gut caeca. The author concluded that in this insect, in contrast to the active transfer in mammals, the amino acids enter into blood probably through "a diffusion gradient established by net movement of water molecules into the hemolymph".

Several S-containing free amino acids like cystine, cysteine, cysteic acid and methionine have been recorded in insect blood. These substances are interrelated in intermediary metabolism (see GILMOUR[80]), and serve as important sources for the sulfhydryl-group in coenzymes and hormones (BARRON[16]; GREENBERG[83]). Adult female mosquitoes are especially rich in methionine sulfoxide (CHEN[31]). According to STEKOL[172] this compound can replace methionine for growth. For the common house fly *Musca domestica*, a tracer study showed that methionine could not be formed from cystine (HILCHEY, COTTY AND HENRY[97]). Taurine was, however, radioactive in flies fed with ^{35}S-cystine. It is interesting to notice that in the ciliate *Tetrahymena geleii*, in contrast to mammals, methionine cannot be formed from homocystine (GENGHOF[79]). The metabolic pathways of these compounds in invertebrates are obviously different from those in higher organisms.

FREE AMINO ACIDS AND TAXONOMY

Paper partition chromatography has been employed for taxonomic purposes. This technique is simple and sensitive for detecting small quantities of biological substances. The question is, however, whether differences in free amino acids do exist among different species. According to FLORKIN[66] the amino acid concentration in the gastropods is in general higher than that in the vertebrates, and that in the crustaceans still higher. Species-specific differences have been recorded in snails (KIRK, MAIN AND BEYER[113]; MICHEJDA AND URBANSKI[135]; KIECOL AND MICHEJDA[110]), Annelids (AUCLAIR, HERLANT-MEEWIS AND DEMERS[10]) and in the eggs of various sea urchins (CHEN AND BALTZER[37]; CHEN[32]). According to DUCHÂTEAU *et al.*[62] the amino acid contents of Molluscs and Annelids are usually higher in the marine forms than in the related fresh-water forms. The same is true for crustaceans (CAMIEN *et al.*[28]). As far as insects are concerned, DUCHÂTEAU AND FLORKIN[59] estimated the free amino acid concentration in such specialized forms like Coleoptera, Hymenoptera and Lepidoptera to be at a higher level than that in the more primitive forms like Odonata and Orthoptera. ROBERTSON[157] investigated the chromatographic patterns in 17 species belonging to Coleoptera, Lepidoptera, Diptera and Hymenoptera. Intergeneric differences were found for *Laemophloeus*, *Acanthoscelides* and *Macrocentrus*, and interspecific difference recorded for a variety of insects. Furthermore, two species of *Laemophloeus* differ in two reducing substances, probably amino sugars. Since the spots described by this author were separated only one-dimensionally and showed a number of overlappings, this work still has a preliminary character.

Chromatographic patterns in various species of Hemiptera *(Triatoma gerstaeckeri, T. infestans)*, Orthoptera *(Periplaneta americana, Blatella germanica, Supella supellectilium)* and Diptera *(Culex molestus, C. fatigans, C. pipiens, Aedes aegypti, Anopheles quadrimaculatus)* have been compared in detail by MICKS[137]. In a later work similar analyses on 29 species of insects and ticks as well as geographic races from nine genera were carried out (MICKS AND GIBSON[140]). In general, different patterns between orders, species of the same genus, and even between races of the same species were noted.

More extensive information is available for various mosquitoes. In an earlier paper CLARK AND BALL[42] stated the absence of tyrosine in *Culex stigmatosoma, C. tarsalis* and *Culiseta incidens*, but the presence of it in *Aedes varipalpus*. The presence of this amino acid was, however, detected in the *Culex* species in a later work of these two authors (CLARK AND BALL[43]; BALL AND CLARK[15]). Aspartic acid was recorded in *C. tarsalis, C. quinquefasciatus*, but not in *C. stigmatosoma, A. varipalpus* and *Culiseta incidens*. Furthermore, cysteic acid was absent in *C. quinquefasciatus* and *Culiseta incidens*, but present in the other species. Asparagine, phenylalanine and hydroxyproline were consistently absent in all mosquito species examined by them.

MICKS AND ELLIS[138] reported that the contents of free amino acids were in general at a high level in *C. quinquefasciatus* and *A. aegypti*. They also observed the absence of aspartic acid in *A. aegypti, A. quadrimaculatus* and *C. salinarius*. In order to detect differences between members of the *pipiens*-complex, MICKS[136] performed both one- and two-dimensional chromatograms using *C. pipiens, C. fatigans* and *C. molestus* as testing materials. Two spots did not appear in the European species (*C. pipiens*) and one spot was absent in *C. fatigans*. Several quantitative differences were also noticed. Unfortunately, no identification of the spots was made, and the nature of these substances is unknown.

In an attempt to detect biochemical differences between different populations of the *Anopheles maculipennis*-complex LEWALLEN[122] undertook chromatographic

References p. 132/135

analysis on *A. occidentalis* and *A. freeborni*. The former is a coastal form, while the latter a continental form. Several spots were identified in adults of the *A. freeborni* population but not in those of the *A. occidentalis*. In a mixed population, the chromatographic pattern was found to be quite homogenous and more similar to the neighbouring *A. occidentalis* group.

Considering the differences in diet, living conditions and other inherent metabolic processes, biochemical differences between individuals of systematically far-related groups are expected. On the other hand, a closer examination of the general amino acid patterns found for the various insects referred to above discloses a high degree of similarity. Small variations in some cases could only be due to differences in the sampling and other technical procedures. Especially if specimens collected from natural populations are employed, there is usually no reliable information on the previous history of such individuals. Without careful control of the metabolic states of these specimens, we do not know whether the detected differences are of real taxonomic characters, or reflect simply individual variations. Paper chromatography is no doubt a useful taxonomic tool, but can be applied only under well-controlled conditions.

FREE AMINO ACIDS AND DEVELOPMENT

Embryonic development

One of the outstanding biochemical properties of insects is their changes in amino acid patterns in the course of development. This is particularly evident in holometabolic type (for references see CHEN[29, 30]). During embryogenesis protein metabolism is especially intensive, and involves mainly the breakdown of yolk materials and their conversion into organ-specific proteins. The analyses of VON DER CRONE-GLOOR[180] showed that the following free ninhydrin-reacting substances are present in the developing egg of *Drosophila melanogaster*: aspartic acid, glutamic acid, glutamine, serine, taurine, cystine, glycine, threonine, alanine, histidine, lysine, arginine, tyrosine, tryptophane, γ-amino-*n*-butyric acid, three peptides and probably proline, valine and leucine. Arginine, β-alanine and γ-amino-*n*-butyric acid appeared only at the end of embryonic development, whereas cystine was present only at the early stage. The total quantities of ninhydrin-positive materials drop to a minimum shortly before hatching. Quantitative determinations on individual components indicate that aspartic and glutamic acid fall off rapidly during development, while alanine and glutamine show a distinct increase, particularly at later developmental period. In the egg hydrolyzates of *D. melanogaster* and *D. viridis*, NAKAMURA *et al.*[142, 143] found that one substance, probably cystine, is present only in the unfertilized egg. For *D. viridis*, up to the stage of germ-band formation, the contents of valine, lysine and isoleucine are quite constant, whereas glutamic acid, aspartic acid, serine, glycine and arginine exhibit characteristic variations (NAKAMURA *et al.*[144]).

For eggs of *Bombyx mori*, DRILHON AND BUSNEL[55] identified four amino acids (glutamic acid, serine, alanine, valine) at the time of fertilization and six amino acids (glutamic acid, serine, alanine, valine, tyrosine, leucine) during diapause. Glycine appeared later, and at the end of incubation five additional amino acids (tryptophane, proline, hydroxyproline, cystine, histidine) could be detected. Furthermore, the

References p. 132/135

presence of methionine sulfoxide, taurine and ethanolaminophosphoric acid in the silkworm egg has been reported (NARUMI *et al.*[145]; SASAKI *et al.*[162]).

SHAW[163] detected in the egg of the grasshopper *Chortophaga viridifasciata*, in addition to other amino acids, methionine sulfoxide, citrulline and ethanolaminophosphoric acid. The author considered that methionine sulfoxide was an oxidation product of methionine formed during the preparation of extracts. Ethanolaminophosphoric acid is probably related to the intermediary metabolism of serine and glycine. Other amino acids of general occurrence such as aminobutyric acid, cystine, taurine and hydroxyproline were not found. There is no qualitative difference of the amino compounds in eggs of different ages, but the concentration appears in general higher in later development.

A detailed study on the S-containing amino acids in developing eggs of the grasshopper *Melanoplus differentialis* has been carried out by FU[74]. The total content of sulfur remained unchanged during embryogenesis. In pre- and post-diapause, methionine and cystine–cysteine showed distinct fluctuations, which obviously reflect metabolic changes between these amino acids and between yolk and embryonic proteins. However, during the period of diapause the values remained strikingly constant. In the late post-diapause a definite drop in the contents of these amino acids was observed, indicating their degradation or destruction through oxidation, and their conversion into –SH or –S–S-containing compounds like coenzymes. The relation between growth and changes in free –SH groups in the grasshopper embryo has been analyzed by NORMAN[146].

In the Moroccan locust *Dociostaurus maroccanus*, LICHTENSTEIN *et al.*[124] demonstrated that glycerine extracts from eggs at diapause were active in hydrolyzing peptone, leucylglycine, leucylglycylglycine, and chloroacetyltyrosine, but neither casein nor gelatine were attacked by such extracts. However, using extracts from eggs in active development and shortly before hatching, digestion of casein was observed. There is thus a distinct difference in the proteolytic enzyme system between eggs at diapause and those at post-diapause.

Larval and pupal development

Growth is no doubt the dominant phenomenon during larval development, while histolysis and histogenesis are the main changes at metamorphosis. The large fluctuations of free amino acids during these developmental periods are apparently related to the morphogenetic processes. In general, the larvae have comparatively higher amino acid concentrations than the pupae or adults. Metabolic changes in free amino acids during post-embryonic development have been reported for *Drosophila melanogaster* (HADORN AND MITCHELL[90]; CHEN AND HADORN[39]; CHEN[36]; BENZ[19]), *Culex pipiens* (CHEN[30]), *Culex quinquefasciatus* and *Aedes aegypti* (MICKS AND ELLIS[139]), *Calliphora augur* (HACKMAN[86]), *Calliphora erythrocephala* and *Phalera bucephala* (AGRELL[1]), *Ephestia kühniella* (CHEN AND KÜHN[41]; *Bombyx mori* (SARLET *et al.*[160, 161]; AMANIEU *et al.*[3]), *Galleria mellonella* (AUCLAIR ET DUBREUIL[8]), *Saturnia pyri* (STAMM AND AGUISSE[171]), *Macrothylacea rubi* (DRILHON[51]) and *Tenebrio molitor* (PATTERSON[149]).

In *D. melanogaster* the total amount of free ninhydrin-positive substances for larvae increases rapidly up to 72 h after egg-laying, remains more or less constant for the next 24 h, and then drops steadily at the time of pupation (HADORN AND STUMM-

References p. 132/135

ZOLLINGER[91]; CHEN[36]). The values per unit body weight show, however, a maximum at 72 h of age (BENZ[19]). As to total concentration in the larval hemolymph, there is a rapid decrease as development proceeds (HADORN AND STUMM-ZOLLINGER[91]; CHEN AND HADORN[39]). For individual amino acids, glutamic acid, alanine, glycine, serine, leucine/isoleucine and valine/methionine decrease steadily, whereas glutamine, threonine, arginine, lysine, cystine, histidine, β-alanine and asparagine exhibit a maximum at 72 h. Tyrosine and proline increase continuously, especially as the time of pupation approaches (CHEN[36].) In the pupal stage the total contents of free ninhydrin-reacting components are rather low and fall off slightly. Quantitative changes of individual amino acids during pupal development have been given by BENZ[19].

In contrast to *Drosophila*, in the mosquito larvae (*Culex pipiens*) both total quantities of ninhydrin-positive substances per unit body weight and total blood concentration remain largely unchanged throughout development[30]. Among all substances identified, tyrosine and proline show a distinct increase, whether the values are expressed per individual or per μg N. The total quantity and total concentration of free amino acids are lower in pupal stage and show a continuous decrease. For *Calliphora augur* (HACKMAN[86]) as well as *C. quinquefasciatus* and *A. aegypti* (MICKS AND ELLIS[139]) it was also found that the pupae contain comparatively less free amino acids than the larvae. MICKS AND ELLIS[139] reported that in the mosquito species studied by them, the amino acid concentration was relatively constant during pupal development. In *C. pipiens* it was further noticed that several amino acids (tyrosine, glutamic acid, glutamine, valine) exhibit a slight increase at the early pupal life[30]. A similar situation was recorded for *Calliphora erythrocephala* (AGRELL[1]) and *Ephestia kühniella* (CHEN AND KÜHN[41]). Such fluctuations reflect most likely the histolytic process at this particular developmental period.

For *Bombyx mori*, DENUCÉ AND ZUBER[47], by paper electrophoresis, observed that during the fifth instar glycine, alanine and serine decrease gradually, histidine increases rapidly, whereas arginine and lysine remain constant. Since no peptide was detected in the hemolymph, the two authors concluded that protein synthesis in this insect takes place in the silk gland. DRILHON AND BUSNEL[54] noted that 10 days after cocoon formation there was again a steady increase of free amino acids. FUKUDA *et al.*[77] reported that at the end of fifth instar glycine, threonine, proline, serine and tyrosine in the hemolymph were used by the silk glands for protein synthesis. Glycine, alanine and serine are known to be the main amino acids of silk fibroin. Tracer studies showed that these three amino acids could be synthesized from pyruvate, although this compound has not been found in the body fluid nor in the silk glands[77, 25].

Another morphogenetic process which should be considered is moulting. In a recent paper ZIELINSKA AND LASKOWSKA[197] identified in the moulting fluid of the fourth instar and prepupae of *Bombyx mori* alanine, arginine, aspartic acid, asparagine, cystine, phenylalanine, glycine, glutamic acid, histidine, leucine, lysine, methionine, proline, serine, threonine, tyrosine, valine, as well as *N*-acetylglucosamine and glucosamine. PASSONEAU AND WILLIAMS[148] reported that the early moulting fluid of the giant silkworm *Platysamia cecropia* contains more proteins and less non-protein N, and the reverse is true for the late moulting fluid. *N*-Acetylglucosamine was also identified. The high proteolytic and chitinolytic activity of such fluid indicates the enzymatic digestion of endocuticle during the process of moulting[105].

The essential amino acids

Nearly all information on the essential amino acids in insects came from studies on insects reared on chemically defined medium. The ten amino acids (arginine, histidine, lysine, tryptophane, phenylalanine, methionine, threonine, leucine, isoleucine, valine), which were shown to be necessary for growth of the rat[158, 192], have been proved to be essential also for *Tribolium confusum*[120, 73, 72], *Chilo simplex*[103], *Apis mellifica*[85], *Trogoderma granarium*[147], *Calliphora erythrocephala*[165] and *Pseudosarcophaga affinis*[101] (see Table II).

TABLE II

ESSENTIAL AMINO ACIDS OF INSECTS

(For references see text)

Amino acids	*Tribolium confusum*	*Chilo simplex*	*Apis mellifica*	*Trogoderma granarium*	*Calliphora erythrocephala*	*Pseudosarcophaga affinis*	*Drosophila melanogaster*	*Phormia regina*	*Blatella germanica*	*Aedes aegypti*
Arginine	+	+	+	+	+	+	+	+	+	+
Histidine	+	+	+	+	+	+	+	+	+	+
Isoleucine	+	+	+	+	+	+	+	+	+	+
Leucine	+	+	+	+	+	+	+	+	+	+
Lysine	+	+	+	+	+	+	+	+	+	+
Methionine	+	+	+	+	+	+	+			+
Phenylalanine	+	+	+	+	+	+	+	+		+
Threonine	+	+	+	+	+	+	+	+		+
Tryptophane	+	+	+	+	+	+	+	+	+	+
Valine	+	+	+	+	+	+	+	+	+	
Alanine									+	
Cystine							+		(+)	+
Glycine							+			+
Proline								+	+*	
Serine									+*	

* Proline and serine appear to be essential only for *Blatella* males[96].

RUDKIN AND SCHULTZ[159] stated, in an abstract form, that *Drosophila* needs the same ten essential amino acids for development as higher animals. In an elaborated study on the amino acid requirements of this insect HINTON, NOYES AND ELLIS[98] pointed out that arginine, isoleucine and glycine are of primary importance. Glycine has especially a growth-promoting effect on *Drosophila* larvae. According to LAFON[117] cystine plays a particular role in metamorphosis.

Ten amino acids also have been determined to be essential for the blowfly *Phormia regina*, but, instead of methionine, proline appears to be indispensable[131, 99]. However, methionine has a strong growth-stimulating effect.

For the cockroach *Blatella germanica* it seems that fewer amino acids are necessary HOUSE[100] reported that only valine, tryptophane, arginine, histidine and most probably cystine were essential. Similarly HILCHEY[96], who worked on the same species, observed that only alanine, serine, leucine, isoleucine, proline and possibly lysine were needed for males, and alanine, lysine, leucine and probably isoleucine for females. It is interesting to notice that, according to this author, this insect is able to synthesize cystine and methionine from inorganic sulfate, and to carry out methylation of amino acids itself. In ruminants it was also found that cystine and methionine could be formed from $Na_2{}^{35}SO_4$ (ref. 24). It is possible that the synthetic ability of these S-containing amino acids is due to the symbiotes in these animals.

For the larval growth of the mosquito *Aedes aegypti*, GOLBERG AND DEMEILLON[81] concluded that ten amino acids were important, but instead of valine they added glycine. Cystine is needed for the emergence of adult mosquitoes. SINGH AND BROWN[168] showed that female adults raised on synthetic diet could lay eggs. The non-essential amino acids can be formed by the mosquitoes, although their contents are comparatively lower than in those raised under "natural" condition[169]. A chemically defined medium for *A. aegypti* has been designed by LEA, DIMOND AND DELONG[118].

In connection with the problem of essential amino acids an interesting piece of work was done by AUCLAIR AND MALTAIS[12], who found that the pea strains which were susceptible to aphid attacks contain three times more arginine, threonine and valine than the resistant strains. Apparently in the latter strains the growth of aphids is inhibited due to inadequate supply of essential amino acids (see AUCLAIR[5]).

Studies on synthetic medium for insect growth has yielded some information about the specific morphogenetic roles of amino acids in addition to their function as protein constituents. For instance, LAFON[117] observed that arginine was essential for pupation of *Drosophila*. According to VANDERZANT AND REISER[178] cystine and glycine were needed for normal pupation of the pink boll-worm. For the flour beetle *Tribolium confusum*, LEMONDE AND BERNARD[120] observed that the pupal life was doubled if lysine was lacking. In the absence of amino acids such as valine, histidine, tryptophane or leucine, pupation did not take place and there was no metamorphosis. In the opinion of FRÖBISCH[73] still another unknown factor is necessary for imaginal differentation.

The earlier work of WILSON on *Drosophila* revealed that proline enhances differentiation[189], tryptophan retards histolysis in the pupae[190], whereas tyrosine, phenylalanine and methionine accelerate moulting[190, 191]. The tripeptide glutathione has most likely a stimulating effect on the larval growth of both *Drosophila*[188] and *Aedes*[168]. A similar effect was found for a certain peptide in the yeast and pancreas extracts[21, 22]. This substance is probably responsible for the growth-promoting effect of yeast extracts found in *Phormia regina*[99]. There is no doubt that different species or even various strains of the same species have different nutritional requirements. The problem of insect nutrition has been reviewed by LIPKE AND FRAENKEL[128].

Contents of soluble proteins

Fluctuations of free amino acids must be related to protein synthesis during development. The patterns of soluble proteins in insect blood have been analyzed by various workers. In the larval blood of *Drosophila melanogaster* at least seven protein fractions, and in that of *Culex pipiens* at least four fractions could be detected[34]. Judging from their mobility and isoelectric point, they seem to be more related to the serum globulins[193, 29]. KRIEG[116] also reached the same conclusion from studies on the blood proteins in *Melolontha* sp., *Neodiprion sertifer* and *Lymantria dispar*. But CLARK AND

BALL[44] reported that the predominant proteins in the insect studied by them probably belong to the albumin group. As a matter of fact, we do not yet know much about the nature of blood proteins in insects. Their electrophoretic patterns are species-specific[46, 179], and it is certainly over-simplified to classify them just as albumin or globulin.

According to CHEN[29, 35] in both *Drosophila* and *Culex* larvae there is a rapid increase in the contents of blood proteins, especially towards the end of larval development. A similar increase has been reported for *Bombyx mori*[52, 195] and *Galleria mellonella*[45]. Using labelled glycine BRICTEUX-GRÉGORIE *et al.*[26] demonstrated that in *Sphinx ligustri* the adult proteins are synthesized directly from free amino acids. This fact points out that the decrease of free amino acids during development is largely due to their incorporation into the proteins. In the mealworm *Tenebrio molitor*, POCHEDLEY[151] observed that the contents of soluble proteins show characteristic changes at various developmental stages. Similar results have been also reported by HELLER[95] for *Deilephila euphorbiae* and by LUDWIG[130] for *Popillia japonica*. Parallel to protein synthesis, variations in RNA contents of *Tenebrio* during morphogenetic development have been detected[149].

EFFECTS OF GENES ON FREE AMINO ACIDS

The genic control of protein metabolism has been repeatedly demonstrated by geneticaı and biochemical studies on both microorganisms[183] and higher animals[2]. In insects, examples are known which show that mutational effects influence to various extents the free amino acid pool. We need only to mention the accumulation of free tryptophane in the *v*-mutant of *Drosophila melanogaster*[82] and in the *a*-mutant of *Ephestia kühniella*[41, 63]. In both cases, the oxidation of tryptophan to kynurenine does not take place. More drastic changes have been reported from studies on lethal factors. This problem has been treated in detail by HADRON[88, 89].

NAKAMURA *et al.*[142] found in *D. melanogaster* the presence of one substance, probably cystine, in the lethal embryos, but not in normal ones. Their further studies on two embryonic lethal factors revealed that the amounts of glutamic acid, aspartic acid, glycine, arginine, threonine and serine show a larger decrease in the lethals, but apparently less in the controls.

From his studies on a melanoma-producing lethal in *Drosophila*, LEWIS[123] reported that the tumor larvae have no cystine. They have, however, a higher concentration of alanine, arginine, glycine, methionine, serine and tyrosine. The concentration of methionine and tyrosine was found to be lower in the heterozygotes.

More detailed studies on the protein metabolism of two lethal factors, *"lethal-translucida" (ltr)* and *"lethal-meander" (lme)*, of *D. melanogaster* have been undertaken by HADORN and his collaborators[88]. The *ltr*-homozygotes accumulate a large amount of hemolymph and stop development in the pupal stage[87]. Biochemical analyses showed that the concentration of free amino acids is much higher in these lethal individuals than in the corresponding normals[90, 91]. The increase, however, is not the same in different amino acids[173]. Since the *ltr*-homozygous larvae have a very low content of blood proteins, it seems that the process of protein synthesis in this mutant is inhibited [193, 29, 34].

The *lme/lme* larvae stop development in the third instar and never pupate[164].

References p. 132/135

Paper chromatographic analyses revealed that in the aged lethal homozygotes the essential amino acids are either entirely absent (valine, leucine, isoleucine, methionine, histidine, arginine, lysine) or greatly reduced (threonine). On the other hand, they accumulate an abnormally high concentration of glycine, especially in later stages[40]. Experiments *in vitro* suggest that a much reduced activity of proteolytic enzymes in the gut is probably responsible for the abnormal protein metabolism in this mutant (for peptides in these two lethal mutants, see paper by MITCHELL AND SIMMONS in this conference).

Recently, chromatographic analyses on three more lethal factors in *Drosophila* have been carried out in HADORN's laboratory. For the factor *"lethal-bluter" (lbl)* BENZ[20] observed that the homozygous *lbl*-individuals in the third larval instar have distinctly less free amino acids than the normal ones. Since the lethals have a thicker cuticle, it is possible that a larger amount of these substances are used for the synthesis of cuticular proteins. For the factor *"lethal-polymorph" (lpm)* it was found that in the larval stage the *lpm*-lethals have an abnormally high concentration of free ninhydrin-reacting components, although the difference becomes less distinct in the prepupal stage. According to BENZ[20] this is probably related to the abnormal formation of muscular proteins in the *lpm*-homozygotes. In a recent paper FAULHABER[64] presented data on the protein metabolism of the mutant *"lethal giant larvae" (lgl)*. Her results indicate that in the *lgl*-lethals there is an excess of free amino acids and a deficiency of blood proteins. The situation is therefore very similar to that found for the factor *"lethal-translucida"*, apparently also due to a reduced protein synthesis.

From all studies just cited, it is evident that the lethal factors have a distinctly specific effect on protein metabolism. Each lethal-mutant has its own free amino acid pattern. As suggested by the work of CHEN[29] and BENZ[20], the targets of the lethal effects vary from case to case; protein formation either in blood, muscle or cuticule is impaired. In the case of *lethal-meander* it seems that mainly the proteolytic enzymes are affected[40]. These facts imply that specific genes are involved in these mutants. However, until more information is available, we do not know whether the observed abnormal protein metabolism results from a direct or indirect effect of the mutational process.

SPECIAL FUNCTIONS AND METABOLIC INTERRELATIONSHIPS OF FREE AMINO ACIDS

In the previous section the developmental roles of amino acids have been mentioned (p. 124). In the following, we shall consider some other specific functions of these substances. The free amino acids are believed to play an active part in adjusting the osmotic pressure and the buffering action of the body fluid[67, 187, 27, 198, 86]. The work of DUCHÂTEAU AND FLORKIN[58] showed that in the marine crustaceans dilution of sea water resulted in a decrease of the amino acid concentration. On the other hand, the fresh-water form, after being transferred into the sea water, showed an increase in the concentration of amino acids, especially that of proline. However, glycine seems to be also effective in osmoregulation[28, 109, 32]. It is known that glycine participates in many activities of the cell[185]. The growth-promoting effect of glycine in *Drosophila* development has been referred to in a previous section (p. 127).

Various studies demonstrated that dietary amino acids have a distinct effect on egg production in insects. Thus, GREENBERG[84] observed that oviposition in *Aedes aegypti* females was increased by isoleucine, but not by a combination of methionine, valine, phenylalanine, tryptophane, threonine, leucine, arginine, histidine and lysine. DIMOND *et al.*[49], who also worked on this mosquito species, reported that for egg-laying the following amino acids had to be included in the diet: arginine, isoleucine, leucine, lysine, phenylalanine, threonine, tryprophane and valine. There was a reduction in the number of eggs laid if histidine, cystine or methionine was eliminated from the diet. These results suggest that yolk proteins are synthesized directly from amino acids. From their study on the house fly, ASCHER AND LEVINSON[4] concluded that protein reserve in the larvae could not be utilized by female adults for egg production.

The presence of γ-amino-n-butyric acid has been detected in a number of insects[13, 39, 41, 20, 64]. According to KATING[108] this unusual amino acid participates in N-assimilation, especially in the transamination of aminocarbonic acid. Recently, MCLENNAN[132] and VAN DER KLOOT AND ROBBINS[177] reported from their studies on the muscle contraction in crayfish, that γ-aminobutyric acid works as transmitter substance of inhibitory neurons. It remains to be investigated whether it has a similar function in insects.

In a recent paper SLOPER[170] identified the presence of a substance rich in protein-bound cystine or cysteine in the neurosecretory system of *Leucophaea maderae.* This finding is of particular interest because, as already mentioned, these amino acids, in addition to their close intermediary relation to methionine, serve as important sources of sulfhydryl groups for the synthesis of coenzymes and hormones (see p. 122).

In their studies on the iodine metabolism in insects, LIMPEL AND CASIDA[126, 127] found that if radioiodide was injected into the cockroach *Periplaneta americana,* most of the iodine was excreted as the iodoamino acid monoiodohistidine. When radioactive iodine was given in the form of monoiodohistidine, di-iodohistidine could be detected in the excreta. These experiments strongly suggest that histidine serves as a detoxicating agent in this insect.

Glutamine is an important component in the hemolymph of insects. It is formed from ammonia and glutamate in the presence of ATP. From a number of experiments on various animals, it is known that this substance participates in transamination[181, 133] and transpeptidation[92, 134]. But it can also be incorporated directly into proteins[18, 167]. In both *Bombyx*[115] and *Schistocerca*[112], glutamine has been shown to be an effective amino donor in transamination reactions (see below). Furthermore, it is involved in the synthesis of uric acid and glucosamine[119, 129]. The latter substance participates in the formation of chitin, and was also found in the protein hydrolyzate of the salivary gland secretion of *Drosophila*[114]. In mammals, evidence also has been presented to show that glutamine serves as a detoxicating agent for a number of aromatic acids and ammonia[184, 182]. Whether it fulfills the same function in insects, is not clear.

The metabolism of amino acids in insects is a complex problem. As in other animals, transamination is a common phenomenon in insect tissues[17]. It involves mainly the formation of glutamate from α-ketoglutarate with aspartate and alanine as amino donors. Transaminase activity has been demonstrated in the silkworm *Bombyx mori*[115, 23, 76]. More recently, KILBY AND NEVILLE[111, 112] presented evidence for the occurrence of transaminase in fat body, gut and Malpighian tubules of the locust *Schistocerca gregaria.* These enzymes, similar to those in mammals, effect transamination of α-keto and α-amino acids. According to these authors the transaminases were present in the mitochondrial and soluble fractions of tissue preparations. Furthermore, they showed that both D- and L-amino acids could be converted to their corresponding keto acids by oxidative deamination. The enzyme involved was most probably L-glutamic acid dehydrogenase. D-amino acid oxidase was also found in the fat body of *Periplaneta americana*, *Blatella germanica* and *Galleria mellonella*[6, 7, 9], but it seems that the intracellular symbionts are at least partly responsible for the enzymatic activity. No doubt microorganisms play an important part in the N-metabolism of insects. In *Aphis brassicae*, TOTH[174] demonstrated that urea or uric acid could be decomposed by symbionts to provide ammonia for other synthetic processes. Recently SEDEE[166], by using ^{15}N, showed that in the blowfly *Calliphora erythrocephala* ammonium nitrate can serve as N-source for new synthesis of amino acids. The transamination processes in this insect have been studied by DESAI AND KILBY[48]. From his work on the German cockroach, AUCLAIR[6, 7] concluded that the mechanism of deamination and transamination in insects is similar to that in mammals. The intermediary metabolism of amino acids in insects has been adequately reviewed by GILMOUR[80].

REFERENCES

1 I. AGRELL, *Acta Physiol. Scand.*, 18 (1949) 247.
2 A. C. ALLISON, *Am. Naturalist*, 93 (1959) 5.
3 M. AMANIEU, G. DUCHÂTEAU, M. FLORKIN AND C. JEUNIAUX, *Arch. intern. physiol. et biochem.*, 64 (1956) 518.
4 K. R. S. ASHER AND Z. H. LEVINSON, *Riv. parassitol.*, 17 (1956) 217.
5 J. L. AUCLAIR, *Can. Entomologist*, 85 (1953) 63.
6 J. L. AUCLAIR, *J. Insect Physiol.*, 3 (1959) 57.
7 J. L. AUCLAIR, *J. Insect Physiol.*, 3 (1959) 127.
8 J. L. AUCLAIR AND R. DUBREUIL, *Can. J. Zool.*, 30 (1952) 109.
9 J. L. AUCLAIR AND R. DUBREUIL, *Can. J. Zool.*, 31 (1953) 30.
10 J. L. AUCLAIR, H. HERLANT-MEEUWIS AND M. DEMERS, *Rev. can. biol.*, 10 (1951) 162.
11 J. L. AUCLAIR AND C. A. JAMIESON, *Science*, 108 (1948) 357.
12 J. L. AUCLAIR AND J. B. MALTAIS, *Rev. can. biol.*, 9 (1950) 332.
13 J. L. AUCLAIR AND J. B. MALTAIS, *Nature*, 170 (1952) 1114.
14 J. L. AUCLAIR AND R. L. PATTON, *Rev. can. biol.*, 9 (1950) 3.
15 G. H. BALL AND E. W. CLARK, *Syst. Zool.*, 2 (1953) 138.
16 E. S. G. BARRON, *Texas Repts. Biol. and Med.*, 11 (1953) 653.
17 E. S. G. BARRON T. N. TAHMISIAN, *J. Cellular Comp. Physiol.*, 32 (1948) 57.
18 J. M. BARRY, *Biochem. J.*, 63 (1956) 669.
19 G. BENZ, *Arch. Julius Klaus-Stiftung Vererbungsforsch. Sozialanthropol. u. Rassenhyg.*, 30 (1955) 498.
20 G. BENZ, *Z. Vererbungslehre*, 88 (1957) 78.
21 M. BEGG, *J. Exptl. Biol.*, 33 (1956) 142.
22 M. BEGG AND F. W. ROBERTSON, *J. Exptl. Biol.*, 26 (1949) 380.
23 B. BHEEMESWAR AND M. SREENIVASAYA, *Current Sci. (India)*, 21 (1952) 253.
24 R. J. BLOCK AND J. A. STEKOL, *Proc. Soc. Exptl. Biol. Med.*, 73 (1950) 391.
25 S. BRICTEUX, T. FUKUDA, A. DEWANDRE AND M. FLORKIN, *Arch. intern. physiol. et biochem.*, 67 (1959) 545.
26 S. BRICTEUX-GRÉGOIRE, W. G. VERLY AND M. FLORKIN, *Nature*, 179 (1957) 678.
27 J. B. BUCK, in K. D. ROEDER, *Insect Physiology*, John Wiley & Sons, Inc., New York, 1953, p. 147.
28 M. N. CAMIEN, H. SARLET, GH. DUCHÂTEAU AND M. FLORKIN, *J. Biol. Chem.*, 193 (1951) 881.
29 P. S. CHEN, *Rev. suisse zool.*, 63 (1956) 216.
30 P. S. CHEN, *J. Insect Physiol.*, 2 (1958) 38.
31 P. S. CHEN, *J. Insect Physiol.*, 2 (1958) 128.
32 P. S. CHEN, *Experientia*, 14 (1958) 369.
33 P. S. CHEN, *Arch. Julius Klaus-Stiftung Vererbungsforsch. Sozialanthropol. u. Rassenhyg.*, 33 (1958) 82.
34 P. S. CHEN, *Rev. suisse zool.*, 66 (1959) 280.
35 P. S. CHEN, *J. Insect Physiol.*, 3 (1959) 335.
36 P. S. CHEN, *11th International Congress Entomology*, 3 (1960) 201.

37 P. S. CHEN AND F. BALTZER, *Nature*, 181 (1958) 98.
38 P. S. CHEN AND C. DIEM, *J. Insect Physiol.*, 7 (1961) 289.
39 P. S. CHEN AND E. HADORN, *Rev. suisse zool.*, 61 (1954) 437.
40 P. S. CHEN AND E. HADORN, *Rev. suisse zool.*, 62 (1955) 338.
41 P. S. CHEN AND A. KÜHN, *Z. Naturforsch.*, 11b (1956) 305.
42 E. W. CLARK AND G. H. BALL, *J. Parasitol.*, 37 (1951) 29.
43 E. W. CLARK AND G. H. BALL, *Exptl. Parasitol.*, 1 (1952) 339.
44 E. W. CLARK AND G. H. BALL, *Physiol. Zoöl.*, 29 (1956) 206.
45 J. M. DENUCÉ, *Z. Naturforsch.*, 12b (1957) 434.
46 J. M. DENUCÉ, *Z. Naturforsch.*, 13b (1958) 215.
47 J. M. DENUCÉ AND H. ZUBER, *Z. Naturforsch.*, 11b (1956) 381.
48 R. M. DESAI AND B. A. KILBY, *Arch. intern. physiol. et biochem.*, 66 (1958) 248.
49 J. B. DIMOND, A. O. LEA, W. F. HAHNERT AND D. M. DELONG, *Can. Entomologist*, 88 (1956) 57.
50 A. DRILHON, *Compt. rend. soc. biol.*, 144 (1950) 224.
51 A. DRILHON, *Compt. rend.*, 234 (1952) 1913.
52 A. DRILHON, *Compt. rend.*, 238 (1954) 2452.
53 A. DRILHON AND R. G. BUSNEL, *Compt. rend. soc. biol.*, 139 (1945) 926.
54 A. DRILHON AND R. G. BUSNEL, *Rev. Ver à Soie*, 1 (1949) 51.
55 A. DRILHON AND R. G. BUSNEL, *Compt. rend.*, 230 (1950) 1114.
56 GH. DUCHÂTEAU AND M. FLORKIN, *Bull. soc. chim. biol.* 37 (1955a) 239.
57 GH. DUCHÂTEAU AND M. FLORKIN, *Arch. intern. physiol. et biochem.*, 63 (1955b) 35.
58 GH. DUCHÂTEAU AND M. FLORKIN, *J. physiol. (Paris)*, 48 (1956) 520.
59 GH. DUCHÂTEAU AND M. FLORKIN, *Arch. intern. physiol. et biochem.*, 66 (1958) 573.
60 GH. DUCHÂTEAU, M. FLORKIN AND H. SARLET, *Arch. intern. physiol. et biochem.*, 60 (1952) 539.
61 GH. DUCHÂTEAU, H. SARLET AND M. FLORKIN, *Arch. intern. physiol. et biochem.*, 60 (1952) 103.
62 GH. DUCHÂTEAU, H. SARLET, M. N. CAMIEN AND M. FLORKIN, *Arch. intern. physiol. et biochem.*, 60 (1952) 124.
63 A. EGELHAAF, *Z. Naturforsch.*, 12b (1957) 465.
64 I. FAULHABER, *Z. Vererbungslehre*, 90 (1959) 299.
65 L. H. FINLAYSON AND D. HAMMER, *Nature*, 163 (1949) 843.
66 W. FLORKIN, in *Vergleichend biochemische Fragen, 6. Kolloquium d. Ges. f. physiol. Chem.*, Springer-Verlag, Berlin, 1956.
67 M. FLORKIN, *Biochemical Evolution*, Academic Press, New York, 1949.
68 M. FLORKIN AND GH. DUCHÂTEAU, *Bull. classe sci. Acad. roy. Belg.*, 28 (1942) 373.
69 A. S. FOX, *Science*, 123 (1956) 143.
70 A. S. FOX, *Z. Vererbungslehre*, 87 (1956) 554.
71 A. S. FOX, C. G. MEAD AND I. L. MUNYON, *Science* 129 (1959) 1489.
72 G. FRAENKEL AND G. E. PRINTY, *Biol. Bull.*, 106 (1954) 149.
73 G. FRÖBISCH, *Z. Vitamin-, Hormon- u. Fermentforsch.*, 6 (1954) 1.
74 Y. Y. FU, *Physiol. Zoöl.*, 30 (1957) 1.
75 T. FUKUDA, *Sanshi Shikenjo Hôkoku*, 13 (1951) 481.
76 T. FUKUDA, *J. Biochem. (Tokyo)*, 44 (1957) 505.
77 T. FUKUDA, J. KIRIMURA, M. MATUDA AND T. SUZUKI, *Nature*, 175 (1955) 1041.
78 H. R. GEIGER, *Rev. suisse zool.*, 68 (1961) 4.
79 D. S. GENGHOF, *Arch. Biochem. Biophys.*, 34 (1951) 112.
80 D. GILMOUR in *Biochemistry of Insects*, Academic Press, New York 1961, p. 238.
81 L. GOLBERG AND B. DEMEILLON, *Biochem. J.*, 43 (1948) 379.
82 M. M. GREEN, *Genetics*, 34 (1949) 564.
83 D. M. GREENBERG, in *Chemical Pathways of Metabolism*, Academic Press, New York, 1954.
84 J. GREENBERG, *J. Nutrition*, 43 (1951) 27.
85 A. P. DE GROOT, *Physiol. Comparata et Oecol.*, 3 (1953) 197.
86 R. H. HACKMAN, *Australian J. Biol. Sci.*, 9 (1956) 400.
87 E. HADORN, *Rev. suisse zool.*, 56 (1949) 271.
88 E. HADORN, in *Letalfaktoren in ihrer Bedeutung für Erbpathologie und Genphysiologie der Entwicklung*, Georg Thieme Verlag, Stuttgart, 1955.
89 E. HADORN, in *Developmental Genetics and Lethal Factors*, Methuen and Co., Ltd., London, 1961.
90 E. HADORN AND H. K. MITCHELL, *Proc. Natl. Acad. Sci. U.S.*, 37 (1951) 650.
91 E. HADORN AND E. STUMM-ZOLLINGER, *Rev. suisse zool.*, 60 (1953) 506.
92 C. S. HANES, F. J. R. HIRD AND F. A. ISHERWOOD, *Biochem. J.*, 51 (1952) 25.
93 J. S. HARINGTON, *Nature*, 178 (1956) 268.
94 F. HATZ, *Helv. Chim. Acta*, 32 (1949) 251.
95 J. HELLER, *Compt. rend. soc. biol.*, 90 (1924) 1360.
96 J. D. HILCHEY, *Contrib. Boyce Thomson Inst.*, 17 (1953) 203.
97 J. D. HILCHEY, V. F. COTTY AND S. M. HENRY, *Contrib. Boyce Thomson Inst.*, 19 (1957) 189.

[98] T. Hinton, D. T. Noyes and J. Ellis, *Physiol. Zoöl.*, 24 (1951) 335.
[99] E. Hodgson, V. H. Cheldelin and R. W. Newburgh, *Can. J. Zool.*, 34 (1956) 527.
[100] H. L. House, *Can. Entomologist*, 81 (1949) 133.
[101] H. L. House, *Can. J. Zool.*, 32 (1954) 351.
[102] F. Irreverre and L. A. Terzian, *Science*, 129 (1959) 1358.
[103] S. Ishii and C. Hirano, *Nôgyô Gijutsu Kenkyûjo Hôkoku*, 5 (1955) 35.
[104] N. Ishimori and M. Muto, *Nippon Sanshigaku Zasshi*, 20 (1951) 128.
[105] Ch. Jeniaux and M. Amanieu, *Experientia*, 11 (1955) 195.
[106] A. Kallhammer, *Naturwissenschaften*, 42 (1955) 126.
[107] W. D. Kaplan, J. T. Holden and B. Hochman, *Science*, 127 (1958) 473.
[108] H. Kating, *Naturwissenschaften*, 41 (1954) 188.
[109] J. L. Kavanau, *J. Exptl. Zool.*, 122 (1953) 285.
[110] T. Kiecol and J. Michejola, *Bull. Soc. amis sci. et lettres Poznan, D*, 1 (1960) 47.
[111] B. A. Kilby and E. Neville, *Biochim. Biophys. Acta*, 19 (1956) 389.
[112] B. A. Kilby and E. Neville, *J. Exptl. Biol.*, 34 (1957) 276.
[113] R. L. Kirk, A. R. Main and F. G. Beyer, *Biochem. J.*, 57 (1954) 440.
[114] M. Kodani, *Proc. Natl. Acad. Sci. U.S.*, 34 (1948) 131.
[115] F. Koide, H. Nagayama and K. Shimura, *Nippon Nôgei-Kagaku Kaishi*, 29 (1955) 987.
[116] A. Krieg, *Naturwissenschaften*, 43 (1956) 60.
[117] M. Lafon, *Ann. physiol. physicochim. biol.*, 15 (1939) 215.
[118] A. O. Lea, J. B. Dimond and D. M. DeLong, *J. Econ. Entomology*, 49 (1956) 313.
[119] L. F. Leloir and C. E. Cardini, *Biochim. Biophys. Acta*, 12 (1953) 15.
[120] A. Lemonde and R. Bernard, *Can. J. Zool.*, 29 (1951) 80.
[121] L. Levenbook, *Biochem. J.*, 47 (1950) 336.
[122] L. L. Lewallen, *Ann. Entomol. Soc. Am.*, 50 (1957) 602.
[123] H. W. Lewis, *J. Exptl. Zool.*, 126 (1954) 235.
[124] N. Lichtenstein, F. S. Bodenheimer and A. Shulow, *Enzymologia*, 13 (1949) 276.
[125] C. H. Li, I. Geschwind and H. M. Evans, *J. Biol. Chem.*, 177 (1949) 91.
[126] L. E. Limpel and J. E. Casida, *J. Exptl. Zool.*, 135 (1957a) 19.
[127] L. E. Limpel and J. E. Casida, *J. Exptl. Zool.*, 136 (1957b) 595.
[128] H. Lipke and G. Fraenkel, *Ann. Rev. Entomol.*, 1 (1956) 17.
[129] D. A. Lowther and H. J. Rogers, *Nature*, 175 (1955) 435.
[130] D. Ludwig, *Physiol. Zoöl.*, 27 (1954) 325.
[131] A. J. McGinnis, R. W. Newburgh and V. H. Cheldelin, *J. Nutrition*, 58 (1956) 309.
[132] H. McLennan, *Naturwissenschaften*, 44 (1957) 116.
[133] A. Meister, *Science* ,120 (1954) 43.
[134] A. Meister, *Physiol. Revs.*, 36 (1956) 103.
[135] J. Michejda and J. Urbanski, *Bull. soc. amis sci. et lettres Poznań*, 14 (1958) 345.
[136] D. W. Micks, *Nature*, 174 (1954) 217.
[137] D. W. Micks, *Ann. Entomol. Soc. Am.*, 49 (1956) 576.
[138] D. W. Micks and J. P. Ellis, *Proc. Soc. Exptl. Biol. Med.*, 78 (1951) 69.
[139] D. W. Micks and J. P. Ellis, *Proc. Soc. Exptl. Biol. Med.*, 79 (1952) 191.
[140] D. W. Micks and F. J. Gibson, *Ann. Entomol. Soc. Am.*, 50 (1957) 500.
[141] H. K. Mitchell, P. S. Chen and E. Hadorn, *Experientia*, 16 (1960) 410.
[142] K. Nakamura, T. Imaizumi and Y. Kitazume, *DISI Bull.*, 25 (1951) 120.
[143] K. Nakamura, Y. Kitazumi, M. Takanuri and T. Imaizumi., *Pap. Coord. Commit. Res. Gent.*, 2 (1951b) 163.
[144] K. Nakamura, T. Imaizumi and M. Takanuri, *DISI Bull.*, 26 (1952) 114.
[145] Y. Narumi, A. Hisao and S. Yoshizumi, *Nippon Sanshigaku Zasshi*, 19 (1950) 530.
[146] C. Norman, *Physiol. Zool.*, 27 (1954) 141.
[147] N. C. Pant, J. K. Nayar and P. Gupta, *Experientia*, 14 (1958) 176.
[148] J. V. Passoneau and C. H. Williams, *J. Exptl. Biol.*, 30 (1953) 545.
[149] D. S. P. Patterson, *Biochem. J.*, 65 (1957) 729.
[150] D. S. Po-Chedley, *Trans. N.Y. Acad. Sci.*, 19 (1956) 19.
[151] D. S. Po-Chedley, *Nature*, 183 (1959) 62.
[152] J. J. Pratt, Jr., *Ann. Entomol. Soc. Am.*, 43 (1950) 573.
[153] J. J. Pratt, Jr., and H. L. House, *Science*, 110 (1949) 9.
[154] R. Raper and J. Shaw, *Nature*, 162 (1948) 999.
[155] L. J. Reed, *J. Biol. Chem.*, 183 (1950) 451.
[156] M. Reiff, *Verhandl. schweiz. naturforsch. Ges.*, 136 (1956) 133.
[157] J. G. Robertson, *Can. J. Zool.*, 35 (1957) 411.
[158] W. C. Rose, *Physiol. Revs.*, 18 (1938) 109.
[159] G. T. Rudkin and J. Schultz, *Anat. Record.*, 99 (1947) 613.
[160] H. Sarlet, Gh. Duchâteau, M. N. Carmien and M. Florkin, *Biochim. Biophys. Acta*, 8 (1952) 571.

161 H. SARLET, GH. DUCHÂTEAU AND M. FLORKIN, *Arch. intern. physiol.*, 60 (1952) 126.
162 S. SASAKI, T. WATANABE AND Y. KONDO, *Nippon Sanshigaku Zasshi*, 26 (1957) 291.
163 E. I. SHAW, *Exptl. Cell Research*, 9 (1955) 489.
164 W. SCHMID, *Z. Vererbungslehre*, 83 (1949) 220.
165 P. D. J. W. SEDEE, *Acta Physiol. et Pharmacol. Neerl.*, 3 (1954) 262.
166 P. D. J. W. SEDEE, *Entomol. Exptl. Appl.*, 1 (1958) 38.
167 J. C. M. SHELDON-PETERS AND J. M. BARRY, *Biochem. J.*, 63 (1956) 676.
168 K. R. P. SINGH AND A. W. A. BROWN, *J. Insect Physiol.*, 1 (1957) 199.
169 K. R. P. SINGH AND D. W. MICKS, *Mosquito News*, 17 (1957) 248.
170 J. C. SLOPER, *Nature*, 179 (1957) 148.
171 M. D. STAMM AND L. AGUISSE, *Rev. españ. fisiol.*, 11 (1955) 63.
172 J. A. STEKOL, in W. D. MCELROY AND H. B. GLASS, *Symposium on Amino Acid Metabolism*, The Johns Hopkins Press, Baltimore, 1955, p. 509.
173 E. STUMM-ZOLLINGER, *Z. Vererbungslehre*, 86 (1954) 126.
174 L. TÓTH, *Zool. Anz.*, 146 (1951) 191.
175 J. E. TREHERNE, *J. Exptl. Biol.*, 36 (1959) 533.
176 H. H. USSING, *Acta Physiol. Scand.*, 11 (1946) 61.
177 W. G. VAN DER KLOOT AND J. ROBBINS, *Experientia*, 15 (1959) 35.
178 E. S. VANDERZANT AND R. REISER, *J. Econ. Entomol.*, 49 (1956) 454.
179 M. VAN SANDE AND D. KARCHER, *Science*, 131 (1960) 1103.
180 U. VON DER CRONE-GLOOR, *J. Insect Physiol.*, 3 (1959) 50.
181 H. WAELSCH, *Advances in Enzymol.*, 13 (1952) 237.
182 H. WAELSCH, *Naturwissenschaften*, 40 (1953) 404.
183 R. P. WAGNER AND H. K. MITCHELL, in *Genetics and Metabolism*, John Wiley and Sons, Inc., New York, 1955.
184 H. WEIL-MALHERBE, *Naturwissenschaften* 40 (1953) 545.
185 S. WEINHOUSE, in W. D. MCELROY AND H. B. GLASS, *Symposium on Amino Acid Metabolism*, The Johns Hopkins Press, Baltimore, 1955.
186 J. WESTLEY, J. J. WREN AND H. K. MITCHELL, *J. Biol. Chem.*, 229 (1957) 131.
187 V. B. WIGGLESWORTH, in *The principles of Insect Physiology*, Methnen and Co., Ltd., London, 1947, p. 230.
188 L. P. WILSON, *Growth*, 3 (1939) 409.
189 L. P. WILSON, *Growth*, 5 (1941) 11.
190 L. P. WILSON, *Growth*, 9 (1945) 341.
191 L. P. WILSON, *Growth*, 10 (1946) 361.
192 K. A. J. WRETLIND, *Acta Physiol. Scand.*, 15 (1948) 304.
193 CH. WUNDERLY AND H. GLOOR, *Protoplasma*, 42 (1953) 273.
194 G. R. WYATT, *Annual Rev. Ent.*, 6 (1961) 75.
195 G. R. WYATT, T. C. LOUGHHEED AND S. S. WYATT, *J. Gen. Physiol.*, 39 (1956) 853.
196 N. YOSHITAKE AND H. ARUGA, *Nippon Sanshigaku Zasshi*, 19 (1950) 530.
197 Z. M. ZIELINSKA AND T. LASKOWSKA, *Acta Biol. Exptl. (Lodz)*, 18 (1958) 209.
198 K. ZWICKY, *Z. Vergleich. Physiol.*, 36 (1954) 367.

AMINO ACIDS AND DERIVATIVES IN DROSOPHILA

HERSCHEL K. MITCHELL AND JOHN R. SIMMONS

California Institute of Technology, Division of Biology, Pasadena, Calif. (U.S.A.)

A number of years ago HADORN AND MITCHELL[1] presented data showing changes in patterns of amino acids and related substances that occur during development in *Drosophila melanogaster*. Two prominent components among the amino acid derivatives were shown to be peptides containing several amino acids and in subsequent work by HADORN, CHEN *et al.*[2–5] at least three more peptides were observed. In addition, FOX[6] reported a peptide difference in male and female Drosophila, and evidence for the existence of peptides among the lipid–amino acid conjugates of the fly was also obtained[7].

It is of importance to point out that nearly all of the observations mentioned above were made by the use of paper chromatography, and, although the method has yielded a wealth of information, it has limitations in resolving power and, at concentrations below the level of overloading the paper, it reveals only the most abundant components to be found in extracts. Consequently when we undertook the work to be summarized here we resorted to ion-exchange chromatography followed by paper chromatography, electrophoresis and other methods suitable for purification of natural substances. Our objectives were, first, to examine in greater detail the pattern of ninhydrin-reacting substances in Drosophila including components present in low concentrations; second, to determine the structures of peptides already observed as well as of others that might be found; and third, to study the metabolism of the peptides particularly in relation to a possible role in protein synthesis. That all these objectives have not been achieved will become evident but some 5 years of exploration has nevertheless, yielded much useful information.

PRELIMINARY OBSERVATIONS

In a study of the occurrence of peptides as normal constituents of tissues it is essential that, during extraction, there is no possibility of significant hydrolysis of protein or existing peptides. Some idea of the magnitude of this problem is illustrated in Fig. 1. As shown, incubation of a larval homogenate even at 0° yields an appreciable increase in ninhydrin-reacting material while a very large increase occurs in a few minutes at 28°. We have therefore made use of extraction at −20° as the first step. It has also been observed that, under some conditions, extraction by heating causes loss of ninhydrin-reacting material presumably due to the presence of reactive amino acid derivatives. We have therefore avoided heating as much as possible. In initial experiments use was made of Dowex-50-4X and a modification of the gradient elution system of THOMPSON[8]. Results from one such experiment are summarized in Fig. 2. This is representative from seven separate runs in which total propanol–water (1 : 1, −20°)

References p. 146

plus boiling-water extract was fractionated with variations in the elution system in the first four. Details of extraction and elution systems are given in the figure and analytical methods are the same as described later. The data are given as μmoles/ml (15-ml fractions) with the white portion on the graph showing ninhydrin equivalents before hydrolysis and the black bars the total amounts after acid hydrolysis. Positions and breadth of bands (for 90% recovery) from fractionation of known amino acids in the same system are indicated by the abbreviations and arrows above the graph. This gives an index of the resolving power of the system since the total moles of amino acid exceeded that of the unhydrolyzed extract.

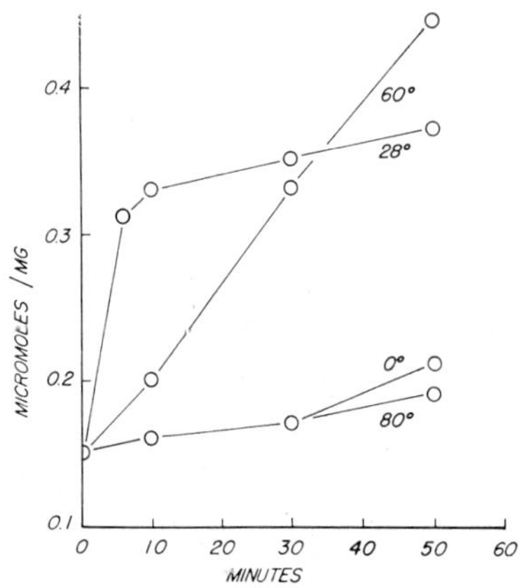

Fig. 1. Instability of amino acid derivatives in larval homogenates. Homogenates of second instar larvae (made in four parts of water at $-0°$) were incubated at temperatures indicated. Samples were spotted directly on paper for chromatography and the spots were dried quickly by suction. Increases observed were general in all amino acid positions and largely at the expense of peptides rather than proteins.

From these experiments and additional ones involving extensive paper chromatography and electrophoresis of each fraction we were able to conclude that the larval extracts contain significant quantities of at least 600 peptides and a considerable variety of other classes of amino acid derivatives. For example, essentially none of the material before fraction 90 represents free amino acids and nearly all of the material from fraction 280–306 represents fairly large mixed peptides. Predominantly acidic peptides come in the region 0–90. Thus, it was clear from secondary fractionations that even most of the minor variations in the amounts of materials in the column fractions were significant. Furthermore, few, if any, fractions contained single pure substances and the necessity for development of more elaborate fractionation methods became evident. Before proceeding to these problems it is again worthy of note that an extract giving the results of Fig. 2 gives, by paper chromatography, a pattern interpretable as representing a mixture of some 20 amino acids plus five or six peptides. Obviously such an interpretation is incomplete.

References p. 146

GENERAL EXTRACTION METHOD

In preparing the extracts already described conditions of −20° and +90° were used and this is essential to avoid enzyme action as shown in Fig. 1. Neither cold-alcohol extracts nor hot-water extracts show such changes though the former do contain substances unstable to heat alone. For these reasons we have retained the use of low-temperature grinding and extraction and have developed a general method that appears applicable to any kind of tissue (Fig. 3). Furthermore, it provides a sharp preliminary fractionation as will be shown (Figs. 4 and 5).

A 10-g sample of 90-h larvae (from egg laying) was washed thoroughly with distilled water and blotted dry with filter paper. The live animals were added quickly to a pre-cooled solution of 30 ml of methanol and 10 ml of water contained in the steel cup of an Omnimixer (Serval). This in turn had been pre-cooled in a dry ice–cellosolve mixture and this was maintained during 10 min grinding at maximum speed. The

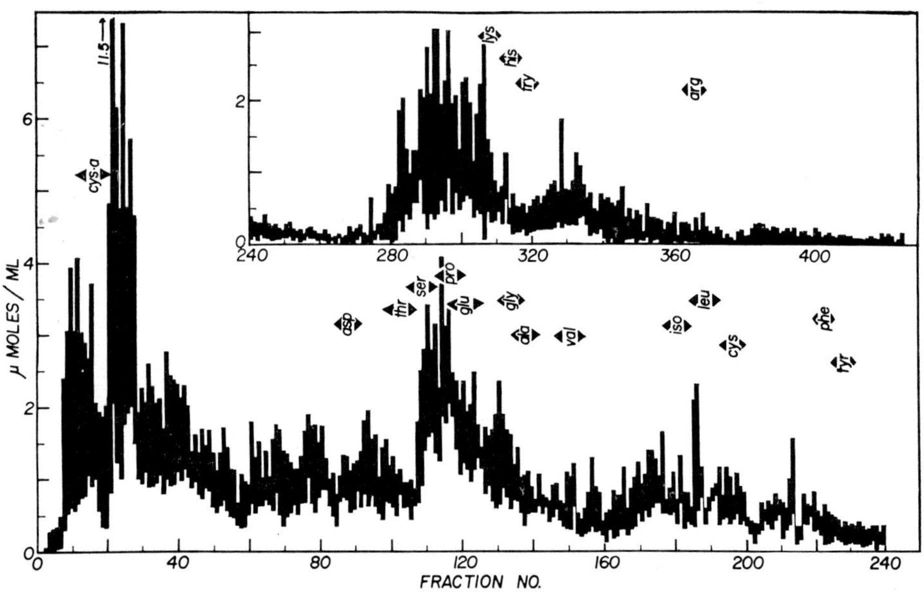

Fig. 2. Column fractionation of an extract of Drosophila larvae. Extract: cold aqueous propanol followed by hot-water extraction of 60 g (wet wt.) of middle third instar larvae. Column: water jacketed (22°) Dowex-50-4X column 2.2 × 45 cm. Equilibrated with starting buffer from acid form and washed with distilled water. Elution: gradient elution with a 100-ml mixing vessel was used starting with ammonium formate (pH 2.49), 0.05 *M* with respect to NH_4^+ in both mixing vessel and solvent reservoir. The gradient was started after collection of 75, 15-ml fractions with the sequence as follows:

Buffer	*pH*	*Molarity* (NH_4^+)	*Change in reservoir at fraction No.*
Ammonium formate	2.9	0.1	75
Ammonium formate	3.3	0.15	142
Ammonium formate	3.65	0.2	191
Ammonium acetate	5.5	0.4	259
Ammonium acetate	6.8	0.6	294
Ammonium acetate	8.0	1.0	359

References p. 146

resulting slurry was poured into a 60-ml medium-porosity glass funnel, pre-cooled at $-20°$ in a deep freezer. Cold 50% methanol was added in batches to a total of 200 ml. The resulting filtrate (Fraction 1, Figs. 3 and 4) was evaporated to a syrup in a rotary vacuum evaporator at 35°. The residue was washed with 30 ml of methanol (100%), 50 ml of ether or chloroform and again with 10 ml of methanol, all at $-20°$ with gravity filtration. The residue was quickly sucked dry with a water pump and the combined filtrates evaporated *in vacuo* to give Fraction 2. The dry residue was added to 30 ml

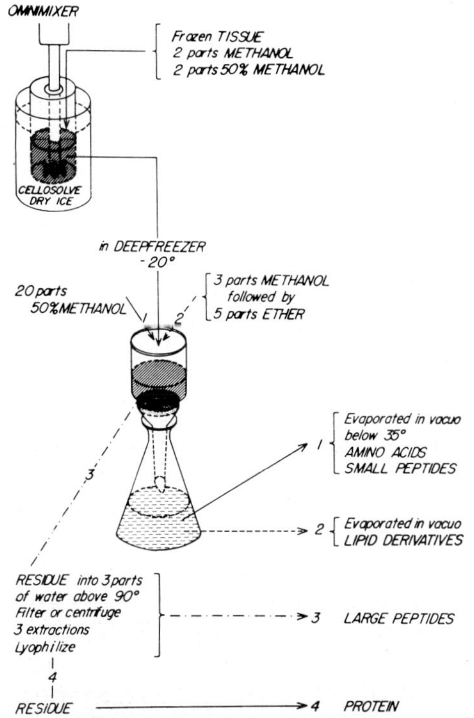

Fig. 3. A diagram showing details of tissue extraction procedure.

of water in an Omnimixer cup contained in water above 90°. After a 3-min mixing (in the hot-water bath) the material was filtered with suction through the original glass funnel. The extraction was repeated twice more and the combined filtrate was lyophilized to give Fraction 3. The residue in the funnel was then extracted twice with 0.1 *M* NaOH by gravity filtration at less than 4°. Protein was precipitated by 10% trichloroacetic acid, washed with trichloroacetic acid and dried or redissolved, as desired, to yield Fraction 4. The method is diagrammed in Fig. 3.

An example of results obtained from an application of the extraction method as detailed above is summarized in Fig. 4. The data were obtained by photometric analyses of paper chromatograms (see analytical methods). Several points are worthy of note. First, before hydrolysis, Fractions 2 and 4 are negligible and Fraction 1 contains three-quarters of the ninhydrin-reacting material. By hydrolysis this fraction increases less than 2-fold indicating a principal content of free amino acids and smaller, more soluble derivatives. On the other hand Fraction 3 increases nearly 10-fold on

Fraction	*Ninhydrin before hydrolysis*		*Ninhydrin after hydrolysis*	
	Dry weight		*Dry weight*	
	μmole/g	*% total*	*μmole/g*	*% total*
1. 50% methanol	210	75	360	17
2. $CHCl_3$–methanol	—	—	10	0.5
3. Water (100°)	69	25	637	31
4. Protein-TCA	—	—	1080	52

Fig. 4. Distribution of amino acids, peptides and other derivatives among the fractions obtained by: (1) 50% methanol at −20°; (2) methanol–ether; (3) hot water; (4) cold alkali. The data are for third instar larvae but those from first instar larvae and embryos are similar.

hydrolysis and its total amino acid content is nearly double that of Fraction 1. With respect to the total balance of amino acids (Fig. 4, last column) it may be observed that about half the amino acids in the larvae are present as protein and, the other half, the pool from which protein could be made, is present largely in the form of peptides and amino acid derivatives.

ANALYTICAL METHODS

In nearly all of the work reported here amino acids and peptides were determined by paper chromatography followed by a photometric determination of the color produced with ninhydrin. Although this method is not as precise as others that could be used it has the advantage of giving additional information besides amounts. That is, in analyzing successive fractions from a column one can see at a glance when changes in composition occur both in unhydrolyzed and hydrolyzed fractions.

In detail the most commonly used procedure was as follows: Aliquots of fractions (3–5 μl) containing 0.02–0.1 μM of ninhydrin-reacting material were pipetted onto sheets (20 × 23 cm) of washed filter paper (Whatman No. 1, washed by descending chromatography for 36 h with 3% aqueous NH_3) with hydrolyzed and unhydrolyzed samples adjacent. Hydrolysis was carried out using 50 μl of sample and 0.1 ml of 6 N HCl in a sealed tube for 16 h at 105°. After hydrolysis the acid was removed *in vacuo* over KOH and P_2O_5. After redissolving in water an appropriate aliquot was spotted beside the unhydrolyzed sample. Each sheet for chromatography was also spotted with 3 or 5 μl of an equal mixture of known amino acids (glutamic acid, glycine, alanine, valine and leucine) at a total concentration of 0.02 M. Chromatograms were then irrigated, four or five at a time in battery jars, by ascending chromatography with n-propanol: 1% aqueous NH_3 (2 : 1). After 8 h the sheets were dried in air and dipped into 1% ninhydrin in acetone. Usually color was allowed to develop in the dark at 35° for 12–16 h. Sheets were then cut in appropriate strips and total color was measured using a Spinco Analytrol. In our experience only proline and hydroxyproline gave seriously low values (less than 50%). Cystine was usually approx. 20% low but most of the other amino acids gave color values within 10% of the maximum (represented by glutamic acid or valine). Peptides occasionally gave peculiar colors and tended to give low values for one terminal amino acid[9]. In spite

References p. 146

of these problems the method is adequate for the present purposes and it is especially informative.

COLUMN CHROMATOGRAPHY OF FRACTIONS SEPARATED BY EXTRACTION

The extraction method described earlier (Figs. 3 and 4) has consistently yielded results that appear to represent a sharp separation of different classes of substances.

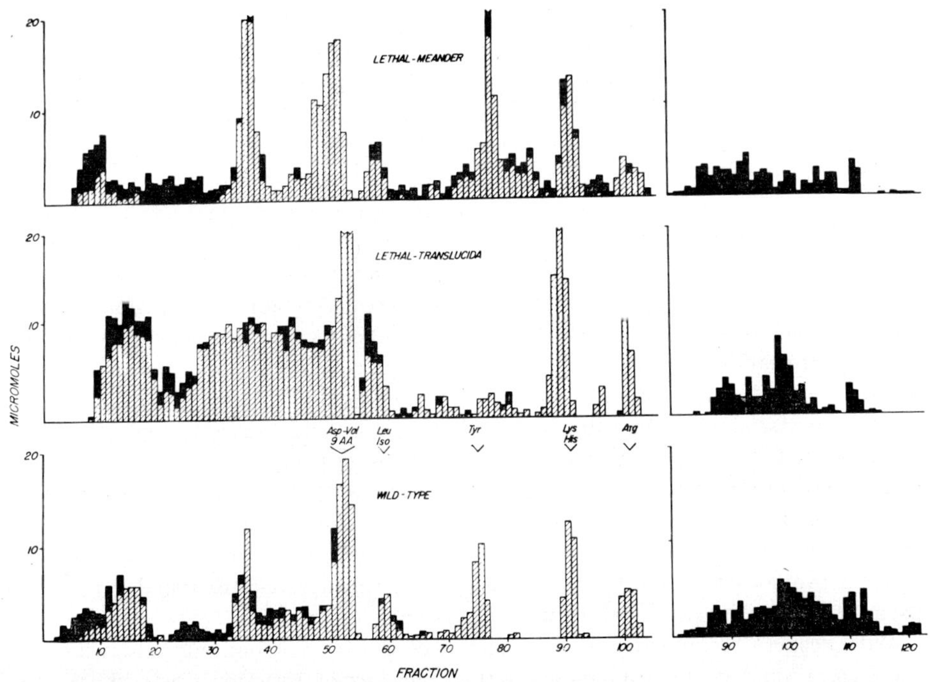

Fig. 5. Patterns of ninhydrin-reacting compounds in Fractions 1 and 3 from wild-type larvae and two mutants. Data on materials from Fraction 1 are shown on the left and from Fraction 3 at the right. Shaded areas show ninhydrin equivalents before hydrolysis and black areas the total after acid hydrolysis. Fraction-3 components gave the ninhydrin reaction before hydrolysis amounting to 2–10% of the total after hydrolysis and in this elution system only negligible amounts of any substance were eluted before tube No. 85. Columns: 1.6 × 15 cm, Dowex-50-4X equilibrated with starting buffer. Elution: linear gradient with an inverted funnel system and the following succession of solvents.

Solvent	*pH*	*Molarity* $(C_2H_5)_3N$	*Change in solvent at fraction*
1. Trimethylamine formate	2.5	0.025	7
2. Trimethylamine formate	3.15	0.05	17
3. Trimethylamine formate	3.5	0.075	39
4. Trimethylamine formate	3.95	0.10	60
5. Trimethylamine acetate	4.55	0.40	78
6. Trimethylamine formate	9.6	1.0	95

9-ml fractions were lyophilized and redissolved for analysis without heating. As shown in the graph for wild type, the common amino acids are all eluted between tubes No. 50 and 110.

References p. 146

Fraction 2 (lipid-soluble) is the most obviously different and since studies on it have been described elsewhere[7] it will not be considered further here. That Fractions 1 and 3 are also distinct and have little if any overlap has been demonstrated in two ways. First, addition of [^{14}C] valine to a homogenate before separation led to a recovery of more than 99.8% of the radioactivity in Fraction 1. Second, as described below, on column chromatography Fractions 1 and 3 contain few if any substances in common. Thus, the use of the preliminary fractionation simplifies further separation and identification of the many components indicated by the data shown in Fig. 2.

The extraction methods described above have been applied to a study of similarities and differences among larvae of wild type Drosophila and two lethal mutants, *lethal translucida (l-tr)* and *lethal meander (l-me)*. Both of the mutants have been shown by CHEN AND HADORN[10, 11] to have abnormalities in protein synthesis. Third instar larvae were used (when the mutant phenotypes are expressed). Fractionations were carried out using Dowex-50-4X but use was made of a linear gradient and a triethylamine buffer. The buffer can be removed sufficiently for chromatography without heating. Details of the system are given with the data in Fig. 5 along with the summary of results. In the figure, elution patterns for Fraction 1 are given on the left where all the common amino acids are eluted between Fraction 48 and 103 (positions of some are shown as an index). Nevertheless a large amount of material including some peptides appears earlier as shown previously (Fig. 2). In contrast, elution patterns of Fraction 3 (shown at the right in Fig. 5) showed negligible quantities of ninhydrin-reacting material prior to tube No. 85. Furthermore, the material obtained after Fraction 85 gave a large increase in amino acids (10–30-fold) on hydrolysis. Unfortunately, in these experiments, the column separations were completed before the analyses and the columns were not continued long enough for complete recovery of the material in Fraction 3. The conclusion is very clear, however, that the extraction method provides a clean separation of materials into Fractions 1 and 3 and thus permits more satisfactory subsequent separations.

With respect to similarities and differences among the mutants and wild type there are a number of considerable interest. Considering Fraction 1 first, the mutant *l-tr* shows a large accumulation of materials all the way from tube No. 5–48 and a striking deficiency in tubes No. 75 and 76. Mutant *l-me* shows accumulations at tubes No. 36, 37 and 78 and a general increase in peptides that appear after tube No. 60. These differences at least (and probably more) appear highly significant since the

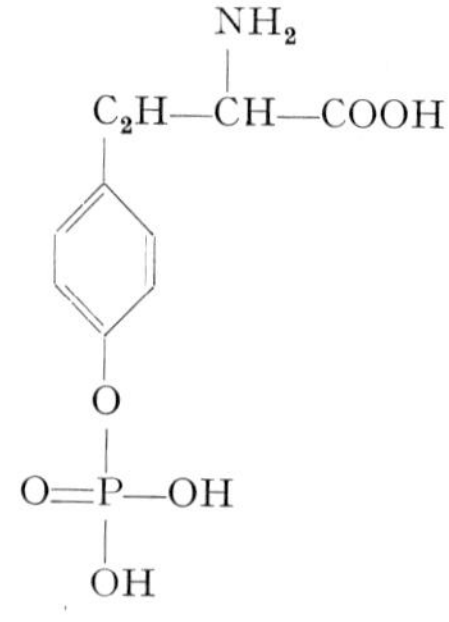

Fig. 6. Tyrosine-*O*-phosphate.

References p. 146

wild-type pattern has been shown to be quite reproducible and the pattern from another mutant *(l-21)* did not differ significantly from that of the wild type.

The differences observed in Fraction 3 are less obvious as presented here than from observations of paper chromatograms. The most striking one is at tubes No. 98 and 99 where mutant *l-me* appeared to be totally deficient in one peptide component (of approx. 20 residues) that is found in wild, *l-tr* and *l-21*.

At present only one substance involved in one of the major differences pointed out above has been identified. This appears, with poor resolution on Dowex-50, in tubes No. 10–18 and accumulation is shown in translucida. This accumulation is associated with the deficiency in *l-tr* at tubes No. 75 and 76 which is the position of tyrosine. The accumulation, which has been noted previously[12] is due to the substance tyrosine-*O*-phosphate (Fig. 6) which has been isolated and identified by comparison with synthetic material. Thus, wild and *l-me* third instar larvae contain about equal amounts of tyrosine and tyrosine phosphate while *l-tr* contains a large amount of the phosphate ester and essentially no free tyrosine.

ACIDIC PEPTIDES

2

	GLU	GLY
a	17	1
b	11	1
c	2	1

	GLU	SER
a	1	1

3

	GLU	GLY	ALA
a	15	1	1
b	1	1	1
c	1	3	1
d	1	1	3

	GLU	GLY	ASP
a	8	1	1

	GLU	GLY	CYS
a	1	1	1
b	2	1	1

	GLU	GLY	SER
a	3	1	1

	GLU	GLY	THR
a	10	1	1
b	8	1	1
c	2	1	4

	GLU	SER	ASP
a	14	1	1

4

	GLU	GLY	ALA	SER
a	7	1	1	1
b	4	1	1	1
c	2	1	1	1
d	1	1	1	1
e	1	2	1	1

	GLU	GLY	CYS	PHE
a	1	1	1	5

	GLU	GLY	ALA	TYR
a	4	1	1	2
b	3	1	1	3
c	1	1	1	3

	GLU	GLY	SER	X
a	4	1	1	THR 1
b	1	1	1	ASP 3

	GLU	GLY	ALA	X
a	14	1	1	CYS 1
b	6	1	1	ASP 1
c	1	1	1	1

5

	GLY	GLY	ALA	SER	CYS
a	2	1	1	1	1
b	1	1	1	1	1

	GLU	GLY	ALA	ASP	PRO
a	6	1	1	1	1
b	3	1	1	1	1

	GLU	GLY	ALA	ASP	X
a	3	1	1	1	CYS 1
b	2	1	1	1	THR 1
c	3	1	3	1	SER 1
d	4	1		1	THR 1 CYS 1
e	6	1 TYR 1 1		1	

	GLU	GLY	THR	ASP	CYS
a	20	1	1	1	1
b	6	1	2	1	1
c	1	1	2	1	1

6

	GLU	GLY	THR	ASP	CYS	PRO
a	1	1	1	1	1	1

Fig. 7. Composition of some acidic peptides from Drosophila larvae. These represent a small proportion of the total number present in larval extracts and not a random sample. The amino acid order given has no significance since only compositions have been determined. Proportions are only approximate for the peptides with many repeats of the same amino acid.

IDENTIFICATION OF PEPTIDES

At the present time we have done relatively little work to determine the composition of the peptides observed and no work on amino acid sequences. There are too many and they are much alike, making the isolation of pure substances a large and tedious program. This is illustrated by the information in Fig. 7 which we present, with some misgivings, since we are aware of the limitations and inaccuracies of the methods

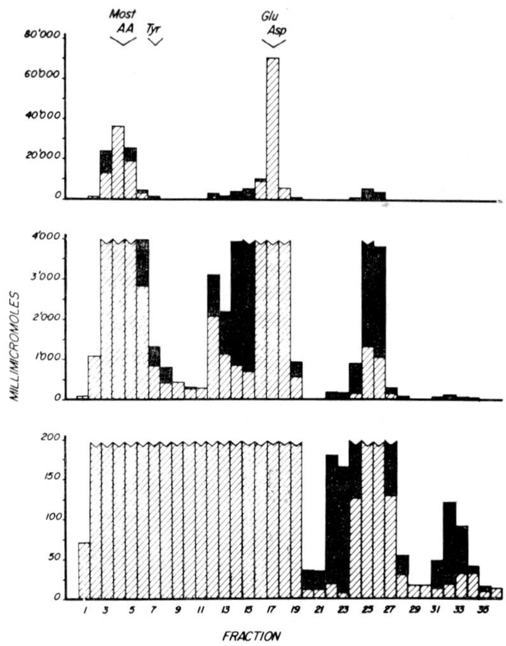

Fig. 8. Amino acids and peptides in rabbit brain. Shaded areas represent quantities before hydrolysis and black areas the increase on hydrolysis. The data are plotted at three levels with 20-fold differences in concentration to show the presence of numerous peptides at relatively low concentrations. Extract: Fraction 1 prepared as in Fig. 3. Column: 1.6 × 16 cm, Dowex-1-8X treated with 4 *M* sodium acetate and washed thoroughly with water. Elution: water to tube No. 5 (10-ml fractions) followed by a gradient produced by introducing 0.5 *M* triethylamine acetate (pH 5.0) into 100 ml of water in a mixing vessel. This system is designed for separation of acidic compounds and elution positions of some known substances are shown in the graph.

used. Samples were taken of acidic components from a column separation (as in Fig. 2) and components were purified further by chromatography and electrophoresis on washed paper. After homogeneity seemed assured samples were hydrolyzed and analyzed by one- and two-dimensional chromatography and by photometric determinations of quantities. Greatest uncertainties are in the ratios when one amino acid is repeated several times and, of course, in original purity.

These results emphasize the separation problem since many of the 42 components listed are very much alike, some series differing only in the number of repeats of glutamic acid. The predominance of non-essential amino acids is also obvious.

References p. 146

It should again be emphasized that these data are not very accurate and that this is not a random sample of the peptides of Drosophila. We expect to provide more extensive and satisfactory information in the near future by use of a Spinco Amino Acid Analyzer although we have not entirely solved the problem of obtaining pure peptides in large numbers.

DISCUSSION

At the outset of this work, some 5 years ago, we expected to find in Drosophila a variety of free amino acids and perhaps as many as ten peptides and other amino acid derivatives as the pool from which proteins could be made. That the situation is nearly 100 times this complex could not have been predicted from earlier work on other organisms although numerous more recent reports have presented evidence for the existence of considerable numbers of peptides in yeast and bacteria (see review by J. HOLDEN in this Symposium). Although these reports and our own qualitative observations have indicated to us that the picture in Drosophila is not unique, we have applied our methods to materials other than fly tissue to a very limited extent.

One example is summarized in Fig. 8 and, although the column elution system was designed only for separation of acidic substances, some interesting conclusions can be drawn. The material placed on the column was obtained by the procedure described here as Fraction 1 (Fig. 4) from a 9-g, fresh rabbit brain. The data are plotted at three levels of concentration to show relative amounts of components. By paper chromatography of such an extract one would see only the components of the two larger peaks shown in the top graph. Nevertheless, as shown in the bottom graph, all the tubes from elution contain one or more components and many are peptides. Therefore, rabbit brain also contains a great variety of ninhydrin-reacting components but the concentrations of most are some 50-fold less than in the Drosophila material. We feel it highly probably that this will be a general observation as more tissues are examined.

In this paper we have confined the discussion largely to methods and a very general description of results. The methods are quite satisfactory as far as they go, but still more systems for separation will be needed to permit total analyses of the many components indicated by the results so far obtained. The results, themselves, demonstrate that there exists in Drosophila a very large and complex pool of peptides and amino acid derivatives which should be considered as potential substrates for synthesis of protein. However, it is a foregone conclusion, that the size and complexity of such a pool is no measure of its metabolic importance.

The extraction method described here is highly satisfactory and it should be applicable to any kind of biological material. The separation of substances by solubility is remarkably clear cut (Fractions 1 and 3 especially) and with the Drosophila material few if any substances appear in more than one fraction in significant quantities. This, of course, is not necessarily true for all materials and each system will require separate consideration. It should also be emphasized that, although low-molecular weight peptides predominate in Fraction 1 and larger ones in Fraction 3, this will always be influenced by composition.

References p. 146

ACKNOWLEDGEMENTS

This work was supported in part by funds from the National Science Foundation (G-4859) and the United States Public Health Service (H-3103). Work on the Drosophila mutants was done under the tenure of a National Science Foundation Senior Fellowship. For this, the senior author acknowledges with pleasure the cooperation and facilities generously provided by Professor E. HADORN at the University of Zurich, Zurich (Switzerland). For assistance in that laboratory we are indebted to Miss R. GLOOR. For assistance during other parts of the program we owe thanks to Mrs. T. SUEOKA, M. TALCOT and Mrs. A. MITCHELL.

REFERENCES

1 E. HADORN AND H. K. MITCHELL, *Proc. Natl. Acad. Sci. U.S.*, 37 (1951) 650.
2 E. HADORN AND E. STUMM-ZOLLINGER, *Rev. suisse zool.*, 60 (1953) 506.
3 P. S. CHEN AND E. HADORN, *Rev. suisse zool.*, 61 (1954) 437.
4 E. STUMM-ZOLLINGER, *Z. Vererbungslehre*, 86 (1954) 126.
5 E. S. BENZ, *Z. Vererbungslehre*, 88 (1957) 78.
6 A. S. FOX, *Physiol. Zool.*, 29 (1956) 288.
7 J. J. WREN AND H. K. MITCHELL, *J. Biol. Chem.*, 234 (1959) 2823.
8 A. R. THOMPSON, *Biochem. J.*, 61 (1955) 253.
9 H. J. BURKHARDT AND H. K. MITCHELL, *Arch. Biochem. Biophys*, 94 (1961) 32.
10 P. S. CHEN AND E. HADORN, *Rev. suisse zool.*, 62 (1955) 338.
11 E. HADORN, *Cold Spring Harbor Symposia Quant. Biol.*, 21 (1956) 336.
12 H. K. MITCHELL, P. S. CHEN AND E. HADORN, *Experientia*, 16 (1960) 1.

METABOLISM OF PEPTIDES IN DROSOPHILA

JOHN R. SIMMONS AND HERSCHEL K. MITCHELL

California Institute of Technology, Division of Biology, Pasadena, Calif. (U.S.A.)

As reported in the previous paper[1], *Drosophila* larvae contain a wide variety of amino acid derivatives, many of them being peptides. This pool of materials is a very large and complex one and many components are present in relatively high concentrations. This system therefore appears to be a very good one for studies on the synthesis and metabolism of peptides and related substances. A good deal of work in this direction has been done[2] using [^{14}C] amino acids and a summary of results obtained so far is given in this report.

METHODS

Most of the methods used here have been described in the previous paper[1]. Injections of *Drosophila* larvae were carried out using an apparatus developed for the purpose (H. K. MITCHELL, unpublished) and by its use we were able to inject large numbers of larvae with a volume of as little as 0.05 μl with an error of approx. $\pm$ 25%. This was determined by direct counts of squashed single larvae injected with [^{14}C] glutamic acid. The error is considerably less in adults or in water drops where there is little or no back pressure. For studies of incorporation of amino acids in time series, larvae were injected and washed thoroughly with water. They were then placed on washed filter paper and squashed with a glass rod as described earlier[3]. Unless otherwise indicated, the solvent used for chromatography was *n*-propanol–1% aqueous ammonia (3 : 1). Radioactive areas on paper chromatograms were determined by use of a paper strip counter (Nuclear Chicago) and a planimeter. Results are reported as percentage of the total radioactivity injected for each sample and the corresponding strip.

RESULTS

Glutamic acid incorporation in direct squash experiments. The results recorded in Table I and Fig. 1 are typical of the pattern found in seven different series in which glutamic acid was injected into third-instar larvae. At the earliest time intervals the major radioactive peak matched the chromatographic position of known glutamic acid very closely. With increasing time from injection there was a marked tendency for this area of greatest radioactivity to shift to higher R_F. When the area was eluted, hydrolyzed and rechromatographed it once more closely matched known glutamic acid in position. It would thus seem that the major chromatographic peak is multi-component in nature, and that the proportion of the various components change with time. All of the radioactive areas showed corresponding ninhydrin-positive reactions, but since the ninhydrin pattern from whole animals is very complex it would be a

References p. 155

mistake to take this as evidence that all of the radioactive components are actually ninhydrin-positive.

On hydrolysis all of the radioactive components yielded glutamic acid, though this was not always the only radioactive component found. All of the radioactive areas,

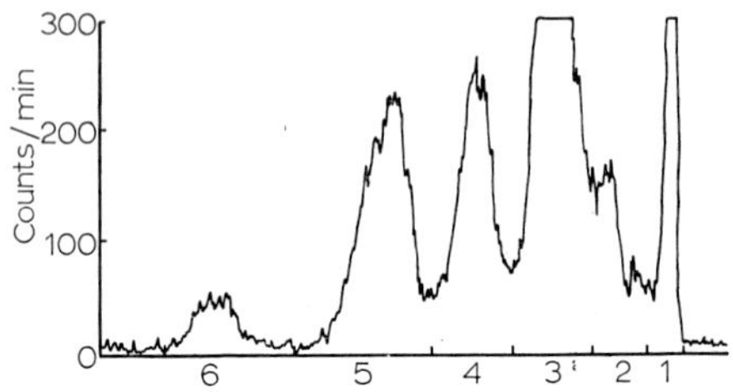

Fig. 1. Distribution of radioactivity resulting from the paper chromatography of squashed whole animals injected with [14C] glutamic acid (4 μC/15 μl). Numbers indicate areas measured by planimeter with 1 being the chromatographic origin and 3 most closely matching the position of known glutamic acid.

when eluted and hydrolyzed, yielded more than one ninhydrin-positive spot following chromatographic separation.

It was found that 4 h of starvation prior to the injection of the [14C] glutamic acid did not appreciably alter the pattern of radioactivity found following chromatography. Increasing the amount of radioactive amino acid injected 4-fold likewise failed to alter the pattern observed.

Leucine incorporation. The results given in Table II and Fig. 2 are examples of those obtained in three experiments in which animals injected with leucine were squashed and chromatographed. As was the case with glutamic acid, the pattern of radioactivity resulting from the injection of leucine was very complex. It differs from

TABLE I

GLUTAMIC ACID INCORPORATION

84-h larvae were injected with a solution of 4 μC/15 μl of L-glutamic acid and three larvae were squashed on Whatman No. 1 paper at the time from injection indicated. After chromatography in the propanol–NH_3 solvent the chromatograms were scanned with the strip counter and the areas determined by planimetry. Radioactivity is reported as the per cent of total radioactivity present. Areas refer to those designated in Fig. 1.

Time (min)	*Total area*	*% of total radioactivity at area*					
		1	*2*	*3*	*4*	*5*	*6*
1	1.855	1.5	5.7	77.8	8.3	6.9	0.0
5	1.955	2.5	9.2	62.0	11.8	12.3	2.4
15	3.885	9.6	9.7	42.7	16.6	17.4	3.9
30	.940	10.0	10.0	32.4	23.1	16.8	7.8
60	2.069	13.5	7.2	30.0	22.6	22.0	4.7

glutamic acid in the more rapid incorporation of radioactivity into compounds remaining at the origin of the chromatogram. Hydrolysis of substances from various areas of the chromatograms resulted in the recovery of material which corresponded chromatographically to leucine, plus other ninhydrin-positive material. However, at 30 min, only about half as much of the non-origin radioactive material hydrolyzed to the amino acid as was the case when glutamic acid was used. On the basis of the

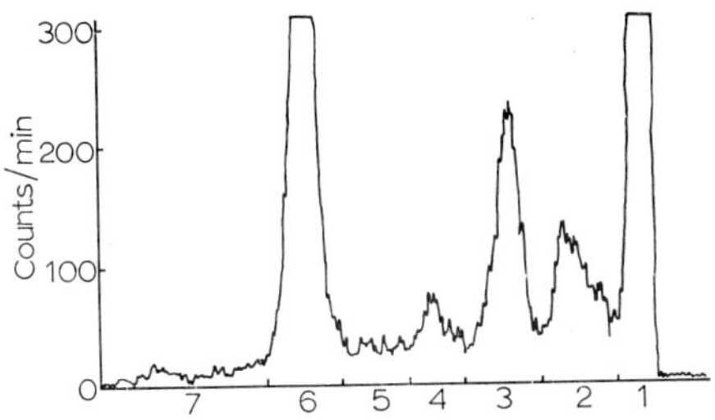

Fig. 2. Distribution of radioactivity resulting from the paper chromatography of squashed whole animals injected with [^{14}C] leucine (10 μC/20 μl). Numbers indicate areas measured by planimeter with 1 being the chromatographic origin and 6 most closely matching the position of known leucine.

direct squash chromatograms of animals which have been injected with leucine it is difficult to draw any conclusions regarding the homogeneity of a given radioactive area. At some stages of the time series recorded in Table II there were well-defined peaks at the chromatographic positions indicated. At both earlier and later times these same areas could not be distinguished from adjacent areas on the basis of radioactivity. It seems probable that different compounds were being measured at different points in the time series. This supposition is supported by the tendency for the point of greatest radioactivity to shift with time, even though it occurs in the same general area.

The leucine pattern is marked by the rapid formation of radioactive material which, with the exception of the substances at the origin, shows little quantitative change

TABLE II

LEUCINE INCORPORATION

84-h larvae were injected with a solution of 10 μC/20 μl of L-leucine and three larvae were squashed on Whatman No. 1 paper at the time from injection indicated. After chromatography in the propanol–NH_3 solvent the chromatograms were scanned with the strip counter and the areas determined by planimetry. Radioactivity is reported as the per cent of total radioactivity present. Areas refer to those designated in Fig. 2.

Time (min)	*Total area*	*% of total radioactivity at area*						
		1	*2*	*3*	*4*	*5*	*6*	*7*
1.5	5.184	0.8	0.7	1.6	1.1	2.3	92.8	0.9
3	5.349	2.1	2.0	4.7	3.9	7.1	79.3	1.1
5	3.886	4.3	1.8	1.4	1.5	6.3	83.1	1.6
10	3.86	11.7	1.2	3.1	3.0	6.4	71.2	3.4
30	4.01	19.4	10.7	4.2	5.7	4.6	53.6	1.8
60	5.985	23.9	3.3	4.6	3.0	6.4	57.7	1.2
140	3.909	54.8	8.3	10.5	3.4	3.2	18.2	1.6

References p. 155

with time. It is also noteworthy that even over a period of 140 min, during which the larvae were not fed, there was not a great loss in the total amount of radioactivity present.

Valine incorporation. The results given in Table III and Fig. 3 are representative of those found in five injection experiments using [^{14}C] valine. At no point during the time series was it possible to distinguish distinct peaks of radioactivity other than at

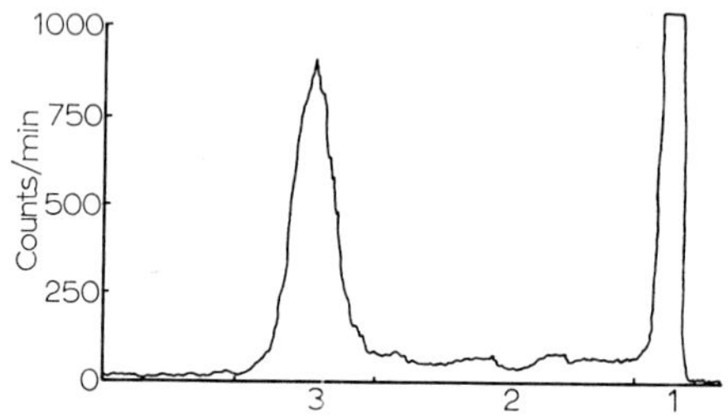

Fig. 3. Distribution of radioactivity resulting from the paper chromatography of squashed whole animals injected with [^{14}C] valine (10 μC/20 μl). Numbers indicate areas measured by the planimeter with 1 being the chromatographic origin and 3 most closely matching the position of known valine.

the origin and in the chromatographic position of valine. There was, however, a continuous zone of radioactivity between these two points. Though the results of the particular experiment recorded in Table III would seem to indicate considerable loss of radioactivity at the end of 60 min. it is felt that this is not the case. In other experiments with valine there was nearly as much total radioactivity remaining at 60 min as was found at earlier times in the series. It is most likely that the difference is attributable to variation in the amount initially injected in this particular experiment.

On elution and hydrolysis of the material from the area between the origin and

TABLE III

VALINE INCORPORATION

84-h larvae were injected with a solution of 10 μC/20 μl of L-valine and three larvae were squashed on Whatman No. 1 paper at the time from injection indicated. After chromatography in the propanol–NH_3 solvent the chromatograms were scanned with the strip counter and the areas determined by planimetry. Radioactivity is reported as the per cent of total radioactivity present. Areas refer to those designated in Fig. 3.

Time (min)	*Total area*	*% of total radioactivity at area*		
		1	*2*	*3*
1.5	2.554	1.6	5.3	93.3
3	1.874	10.5	10.6	78.8
5	1.873	16.5	10.1	73.3
10	2.159	14.5	9.3	76.4
15	1.70	30.4	7.6	62.0
30	1.797	35.4	10.4	54.1
60	1.221	54.1	14.7	31.0

References p. 155

valine, radioactive valine was recovered as were other radioactive components. The ninhydrin pattern of the hydrolyzed material was complex and it is certain that amino acids in bound form were present prior to hydrolysis. As was the case with leucine, radioactive components other than the origin peak increase very rapidly initially and then show a much reduced rate of increase during the rest of the series.

Attempts were made to alter the incorporation patterns found following injection of radioactive amino acids by using compounds which had been found to act as inhibitors of protein synthesis in other systems[4,5]. Among a number of substances tested

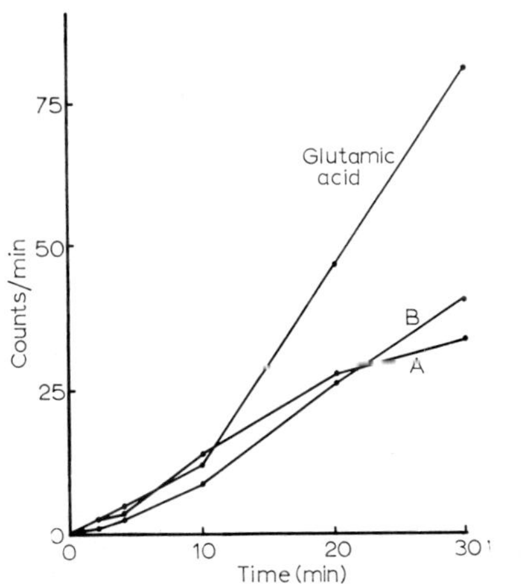

Fig. 4. Larvae were injected with either chromatographic component A or B from Fraction I or with known glutamic acid. All three of the injected materials had an approximate specific activity of 70 000 counts/min/μmole of glutamic acid present. Protein samples were obtained from injected larvae by homogenizing the animals in hot 80% ethanol. The insoluble material from this homogenization was defatted and washed. It was then extracted with 0.1 *N* NaOH. Trichloroacetic acid insoluble material was then separated from the NaOH extract and following an additional washing radioactivity was determined and protein was assayed by the method of LOWRY *et al.*[6]

were chloramphenicol and *p*-fluorophenylalanine. Neither of these compounds nor any of the others tested was effective in altering the incorporation pattern, nor did they reduce radioactivity incorporated into protein extracts. Potential inhibitors were injected both prior to and simultaneously with the [^{14}C] amino acid being tested.

Reinjection of isolated fractions. While a primary aim of the investigation on the low molecular-weight amino acid-containing compounds of *Drosophila* has been to determine their possible implication in protein synthesis, the complexity of the system has made this problem difficult. The results of one set of experiments in which partially purified fractions obtained from larvae injected with [^{14}C] glutamic acid are given here. The chromatographic components A and B were obtained from Fraction I of the extraction procedure given in the preceding paper. A and B were then separated from the bulk of the Fraction-I material by chromatography with propanol–water (3 : 1). Component A moves to the same chromatographic R_F as glutamic acid does in the solvent used. On hydrolysis A gave three spots, probably glycine and lysine in addition to glutamic acid, on the basis of chromatographic behavior. The glutamic acid

References p. 155

concentration was about 3 times that of either of the other amino acids. The overall increase in ninhydrin reactivity following hydrolysis was about 2-fold. Component B separated cleanly from glutamic acid on chromatography in the propanol–water solvent. Hydrolysis followed by two-dimensional chromatography revealed 4 ninhydrin-positive spots, probably lysine, glycine and aspartic acid in addition to glutamic acid. The intensity of the ninhydrin color of the glutamic acid spot was about 4 times that of lysine and aspartic acid and about twice that of the glycine. The overall increase on hydrolysis was again about 2-fold.

The results obtained by injecting A and B and then determining protein-specific

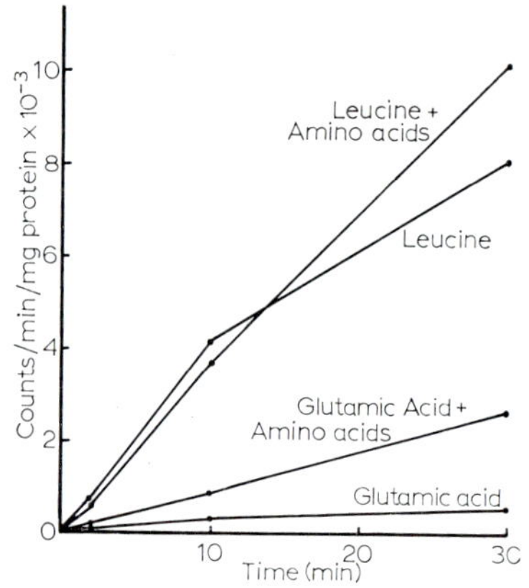

Fig. 5. 84-h larvae were injected with a solution (3 μC/10 μl) of the radioactive amino acid, with and without added essential amino acids. Ten larvae were injected for each time interval and the specific activity of an extracted protein preparation was determined.

activity are given in Fig. 4, and are compared with the result of a control experiment. The results leave little doubt that both A and B are different from free glutamic acid with respect to incorporation into larval protein. Whether the incorporation proceeds directly from the injected component A and B or first is hydrolyzed to the level of free glutamic acid is not determinable with the limited information now available.

Incorporation of amino acid into protein. The limited amount of information obtainable from paper chromatograms of squashed larvae indicated that most of the radioactive material found at the chromatographic origin is protein though there is also some large peptidic material present. To obtain a more direct measure of incorporation of amino acids into larvae, extracts were made and the specific radioactivity of proteins was determined. The results of such experiments using two [^{14}C] amino acids are shown in Fig. 5. The findings are in good agreement with those of the direct squash experiment. The slow incorporation of glutamic acid into protein is typical of the non-essential amino acids (glutamic acid, aspartic acid, glycine and alanine) which were tested while the essential amino acids, leucine and valine both show more rapid incorporation into protein. The addition of an amino acid mixture (acid hydrolyzate of *Drosophila* protein) to the injection mixture enhanced the incorporation of both glutamic acid and leucine into protein.

References p. 155

Column fractionation. The work with injected larvae that were squashed directly on paper and then chromatographed made it clear that the system being studied was very complex. It was hoped that greater resolution could be obtained through the use of ion-exchange chromatography. After the injection of 7000 *Drosophila* larvae (approx. 5 g) with a mixture of L-[^{14}C] glutamic acid and L-[^{14}C] leucine enough uninjected animals were added to give a total weight of 51 g. The larvae were then ground with dry ice at $-80°$ in an Omni-Mixer (Servall). The resulting powder was placed on a sintered glass funnel and extracted with 900 ml of chloroform–methanol (2 : 1)

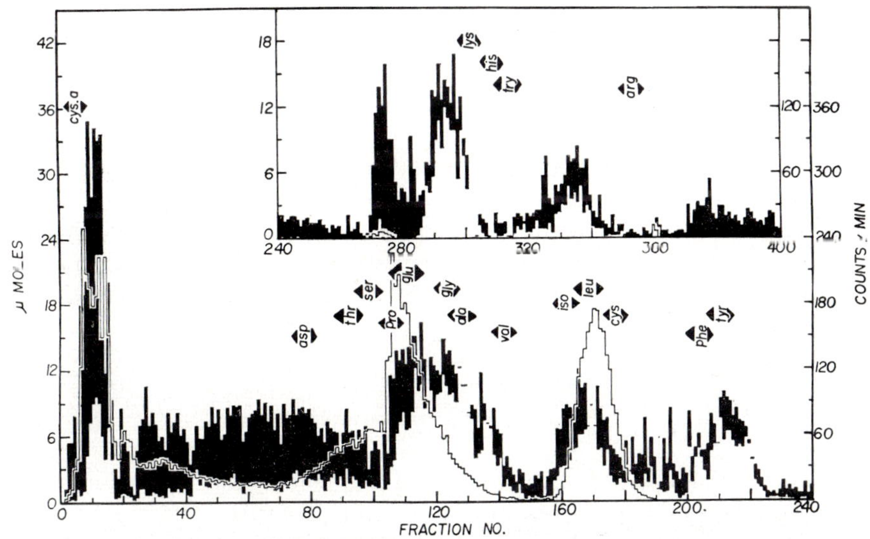

Fig. 6. Elution diagram resulting from the ion-exchange chromatography of a larval extract. The before-hydrolysis concentration of ninhydrin-positive material is indicated by the height of the open bars and the after-hydrolysis concentration by that of the black bars. Radioactivity is shown by the continuous black line. The fractions in which known amino acids were eluted are indicated. This information was obtained by separating a known amino acid mixture under the same conditions used in this experiment. Introduction of eluting solvents was started with ammonium formate buffer (pH 2.49, 0.05 *M* with respect to ammonium). After the collection of Fraction 11, elution was continued to Fraction 75 with this buffer. At this point a pH and concentration gradient was started by slowly introducing 0.1 *M* ammonium formate buffer at pH 2.9 into the solvent reservoir (100 ml) containing the first buffer. Rapid mixing was assured by mechanical stirring. The addition of the pH-2.9 buffer continued until the introduction of the next buffer started. The other buffers, and the fraction at which their introduction into the mixing reservoir was started, were as follows:

Buffer	*pH*	*Molarity*	*Fraction*
Ammonium formate	3.3	0.15	142
Ammonium formate	3.65	0.2	191
Ammonium acetate	5.5	0.4	259
Ammonium acetate	6.8	0.6	294
Ammonium acetate	8.0	1.0	359

Fractions eluted from the column were analyzed by the methods previously given.

References p. 155

at −20°. The extract was washed with water according to Folch's procedure[7] and after removal of chloroform and methanol the aqueous phase was retained. The residue from the chloroform–methanol extract was dropped into boiling water while still frozen and was homogenized for 5 min at 100°. The homogenate was then centrifuged at 15 000 rev./min for 20 min in the 30 rotor of the Spinco Model L ultracentrifuge. The supernatant was decanted and the water extraction of the residue was then twice repeated. The combined water extracts (170 ml) were added to the aqueous phase from the wash of the chloroform–methanol extract and the whole was placed on an ion-exchange column for fractionation. Fractionation was carried out on a Dowex-50-X4, 22 mm × 45 cm column. The results of the fractionation of the extract are shown in Fig. 6.

As indicated previously[1] the products of the column fractionations are very complex. With respect to radioactive components this is particularly true with glutamic acid. Glutamic acid was widely distributed in the column fractions in bound form. When radioactive material (not in the region of free glutamic acid) was isolated from column fractions by additional purification steps and then hydrolyzed it was found that most contained radioactive glutamic acid and other non-radioactive amino acids. It would thus seem that glutamic acid was incorporated into a wide variety of peptidic materials. The amount of leucine in such material was less and not enough was found in any given fraction to do more detailed analysis. This is apparently a reflection of smaller pools of leucine-containing material.

DISCUSSION AND SUMMARY

The data presented here permit the conclusion that amino acids injected into *Drosophila* larvae are very rapidly incorporated into peptides of various sizes. Subsequently they enter into proteins, but the results do not give information as to the mechanisms by which the syntheses take place. In initial stages, mixed peptides containing [^{14}C] glutamic acid are incorporated into proteins at about the same rate as glutamic acid itself but whether the peptides are incorporated directly or are first hydrolyzed remains an open question.

It is of interest to note that glutamic acid (glycine, alanine and aspartic acid behave similarly) goes very rapidly into peptides and relatively slowly into proteins. In contrast leucine and valine, which are essential amino acids to *Drosophila*, appear in relatively few peptides and appear to go more directly and rapidly into proteins. However, this may be only a reflection of pool sizes for essential and non-essential amino acids. Much greater accumulation of peptides would be expected with amino acids (such as glutamic acid and glycine) that can be synthesized by the tissues of the larvae.

It is of special interest to note that in this *in vivo* system, chloramphenicol and fluorophenylalanine had no influence on the rate of incorporation of amino acids into protein even when saturated solutions were injected. So far no inhibitors have been found which specifically inhibit protein synthesis in the larvae.

The methods described here, provide an adequate background for further work on isolation of pure peptides and further studies on their metabolism both *in vivo* and *in vitro* (*cf.* ref. 1).

References p. 155

ACKNOWLEDGMENTS

During this work one of us (JRS) was supported by a National Science Foundation Pre-Doctoral Fellowship. We are also indebted to the National Science Foundation for additional support in the form of a Research Grant (G-4859).

REFERENCES

1 H. K. Mitchell and J. R. Simmons, This Symposium, p. 136.
2 J. R. Simmons, *Doctoral Thesis*, Biology Division, California Institute of Technology, Pasadena, Calif., 1960.
3 E. Hadorn and H. K. Mitchell, *Proc. Natl. Acad. Sci. U.S.*, (1951) 650.
4 D. W. Woolley, in *A Study of Antimetabolites*, J. Wiley and Sons, New York, 1952.
5 W. Shive and C. G. Skinner, *Ann. Rev. Biochem.*, 27 (1958) 643.
6 O. H. Lowry, N. J. Rosebrough, A. L. Farr and R. J. Randall, *J. Biol. Chem.* 193 (1951) 265.
7 J. Folch, M. Lees and G. H. Stanley, *J. Biol. Chem.*, 226 (1957) 497.

DISCUSSION

Chairman: NORMAN H. HOROWITZ

GUROFF: DR. MITCHELL, I wonder if you could comment on the biosynthesis of tyrosine phosphate and what its function might be?

MITCHELL: I have some evidence that tyrosine phosphate is combined in peptides and perhaps a bit in some proteins. We have speculated on the possibility that the substance is concerned in blood coagulation as tyrosine sulfate is in mammals. However, there is no evidence for this so far.

GUROFF: I wonder if it could have any relation to the serine phosphate found in mammalian tissue?

MITCHELL: The tyrosine derivative is much more labile than serine phosphate. This is a phenol ester and should have a higher energy bond, so it is, I think, in an entirely different category so far as synthesis and reactions are concerned.

HALVORSON: We have made some observations similar to yours while working with peptides attached to yeast ribosomes. These peptides are rapidly labeled during a pulse labeling experiment. I wonder if you looked at the microsomal and ribosomal fractions? Is it possible that some of the peptides you have observed result from breakdown of such cell structures?

MITCHELL: We do not know whether the peptides come from direct synthesis or from degradation of something more complex. Using radioactive amino acids these peptides are labeled much faster than proteins. The label in peptides goes into protein, but our present evidence does not preclude peptide hydrolysis before protein synthesis. So far as viteosomes are concerned, preparation from larvae contain the larger, more complex peptides. However, the procedure for their preparations requires a good deal of time and sufficient reactions can occur even at 0° to give a misleading picture.

HALVORSON: Is it feasible to use diisopropylfluorophosphate during rupture of the larvea to minimize proteolytic activity?

MITCHELL: I do not think we need to worry about proteolytic action at $-80°$. Nothing happens on incubation of low temperature extracts prepared as I have described. I do not think the use of an inhibitor is necessary when making extracts in this way.

HOROWITZ: I want to return to the point that DR. HALVORSON raised; namely, have you done any experiments to find out how soon the label appears in ribosomes?

MITCHELL: No. We have done no labeling experiments with isolated ribosomes. Our intent has been to try and find the system which will synthesize protein from peptides but which will not incorporate amino acids. If peptides are used directly for making proteins, one should be able to find such a system.

HOROWITZ: I thought perhaps what DR. HALVORSON had in mind was the possibility that the injected amino acids enter the ribosomes rapidly where they are attached to a template, and that your treatment removes them as peptides.

MITCHELL: We have not done an experiment directly on this point. I think there would not be enough ribosomes to carry all of the peptides that are present. However, I have not made the necessary calculations.

HENDLER: In some of your earlier work you described a large lipid–amino acid pool in larvae which seemed to decrease during development. In short-term incubations did you determine whether there was isotope uptake in that fraction, and, if so, how did it compare with the uptake into the peptide fraction?

MITCHELL: The lipid fraction is labeled very quickly but there is no subsequent increase after even the *earliest samples* we have taken.

HENDLER: You mean ahead of your peptide fraction?

MITCHELL: Yes, but we have done no work on measuring turnover of label in the lipid fraction. WREN has found that a fresh, carefully prepared lipid fraction will firmly bind added amino acids. Thus, some artifact of incorporation can be created, but this is a much smaller amount of amino acid than is found normally in lipid preparations. I do not know whether the rapid incorporation of labeled amino acid into the lipid fraction means anything or not and we have done no work to establish its significance.

CHRISTENSEN: I have a question with regard to the possibility that proteins may be split under the conditions of extraction. Although your primary extraction is made at a very low temperature, some of the subsequent steps are carried out at higher temperatures, and these are the ones that yield your peptides. Perhaps the greatest danger occurs in the finite interval that must occur in passing from −20° to 90°.

Mitchell: The material at −20° is plunged directly into water that is boiling. That is the best I can do. We have done extractions at a little below zero, and successively at different temperatures, and we have obtained different increments and different sizes of products at these different temperatures. On the average, the size of what is extracted is larger as the temperature is increased.

E. Roberts: Is there any possibility that these peptides are produced by methanolysis of bigger units?

Mitchell: I think that it would be necessary to have acyl amino acids or acyl peptides in order to get such reactions. You surely would not get it just from a peptide bond, and in any case hydrolysis would occur more rapidly than methanolysis in the 50 per cent solution.

Eagle: I have not so much a question as a comparison. In cultured animal cells, using cold TCA (trichloroacetic acid) to extract the amino acid pool, we are not plagued by this large amount of peptides. The TCA extract represents about 10 per cent of the total cell nitrogen. Of that, about half is in free amino acids, and the other half in peptides which release amino acids on hydrolysis. I rather suspect this is a difference in biological material, rather than in method of preparation.

L. Miller: I presume that the larvae have a functional gastro-intestinal tract and I wonder, therefore, how you eliminate this gastro-intestinal contribution to the peptide "gemisch"?

Mitchell: You can take the tract out. We have done this previously with small samples and the amounts of amino acids and peptides formed in the tract is negligible.

E. Roberts: I do not think we have ever succeeded with paper chromatographic methods in finding any real evidence for occurrence of considerable amounts of peptides in brain extracts. Dr. Tallan has used the column-chromatographic methods extensively in brain extracts and I wonder if he has found peptidic material in his extracts to any large extent.

Tallan: We have never hydrolyzed the effluent from a chromatogram of brain, in the manner that Dr. Mitchell described. However, when we hydrolyzed various portions of the effluent from Dowex-50 chromatograms of human urine, we did find a liberation of amino acids; glycine and glutamic acid were present in the largest amounts, but we also found aspartic acid, threonine, alanine, valine, isoleucine and leucine.

Christensen: With regards to Dr. Tallan's comment about bound glycine in the urine, this appears largely to be hippuric acid, which we have also been able to detect in the liver. In general, we must always be suspicious that amino acid conjugates found in tissue extracts or body fluids may not be peptides at all.

L. Miller: That brought to mind another perhaps totally extraneous observation; that is, in working with dogs years ago, we found that we could get enormous increases in urinary peptides by infusing intravenously enzymatic digests of protein. Are these *Drosophila* larvae being fed a completely defined fodder or are they being fed a partially digested undefined diet containing peptides which may not be particularly appealing to them and which are consequently accumulating somewhere in the back-waters, if you will, of the extracellular fluid?

Mitchell: Our larvae are fed whole, live yeast.

L. Miller: And whole live yeast have no peptides at all that are comparable to those you find in the larvae.

Mitchell: In the amount present, no. Whole live yeast contain peptides, but the amounts present in the intestinal tract are negligible.

L. Miller: Of course, one would assume that in the process of transforming whole live yeast to whole live larvae there are no peptides formed excepting those that the beasts had a fancy for; that is, metabolically speaking?

Mitchell: Well, only the larvae could answer that.

Halvorson: I have had some experience with protein breakdown in starving yeast cells and I would like to have one point clarified. If you look at non-starving larvae, do you see any difference in the peptide content as compared with those that you place under starvation conditions?

Chen: There is some increase of peptides, but not those originally present.

V. COMPARATIVE, DEVELOPMENTAL AND EVOLUTIONARY ASPECTS

FREE AMINO ACIDS IN INVERTEBRATES: A COMPARATIVE STUDY OF THEIR DISTRIBUTION AND METABOLISM

JORGE AWAPARA

Biology Department, Rice University, Houston, Texas (U.S.A.)

INTRODUCTION

Tissues of both vertebrates and invertebrates contain a number of nitrogenous substances that can be easily extracted. This fraction is relatively small as compared with the protein fraction of tissues. However, it contains all the free amino acids and many compounds which are derived from them. The free amino acids were discovered during the study of protein digestion in mammals and the factors which affect their concentration in various organs were carefully studied. Free amino acids were also discovered in the non-protein fraction of tissues from invertebrates. Many analyses were made of the non-protein fraction from all sorts of animal tissues and from time to time a new nitrogenous substance would be found for which no metabolic role could be ascribed. ACKERMANN has isolated and identified a very large number of nitrogenous compounds from invertebrates, many of them obviously derived from amino acids. The tedious but rigorous methods of isolation used could not be applied to the analysis of large numbers of specimens and any comparative study would be almost impossible. The development of microbiological methods of assay for amino acids made possible the study of many specimens and these methods were successfully applied to the analysis of free amino acids in the non-protein fraction of tissues from mammals as well as invertebrates. The interest in the free amino acids in invertebrates was limited to their possible role as osmoregulators. It is true that the concentration of free amino acids in invertebrates is much higher than in mammals and in marine invertebrates higher than in terrestrial or fresh water invertebrates. The question has not been resolved and the study of amino acids as osmoregulators is being followed actively in many laboratories.

Microbiological assay is a good and rapid method to measure qualitatively those amino acids that can be measured by this method; it will not reveal the existence of new amino acids or compounds such as taurine which are closely related to them. Taurine is present in many invertebrates in very high concentration and failure to measure this compound would neglect an important factor in osmoregulation; in some marine invertebrates taurine comprises over one-third of the total non-protein nitrogen fraction. This raises the possibility that other nitrogenous compounds could exist which could contribute in large measure to osmoregulation. This problem is eliminated when the analysis of the non-protein fraction is carried out by chromatographic methods. Paper chromatography is simple and has a high resolving power; it lends itself to the analysis of many samples and the quantities of tissues needed are relatively small. Paper chromatography has been used extensively in the analysis of free amino acids in mammalian tissues. We know that many new compounds were discovered in tissues of mammals by using paper chromatography; it is both a survey method and a method easily adapted to rough quantitation. Many invertebrates have been thus analyzed and new compounds detected as will be discussed later. The method has been used to establish patterns of ninhydrin-reactive substances in the hope to find a method for classification and speciation but the results have not been too satisfactory. Paper chromatography is now a well-established laboratory method but it has not been exploited to the limit of its potentialities in the study of amino acids in invertebrates. In the past, this work was in abeyance for want of a good analytical method. In the present, the problem is one of sampling. Invertebrates are collected randomly from untested environments; there is no knowledge of their physiological condition and often there is no precise knowledge of their feeding habits. In spite of these limitations the analysis of the non-protein fraction from tissues of invertebrates has been revealing and stimulating. Clues have been obtained that can be used to search for new enzymes, new metabolites and perhaps some unique metabolic sequences. It should be easy to maintain stocks of invertebrates in the laboratory where conditions could be controlled. Then a number of interesting problems could be solved. Invertebrates appear to contain free amino acids in extremely high concentration as compared with mammals. One could explain this extremely high concentration of free amino acids if one could demonstrate very active proteolysis in these organisms, but this could not explain the uneven distribution of some amino acids; for example, it could not explain the fact that in many crustaceans over one-third of the amino acid nitrogen is made up of glycine. It could not explain the fact that in some echinoderms glycine is the only detectable amino acid next to taurine. The answer to this uneven distribution might be related to the rates of formation and utilization of these compounds; or it could be related to the ability of invertebrate cells to concentrate nitrogenous substances by some unknown mechanism. This is true in the case of the mussel *Mytilus edulis* which contains taurine in a concentration of approx. 2% of its fresh weight. *M. edulis*, however, forms taurine from its precursors at a very slow rate. This invertebrate must possess a mechanism to maintain this high concentration of taurine in the cells.

There are few papers published related to nitrogenous compounds of invertebrates; a few have been selected and will be discussed in relation to the role of amino acids in osmoregulation. Others have been selected as interesting reports on either new nitrogenous compounds or compounds which have been studied in relation to the metabolism of amino acids in invertebrates.

FREE AMINO ACIDS IN INVERTEBRATES

The available information on the content of amino acids in invertebrates has been obtained mostly by means of microbiological assays or by means of paper chromatography; in the latter case many other substances which react with ninhydrin are usually detected and reported under the heading "amino acids". For the present discussion, "free amino acids" will mean α-amino-carboxylic amino acids and not all compounds detected with ninhydrin in paper chromatograms or in the effluents from ion-exchange resin columns. Often the composition of a whole animal is given, simply because the animal is too small to separate its component organs. In many papers this is not specified and comparisons are difficult.

References p. 174/175

One fact emerges clearly from most of the information reviewed: amino acids in their free state exist in much greater concentration in the tissues of invertebrates than in those of vertebrates. Comparison of organs from a lobster to those of a cat may have little meaning, but such a comparison will be shown in Table I to emphasize the enormous difference in composition. Invertebrates, as said before, have very large amounts of free amino acids in their tissues. The data in Table I were obtained from a paper by CAMIEN *et al.*[1] who analyzed lobster tissue by microbiological methods and from a paper by TALLAN, MOORE AND STEIN[2] who analyzed cat organs by their

TABLE I

FREE AMINO ACID CONTENT OF TISSUES FROM THE CAT AND THE LOBSTER *(Homarus vulgaris)*
Values are expressed in mg/100 g wet wt.

	Lobster			*Cat*					
				Liver		*Muscle*		*Kidney*	
Amino acid	*Regular*	*Dialyzed*	*Unhydrolyzed*	*Extract*	*Hydrolyzed*	*Extract*	*Hydrolyzed*	*Extract*	*Hydrolyzed*
Glycine	1025	892	1025	9.1	54.0	6.7	17.2	14.4	36.0
Proline	728	707	728	2.6	L	3.2	8.5	4.6	5.3
Arginine	778	830	778	0.2	L	2.7	3.0	1.2	1.4
Glutamic acid	267	267	44	66	241	36.2	156	137	174
Alanine	133	133	133	16.5	16.9	24.7	24.3	20.7	19.4
Aspartic acid	12	13	4.6	11.6	18.2	3.9	9.8	7.3	14.3
Lysine	23	25	24	3.6	4.1	5.5	9.6	3.7	9.1
Threonine	8.6	8.1	9.1	3.1	3.7	3.9	4.6	3.6	4.1
Valine	22	22	21	4.3	3.4	2.3	2.9	6.2	3.8
Isoleucine	—	—	—	1.7	0.8	1.7	1.6	2.3	1.7
Leucine	9.3	8.9	8.9	3.6	4.5	2.3	2.6	3.2	3.2
Histidine	7.7	8.3	9.7	9.1	13.4	3.6	103	2.7	4.0
Methionine	11	12	12.0	0.9		0.4		1.1	
Phenylalanine	5.4	5.0	5.2	1.8	2.5	1.0	1.6	1.6	1.7
Tyrosine	1.6	1.8	11	2.1	1.3	0.8	1.1	1.8	2.3

ion-exchange method. Most amino acids are in higher concentration in the lobster muscle, but glycine is particularly abundant in the lobster as it is in other related species which will be discussed later. The sum of all the amino acids in lobster muscle is more or less ten times greater than the sum of all the amino acids in cat muscle. Mammalian organs vary in total free amino acid nitrogen concentrations from 10 to perhaps 50 mg/100 g wet wt.[3]. In invertebrates the total free amino acid concentration is much higher. A comparison of the concentration of free amino acids in the muscle of the rooster with muscle of the lobster is given in Table II taken from a paper by FREDERICQ, BACQ AND FLORKIN[4]. The distribution of nitrogenous compounds in aqueous extracts of lobster muscle and hepatopancreas is given in Table III (ref. 5). Again one can see that free amino acids make up a substantial fraction of the total non-protein nitrogen of the animal. High values for non-protein nitrogen are not characteristic of marine animals. In a comparison of values between marine invertebrates and fishes, it was found that invertebrates (four species analyzed) contain from 210–370 mg of α-amino acid nitrogen whereas fishes contain only 9–72 mg of

References p. 174/175

α-amino nitrogen per 100 g of fresh tissue[6, 7]. Intracellular free amino acids are more abundant in marine than in fresh water forms[1, 8]. This was shown at least to be the case in a comparative study of the North Sea lobster *(Homarus vulgaris)*, the spider crab *(Maia squinado)*, the chinese crab *(Eriocheir sinensis)* and the crayfish *(Astacus fluvialis)* and in a comparative study of tissues of lamellibranch molluscs and some worms. A series of experiments were carried out by DUCHÂTEAU *et al.*[9] to define the

TABLE II

FREE AMINO ACIDS IN LOBSTER MUSCLE AND ROOSTER MUSCLE
Values are expressed in mg/100 g wet wt.

	Rooster	*Lobster*
Alanine	5.2	133.0
Arginine	1.3	178.0
Glycine	12.3	1025.0
Leucine	1.6	9.3
Lysine	2.5	23.0
Methionine	1.0	11.0
Phenylalanine	1.3	5.4
Proline	4.0	728.0
Threonine	2.8	8.6
Valine	1.3	22.0
Total	33.3	2743.3

conditions under which the concentration of free amino acids vary in relation to environment; from a study with the crab *Carcinus maenas* L. they concluded that free amino acid concentration decreases as the salinity of the medium decreases. When living in sea water, the crab muscle had a total of 2940 mg/100 g of amino acids (the sum of 15 amino acids measured microbiologically in the muscle). When sea water was diluted 1 : 1 and the animals adapted to the new environment for 2

TABLE III

THE NITROGENOUS COMPOUNDS PRESENT IN AQUEOUS EXTRACTS OF LOBSTER MUSCLE AND HEPATOPANCREAS*
Values are expressed in mg/100 g wet wt.

Tissue	*Total N*	*α-Amino N*	*Trimethyl amine oxide*	*Betaine N*	*Glutamine amide*	*Volatile base N*
Muscle	820	358	100	92	29	10
Muscle	726	306	104	96	30	8
Muscle	805	280	106	—	—	—
Muscle	749	369	110	—	26	14
Hepatopancreas	628	248	17	91	12	30
Hepatopancreas	625	233	21	85	19	16

* From a paper by KERMACK, LEES AND WOOD[5].

References p. 174/175

weeks they showed a decrease of approx. 40% in the concentration of free amino acids in muscle. The decrease was not equal for all of the amino acids measured but some decreased more than others. The greatest change observed was in the concentration of alanine, arginine, aspartic acid, glutamic acid and proline.

Other invertebrates studied are some parasitic worms. The concentration of free amino acids in parasitic worms is not different from mammalian tissues. The total free α-amino nitrogen found for three species of anoplocephalid cestodes was measured by CAMPBELL[10]. He reports values for non-protein carboxyl nitrogen of 16, 13 and 10 mg/100 g, which is actually on the lower range for mammalian organs, and certainly very low when compared with marine crustacea like the lobster (Table IV). Many ideas have been proposed to explain the extremely high concentration of free amino acids in invertebrates. High concentrations of amino acids could be osmoregulatory. They could be products of protein breakdown and excretory products, but this would not explain easily the enormous difference in patterns which exist between one species and another. Dietary factors must be also considered but again difficulties would arise in explaining for example the extremely high concentration of glycine in some crustacea. This deserves further discussion. Glycine makes up about one-third of the total free amino acids in the lobster. Three other crustacea studied by us[11] vary in glycine content considerably; in the Gulf Coast shrimp *(Penaeus aztecus)* glycine makes up also one-third or more of the free amino acid nitrogen (450 mg of glycine/100 g), but in the hermit crab *(Clibinarius vittatus)* the concentration of glycine is only 45 mg/100 g. In the crab *Pagurus pollicaris* the concentration of glycine is

TABLE IV

NITROGEN DISTRIBUTION IN *M. expansa*, *T. actinoides* AND *C. perplexa*
Values are expressed in mg/100 g wet wt.

Species	*Total N*	*Non-protein N*	*Non-protein carboxyl N*	*Protein N*	*Protein carboxyl N*
M. expansa	544	88	35	414	321
Range	(425–767)	(70–112)	(24–53)	(254–621)	(198–501)
% of total N	—	16.2	6.4	76.1	59
T. actinoides	1296	164	81	896	756
Range	(1159–1501)	(140–177)	(58–105)	(630–1081)	(530–961)
% of total N	—	12.7	6.3	69.1	58.3
C. perplexa	1291	124	61	978	830
Range	(1217–1386)	(110–136)	(52–71)	(846–1077)	(685–885)
% of total N	—	9.6	4.7	75.8	64.3

intermediate between the two other crustacea, namely 150 mg/100 g. In all three, glycine was the most abundant of the α-amino acids. Comparisons of amino acids between various species have been attempted as means of speciation but with little success. GIORDANO, HARPER AND FILICE[12, 13] compared the amino acid composition of blood from two taxonomically related species, the sea urchin and the starfish, and found that the composition was dissimilar. I do not believe that the study of patterns can be used for this purpose. Invertebrates are in an evolutionary sense very advanced and already possess all the biochemical characteristics of the mammals. Amino acids

References p. 174/175

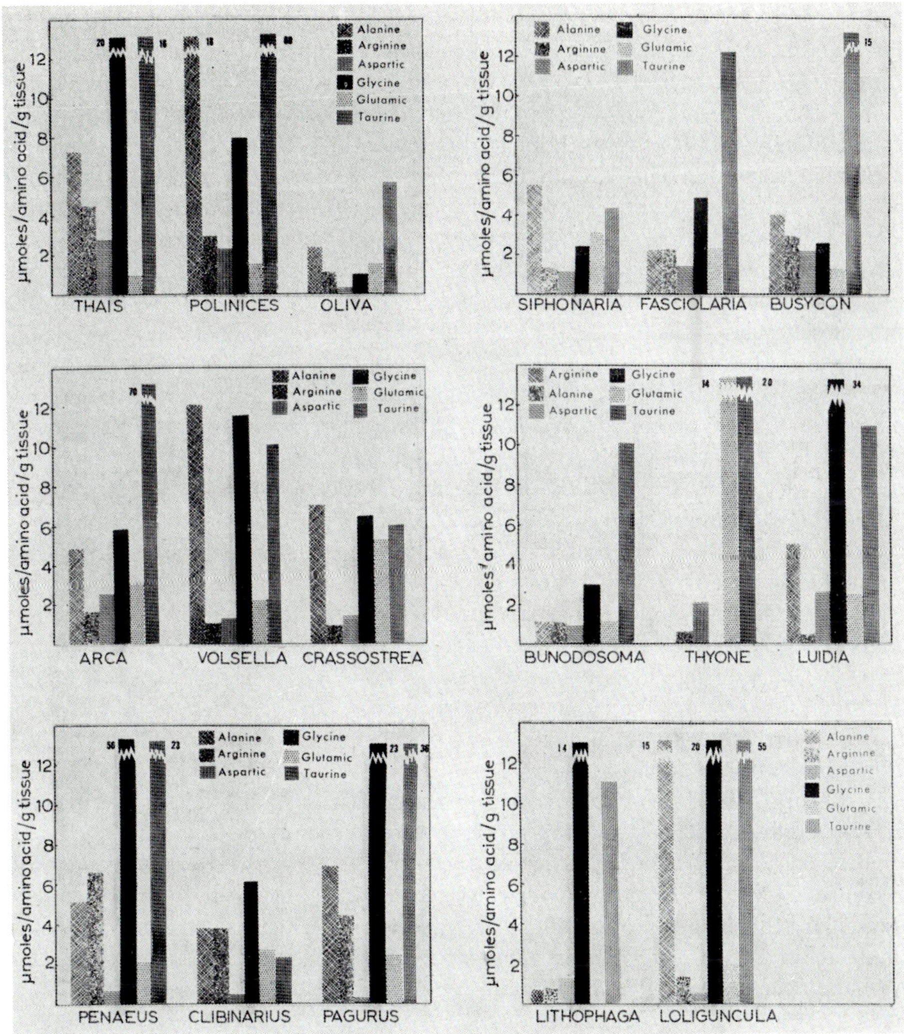

Fig. 1. Amino acid contents of marine invertebrates.

are metabolized and interconverted in similar ways as in the mammal[14]. Some species may possess unique characteristics which deserve further study but no more. The abundance of glycine was mentioned; in other species, particularly among the molluscs, taurine is present in very high amounts as will be shown later. In some other species, other amino acids are abundant; in the hemolymph of the Orthoptera *Anacridium aegyptium* three amino acids accounted for 50% of the total free amino acid nitrogen, namely glycine, proline and glutamic acid[15]. In a survey of 17 different marine invertebrates SIMPSON *et al.*[11] found that most of the free α-amino acid nitrogen was distributed in a few amino acids, namely the three amino acids which form a link with the citric acid cycle: alanine, aspartic acid and glutamic acid (glutamine) and also glycine. Taurine was abundantly present in all marine forms (Fig. 1). The aromatic amino acids as well as the branched chain amino acids were present in small amounts. Proline was found in most of them in higher concentration than in most mammalian organs. ACKERMANN has analyzed a large number of invertebrates and invariably found large quantities of glycine in most of them. In the mussel *Mytilus edulis* he found a number of bases but glycine betaine was the most abundant of all [16]. Since that time, ACKERMANN *et al.* have isolated glycine betaine from other invertebrates along with a number of new nitrogenous compounds. From the king crab *(Limulus polyphemus)* they obtained nearly all known α-amino acids and a number of other related substances like glycine betaine, choline, ergothioneine and others[17]. From the sea snail *Patella sp.* he isolated also glycine betaine, choline, arginine, taurine and a number of other metabolites, some closely related to amino acids[18]. From a tunicate *Cionia intestinalis* they isolated glycine and taurine[19] and from the marine worm *Nereis virens* he isolated glycocyamine, choline, lysine, leucine, tyrosine and α-alanine[20]; he also showed the presence of large amounts of glycine. It is in fact tempting to ascribe an important function to glycine and its various derivatives: glycocyamine, glycine betaine and choline. It would be very interesting to study the biogenesis and catabolism of glycine in invertebrates where this amino acid is abundantly present. The occurrence of glycine betaine and choline in most invertebrates suggest active transmethylation, a reaction which has not been studied extensively among invertebrates. In addition to those methylated derivatives a new and unique one was found by ACKERMANN in a giant sponge, namely taurobetaine. Although taurobetaine has not been discovered in any other organism it shows that methylations must occur even in the lowly sponge and with methyl acceptors not recognized until now such as taurine. The role of taurobetaine is not known. Other unique betaines have been known before the discovery of taurobetaine; one such is butyrobetaine.

To sum up: There are some unique features in the distribution of non-protein nitrogen in invertebrates. Much of the non-protein nitrogen is present as α-amino acids and some derivatives. The concentration of free amino acids in invertebrates is very high as compared to the free amino acids of vertebrates. A marine environment does not account for this high concentration of free amino acids since fishes do not have more amino acids in the free form than mammals do. Whereas in mammalian organs there is a more even distribution of free amino acids, in invertebrates' organs some amino acids make up the largest portion of the free amino acid nitrogen. Glycine and, as will be discussed later, taurine are found in nearly all species of invertebrates and in very large amounts. Glycine betaine is also found in the tissues of many invertebrates suggesting that it is formed by transmethylation in which glycine participates. High levels of free amino acids, it has been suggested, is a manifestation of osmoregulatory mechanisms in some animals. Dietary factors could affect the distribution of free amino acids and related compounds but cannot account for a relatively uniform and unique pattern of distribution such as found in the crustacea discussed.

From the standpoint of comparative biochemistry it would be more profitable to study taxonomically related species under uniform laboratory conditions. The content and types of amino acids which are found in the invertebrates have given us many clues to fascinating problems. The next step is to study metabolic sequences and to compare rates of reactions in different species. The comparative study on urea formation by BROWN AND COHEN[21] is a good example of the interesting consequences of this type of study.

References p. 174/175

OCCURRENCE OF COMPOUNDS RELATED TO α-AMINO ACIDS IN INVERTEBRATES

Many compounds of nitrogen have been isolated from invertebrates; in some instances an important role has been ascribed to them as in the case of arginine phosphate which replaces creatine phosphate in a number of invertebrates; in other instances no role has been assigned to the various compounds discovered in invertebrates. This is the largest group and it appears that the more widely distributed the compound the less is known about its role. Such compounds as taurine for example are found in both vertebrates and invertebrates in high concentrations but no particular role has been assigned to it. A partial list of compounds of nitrogen isolated from invertebrates follows:

Taurine

Taurine was isolated from extracts of the clam *Mytilus edulis* by KELLY[22]. *M. edulis* contains very large quantities of taurine: as much as 4% of its fresh weight. HENZE[23]

$$H_2N—CH_2—CH_2—SO_3H$$

found it in the octopus; KOSSEL AND EDLBACHER[24] found it in some echinoderms;

TABLE V

TAURINE CONTENT IN VARIOUS MOLLUSCS

Mollusca	*Environment*	*Taurine*
Gastropoda		
Lymnaea palustris	Fresh water	—
Marisa cornuarietis	Fresh water	—
Pomacea bridgesi	Fresh water	—
Rumina decollata	Terrestrial	—
Otala lactea	Terrestrial	—
Mesodon thyroidus	Terrestrial	—
Bulimulus alternatus	Terrestrial	—
Murex fluvescens	Marine	+
Littorina irrorata	Marine	+
Oliva sayana	Marine	+
Polinices duplicata	Marine	+
Busycon perversum	Marine	+
Siphonaria lineolata	Marine	+
Fasciolaria distans	Marine	+
Thais haemastoma haysae	Marine	+
Pelecypoda		
Anadonta grandis	Fresh water	—
Quadrula quadrula	Fresh water	—
Lampsilis sp.	Fresh water	—
Elliptio sp.	Fresh water	—
Rangia cuneata	Brackish-fresh water	+
Brachiodontes recurvus	Brackish-marine	+
Crassostrea virginica	Brackish-marine	+
Donax variabilis	Marine	+
Venus mercenaria	Marine	+
Dosinia discus	Marine	+
Arca incongrua	Marine	+
Arca compechiensis	Marine	+
Noetia ponderosa	Marine	+
Cephalopoda		
Loliguncula brevis	Marine	+

References p. 174/175

MENDEL[25] in the muscle of a number of invertebrates, OKUDA[26] in some cephalopods and SCHMIDT AND WATSON[27] in the abalone *(Haliotus)*. SIMPSON, ALLEN AND AWAPARA[11] compared the taurine concentration of some marine molluscs with the taurine concentration of fresh water and terrestrial molluscs. As shown in Table V, none of the fresh-water or terrestrial molluscs had taurine, at least not enough to be detected by paper chromatography. All the marine molluscs had taurine and some in very high concentration (Table VI). In *Polinices duplicata* the concentration of taurine was as

TABLE VI

TAURINE CONCENTRATION IN SOME INVERTEBRATES
Values are expressed in μmoles/g fresh wt.

	Taurine concentration
Penaeus aztecus	23
Clibinarius vittatus	2
Pagurus pollicaris	36
Siphonaria lineolata	5
Fasciolaria distans	12
Busycon perversum	15
Thais haemastoma	16
Polinices duplicata	60
Oliva sayana	6
Arca umbonata	70
Volsella demisus	10
Crassostrea virginica	6
Lithophage bisulcata	11
Loliguncula brevis	55

high as 60 μmoles/g of fresh tissue. ALLEN AND AWAPARA[28] compared the rate of formation of taurine from cysteine in two molluscs: *Rangia cuneata,* a mollusc that lives in brackish water and is known to have traces of taurine and *Mytilus edulis,* a marine mollusc which as indicated before has a very high amount of taurine. Methods were developed to inject in the tissues of these molluscs small volumes of solution of [^{35}S]cysteine or methionine. Analyses of all ^{35}S-labeled compounds were then performed after various time-intervals. They found that taurine is formed from both cysteine and methionine. Sulfate was formed in both animals. Other intermediates detected were cysteic acid and cysteine sulfinic acid. Taurine formation from methionine, cystine or cysteine occurred by the same steps described for the rat[29]. Differences in rates of formation were observed but not sufficiently different to account for the discrepancy in their taurine content. We postulated that taurine is in fact formed in both organisms but for unknown reasons *R. cuneata* loses its taurine to the medium whereas *M. edulis* concentrates it in the muscle. A similar situation was observed by AWAPARA[30] who showed that taurine is concentrated and held for long periods of time by the heart muscle of the rat, but other organs did not concentrate taurine nor hold it for long periods of time. In one experiment we showed that *R. cuneata* cannot hold injected taurine. This is shown in Fig. 2. A similar experiment was not possible for *M. edulis* because the high concentration of endogenous taurine prevented us

References p. 174/175

from measuring injected taurine accurately. However, proof of its retention is furnished by its presence in such large amounts. We do not know with certainty whether taurine is formed by decarboxylation of cysteic acid or oxidation of hypotaurine. The latter intermediate, first found in rat livers by AWAPARA[31] has also been found in invertebrates[32, 33]. There is no doubt that taurine is a metabolic product and not an inert dietary component accumulated in the tissues of some animals. The role of this

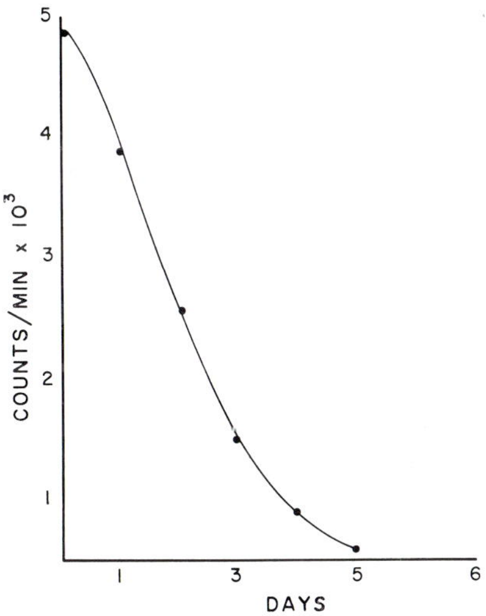

Fig. 2. Concentration of taurine in *R. Cuneata*.

simple compound remains obscure. In some species it serves as a precursor to other compounds for which a possible role can be ascribed.

Taurocyamine

Taurocyamine was found in polychaete worms by THOAI *et al.*[34], by ROBIN[35], ACKERMANN[36], GRIFFITHS, ENNOR AND MORRISON[37], and ABBOTT AND AWAPARA[38]. ROBIN[35] postulated that taurocyamine is formed from taurine but no evidence of its formation was shown. SCHRAM AND CROKAERT found taurocyamine in urine from human beings

$$H_2N{-}\overset{\overset{\displaystyle NH}{\|}}{C}{-}NH{-}CH_2{-}CH_2{-}SO_3H$$

and rats[39]. ACKERMANN[40] studied taurocyamine as a possible precursor of asterubin, the N,N-dimethyl derivative; asterubin was first isolated by ACKERMANN from two species of starfish: *Asterias ruben* L. and *Asterias glacialis* L. The process of biological methylation of taurocyamine was studied by ACKERMANN but he failed to show methylation of taurocyamine in the dog. Taurocyamine was then found as the phosphate in the polychaete worms which contain free taurocyamine. The formation of taurocyamine from [^{35}S]taurine was then investigated by ABBOTT AND AWAPARA in the lugworm *Arenicola cristata* Stimpson[38]. The administration of [^{35}S]methionine to

References p. 174/175

this animal gave rise to a series of ^{35}S-labeled intermediates which were separated by chromatography in ion-exchange columns, and identified. The metabolism of methionine and cysteine in *A. cristata* is the same, at least qualitatively, as in the mammal: methionine → homocysteine → cysteine → cysteine sulfinic acid → taurine and/or sulfate. The formation of taurocyamine could be shown *in vivo* but only when large amounts of very radioactive taurine were administered; it was separated by chromatography in ion-exchange resins as shown in Fig. 3, where the curve representing taurocyamine concentration measured colorimetrically overlaps the curve obtained by plotting values for radioactivity. A transamidination reaction with taurine as the acceptor could not be shown with a number of preparations from the lugworm. *N*-phosphoryl-taurocyamine is formed by a transphosphorylation from ATP and catalyzed by the enzyme taurocyamine phosphoryl transferase[41].

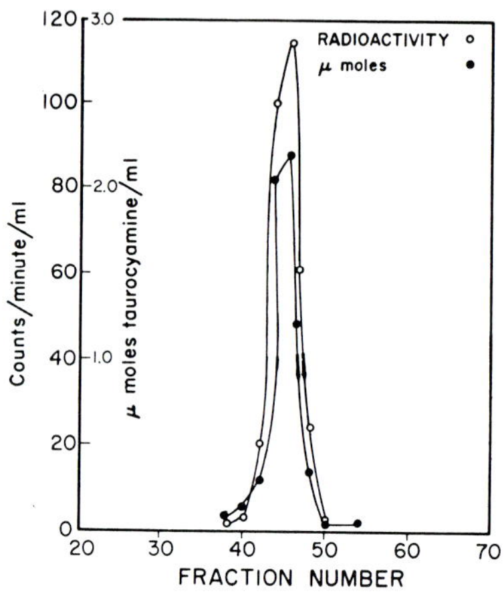

Fig. 3

Other taurine derivatives

In addition to taurocyamine and asterubin, a new taurine derivative has been reported by Ackermann[42], namely taurobetaine. The compound was obtained from a giant siliceous sponge found along the coast of Greece. This is the first time that such a derivative of taurine has been reported to exist in nature. Two other *N*-methylated taurines have been reported previously but not in animals. Mono- and dimethyltaurine have been obtained from red algae[43, 44]. This observation is interesting because it shows that taurine is not only widely distributed but it participates in a number of reactions, if we assume that all three methylated taurines are formed by a methylation reaction similar to that observed in mammals. No role has been assigned to monomethyltaurine, dimethyltaurine or taurobetaine. Since we have shown already that methionine is demethylated by some invertebrates it would be very interesting to study transmethylation in these organisms in which methylated derivatives appear.

References p. 174/175

β-Aminoisobutyric acid

β-Aminoisobutyric acid was identified in extracts from *Mytilus edulis* by AWAPARA AND ALLEN[45] and in a number of flatworms by CAMPBELL[96]. The same flatworms contained β-alanine indicating that the two β-amino acids were formed probably from pyrimidines as in the mammal[47–49]. CAMPBELL[50] investigated pyrimidine meta-

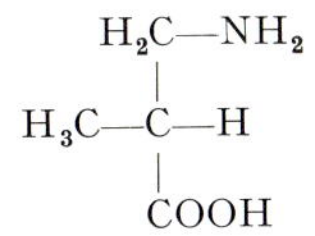

bolism in the parasitic flatworms *Hymenolepsis diminuta* and showed that uracil and thymine are degraded to β-alanine and β-aminoisobutyric acid respectively. The degradation of [2-^{14}C]uracil resulted in the formation of radioactive dihydrouracil, carbamyl-β-alanine, carbon dioxide and in addition α[1-^{14}C]alanine, some unidentified compounds and an organic acid which was tentatively identified as succinic acid. The presence of β-aminoisobutyric acid in the clam *M. edulis* also indicates that pyrimidines could break down in marine invertebrates by a similar route. Evidence for this has been obtained in our laboratory by CAMPBELL AND ALLEN. They showed formation of β-alanine in *Rangia cuneata* after injection of uracil, carbamyl-β-alanine and dihydrouracil.

Lombricine

The isolation of lombricine from the earthworm *Lumbricus terrestris* was first described by VAN THOAI AND ROBIN[51]. ROSENBERG AND ENNOR[52] described a simple method to isolate lombricine; they also found that the earthworm contained a phosphodiester of serine and ethanolamine; this phosphodiester (2-aminoethyl-2-amino-

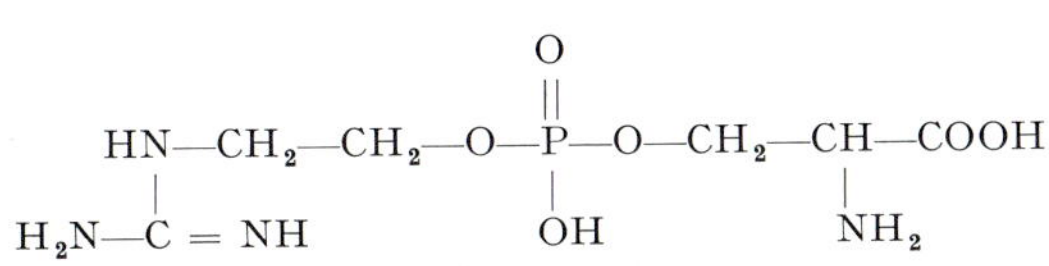

2-carboxyethyl hydrogen phosphate) was postulated as a precursor of lombricine. Such a phosphate ester was not new in the animal kingdom for ROBERTS AND LOWE[53] had already found it in the muscle of turtles. The serine in the ester from turtles was reported to be L-serine whereas in lombricine serine exists as the D-enantiomorph[54]. More work by ENNOR *et al.* permitted them to isolate a sufficient quantity of the phosphodiester of serine and ethanolamine to identify it by chemical and physical means. The serine portion of the molecule was D-serine. This fact provided sufficient evidence to favor its precursor role in the formation of lombricine.

The biosynthesis of lombricine was finally demonstrated by ROSSITER, GAFFNEY, ROSENBERG AND ENNOR[55]. Labeled precursors were administered to earthworms *(Megascolides cameroni)*: $^{32}P_i$, [1 : 2-^{14}C]ethanolamine, [3-^{14}C]serine and L-[^{14}C]-amidinoarginine. Lombricine was obtained by chromatography on filter paper and radioactivity determined; the serine ethanolamine phosphodiester was also separated by the same procedure and the specific activity determined. From the results obtained they concluded that the precursor of lombricine is ethanolamine serine phosphodiester which is converted to lombricine by a transamidination reaction in which arginine supplies the amidino group.

Histamine

Histamine is found in mammalian organs and it is formed by the decarboxylation of histidine; it has been reported that histamine is present in crustacea[56], in various parts of the bee[56] and also in bee poison and in coelenterates[56]. Most of these claims are based on biological assays and furnish proof only of the presence, in those organisms, of a histamine-like factor. ACKERMANN AND LIST[57] isolated histamine from the giant siliceous sponge *(Geodia gegas)* in substantial amounts so that chemical identification of the base was possible. They calculated that this organism contains approx. 45 mg of histamine/kg. Histamine has been found in the posterior salivary gland of a number of octopoda and HARTMAN *et al.*[58] reported that it is formed in that organ. They measured a number of amino acid decarboxylases in the posterior salivary glands of *Octopus apollyon,* and *Octopus bimaculatus* and found along with other amino acid decarboxylases an active histidine decarboxylase, thus giving proof that histamine is formed in the salivary gland of octopoda.

Octopine

This unique compound was first isolated from the muscle of octopus[59] and later identified in some cephalopods[60, 61] and in some lamellibranches[62, 62]. Many compounds of unique configuration like octopine have been discovered in invertebrates but only in a few cases has their biosynthesis been elucidated. In the case of octopine, due to the

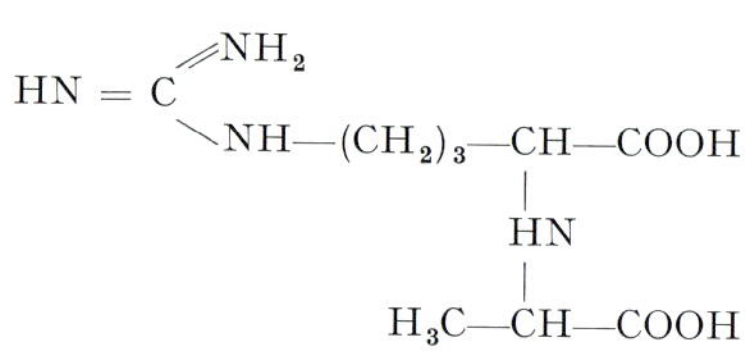

efforts of THOAI AND ROBIN[64] the mechanism of its formation has been clarified. It is formed from arginine and pyruvic acid by a process of reductive condensation involving a DPN-dependent dehydrogenase. They observed also that the enzymes are present only in muscles of animals containing octopine. The role ascribed to this compound by THOAI AND ROBIN is one of terminal receptor of pyruvic acid in glycolysis.

Other amines

Invertebrates contain a number of amines in addition to histamine. Most of them are closely related to amino acids which would indicate that amino acid decarboxylases must be active in these organisms. Many amines have been detected by paper chromatography. ERSPAMER *et al.*[65–67] analyzed by this method extracts from the salivary glands of octopoda and found enteramine, tyramine, octopine, histamine and other similar compounds. WELSH[68] and WELSH AND MOORHEAD[69] surveyed the distribution

References p. 174/175

of 5-hydroxytryptamine in tissues of invertebrates. The latter is formed by the decarboxylation of 5-hydroxytryptophane. Spermine has been isolated by ACKERMANN[70] from the tunicate *Ciona intestinalis*.

An interesting compound, N-acetyltyramine has been isolated from the silkworm *Bombyx mori* by BUTENANDT *et al.*[71].

SIMPSON[72] using a method for separating amines from amino acids[73] detected a number of amines in extracts of all 17 marine species studied but none were identified.

WHAT IS THE MAJOR SOURCE OF FREE AMINO ACIDS IN AQUATIC INVERTEBRATES?

In Figs. 1–5 are shown chromatograms of extracts from whole animals. It is immediately apparent that there are great differences in patterns. As stated in the preceding discussion, glycine was very abundant in crustaceans and taurine in marine molluscs. This rather uneven distribution cannot be explained on the basis of diet. Many of the organisms studied have similar feeding habits and were obtained from the same area; yet, their free amino acid pattern is quite different. Marine gastropods have abundant quantities of taurine, but their terrestrial counterparts have none. In Fig. 7 one can see that the land snails have a relatively similar pattern. Glycine is present in very small amounts or not present; taurine is absent; glutamic acid is present in relatively higher concentration and so is aspartic acid. There are no detectable amounts of branched chain or aromatic amino acids (in the range studied which is 50 mg of tissue per chromatogram). The land snail *Otala lactea* was analyzed after 6 months of starvation and compared with specimens of fed snails. The chromatograms are shown in Fig. 7 and it is apparent that the differences are relatively small. In the same Fig. are shown chromatograms of a herbivorous snail. The carnivorous snail *Euglandina singlyana* feeds on the herbivorous snail *Bulimulus*. There is relatively little difference in the two patterns.

It should be interesting to study sulfur metabolism in land snails. The pathway of methionine and cysteine metabolism are similar in nearly all organisms studied and there is ample evidence to show that taurine is formed in the marine gastropods. Land gastropods must metabolize sulfur in a different manner and it should be studied.

Basically the biochemical events occurring in invertebrates must be the same or very similar to those occurring in the mammal. But the differences which one finds, however small, are interesting clues to biochemical evolution. Amino acids are an important group of compounds and the patterns, or "finger prints" as ROBERTS calls them are expressions of their metabolic activities. I believe that there is sufficient evidence to support the view that these "finger prints" are not fortuitous, that they do not represent dietary habits, but that they represent a picture of the metabolic activities of these organisms. They represent the balance between formation or accumulation of these substances, and disposition of them by catabolic reactions or by some excretory process. It has been suggested that free amino acids maintain osmotic balance in marine invertebrates. This is probably true, but this is only one function. Is this an adaptation to a marine environment? The question needs to be answered. Before it can be answered we need to know far more about metabolism of amino acids in invertebrates.

References p. 174/175

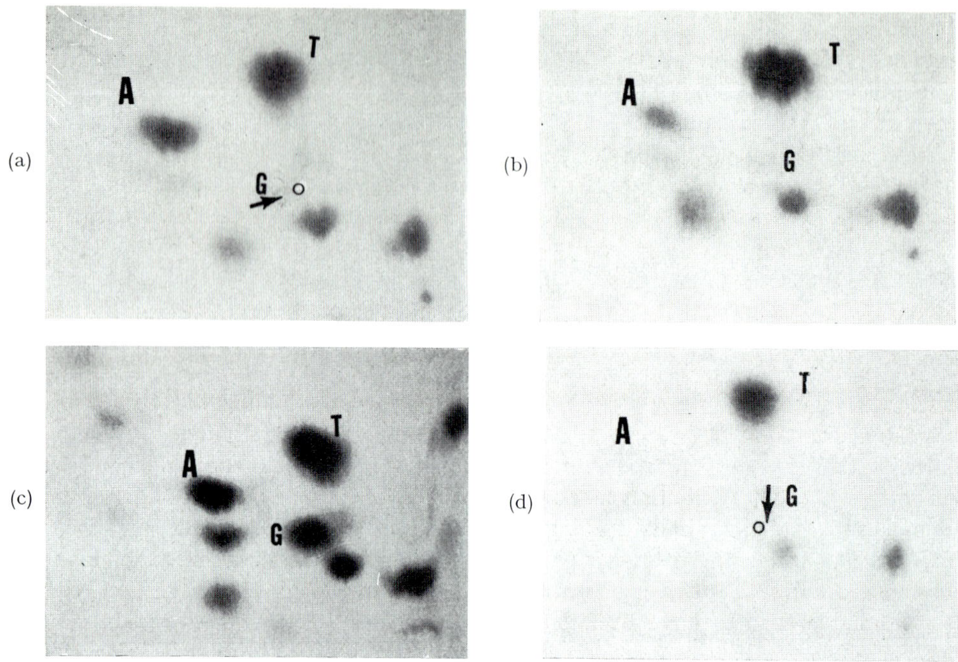

Fig. 4. Marine gastropods. (a) Moon shell *(Polinices duplicatus)*; (b) Left handed whelk *(Busycon perversum)*; (c) Oyster drill *(Thais haemostoma)*; (d) Slipper shell *(Crepidula fornicata)*. G = Glycine, T = Taurine, A = Alanine.

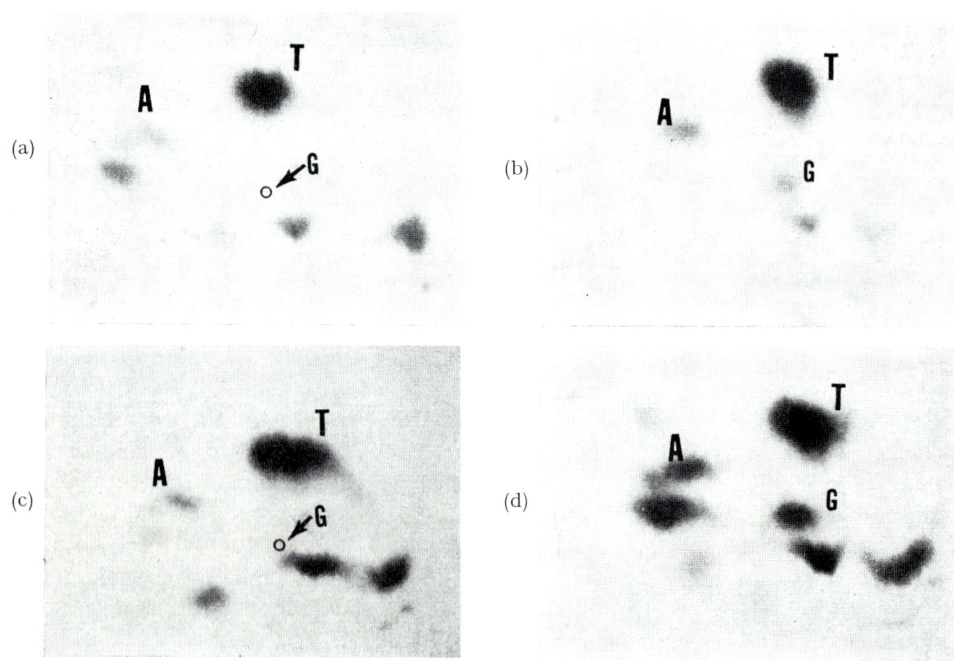

Fig. 5. Pelecypods. (a) Incongruous ark *(Arca incongrua)*; (b) Oyster *(Crassostrea virginica)*; (c) Bloody ark *(Arca campechiensis)*; (d) Ponderous ark *(Noetia ponderosa)*. G = Glycine, T = Taurine, A = Alanine.

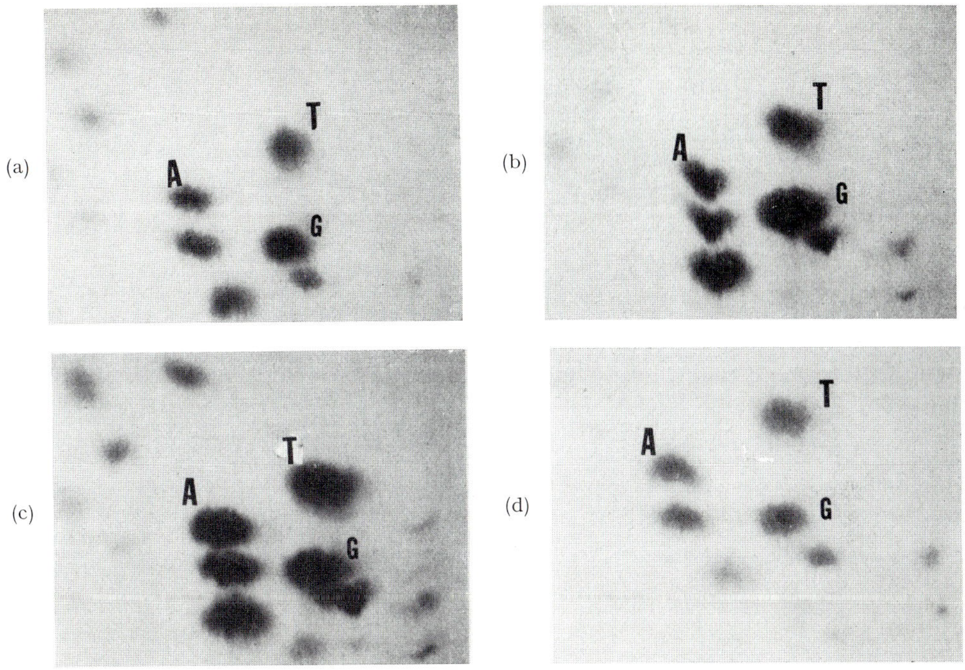

Fig. 6. Marine crustacea. (a) Stone crab *(Neopanope taxana)*; (b) Pistol shrimp *(Crangon heterochelis)*; (c) Hermit crab *(Pagurus pollicaris)*; (d) Hermit crab *(Clibinarius vittatus)*. G = Glycine, T = Taurine, A = Alanine.

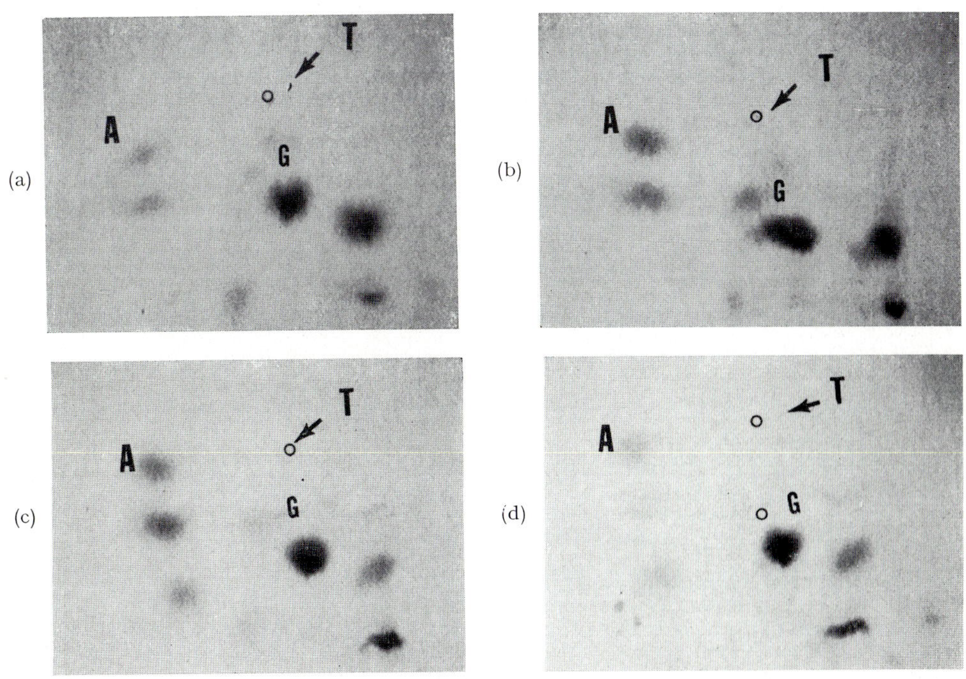

Fig. 7. Land Gastropods. (a) Land snail *(Otala lactea*, fed*)*, (b) *Otala lactea* (starved for 6 months), (c) Herbivorous land snail *(Bulimulus alternatus)*, (d) Carnivorous land snail *(Euglandina singlyana)*. G = Glycine, T = Taurine, A = Alanine.

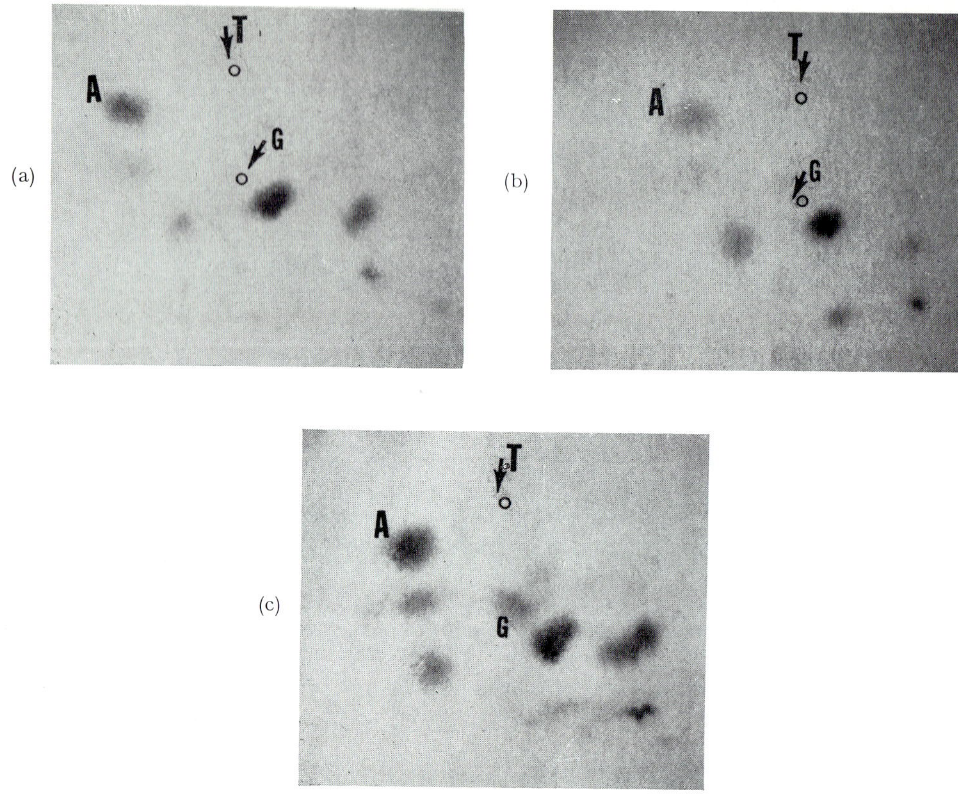

Fig. 8. Fresh water and land molluscs. (a) Fresh water snail *(Helisoma trivolvis)*, (b) Fresh water clam (*Elliptio*, sp), (c) Slug *(Limax flavus)*. G = Glycine, T = Taurine, A = Alanine.

REFERENCES

1 M. N. Camien, H. Sarlet, G. Duchâteau and M. Florkin, *J. Biol. Chem.*, 193 (1951) 881.
2 H. H. Tallan, S. Moore and W. H. Stein, *J. Biol. Chem.*, 211 (1954) 927.
3 F. Friedberg and D. M. Greenberg, *J. Biol. Chem.*, 168 (1947) 405.
4 H. Fredericq, Z. M. Bacq and M. Florkin, *Arch. intern. physiol. et biochem.*, 62 (1954) 205
5 W. O. Kermack, H. Lees and J. D. Wood, *Biochem. J.*, 60 (1955) 424.
6 N. K. Velankar and T. K. Govindan, *Current Sci. (India)*, 26 (1957) 285.
7 N. K. Velankar and T. K. Govindan, *Proc. Indian Acad. Sci., B*, 47 (1958) 202.
8 Gh. Duchâteau, H. Sarlet, M. N. Camien and M. Florkin, *Arch. intern. physiol. et biochem.*, 60 (1952) 124.
9 Gh. Duchâteau, M. Florkin and Ch. Jeuniaux, *Arch. intern. physiol. et biochem.*, 67 (1959) 489.
10 J. W. Campbell, *Exptl. Parasitol.*, 9 (1960) 1.
11 J. W. Simpson, K. Allen and J. Awapara, *Biol. Bull.*, 117 (1959) 371.
12 M. F. Giordano, H. A. Harper and F. P. Filice, *Wasmann J. Biol.*, 8 (1950) 1.
13 M. F. Giordano, H. A. Harper and F. P. Filice, *Wasmann J. Biol.*, 8 (1950) 129.
14 K. Ramamoorthi, *J. Animal Morphol. and Physiol.*, 5 (1958) 34.
15 C. A. Benassi, G. Colombo and G. Peretti, *Experientia*, 15 (1959) 457.
16 D. Ackermann, *Z. Biol.*, 74 (1921) 67.
17 D. Ackermann and P. H. List, *Z. physiol. Chem., Hoppe-Seyler's*, 313 (1950) 30.
18 D. Ackermann and R. Janka, *Z. physiol. Chem., Hoppe-Seyler's*, 298 (1954) 65.

[19] D. Ackermann and R. Janka, *Z. physiol. Chem., Hoppe-Seyler's*, 296 (1954) 283.
[20] D. Ackermann, *Z. physiol. Chem., Hoppe-Seyler's*, 299 (1955) 186.
[21] G. W. Brown, Jr., and P. P. Cohen, *Biochem. J.*, 75 (1960) 82.
[22] A. Kelly, *Beitr. Zeitschr. Chem. Physiol. Pathol.*, 5 (1904) 377.
[23] M. Henze, *Z. physiol. Chem., Hoppe-Seyler's*, 43 (1905) 477.
[24] A. Kossel and S. Edlbacher, *Z. physiol. Chem., Hoppe-Seyler's*, 94 (1915) 1951.
[25] L. B. Mendel, *Beitr. Zeitschr. Chem. Physiol. Pathol.*, 5 (1904) 582.
[26] Y. Okuda, *Chem. Abstr.*, 14 (1920) 2827.
[27] C. L. A. Schmidt and T. Watson, *J. Biol. Chem.*, 33 (1918) 499.
[28] K. Allen and J. Awapara, *Biol. Bull.*, 118 (1960) 173.
[29] J. Awapara and W. J. Wingo, *J. Biol. Chem.*, 203 (1953) 189.
[30] J. Awapara, *J. Biol. Chem.*, 225 (1957) 877.
[31] J. Awapara, *J. Biol. Chem.*, 203 (1953) 183.
[32] S. Schibuya and S. Shuniji, *Nature*, 180 (1957) 549.
[33] S. Ouchi, *J. Biochem.*, 46 (1959) 765.
[34] N. V. Thoai, J. Roche, and Y. Robin, *Biochim. Biophys. Acta*, 11 (1953) 593.
[35] Y. Robin, *D. Sc. Thesis*, College de France, 1954.
[36] D. Ackermann, *Z. physiol. Chem., Hoppe-Seyler's*, 302 (1955) 80.
[37] D. E. Griffiths, A. H. Ennor and J. F. Morrison, cited in A. H. Ennor and J. F. Morrison, *Physiol. Revs.*, 38 (1958) 631.
[38] W. Abbott and J. Awapara, *Biol. Bull.*, 119 (1960) 357.
[39] E. Schram and R. Crokaert, *Bull. soc. chim. biol.*, 39 (1957) 561.
[40] D. Ackermann, *Z. physiol. Chem., Hoppe-Seyler's*, 239 (1936) 231.
[41] N. V. Thoai, *Bull. soc. chim. biol.*, 39 (1957) 197.
[42] D. Ackermann, *Z. physiol. Chem., Hoppe-Seyler's*, 232 (1935) 206.
[43] B. Lindberg, *Acta Chem. Scand.*, 9 (1955) 1093.
[44] B. Lindberg, *Acta Chem. Scand.*, 9 (1955) 1323.
[45] J. Awapara and K. Allen, *Science*, 130 (1959) 1250.
[46] J. W. Campbell, *Biol. Bull.*, 119 (1960) 75.
[47] K. Fink, *J. Biol. Chem.*, 218 (1956) 9.
[48] K. Fink, R. E. Cline, R. B. Henderson and R. M. Fink, *J. Biol. Chem.*, 221 (1956) 425.
[49] S. Grisolia and S. S. Cardoso, *Biochim. Biophys. Acta*, 25 (1957) 430.
[50] J. W. Campbell, *Biochem. J.*, 77 (1960) 105.
[51] N. V. Thoai and G. Robin, *Biochim. Biophys. Acta*, 14 (1954) 76.
[52] H. Rosenberg and A. H. Ennor, *Biochem. J.*, 73 (1959) 521.
[53] E. Roberts and T. P. Lowe, *J. Biol. Chem.*, 211 (1954) 1.
[54] I. M. Beatty, D. I. Magrath and A. H. Ennor, *Nature*, 183 (1959) 591.
[55] R. J. Rossiter, T. Gaffney, H. Rosenberg and A. H. Ennor, *Nature*, 185 (1960) 383.
[56] M. Guggenheim, *Die Biogene Amine*, S. Karger, New York, 1951, p. 451.
[57] D. Ackermann and P. H. List, *Z. physiol. Chem., Hoppe-Seyler's*, 308 (1957) 274.
[58] W. J. Hartman, W. G. Clark, S. D. Cyr, A. L. Jordon and R. A. Leibhold, *Ann. N. Y. Acad. Sci.*, 90 (1960) 637.
[59] K. Morizawa, *Acta Schol. Med. Univ. Imp. Kioto*, 10 (1927) 281.
[60] D. Ackermann and M. Mohr, *Z. physiol. Chem., Hoppe-Seyler's*, 250 (1937) 249.
[61] J. Roche, N. V. Thoai, Y. Robin, I. Garcia and J. L. Hatt, *Compt. rend. soc. biol.*, 146 (1952) 1899.
[62] M. Mayeda, *Acta Schol. Med. Univ. Imp. Kioto*, 18 (1936) 218.
[63] E. Moore and D. W. Wilson, *J. Biol. Chem.*, 119 (1937) 573.
[64] N. V. Thoai and Y. Robin, *Biochim. Biophys. Acta*, 35 (1959) 446.
[65] V. Erspamer, *Experientia*, 6 (1950) 348.
[66] V. Erspamer and B. Asero, *J. Biol. Chem.*, 200 (1953) 311.
[67] V. Erspamer and V. Boretti, *Arch. intern. pharmacodynamie*, 88 (1951) 296.
[68] J. H. Welsh, *Federation Proc.*, 13 (1954) 162.
[69] J. H. Welsh and M. Moorhead, *J. Neurochem.*, 6 (1960) 146.
[70] D. Ackermann and R. Janka, *Z. physiol. Chem., Hoppe-Seyler's*, 296 (1954) 279.
[71] A. Butenandt, U. Gröschel, P. Karlson and W. Zillig, *Arch. Biochem. Biophys.*, 83 (1959) 76.
[72] J. W. Simpson, *M. A. Thesis*, Rice University, 1959.
[73] J. Awapara, O. Graham and V. E. Davis, *J. Chromatog.*, 3 (1960) 11.

INVITED DISCUSSION

FREE AMINO ACIDS OF MARINE INVERTEBRATES

JAMES S. KITTREDGE, DAISY G. SIMONSEN, EUGENE ROBERTS AND BOHDAN JELINEK

Department of Biochemistry, Medical Research Institute, City of Hope Medical Center, Duarte, Calif. (U.S.A.)

We would like to make a brief addendum to Dr. Awapara's remarks.

Paradoxically, the comparative biochemistry of marine invertebrates has been handicapped by two very fortunate circumstances. The first is man's position on the evolutionary scale, of which the anthropocentric concentration on the Chordata is an understandable consequence. The second is the basic similarity of modes of intracellular metabolism in all forms of life. This similarity, without the recognition of which biochemistry today probably would resemble pre-LINNAEAN zoology, has permitted biochemists to skip over the entire evolutionary scale and to utilize the microorganisms as an expedient tool in biochemical investigations.

All of us are on the alert for deviations from this similarity such as occur in mutants of microorganisms or are reflected in some organ differences in mammals. Few have attempted to utilize the wealth of "natural mutants" available. In fact, most biochemists are only vaguely aware of the extent of this plethora. There are 17 phyla of animals, yet we commonly utilize less than a dozen species of one phylum. How many

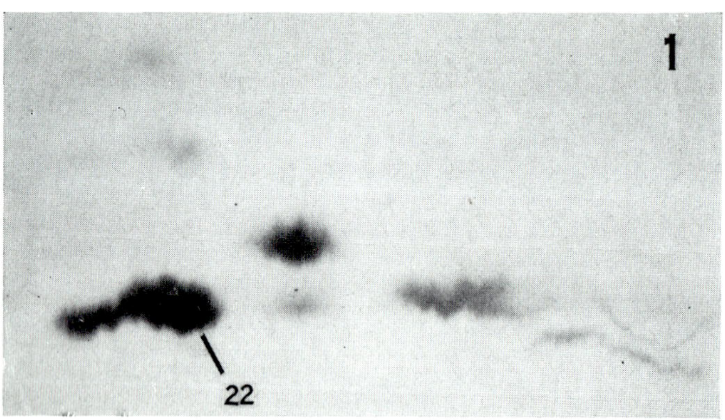

Fig. 1. A dinoflagellate, *Gonyaulax polyhedra*. γ-Aminobutyric acid, 22.

excursions into "comparative biochemistry" have been limited to this rather small phylum? Of the other 16 phyla, six are confined exclusively to the sea and five other phyla are almost entirely marine. Why the diversity of marine forms? Probably the extremely uniform environmental conditions in the sea have permitted the development and survival of many divergent branches of the evolutionary experiments, although only a few have mastered the rigors of terrestrial life.

References p. 186

Primarily as a guide for more detailed investigations, we have conducted a preliminary survey of the free amino acids of the major marine phyla. All of the specimens were collected on the coast of San Diego county (Calif., U.S.A.), during the months of July and August. In the following discussion the designation "free amino acids" will be applied to those ninhydrin-reactive materials which are extractable by 80% ethanol and which can be detected on two-dimensional paper chromatograms. An overall examination of the results confirms and expands the range of previous observations[1–15].

High taurine concentrations were observed in many species. *Mytilus edulis*, the

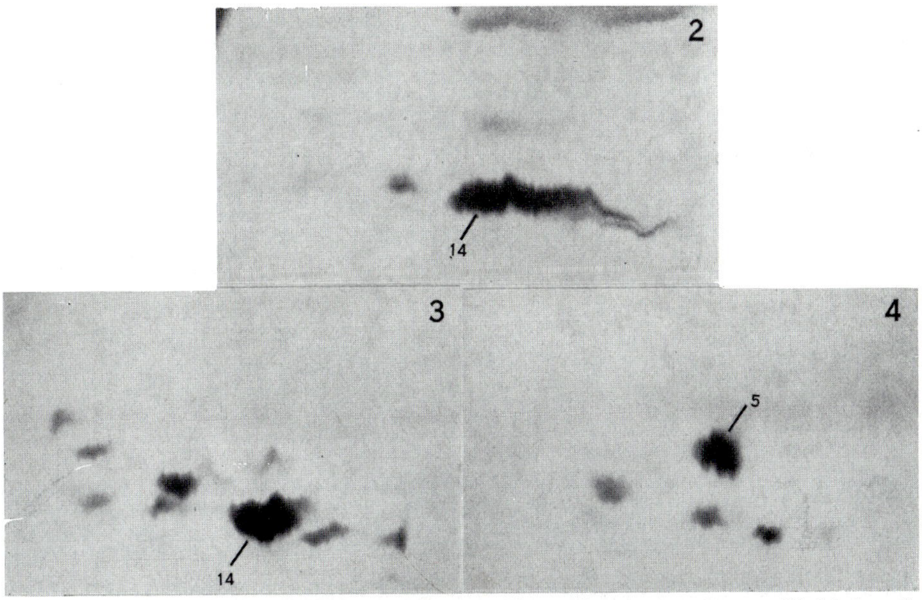

Figs. 2–4. Fig. 2, a Siliceous sponge, *Stelletta* sp; Fig. 3, Calcareous sponges, *Rhabdodermella nuttingi*; Fig. 4, *Xestospongia vanilla*. Taurine, 5; glycine, 14.

mussel, has been cited as the classic example of a marine organism with a high taurine concentration[12, 16]. The high taurine concentration in the abalone, *Haliotis*, was observed at an equally early date[17]. Both of these species, and the octopus which also has been proposed as a commercial source of taurine[18, 19], exhibited the expected high levels of this amino acid. However, the relative content of taurine is even higher in some other species, *e.g.*, *Renilla kollikeri*, the sea pansy. As observed by previous investigators[5, 8, 12], in many species the high taurine concentration was paired with an equally high glycine content. In some instances glycine alone was dominant[7–12], and occasionally alanine was a major component. The expected high arginine concentrations[5, 12] were evident throughout the Crustacea. This probably reflects the breakdown of the phosphogen arginine phosphate.

Little phylogenetic correlation is to be expected in such an extensive survey. Here we may profitably reflect on the observations of botanical biochemists who have found that phylogenetic correlations usually become more evident in intensive investigations rather than extensive surveys. Another frequent observation is that the parallel

References p. 186

evolution of similar metabolic characteristics is not at all uncommon[21]. Often there was found to be a greater difference between the free amino acid patterns of the various tissues of a given species than between the same tissue of related species[4]. An example of this was the observation of a marked similarity of the free amino acid patterns of the radular muscles throughout a number of the Gastropoda (snails). An exception to the lack of distinctive phylogenetic patterns is the occasional occurrence of a specific compound. Where the ability to synthesize a new compound occurred early in the evolution of a major group, this capacity may be maintained throughout its evolution and morphological diversification.

At present we are examining two new amino acids occuring in the Coelenterata. A point of taxonomic interest is the marked difference in the concentration of these two new compounds, relative to the other free amino acids, in the two closely related sea

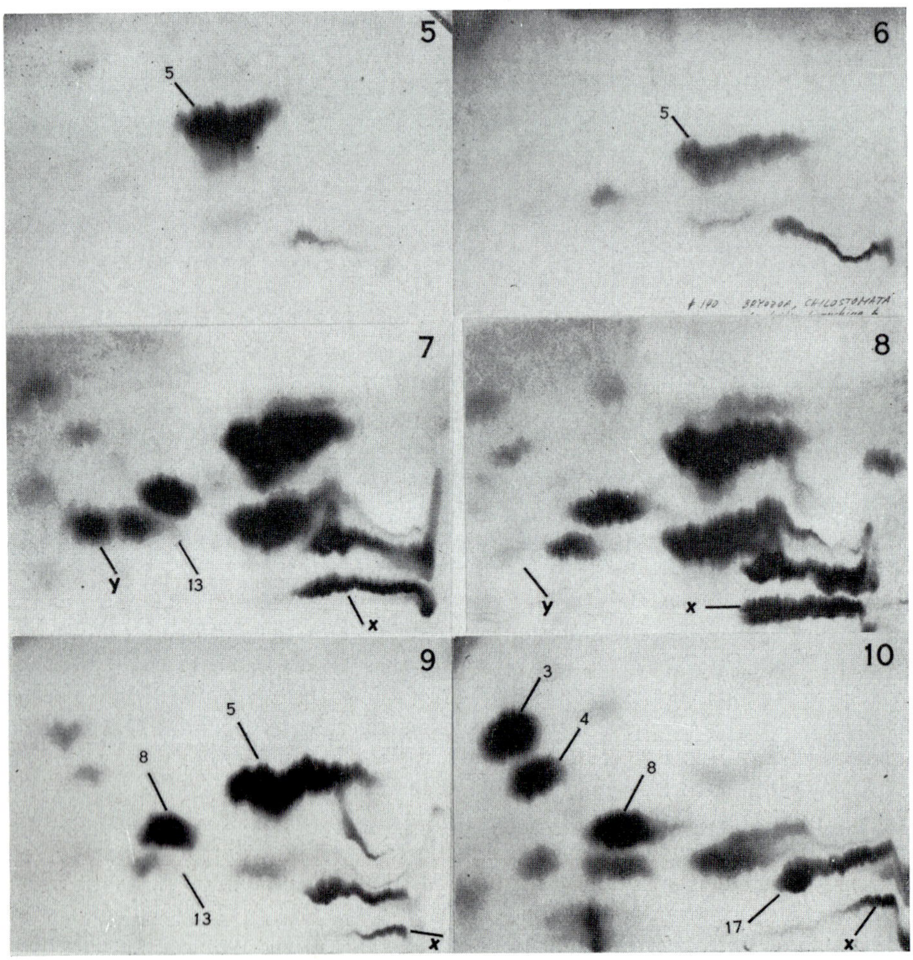

Figs. 5–10. Fig. 5, the sea pansy, *Renilla kollikeri*; Fig. 6, a Bryozoan, *Chilostomata* sp; Fig. 7, sea anemones, *Anthopleura elegantissima*; Fig. 8, *A. elegantissima*, hydrolyzed extract; Fig. 9, *Metridium senile*; Fig. 10, *Corynactis californica*. Leucines, 3; valine, 4; taurine, 5; alanine, 8; glutamine, 13; glutamic acid, 17; unknown amino acids, X and Y.

References p. 186

anemones, *Anthopleura elegantissima* and *A. xanthogrammica*. Many taxonomists of the Actiniaria consider these two to be the same species[21], *A. xanthogrammica* being the solitary form and *A. elegantissima* the aggregate form. The free amino acid patterns provide some biochemical evidence supporting the validity of their taxonomic segregation.

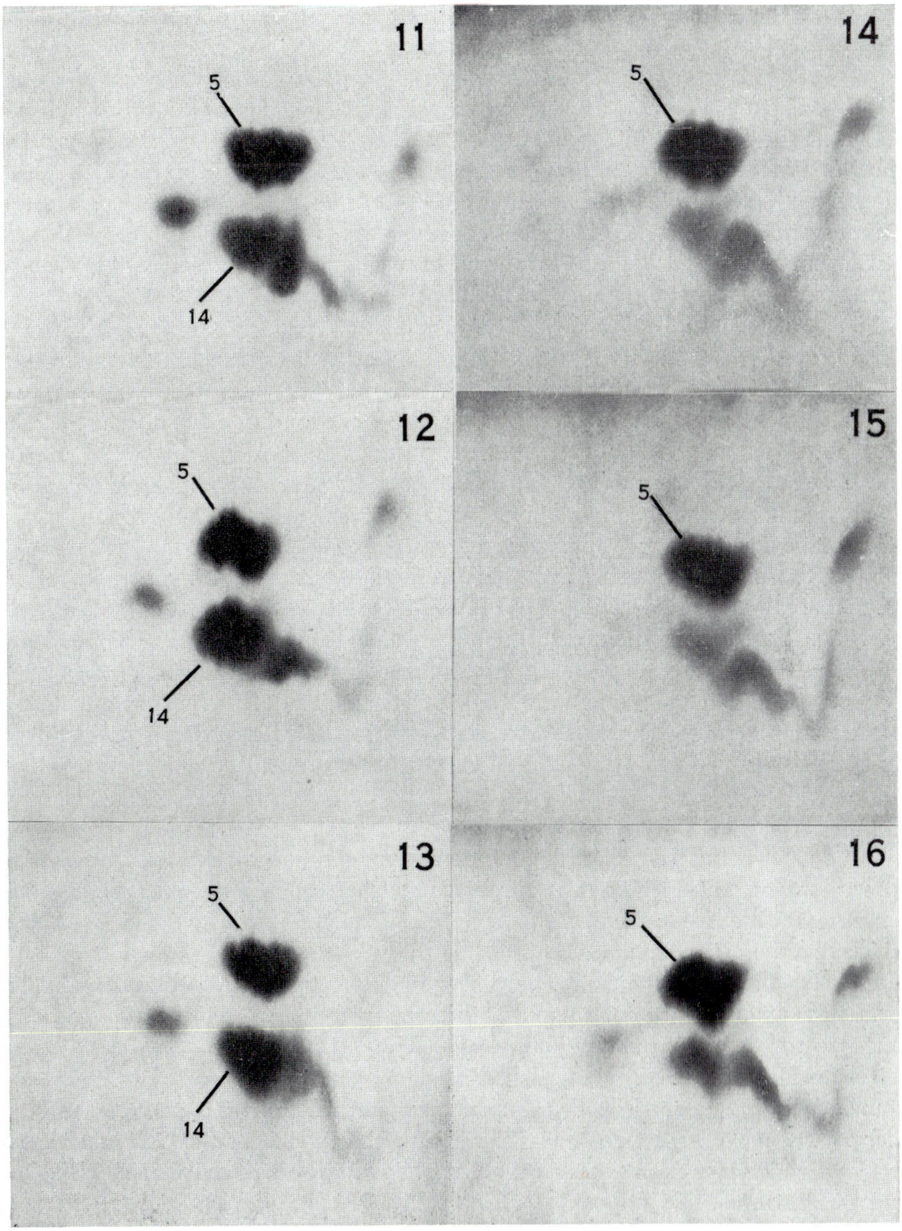

Figs. 11–16. A brachiopod, *Terebratella transversa*. Figs. 11–13, muscle; Figs. 14–16, lophophore and gills. Taurine, 5; glycine, 14.

References p. 186

Below we will present a few of our observations. All of the chromatograms, except where noted, were run with applications equivalent to 75 mg of fresh weight of tissue.

Gonyaulax polyhedra, a dinoflagellate (Fig. 1). Many members of this class, the Mastigophora, have the capacity to carry on photosynthesis and are included among the plants by the biochemists, and since many are heterotrophic and all are motile, the subclass Phytostigmata is placed among the protozoa by the zoologists[22]. The high concentration of γ-aminobutyric acid, an amino acid occurring frequently in plants[20], but only in the central nervous system of animals[23], suggests a stronger relationship to the plants.

Chromatograms of a Siliceous sponge, *Stelletta* sp. (Fig. 2) and two Calcareous sponges, *Rhabdodermella nuttingi* (Fig. 3) and *Xestospongia vanilla* (Fig. 4), show high glycine in the Siliceous sponge and one Calcareous sponge, but a high taurine concentration in the other Calcareous sponge.

The chromatograms in Figs. 5 and 6 show two examples of relatively high taurine contents, the sea pansy, *Renilla kollikeri* and the bryozoan, *Chilostomata* sp.

Fig. 7 is presented to show the two new compounds found in *Anthopleura (Cribina) elegantissima*, a sea anemone, which we are investigating. The chromatogram of an acid hydrolyzed extract (Fig. 8) shows that spot "Y" disappeared and an increase occurred in the spot "X", suggesting that "Y" may contain a bound form of "X". Chromatograms of extracts of two other sea anemones, *Metridium senile* (Fig. 9), and *Corynactis californica* (Fig. 10), both showed some "X". *M. senile* contains a preponderance of taurine and alanine, and *C. californica* shows high concentrations of glutamic acid, alanine, valine and the leucines to be present, but a relatively low level of taurine. Glutamine was absent from the extracts of *A. elegantissima* and *M. senile.*

In Figs. 11–16 chromatograms are shown of the free amino acids of the brachiopod, *Terebratella transversa*. Figs. 11–13 are the results from muscle tissue, Figs. 14–16 extracts of the lophophore and gills. The reproducibility of the pattern for a given tissue from several specimens, as well as the contrast between different tissues of the same species is evident. The muscle shows high taurine and glycine concentrations, while the preponderance of taurine is evident in the lophophore and gills. Glutamine is low or absent in this species also.

Fig. 17 shows the pattern of free amino acids found in a sabellid worm, *Sabellaria* sp. with an interesting unknown spot above taurine. The peanut worm, *Dendrostoma zostericolum* (Fig. 18), exhibited high concentrations of aspartic acid and a substance which may be ethanolamine phosphate or the phosphodiester of ethanolamine and serine, also a small amount of tyrosine-*O*-sulfate and several unidentified ninhydrin-reactive substances. Chromatograms of the gooseneck barnacle, *Mitella polymerus* (Figs. 19–21), exhibited two unknown spots above taurine.

The results in Fig. 22–25 are presented to show an ecological correlation. Chromatograms of extracts of the isopod, *Ligyda occidentalis*, which lives in the spray zone on rocky cliffs above high tide level (Fig. 22), of the beach hopper, *Orchistoidea californiana*, which occupies a similar niche at the high tide level of sandy beaches (Fig. 23), and two terrestrial Arthropoda, the tarantula, *Acanthrophrynus coronatus* (Fig. 24), and the scorpion, *Hadrurus hirsutus* (Fig. 25) are shown for comparison. Extremely high concentrations of several free amino acids were found in the two marine Arthro-

References p. 186

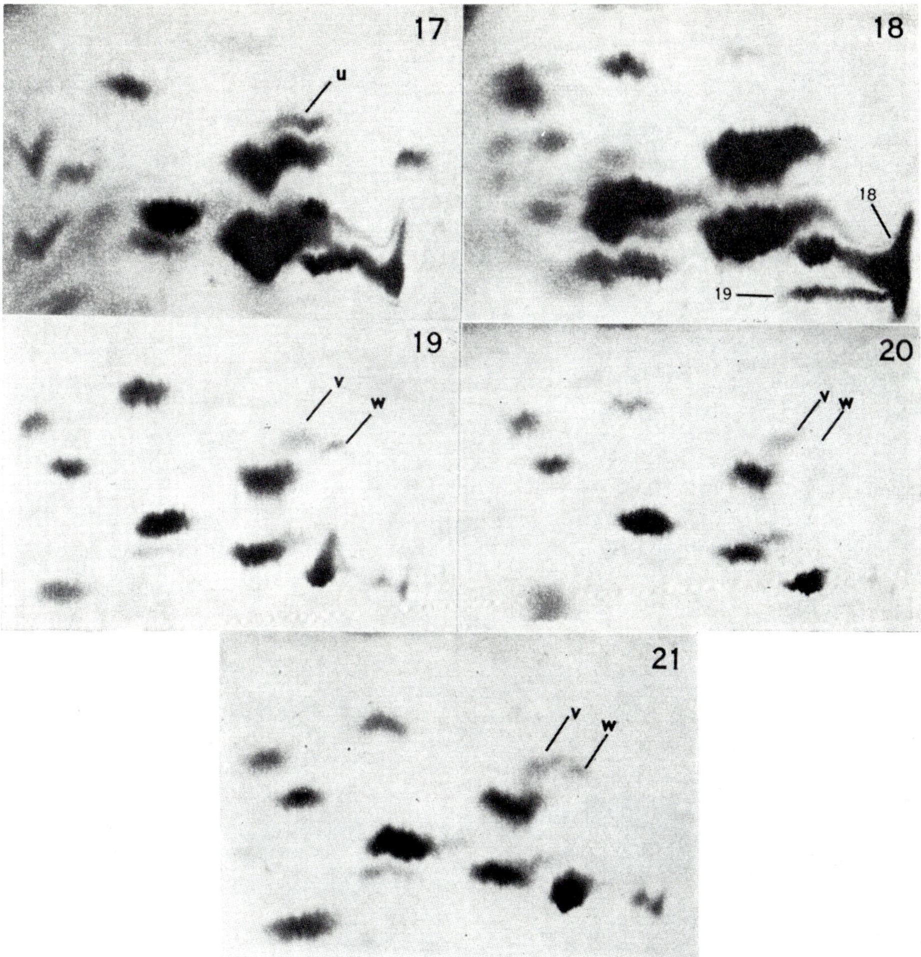

Figs. 17–21. Fig. 17, a sabellid worm, *Sabellaria* sp; Fig. 18, a peanut worm, *Dendrostoma zostericolum*; Figs. 19–21, a gooseneck barnacle, *Mitella polymerus*. Aspartic acid, 18; ethanolamine phosphate or the phosphodiester of ethanolamine and serine, 19; tyrosine-*O*-sulfate, 27; unknown amino acids, U, V, and W.

poda which occupy "semiterrestrial" niches, their amino acid patterns more closely resembling those of terrestrial Arthropoda.

Examples of chromatograms of extracts from representative marine Crustacea are those of a shrimp, *Spirontocaris picta* (Fig. 26), a crab, *Pachygrapsus crassipes* (Fig. 27), and the spiny lobster, *Panulirus interruptus* (nerve) (Fig. 28). All show high levels of taurine, glycine and arginine, the shrimp also exhibiting tyrosine-*O*-sulfate and the crab and lobster high lysine concentrations. In addition, the chromatogram from the lobster nerve shows aspartic acid to be a major free amino acid in this tissue. The crab selected, *P. crassipes*, occupies an ecological niche intermediate between the "semiterrestrial" *L. occidentalis* and *O. californiana* and the subtidal *S. picta* and *P. interruptus*. It prefers to remain above the water level around tide pools. The free amino acid pattern is suggestive of this intermediate position.

References p. 186

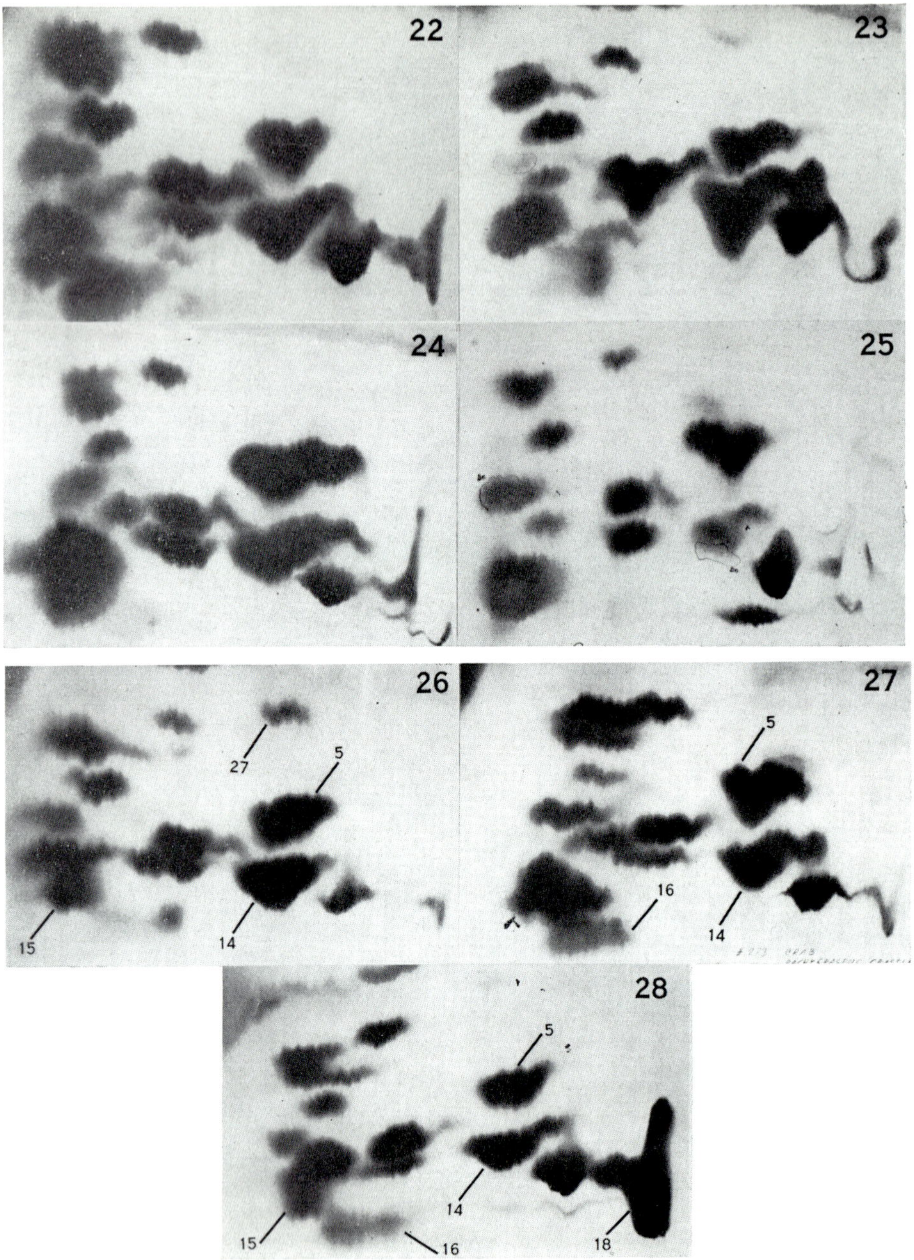

Figs. 22–28. Fig. 22, an isopod, *Ligyda occidentalis*; Fig. 23, a beach hopper, *Orchistoidea californiana*; Fig. 24, a tarantula, *Acanthrophrynus coronatus*; Fig. 25, a scorpion, *Hadrurus hirsutus*. Fig. 26, a shrimp, *Spirontocaris picta*; Fig. 27, a crab, *Pachygrapsus crassipes*; Fig. 28, a spiny lobster, *Panulirus interruptus*. Taurine, 5; glycine, 14; arginine, 15; lysine, 16; aspartic acid, 18; tyrosine-*O*-sulfate, 27.

References p. 186

Chromatograms of extracts from various tissues of the green abalone, *Haliotis fulgens*, illustrate the differences between the various tissues in this species. The white muscle (Fig. 29) had large amounts of taurine and arginine. The male gonadal tissues (Fig. 30) were extremely rich in taurine and glycine, while glutamine was not evident at all on the chromatograms. Taurine was the major ninhydrin-reactive constituent detected in the digestive diverticulum (Fig. 31).

Amino acid patterns of various tissues of the octopus, *Octopus bimaculatus*, in this case illustrating considerable constancy from one tissue to another (Figs. 32–34), show a high level of taurine and the presence of an unidentified compound near glutamic acid which gives a yellow color with ninhydrin. Other tissues showed different patterns: the brain and eyes had high levels of aspartic acid (which also was high in the lobster nerve); the gonads of glutamine; the ink glands of valine, the leucines and tyrosine; the digestive glands of arginine.

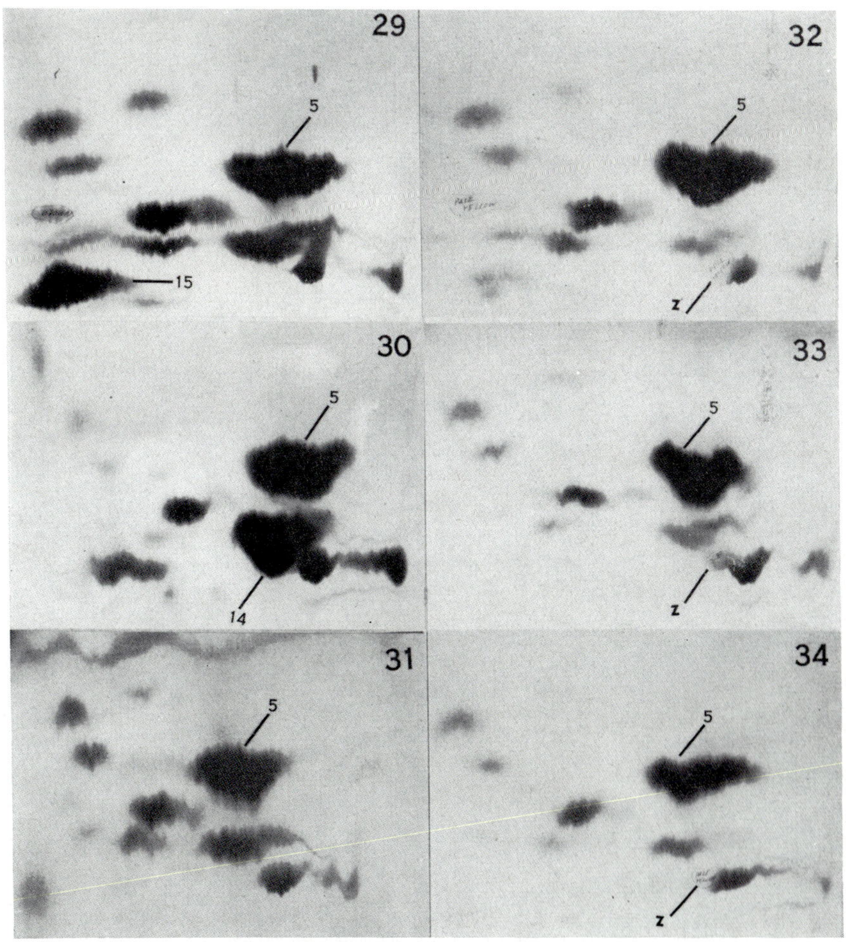

Figs. 29–34. Green abalone, *Haliotis fulgens*. Fig. 29, muscle; Fig. 30, male gonadal tissues; Fig. 31, digestive diverticulum. Octopus, *Octopus bimaculutus*. Fig. 32, leg muscle; Fig. 33, gill; Fig. 34, excretory tissue. Taurine, 5; glycine, 14; arginine, 15; unknown amino acid, Z.

References p. 186

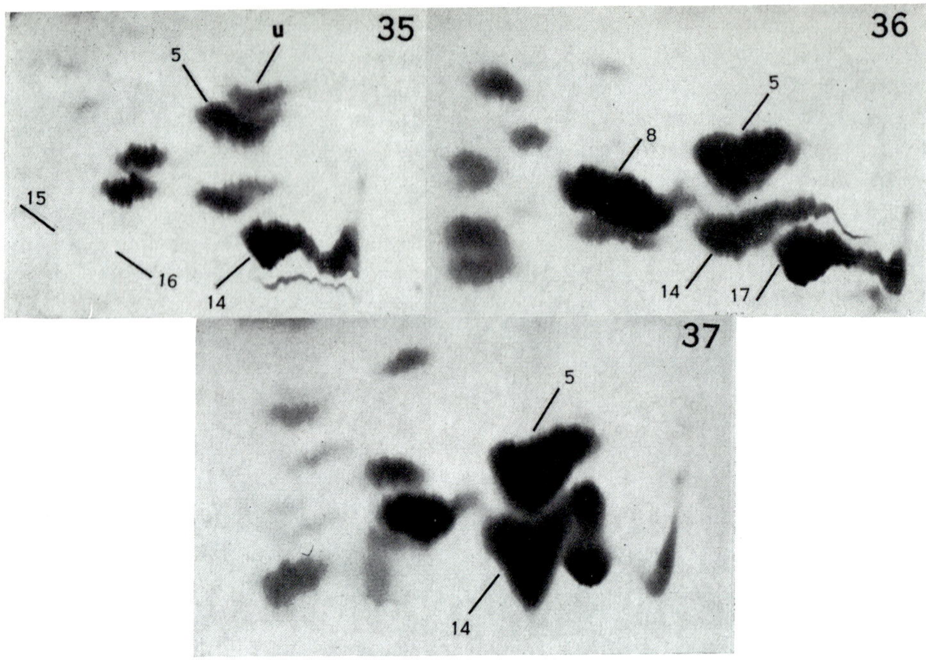

Figs. 35–37. Fig. 35, a bubble shell, *Bullaria gouldiana*; Fig. 36, a periwinkle, *Littorina planaxis*; Fig. 37, a mussel, *Mytilus californianus*. Taurine, 5; alanine, 8; glycine, 14; arginine, 15; lysine, 16; glutamic acid, 17; unknown amino acid, U.

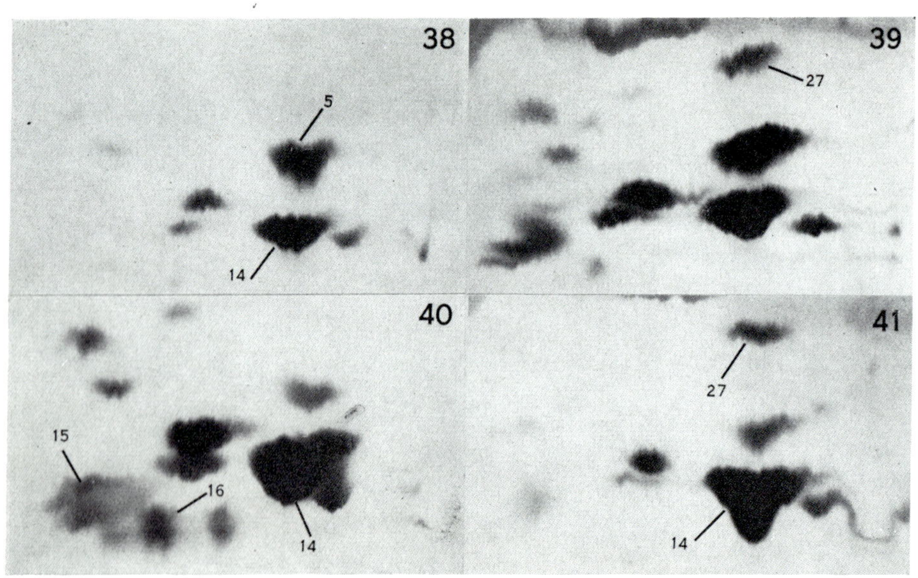

Figs. 38–41. A sea urchin, *Strongylocentrotus purpuratus*. Fig. 38, Aristotle's Lantern muscles; Fig. 39, intestinal tract; Fig. 40, gonads. Fig. 41, a starfish, *Pisaster ochraceus*. Taurine, 5; glycine, 14; arginine, 15; lysine, 16; tyrosine-*O*-sulfate, 27.

References p. 186

Other chromatograms of extracts of tissues from Mollusca are shown in Figs. 35–37. In the bubble shell, *Bullaria gouldiana* (Fig. 35), the absence of the basic amino acids and the predominance of glutamic acid was notable. Also there is a strong unknown spot above taurine. In the periwinkle, *Littorina planaxis* (Fig. 36), the other dominant inhabitant of the high spray-zone niche, the concentrations of taurine, glycine, alanine and glutamic acid were found to be high. The periwinkle combats desiccation by sealing off the aperture of its shell with its operculum and with a mucoid secretion. The mussel, *Mytilus californianus* (Fig. 37), possesses both a high taurine and glycine content.

Chromatograms of tissues from typical Echinodermata are shown in Figs. 38–41. In the sea urchin, *Strongylocentrotus purpuratus*, the muscles of Aristotle's Lantern had high levels of taurine and glycine (Fig. 38). The intestinal tract contained considerable tyrosine-*O*-sulfate (Fig. 39). The ripe gonads contained large amounts of glycine and the basic amino acids (Fig. 40). Glycine was the major constituent seen on the chromatograms of the free amino acids of the tube feet and associated muscles of the starfish, *Pisaster ochraceus* (Fig. 41). Also a considerable quantity of tyrosine-*O*-sulfate was detected.

Studies also were made of some lower Chordata. The major constituents found on chromatograms of the red gland of the sea squirt, *Ciona intestinalis*, were taurine, glycine and alanine (Fig. 42). A striking feature of the ripe female gonads from the same species was the large amount of histidine (Fig. 43). The final chromatogram is of an extract of the amphioxus, *Branchiostoma californiense* (Fig. 44). Even when the amount of extract was reduced to that corresponding to 25 mg of fresh tissue, the chromatogram was virtually swamped with glycine.

We hope that this brief survey, in addition to providing a few aspects of free amino

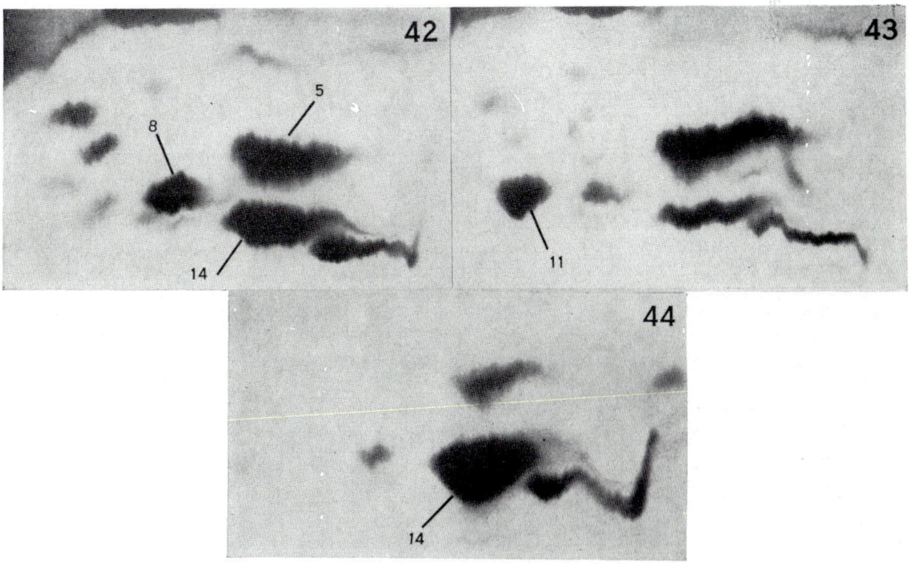

Figs. 42–44. A sea squirt, *Ciona intestinalis*. Fig. 42, red gland; Fig. 43, female gonads. Amphioxus, *Branchiostoma californiense*. Fig. 44, extract equivalent to 25 mg tissue wt. Taurine, 5; alanine, 8; histidine, 11; glycine, 14.

References p. 186

acid distribution in marine invertebrates, illustrates the necessity of considering several criteria in any approach to comparative biochemistry. (1) Adequate taxonomy. (2) Phylogenetic relations of specimens. (3) Ecological influences. (4) Developmental state. (5) Comparative morphology. (6) Sampling of individual tissues. (7) Biochemical techniques that will permit the recognition of new factors.

ACKNOWLEDGEMENT

This work was supported in part by grant No. 3001 (00) from the Office of Naval Research.

ADDENDUM

The new amino acid occurring in the Coelenterata ("X", Fig. 7, 8, 9 and 10) has now been identified as 2-aminoethyl phosphonic acid and "Y" has tentatively been identified as its glycerol ester. This phosphonic acid, previously isolated from the hydrolyzed "proteolipid-like" material extracted from sheep rumen ciliates[24], is the first compound containing a phosphorus to carbon bond reported from biological material.

REFERENCES

1 M. N. Camien, H. Starlet, G. Duchâteau and M. Florkin, *J. Biol. Chem.*, 193 (1951) 881.
2 G. G. J. Deffner and R. E. Hafter, *Biochim. Biophys. Acta*, 32 (1959) 362.
3 G. Duchâteau and M. Florkin, *Arch. intern. physiol.*, 62 (1954) 487.
4 M. Florkin, *Bull. soc. zool. France*, 62 (1954) 487.
5 W. O. Kermack, H. Lees and J. D. Wood, *Biochem. J.*, 60 (1955) 424.
6 S. Konosu, T. Akiyama and T. Mori, *Nippon Suisangaku Kaishi*, 23 (1957) 561.
7 S. Konosu, T. Akiyama and T. Mori, *Nippon Suisangaku Kaishi*, 23 (1957) 566.
8 P. R. Lewis, *Biochem. J.*, 52 (1952) 330.
9 L. Messina, *Pubbl. staz. zool. Napoli*, 30 (1957) 127.
10 S. E. Severin and A. A. Dikanova, *Biokhimiya*, 25 (1960) 1012.
11 W. Simidu, H. Sigeyuki, S. Sibata and K. Takeda, *Bull. Japan Soc. Sci. Fisheries*, 19 (1953) 871.
12 J. W. Simpson, K. Allen and J. Awapara, *Biol. Bull.*, 117 (1959) 317.
13 Y. Suzuki, *Tôhoku J. Agr. Research*, 6 (1959) 85.
14 N. K. Velankar and T. K. Govindan, *Proc. Indian Acad. Sci., B*, 47 (1958) 202.
15 N. K. Velankar and K. Mahadeva Iyer, *J. Sci. Ind. Research (India)*, 20C (1961) 64.
16 A. Kelly, *Beitr. Chem. Physiol. Pathol.*, 5 (1904) 377.
17 L. B. Mendel, *Beitr. Chem. Physiol. Pathol.*, 5 (1904) 582.
18 C. L. A. Schmidt and T. Watson, *J. Biol. Chem.*, 33 (1918) 499.
19 R. Kinosita, personal communication.
20 L. Fowden, *Biol. Revs. Cambridge Phil. Soc.*, 33 (1958) 393.
21 E. F. Ricketts and J. Calvin, in *Between Pacific Tides*, Stanford University Press, Stanford, Calif., revised edition, 1948, p. 30.
22 L. H. Hyman, in *The Invertebrates: Protozoa through Ctenophora*, Vol. 1, McGraw-Hill Book Co., Inc., New York, 1940, p. 85.
23 E. Roberts, I. P. Lowe, L. Guth and B. Jelinek, *J. Exptl. Zool.*, 138 (1958) 313.
24 M. Horiguchi and M. Kandatsu, *Bull. Agr. Chem. Soc. Japan*, 24 (1960) 565.

INVITED DISCUSSION

LOMBRICINE AND SERINE ETHANOLAMINE PHOSPHODIESTER

A. H. ENNOR AND H. ROSENBERG

Department of Biochemistry, John Curtin School of Medical Research, Australian National University, Canberra (Australia)

Both serine ethanolamine phosphodiester (SEP, I) and lombricine (II) have been discovered only recently and are compounds of more than usual interest. SEP is now known to exist in nature in both its isomeric forms; these have different and peculiar distributions and demonstrably different functions, although the importance and function of the L-isomer is as yet unknown.

L-SEP owes its recognition to the work of ROBERTS AND LOWE[1], who detected it as a ninhydrin-positive spot on a paper chromatogram of an alcoholic extract of river turtle *(Pseudemys elegans)* muscle. Sufficient material was isolated from paper chromatograms to permit acid degradative studies as a result of which serine, serine phosphate, ethanolamine, ethanolamine phosphate and inorganic phosphate were detected as products. ROBERTS AND LOWE[1] identified the serine moiety as the L-isomer and suggested the structure:

$$\begin{array}{c} \mathrm{O} \\ \| \\ \mathrm{H_2N{\cdot}CH_2{\cdot}CH_2{\cdot}O{\cdot}P{\cdot}O{\cdot}CH_2{\cdot}CH{\cdot}COOH} \\ \quad\quad\quad | \quad\quad\quad\quad | \\ \quad\quad\quad \mathrm{OH} \quad\quad\quad \mathrm{NH_2} \end{array}$$

I

which was subsequently confirmed by synthesis[2]. Further investigations indicated that L-SEP was present also in snake and alligator muscle[3]. Intracardiac injection of turtles with $^{32}P_i$ led to labelling of ethanolamine phosphate and L-SEP in several organs, and autoradiography indicated that isotope incorporation into ethanolamine phosphate and SEP was greatest in the kidney and heart. Much less isotope was incorporated in SEP isolated from the red and white muscles where the absolute amounts of SEP were believed to be greatest. Attempts at the *in vitro* incorporation of ^{32}P into L-SEP in a variety of organs were unsuccessful although a small amount of isotope was incorporated into L-SEP in turtle erythrocytes after 12 h incubation at 37° (ref. 4). The apparent slow rate of synthesis of L-SEP gave rise to the suggestion that the compound was not synthesized from small molecular-weight precursors[3, 4].

The interest of the present authors in the work of ROBERTS and his group arose from the structural similarity of SEP and lombricine to which attention had been drawn by ENNOR AND MORRISON[12]. Lombricine was discovered by the French group[5, 6] in Roche's laboratory and was isolated in low yield from the earthworm *(Lumbricus terrestris)*. On the basis of degradative studies[5,6] the compound was assigned the following structure:

References p. 192/193

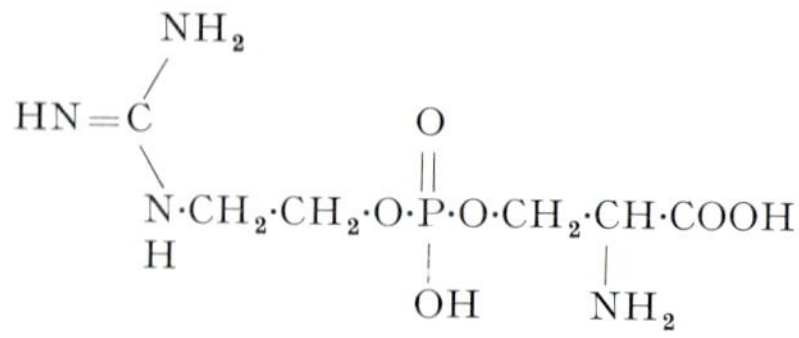

II

Subsequent studies in this laboratory yielded a method[7] for the isolation of chemically pure material in high yield from earthworms and provided ample material for chemical examination. Yields (1.1 g/kg) of lombricine in almost quantitative amounts are now possible by improvements in the isolation procedure[8]. Chemical studies undertaken with a view to a final determination of the structure of lombricine showed that the serine moiety possessed the D-configuration[9, 14] and provided the first unequivocal proof for the presence of a D-amino acid in animal tissue. Chemical synthesis of DL- and L-lombricine was then achieved[10] and was followed by synthesis of the D-isomer which was shown to be identical with D-lombricine isolated from natural sources[11]. Confirmation of the proposed structure of the compound was thus provided and was followed by a search for the biological precursor which, it had been suggested, might be SEP (ref. 12).

The presence of SEP was detected[7] in earthworm extracts but the amounts obtained were sufficient only to permit chromatographic examination of the products of acid hydrolysis. A re-examination[8] of the problem resulted in the almost quantitative isolation of SEP from earthworms and permitted positive identification of the serine component. It was found that this had the D-configuration. The presence of D-SEP in the earthworm, although consistent with the hypothesis[12] that it might be the biological precursor of lombricine, could not be regarded as proof of it. Convincing evidence that D-SEP was indeed the precursor of D-lombricine was however afforded by the demonstration[13] that the oral administration of [^{14}C]amidine-labelled arginine to an earthworm was followed by the appearance of 94% of the label in the guanidinoethanol moiety of D-lombricine. It would seem, then, that the amidine group of lombricine is derived from arginine as a consequence of a transamidinase reaction. The unequivocal demonstration of such a reaction has not yet been achieved in an *in vitro* system because of the large amounts of arginase which are present in extracts of earthworm tissue and the impossibility of separating this enzyme from transamidinase. However such a demonstration would appear unnecessary in view of the appearance in lombricine of the ^{14}C-labelled amidine group from arginine. This result also makes it unnecessary to postulate any other biosynthetic pathway for the formation of lombricine as THOAI[28] has done particularly as this was based on a failure to find SEP in earthworms.

Some attention has been paid to the nature of the possible precursors of D-SEP and D-lombricine in the earthworm. $^{32}P_i$ when given orally is incorporated into both D-SEP and D-lombricine at a rate which is about equal to that at which the isotope appears in phospholipids[15]. The oral administration of [1,2-^{14}C]ethanolamine, DL-[3-^{14}C]serine, L-[^{3}H]serine or D-[^{3}H]serine is also followed by incorporation of isotope into both D-SEP and D-lombricine[16]. In these experiments the specific radioactivity of both D-SEP and D-lombricine was greater in the viscera than in the muscle, indicating that biosynthesis occurred in some internal organ. The incorporation of

References p. 192/193

both D- and L-[^{3}H]serine into D-SEP and D-lombricine was taken[16] as indicating a change from the L- to the D-isomer, although it was not possible to determine whether this change took place before, or at the moment of incorporation. Administration of DL-[3-^{14}C]serine resulted in radioactivity appearing in both the guanidinoethanol and serine parts of the molecule, as would be expected if a decarboxylation of serine took place[16].

The presence of D-serine in SEP and lombricine isolated from earthworms and the incorporation of isotopically labeled D-serine into both of these compounds suggested the possible presence of D-serine in the free amino acid pool of the earthworm. D-serine has been isolated[17] from perchloric acid extracts of earthworms and in a more detailed investigation[18] unequivocal evidence was adduced to show that the D-isomer could not have arisen as a result of racemization induced by the techniques and solutions used in the isolation procedure. There can be no doubt, therefore, that in the earthworm D-serine has a biological origin, although the precise nature of this origin is a matter for conjecture. It has been reported[5] that lombricine undergoes enzymatic degradation and although no evidence was adduced to support this contention, it would provide a possible explanation for the presence of free D-serine in the free amino acid pool. However, experiments[18] by the present authors have failed to confirm the reported lability of lombricine in earthworm homogenates. Experiments designed to give a definitive answer to the question of the origin of D-serine in the earthworm are in progress.

The importance of lombricine in the earthworm lies in the fact that it is the guanidino base of the phosphagen, *N*-phosphoryllombricine (PL). PL was first isolated from earthworms as an impure compound by THOAI *et al.*[19, 20] but has since been obtained in good yield from natural sources as chemically pure barium, magnesium and ammonium salts[21]. The isolation method involves the use of ion-exchange techniques and gives practically quantitative recoveries.

N-phosphoryllombricine of the structure

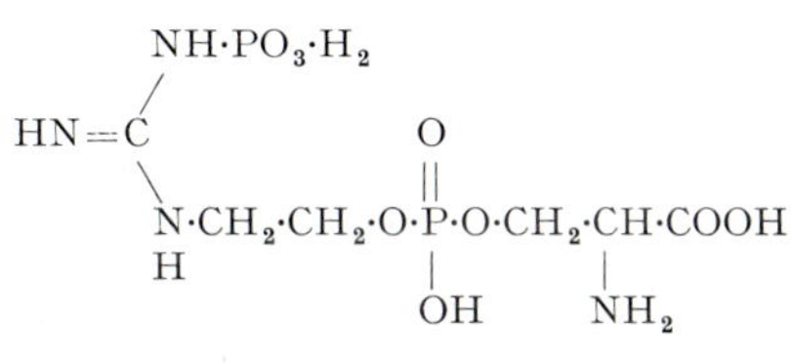

has been synthesized[22] and shown to be identical with the naturally occurring compound[23]. Proof that this compound acts as a phosphagen has been afforded by the demonstration[24] of an enzyme which catalyzed the reaction:

$$\text{Lombricine} + \text{ATP} \rightleftarrows \text{PL} + \text{ADP}$$

The enzyme was activated by Mg^{2+}, Co^{2+} and Mn^{2+}, but not by Ca^{2+} or Sr^{2+}, and catalysed the phosphorylation of both the naturally-occurring D-lombricine and the synthetic L-isomer. The preparation examined also catalysed the phosphorylation of taurocyamine but whether this was due to a lack of specificity or to possible contamination with taurocyamine phosphoryltransferase (TPT) is not known with certainty. Somewhat similar observations have been made by PANT[25], who used extracts of fresh and acetone-dried whole worms. Highly purified preparations have now been obtained and this problem, together with the properties of the enzyme, is under investigation.

Nothing is known of the *in vivo* reactions which lead to the synthesis of either L- or D-SEP other than from the *in vivo* experiments referred to above, but it seemed likely to us that similar reactions would be involved in the biosynthesis of both isomers. Because of the technical problems in working with earthworms, initial interest has therefore centred on L-SEP.

A survey of the distribution of SEP in nature has been partially completed and the compound has been detected only in the earthworm, where it is in the form of the D-isomer, and in birds, fishes, reptiles and amphibia in which it is present as the L-enantiomorph. Representative species within the following groups have been examined and no trace of SEP has been found: protozoa (one), porifera (one), coelenterata (one), ctenophora (one sp.), chaetognatha (one sp.), platyhelminthes (one sp.), nematoda (one sp.), annelida (four polychaetes, one hirudonea), arthropoda (seven sp.), mollusca (nine sp.), echinodermata (five sp.), urochordata (one sp.), mammalia (ten sp. incl. representatives of monotremes, marsupials and eutherians). The rather sharply defined distribution of the L-isomer suggests that it may have some evolutionary significance the nature of which is at present obscure.

Preliminary experiments employing the intraperitoneal or intravenous injection of $^{32}P_i$ have shown that isotopic incorporation into L-SEP isolated from the muscles is most rapid in the bird and the preliminary experiments reported here have been carried out with 16-week-old chickens.

The amount and specific radioactivity of L-SEP present in various organs of the chicken has been determined after intravenous injection of $^{32}P_i$. Considerably higher concentrations of SEP are present in the kidney and small intestine than in any other organ (Table I) while the specific radioactivity of the isolated material is highest in the kidney and plasma. These latter results suggest that L-SEP is synthesized in the kidney and transported via the blood stream to other organs.

TABLE I

THE DISTRIBUTION AND LABELING OF SEP IN THE CHICKEN FOLLOWING THE INTRAVENOUS INJECTION OF $^{32}P_i$

Tissue	*SEP content (µmoles/100 g)*	*SEP S.A.–counts/min/µmole*		
		70 min	*10 min*	*100 min*
Kidney	240	1150	560	3780
Liver	0.7	310	137	1270
Pancreas	31	117		
Duodenum	190	202		
Jejunum	190	328	152	1920
Ileum	120	805		
Spleen	11	310		
Brain	6.3	274		
Heart	10	515		
Muscle	1.3	224	155	2760
Plasma	0.3	1190		
Erythrocytes	0.9	970		

References p. 192/193

The high concentration of SEP in the small intestine compared with that in other tissues (apart from kidney) is of interest and may well provide a clue to the functional significance of this compound. Thus from the structural similarity of SEP with the phosphatides it is a temptation to suggest that it may be concerned with absorption phenomena. The results do not exclude the possibility of SEP biosynthesis in the small intestine but *in vitro* experiments with avian intestinal mucosa have failed to show any incorporation of $^{32}P_i$ into SEP under conditions which have been successful with avian kidney homogenates.

Thus incorporation of $^{32}P_i$ has been achieved with chicken-kidney homogenates

TABLE II

In vitro INCORPORATION OF $^{32}P_i$ INTO SEP IN CHICKEN-KIDNEY HOMOGENATE

Incubation mixtures contained: Mg^{2+}, 100 μmoles; ethanolamine, 50 μmoles; serine, 50 μmoles; ADP, 50 μmoles; CTP, 5 μmoles; K-succinate, 100 μmoles; *N*-ethyl–morpholine buffer (pH 7.2), 500 μmoles; kidney homogenate (3 g tissue) in a total vol. of 15 ml. Shaken for 2 h at 37°.

Conditions	*Phosphorus esterified* (%)	*SEP in mixture after incubation*		
		μmoles	*counts/min*	*S.A.*
Complete mixture	4.9	0.97	779	804
Complete mixture + DNP (3×10^{-4} *M*)	2.5	0.66	0	0
Complete mixture less CTP	—	0.92	452	491

prepared in phosphate or *N*-ethylmorpholine buffer and supplemented by a number of possible cofactors and precursors (Table II).

The lack of new synthesis of SEP in the presence of dinitrophenol and its stimulation by CTP (Table II) show that the reaction is energy-dependent and that a cytidine derivative may be involved in the synthesis. The nature of the reactions involved and the cofactors necessary are now under investigation and will be reported elsewhere.

SUMMARY AND CONCLUSIONS

The discovery of L- and D-SEP and of D-lombricine has revealed a number of fascinating problems which are gradually nearing solution. It is abundantly clear that some at least of the functions of the L- and D-enantiomorphs of SEP are quite different, although the possibility of the similarity of some functions cannot be discarded. Doubtless further work will enable this question to be resolved.

The prime function of D-SEP is to act as a precursor of lombricine which is the guanidino base of the phosphagen, *N*-phosphoryllombricine, in the earthworm. It is known that the transfer of an amidino group from arginine to an acceptor results in the formation of a compound which is metabolically inert as far as its participation in intermediary metabolism is concerned. It is tempting, therefore, although it may nevertheless be sophistry, to suggest that D-lombricine functions not only as a phosphoryl-group acceptor, but also as the ultimate "sink" for the disposal of a D-amino acid the presence of which might conceivably lead to some derangement of normal metabolic mechanisms. Such a teleological explanation gains support, although not necessarily

References p. 192/193

credence, from the fact that D-amino acid oxidase could not be detected in the earthworm by the present authors.

The function of L-SEP presents a different problem to which there is, as yet, no categorical answer. The site of biosynthesis is now clearly defined in the bird, where it seems to be restricted to the kidney. The results of isotopic experiments indicate that L-SEP, once synthesized in the kidney, is then transported via the blood stream to the various organs where it appears in widely varying concentrations. The compound appears in the highest concentration in the small intestine and this fact, together with its chemical structure, suggests that it may be involved in absorption mechanisms. Work now in progress should lead to clarification of this point and until this is completed it would seem fruitless to hypothesize further as to its possible functions.

The mechanisms by which L- and D-SEP are biosynthesized are obscure at present and while some advance has been made with L-SEP a study of D-SEP biosynthesis has not yet been attempted. Elucidation of the reactions involved in L-SEP synthesis is a desirable, if not a necessary, prerequisite before this can be initiated. The preliminary experiments reported here point to a possible involvement of CTP in SEP biosynthesis. This and the structural similarity of this compound with the phosphatides suggests that the mechanism of its biosynthesis may follow pathways similar to those involved with these compounds. Thus the final step may be envisaged as involving the condensation of cytidine diphosphoethanolamine and free serine, a mechanism compatible with what is known of the biosynthetic pathways leading to phosphatidylserine and phosphatidylethanolamine[26, 27]. It should be remembered that an additional problem arises in the case of D-SEP in that the origin of the D-serine moiety must be determined. The difficulty in isolating, for experimental purposes, different organs in the earthworm complicates the investigation, and initial experiments with crude extracts and suspensions of earthworm viscera have failed to demonstrate the presence of an amino acid racemase. D-serine, however, could also be formed by a number of pathways not involving racemization, and the use of preparations of isolated organs from giant earthworms, common in Australia, should allow a detailed investigation into the origin of the D-isomer.

ACKNOWLEDGEMENT

This work has been partially supported by a generous grant from the Rockefeller Foundation.

REFERENCES

1 E. ROBERTS AND I. P. LOWE, *J. Biol. Chem.*, 211 (1954) 1.
2 E. E. JONES AND D. LIPKIN, *J. Am. Chem. Soc.*, 78 (1956) 2408.
3 P. AYENGAR AND E. ROBERTS, *Federation Proc.*, 16 (1957) 147.
4 P. AYENGAR AND E. ROBERTS, *Proc. Soc. Exptl. Biol. Med.*, 103 (1960) 811.
5 Y. ROBIN, *D. Sc. Thesis,* College de France, 1954.
6 N. V. THOAI AND Y. ROBIN, *Biochim. Biophys. Acta*, 14 (1954) 76.
7 H. ROSENBERG AND A. H. ENNOR, *Biochem. J.*, 73 (1959) 521.
8 A. H. ENNOR, H. ROSENBERG, R. J. ROSSITER, I. M. BEATTY AND T. GAFFNEY, *Biochem. J.*, 75 (1960) 179.
9 I. M. BEATTY, D. I. MAGRATH AND A. H. ENNOR, *Nature*, 183 (1959) 591.
10 I. M. BEATTY AND D. I. MAGRATH, *Nature*, 183 (1959) 591.
11 I. M. BEATTY AND D. I. MAGRATH, *J. Am. Chem. Soc.*, 82 (1960) 4983.
12 A. H. ENNOR AND J. F. MORRISON, *Physiol. Revs.*, 38 (1958) 631.
13 R. J. ROSSITER, T. GAFFNEY, H. ROSENBERG AND A. H. ENNOR, *Nature*, 185 (1960) 383.

14 I. M. Beatty, A. H. Ennor, H. Rosenberg and D. I. Magrath, *J. Biol. Chem.*, 236 (1961) 1028.
15 T. J. Gaffney, R. J. Rossiter, H. Rosenberg and A. H. Ennor, *Biochem. Biophys. Acta*, 42 (1960) 218.
16 R. J. Rossiter, T. J. Gaffney, H. Rosenberg and A. H. Ennor, *Biochem. J.*, 76 (1960) 603.
17 H. Rosenberg and A. H. Ennor, *Nature*, 187 (1960) 617.
18 H. Rosenberg and A. H. Ennor, *Biochem. J.* 79 (1961) 424.
19 N. V. Thoai, J. Roche, Y. Robin and N. V. Thiem, *Compt. rend. soc. biol.*, 147 (1953) 1670.
20 N. V. Thoai and Y. Robin, *Biochim. Biophys. Acta*, 14 (1954) 76.
21 H. Rosenberg and A. H. Ennor, unpublished observations, 1960.
22 I. M. Beatty, A. H. Ennor and D. I. Magrath, *Nature*, 188 (1960) 1026.
23 I. M. Beatty, A. H. Ennor and D. I. Magrath, unpublished observations.
24 H. Rosenberg, R. J. Rossiter, T. J. Gaffney and A. H. Ennor, *Biochim. Biophys. Acta*, 37 (1960) 385.
25 R. Pant, *Biochem. J.*, 73 (1959) 30.
26 E. P. Kennedy and S. B. Weiss, *J. Biol. Chem.*, 222 (1956) 193.
27 G. Hübscher, R. R. Dils and W. F. R. Pover, *Biochim. Biophys. Acta*, 36 (1959) 518.
28 N. V. Thoai, in *Biochimie Comparée des Acides Amines Basiques*, Editions Du Centre National De La Recherche Scientifique, Paris, 1960, p. 297.

DISCUSSION

Chairman: NORMAN H. HOROWITZ

AWAPARA: I would like to ask DR. ROSENBERG a question on the formation of ethanolamine. So far as I know, no one has yet shown a serine decarboxylase. It is very apparent from *in vivo* evidence using double labeled material that serine gives rise to ethanolamine. There is no doubt then that serine first combines in some form, perhaps in nucleotides similar to those you have mentioned, and that it can then be decarboxylated. It is quite appalling, however, that the so-called serine decarboxylase has not been demonstrated. Do you have any information on that?

H. ROSENBERG: I believe a very recent communication has shown that phosphatidyl serine and ethanolamine are interconvertible; perhaps someone has seen this paper.

ROUSER: What happens, presumably, is that phosphatidyl ethanolamine undergoes an exchange with serine to make phosphatidyl serine which is then decarboxylated back to phosphatidyl ethanolamine. In this way ethanolamine is released, and in the cycle serine is converted (decarboxylated) to ethanolamine while combined in a phospholipid.

GUROFF: In our laboratory GIBSON AND WILSON were able to show that serine phospholipids could be decarboxylated and methylated all the way to the choline phospholipid in liver preparations. I think this has also been done in DR. GREENBERG's laboratory.

LOFTFIELD: I would just like to ask either of the marine invertebratologists here whether the correlation between taurine content and the nature of the beast is related to whether they are in marine conditions. You said some brackish water animals had very low taurine, and I was wondering whether they would begin to accumulate taurine, if you put them into isotonic salt. In other words, is the possession of taurine only a reflection of the effect of the environment on their capacity to retain this substance?

AWAPARA: We tried to do this but, unfortunately, the animal died very quickly. The only one tried was *Rangia cuneata*, a clam, which lives in brackish water. We managed to put it in much higher concentration of salt water, but the evidence was very erratic. I believe that the idea is correct.

KITTREDGE: Just a point on the ionic strength of the medium. In attempting to force the synthesis of taurine from radioactive sulfate, we took abalone and brought them rather rapidly down to about half the normal ionic concentration of their environment, and then back up to higher than normal concentrations. We found, as DR. AWAPARA mentioned, that marine invertebrates retain their taurine very tenaciously. The tissues would become flooded with excess water during exposure to sea water with low ionic concentration, but in no case could we drastically change the tissue content of taurine or force synthesis of new taurine from sulfate.

H. ROSENBERG: I would just like to add here that we find quite a lot of taurine in the earthworm, which is a terrestrial polychaete, and there is also quite a lot in the bird. In fact, taurine can be easily crystallized from concentrates of chicken muscle.

SCHREIER: DR. AWAPARA talked about the puzzles of taurine. I should add another puzzle to this problem which is that human babies excrete taurine only for about three to six months, and prematures do not excrete it until they are about ten to fourteen days old.

SEGAL: In relation to DR. SCHREIER's comment on the excretion of taurine in the newborn, which then ceases, DR. ROSENBERG and I have measured taurine excretion in the Sprague-Dawley rat and found that a great deal of the amino nitrogen coming out on the urine daily is in the form of taurine. Some years ago there was a very nice report in the analysis of biological fluids by chromatographic techniques published by the University of Texas. This monograph included a paper which showed that there is a marked difference in taurine excretion by various strains of white rats.

VI. VERTEBRATES

THE FREE AMINO ACIDS OF BODY FLUIDS AND SOME HEREDITARY DISORDERS OF AMINO ACID METABOLISM

R. G. WESTALL

Medical Unit, University College Hospital Medical School, London (Great Britain)

THE FREE AMINO ACIDS OF BODY FLUIDS

It is intended in this article to summarize, as far as possible, our present knowledge of the content of the free amino acids in the body fluids. It cannot be, in the space allowed, a complete historical account and hence the results obtained by workers using modern methods of analysis take precedence over those results obtained by the earlier investigators who were, in many cases, the pioneers in the field. Also, from the terms of reference, the article should cover all the vertebrates but one is forced to concentrate on mammals and especially on the human, since, not unnaturally much of the work has been carried out on our own species. Only comparatively recently have reliable and accurate analyses of the amino acid content of the blood plasma and urine of normal healthy individuals become available although a great deal of work had already been done on these body fluids in various disease states.

Methods

The occurrence of free amino acids in blood and urine has been known for many years and a number of different methods have been devised to measure them quantitatively. The earlier methods were based upon the measurement of the amino groups by titration in the presence of formalin[1]. Later, this method was modified by the substitution of ethanol for formalin[2]. Neither these methods nor that proposed by VAN SLYKE[3], which measured volumetrically the N_2 gas liberated by the amino acids on treatment with nitrous acid were entirely satisfactory although a modified gasometric method devised by VAN SLYKE, MACFADYEN AND HAMILTON[4] proved to be much more reliable. However, these methods only measured the total amino acids present. A number of methods were available for measuring individual amino acids usually based on a specific chemical reaction which yielded a coloured end product which could be measured in a colorimeter. At this time (the late 1930's) there were no practical ways of estimating the series of aliphatic mono-amino, mono-carboxylic acids (glycine, alanine etc.). This deficiency was, to a large extent, made up by the development of the microbiological methods using mutant strains of bacteria but these techniques still have their limitations and are time consuming.

Certainly the greatest advance in our knowledge of the distribution of the amino acids in biological systems has arisen since the invention of the technique of paper chromatography[5]. The potential of this method was not immediately recognized by many workers but DENT[6] quickly realized its value and used it for screening urine and plant extracts for amino acids. Paper chromatography gives a semi-quantitative measurement of the amino acids and attempts are still being made currently to improve the quantitation. At the present time, the elegant method of ion-exchange column chromatography devised by MOORE AND STEIN in its latest partly automatic form[7] provides the most accurate means of estimating the commonly occurring amino acids. Nevertheless, paper chromatography with its simple apparatus, rapid results and adaptability will remain a valuable technique in biological research for a long time.

The body fluids

Blood plasma. Studies on the amino acids present in blood plasma or serum have been carried out by numerous workers who have been interested in the metabolic fate of these substances in the body. The systemic blood which replenishes its store of amino acids from the gut via the portal system and the liver can be regarded, in a general way, as the medium from which the tissue cells, by way of the extracellular fluid, draw their nutriment and into which they discard the unwanted by-products. Since a proportion of the cellular protein is being continuously broken down and new protein is being synthesized it might be expected that the nutrient medium would have an optimum composition especially with regard to those amino acids which are protein constituents. In fact the amino acid composition of the plasma, in the fasting condition, has a fairly constant pattern which simulates approximately the amino acid composition of the tissue proteins. The plasma amino acid level rises after a protein meal[8] but it slowly drops back to the fasting level within approx. 4 h. Those amino acids which are protein constituents, and this, of course, includes those which are essential to our diet and which we are unable to synthesize ourselves in sufficient quantities, if at all, are only excreted in the urine in amounts of the order of 1–2% of the dietary intake. Glycine and histidine, which are not so essential, are found in the urine in larger amounts (up to 20%). Hence it follows that since in good nutritional status the intake of the essential amino acids exceeds the amount required for protein replacement, in the adult, these amino acids in excess must be used in other ways. That this is true is well exemplified in the two metabolic diseases phenylketonuria and maple-syrup-urine disease where in the first case phenylalanine and in the second instance the branched chain amino acids leucine, isoleucine and valine cannot be further metabolized at the required rate due to an insufficiency of specific enzymes and these amino acids build up to a high concentration in the body fluids. On the other hand, when the nutritional status is poor and the individual is in negative nitrogen balance there is a tendency for the plasma amino acid level to fall. For a while this fall in level is prevented from going too far by the utilization of endogenous supplies of expendable tissue protein but once this store has been used up the plasma amino acid level begins to fall seriously. This is the state of affairs frequently found in cases of kwashiorkor and other forms of protein malnutrition where the plasma amino acids may fall to as low as one-fifth of the normal concentration.

A two-way paper chromatogram showing the amino acid pattern in human adult blood plasma is shown in Fig. 1. For this chromatogram a sample of 625 μl of ultra-filtered and desalted plasma was used as suggested by DENT[9]. A number of quantitative analyses, performed by various workers, are given in Table I and a composite range of values is given in the final column.

Aarpt from those amino acids which are utilized directly for protein synthesis

References p. 217/219

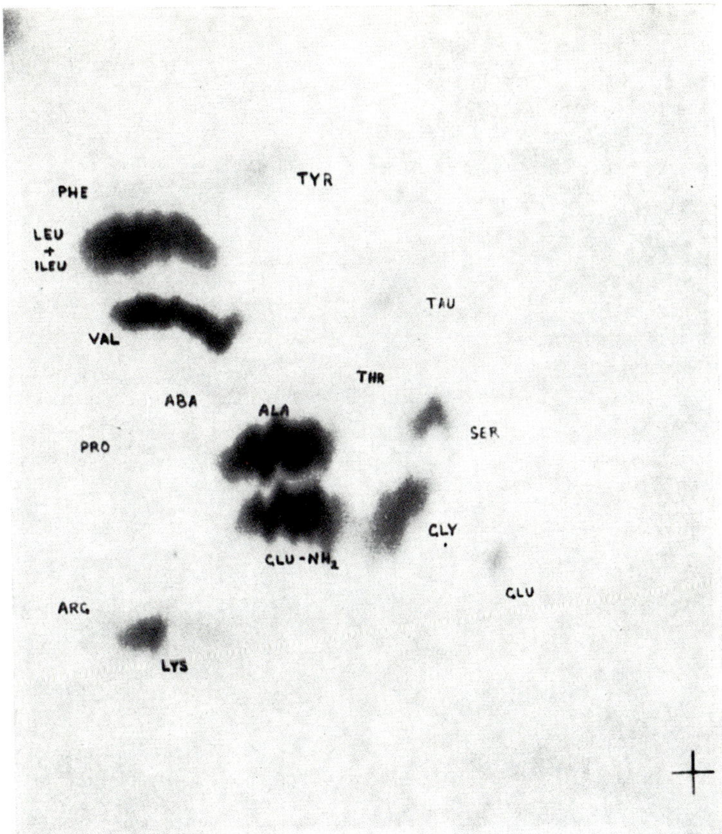

Fig. 1. A paper chromatogram of normal human blood plasma (625 μl).*

there are a number of others, usually found in the plasma in lower concentration, which are normally present. The most conspicuous of these are taurine, ornithine, α-aminobutyric acid and citrulline. There are others also which are found abundantly in urine but which have plasma clearances which are high, or are suspected to be high, and their concentration in the plasma is extremely low. The 1- and 3-methylhistidines, β-aminoisobutyric acid as well as the dipeptides carnosine and anserine are in this category.

With regard to the blood plasma amino acids of other vertebrates very little is

* Illustrations of paper chromatograms: general remarks. +, marks the point of application of the sample. 1st solvent (from right to left) water-saturated phenol used in an ammoniacal atmosphere (several drops of concentrated ammonia solution put in the tank before sealing). 2nd solvent (from bottom to top) lutidine (commercial grade, a mixture of the 2,4- and 2,5-dimethyl-pyridines). The solvent is made up by mixing 2.2 parts of lutidine with 1 part of water (v/v). *Key to the abbreviations on the chromatograms*: ABA, α-aminobutyric acid; ALA, alanine; ARG, arginine; ASA, argininosuccinic acid; ASA-B, anhydride form of ASA; ASP, aspartic acid; BAIB, β-aminoisobutyric acid; CIT, citrulline; CYS-A, cysteic acid; CYSTA, cystathionine; ETH-NH_2, ethanolamine; ETH-PO_4, ethanolamine phosphate; GLU, glutamic acid; GLU-NH_2, glutamine; GLY, glycine; HIS, histidine; ILEU, isoleucine; LEU, leucine; LYS, lysine; METH, methionine; ORN, ornithine; PHE, phenylalanine; PRO, proline; SER, serine; TAU, taurine; THR, threonine; TYR, tyrosine; VAL, valine.

References p. 217/219

TABLE I

THE FREE AMINO ACIDS OF HUMAN ADULT BLOOD PLASMA

Values are expressed in mg/100 ml.

	STEIN AND MOORE[94] *average*	EVERED[95] *average*	HEIR AND BERGEIM[96] *average*	HARPER *et al.*[97] *average*	DOOLAN *et al.*[98]	MCMENAMY *et al.*[99]	*Range of values*
Alanine	3.4	2.9		3.96	2.5	2.85	2.4–7.6
α-Aminobutyric acid	0.3	0.14				0.21	0.1–0.3
Asparagine	0.58						
Aspartic acid	0.03	0.13		0.33			0.01–0.3
Arginine	1.51	1.55	2.32	2.26	2.8	1.18	1.2–3.0
Citrulline	0.5						
Cystine	1.18*	0.85		3.02	2.16		0.8–5.0
Glutamic acid	0.70	0.94		0.89			0.5–1.2
Glutamine	8.3			7.51	7.70	6.57	4.6–9.7
Glycine	1.54	1.47		2.91	2.01	1.76	0.8–5.4
Histidine	1.15	0.90	1.42	2.11	1.37	1.37	0.8–3.8
Isoleucine	0.89	0.90	1.66	2.00	1.21	0.75	0.7–4.2
Leucine	1.69	1.34	2.03	2.48	1.90	1.04	1.0–5.2
Lysine	2.72	1.97	2.97	3.68	2.40	2.09	1.4–5.8
Methionine	0.38	0.40		0.57	0.50	0.42	0.2–1.0
1-Methylhistidine	0.11						
3-Methylhistidine	0.08						
Ornithine	0.72				0.77		0.6–0.8
Phenylalanine	0.84	0.77	1.40	1.99	1.55	0.93	0.7–4.0
Proline	2.36	1.95		2.61	1.71	1.63	1.5–5.7
Serine	1.12			1.39	1.50		0.3–2.0
Taurine	0.55	0.77				0.12	0.2–0.8
Threonine	1.39	1.43	2.08	2.06	1.51	1.61	0.9–3.6
Tryptophane			1.11	1.74	1.20	0.43	0.4–3.0
Tyrosine	1.03	0.97	1.50	1.32	0.90	1.27	0.8–2.5
Valine	2.88	2.30	2.89	3.23	2.90	2.04	1.9–4.2

* About one quarter of this amount is cysteine[93].

known. The cat seems to have a similar plasma amino acid pattern to the human except for the occurrence of a detectable amount of felinine[10]. Any differences that exist between the amino acid patterns of males and females are smaller than those that occur between individuals and are probably not significant. It is, however, interesting to note that in pregnancy the plasma amino acid concentration in the foetus is higher than in the maternal blood[11, 12].

Urine. Since urine is so readily accessible it is not surprising that it has been more extensively studied than any other body fluid. In recent years, since the first application of the method of paper chromatography to urine analysis by DENT[6], probably many tens of thousands of urine samples have been screened for amino acids. In many medical centres it has become a routine procedure and serves as a useful aid to the diagnosis of certain diseases.

References p. 217/219

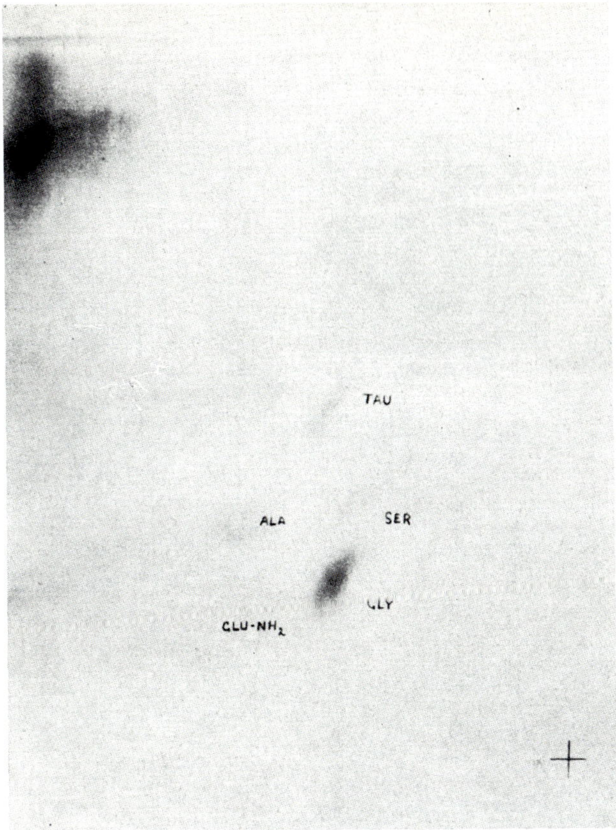

Fig. 2. A paper chromatogram of normal human urine (sample taken contained 250 μg of total nitrogen).

The urine amino acid excretion of the healthy human adult conforms to a fairly standard pattern which is only changed by disease or by a gross alteration in dietary intake. A paper chromatogram of a normal urine (the amount used contained 250 μg of total N) after treatment with ninhydrin to show up the main amino acids, is illustrated in Fig. 2. Note the marked difference from the plasma pattern due to the intervention of the kidney. It will be seen that glycine is the most abundant amino acid excreted followed by histidine, taurine, glutamine, alanine and serine. This general pattern does not alter greatly from individual to individual and such changes that are usually observed are quite subtle. Further, the amino acid excretion pattern of the individual, after childhood, and providing he remains healthy, is constant over many years. However, one occasionally finds a variant from this pattern in a healthy person and an instance of this is seen in those people who excrete excessive amounts of β-aminoisobutyric acid[13, 14]. A number of recent quantitative analyses of the amino acid content of normal human urine carried out by several investigators are set out in Table II. Reference to this list shows that glycine, taurine, 1-methylhistidine and histidine are the only amino acids excreted in amounts exceeding 100 mg/day. All of the remaining protein-derived amino acids are found in the urine and are excreted

References p. 217/219

TABLE II

THE FREE AMINO ACIDS OF NORMAL HUMAN ADULT URINE

Values are expressed in mg/24 h.

	STEIN[100], average of 6 normal males	EVERED[95], average of 3 normal males	SOUPART[31], average of 6 normal males	SOUPART[31], average of 9 normal females	Range of normal males	Range of normal females
Alanine	46	28	22	24	5– 71	9– 44
β-Alanine			6	3	3– 10	2– 9
α-Aminoadipic acid			8	4	5– 13	0– 13
β-Aminoisobutyric acid		13	22	29	6– 37	10– 52
Aspartic acid	<10		8	4	3– 29	2– 11
Arginine	<10		6	4	0– 14	0– 11
Cystine	10	9	14	6	3– 33	0– 13
Glutamic acid	<10	13			7– 40	
Glutamine			73	62	40–103	43– 88
Glycine	132	109	104	142	53–200	67–312
Histidine	216	97	138	128	20–320	79–208
Isoleucine	18	13	15	10	5– 30	5– 20
Leucine	14	8	11	9	5– 25	2– 16
Lysine	19	12	7	8	0– 48	0– 16
Methionine	18	6	7	5	5– 11	3– 12
1-Methylhistidine	180	22	73	65	9–210	26–155
3-Methylhistidine		40	65	48	33– 87	30– 69
Ornithine			1	2	0– 4	0– 11
Phenylalanine	18	14	13	13	8– 31	6– 41
Proline	<10					
Serine	43		42	37	25– 75	22– 61
Taurine	156	59	123	87	35–300	27–161
Threonine	28		17	23	2– 50	5– 33
Tyrosine	35	23	19	15	7– 50	9– 26
Valine	10	5	10	6	4– 17	0– 30

in amounts of between 5 and 100 mg/day. Whilst, ornithine and 3-methylhistidine[15] are also present in similar amounts.

The complexity of the mixture of amino acids which can be detected in normal human urine is such that although all the amino acids in Table II are present so also are a large number of others occurring in amounts of 5–10 mg/day or less. Evidence for the occurrence of these substances which react with ninhydrin to give a coloured product has been provided by workers who have analysed urine concentrates by a variety of methods including the use of ion-exchange resins[16–20]. Many of these substances are as yet uncharacterized and the number of these substances seems almost limitless as new methods lower the threshold of detection. The following amino acids have also been detected in urine: β-alanine, α-aminobutyric acid[21], γ-aminobutyric acid (GABA), α-aminoadipic acid[22], α-aminolevulinic acid[23], γ-guanidinobutyric acid[24], phenylacetylglutamine[25] and sarcosine[18]. In addition the following substances, not strictly amino acids, give blue-coloured spots with ninhydrin on paper chromatograms and may be often detected in urine: ethanolamine[26, 27] and phosphoethanolamine[28, 29].

There is apparently no appreciable difference in the amino acid excretion pattern between men and women[30, 31]. Several groups of workers have maintained that certain differences do exist but on the whole the variations are so subtle as to be insignificant. It is interesting, however, to note that there is an increase in the excretion of histidine during pregnancy[32, 33].

It has been known for many years that infants excrete more α-amino nitrogen than adults relative to their body weights[34]. Paper-chromatographic studies of the urine of the newborn has confirmed this observation and has shown that certain amino acids are excreted preferentially[28, 35, 36]. The most striking feature of the urine amino acid pattern of the newborn is the increased excretion of proline and hydroxyproline which commences on the 5–6th day and persists for about 5–6 months, after which time they are rarely, if ever, observed on paper chromatograms of urine from healthy people. The other amino acids which are involved in the increased aminoaciduria of the newborn are β-aminoisobutyric acid, lysine and taurine. Ethanolamine is also excreted in increased amount[36].

The amino acid excretion of other mammals. The excretion pattern of the amino acids which is relatively constant within species can vary quite markedly between species. DATTA AND HARRIS[37] made a study of a number of different vertebrate species with respect to their amino acid excretion and the results are reproduced in Table III. It is noticeable that the carnivores excrete large amounts of 1-methylhistidine which is almost certainly derived from the dietary muscle tissue which is an abundant source of this amino acid. The domestic cat, and to a lesser extent the ocelot, is unique in excreting the amino acid felinine[38] (S-hydroxyisopentanylcysteine, previously known as cat-spot). The Kenya genet excretes large amounts of cystine but does not suffer from stone formation as does the human cystinuric patient when excreting similar

TABLE III

THE PRINCIPAL AMINO ACIDS DETECTED BY PAPER CHROMATOGRAPHY IN THE URINES OF A NUMBER OF ANIMALS[37]*

	Glycine	*Alanine*	*Glutamine or glutamic acid*	*Taurine*	*"Cat-spot"*	*Cystine*	*Methyl-histidine*
Domestic cat	+	+	—	—	++++	—	++++
Lion	+	+	+	+	—	—	+++
Tiger	+	+	—	+	—	—	+++
Ocelot	+	+	+	—	++	—	+++
Puma	+	+	—	++	—	—	++
Genet	+	+	—	+	—	++++	+
Binturong	—	—	—	+++	—	—	—
Dog	+	+	+	+++	—	—	+++
Rabbit	+	+	+	—	—	—	—
Rat	+	+	—	++	—	—	—
Mouse	—	—	—	++++	—	—	—
Guinea-pig	+	+	+	—	—	—	—
Horse	+	+	+	+	—	—	—
Cow	+	—	—	—	—	—	—

* Reproduced by permission of the Physiological Society, London.

References p. 217/219

concentrations of this amino acid. In the dog, rat and mouse the excretion pattern is dominated by taurine.

Cerebrospinal Fluid. The amino acid concentration in cerebrospinal fluid (C.S.F.) is much lower than in blood plasma as can readily be seen in the chromatogram illustrated in Fig. 3 when compared with that of plasma (Fig. 1) using a 625-μl sample in

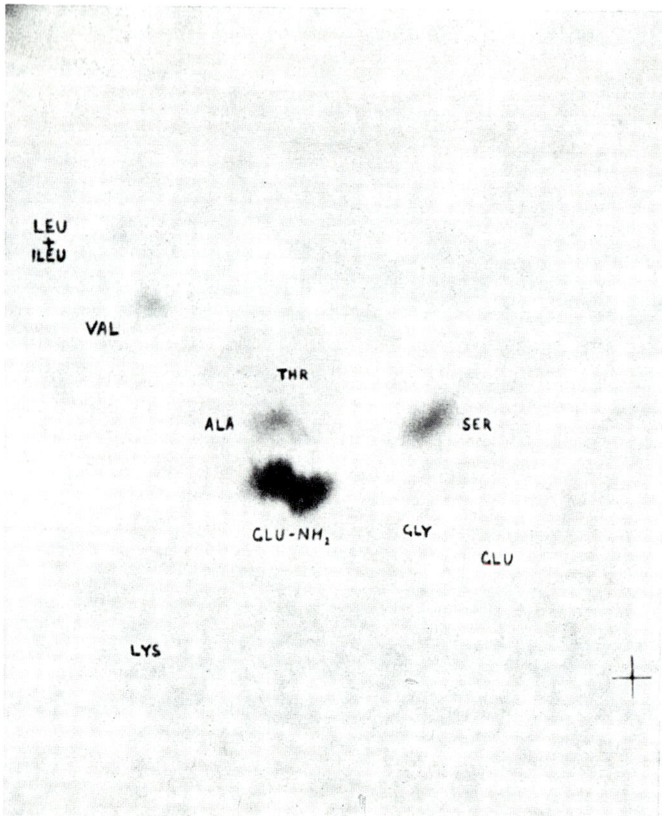

Fig. 3. A paper chromatogram of normal human cerebrospinal fluid (625 μl).

each case. Glutamine stands out as the most conspicuous spot and it is present in about the same concentration as in plasma. From investigations carried out by a number of workers[39–41] it is apparent that most, if not all, of the amino acids which occur in blood can be found in C.S.F. although in a lowered concentration. This cannot mean that the C.S.F. derives its amino acids from the blood by passive diffusion through the blood-brain barrier as the two amino acid patterns are quite different. That the pattern should be usefully different is perhaps not surprising since the C.S.F. serves as a fluid medium for predominantly neurological tissues with their very different metabolism. For instance, GABA which is abundantly found in brain[42] is often present in C.S.F. whereas it can be barely detected in the blood. So far,

comprehensive quantitative estimations of the concentration of the various amino acids present in normal human C.S.F. have not been reported but WALKER, TELLES AND PASTORE[40] quote the following mean figures from a series of 26 normal C.S.F. samples: aspartic acid, 0.4; glutamic acid, 0.6; serine, 1.6; glycine, 1.0; threonine, 1.3; glutamine, 7.1; alanine, 1.1; tyrosine, 0.8; valine, 0.7; phenylalanine, 2.1; and leucine, 1.4 μg/ml.

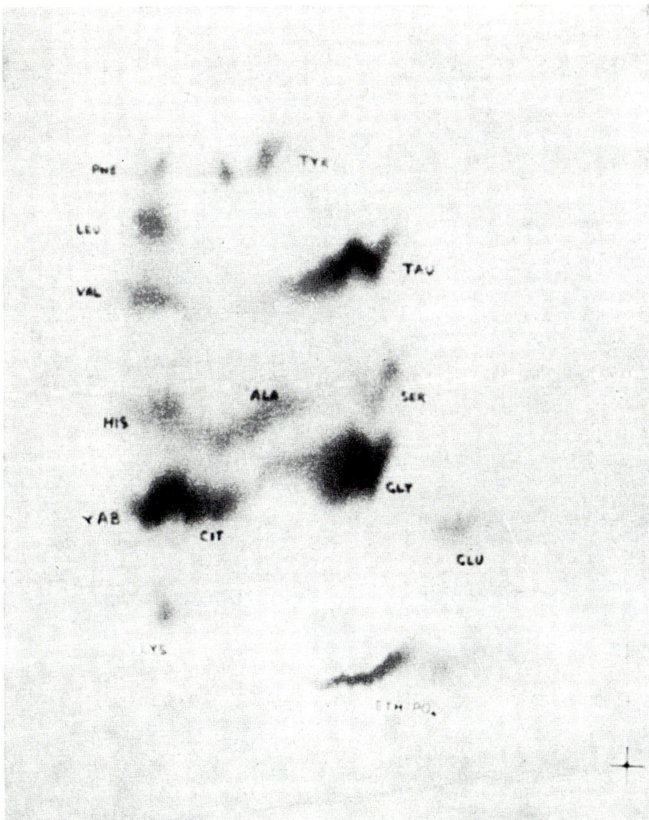

Fig. 4. A paper chromatogram of normal human saliva (2.0 ml).

Saliva. The amino acids of whole saliva have been investigated by many workers (see bibliography by ROSE AND KERR[43]). A paper chromatogram obtained from running a 2-ml sample of whole saliva, after ultrafiltration and electrolytic desalting, is shown in Fig. 4. Once again, it can be seen that the pattern given by the amino acid spots is quite different from that given by other body secretions. In particular, GABA seems to be present although BERRY AND CAIN[44] decided that the spot that they found in a similar position to that taken up by GABA was, in fact, something else. There is also a well-defined spot due to ethanolamine phosphate. Some further interesting patterns were obtained by ROSE AND KERR[43] who ran paper chromatograms on secretions, obtained by cannulation, from the parotid and submaxillary glands. They found that the concentration of ethanolamine phosphate in these secretions was some four to five times higher than in whole saliva. This is curious since it is estimated that the secretions from these two glands makes up approx. 75% of the volume of the whole saliva[45]. ROSE AND KERR[43] concluded that there must be an enzyme present in whole saliva capable of hydrolysing ethanolamine phosphate and which must be absent in the parotid and submaxillary gland secretions. This explanation is supported by an observation by CHAUNCEY, LIONETTI, WINER AND LISANTI[46], who reported that

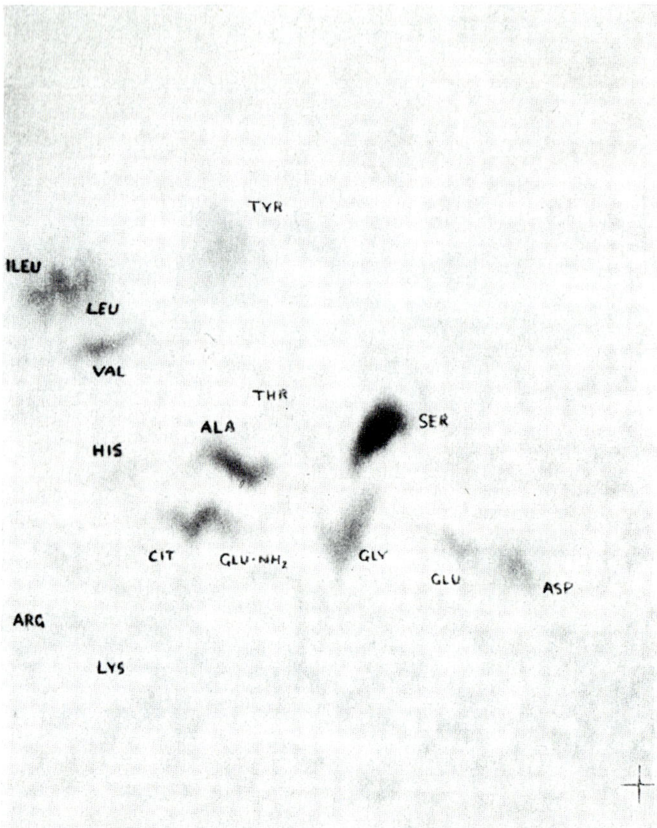

Fig. 5. A paper chromatogram of normal human sweat (125 μl).

alkaline phosphatase activity was present in whole saliva. The GABA which is present in whole saliva seems to come from some source other than from the parotid or submaxillary glands[43].

Sweat. The presence of amino acids in sweat has been known for over 50 years when EMBDEN AND TACHAU[47] managed to isolate serine from human sweat. Fig. 5 shows a paper chromatogram of 125 μl of human sweat and it can be noted that serine is probably the most abundant amino acid. A similar paper chromatogram to that illustrated was published by ROTHMAN AND SULLIVAN[48] who obtained their samples by wiping down their subjects with water-soaked cotton 24 h after bathing. It is

References p. 217/219

difficult, from their results, to estimate the amounts of the amino acids which occur in the perspiration. However, if droplets of sweat are collected during exercise then the amino acid concentration in the sample is found to be about five times that of blood plasma. The amino acid pattern of sweat differs from that of other body fluids in having a high concentration of citrulline, a substance which occurs in other fluids

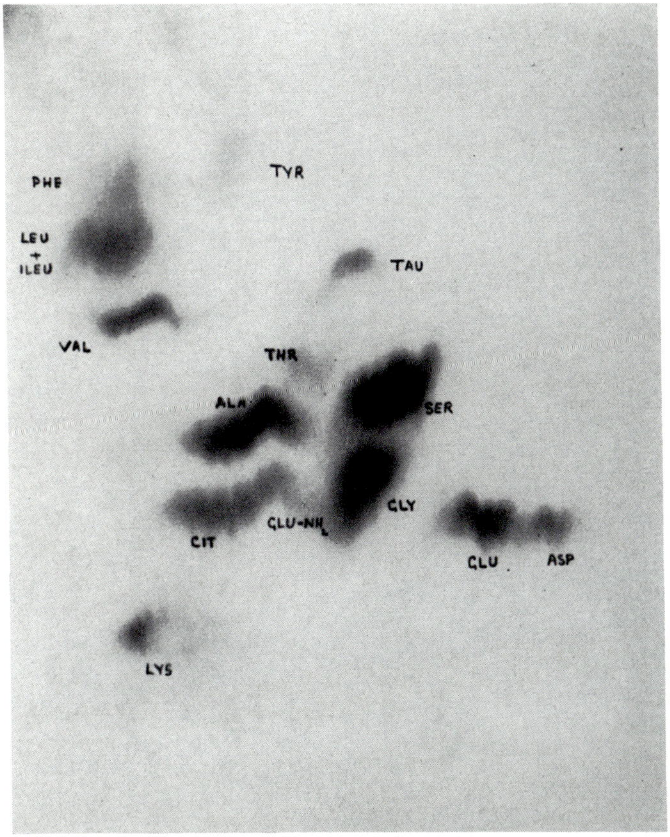

Fig. 6. A paper chromatogram of human tears (250 μl).

but usually only in trace amounts. Aspartic acid is also found in greater concentration in sweat than in the other body fluids.

Tears. As far as I am aware very little is known about the amino acids in tears. For the purpose of this article a paper chromatogram was prepared using 250 μl of ultra-filtered and electrolytically desalted lacrimal fluid and this is illustrated in Fig. 6. It will be seen that the amino acid pattern is somewhat similar to that of sweat except that the citrulline spot is not as prominent. The amino acid concentration is higher than in blood plasma but does not quite reach that of sweat.

References p. 217/219

HEREDITARY DISORDERS OF AMINO ACID METABOLISM

A number of hereditary diseases have been described which exhibit an increased aminoaciduria. These diseases will not be discussed from the standpoint of genetics or of more than passing references to certain clinical features but they will be used as instances of the various ways that the handling of the amino acids by the body may be altered.

The aminoacidurias have been divided into two groups[49]: the renal type where

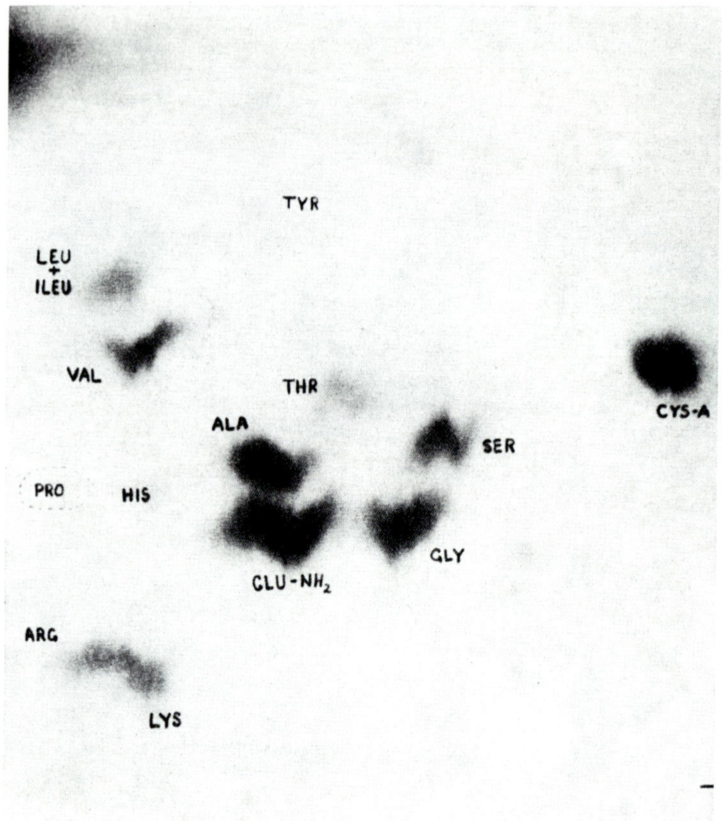

Fig. 7. A paper chromatogram of the urine from a child with cystinosis.

the amino acid pattern of the blood plasma is normal and the aminoaciduria is due to a defect in renal tubular reabsorption which may be selective for certain amino acids only or which may be more generalized due to more widespread damage to the kidney. The second type is known as the "overflow" aminoaciduria. Here the kidney function is normal and the increased excretion is due to a high plasma concentration of certain amino acids. These may be normal constituents, for which there is a low renal clearance, or they may be amino acids which are not usually found in plasma and for which the renal clearance is high. This latter sub-type has been called a "no threshold"-aminoaciduria[50].

The renal aminoacidurias

The Fanconi syndrome. Under this heading is grouped a number of diseases, affecting both adults and children, which are characterized by a marked aminoaciduria, chronic acidosis, renal glycosuria, proteinuria and electrolyte disturbances. Evidence has been produced[51] that there is a morphological abnormality of the proximal kidney tubule in some of these cases. This may well be the reason for the reduced tubular reabsorption of the amino acids. Fig. 7 shows the amino acid pattern of the urine of a

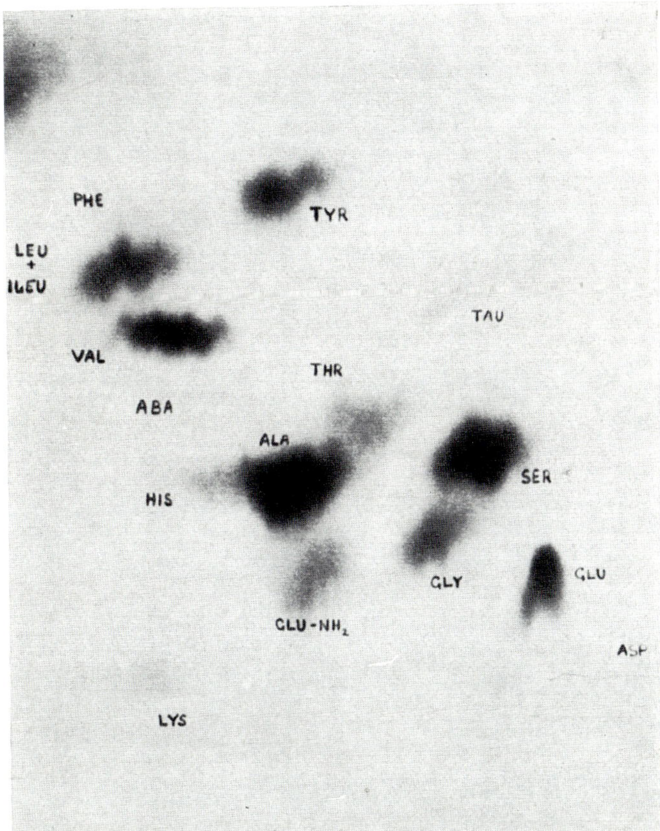

Fig. 8. A paper chromatogram of the urine from a patient with Hartnup disease.

child with cystinosis which is one of the commoner types of the Fanconi group. In these patients there is also a widespread deposition of cystine crystals in the body tissues[52]. This aminoaciduria is of a general type and shows a pattern which is markedly similar to that of blood plasma (Fig. 1).

Hartnup disease. This condition was first described in 1956 by BARON, DENT, HARRIS, HART AND JEPSON[53]. They reported on a family in which four out of eight children were affected and they believed the disease to be hereditary. The discovery of a num-

References p. 217/219

ber of further cases has supported this belief. In these cases there is a constant aminoaciduria, an intermittent pellagra-like rash, occasional attacks of cerebellar ataxia and some degree of mental deficiency. The gross aminoaciduria is thought to be mainly, if not entirely, of renal origin as the plasma amino acid levels are not raised and in fact seem to be, in respect of serine, threonine and glutamine, lower than usual[50]. The urinary excretion, shown in Fig. 8, is somewhat similar to that seen in the Fanconi syndrome (Fig. 7), but apart from differences in the concentration of several

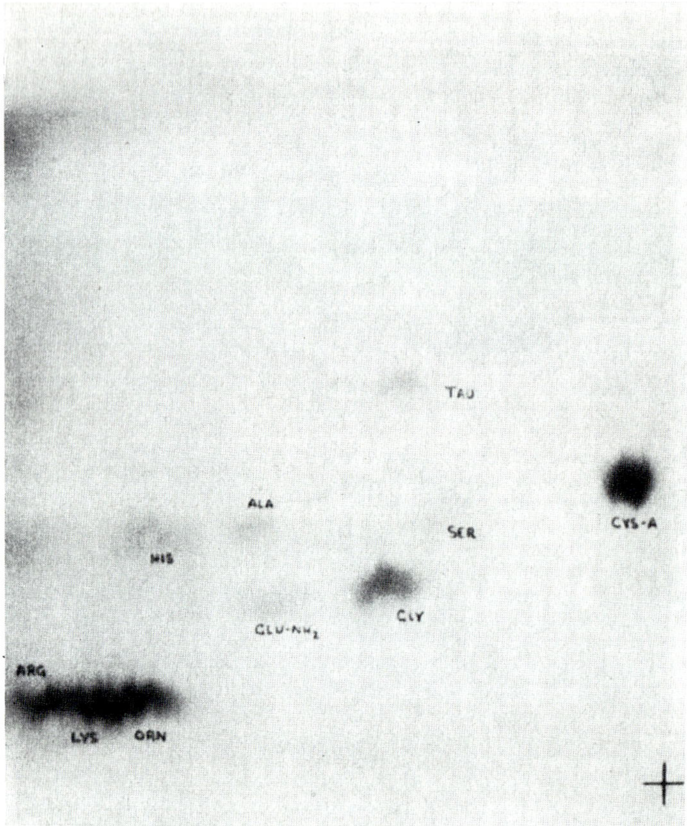

Fig. 9. A paper chromatogram of the urine from a patient with cystinuria.

of the amino acids proline is always absent and asparagine is more prominent in the Hartnup pattern. Another outstanding biochemical abnormality in this disease is the increased excretion of the indole derivatives indolylacetic acid and indolylglutamine[54] which are considered to be derived from tryptophane. No single hypothesis has yet been put forward which could account for the various biochemical abnormalities found in these cases. It has been suggested[55] that the primary cause might be a block in tryptophane metabolism and that the aminoaciduria is a secondary effect. On the other hand the primary cause might be an error in the amino acid

References p. 217/219

transport mechanisms affecting tubular reabsorption and that tryptophane is involved secondarily.

Cystinuria. It has been known for 150 years that certain people form cystine stones in the urinary tract and that the disorder was familial. Further investigations, carried out since the advent of chromatography, have shown that not only cystine is excreted in larger amounts than usual but there is also an increased excretion of lysine

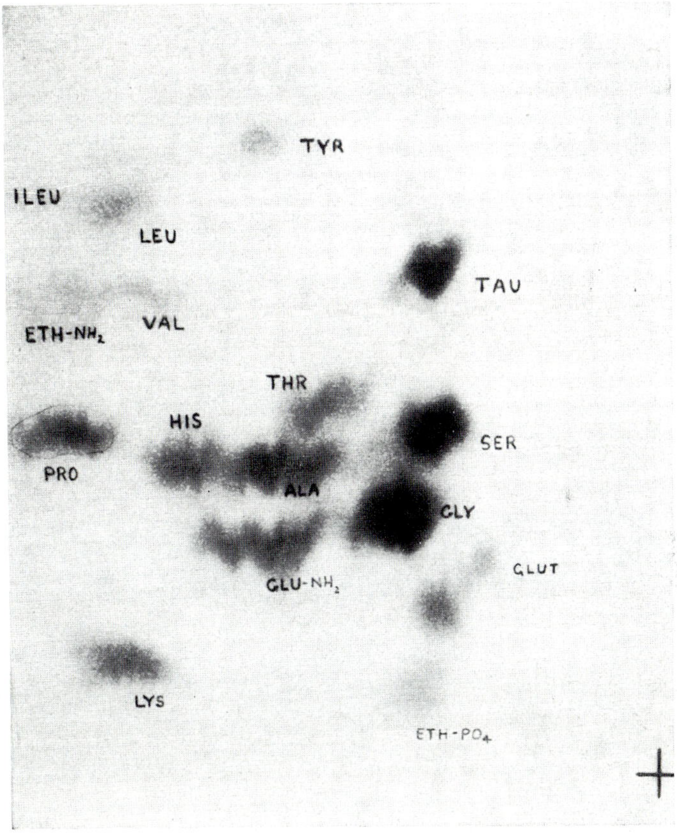

Fig. 10. A paper chromatogram of the urine of a child with galactosaemia.

arginine, and ornithine[56, 57]. But for the fact that cystine has a low solubility in water and urine and consequently tends to form calculi when excreted in a supersaturated concentration these patients would remain healthy as the daily loss of these amino acids is easily compensated for in the dietary intake. The urine chromatogram of a typical cystinuric patient is shown in Fig. 9. The cystine is oxidized before chromatography as the resulting cysteic acid behaves more reliably than the parent substance which tends to streak. The plasma amino acid levels are normal in these subjects but the cystine level tends to be slightly low[58]. The metabolic error in this disease is apparently an inability, partial or complete, to reabsorb cystine from the glomerular

References p. 217/219

filtrate. Dent *et al.*[58] found that the rate of clearance of cystine in cystinuric patients was about the same as the rate of glomerular filtration. The renal clearances of lysine, arginine and ornithine are also raised[59, 60]. It seems therefore that in this disease there is an error in the transport of these four amino acids through the lumen of the renal tubule. That these amino acids should be involved, and no others, may be because they have a certain similarity of chemical structure, for instance a similar spacing between the two amino groups. This was suggested by Dent and Rose[56].

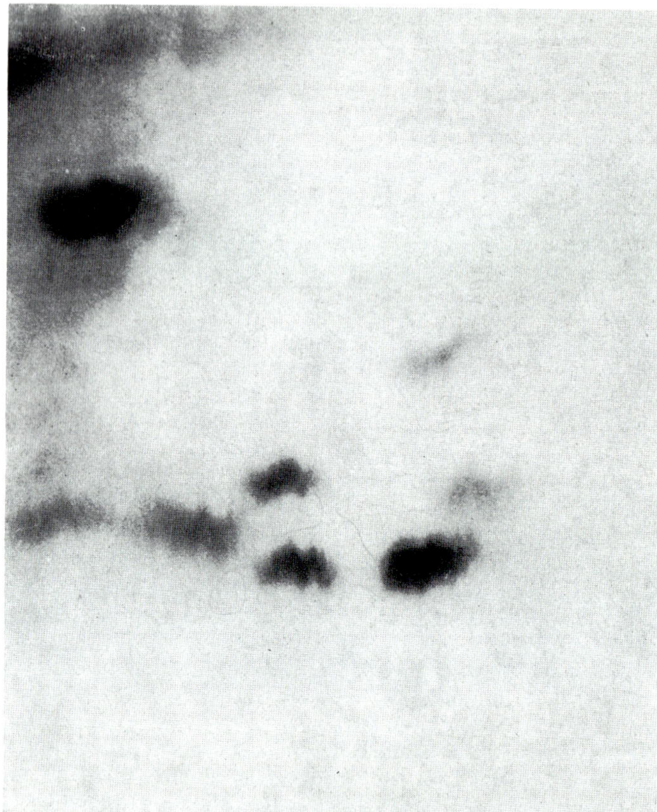

Fig. 11. A paper chromatogram of the urine of a patient with phenylketonuria.

Further evidence that these amino acids do, in fact, share the same transport mechanism has been obtained by Robson and Rose[61] and Smeenk[62]. They found that a considerably increased excretion of cystine and, indeed, of the other two amino acids, could be induced artificially in normal individuals by the infusion, by vein, of lysine and of arginine.

Galactosaemia. It has been found that certain babies cannot tolerate milk. Infants with this defect, which has been found to be hereditary, appear normal at birth but

References p. 217/219

after a few days on a milk diet begin to vomit and fail to thrive. If the milk diet is continued other clinical features such as abdominal distension, hepatomegaly and possibly jaundice may show. There is also a gross-generalized aminoaciduria, shown in Fig. 10, and an excessive excretion of galactose in the urine. These children are unable to metabolize galactose, which is derived from the lactose in the milk, in the normal way. Enzyme studies on the red blood cells and on liver biopsy samples from affected patients have established that they lack an enzyme necessary for the conversion of galactose to glucose 1-phosphate[63]. This enzyme has since been tracked down to galactose 1-phosphate uridyl transferase[64]. Fortunately, if the source of

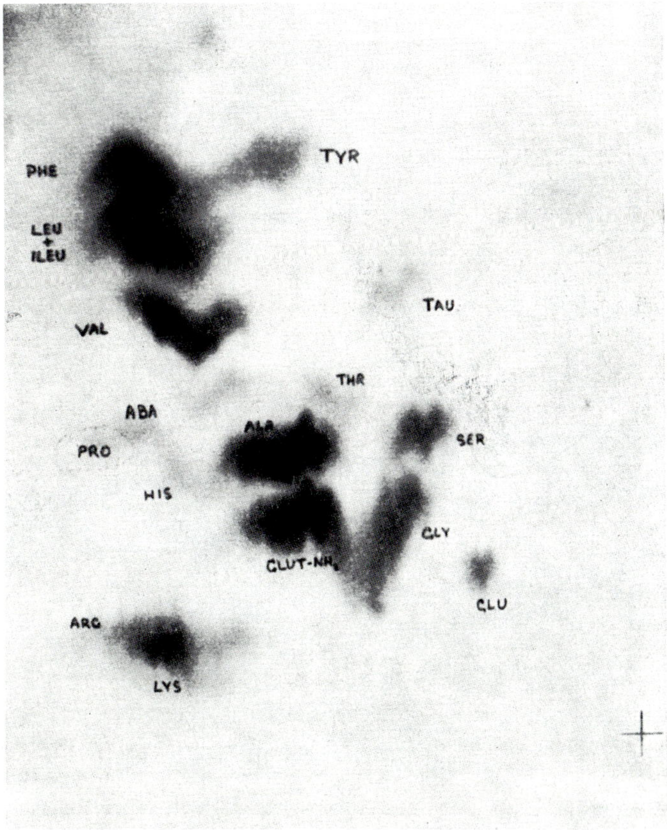

Fig. 12. A paper chromatogram of the blood plasma from the phenylketonuric patient.

galactose is removed from the diet then the infants recover but, however, if the exposure to galactose has been prolonged some of the damage is irreversible. With the removal of galactose from the diet the aminoaciduria clears up but reappears on restoring the galactose. It is therefore considered, in this disease, that the aminoaciduria is of a secondary nature only and may be due to some toxic effect on the kidney tubule caused either by galactose itself or by one of its metabolites.

References p. 217/219

The "overflow" aminoacidurias

Phenylketonuria. In 1934, FOLLING[65] observed that there was an excessive excretion of phenylpyruvic acid in the urine of certain mentally defective patients. Further studies established that the disease was hereditary and that the defect seemed to be handed on as a typical Mendelian recessive character. It is now known that phenylpyruvic acid is not the only metabolite which is excreted excessively in these cases; phenylacetic acid, phenyllactic acid and phenylacetylglutamine are also impli-

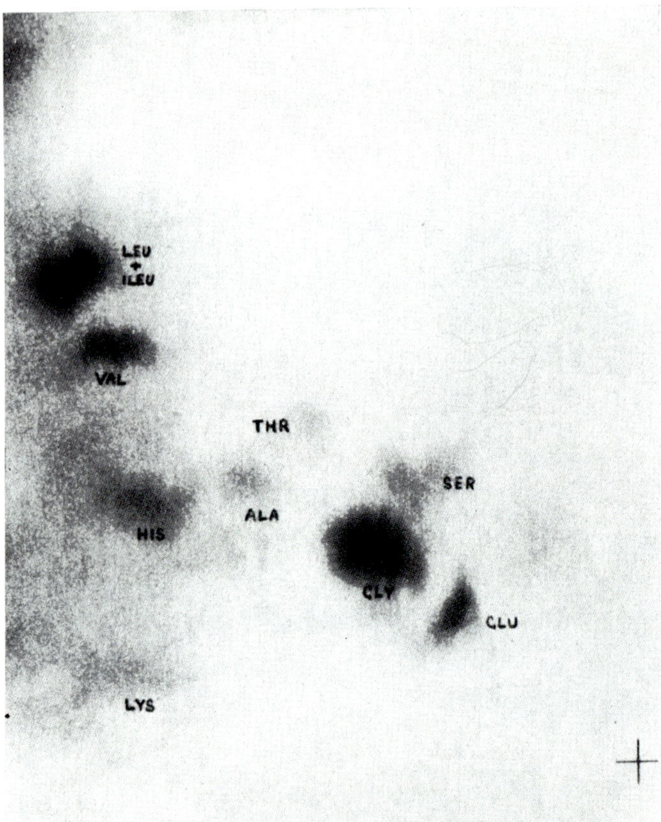

Fig. 13. A paper chromatogram of the urine of a child with maple-syrup-urine disease.

cated[65a, 66, 67]. Phenylalanine, the immediate precursor of phenylpyruvic acid, is also present in the urine in high concentration (Fig. 11) and, indeed, is also found in excessive amount in the blood plasma (Fig. 12) and the C.S.F.[68–71]. The site of the metabolic lesion in this disease is at the conversion of phenylalanine to tyrosine and these patients lack the phenylalanine hydroxylating enzyme which catalyses this step[72, 73]. Hence in this disease, where one of the major pathways of phenylalanine metabolism is blocked, this amino acid tends to accumulate in the body tissues and fluids. Some of the phenylalanine is deaminated to form phenylpyruvic acid and thence, by further reactions, to other derivatives all of which, having a high renal clearance, are readily excreted in the urine. However, phenylalanine itself is almost all reabsorbed by the kidney tubule and so cannot be excreted in sufficient amounts to lower the plasma level to normal limits. This can only be done by limiting the intake of phenylalanine in the diet. There are certain other biochemical abnormalities associated with the disease which have yet to be explained, one of which is the reason for an apparent disturbance of tryptophane metabolism leading to an increased urinary excretion of indolylacetic acid and indolyllactic acid.

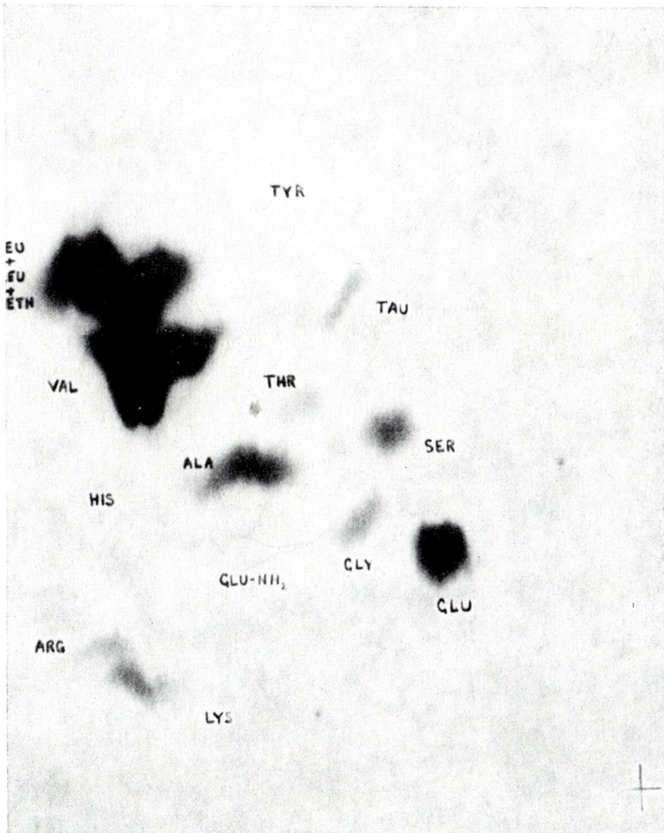

Fig. 14. A paper chromatogram of the blood plasma from the maple-syrup-urine disease case.

Maple-syrup-urine disease. In 1954, MENKES, HURST AND CRAIG[74] described a family in which four out of six children were severely affected with neurological disturbances and failed to survive longer than a few months. They drew attention to the fact that the urine of the children had a strange smell like that of maple syrup. In 1958, WESTALL, DANCIS, MILLER AND LEVITZ[75] reported a further case and showed that the urine contained abnormally high amounts of the branched chain animo acids leucine, isoleucine and valine (Fig. 13). The plasma levels of these amino acids were also markedly elevated (Fig. 14) as also was the level of methionine whilst the cystine

References p. 217/219

was barely perceptable. Soon after it was found that the keto acid analogues of the branched amino acids *viz.* α-ketoisocaproic, α-keto-β-methylvaleric and α-ketoisovaleric acids were excreted in the urine in excessive amounts[76,101,102]. There is, therefore, a similar state of affairs in this disease as exists in phenylketonuria except that the amino acids involved are different and in this newer disease the effects tend to be more lethal. Further cases of this disease are frequently being found but so far the cause of the metabolic error has not been proved. It is strange that these three amino

Fig. 15. A paper chromatogram of the urine from a patient with argininosuccinic aciduria.

acids leucine, isoleucine and valine, and possibly also methionine, all essential amino acids to man, should all be involved in the metabolic disturbance. DENT AND WESTALL[77] carried out a series of dietary experiments on a child with this disease during which the intake of the branched amino acids were limited to the minimum requirements for growth. The analogous keto acids were no longer excreted and the characteristic smell disappeared from the urine. We wondered whether perhaps only one of the branched amino acids was truly implicated and that the others were involved by some form of transport inhibition. However by feeding the limited amino acids back to the diet, in normal amounts, separately and one at a time, we were unable to find the culprit. The general opinion of those who have studied the biochemical aspects of the disease is that the metabolic block is at the stage of decarboxylation of the branched ketoacids where the chain is shortened by one C atom and the residue condenses with coenzyme A. This seems to be the most likely stage since later in the series of degradations the pathways of the branched acids diverge[78].

Fig. 16. A paper chromatogram of the urine of a typical β-aminoisobutyric acid excretor.

Argininosuccinic aciduria. In 1958, ALLAN, CUSWORTH, DENT AND WILSON[79] described a family in which two children, both mentally defective, excreted a large amount of an unknown amino acid-like substance in the urine (Fig. 15). It was, however, absent from the urine of the parents and of two other apparently normal siblings. The unknown substance could be detected in the plasma of the affected children and also in the C.S.F. where, curiously enough, the concentration was over twice as high as in the plasma. Later, the strange substance was isolated from the urinary source and characterized as argininosuccinic acid (ASA)[80].

ASA was postulated to occur as an intermediate in the ornithine cycle[81]. It has been found in plant tissue[82] and in mould[83] but it had not been found previously to occur naturally in mammalian tissues or fluids. ASA can be synthesized *in vitro* by

References p. 217/219

the condensation of citrulline and aspartic acid in the presence of ATP and an enzyme present in liver tissue. It can also be prepared similarly by the action of argininosuccinase (ASase), which will reverse its usual action if the substrate is loaded with arginine and fumaric acid[84, 85]. RATNER *et al.*[84] found ASase activity in liver and kidney and lately WALKER[86] has found similar activity in a number of other body organs including the brain.

We can only speculate, at this stage, as to where the metabolic block in this disease

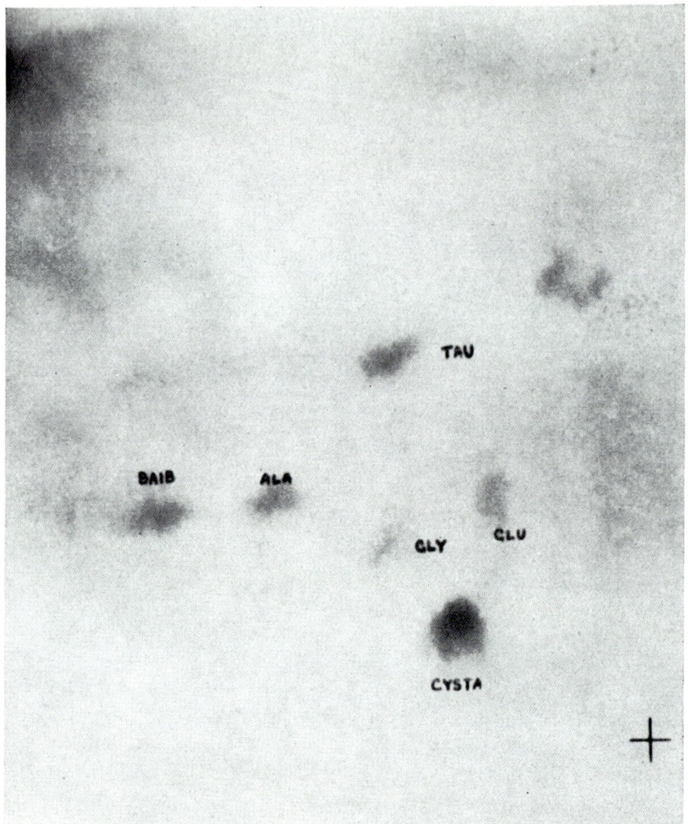

Fig. 17. A paper chromatogram of the urine of the only known case of cystathioninuria.

might be found. The most obvious site would be at the point of breakdown of ASA into arginine and fumaric acid and this reaction is catalysed by the enzyme ASase. We have recently been able to show that ASase is present in normal red blood cells. On testing the red cells of the affected children we found that the ASase activity was reduced to one-fifth or less[87]. If this lowered activity in the red cells reflects a similar condition in the other cells in the body where ASA is being synthesized then it is likely that ASA will accumulate, flood out of the cells into the extracellular system and be excreted in the urine. If there was no ASase at all in their tissues it is difficult to see how they could make urea, as in fact they do. Perhaps in such a large organ as

References p. 217/219

the liver even the reduced concentration of ASase is sufficient to enable some of the ASA to be further metabolized to urea.

β-Aminoisobutyric-aciduria. In the course of routine studies of the amino acids in urine by paper chromatography it was observed that a number of apparently normal healthy people, about one in twenty, excreted as much β-aminoisobutyric acid (BAIBA) as they did glycine[13, 14]. The amount of BAIBA excreted (Fig. 16) remained constant for the individual, was uninfluenced by diet, and from family studies it appeared to be under genetic control. The frequency of incidence of "high excretors" varies with the racial groups and are highest in the Mongolian races and in the American Indians[88, 89]. BAIBA seems to be present in most human urine samples but usually only in trace amounts. The blood plasma level of BAIBA is barely detectable and it is probably one of those substances which are not reabsorbed at all by the kidney tubule[89]. It is thought that BAIBA arises from the breakdown of DNA via thymine[90] but there is also an alternative source arising from the degradation of valine[91]. The most likely explanation, at present, for the high excretion of BAIBA by some people is that they have a partial block in its further metabolism.

Cystathioninuria. Another interesting form of aminoaciduria has recently been discovered by HARRIS[55] during a systematic screening of the urines of patients in a mental deficiency institution. This patient excreted approx. 500 mg/day of cystathionine (Fig. 16) and the output was markedly increased by feeding methionine. Cystathionine is a postulated intermediate in the series of reactions taking place in the conversion of methionine to cystine. This substance, like argininosuccinic acid and β-aminoisobutyric acid in the previous diseases, is probably an intracellular substance only and rarely reaches the extracellular fluids. Cystathionine has recently been found in brain tissue[92]. Although no other evidence is available it seems likely that this condition is caused by a lack or shortage of the enzyme responsible for cleaving cystathionine to form cysteine and homoserine.

REFERENCES

1 S. P. L. SORENSEN, *Biochem. Z.*, 7 (1907) 45.
2 F. W. FOREMAN, *Biochem. J.*, 14 (1920) 451.
3 D. D. VAN SLYKE, *J. Biol. Chem.*, 9 (1911) 185.
4 D. D. VAN SLYKE, D. A. MACFADYEN AND P. HAMILTON, *J. Biol. Chem.*, 141 (1941) 671.
5 R. CONSDEN, A. H. GORDON AND A. J. P. MARTIN, *Biochem. J.*, 38 (1944) 224.
6 C. E. DENT, *Lancet*, i (1947) 637.
7 D. H. SPACKMAN, W. H. STEIN AND S. MOORE, *Anal. Chem.*, 30 (1958) 1190.
8 D. D. VAN SLYKE AND C. M. MEYER, *J. Biol. Chem.*, 12 (1912) 399.
9 C. E. DENT, in S. C. DYKE, *Recent Advances in Clinical Pathology*, Churchill, London, (1951) p. 238.
10 H. H. TALLAN, S. MOORE AND W. H. STEIN, *J. Biol. Chem.*, 211 (1954) 927.
11 H. N. CHRISTENSEN AND J. A. STREICHER, *J. Biol. Chem.*, 175 (1950) 95.
12 H. R. CRUMPLER, C. E. DENT AND O. LINDAN, *Biochem. J.*, 47 (1950) 223.
13 H. R. CRUMPLER, C. E. DENT, H. HARRIS AND R. G. WESTALL, *Nature*, 167 (1951) 307.
14 K. FINK, R. B. HENDERSON AND R. M. FINK, *Proc. Soc. Exptl. Biol. Med.*, 78 (1951) 135.
15 H. H. TALLAN, W. H. STEIN AND S. MOORE, *J. Biol. Chem.*, 206 (1954) 825.
16 M. E. CARSTON, *J. Am. Chem. Soc.*, 74 (1952) 5954.
17 P. BOULANGER, G. BISERTE AND F. COURTOT, *Bull. Soc. Chim. Biol.*, 34 (1952) 366.
18 R. G. WESTALL, *Biochem. J.*, 60 (1955) 247.
19 H. HANSON AND S. FITTKAU, *Z. physiol. Chem., Hoppe-Seyler's*, 313 (1958) 152.
20 O. R. JAGENBURG, *Scand. J. Clin. and Lab. Invest., Suppl. No. 43*, 1959.
21 C. E. DENT, *Federation Proc.*, 6 (1947) 390.

[22] P. Boulanger and G. Biserte, *Compt. rend.*, 232 (1951) 1451.
[23] D. Mauzerall and S. Granick, *J. Biol. Chem.*, 219 (1956) 435.
[24] F. Irrevere, R. L. Evans, A. R. Hayden and R. Silber, *Nature*, 180 (1957) 704.
[25] W. H. Stein, A. C. Paladini, C. H. W. Hirs and S. Moore, *J. Am. Chem. Soc.*, 76 (1954) 2848.
[26] J. M. Luck and A. Wilcox, *J. Biol. Chem.*, 205 (1953) 859.
[27] H. K. Berry, *Am. J. Phys. Anthropol.*, 11 (1953) 559.
[28] F. Souchon, *Z. ges. exptl. Med.*, 118 (1952) 219.
[29] G. Reuter, *Inaug. Dis.*, Kiel, 1952.
[30] E. S. Nasset and R. H. Tully, *J. Nutrition*, 44 (1951) 477.
[31] P. Soupart, *Clin. Chim. Acta*, 4 (1959) 265.
[32] C. Voge, *Brit. Med. J.*, 2 (1929) 829.
[33] P. Soupart, *Acta Clin. Belg.*, 9 (1954) 318.
[34] S. Simon, *Kinderheilk.*, 2 (1911) 1.
[35] H. Bickel, *Thesis*, University of Birmingham, England, 1952.
[36] L. I. Woolf and A. P. Norman, *J. Pediatrics*, 50 (1957) 271.
[37] S. P. Datta and H. Harris, *J. Physiol.*, 114 (1951) 39 P.
[38] R. G. Westall, *Biochem. J.*, 55 (1953) 244.
[39] J. Logothetes, *J. Neurol. (Minneapolis)*, 5 (1955) 767.
[40] B. S. Walker, N. C. Telles and E. J. Pastore, *A. M. A. Arch. Neurol. Psychiat.*, 73 (1955) 149.
[41] H. G. Knauff, *Nature*, 182 (1958) 937.
[42] E. Roberts and S. Frankel, *Federation Proc.*, 9 (1950) 219; *J. Biol. Chem.*, 187 (1950) 55.
[43] G. A. Rose, A. C. Kerr, *Quart. J. Exptl. Physiol.*, 43 (1958) 160.
[44] H. K. Berry and L. Cain, *Univ. Texas Publ., No. 5109* (1951) 71.
[45] L. H. Schneyer and L. K. Levin, *J. Appl. Physiol.*, 7 (1955) 609.
[46] H. N. Chauncy, F. Lionetti, R. A. Winer and V. F. Lisanti, *J. Dental Research*, 33 (1954) 321.
[47] G. M. Embden and H. Tachau, *Biochem. Z.*, 28 (1910) 230.
[48] S. Rothman and M. B. Sullivan, *J. Invest. Dermatol.*, 13 (1949) 319.
[49] C. E. Dent, *Biochem. J.*, 41 (1947) 240.
[50] D. C. Cusworth and C. E. Dent, *Biochem. J.*, 74 (1960) 550.
[51] R. D. Clay, E. M. Darmady and M. Hawkins, *J. Pathol. Bacteriol.*, 65 (1953) 551.
[57] H. Bickel, W. C. Smallwood, J. M. Smellie, H. S. Baar and E. M. Hickmans, *Acta Paediat.*, Suppl. 90, 42 (1952).
[53] D. N. Baron, C. E. Dent, H. Harris, E. W. Hart and J. B. Jepson, *Lancet*, ii (1956) 421.
[54] J. B. Jepson, *Biochem. J.*, 64 (1956) 14 P.
[55] H. Harris, in *Human biochemical genetics*, Cambridge University Press, 1959, p. 92.
[56] C. E. Dent and G. A. Rose, *Quart. J. Med.*, 20 (1951) 205.
[57] W. H. Stein, *Proc. Soc. Exptl. Biol. Med.*, 78 (1951) 705.
[58] C. E. Dent, B. Senior and J. M. Walshe, *J. Clin. Invest.*, 33 (1954) 1216.
[59] V. Arrow and R. G. Westall, *J. Physiol. (London)*, 142 (1958) 14.
[60] P. D. Doolan, H. A. Harper, M. E. Hutchin and E. L. Alpin, *Am. J. Med.*, 23 (1957) 416.
[61] E. B. Robson and G. A. Rose, *Clin. Sci.*, 16 (1957) 75.
[62] D. Smeenk, unpublished results.
[63] V. Schwartz, L. Goldberg, G. M. Komrower and A. Holzel, *Biochem. J.*, 62 (1956) 34.
[64] K. J. Isselbacher, E. P. Anderson, K. Karahashi and H. M. Kalckar, *Science*, 123 (1956) 635.
[65] A. Folling, *Z. physiol. Chem., Hoppe-Seyler's*, 227 (1943) 169.
[65a] L. I. Woolf and D. G. Vulliarmy, *Arch. Disease Childhood*, 26 (1951) 487.
[66] E. A. Zeller, *Helv. Chim. Acta*, 26 (1943) 169.
[67] M. Dann, E. Marples and S. Z. Levine, *J. Clin. Invest.*, 22 (1943) 87.
[68] E. Borek, A. Brecher, G. A. Jervis and H. Waelsch, *Proc. Soc. Exptl. Biol. Med.*, 75 (1950) 86.
[69] G. A. Jervis, R. J. Block, D. Bolling and E. Kanze, *J. Biol. Chem.*, 134 (1940) 105.
[70] H. Bickel, J. Gerrard and E. M. Hickmans, *Acta Paediat.*, 43 (1954) 64.
[71] M. D. Armstrong and N. L. Low, *Proc. Soc. Exptl. Biol. Med.*, 94 (1957) 142.
[72] G. A. Jervis, *Proc. Soc. Exptl. Biol. Med.*, 82 (1953) 514.
[73] H. W. Wallace, K. Moldave and A. Meister, *Proc. Soc. Exptl. Biol. Med.*, 94 (1957) 632.
[74] J. H. Menkes, P. L. Hurst and J. M. Craig, *Pediatrics*, 14 (1954) 462.
[75] R. G. Westall, J. Dancis, S. Miller and M. Levitz, *Federation Proc.*, 17 (1958) 334.
[76] D. Y. Mackenzie and L. I. Woolf, *Brit. Med. J.*, 1 (1959) 90.
[77] C. E. Dent and R. G. Westall, *Arch. Disease Childhood*, 36 (1961) 259.
[78] M. J. Coon and S. Gurin, *J. Biol. Chem.*, 180 (1949) 1159.

79 J. D. ALLAN, D. C. CUSWORTH, C. E. DENT AND V. WILSON, *Lancet*, i (1958) 182.
80 R. G. WESTALL, *Proc. Intern. Congr. Biochem., 4th Congr., Vienna*, 13, 34 (1958) 168.
81 S. RATNER AND A. PAPPAS, *J. Biol. Chem.*, 179 (1949) 1183.
82 D. C. DAVISON AND W. H. ELLIOTT, *Nature*, 169 (1952) 313.
83 J. R. S. FINCHAM AND J. B. BOYLEN, *Biochem. J.*, 61 (1955) 23.
84 S. RATNER, B. PETRACK AND O. ROCHOVANSKY, *J. Biol. Chem.*, 204 (1953) 95.
85 J. B. WALKER, *Proc. Natl. Acad. Sci. U.S.*, 38 (1952) 561.
86 J. B. WALKER, *Proc. Soc. Exptl. Biol. Med.*, 98 (1958) 7.
87 R. G. WESTALL AND S. TOMLINSON, *Proc. Intern. Congr. Biochem., 5th Congr. Moscow 16*, 103 (1961) 366.
88 S. M. GARTLER, I. L. FIRSCHEIN AND T. GIDESPOW, *Acta Genet.*, 6 (1957) 435.
89 H. E. SUTTON, in *The Metabolic Basis of Inherited Disease*, McGraw-Hill, New York, 1960, p. 803.
90 K. FINK, R. E. CLINE, R. B. HENDERSON AND R. M. FINK, *J. Biol. Chem.*, 224 (1956) 425.
91 W. G. ROBINSON, R. NAGLE, R. K. BACHAWAT, F. P. KUPIECKI AND M. J. COON, *J. Biol. Chem.*, 224 (1957) 1.
92 H. H. TALLAN, S. MOORE AND W. H. STEIN, *J. Biol. Chem.*, 230 (1958) 707.
93 M. P. BRIGHAM, W. H. STEIN AND S. MOORE, *J. Clin. Invest.*, 39 (1960) 1633.
94 W. H. STEIN AND S. MOORE, *J. Biol. Chem.*, 211 (1954) 915.
95 D. F. EVERED, *Biochem. J.*, 62 (1956) 416.
96 S. W. HEIR AND O. BERGEIM, *J. Biol. Chem.*, 163 (1946) 129.
97 H. A. HARPER, M. E. HUTCHIN AND J. R. KIMMEL, *Proc. Soc. Exptl. Biol. Med.*, 80 (1952) 768.
98 P. D. DOOLAN, H. A. HARPER, M. E. HUTCHIN AND W. W. SHREEVE, *J. Clin. Invest.*, 34 (1955) 1294.
99 R. H. McMENAMY, C. C. LUND AND J. L. ONCLEY, *J. Clin. Invest.* 36 (1957) 1672.
100 W. H. STEIN, *J. Biol. Chem.*, 201 (1953) 15
101 J. H. MENKES, *Pediatrics*, 23 (1959) 348.
102 J. DANCIS, M. LEVITZ, S. MILLER AND R. G. WESTALL, *Brit. Med. J.*, 1 (1959) 91.

FREE AMINO ACIDS OF BLOOD AND URINE IN THE HUMAN

PIERRE SOUPART

Department of Biochemistry and Nutrition, Faculty of Medicine, University of Brussels, Brussels (Belgium)

Interest in free amino acids of blood and urine in man may arise from many sources. For instance, physicians and more specially pediatricians might be interested in amino acid excretion, which exhibits disturbances related to several pathological conditions in children. Kidney physiologists or obstetricians might be interested in urinary amino acids, as was the case at a time, when histidine excretion measurement was thought to furnish the basis of a "simple chemical test" for pregnancy diagnosis. Owing to kidney function, or dysfunction, free amino acids of urine are necessarily related to plasma free amino acid levels. More recently, many important studies in the field of cancer (leukemia) and also in that of radiation therapy or injury effects, as well as studies on free amino acids available for protein synthesis, have reinforced the interest of research workers in free amino acid composition of body fluids and its possible alterations. As our knowledge of free amino acids in body fluids has developed chronologically, and it has been particularly true in the case of urine amino acids, a striking observation has arisen, namely, that much information was gathered in the study of aminoaciduria in disease long before an exact picture of the normal situation regarding amino acid excretion became available. Despite the number of invaluable data collected in this way, it has however often led to misleading interpretation of the pathological findings, this being mainly due to the very nature of methods used for amino acid determination. It is only recently that a rather exact and complete account of amino acid excretion has been given for normal subjects[1, 1a]. There was therefore a need for thorough revision of our knowledge of free amino acids in normal and pathological body fluids.

The purpose of this report is threefold. First, it will describe, qualitatively and quantitatively, the free amino acid composition of normal human blood plasma and urine, both in adults and children. Secondly, as free amino acids of blood plasma are necessarily closely related to free amino acids of tissue cells and blood cells, this report will describe a tentative model of quantitative distribution of free amino acids between blood plasma and the various types of blood cells. Thirdly, as this report is also intended to be a progress report, some problems of more limited scope but currently under examination, will be dealt with which have arisen from observations made during studies on intracellular free amino acid composition as well as studies on radiation and protein-malnutrition effects. As will be seen, radiation and protein-malnutrition effects seem to share some particular features, at least as regard to two amino acids, namely taurine and β-aminoisobutyric acid. In the light of these recent developments, a provisional classification of free amino acid excretion patterns in pathological disorders will also be submitted. Before closing these introductory remarks, it seems advisable to stress the point that what is referred to in this report

References p. 261/262

as "free amino acids" is by no means a simple and logical concept but a conventional and provisional one, depending largely on the nature of the method used for the purpose of amino acid separation and determination. The author's laboratory has specialized very early (1950) in the analytical use of ion-exchange chromatography and accordingly all experimental data submitted below have been obtained by one or other variant of the MOORE AND STEIN ion-exchange chromatographic method. The group of substances referred to here as "free amino acids" has thus to be defined in terms of ion-exchange analysis. It includes free amino acids, substituted free amino acids or amino acid derivatives in which the amino group stays free, enabling them to react with ninhydrin. Examples of such substances are taurine, glutamine, or asparagine (the last two being amino acids combined with ammonia) or tyrosine-*O*-sulphate[2], an amino acid combined with sulfuric acid by a linkage which does not involve the amino group. The group also includes alcohol amines which are other types of amino acid derivatives in which the amino group remains free, for instance ethanolamine, which corresponds to decarboxylated serine. Chromatographic separation on ion-exchange columns occurs when unhydrolyzed urine is used and these substances are detected by means of the ninhydrin reaction. Additional peaks show up on elution curves obtained from unhydrolyzed urine, comprising urea, ammonia, probably also small amounts of peptides (BOULANGER AND BISERTE[3, 4]; WESTALL[5]) and eventually still unidentified amino acids but, for obvious reasons, these substances are not included in the group referred to here as "free amino acids". On the other hand, the term "aminoaciduria" has not been consistently used with the same meaning. In the past 10 years, according to certain clinical biochemists, it referred to pathological conditions only, just as the word uremia refers to a disorder and not to blood-urea level in general. It is also used in a broader sense, referring to amino acid excretion in general, whether it is normal or not, in the same way as the word glycemia is used. In this report, the term "aminoaciduria" will be used in the latter sense since there is a significant output in normal conditions and since there are normal conditions in which amino acid output may be different than usual, although the condition is not pathological, an instance of which is found in the hyperaminoaciduria which occurs in the course of normal pregnancy[6].

MATERIAL AND METHODS

Ion-exchange chromatography

Ion-exchange chromatographic method, as worked out by S. MOORE AND W. H. STEIN, is presently the most satisfactory method for amino acid analysis. According to the chronological development of this method, four different procedures may be distinguished:

a) 1951 method on Dowex-50-X8 columns[7], which is suitable for protein-hydrolyzates analysis.

b) 1954 method on Dowex-50-X4 150-cm column[8, 9], which is particularly adapted to complex physiological fluids analysis.

c) 1958 method on Amberlite-IR-120 column[10], also well adapted to physiological fluids analysis but characterized by an increased speed of operation.

d) 1958 completely automatic procedure on Amberlite-IR-120-X8 150 and

References p. 261/262

50-cm columns[1], which is more rapid and less laborious than the techniques employed heretofore.

Data presented in this report have been obtained by use of the 1954 technique, referred to here as the "manual" procedure, as well as by use of the 1958 "automatic" procedure. With regard to blood and urine amino acid analysis, the advantages of the 1954 procedure have been discussed at length elsewhere[1]. The overall precision of these two ion-exchange methods is ± 5%, except when the amounts of some of the amino acids contained in the sample are very small, in which cases the relative precision becomes less satisfactory.

However, there are three amino acids for which recovery is low. First, methionine yield is consistently 90% of theoretical, and has to be corrected for this loss. Secondly, tryptophane is destroyed to a variable but large extent (40–60%) during the chromatographic process. As it seems to be present as traces only in the free state in biological fluids and owing to the loss during chromatographic analysis, no value for tryptophane concentrations will be reported here. Thirdly, part of the free glutamic acid escapes determination because of partial cyclization into pyrrolidone carboxylic acid which does not react with ninhydrin. Values given below for free glutamic acid are thus to be taken as minimal.

On the other hand, values reported for glutamine are somewhat too high; the glutamine peak, in the case of blood plasma for instance, also contains asparagine in the proportion of about one tenth of the glutamine present[12]. When urine is examined this peak might also contain sarcosine if Dowex 50-X4 is used but is completely free from it if chromatographed on Dowex-50-X5 (manual procedure). When the automatic procedure is used, sarcosine, if present, is also completely separated from the mixture of glutamine and asparagine.

Manual procedure (MOORE AND STEIN's 1954 technique). Most of the data submitted here have been obtained by use of the MOORE AND STEIN 1954 procedure, ion-exchange chromatography on Dowex-50-X4 columns[8,9] associated with an automatic siphon device for automatic collection instead of the original drop-counting procedure[13]. The main drawback of the manual procedure lies in the fact that it is rather time- and labour-consuming, a complete analysis requiring about 1000 individual colorimetric determinations at a rate not exceeding 200 analyses per day.

Automatic procedure. This completely automatic procedure, described by SPACKMAN, STEIN AND MOORE in 1958[11], is by far the most convenient method and, in our opinion, should definitely become the standard method for any amino acid investigation. Notwithstanding the heavy initial investment involved and the fact that its operation still requires much technical skill, the economy of labour it allows for and the precision of the results it furnishes are well worth the investment. We have had the opportunity to verify at length the assessment of its authors[14], that the greatest value of the automatic recording equipment resides in the fact that it acts like an objective and constant observer of the full course of a chromatographic experiment. From such observation valuable and unexpected information has often been gained. Some minor technical modifications have been introduced in the original procedure which allow for greater ease in continued operation and maintenance[15]. Such a machine has been continuously operated in the department since June 1959 and a good deal of the data submitted here have been obtained by its use.

References p. 261/262

Sample preparation

Blood plasma. Fasting blood samples (50 ml) are collected in a plastic centrifuge tube, directly from a silicone-coated (Arquad 2 C, Armour and Co.) large gauge needle, on dry powdered anticoagulant (sodium polyethanolsulfonate, "Liquoide" Roche, F. Hoffmann, La Roche and Co, Basel, Switzerland). An alternative procedure is also described below. The plasma is immediately separated by centrifugation at 5000 × *g* for 10 min. Two 10-ml aliquots of plasma, numbered 1 and 2 according to order of sampling, are deproteinized at once by 50 ml of 1% picric acid solution, according to the method of HAMILTON AND VAN SLYKE[16], and centrifuged. From the supernatant of each sample, two 25-ml aliquots are separately processed by the desalting method described by STEIN AND MOORE[12], if the automatic procedure is to be used. In the case of the manual procedure a 50-ml aliquot is required. After desalting, concentration and adjustment to pH 2.0, the two samples, equivalent each to 4.16 ml of the original plasma, will be suitable one for acidic and neutral amino acid separation and the other for basic amino acid separation in the double-stage automatic procedure. Aliquots prepared from sample No. 1 will be preferentially used for analysis since, being taken from the upper layer of plasma, they are probably free of light blood cells like platelets if any of these had failed to sediment. Recently, BRIGHAM, STEIN AND MOORE have described a new procedure for blood-samples preparation which allows for the separate determination of cysteine and cystine concentrations in blood plasma[17], but this procedure has still not been applied in our investigations.

Urine. As a rule urine analyses are carried out on aliquots of quantitative 24-h collections, irrespective of diet. In some instances 24-h collections have been fractionated in sub-collections of variable length of time when desirable for location of a peak excretion. During the 24-h collection, urines are collected in a one-gallon bottle containing 5 ml of toluene and 5 ml of chloroform as preservative. When collection is completed, total volume is recorded and an aliquot sufficient for repeated analyses is stored in a plastic vial and kept in the deep-freezer at −35° pending analysis. In some pathological specimen, when proteins were present, urine samples have been deproteinized by the method used for plasma samples[12]. 1 or 2 ml of urine brought to pH 2.0 by HCl addition are used for analysis depending upon the 24-h urine output, the latter one being smaller or greater than 1 l.

Blood cells and plasma. Separation of the four constituents of blood in a pure form is by no means an easy task, especially when dealing with small volumes of blood. The following procedure, slightly modified of that of TRUHAUT[18], has been used, but, as will clearly be seen from discussion of analytical results, suffers from certain drawbacks, which are to be avoided if possible:

1) 100-ml blood collection in unwettable centrifuge tube, using 5 ml of a 2% disodium EDTA solution in saline as an anticoagulant.

2) Separation of plasma by undelayed centrifugation at 5000 × *g* for 45 min in the refrigerated centrifuge at 4°. Deproteinization of two 10-ml aliquots of plasma is performed as described under *Blood plasma.* A correction factor, owing to dilution of the blood by EDTA solution has to be calculated.

3) Dispersion of the cell sediment, breaking thoroughly the buffy coat, in an equal volume of a buffered solution at pH 7.3* containing 2% polyvinylpyrrolidone (Société parisienne d'expansion chimique, Paris, France). Spontaneous sedimentation of the erythrocytes for 60 min, the centrifuge tube being placed at 45° angle in a 37° water bath. Collection of the supernatant, which contains leucocytes and platelets, and storage in a plastic centrifuge tube in the ice box at 4°. This spontaneous sedimentation is repeated twice, all supernatants being poured together.

4) Two washings of the erythrocytes with saline and sampling of two aliquots of packed red cells, respectively of 4 and 8 g, which are then lysed by 6 and 12 ml of distilled water. Each of the lysates is deproteinized by 5 vol. of 1% picric acid solution by the method used for plasma.

5) Separation of the leukocytes from the platelets in the polyvinylpyrrolidone supernatant by means of repeated centrifugations of short duration: 1.5 min at less than 500 $\times$ g in the refrigerated centrifuge at 4°. This procedure, which may be repeated three to four times, is controlled by checking the supernatant and sediment contents by examination in phase microscopy after each step. Only the first and second sediments, which contain mainly leucocytes but a very few platelets and red cells, are collected and quantitatively washed in another centrifuge tube, previously weighed, with pH 7.3 buffered isotonic solution *without* polyvinylpyrrolidone. The washing of the leucocytes sediment is repeated twice, centrifuging each time for 2.5 min at less than 500 $\times$ g at 4°. The supernatant is then poured off, drained thoroughly with filter paper and the weight of the wet sediment is recorded. The cells are lysed in 5 ml of distilled water, the solution is quantitatively transferred in a medium-size tissue grinder and the final volume made up to 10 ml with the washings. The solution is then ground to insure complete rupture of the cells, and deproteinized by 5 vol. of 1% picric acid solution. After centrifugation, the supernatant is deep-frozen pending desalting and analysis.

6) The supernatant obtained in 5), which contains platelets is centrifuged at 5000 $\times$ g for 30 min at 4° in a plastic centrifuge tube previously weighed. The sediment is washed twice in the same way in pH 7.3 isotonic buffered solution without polyvinylpyrrolidone, and the last sediment is processed the same way as the leucocyte sediment.

After desalting to remove picric acid[12], all deproteinized extracts are concentrated under reduced pressure in a Craig rotary evaporator[19] to a volume of approx. 1 ml. The extracts of the 4 and 8 g of red cells are brought to pH 2.0 and quantitatively transfered to 150 and 50-cm ion-exchange columns, respectively. As glutathione is eluted in the range just in between those of threonine–glutamine and glutamic acid an alternative procedure is to transform it into the glutathione S-sulfonate, by means of treatment by 0.5 M sodium sulfite solution as advised by MOORE, SPACKMAN AND STEIN[10]. This substance will be eluted with the front of the eluent and will appear at the very beginning of the elution curve, before glycerophosphoethanolamine (GPE). Owing to the amount of glutathione usually present in 4 g of packed red cells, 1.5 ml of the 0.5 M solution of sodium sulfite is sufficient for complete transformation into glutathione S-sulfonate under conditions des-

* Buffered isotonic solution at pH 7.3. NaCl, 7.65 g; KCl, 0.20 g; sodium acetate, 1.50 g; NaH_2PO_4, 0.05 g; KH_2PO_4, 0.10 g; $NaHCO_3$, 0.70 g; dextrose, 1.00 g; distilled water 1000 ml.

cribed[10]. The extracts of leucocytes and platelets are concentrated to a few ml, then quantitatively transferred in a 10-ml volumetric flask, brought to pH 2.0 and made up to volume. Five ml samples are used respectively for 150 and 50 cm ion-exchange columns analysis. In the case of leucocytes and platelets, samples of a size suitable for each of the two steps of the complete chromatographic analysis are equivalent extracts of 100 mg of cells. No attempt at subfractionation of the different types of leucocytes has been made here.

Methods for expressing data on free amino acid content or excretion

This important point has been discussed at length elsewere[1]. Here data will be expressed mainly on a molar basis, which is more suitable for the sake of comparison. Urine free amino acids will be referred to as 24-h outputs.

FREE AMINO ACIDS OF URINE AND BLOOD PLASMA

Normal aminoaciduria in healthy adults

Qualitative composition. A typical elution curve obtained by the MOORE AND STEIN 1954 manual procedure using unhydrolyzed urine from a normal adult female is to be found in a review published in 1959 by BIGWOOD *et al.*[1]. Fig. 1 shows the automatically recorded elution curve obtained by use, in our department, of the SPACKMAN, MOORE AND STEIN automatic procedure[11] and is in complete agreement with that published by these authors[14]. Although the amino acid composition illustrated by Fig. 1 is that of urine obtained from an untreated leukemic patient, it is given in this section in order to avoid duplication of figures since it will also be used in another section of this report; moreover the amino acid pattern was in this case quite normal, both qualitatively and quantitatively. The analytical methods used here

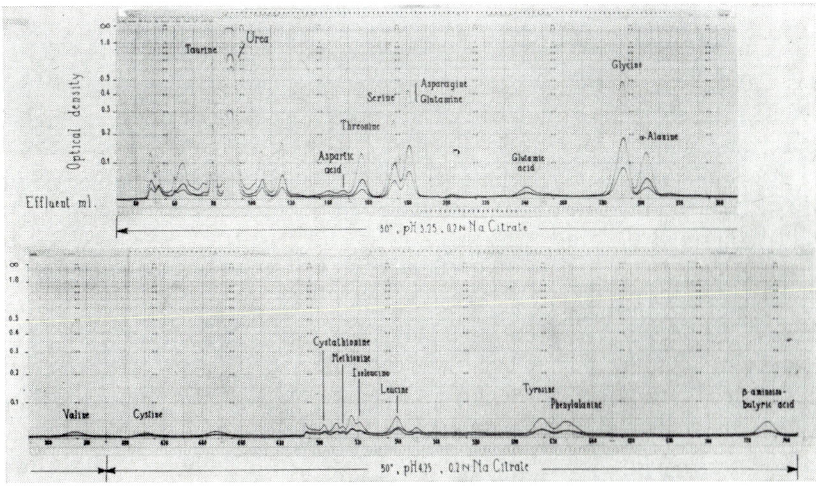

Fig. 1a. Chronic lymphocytic leukemic patient's urine. Determination of acidic and neutral amino acids on a 150-cm column (automatic procedure). Free amino acid excretion in this case is normal, both qualitatively and quantitatively.

enable one to recognize in normal urine more than 40 different ninhydrin-reacting substances when the manual procedure is used, and up to 50 or more when the automatic procedure is applied. Of those 50 ninhydrin-reacting substances 29 form the group of well-identified free amino acids, including taurine, asparagine, and glutamine. Under the experimental conditions used, ninhydrin-reacting peptides containing up to eight amino acids residues should show up as well-defined peaks if their color factor allows for it. The remaining unidentified peaks correspond to unknown amino acids or to amino acid derivatives or to peptides. They appear as very small peaks either because they are excreted in minute amounts or because their color

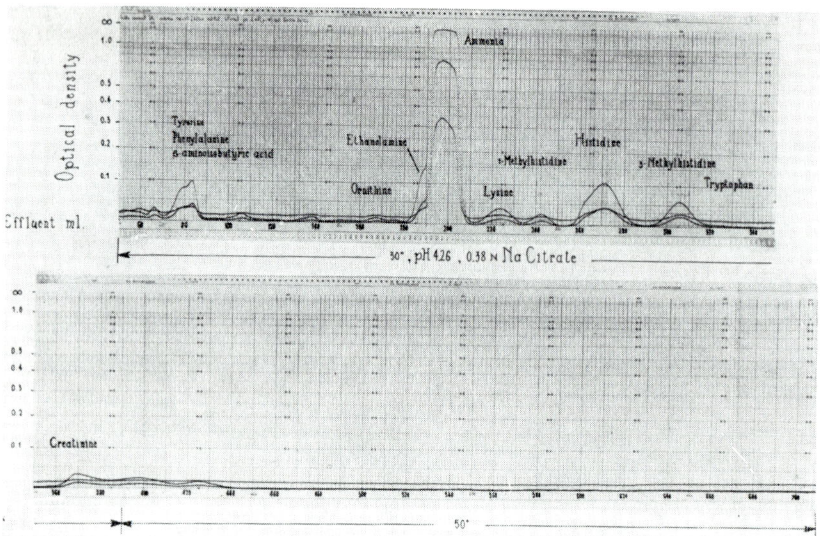

Fig. 1b. Chronic lymphocytic leukemic patient's urine. Determination of basic amino acids on a 50-cm column (automatic procedure). Free amino acid excretion in this case is normal, both qualitatively and quantitatively.

factor is very low. Many of them are acid-stable. Acid hydrolysis of the urine samples produces a marked increase in some of the amino acids of the group of identified substances and a minor increase in almost all of the other amino acids of the same group. The bulk excretion of combined amino acids is chiefly composed of substances such as hippuric acid or phenylacetylglutamine and even some peptides, although the presence of the latter in urine is still a matter for debate. WESTALL[5], and BOULANGER *et al.*[3,4] claim that there is evidence for the presence of peptides in normal urine, but they must be present in very minute amounts since these authors had to process considerable volumes of urine to recognize the presence of these peptides. The term "combined amino acids" refers in our opinion to substances in which the amino acid is linked to another substance by its amino group. We therefore include compounds such as glutamine and asparagine or still others, such as tyrosine-*O*-sulphate[2] in the group of free amino acids.

Quantitative composition. Among the 29 free amino acids excreted in normal urine, 11 have 24-h outputs which range from approx. 10–300 mg. For these amino acids

References p. 261/262

the precision of the determination is of the order of ± 5% when the manual procedure is used. For amino acids whose 24-h outputs average 10 mg, the relative error may rise to 10% or more. When determinations are made by use of the automatic procedure the precision which can be attained is of the order of 2 or 3%, even for amino acids which have low urinary outputs.

Other amino acids are regularly present in urine but in very minute amounts, lower than 10 mg/day. Still others are occasionally present but in non-measurable amounts or traces.

Quantitative data for 23 amino acids in a group of 15 normal healthy adults, six males and nine females, are shown in Table I, drawn from a paper by SOUPART[20]. Fig. 2 shows, on a molar basis, that nine of these amino acids, listed in decreasing order of average excretion, add up to approx. 85% of the total amino acid excretion. In this group, approximate values are given only for glutamine which corresponds to an excretion of at least 300 μmoles or approx. 50 mg/day. The remaining 14 amino acids form only 15% of the total excretion. Differences in the average figures of the latter are therefore not to be taken as significant. On a normal protein diet, for instance, a supply of protein varying between 70 and 125 g/day, total urinary nitrogen may vary from 9 to 18 g/24 h, of which approx. 3% in the average is

TABLE I

DAILY URINARY OUTPUT OF 23 FREE AMINO ACIDS IN 15 NORMAL ADULT SUBJECTS

Nine women and six men. Values expressed in μmoles/24 h (SOUPART[20]).

A. 9 amino acids (85–90% of total)	*Range*		*Average*
Glycine	710–4160		1687
Taurine	220–1850		812
Histidine	130–1370		790
1-Methylhistidine	130– 930		433
Glutamine*		At least	350
Serine	310– 620		374
3-Methylhistidine	180– 520		323
Alanine	60– 500		257
β-Aminoisobutyric acid	0– 500		252
Total		At least	5278 μmoles
*B. 14 amino acids***, excreted in small amounts (0–300 μmoles) in the average			65 μmoles
	65 × 14 =		910 μmoles
	Total A+B	Approx.	6200 μmoles

A = at least 85% of total aminoaciduria.
B = no more than 15% of total aminoaciduria.

* Minimal values (see p. 222); glutamic acid added to glutamine on a molar basis since freshly voided urine does not contain glutamic acid.

** In decreasing order of average daily output: threonine, tyrosine, phenylalanine, isoleucine, leucine, cystine 1/2, lysine, valine, β-alanine, aspartic acid, methionine, α-aminoadipic acid, arginine and ornithine.

amino nitrogen (free and combined, irrespective of the alpha or other position of the amino group). SOUPART[20] has found that values for the total amino acid excretion, calculated from the column data in his group of 15 normal adults, ranged from 0.48 to 1.11 g/24 h in females and from 0.35 to 1.18 g/24 h in males, the average figures being 0.73 and 0.80, respectively. The average figure found by STEIN[22] for men is 1.1 g/24 h. When amino acid excretion is calculated from SOUPART's data[20] in terms of α-amino nitrogen, it appears that values range from 59 to 132 mg/24 h in females and from 41 to 133 mg/24 h in males. The average figures are 87 and 91 mg respectively of α-amino nitrogen. These results are in very good agreement with the value of 1% of total nitrogen usually reported in the literature.

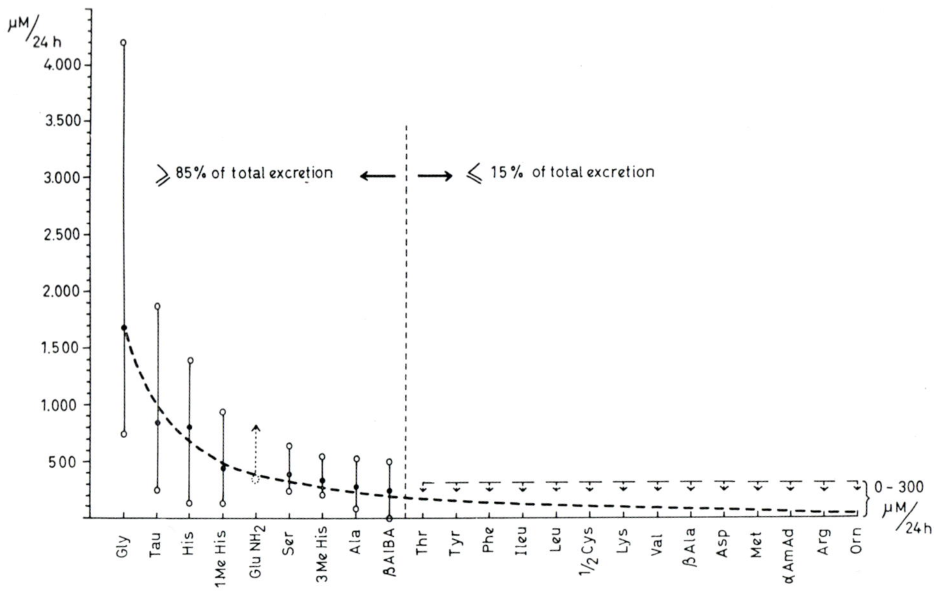

Fig. 2. Aminoaciduria in normal healthy adults of both sexes. Open circles show normal variation range of individual daily outputs. Filled circles indicate average outputs. For glutamine, only approximate average value is given (see p. 222).

Factors affecting free amino acid excretion in normal healthy adults. There is rather little information to be found on this point. As a general remark, it seems that free amino acid excretion varies within a narrow range in a given individual if followed in the course of time. There are of course some exceptions to this rule, namely 1-methylhistidine, the excretion of which is correlated to meat consumption of the individual[22]. Moreover taurine and β-aminoisobutyric acid excretion are increased when fasting for a sufficient duration is prescribed to a normal individual. On the other hand, there are rather large differences in free amino acid excretion when different individuals are compared. Among other factors which may affect free amino acid excretion in normal healthy adults the following may be listed:

1) Menstrual cycle in females seems to influence the excretion of some of the free amino acids. This is particularly the case with histidine excretion. SOUPART[6] has described the evolution of histidine output in the course of normal menstrual cycles in a given individual, whose amino acid excretion has been repeatedly investigated at various intervals during a period extending through 6 consecutive years. There was found regularly that a maximum histidine output, of highly reproducible magnitude, occurs at a time when there was maximum estrogenic activity during the course of menstrual cycles, namely around mid-cycle. Observations made at 6-years interval in the same individual were perfectly superposable as regard to lower and higher urinary histidine output during the cycles, and also as regard to the occurrence of the maximum output at mid-cycle. These observations have been cross-checked by use of two different ion-exchange chromatographic methods and a specific enzymatic decarboxylation method for histidine determination[23].

2) Normal pregnancy. This physiological circumstance has been extensively investigated by SOUPART[6] during the past 6 years. Normal pregnancy is characterized by a generalized hyperaminoaciduria, affecting essentially the following amino acids listed in decreasing order of hyperexcretion during pregnancy: histidine, glycine, threonine, serine, alanine, tyrosine, phenylalanine, isoleucine and arginine. Unfortunately, although very typical of pregnancy in an individual whose usual amino acid excretion is well known, this hyperaminoaciduria, which appears very early after fecondation has occurred, cannot be used as a diagnostic test for pregnancy because of the very large variations of normal excretion levels among different individuals. This has been demonstrated beyond any doubt by SOUPART[6]. The mechanism of this phenomenon will be dealt with in one of the following sections.

3) Effects of some hormonal factors. In experiments designed to assess an interpretation of the hyperaminoaciduria occurring during normal pregnancy, SOUPART[6] has investigated the effects on amino acid excretion of large doses of estrogens (estradiol benzoate) and of dihydrocortisone administered to patients suffering of prostate carcinoma. In those instances, and specially after dihydrocortisone, a hyperaminoaciduria of the type observed in pregnancy has been induced in the patient, who was formerly normal in this respect. Adrenal hormones are known to increase definitely the glomerular filtration rate and this may be responsible for the hyperaminoaciduria observed in the patient as well as that observed in the course of pregnancy, since this latter state is known to comprise hyper-hormonal activities.

4) Effects of high protein diet. It seems that free amino acid excretion is not, or rather poorly influenced by overdosage of protein in the diet. Even a very large supply of 3–4 g of protein/kg of body weight does not seem to affect significantly the free-amino acid urinary outputs. But, as already stated, there is an exception, namely in the case of 1-methylhistidine, the excretion of which is correlated not to the total protein supply but to that part of the supply which consists of meat.

Normal aminoaciduria in healthy children

Normal newborns show a higher total aminoaciduria, when related to total nitrogen output, than do normal healthy adults. This phenomenon is even more pronounced in premature babies. Such a situation might be the result of less complete reabsorption of free amino acids by immature renal tubular mechanisms. The relative amount of α-amino nitrogen excreted in urine is 1% of total nitrogen in adults, approx. 3% in newborns and more than 4% in premature babies[20, 24, 25]. This situation is normalized during the first year of life, but until the age of 2 years,

References p. 261/262

TABLE II

DAILY URINARY OUTPUT OF 26 FREE AMINO ACIDS IN 15 NORMAL CHILDREN

Seven girls and eight boys, white and black; age range 9 months–2 years. Values are expressed in μmoles/24 h (Dr. H. VIS[1a]).

A. 10 amino acids (87% of total)	*Range*	*Average*
Glycine	143– 561	370
Histidine	99– 533	300
Glutamine*	>12–>393	>170
Alanine	64– 164	107
Serine	65– 120	95
Lysine	10– 137	70
1-Methylhistidine	0– 250	52
Taurine	18– 112	52
Threonine	26– 74	48
β-aminoisobutyric acid	0– 93	46
Total		⩾1310 μmoles
B. 9 amino acids (up to approx. 12% of total)		
Phenylalaline, 3-methylhistidine, tyrosine, ornithine, leucine, isoleucine, methionine sulfoxides, cystine $^1/_2$, aspartic acid.		
Average output from 30–10 μmoles: 9 × 20 =		180 μmoles
C. 8 amino acids (less than 1% of total)		
Arginine, valine, methionine, α-aminobutyric acid, γ-aminobutyric acid, β-alanine, α-aminoadipic acid, proline.		
Average output 3 μmoles:	8 × 3 =	24 μmoles
Grand total	Approx.	1514 μmoles

* See footnote * in Table I.

the free amino acid excretion pattern seems to be different from that found in adults. Table II gives free amino acid excretion data, in the same manner as those given for adults in Table I. These data are borrowed from a study done in the department by Dr. H. VIS[1a].

Although the absolute amounts of free amino acids excreted by this group of normal children (eight boys and seven girls, ages ranging from 9 months to 2 years) are very different in size from those found in adults, a convenient way of comparison is to express individual free amino acid average excretions in both cases in percent of the total average molar amount excreted in 24 h. Such a comparison is shown in Fig. 3, for two groups of adults and children of comparable size and sex partition (data of Table I and Table II). To be noticed is that the average excretion of histidine, glutamine, alanine, lysine and ornithine, are higher in children of this age group than they are in adults. On the other hand, the reverse situation is apparent with regard to taurine, 1- and 3-methylhistidine, valine, β-alanine and α-aminoadipic acid. These observations are of special interest in light of our present knowledge of metabolism of some of the involved amino acids, namely taurine, 3-methylhistidine and lysine. In the case of taurine, which is one of the end-products of sulfur amino acid metabolism, a rather low excretion is not surprising as sulfur retention is an essential requisite of growth. As regard to 3-methylhistidine, there seems to be some link between urinary output of this amino acid and that of creatine. During all the time when there is a urinary creatine output, *i.e.* before the age of puberty, there is a rather low excretion of 3-methylhistidine. When this critical period of life is over, creatinuria disappears and urinary output of 3-methylhistidine, as well as that of taurine, tends to be elevated (Dr. H. VIS[1a]).

The case of lysine excretion is difficult to interpret owing to our present state of knowledge. In fact it is very surprising to observe that in infants 9 months–2 years

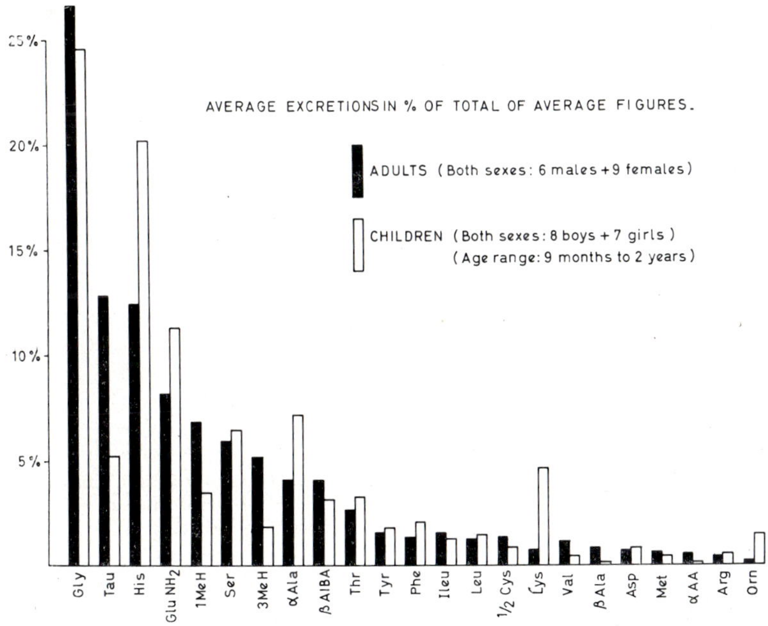

Fig. 3. Aminoaciduria in normal adults and normal children. Average excretion in percents of total average figures. Notice that the situation in children and adults is different.

old the excretion of lysine is relatively much higher than it is in adults, when one takes into account that specific needs for this indispensable amino acid are so important for growth.

It has been demonstrated very recently (April 1961) that homocitrulline is a normal constituent of children's urine[80].

Normal free amino acid levels in plasma

Qualitative composition. A typical elution curve obtained by the MOORE AND STEIN 1954 manual procedure from deproteinized human blood plasma will be found in a paper by SOUPART[6]. Fig. 4 shows the automatically recorded elution curve obtained in our department by use of the SPACKMAN, MOORE AND STEIN automatic procedure[11] which is in complete agreement with that published by these authors[14]. Although Fig. 4 represents the free amino acid composition of plasma in an untreated leukemic patient, it is given here for the same reasons as Fig. 1. When compared with urine amino acid composition, the free amino acid composition of plasma appears much simpler and the picture is more clear-cut. There are a small number of unidentified ninhydrin-reacting substances and only a few remarks are needed. There are two unknown components, constantly present, eluted in the range between that of urea and that of methionine sulfoxides. The glutamine peak is, in fact, a mixture containing asparagine as well, to the extent of

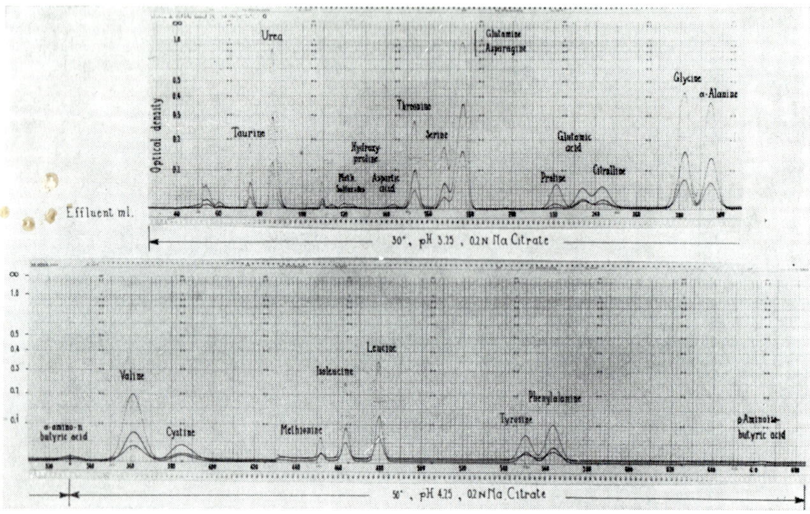

Fig. 4a. Amino acid composition of protein-free human blood plasma from a chronic lymphocytic leukemic patient. Determination of acidic and neutral amino acids on a 150-cm column. Free amino acid plasma levels are normal in this case.

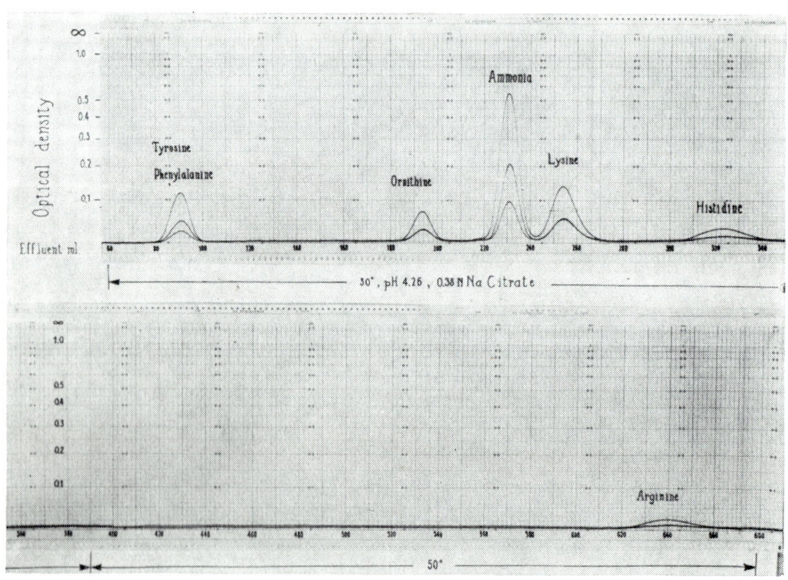

Fig. 4b. Amino acid composition of protein-free human blood plasma from a chronic lymphocytic patient. Determination of basic amino acids on a 50-cm column. Free amino acids plasma levels are normal in this case.

TABLE III

FREE AMINO ACID PLASMA LEVELS IN NORMAL FASTING ADULTS

Literature data. Values are expressed in μmoles/l.

Amino acid	Ion-exchange chromatography I	II	III	IV	V	VI	Range for groups I–VI	Paper chromatography VII	VIII	Microbiol. method IX
	SOUPART[26]	STEIN AND MOORE[12]	CHRISTENSEN[27]	FRAME[28]	CUSWORTH AND DENT[29]	EVERED[30]	(30 subjects)	McMENAMY[31]	SERINI *et al.*[32]	DOOLAN *et al.*[33]
α-Alanine	300–380	338–472	213–270	213–382	260–400	337–416	213–472	350–450	220–440	281
Glycine	232–283	179–231	280–587(?)	200–320	240–290	200–267	179–587	—	250–610	267
Valine	168–216	203–317	—	171–239	200–280	197–316	168–317	180–225	130–430	248
Threonine	137–194	102–145	109–160	76–147	110–170	101–143	76–194	150–180	100–150	126
Proline	103–210	160–290	165–183	122–226	150–230	157–287	103–290	150–220	40–260	148
Serine	115–164	96–119	133–162	76–124	114–133	95–124	76–164	—	80–170	143
Lysine	105–155	172–207	120–151	130–205	144–164	171–205	105–207	180–230	—	164
Leucine	78–115	108–176	84–115	84–122	90–130	107–176	78–176	—	—	145
Taurine	80–100	33– 66	32– 58	32–138	63–104	32– 64	32–138	—	—	—
Cystine $^{1}/_{2}$ + Cysteine	70–100	90–108	—	—	—	83–108	70–108	—	60–150	90
Arginine	40–102	70–111	44– 72	66– 92	60–140	69–115	40–140	55– 80	—	161
Histidine	32– 92	51– 96	59– 84	65– 90	77– 97	52– 97	32– 97	70–110	140–330	88
Isoleucine	40– 63	53– 98	52– 60	46– 61	47– 68	54– 99	40– 99	—	—	92
Phenylalanine	38– 67	42– 58	47– 48	42– 73	48– 55	42– 61	38– 73	45– 60	20–100	94
Ornithine	30– 64	47– 61	—	45– 61	31– 58	45– 61	30– 64	70–100	—	—
Glutamic acid	(27– 50)	(29– 78)	(27– 40)	(20– 40)	(40– 90)	(27– 81)	(20– 90)	(30)	0– 30	—
Glutamine	(140–370)	(570)	—	—	(180–310)	—	(140–570)	450–550	280–450	527
Tyrosine	22– 58	45– 80	39– 48	33– 66	35– 49	44– 83	22– 83	60– 80	20– 60	44
Methionine	11– 26	22– 29	21– 25	13– 30	20	20– 27	11– 30	18– 26	—	34
α-Aminobutyric acid	14– 21	21– 34	16– 35	10– 30	0	20– 34	10– 35	22– 34	0–100	—
Aspartic acid	2– 8	1– 5	9– 11	1– 7	7	Tr	1– 11	—	—	—
β-Aminoisobutyric acid	Tr	—	—	—	—	—	Tr	—	—	—
1-Methylhistidine	Tr	2– 10	—	—	—	—	0– 10	—	—	—
3-Methylhistidine	Tr	2– 8	—	—	—	—	0– 8	—	—	—
β-Alanine	Tr	—	—	—	—	—	—	—	—	—
α-Aminoadipic acid	Tr	—	—	—	—	—	—	—	—	—
Tryptophane	—	—	—	—	—	Tr	Tr	(18)	—	6

References p. 261/262

about one-tenth of the total amount of ninhydrin-reacting material forming the peak, as shown by other methods[12]. There are two small unidentified peaks regularly present following that of leucine. Sometimes traces of β-alanine, 1- and 3-methylhistidine and tryptophane are found. Creatinine is never apparent in plasma chromatograms because of its very low color factor. Some of the amino

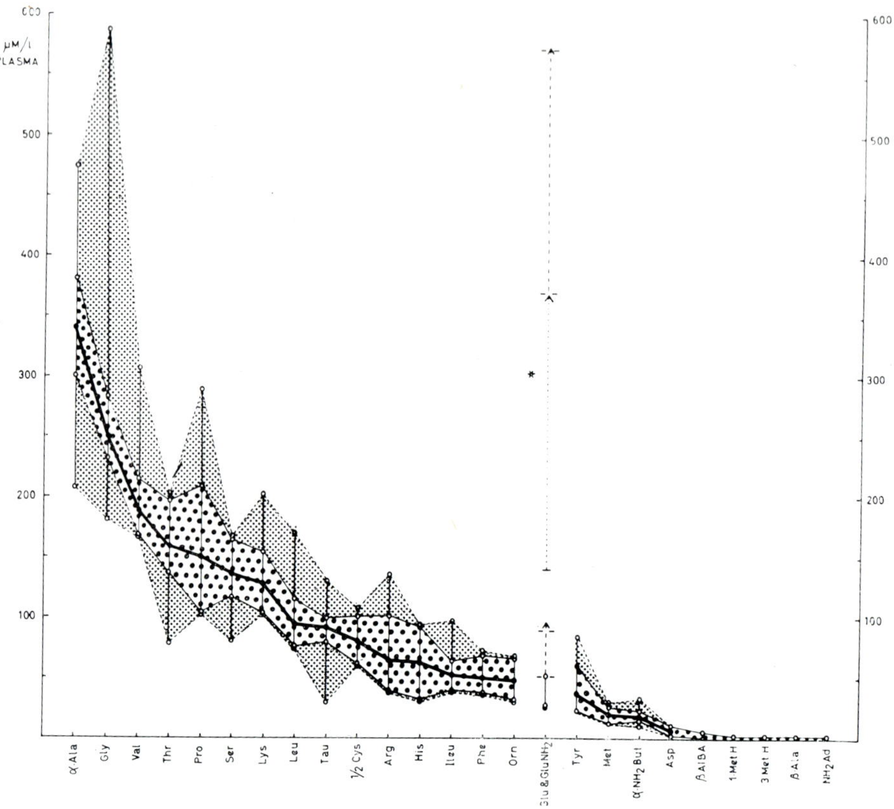

Fig. 5. Variation range and average values for normal free amino acid plasma levels. Literature data (shadowed area) compared to SOUPART's data (filled circles area). Values given for glutamic acid and glutamine (*) are minimal values as indicated by arrow pointing upwards (see p. 222). Fine doted line shows the range observed by SOUPART, large doted line that observed by others. Glutamine values are approximate since about one tenth of the glutamine peak is made of non-separated asparagine.

acids present in plasma are never found in urine by this method, namely hydroxyproline, proline and citrulline.

Quantitative composition

a) In normal healthy adults. In Table III, data from the literature relative to free amino acids in blood plasma have been grouped. Grouping has been done for data obtained through use of ion-exchange chromatography (Groups 1–6)[12, 26–30] on the first hand, and for data obtained by use of other methods on the other hand[31–33]. The decreasing order of free amino acid concentration described in plasma by

SOUPART[26] has been adopted for the whole arrangement of the data. An additional column is given, grouping the overall ranges of concentrations as measured by the ion-exchange chromatographic method. Fig. 5 shows graphically how SOUPART's data fit with the overall range of concentrations described by similar methods for 30 normal adult subjects. Fig. 5 thus gives a general picture of our present knowledge of free amino acid levels in normal plasma. As can be seen, all data obtained by ion-

TABLE IV

FREE AMINO ACID PLASMA LEVELS IN 20 NORMAL FASTING CHILDREN

15 coloured and five white; 9 months–2 years old. Values are expressed in μmoles/l (Dr. H. VIS[1b])

	Range	*Average*
Alanine	99–313	219
Glycine	56–308	170
Valine	57–262	127
Threonine	33–128	60
Proline	51–185	115
Serine	24–172	92
Lysine	45–144	87
Leucine	45–155	75
Taurine	19– 91	49
Arginine	11– 65	31
Histidine	24–112	64
Isoleucine	26– 94	44
Phenylalanine	23– 69	40
Ornithine	10–107	40
Glutamine	46–290	135
Tyrosine	11–122	45
Methionine	3– 29	21
α-Aminobutyric acid	0– 17	5
Aspartic acid	0– 9	2
β-Aminoisobutyric acid	0– 22	5
1-Methylhistidine	0	0
3-Methylhistidine	0	0
Cystine $^1/_2$	0– 40	4
β-Alanine	0	0
α-Aminoadipic acid	0	0

exchange method are in fairly good agreement despite the various origins of the subjects investigated. 26 well-identified free amino acids are found in human blood plasma, of which 20 are present in measurable amounts, the other six being present only as traces. Fig. 5 also shows amino acids whose concentrations fluctuate within a relatively large range and those for which the range of variation is narrower. As can be seen, variation ratios of normal concentrations may range between $^1/_2$ and $^1/_3$.

b) In normal healthy children. Data presented here have been collected by Dr. H. VIS in a composite group of 20 children, boys and girls: five Western-European and 15 Central-African children, age range 9 months–2 years. Quantitative data are grouped in Table IV. Fluctuation ranges and average values are visualized in Fig. 6. There is no race-linked difference to be noticed. The plasma free amino acid concentration pattern of children is about the same as that found in adults. The decreasing

References p. 261/262

order of the average concentrations is quite similar in the two groups. On comparison of Figs. 5 and 6, the following differences may be distinguished: (1) average concentrations are in general a bit lower in infants than in adults; (2) the range of variation tends to be larger in infants than in adults; (3) there are some amino acid concentrations in infants which differ markedly from what they are in adults, namely that of half cystine plus cysteine and that of glutamic acid and glutamine.

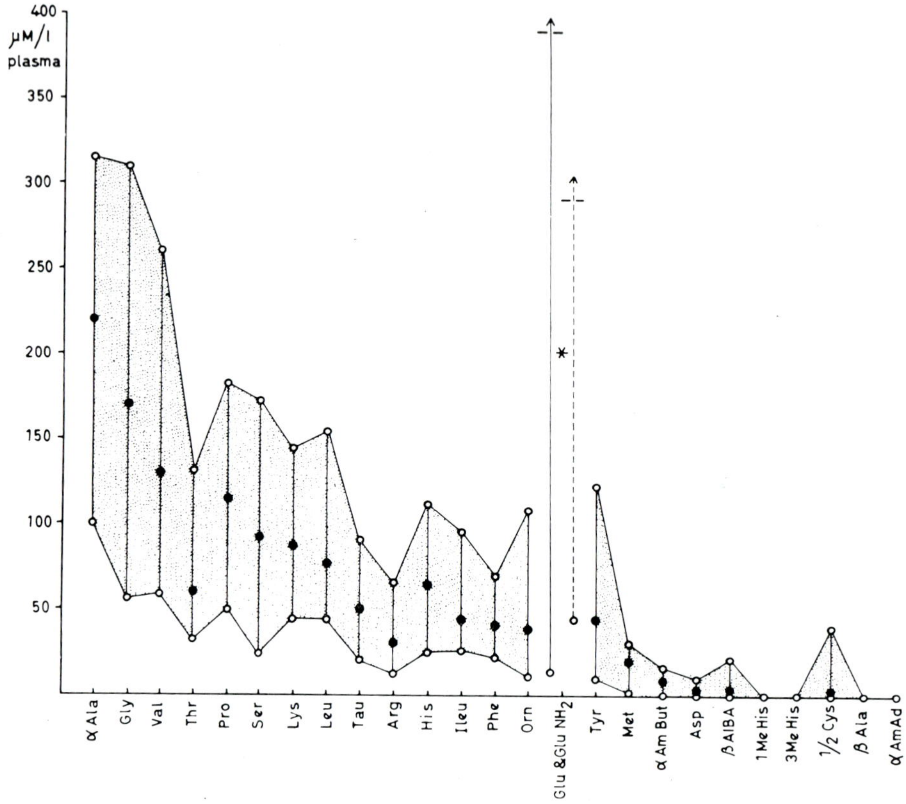

Fig. 6. Free amino acid plasma levels in normal children. Open circles indicate variation range observed and filled circles average values. For cystine, glutamine and glutamic acid see text*. Straight line shows glutamic acid levels variation range and dotted line that of glutamine. Both groups are made of minimal approximate values (see p. 222).

Half cystine plus cysteine being of medium abundance (70–108 μmoles/l of plasma) in adults, this amino acid is present only as traces in children's plasma*. *Glutamic acid and glutamine* values found have to be considered as a special case, owing to the causes of error which interfere with their determination as previously indicated. The concentrations given here are thus to be taken as minimal values. In plasma the

* In the present data, cysteine has been oxidized into cystine prior to analysis because cysteine does not react with ninhydrin. It is only recently (1960) that a method has been worked out by BRIGHAM, MOORE AND STEIN[17] which affords for the separate determination of cysteine and cystine, the former under the form of carboxy-*S*-methylcystine, by means of ion-exchange chromatographic analysis.

glutamic acid level is much more elevated in children than in adults, in contrast with all other concentrations which, as a rule, tend to be lower in children.

Apart from these data, collected by Dr. H. VIS, there are rather few others to be found in the literature at the present time. WESTALL *et al.*[34] have given normal ranges, but they include values given by STEIN AND MOORE for adults[12]. Such is also the case for values given by IBER *et al.*[35]. HUISMAN[36] on the other hand, refers only to the child.

Factors affecting free amino acid levels in plasma of normal adults. As in the case of free amino acid urinary excretion, there is rather little reliable information to be found on this point up to date. Among the factors which may affect the plasma levels, the following may be listed: menstrual cycle and normal pregnancy, which are in fact ultimately controlled by endocrine factors, effects of hormonal factors and effects of high protein diet.

1) Menstrual cycle. SOUPART[26] has studied fasting free amino acid plasma levels in a group of five healthy females, at the beginning of both follicular and luteal phases of their cycle. Analysis of the results showed that the shift from follicular to luteal phase induces a general lowering of the average plasma levels for all amino acids. The difference between average levels was statistically significant for the following amino acids: proline and threonine ($P < 0.05$), serine and lysine ($P < 0.01$), and alanine ($P < 0.001$). This variation of plasma levels seems at first to characterize a systematic phenomenon, involving all amino acids, but the fact that the lowering of average concentrations is statistically significant only for five of them strongly suggests that it reflects a more specific phenomenon, not yet understood, but related in some way to the physiological rebuilding involved in the menstrual cycle.

2) Normal pregnancy. As previously stated (see p. 229) normal pregnancy is accompanied by the hyperexcretion of the following amino acids: histidine, glycine, threonine, serine, alanine, tyrosine, phenylalanine, isoleucine and arginine. Glutamine should also be added to this group, notwithstanding the approximation which affects its determination (see p. 222). SOUPART[6] has compared free amino acid levels in a group of normal pregnant females to those found in a group of five non-pregnant individuals. Although statistical analysis was not practicable in this study, owing to the lack of homogeneity of the group of pregnant females investigated with regard to the age of pregnancy, it appears that at any age of pregnancy studied plasma free amino acid levels were of the same order of magnitude, or even lower, than those observed in the luteal phase of menstrual cycle; the only two exceptions to this rule are glutamic and aspartic acids. Their concentration is definitely higher in pregnant than in non-pregnant women. Interpretation of the effect of pregnancy on free amino acid plasma levels will be found in the section *Renal handling of free amino acids.*

3) Effects of some hormonal factors. Administration of large doses of estrogens (estradiol benzoate) does not influence free amino acid plasma levels, except that of glycine, which is lowered at the same time that glycine is hyperexcreted in urine[6]. Concentrations of almost all amino acids present in plasma are decreased when large doses of hydrocortisone are given, when compared with plasma levels of a control period[6]. This lowering of free amino acid plasma levels is quite similar to that observed during pregnancy.

4) High protein diet. Effects of high protein diet on free amino acid plasma levels have been studied by STEIN *et al.*[37] and by FRAME[28] by means of ion-exchange chromatography. From the results of STEIN's experiment it was concluded that the increase in free amino acid plasma levels, observed 2 h after feeding 50 g of casein to a normal subject, reflects to some degree but not in detail the amino acid composition of the casein fed. Such a protein meal does not lead to more than a 2-fold increase of concentration for the involved amino acids. Similar but more detailed conclusions are drawn from the experiment of FRAME[28], in which plasma free amino acid levels were determined in peripheral blood, in a group of six normal subjects, at intervals up to 8 h after a high protein meal. Most of the amino acids increased

TABLE V

BLOOD AMINO ACID CLEARANCES

Literature data. Values are expressed in ml/min, corrected or non-corrected for 1.73 sqm.

	SOUPART[26] *5 subjects (♀)*	CHRISTENSEN[27] *3 subjects*	EVERED[30] *3 subjects*	CUSWORTH AND DENT[29] *4 subjects*	ARROW AND WESTALL[38] *10 subjects*	DOOLAN[33] *4 subjects*	SERINI *et al.*[32] *5 adults*
Histidine	2.0–16.3	4.5–4.9	1.1–7.5	4.7 – 9.7	1.4 –15.0	4.6 –8.5	5.2–8.7
Taurine	3.3–12.4	3.4–3.5	1.6–6.1	1.7 –14.0	0.2 –20.0	—	—
Glycine	2.5– 6.7	3.2–5.6	1.5–5.7	2.7 – 5.8	1.8 –19.0	4.0 –6.3	1.0–7.2
Serine	1.0– 3.7	1.8–2.9	—	1.9 – 3.1	1.5 – 5.6	1.2 –2.1	1.7–3.3
Tyrosine	0.6– 2.3	0.8–1.3	0.9–1.5	1.0 – 1.7	0.7 – 3.8	1.2 –2.4	0.5–4.0
Phenylalanine	0.4– 2.0	0.7–1.1	0.8–1.5	0.7 – 1.4	0.5 – 1.9	0.2 –1.3	0.4–2.0
Methionine	0.7– 7.6	0.9–1.3	0.7–1.6	1.1	0.2 – 3.9	0 –0.4	—
Threonine	0.4– 2.2	0.7–1.1	—	0.8 – 1.5	0.7 – 3.5	0.5 –1.4	0.4–1.9
Isoleucine	0.7– 3.3	1.0–2.0	0.3–1.7	0.2 – 1.0	0.3 – 2.5	0 –0.9	—
Aspartic acid	1.9–18.0	0.6–0.7	—	2.4	—	—	—
Arginine	0.4– 2.5	0.4–0.5	—	0.2 – 0.8	1.4	0 –2.5	—
Lysine	0.1– 0.8	0.2–0.5	0.2–0.5	0.2 – 1.9	—	0.2 –0.4	—
Alanine	0.2– 0.9	0.4–0.7	—	0.3 – 0.9	0.2 – 1.2	0.3 –1.0	0.7–1.9
Cystine	—	0.2–0.5	0.2–0.5	0.7 – 2.9	—	0.1 –0.3	3.6–6.7
Leucine	0.2– 0.9	0.3–0.3	0.3–0.5	0.2 – 0.9	0.3 – 2.2	0.2 –0.4	—
Valine	0.2– 0.4	—	0.1–0.2	0.04– 0.1	0.04– 0.5	0.04–0.2	0.1–0.4

in concentration after the meal, but the rise did not parallel the relative amino acid composition of the food and the relative distribution of the different amino acids in the plasma differed from that in the fasting state. The behavior of the observed concentrations was not uniform among the different amino acids. Some of them increased in concentration, namely valine, leucine and isoleucine, which rose markedly after the meal and, even after 8 h, were higher than in the fasting state. Others, namely alanine and glycine, fell to levels half the fasting values in 6–8 h.

Renal handling of free amino acids

It has become usual to evaluate the excretory function of the kidney in terms of the filtration–reabsorption theory by means of calculating clearances for the substances excreted. It is also done for amino acids. When the available data on individual amino acid clearances are considered, it appears clearly that as a rule the

References p. 261/262

amino acid clearance is very low, although there is a considerable individual variation as shown by Table V (refs. 26, 27, 29, 30, 32, 33, 38). According to their average clearance values, the amino acids may be distributed into three more or less well defined groups:

1) Clearance lower than 1 ml/min: lysine, ornithine, valine, alanine and leucine.

2) Clearance comprised between 1 and 2 ml/min: arginine, phenylalanine, tyrosine, isoleucine, serine and methionine.

3) Clearance higher than 2 ml/min but not higher than 20 ml/min: glycine, aspartic acid, taurine and histidine.

It appears therefore that the reabsorption is accomplished at various degrees of efficiency for the different amino acids.

TABLE VI

ESTIMATED EFFICIENCY OF FREE AMINO ACID REABSORPTION BY THE KIDNEY

Amino acid	*Percent of filtered load reabsorbed**		*Filtered load***
	Range	*Average*	
Histidine	87.3 –98.4	92.5	20
Taurine	90.2 –97.4	94.2	22
Aspartic acid	87.2 –98.6	95.4	1
Glycine	94.7 –98.1	96.0	38
Methionine	94.12–99.34	96.8	6
Serine	97.10–99.20	98.2	29
Isoleucine	97.46–99.47	98.55	14
Tyrosine	98.21–99.56	98.64	14
Threonine	98.34–99.72	99.07	38
Phenylalanine	98.44–99.67	99.26	17
Arginine	99.15–99.72	99.43	22
Leucine	99.31–99.83	99.53	25
Alanine	99.29–99.84	99.54	58
Valine	99.65–99.87	99.76	44
Ornithine	99.77–99.87	99.83	12
Lysine	99.77–99.97	99.89	38
Proline	100	100	34

* For method of calculation see text.
** Relative to that of aspartic acid taken as unity.

A quantitative evaluation of the efficiency of individual amino acid reabsorption may be obtained in the following way: assuming that the glomerular filtrate amounts in the average to a volume of 183 l in 24 h[39], that the concentration of the free amino acids is the same in the glomerular filtrate and in plasma and that the free amino acid plasma levels remain reasonably constant during a 24-h urine collection period, the percentage of the reabsorption may be approximately calculated for each of the amino acids. Such calculation has been done from data collected by SOUPART in a group of five normal subjects[26]. Results are shown in Table VI and Fig. 7, in which the calculated percentage range of reabsorption for each amino acid is compared to the individual filtered load, clearly shows that there is no relationship in normal conditions between the amounts of amino acid filtered and their reabsorption. In this figure, the amounts of amino acid filtered are expressed relatively to that of aspartic acid, assuming once again that individual amino acid concentrations are the same in the glomerular filtrate and in plasma.

For alanine, which is the most abundantly filtered amino acid, 99.54% is reabsorbed in the average, and this percentage varies little among the group of individuals investigated. At the opposite, aspartic acid, which is scarcely present in plasma, and thus filtered in rather small amount, is less effectively reabsorbed. In the

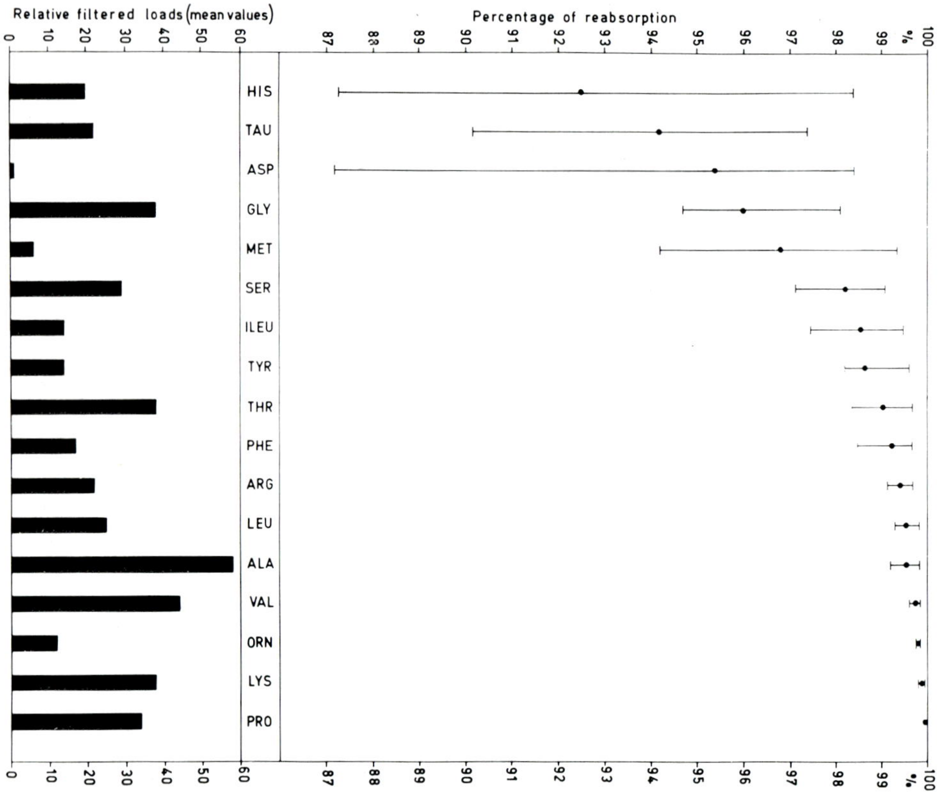

Fig. 7. Estimated efficiency of free amino acid reabsorption by the kidney, as compared with relative filtered loads.

average, the less well-reabsorbed amino acids are histidine, taurine, aspartic acid and methionine, although some individuals distinguish themselves by fairly good reabsorption of these amino acids as shown by Table VI. Reabsorption of a few of the amino acids present in plasma has still not been correctly evaluated, in most cases owing to technical reasons which have been discussed previously. 1- and 3-methylhistidine, BAIBA, and β-alanine are generally considered as not being reabsorbed at all, since they are excreted in non-negligible amount, although they are only present in plasma as very faint traces when detectable. Proline, on the other hand, one of the most abundant amino acids in the plasma, is never found in normal urine. It is generally assumed that its reabsorption is complete. With this particular case, another question arises: is proline so completely reabsorbed that not a trace is to be found in normal urine, even when such a high resolution-power instrument as the automatic recording ion-exchange chromatographic machine is used, or does it not pass into the glomerular filtrate? It must be recalled here that the use of the automatic recording procedure has led to the recognition of a faint trace of hydroxyproline in normal human fasting plasma[14] heretofore not detected by any other analytical procedure.

The clearance concept, as applied to amino acids, relies on the assumption that they all pass equally easily into the glomerular filtrate and are present there at approximately the same concentration as in plasma. For most of the amino acids this might be the case, but nothing whatever tells that their concentration in the glomerular filtrate is not significantly different from what it is in the plasma. A Donnan effect in the distribution of amino acids between plasma and glomerular filtrate does not seem to be taken into account although amino acids are electrolytes. Except in one instance, it is not known if protein binding of amino acids occurs in the plasma, in which case the filterability of these solutes would be reduced to some yet unknown extent.

It must be born in mind that what is called free amino acid concentration in the plasma are values determined on deproteinization filtrates, a procedure which might well be able to break weak bonds between protein and amino acids if such a protein binding does really occur. Protein binding seems to be a reversible reaction, determined by the specific nature of the solute, by the concentration of the solute, by the concentration of protein as well as by H^+ concentration, temperature or other factors.

Up to now, only McMenamy *et al.*[31, 40] claim that there is evidence for such a protein binding, especially in the case of tryptophane which, they say, is largely bound to albumin.

The presence and extent of protein binding of amino acid should be determined by ultrafiltration experiments before calculating the filtered load of such solutes. Whatever its response, such an experiment would be worth carrying out by means of reliable analytical methods. It will presumably show that some of the amino acids do not pass freely through the dialyzing membrane, as suggested by a former study done by McMenamy *et al.*[40]. As it is, the clearance concept applied to renal handling of free amino acids has already proved useful since it is used as the basis for interpretation of the mechanism of abnormal aminoaciduria and has led to Dent's classification of abnormal types of aminoaciduria in disease[41]. It has also led to interpretation of some non-pathological situation such as the hyperaminoaciduria that occurs in the course of normal pregnancy as a result of increased glomerular filtration rate induced by hormonal hyperactivities[6].

However, it seems that more knowledge has still to be gained, especially in regard to amino acid glomerular filtration and the very nature of the reabsorption mechanism, before full understanding of the process will be achieved.

BLOOD POOLS OF FREE AMINO ACIDS

Animal tissues contain much greater amounts of free amino acids than the corresponding blood plasma. This has been demonstrated by Tallan *et al.* who published results of a detailed quantitative study on free amino acid composition of cat tissues by means of ion-exchange chromatography[42]. A similar study dealing with invertebrates' tissues by means of microbiological methods[43] has been done by DUCHÂTEAU AND FLORKIN, among their numerous studies of amino acids in invertebrates.

As pointed out by MCMENAMY *et al.*[31] "understanding of the role of the blood in amino acid transport and metabolism requires information not only about the plasma concentration and turnover of these metabolites but also about their distribution between the plasma and the formed elements of the blood". In man, as already stated, while the free amino acid content of the blood plasma has been extensively investigated, there are rather few studies to be found on free-amino acid content of blood cells, except that of IYER on erythrocytes[44], that of NOUR-ELDIN AND WILKINSON[45] and of ROUSER[46] on leucocytes, and the very recent one of MCMENAMY *et al.*[31], the latter study dealing with the simultaneous determination of unbound amino acids in plasma, erythrocytes, leucocytes and urine of normal subjects. Two other papers, dealing with a limited number of amino acids only, have also been published by CHRISTENSEN *et al.*[47] and by JOHNSON AND BERGEIM[48]. SCHRAM AND HUBINONT have communicated preliminary results obtained by ion-exchange chromatography on the distribution of free amino acids between erythrocytes, leucocytes and plate-

TABLE VII

FREE AMINO ACID CONCENTRATION IN PLASMA, ERYTHROCYTES, LEUCOCYTES AND PLATELETS

Adult man. Values are expressed in μmoles/100 g of wet wt.*.

Amino acid	*Plasma***	*Erythrocytes*		*Leucocytes*	*Platelets*
		non-corrected for diffusion	*corrected losses****		
Taurine	5	3	3.6	2603	2100
Lysine	20	5	13.0	236	122
Glycine	22	34	37	508	365
Alanine	42	21	35	661	270
Valine	27	2	33	375	150
Leucine	14	2	40	630	188
Isoleucine	7	0.3	4	290	120
Serine	12	11	15	510	365
Threonine	13	4	16	340	155
Phenylalanine	5	0.5	4	248	85
Tyrosine	6	1	5	197	78
Histidine	8	3.3	14	63	31
Arginine	10	0	0	33	53
Aspartic acid	0.2	33	37	350	270
Glutamic acid§	(5)	(32)	(32)	(736)	(316)
Methionine	3	0	Traces	175	38
Cystine $^1/_2$	11	0	0	37	0
Proline	22	2	17	210	102

* Corrected for water content: plasma 92 %, erythrocytes 65 %, leucocytes 75 %, platelets 87 %.
** Average concentration according to STEIN AND MOORE[12].
*** Correction for diffusion losses based on results shown in Table IX.
§ Approximate values.

lets[49]. The free amino acid content of plasma, erythrocytes, leucocytes and platelets is currently under investigation in our department by means of automatic ion-exchange chromatography. The data already collected by SOUPART AND SCHRAM (unpublished) draw attention to some problems that arise from methods used for single species cell preparations in view of amino acid analysis. Results of a typical analysis of formed elements prepared as described in MATERIAL AND METHODS, are presented in Table VII, and graphically shown in Fig. 8. As can be seen, free amino acid concentrations differ considerably in the four compartments of the blood

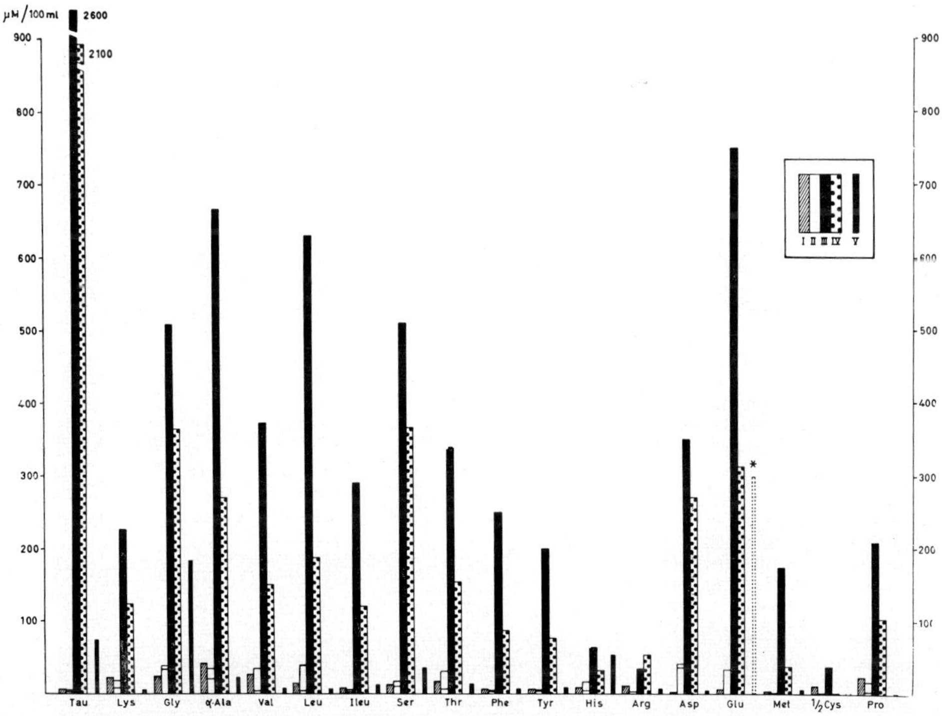

Fig. 8. Free amino acid concentration in plasma (I), erythrocytes (II, the lower part of each bar indicates what is left after diffusion losses, see text), leucocytes (III), platelets (IV) and urine (V). Glutamic acid (*) is present in freshly voided urine only in the form of glutamine.

reservoir, except those of erythrocytes in which the concentrations of a majority of amino acids are similar to those found in the plasma. It is also to be noticed that when blood cells are separated by such a method as the one used in this experiment, some amino acids are allowed to diffuse out of the erythrocytes, which does not seem to be the case with leucocytes and platelets, or at least not to the same degree. Differences in concentration may be appreciated by expressing individual amino acid concentration relatively to that found in the plasma taken as unity. This has been done in Table VIII, which emphasizes the following facts:

a) Leucocytes. Free amino acid concentrations are the highest in this reservoir. Of special interest is the case of taurine which is several hundred times more concentrated in leucocytes than in plasma. For all other amino acids, without exception, concentrations are 4–60 times higher than in the surrounding plasma. In the case of aspartic and glutamic acids, concentration ratios may be somewhat in error since aspartic acid is present in a very small amount in plasma and glutamic acid might have been produced in part by glutamine decomposition.

b) Platelets. The situation in this reservoir is similar to that found in leucocytes, when compared to plasma levels. Comparison of analytical data given in Table VII for leucocytes and platelets shows concentration differences between these two

TABLE VIII

RELATIVE CONCENTRATION OF FREE AMINO ACIDS IN ERYTHROCYTES, LEUCOCYTES AND PLATELETS*

Amino acid	*Erythrocytes*	*Leucocytes*	*Platelets*
Taurine	0.72	521	420
Lysine	0.65	20	6
Glycine	1.70	24	17
Alanine	0.83	16	6
Valine	1.20	14	5
Leucine	3.00	45	13
Isoleucine	0.60	41	17
Serine	1.25	43	30
Threonine	1.23	26	12
Phenylalanine	0.80	50	17
Tyrosine	0.83	33	13
Histidine	1.75	8	4
Arginine	Traces	—	—
Aspartic acid	(185)	(1750)	(1350)
Glutamic acid	(6)	(147)	(63)
Methionine	—	58	13
Cystine $^1/_2$	—	4	—
Proline	0.80	10	5

* Plasma concentration taken as unity. Brackets indicate imprecise values.

types of cells which cannot be accounted for on the basis of analytical errors in amino acid determination. It is concluded therefore that the platelet preparation was not contaminated to a significant extent by leucocytes. As in the case of leucocytes, taurine is several hundred times more concentrated in platelets than in plasma. Taurine has been shown to derive from cysteine in platelets by means of tracer studies[50]. It appears either that sulfur metabolism must be specially active in those compartments, or that taurine has a special meaning as an electrolyte since it has been calculated that taurine equivalents could balance approx. 70% of the total Na^+, K^+, and Ca^{2+} content of platelets[51].

Errors in concentration measurements of free amino acids in leucocytes and platelets are unavoidable, owing to the method used for the separation of the cells, since the sediment obtained is made of wet cells and is weighed as such after as complete as possible drainage of the excess washing solution. Nevertheless, these results clearly indicate that free amino acid concentrations in leucocytes and platelets

References p. 261/262

TABLE IX

ERYTHROCYTES FREE AMINO ACID CONTENT AT DIFFERENT STAGES OF THE SEPARATION PROCESS

Normal adult woman. Values are expressed in μmoles/100 g water.

Amino acid	*Plasma*	*Erythrocytes*				*Non-washed erythrocytes/plasma ratio*
		Non-washed	*After PVP treatment*	*After 1st NaCl washing*	*After 2nd NaCl washing*	
Phosphoserine	0.44	7.85	6.16	—	37.9	17.84
Taurine	3.71	3.85	3.23	—	4.62	1.04
Methionine sulfoxides	0.55	1.54	—	—	0.92	2.80
Aspartic acid	0.11	21.25	19.40	19.60	18.63	193.18
Threonine	9.05	9.55	2.16	2.20	2.77	1.06
Serine	8.50	7.24	6.00	6.10	6.01	0.85
Glutamine	12.50	46.80	32.80	35.70	34.19	3.75
Proline	23.00	12.47	2.00	1.50	1.10	0.54
Glutamic acid	2.00	23.10	26.80	24.20	23.41	11.55
Citrulline	1.85	0.31	—	0.15	—	0.17
Glycine	18.20	31.88	28.30	28.20	28.34	1.75
Alanine	26.30	27.10	26.30	19.70	15.40	1.03
α-Aminobutyric acid	1.40	1.23	—	—		0.88
Valine	18.50	14.17	2.60	2.50	0.92	0.77
Cystine $^1/_2$	2.40	Traces	—	—	—	—
Methionine	2.00	1.85	Traces	0.50	0.40	0.93
Isoleucine	6.80	6.01	1.10	0.90	0.46	0.88
Leucine	11.40	10.01	2.16	1.70	0.49	0.88
Tyrosine	5.90	7.39	2.00	1.69	1.54	1.25
Phenylalanine	5.90	5.24	1.39	1.08	0.62	0.89
β-alanine	0	0.77	—	—	—	—
β-aminoisobutyric acid	Traces	Traces	—	—	—	—
Ornithine	6.04	13.27	4.79	—	2.63	2.20
Lysine	13.25	12.92	4.99	—	3.66	0.92
1-Methylhistidine	0	1.28	0.98	—	Traces	—
Histidine	7.05	7.53	2.79	—	1.66	1.07
3-Methylhistidine	Traces	Traces	—	—	—	—
Arginine	5.32	2.00	0	—	0	0.39
Urea	550	517	52.70	—	14.30	0.94
Ammonia	56.89	88.81	76.85	—	76.58	1.56

are much higher than in the surrounding plasma, and they agree reasonably with those published by SCHRAM AND HUBINONT[49] and by MCMENAMY *et al.*[31] although in the latter study, the methods used for amino acid determination and preparation of the samples were quite different from those used in the present study. Although we have no evidence that free amino acids could have leaked away from leucocytes and platelets during the cell separation process we have used, this possibility must be kept in mind since, as will be seen, it actually happens to a large extent with erythrocytes when plasma is washed off.

c) Erythrocytes. To obtain separation of the leucocytes and platelets from the erythrocytes, use was made of spontaneous sedimentation of the red blood cells in an isotonic (pH 7.3) buffered solution containing 2% polyvinylpyrollidone. This process was repeated twice, the erythrocytes were packed by centrifugation and the red cell sediment resuspensed in saline and also centrifuged twice. Values obtained from the last sediment, after lysis and picric acid deproteinization of the lysate, are shown in Table VII under the heading *non-corrected for diffusion*. Concentrations listed in this column suggest that most of the amino acids contained by red cells have been lost by diffusion during the preparative steps. This has been controlled in the following experiment: a sample has been collected after each step involving erythrocytes, in order to assess the extent and moment of loss by diffusion. The results of these experiments are shown in Table IX, and the following facts can be emphasized:

1) In erythrocytes which have not been washed, free amino acids may be divided in three groups according to their respective concentration. In the first group there are 13 free amino acids; their concentration is similar to that observed in the plasma (*i.e.* amino acids having an erythrocytes/plasma concentration ratio close to unity, comprised between 0.77 and 1.25). It includes taurine, threonine, serine, alanine, α-aminobutyric acid, valine, methionine, isoleucine, leucine, tyrosine, phenylalanine, and histidine. Such is also the case for urea. The second group includes seven free amino acids or amino acid derivatives, which have an intracellular concentration significantly higher than that found in plasma, namely: aspartic acid, phosphoserine, glutamic acid, glutamine (and asparagine), ornithine, methionine sulfoxides and glycine. Apparently, such is also the case for ammonia. In the third group are found three amino acids having an intracellular concentration significantly lower than that found in the plasma, namely: citrulline, arginine and proline. One must be cautious when including proline in this group since the presence of glutathione in its elution range renders its determination difficult, unless glutathione is converted into glutathione S-sulfonate which will be eluted with the front of the eluent, but this has not been done in this particular experiment. It is possible therefore that proline has to be included in the first rather than in the third group. The finding of a low arginine concentration associated to a high ornithine concentration is not surprising since erythrocytes are known to contain arginase[52].

2) After the three-stage process for erythrocytes spontaneous sedimentation in buffered isotonic solution containing 2% polyvinylpyrollidone, it can be seen that some of the amino acids are easily lost by diffusion (threonine, proline, valine, isoleucine, leucine, tyrosine, phenylalanine, ornithine, lysine and histidine; and also urea), while others diffuse rather slowly into the amino acid-free washing solution (phosphoserine, taurine, aspartic acid, serine, glutamine (and asparagine), glutamic acid, glycine, alanine, cystathionine, β-alanine and 1-methylhistidine; and also ammonia). Arginine has completely disappeared presumably as a result of continued arginase activity.

3) The two washings in saline solution introduce a new factor since the solution used, although isotonic, is not buffered. After each of these two steps, the intracellular concentration of taurine, aspartic acid, glutamine (and asparagine), glutamic acid, glycine, alanine and also of ammonia, is not impressively lowered, whilst other amino acid concentrations continue to drop progressively or even abruptly, which drop might presumably be accounted for by the fact that pH conditions are modified and that volume contraction might have been induced.

From such an observation many questions arise. Why is an amino acid such as serine apparently so firmly retained in the red cell when threonine leaks away so easily, although they are so similar in structure? Why are amino acids of small molecular weight such as glycine and alanine kept within the erythrocytes when their molecular weight, so close to that of urea, should allow them presumably to pass rather easily through the erythrocyte membrane? The case of aspartic and glutamic acids might probably be explained by the fact that erythrocytes contain glutamic–oxalacetic and glutamic–pyruvic transaminase systems.

From the data presented in Table VII, one may try to figure out a tentative quantitative model for distribution of free amino acids between the different compartments of the whole blood reservoir. Comparison of the free amino acid concentrations in the three types of formed elements to those found in plasma is impressive, but does not reflect what amounts are actually present. Such an evaluation may be obtained by assuming theoretical values for the different compartments, for instance, a mean whole blood volume of 6 l, an hematocrit value of approx. 50%, leucocytes and platelets volumes of respectively 1 and 0.5% of the whole blood volume. When such calculation is made, as shown in Table X, one can roughly evaluate the relative total content of the four compartments of the blood in the following way:

TABLE X

TENTATIVE DISTRIBUTION OF FREE AMINO ACIDS IN THE FOUR COMPARTMENTS OF THE BLOOD RESERVOIR IN MAN

Amino acid	*Blood-plasma (μmoles*)*	*Erythrocytes (μmoles**)*	*Leucocytes (μmoles**)*	*Platelets (μmoles**)*	*Urinary output (μmoles/24 h***)*	*Plasma clearance time§*
Taurine	120	95	1200	560	990	3 h
Lysine	548	264	103	34	50	11 days
Glycine	666	740	266	133	1380	12 h
Alanine	1124	750	337	112	240	5 days
Valine	726	700	171	43	90	9 days
Leucine	382	760	305	76	80	5 days
Isoleucine	229	75	114	38	110	2 days
Serine	286	230	238	95	400	17 hr
Threonine	336	336	168	42	150	2 days
Phenylalanine	152	75	121	30	80	2 days
Tyrosine	166	105	83	28	110	1.5 day
Histidine	226	300	32	9	900	6 h
Arginine	258	60	14	15	30	8.5 days
Aspartic acid	7	710	150	75	60	3 h
Glutamic acid	(136)	(612)	(340)	(68)	(500)	?
Methionine	67	0?	67	7	50	1 day
Cystine $^1/_2$	292	0?	16	0?	120	2.5 days
Proline	608	350	86	43	0	—
Total	6369	6162	3811	1408	5340	—
% of whole blood volume	± 50 %	± 50 %	1 %	0.5 %	—	—

* Average values calculated from STEIN AND MOORE data[12].
** Calculated from data shown in Table VII.
*** Average values in man according to SOUPART[20].
§ Urinary daily output divided by amount contained in plasma, expressed in h or days. Indicates time necessary for loss by urinary excretion of the amount normally present in plasma.
* and ** calculated for a theoretical whole blood volume of 6 l.

References p. 261/262

Leucocytes. Although no distinction can be made here between the different varieties of white blood cells, it appears that leucocytes as a whole contain about ten times more free amino acids than the plasma compartment, although the latter has a volume 50 times greater than that occupied by the circulating leucocytes under normal conditions. Taurine, aspartic acid and glutamic acid content is remarkable in this respect.

Platelets. Although half the volume of the circulating leucocytes this compartment has a very high content in free amino acids.

Erythrocytes. When non-washed red cells are considered it may be seen that its capacity is lower than that of the plasma with regard to taurine, lysine, alanine, isoleucine, serine, phenylalanine, tyrosine, methionine, cystine and proline; higher than that of plasma as regard to glycine, leucine, histidine, aspartic and glutamic acids (especially the latter two); and about the same as plasma as regard to valine and threonine.

When considering all the above data, it must be borne in mind that it is only an approximate evaluation, bound to several limitations, namely the method of preparation of the formed elements which has been used, the fact that the phenomenon described reflects complex situations, depending upon continuing intrinsic metabolism during the separation process which requires at least several hours to be completed, and also depending upon secondary factors, such as volume contraction, which might be involved in *in vitro* experiments.

AMINOACIDURIA IN VARIOUS CONDITIONS

Since a general review on aminoaciduria has been published by BIGWOOD *et al.*[1] giving a complete and detailed account of the literature up to August 1958, some new knowledge has been gained which leads to reconsideration of the classification of types of aminoaciduria in diseases[1a]. This section is devoted to data recently collected with regard to aminoaciduria in various conditions in man: the problem of "β-aminoisobutyric acid excreters", the free amino acid excretion in protein malnutrition and some observations on the effects of X-irradiation on free amino acids of blood plasma and urine in certain types of malignant diseases. By way of conclusion a provisional classification of hyperaminoaciduria in diseases will be submitted.

β-aminoisobutyric acid excreters

As shown in Table I, β-aminoisobutyric acid (BAIBA) is excreted in rather small amounts by healthy adults: from 0–52 mg/day, in the average 26 mg/24 h (or 0–500 μmoles, average 252 μmoles/24 h). This excretion range has been found in a group of 24 white adult subjects and the individual values were uniformly distributed among the excretion range.

In the child, up to 2 years old, BAIBA excretion was found to vary between 0 and 12 mg/24 h in a group of ten young white normal subjects. "BAIBA excreters" have been so called after a study done by HARRIS in 1953 (ref. 53) in a group of 345 healthy white people of the British Isles. HARRIS found that approx. 10% of the subjects investigated excreted abnormally large amounts of BAIBA, and he concluded that this frequency of excreters could be taken as a hereditary character of the white, being monogenic and recessive[54, 53]. According to GARTLER *et al.*[54] BAIBA output in those excreters could amount up to 70 mg/day. These data have been obtained by means of paper chromatography techniques. By similar methods, LINDAN[55] in Gambia, assigned a frequency of some 25% of BAIBA excreters as a genetic character, here too hereditary and recessive, to the colored population (Central African tribes) living in the area where he performed his study. In other population groups, namely among Asians, the percentage of BAIBA excreters is said to be still higher[56].

Excretion was said to be independant of age, sex and diet, but was considered to be under the control of more than one gene. The distribution curve is bimodal; among Mongoloid subjects 60% are claimed to be BAIBA excreters whereas among the Negroid type of individuals the frequency is of the order of 40% and this percentage drops to 10% among Caucasians[57]. Originally it was considered to be a renal type of aminoaciduria[30] but this has become doubtful; DENT[41] includes it rather in his third category of aminoaciduria process, "no threshold" category, the threshold being very low if there is one. The point to be stressed here is twofold: first, the analytical method used in these studies is questionable from the quantitative point of view; and secondly, it seems that there could be an underlying factor, responsible for BAIBA excretion, that probably escaped the scrutiny of former investigators.

From the quantitative point of view, it seems very hazardous to judge the amount of BAIBA excreted from semi-quantitative results furnished by paper chromatography techniques, especially when an amino acid excreted in minute amounts is concerned. Moreover, in many cases, the daily BAIBA output was estimated from 6-h urine collections[58], a procedure which is to be considered as definitely unsafe. If the frequency of 10% of BAIBA excreters, indicated by HARRIS as a genetic character of the white population of the British Isles, was to be fully accepted, the odds would have been that at least two or more BAIBA excreters should show up in the group of 34 white normal subjects, 24 adults and ten children picked at random, whose aminoaciduria is described in the section FREE AMINO ACIDS IN URINE AND BLOOD PLASMA. None of them was abnormal in this respect. Of course, it is realized that this set of observations is of too small a size to permit a conclusion. A similar observation has been made by Dr. H. VIS[1a], who extensively studied aminoaciduria in chronic protein deficiency among Central-African populations by means of ion-exchange chromatography. VIS found only two or three cases in a group of 45 subjects of various ages, whose BAIBA excretion could be considered as definitely abnormal, hence approx. 4–6%, a frequency much lower than the 25% assigned by LINDAN to such a population group. The previously unsuspected factor underlying here, as will be seen in the next section is that BAIBA may be considerably increased owing to chronic protein deficiency, a condition which is unfortunately not rare in underdeveloped countries. Impaired nutritional conditions probably have more bearing on the frequency of hyperaminoaciduria than any genetic character.

Free amino acid excretion in protein malnutrition

Low protein content of the available diet of populations living in tropical underdeveloped areas leads primarily to a degeneration of certain organs, namely those glandular ones whose external secretions are protein-rich in physiological conditions. As a result of this tissue breakdown a hyperaminoaciduria develops*, which affects all urinary amino acids and specially taurine and BAIBA. When protein deprivation is slight only taurine and BAIBA excretion are involved. It is in the more severe cases, when the kidney itself is affected, that hyperaminoaciduria extends to all urinary amino acids. As shown by VIS *et al.*[59, 60] adequate therapy, consisting in a protein-rich diet, mainly constituted of milk proteins which allow for a high supply of indispensable amino acids and especially of lysine, is highly effective in these patients. Aminoaciduria is enhanced at first but very soon, in a week or two, and in 24 h in the case of taurine, the urinary amino acid pattern returns to normal, except for BAIBA, which remains elevated in urine for a longer period, even for 2 months after otherwise succesful treatment[61].

Fasting itself, provided it is of some duration, has been observed to produce a similar hyperexcretion of taurine and BAIBA[62, 81]. As will be seen in the next section, similar findings are also typical of X-irradiation.

Effects of X-irradiation on free amino acids in urine and plasma in certain types of malignant diseases

Studies of patients accidentally injured by nuclear radiations have drawn attention, among other features, to the hyperaminoaciduria that follows the exposure. Such was the case for the Los Alamos and more recently for the Yugoslavian accidents which effects have been described extensively by EMPELMAN *et al.*[63] and by JAMMET *et al.*[64], respectively. Aside from such undesirable events, there is a much more common occurrence, namely the therapeutic irradiation of a cancerous patient. Two such instances** have been investigated by SOUPART[65, 66] by means of ion-exchange chromatography and typical observations are described below.

In a case of chronic lymphocytic leukemia plasma and urine amino acid analyses were performed before and during treatment which consisted in this case of 50-r X-rays external doses delivered at first on the very enlarged spleen area only and afterwards on very large lymph nodes located in the neck. Free amino acid urinary pattern and blood plasma composition in this patient before treatment are illustrated by Fig. 1 and Fig. 4, respectively. Both of them were quite normal according to the standards given in the section FREE AMINO ACIDS IN URINE AND BLOOD PLASMA (p. 225). Following local X-rays therapy, the main feature of the urinary amino acid pattern was an increased output of both taurine and BAIBA, all other urinary amino acids remaining apparently unaffected, except for an increased methionine sulfoxide excretion and the very transient appearance of an unusual component, eluted in the range of methionine sulfone.

Fig. 9 shows the daily excretion of BAIBA during the treatment as compared to

* According to very recent work[89] hyperaminoaciduria which develops in such protein deficiencies, as well as in rickets, might be a reflection of a trend tc adaptation of the intracellular pool of free amino acids to electro-osmotic variations. Aminoaciduria under normal conditions might have the same meaning.

** The opportunity to study these cases has been provided by courtesy of Dr. H. J. TAGNON, Head of Internal Medicine Department. Institut Jules Bordet, Brussels University, Cancer Treatment and Research Center, Brussels (Belgium).

References p. 261/262

its excretion during the control period. The excretion of this amino acid rises up to four times the control value. The arrows indicate the time of X-irradiations.

Fig. 10 shows the free and combined taurine excretion pattern after irradiation. The excretion of the free, as well as that of the combined form, immediately after the first irradiation, rises up to ten times the control values, and this is more than twice the highest value recorded in the normal up to now. But subsequent irradiations seem to be less active on taurine excretion.

This increased urinary excretion of both taurine and BAIBA is attributable to plasma overloading, since plasma levels of these amino acids are significantly raised

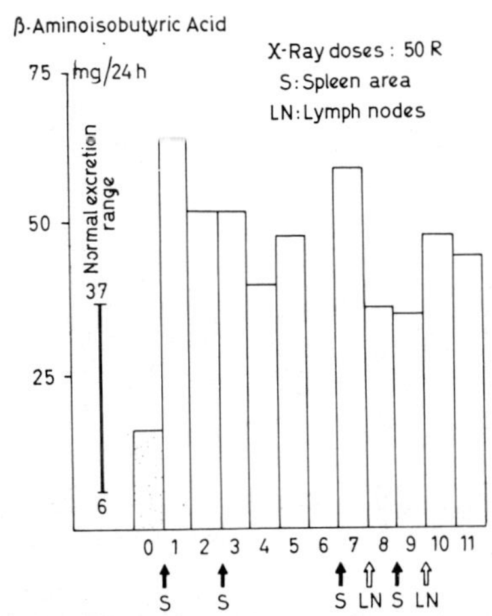

Fig. 9. β-Aminoisobutyric acid excretion in a chronic lymphocytic leukemic patient after therapeutic X-irradiation (o, control excretion).

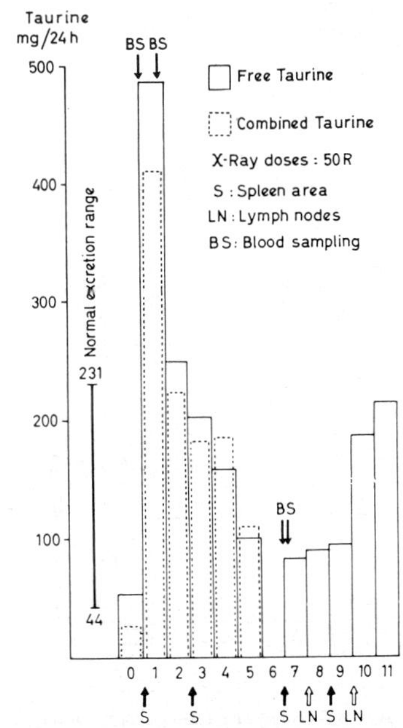

Fig. 10. Free and combined taurine excretion in a chronic lymphocytic leukemic patient after therapeutic X-irradiation (o, control excretion).

shortly after irradiation. Increased renal function (*i.e.* increased glomerular filtration rate, for instance) and/or lowering of renal tubular reabsorption do not seem to be involved in this particular case, since none of the other amino acids is excreted in higher amount than usual.

BAIBA is known to be released in abnormally high amount as a result of cell breakdown initiated by X-irradiation. This amino acid is in fact, one of the end products of DNA catabolism. Moreover, DNA is presumably particularly abundant in this case.

The origin of the taurine excreted under the influence of X-irradiation might be slightly different. As shown in the section BLOOD POOLS OF FREE AMINO ACIDS (p. 241), taurine is a typical intracellular component, the concentration of which is remarkably higher in leucocytes and platelets than it is in the surrounding plasma. Taurine also is one of the end products of the oxidative metabolism of the sulfur amino acids cystine and cysteine. The increased amount of free and combined taurine excreted, as well as that of total sulfate which shows a similar pattern of excretion, may well be due in part to increased oxidation of cystine and cysteine. The fact that repeated X-irradiation of the patient seems to be unable to produce a sustained high taurine excretion may perhaps be due to temporary exhaustion of oxidizable sulfur amino acids in the cells. To understand the extra taurine excretion after X-irradiation it seems that some factor other than oxidation of amino acids has to be taken into account. This was strongly suggested by the second observation to be described hereafter.

As taurine is essentially an intracellular component[67], this amino acid, pre-existing

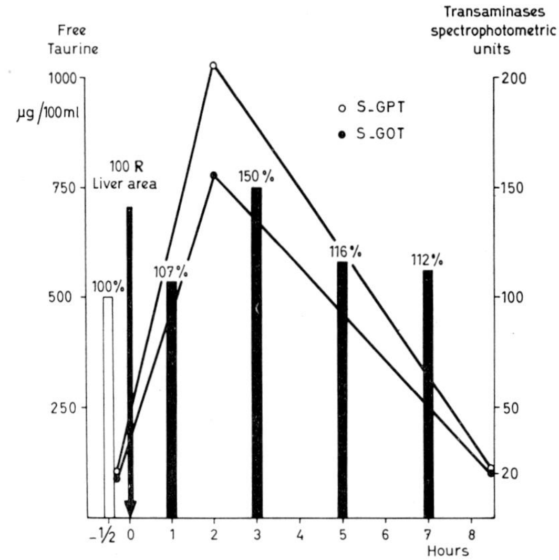

Fig. 11. Free taurine plasma levels before and after X-irradiation on liver area of a case of Hodgkin's disease, as compared with serum-transaminase activities taken as an index of cell injury.

or formed under the influence of X-rays, probably finds its way out of the cells to appear in increased amount in the plasma. This suggests that, aside from their oxidative power, X-rays might also influence the permeability of the cell, leading to the exit of normally enclosed components. Accordingly, the free taurine plasma concentration as well as glutamic-oxalacetic and glutamic-pyruvic transaminase activities in plasma have been determined before and at intervals after the patient (a case of Hodgkin's disease) received 100 r external dose of X-rays on the liver area. As shown in Fig. 11, the maximum increase of free taurine plasma concentration coincides with a transient but definitely abnormal transaminase-activity level, taken here as an index of hepatic and/or Hodgkin cell injury.

Such a finding, in the author's opinion, strongly suggests that the oxidative effects of X-rays are also accompanied by a specific effect on the permeability of the cell, namely on the enzymatic mechanisms which may be involved to sustain a high intracellular content of taurine in normal conditions. Direct evidence, supporting this opinion, has been afforded by SCHRAM[51] in this department, by total irradiation of rats in which taurine was labeled by means of daily injections of 1 mg of [^{35}S]-

taurine for 3 days prior to exposure. This experiment is based on the fact that intracellular taurine is characterized by a rather low turnover rate, contrary to its extracellular fraction, which is negligible with regard to the amount excreted. In those conditions, one may assume, either that if the specific activity of excreted taurine after irradiation does not vary or increases, that it came at least partly from the intracellular compartment; or, if the specific activity of excreted taurine after irradiation is decreased (total radioactivity being constant), that it came from its precursors. 48 h after the last labeled-taurine injection, the test animals were irradiated, receiving a single 500 r external dose of X-rays, and at the same time control animals were submitted to mock irradiation. Within the first 24 h following that treatment, taurine excretion in the test animals increased by 75% as compared to controls, and its specific radioactivity remained unchanged. Taurine urinary hyperexcretion may thus be attributed to escape from the cells where this amino acid is highly concentrated, rather than to excessive oxidation of sulfur amino acids.

It seems that, as far as taurine and BAIBA are concerned, protein malnutrition as well as radiation effects share some properties, which are to increase BAIBA formation from DNA and to injure some sensitive mechanisms, probably enzymatic in nature, which sustain high taurine intracellular concentration in physiological conditions.

Provisional classification of aminoaciduria in diseases

Attractive in its simplicity, DENT's classification[41] of abnormal types of aminoaciduria is based on the blood amino acid clearance concept. Pathological conditions are divided in three categories: (a) "overflow" aminoaciduria in which the blood level of the amino acid concerned is definitely raised above normal; (b) "renal" aminoaciduria in which case the blood level is either normal or below normal, and yet the urinary excretion is above normal (tubular reabsorption deficiency); (c) "no-threshold" aminoaciduria in which the condition is attributed to an extra-renal disturbance of the metabolism of an amino acid with a high blood clearance, and in which, on account of this latter circumstance, the corresponding blood level may stay normal or is hardly increased.

In 1958, when preparing a general review on aminoaciduria with BIGWOOD *et al.*[1], we strongly felt that our knowledge of blood amino acid levels and clearances was still insufficient to fill simply the three theoretical categories of DENT with the available data. Today we still adhere to this opinion but owing to some recent advances we feel the need for a revision[1a] of the provisional classification published in 1958. Accordingly we propose a renewed but still provisional classification* which will distinguish two main types of abnormal aminoaciduria: (1) a common type of aminoaciduria, involving more or less all amino acids excreted in urine; (2) a specific type of aminoaciduria, involving a limited number of urinary amino acids.

An extensive reference list is to be found in the review published by BIGWOOD *et al.*[1].

---

* Based mainly on the views of DR. H. VIS; see footnote page 250.

New provisional classification of aminoaciduria in pathological disorders

Aminoaciduria of the common type, involving all urinary free amino acids. This type of pathological picture may be due either to congenital or acquired conditions. It is not, at first, generalized to all amino acids, but starts with hyperexcretion of the amino acids which are the less well reabsorbed in physiological conditions. When the disorder extends the hyperaminoaciduria progressively affects all amino acids. There is thus only a matter of degree which could distinguish the condition at its starting point and the situation when the condition has fully developed. Three sets of circumstances may lead to such amino acid excretion patterns:

(1) Increased blood plasma amino acid concentrations, resulting in a true "overflow" aminoaciduria. The urinary amino acid excretion pattern seems to be related to the type of cells specifically involved in the disorder, since tissues differ in their relative free amino acid composition. (2) The unusual process may not involve increased plasma amino acid concentration but, on the contrary, may result in lower plasma concentration, the intact reabsorption mechanism being overloaded by an increased glomerular filtration rate. Such an occurrence is found for instance in normal pregnancy. (3) Free amino acid levels in plasma are normal or even lowered, the glomerular filtration rate is normal, but there is still a generalized hyperaminoaciduria which is attributable to an impaired reabsorption function, as is the case in the so-called "De Toni-Debré-Fanconi syndrome", among the congenital disorders, and possibly that of vitamin D deficiency, among acquired ones.

These three processes do not depart very far from the three main categories of DENT's classification, although they do not show up in full development at first.

Specific types of aminoaciduria involving a limited number of free amino acids. Here too, two different types of pathogenic mechanisms must be recognized: congenital and acquired conditions. Among the first category of situations one may rank genetic metabolic blocks, namely aromatic acid metabolic disturbances and essential cystinuria, as well as "maple-sugar-urine disease". Among acquired disorders, all situations are characterized by a common feature, there is always a tissue breakdown to be found resulting in taurine and BAIBA hyperexcretion. It is realized that the main distinction submitted here is only valid as a working hypothesis since in most cases, either sound information is not available or the available data have been provided by means of methods actually considered as unreliable.

After these preliminary statements, one may reconsider a provisional but practical classification on the following basis:

1) The anomaly is a congenital metabolic block, involving one or more amino acids and resulting in a hyperaminoaciduria of the generalized type.

2) The anomaly results from a defect of the proximal convoluted tubule. The process may be either congenital or acquired and the resulting hyperaminoaciduria is still of the common generalized type.

3) The primary pathogenic factor is congenital in nature but may also involve metabolic disturbance of substances other than amino acids. It results, nevertheless, in a secondary hyperaminoaciduria of the common generalized type.

4) The anomaly results from an acquired condition, inducing either a generalized hyperaminoaciduria of the common type or an hyperexcretion of a limited number

References p. 261/262

of free amino acids, related to a common pathological process which may be due to distinct causative factors.

The proposed practical classification may now be developed as follows:

A. Aminoaciduria resulting from a congenital metabolic block involving one or more amino acids

According to the "one enzyme–one gene" theory if, in an enzymatically driven sequence of reactions, one particular enzyme is missing, there is a metabolic block at the stage of the missing enzyme as a result of the genetic defect. If no alternative metabolic pathway is available, there will be an accumulation of the precursors which may result in harmfull effects on the tissues and in urinary hyperexcretion of these precursors. In this category we classify the following types of aminoaciduria which are very infrequently met:

1) Lesion of phenylalanine metabolism. In phenylpyruvic oligophrenia there is an increased level of plasma phenylalanine and hyperexcretion of this amino acid as well as of its derivatives (phenylpyruvic, phenyllactic, phenylacetic, *p*-phenolacetic, *p*-phenollactic acids and phenylacetylglutamine). Oxidation of phenylalanine into tyrosine is impaired and tryptophane metabolism seems also to be involved since there is a hyperexcretion of indoleacetic, indolelactic and indoxylacetic acids. Phenylalanine clearance is said to be normal.

2) Lesion of tyrosine metabolism. It is not well established whether alcaptonuric urines show evidence of abnormal amino acid excretion. According to PELC AND VIS[68], there is an increased tyrosine and phenylalanine excretion as well as the appearance of an unidentified unhydrolyzable constituent, yielding a yellow color on reaction with ninhydrin, which presumably is homogentisic acid. One case of tyrosinosis has been described by MEDES in 1932 in a patient suffering from myasthenia gravis, who excreted large amounts of *p*-phenolpyruvic acid. It was noticed that phenylalanine intake produced tyrosine hyperexcretion w thout alcaptonuria. The reality of an authentic biochemical lesion has been questioned since it is well established now that pyruvic acid hyperexcretion may result from acquired hepatic insufficiency, from malignant growths and from vitamin C deficiency (see p. 258 and 259).

3) Lesion of tryptophane metabolism. In Hartnup syndrome, described by BARON *et al.*[69], there is a more or less generalized hyperaminoaciduria associated with indican as well as indoleacetic acid hyperexcretion. The biochemical lesion concerned here impairs tryptophane conversion into nicotinic acid.

4) Lesion of valine, leucine and isoleucine metabolism. There is an "overflow" aminoaciduria involving valine, leucine and isoleucine present in a syndrome known as "maple-sugar-urine disease"[70, 71] in which urine smells like maple syrup. Urine also contains large amounts of the corresponding α-keto acids. The metabolic block has been definitely located by DANCIS *et al.*[72] who showed that desamination of the three branched-chain amino acids normally occurs but that the resulting keto acids cannot undergo further degradation by oxidative decarboxylation.

5) Lesion of phosphoethanolamine metabolism. In a condition known as hyperphosphatasia, there is a metabolic block bearing on phosphoethanolamine hydrolysis, resulting in an accumulation of this intermediate metabolite and its urinary hyperexcretion.

6) Lesion of arginine metabolism. ALLAN *et al.*[73] have described an abnormal excretion of arginine–succinic acid[74] in children suffering from severe mental deficiency, probably hereditary in nature.

7) Lesion of glycine metabolism. CHILDS *et al.*[32] have discribed very recently a new disorder which can be listed here: *idiopathic hyperglycinemia and glycinuria.* The syndrome itself comprises growth and mental disturbances, anorexia, vomitings and convulsions. This clinical condition becomes worse when threonine, leucine or isoleucine are fed.

B. Aminoaciduria resulting from a congenital or acquired defect of the proximal convoluted tubule

1) Essential cystinuria. This condition is characterized by hyperexcretion of the "renal" type of cystine, lysine, arginine and ornithine exclusively, with an eventual formation of cystine stones in the urinary tract. No other amino acid is involved in this type of aminoaciduria. As plasma concentrations of the above amino acids are not elevated, the condition is thought to be due to a reabsorption defect of a congenital nature. After infusion of any amino acid of this group, excretion of all four amino acids is found to be increased. Since there are other occurrences of cystinuria not to be confused with the present condition, it is advisable to characterize it as "essential cystinuria".

2) De Toni-Debré-Fanconi syndrome. There is an aminoaciduria of the common type, involving all amino acids. All amino acids have increased clearances and the syndrome, still not fully understood, may be accompanied by cystinosis. In the later case, the condition is sometimes called "Lignac-Fanconi" since LIGNAC has been the first to describe cystinosis. It has been shown experimentally, that high doses of cystine produce albuminuria, gluciduria and a generalized cystinosis in dogs.

3) Cystinosis. In this condition cystine accumulates in very large amounts in tissues and crystallizes there, forming important cystine deposits in the reticuloendothelial tissue of the liver, spleen, bone marrow, cornea and kidney, chiefly in the renal tubules. Plasma concentration of cystine is normal and there may be a hypercystinuria, although this does not seem to be the rule. It is still not known if eventual hypercystinuria is accompanied by hyperexcretion of lysine, arginine and ornithine, and the condition is therefore to be considered momentarily as distinct from essential cystinuria. Cystine deposits found in tissues suggest a metabolic block as the origin of the condition but on the other hand, when cystinuria is present in cystinosis, a renal tubular lesion is thought of. As cystinosis cases are scarce, correct classification of this condition must await further investigation.

4) Heavy metal intoxication. Toxic effects of heavy metals have been considered to depress tubular reabsorption mechanisms, probably by means of enzyme inhibition. Accumulation of malic acid formed by the Krebs cycle has been claimed to depress the reabsorption mechanism, a situation which could result in a generalized aminoaciduria.

5) Glycinuria in oxalic lithiasis. DE VRIES *et al.*[83] have discribed hyperglycinuria in a patient with oxalic lithiasis. The condition seems to be a family one since it has been observed in 4 other members of the same family.

6) Hyperaminoaciduria in lipoidic nephrosis. HOOFT AND HERPOL[84], WOOLF AND

McGiles[85], Tegelaers and Tiddens[86] Standbury and McGaulay[87], Hooft and Vermassen[88], have described hyperaminoaciduria in patients with lipoidic nephrosis and also under some special conditions of treathment with protein and ACTH.

C. Hyperaminoaciduria secondary to some other congenital metabolic disturbances

In those cases, the aminoaciduria is of the common type, generalized and not specific to the etiologic factor of the metabolic disorder. It is found in the following conitions:

1) Wilson's disease. Copper metabolism is primarily concerned. The aminoaciduria is secondary to a reabsorption defect and many amino acids are involved among which proline, citrulline and cystine deserve special mention. There is neither hyperlysinuria nor hypertaurinuria and the plasma levels are close to the lower limit of their normal distribution range.

2) Galactosemia. It is accompanied by a generalized aminoaciduria resulting from an impairment of tubular reabsorption due to galactose accumulation in the tubular cells. As a secondary factor, some degree of hepatic insufficiency may complicate the picture by addition of a true "overflow" aminoaciduria to the renal one. In this case, all amino acids are hyperexcreted, including taurine and BAIBA. These troubles disappear when galactose is excluded from the diet.

3) Levulosemia. It leads to a picture very close to that produced by galactosemia.

4) Lowe syndrome. This syndrome is characterized by glaucoma, mental deficiency, thermoregulation unbalance and acidosis due to insufficient ammonia formation in the kidney. It is also associated with hyperaminoaciduria of the common type and the plasma amino acid levels are within normal limits.

5) To the above categories of "renal" aminoaciduria may also be added the abnormal excretion found in progressive muscular dystrophy, in gargoylism, in some types of coproporphyrinurias and in infantile congenital cirrhosis. Unfortunately information on such disturbances as regard to amino acid excretion is still scarce and difficult therefore to interpret.

D. Hyperaminoaciduria secondary to acquired metabolic disturbances

In this group of conditions, the aminoaciduria may be either of the common generalized type, or of the specific type involving only a few amino acids. Different etiologic factors, by means of a common initial pathogenic process may result in the same urinary amino acid pattern. The conditions listed below are of a much more common occurrence than those which are genetically determined.

1) Conditions having in common profound general disturbances, excessive cellular catabolism, associated or not to a nutrition deficiency and a negative nitrogen balance. The common feature shared by all those conditions is taurine or taurine and BAIBA hyperexcretion. Owing to adequate therapy the hyperexcretion is reversible, rapidly as regard to taurine, more slowly when BAIBA is concerned. Rubini *et al.*[75] and Fink *et al.*[76] have shown that BAIBA metabolism is closely related to that of nucleic acids and particularly to that of thymine. Excessive tissue breakdown leads to BAIBA hyperexcretion, but its plasma level hardly rises to more than detectable traces since it has a clearance near that of inulin. Taurine hyperexcretion may be either a consequence of increased oxidation of sulfur amino acids or of abnormal release of this amino acid from inside the cell where it is maintained normally at higher concentration than that of the extracellular fluid. In this group of hyperaminoaciduria may be listed:

a) Chronic protein deficiency of the kwashiorkor type (see p. 250).

b) Fasting effects in a normal subject (see p. 250).

c) Radiation effects, even of moderate dosage such as in usual therapeutic X-irradiation (see p. 250).

d) Miscellaneous conditions: shock, hepatic coma, vitamin C deficiency as well as pernicious anemia[77]. In hepatic disorders, the condition may lead progressively to a generalized aminoaciduria due to general increase of all plasma levels and is correlated to the severity of the clinical state.

2) Aminoaciduria occurring in liver disorders. That functional liver insufficiency increases the ratio of amino nitrogen to total nitrogen in urine has long been known, but it is only noticeable in very severe involvement of the liver. First investigation of urinary amino acids in such condition, by means of microbiological methods, led 15 years ago to the recognition of an increased urinary excretion involving cystine, histidine and tryptophane in acute and subacute cirrhosis, acute yellow atrophy, viral hepatitis and hepatic coma. Frequency and intensity of the hyperaminoaciduria are closely correlated to the severity of the condition. When such diseases were investigated by means of ion-exchange chromatography it appeared that hyperexcretion also involved methionine, lysine, phenylalanine, tyrosine, glycine, alanine, BAIBA, 1-methylhistidine, taurine and ethanolamine. In this way it was demonstrated that the urinary amino acid pattern was that of a hyperaminoaciduria of the generalized common type, characterized by progressive installation, the speed of which was related to the extent of liver injury. Frequently, but not as a rule, blood plasma levels of some of the amino acids are found to be increased. It has been observed in some cases that the increased urinary output of an amino acid corresponded to a high clearance, whereas in other cases the clearance was low, but this point still calls for further investigation.

3) Aminoaciduria occurring in vitamin D deficiency. JONXIS AND HUISMAN were the first to describe a hyperaminoaciduria in rickets. The aminoaciduria* progressively develops into a pattern of the generalized common type apparently without abnormalities of the plasma levels. This situation is easily corrected by adequate therapy.

4) Aminoaciduria occuring in vitamin C deficiency. Scurvy is accompanied by a progressive aminoaciduria of the generalized common type. It was thought at one time that tyrosine and phenylalanine were specifically involved, but no evidence supporting this opinion has been found in spite of extensive investigation done by H. VIS (unpublished data). On the other hand, what VIS found as a typical, constant and very early character is free proline and hydroxyproline excretion, a highly specific finding since those amino acids are never found, even as traces, in normal urine, associated to a hyperexcretion of taurine and BAIBA, initiated by the process already described (see p. 251).

In vitamin C deficiency interstitial tissue and more precisely collagen fibers are especially involved, which are rich in proline and hydroxyproline. These two amino

* See footnote page 250.

acids are also excreted in combined form in scurvy. When the deficiency becomes more pronounced the picture extends to a generalized aminoaciduria. As in the case of vitamin D deficiency, adequate specific therapy easily corrects the situation.

5) *Aminoaciduria in some types of intoxication.* Heavy metal (Pb, Hg, U, Cd) intoxication may lead to a hyperaminoaciduria but its effects on plasma levels still have not been investigated. If the pathogenic process involves the proximal convoluted tubule, the picture is that found in the "De Toni-Debré-Fanconi syndrome". When the process is an endogenous one and when copper is concerned, the picture is that of Wilson's disease, but in this case the process is congenital in nature.

An experimental "De Toni-Debré-Fanconi syndrome" may be induced in rats by intoxication with malic acid and presumably its cause might be a lowering of the supply of energy needed for reabsorption, as a result of impaired Krebs cycle function in the tubular cells. Other toxic substances, such as lysol, oxalic acid and *p*-aminosalicylic acid have been charged with inducing hyperaminoaciduria.

A TECHNICAL IMPROVEMENT FOR THE CONTINUOUS COUNTING OF LABELED AMINO ACIDS COUPLED WITH ION-EXCHANGE CHROMATOGRAPHY

The opportunity is taken here to present a new and important technical development brought about by SCHRAM AND LOMBAERT in our department[78]. It consists of

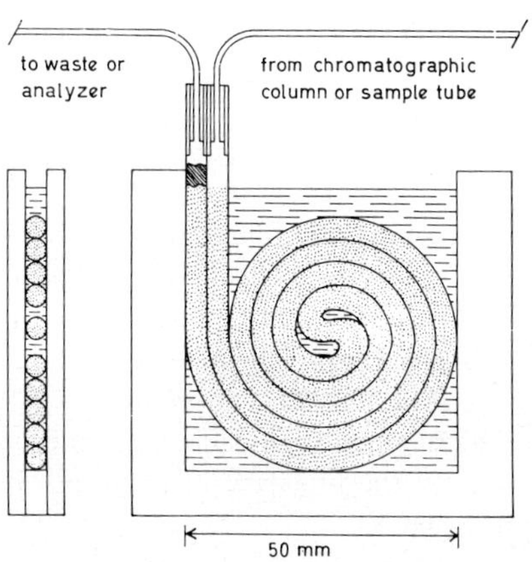

Fig. 12. Anthracene cell for the scintillation counting of aqueous solutions.

a constant volume cell (approx. 1 ml) filled with calibrated anthracene particles[79] which may be inserted in the line of an ion-exchange chromatography automatic machine, by means of special connectors[15], without producing even the least distortion of the eluted peaks thanks to absence of mixing. It is used for scintillation counting of aqueous solution. The counting cell itself is made of transparent polyethylene tubing, of approx. 60 cm length filled with anthracene powder and bent into a spiral inserted into a flat lucite vial filled with silicone oil, as shown in Fig. 12.

As the size of anthracene crystals influences both flow resistance and efficiency, the mesh size choosen has to be a compromise: 300 μ mesh size were selected for ^{14}C and 105 μ particles for tritium. This counting cell may be used in three different

References p. 261/262

ways: with a single refrigerated photomultiplier (for ^{14}C only); with photomultipliers applied on both faces of cell and with photomultipliers placed at right angle, the same face of the cell facing both tubes with a prism interposed as a light-guide. The device fits, by means of a special adapter when necessary, in most commercial units currently used for the liquid scintillation counting of weak beta emitters. It

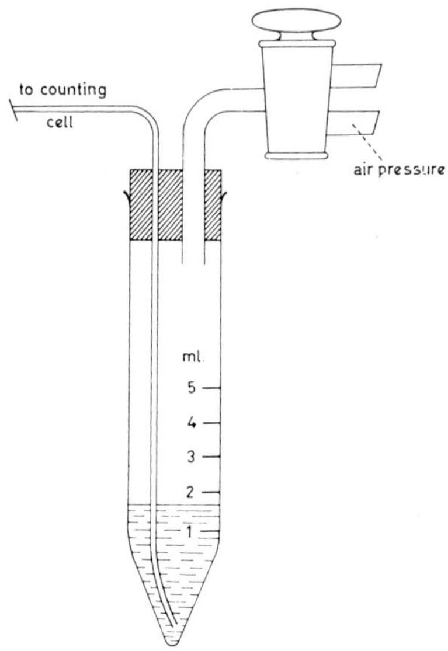

Fig. 13. Device for the counting of individual samples.

may also be used for counting of individual samples by means of a device shown in Fig. 13. When the sample is counted, the stopper is put onto the next tube after wiping off the immersed part of the Teflon tubing with a piece of filter paper taking care not to introduce air bubbles. Due to the absence of mixing, rinsing with distilled water between successive samples is not necessary. The apparatus is calibrated with a standard solution which is run with each set of measurements. The reproducibility of the results as well as the linearity of response for solutions of increasing radioactivity are shown in Table XI.

TABLE XI

LINEARITY AND REPRODUCIBILITY OF SCINTILLATOR RESPONSE FOR ^{14}C

	Radioactivity (dilution factor)	*Counts/min*	*Yield (%, versus solution 3)*
1)	1.000 (stock solution)	28 895	100.6
2)	0.332	9 478	99.4
3)	0.201	5796–5789–5768 5765–5774–5756 Mean = 5775	100.0
4)	0.1004	2851	98.9

References p. 261/262

This method appears suitable for the counting of chromatographic effluents and almost all biological substances which may be brought into solution in an acid, neutral or alkaline aqueous medium.

Insoluble biological materials (tissues) may also be counted after mineralization. Carbon dioxide may be counted after trapping in sodium hydroxide.

When using a fast electronic circuit and unselected E.M.I. photomultipliers (9514 S) the efficiency attained with the finer anthracene powder is 2% for tritium and 55% for ^{14}C. The corresponding background amounts to 60 counts/min when operating at room temperature.

REFERENCES

[1] E. J. Bigwood, R. Crokaert, E. Schram, P. Soupart and H. Vis, *Advances Clin. Chem.*, vol. 2, Academic Press, New York, 1959.
[1a] E. J. Bigwood, E. Schram, P. Soupart and H. Vis, *Exposés ann. biochim. méd.*, 23 (1961) 9.
[2] H. H. Tallan, S. T. Bella, W. H. Stein and S. Moore, *J. Biol. Chem.*, 217 (1955) 703.
[3] P. Boulanger and G. Biserte, *Exposés ann. biochim. méd.*, 18 (1956) 123.
[4] P. Boulanger, G. Biserte, P. Paysant and T. Schoonaert, personal communication.
[5] R. G. Westall, *Biochem. J.*, 60 (1955) 247.
[6] P. Soupart, *Ann. soc. roy. sci. med. et nat. Bruxelles*, 12 (1959) 33.
[7] S. Moore and W. H. Stein, *J. Biol. Chem.*, 192 (1951) 663.
[8] S. Moore and W. H. Stein, *J. Biol. Chem.*, 211 (1954) 893.
[9] S. Moore and W. H. Stein, *J. Biol. Chem.*, 211 (1954) 907.
[10] S. Moore, D. H. Spackman and W. H. Stein, *Anal. Chem.*, 30 (1958) 1185.
[11] D. H. Spackman, W. H. Stein and S. Moore, *Anal. Chem.*, 30 (1958) 1190.
[12] W. H. Stein and S. Moore, *J. Biol. Chem.*, 211 (1954) 915.
[13] E. Schram and E. J. Bigwood, *Anal. Chem.*, 25 (1953) 1424.
[14] S. Moore, D. H. Spackman and W. H. Stein, *Federation Proc.*, 17 (1958) 1107.
[15] E. Schram and R. Lombaert, *Anal. Chem.*, 33 (1961) 1134.
[16] P. B. Hamilton and D. D. van Slyke, *J. Biol. Chem.*, 150 (1943) 231.
[17] M. P. Brigham, W. H. Stein and S. Moore, *J. Clin. Invest.*, 39 (1960) 1633.
[18] R. Truhaut, personal communication.
[19] L. C. Craig, J. D. Gregory and W. Hausmann, *Anal. Chem.*, 22 (1950) 1462.
[20] P. Soupart, *Clin. Chim. Acta*, 4 (1959) 265.
[21] W. H. Stein, *J. Biol. Chem.*, 201 (1953) 45.
[22] S. P. Datta and H. Harris, *Nature*, 168 (1951) 296.
[23] P. Soupart, *Clin. Chim. Acta*, 3 (1958) 349.
[24] J. P. Dustin, *Exptl. Med. Surg.*, 12 (1954) 233.
[25] J. P. Dustin, S. Moore and E. J. Bigwood, *Metabolism*, 4 (1955) 75.
[26] P. Soupart, *Clin. Chim. Acta*, 5 (1960) 235.
[27] J. P. Christensen, J. W. Date, F. Schønheyder and K. Volqvartz, *Scand. J. Clin. Lab. Invest.*, 9 (1957) 54.
[28] E. G. Frame, *J. Clin. Invest.*, 37 (1958) 1710.
[29] D. C. Cusworth and C. E. Dent, *Biochem. J.*, 74 (1960) 550.
[30] D. F. Evered, *Biochem. J.*, 62 (1956) 416.
[31] R. H. McMenamy, C. C. Lund and D. F. H. Wallach, *J. Clin. Invest.*, 39 (1960) 1675.
[32] M. C. Serini, H. McNamara, M. Sibuya, N. Kretchmer and H. H. Barnett, *Pediatrics*, 15 (1955) 575.
[33] P. D. Doolan, H. A. Harper, M. E. Hutchin and W. W. Shreeve, *J. Clin. Invest.*, 34 (1955) 1247.
[34] R. G. Westall, E. Roitman, C. de la Peña, H. Rasmussen, J. Cravioto, F. Gomez and E. Holt, Jr., *Arch. Disease Childhood*, 33 (1958) 499.
[35] F. L. Iber, H. Rosen, S. M. Levenson and T. C. Chalmers, *J. Lab. Clin. Med.*, 50 (1957) 417.
[36] T. H. J. Huisman, *Pediatrics*, 14 (1954) 245.
[37] W. H. Stein, A. G. Bearn and S. Moore, *J. Clin. Invest.*, 33 (1954) 410.
[38] V. K. Arrow and R. G. Westall, *J. Physiol. (London)*, 142 (1958) 141.
[39] H. W. Smith, *Principles of renal physiology*, Oxford University Press, New York, 1956, p. 30.
[40] R. H. McMenamy, C. C. Lund and J. L. Oncley, *J. Clin. Invest.*, 36 (1957) 1672.
[41] C. E. Dent, *Scand. J. Clin. and Lab. Invest.*, 10, suppl. 13 (1957) 122.
[42] H. H. Tallan, S. Moore and W. H. Stein, *J. Biol. Chem.*, 211 (1954) 927.
[43] G. Duchâteau and M. Florkin, *Arch. intern. physiol.*, 62 (1954) 487.

[44] G. Y. N. Iyer, *Indian J. Med. Research*, 45 (1957) 213.
[45] F. Nour-Eldin and J. F. Wilkinson, *Brit. J. Haematol.*, 1 (1955) 358.
[46] G. Rouser, in *The Leukemias: Etiology, Pathophysiology and Treatment*, J. W. Rebuck, F. H. Bethell and R. W. Monto, Eds., Academic Press, New York, 1957, p. 361.
[47] H. N. Christensen, P. F. Cooper, Jr., R. D. Johnson and E. L. Lynch, *J. Biol. Chem.*, 168 (1947) 191.
[48] C. A. Johnson and O. Bergeim, *J. Biol. Chem.*, 188 (1951) 833.
[49] E. Schram and P. O. Hubinont, *7th European Congr. Haematology*, London, 1959, Abstr. Communs., No. 159.
[50] J. Frendo, A. Koj and J. M. Zgliczynski, *Nature*, 183 (1959) 685.
[51] E. Schram, *Arch. intern. physiol. et biochem.*, 69 (1961) 101.
[52] G. C. Covolo and R. West, *J. Clin. Endocrinol.*, 7 (1947) 325.
[53] H. Harris, *Ann. Engenics*, 18 (1953) 43.
[54] S. M. Gartler, *Am. J. Human Genet.*, 8 (1956) 120.
[55] O. Lindan, *Rep. 2nd Inter-african (C.C.T.A.) Conf. on Nutrition*, Gambia, 1952, Colonial Office Publication, London, Her Majesty's Stationary Office (1954).
[56] J. de Groucny and H. Eldon-Sutton, *Am, J. Human Genet.*, 9 (1957) 76.
[57] S. M. Gartler, I. L. Firschein and B. S. Kraus, *Am. J. Human Genet.*, 9 (1957) 200.
[58] S. M. Gartler, *Arch. Biochem. Biophys.*, 80 (1959) 400.
[59] H. Vis, R. Dubois, H. Loeb, M. Vincent and E. J. Bigwood, *Ann. soc. belge méd. trop.*, 38 (1958) 991.
[60] R. Dubois, H. Vis, H. Loeb and M. Vincent, *Helv. Paediat. Acta*, 14 (1959) 13.
[61] H. Vis and E. de Mayer, *4th Internatl. Congr. Clinical Chemistry*, Edinburgh, August 1960, Abstracts of Papers, p. 90.
[62] H. K. Berry, *Metabolism*, 9 (1960) 373.
[63] L. H. Hempelman, H. Lisco and S. G. Hoffman, *Ann. Internal Med.*, 36 (1952) 444.
[64] H. Jammet, G. Mathé, B. Pendic, J. F. Duplan, B. Maupin, R. Latarjet, D. Kalic, L. Schwarzenberg, Z. Djubic and J. Vigne, *Rev. franç. études clin. biol.*, 4 (1959) 210.
[65] P. Soupart, *4th Internatl. Congr. Clinical Chemistry*, Edinburgh 1960, Abstracts of Papers, p. 29.
[66] P. Soupart, *5th Internatl. Congr. Biochemistry*, Moscow 1961. Abstracts of Communications, 364.
[67] E. Schram, *Arch. intern. physiol. et biochem.*, 68 (1960) 698.
[68] S. Pelc and H. Vis, *Acta Neurol. Psychiat. Belg.*, 10 (1960) 905.
[69] D. N. Baron, C. E. Dent, H. Harris, E. W. Hart and J. B. Jepson, *Lancet*, ii (1956) 421.
[70] J. H. Menkes, P. L. Hurst and J. M. Craig, *Pediatrics*, 14 (1954) 426.
[71] R. G. Westall, J. Dancis and S. Miller, *Am. J. Diseases Children*, 94 (1957) 571.
[72] J. Dancis, M. Levitz and R. G. Westall, *Pediatrics*, 25 (1960) 72.
[73] J. D. Allan, D. C. Cusworth, C. E. Dent and V. K. Wilson, *Lancet*, i (1958) 182.
[74] R. G. Westall, *Biochem. J.*, 77 (1960) 135.
[75] J. B. Rubini, E. P. Cronkite, V. P. Bond and T. M. Fliedner, *Proc. Soc. Exptl. Biol. Med.*, 100 (1959) 130.
[76] K. Fink, R. E. Cline, R. B. Henderson and R. M. Fink, *J. Biol. Chem.*, 221 (1956) 425.
[77] J. A. Weaver and D. W. Neill, *Lancet*, i (1954) 1212.
[78] E. Schram and R. Lombaert, *Anal. Biochem.* 3 (1962) 68.
[79] E. Schram and R. Lombaert, *Arch. intern. physiol. biochim.*, 68 (1960) 845.
[80] Th. Gerritsen, S. H. Lipton, F. M. Strong and H. A. Waisman, *Biochem. Biophys. Research Communs.*, 4 (1961) 379.
[81] M. Sandler and C. M. B. Pare, *Lancet*, i (1954) 494.
[82] B. Childs, W. L. Nyham, M. Borden, L. Bard and R. E. Cooke, *Pediatrics*, 27 (1961) 522, 539.
[83] A. Devries, S. Kochva, J. Lazebnik, M. Frank and M. Djaldetti, *Am. J. Med.*, 23 (1957) 408.
[84] C. Hooft and J. Herpol, *Acta Paediat.*, 48 (1959) 135.
[85] L. I. Woolf and H. McGiles, *Acta Paediat.*, 45 (1956) 489.
[86] W. H. H. Tegelaers and H. W. Tiddens, *Helv. Paediat. Acta*, 10 (1955) 269.
[87] S. W. Standbury and D. McCaulay, *Quart, J. Med.*, 26 (1957) 7.
[88] C. Hooft and A. Vermassen, *Ann. Paediat.*, 1 (1958) 190.
[89] H. Vis, *Rev. Franç. Études Clin. et. Biol.*, 7 (1962) 492.

THE BEHAVIOR OF AMINO ACIDS IN BODY FLUIDS DURING DEVELOPMENT AND GROWTH: PHYSIOLOGY AND PATHOLOGY

K. SCHREIER

Children's Hospital, University of Heidelberg, Heidelberg (Germany)

Prior to the development of reliable methods for the determination of single amino acids, German pediatricians had found that the amino-nitrogen content in the urine of the newborn and especially of prematures is definitely higher than that found during later life. For several decades this finding has evoked a discussion as to whether the increased amino acid loss is of renal origin or whether it is due to the

TABLE I

THE MATURITY OF NEWBORNS AND THE LEVELS OF DIFFERENT PROTEIN METABOLITES IN BLOOD ACCORDING TO LICHTENSTEIN

Averages value in mg/100 ml

Birth weight	*Non-protein N*	*Urea-N*	*Creatinine*	*Amino-nitrogen*	*Uric acid*
500–1000 g	27.44	11.07	5.05	9.06	2.40
1000–1500 g	30.27	13.42	5.19	8.31	3.12
1500–2000 g	33.03	13.72	4.96	8.02	3.25
2000–2500 g	33.55	14.01	4.86	7.32	3.65
2500–3000 g	32.41	13.86	4.33	7.22	3.55
3000 g and above	35.05	14.93	4.71	6.48	3.70

immaturity of the liver metabolism. We are now in a position to present a more or less clear picture of the behavior and the metabolic processes of the amino acids in the pre- and post-natal periods in health and disease.

A. BLOOD

RABINOWITSCH[1] and 5 years later MORSE[2] and VAN SLYKE AND MEYER[3] were apparently the first to ascertain an increased amino-nitrogen content in the blood of the umbilical cord. This finding has been confirmed by several workers (*cf.* refs. 4–15). The most profound study was done by LICHTENSTEIN[16]. He showed clearly that the amino-nitrogen content depends on the maturity of the newborn as Table I demonstrates. According to POMMERENKE[17], the amino-nitrogen level, like that of the other non-protein-nitrogen substances, is slightly higher in the cord arteries than in the cord vein. The results of this early research period have recently been extended through the use of contemporary methods (see for example, the studies of

References p. 279/283

CRUMPLER *et al.*[18], DUSTIN *et al.*[19, 20], SERENI *et al.*[21], and others[22–25]). In Table II the average value of 13, mainly microbiologically determined amino acids in the cord plasma[26] is compared with the values obtained from children and adults by the same technique. It is evident that the single amino acids are not increased proportionately.

It should be noted that the results obtained so far in studies of mammals demonstrate the same trend. CHRISTENSEN *et al.*[27] have found an increase of amino acids in the fetal circulation of various species of mammals which is two to five times that of the maternal value. The same is true for glutathione[28].

TABLE II

AMINO ACID VALUES IN ADULTS, INFANTS, AND CORD BLOOD

Average values in mg %

Amino acid	*Adults*	*Infants*	*Cord blood*
Alanine	3.8	3.9	4.8
Arginine	2.1	2.2	3.3
Glycine	2.8	2.6	3.4
Histidine	1.7	1.8	3.4
Isoleucine	1.6	1.7	2.3
Leucine	2.0	2.3	2.5
Lysine	2.8	2.4	8.1
Methionine	0.35	0.3	0.5
Phenylalanine	1.6	1.6	2.3
Threonine	2.0	2.3	2.8
Tryptophane	1.1	0.8	1.7
Tyrosine	1.4	1.6	2.3
Valine	3.0	3.2	4.9

As one would expect from the data of LICHTENSTEIN[16], during the intra-uterine period, the free amino acid pool is not constant. In Fig. 1 the content of free amino acids in brain of rabbit fetus, during the last days of gestation, is shown. This curve was obtained by plotting the values of free amino acids in brain of fetus which received ^{14}C-labeled lysine intraperitoneally while *in utero*.

It is evident that a decrease in the specific activity of the deproteinized body fluids occurs, reaching its lowest value at the time of birth. There is no doubt, therefore, that the placenta is able to concentrate single amino acids in the direction of the fetus. This ability necessitates an active transport mechanism, although it could be also an accumulation mechanism akin to that for vitamin C. Almost nothing is known of the essentials of this process. PAGE *et al.*[29] have been able to demonstrate that this mechanism is stereo-specific by showing that the physiological L-amino acids are transported much faster through the placenta than the D-isomers. Furthermore, a certain concurrence of the single amino acid is very probable. One can also accept as possible the fact that a saturation effect exists[30].

It is evident that the concentration capacity of single cells for amino acids limits the growth intensity of the organ[31]. This has not only been shown for normal embryonic cells, but ZAMECNIK *et al.*[32] have found a higher concentration of free amino acids in certain malignant tissues as compared to healthy structures. The

References p. 279/283

same is true for several amino acids in respect to leukocytes of patients with leukemia[33–35].

In addition to the "placental pump mechanism" (DENT), there are theoretically several other ways to satisfy the high protein demand of the fetus. Very briefly, the following possibilities exist[36]:

1. Proteolytic enzymes of the placenta could split maternal proteins into amino acids. This is unlikely since active proteolytic enzymes have not yet been detected (*cf.* refs. 37, 38).

2. The placenta could synthesize proteins destined for the fetus. No reliable studies bearing on this point seem to have been done.

3. Some proteins of the mother are transported intact through the placenta. Extensive literature exists concerning this problem (see refs. 39–42). From these studies one may conclude, that practically only γ-globulin can pass through the placenta. Despite the fact that albumin has a much smaller molecular size, it permeates the placental membrane to less than 10% of the values for γ-globulin.

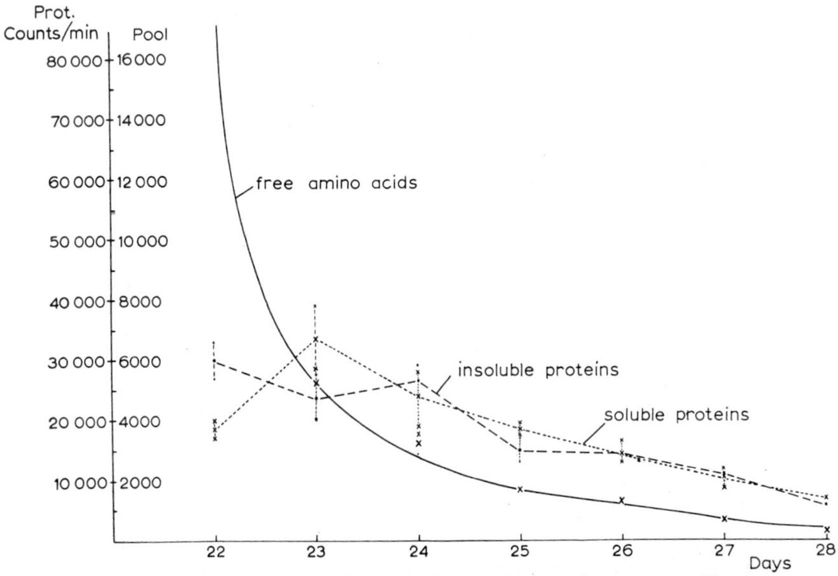

Fig. 1. Specific activity of proteins and free amino acids in brain of rabbit fetus following intraperitoneal administration of ^{14}C-lysine.

After cutting the cord of rats and rabbits a very steep decline of the incorporation rate of labeled amino acids takes place (Fig. 2).

I have not been able to find corresponding studies in the human newborn. But the changes of amino acids in blood and urine indicate that the reaction proceeds in the same way.

In later life the regulation of amino acid levels in body fluids is achieved by:

1. *Homeostatic processes.* These are governed unquestionably by the liver with its great capacity to concentrate and incorporate amino acids. The existence of other as yet poorly understood regulative processes, which are responsible for the degree of the reaction, is demonstrated, for instance, by the very different breakdown rates

References p. 279/283

after a comparable surgical trauma depending on the nutritional state of the individual[43].

2. *Hormones and central mechanisms.* For a short time after the start of feeding, we have a moderate hyperaminoacidemia in fasting prematures and most full-term newborns. A definite stabilization of the behavior of amino acid metabolism, as demonstrated in children and adults, is reached around the 3rd post-natal month. It is noteworthy that at about the same time the activity of the blood transaminases declines to the adult level[44]. I should like to emphasize that individual variation in respect to the regulation of the blood levels of amino acids is great.

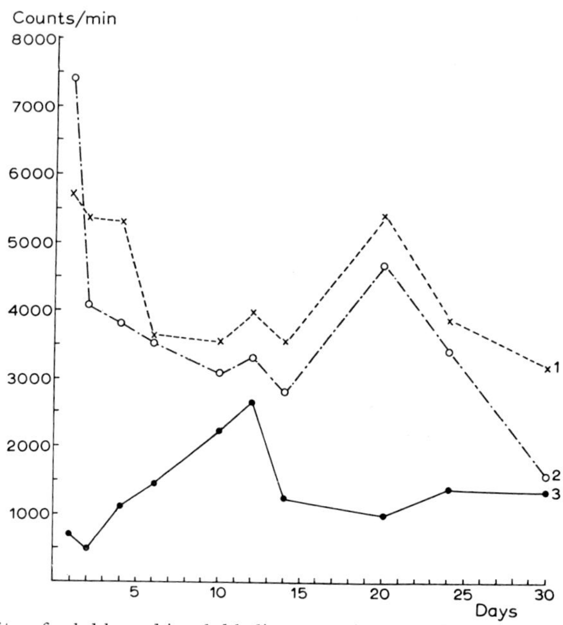

Fig. 2. Specific activity of soluble and insoluble liver proteins and of total nucleic acids of rabbits, 1–30 days of age. Counts/min/10 mg, insoluble proteins: 1; soluble proteins: 2; total nucleic acids: 3.

Until the 6th year of life a lowering of the different amino acid values in the fasting blood often occurs (*e.g.*, KIRBY[45]; SCHREIER[100]). But the physiological and individual variations of the single amino acid values are so large that a decrease cannot be ascertained statistically. The trend toward a hypoaminoacidemia is unequivocal.

Since the protein synthesis rate of the thriving premature infant proceeds with an increased activity, one would actually expect a *decrease* of the amino acids in the body fluids, particularly since the concentration capacity of juvenile cells for amino acids is exceptionally good.

As a cause for the hyperaminoacidemia in prematures one would first of all presume that a diminished catabolic activity, especially in the oxidizing enzymes, occurs. It has been shown recently that particularly those enzymes which oxidize the cyclic amino acids have a very low activity during the post-natal period (see below). Of course, an increased proteolysis would lead to higher levels in blood too; but we have no evidence for such activity. In fact, catheptic activity has been found to be

References p. 279/283

lower in fetal as compared to maternal tissue[46]. COHEN AND HEKHUIS[47] and BEATON *et al.*[48] stress the fact, that glutamic-oxalacetic and glutamic-pyruvate transaminase activity vary inversely with proteosynthetic activity. Indeed, in prematures the transaminase activity is lower than in full-term newborns[44].

Possibly the higher permeability of all cell membranes contributes to the hyperaminoacidemia, especially since the lowering of pH due to the metabolic and respiratory acidosis increases it further. It has been shown that such pH-dependent changes of the cell membrane observed post-natally also have an influence in later life on the amino acid content of the tissue. For instance, in potassium deficiency the level of basic amino acids is increased, whereas that of monoaminodicarboxylic acids is decreased[49, 50].

I assume that the higher turnover rate of almost all body proteins during early youth, as we have shown to be the case in humans, rats and rabbits, necessitates an increase in their breakdown to amino acids, and, therefore, an increase of the free amino acid pool. In support of this concept, one may cite the fact that in later life diseases which increase the protein turnover rate (*e.g.*, leukemias, surgical trauma, etc.) usually show a hyperaminoacidemia[35, 51–53]. It should be mentioned that this higher rate of turnover in youth has a parallel in the increased "minimal nitrogen turnover" and in the basal metabolic rate[54].

It is not likely that STH (somatropic hormone) is the reason for post-natal hyperaminoacidemia, since STH increases the permeability of the cell membrane to all amino acids including the non-physiological amino acids so far studied, and leads in this way to a *decline* of amino acids in the blood[55–57]. Furthermore, the incorporation of amino acids into proteins is increased[58].

It is well known that the so-called "birth stress" leads to excretion of steroid hormones, which function in various ways. Among other effects they show a marked influence on the cell membrane[59] and on amino acid turnover, thereby increasing the amino acid values in blood[60–62]. The instability of blood amino acid values, with their tendency to increase, usually continues much longer than one would expect, when the after-effects of birth trauma are considered. Similarities to the changes observed after surgical trauma or severe burns nevertheless are evident. It may be noteworthy in this connection that the turnover rate of [^{35}S]methionine-labeled body proteins is increased significantly after an abdominal operation in rats[63].

In addition to these two hormone groups (STH and steroids), insulin apparently also plays an essential role in protein metabolism in the early period of life. It appears rather certain (KORNER AND MANCHESTER) that this hormone has a direct influence on amino acids, independent of any effect on carbohydrate metabolism[64]. The most plausible interpretation of all available findings at this time is that insulin has a positive effect on the permeation of amino acids into cells.

In well-controlled cases of diabetes mellitus in adults and children the amino acid levels in the plasma are within the normal range. In coma diabeticum, however, several essential and non-essential amino acids, such as leucine, isoleucine, valine, lysine, phenylalanine, etc., are elevated. The degree of hyperaminoaciduria depends not only on the plasma value of the amino acid, but also on the condition of the kidneys and several extrarenal factors[65]. Our results have been fully verified by CZYZYK[66]. Diabetic subjects are said to have a disturbance in tryptophane metabolism[67] (see also refs. 68, 69). As indicated before, the liver plays an extremely important role in the homeostasis of blood amino acid levels. Especially during the unstable post-natal regulatory period, anoxemic toxic and viral diseases which damage the liver tissue provoke a very marked hyperaminoacidemia and, of course, a hyperamino-aciduria as well due to the so-called "overflow" mechanism. In later life the changes usually are less pronounced. In agreement with results of various authors (*cf.* refs. 70–78) in studies mainly on adults we have found in children, first of all, a disturbance in the turnover of methionine, as well as of lysine, arginine and the cyclic amino acids. The cause for the increased concentration of these amino acids in the body fluids in cases of several liver disorders appears to be a disturbance in deamination and desulfuration and in the anabolic capacity of this organ. Besides this, one should not forget the hepato–renal relations (*i.e.* tubular damage, *cf.* ref. 79).

TABLE III

SOME AMINO ACIDS IN SERUM OF INFANTS WITH ENTERITIS

Method: microbiological assay

Average values: μg/ml

Amino acid	*Mild form*	*Severe form*	*Toxic enteritis*
Isoleucine	21.5	22.6	35.6
Leucine	27.3	31.0	64.3
Methionine	5.3	6.7	14.8
Phenylalanine	26.8	27.4	33.3
Tryptophane	12.6	16.9	20.6
Tyrosine	18.6	21.2	22.0
Valine	31.8	42.8	51.3

Infants having severe enteritis demonstrate a special form of hyperaminoacidemia and hyperaminoaciduria. FINKELSTEIN fittingly called the severe forms of infantile dehydration "coma trophopathicum", emphasizing the similarity with coma hepaticum.

We have made an extensive study of the amino acids of the blood and urine from infants of varying ages who had enteritis in degrees ranging from mild to very severe. In those cases studied, we found slight to extreme alterations in amino acid levels, correlating with the clinical picture (Table III). Our findings have been confirmed by KLEINBAUM[80] and LOEB AND ENGELEN[81] who found disturbances in amino acid metabolism especially in the "toxic-infectious" form of severe enteritis.

The degree of the hyperaminoaciduria is influenced greatly by the accompanying kidney damage which is very severe in some cases [82–84].

There exist only a few investigations on the effect of the basic nutritional substances on the amino acid levels of plasma[85–87]; this is especially true for infants and children. The determination of amino acids in the serum of children[88] following a protein-rich meal showed a rise in the individual amino acids which, as in adults[89–93] did not correspond to the quantity of the acids in the meal. The behavior of amino acids in the post-prandial period is mainly influenced by the rate of breakdown of ingested proteins, the quantity of amino acids released and the rate of absorption of each individual amino acid[94]. This resorption rate is, of course, influenced by the concentration of amino acids in the chyme and by the presence of saccharides and perhaps of fats. In our studies the blood levels of lysine, tryptophane, isoleucine and valine showed a distinctive increase after ingestion of meat or casein. These results agree quite favorably with those obtained by FRAME[93] in his examination of adults. On the other hand, we should note that several amino acids decrease after a protein load; for instance alanine and glycine[95]. Apparently their use in protein synthesis is greater than the supply. The ingestion of individual amino acids[95–97] leads to appreciably higher peaks[98] than the same quantity of amino acid supplied as protein or in protein hydrolyzates.

According to JACCOTTET *et al.*[99] in prematures the maximum of amino-nitrogen after oral ingestion of hydrolyzates is reached earlier than in full-term babies. Low-value-proteinhydrolyzates change the amino acid levels in the sense of a temporary imbalance.

The ingestion of monosaccharides decreases the amino-nitrogen content of the plasma[101, 102, 240]. The decline in the blood levels of the individual amino acids varies[103] and is especially impressive with isoleucine[88]. ALBANESE *et al.*[6] claim to have observed a decrease in tryptophane elimination in urine following a glucose load, while the other amino acids remained unchanged. This is in contradiction to the experimental results of BUTLER AND SARETT[104] who reported that tryptophane excretion is more or less independent of the carbohydrate content of food. MUNRO AND THOMSON[103] also have been unable to detect an influence of monosaccharides on tryptophane elimination. Fat has no certain influence on the levels of most amino acids.

In starvation or in the so-called absolute protein-minimum stage the amino-nitrogen content and the levels of the amino acids decrease until only shortly *ante exitum*. At this time the levels increase above the physiological range. STELGENS AND BOTHE[105] have found in two infants on a long standing protein-free diet that a relative rise in amino acid values in urine occurs. Animal experiments have shown, that the content of free amino acids in the liver is increased during starvation[106]. Our studies with ^{14}C-labeled lysine in young rats confirm these findings and revealed that the incorporation rate of [^{14}C]lysine in protein shortage is increased in the visceral organs, especially in the liver, while the incorporation into other structures, for instance, the erythrocytes, is sharply decreased. From this one may conclude that in extreme protein deficiency there is a shunting of protein metabolism to the benefit of essential synthetic processes in the liver, kidneys, etc. This effect is so pronounced, that in newborn rabbits, even shortly before death, labeled lysine is incorporated at a high rate into liver proteins.

We do not believe that the hypoaminoacidemia observed in severe nutritional dystrophy (kwashiorkor etc.) is due to enzyme inactivity but rather that it results from an increased utilization of amino acids for energy-yielding processes and partly through a higher uptake by the liver.

The kidney has only an insignificant influence on the amino acid content of the blood. Even in the final stages of nephrosclerosis, we could like many other authors, find no significant changes in most amino acids. Only alanine and glycine were found to be increased in several cases of uremia[95,107]. A hypoaminoacidemia is present in the acute stage of nephrotic syndrome[108, 109]. Only thriving infants, after recovery from severe malnutrition, show a similar decrease of the amino acids in blood.

The metabolism and the handling of the aromatic amino acids during the post-natal period deserves a somewhat detailed discussion. In 1941 LEVINE *et al.*[110] had detected, that prematures on a protein-rich diet lacking ascorbic acid excreted breakdown products of phenylalanine and tyrosine in considerable amounts. By loading of these infants with one or both amino acids the tyrosyluria could be greatly increased. SEALOCK AND SILBERSTEIN[111] had made corresponding observations on guinea-pigs with scurvy (1940). In 1901 STOKVIS[112] observed a darkening of the urine of scorbutic human beings. In the meantime several teams have studied the problems of the influence of vitamin C on the handling of amino acids in the kidney and particularly on the metabolism of aromatic amino acids[113–116].

The addition of ascorbic acid to the diet resulted always in the disappearance of *p*-hydroxyphenylpyruvic acid and related compounds It also was observed that in scorbutic animals and prematures pteroylglutamic acid was able to decrease the tyrosyluria[117–119]. Furthermore, ACTH decreased the excretion of aromatic keto acids[115, 120]. The mode of action of folic acid and ACTH is still obscure. The inconsistent effect may be due to an augmented non-specific protein synthesis or an influence on the synthesis of a specific apo-enzyme. Initially it was assumed that the oxidation of *p*-hydroxyphenylpyruvate requires ascorbic acid as an essential cofactor[112]. The fact that vitamin C can be replaced by substances like dichlorophenylindophenol[122] supports the interpretation that these compounds provide an appropriate oxidation–reduction potential required for activity of the *p*-hydroxyphenylalanine oxidase.

The studies of KRETCHMER and his team[123, 124] cast a new light on the difficulties which prematures experience with the metabolism of tyrosine. They showed that the activity of the tyrosine-oxidizing system, the tyrosine transaminase and phenylalanine hydroxylase in the liver of prematures has only $^1/_{10}$–$^1/_{30}$ of the adult organ activity. The same is true for fetal rats. The *in vitro* addition of large amounts of ascorbic acid had no stimulatory influence on the tyrosine oxidizing enzyme system. It seems likely that the apoenzyme itself is lacking.

p-Hydroxyphenylpyruvate oxidase on the other hand seems to be present in fairly great amounts in tissue of premature and fetal rodents but the enzyme requires very large amounts of ascorbic acid for activation. The ascorbic acid might be necessary to remove an inhibitor of the enzyme[123]. Phenylalanine hydroxylase consists of two fractions, of which one (I) is missing in immature liver (see also ref. 125).

The changes of tyrosine and phenylalanine metabolism in prematures differ therefore from those of the scorbutic adult. The activities of the individual enzymes maturate with different speed. Tyrosine transaminase, for example, is active soon after birth, producing *p*-hydroxyphenylpyruvate which accumulates due to the shortage of ascorbic acid thus leading to tyrosyluria. According to AURICCHIO *et al.*[126] some enzymes of tryptophane turnover also are lacking or insufficiently active in the post-natal period. Besides these acquired anomalies in the turnover of aromatic amino acids MEDES[127], FELIX *et al.*[128] and SAKAI AND KITAGAWA[129] have found changes which some include with the inborn errors of metabolism. In my opinion, there is no convincing evidence for an inborn lack of an enzyme in "tyrosinosis".

BLOXAM[130] has found an increased excretion of tyrosine metabolites in apparently

References p. 279/283

healthy infants provided with enough dietary vitamin C. ROBINSON AND SMITH[131] even claim that "many" adults excrete *p*-hydroxyphenylpyruvic acid and *p*-hydroxyphenyllactic acid in considerable amounts under different pathological conditions, while GROS AND KIRNBERGER[132] found the two acids only in liver disorders. AMMON AND HENNING[133] detected a small amount of *p*-hydroxyphenylpyruvate in urine of every healthy individual examined.

B. URINE

In infancy, as in adulthood, the excretion of amino acid in urine depends, among other things, on the following factors: (1) age; (2) nutrition (protein content, fluid and vitamins); (3) hormones; (4) function of the liver; (5) function of the kidneys; (6) presence of malignant diseases; (7) different poisons; (8) inborn errors of metabolism (see WESTALL's paper p. 195).

The simplification in the detection of amino acids achieved by modern methods[134], especially paper chromatography[135a,b c,d], has led to a very broad and sometimes non-critical use of these procedures in the clinical laboratory. Therefore, published values determined by the various methods differ considerably. Additional sources of variation are found in the reference values used, such as body weight, volume of 24-h urine specimen, total nitrogen, and especially creatine. The creatine index particularly should not be used in the first post-natal days, since the excretion of this substance shows marked fluctuations.

1. Age

PFAUNDLER[136], the famous pediatrician from Munich, discovered (1900) with an inappropriate method the principle of hyperaminoaciduria in young infants. Using SORENSEN's formol titration method, his pupil SIMON[137] showed that newborns eliminate up to 10% of the total nitrogen excretion as amino nitrogen. To be sure, this value is too high due to procedural limitations. The relationship to the adult urine values, however, is principally correct and agrees with the findings of recent years. BARLOW AND MCCANCE[22] showed that this ratio is especially high in the first 24 h and decreases somewhat in the next few days. This has been confirmed many times[138–144].

The amino-nitrogen elimination, expressed in mg/day, increases in the 1st week of life approximately parallel to the increase of urine volume[145–148]. The total elimination reaches its maximum at about the 6th week of life. As is evident from Fig. 3, the amino-nitrogen/total-nitrogen quotient differs from this trend in that it shows a precipitous decline about the 14th day of life[149]. In the first few days after birth the amino-N elimination in prematures is often lower than, or at least not higher than, that of full-term infants. In the ensuing weeks, however, almost all amino acids show an increased excretion. This has been reported by GOEBEL[8] in respect to amino-nitrogen.

The large quantities of taurine found immediately after birth in the urine of the bladder and in the blood of the umbilical cord[150], are obviously a characteristic of fetal metabolism. This can result either from the immaturity of the taurine degradation mechanism, or from an increased synthesis. On the other hand, it cannot be properly explained why taurine is excreted by full-term infants in greater quantity than by prematures. The large range of variation of lysine and cystine excretion in the post-natal period probably results from the different stages of maturity of the transport mechanism in the tubuli of the kidney. Basically, the excretion of lysine is quickly reduced during the first month of life. It is also noteworthy that ethanolamine is initially eliminated in the urine in large quantities, but about the third month of life, except in cases of severe liver damage and occasional cases of leukemia, its presence can no longer be demonstrated. O'BRIEN *et al.*[151] believe they have demonstrated methylglycine in early urines.

By the 5th or 6th day of life the amino acid pattern has changed to the extent that the elimination of taurine, cystine and ethanolamine is decreased sharply. Proline and hydroxyproline have become quantitatively the principal amino acids. Young rats also excrete more hydroxyproline than adult animals[152]. The elimination of the majority of the other amino acid is more or less increased. This is especially true of glutamine[153]. Some urines even contain appreciable quantities of aromatic amino acids. The behavior of amino acids in urine after birth coincides in most respects with

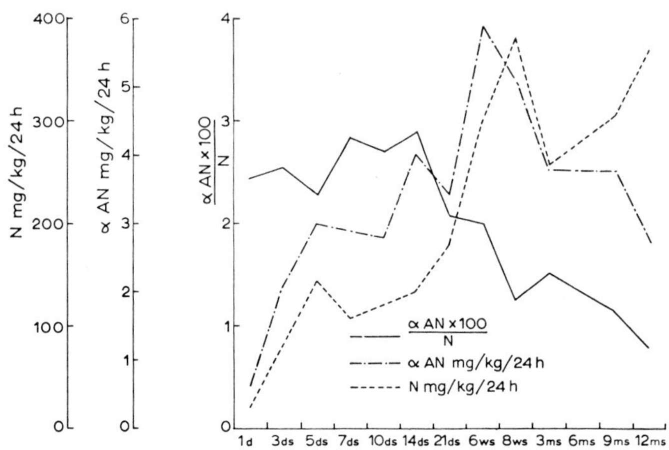

Fig. 3. α-Amino-nitrogen and non-protein nitrogen in mg/k/24-h and as amino-quotient (taken from Bickel: *Physiolog. Entwicklung des Kindes*).

reactions of other processes related to protein metabolism, for example, with the activity curves of some coagulation factors[154]. The extent to which the hyperaminoaciduria of prematures and full-term infants is of renal origin can be definitely resolved by clearance studies. Table IV presents our values[155] from studies with infants and those of SERENI *et al.*[21] compared to adult values[156,281] (see also ref. 157). It is evident that an immaturity of the tubule function exists in infants. The relative hyperaminoaciduria of infants is sometimes even more marked than paper-chromatographic studies would suggest since the glomerular filtration rate in some prematures is severely decreased.

In premature infants the hyperaminoaciduria is, in most cases, demonstrable for a longer period than in full-term infants. Furthermore, the excretion of the characteristic amino acids, proline and hydroxyproline, begins later. The essentials of prolinuria and hydroxyprolinuria in infancy are not yet properly understood because the

References p. 279/283

biosynthesis of hydroxyproline is still unclear[5]. We know that proline can be synthesized from glutamic acid *in vivo* and *in vitro*[158, 159] but until the present the direct oxidation to hydroxyproline could not be ascertained. Apparently the formation of hydroxyproline is intimately connected with collagen synthesis[160–162]. GERBER *et al.*[163] speculate that two different metabolic pools of collagen and of prolin derivatives exist. Free hydroxyprolin is not incorporated into cartilage[164]. One may postulate that the increased excretion of this amino acid is due to a higher turnover rate of the growing collagen in infancy, especially since syndromes with disturbances of cartilage synthesis, like Marfan's disease, are said to show a hydroxyprolinuria.

TABLE IV

CLEARANCE VALUES OF HUMAN ADULTS, INFANTS AND PREMATURES

Amino acid	*Adults*			*Infants*		*Prematures*	
	Doolan[156]	*Evered*[281]	*Sereni*[21]	*Sereni*[21]	*Schreier*[155]	*Sereni*[21]	*Schreier*[155]
Histidine	6.38	4.70	5.38	10.00	7.96	10.20	7.91
Leucine	0.37	0.40	0.43**	0.50**	0.40	0.74**	0.70
Lysine	0.22	0.30	1.65*	10.87*	1.23	7.09*	1.44
Threonine	0.79	0.87	1.22	4.13	1.01	6.00	3.18
Tryptophane	1.37	—	—	—	0.97	—	1.58
Tyrosine	1.65	1.23	1.57	2.64	1.63	1.47	1.27
Creatinine	—	—	—	92.30***	69.40	45.60***	39.58

* Lysine–arginine clearance values.
** Leucine–isoleucine clearance values.
*** Method Hare.

In evaluating the extent of infantile hyperaminoaciduria it must be born in mind that the total N elimination in early infancy is only approx. 0.075 g/kg/day as compared with approx. 0.175 in the adult[165]. Parallel to this the glomerular filtration rate, the urea clearance and the osmolar concentration are reduced in the full-term infant[167]. A leaking of plasma into the interstitium also occurs. This displacement is especially impressive in prematures[166].

The inability to secrete a urine as concentrated as that produced by adults under comparable conditions is not to be attributed to a reduced response to the antidiuretic hormone[167–170]. In addition to the tremendous difference in nutritional composition, the inability to produce an interstitial ionic hypertonia also must be taken into consideration[171]. Protein-rich feedings produce an increase in the concentration capacity of the infantile and adult kidney[172].

The length of the loops of Henle, which is now considered as a counter-current distribution system, is at birth very short but extends itself rapidly. The tubuli likewise have not reached the definite morphological structure. I should like to emphasize particularly the fact that the maturation of the kidney in respect to amino acid handling shows a great individual variation. In some children this organ reaches the functional capacity of the adult apparently within a few months, while in others it takes up to 2 years to reach this stage of development. Besides the immaturity of the reabsorption mechanism, a temporary hypoxia of the kidney may be partly responsible for the extreme hyperaminoaciduria in individual cases after birth. Furthermore the shortage of energy donors in the immediate postnatal period must be considered.

It is remarkable indeed that the excretion of total amino-nitrogen and of single amino acids shows such a constancy during almost the entire life span. Only in very old people can one again find an increased excretion of some amino acids. At least in children a day-night rhythm of amino acid elimination has been detected. During the night the amino acid concentration in urine is higher. Since the urine volume is smaller the total amount is not elevated. The minimum of amino acid excretion occurs in the early morning hours.

2. *Nutrition*

a) Protein content

It is not quite understandable why opinions differ as much as to whether the protein content of the food has an influence on amino acid elimination in the urine[173, 174], since it has been known for more than 50 years that some correlation exists (*cf.* PETERS AND VAN SLYKE[174a]). Some authors[175, 177] deny it almost entirely. The increase of amino-nitrogen elimination in the first days of life and the rise of the amino-nitrogen/total-nitrogen quotient from about 5–10% clearly shows this influence. In later life, amino acid excretion during ingestion of a normal diet increases by about 50–60% in comparison with the fasting state[88, 176]. After a high protein diet those amino acids which have a low T_m value are excreted in higher amounts (histidine, threonine, lysine, tyrosine, etc., *e.g.* HUISMAN[184,186]). A high protein diet has an unusually marked influence if administered to an organism deprived of proteins[178]. This we verified in severe cases of nutritional dystrophy of infancy. When these patients received an adequate amount of milk a very impressive hyperaminoaciduria resulted.

It needs not to be emphasized that no *direct* relation exists between protein content of the food and amino acid excretion in urine, since several organs are inserted in the circuit. THURAU[271] has found that the urine of infants fed human milk (protein-poor) contained less amino acids than that of infants fed on cow's milk formula. This has been confirmed by SCHMIDT[179]; however, DUSTIN[19] did not observe great differences.

In acute gastroenteritis in infancy a slight or very marked hyperaminoaciduria, parallel to the severity of the disease, can be found[80, 180, 181] (see also p. 268). In chronic nutritional disorders usually no significant change of the amino acid content of urine is detectable. In a few cases of celiac syndrome and mucoviscidosis a slight increase can be found[145, 182]. CHOREMIS *et al.*[183] found a hyperaminoaciduria in infants with "thirst-fever" which is difficult to interpret.

b) Fluid

The dependence of the amount of amino acid excretion on diuresis was discussed ardently a few years ago mainly in the German medical literature. In older children and in adults no dependence does exist[178, 179, 184, 185]. It is maintained that under normal conditions, that is when the tubule function is intact, the adult kidney is able to reabsorb amino acids almost quantitatively, even under great variation of fluid intake. However, in young infants the rate of amino acid excretion increases more or less proportionately to the urine volume.

c) Vitamins

The hyperaminoaciduria of vitamin C deficiency was found by JONXIS[186] and has been confirmed by most of the other workers in the field[145, 187–189]. In our own study, all ten children, even those demonstrating early prescurvy symptoms, showed an increased excretion of glutamic acid, aspartic acid, serine, threonine and some other acids. The plasma values of all the amino acids were within normal limits in such infants[186]. In scorbutic guinea-pigs RANGNEKER[190] found an increase of phenylalanine, lysine and histidine in blood.

ARMBRUSTER[191] claimed that vitamin C in relatively high doses (100 mg or more) causes an increase in the amino acid content of the urine (this publication is not well documented). A hyperaminoaciduria is also found in rickets, but not with the same regularity as with vitamin C deficiency[189, 192,199]. Only about half of our 49 rickets cases have shown an elevated excretion rate. The plasma values were also found to be normal. JONXIS found an increased excretion of some amino acids in some blood relatives of children with rickets. This leads one to wonder whether peculiarities of tubular resorption of hereditary origin are responsible, in addition to the vitamin D deficiency, for this ricketsial hyperaminoaciduria. Vitamin E deficiency in rabbits also produced a hyperaminoaciduria[200]. This is of some interest, since in human and animal progressive muscular dystrophy an increased excretion of some amino acids occasionally has been found[7, 201–203]. Vitamin A deficiency does not produce any significant change in the amino acid content of the serum or various organs of the rat[4]. A shortage of vitamin B_6 and pantothenic acid on the other hand is said to increase the level of arginine, methionine and tryptophane, whereas riboflavin deficiency leads to a higher level of valine in animals[204]. It is not clear whether the hyperaminoaciduria described by HANSEN *et al.* in infants deprived of pyridoxine is due only to the vitamin deficiency. Vitamin B_{12} deficiency according to GOKHALE AND PUNEKAR[205] in rats produces a hyperaminoaciduria.

3. Hormones

Several hormones exhibit a significant influence on protein metabolism (*cf.* p. 267). The effect of glucocorticoides and ACTH on amino acid elimination by the kidney has been studied thoroughly, and found to produce a relatively intense hyperaminoaciduria[206–209]. A marked decrease in urinary amino acid elimination occurs after application of STH (ref. 210), as well as after effective doses of anabolic testosterone derivatives. As one might expect, insulin likewise produces a definite decrease in amino acid excretion.

Central regulation is a field of amino acid metabolism which has hardly been studied. Better knowledge here would be valuable even for clinical purposes, since we have found that not only infants with well-known inborn errors of metabolism (*e.g.* phenylketonuria, maple-sirup-disease, etc.) but also children with common cerebral defects and mental retardation frequently show an increased amino acid excretion. In our material, of 50 cases 60% had at least a slight increase and approx. 25% a striking hyperaminoaciduria of the general type (see also refs. 211, 212). Two cases of so-called "myoclonus epilepsia" showed a very pronounced increase in almost all the usual amino acid found in urine. This instability of amino acid handling in cerebrally damaged children has a parallel in their difficulties with the regulation

References p. 279/283

of Na and H_2O in the body fluid[213, 214]. Virus infections and fever due to other etiology, apparently disturb the regulation of water and amino acid distribution.

4. Function of the liver

Studies on the excretion of amino acids in urine by adult patients with mild to severe liver diseases have been published in great numbers[215–222]. The various hepatopathies in childhood are equally well studied[223]. Less is known concerning the influence of liver disorders in the post-natal period. To summarize, one can state that in chronic hepatopathy and especially in liver cirrhosis no significant hyperaminoaciduria is detectable. After a protein load, methionine, histidine, glycine etc. are usually excreted in greater amounts by these patients than by healthy individuals. In acute hepatitis, a hyperaminoaciduria of the overflow type is present in all age groups and particularly in infants. This type is modified only by the functional state of the kidneys. In acute yellow atrophy of the liver, the high excretion of tyrosine and leucine is one of the oldest known biochemical signs of this disorder (FRERICHS).

5. Function of the kidneys

Pathological alterations of glomerula generally produce no marked changes in amino acid excretion[224, 225]. In relatively compensated cases of the nephrotic syndrome of childhood one does not find a significant hyperaminoaciduria[225, 226a]. However, in the acute phase mainly leucine, tyrosine, phenylalanine, arginine and tryptophane are excreted in higher amounts than normally[219, 227–230]. A pronounced hyperaminoaciduria is often the biochemical sign of disturbances of tubular function. Especially the so-called tubular necrosis is characterized by an excessive amount of amino acids in the urine[231]. When only a part of the nephrons is involved, the final urine shows only a slight or moderate hyperaminoaciduria; this is due to dilution by addition of normal urine. If inflammatory or degenerative changes lead to a sclerosis and hyalinization of the nephron the hyperaminoaciduria may disappear completely as is the case in prematures, when the effective renal plasma flow is so much decreased, that a pseudonormal amino acid excretion results. A familiar disorder corresponding to renal glucosuria has been described for amino acids[232, 233].

6. Presence of malignant diseases

More or less marked hyperaminoacidurias have been reported in various malignancies. Most of these studies deal with individual cases in which no autopsy was performed. Therefore, it is impossible to say whether or not the liver or the kidney was involved. Only in leukemia have many cases been studied thoroughly. We have done follow-up studies on 20 cases using paper-chromatographic and to some extent microbiological methods. The results were extremely variable. Later stages of the disease exhibited a massive excretion of β-aminoisobutyric acid (also found by others[234]) and frequently an increase of alanine, glycine, phenylalanine, arginine, and other amino acids. β-Aminoisobutyric acid on the other hand has been found in the urine of approx. 5% of the normal population[235–237].

7. *Different poisons*

To conclude this brief summary I should like to mention hyperaminoaciduria caused by various poisons. This is true for lead, cadmium, uranium and other heavy metals[184, 238, 239]; for sodium fluoride, dinitrobenzene, lysol, maleic acid[240, 241, 244], succinic acid, γ-butyrobetaine[242], oxalic acid[243] and also for amino acid antagonists[245]. Large amounts of amino acids particularly cystine, methionine and lysine lead not only to high excretion rates of the respective amino acid, but also of many others[98, 246–248]. This is apparently due to a competitive or non-competitive inhibition of the transport mechanism. Some amino acids may even damage the tubuli irreversibly[249]. According to VAN CREFELD[250] toxic amounts of vitamin D produced in one child a severe hyperaminoaciduria. In our three cases and also in those of WEWALKA[251] no significant increase of the amino acids in urine could be detected.

Severe shock syndromes, for instance, those due to burns[252, 253], surgical trauma[43] and accidents[254–256], evoke a temporary hyperaminoaciduria. The same is true for severe infectious diseases[258, 259]. The results of amino acid studies in the different virus and bacterial infections vary greatly. This is to be expected, since so many divergent factors may play a role in influencing amino acid excretion.

TABLE V

FREE AMINO ACIDS IN LIQUOR CEREBROSPINALIS OF HEALTHY HUMAN BEINGS

Average values in mg%

AUTHOR	MÜTING *et al.*[265]	TORRE *et al.*[280]	LOGOTHETIS[263]	WALKER *et al.*[276]	HUISMAN[279]	SOLOMON *et al.*[273]
Method	*Paper chromatography*				*Column chromatography*	*Microbiol.*
Aspartic acid	0.13	—	0.11	0.04	0.10	—
Glutamic acid	0.27	0.50–1.76	0.09	0.06	3.35	—
Lysine	0.36	—	0.17	—	0.20	0.28
Arginine	0.20	0.07–0.13	0.16	—	0.10	0.60
Histidine	0.26	0.07–0.55	—	—	0.05	0.17
Tyrosine	0.18	0.10–0.90	0.16	0.08	0.10	0.20
Tryptophane	0.20	0.20–0.33	—	—	—	—
Phenylalanine	0.19	0.07–0.15	0.18	0.21	0.15	0.19
Hydroxyproline	0	—	—	—	—	—
Proline	0.27	—	—	—	—	—
Cystine	0.24	0.10–0.55	—	—	0.05	0.18
Methionine	0.13	—	—	—	0.05	0.04
Leucine	0.23	0.05–0.42	0.14	0.14	0.05	0.14
Isoleucine	0.12	—	—	—	—	0.095
Valine	0.36	0.05–0.41	0.12	0.07	0.15	0.21
Glycine	0.39	0.05–0.18	0.09	0.10	0.10	—
Alanine	0.34	0.05–0.17	0.15	0.11	0.15	—
Serine	0.28	—	0.22	0.16	0.10	—
Threonine	0.18	—	0.12	0.13	0.30	0.28
Taurine	0.32	—	—	—	—	—
Glutamine	0.45	—	0.30	0.71	—	—
α-Aminobutyric acid	traces	—	0.02	—	—	—
γ-Aminobutyric acid	traces	—	0.04	—	—	—

C. LIQUOR CEREBROSPINALIS

Of all the field of amino acid metabolism found in the literature the most controversial data concern liquor cerebrospinalis[260–273]. This arises from the fact that the amino acid content is much lower than that of the plasma and therefore in unconcentrated liquor the amino acid levels may be below the sensitivity of the methods employed.

I doubt that the immense differences in values for single amino acids reported by different authors[2] are really due to biological variations. It is much more probable that they are caused by losses during the desalting procedure and other methodological inadequacies. It has been claimed that differences exist between the spinal liquor and that of the brain ventricles[274]. The data available in the literature are compiled in Table V. It was to be expected that all amino acids found in the other body fluids also would be detected in the liquor. KNAUFF *et al.*[274] have found regularly 25 ninhydrin-positive substances of which 21 were amino acids. The great differences from the blood values demonstrate that the brain has its own way of handling amino acids. Since the reported values are so variable, no characteristic pattern of amino acid in the liquor for age groups or for any disease can be defined. Only in phenylketonuria is the phenylalanine level increased. Liver coma and presumably uremia change the amino acid content of the liquor in a non-specific way. It must be expected that increased permeability of the blood–brain barrier will increase the amino acids in the liquor on account of the high amino acid gradient between blood and liquor.

D. ASCITES AND OTHER EXUDATES AND TRANSUDATES

I was able to find two publications on amino acid content of trans- and exudates of the pleura and abdominal cavity[275, 278]. The amino acid levels in both types of effusion were of the same magnitude as in the blood plasma. It seems likely that the amino acids permeate passively from the blood into the exudates.

E. PERSPIRATION (SWEAT)

Amino acids also have been found regularly in sweat[268,277]. To my knowledge quantita tive values have not yet been published. Arginine, citrulline and some unidentified spots in the position of tyrosine have always been found in the paper chromatogram of sweat.

F. FECES

Feces of infants fed human milk contain a greater amount of amino acids than those fed on cow milk formulas[270–271]. This, of course, is due to differences in the bacterial flora. *Escherichia coli* incorporates amino acids not only with greater activity than lactobacilli, but also deaminates them more vigorously.

REFERENCES

[1] K. N. Rabinowitsch, *C. R. Soc. Biol. (Paris)*, 66 (1914) 457.
[2] A. Morse, *Compt. Rend Soc. Diol.*, 28 (1917) 199.
[3] D. D. Van Slyke and G. M. Meyer, *J. Biol. Chem.*, 12 (1912) 399.
[4] M. Abdulnabi and J. G. Bieri, *Texas Rep. on Biol. and Med.*, 2 (1953) 181.
[5] E. Adams and A. Goldstone, *J. Biol. Chem.*, 235 (1960) 3492, 3499, 3504.
[6] A. A. Albanese, V. Irby, J. E. Frankston and M. Lein, *Am. J. Physiol.*, 150 (1947) 389.
[7] S. R. Ames and H. A. Risley, *Proc. Soc. Exptl. Biol. Med.*, 68 (1948) 131.
[8] F. Goebel, *Z. Kinderheilk.*, 34 (1922) 94.
[9] F. Goebel, *Z. Kinderheilk.*, 38 (1924) 27.
[10] K. Hellmuth, *Arch. Gynäkol.*, 118 (1923) 18.
[11] K. Hellmuth, *Arch. Gynäkol.*, 123 (1925) 57.
[12] F. L. Meyer and H. Rietschel, *Biochem. Z.*, 3 (1907) 31.
[13] E. Müller, H. Steudel and J. Ellinghaus, *Arch. Kinderheilk.*, 76 (1925) 272.
[14] E. D. Plass and C. W. Matthews, *Bull. Johns Hopkins Hosp.*, 36 (1925) 393.
[15] H. Rietschel and L. Langstein, *Biochem. Z.*, 1 (1906) 75.
[16] A. Lichtenstein, *Z. Kinderheilk.*, 51 (1931) 748.
[17] W. T. Pommerenke, *J. Clin. Invest.*, 15 (1936) 485.
[18] H. R. Crumpler, C. E. Dent and O. Lindau, *Biochem. J.*, 47 (1950) 223.
[19] J. P. Dustin in F. Linneweh, *Die Physiologische Entwicklung des Kindes*, Springer Verlag, Berlin, 1959, p. 232–240.
[20] J. P. Dustin, S. Moore and E. J. Bigwood, *Metabolism*, 4 (1955) 75.
[21] F. Sereni, H. McNamara, M. Shibuya, N. Kretchmer and H. L. Barnett, *Pediatrics*, 15 (1955) 575.
[22] A. Barlow and R. A. McCance, *Arch. Dis. Childh.*, 24 (1948) 225.
[23] S. Lo Bianco, *Lattante*, 24 (1953) 781.
[24] H. Loeb, M. Engelen and R. Van Geffel, *Acta Paediat. Belg.*, 10 (1956) 145.
[25] C. Rivier and M. R. Jeanneret, *Helv. Paediat. Acta*, 11 (1956) 489.
[26] K. Schreier and H. Steg, *Z. Kinderheilk.*, 68 (1950) 563.
[27] H. N. Christensen and J. S. Streicher, *J. Biol. Chem.*, 175 (1948) 95.
[28] A. B. Czeinberg, C. Rauert, A. Sheba, J. Adam, M. Halbrecht, S. Rikover, S. Wishnievsky and E. Rabau, *J. Clin. Invest.*, 37 (1958) 1936.
[29] E. W. Page and M. B. Glendening, *Obstet. and Gynecol.*, 5 (1955) 781.
[30] J. Dancis and M. Shafran, *J. Clin. Invest.*, 37 (1958) 1093.
[31] H. N. Christensen, *Geriatrics*, 14 (1959) 429.
[32] P. C. Zamecnik, J. D. Frantz, R. B. Loftfield and M. L. Stephanson, *J. Biol. Chem.*, 175 (1938) 299.
[33] G. Rouser, in *The Leukemias, Etiology, Pathophysiology and Treatment*, Academic Press, New York, 1957, p. 361.
[34] F. Nour-Eldin and J. F. Wikinson, *Brit. J. Haematol.*, 1 (1955) 358.
[35] R. H. McMenamy, C. C. Lund and D. F. H. Wallach, *J. Clin. Invest.*, 39 (1960) 1688.
[36] J. Dancis, N. Braverman and J. Lind, *J. Clin. Invest.*, 36 (1957) 398.
[37] W. Bickenbach and H. Rupp, *Z. Geburtsh. u. Gynäkol.*, 103 (1932) 170.
[38] G. Stark and G. Siebert, *Arch. Gynäkol.*, 185 (1954) 50.
[39] R. Doerr, *Antikörper in die Immunitätsforschung*, Vol. I, Springer Verlag, Berlin, 1947.
[40] W. A. Hemmings and F. W. Brambell, *Brit. Med. Bull.*, 17 (1961) 96.
[41] F. Nordbring, *Studies on antibodies and immuno-globulines in colostrum from different species and their transference to the newborn infant and animal*, Almquist and Co., Uppsala, 1958.
[42] B. Vahlquist, in F. Linneweh, *Die physiologische Entwicklung des Kindes*, Springer Verlag, Berlin, 1959, p. 362.
[43] E. Holder and K. Schreier, *Arch. Klin. Chir.*, 294 (1960) 394.
[44] U. Stave, *Z. Kinderheilk.*, 81 (1958) 472.
[45] B. H. Kirby, *Ann. J. Phys. Anthrop.*, 11 (1953) 559.
[46] R. Merten, *Ergeb. inn. Med. u. Kind. Neue Folge*, 2 (1951) 49.
[47] P. P. Cohen and G. L. Hekhuis, *Cancer Res.*, 1 (1941) 620.
[48] G. H. Beaton, D. M. Curry and M. J. Veen, *Arch. Biochem.*, 70 (1957) 288.
[49] M. Jacobellis, G. E. Griffin and E. Muntroyler, *Proc. Soc. Exp. Biol. Med.*, 96 (1957) 64.
[50] I. K. Brandt, V. A. Matalaka and J. T. Combs, *Am. J. Phys.*, 199 (1960) 39.
[51] K. Sandiford, W. M. Boothly and H. Z. Griffin, *J. Biol. Chem.*, 55 (1923) 23.
[52] J. J. Kelley and H. A. Waisman, *Blood*, 12 (1957) 635.
[53] G. Y. N. Jyer, *J. Lab. Clin. Med.*, 54 (1954) 229.
[54] K. Schreier, U. Porath, M. Schneider, M. Weber, K. Mehra and H. Plückthun, *Clin. Chim. Acta*, 6 (1961) 295.

[55] J. C. Beck, E. E. McGarry, I. Dyrenfurth, R. O. Morgen, E. D. Bird and E. H. Venning, *Metabolism*, 9 (1960) 699.
[56] H. Gershberg, *J. Clin. Endocrinol. and Metabolism*, 20 (1960) 1107.
[57] T. R. Riggs and L. M. Walker, *J. Biol. Chem.*, 235 (1960) 3603.
[58] O. H. Gaebler, R. Glovinsky, H. Lees, D. Kurrie and H. C. Choitz, *Endocrinology*, 65 (1959) 283.
[59] W. Wilbrandt, *Schweiz. med. Wschr.*, 89 (1959) 363.
[60] D. E. Buetow and B. H. Levedahl, *Arch. Biochem. Biophys.*, 86 (1960) 34.
[61] A. Butenandt, H. Günther and F. Turba, *Z. physiol. Chem. Hoppe-Seyler's* 322 (1960) 28.
[62] J. B. Derrick and A. P. Hauley, *Canad. J. Biochem. Physiol.*, 35 (1957) 1005.
[63] K. Schreier and W. Lesoine, *Strahlentherapie*, 102 (1957) 209.
[64] W. D. Lotspeich, *J. Biol. Chem.*, 179 (1949) 175.
[65] K. Schreier and V. Szybko, *Klin. Wochschr.*, 17/18 (1953) 430.
[66] J. Czyzyk, *Z. inn. Med.*, 16 (1961) 196.
[67] M. H. Wiseman, N. Kalant and M. M. Hoffmann, *J. Lab. Clin. Med.*, 52 (1958) 27.
[68] E. G. McDaniel, J. M. Hundley and W. H. Serbell, *J. Nutrition*, 59 (1956) 407.
[69] D. A. Rosen, G. D. Maengwyn-Davies, B. Becker, H. H. Stone and J. S. Friedenwald, *Proc. Soc. Exptl. Biol. Med.*, 88 (1953) 321.
[70] L. Heim and V. Kaffanke, *Ann. Paediat.*, 185 (1955) 348.
[71] F. L. Iber, H. Rosen, S. M. Levenson and T. C. Chambers, *J. Lab. Clin. Med.*, 50 (1957) 417.
[72] H. G. Knauff, H. Selmair and A. Reitlinger, *Klin. Wochschr.*, 38 (1960) 812.
[73] H. G. Knauff and F. Windsheimer, *Arch. Exp. Path. u. Pharmak.*, 239 (1960) 442.
[74] S. M. Mellinkoff, D. J. Jensen and M. Frankland, *Arch. Int. Med.*, 94 (1954) 604.
[75] S. M. Mellinkoff, M. Frankland, D. Boyle and M. Greipel, *J. Appl. Physiol.*, 8 (1956) 535.
[76] S. M. Mellinkoff, M. Frankland, M. Greipel, H. N. Shibata and W. J. Dixon, *Gastroenterology*, 32 (1957) 592.
[77] S. M. Mellinkoff, T. B. Neynolds, M. Frankland and M. Greipel, *Gastroenterology*, 36 (1059) 780.
[78] D. Müting and V. Wortmann, *Deut. Med. Wochschr.*, 81 (1956) 1853.
[79] H. W. Smith, *The Kidney, Structure and Function in Health and Disease*, Oxford University Press, 1951.
[80] H. Kleinbaum, *Mschr. Kinderheilk.*, 103 (1955) 26.
[81] H. Loeb and M. Engelen, *Arch. franç. ped.*, 14 (1957) 7.
[82] P. L. Calcagno and M. J. Rubin, *Pediatrics*, 7 (1951) 328.
[83] O. Kerpel-Fronius, F. Varza, K. Kun and J. Vonoczky, *Acta Med. Acad. Sci. Hung.*, 5 (1959) 27.
[84] J. Black, *Arch. Dis. Child*, 34 (1959) 158.
[85] R. D. Eckhardt and C. S. Davidson, *J. Biol. Chem.*, 177 (1949) 687.
[86] L. M. Henderson, P. E. Schurr and C. A. Elvehjem, *J. Biol. Chem.*, 177 (1949) 815.
[87] J. B. Kirsner, A. L. Shefener and W. L. Palmer, *J. Clin. Invest.*, 28 (1949) 716.
[88] K. Schreier and H. Remsperger, *Biochem. Z.*, 322 (1952) 298.
[89] A. E. Denton and C. A. Elvehjem, *J. Biol. Chem.*, 206 (1954) 449.
[90] E. S. Wynne and C. L. Scott, *Am. J. Clin. Nutr.*, 4 (1956) 275.
[91] W. H. Stein, *J. Biol. Chem.*, 201 (1953) 45.
[92] L. W. Charkey, A. K. Kano and D. F. Hougham, *J. Nutrition*, 55 (1955) 469.
[93] E. G. Frame, *J. Clin. Invest.*, 37 (1958) 1710.
[94] K. Guggenheim, S. Halvey and N. Friedmann, *Arch. Biochem. Biophys.*, 91 (1960) 6.
[95] K. Schreier and V. Bergmann, *Klin. Wochschr.*, 29 (1951) 285, 608.
[96] K. Schreier and H. Schönsee, *Deut. med. Wochschr.*, 77 (1952) 418.
[97] B. F. Steele and C. B. Le Bovit, *J. Nutrition*, 45 (1951) 235.
[98] H. Kamin and P. Handler, *Am. J. Physiol.*, 164 (1951) 654.
[99] M. Jaccottet, E. Juillard and C. Piguet, *Med. Surg.*, 12 (1954) 222.
[100] K. Schreier and H. Plückthun, *Biochem. Z.*, 320 (1949/50) 447.
[101] E. G. Schmidt and J. S. Eastland, *J. Lab. Clin. Med.*, 21 (1935) 1.
[102] C. H. Greene, K. Sandiford and H. Ross, *J. Biol. Chem.*, 58 (1923) 845.
[103] H. N. Munro and W. S. T. Thomson, *Metabolism*, 2 (1953) 354.
[104] R. E. Butler and H. P. Sarett, *J. Nutrition*, 35 (1948) 539.
[105] P. Stelgens and R. Bothe, *Z. Kinderheilk*, 80 (1957) 26.
[106] O. Wiss, *Helv. Chim. Acta*, 31 (1948) 2148.
[107] H. E. Bock and W. Gerok, *Deut. med. Wochschr.*, 86 (1961) 704.
[108] H. Plückthun and K. Schreier and H. Hauss, *Klin. Woschr.*, 31 (1953) 558.
[109] L. E. Farr and D. A. MacFayden, *Am. J. Dis. Child*, 59 (1940) 782.
[110] S. Z. Levine, *Harvey Lecturen*, 47 (1946) 303.

111 H. R. SEALOCK AND H. E. SILBERSTEIN, *J. Biol. Chem.*, 135 (1940) 251.
112 B. J. STOKVIS, *Ned. Tijdschr. Geneesk.*, (1902) 678.
113 W. F. ROGERS, JR. AND F. GARDNER, *J. Clin. Invest.*, 28 (1949) 806.
114 J. E. MORRIS, E. R. HARPER AND A. GOLDBLOOM, *J. Clin. Invest.*, 29 (1950) 325.
115 D. H. BASINSKI, H. A. FREUND AND D. WILLIAMS, *J. Biol. Chem.*, 197 (1852) 433.
116 H. A. PAINTER AND S. S. ZILVA, *Biochem. J.*, 41 (1947) 511.
117 C. W. WOODRUFF, M. E. CHERRINGTON, A. K. STOCKELL AND W. J. DARBY, *J. Biol. Chem.*, 178 (1949) 861.
118 W. D. SALMON, *J. Lab. Clin. Med.*, 56 (1960) 673.
119 H. M. NITOWSKY *et al.*, *Proc. Soc. Exp. Biol. Med.*, 81 (1952) 62.
120 S. Z. LEVINE, E. MARPELES AND H. H. GORDON, *J. Clin. Invest.*, 20 (1941) 199.
121 W. E. KNOX AND M. LE MAY-KNOX, *Biochem. J.*, 48 (1951) 686.
122 B. N. LA DU AND D. M. GREENBERG, *J. Biol. Chem.*, 190 (1951) 245.
123 N. KRETCHMER, *Pediatrics*, 23 (1959) 606.
124 N. KRETCHMER, S. Z. LEVINE AND H. MCNAMARA, *J. Clin. Invest.*, 2 (1956) 236.
125 S. KAUFMANN, *J. Biol. Chem.*, 230 (1958) 2.
126 S. AURICCHIO, E. GUAGLIARIELLO AND A. RUBINO, *Helv. paediat. Acta*, 15 (1960) 479.
127 A. G. MEDES, *Biochem. J.*, 26 (1932) 917.
128 K. FELIX, G. LEONHARDI AND I. GLASENAPP, *Z. physiol. Chem.*, 287 (1951) 141.
129 K. SAKAI AND T. KITAGAWA, *Jikeikai Med. J.*, 4 (1957) 1, 11; 6 (1959) 15.
130 H. R. BLOXAM, M. G. DAY, N. K. GIBBS AND L. I. WOOLF, *Biochem. J.*, 77 (1960) 320.
131 R. ROBINSON AND P. SMITH, *Nature*, 189 (1961) 323.
132 H. GROS AND H. J. KIRNBERGER, *Klin. Wochschr.*, 32 (1959) 645.
133 R. AMMON AND U. HENNING, *Z. physiol. Chem.*, 30b (1957) 214.
134 M. S. DUNN, M. N. CAMIEN, S. SHANKMAN AND H. BLOCK, *Arch. Biochem.* 13 (1947) 207.
135a J. AWAPARA AND Y. SATO, *Clin. Chim. Acta*, 1 (1956) 75.
b C. E. DENT, *Biochem. J.*, 43 (1948) 169.
c C. GIDDEY, *Schweiz. Med. Wochschr.*, 83 (1953) 431.
d H. A. HARPER, M. E. HUTCHIN AND J. R. KIMMEL, *Proc. Soc. Exp. Biol. Med.*, 80 (1953) 768.
136 P. M. PFAUNDLER, *Z. phys. Chem.*, 30 (1900) 75.
137 S. SIMON, *Z. Kinderheilk.*, 2 (1911) 1.
138 D. J. FOWLER, P. M. NORTON, M. W. CHEUNG AND E. L. PRATT, *Arch. Biochem. Biophys.*, 68 (1957) 452.
139 B. CHILDS, *Proc. Soc. Exptl. Biol. Med.*, 81 (1952) 225.
140 R. HADLICH AND P. GROSSER, *Jahrb. Kinderheilk.*, 73 (1911) 421.
141 F. W. SCHULZ, *Jahrb. Kinderheilk.*, 72 (1910) 94.
142 F. SOUCHON, *Z. ges. exptl. Med.*, 118 (1952) 218.
143 F. SOUCHON AND G. GRUNAU, *Klin. Wochschr.*, 30 (1952) 663.
144 F. SOUCHON AND G. GRUNAU, *Arch. Kinderheilk.*, 144 (1952) 143.
145 H. BICKEL AND F. SOUCHON, *Arch. Kinderheilk.*, *Beih.*, 31 (1955) 1, 151.
146 J. STRÖDER, H. GRÜNHOFER AND W. ENGEL, *Ann. paediat.*, 177 (1951) 304.
147 J. STRÖDER, H. GRÜNHOFER AND L. HEIM, *Ann. paediat.*, 180 (1953) 52.
148 H. BERGER, *Schweiz. med. Wochschr.*, 24 (1956) 86.
149 H. BICKEL, in F. LINNEWEH, *Die Physiologische Entwicklung des Kindes*, Springer Verlag, Berlin, 1959, p. 240.
150 B. L. J. WOOLF AND A. P. NORMAN, *J. Pediat.*, 50 (1957) 271.
151 D. O'BRIEN, J. BERGSTEDT, J. BUTTERFIELD, F. IBBOTT AND L. LUBCHENCO, *Acta Peadiat.*, 49 (1960) 258.
152 K. J. KIVIRIKKO AND M. LIESMAA, *Acta endocrinol.*, 27 (1958) 441.
153 N. F. TOLKACHEVSKAYA, *Clin. Chim. Acta*, 4 (1959) 807.
154 J. J. MURPHY, *Ann. Paediat.*, 196 (1961) 122.
155 K. SCHREIER, R. ITTENSOHN, U. HANS AND W. SIEVERS, *Z. Kinderheilk.*, 79 (1957) 165.
156 P. D. DOOLAN, H. A. HARPER, M. E. HUTCHIN AND E. L. ALPEN, *J. Clin. Invest.*, 34 (1955) 1247.
157 G. STALDER, H. VETTERLI-BUCHNER AND H. BERGER, *Klin. Wochschr.*, 38 (1960) 472.
158 A. MEISTER, *Biochemistry of the Amino Acids*, Academic Press, New York, 1957, p. 99.
159 H. J. VOGEL AND B. D. DAVIS, *J. Am. Chem. Soc.*, 74 (1952) 109.
160 M. R. STETTEN, *J. Biol. Chem.*, 181 (1949) 31.
161 N. M. GREEN AND D. A. LOWTHER, *Biochem. J.*, 71 (1959) 55.
162 M. ZIFF, A. KIBRICK, E. DRESNER AND H. J. GRIBRTZ, *J. Clin. Invest.*, 35 (1956) 579.
163 G. B. GERBER, G. GERBER AND K. J. ALTMAN, *Nature*, 185 (1960) 767.
164 D. J. PROCKOP, C. MITOMA AND A. SJOEDEDSMA, *Fed. Proc.*, 18 (1959) 434.
165 R. A. MCCANCE, *Arch. Dis. Childhood.*, 34 (1959) 361, 459.
166 A. C. L. CLARK AND D. GAIRDNER, *Arch. Dis. Childhood*, 35 (1960) 352.

[167] H. L. Barnett and J. Vesterdal, *J. Ped.*, 42 (1953) 98.
[168] R. G. Ames, *Pediatrics.*, 12 (1953) 272.
[169] H. Heller, *Arch. Dis. Childhood*, 26 (1951) 195.
[170] D. L. Calcagno, M. J. Rubin and D. H. Weintraub, *J. Clin. Invest.*, 33 (1954) 91.
[171] C. M. Edelmann and H. L. Barnett, *J. Ped.*, 56 (1960) 154.
[172] F. H. Epstein, C. R. Kleeman, S. Pursel and A. Hendrikx, *J. Clin. Invest.*, 36 (1957) 635.
[173] S. W. Hier, *Trans. N.Y. Acad. Sci.*, Ser. II, 10 (1948) 280.
[174] H. Kalant and H. Ducci, *J. Clin. Endocrinol. and Metabolism*, 15 (1955) 481.
[174a] J. P. Peters and D. D. Van Slyke, *Quantitative clin. chem., Interpretations*, Baltimore, 1 (1946).
[175] M. A. Wharton and M. B. Patton, *J. Am. Dietet. Assoc.*, 29 (1953) 762.
[176] R. H. McMenamy, C. C. Lund, G. J. Neville and D. F. H. Wallach, *J. Clin. Invest.*, 39 (1960) 1675.
[177] G. J. Gabuzda, J. R. Eckhardt and C. S. Davidson, *J. Clin. Invest.*, 31 (1952) 1015.
[178] H. Berger and I. Antener, *Intern. Z. Vitaminforsch.*, 28 (1957) 17.
[179] G. W. Schmidt, *Ann. Paediat.*, 180 (1953) 52.
[180] K. Schreier, *J. Ped.*, 46 (1955) 86.
[181] H. V. R. Dubois, H. Loeb, M. Vincent and E. J. Bigwood, *Ann. Soc. Belg. de Med. Trop.*, 38 (1958) 5.
[182] H. Sochor, *Oesterr. Z. Kinderheilk.*, 7 (1952) 264.
[183] K. Choremis, C. Damelaton, F. Maoumis, B. Basti, P. Lapatsanin and K. Kiossoglon, *Helv. paediat. Acta*, 14 (1959) 44.
[184] T. H. Huisman, *Voeding*, 19 (1958) 138.
[185] G. W. Schmidt, *Ann. paediat.*, 180 (1953) 52.
[186] H. J. P. Jonxis and T. H. J. Huisman, *Pediatrics*, 14 (1954) 238.
[187] G. Biserte, A. Breton and G. Fontaine, *Arch. Franç. Ped.*, 12 (1955) 988.
[188] J. P. Dustin and E. J. Bigwood, *Proc. Nutrition Soc.*, 12 (1953) 293.
[189] T. H. Huisman, *Pediatrics*, 14 (1954) 245.
[190] P. V. Rangneker and L. P. Dugal, *Canad. J. of Biochem. and Phys.*, 36 (1958) 25.
[191] W. Armbruster, R. Schäfer and K. Schäfer, *Schweiz. Med. Wochschr.*, 87 (1957) 147.
[192] H. Berger and G. Stalder, *Ann. Paed.*, 186 (1956) 163.
[193] H. Berger, J. R. Werber, L. Antener and U. Pfändler, *Intern. Z. Vitamin-forsch.*, 26 (1955) 96.
[194] H. Berger and G. Stalder, *Ann. Paediat.*, 186 (1956) 163.
[195] T. H. Huisman, P. A. Smith and J. H. P. Jonxis, *Lancet*, 1015 (1952) 17.
[196] J. H. P. Jonxis, *Helv. Paediat. Acta*, 10 (1955) 245.
[197] J. H. P. Jonxis and T. H. J. Huisman, *Lancet*, II (1953) 428.
[198] J. H. P. Jonxis and T. H. J. Huisman, *Lancet*, II (1954) 513.
[199] P. Royer, A. Spahr and S. Gerbeaux, *Seamaine*, 32 (1956) 254.
[200] J. S. Dinning, K. W. Cosgrove, C. D. Fitch and P. L. Day, *Proc. Soc. Exptl. Biol. and Med.*, 91 (1956) 632.
[201] W. H. Blahd, A. Bloom and W. Drell, *Proc. Soc. Exptl. Biol. and Med.*, 90 (1955) 704.
[202] K. E. Hurdey and R. J. Williams, *Arch. Biochem. Biophys.*, 54 (1955) 384.
[203] H. Schönenberg, *Klin. Wochschr.*, 33 (1955) 513.
[204] L. R. Richardson, L. G. Blaylock and C. M. Lyman, *J. Nutrition*, 49 (1953) 21.
[205] S. K. Gokhale and B. D. Punekar, *Ann. Biochem.*, 19 (1959) 159.
[206] E. C. Brodie, E. B. Wallraff, A. L. Borden, W. P. Holbrook, C. A. L. Stephens, D. F. Hill, L. J. Kent and A. R. Kremmerer, *Proc. Soc. Exptl. Biol. Med.*, 285 (1950) 75.
[207] D. Grob, *J. Clin. Invest.*, 31 (1952) 639.
[208] D. F. Hill, L. J. Kent and A. R. Kremmerer, *Proc. Soc. Exptl. Biol. Med.*, 285 (1950) 75.
[209] E. Martin, G. Milhaud and J. P. Doret, *Exptl. Med. Surg.*, 12 (1954) 249.
[210] C. G. Li, I. Geschwind and H. M. Evans, *J. Biol. Chem.*, 177 (1949) 91.
[211] H. Porter, *J. Lab. Clin. Med.*, 34 (1949) 1623.
[212] H. E. Thelander and R. Imagawa, *J. Pediat.*, 49 (1956) 15.
[213] F. H. Epstein and H. Levitin, *J. Clin. Invest.*, 38 (1959) 1001.
[214] M. Goldberg and J. S. Handler, *New Engl. J. Med.*, 263 (1960) 1037.
[215] R. Barow and F. Hartmann, *Dt. Arch. klin. Med.*, 260 (1956) 203.
[216] M. S. Dunn, S. Akawie, H. L. Yeh and H. Martin, *Clin. Invest.*, 3 (1950) 302.
[217] D. Y. Hsia and S. S. Gellis, *J. Clin. Invest.*, 33 (1954) 1603.
[218] W. Löffler and W. Gerock, *Ärztl. Forschung*, 9 (1955) 403.
[219] J. D. Lyttle, E. Goettsch, D. M. Greeley and W. M. Grim, *J. Clin. Invest.*, 22 (1943) 169.
[220] C. E. Dent and J. M. Walshe, *Ciba Found. Symp. on Liver Diseases Blakiston Philadelphia 1951*.
[221] J. M. Walshe, *Quart. J. Med.*, 22 (1953) 483.
[222] C. Wu, J. L. Bollmann and H. R. Butt, *J. Clin. Invest.*, 34 (1955) 845.

223 K. Schreier and H. G. Sattelberg, *Deut. Med. Wochschr.*, 76 (1951) 868.
224 W. Latheim, D. Baker and S. Bradley, *Amer. J. Med.*, 18 (1955) 249.
225 P. F. Salisbury, M. S. Dunn and E. A. Murphy, *J. Clin. Invest.*, 36 (1957) 1227.
226 C. Hooft and J. Herpol, *Acta Paediat. Scand.*, 48 (1959) 135.
226a H. Plückthun, K. Schreier and H. Hauss, *Klin. Wochschr.*, 31 (1953) 558.
227 P. Careddu, *Bull. Soc. ital. Biol. sper.*, 28 (1952) 1194.
228 L. Levi and S. Lo Blanco, *Medicina (Roma)*, 5 (1955) 681.
229 L. Levi and S. Lo Blanco, *Z. Kinderheilk.*, 59 (1957) 379.
230 W. W. Shreeve, M. E. Hutchin, H. A. Harper, C. D. Miller and P. D. Doolan, *Proc. Soc. Exptl. Biol. Med.*, 88 (1955) 510.
231 D. Emalie-Smith, J. H. Hohnstone, M. B. Thomas and K. G. Lowe, *Clin. Sci.*, 15 (1956) 171.
232 E. Juillard and C. Pignet, *Ann. paediat. (Basel)*, 183 (1954) 257.
233 J. Luder and W. Sheldon, *Arch. Dis. Childh.*, 30 (1955) 160.
234 K. Fink, R. B. Henderson and R. M. Fink, *Proc. Soc. Exptl. Biol. Med.*, 78 (1951) 135.
235 S. M. Gartler, *Am. J. Human Gen.*, 8 (1956) 120, 126.
236 H. R. Crumpler, C. E. Dent, H. Harris and R. G. Westall, *Nature* 167 (1951) 307.
237 J. DeGromly and H. Eldon-Sutton, *Am. J. Human Gen.*, 9 (1957) 76.
238 J. J. Chisholm, H. C. Harrison, W. R. Eberlein and E. E. Harrison, *Am. J. Dis. Child.*, 89 (1955) 159.
239 V. K. Wilson, M. L. Thomas and C. E. Dent, *Lancet*, II (1954) 66.
240 H. E. Harrison and H. C. Harrison, *Science*, 120 (1954) 606.
241 S. Angielski and J. Rogulski, *Acta Biochem. Polonica*, 25 (1959) 184.
242 E. A. Hosein, M. Smart and K. Hawkins, *Canad. J. Biochem. Physiol.*, 38 (1960) 837.
243 D. Emalie-Smith, J. H. Hohnstone, M. B. Thomas and K. G. Lowe, *Clin. Sci.*, 15 (1956) 171.
244 A. G. Spencer and G. T. Franglen, *Lancet*, 262 (1952) 190.
245 R. G. Westall, E. Roitmann, D. de la Plena, H. Rasmussen, J. Chravioto, F. Gemez and L. E. Holt, *Arch. Dis. Child.*, 33 (1958) 17?.
246 D. P. Doolan, H. Harper, H. E. Hutchin and E. L. Alpen, *J. Clin. Invest.*, 35 (1956) 888.
247 E. B. Robson and G. A. Rose, *Clin. Sci.*, 16 (1957) 75.
248 K. Schreier, *Z. Kinderheilk.*, 82 (1959) 634.
249 U. Stave, *Z. Kinderheilk.*, 78 (1956) 275.
250 S. Van Crefeld and P. Arons, *Ann. Paediat.*, 182 (1954) 191.
251 F. Wewalka, *Acta hepat. (Hamburg)*, 41 (1956) 37.
252 C. H. Eades, R. L. Pollack and J. D. Hardy, *J. Clin. Invest.*, 34 (1955) 1756.
253 G. L. Nardi, *J. Clin. Invest.*, 33 (1954) 847.
254 S. M. Levenson, J. M. Howard and H. Rosen, *Surg. Gynec. Obstet.*, 101 (1955) 35.
255 K. Peter, *Studies of the growth and development of the kidney*, Jena Gustav Fischer (1927).
256 C. M. B. Pare and M. Sandler, *Lancet*, I (1954) 702.
257 H. Kleinbaum, *Mschr. Kinderheilk.*, 103 (1955) 26.
258 M. Kulonen and E. Kulonen, *Scan. J. of Clin. u. Lab. Invest.*, 12 (1960) 1.
259 G. B. Ansel and D. Richter, *Biochem. J.*, (1954) 70.
260 H. M. Häkkinen and E. Kulonen, *Biochem. J.*, 78 (1961) 588.
261 P. Kemali and G. Porcellati, *Boll. Soc. Biol. Sper.*, 28 (1952) 1280.
262 P. Kemali and G. Porcellati, *Boll. Soc. Ital. Biol. Sper.*, 29 (1953) 1182.
263 L. Logothetis, *Neurology (Minneap.)*, 5 (1955) 767.
264 S. Ludewig, *Arch. Neurol. Psychiat.*, 70 (1953) 268.
265 D. Müting and K. N. Shivaram, *Z. physiol. Chem.*, 317 (1959) 34.
266 M. H. Pond, *J. Mental Sci.*, 96 (1950) 1048.
267 F. Reitman, *J. Mental Sci.*, 100 (1954) 149.
268 J. Kral, J. M. Hais and A. Zenisch, *Casop. Lék. Cesk.*, 92 (1953) 62.
269 H. Schönenberg, *Z. Kinderheilk.*, 73 (1953) 8; 75 (1955) 33.
270 C. A. C. Ross, *Lancet*, i (1950) 716.
271 R. Thurau, *Arch. Kinderheilk. Beiheft*, 33 (1956) 43.
272 H. Schönenberg, *Ärztl. Forschung*, 9 (1955) 33.
273 J. D. Solomon, S. W. Hier and O. Bergeim, *J. Biol. Chem.*, 171 (1947) 695.
274 H. G. Knauff, W. Mialkowsky and H. Zickgraf, *Z. Klin. Med.*, 155 (1959) 483.
275 J. P. Ellis, W. W. Nowinski and J. G. Bieri, *Texas Repts. Biol. Med.*, 11 (1953) 231
276 B. S. Walker, N. C. Telles and E. J. Pastore, *Arch. Neurol. Psychiat.*, 73 (1955) 149.
277 J. T. Clarke, E. Elian and H. Schwachman, *A.M.A. J. Diseases Children*, 181 (1961) 491.
278 H. G. Knauff, H. Selmair and H. Zickgraf, *Z. physiol. Chemie Hoppe-Seyler's*, 318 (1960) 73.
279 T. H. Huisman, *Ned. Tijdschr. Geneesk.*, 99 (1955) 3357.
280 M. Torre, T. Scarzella and A. Zanalda, *Boll. Soc. ital. Biol. sper.*, 29 (1953) 196.
281 D. F. Evered, *Biochem. J.*, 62 (1956) 416.

FREE AMINO ACIDS IN ANIMAL TISSUE

EUGENE ROBERTS AND DAISY G. SIMONSEN

Department of Biochemistry, Medical Research Institute, City of Hope Medical Center, Duarte, Calif. (U.S.A.)

Early in our studies of the nitrogen metabolism of normal and neoplastic tissues[1] it appeared desirable to study the pools of non-protein amino acids and related substances. However, as in almost every biochemical field, progress was slow and curiosity limited until methods became available which enabled a large number of determinations to be made in a reasonably short period of time. The development of two-dimensional paper-chromatographic procedures, by which it was possible to detect microgram quantities of substances for which other adequate microanalytical procedures were not available[2–5], made it feasible to survey rapidly the distribution of free or loosely bound amino acids and other ninhydrin-reactive substances in a variety of animal tissues. The paper-chromatographic procedures furnished tools which were ideally suited for giving simultaneous information rapidly about the maximal number of ninhydrin-reactive constituents, and although often employed in a semi-quantitative fashion, could give valuable hints about the presence of new materials and could indicate which substances should be studied further in particular biological situations. Column-chromatographic methods have also been applied extensively, but the procedures, although quantitative, are more time consuming and allow fewer samples to be examined (see ref. 6 and paper by WINITZ in this Symposium for review of analytical methods).

As the work developed it became apparent that the perceptions of the patterns of the spots of the different constituents on the chromatograms were more meaningful to most workers than the same data given in lists of names of the constituents followed by numbers designating the amounts, or in bar graphs, etc. Most human computers seem to be able to store and retrieve the patterned, pictorial information more effectively and to relate it to metabolic events more rapidly than they can the numerical representation of the same information. One might wonder whether the transmittal of some types of numerical information about multivariant situations, in general, from non-human computers to human recipients could be made more effective if the information were transformed into quantitative, pictorial patterns resembling those seen on paper chromatograms!

SOME GENERAL REMARKS ABOUT DATA FROM EXTENSIVE SURVEYS OF FREE AMINO ACIDS OF TISSUES

The first application of the two-dimensional paper-chromatographic technique to protein-free extracts prepared from mammalian tissues immediately after removal

References p. 348/349

from the animal showed that the free amino acid pattern of the transplantable squamous cell carcinoma was completely different from that found in epidermis of normal adult and newborn mice, in epidermis made hyperplastic by application of carcinogen, or in non-malignant papillomata[7]. The extension of these observations to other tissues and types of tumors showed that *each tissue of the healthy adult mouse of a particular strain has a characteristic distribution of free amino acids, while quite similar patterns of free amino acids are found in many different types of transplanted and spontaneous tumors*[8]. The latter results indicate (see ref. 9 for similar data on

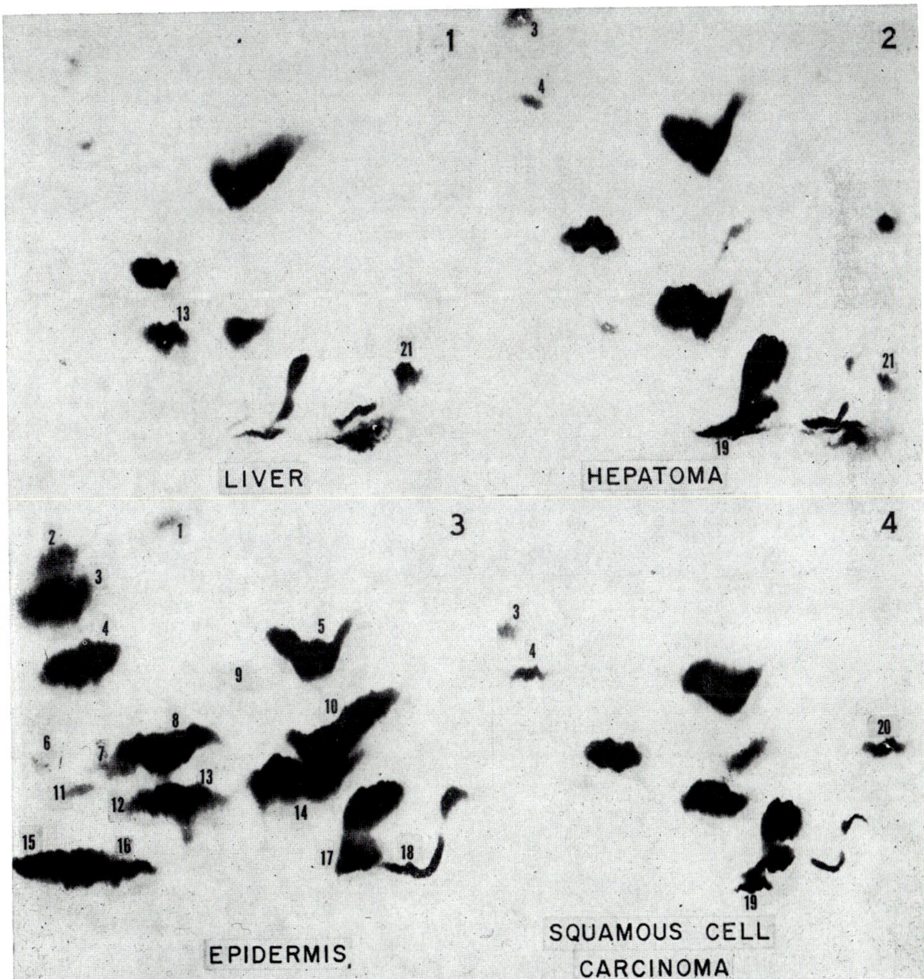

Figs. 1–4. Comparison of free amino acid patterns of mouse liver (Fig. 1) with those found in a transplantable hepatoma (Fig. 2) and epidermis (Fig. 3) and squamous cell carcinoma (Fig. 4). Extracts obtained from 75 mg of fresh weight of tissue were employed for descending two-dimensional chromatography (phenol, right to left; lutidine, bottom to top). Constituents on chromatograms: tyrosine, 1; phenylalanine, 2; leucine and isoleucine, 3; valine, 4; taurine, 5; proline, 6; hydroxyproline, 7; alanine, 8; threonine, 9; serine, 10; histidine, 11; glycerylphosphorylethanolamine and/or β-alanine, 12; glutamine, 13; glycine, 14; arginine, 15; lysine, 16; glutamic acid, 17; aspartic acid, 18; ethanolamine phosphate, 19; cystine (cysteic acid), 20; glutathione, 21.

References p. 348/349

the rat) that steady-state concentrations of small molecules reflect the enzymatic differentiation of normal tissues and the similarities found in tumors agree with GREENSTEIN's generalization[10] based on enzyme assays: "No matter how or from which tissues tumors arise, they more nearly resemble each other chemically than they do normal tissues or than normal tissues resemble each other". Typical results in Figs. 1–4 show comparisons of chromatograms of extracts of epidermis and normal mouse liver with those obtained from equal fresh weights of a transplanted squamous-cell carcinoma and hepatoma, respectively.

Results obtained in a study of extracts of tissues from human, monkey, dog, cat, guinea-pig, rabbit, rat, mouse, opossum, chicken, alligator, snake, turtle, frog, salamander, and a wide variety of marine organisms in our laboratory by this technique[1] have shown that *in a given species at a particular stage of development each*

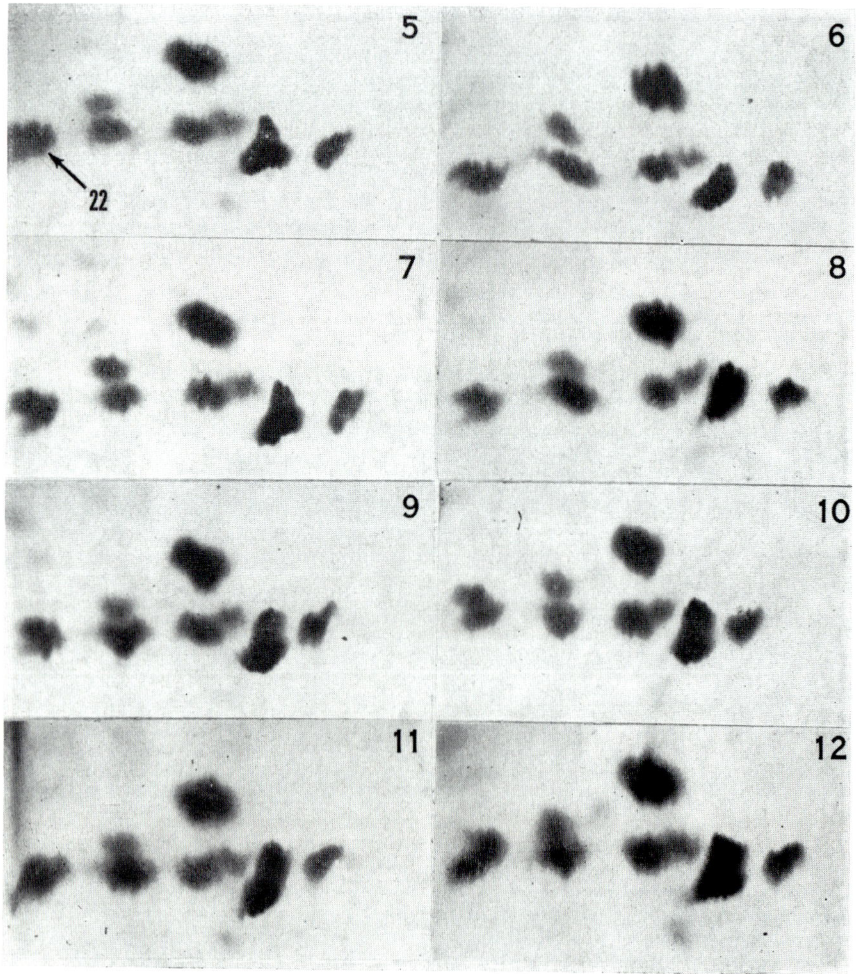

Figs. 5–12. Chromatograms showing constancy of patterns of free amino acids in extracts of brains (37.5 mg) of eight Swiss mice. γ-Aminobutyric acid, 22.

normal tissue, including every type of blood cell, has a distribution of easily extractable ninhydrin-reactive constituents which is characteristic for that tissue. In healthy organisms the patterns are specific and consistent so that with some experience it is possible to determine the tissue studied upon inspection of the chromatogram. A remarkable constancy of distribution of ninhydrin-reactive constituents was observed from one animal to another in the case of adult mice and rats of a particular strain. This is illustrated in Figs. 5–12 in which are shown chromatograms made of extracts of brain from mice of the Swiss strain of both sexes ranging between 16.9 and 18.5 g in weight. A similar constancy in amino acid distribution was found in

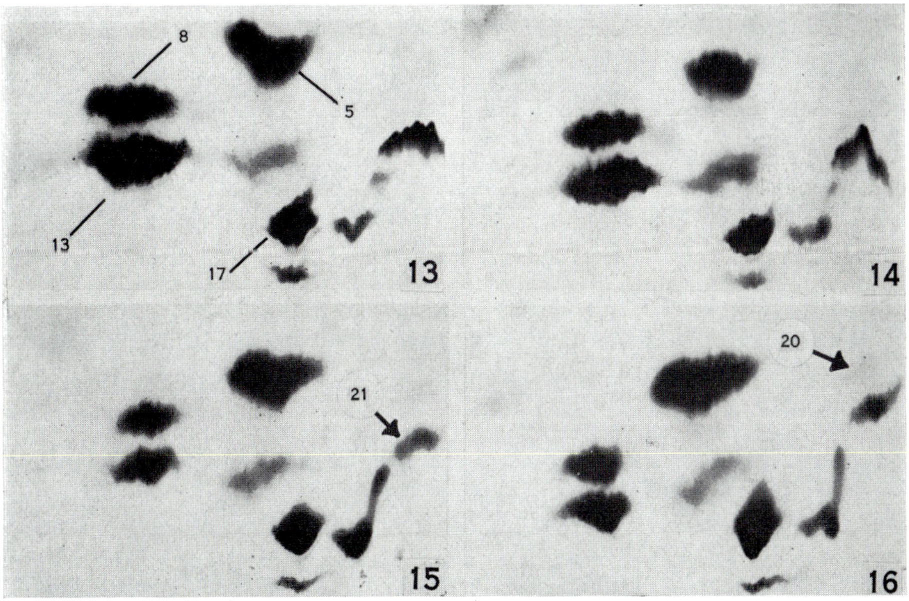

Figs. 13–16. Amino acids in extracts of the myocardium of the different chambers of dog heart (75 mg). Fig. 13: left ventricle. Fig. 14: right ventricle. Fig. 15: left auricle. Fig. 16: right auricle. Taurine, 5; alanine, 8; glutamine, 13; glutamic acid, 17; cystine (cysteic acid), 20; oxidized (H_2O_2) glutathione, 21.

the brains of young adult Sprague-Dawley rats and in other tissues of these and other species.

Reproducible differences in amino acid pattern can be found even within one organ as illustrated in the patterns of the auricles and ventricles of the dog heart (Figs. 13–16)[11]. The ventricular pattern in a number of dogs differed from that found in the auricles, the glutamine and alanine levels of the ventricle being found to be somewhat higher and the taurine level generally somewhat lower than that found in auricular tissue. The chief ninhydrin-reactive constituents found in the extracts of dog heart were glutamine, alanine, taurine and glutamic acid, the same amino acids that were prominent in the extracts of other mammalian hearts. The above point is illustrated further by the demonstration in Figs. 17–22 of the variations in the amounts of ninhydrin-reactive constituents noted in different parts of rat brain, the most marked difference being found in the content of γ-aminobutyric acid

(GABA) which appears to have a unique occurrence in the central nervous system (see ref. 12). Other constituents which have been detected on chromatograms of alcoholic extracts of brain and spinal cord have also been found to occur in varying concentrations in the other tissues. A recent survey of 32 areas of monkey brain has also revealed regional differences in distribution of ninhydrin-reactive constituents. However, only minimal differences were apparent when different areas

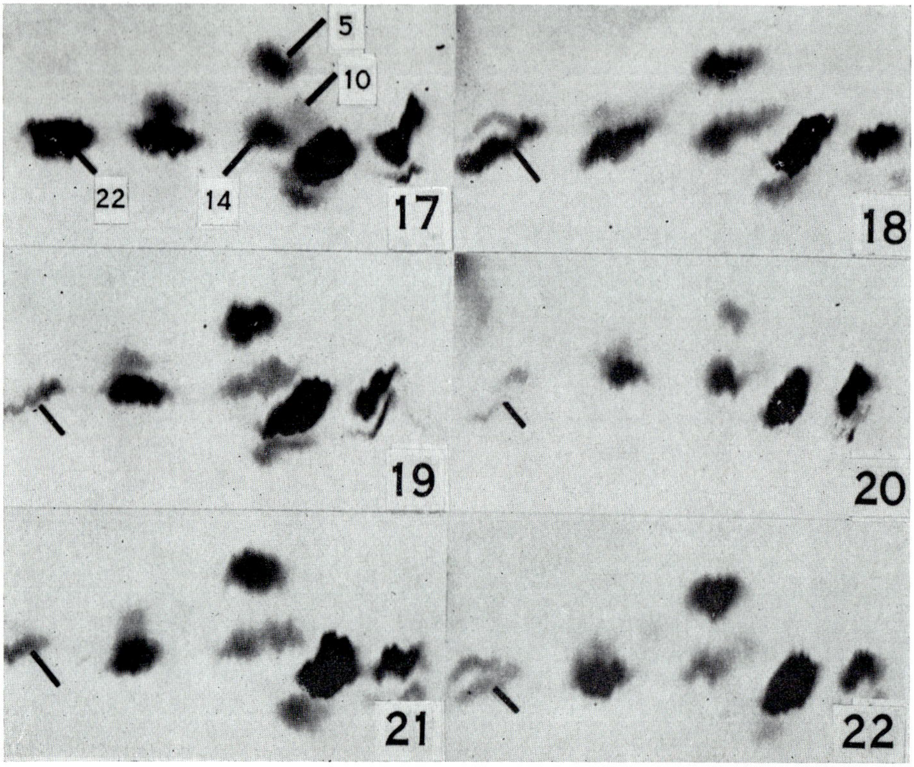

Figs. 17–22. Chromatograms of extracts obtained from 20 mg of original fresh weight of different areas of rat brain; Fig. 17: corpora quadrigemina. Fig. 18: diencephalon. Fig. 19: cerebral hemispheres. Fig. 20: pons and medulla. Fig. 21: cortex. Fig. 22: cerebellum. Taurine, 5; serine, 10; glycine, 14; γ-aminobutyric acid, 22.

of cortical gray matter were compared with each other. Some recent values have been published (Table I) for various areas of human brain determined by column chromatography[13] and for areas of brains of other species[15, 16].

WHAT IS THE INTRACELLULAR DISTRIBUTION OF THE FREE AMINO ACIDS?

The various tissues contain much greater concentrations of a number of the detectable constituents than are found in whole blood or plasma. The intracellular patterns do not reflect the composition of the plasma. The exact mode of intracellular retention of large amounts of easily extractable amino acids and other compounds is not known. Relatively gentle procedures such as homogenization in cold 80% alcohol, heat coagulation and dialysis, or deproteinization with cold trichloroacetic and perchloric acids yield extracts which give virtually identical chromatograms. Similar patterns of free amino acids were found in whole cells, ground cytoplasm, and cell nuclei prepared from livers of fed Wistar rats by the BEHRENS technique, which utilizes organic solvents[17]. All of the numerous attempts in our laboratory to analyze free amino acids in cell particulates of liver and tumor cells prepared in aqueous media

TABLE I

CONCENTRATIONS OF FREE AMINO ACIDS IN VARIOUS AREAS OF 19-YEAR OLD HUMAN BRAIN[13]*

Values expressed in mg/100 g wet wt.

Constituent	*Frontal lobe*	*Corpus callosum*	*Caudate nucleus*	*Globus pallidus*	*Thalamus*	*Hypothalamus*	*Medulla*
Glycerophosphoethanolamine	22.5	23.4	24.9	16.6	23.6	24.8	50.7
Phosphoethanolamine	16.5	—	14.6	4.6	10.8	5.4	2.8
Taurine	18.1	11.9	15.1	14.1	9.2	14.9	13.5
Aspartic acid	38.2	10.7	26.8	33.7	38.8	39.2	16.2
Threonine	6.3	6.4	9.1	18.0	12.4	9.9	10.0
Serine	12.7	11.0	23.2	18.9	18.3	15.7	22.9
Glutamic acid	154.2	61.5	180.5	146.2	135.8	103.5	94.0
Glycine and alanine	34.3	22.6	39.2	49.8	42.7	36.4	54.4
Cystathionine	1.7	9.5	2.3	—	—	3.3	5.3
Isoleucine	2.4	3.6	4.0	—	—	1.8	7.8
Leucine	7.3	5.2	11.4	9.4	9.3	11.6	15.6
γ-Aminobutyric acid	21.0	7.2	27.4	70.5	39.5	48.0	22.0

* Method of MOORE AND STEIN[14].

ordinarily considered suitable for the metabolic study of cell fractions have resulted in the virtually complete liberation of free amino acids into the suspending medium. Typical results are shown in Figs. 23–28. Livers were removed from male rats which had been fasted for 24 h. The livers were homogenized either in 0.25 M sucrose alone or in a medium containing 0.25 M sucrose, 0.1 M potassium phosphate buffer, and 0.002 M versene at pH 7.38. All operations were performed at 0°. Smears made of the homogenates revealed that at least 99% of the cells had been disrupted. The homogenates were centrifuged for 30 min at 105 000 $\times g$ in the Spinco Model L ultracentrifuge. The supernatant fluid was removed and the precipitates were extracted with 80% alcohol in the usual fashion for chromatography. Chromatograms prepared from the original fresh homogenate and from the residue which contained all the particulate matter (microsomes, mitochondria, nuclei and cell debris) revealed that virtually all of the free or easily extractable amino acids had been removed into the suspending medium. Similar experiments were performed in which the suspending medium contained 10% polyvinylpyrrolidone and 20% sucrose. The latter experiments were performed both at 0° and 25°. In each instance the free amino acids were found largely in the supernatant medium.

It has not yet been possible to relate the extractability of the small molecular-weight ninhydrin-reactive constituents of tissues by the above procedures to their state in living tissue. For this reason they have been defined as "free or easily extractable". Insofar as virtually identical quantities of these constituents have been obtained from the same tissues by the different procedures employed, it appears unlikely that they arise from larger molecules by the rupture of covalent bonds. The amino acids may exist in the free form in the cytoplasm and nucleoplasm or may be held by adsorptive forces at interphases and surfaces. Loose bonds may exist between these constituents and proteins, nucleic acids and lipids or complexes of these materials; or sequestration may take place in intracellular structures. It is also possible that the amino acids may exist in the cell in the form of extremely labile derivatives or complexes which are dissociated by the extraction procedures

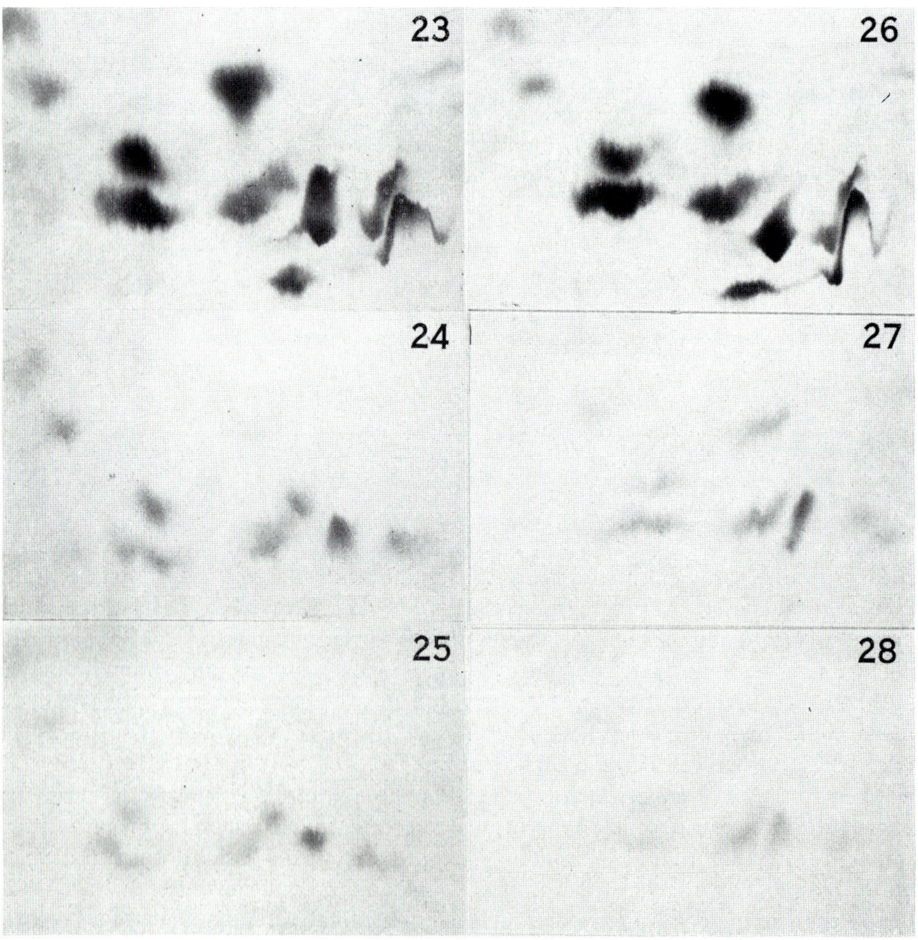

Figs. 23–28. Free amino acids of whole rat liver homogenates (75 mg) Figs. 23, 26: prepared in 10% polyvinylpyrrolidone and 20% sucrose. Figs. 24, 27: extracts of residues from homogenates of 75 mg of fresh weight of tissue prepared at 0° and 25°, respectively. Figs. 25, 28: residues washed once with original volumes of suspending medium at 0° and 25°, respectively and centrifuged at these temperatures.

References p. 348/349

which have been employed. Likewise, even if these small molecular-weight constituents were shown to be associated in loose linkage with a subcellular component or structure when procedures are employed which disrupt cell structure but do not degrade complex molecules and organelles, it could not be assumed that such an association exists *in vivo*. When the cell structure is destroyed myriad new opportunities for interactions between cellular constituents may arise which do not exist in the living cell. We have recently obtained experimental evidence which indicates the possibility of the existence of a physical binding of GABA in brain which is

TABLE II

UPTAKE OF [1-^{14}C]-γ-AMINOBUTYRIC ACID INTO SEDIMENTS OF HOMOGENATES OF BRAIN AND OTHER TISSUES[18]

Expt. No.	*Tissue*	*% of total volume in sediment* (A)*	*% of total counts in sediment (B)*	*B/A*
1	Brain	10.0	26.0	2.60
	Liver	9.0	5.9	0.65
	Heart	10.4	7.4	0.71
	Lung	7.4	5.9	0.80
	Spleen	8.6	5.7	0.66
	Kidney	9.9	7.9	0.80
2	*Brain:*			
	a. Original sediment	10.1	37.8	3.74
	b. 1 × washed sediment		29.9	2.96
	c. 2 × washed sediment		32.3	3.20
	Heart:			
	a. Original sediment	9.8	8.3	0.85
	b. 1 × washed sediment		1.0	0.10
	c. 2 × washed sediment		0.8	0.09
3	Mouse brain acetone powder**	15.7	12.7	0.81

* Derived from residue weight assuming a density of unity for both the sediment and suspending medium. The suspending medium had the following composition: 0.154 *M* NaCl, 10.4 parts; 0.154 *M* $MgSO_4$, 0.1; 0.25 *M* glucose, 0.3; 0.11 *M* sodium phosphate buffer (pH 7.2), 1.2. 0.1 mg of [1-^{14}C]GABA (2.7 mC/m*M*) was employed in 10 ml of the above medium. The sediments in Expt. 2 were resuspended in isotope-free medium.

** 100 mg of acetone powder containing high levels of L-glutamic acid decarboxylase and γ-aminobutyrate–α-ketoglutarate transaminase activities was suspended in 2.5 ml of medium and incubated for 70 min.

not enzymatic (Table II)[18]. The experiments were performed at 0–4° under conditions which did not permit metabolism of the added [1-^{14}C]GABA to take place. To weighed portions of freshly dissected rat brain, lung, heart, spleen, kidney and liver were added 9 vols. of incubation medium containing the labeled amino acid. Tissues were homogenized, incubated with shaking for 50 min at 0°, and were then centrifuged for 30 min at 23 000 × *g* at 0°. The supernatant fluid was poured off, the residue weighed and resuspended in the original volume of fresh medium not containing isotope, and suitable aliquots of the original homogenate and resuspended residue were counted in the scintillation counter. Results show that GABA is bound to the brain sediment but not to that of other tissues or to brain acetone powder. The radioactivity in the sediment from heart homogenate was washed out by resuspension in the medium but most of that in the original sediment from brain remained through two washings. Similar experiments employing equilibrium dialysis have shown that glutamic acid and possibly other amino acids may also be bound by brain preparations. Attempts to solubilize the binding material in brain so far have failed. Work is in progress to delineate further the properties of the binding material and to purify the factor or factors involved. Thus, at least in the case of brain, there is some evidence for the existence of substances in tissues with specific binding capacities for small molecules, but it cannot be determined at present whether or not any considerable proportion of the total intracellular content of any of the easily extractable amino constituents actually exist in this form, nor can it be said that such binding, where found, has any relationship to the physiological effects which some of the constituents may exert in various biological test systems.

ARE THE FREE OR EASILY EXTRACTABLE AMINO ACIDS ARTIFACTS RESULTING FROM AUTOLYTIC CHANGES?

One of the questions which was raised early in the work on free ninhydrin-reactive substances was whether these easily extractable constituents actually were normal cell constituents or whether they represented autolytic artifacts, since the quantities found were very small by comparison with the amounts of protein-bound amino acids occurring in the tissue samples studied. In order to minimize this possibility in our laboratory tissues are fixed immediately after excision. A number of the substances which occur prominently in one tissue or another such as taurine, GABA, carnosine and anserine, ethanolamine phosphate and glutathione are not normal constituents of proteins. However, recently a new nucleotide-peptide has been isolated from bovine liver which contains β-alanine, cysteic acid and taurine as well as other amino acids[19]. Patterns of amino acids observed after hydrolysis of whole homogenates of tissues were never found to resemble those obtained in the protein-free extracts (see example in ref. 20). Furthermore, no evidence for the extensive occurrence of peptides other than glutathione was found in the extracts (see also ref. 21), the presence of which would be expected to a considerable extent if catheptic activity had played a prominent role in the origin of the detectable constituents. Fig. 29 shows a typical pattern of a picric acid filtrate of the ventricle of dog heart and Fig. 30 shows the changes produced in this pattern upon hydrolysis with 6 N HCl in a sealed tube for 24 h (ref. 11). The changes noted on hydrolysis were the complete disappearance of glutamine, glutathione and ethanolamine phosphate with concomitant increases in the concentrations of glutamic acid, glycine, cysteine (detected as cysteic acid) and ethanolamine. In addition, β-alanine and histidine, the products of hydrolysis of carnosine, were detected. The latter substance, which gives a yellow-brown ninhydrin color, was below the level of detection on the chromatogram of the unhydrolyzed extract. It is noteworthy that the content of alanine and taurine remained essentially constant and that most of the other amino acids did not even become detectable after acid hydrolysis, indicating that, at most, only small amounts of diverse peptides may be present in these extracts. Similar results were obtained in many instances in which protein-free extracts of different tissues of various species were compared with the hydrolysates of the extracts. It is, of course, likely that small amounts of peptidic material would escape detection by the procedures employed.

An extensive series of experiments then showed that the amino acid patterns observed in the freshly fixed tissues are in no way related to those found after autolytic changes have been allowed to take place[22]. Brain, muscle, liver and kidney were removed from normal Sprague-Dawley rats or those which have been hypophysectomized or adrenalectomized, as well as from tumor-bearing hosts. The Walker carcinoma was also studied under the conditions to be described. Results will be described in this report only for tissues of the normal rats. In order to insure that

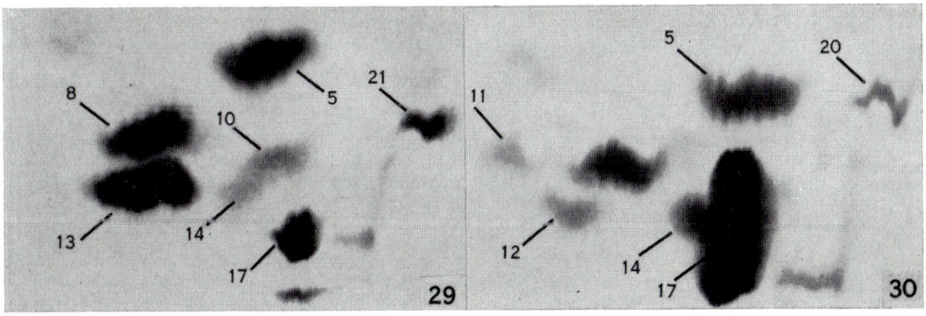

Figs. 29, 30. Influence of acid hydrolysis on distribution of constituents in protein-free extract of left ventricle of dog heart (75 mg). Fig. 29: unhydrolyzed extract. Fig. 30: same aliquot of extract after hydrolysis with 6 *N* HCl in a sealed tube for 24 h. Taurine, 5; alanine, 8; histidine, 11; β-alanine, 12; glutamine, 13; glycine, 14; glutamic acid, 17; cystine (cysteic acid), 20; glutathione (oxidized with H_2O_2), 21.

all changes observed would be attributable to enzymes found in the tissues, care was taken to exclude microorganisms. In an effort to avoid mixing substrate and enzymes not ordinarily in contact with each other, pieces of intact tissue were used rather than homogenates, minces, or slices. The tissues were removed observing sterile precautions and samples were placed in individual sterile weighing bottles of known weight. After weighing, samples were placed in an incubator at 38° for varying periods of time. Upon removal from the incubator, cultures were made in thioglycolate broth from a small particle of each sample and the remainder of the sample was then worked up for chromatography in the usual manner. Results were used only for samples showing no microbial contamination. Virtually identical results were obtained for tissues of normal rats as were found for their hypophysectomized or adrenalectomized litter mates. The results at various time intervals after removal of the samples from the animals showed that the pattern of proteolytic breakdown is characteristic for each tissue.

The results for various intervals of autolysis of brain up to 96 h after removal of the samples from the animals showed that extensive overall proteolysis had not taken place (Figs. 31–38). The amino acids were not liberated in the same relative amounts as during acid hydrolysis of brain protein. The changes in free amino acid content probably were the results of some processes by which the individual amino acids were formed or liberated and used during the period of observation. Since GABA is found chiefly in the easily extractable form in brain, the relatively large

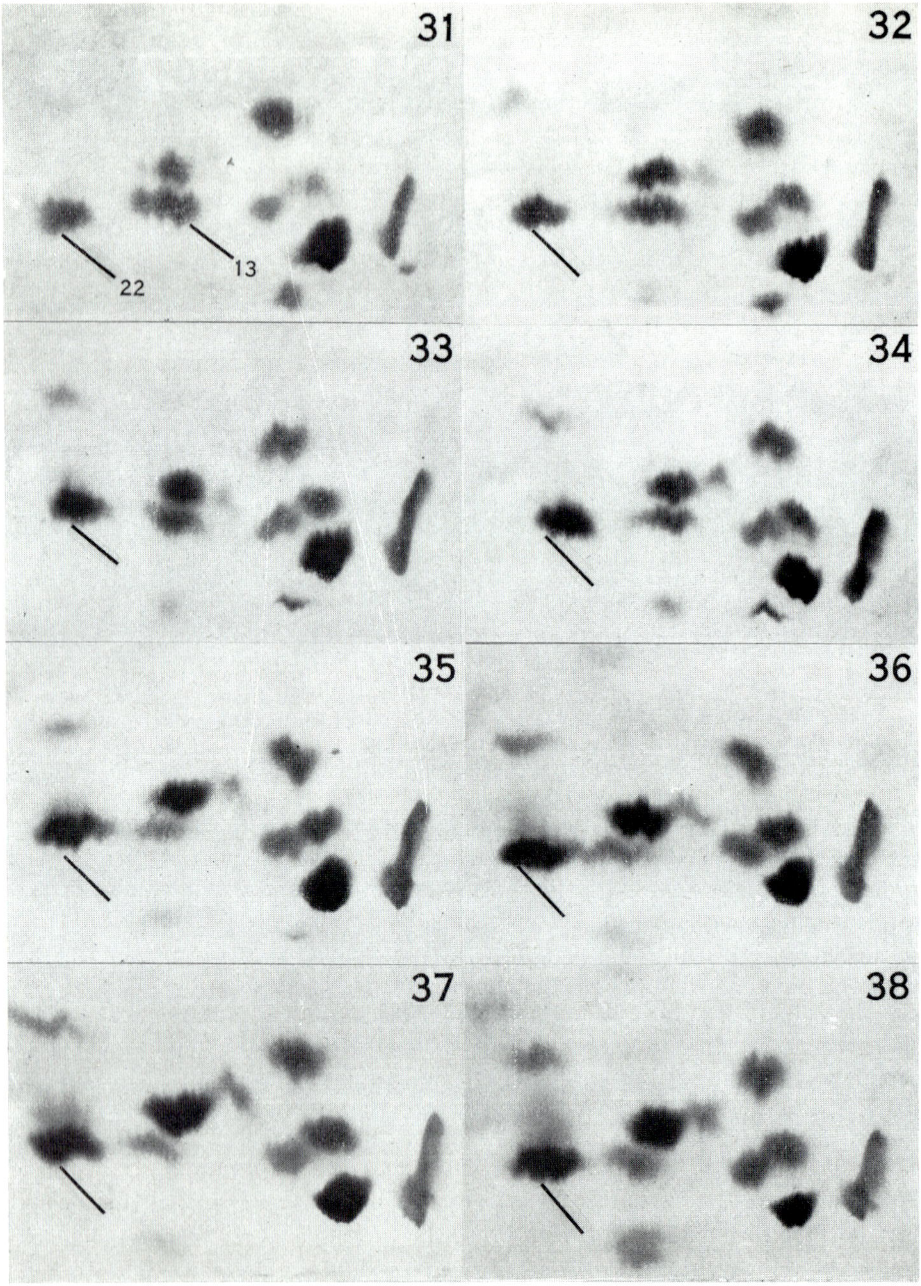

Figs. 31–38. Paper chromatograms of extracts obtained from 37.5 mg of original fresh weight of rat brain at various times of sterile autolysis. Fig. 31: control. Fig. 32: 4 h. Fig. 33: 8 h. Fig. 34: 12 h. Fig. 35: 24 h. Fig. 36: 48 h. Fig. 37: 72 h. Fig. 38: 96 h. Glutamine, 13; γ-aminobutyric acid 22.

increase in content of this amino acid can probably be attributed to continued *post mortem* activities of the glutamic acid decarboxylase, the enzyme which forms GABA and which can function anaerobically, and the cessation of the tricarboxylic acid cycle which must furnish α-ketoglutarate, one of the substrates of the γ-aminobutyrate transaminase system, probably the major system by which GABA is utilized in brain. Glutamine decreased in content between 8 and 12 h and was no longer detectable after 24 h. It was probably decomposed into glutamic acid and ammonia by the action of cerebral glutaminase. The increases noted in the aspartic acid content could be attributable to formation by transamination from oxalacetate or by breakdown of *N*-acetylaspartic acid, which is present in large amounts in brain[23]. The increase in alanine could possibly result from transamination of pyruvate with a number of the available amino acids. Proteolytic activity probably accounts for the relatively small increases which were observed in the contents of threonine, lysine, valine, leucine and isoleucine, tyrosine and histidine. Ethanolamine phosphate was destroyed slowly, probably by action of phosphomonoesterases present in brain.

One of the chief points of interest in the study with the Walker carcinoma was to ascertain whether the liberation of free glutamine, which has been found to occur during the treatment of a variety of animal tumors with agents which produce regression[24], would also take place under conditions of sterile autolysis. The results (Figs. 39–46) showed that only a small amount of glutamine appeared during the first 24 h of autolysis in the samples of tumors studied, while valine, the leucines, tyrosine and some of the other amino acids were liberated from the tissue proteins at a rapid rate. At later time intervals more glutamine appeared. The changes, which occur after treatment with chemical agents[24] or during regression resulting from genetically determined resistance of the host[25], do not appear to be the same as those which occur when the tumor is allowed to autolyze under sterile conditions.

In Figs. 47–58 are shown comparisons of the chromatograms obtained from freshly dissected samples of liver, kidney, muscle and brain with those obtained at 8 and 24 h after sterile autolysis. It is seen that in the case of liver extensive overall proteolysis had taken place rather rapidly, while the brain showed relatively little liberation of ninhydrin-reactive constituents. The kidney and muscle were intermediate in their rates of breakdown. An interesting finding in the case of the liver was that free arginine was not detected at any time but only ornithine, reflecting the great amount of arginase activity in the liver. In the kidney considerable ornithine was formed but arginine was also detected, while in the case of muscle there was relatively little if any ornithine formed and the free arginine had considerably increased as a result of the autolytic process. The amounts of ornithine found are in keeping with the relative arginase activities in these tissues, which in mice were found to be for liver, kidney and muscle, respectively, 300, 6 and 1 (ref. 26). Another interesting finding is that in none of the tissues were there detectable changes in the level of taurine at any of the periods of autolysis which were observed. In liver, kidney, and muscle relatively rapid increases took place in the content of free valine, leucine and isoleucine, and tyrosine, whereas relatively small changes took place in some of the other detectable constituents. In particular, little increase if any took place in the content of glycine in the muscle during the period of observation at a time during which considerable increases in the above free amino acids as well as lysine and arginine were observed. It is apparent from cursory examination

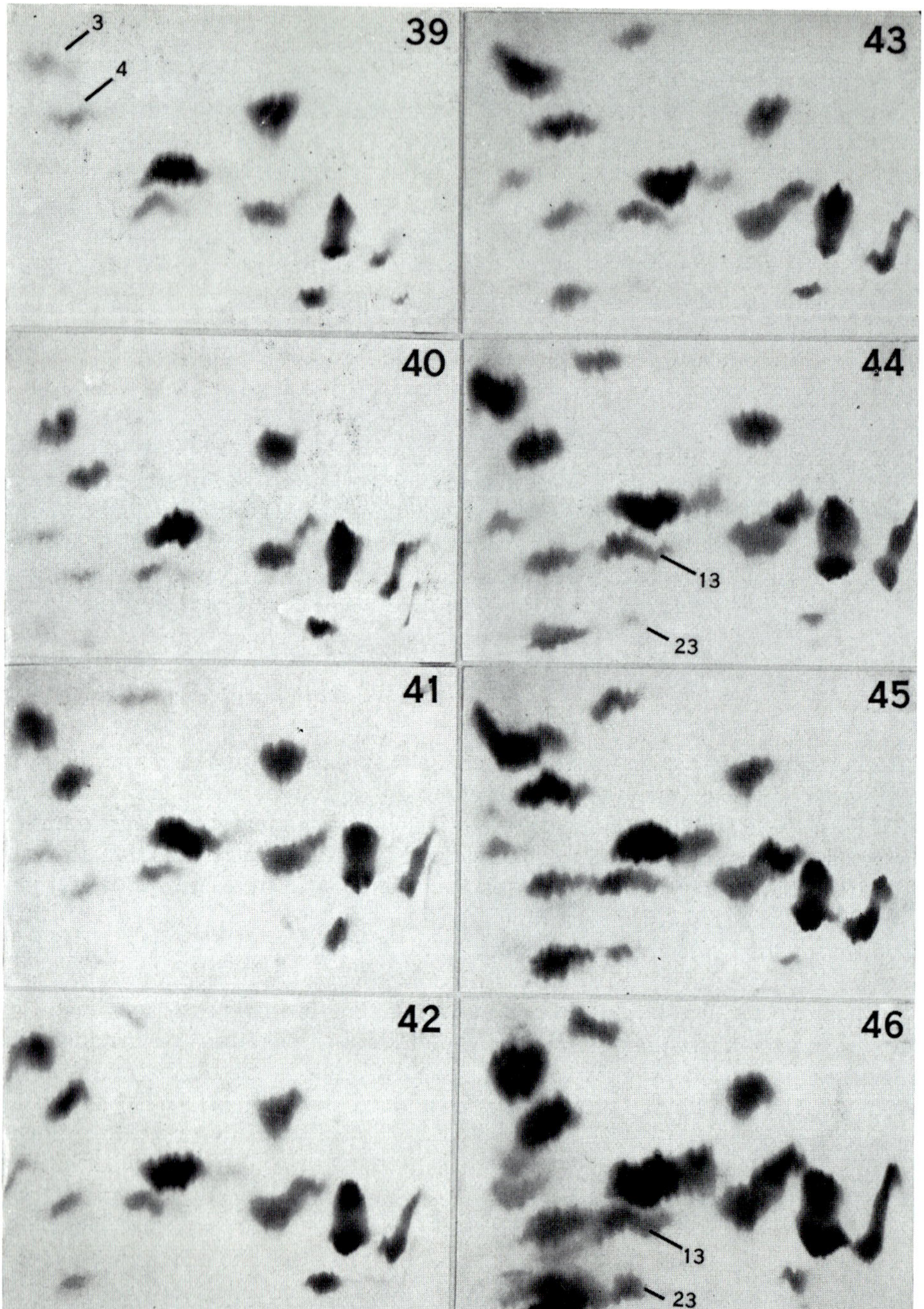

Figs. 39–46. Changes in free amino acids of samples (37.5 mg) of Walker carcinoma undergoing sterile autolysis. Fig. 39: immediately after removal. Figs. 40–46: samples extracted after 4, 8, 12, 24, 48, 72 and 96 h, respectively, of sterile autolysis. Leucine and isoleucine, 3; valine, 4; glutamine, 13; ornithine, 23.

of the chromatograms in Figs. 47–58 that there is a pattern of autolytic change which is characteristic for each tissue. Separate experiments showed that in no case were the amino acids liberated in the same relative amounts during autolysis as

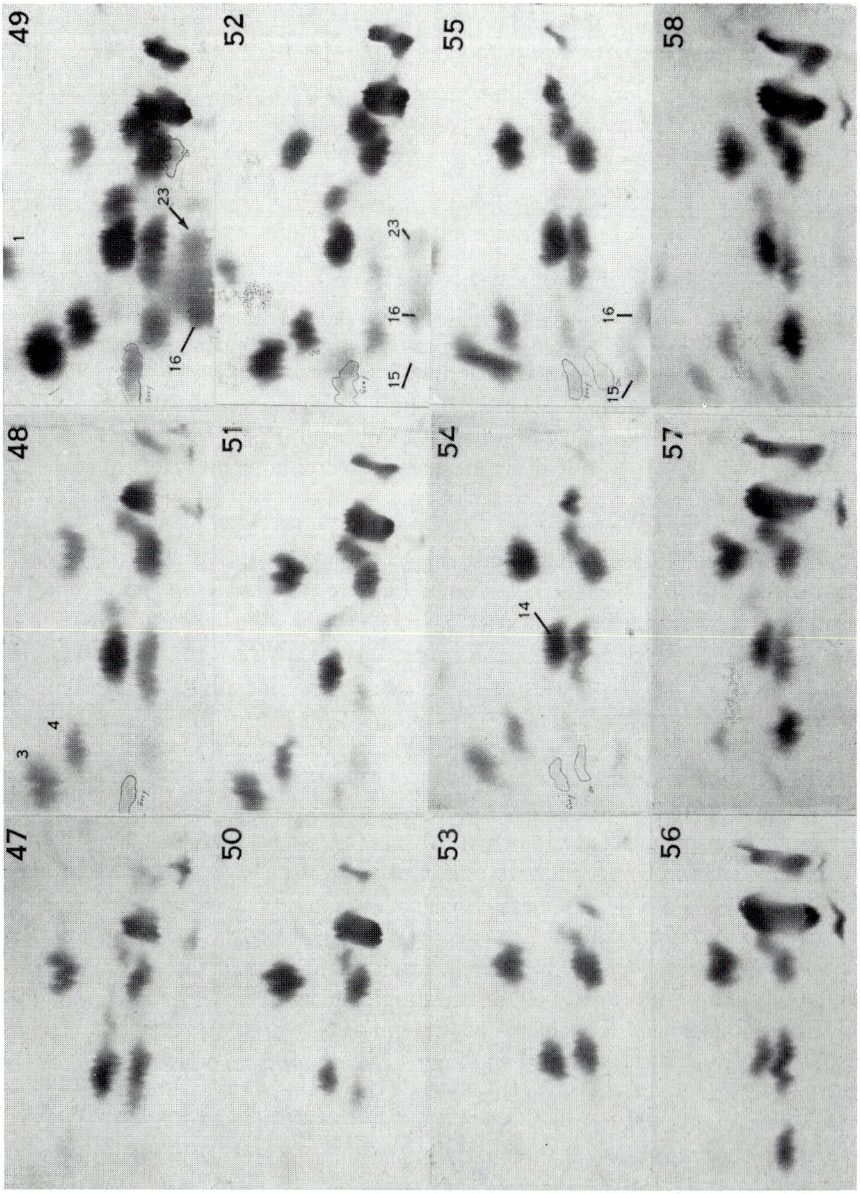

Figs. 47–58. Changes in free amino acids during sterile autolysis of liver, kidney, muscle and brain. Aliquots corresponding to 37.5 mg of fresh weight of tissue were employed. For each tissue the numbers are in the following order: control, 8 h autolysis, and 24 h autolysis. Figs. 47–49: liver, Figs. 50–52: kidney. Figs. 53–55: muscle. Figs. 56–58: brain. Alanine, 14; arginine, 15; lysine, 16; ornithine, 23.

References p. 348/349

during acid hydrolysis of the tissue proteins. Hydrolysis of the protein-free extracts at all stages of autolysis in the various tissues revealed little liberation of peptidic material.

From the above discussion and results presented, it appears highly unlikely that the amino acids and other ninhydrin-reactive constituents which are found in extracts of fresh tissues arise by autolytic and proteolytic breakdown of the proteins after excision of the tissue. The results indicated that the changes in these constituents probably occur relatively slowly enough so that precautions such as freezing of tissue prior to dissection of specimens does not appear to be necessary.

CHANGES IN THE CONTENT OF EASILY EXTRACTABLE NINHYDRIN-REACTIVE CONSTITUENTS DURING DEVELOPMENT

Although each tissue of a mature organism of a particular species has a characteristic distribution of free or easily extractable amino acids and related substances, changes in the content of these constituents per unit of fresh weight may take place at all stages of development until the final adult functional and structural patterns are laid down. Tissues from the frog, chick embryo (including yolk, albumen and membranes), salamander, mouse and other organisms have been analyzed for their content of free amino acids at different stages of development and compared with adult tissues.

It was found that ovarian eggs of *Rana pipiens* contained higher concentrations of free amino acids than shed unfertilized eggs from which the jelly coat had been removed. After fertilization there was an increase in the content of free glycine at the two-cell stage, while taurine was first detected in the late gastrula state. In other stages of the development of *Amblystoma punctatum* the pattern of free amino acids of the yolk was different from that of the neural plate and ectoderm. The free amino acid pattern of the blastodisc of the unincubated fertilized chicken egg was very similar to that of the yolk. However, after the 1st day of incubation the blastodisc, made up of adhering yolk and embryonic tissue, was found to contain taurine, an amino acid which was not detected in the yolk at any stage. In the case of the chick embryo, alcoholic extracts made from 150 mg of fresh weight of albumen showed only traces of freely extractable ninhydrin-reactive material, possibly peptidic. Only at 16 days of incubation did small amounts of some free amino acids appear (Fig. 59) and remain at similar concentrations to 20 days. In Fig. 60 is shown a chromatogram from the hydrolysate of the mixed proteins of the albumen. The finding of a remarkably small pool of free amino acids at the time that rapid utilization of the albumen is taking place by the embryo suggests that the albumen proteins may be transported out of the albumen as intact units, or at least as relatively large polypeptide units, for use elsewhere in the embryo. The changing pattern in the yolk as a function of time of incubation of the chick embryo is shown in Figs. 61–64. As in the case of the albumen, at no time did the free amino acids reflect the amino acid composition of the mixed yolk proteins.

In Figs. 65–68 are shown the amino acid patterns in the liver of the tadpole and adult bullfrog. Particularly noteworthy was the progressive increase in the taurine content with the stage of development, although other changes, both increases and decreases, in the content of various constituents were noted as well. Examination of

References p. 348/349

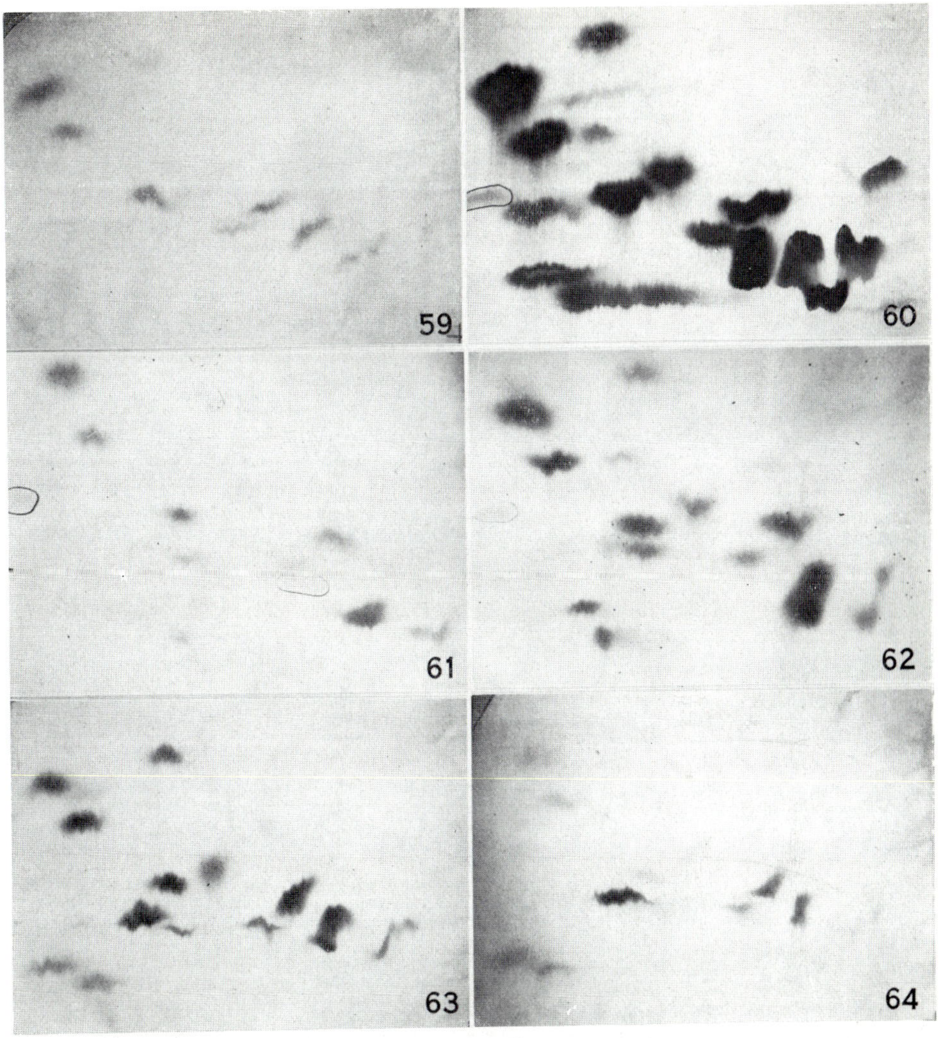

Figs. 59–64. Fig. 59: chromatogram of extract corresponding to 150 mg of fresh weight of albumen from chick embryo at 18 days of incubation. Fig. 60: chromatogram of acid hydrolysate of 1.5 mg of the dried, alcohol-extracted albumen. Figs. 61–64: chromatograms of alcoholic extracts of 150 mg of yolk before incubation (Fig. 61) and at 3, 16 and 18 days of incubation (Figs. 62, 63, 64, respectively).

extracts of fetal mouse liver showed that glutamine was not detectable until the 19th day of gestation, at which time the pattern of amino acids only differed from that found in the adult mouse liver by having less glutamine and glycerylphosphorylethanolamine (Figs. 69–72).

Figs. 73–75 show chromatograms of extracts of heart obtained from 16-day fetal mice (Swiss), 1-day-old mice and adult mice, respectively. Extracts of the heart of fetal mice between 16 and 19 days of gestation (results shown only for 16 days) and of newly born mice contained higher concentrations of a number of the amino acids

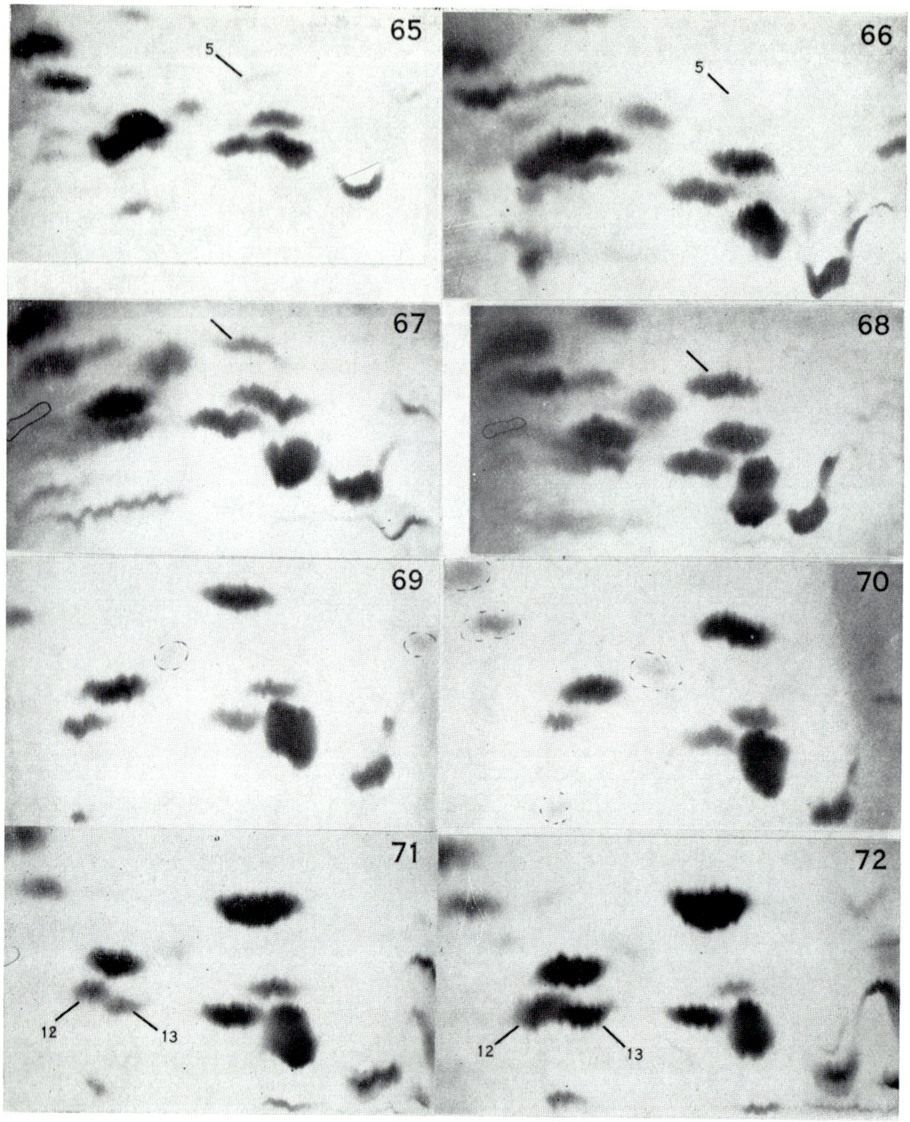

Figs. 65–68. Amino acid patterns in the liver (75 mg) of the tadpole and adult bullfrog *(Rana catesbeiana)*. Fig. 65: 3.5 cm tadpole. Fig. 66: 4 cm tadpole. Fig. 67: 7 cm tadpole. Fig. 68: adult frog.

Figs. 69–72. Amino acid patterns of mouse liver (75 mg) at different stages of development. Figs. 69, 70: 16th- and 18th-day fetuses, respectively. Figs. 71, 72: 19th-day fetus and adult mouse, respectively. Taurine, 5; β-alanine and/or glycerylphosphorylethanolamine, 12; glutamine, 13.

than did the extracts of the adult heart, and hydroxyproline and asparagine were detected only in the younger tissues. Extracts of fetal and 1-day-old mouse heart contained relatively large amounts of glycine and serine. These amino acids were present in low concentrations in extracts of adult tissue. A relatively low level of glycine and serine is characteristic of adult mammalian and reptilian heart. Glutamine was present in higher concentration in newborn and adult heart than in fetal heart. As in hearts of other mammalian species[11], glutamine, alanine, taurine and

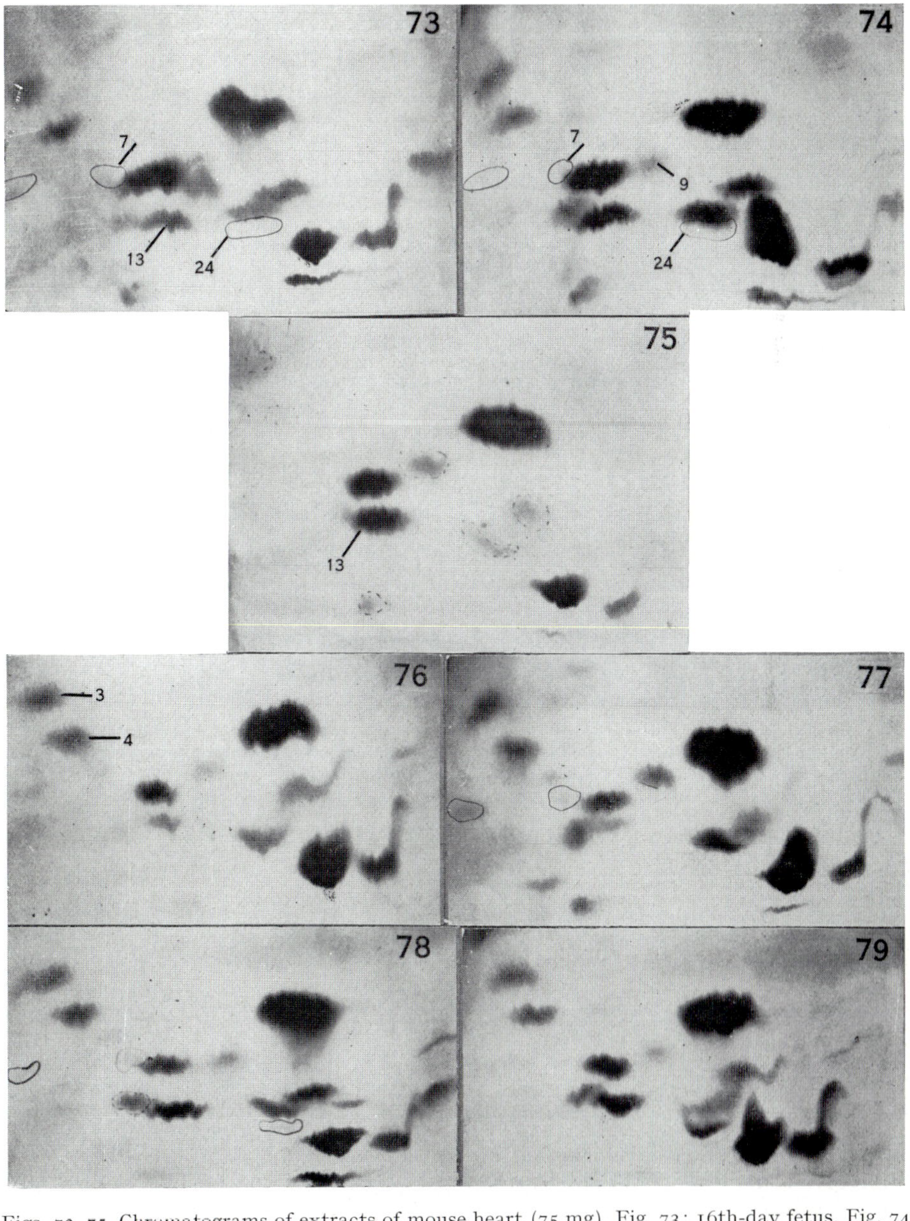

Figs. 73–75. Chromatograms of extracts of mouse heart (75 mg). Fig. 73: 16th-day fetus. Fig. 74: 1-day old mouse. Fig. 75: adult mouse (6-months old). Leucines, 3; valine, 4; hydroxyproline, 7; threonine, 9; glutamine, 13; asparagine, 24.

Figs. 76–79. Amino acid patterns of chick heart. Fig. 76: 14th-day embryo. Fig. 77: 16th-day embryo. Fig. 78: 17th-day embryo. Fig. 79: young hen.

glutamic acid were the amino acids present in the highest concentrations. Examinations of extracts of cardiac muscle from Barred Rock chick embryos of between 5 and 19 days of incubation and from adult chickens of the same strain revealed that marked changes take place during the period of development. Figs. 76–79 illustrate the patterns observed on chromatograms of cardiac muscle of 14-, 16- and 17-day chick embryos and of the adult chicken, respectively, changes being noted in a number of the detectable constituents. Valine and hydroxyproline were detected in the embryo hearts, but the latter amino acid decreased below the level of detec-

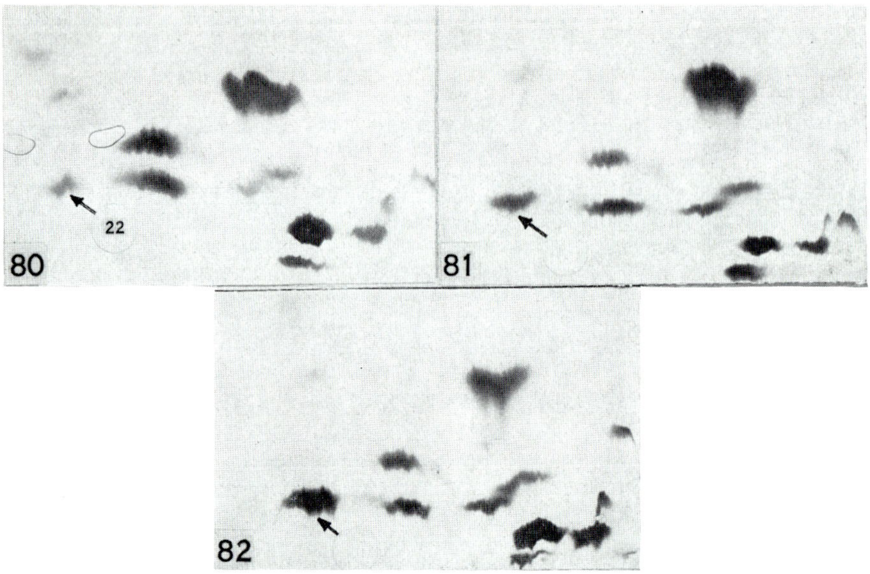

Figs. 80–82. Chromatograms of extracts (75 mg samples) of brain of a 16th-day fetal mouse, 1-day old mouse and an adult mouse, respectively (Figs. 80–82); γ-Aminobutyric acid, 22, and the unlabeled arrows.

tion in extracts of hearts of adult chickens. The extracts of the adult chicken heart showed larger amounts of glycine, serine, valine and the leucines than were found in the mouse and rat hearts and the hearts of other mammalian species.

Figs. 80–82 show chromatograms of extracts of brain obtained from a 16-day fetal mouse, 1-day-old mouse and an adult mouse, respectively. On examination of a large number of brains from mice of intermediate ages, it was found that the content of GABA increased progressively with age up to approx. 30 days after birth. The progressive increase in content of GABA with development is also illustrated on the chromatograms (Figs. 83–86) from extracts of pooled brains of tadpoles and adult bullfrog. Patterns of amino acids in the bullfrog tadpole, 7 cm long (Fig. 85), and at subsequent larval stages were similar to those found in the adult bullfrog. Taurine was noted only in the brain of the adult frog, none being present in the earlier stages (Fig. 86). Examples of chromatograms prepared from tissues of a series of chick embryos at all stages between 4 days of incubation and hatching are shown in Figs. 87–90. GABA was first detectable in the brain of the chick embryo on the 4th day of incubation (Fig. 87) and increased in concentration with the increasing

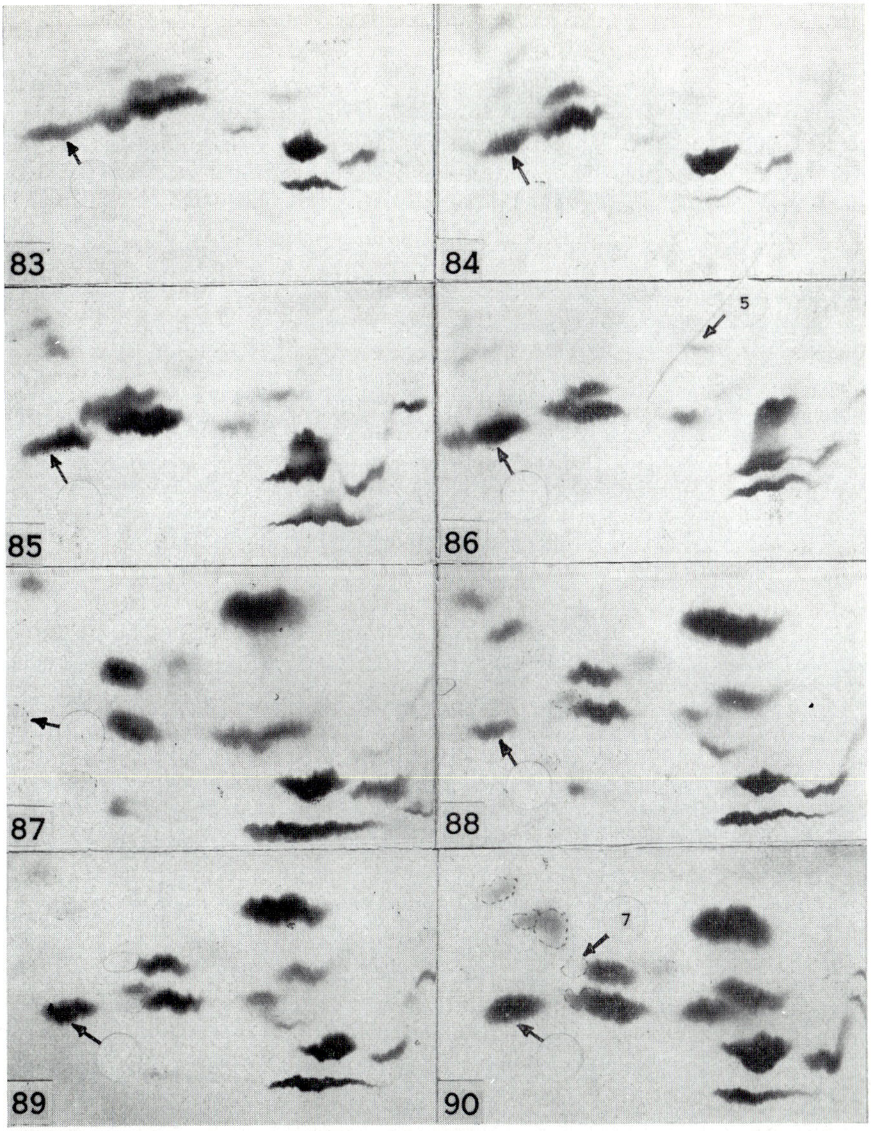

Figs. 83–90. Fig. 83: brains of 3.5-cm tadpole. Fig. 84: 4-cm tadpole. Fig. 85: 7-cm tadpole. Fig. 86: adult frog *(Rana catesbeiana)*. Fig. 87: brains of chick embryos on 4th day. Fig. 88: 8th day. Fig. 89: 13th day. Fig. 90: 17th day of incubation. Taurine, 5; hydroxyproline, 7; γ-aminobutyric acid, 22. Unlabeled arrows point to γ-aminobutyric acid.

age of the embryo. As in the case of the previously discussed species, GABA was the only detectable constituent that showed progressive changes during the course of development. The results in Figs. 91–98 show the changes in different parts of the chick embryo brain between 15 and 21 days of incubation. There was relatively large quantities of GABA early in development in the optic lobes (Figs. 95, 96) and the diencephalic structures (Figs. 97, 98) while the cerebellum (Figs. 91, 92) and the hemispheres (Figs. 93, 94) showed relatively low levels of this amino acid. From inspection of the chromatograms prepared from the extracts of the different cerebral areas either at 15 or 21 days of incubation it is apparent that the most marked difference in any detectable ninhydrin-reactive constituent from one area to another is in the content of GABA. The progressive increase of GABA with the age of the embryo is illustrated for the optic lobes of the chick in Figs. 99–106. The decrease in content of ethanolamine phosphate with age also is noteworthy.

Quantitative determinations of GABA and the activity of glutamic acid decar-

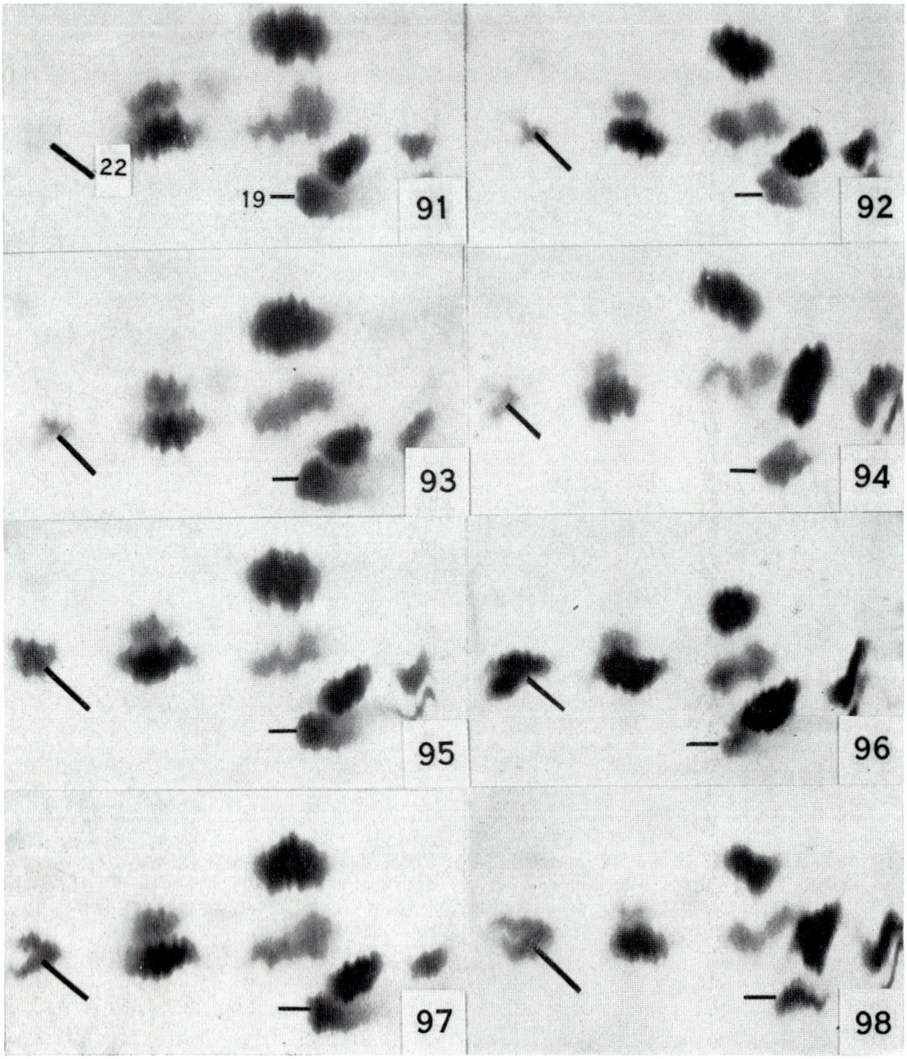

Figs. 91–98. Amino acids in extracts of chick embryo brain at 15 and 21 days of incubation (20 mg equivalents). Fig. 91: cerebellum, 15 days. Fig. 92: cerebellum, 21 days. Fig. 93: hemispheres, 15 days. Fig. 94: hemispheres, 21 days. Fig. 95: optic lobes, 15 days. Fig. 96: optic lobes 21 days. Fig. 97: diencephalon, 15 days. Fig. 98: diencephalon, 21 days. γ-Aminobutyric acid, 22; ethanolamine phosphate, 19.

boxylase, the enzyme which forms GABA from glutamic acid, previously had been correlated with cytological changes in the developing mouse brain[27]. A recent study has been made of the changes of the enzymes related to the metabolism of GABA and of the content of this amino acid in the optic lobe as a function of age of the embryo and the newly hatched chick[28].

From the above results as well as from many data from our own and other laboratories (refs. 28–30 and others) which have not been presented it appears that *although*

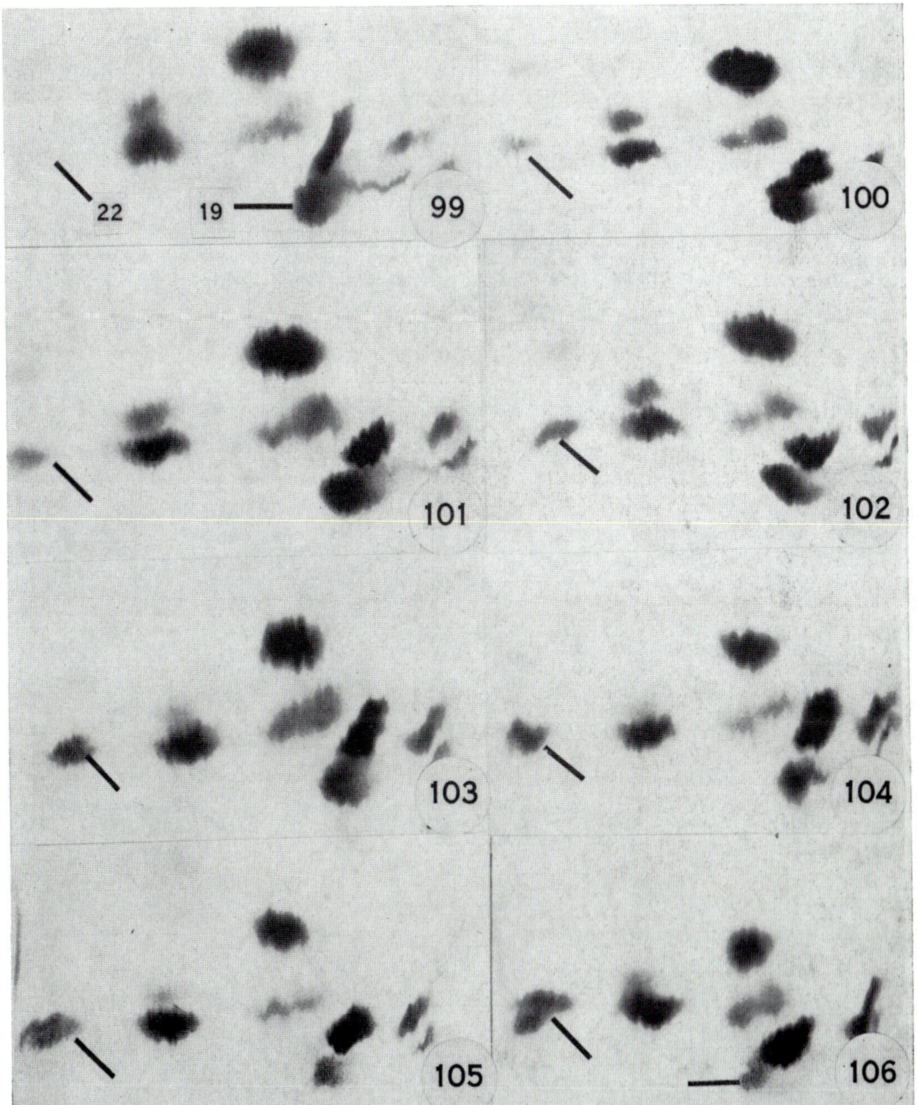

Figs. 99–106. Amino acid patterns of extracts of optic lobes of chick embryo brain (20 mg equivalents) at various times of incubation. Fig. 99: 6 days. Fig. 100: 8 days. Fig. 101: 10 days. Fig. 102: 13 days. Fig. 103: 17 days. Fig. 104: 19 days. Fig. 105: 20 days. Fig. 106: 21 days. γ-Aminobutyric acid, 22; ethanolamine phosphate, 19.

References p. 348/349

certain aspects of the amino acid pattern of tissues become established at a given stage in development, one or more of the amino acids may still continue to change in concentration before the adult pattern is established. To date only in the case of GABA has the content of an amino acid been related to the increase in activity of the enzyme which forms this amino acid.

RELATIVE CONSTANCY OF THE FREE AMINO ACID PATTERNS OF TISSUES: EXPERIMENTAL ALTERATIONS

At any time, the concentration of a particular constituent in a cell must be, at the least, the function of the rate of entry, exit, formation, utilization and possible

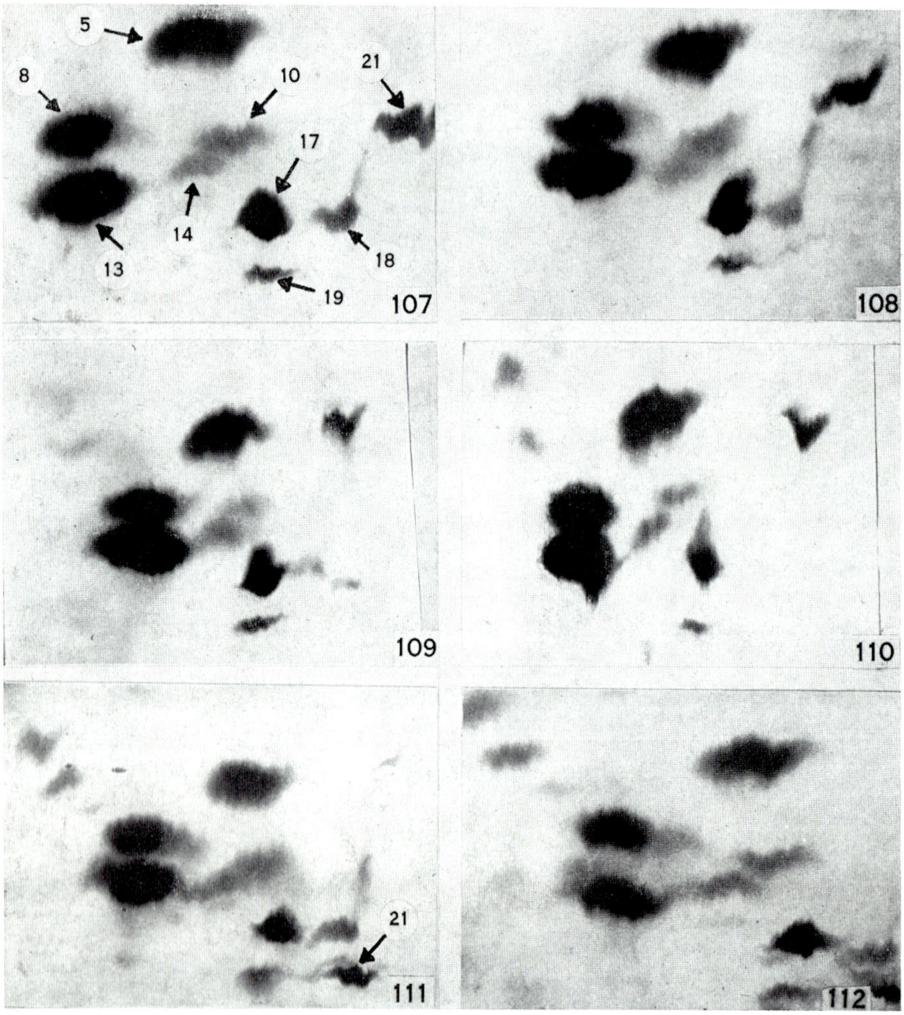

Figs. 107–120. Free amino acids of left ventricle of dog heart (75 mg) at various times after ligation of the anterior descending coronary artery. The odd numbered figures show the uninfarcted areas and the even numbered figures the infarcted areas. Figs. 107, 108: 2 h. Figs. 109, 110: 4 h. Figs. 111, 112: 8 h.

References p. 348/349

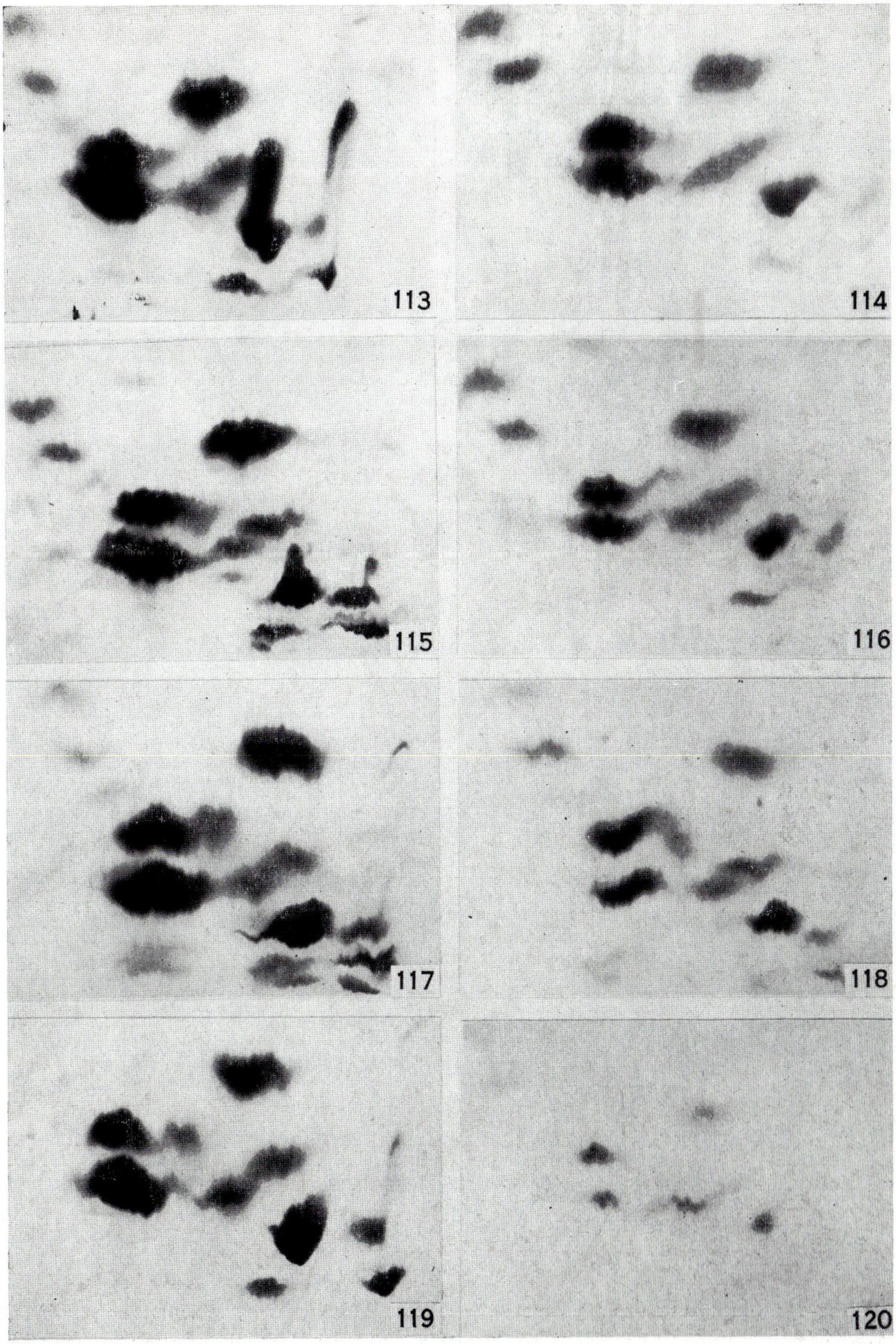

Figs. 113, 114: 16 h. Figs. 115, 116: 24 h. Figs. 117, 118: 48 h. Figs. 119, 120: 5 days. Taurine, 5; alanine, 8; serine, 10; glutamine, 13; glycine, 14; glutamic acid, 17; aspartic acid, 18; ethanolamine phosphate, 19; oxidized glutathione (H_2O_2), 21 (in Fig. 107) and unoxidized glutathione, 21 (in Fig. 111). See also legend p. 306.

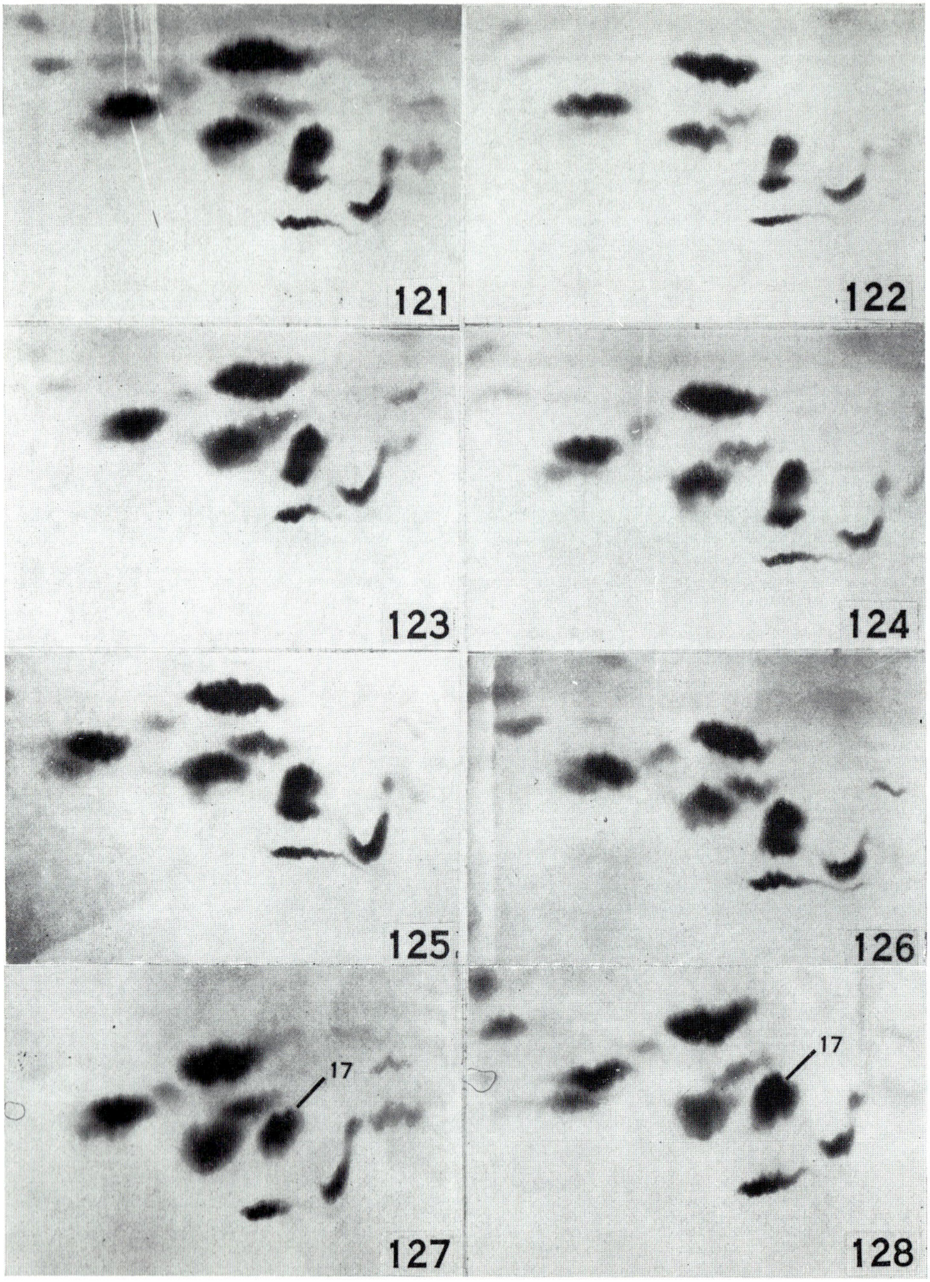

Figs. 121–132. Chromatograms of extracts of solid form of tumor C1498. Aliquots corresponding to 75 mg of fresh weight of tissue were employed in each case. The results in Figs. 121, 123, 125, 127, 129 and 131 are from extracts of tumors grown in C57BL/10-H-2^b (susceptible) mice at 4, 6, 8, 10, 12 and 14 days, respectively, after transplantation. The results in Figs. 122, 124, 126, 128, 130 and 132 are from extracts of tumors grown in C57BL/10-H-2^d (resistant) mice at 4, 6, 8, 10, 12 and 14 days, respectively, after transplantation. Glutamine, 13; glutamic acid, 17.

References p. 348/349

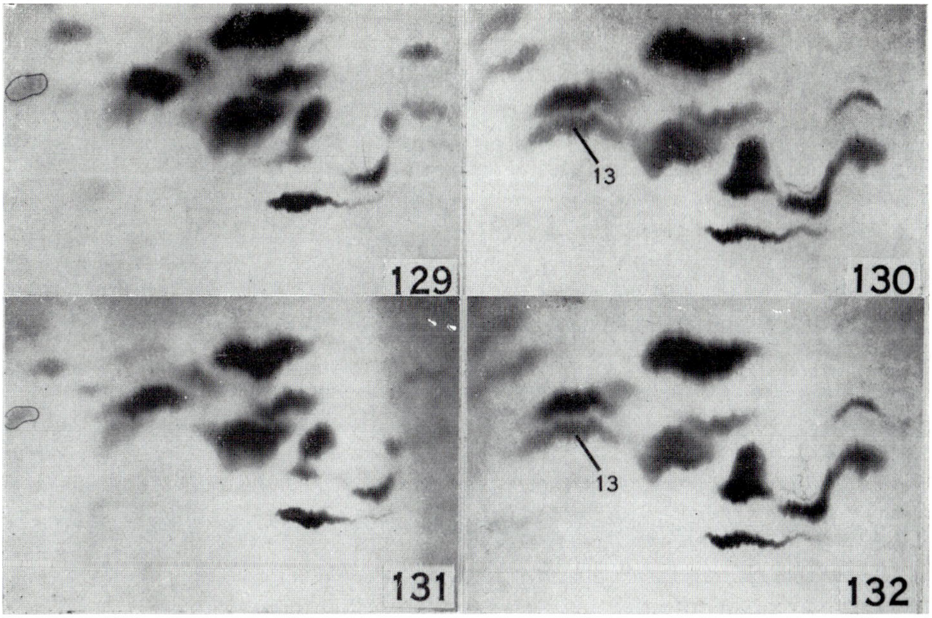

Figs. 129–132. For legend see p. 308.

special retention factors related to adsorption, sequestration in organelles, etc. In the course of attempting to elucidate some of the mechanisms which might be of importance in regulating the patterns of the freely extractable ninhydrin-reactive constituents of tissues, a variety of physiological changes were attempted. It was found that the distribution of free amino acids and related substances in the tissues of the various animals studied are remarkably constant in the face of a variety of induced physiological changes and that only cell death could lead to a virtually complete loss of these constituents from a tissue.

The influence of myocardial infarction on free amino acids of dog heart

It was of particular interest to follow the changes in free amino acids in cells of a tissue undergoing progressive irreversible damage and to attempt to correlate the changes in the patterns of these constituents with the cellular changes observed. In one such study the free or easily extractable amino acids were determined chromatographically in the hearts of dogs at various time intervals up to 6 days after the production of experimental myocardial infarction by ligation of the anterior descending coronary artery[31]. Comparisons were made of the infarcted and non-infarcted areas of the left ventricle.

Typical results are shown in Figs. 107–120. For each time interval after ligation it was possible to compare the infarct with a grossly normal area of the same ventricle. Numerous comparisons were made of normal and infarcted areas of hearts of different dogs. The experimental procedure did not appear to produce profound effects on the distribution of free amino acids in those areas of myocardium which were not affected by ligation, and no significant histological changes were seen in these areas. A comparison of chromatograms of the infarcted area with a normal area from the ventricle of each dog showed little change in 30 min and 2 h after ligation (Figs. 107, 108). At these periods there was marked capillary congestion with minimal histological changes. In the 4-h experiment (Figs. 109, 110), in which slight pericarditis was noted but minimal gross abnormality of the muscle fibers in the section studied, there was a decrease in aspartic acid by comparison with the control, but otherwise there was no change in amino acid pattern. At the 8-h interval (Figs. 111, 112) there was a decrease in glutathione content as well as in aspartic acid. The contents of the other detectable constituents remained unchanged. At this time segmentation of the muscle fibers, necrosis, loss of cross-striations, and extravasation of red corpuscles and polymorphonuclear leukocytes into the infarcted area were noted. Further changes had occurred in the amino acid pattern at 16 h after ligation (Figs. 113, 114). A marked decrease in content of taurine was noted as well as decreases in the levels of alanine, glutamine, glutamic and aspartic acids, glutathione and ethanolamine phosphate. Further significant changes appeared to have taken place at 24 and 48 h (Figs. 115–118). A sample obtained from a dog surviving the ligation for 5 days (Fig. 120) showed a great loss of all of the constituents in the infarct by comparison with the control area (Fig. 119). Losses of carnosine from the affected areas also occurred, the quantities of β-alanine and histidine detected on chromatograms of acid-hydrolyzed extracts being decreased markedly at 16 h after ligation and at later time intervals.

From this study it is apparent that extensive damage in the heart muscle did not result in the immediate loss of the easily extractable ninhydrin-reactive constituents, but rather that this loss took place over a considerable period of time. At the time that marked changes in amino acid content became apparent the degree of damage to the myocardium was of such an extent that probably many of the cellular constituents of both higher and lower molecular weight were lost. However, not all injury to muscle results in the loss of amino acids. Determinations of free amino acids and related substances in extracts of muscle from normal and vitamin E deficient (dystrophic) rabbits have shown that in a severely dystrophic rabbit a number of the amino acids in the picric acid filtrates were elevated, but that the contents of glycine, histidine, 1-methylhistidine, carnosine and anserine were markedly reduced[32].

Free amino acids of tumors grown in resistant and susceptible hosts

The results to be reported in this section again serve to emphasize the stability of those properties of cells which are related to the retention of relatively high levels of the easily extractable ninhydrin-reactive constituents. Tumor cells have been shown to have the capacity to retain much larger concentrations of the amino constituents than are found in extracellular fluid of the tumors even when severe cellular damage was produced as judged by cytological criteria.

Determinations were made by two-dimensional paper chromatography of free amino acid patterns in tumor C1498 in both solid and ascites forms at various times after transplantation into C57BL/10-H-2^b mice, a strain in which the tumor grows progressively and kills the animals, and into C57BL/10-H-2^d mice, a subline which

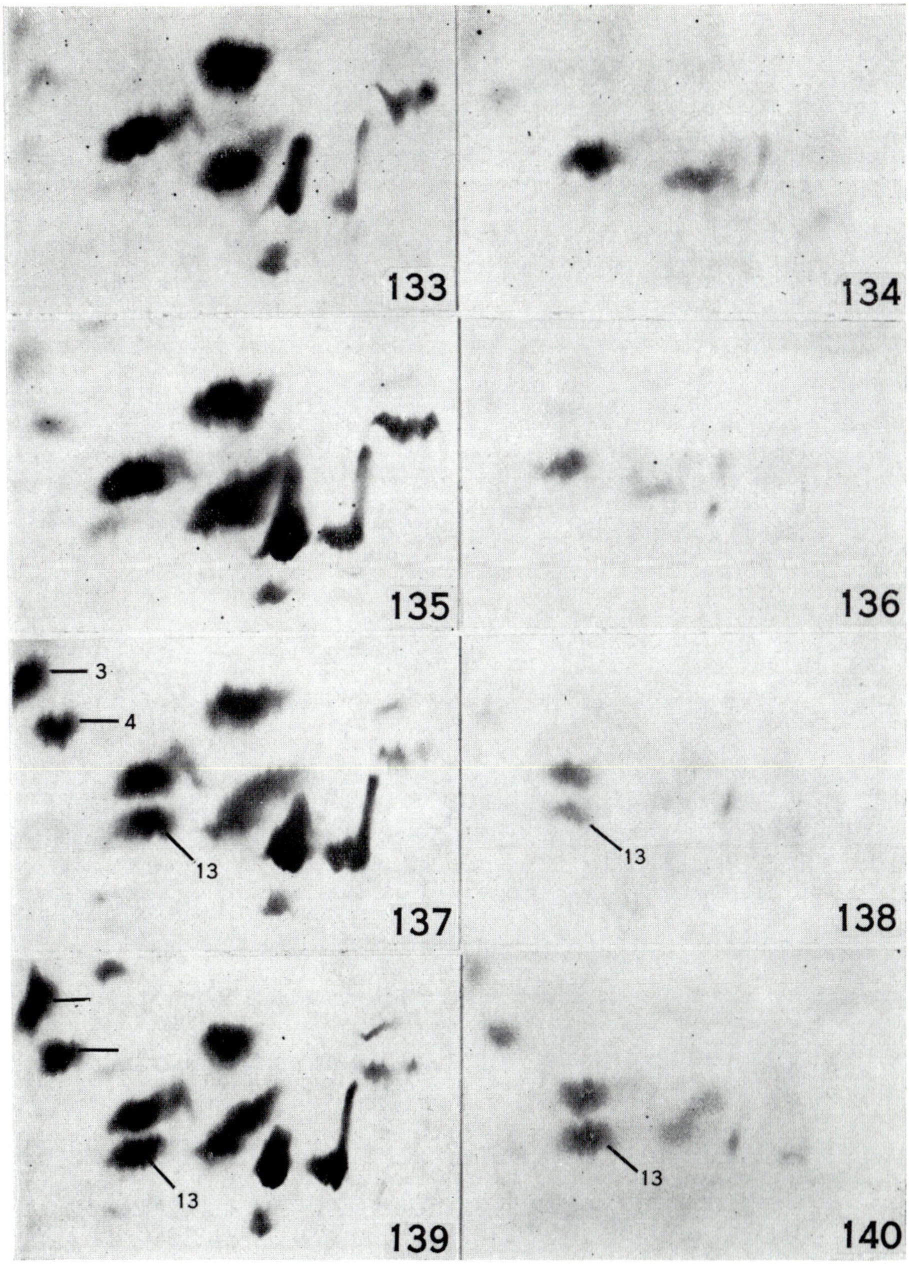

Figs. 133–140. Chromatograms of extracts of cells and fluid of ascites form of tumor C1498. Aliquots of extracts corresponding to 75 μl of packed cells or 100 μl of ascitic fluid were employed for chromatography. Figs. 133 and 134 are from cells and fluid in susceptible mice at 7 days after transplantation. Figs. 135, 137 and 139 are chromatograms of extracts of cells and Figs. 136, 138 and 140 are chromatograms of extracts of ascitic fluid in the resistant mice at 6, 8 and 9 days, respectively, after transplantation. Leucines, 3; valine, 4; glutamine, 13.

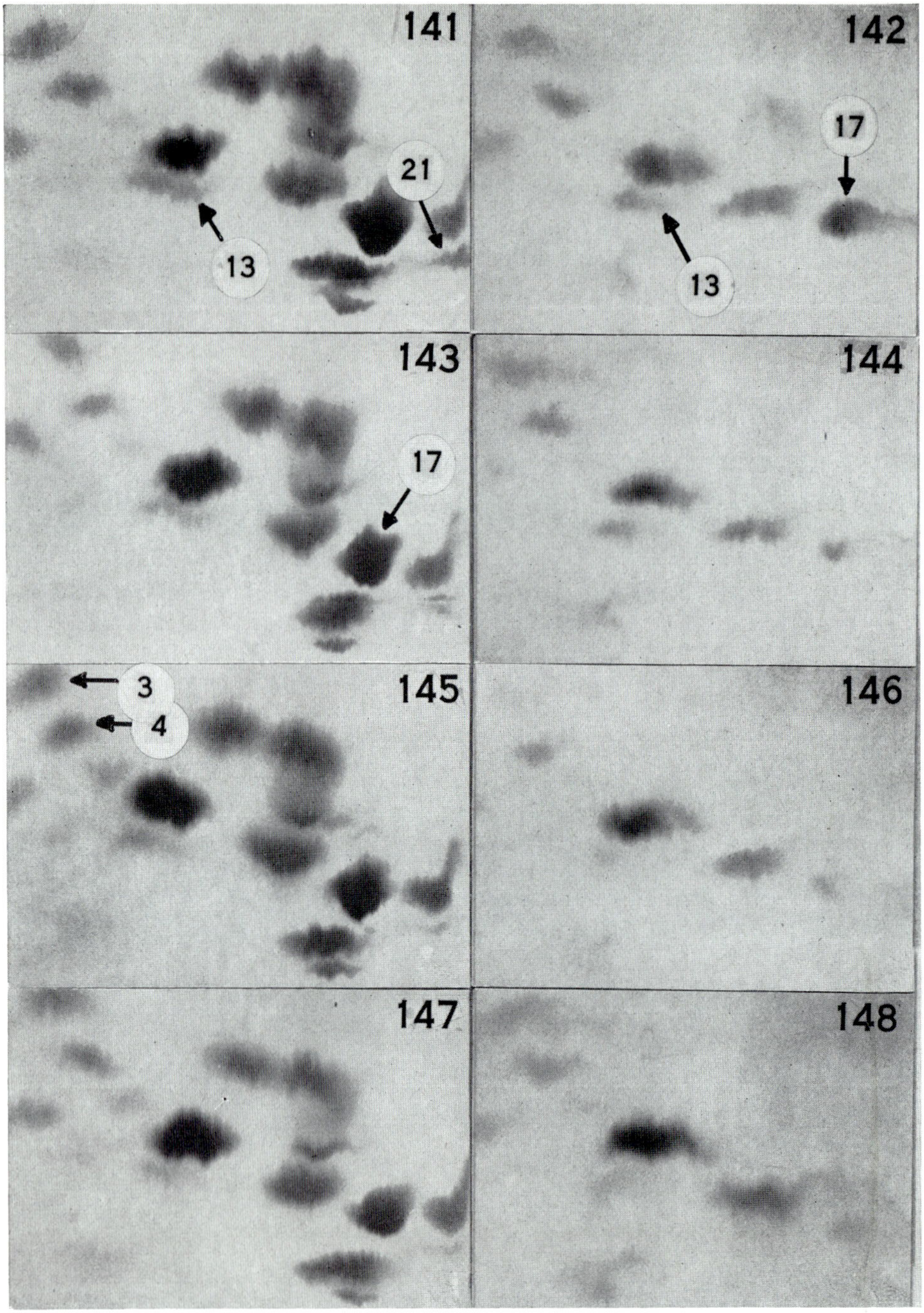

Figs. 141–148. Chromatograms of extracts of cells and fluid at various times after transplantation of the Yoshida sarcoma into J-strain rats. Chromatograms 141, 143, 145 and 147 are from 75 μl of cells and 142, 144, 146 and 148 are from 150 μl of ascitic fluid obtained at 4, 5, 6 and 7 days, respectively, after transplantation. Leucines, 3; valine, 4; glutamine, 13; glutamic acid, 17; glutathione, 21.

References p. 348/349

differs from the former by a single histocompatability gene and in which the tumor grows initially and then regresses[25]. The amino acid patterns of the solid form of the tumor were virtually identical in both sublines for the first 8 days after transplantation. Subsequently the patterns diverged. The tumors in the resistant strain showed appearance of free glutamine while this amino acid was not detected at any time on the chromatograms of extracts from tumors grown in the susceptible subline. Also, there were relative increases in the content of free glutamic acid in the tumor in the resistant subline (Figs. 121–132). Similar observations were made with an ascites form of tumor C1498 grown in the above two sublines of mice. The results showed that glutamine eventually appeared in the cells and ascitic fluid in the resistant strain but was not detectable in either the cells or the fluid of the susceptible mice at any time (Figs. 133–140).

A detailed study was then made of the cytological characteristics of the Yoshida ascites tumor cells and of the content of the ninhydrin-reactive constituents detected on two-dimensional paper chromatograms from extracts of cells and ascitic fluid at various times after transplantation into susceptible rats (J strain) (Figs. 141–148), in which the tumor grows progressively and kills the animal in approx. 14 days, and into a resistant rat strain (Wistar) (Figs. 149–156) in which the tumor grows initially but regresses completely within 8 days[24]. At all times after transplantation the content of the free or easily extractable glutamine was higher in the cells grown in the resistant rats. The chemical findings, generally, were similar to those observed in the experiments with the mice bearing the C1498 leukemia.

Free amino acids of tumors after treatment with cytotoxic agents

A detailed analysis was made of the sequential changes in the amino acid patterns of the extracts of the cells and fluid of the Yoshida sarcoma in rats after the administration of sarkomycin, nitromin, or crude podophyllin, and of the Ehrlich ascites tumor in mice after the injection of maleuric acid, sarkomycin and E-39 (2,5-di-*n*-propoxy-3,6-bis-ethyleniminobenzoquinone)[24]. These agents alone and in combination produced different types of cytological damage and an attempt was made to correlate the type and extent of histologically observable abnormality produced with the chromatographic patterns found.

Sarkomycin (active material, 2-methylene-3-oxocyclopentanecarboxylic acid) is a weak antibiotic which produces both nuclear and cytoplasmic damage in ascites tumor cells. The effects upon the Yoshida tumor grown in a Wistar rat are shown in Figs. 157–168. After removal of the control sample (Figs. 157–159), 1 ml of physiological saline containing 50 mg of sarkomycin was injected intraperitoneally into a rat on the 5th day after intraperitoneal inoculation of the Yoshida tumor. Samples of tumor were removed at intervals up to 300 min, at which time the intraperitoneal cellular reaction became marked, and no further cytological or chemical studies were made. The control sample showed a high mitotic index (Fig. 157) and the chromatograms of the cells (Fig. 158) and fluid (Fig. 159) showed typical patterns of free amino acids for this tumor grown in Wistar rats, small amounts of glutamine being noted in the cells but not in the fluid. 30 min after the injection, most of the tumor cells showed cytoplasmic blebbing and chromosomal abnormalities (Fig. 160) while the cells showed a greatly elevated level of free glutamine (Fig. 161), a small

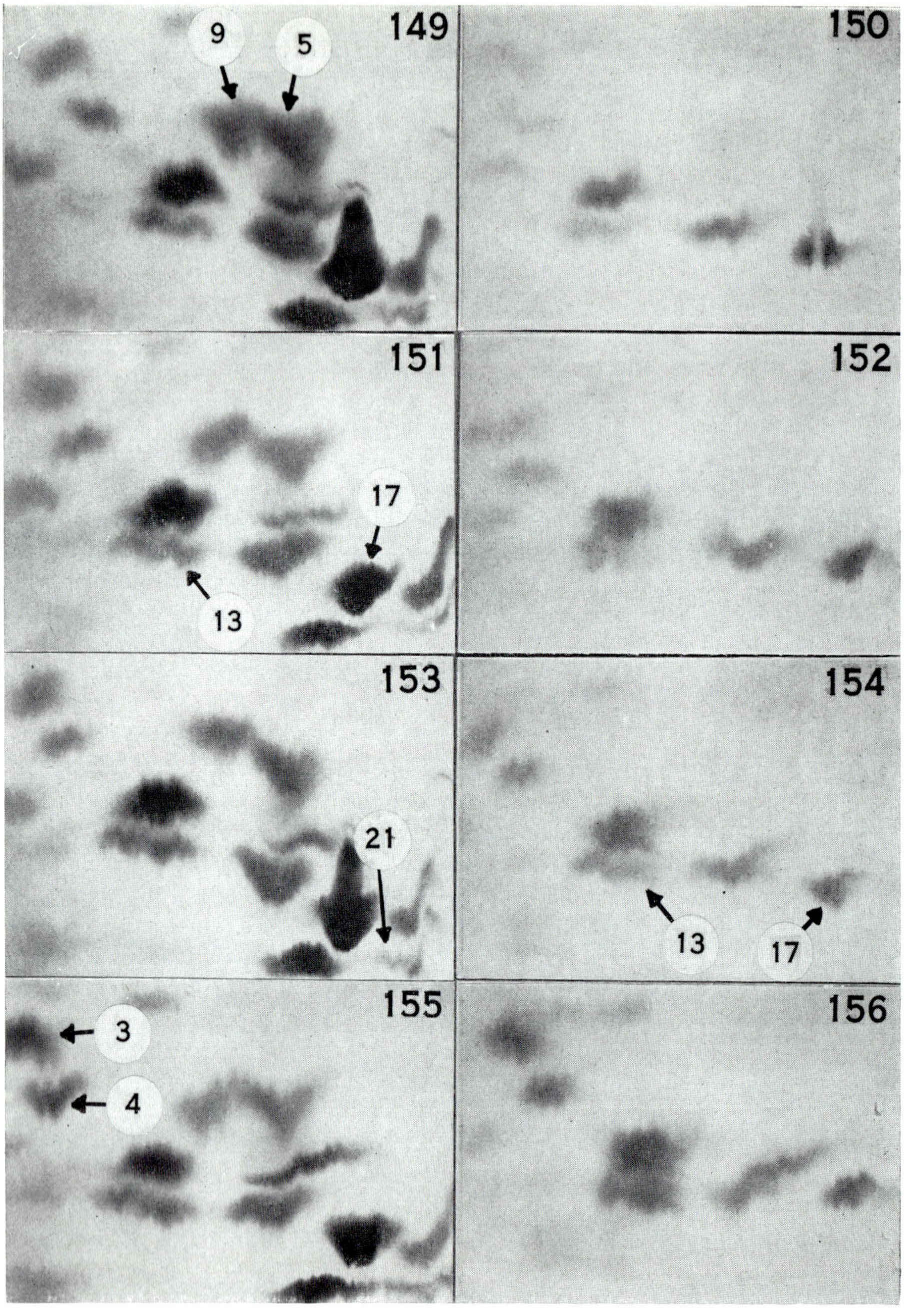

Figs. 149–156. Chromatograms of extracts of cells and fluid at various times after transplantation of the Yoshida sarcoma into Wistar rats. Figs. 149, 151, 153 and 155 are from 75 μl of cells and Figs. 150, 152, 154 and 156 are from 150 μl of fluid obtained at 4, 5, 6 and 7 days, respectively, after transplantation. Leucines, 3; valine, 4; taurine, 5; threonine, 9; glutamine, 13; glutamic acid, 17; glutathione, 21.

References p. 348/349

quantity of which also appeared in the ascitic fluid (Fig. 162). The cytoplasmic and nuclear damage increased with time (60-min cells, Fig. 163; 300-min cells, Fig. 166), no normal mitotic cells being observed at 300 min. Observations with phase microscopy showed that damage also occurred to the mitochondria after treatment with sarkomycin, the fine granules or thread-like shapes in the untreated tumor cells changing progressively after treatment into rounded forms. In spite of the extensive damage found in the cells the only remarkable changes in the chromatograms of the cells (Figs. 164 and 167) and of the fluid (Figs. 165, 168) were increases in the glutamine levels. With the exception of the glutamine content there was a truly remarkable similarity between the patterns found in the cells and fluid after treatment and those of the controls.

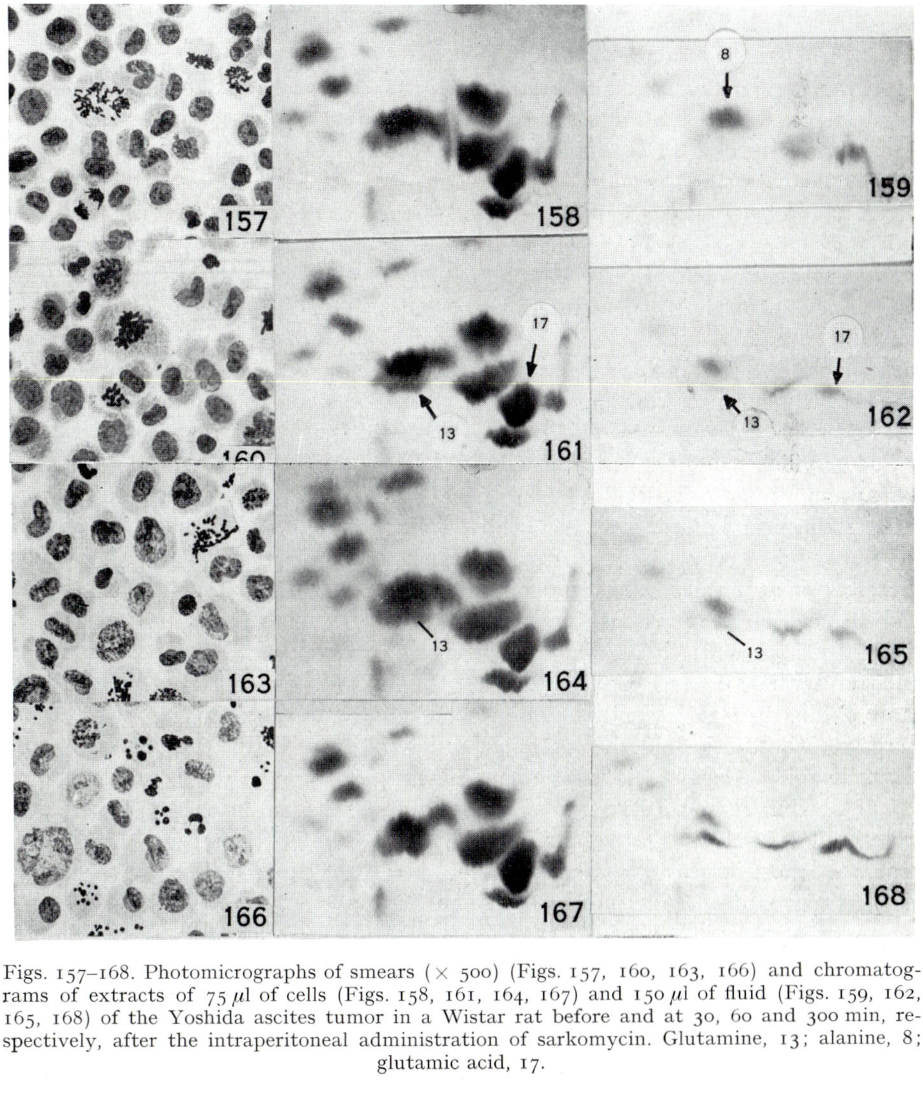

Figs. 157–168. Photomicrographs of smears (× 500) (Figs. 157, 160, 163, 166) and chromatograms of extracts of 75 μl of cells (Figs. 158, 161, 164, 167) and 150 μl of fluid (Figs. 159, 162, 165, 168) of the Yoshida ascites tumor in a Wistar rat before and at 30, 60 and 300 min, respectively, after the intraperitoneal administration of sarkomycin. Glutamine, 13; alanine, 8; glutamic acid, 17.

References p. 348/349

Results similar to those above (Figs. 169–180) were obtained when 3.5 mg of sarkomycin was administered to 35-g mice 48 h after inoculation with the Ehrlich ascites tumor. Even at 5 min after the injection both cytoplasmic and nuclear damage were evident (Fig. 172) and glutamine, which was not detected in the control cells or fluid, appeared (Figs. 173, 174). There was a marked decrease in the content of glutathione in the tumor cells. Valine and the leucines increased in amounts, and a new ninhydrin-reactive material, possibly an addition product of glutathione and sarkomycin, appeared to the left of glutamic acid (spot X) and persisted through the period of study. The chromatographic results obtained at the 10- and 20-min intervals (results for 10 min, Figs. 173, 174) were similar to those found at the 5-min period with the exception that taurine appeared in the fluid in considerably larger amounts. However, the cytological damage progressed rapidly (Fig. 175). The cells in the 40-min sample (Fig. 178) showed extensive cytoplasmic and nuclear damage. In many cells disintegration of nucleus and cytoplasm appeared to be occurring simultaneously. The ascitic fluid (Fig. 180) showed elevations in the levels of glutamine, taurine, and aspartic acid over the control. At this time, however, the badly damaged cells (Fig. 179) showed markedly elevated levels of glutamine, valine, leucine and isoleucine, tyrosine and lysine over the controls, while the other ninhydrin-reactive constituents were unchanged. It is truly remarkable that cells showing the degree of destruction observed in Fig. 178 should be able to retain high levels of small molecules such as amino acids. The findings are certainly not in keeping with these constituents being in free solution in the cytoplasm or nucleoplasm and being maintained intracellularly by intact membranes. Separate experiments with ^{14}C-labeled glutamine gave results in keeping with the interpretation that the effect of sarkomycin resulting in the increased content of easily extractable glutamine in the tumor cells is not necessarily related to an altered ability of the tumor cells to take up exogenous glutamine and to convert it to glutamic acid[24].

Podophyllin is another material which produces both cytoplasmic and nuclear damage in tumor cells. Podophyllin (10–15 mg/kg) was administered to J-strain rats bearing the Yoshida sarcoma. Extensive and progressive damage occurred to the tumor cells even within 1 h after the administration of the drug, many irregular processes or atypical ameboid protrusions appearing in the cytoplasm of the cells and all cells in metaphase showing chromosome clumping or other degenerative changes such as formation of deformed, rounded, or bizarre bodies. The chromatograms of both extracts from the cells and the fluid for the control sample and for those obtained at 30, 45 and 60 min after injection are shown in Figs. 181–188. In the case of the ascitic fluid, the post-injection samples showed the presence of taurine, glutamic acid, aspartic acid, and ethanolamine phosphate in small but detectable amounts, while these constituents were not noted on the chromatograms of the control sample of fluid, suggesting a slight leakage of some of the intracellular constituents. At most, traces of glutamine were detected in the extracts of fluid or cells at the times studied. The most remarkable thing about the above findings is the constancy of the pattern of the easily extractable amino acids of the tumor cells even at 60 min after treatment (compare Fig. 187 with the control in Fig. 181) at a time when marked cytological damage had occurred. A similar constancy in amino acid distribution in cells and fluid was found in a serial study of the Yoshida tumor after treatment with nitromin (methyl-bis-(β-chloroethyl)amine-N-oxide hydro-

References p. 348/349

chloride), an effective agent which acts largely at the level of the nucleus. It is apparent that the properties of these cells which are responsible for maintaining

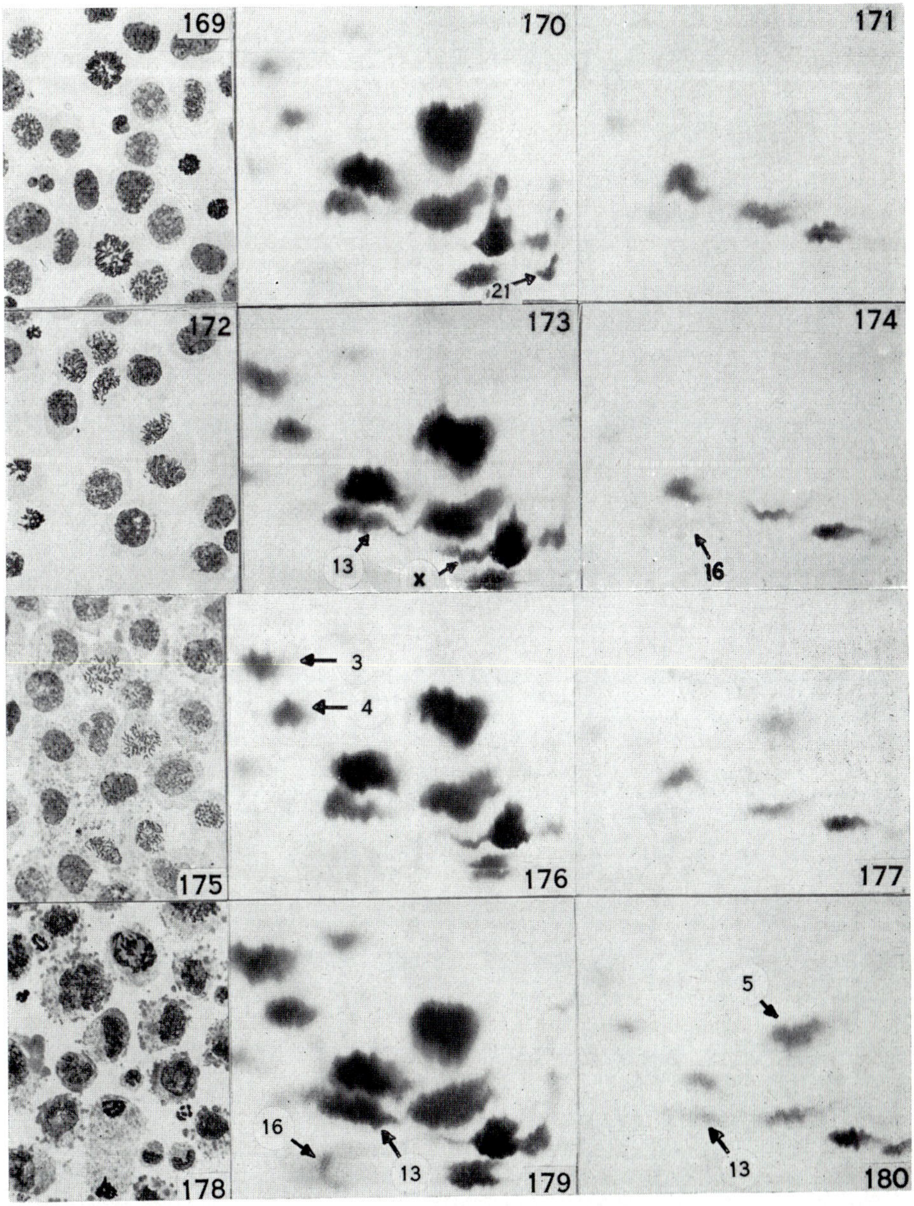

Figs. 169–180. Photomicrographs of smears (× 500) (Figs. 169, 172, 175, 178) and of chromatograms of extracts of 75 μl of cells (Figs. 170, 173, 176, 179) and 150 μl of fluid (Figs. 171, 174, 177, 180) of the Ehrlich ascites tumor grown in A/He mice before and at 5, 10 and 40 min, respectively, after the intraperitoneal administration of sarkomycin. Leucines, 3; valine, 4; taurine, 5; glutamine, 13; lysine, 16; glutathione, 21; X, possibly an addition product of sarkomycin and glutathione.

high intracellular concentrations of these constituents in relatively constant proportions can be resistant to the action of at least some severely cytotoxic agents.

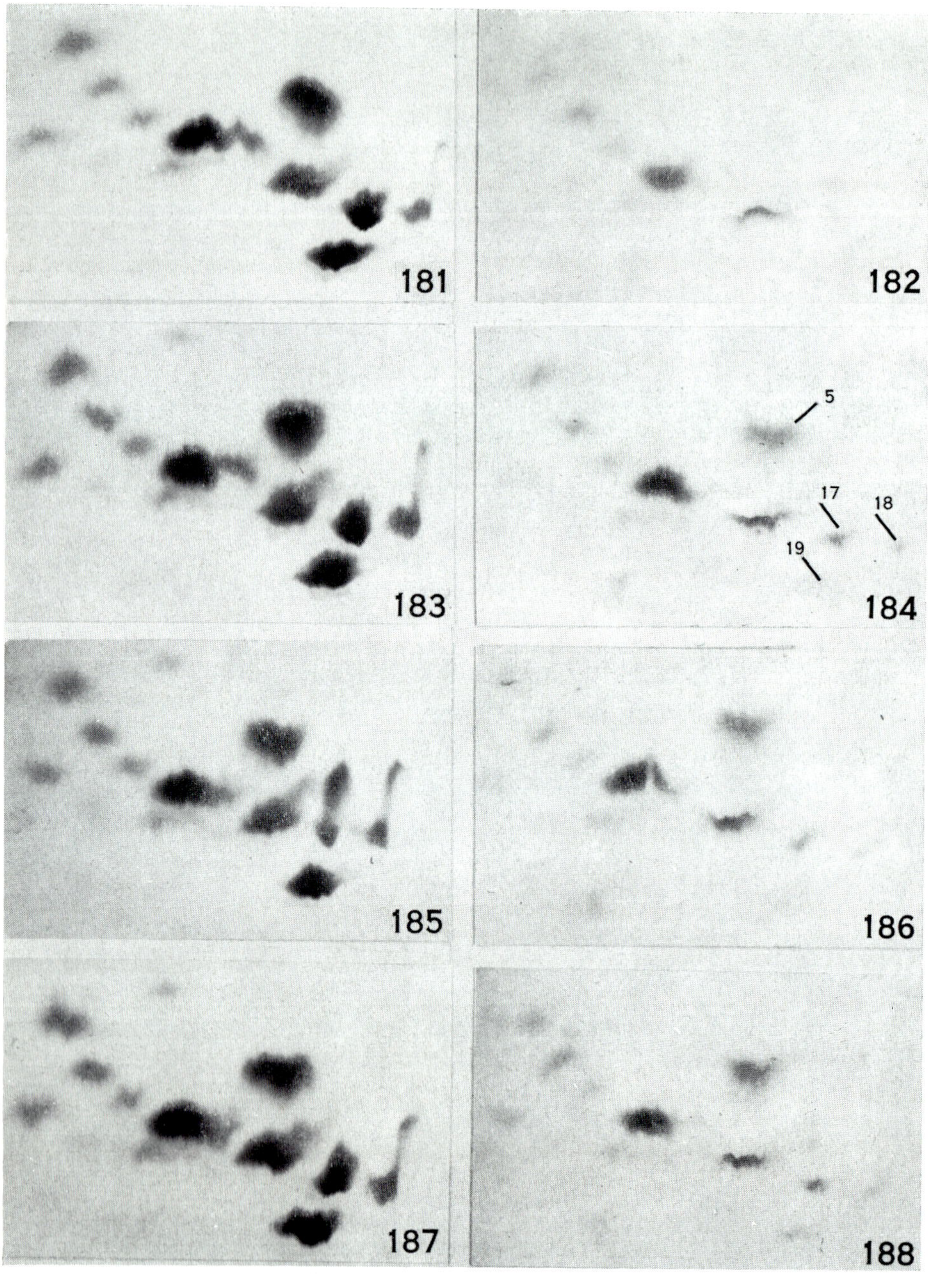

Figs. 181–188. Effect of podophyllin on free amino acid distribution in Yoshida sarcoma cells. Control cells (75 μl) and fluid (150 μl) (Figs. 181, 182) and at 30 (Figs. 183, 184), 45 (Figs. 185, 186), and 60 min (Figs. 187, 188) after injection of podophyllin. Taurine, 5; glutamic acid, 17; aspartic acid, 18; and ethanolamine phosphate, 19.

References p. 348/349

Changes in amino acids of tumor cells exposed to high extracellular levels of glutamine and glutamic and α, γ-diaminobutyric acids

It is of interest to compare the effects on amino acids of ascites tumor cells of high extracellular concentrations of a basic, acidic and neutral substance. Because of the virtual absence of free glutamine[1, 7, 8, 33] from tumor cells it was easy to follow the changes of this amino acid after administration to animals bearing ascites tumors[34]. When 50 mg of glutamine was injected intraperitoneally into rats bearing the Yoshida sarcoma, a serial study of extracts of cells and fluid (Figs. 189–202) showed there to be a rapid uptake of glutamine by the cells. At the 2-min period the fluid (Fig. 192) showed a large quantity of free glutamine. The content of glutamine in the cells and fluid decreased progressively after the 5-min period (Figs. 193, 194), the levels in

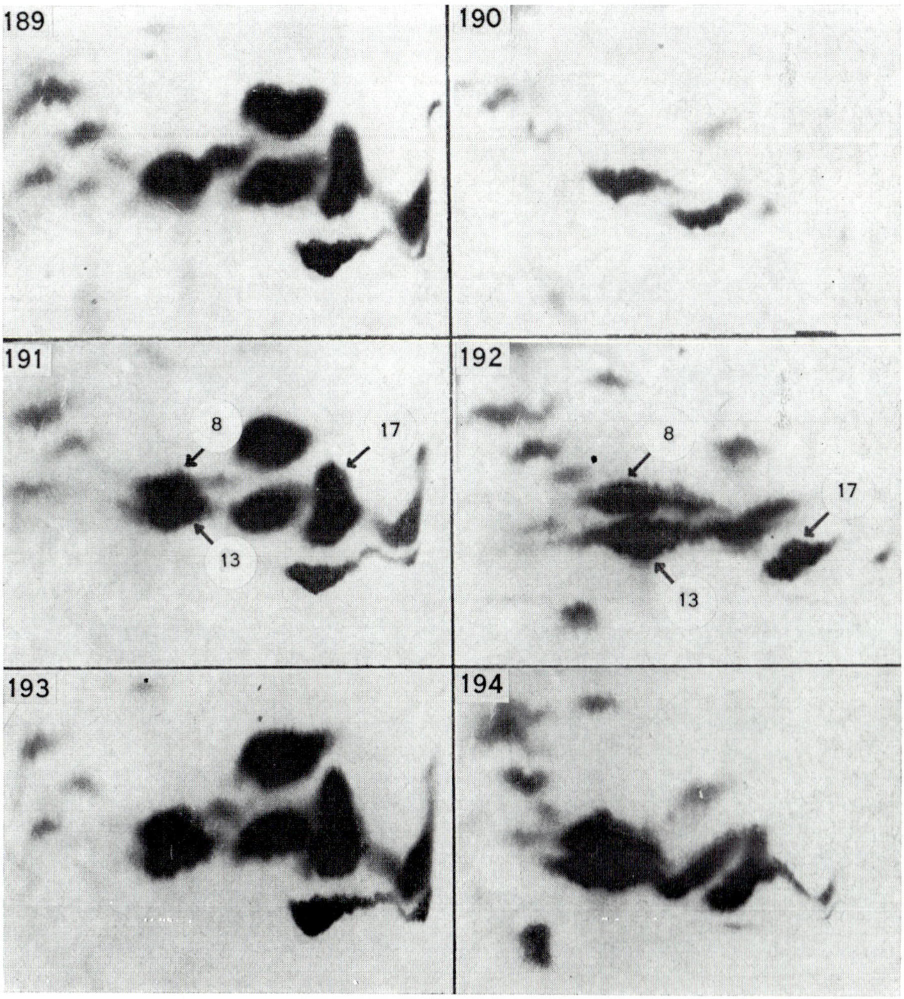

Figs. 189–194. For legend see p. 320.

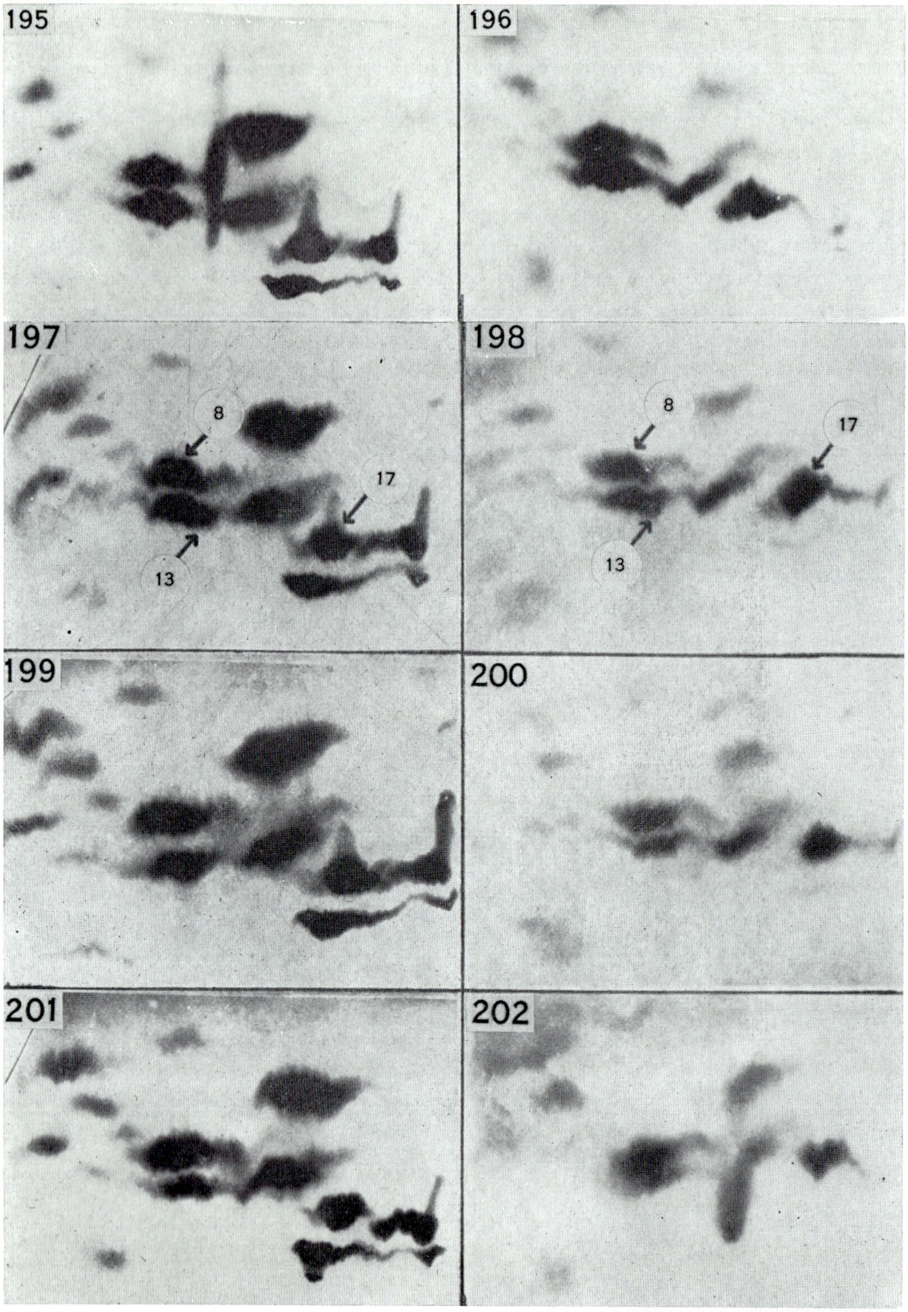

Figs. 189–202. Chromatograms of extracts of cells and fluid before and at various times after intraperitoneal injection of 50 mg of L-glutamine into a rat bearing the Yoshida sarcoma. Aliquots of extracts corresponding to 100 μl of packed cells or 200 μl of ascitic fluid were employed for chromatography. The odd-numbered figures between 189 and 202 are chromatograms of extracts of cells and the even-numbered figures are chromatograms of extracts from fluid obtained before and at 2, 5, 15, 25, 40 and 55 minutes, respectively, after injection of the glutamine. The vertical smear on Fig. 195 is an artifact. Glutamine, 13; alanine, 8; glutamic acid, 17.

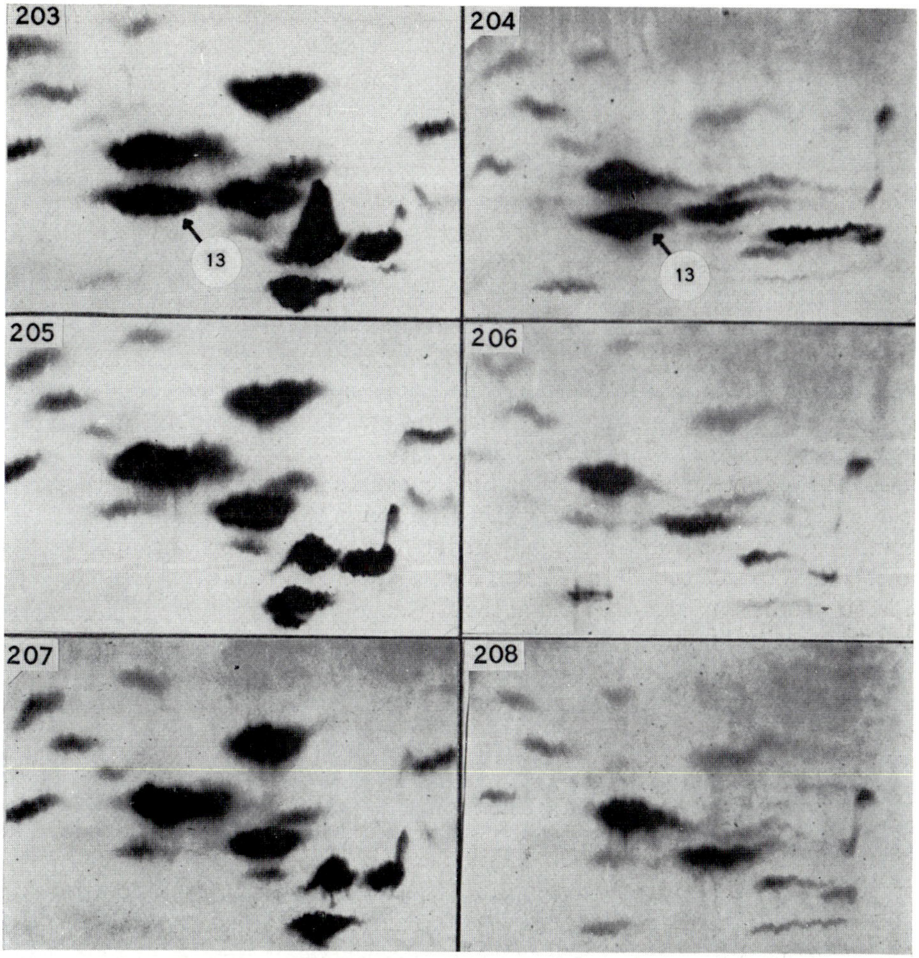

Figs. 203–208. Chromatograms of extracts of cells and fluid at $^1/_2$ h (Figs. 203, 204, respectively), $1^1/_2$ h (Figs. 205, 206, respectively), $2^1/_2$ h (Figs. 207, 208, respectively) after intraperitoneal injection of 50 mg of L-glutamine under same conditions as in Figs. 189–202. Glutamine, 13.

the fluid decreasing at a more rapid rate than in the cells. Much larger quantities of glutamine were found in the cells obtained at 40 min (Fig. 199) and 55 min (Fig. 201) after the injection of glutamine than in the fluid. Throughout this experiment the size and morphology of the tumor cells was perfectly normal. The results of an experiment in which complete disappearance of the injected glutamine from both cells and fluid had taken place within $1^1/_2$ h after the administration of glutamine are shown in Figs. 203–208. These results proved conclusively that glutamine enters the tumor cells rapidly and then disappears.

An interesting finding in this experiment was that after the injection of glutamine there was a general increase in most of the detectable ninhydrin-reactive constituents in the ascitic fluid (Figs. 192, 194, 196, 198), notable increases occurring in contents of alanine, glycine, glutamic acid, serine, threonine and lysine. The levels of most of these amino acids returned toward control levels as the glutamine content fell. The taurine level did not appear to be altered markedly during the experiment. The glutamic acid probably arose from the glutamine itself, since experiments with DL-[2-^{14}C]glutamine showed there to be a rapid cellular uptake of glutamine and a major conversion of the glutamine to glutamic acid[24] as well as rapid appearance of isotope in succinate, aspartate, glutathione, and pyrrolidone carboxylic acid. In the case of the other amino acids mentioned, the changes observed may possibly be attributed to the displacement of the amino acids from the cells and/or tissues into the fluid with a subsequent re-entry into the cells or removal from the fluid by the circulation when the content of glutamine fell. Rapid uptake and metabolism of glutamine has recently also been reported for Ehrlich ascites tumor cells[35]. A more detailed exposition of the possible significance of glutamine in tumor metabolism is not germane to the present discussion and has been given elsewhere [1]. In contrast to the findings with glutamine, chromatograms of serial samples obtained after the injection of 50 mg of L-glutamic acid into a rat bearing the Yoshida tumor showed no evidence of uptake of glutamate by the tumor cells (Figs. 209–216). The glutamic acid level of the ascitic fluid fell rapidly after the injection, virtually none of this amino acid being detected in the fluid at the 90-min period. Concentrations of the other amino acids detectable in the fluid were not altered perceptibly by the presence of large quantities of glutamic acid in the fluid, and no glutamine was detected. An experiment with DL-[2-^{14}C]glutamic acid was in agreement with the interpretation that glutamic acid does not enter the cells readily[34]. Interestingly, the cytological observations suggested that the injection of glutamic acid had a slight but definite enhancing effect on the growth of the Yoshida sarcoma cells. The results with glutamine and glutamic acid are in keeping with the finding of a ready permeability to glutamine and a *relative* impermeability to glutamic acid of animal tissues in general[36].

Our interest was focused on α, γ-diaminobutyric acid because of the intense accumulation of this substance by Ehrlich ascites tumor cells *in vitro* (see ref. 37 for discussion). Under the latter experimental conditions almost all of the potassium is replaced by the amino acid and there is an entry of chloride and water causing the cells to swell. It was of interest to determine whether the extensive intracellular changes taking place during uptake of this substance would be associated with changes in content of any of the detectable ninhydrin-reactive constituents. Mice of the C57 black strain bearing the Ehrlich ascites tumor were injected intraperitoneally with 20 mg of DL-α, γ-diaminobutyric acid on the 5th day after transplantation of the tumor. Samples were taken at 5, 10 and 20 min and 1, 2 and 24 h after the injection (Figs. 217–230). Uptake of the amino acid into the cells occurred readily, maximal levels being attained at 1 h (Fig. 225). At 24 h (Fig. 229), traces, at most, of the diaminobutyrate were detected in the cells and none in the fluid (Fig. 230). Relatively large amounts of glutamic acid appeared in the ascitic fluid at 5 min (Fig. 220) after the injection, the levels decreasing progressively thereafter. This amino acid was barely detectable in the control sample of ascitic fluid (Fig. 218). At 1 h (Fig. 225) an extraordinarily large amount of taurine appeared in the fluid and decreased in subsequent samples studied. Since neither the glutamic acid nor taurine levels of the tumor cells showed marked alteration during the course of the experiment, it is possible that release of these amino acids took place from tissues other than the tumor cells. With the exception of the presence of glutamine

References p. 348/349

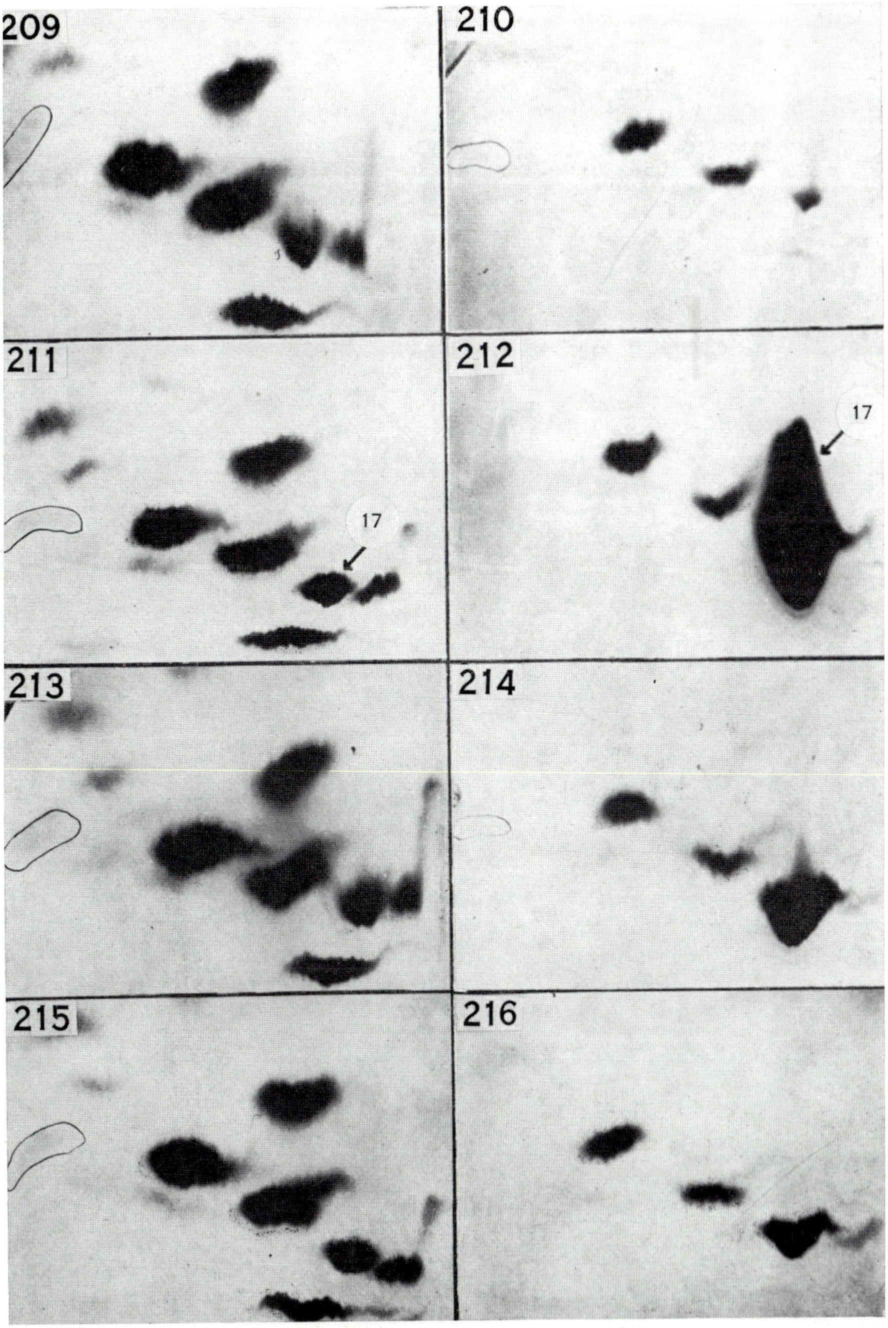

Figs. 209–216. Chromatograms of extracts of Yoshida tumor cells and fluid before (Figs. 209, 210, respectively) and 5, 10 and 30 min (Figs. 211–216) after the intraperitoneal injection of 50 mg of L-glutamic acid. Glutamic acid, 17.

References p. 348/349

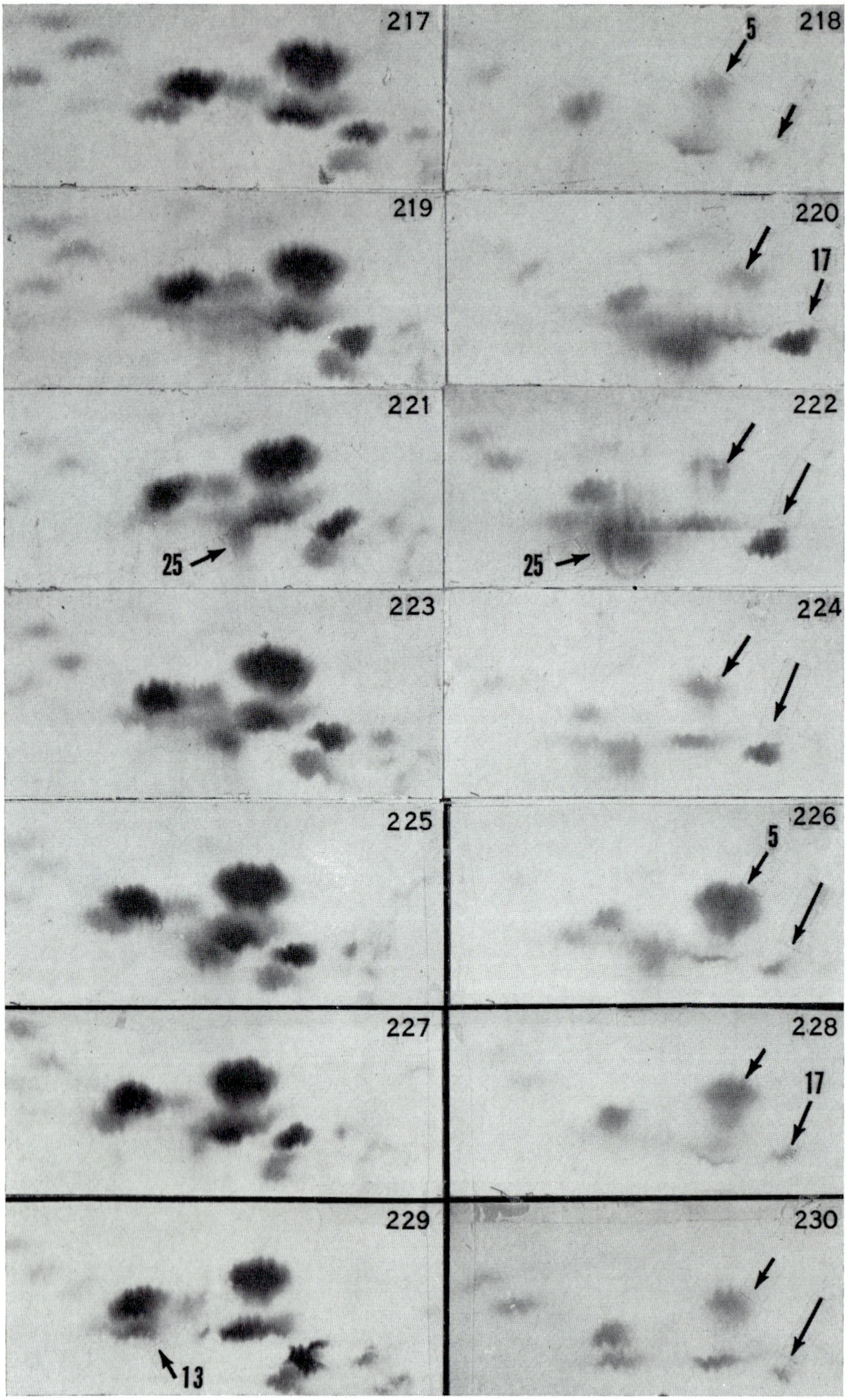

Figs. 217–230. For legend see p. 325.

in both the cells and fluid of the 24h sample, no changes were observed in the distribution of the amino acids normally found in the tumor cells. Cytological observations made during the first 4 h after the injection of the diaminobutyrate showed there to be a marked diminution in a number of mitotic cells and some chromosomal abnormalities characterized by the presence of clumping of chromosomes. The intense swelling accompanying uptake of the diaminobutyrate *in vitro*[37] was not evident in our *in vivo* study. This possibly may be attributable to the different degrees of protection against the swelling afforded by the protein-rich ascitic fluid and the buffer medium employed in the *in vitro* studies. It is probably allowable to assume that intracellular ionic changes took place in our experiments which were similar to those observed *in vitro*.

Thus, in addition to withstanding the effects of cytotoxic agents which affect all microscopically observable cellular structures (previous section), the mechanisms which control intracellular amino acid concentrations are also able to adapt to the uptake of large amounts of a neutral substance which is metabolized rapidly (glutamine) or of a basic substance which produces alterations in intracellular ionic and water balance (α,γ-diaminobutyrate).

Amino acid changes in tissues during induced growth or atrophy

Liver at various times after partial hepatectomy. The reader is referred to recent reviews on liver regeneration for details of the sequential changes in many variables that have been studied[38, 39]. The examination of the free amino acids of regenerating liver at various times after partial hepatectomy gave an opportunity to study these constituents in a tissue exhibiting a high rate of growth. It was found that for 20 h after partial hepatectomy the total amino acid content of picric acid filtrates of liver did not exceed those found in animals which had been subjected to laparotomy[40]. Between 20 and 30 h there was a general rise in amino acids, which persisted for 16 days. Also a fall in glutamine content and an increase in glutathione content were observed. In a more recent study determinations were made by photometric estimation of quantities of the amino acids on two-dimensional paper chromatograms of the amino acid patterns of plasma and of liver of control and partially hepatectomized rats at 6 and 24 h after surgery[41]. No significant changes in total free amino nitrogen or in individual free amino acids were found in whole blood or plasma. In both the 6 and 24h liver samples there were reported to be marked increases in content of aspartic and glutamic acids, lysine and ethanolamine phosphate and decreases in glutamic acid and taurine by comparison with unoperated controls. Glutathione also was higher in the 24h sample of liver in the hepatectomized animal than in the control. The concentrations of a number of other constituents were not affected by the procedure.

Measurements similar to those above were also performed in our laboratories[42]

Figs. 217–230. Amino acids of Ehrlich ascites tumor cells (odd numbers 217–229) and ascitic fluid (even numbers 218–230) before and at various time after injection of α,γ-diaminobutyric acid. Figs. 217, 218: control sample. Figs. 219, 220: 5 min. Figs. 221, 222: 10 min. Figs. 223, 224: 20 min. Figs. 225, 226: 1 h. Figs. 227, 228: 2 h. Figs. 229, 230: 24 h. Taurine, 5; glutamine, 13 glutamic acid, 17; α,γ-diaminobutyric acid, 25.

References p. 348/349

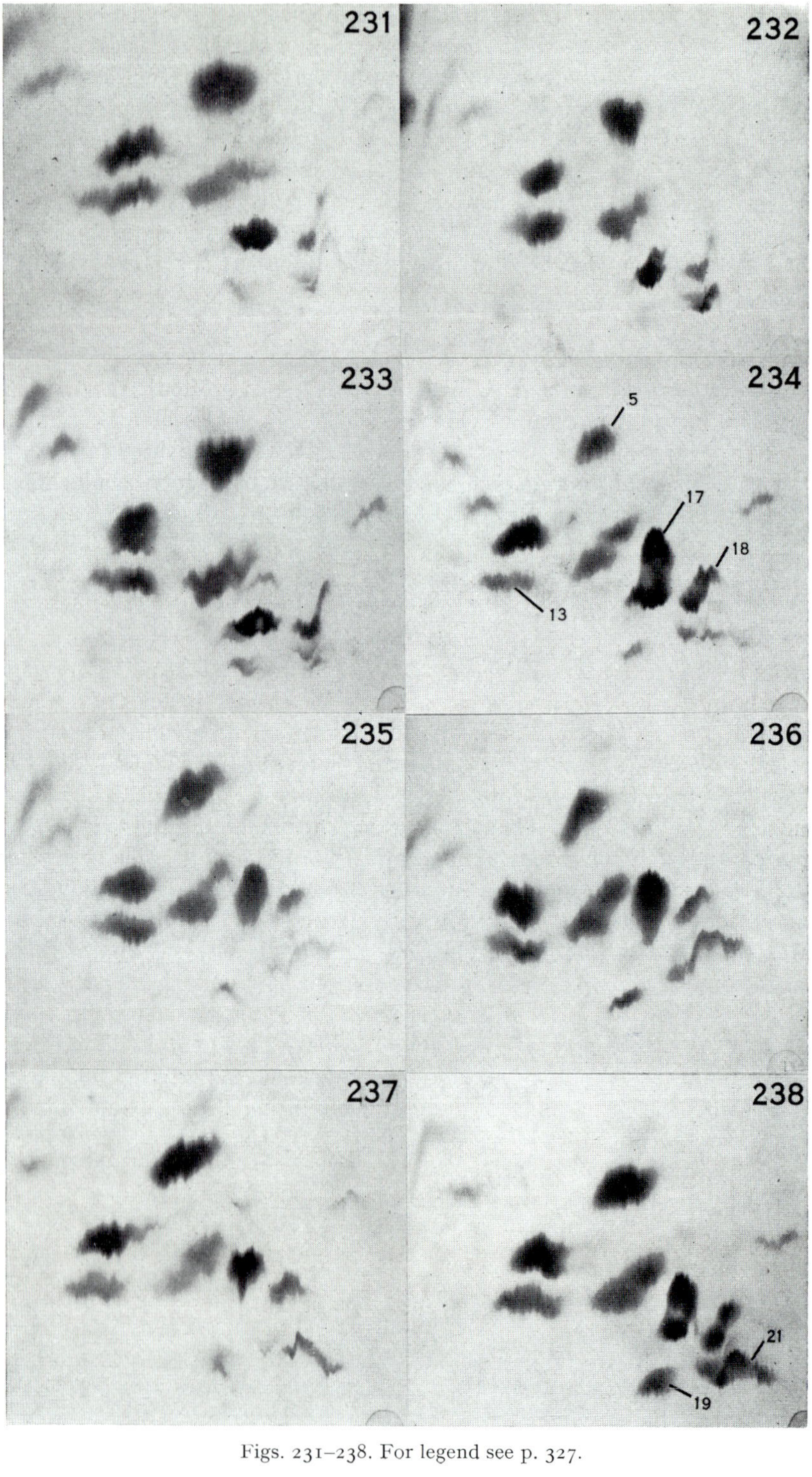

Figs. 231–238. For legend see p. 327.

on livers of rats at 12 h and 2 and 6 days after laparotomy (controls) and at 12 h, and 1, 2, 3, 6 and 9 days after removal of two-thirds of the liver mass. Chromatograms of extracts prepared from livers of unoperated rats are shown in Figs. 231 and 232 and comparisons of livers of laparotomized and hepatectomized animals at 12 h and 2 and 6 days are shown in Figs. 233–238. The surgical procedure involved in the laparotomy produced virtually no observable effect on the distribution of detectable ninhydrin-reactive constituents. However, even at 12 h after partial hepatectomy (Fig. 234) there was a marked increase in the content of glutamic acid, a small relative increase in aspartic acid, and notable decreases in taurine and glutamine. The extract from the liver sample obtained at this time was the only one which showed the presence of α-amino-n-butyric acid. At 1–6 days after partial hepatectomy there was an elevation in the content of ethanolamine phosphate and glutathione as well as of glutamic and aspartic acids, as reported previously[41]. In addition, serine was higher than in the controls during this period. At no time was an increase in lysine detected. The taurine and glutamine levels returned to normal by the 6th day. The extracts obtained at 9 days from the livers of the experimental animals were normal in all respects.

The relative increases in glutamic acid and decreases in glutamine contents observed during the most active phases of regeneration (up to 3 days) might be attributed to a greatly increased rate of utilization of the glutamine amide nitrogen for reactions in which the amide group is used specifically, such as purine[43], glucosamine[44], and diphosphopyridine nucleotide[45] biosyntheses. The increase in ethanolamine phosphate may reflect the decreased rate of synthesis of phospholipids observed after partial hepatectomy[46, 47]. The lowered taurine level may result from the decreased availability of cystine for synthesis of taurine, the function of which in the free form is still not known, because of the increased rate of utilization of cystine for the synthesis of protein and glutathione.

Considering the multitude of major physiological and biochemical alterations which take place in liver during regeneration after partial hepatectomy[38, 39], it is remarkable that relatively few changes were observed in the distribution of easily extractable ninhydrin-reactive constituents.

Effect of hormones on prostate and uterus. Orchiectomy produces atrophy of the prostate and ovariectomy acts similarly on the uterus. Concentrations of free amino acids were found by paper chromatography to decrease profoundly in the prostate after castraction and atrophy, and normal levels of these constituents were restored by administration of testosterone propionate[48, 49]. Treatment of normal male rats with testosterone, progesterone, or estradiol produced little effect on amino acid distribution of the prostate. A similar study was performed recently[50] in which column-chromatographic determination was made of ninhydrin-reactive constituents in tungstate filtrates of uteri of ovariectomized rats before and at 4 and 24 h after the administration of estradiol. With the exception of a decrease in taurine

Figs. 231–238. Amino acids in liver (75 mg) at various times after partial hepatectomy. Figs. 231, 232: unoperated controls. Fig. 233: laparotomized, 12 h. Fig. 235: 2 days. Fig. 237: 6 days. Fig. 234: partially hepatectomized, 12 h. Fig. 236: 2 days. Fig. 238: 6 days. Taurine, 5; glutamine, 13; glutamic acid, 17; aspartic acid 18; ethanolamine phosphate, 19; glutathione, 21.

References p. 348/349

content, there were marked increases in the levels of all of the detectable constituents at 24 h after estradiol injection. Unfortunately, data for uteri of normal rats were not included. However, since the highest level of amino acid content was observed at a time when the dry weight had attained the maximal level, it is unlikely that

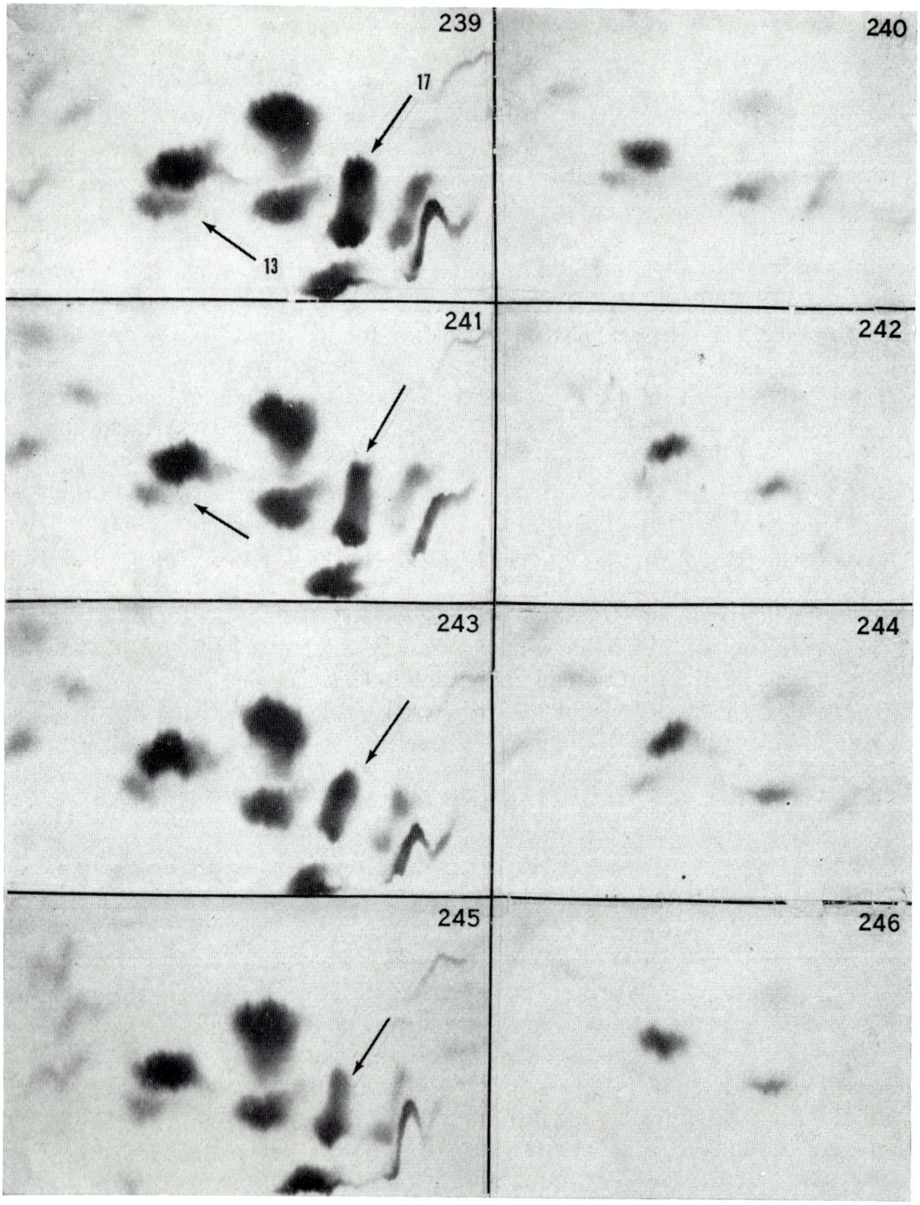

Figs. 239–246. Free amino acids of cells (75 μl) and fluid (150 μl) of the ascites form of the Murphy lymphosarcoma at different times after implantation into rats. Figs. 239, 241, 243, 245: chromatograms of extracts of cells. Figs. 240, 242, 244, 246: corresponding chromatograms of extracts of ascites fluid obtained at 2, 3, 4 and 8 days, respectively, after transplantation. Glutamine, 13; glutamic acid, 17.

there was a direct relationship between the tissue synthesis and the size or composition of the amino acid pool.

In both prostate and uterus the appropriate sex hormones are essential for the

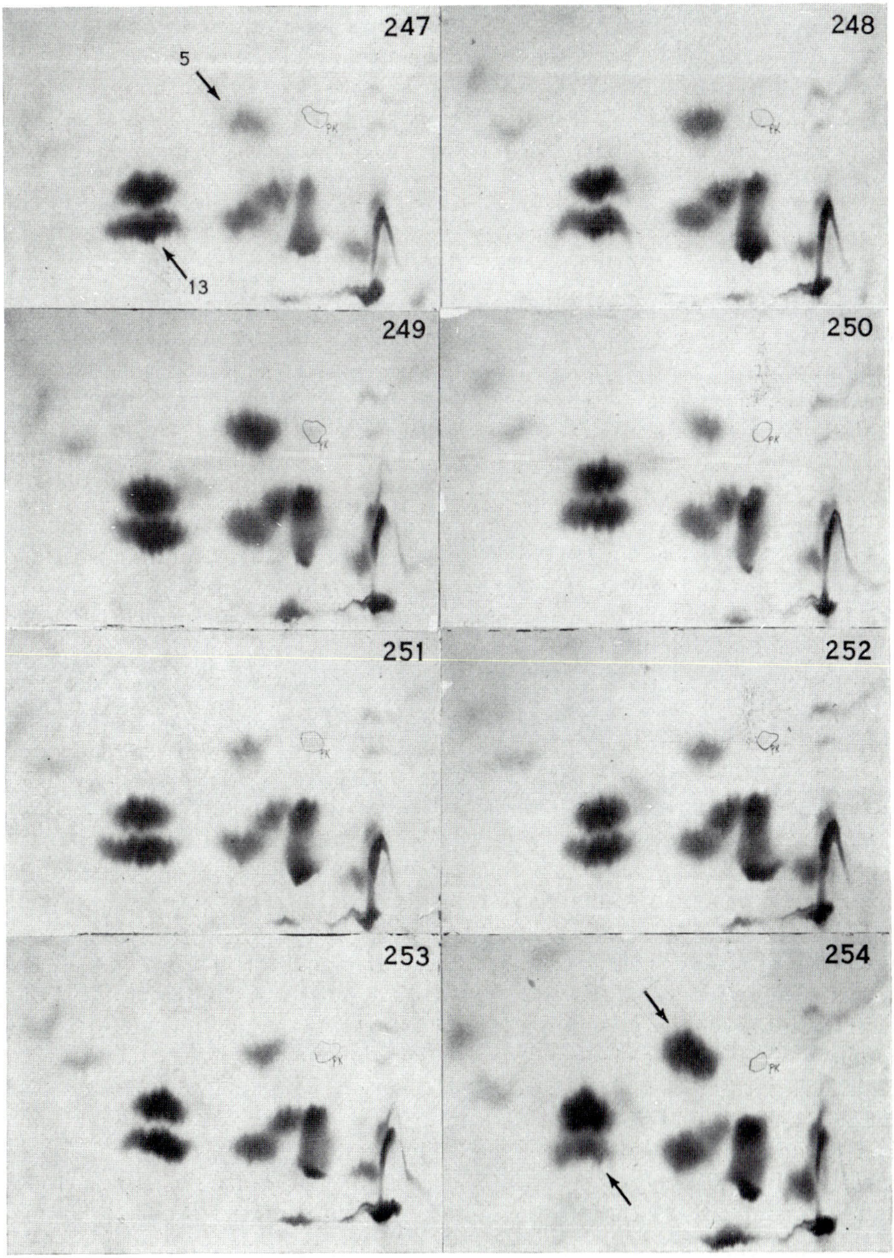

Figs. 247–254. Free amino acids of liver (75 mg) of rats bearing the Murphy lymphosarcoma (ascites) at different times after transplantation of the tumor. Fig. 247 was from a control animal and Figs. 248–254 were from animals at 6 h and 1, 2, 3, 4, 5 and 8 days, respectively, after transplantation. Taurine, 5; glutamine, 13.

integrity of many cell properties, one of which is the maintenance of characteristic concentrations of easily extractable ninhydrin-reactive constituents. The increases found in amino acid concentrations in the above experiments on the administration of hormones to castrated animals are probably not correlated directly with the process of regrowth of the atrophied tissue, but rather reflect the restoration of a metabolic balance normal for the tissues involved which is necessary for the maintenance of normal size and function of these tissues. There is at present no evidence that the *rate-limiting reactions or processes* through which the sex hormones exercise their control are directly related to any aspects of amino acid metabolism.

Free amino acids of tumor cells and host tissues during progressive growth of tumors

During progressive tumor growth there may be widespread alterations in the metabolism of the tumor-bearing host. Changes may occur in enzyme activities and even structure of the tissues, and in nutrition, hormonal balance, and the composition of the blood. Particularly great disturbances would be expected to take place in the metabolism of the host when the dietary and endogenous sources of nitrogen and energy become sufficiently limiting so that they are inadequate to supply both the needs of the tumor for growth and the host for maintenance and the available building blocks and energy are pre-empted by the neoplasm for its own growth at the expense of the normal tissues[51, 52]. It was anticipated that some changes might be observed in the patterns of easily extractable ninhydrin-reactive constituents in the tissues of the tumor-bearing animals during progressive growth of various tumors. Studies were made of extracts of liver, kidney, muscle, and brain as well as of tumor cells at various times after implantation of ascites tumors into mice and rats and during the growth of a solid tumor in rats[53].

Results obtained in the case of rats bearing the Murphy ascites lymphosarcoma are shown in Figs. 239–278. 12 male Sprague-Dawley rats, 45–50 days old, received intraperitoneal transplants consisting of 1.7×10^7 tumor cells. Tumor-bearing animals were sacrificed at 6 h and at 1, 2, 3, 4, 5 and 8 days after transplantation, at which time the experiment was terminated. Suitable controls were obtained at all times. Small amounts of glutamine were detectable in the tumor cells at 48 h (Fig. 239) after transplantation of the tumor, possibly because of the presence of a small number of leucocytes at this time, but this amino acid was not detectable at subsequent time intervals. During the progressive growth of the tumor there was a decrease in the content of glutamic and aspartic acids and glutathione with little or no change in the other observable constituents in the tumor cells (Figs. 241, 243, 245). The pattern of amino acids in the fluid was essentially unchanged during the 8 days of observation (Figs. 240, 242, 244, 246). The amino acid patterns of the liver during tumor growth were remarkably constant throughout the first 7 days of observation (Figs. 247–253), with the exception of the taurine level, which varied from one sample to another. In the liver of a terminal animal (Fig. 254) there was an increase in contents of ethanolamine phosphate and glutamic acid and a decrease in glutamine by comparison with the controls. The only change found in the kidney during the course of the experiment (Figs. 255–262) was a decrease in free glutamine level at 8 days (Fig. 262). The amino acid patterns in samples of leg muscle of the tumor-bearing rats were normal for 5 days (Figs. 263–269), but decreases in glutamine, glutamic acid, alanine, glycine and serine were observed by comparison with the controls at 8 days (Fig. 270). No changes were found in the patterns of amino acids of the brains of tumor-bearing animals at any time (Figs. 272–277) nor were any differences found between the brains of control (Fig. 271) or tumor-bearing animals.

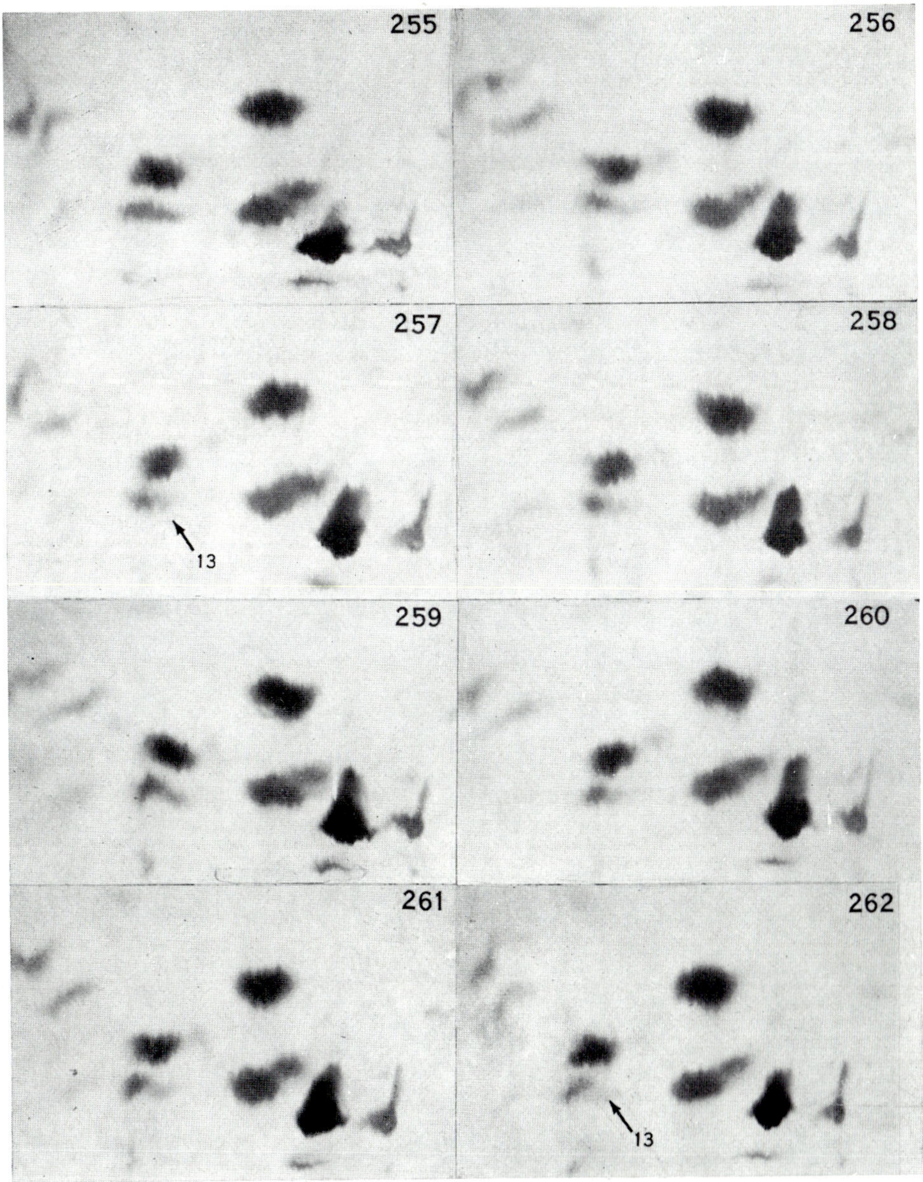

Figs. 255–262. Fiee amino acids of kidney (75 mg) of rats bearing the Murphy lymphosarcoma (ascites) at different times after transplantation of the tumor. Fig. 255 was from a control animal and Figs. 256–262 were from animals at 6 h and 1, 2, 3, 4, 5 and 8 days, respectively, after transplantation. Glutamine, 13.

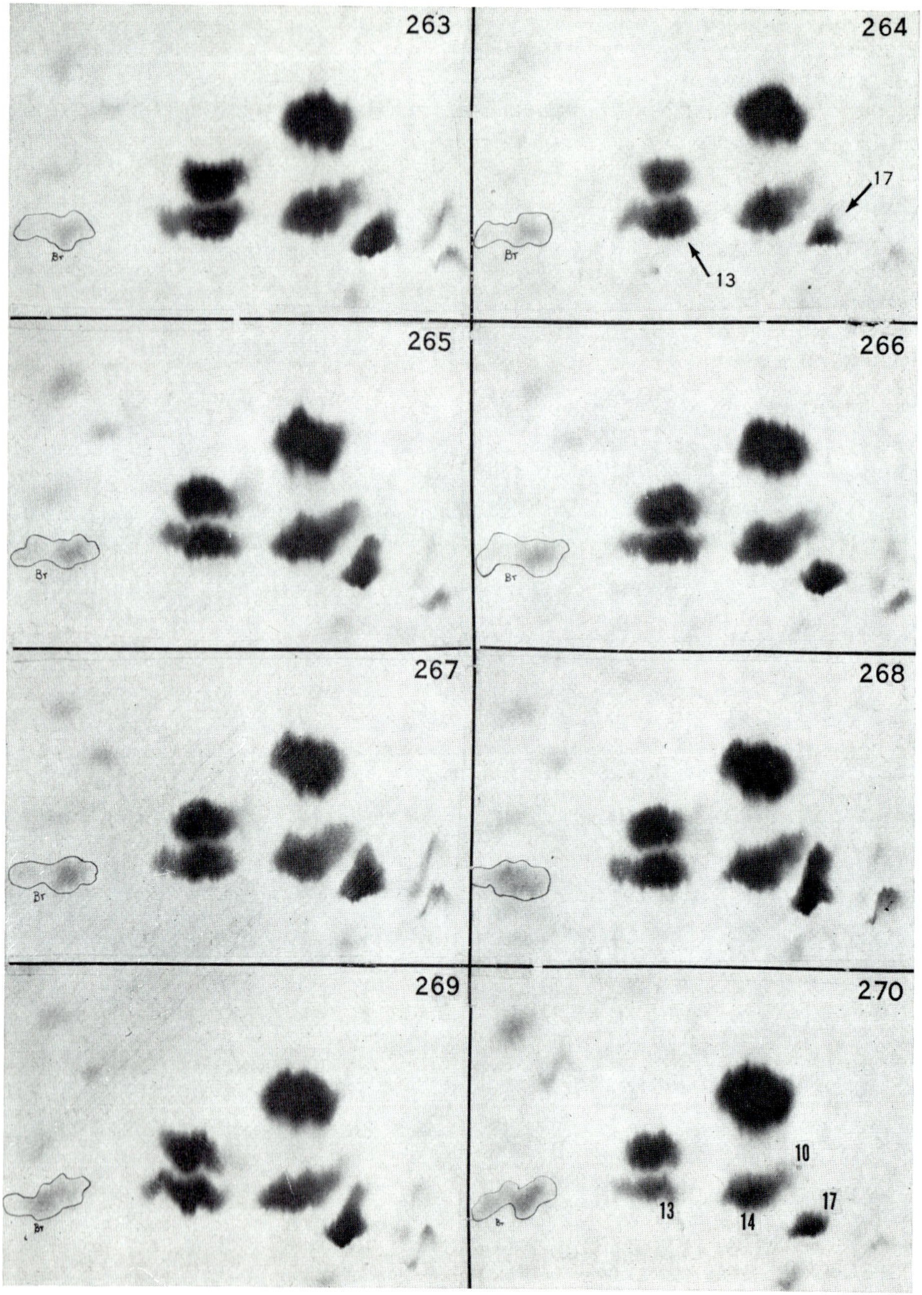

Figs. 263–270. Free amino acids of muscle (75 mg) of rats bearing the Murphy lymphosarcoma (ascites) at different times after transplantation of the tumor. Fig. 263 was from a control animal and Figs. 264–270 were from animals at 6 h and 1, 2, 3, 4, 5 and 8 days, respectively, after transplantation. Serine, 10; glutamine, 13; glycine, 14; glutamic acid, 17; Br, brown spot due to carnosine and anserine.

Studies similar to the one above were made in C57 black/dba female mice implanted with 5×10^6 Ehrlich ascites tumor cells. Samples of tumor, liver, muscle, kidney, and brain were studied from 48 h after transplantation until 11 days, the average life span of the tumor-bearing animals. The amino acids in tumor cells and ascitic fluid were remarkably constant as were the patterns found in the tissues of the tumor-bearing mice. The only changes observed were decreases in glutamine levels of both muscle and liver at 8 and 11 days after transplantation. Similarly, tissues and tumors were studied in Wistar rats (6–7 weeks of age, 135–142 g) 6 days after inoculation into the axillary region with 1 ml of 20% suspension of sterile Walker tumor. Tissues of littermates were used for control determination. No

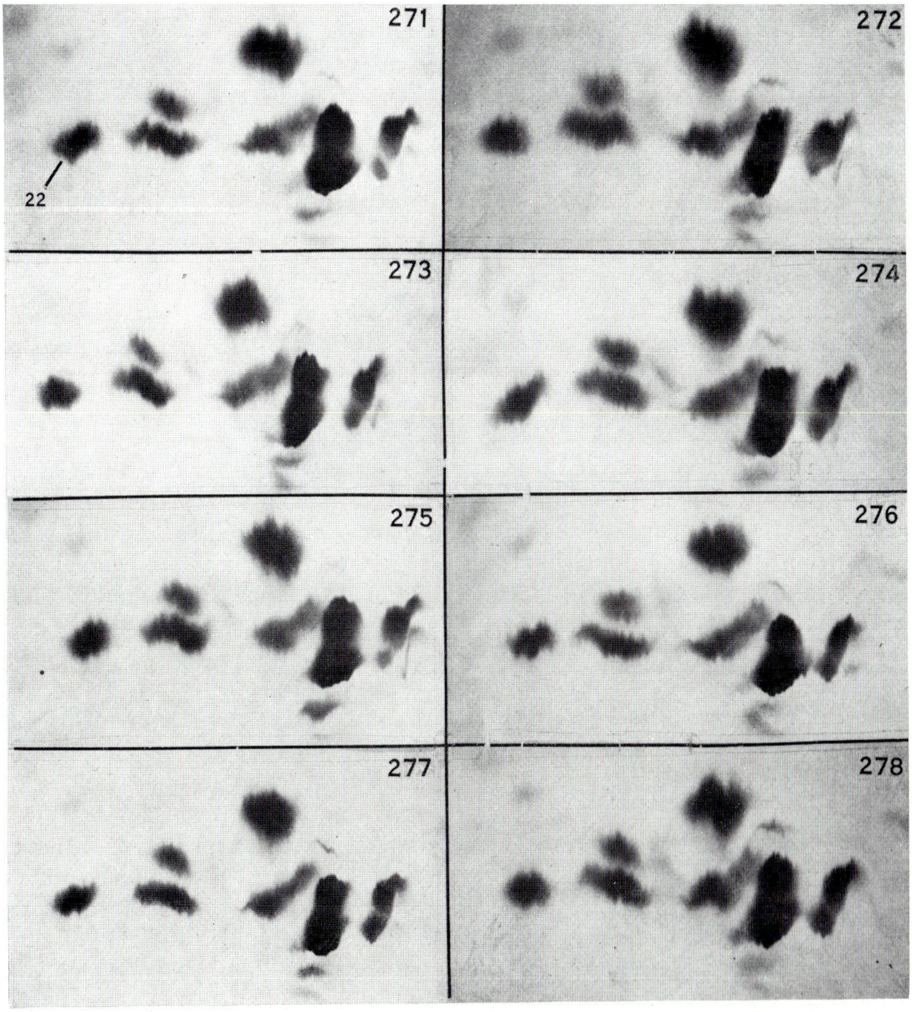

Figs. 271–278. Free animo acids of brain (75 mg) of rats bearing the Murphy lymphosarcoma (ascites) at different times after transplantation of the tumor. Fig. 271 was fiom a control animal and Figs. 272–278 were from animals at 6 h and 1, 2, 3, 4, 5 and 8 days, respectively, after tiansplantation. γ-Aminobutyric acid, 22.

References p. 348/349

differences were found between the tissues of the control and tumor-bearing animals.

The above observations show clearly that the tissues of rodents bearing transplantable tumors maintain a remarkably constant pattern of easily extractable ninhydrin-reactive constituents during progressive and eventually fatal growth of the neoplasms. Only in the terminal stages do there appear to be alterations, the most frequently observed change being a decrease in content of free glutamine. It is interesting that plasma-glutamine levels generally were lower and in some instances glutamic and aspartic acid and alanine concentrations were higher than normal in patients with acute myeloblastic and lymphoblastic and chronic lymphocytic leukemias and in the body fluids of patients with carcinoma of the breast, Hodgkin's disease, lymphosarcoma and reticulum cell sarcoma[54]. In some patients treatment with chemotherapeutic agents tended to cause an increase in plasma glutamine toward normal levels. Decreases in glutamine and increases in glutamic acid content in plasma were also observed in tumor-bearing rats[55].

Influence of dietary variations on free amino acids

Effects of starvation and dehydration. A paper-chromatographic study was made of the effects of starvation for 9 days on the content of free amino acids in liver, muscle and blood and on urinary excretion, comparing the results with those obtained in rats fed a 12% casein diet[56] (see also for pertinent previous references). The fasted animals lost approx. 35% of their initial weight and on a fresh weight basis there was a decrease in plasma of 7.6% in protein nitrogen and 21.0% in total α-amino nitrogen (taurine not included) by comparison with the controls, the corresponding decreases for liver being 5.9 and 45.4% and for muscle 13 and 18%, respectively. Thus, at the time the analyses were performed a considerable loss in tissue protein already had taken place. The most marked changes in *relative* distribution of ninhydrin-reactive constituents in plasma were increases in valine, leucine and isoleucine (determined together), and taurine contents and a decrease in serine. Similar changes in these constituents were found in the liver. In addition, the liver showed a relative increase in aspartic acid and a decrease in glutamine content. In the muscle there were reported to be relative decreases in glycine and serine and increases in valine and taurine (results for the leucines not being given). Even on the 1st day of the fast there was a remarkable decrease in the excretion of all the detectable urinary amino acids other than taurine, which increased approx. 1.5-, 7- and 10-fold over the control period on the 1st, 3rd and 5th days of the fast, respectively. The increased quantities of valine and the leucines, essential amino acids for the rat, in the tissues and plasma must have originated in the breakdown of tissue protein in the fasted animals. The finding that a general increase in amino acids of tissues and blood does not occur suggests that for most of the constituents of tissue proteins the rates of degradative metabolism balance the release of these constituents from bound form. This does not seem to be the case for valine and the leucines. The marked increase in tissue, plasma and urinary taurine indicates that this substance is a major end-product of cystine and methionine catabolism in the severely fasted animal.

It was of interest to us to determine whether the amino acid patterns of tissues could be altered during relatively short periods of starvation or dehydration during

References p. 348/349

which appreciable loss in weight occurred. In one study one group of young Sprague-Dawley rats (150–160 g) was deprived of food but given water, while another was given food *ad libitum* but no water. Dehydrated rats were sacrificed at 24, 48 and 68 h. Results are shown for the starved animals only for the 24- and 48-h periods because these animals died before 68 h. Between the 48- and 68-h period all experimental animals had lost up to 30% of their original body weight. The amino acid patterns in brain were not altered in any detectable fashion either by dehydration or starvation (Figs. 279–284). The kidneys of rats showed slight increases in contents of glutamic and aspartic acids and glycerylphosphorylethanolamine during dehydration and small increases in cystine, aspartic acid and glycerylphosphorylethanolamine and a decrease in glutamine during inanition (Figs. 285–290). In the liver (Figs. 291–296) dehydration resulted in increases in taurine and threonine levels at all times and in the appearance of an unidentified ninhydrin-reactive substance above taurine. In the sample obtained after 68 h of dehydration there

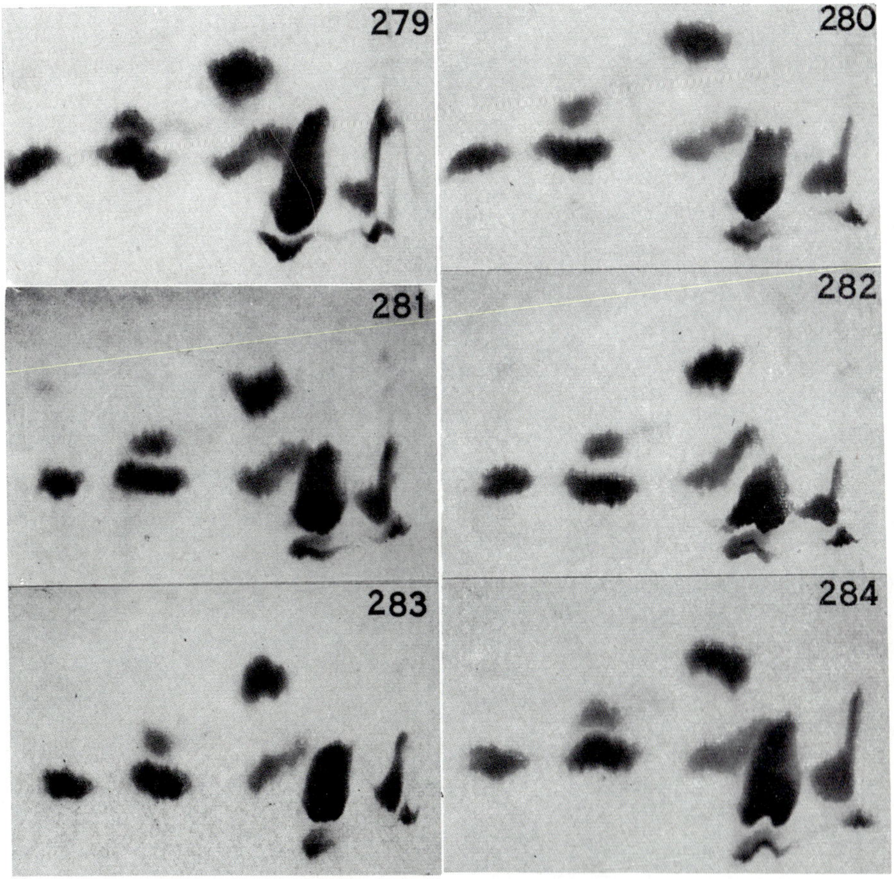

Figs. 279–284. Amino acids of brains (75-mg aliquots) of dehydrated and starved rats. Fig. 279: control. Figs. 280, 282, 284: dehydrated for 24, 48 and 68h, respectively. Fig. 281, 283: starved for 24 and 48h, respectively.

appeared to be an increase in glutamic acid content. Upon starvation there were decreases in glycine, serine and taurine contents and increases in glutamic and aspartic acids. Without more extensive data the decrease in taurine cannot be considered to be significant, since in our experience inexplicable variations often occur in taurine levels from the liver of one rat to another. The only changes observed in the muscle were slight decreases in glutamine, glycine and serine contents in both starved and dehydrated animals (Figs. 297–302). In another study with older rats (300–350 g) it was found that prolonged (6 days) starvation did not alter the pattern of free amino acids significantly either in heart (Figs. 303, 304) or in skeletal muscle (Figs. 305, 306). It has been reported that the taurine content of rat heart remains constant in starvation or when the animals are fed a protein-free diet[57].

When one considers the magnitude of the biochemical and physiological changes which occur in the organism as a whole and at the tissue level during starvation and

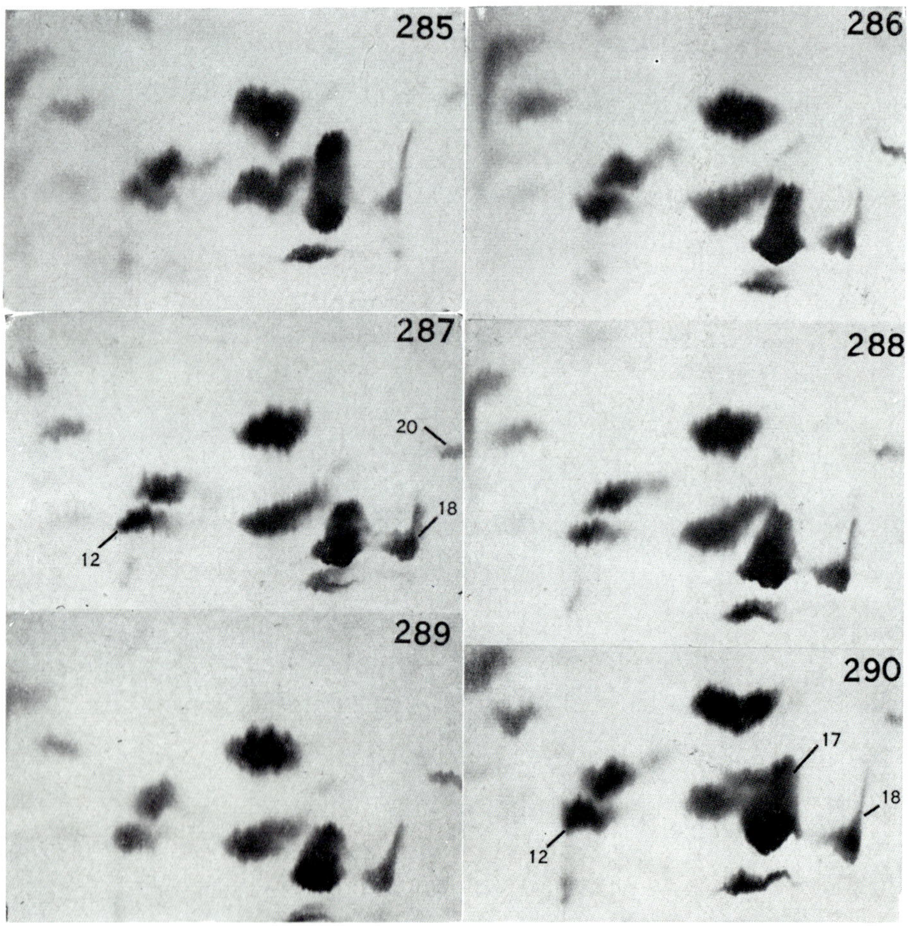

Figs. 285–290. Effect of dehydration and inanition on free amino acids of kidney (75 mg). Fig. 285: control. Figs. 286, 288, 290: dehydrated for 24, 48 and 68 h, respectively. Figs. 287, 289: starved for 24 and 48 h, respectively. Glycerylphosphorylethanolamine and/or β-alanine, 12; glutamic acid, 17; aspartic acid, 18; cystine (cysteic acid), 20.

dehydration it is remarkable that so few, relatively minor, changes are produced in the free amino acids. There must be extremely sensitive mechanisms for optimalizing the kinetics of the various processes by which the relatively constant steady-state concentrations of these constituents are maintained.

Effects of potassium deficiency. It was found that lysine could replace the loss of potassium in muscle of potassium-deficient rats to the extent of 8–40%, approximately two thirds of the deficit being made up by sodium[58]. A paper-chromatographic study was made of the amino acid distribution of extracts of skeletal muscle, diaphragm, kidney and liver from rats kept on a potassium-deficient ration for 35 days[59]. In the skeletal muscle and diaphragm marked increases were noted in the contents of lysine and arginine and losses were observed in aspartic and glutamic acids. A similar effect was observed in normal or potassium-deficient rats that had received desoxycorticosterone acetate. The amino acid patterns returned to normal

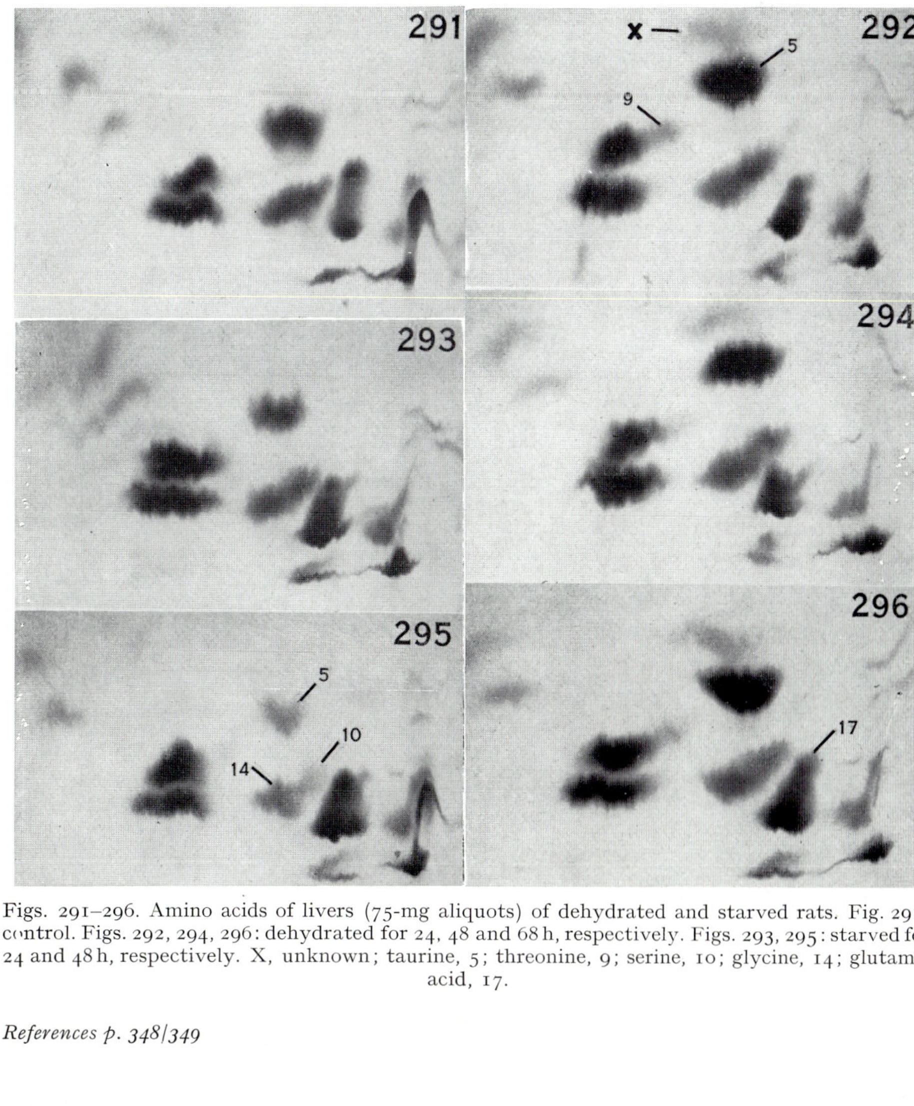

Figs. 291–296. Amino acids of livers (75-mg aliquots) of dehydrated and starved rats. Fig. 291: control. Figs. 292, 294, 296: dehydrated for 24, 48 and 68 h, respectively. Figs. 293, 295: starved for 24 and 48 h, respectively. X, unknown; taurine, 5; threonine, 9; serine, 10; glycine, 14; glutamic acid, 17.

References p. 348/349

within 24 h after restoration of potassium. In the kidney potassium deficiency resulted only in increases in lysine and arginine, and no changes were noted in the liver. However, no changes in any of the detectable constituents were found in extracts of skeletal muscle, left ventricle and kidney of severely potassium-deficient dogs[60].

The above studies show that although some compensatory changes may occur in the free amino acid of some tissues of some species when disturbances are induced in the distributions of the major electrolytes, these changes are not obligatory. The above results emphasize further the stability of the mechanisms which are conservative of the characteristic steady-state concentrations of the free or easily extractable amino acids.

Vitamin A deficiency. A recent study of amino acid patterns in tissues of vitamin

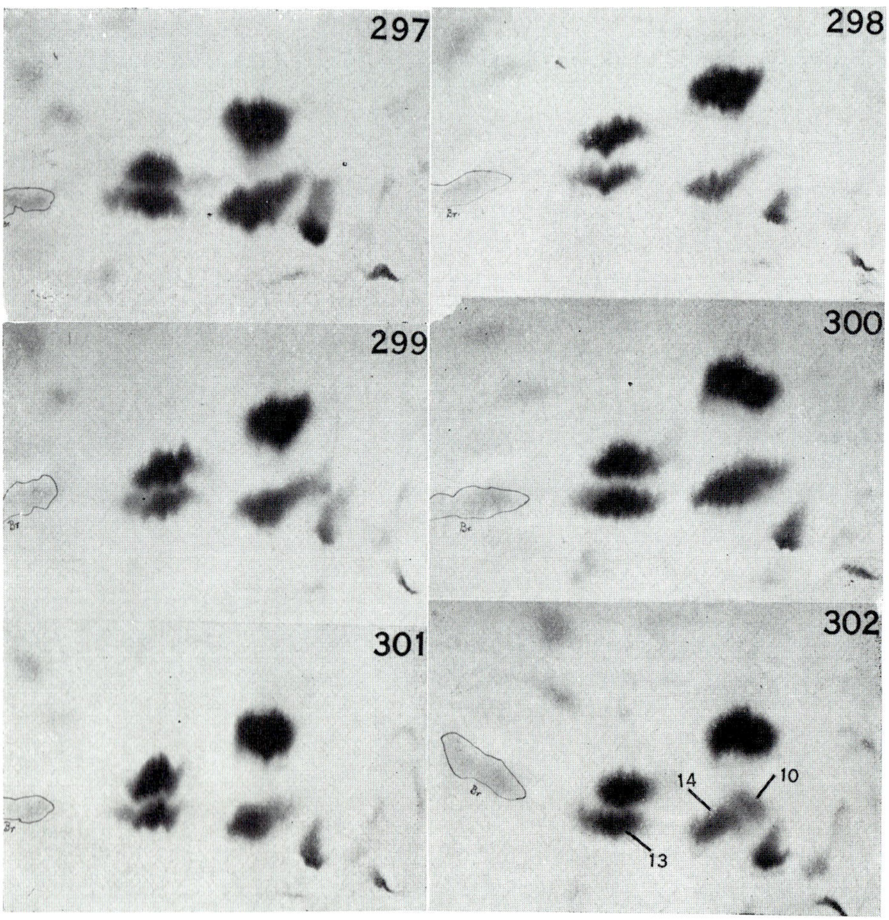

Figs. 297–302. Effect of dehydration and inanition on free amino acids of muscle (75 mg). Fig. 297: control. Figs. 298, 300, 302: dehydrated for 24, 48 and 68 h, respectively. Figs. 299, 301: starved for 24 and 48 h, respectively. Serine, 10; glutamine, 13; glycine, 14; Br, brown spot of carnosine and anserine.

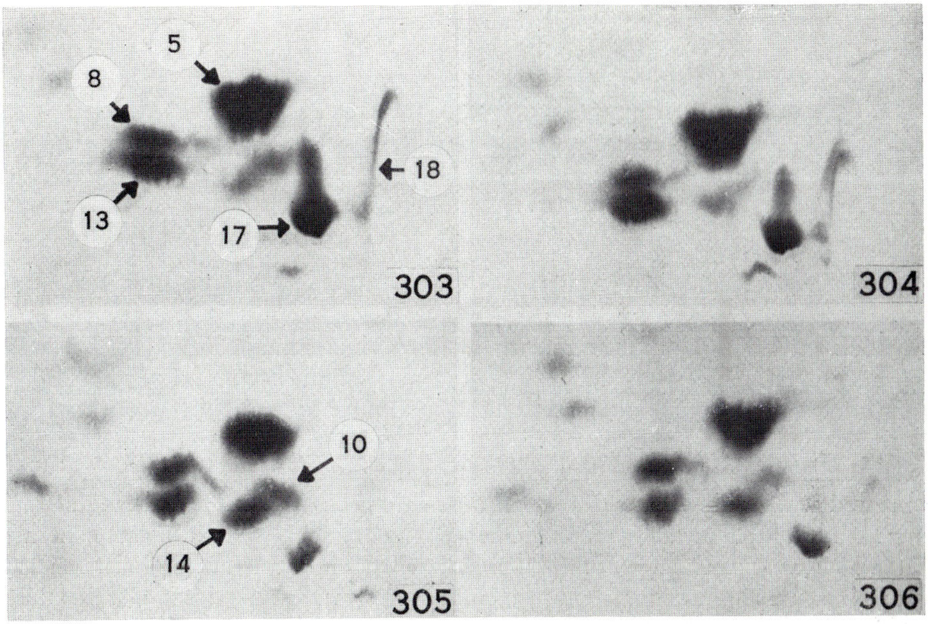

Figs. 303–306. Chromatograms of extracts of rat heart and muscle (75 mg). Fig. 303: normal rat heart. Fig. 304: rat heart after 6 days inanition. Fig. 305: normal rat leg muscle. Fig. 306: rat leg muscle after 6 days inanition. Taurine, 5; alanine, 8; serine, 10; glutamine, 13; glycine, 14; glutamic acid, 17; aspartic acid, 18.

A-deficient rats[61] was chosen as an example of the effect of a vitamin deficiency on these constituents. Quantitative colorimetric determinations of 15 of the constituents were made by paper-chromatographic procedures (Table III) in extracts of

TABLE III

AMINO ACID PATTERNS IN THE TISSUES OF VITAMIN A-DEFICIENT AND PAIR-FED RATS[61]

Amino acids	*Average values for seven groups of rats*					
	Blood (µg/ml)		*Liver (µg/g)*		*Kidneys (µg/g)*	
	*A**	*B*	*A*	*B*	*A*	*B*
Aspartic acid	52	51.6	353	355	995	989
Glutamic acid	29	28.2	353	358	249	240
Serine	62.3	60.0	504	506	1094	1108
Glycine	30.4	31.7	498	498	751	749
Cystine	147.7	147.7	195	191	186	182
Lysine	77.4	79.0	574	579	723	723
Histidine	52.1	53.1	244	243	567	563
Arginine	113	110.7	440	439	1770	1765
Alanine	49	50.2	1175	1168	1756	1753
Methionine	17.7	18.5	104	106	96	93
Valine	28.9	28.8	396.4	395	253	256
Leucines	16.4	16.2	534.0	536	942	939
Threonine	40.0	64.0	1469	2089	2453	3941
Tyrosine	104.5	60.8	1032	808	909	699
Phenylalanine	88.6	49.8	1469	965	1894	1579

* A, deficient rats; B, pair-fed normals.

blood, liver and kidneys of severely vitamin A-deficient rats and pair-fed controls. There was no effect of the deficiency on the content of 12 of the constituents. However, in blood and the two tissues examined there were significant increases in contents of tyrosine and phenylalanine and decreases in threonine in the deficient animals. Although the significance of the specific changes is not clear, these results serve to show that a deficiency of a dietary essential sufficiently severe to cause anorexia and loss in weight may result in relatively small changes in distribution of free amino acids.

Changes in tissue amino acids in ammonia intoxication

Ammonia intoxication produces remarkably few changes in amino acids of poisoned animals. In an acute study $LD_{99.9}$ doses of ammonium acetate (10.8 mmoles/kg) were injected into young rats and at death (approx. 15 min), quantitative paper-chromatographic determinations were made of alcohol extractable amino acids in blood, kidney, muscle, liver, heart, spleen, pancreas, liver and testes[62] and were compared with those from suitable controls. In the liver significant changes were found only in aspartic acid (1200% increase) and alanine (500% increase) contents of the 18 constituents detected. The only change noted in the other tissues studied was an increase in glutamine in muscle, brain and testes. Smaller doses of ammonium chloride (3.7 mmoles/kg) doubled levels of glutamine in the brains of injected rats[63]. In dogs receiving bilateral carotid infusion of lactate–Ringer's solution containing 1% ammonium hydroxide the only significant change found in cerebral free amino acids was an increasing content of glutamine with the time of infusion[64]. Dogs in which coma was produced by rapid infusion of the ammonium salt showed virtually identical patterns with those of controls. The above results, together with the demonstration of an extremely rapid conversion of intracerebrally administered L-[U-^{14}C]-glutamic acid to glutamine in mice[65], are compatible with the interpretation that in brain there is an extremely rapid incorporation of ammonia into the amide group of glutamine *in vivo* by the known synthetic pathway involving glutamic acid, ammonia and adenosine triphosphate, by an enzyme known to be at a high level in brain. In no case was there found to be a decrease in the content of glutamic acid. Although it appears likely that the accumulation of glutamine occurs because the combined rates of degradation and exit of glutamine from brain are exceeded by the rate of formation when the blood level of ammonia is high, the possibility must be considered that inhibition of some normal pathway of glutamine utilization or breakdown may also contribute to the elevated levels of cerebral glutamine.

Influence of hormonal changes on free amino acids

An extensive series of studies was performed in which comparisons were made of amino acids of tissues of normal rats with littermate controls in which there was experimental impairment of function or removal of one of the major endocrine glands. The results to be reported further confirm the remarkable stability of the free amino patterns in tissues of an organism in which great physiological changes have been produced and a variety of adaptations are taking place (see refs. 66 and 67 for pertinent references).

References p. 348/349

Diabetes and hypoglycemia. Six Long-Evans rats approx. 200 g in weight were fasted for 48 h and then injected intravenously with 40 mg/kg of alloxan. All the treated animals exhibited a glycosuria and elevated blood-sugar levels. The diabetic animals and the littermate controls were sacrificed 1 month after the injection and the easily extractable ninhydrin-reactive constituents were determined. The patterns found in the diabetic animals were virtually identical with those found in the controls, the differences between the patterns in normal and diabetic littermates being no greater than those observed from one normal animal to another. Typical results

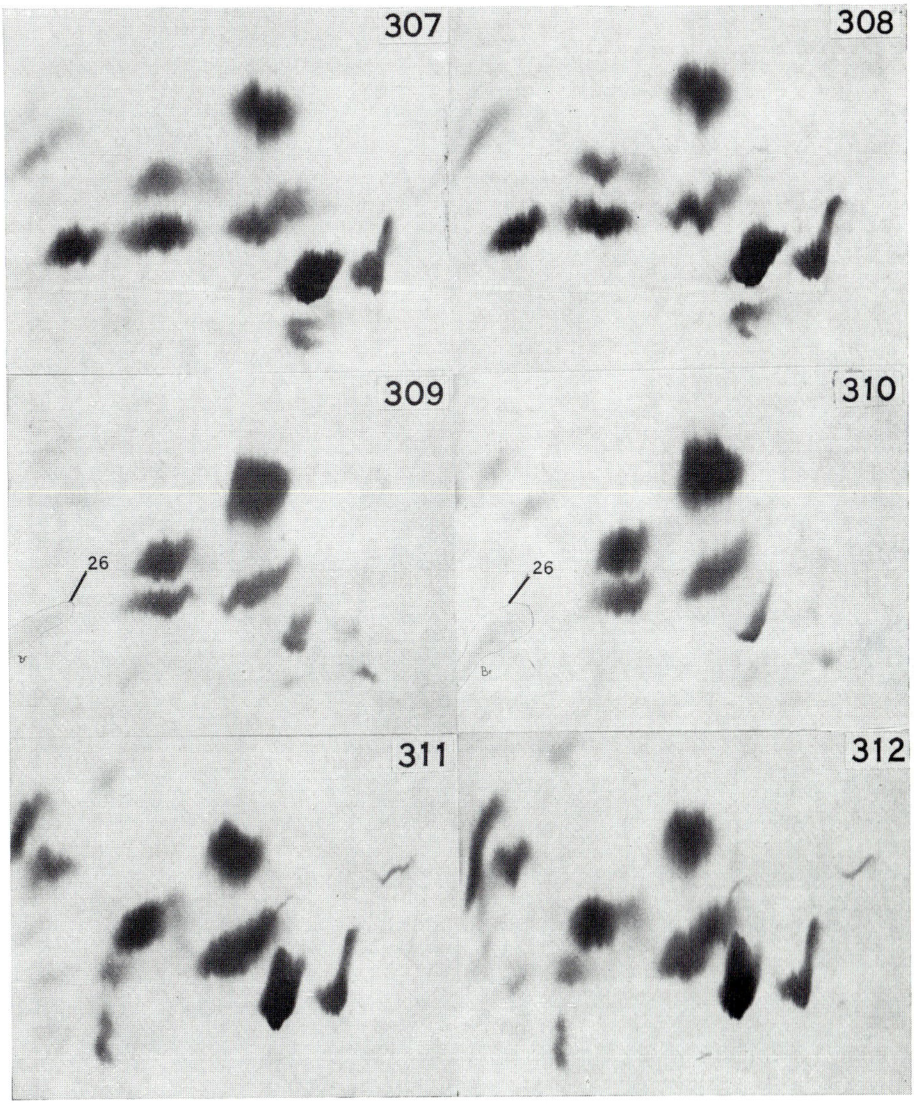

Figs. 307–312. Amino acids of extracts from 75 mg samples of brain (Fig. 307), muscle (Fig. 309) and kidney (Fig. 311) of normal rats and brain (Fig. 308), muscle (Fig. 310) and kidney (Fig. 312) of diabetic littermates. Carnosine and anserine, 26.

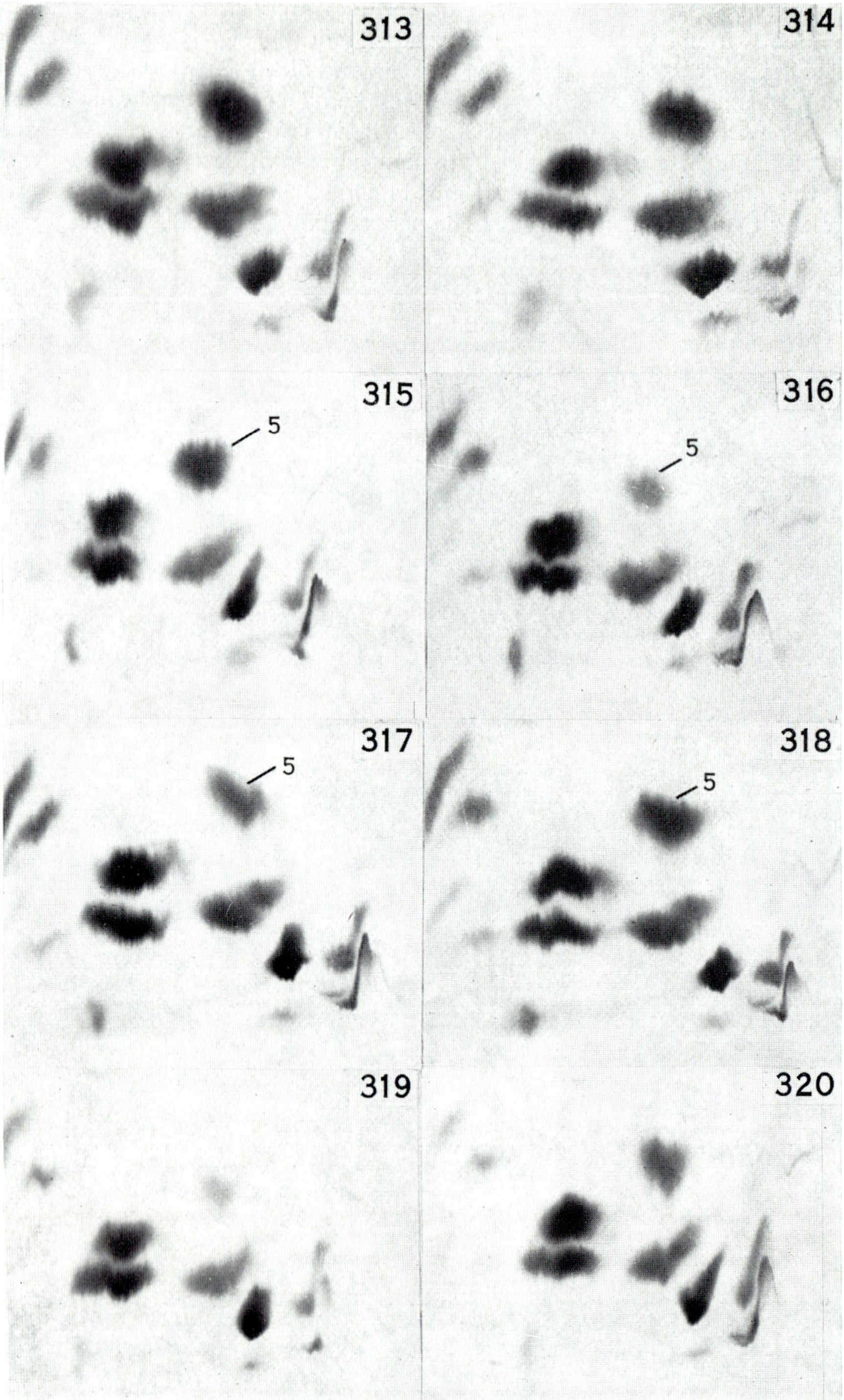

Figs. 313–320. Free amino acids of livers (75 mg samples) of normal rats (Figs. 313, 315, 317, 319) and diabetic littermates (Figs. 314, 316, 318, 320). Taurine, 5.

for single representative diabetic and control animals are shown for muscle, kidney and brain in Figs. 307–312. Recently a column-chromatographic study was reported of the content of 16 constituents in extracts of livers of four Wistar rats, made diabetic by the intravenous injection of 50 mg/kg of alloxan, and two controls[68]. There were found to be decreases in the contents of aspartate, alanine, serine and threonine in the livers of the diabetic animals so that the ratios of the control to diabetic values were as follows: aspartate, 5.6; alanine, 3.8; serine, 5.2; threonine, 6.5. Such changes were not found in our experiments. Results are shown for the livers of four pairs of rats in Figs. 313–320. The contents of free amino acids were not altered in any detectable manner in the livers of the diabetic animals. There appeared to be a random fluctuation of taurine content, which has been mentioned before. The above results show that the severe deficiency in carbohydrate metabolism accompanying diabetes and the secondary metabolic disturbances which result need not be accompanied by marked changes in steady-state concentrations of the free amino acids, although the concentrations of alanine and aspartic and glutamic acids, amino acids which can be formed from intermediates of carbohydrate metabolism, and glutamine and GABA, which are made from glutamic acid, could be regulated in part by availability of carbon from carbohydrate intermediates. Indeed, in extracts of brains of rats in hypoglycemic coma produced by insulin there were found decreases in alanine and glutamic acid contents and an increase in aspartic acid[69–71], the changes in the latter two amino acids being isomolar. These results could be explained on the basis of decreased availability of glucose to brain tissue because of the hypoglycemia. There would be less pyruvate available for alanine formation and the lowered level of acetyl coenzyme A could result in decreasing the rate of condensation of acetyl-CoA with oxalacetate to form citrate, thus allowing the transamination of glutamic acid with oxalacetate, forming aspartate and α-ketoglutarate, to proceed at a more rapid rate. Fluoroacetate, which blocks the tricarboxylic acid cycle prior to the α-ketoglutarate oxidase step, produced a reduction in the contents of both glutamic acid and aspartic acids in the brains of rats[71]. Although the rates of incorporation of amino acids into protein may be decreased in tissues of diabetic animals[72, 73], this would not appear to be related to changes in the total extractable pool.

Effect of thyroidectomy and injection of thyroxin. It has long been known that the thyroid hormone plays an important role in the regulation of protein metabolism, an absence of the hormone resulting in retardation of growth and development in immature animals and disturbances in nitrogen metabolism as well as other aspects of tissue function in mature animals[74] (and see ref. 75 for further pertinent references). Recent work with cell-free homogenates of rat liver showed that daily pretreatment of rats with 100 μg of sodium L-thyroxin for an average period of 10 days resulted in an increased *in vitro* incorporation of amino acids over that found in homogenates of livers of normal controls and that there was a reduction of this rate in livers of thyroidectomized animals[75]. The results suggest the possibility that thyroxin may play a role in linking oxidative phosphorylation to protein synthetic processes.

We have made a paper-chromatographic study of the tissues of thyroid-injected and thyroidectomized rats to determine whether the physiological changes produced by these treatments would be reflected by any alterations in the patterns of the extractable amino acids. The injection of 10 μg of thyroxin daily for 15 days into

References p. 348/349

suitably paired young adult Sprague-Dawley rats was employed to induce the hyperthyroid state. Typical results in Figs. 321–326 show that no changes, whatsoever, had taken place in the amino acid distribution in liver, muscle, or kidney. Experiments were then performed with Sprague-Dawley rats that were surgically thyroidectomized at 40 days of age and were sacrificed together with littermate controls at 5, 9, 17, 24, 29, 38 and 45 days, respectively, after surgery. Extracts of brain, liver, kidney and muscle were chromatographed in the usual fashion. The only consistently reproducible difference found between the experimental and control groups was an increase in the content of ethanolamine phosphate in the extracts of the kidneys of the thyroidectomized animals. An impression was also gained that the taurine level was elevated in liver of the thyroidectomized rats.

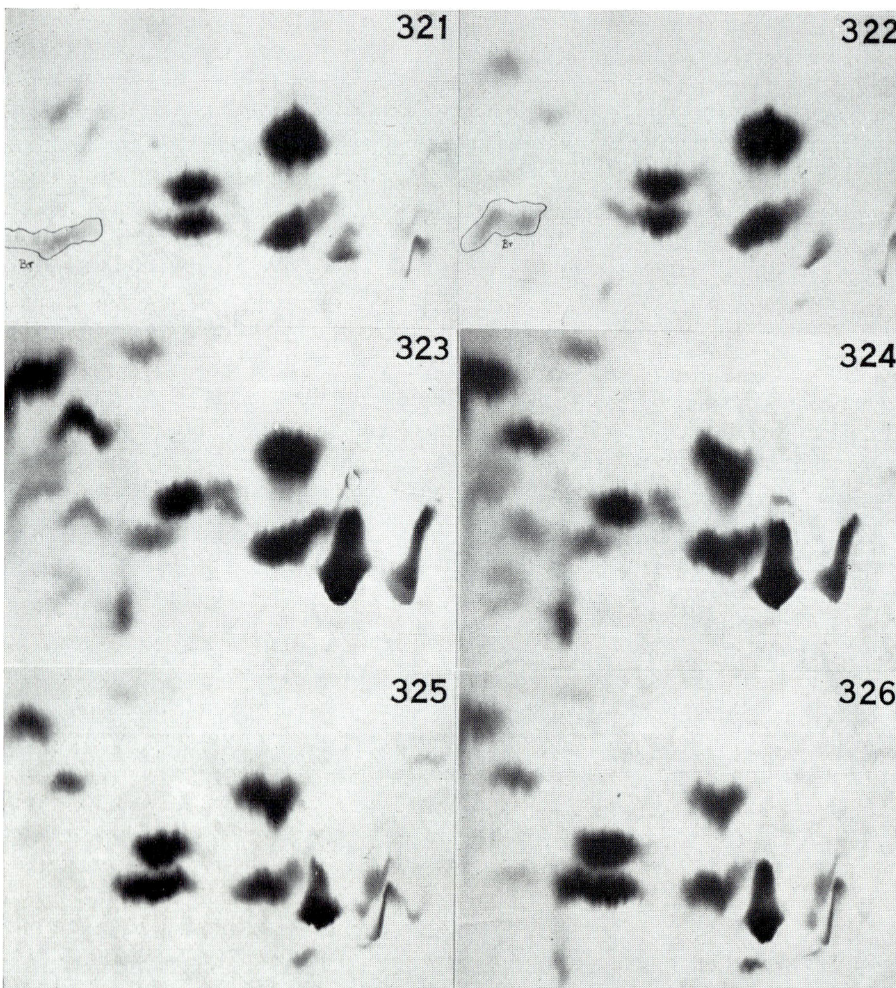

Figs. 321–326. Failure of thyroxin injection to affect free amino acids (75-mg samples) of muscle (Fig. 321: control. Fig. 322: thyroxin), kidney. (Fig. 323: control. Fig. 324: thyroxin), or liver (Fig. 325: control. Fig. 326: thyroxin).

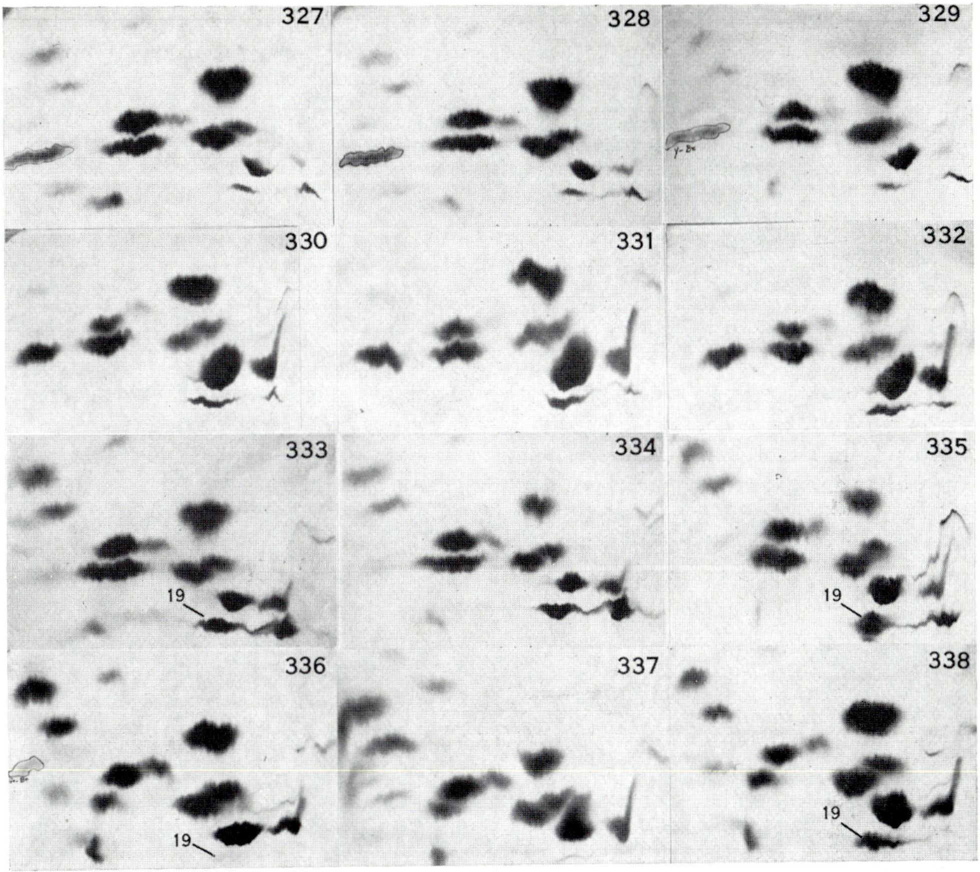

Figs. 327–338. Effect of adrenalectomy and hypophysectomy on amino acids of muscle, brain, liver and kidney (aliquots corresponding to 75 mg of fresh weight of tissue). The order of the numbers is control, adrenalectomized and hypophysectomized. Figs. 327–329: muscle. Figs. 330–332: brain. Figs. 333–335: liver. Figs. 336–338: kidney. Ethanolamine phosphate, 19.

It was of interest then to study the effects of early deprivation of thyroid which results in the cretinoid rat. One member of each of three pairs of littermate rats of the Birmingham strain was thyroidectomized at birth by a single subcutaneous injection of 100 μC of carrier-free ^{131}I in the form of sodium iodide dissolved in normal saline*. A single subcutaneous injection of sodium iodide (^{128}I) was given to littermate controls[76]. All animals were sacrificed at 32 days of age. There were no significant changes in the patterns of the livers of the thyroidectomized animals with the exception of an elevation of taurine content in each instance studied. In neither muscle nor kidney were there any readily detectable differences between the two groups. Particular attention was paid to an examination of the extracts of whole cerebral cortex and the gray matter of the cerebral cortex because structural and electroencephalographic abnormalities occur in the brains of the thyroidectomized animals

* Animals were prepared and tissues sent to us by Dr. J. T. Eayrs, Department of Anatomy, University of Birmingham, Birmingham (Great Britain).

and the behavioral development is retarded (see ref. 77 for summary). In all cases, the chromatographic patterns of the brain tissue of the control and experimental animals were indistinguishable.

Thus, although profound effects are produced by thyroxin excess or deficiency in animals at virtually all levels of observation only minimal changes are produced in the free amino acids of the tissues which have been studied.

Hypophysectomy and adrenalectomy. Sprague-Dawley rats either were hypophysectomized or adrenalectomized at 40 days of age. The hypophysectomized rats and unoperated littermates were sacrificed at 12 days after surgery, the operated animals weighing 30–50% less than the controls. The bilaterally adrenalectomized rats were sacrificed at 6 and 11 days and weighed 10–30% less than the normal controls. Typical results for liver, kidney, muscle and brain are shown in Figs. 327–338. The variations in taurine content in the livers were no greater than those observed from one control to another. There was an increase in content of ethanolamine phosphate in kidneys and livers of hypophysectomized animals, as was the case for thyroidectomized animals. Virtually no other significant changes were seen in the amino acid patterns of the tissues of the experimental animals. The results obtained in these groups of animals are in keeping with the previously observed stability of the steady-state concentrations of the free amino acids in tissues of animals subjected to various physiological stresses.

DISCUSSION

The relative constancy of the concentrations of easily extractable ninhydrin-reactive constituents found in the tissues of mature animals under a variety of physiological conditions suggests that the concentrations of these substances are regulated by remarkable biological servo-mechanisms which involve not only the coordination of the rates of a variety of complex biosynthetic and degradative pathways but also the continuous adjustment of the rates of entry into cells and exit from cells and the movements between intracellular sites and organelles. The characteristic amino acid patterns are largely maintained in the tissues when changes take place in the external environment of the animal. This was illustrated in the present report primarily in the discussion of experiments in which it was found that only small changes in tissue amino acids were produced by starvation, dehydration, or potassium or vitamin A deficiencies. The characteristic distribution of the amino acids and related substances were retained by the tissues of organisms even when major disturbances were produced in the homeostatic mechanisms of the animal as a whole. Thus, thyroxin injection, thyroidectomy, hypophysectomy, adrenalectomy, production of diabetes and the induction of progressive, eventually fatal growth of tumors produced only small changes in the free amino acid patterns of liver, kidney, muscle and brain. A number of the above experimental alterations is known to be accompanied by some significant changes in enzyme activities and in content of various proteins, nucleic acids, lipids and polysaccharides in at least some of the tissues studied. Therefore, the constancy of the amounts of the small molecular weight substances with which we have been concerned may be maintained even when some changes take place in the macromolecular composition of the intracellular environment. Nor do the patterns of amino acids appear to depend on intactness of micro-

References p. 348/349

scopically observable intracellular structures, since tumor cells which were affected severely by cytotoxic agents were still found capable of maintaining high concentrations of the ninhydrin-reactive substances. Particularly striking was the failure to produce any change, whatsoever, in the amino acid pattern of Yoshida sarcoma cells after treatment with podophyllin at a time when severe damage to individual cells was evident, as reflected in abnormalities at the cell surface as well in mitochondrial and nuclear structures. Even sudden and complete cessation of the circulation to a portion of the left ventricle of the dog achieved by ligation of the descending coronary artery resulted in only a relatively slow loss of the high concentrations of some of the free amino acids found in the normal tissue, the decrease becoming considerable between 8 and 16 h after ligation when the histologically observed damage to the myocardium was of such an extent that many of the cellular constituents of both high and low molecular weight undoubtedly were lost from the injured tissue.

The data presented suggest that the steady-state concentrations of the various detectable constituents are regulated largely by their own separate metabolic servomechanisms. Although the mechanisms for the regulation of amounts of the individual substances may possibly interact with one another at one point or another, there are some examples of changes produced in the concentration of one constituent without marked effects occurring in any of the others, even though known metabolic relations exist between the constituents which change and those that remain essentially constant. In a following communication (BAXTER AND ROBERTS, this Symposium) will be discussed the specific elevations of GABA in brain and β-alanine in other tissues by administration of hydroxylamine or aminooxyacetic acid to rats. Administration of lethal doses of ammonium acetate to rats produced large increases only in free aspartic acid and alanine levels in liver and in glutamine in muscle, testes and brain. Remarkable increases in glutamine content, and no other consistent changes, were found in the brains of dogs after the bilateral intracarotid infusion of lactate–Ringer's solution containing ammonium hydroxide. Early in regression of tumors changes in glutamine content were found and not in other amino acids. A number of experimental procedures discussed in previous sections were found to result in changes in only a few of the large number of the detectable constituents, *e.g.* the changes found in muscle of potassium-deficient rats and serum and tissues of vitamin A-deficient rats. The above observations point to the relatively independent regulation of the *amounts* of a number of the substances which can be shown in other types of experiments to be closely related to each other metabolically. Thus, the quantity of glutamic acid, the precursor of GABA in brain, remains essentially unchanged whether the amount of GABA is increased by blocking its utilization with hydroxylamine or aminooxyacetic acid or decreased by inhibiting its formation with thiosemicarbazide (see BAXTER AND ROBERTS, this Symposium). Since the glutamine synthetase activity in brain is much greater than glutamic dehydrogenase it might have been expected that the great accumulation of glutamine occurring in brain during rapid infusion of ammonium salts would have been accompanied by a decrease in the level of glutamic acid, the precursor of glutamine, or at least by a decrease in aspartic acid or some other amino acid which can transaminate with α-ketoglutarate to form glutamic acid. Such changes were not found. The above considerations emphasize that although our knowledge of possible metabolic

References p. 348/349

transformations in a tissue may be extensive our comprehension of the kinetics of these processes within living cells is negligible. We are thus faced with a situation in animal cells in which a variety of experimental conditions have been employed which probably have had great effects on *turnover* rates of the free amino acids while having minimal effects on their *steady-state concentrations*. If the above interpretation is correct, further analyses of the servo-mechanisms involved should be carried out by studying intensively the rates of the various processes involved in the maintenance of the concentration of single constituents and not by trying to find a general answer which would fit all cases. Perhaps the most favorable substances with which to begin would be those that have the smallest number of alternate metabolic pathways within the cells employed for study.

In this report no attempt has been made to cover the entire literature dealing with the occurrence of free amino acids in animal tissues. Those aspects of our own work and that of others were discussed which could be woven into a meaningful pattern without infringing on the material to be reported by other speakers at the Symposium.

ACKNOWLEDGEMENTS

This investigation was supported in part by research grant C-2568 from the National Cancer Institute, and grant B-1615 from the National Institutes of Neurological Diseases and Blindness, National Institutes of Health; grant 3001 (00) from the Office of Naval Research; and a grant from the National Association for Mental Health.

REFERENCES

[1] E. ROBERTS AND D. G. SIMONSEN, in J. T. EDSALL, *Amino Acids, Proteins and Cancer Biochemistry*, Academic Press, New York, 1960, p. 121.
[2] R. CONSDEN, A. H. GORDON AND A. J. P. MARTIN, *Biochem. J.*, 38 (1944) 224.
[3] C. E. DENT, *Biochem. J.*, 41 (1947) 240.
[4] C. E. DENT, *Biochem. J.*, 43 (1948) 169.
[5] C. E. DENT, W. STEPKA AND F. C. STEWARD, *Nature*, 160 (1947) 682.
[6] J. P. GREENSTEIN AND M. WINITZ, in *Chemistry of the Amino Acids*, John Wiley and Sons, Inc., New York, 1961, p. 1366.
[7] E. ROBERTS AND G. H. TISHKOFF, *Science*, 109 (1949) 14.
[8] E. ROBERTS AND S. FRANKEL, *Cancer Research*, 9 (1949) 645.
[9] J. AWAPARA, A. J. LANDUA AND R. FUERST, *Biochim. Biophys. Acta*, 5 (1950) 457.
[10] J. P. GREENSTEIN, in *Biochemistry of Cancer*, Academic Press, New York, 1954, p. 362.
[11] E. ROBERTS, I. P. LOWE, M. CHANIN AND B. JELINEK, *J. Exptl. Zool.*, 135 (1957) 239.
[12] E. ROBERTS AND E. EIDELBERG, in C. C. PFEIFFER AND J. R. SMYTHIES, *International Review of Neurobiology*, Academic Press, New York, 1960, p. 279.
[13] N. OKUMURA, S. OTSUKI AND A. KAMEYAMA, *J. Biochem. (Tokyo)*, 47 (1960) 315.
[14] S. MOORE AND W. H. STEIN, *J. Biol. Chem.*, 211 (1954) 893.
[15] N. OKUMURA, S. OTSUKI AND N. FUKAI, *Acta Med. Okayama*, 13 (1959) 27.
[16] N. OKUMURA, S. OTSUKI AND T. AOYAMA, *J. Biochem. (Tokyo)*, 46 (1959) 207.
[17] A. L. DOUNCE, G. H. TISHKOFF, S. R. BARNETT AND R. M. FREER, *J. Gen. Physiol.*, 33 (1950) 629.
[18] K. SANO AND E. ROBERTS, *Biochem. Biophys. Research Communs.*, 4 (1961) 358.
[19] D. R. WILKEN AND R. G. HANSEN, *J. Biol. Chem.*, 236 (1961) 1051.
[20] E. ROBERTS, S. FRANKEL AND P. J. HARMAN, *Proc. Soc. Exptl. Biol. Med.*, 74 (1950) 383.
[21] H. N. CHRISTENSEN, J. T. ROTHWELL, R. A. SEARS AND J. A. STREICHER, *J. Biol. Chem.*, 175 (1948) 101.
[22] D. G. SIMONSEN, E. ROBERTS AND R. J. JORDAN, to be published.
[23] H. H. TALLAN, *J. Biol. Chem.*, 224 (1957) 41.
[24] E. ROBERTS, K. K. TANAKA, T. TANAKA AND D. G. SIMONSEN, *Cancer Research*, 16 (1956) 970.
[25] E. ROBERTS AND P. R. F. BORGES, *Cancer Research*, 15 (1955) 697.

26 E. Roberts, *J. Biol. Chem.*, 176 (1948) 213.
27 E. Roberts, P. J. Harman and S. Frankel, *Proc. Soc. Exptl. Biol. Med.*, 78 (1951) 799.
28 B. Sisken, K. Sano and E. Roberts, *J. Biol. Chem.*, 236 (1961) 503.
29 E. M. Deuchar, *J. Embryol. Exptl. Morphol.*, 4 (1956) 327.
30 J. L. Kavanau, *Exptl. Cell Research*, 7 (1954) 530.
31 M. Chanin, E. Roberts and A. Goldman, *Circulation Research*, 4 (1956) 713.
32 H. H. Tallan, *Proc. Soc. Exptl. Biol. Med.*, 89 (1955) 553.
33 S. Kit and J. Awapara, *Cancer Research*, 13 (1953) 694.
34 E. Roberts and T. Tanaka, *Cancer Research*, 16 (1956) 204.
35 E. Kvamme and G. Svenneby, *Cancer Research*, 21 (1961) 92.
36 P. Schwerin, S. P. Bessman and H. Waelsch, *J. Biol. Chem.*, 184 (1950) 37.
37 H. N. Christensen, in W. D. McElroy and B. Glass, *Amino Acid Metabolism*, The Johns Hopkins Press, Baltimore, 1955, p. 63.
38 R. D. Harkness, *Brit. Med. Bull.*, 13 (1957) 87.
39 K. Weinbren, *Gastroenterology*, 37 (1959) 657.
40 H. N. Christensen, J. T. Rothwell, R. A. Sears and J. A. Streicher, *J. Biol. Chem.*, 175 (1948) 101.
41 V. Ferrari and R. D. Harkness, *J. Physiol.*, 124 (1954) 443.
42 K. Sano and E. Roberts, to be published.
43 D. A. Goldthwait, G. R. Greenberg and R. A. Peabody, *Biochim. Biophys. Acta*, 18 (1955) 148.
44 L. F. Leloir and C. E. Cardini, *Biochem. Biophys. Acta*, 20 (1956) 33.
45 J. Preiss and P. Handler, *J. Biol. Chem.*, 233 (1958) 493.
46 K. K. Tsuboi, H. O. Yokoyama, R. E. Stowell and M. E. Wilson, *Arch. Biochem. Biophys.*, 48 (1954) 275.
47 E. Chargaff, K. B. Olson and P. F. Partington, *J. Biol. Chem.*, 134 (1940) 505.
48 H. N. Marvin and J. Awapara, *Proc. Soc. Exptl. Biol. Med.*, 72 (1949) 93.
49 J. Awapara, *Texas Repts. Biol. Med.*, 10 (1952) 22.
50 S. M. Kalman and M. E. Lombrozo, *J. Pharmacol. Exptl. Therap.*, 131 (1961) 265.
51 G. B. Mider, *Ann. Rev. Med.*, 4 (1953) 187.
52 M. Rechcigl Jr., F. Grantham and R. E. Greenfield, *Cancer Research*, 21 (1961) 238.
53 D. G. Simonsen and E. Roberts, to be published.
54 G. Rouser, in *The Leukemias*, Academic Press, New York, 1957, p. 361.
55 J. M. White, G. Ozawa, G. A. L. Ross and E. W. McHenry, *Cancer Research*, 14 (1954) 508.
56 C. Wu, *J. Biol. Chem.*, 207 (1954) 775.
57 J. Awapara, *Proc. Soc. Exptl. Biol. Med.*, 90 (1955) 435.
58 R. E. Eckel, C. E. Pope and J. E. C. Norris, *Arch. Biochem. Biophys.*, 52 (1954) 293.
59 M. Iacobellis, E. Muntwyler and C. L. Dodgen, *Am. J. Physiol.*, 185 (1956) 275.
60 M. Iacobellis, G. E. Griffin and E. Muntwyler, *Proc. Soc. Exptl. Biol. Med.*, 96 (1957) 64.
61 P. Malathi, P. S. Sastry and J. Ganguly, *Nature*, 189 (1961) 660.
62 J. P. du Ruisseau, J. P. Greenstein, M. Winitz and S. M. Birnbaum, *Arch. Biochem. Biophys.*, 68 (1957) 161.
63 Y. Tsukada, G. Takagaki, S. Sugimoto and S. Hirano, *J. Neurochem.*, 2 (1958) 295.
64 B. Eiseman, H. Osofsky, E. Roberts and B. Jelinek, *J. Appl. Physiol.*, 14 (1959) 251.
65 E. Roberts, M. Rothstein and C. F. Baxter, *Proc. Soc. Exptl. Biol. Med.*, 97 (1958) 796.
66 J. Tepperman and H. M. Tepperman, *Pharmacol. Revs.*, 12 (1960) 301.
67 W. E. Knox, V. H. Auerbach and E. C. C. Lin, *Physiol. Revs.*, 36 (1956) 164.
68 E. Kirsten, R. Kirsten, H. J. Hohorst and T. Bücher, *Biochem. Biophys. Research Communs.*, 4 (1961) 169.
69 O. Nikichi, O. Saburo and N. Hiroyuki, *J. Biochem. (Tokyo)*, 46 (1959) 247.
70 R. O. Cravioto, G. Massieu and J. J. Izquierdo, *Proc. Soc. Exptl. Biol. Med.*, 78 (1951) 856.
71 R. M. C. Dawson, *Biochem. Biophys. Acta*, 11 (1953) 548.
72 M. E. Krahl, *J. Biol. Chem.*, 200 (1953) 99.
73 M. G. Kritsman, A. S. Konikova, D. G. Stepanyan and L. M. Pyatigorskaya, *Biokhimiya*, 16 (1951) 246.
74 R. W. Rawson and J. E. Rall, in G. Duncan, *Diseases of Metabolism*, 3rd edition, W. B. Saunders Co., Philadelphia, 1952, p. 957.
75 L. Sokoloff and S. Kaufman, *J. Biol. Chem.*, 236 (1961) 795.
76 J. T. Eayrs and G. Horn, *Anat. Record*, 121 (1955) 53.
77 J. T. Eayrs, in D. B. Tower and J. P. Schadé, *Structure and Function of the Cerebral Cortex*, Elsevier Publishing Co., Amsterdam, 1960, p. 44.

FREE AMINO ACIDS IN THE BLOOD OF MAN AND ANIMALS

I. METHOD OF STUDY AND THE EFFECTS OF VENIPUNCTURE AND FOOD INTAKE ON BLOOD FREE AMINO ACIDS

GEORGE ROUSER, BOHDAN JELINEK, ARTHUR J. SAMUELS AND KEIJI KINUGASA*

Departments of Biochemistry and Medicine, City of Hope Medical Center, Duarte, Calif. (U.S.A.)

Our interest in the *free* or easily extractable amino acids and related ninhydrin positive substances of cells and body fluids stems to a great extent from the fact that free amino acids appear to be precursors of protein while preformed proteins or peptides are either of minor importance or not involved in protein synthesis. The free amino acid pool and the regulatory mechanisms involved in its maintenance thus bear an important relationship to cell growth and reproduction. The relationships might be altered in pathological states, such as leukemia, in which there is a disturbance of growth. The possibility that the metabolism of individual amino acids might be different in the leukemias and that differences might be utilized in controlling these conditions constituted another reason for the present studies. The influence of cytotoxic drugs on amino acid metabolism has been studied in an effort to define, at least in part, the mode of action of the drugs. Elucidation of changes produced by drugs may be useful in the development of new measures for the control of the leukemic process.

These investigations are extensions of previous studies[1–3]. The objectives were: (1) To determine the nature of the constituents in the free amino acid pools of plasma and cells; (2) to define relationships existing between plasma and cellular free amino acids; (3) to detect deviations from normal in the leukemias; and (4) to determine the effects on free amino acid levels of cytotoxic drugs commonly administered to patients with leukemia.

A number of earlier reports on blood amino acid nitrogen in various pathological states involving the hematopoietic system were available when we began our studies. GREENE AND CONNOR[4] and LUCK[5] had reported that amino acid nitrogen was elevated in polycythemia, and SCHMIDT[6] reported that the amino acid nitrogen content of the blood was frequently moderately increased in leukemia. OKADA AND HAYASHI[7] had also found a moderate increase of amino acid nitrogen in leukemia, particularly of the myelogenous type.

BEATON *et al.*[8] reported an elevation of plasma glutamic acid levels in malignancy. They noted that total amino nitrogen in the blood, as determined by a modification of the Folin method, was not significantly elevated above normal. Subsequently, WHITE *et al.*[9, 10] observed an increase in the plasma glutamic acid level in other conditions and showed that there was a rise in the free glutamic acid concentration

* Present address: Professor OKINAKA's Clinic School of Medicine, University of Tokyo, (Japan).

References p. 447/448

of eggs containing growing tumors. The rise in glutamic acid was accompanied by a decrease in glutamine. The same investigators[11] reported that 2 strains of rats in which tumors were produced by methyl cholanthrene or by implantation had increased levels of free glutamic acid in the blood plasma.

More recently ROBERTS AND FRANKEL[12] reported that free glutamine was not present in ascites tumor cells of mice. ROBERTS AND TANAKA[13] later showed that glutamine was absent from both Yoshida ascites tumor cells and the ascitic fluid, and that glutamine injected into the ascitic fluid in mice was metabolized rapidly by the tumor cells. Injected glutamic acid was not metabolized rapidly, apparently because of limited permeability to this amino acid as shown by studies with DL-glutamic acid-2-^{14}C.

It was evident at the beginning of the studies in humans that a great deal of variability was to be expected since there were many conflicting reports in the older literature. Reports that have appeared since the beginning of these studies have emphasized again the variability. Thus, while we reported a rise in plasma glutamic acid and a tendency to lower glutamine in some leukemias[2], WAISMAN[14] failed to find such changes, although subsequently KELLEY AND WAISMAN[15] using a different method did find an elevation of the plasma glutamate level in their leukemia patients. NOUR-ELDIN AND WILKINSON[16] found the fasting plasma amino nitrogen concentrations to be within normal limits in patients with different types of leukemia before and after treatment, while IYER[17] confirmed the rise in plasma glutamic acid and also observed decreases in the levels of glutamine and some of the other amino acids. SASSENRATH AND GREENBERG[18] failed to observe any differences in plasma amino acids of rats bearing the Walker carcinoma and could not find any distinct pattern of change in the free amino acids of tissues. Later, WU AND BAUER[19] reported that rats bearing the Walker carcinoma showed no change in the concentration of the free amino acids in plasma or tissues when the tumors were small, but as the tumors grew larger these investigators noted that the concentrations of most free amino acids were increased in plasma and liver and decreased in muscle. They further observed that glutamine was the only compound whose concentration was consistently decreased in plasma, liver, and muscle. These investigators reported a definite elevation of the plasma glutamic acid level in animals bearing both medium and large sized tumors.

BERNARD *et al.*[20] reported highly variable results in their studies of human leukemias. The blood plasma amino nitrogen was not regularly found to be above normal and was, in fact, frequently below the normal level. These studies were complicated by the fact that the patients were receiving various drugs. As will be shown in subsequent papers in this series, some rather marked changes can be brought about by drugs. In agreement with the work of WU AND BAUER[19], LUTZ *et al.*[21] reported that blood amino acid values (determined by the Folin method) of rats with subcutaneous tumors remain relatively unchanged until the terminal stage when a significant increase is observed. MCMENAMY *et al.*[22] in a paper-chromatographic study of 32 plasma specimens from 23 patients with various types of leukemia, both treated and untreated, were unable to find significant differences in plasma glutamic acid and glutamine levels. The very small number of samples and the inclusion of treated patients in their investigation emphasizes the difficulties in drawing conclusions from limited data.

References p. 447/448

The marked variability in the results reported from different laboratories appears to be due in part to the relatively small number of samples examined so that a broad picture of the factors that influence the free amino acid levels in cells and body fluids was not obtained. The large number of samples to be examined in our own studies precluded the use of any but simple, rapid methods. It is for this reason that we chose *semi-quantitative* examination by paper chromatography. When samples are carefully prepared under controlled conditions and frequent comparisons are made between samples and known mixtures, reproducible results can be obtained and patterns or levels of free amino acids may be determined without giving the levels a numerical designation. Such studies should include photographs of the paper chromatograms so that the reader may make his own comparisons and draw his own conclusions. Paper chromatography was selected in preference to column-chromatographic procedures because the column procedures are very time consuming, there is usually some fraction overlap, glutamine is not recovered quantitatively from columns presently in use, and glutamic acid values may be in error. As we were particularly interested in possible changes in glutamic acid and glutamine, the latter objections are particularly important ones.

The present report describes the general methods of study used in our investigations in which 2385 two-dimensional paper chromatograms were prepared from 34 blood samples and 18 urine samples from normal individuals; 12 blood and 6 urine samples from patients with acute leukemia; 174 blood samples, 23 urine samples, and 15 bone marrow specimens from patients with chronic lymphatic leukemia; 199 blood samples and 20 urine specimens from patients with chronic granulocytic leukemia; 59 blood samples and 70 tissue samples from rabbits; and 7 blood samples from leukemic and normal dogs.

METHODS

Separation of blood cells

Blood from fasted individuals (12–18 h since the last meal) was drawn into heparin (0.5 mg/10 ml of blood) and centrifuged immediately. The following centrifugation technique was used for high count bloods. Blood, usually 10–12 ml, was placed in a 12-ml graduated centrifuge tube and the speed of centrifugation gradually increased to 500 $\times$ *g* over a period of 1–3 min, and centrifugation was continued at this speed for 7 min. After the initial spin, platelet-rich plasma overlying the white cells and red cells was withdrawn and transferred to another tube for recentrifugation. The white cell layer was next withdrawn with a small amount of blood plasma and placed in another 12-ml centrifuge tube for recentrifugation. Erythrocytes were withdrawn by inserting a needle to the bottom of the layer and aspirating the lower two-thirds of the cell mass (to minimize contamination with white cells) and placed in another 12-ml tube for recentrifugation. All transfers were made with a syringe equipped with a stainless steel needle 4 in. long that was rinsed with saline after each use.

The relatively large white cell layer obtained from high count leukemias was recentrifuged at 1300 $\times$ *g* for 7–10 min in a 12-ml centrifuge tube. The packing at this force gave reproducible volume measurements for white blood cells. Where the mass of leukocytes was small, a series of 10-ml tapered tubes was used for recentrifu-

References p. 447/448

gation of white cells at 1300 × g. The tapered portions of the tubes had volumes of 0.1, 0.4, or 1 ml and were calibrated in small units. These tubes facilitate accurate volume determinations and withdrawal of contaminating platelets and/or red cells.

Erythrocyte samples were recentrifuged at 1300 × g for 10 min, and plasma and any residual leukocytes were aspirated before the cells were extracted. The red cell samples were not washed since control studies indicated losses of amino acids into saline or buffer washes. Generally, plasma and cell samples were extracted with alcohol within 30–40 min from the time blood was withdrawn.

The isolation of white blood cells from normal individuals was accomplished with a unit of blood centrifuged in 500-ml bottles. The blood was spun at about 1300 × g to pack leukocytes, the cells were pipetted off, and recentrifuged in 12-ml tubes. After the second centrifugation white cells could be separated from most of the contaminating erythrocytes. A third centrifugation (1300 × g) gave white cells largely free of erythrocytes. Cells prepared in this way were over 95% neutrophilic polymorphonuclear leukocytes.

Extraction and preparation of samples for chromatography

Samples were treated with 3 volumes of 95% ethanol and the precipitate that developed after vigorous shaking was filtered off on a sintered glass filter. The residue was extracted 3 times with small portions of 80% ethanol. The combined ethanol solutions obtained from white cells and blood platelets were applied directly to filter paper for chromatography.

The ethanolic extracts from blood plasma, erythrocytes, and urine were evaporated to dryness in 20–50 ml beakers under infrared heat lamps in a current of air from a fan. The beakers were placed on bright aluminum foil to prevent overheating of the samples. Next, the samples were dialyzed to remove lipid. The dry solids were treated with a measured volume of distilled water, transferred to well-washed Nojax cellulose casing and dialyzed against a measured volume of water for 4 hours at room temperature with constant shaking. The dialysis tubing was first washed by boiling twice for 30 min in distilled water to remove impurities that would otherwise interfere with the chromatographic examinations. The true aliquot size of the dialysate was determined from the ratio of water volumes inside and outside the bag. The dialysate was then desalted in a Reco electrolytic desalter with a 10 ml well at 20 mV for 5 to 10 min. Usually samples equivalent to 3 ml of original plasma, erythrocytes, or urine were treated for 5 min. The desalted extract was then evaporated to dryness as described for ethanol extracts.

Paper chromatography

Untreated sheets of Whatman No. 1 filter paper for chromatography ($18^1/_4 \times 22^1/_2$ in.) were used. The dry solids from the desalting step were taken up in a measured volume of water and an equivalent of from 0.1–1.0 ml of original sample (plasma, urine, erythrocytes) applied to one corner of the paper. A spot 3 cm in diameter is convenient. When hydrogen peroxide was used to oxidize sulfur-containing amino acids, a solution of 30% hydrogen peroxide (Superoxol) was spotted over the point of application of the sample. Rapid evaporation of the applied solutions was achieved by a current of air from a hair dryer.

Aliquots for platelets and white cells are expressed on a wet weight basis assuming 1 ml of packed cells (1300 × *g*) to be equivalent to 1 gram. Erythrocyte aliquots are expressed as milliliters of packed cells (1300 × *g*).

Filter paper chromatography was carried out using water-saturated phenol in an ammonia atmosphere in the first dimension, followed by 2,4-lutidine saturated with water in the second dimension. The chromatograms were sprayed with ninhydrin dissolved in *n*-butanol (1 g/l) while still slightly moist with lutidine and allowed to develop at room temperature before they were photographed. Lutidine increases the color obtained with ninhydrin, and in its presence the proline spot is blue rather than yellow when fully developed.

Identification of compounds

The tentative identifications of compounds based on migrations in phenol and lutidine, behavior after hydrogen peroxide treatment, and stability to 6 *N* hydrochloric acid in a sealed tube at 100° for 24 h was confirmed as follows. Several two-dimensional chromatograms were prepared in the usual manner. The dry papers were heated at 105° for 30–60 min and amino acid spots visualized by their fluorescence under ultraviolet light. The spots were marked and eluted with water or a dilute sodium carbonate solution. When such eluates are evaporated to dryness in a small beaker, crystals can usually be obtained in the center of the beaker that are largely free of coloration from lutidine. A comparison of these crystals with those obtained from authentic compounds obtained in a similar manner can give valuable information as to the nature of the material.

Dinitrofluorobenzene derivatives were prepared from the compounds eluted from paper and, after removal of excess dinitrofluorobenzene by extraction with diethyl ether, the derivatives were applied to washed strips of Whatman No. 1 paper alongside authentic compounds treated in the same manner. The paper strips were washed by capillary descent in 2 *N* acetic acid prior to use. Chromatographic comparisons were carried out in: (1) dioxane–concentrated ammonia (4 : 1, v/v); (2) 2,4-lutidine–water (11 : 1, v/v); (3) isobutyric acid saturated with water; (4) 1 *N* aqueous hydrochloric acid; (5) 5% potassium dihydrogen phosphate in water layered with an equal volume of isoamyl alcohol; (6) 5% potassium dihydrogen phosphate in water adjusted to pH 7.0 with concentrated ammonia and layered with an equal volume of isoamyl alcohol; and (7) 5% disodium hydrogen phosphate in water layered with an equal volume of isoamyl alcohol. Solvents 5 to 7 are used only for ascending paper chromatography.

The differentiation of glutathione, cysteine and cystine (as cysteic acid), and cysteinylglycine in leukocyte samples was achieved as follows: With some batches of filter paper the spot applied to paper could be warmed with air from a hair dryer to oxidize cysteine and cysteinylglycine without oxidizing glutathione. With other batches of paper it was necessary to treat with 30% hydrogen peroxide to obtain oxidation. The differentiation of cysteic acid and cysteinylglycine (probably as the sulfonic acid diketopiperazine) was facilitated by observing the color changes of the spots after spraying with ninhydrin. Cysteinylglycine was first yellow, then blue-green, and finally purple. The cysteic acid spot was purple at all times. Authentic cysteinylglycine was prepared by the hydrolysis of glutathione (250 mg) in 1 *N* hydrochloric acid (10 ml) in a boiling water bath for 1 h.

References p. 447/448

RESULTS

Limitations of the method

The limitations of the method are related largely to the presence in some samples of interfering substances of unknown nature not removed in the sample preparation, and to the limits of detectibility set by the ninhydrin reagent. In plasma and erythrocytes distinct spots of phenylalanine, tyrosine, histidine, arginine, and lysine are not usually observed except at aliquot sizes of the order of 0.5–1 ml. Phenylalanine and tyrosine are present in relatively small amounts while the ninhydrin color for histidine is somewhat less than for other compounds. The arginine and lysine spots are frequently poorly formed due to interference of migration from other substances.

The overlap of spots may prevent the recognition of some compounds. Thus, leucine and isoleucine migrate together and the glycine and asparagine spots may overlap and merge to varying degrees with the serine spot in samples, even though these compounds in a standard mixture are usually well separated (see Figs. 1–10). A good deal of variation has been observed with paper obtained at different times. With some batches of paper glycine, serine, and asparagine were separated completely from each other. On occasion paper was obtained that had the interesting quality of allowing wide separations of the pairs glycine–serine and alanine–threonine by virtue of a great increase in the migration of threonine and serine in lutidine without alteration of the migration characteristics of other amino acids.

The variability of the filter paper gave rise to other limitations. Cysteine and cystine must be oxidized before they can be visualized on chromatograms. When samples are treated with hydrogen peroxide, variable destruction of amino acids may take place. This destruction may be very great on some filter papers and is particularly noticeable with plasma samples. The destruction of amino acids by peroxide was not as extensive with erythrocyte, leukocyte, or urine samples. The phenomenon made it impossible to examine some plasma samples for the sulfur amino acids.

The various steps in the procedure for preparation of the samples were checked with authentic compounds and mixtures of standards and no alterations of any of the known substances were observed after dialysis or desalting.

Compounds seen on two-dimensional paper chromatograms from blood and urine samples

Table I gives the composition of an artificial plasma amino acid mixture prepared from pure substances with amounts in general to correspond to average values from the literature. Some adjustments were made upwards or downwards from the literature values. Figs. 1–10 show the paper-chromatographic results obtained with different aliquot sizes of the artificial plasma amino acid mixture. These chromatograms show the variations in spot size and color intensity with different amounts of known substances and are of value for comparison with sample runs. Evidently the literature values are similar to the levels frequently seen in our studies.

The free amino acids seen in plasma are: leucine plus isoleucine, valine, proline, alanine, glutamine, threonine, glycine, serine, taurine, glutamic acid, and cystine–cysteine (as cysteic acid after peroxide oxidation). Aspartic acid is frequently observed and at times α-amino n-butyric acid is also observed. Very occasionally,

References p. 447/448

particularly in leukemia, ethanolamine-O-phosphate may be seen. Phenylalanine, tyrosine, histidine, arginine, and lysine are not always observed for the reasons described above. Asparagine cannot always be visualized readily as it tends to migrate with glycine. The identification of the spots seen on paper chromatograms of plasma samples were confirmed by the methods outlined above. Upon hydrolysis with 6 *N* hydrochloric acid for 18 h at 100° in a sealed tube, the disappearance of glutamine and asparagine with the appearance of increased amounts of glutamic and aspartic acids were observed in all samples.

The substances identified on paper chromatograms prepared from different types of white blood cells and platelets were: leucine plus isoleucine, valine, proline, alanine,

TABLE I

ARTIFICIAL PLASMA AMINO ACID MIXTURE

Amino acid	*mg/100 ml*	*Amino acid*	*mg/100 ml*
Alanine	3.6	Lysine	2.6
α-Amino *n*-butyric acid	0.2	Methionine	0.5
Arginine	1.8	Ornithine	0.7
Asparagine	1.0	Phenylalanine	1.2
Aspartic acid	0.1	Proline	1.9
Cysteic acid	1.5	Serine	1.2
Glutamic acid	0.8	Taurine	1.2
Glutamine	10.0	Threonine	1.8
Glycine	2.0	Tryptophan	1.2
Histidine	1.3	Tyrosine	1.0
Isoleucine	1.7	Valine	2.7
Leucine	2.0		

threonine, glutamine, glycine, serine, glutamic acid, aspartic acid, taurine, glutathione and ethanolamine-O-phosphate. Cysteic acid and cysteinylglycine were observed after oxidation. Cysteinylglycine was not seen regularly in platelets. Some samples were observed to contain traces of other compounds such as α-amino *n*-butyric acid.

The amino acids seen regularly in erythrocyte samples were aspartic and glutamic acids, glycine, alanine, serine, threonine, glutamine, leucine plus isoleucine, valine and glutathione. Taurine was present in some samples and occasionally traces of ethanolamine-O-phosphate were observed. Small amounts of uncharacterized substances were frequently observed.

Chromatograms prepared from urine samples showed glycine, serine, alanine, glutamine, histidine, and taurine as readily visualized and identified spots and valine, the leucines, phenylalanine, and tyrosine were usually seen despite some distortion of this area of the chromatogram due to the presence of ninhydrin negative material. The recognition of tyrosine-O-sulfate, 1-methylhistidine, and 3-methylhistidine as constituents of all samples was possible with aliquots equivalent to 0.5 ml or more of urine. The exact correspondence of migration of added authentic tyrosine-O-sulfate (obtained from H. H. TALLAN) with the spot on urine chromatograms giving rise to tyrosine and sulfate on hydrolysis is shown in Figs. 11 and 12.

References p. 447/448

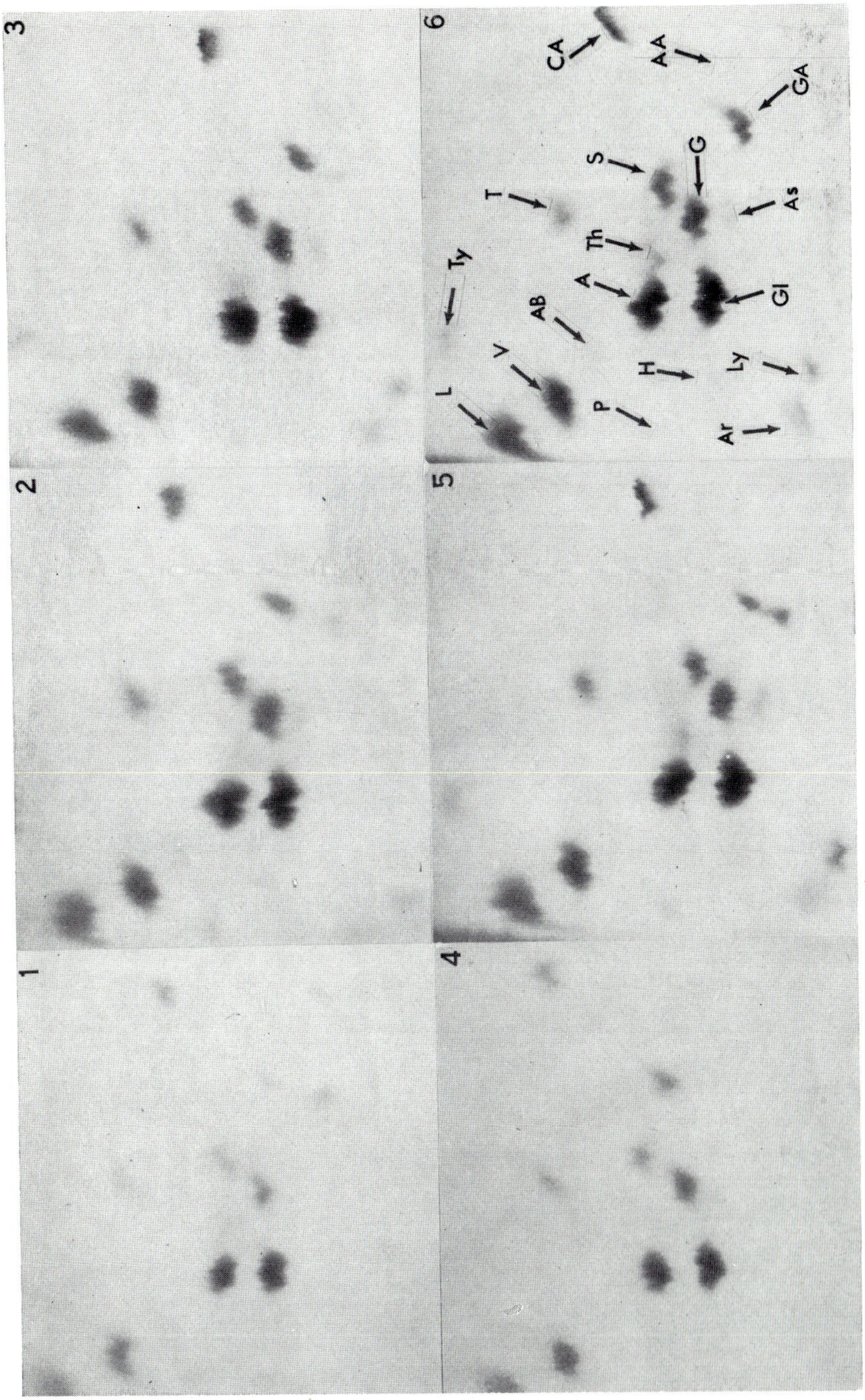

Figs. 1–6. Prepared from artificial plasma amino acid mixture (Table I). Aliquots of 0.1, 0.2, 0.3, 0.1, 0.25, 0.3 ml, respectively used. Some compounds barely detectable in Fig. 6 are more prominent in Figs. 7–10. For abbreviations see p. 369.

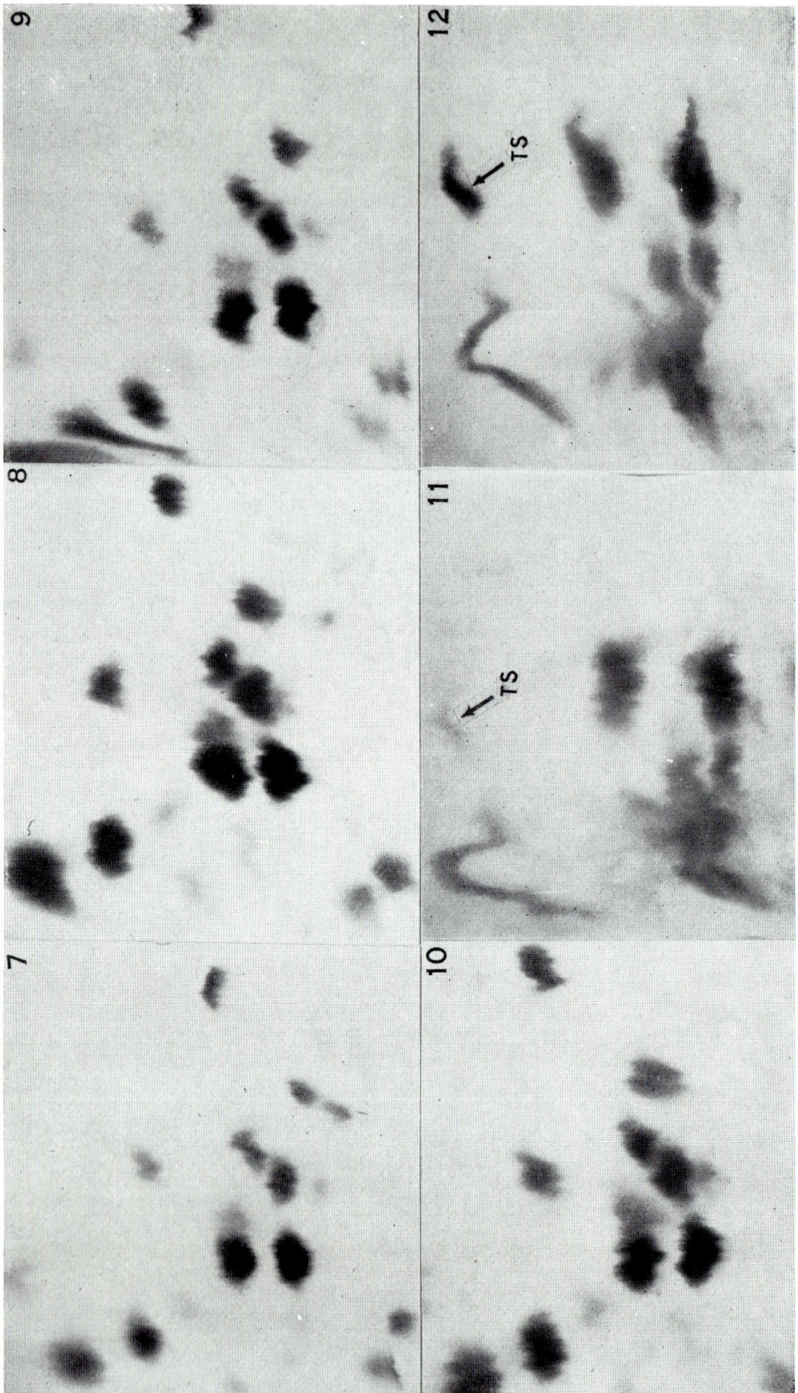

Figs. 7–10. Prepared from artificial plasma amino acid mixture (Table I). Aliquots of 0.50, 1.0, 0.5, and 1.0 ml, respectively used

Figs. 11, 12. Prepared from a normal urine sample (equivalent of 0.5 ml of urine). 80 μg of tyrosine-O-sulfate added to the sample in Fig. 12. Note the exact coincidence of movement of authentic tyrosine-O-sulfate and compound in urine (Fig. 11). For abbreviations see p. 369.

In vitro incubation studies with blood from chronic lymphatic leukemia patients

In vitro incubations were performed to determine what arteifacts might arise during the processing of blood samples. Incubation of whole blood from untreated patients with chronic lymphatic leukemia disclosed that marked changes in the amino acid composition of plasma, leukocytes, and erythrocytes took place at 37°. The complete degradation of glutamine in cells and plasma was observed, and a marked increase of glutamic acid and alanine was observed in both plasma and cells. A similar study of whole blood from a patient with a high leukocyte count with 0.7 mg/ml of glutamine added followed by incubation at 37° with shaking at 120 cycles/min demonstrated that glutamine was completely metabolized in a little over 4 h. Incubations of white cells in plasma and red cells in plasma were then carried out to determine the relative metabolic activity of the leukocytes and erythrocytes. 15 *in vitro* incubations in all were carried out with chronic lymphatic and chronic granulocytic leukemia blood samples.

One study with added glutamine was carried out as follows. 50 ml of blood was drawn into heparin from a patient with chronic lymphatic leukemia and cells and plasma separated as described above. Glutamine (0.25 mg/ml) was added to plasma. 18 ml of plasma was then mixed with 18 ml of erythrocytes, and 17 ml of the plasma was mixed with 2.7 ml of packed leukocytes (1300 × *g*). The incubations were carried out in a Warburg bath at 37° with shaking at 120 cycles/min in siliconed Erlenmeyer flasks in an atmosphere of air. Samples were withdrawn at the following times after mixing: 5, 10, 30, 45, 60, 90, 120, 200, and 240 min. An aliquot of 1.5–2.0 ml of the plasma plus leukocyte mixture was removed each time, and 3.0–3.3 ml of the red cell plus plasma mixture was removed for each sample. The samples were spun and the volume of plasma and packed cells was noted in each case. Approximately 5 min was required for satisfactory separation of cells and plasma at 1300 × *g*. The plasma was aspirated as completely as possible and the cells were treated (without washing) with alcohol for extraction. The last trace of plasma above the packed cells was removed with a small swab.

During the course of the erythrocyte incubation a trace of hemolysis was evident first at 160 min, but was still very slight at 240 min. The pH of the plasma at the beginning of the erythrocyte incubation was 7.5 and rose to 8.1 at the end of the incubation. The pH of plasma rose from 7.5–8.3 in the leukocyte incubation flask. At the end of the erythrocyte incubation the erythrocyte volume had increased from an initial value of 45% to 47% (volumes determined at 1300 × *g* for 5 min). The initial packed volume of leukocytes (1300 × *g*, 5 min) was 12%. The cells swelled gradually to a value of 14.5% at the end of the incubation (21% increase).

Figs. 13–16 show the plasma samples and Figs. 17–20 the leukocyte samples obtained at the beginning of the incubation and after 60, 90, and 240 min. The complete absence of glutamine from the white cells was first evident at 60 min (Fig. 18) and the complete disappearance from plasma was evident at 90 min (Fig. 15). White cells lost taurine to the plasma and large increases of alanine and glutamic acid with lesser increases in aspartic acid and some of the other amino acids are apparent in plasma.

From this study it is evident that the lymphocytes from chronic lymphatic leuke-

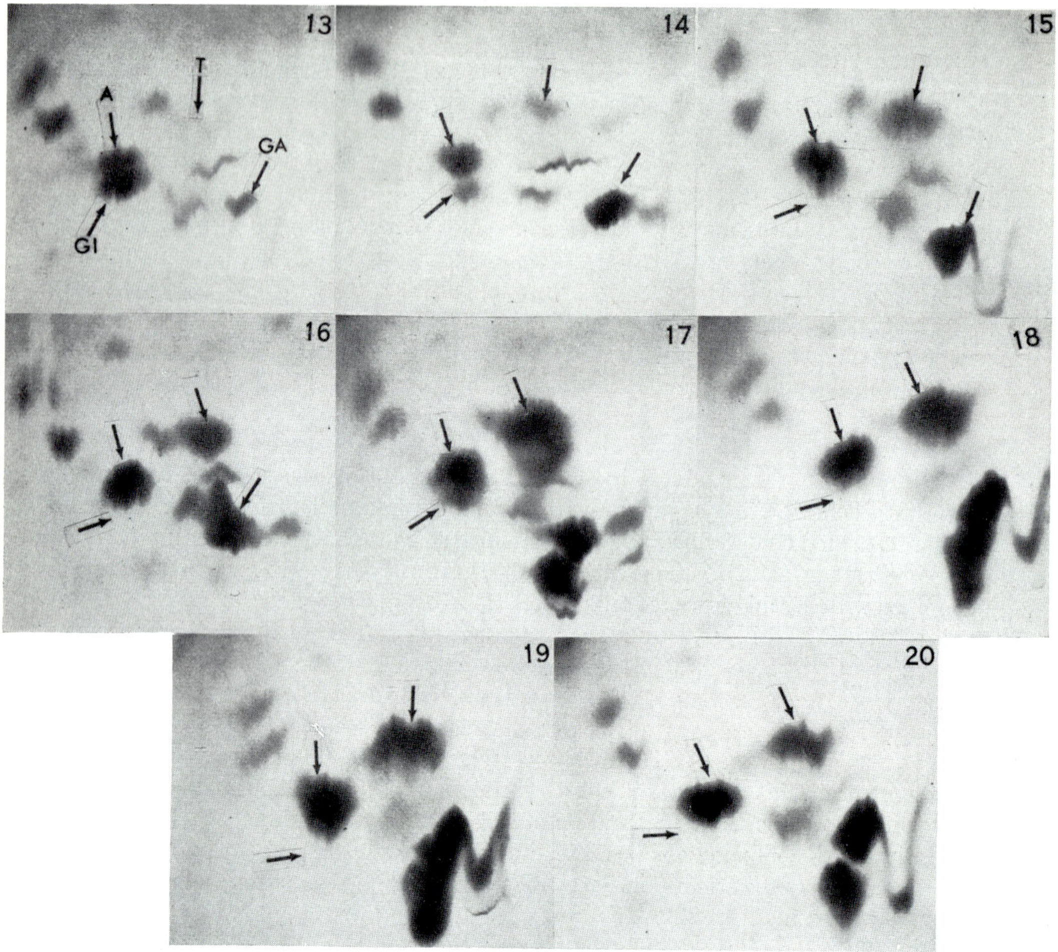

Figs. 13–20. Plasma and leukocyte samples from an *in vitro* incubation study (see text). Figs. 13–16 prepared from the equivalent of 0.25 ml of plasma samples removed at zero time, 60, 90, and 240 min, respectively. Figs. 17–20, the leukocyte samples removed at the same time intervals. Glutamine, Gl; Glutamic acid, GA.

mia patients degrade glutamine and produce alanine and glutamic acid. When blood is allowed to stand prior to processing, a variable increase in alanine and glutamic acid and a decrease in glutamine may be seen in plasma, although this should be almost undetectable by the standard procedure adopted.

Figs. 21–24 show the plasma and Figs. 25–28 the erythrocyte samples removed at the beginning and after 30, 120 and 240 min of incubation. Glutamine was not degraded, although there was a marked increase in both the aspartate and glutamate levels in erythrocytes and a marked increase in the glutamate level of plasma. The major artifacts that would be introduced from erythrocytes by allowing blood from chronic lymphatic leukemia patients to stand *in vitro* would be an elevation of the plasma glutamic acid level with increases in the dicarboxylic acids in the eryth-

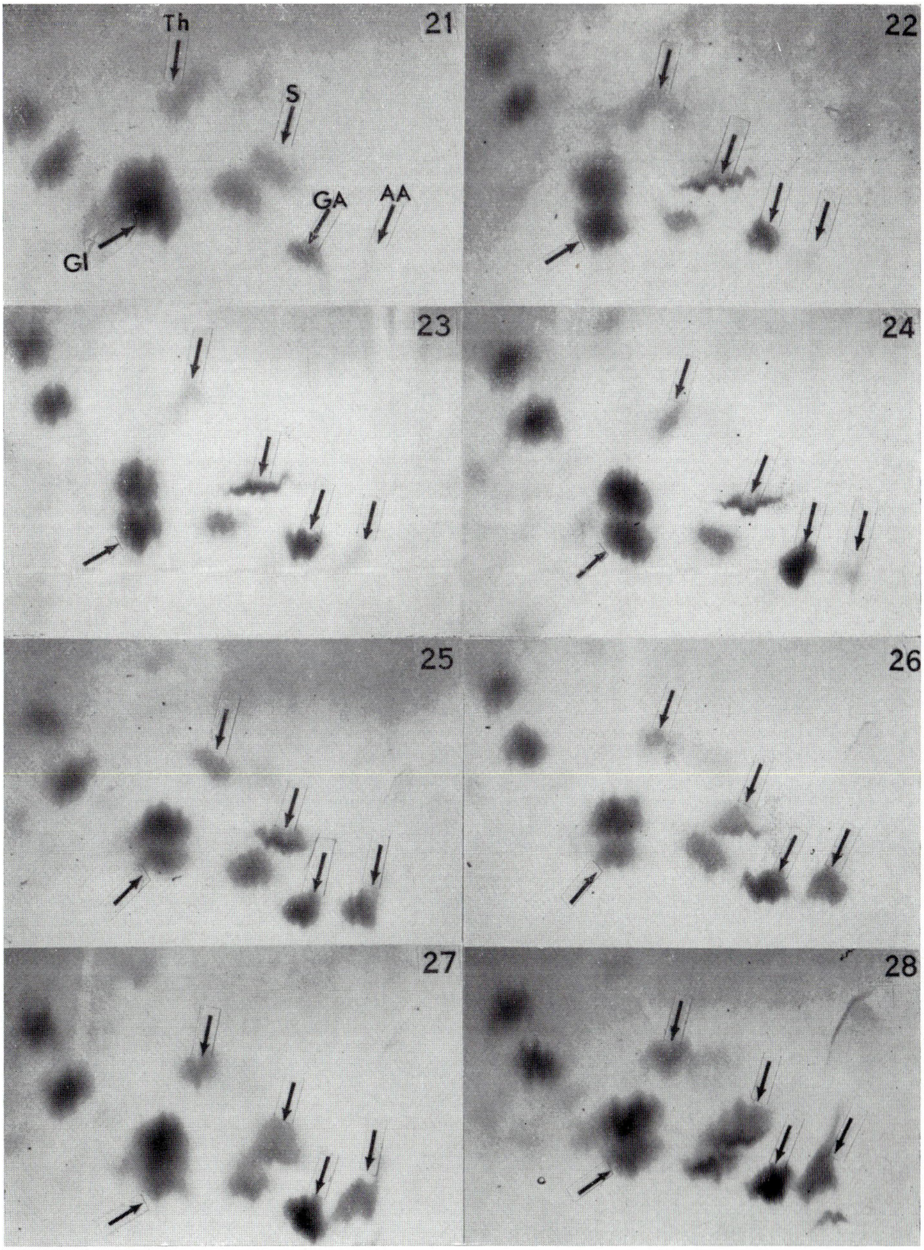

Figs. 21–28. Chromatograms prepared from samples removed from an *in vitro* incubation study of erythrocytes in plasma (see text). 0.25 ml of plasma and red cells used. Figs. 21–24 from plasma and Figs. 25–28 from erythrocytes removed at zero time, and after 30, 120, and 240 min of incubation. The uppermost arrows point to threonine (left) and serine (right) that separated widely on this particular batch of filter paper. The lower row of arrows point to (from left to right) glutamine, glutamic acid, and aspartic acid. Note the increase of the acidic amino acids in both plasma and cells during the incubation. For abbreviations see p. 369.

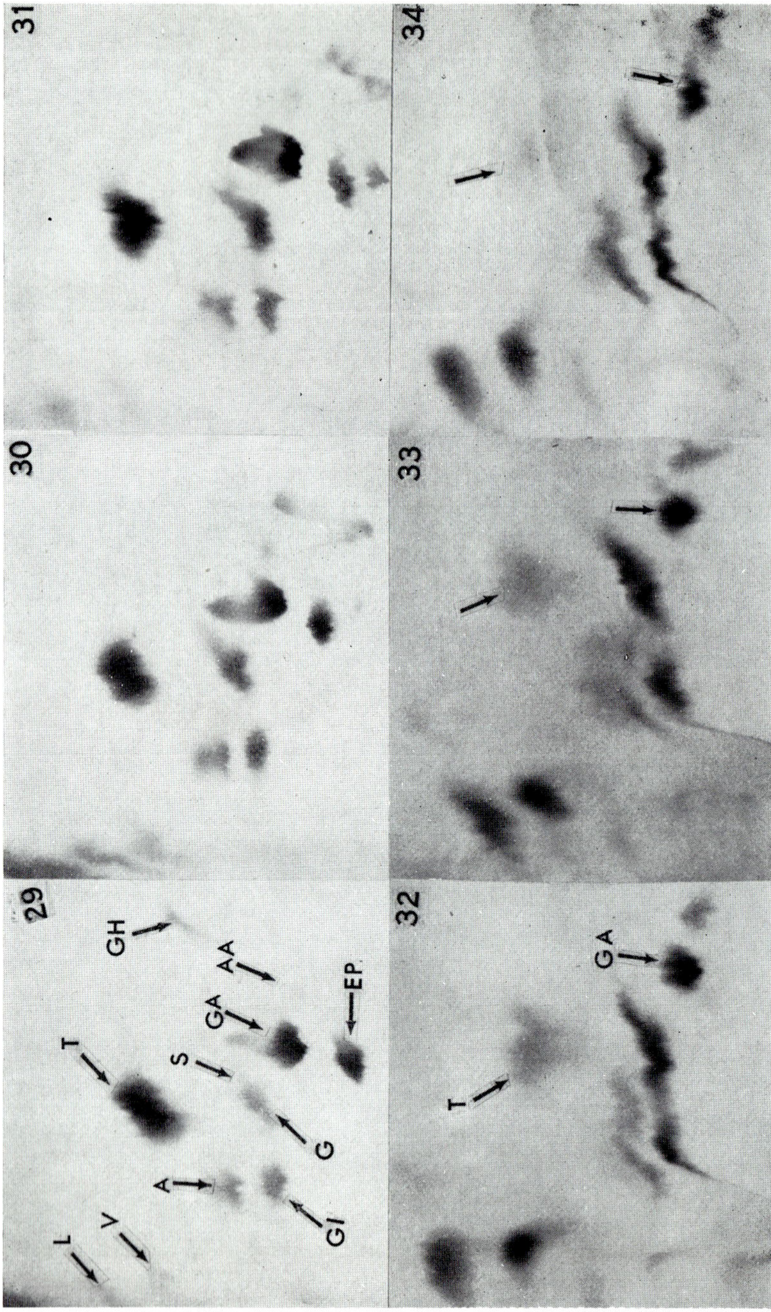

Figs. 29–31. Extract of 50 mg of leukocytes from a patient with chronic granulocytic leukemia after incubation for 1, 2, and 3 h. Note the great similarity of the chromatograms and the relative lack of change during the incubation.

Figs. 32–34. Equivalents of 0.3 ml of plasma from a patient with chronic granulocytic leukemia to illustrate the venipuncture response. Blood samples drawn at 15-min intervals. The arrows point out the fall in the levels of taurine (upper) and glutamic acid (lower).

For abbreviations see p. 369.

rocytes, although such changes are too small to be apparent with the procedure finally adopted for the separation of plasma and cells.

In vitro incubation studies with blood from patients with chronic granulocytic leukemia

Studies similar to those carried out with blood from chronic lymphatic patients were done with blood from chronic granulocytic leukemia patients, and in addition an instructive series of incubations (a total of 7 *in vitro* studies) were done where blood, after being withdrawn in the usual manner, was allowed to stand in a centrifuge tube at room temperature (about 25°) for periods up to 4 h. The blood was then centrifuged and the cells and plasma separated in the usual manner. It was found that leukocyte free amino acid patterns did not change. Glutamine was not degraded and glutamic and aspartic acids were not increased in amount. Figs. 29–31 show white cells recovered from whole blood allowed to stand for 1, 2, and 3 h at room temperature without shaking. No definite changes were observed.

In contrast to the very consistent pattern obtained from the leukocytes, it was found that plasma and red cell samples might show large changes related to the fragmentation of platelets and perhaps white cells, and to the increased difficulty of removal of the leukocytes from the erythrocyte mass after the blood had stood for several hours. From these studies it was concluded that blood from patients with chronic granulocytic leukemia should be processed immediately, but that artefacts would be unlikely with the procedure adopted for routine work.

The fall of plasma free amino acids following withdrawal of blood samples

Studies were carried out to demonstrate the reproducibility of the blood free amino acid picture over the period of one day. It became apparent that a fall of plasma and cell amino acids could follow a single venipuncture. This decrease of amino acids has been observed as early as 10 min after a single venipuncture, and appears to reach a maximum within 30 min with a return to or near the initial level in about 1 h. The response may not be observed when venipunctures are spaced at intervals of 1 h or more. The response has been observed in normal and leukemic man and in rabbits and is termed the venipuncture response.

The lowering of the free amino acids of the blood plasma in response to venipuncture has been observed to be most marked for taurine and glutamic acid, although the levels of other amino acids may decrease. Figs. 32–34 illustrate the nature of the response of plasma free amino acids in a chronic granulocytic leukemia patient (H. Gol. ♂) with a white count of 450 000/mm³. The three samples were drawn at 15-min intervals. There was a striking decrease in both taurine and glutamic acid as indicated by the arrows in Figs. 32–34. A clear demonstration of the venipuncture response in a rabbit is shown in Figs. 35–39. 15 min after the first blood sample was drawn a decrease in most of the amino acids of plasma was observed. A return toward control levels was evident in the 30-min sample.

A patient with chronic granulocytic leukemia (D. Fol., a woman with a leukocyte count of 157 000 cells/mm³) who had not received any form of treatment for the disease was sampled over a period of 4 h beginning at 9.30 in the morning. Just after the first sample was taken the patient drank 400 ml of distilled water. The same

References p. 447/448

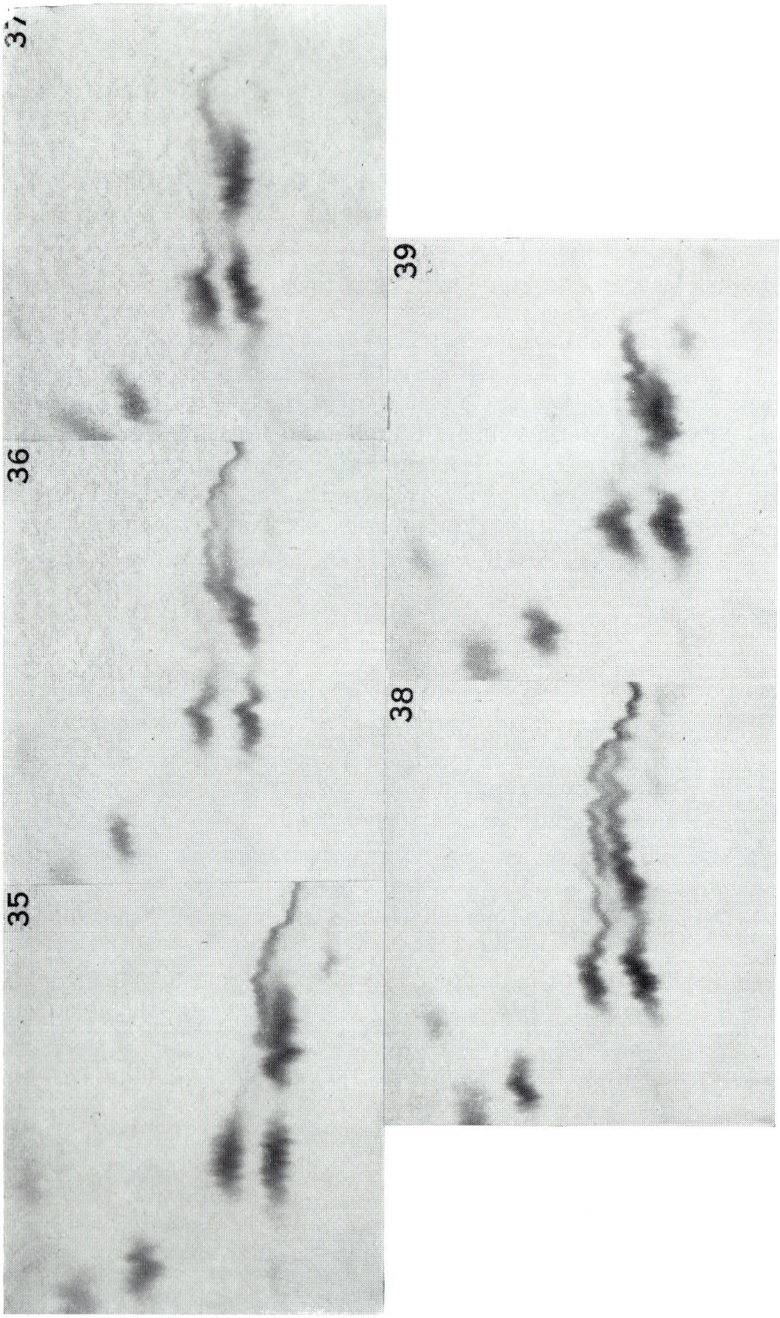

Figs. 35–39. Equivalents of 0.3 ml of plasma from a rabbit to illustrate the venipuncture response Samples removed at zero time, 15 min, 30 min, 1 h and 2 h. Note the marked decrease of all free amino acids in the 15-min sample (Fig. 36).

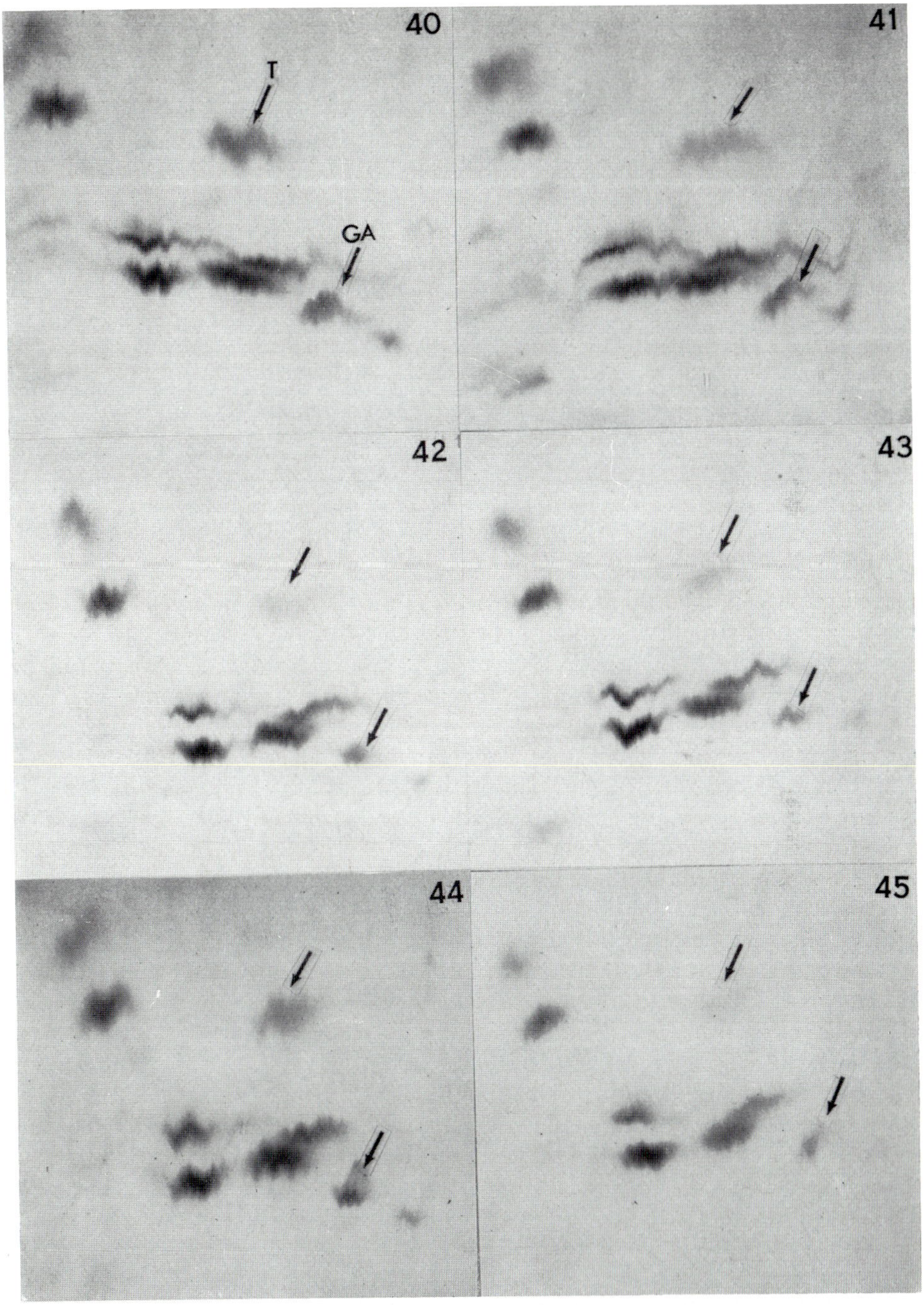

Figs. 40–45. Response of plasma free amino acids of patient (D. Fol.). Equivalents of 0.3 ml of plasma removed at zero time, 30 min, 1, 2, 3, and 4 h, respectively, after ingestion of water. Note changes indicated by arrows. Taurine, T; Glutamic acid, GA.

procedure used in glutamine ingestion studies (part VI of this series) was followed except that glutamine was not given.

Figs. 40–45 show the plasma samples obtained at zero time, 30 min, 1, 2, 3, and 4 h. The changes in the blood plasma levels were of particular interest as they were most marked at 1 h. This appears to be associated largely with the ingestion of water. A drop in the taurine and glutamic acid levels is evident. Figs. 46–51 show the free amino acids of the erythrocytes at the same time intervals recorded for plasma, Some slight fluctuations of the free amino acids of the erythrocytes can be observed. although the response seen in the plasma is not evident. We have observed on many occasions that changes may take place in plasma free amino acid levels that are not reflected in the erythrocyte levels. Figs. 52–57 show the free amino acid levels of leukocytes from the same blood samples. A very distinct drop in both taurine and glutamic acid in leukocytes was evident in the 1-h sample. Small differences in other amino acid levels can be seen as well. Leukocytes showed the changes seen in plasma. Other examples of the venipuncture response are presented in part VI of this series.

The effects of food intake on blood free amino acids

The effect of food upon the free amino acids of blood plasma and blood cells was of interest as some patients were known or suspected to have eaten prior to the withdrawal of a routine blood sample. Patient H. Gol., a man with chronic granulocytic leukemia who had received 40 mg of dimethylmyleran 4 months previously, was studied. The patient's control plasma showed an increase of free amino acids above the level of normal individuals (also observed prior to any form of treatment) at the time the study was carried out. This relatively high level of blood amino acids appeared to be characteristic for this patient in the untreated state. The patient appeared at the out-patient clinic at 9 AM and control blood samples were drawn at 9.15 and 9.35 AM. The patient was then allowed to eat his usual meal consisting of 2 medium-sized scrambled eggs, 1 small fried ground beef patty, 2 pieces of toast without butter, a small portion of potatoes, and black coffee. Food was consumed between 10.05 and 10.20 AM. and another blood sample was taken at 10.30, followed by samples at 11.30, 12.30, 1.30 and 2.30. Figs. 58–63 illustrate the findings in blood plasma and Figs. 64–69 the findings with erythrocytes at the times stated except that the blood sample drawn at 1.30 is not shown.

Except for a small drop of taurine in the second sample prior to eating (venipuncture response), the plasma amino acid pattern remained very nearly the same throughout the study. No marked changes of free amino acids occurred as a result of the ingestion of food. The erythrocyte free amino acid levels did not change. Figs. 70–75 illustrate the findings with the leukocyte samples at the same time intervals for plasma and erythrocytes. The chromatograms were very similar and it is evident that the consumption of food did not change the free amino acid pattern of leukocytes.

The results of this study are in agreement with the observations made during the course of routine examinations of patients where an occasional known food intake could not be correlated with a change in the free amino acid levels of blood plasma or cells (when compared to samples taken on other occasions with the patient in the fasting state).

References p. 447/448

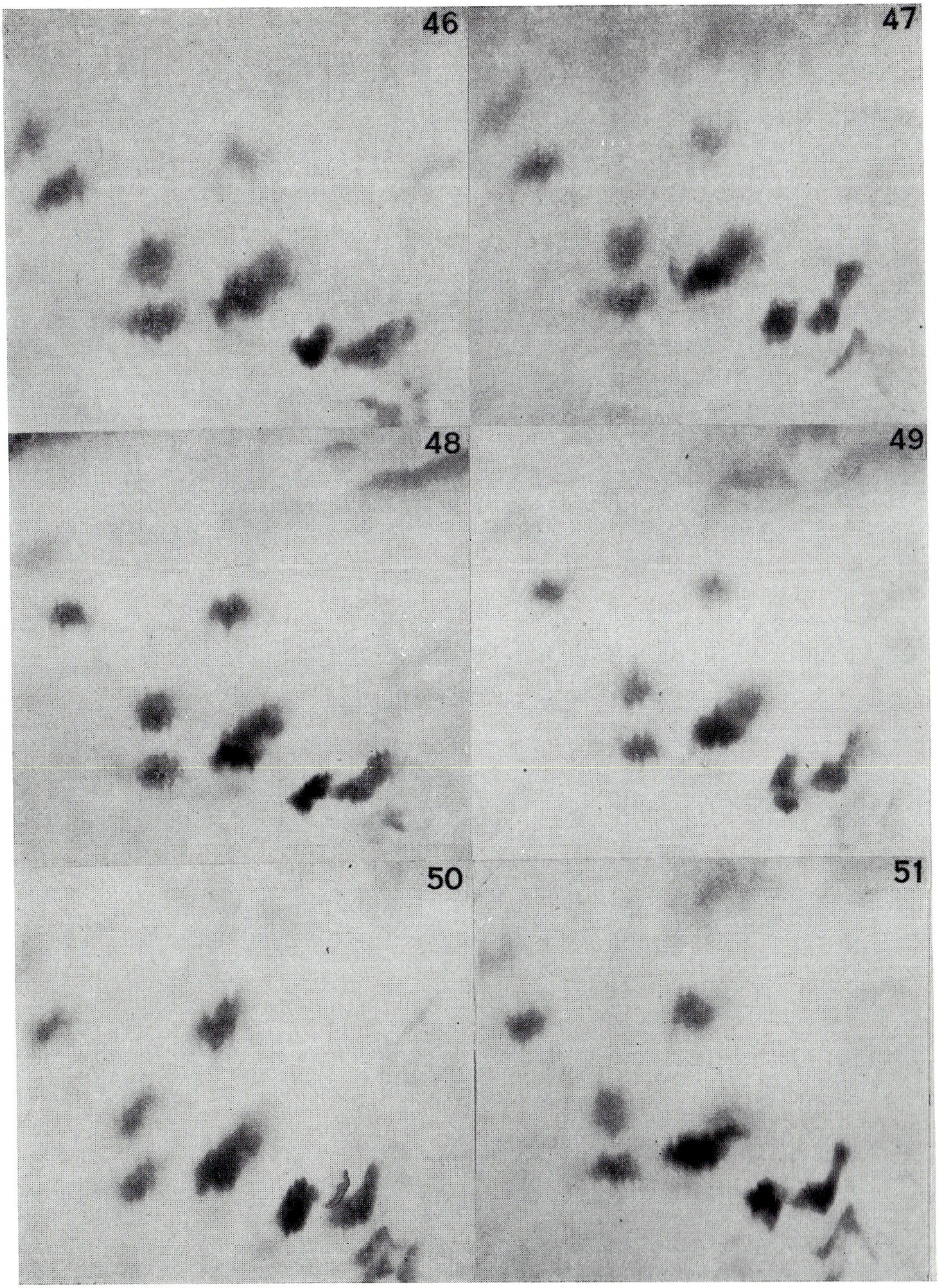

Figs. 46–51. Equivalents of 0.3 ml of packed erythrocytes from patient D. Fol. (see legend Figs. 40–45). Erythrocyte free amino acids did not change.

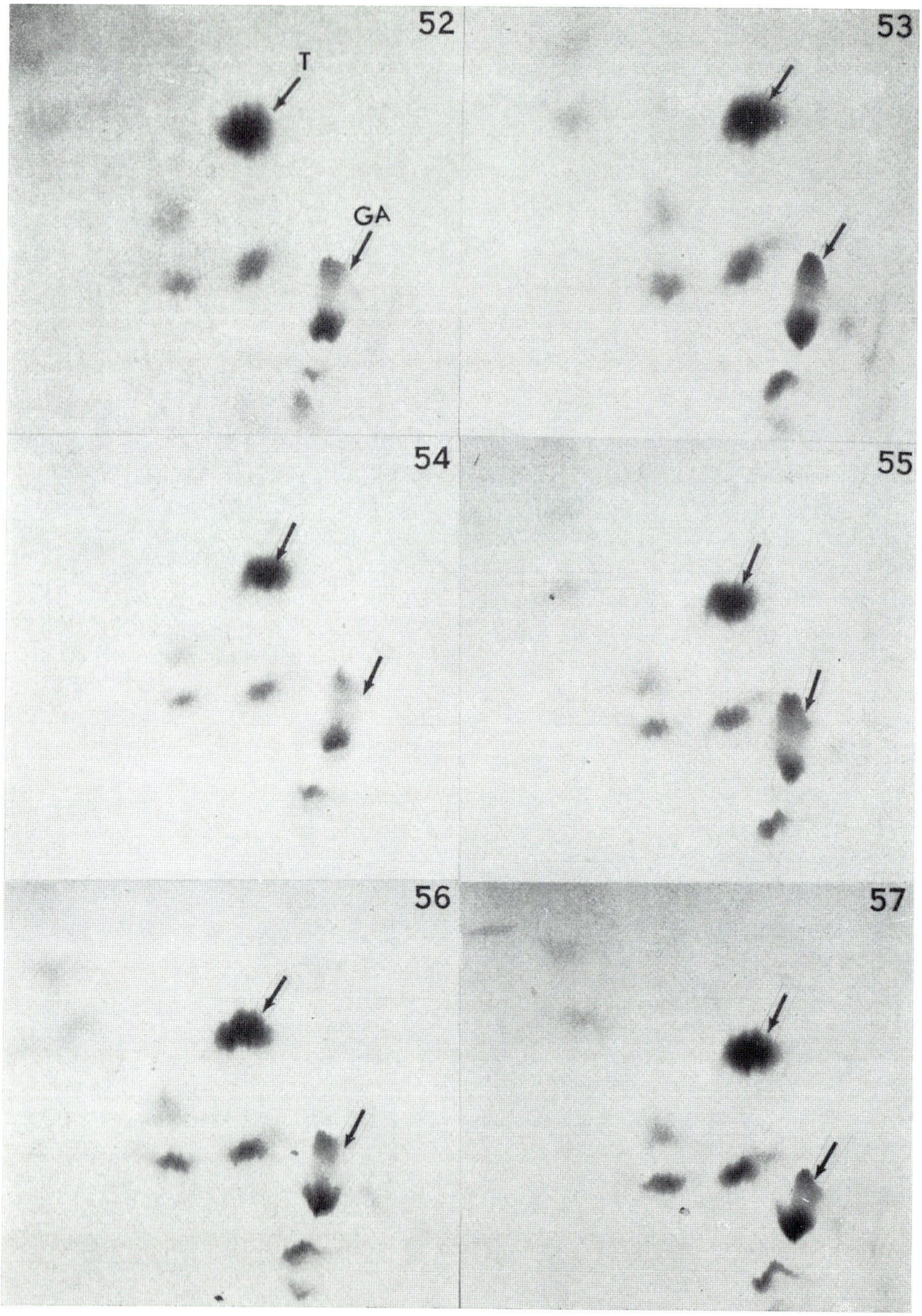

Figs. 52–57. Extracts from 50 mg of leukocytes from patient D. Fol. (see legend Figs. 40–45). There was a marked fall of all the free amino acids in the 1-h sample (Fig. 54). Taurine, T; Glutamic acid, GA.

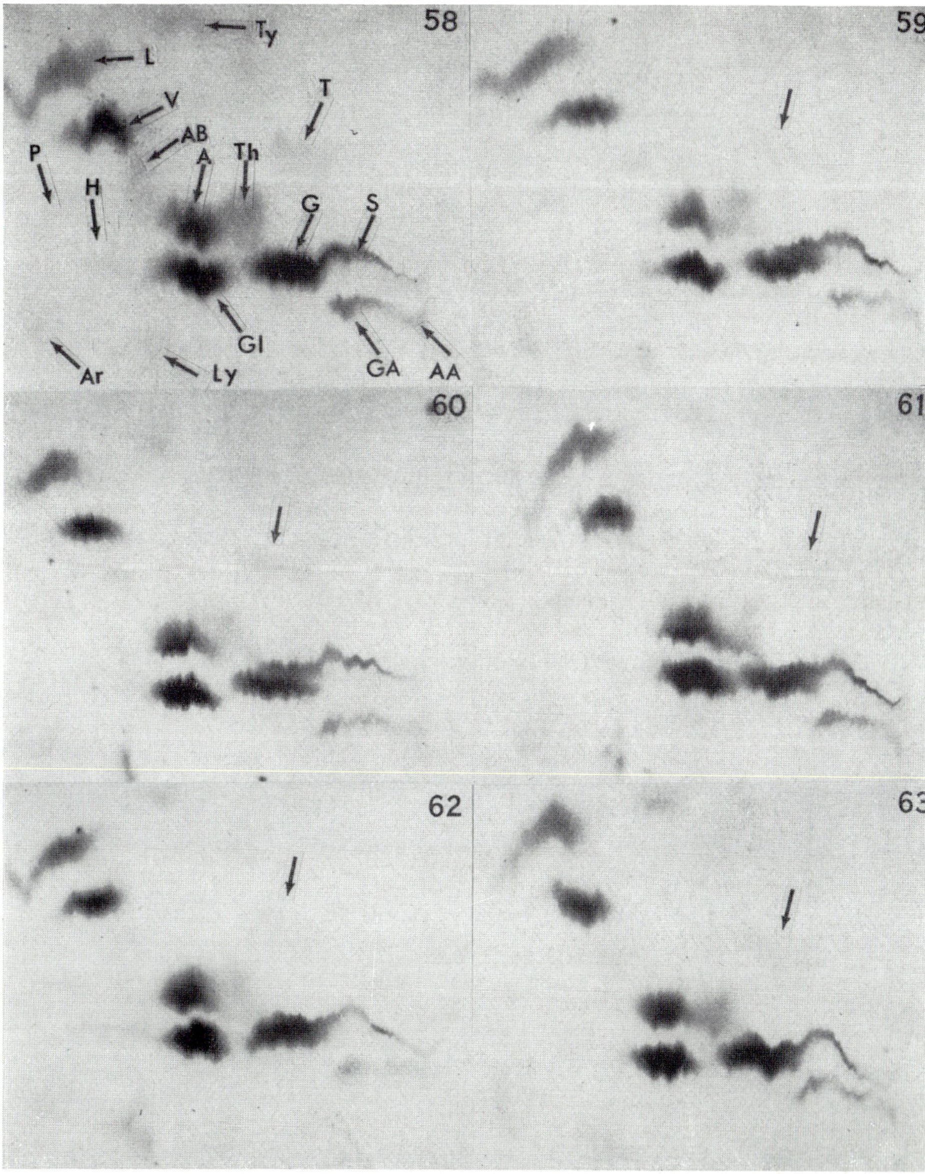

Figs. 58–63. Equivalents of 0.3 ml of plasma from H. Gol. with chronic granulocytic leukemia to illustrate the absence of changes after ingestion of food. Samples removed at 9.15, 9.35, 10.30, 11.30 AM, 12.30 and 2.30 PM, respectively. The arrow to taurine indicates the typical fall of taurine following venipuncture.

Abbreviations: alanine, A; α-amino *n*-butyric acid, AB; arginine, Ar; asparagine, As; aspartic acid, AA; cysteic acid, CA; cysteinylglycine, CG; ethanolamine-O-phosphate, EP; glutathione, GH and GO (oxidized); glutamic acid, GA; glutamine, Gl; glycerylphosphorylethanolamine, GPE; glycine, G; histidine, H; leucines, L; lysine, Ly; methionine, M; 1-methylhistidine, 1-MH; 3-methylhistidine, 3-MH; proline, P; serine, S; taurine, T; threonine, Th; tyrosine, Ty; tyrosine-O-sulfate, TS; valine, V.

Samples applied to the lower right hand corner of each chromatogram; phenol/ammonia was used for development in the horizontal direction and lutidine/water in the vertical direction.

References p. 447/448

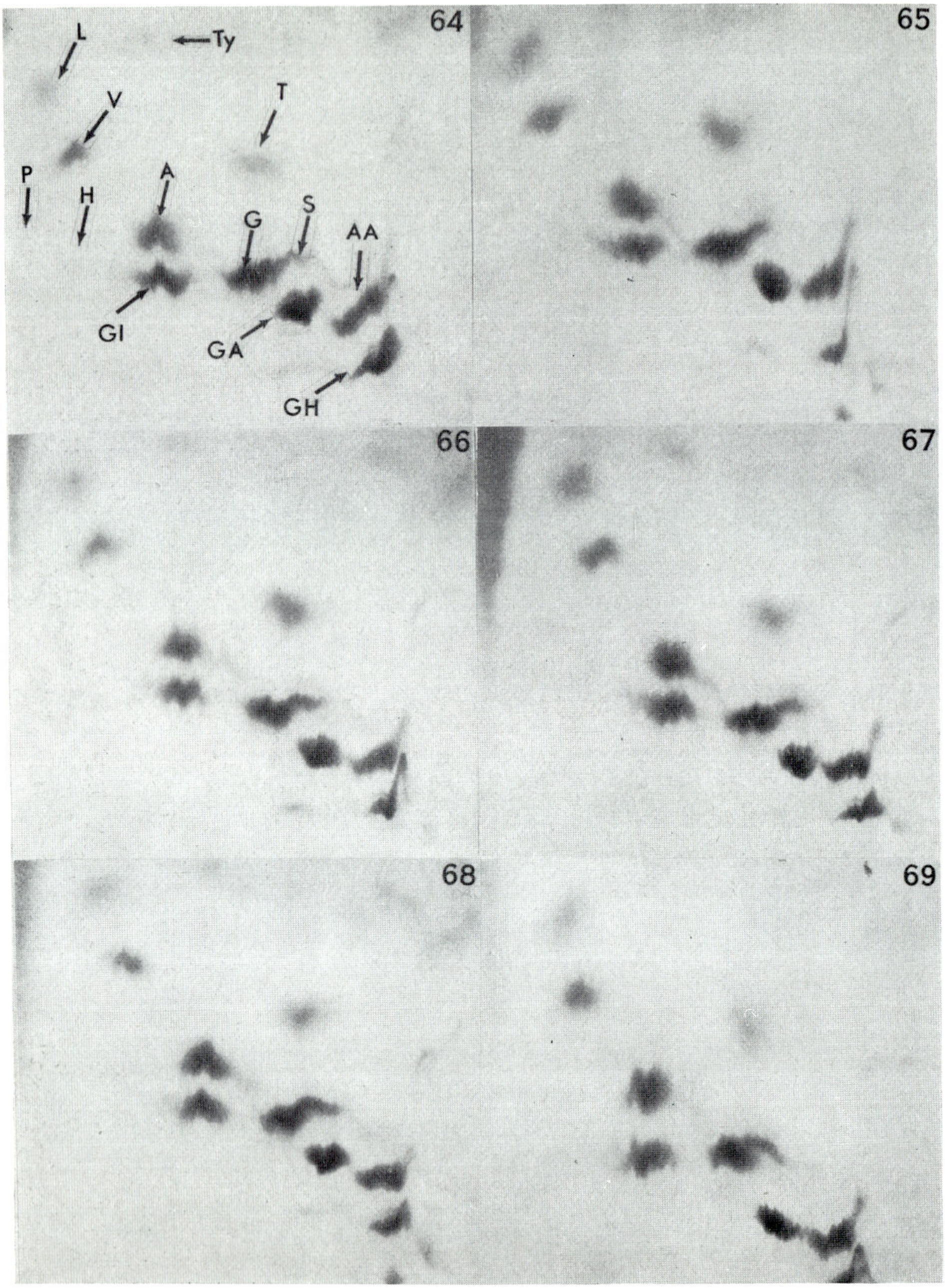

Figs. 64–69. Equivalents of 0.3 ml of packed erythrocytes from H. Gol. after food intake (see legend Figs. 58–63). Note the relative constancy of the erythrocyte free amino acid pool. For abbreviations see p. 369.

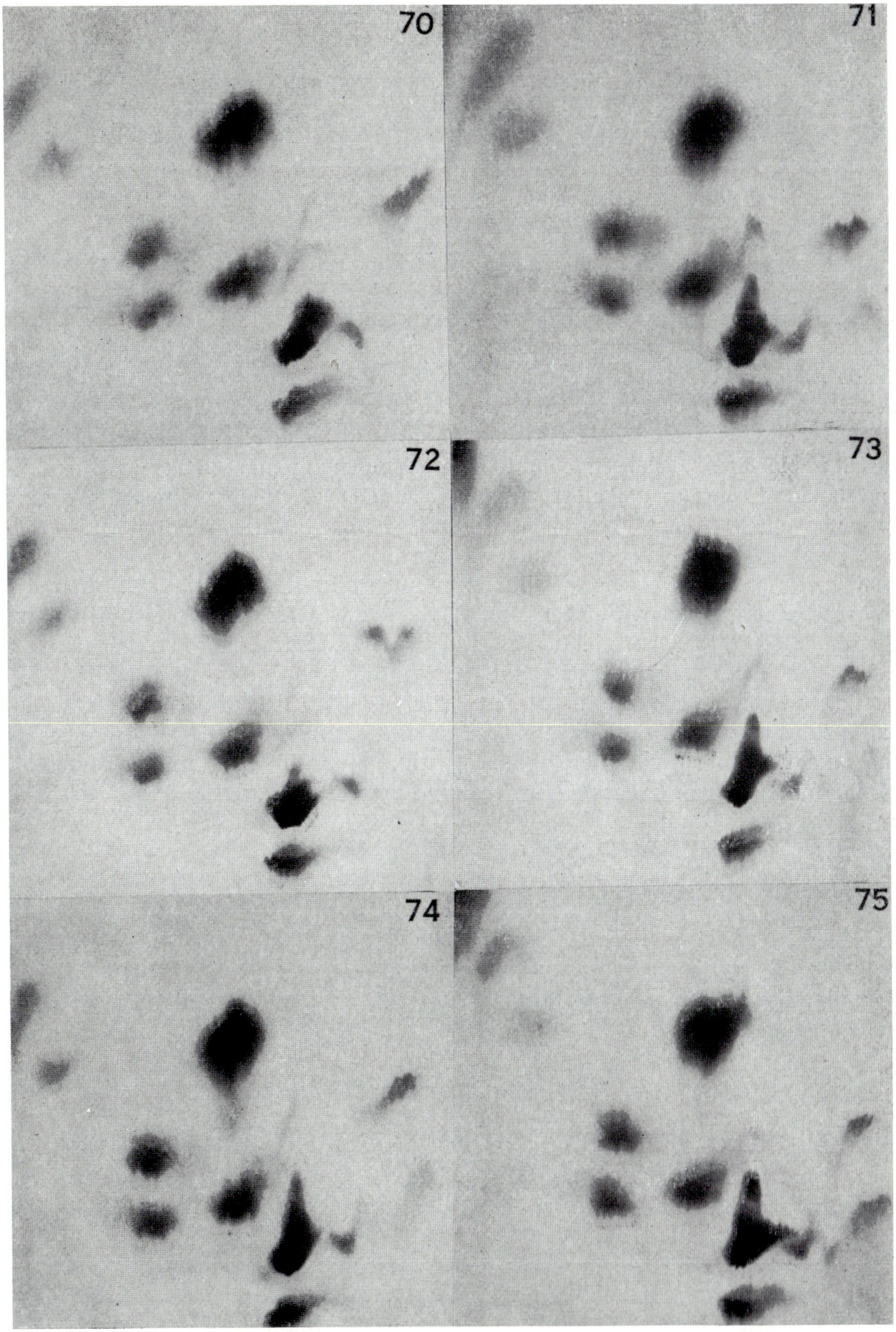

Figs. 70–75. Illustrate the findings with extracts of 50 mg of leukocytes from H. Gol. after food intake (see legend Figs. 58–63).

Effects of fasting and food intake on the free amino acids of plasma and erythrocytes in rabbits

The effects of fasting and food intake in rabbits were studied particularly in connection with simultaneous investigations on the effects of nitrogen mustard. The results are presented in part IV of this series (Figs. 157–162). The plasma samples showed minimal variations and it can be concluded that there are no major changes produced by a short fast or normal food consumption.

CONCLUSIONS

The data demonstrate a good degree of reproducibility of the chromatographic method and point out a number of difficulties and pitfalls that may be encountered in studies of this type. Special considerations for the handling of blood from patients with leukemia have been disclosed by *in vitro* incubation studies.

A marked decrease in some or all amino acids in blood plasma, and to a lesser extent in blood cells, has been found to follow venipuncture in both man and animals and should be considered in all studies of free amino acids of blood. Previous investigators have failed to note the response and it is possible that this failure is due to the fact that blood samples are usually withdrawn at hourly intervals, whereas the venipuncture response occurs within a few minutes and the control level is generally reached within one hour. Furthermore, all the amino acids are seldom involved in the response and the determination of total amino acid nitrogen or determination of a few amino acids may not disclose the changes that frequently take place.

The study of the effects of eating have disclosed that little or no change is to be expected as a result of the intake of food in the morning in the normal manner for the patient. Similar results were obtained with rabbits. No consistent effects due to fasting for 23 or 33 h or eating were observed upon the free amino acids of plasma and erythrocytes of rabbits.

References p. 447/448

FREE AMINO ACIDS IN THE BLOOD OF MAN AND ANIMALS

II. NORMAL INDIVIDUALS AND PATIENTS WITH CHRONIC GRANULOCYTIC LEUKEMIA AND POLYCYTHEMIA

GEORGE ROUSER, KEITH KELLY, ARTHUR J. SAMUELS, BOHDAN JELINEK AND DOROTHY HELLER

Departments of Biochemistry and Medicine, City of Hope Medical Center, Duarte, Calif. (U.S.A.)

Part I of this series presents the methods used for the study of the free amino acids of blood and urine and the results of various control studies. This report presents the findings for normal individuals and patients with diseases affecting the myeloid series of cells.

General nature of the free amino acid pools in leukemia as compared to normal

After the examination of 199 blood samples, 20 urine samples, and 15 bone marrow specimens from 16 patients with chronic granulocytic leukemia it was concluded that the free amino acids of plasma, cells, and urine of untreated leukemic patients were qualitatively similar to those from normal individuals. No new compounds were observed in leukemic patients except after treatment. The amino acid levels of the blood in leukemia and polycythemia were observed to deviate on occasions from the levels observed in normal individuals.

Blood plasma

Approximately one-half of the plasma samples from the leukemic patients examined were within the normal range. This includes samples taken both before and after drug administration. Drugs may bring about a variety of changes in plasma free amino acids (see parts IV and V). Deviations from the normal free amino acid patterns for untreated patients were of two general types: an elevation of the levels of most of the plasma amino acids, or the distinct elevation of the level of glutamic acid only. Illustrations of the maximum deviations from normal will be emphasized here. Numerous additional examples of findings for normals and leukemics are presented in other parts of this series.

Fig. 76 shows the free amino acids found in blood plasma of a normal male (E. Eme.). The sample was chosen for the present illustration because it shows one of the highest levels of glutamic acid that we have observed in a normal individual. This is to be contrasted with the plasma sample shown in Fig. 77 obtained from a patient (E. McG., ♀) with chronic granulocytic leukemia that shows an elevation of the plasma glutamic acid level. The patient had a 35 000/mm^3 white cell count and had been treated with myleran until 2 weeks prior to examination. At the time of examination the patient's count was rising and the blood contained many immature cells (49% polymorphonuclears, 26% myelocytes, 5% promyelocytes, 16% myeloblasts, 1% eosinophils, 1% monocytes, and 2% lymphocytes).

Fig. 78 illustrates the plasma free amino acid pattern from a patient whose plasma

References p. 447/448

showed an elevation of all of the amino acids with a particularly large elevation of the glutamic acid level. This patient (L. Gre. ♀) with polycythemia of several years standing had not been treated (other than periodic blood withdrawal) and at the time of examination had a white count of 30 000/mm³ and a thrombocytosis. Almost all (95%) of the white cells in the peripheral blood were neutrophilic polymorphonuclear leukocytes. The more extreme cases shown in Figs. 77 and 78 are to be compared with the results from other patients (see parts V and VI). A rather marked elevation of the plasma glutamic acid level is shown for a patient (H. Gol. ♂) with granulocytic leukemia in the first part of this series (Figs. 32–34).

Aspartic acid is not detectible on the chromatograms shown in Figs. 76–78. Aspartic acid was not observed on a number of chromatograms from normal plasma, and was frequently absent from the plasma samples obtained from patients with chronic granulocytic leukemia, particularly after drug administration with a reduction in the leukocyte count (see part V). Aspartic acid in normal plasma is illustrated in part VI (Figs. 268–277, 304–308 and 314–318).

Platelets and leukocytes

Fig. 79 shows a normal platelet free amino acid pattern (subject E. Eme.), while Figs. 80 and 81 show the platelet free amino acids from the two patients (E. McG. and L. Gre.) whose plasmas were presented in Figs. 77 and 78. Fig. 82 shows the leukocyte free amino acid pool from the same normal individual (E. Eme.), while Fig. 83 (patient E. McG.) and Fig. 84 (patient L. Gre.) are chromatograms from leukocytes of the patients whose plasma and platelet free amino acids are illustrated (Figs. 77, 78, 80, 81). By comparison of Figs. 79 and 82 (as well as 80 and 83, and 81 and 84) it is evident that blood platelets and neutrophilic polymorphonuclear leukocytes from the same blood samples have similar free amino acid patterns. The platelets and leukocytes from patients with chronic granulocytic leukemia tend to have somewhat higher glutamic acid and somewhat lower taurine levels than normal platelets and leukocytes.

Figs. 85–87 show the free amino acid pools of three different preparations of neutrophilic polymorphonuclear leukocytes isolated from normal human blood. These preparations are shown at a lower aliquot size (equivalent to 50 mg of packed leukocytes).

Fig. 88 presents the free amino acid pool of leukocytes obtained from a bone marrow puncture of patient E. McG. whose peripheral blood leukocyte free amino acids are shown in Fig. 83. Note the increased amount of ethanolamine phosphate in particular as compared to the peripheral blood cells. The glutamic and aspartic acid levels are also higher while the taurine level is reduced in cells from bone marrow. Most bone marrow specimens large enough to examine by paper chromatography contained so much venous sinus blood that the cell populations and chromatographic results were similar to those from peripheral blood.

Fig. 89 shows the free amino acid pool in an almost pure preparation of myelocytes. The blood sample was drawn from a patient with a white cell count of 125 000/mm³. The white cell preparations from the patient were 92% or more myelocytes. Only a few polymorphonuclear leukocytes and myeloblasts were seen in the peripheral blood. Note the higher level of ethanolamine phosphate in particular in the myelocytes as compared to neutrophilic polymorphonuclear leukocytes from leukemia

References p. 447/448

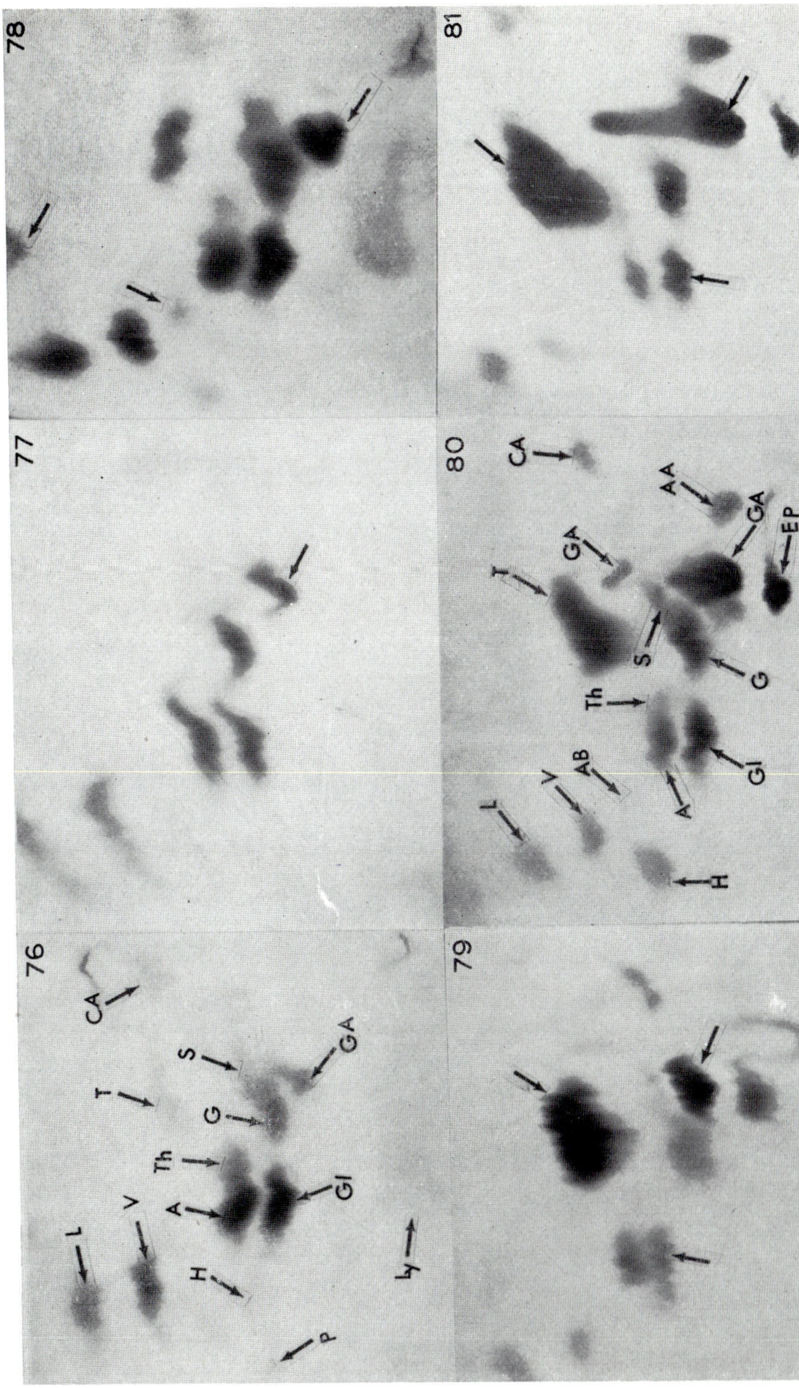

Figs. 76–78. Extracts of 0.5 ml of plasma from a normal individual, a patient with chronic granulocytic leukemia, and a patient with polycythemia vera, respectively. Note the very high levels of glutamic acid in plasma from the patients and the marked elevation of total free amino acids in the plasma of the patient with polycythemia (Fig. 78, the upper 2 arrows point to tyrosine and α-amino n-butyric acid that are clearly visible when all free amino acids are increased).

Figs. 79–81. Extracts of 100 mg of normal human blood platelets, 100 mg of blood platelets from a patient with chronic granulocytic leukemia, and 160 mg of blood platelets from a patient with polycythemia, respectively. For abbreviations see p. 369.

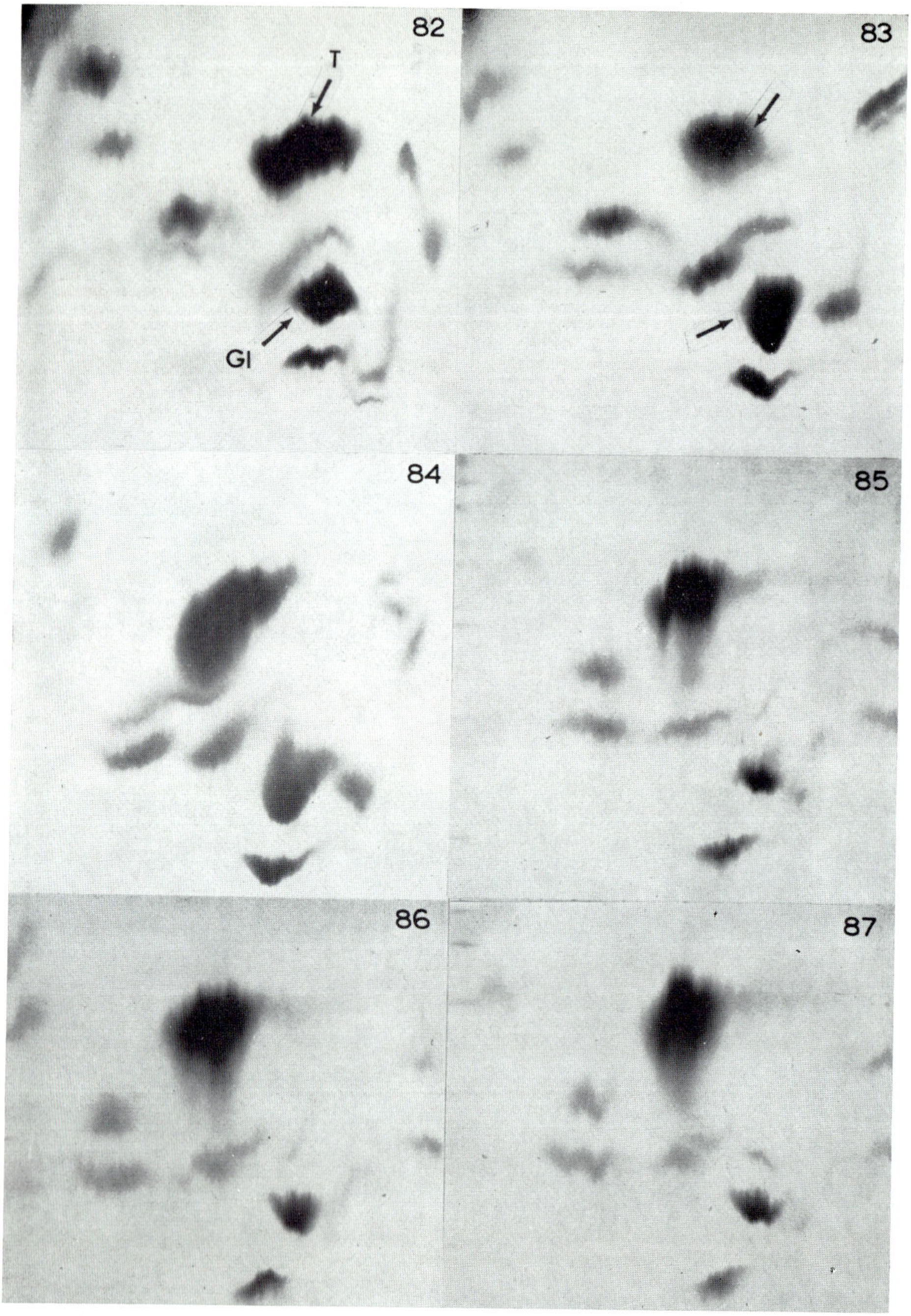

Figs. 82–87. Extracts of 100 mg of normal human leukocytes, 100 mg of leukocytes from a patient with chronic granulocytic leukemia, 160 mg of leukocytes from a patient with polycythemia, and extracts from 50 mg of leukocytes from 3 normal individuals (samples treated with hydrogen peroxide prior to paper chromatography). Note the lower taurine and higher glutamic acid levels in the leukemic cells.

References p. 447/448

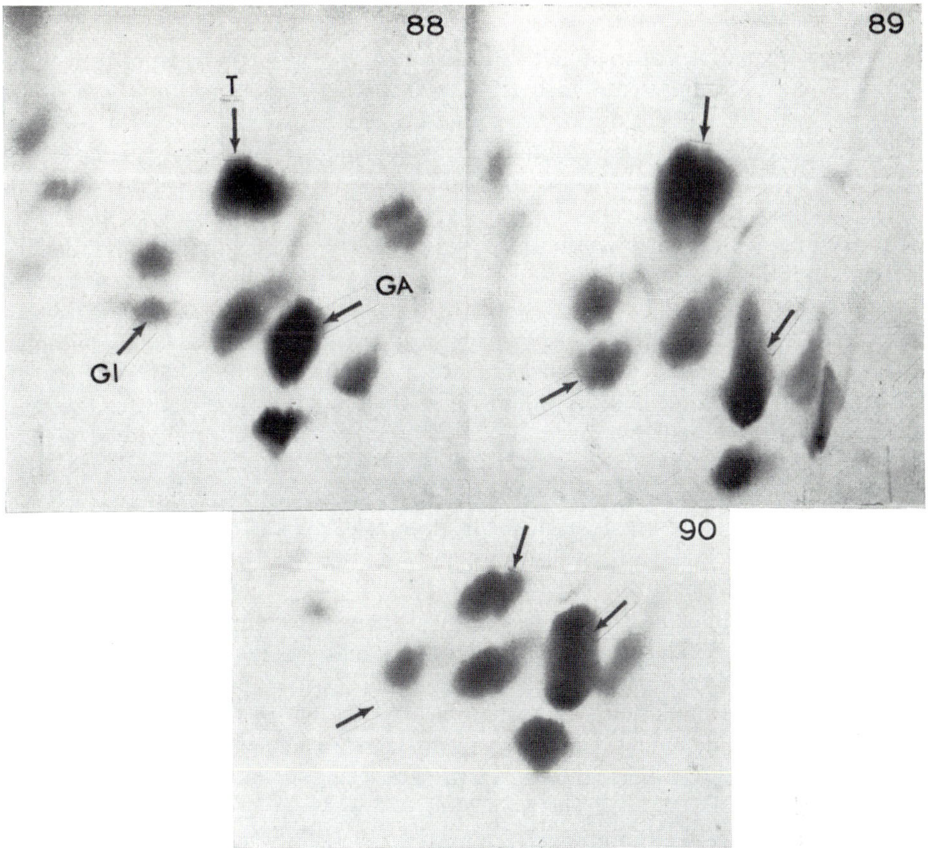

Figs. 88–90. Extracts of 100 mg of leukocytes from the bone marrow of a patient with chronic granulocytic leukemia (sample oxidized with peroxide), 100 mg of myelocytes removed from the peripheral blood of a patient with chronic granulocytic leukemia, and 50 mg of myeloblasts removed from the blood of a patient with acute myeloblastic leukemia. Note the marked differences in the free amino acid patterns of the different cell types. The upper arrow points to taurine, the one at the lower left to glutamine, and the one to the lower right to glutamic acid.

patients or normal individuals. There was an overall increase in the free amino acids in these cells as compared to the more mature cells and the relative distribution of constituents is different. It has been pointed out previously that myelocytes tend to have a large free amino acid pool and the highest levels of glutamine of the various cells in the myeloid series[2].

The free amino acid pool of a preparation of almost pure myeloblasts from a patient with acute myeloblastic leukemia is shown in Fig. 90. This sample was removed from peripheral blood and is notable for the extremely low level of glutamine, the moderately low level of taurine, and the very high levels of glutamic acid and ethanolamine-O-phosphate. Glutathione and cysteic acid have not been observed in myeloblast preparations. Myeloblasts appear to be unique in this respect as the other cells of the myeloid series contain both glutathione and cysteic acid (from cysteine and cystine) in readily detectable amounts.

References p. 447/448

Erythrocyte free amino acids

A number of examples of the free amino acid pools of erythrocytes in chronic granulocytic leukemia are presented in parts I, V and VI of this series. Figs. 91–93 show three erythrocyte samples drawn from the same normal individual on the same day. The cells shown in Fig. 92 were obtained from a blood sample drawn 30 min after the cells shown in Fig. 91, while cells shown in Fig. 93 were obtained from blood drawn 1 h after the blood sample shown in Fig. 92. The aliquot size is equivalent to 0.5 ml of packed red cells and shows quite clearly the free amino acid pool constituents of normal erythrocytes listed in the first paper of this series. Erythrocytes contain more free glutamate and aspartate than plasma. Note that Figs. 91–93 show the presence of α-amino *n*-butyric acid, proline, histidine, arginine, and ethanolamine-O-phosphate in normal erythrocytes. The marked similarity of the free amino acid pool of erythrocytes obtained from blood samples drawn on the same day demonstrates the reproducibility of the methods employed. It was observed that the same individual, whether normal or leukemic, showed some fluctuations of the erythrocyte free amino acid pool constituents from one day to the next, and that differences were more marked when samples were spaced at weekly or monthly intervals. Taurine is a highly variable constituent of both plasma and erythrocytes, and many erythrocyte samples do not show the presence of the small amount of ethanolamine-O-phosphate seen in other samples.

No consistent differences could be observed between erythrocyte free amino acid pools of normal individuals and patients with granulocytic leukemia. The variability of the free amino acid pools of both groups makes such comparisons difficult. It appeared that the erythrocyte free amino acid pools tended to be greater when the plasma levels of free amino acids were elevated in granulocytic leukemia. This was not, however, an invariable finding. Erythrocytes do not always respond to amino acid pool changes of plasma. The failure of erythrocytes to reflect plasma changes was observed after drug therapy and amino acid ingestion and will be discussed later.

Atypical findings

We have generally observed deviations from the more or less typical findings with plasma and cells of untreated patients after administration of cytotoxic drugs. Figs. 94–96 illustrate the findings in plasma, erythrocytes, and leukocytes of a blood sample from a patient (R. San.) with chronic granulocytic leukemia diagnosed 5 years prior to the time of sampling. The patient had received extensive myleran and X-ray therapy and 3 days before the blood sample was drawn had received 400 ml of fresh whole blood. The plasma sample (Fig. 94) from this patient had an extremely low glutamic acid level. A reduction in plasma glutamate occurs regularly after drug therapy and is considered in detail in the discussion of the effects of myleran and dimethylmyleran (part V). The erythrocyte free amino acid pool (Fig. 95) is of considerable interest in that glutathione is far above the levels seen in untreated patients. Fig. 96 shows leukocyte free amino acids that are distinctly different from the usual cell population of this type (the patient had a leukocyte count of 142 000/mm^3 with 42% polymorphonuclear leukocytes, 28% myelocytes and metamyelocytes, 7% promyelocytes, 3% myeloblasts, 5% eosinophils, and 15% basophils). The leukocyte ethanolamine-O-phosphate level is very high while the taurine level is lower than the level usually encountered. Erythrocytes may contain taurine when it is not observed in the blood plasma (Figs. 94 and 95). This

References p. 447/448

lack of correspondence between plasma and red cell taurine has also been observed in normal individuals.

Special findings in chronic granulocytic leukemia with a predominance of neutrophilic polymorphonuclear leukocytes in the peripheral blood

Two patients were studied who had an unusual peripheral blood picture. One had 90% and the other 95% neutrophilic polymorphonuclear leukocytes in the peripheral blood with only a few myelocytes and lymphocytes making up the remainder of the cell population. The free amino acid pool constituents in one of these patients has been described[2]. It was pointed out that this patient, as well as a patient with reticulum cell sarcoma and granulocytosis, showed very low leukocyte glutamine with very high glutamic acid and reduced taurine levels as compared to normal leukocytes. Another patient was found to show very nearly the same free amino acid pool changes. The patient (L. Gre.) was examined on a previous occasion in a polycythemic phase (Figs. 78, 81, 84). Another sample was withdrawn when the patient was in a leukemic phase. The leukocyte count was 55 000/mm^3 and the erythrocyte count had fallen to $4 \cdot 10^6$ cells/mm^3. The peripheral blood leukocytes were 95% neutrophilic polymorphonuclear leukocytes. The patient terminated three months later with a picture of chronic granulocytic leukemia. Fig. 103 shows the pool from leukocytes of the blood of a normal individual (R. Lea.) and Fig. 104 the leukocytes obtained from the patient. The arrows in the figures point to the characteristically low glutamine and high glutamic acid and aspartic acid levels of the leukemic cells. The leukocyte samples from these two individuals were obtained on the same day and processed together so that variations due to the methods employed are minimal. The plasma of the patient showed a high glutamic acid level similar to that illustrated in Fig. 78, while the erythrocyte free amino acids were not clearly different from normal.

That the findings in patients with high neutrophilic polymorphonuclear leukocyte counts is characteristic for this blood picture is indicated by the results from normal and leukemic dogs. Figs. 97–99 were prepared from the plasma, erythrocytes, and leukocytes, respectively, of a normal dog, while Figs. 100–102 were prepared from plasma obtained from a leukemic dog with a white cell count of 160 000/mm^3, 95% of which were neutrophilic polymorphonuclear leukocytes. The leukemic dog remained markedly anemic despite frequent blood transfusions. At autopsy the animal showed findings expected for granulocytic leukemia. The marked increase of glutamic acid in blood plasma of the leukemic animal is clearly shown, while the aspartic acid level is nearly normal. Fig. 101 shows the marked increase of glutamic acid in the erythrocytes and the almost complete absence of glutamine from the red cells of the leukemic dog in contrast to findings with erythrocytes of the normal dog (Fig. 98). Fig. 102 shows the marked increase of glutamic acid, and the distinct increase in aspartic acid in the neutrophilic polymorphonuclear leukocytes of the leukemic dog as compared to the morphologically similar cells from the normal dog. Glutamine is not detectable on the chromatogram from the leukemic dog. It is to be noted that the chromatogram shown in Fig. 99 was developed with lutidine–water for a slightly longer time than that shown in Fig. 102 so that the spots are spread out more in the vertical direction.

The results with the leukemic patients and the leukemic dog with large numbers

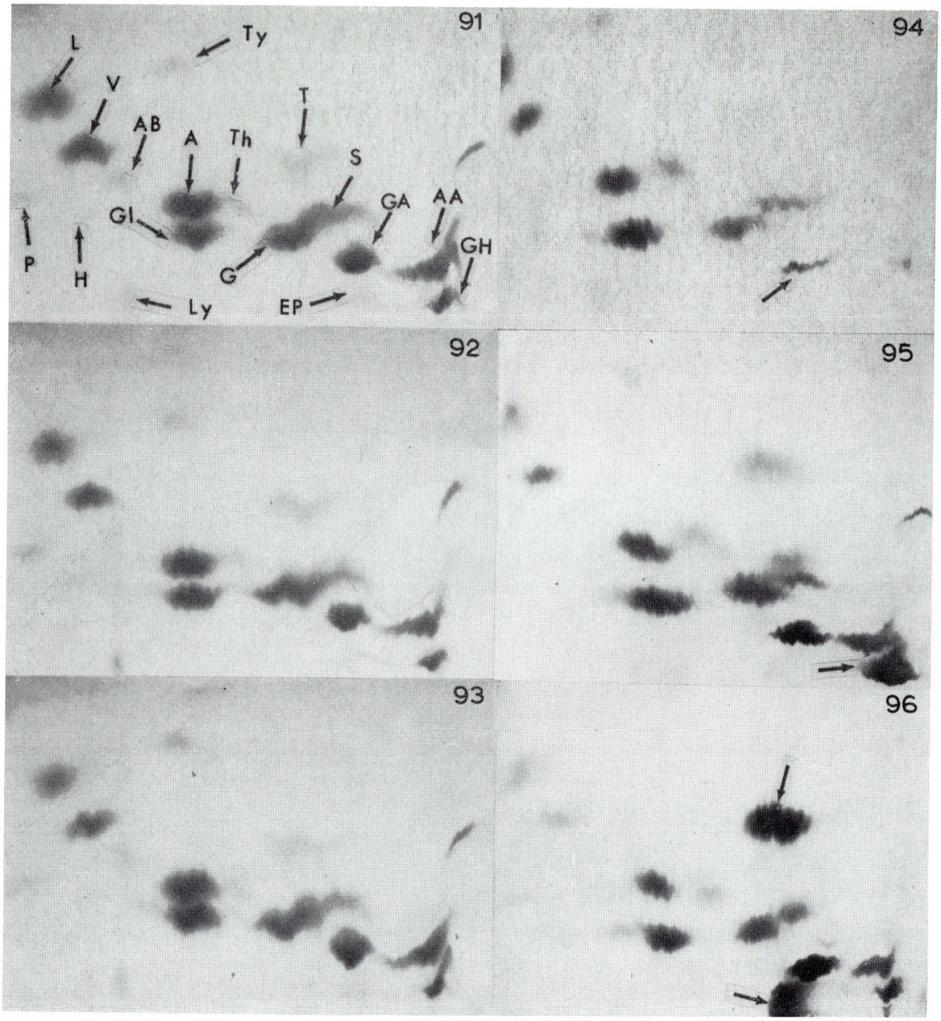

Figs. 91–93. Extracts equivalent to 0.5 ml of normal human red cells from the same individual at different times (see text). Note the marked uniformity of the chromatograms.
Figs. 94–96. Illustrate findings for plasma, erythrocytes, and leukocytes from a patient with chronic granulocytic leukemia after drug therapy. Extracts from 0.3 ml of plasma and erythrocytes and 50 mg of leukocytes were used. Note the extremely low level of glutamic acid in blood plasma (arrow, Fig. 94), the very high level of glutathione in erythrocytes (arrow, Fig. 95), and the relatively low level of taurine (upper arrow) and high level of ethanolamine-O-phosphate (lower arrow) in leukocytes (Fig. 96). For abbreviations see p. 369.

of neutrophilic polymorphonuclear leukocytes illustrate the marked differences that may be observed for individuals with granulocytic leukemia with different blood pictures. Some of these differences have been pointed out previously[2].

Urine

Figs. 105–107 illustrate the findings in urine from a normal male (subject R. Car.). These 24-h urine collections were obtained on three consecutive days. The urine

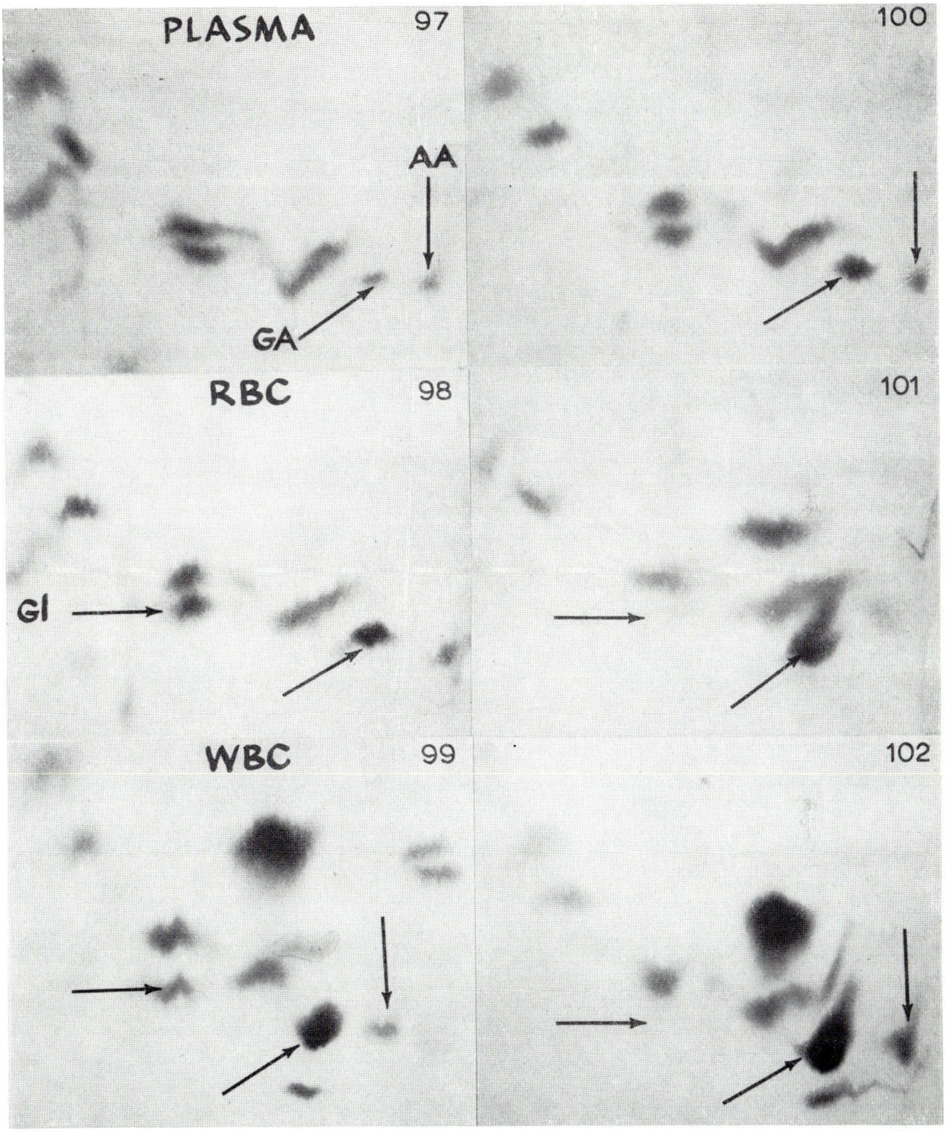

Figs. 97–102. Illustrate the findings in the blood of normal and leukemic dogs. Figs. 97–99 from extracts of 0.3 ml of plasma and erythrocytes, and 50 mg of leukocytes from a normal dog. Figs. 100–102 from extracts of 0.3 ml of plasma and erythrocytes and 50 mg of leukocytes from a dog with granulocytic leukemia. The arrows indicate the marked elevation of glutamic acid in the plasma, erythrocytes, and leukocytes and the increase in aspartic acid in plasma and leukocytes of the leukemic dog. Note the extremely low level of glutamine in erythrocytes and leukocytes from the leukemic dog.

volumes were 1230 ml, 855 ml, and 740 ml on the first, second, and third days, respectively. For each paper chromatogram an equivalent of 0.5 ml of urine was applied to paper. Aside from the gradual decrease in the content of 1-methylhistidine, free amino acids remained surprisingly constant over the three day period. Fig. 108

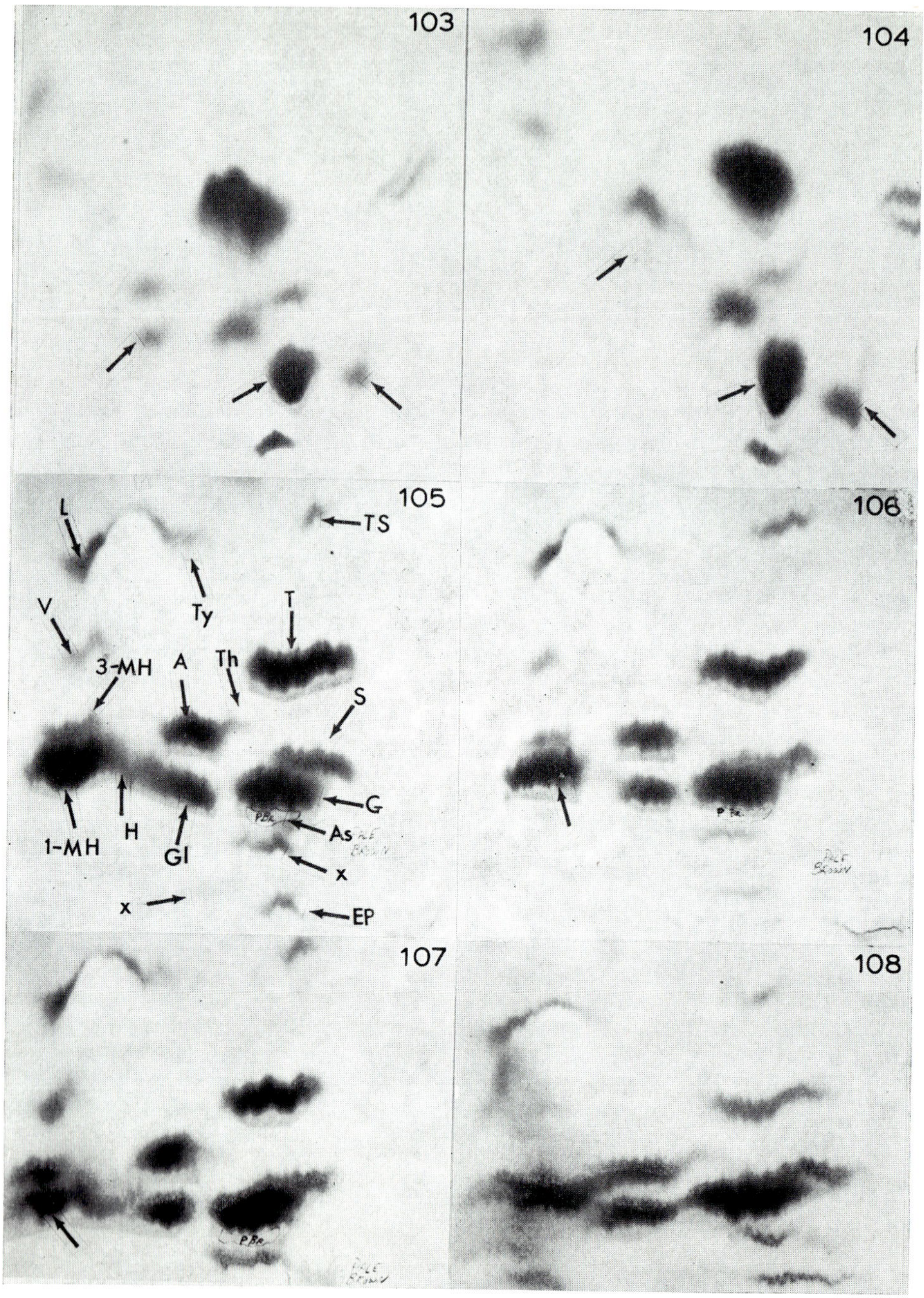

Figs. 103 and 104. Compare the free amino acid pools of human normal and leukemic neutrophilic polymorphonuclear leukocytes (extracts from 100 mg). The arrow to the left in each figure is to glutamine (low in leukemic cells); middle arrow to glutamic acid (higher in leukemic cells); and arrow to the right to aspartic acid (also higher in leukemic cells).

Figs. 105–108. Illustrate the findings with normal urine samples (extracts of 0.5 ml of urine in each case). Figs. 105–107 from the same individual on 3 consecutive days (urine volumes 1230, 855, and 740 ml). Note steady fall in level of 1-methylhistidine and the general uniformity of the chromatographic results. For abbreviations see p. 369.

illustrates the findings with another normal individual (A. Knu.). Evidently the normal urine pattern is moderately constant, but large differences in the excretion of taurine and the methylhistidines are noted in particular.

Fig. 109 shows a fairly typical urine sample from a chronic granulocytic leukemia patient (J. Hay.), while Fig. 110 shows a urine sample from a patient with polycythemia (L. Gre.). No consistent differences were noted for the untreated patients as compared to normal controls even when plasma levels deviated from normal levels. The patients showed a somewhat greater variability, but the cause of these variations was not ascertained since an extensive study of the urine was not undertaken.

It was felt that the uncertainties related to the method of studying urine by paper chromatography were too great to warrant an extensive investigation. The major problem is one of choice of sample size. Frequently sample size is controlled by applying an amount of urine that contains a constant amount of total nitrogen, urea nitrogen, or creatinine. These methods presuppose that free amino acid excretion is related to the excretion of these other substances, an assumption that does not seem justified to the present investigators. We have observed that most urine samples (excluding obvious diuresis with very dilute urine) could be spotted with a constant aliquot (such as 0.5 ml) to give useful comparative results despite differences in urine volume. This is well illustrated by the findings shown in Figs. 105–107. Moderate variations in urine volume do not appear to make major differences in the concentration of any particular free amino acid. The variations and uncertainties make it difficult, if not impossible, to demonstrate mild amino acidurias or relatively small deviations of a few amino acids from the normal levels by paper chromatography. Gross changes can be observed as have been reported for certain types of metabolic diseases. Such gross deviations have not been encountered in patients with leukemia.

The urine may show a sequence of progressive changes when therapy is instituted, or when therapy is stopped. Figs. 111–114 show paper-chromatographic results obtained from a patient (E. McG.) with chronic granulocytic leukemia who had been treated with myleran over an extended period until toxic symptoms developed. The leukocyte count fell for a time to 500/mm^3 and the patient suffered a loss of hair. The drug was withdrawn one week before the first urine sample was obtained. Figs. 111–114 show the free amino acids in urine samples obtained at weekly intervals. The leukocyte count at the time of the first urine sample was 25 000/mm^3 and changed to 63 000, 39 555, and 103 000 on the days when the other three urine samples were obtained. The rise in the leukocyte count after withdrawal of the drug was accompanied by a steady decrease of the excretion of taurine and cystine–cysteine without any marked changes in levels of other amino acids except the methylhistidines that appeared to decrease with time. Although this is evidently associated with the withdrawal of drug in this particular case, the variations in the excretion of both taurine and cystine–cysteine (as cysteic acid on the chromatograms) are sufficiently great in both normal individuals and untreated patients with chronic granulocytic leukemia that distinct deviations from the normal levels are difficult to ascertain with accuracy.

On the basis of the relatively small number of urine samples examined in the present study, it is our general impression that the excretion of cystine–cysteine may be somewhat increased at certain stages in chronic granulocytic leukemia while both taurine and cysteic acid appear less frequently on the chromatograms in other stages of the disease. An extensive examination of urine samples from patients studied at

References p. 447/448

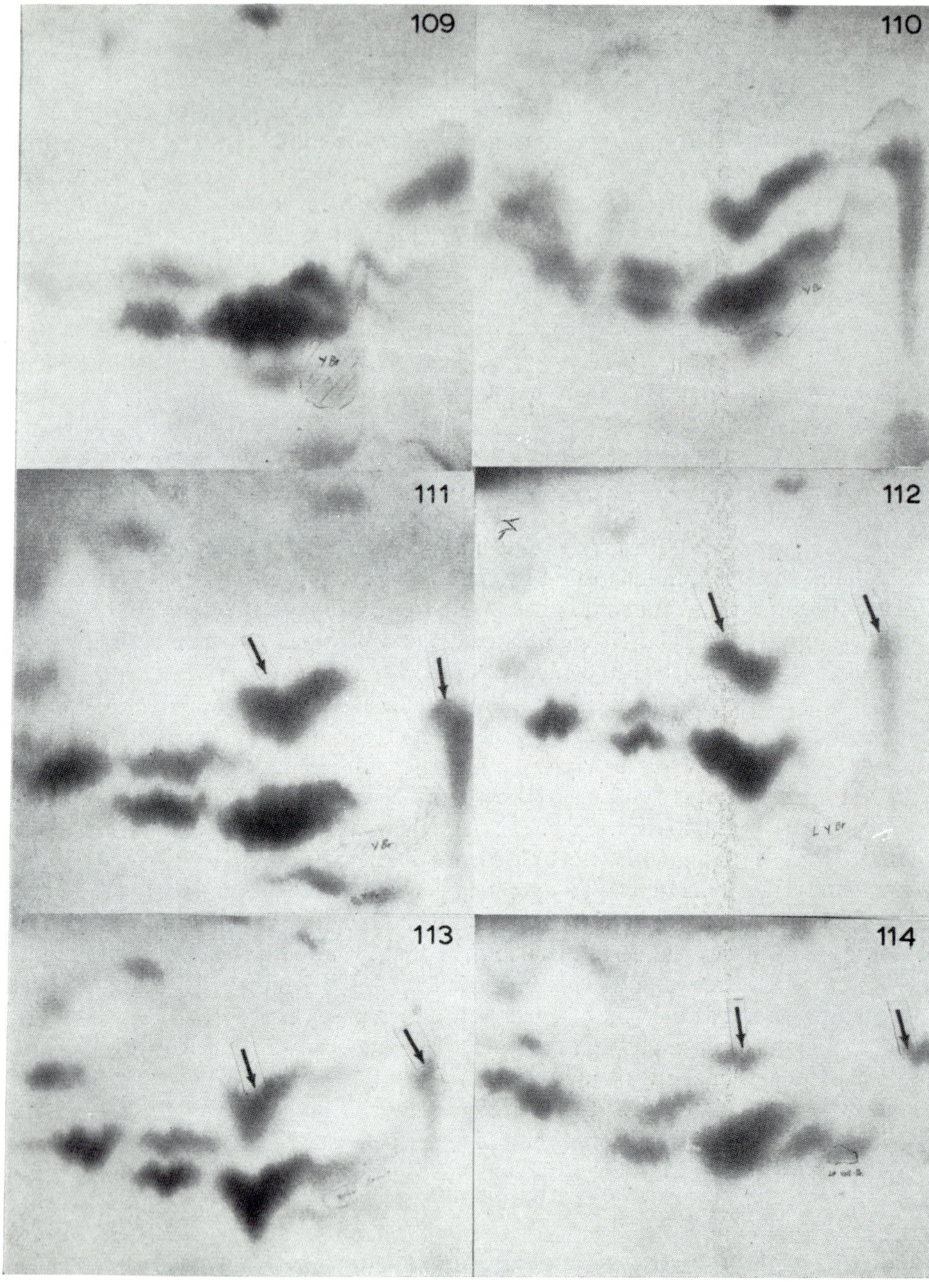

Figs. 109–114. Extracts equivalent to 0.5 ml of uiine. Fig. 109 from a patient with chronic granulocytic leukemia: Fig. 110 from patient with polycythemia; Figs. 111–114 from samples obtained at weekly intervals from a patient with chronic granulocytic leukemia who was recovering from toxic effects of myleran. Note the steady decline in tauiine (ariow to left) and cysteic acid (arrow to right).

very frequent intervals would be necessary to determine the exact nature and significance of these changes.

DISCUSSION

The findings in patients with chronic granulocytic leukemia and the patient with polycythemia make it apparent that there are certain deviations from the normal free amino acid patterns in plasma and cells. These deviations are variable and the free amino acid levels may be well within the normal range on many occasions. As has been noted above, there is a definite tendency for patients with chronic granulocytic leukemia and polycythemia to show elevations of all free amino acids on some occasions and on other occasions the plasma glutamic acid level may be elevated without any appreciable change in the other free amino acids. Many of the plasma samples had a glutamine level that was at the lower limit seen in any normal individual on a number of occasions, and the impression is thus gained that, except when there is an elevation of total free amino acids, glutamine tends to be somewhat reduced in the plasma of patients with granulocytic leukemia. Quantitative values are also in agreement with this concept (see Figs. 377 and 378 of part VI).

The most marked and consistent deviations from normal were observed in three humans and a dog that had in common the somewhat unusual finding of 90% or more neutrophilic polymorphonuclear leukocytes in the peripheral blood. In these cases the plasma glutamic acid level was distinctly elevated; leukocytes and platelets showed the elevation of glutamic acid and a very low glutamine level. The erythrocytes from the three patients did not show the glutamine and glutamic acid changes seen in the plasma, leukocytes, and platelets, but the dog erythrocytes showed the changes clearly. The urine did not appear to reflect the changes seen in the blood. It is to be noted that, although rare, granulocytic leukemia has been encountered in dogs by other investigators[23, 24].

It is of particular interest that where morphologically similar (and almost identical) leukocytes could be compared from normal individuals and patients with granulocytic leukemia and polycythemia, the leukemic cells proved to be distinctly different from normal. Leukocytes from other types of leukemia may be abnormal, but it is not possible to obtain morphologically similar cells from normal individuals for direct comparison. That the changes seen in the polymorphonuclear leukocytes are not specific for leukemia is indicated by the fact that similar findings were obtained in the plasma, leukocytes, and erythrocytes of a patient with reticulum cell sarcoma[2]. The finding of an elevated plasma glutamic acid level is in keeping with previous reports of WHITE *et al.*[9–11], and the low glutamine level in some of the white cells is in keeping with the findings of ROBERTS AND FRANKEL[12] with transplantable tumor cells.

It is perhaps surprising that patients with such large increases in the myeloid series of cells do not always show distinctly abnormal plasma free amino acid levels. The myeloid system is one of the major organs of the body in these patients. The absence of such changes can be explained in part by *in vitro* studies (reported in part I). Very little change in the free amino acid pool of myeloid cells is noted when blood is allowed to stand for as long as 4 h at room temperature. This indicates that the cells do not degrade free amino acids rapidly. The major tendency to produce a deviation from the normal amino acid levels of plasma would be by the uptake of amino acids by the cells or the metabolism of other substances. This uptake is limited and the host can readily reach a balance close to that of the normal individual.

It seems probable that the deviations that have been observed from the normal levels on occasions in untreated patients represent changes in the equilibrium established between the metabolism of myeloid cells and the remainder of the body, and that, through various metabolic adaptations on the part of the host, the changing conditions produced by growth and metabolism of the leukemic cells is counterbalanced with a return to the normal level. When the host is unable to balance the changes, marked deviations and death appear to follow. Normal or nearly normal levels of free amino acids appear to be maintained in many of the patients until shortly before death when the levels may rise. This has been observed by other investigators in experimental animals bearing tumors[19]. The finding that plasma and cellular glutamic acid levels are frequently elevated and cellular glutamine levels may be quite low in some types of leukemia points to the metabolism of these compounds as being of considerable importance. The fact that aspartic acid tends to be elevated along with glutamic acid in leukocytes and platelets suggests that transamination to form these dicarboxylic acids may be increased in leukemia.

The findings in patients with a generalized increase in myeloid cells illustrate certain basic relationships existing between cells and plasma. Erythrocytes may fail to show an elevation seen in plasma, although a more direct correspondence has been observed with dog erythrocytes. A generalized increase in plasma amino acids is associated with an overall elevation of leukocyte and platelet free amino acids without appreciable alteration of the characteristic relative amounts of the free amino acids. Urine, like erythrocytes, may not reflect the changes seen in plasma.

One substance of special interest is taurine. Since taurine is a metabolic end-product (not metabolized further), taurine levels can be used to deduce some relationships between the levels of plasma and cellular free amino acids. The plasma of patients with granulocytic leukemia may show a characteristically high or low taurine level as compared to normal individuals or other patients with leukemia. This characteristic plasma level is reflected in a comparable leukocyte level. Taurine was not detected on chromatograms from plasma of some of the patients with chronic granulocytic leukemia who were studied over a period of many months. The leukocytes and platelets of these individuals contained approximately one-half as much taurine as leukocytes obtained from patients whose plasmas showed a distinct spot for taurine. The erythrocytes from patients whose plasma does not contain taurine are usually devoid of taurine. Evidently leukocytes concentrate taurine and contain more taurine when plasma taurine is highest. The basis for these differences in the taurine levels of plasma is not known, although it does appear to be characteristic for a patient and may change during the course of the disease. It probably represents a decreased rate of formation of taurine.

The general lack of correspondence of the erythrocyte and plasma levels of amino acids is perhaps to be expected upon the basis of a relatively poor permeability for amino acids shown by erythrocytes[26]. When normal humans ingest glycine or alanine the plasma levels rise, but the levels in erythrocytes do not rise as rapidly and equilibration takes place over a period of several hours[19, 25]. A similar slow equilibration of plasma and erythrocyte glutamine has been observed (see part VI). Dog erythrocytes appear to equilibrate more rapidly with plasma since HANDLER *et al.*[27] observed a rapid equilibration of glycine between dog plasma and erythrocytes, and we have observed that these cells reflect plasma changes in dog myeloid leukemia.

The relationship between plasma amino acids and platelet and leukocyte amino acids has not been studied by other investigators. Our findings indicate that both leukocytes and platelets tend to have higher free amino acid pools when they are obtained from plasma with higher free amino acid levels. This suggests that these cells have the ability to concentrate certain amino acids. The fact that myeloid cells do not appear to metabolize free amino acids rapidly on standing *in vitro* (see part I) suggests that the free amino acid pool of the leukocyte may be an accurate reflection of the ability of these cells to concentrate amino acids such as leucine, isoleucine, valine, and other amino acids that are not synthesized within the cells. Glutamine ingestion studies have definitely demonstrated that both leukocytes and platelets can concentrate glutamine from plasma. The glutamine levels in these formed elements of the blood were observed to increase when the plasma level increased and to fall when the plasma level returned to the control level (see part VI).

The present findings are to be contrasted with some of the reports in the literature[15, 16, 20, 22]. In our opinion the highly variable findings reported by other investigators are related to the variations encountered in leukemia and in the small number of samples examined by these investigators. Previous studies (other than amino nitrogen determinations) have been carried out on single samples from each patient and variability cannot be determined in this way. The practice of grouping treated and untreated patients in the same category for determination of free amino acid levels is quite unjustified, and evidence of marked changes as a result of drug therapy are presented in parts IV and V.

The information obtained in our investigations of chronic granulocytic leukemia and polycythemia do not appear to lend support to the concept advanced by CHRISTENSEN AND HENDERSON[28] and subsequently made into a general theory of cancer by WISEMAN AND GHADIALLY[29]. Their thesis is that the selective advantage of tumor cells that allows them to grow and reproduce in the host without the control normally exerted by the host is an increased ability to capture free amino acids. Although it is true that the leukocytes, platelets, and to a lesser extent erythrocytes may have increased free amino acid pools in some cases of leukemia, this appears to be associated with similar changes in the plasma levels and the findings are variable. The simplest interpretation of the present findings is that the uptake of free amino acids by leukemic leukocytes is not particularly different from the uptake by normal cells, at least where morphologically similar cells from normal individuals and patients with leukemia can be compared. It is true that the myelocytes do contain larger concentrations of free amino acids than the neutrophilic polymorphonuclear leukocytes of either normal humans or patients with leukemia. The significance of this cannot be judged as morphologically similar cells of normal individuals are not available for study. The fact that myeloblasts contain very large amounts of taurine, glutamic acid, and ethanolamine phosphate, but smaller amounts of some of the other free amino acids demonstrates that the large free amino acid pool of myelocytes is not a characteristic of all immature cells. All of these findings indicate that the free amino acid pools as studied in the present investigations are not directly related to the basic underlying cause of the disease. The free amino acid pools are related to leukocyte metabolism, permeability, and concentrative ability that differ depending upon the type of cell.

References p. 447/448

FREE AMINO ACIDS IN THE BLOOD OF MAN AND ANIMALS

III. CHRONIC LYMPHATIC AND ACUTE LEUKEMIAS

GEORGE ROUSER, ARTHUR J. SAMUELS, DOROTHY HELLER AND BOHDAN JELINEK

Departments of Biochemistry and Medicine, City of Hope Medical Center, Duarte, Calif. (U.S.A.)

Previous parts of this series present the general methods of study of free amino acids of blood and the results obtained with untreated patients with chronic granulocytic leukemia and polycythemia. This part presents the results obtained from the study of untreated patients with chronic lymphatic leukemia and acute leukemia.

PATIENTS STUDIED

Eight patients with chronic lymphatic leukemia were studied. 4 of these patients had no previous form of treatment. The other 4 patients had been treated at some time prior to our studies. All of the patients had high lymphocyte counts and other features typical of chronic lymphatic leukemia. As the platelet levels were very low in these patients, no platelet samples were examined in the present series. Three of the patients with chronic lymphatic leukemia were studied on numerous occasions. One was studied for a period of 6 months, another for a period of 1 year, and 1 patient was followed for 2 years. These longer term studies were in connection with the effects of administration of various drugs that will be discussed in part IV.

A total of 174 blood samples, 45 bone marrow aspirates, and 23 urine samples were examined from patients with chronic lymphatic leukemia. Five patients with acute leukemia were studied. These studies were generally less satisfactory than the chronic leukemia studies because of the necessity for prompt clinical intervention with blood transfusions and other forms of therapy that made it impossible to obtain adequate pretreatment examinations. Some deviations from the normal free amino acid levels were observed in blood from these patients. Since the changes caused by therapy could not be studied separately, only illustrative examples will be given of findings that indicate how extensive the deviations from normal may be in acute leukemia.

RESULTS AND DISCUSSION

Figs. 115 and 116 illustrate fairly typical findings in plasma and erythrocytes, respectively, from an untreated patient (R. Mon. ♂) with chronic lymphatic leukemia. A number of other examples are given in parts IV and VI. No absolutely reproducible differences from normal were found for plasma or erythrocytes from untreated patients. The untreated chronic lymphatic leukemia patients showed greater fluctuations from one sample to the next than was observed for normal individuals. A consistent finding was a plasma glutamine level at the lower limit of the normal range as was observed for chronic granulocytic leukemia. The plasma glutamate level was frequently above the highest level seen in normal individuals as noted also for patients with chronic granulocytic leukemia.

One difficulty encountered early in the studies of chronic lymphatic leukemia was the marked effect of nitrogen mustard upon the free amino acids of blood and urine. Initially, patients who had been treated with small doses of drugs inadequate to produce any demonstrable clinical or hematological change were included in the untreated patient category, particularly when the drug had been given several weeks prior to the withdrawal of blood for free amino acid studies. It was subsequently found that this method of examination was not suitable. The cytotoxic drug frequently had marked effects on the free amino acid levels of blood even when there were no observable clinical or hematological effects. These changes were evident several weeks after even a small dose of nitrogen mustard (see part IV). The failure to appreciate this difference accounts for the earlier report from this laboratory[2] to the effect that glutamine might disappear from the blood of chronic lymphatic leukemia patients. At this time it was not realized that small amounts of nitrogen mustard could produce this change. Subsequently, the effect of the drug was appreciated[30].

Leukocyte samples obtained from the blood and bone marrow of some of the patients with chronic lymphatic leukemia had very low levels of glutamine without any form of treatment. A very low glutamine level is illustrated in Fig. 117. The chromatogram was prepared from an extract of leukocytes removed from the bone marrow of a patient (A. Sil. ♀) with chronic lymphatic leukemia without previous treatment. Fig. 118 illustrates the opposite type of finding where more glutamine, as well as a large amount of glutathione, is evident in the alcohol extracts of lymphocytes (patient E. Sch. ♀). Fig. 119 shows the same extract after treatment with hydrogen peroxide to oxidize glutathione completely. Fig. 120 illustrates the presence of an intermediate amount of glutamine in the lymphocytes of the same patient on a different occasion. Fig. 121 shows the free amino acid pattern of lymphocytes obtained from still another patient (M. Blo. ♂) where the glutamine level was quite low. The free amino acids in the leukocytes of this patient with chronic lymphatic leukemia resembled to a marked extent the free amino acid pool observed in lymphoblasts from a patient with acute lymphoblastic leukemia (Figs. 130, 131). Fig. 122 shows the free amino acid pattern of lymphocytes obtained from another patient (A. Sil. ♀) where glutamine was much higher, and it is to be noted that the free amino acid patterns of the lymphocytes from the patients shown in Figs. 121 and 122 are different in several respects. Patient M. Blo. shown in Fig. 121 followed a more acute clinical course. It is not surprising that the free amino acid levels of leukocytes from this patient resembled the pattern seen in the acute lymphoblastic phase to be discussed below.

The plasma glutamine levels in the patients with chronic lymphatic leukemia were frequently at, or slightly below, the lower limit seen in the normal individuals, but it is not possible to state that plasma glutamine levels are characteristically lower than normal in chronic lymphatic leukemia. There appears to be a definite tendency toward lower plasma glutamine levels, particularly for patients with the more progressive form of the disease.

Most of the findings in the urine from patients with chronic lymphatic leukemia were within the normal range as illustrated in parts IV and VI. Fig. 123 shows an interesting urine pattern that was obtained on three different occasions from the same patient (M. Kle. ♀). The patient excreted a large amount of glycine with lesser amounts of other amino acids (compare with normal urine patterns of Figs. 105–108, part II). This finding was not characteristic of the disease, but rather was related to the urinary excretion characteristics of this particular patient.

As noted above, a small number of patients with acute leukemia were studied, but pretreatment findings were difficult to obtain. Figs. 124 and 125 show the free amino acid patterns of leukocytes obtained on two separate occasions from a patient with the diagnosis of monocytic leukemia (S. Nie. ♂). It is to be noted that the glutamine level of these cells was higher than seen in cells from patients with acute lymphoblastic and myeloblastic leukemia (Figs. 130–132). The cells from the monocytic leukemia patient had a distinctive amino acid pattern that is readily differentiated from the patterns seen in acute lymphoblastic and acute myeloblastic leukemia as can be seen by the comparison of Figs. 124 and 125 with Figs. 130, 131, and 132.

A good illustration of the marked deviations that may be seen in plasma samples from some of the patients with acute leukemia is illustrated in Fig. 126. The plasma sample for this chromatogram was obtained from patient S. Nie. whose leukocytes are shown in Figs. 124 and 125. Marked elevations of the leucines, valine, glycine, and lysine are apparent. Fig. 127 illustrates the abnormal urine findings in the same patient (S. Nie.), and it is to be noted that there was a marked increase in both the leucine and valine areas on the chromatograms and a very large amount of histidine. In this patient there appeared to be an overall aminoaciduria. The free amino acid pool of the erythrocytes of patient S. Nie. are shown in Fig. 128. Spots to the left of alanine and glutamine on the chromatogram were not seen in normal individuals and have seldom been seen in other patients.

The erythrocytes from a patient (A. May. ♀) with acute lymphoblastic leukemia are illustrated in Fig. 129. The chromatogram is of particular interest in that glutamic and aspartic acids and glutathione are very high compared to the levels of other

---

Figs. 115 and 116. Extracts of 0.3 ml of plasma and erythrocytes from a patient with chronic lymphatic leukemia.

Figs. 117–122. Prepared from leukocyte extracts from patients with chronic lymphatic leukemia. Fig. 117 from extract of 100 mg of leukocytes from bone marrow to show the relatively low glutamine level in the cells. Fig. 118 from extract of 100 mg of lymphocytes from peripheral blood of another patient to show the free amino acid pools of lymphocytes obtained from peripheral blood. Fig. 119, the same sample shown in Fig. 118 after oxidation with hydrogen peroxide (the arrow points to the increase in oxidized glutathione). Fig. 120, from an extract of 100 mg of lymphocytes from a bone marrow aspirate from the patient whose peripheral blood leukocytes are shown in Figs. 118 and 119 (arrow to glutamine). Fig. 121 prepared from an extract of 100 mg of lymphocytes from peripheral blood of another patient to show the marked similarity of the lymphocyte free amino acid pattern in this case to the pattern seen with lymphoblasts from acute lymphoblastic leukemia (Figs. 130, 131). Arrows from above down to taurine, glutamine, glutamic acid, and ethanolamine-O-phosphate. Fig. 122 from an extract of 100 mg of lymphocytes from the peripheral blood of another patient to show the variability in the free amino acid pool of the morphologically similar cells (arrow to glutamine).

Fig. 123. From an extract of 0.5 ml of urine from a patient with chronic lymphatic leukemia that illustrates the unusually large amount of glycine excreted.

Fig. 124 and 125. From extracts of 165 mg and 100 mg of monocytes removed from the peripheral blood of a patient with monocytic leukemia.

Fig. 126. Shows the marked alterations in free amino acids of plasma (extracts of 0.5 ml) from the patient with monocytic leukemia. The arrows point to the high levels of (from above down) the leucines, valine, glycine, and lysine. For abbreviations see p. 369.

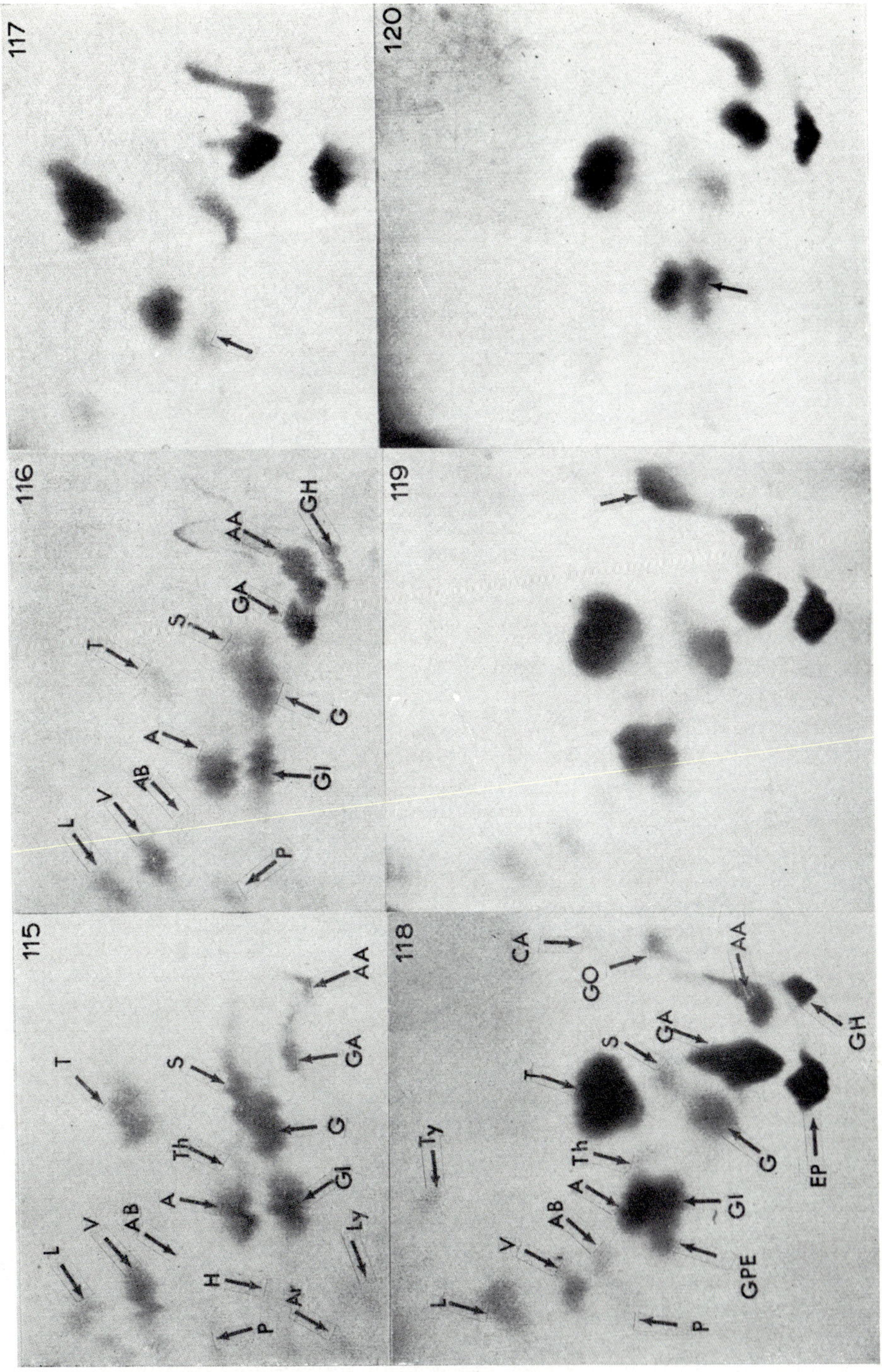

Figs. 115–120. For legend see p. 390.

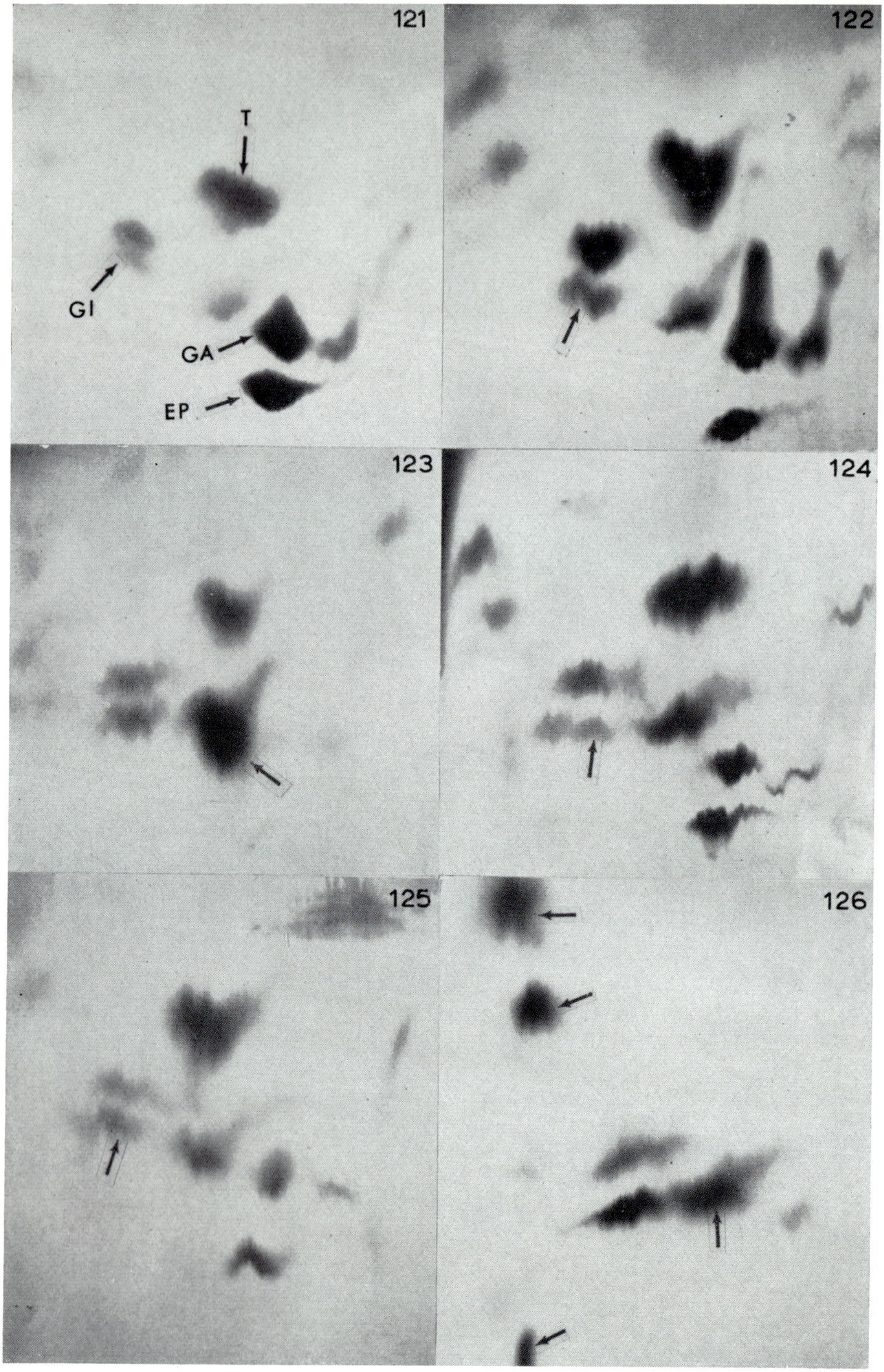

Figs. 121–126. For legend see p. 390.

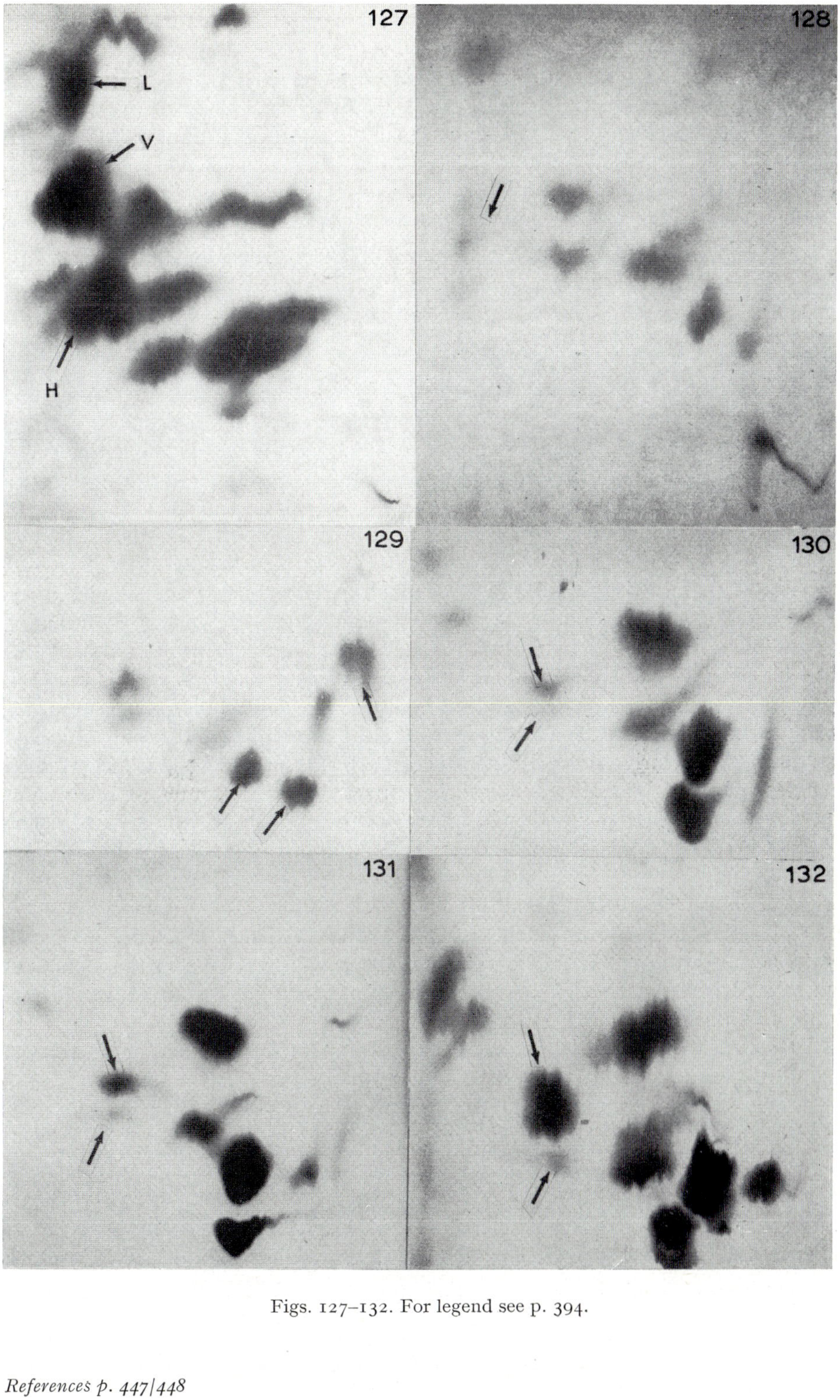

Figs. 127–132. For legend see p. 394.

compounds and glutamine is relatively low. This pattern was not characteristic either for this particular patient or for the disease in general.

Figs. 130 and 131 illustrate the free amino acid pools of lymphoblasts obtained on two different occasions from a patient (R. Can. ♂) with acute lymphoblastic leukemia. Of particular interest is the extremely low level of glutamine and the very low level of alanine compared to other types of normal or leukemic cells. As shown in Fig. 132, myeloblasts obtained from a patient (C. Ver. ♂) with acute myeloblastic leukemia are similar to the lymphoblasts in having an extremely low glutamine level but differ in that the alanine level is very high in myeloblasts. The myeloblasts have a larger free amino acid pool and glutathione is undetectable in these cells.

CONCLUSIONS

It can be concluded that untreated patients with chronic lymphatic leukemia tend to show a lower than normal plasma glutamine level and on occasions an elevated plasma glutamate level. Other free amino acids are usually within the normal range in the untreated patients.

No characteristic changes specific for chronic lymphatic leukemia or consistent differences from normal were observed for plasma, red cell, or urine free amino acid levels.

The most characteristic findings were with respect to the free amino acids of the leukocytes. Leukocytes from some patients showed very low levels of glutamine. The glutamine level was particularly low in cells obtained from bone marrow aspirates. The leukocytes from different patients showed significantly different free amino acid patterns. This is in spite of the fact that the cells were similar morphologically.

Although marked deviations from the normal patterns were observed in plasma, erythrocytes, and urine of patients with acute leukemia, the significance of these changes could not be assessed as the patients required treatment. These studies were not pursued to the point where definite conclusions could be drawn. It is evident, however, that both lymphoblasts and myeloblasts from patients with acute lymphoblastic and myeloblastic leukemia show very low levels of glutamine and that the peripheral blood leukocytes obtained from patients with acute monocytic, acute lymphoblastic, and acute myeloblastic leukemia are distinctly different and can be recognized readily by their free amino acid patterns. The cells from the monocytic leukemia patient contained much more glutamine than cells from acute lymphoblastic and myeloblastic leukemias.

It is difficult to compare the findings of the present study with those of McMenamy

---

Fig. 127. Shows the free amino acids from 0.5 ml of urine of the patient with monocytic leukemia to show the marked elevations of the leucines, valine, and histidine (indicated by arrows from above downward).

Figs. 128 and 129. From extracts of 0.5 ml and 1.0 ml of erythrocytes from the patient with monocytic leukemia and a patient with an acute blastic leukemia to show some of the deviations from normal. The arrow in Fig. 128 points to uncharacterized compounds, while the arrows in Fig. 129 indicate, from left to right, the relatively high level of glutamic acid, aspartic acid, and glutathione.

Figs. 130–132. From extracts of 100 mg of leukoyctes from a patient with acute lymphoblastic leukemia (130 and 131) and myeloblasts from peripheral blood of a patient with acute myeloblastic leukemia (Fig. 132). For abbreviations see p. 369.

et al.[22] who examined a small number of patients with various types of leukemia by paper chromatography. Their methods for the separation of plasma and cells have not been satisfactory in our hands (see part I) and the grouping together of all the different forms of leukemia, both treated and untreated, for consideration is totally unjustified in our experience. One surprising finding in view of the results from the present studies was the inability of McMenamy *et al.* to observe a characteristic free amino acid pattern for each type of leukocyte. As pointed out in the present paper, cells from lymphocytic, myeloblastic, and monocytic leukemias could be differentiated readily on the basis of their free amino acid patterns and illustrations of the characteristic patterns observed in other forms of leukemia are presented in other parts of this series. This difference is presumably due to the very small number of samples examined by the other investigators and to differences in the methods employed.

References p. 447/448

FREE AMINO ACIDS IN THE BLOOD OF MAN AND ANIMALS

IV. EFFECTS OF METHYL(BIS)β-CHLOROETHYLAMINE (NITROGEN MUSTARD), 4-[p-BIS(2-CHLOROETHYLAMINOPHENYL)] BUTYRIC ACID (CHLORAMBUCIL) AND PHENYLHYDRAZINE

GEORGE ROUSER, ARTHUR J. SAMUELS, KEIJI KINUGASA*, BOHDAN JELINEK AND DOROTHY HELLER

Departments of Biochemistry and Medicine, City of Hope Medical Center, Duarte, Calif. (U.S.A.)

This report presents the results of studies on the effects of nitrogen mustard and chlorambucil on the free amino acid levels of blood of patients with chronic lymphatic leukemia, and the effects of nitrogen mustard and phenylhydrazine on the blood and tissue free amino acids of rabbits.

GENERAL METHODS OF STUDY

The general methods for the separation of the blood constituents, extraction, and chromatographic examination are described in part I. Two carefully controlled studies were carried out in two different patients to determine the effects of nitrogen mustard and chlorambucil. The first patient (G. Cap. ♂) had been followed for several months prior to treatment and was then given 0.7 mg of nitrogen mustard intravenously on 4 consecutive days and the plasma and cell levels of free amino acids were followed for 3 months.

A patient with chronic lymphatic leukemia (W. Ric. ♂) was studied for the effects of chlorambucil after the oral administration of 2 mg/day for 1 week (total dose 14 mg). The free amino acids of plasma and blood cells were then followed for about 2 months.

The effects of nitrogen mustard on blood and tissue free amino acid levels of rabbits was studied with male and female New Zealand white rabbits weighing from 2.4–2.9 kg. Several control studies were necessary. One control animal was subjected to venipuncture at frequent intervals over a one day period. The results of this study are presented in part I of this series. The effects of fasting over the longest period of time (about 33 h) that any mustard-treated animals went without food was studied in a second animal. Two additional animals were fed from 12 to 4 PM while blood samples were drawn at 10 AM. These animals were followed for 4 days in this fashion to serve as controls for mustard-treated animals that were examined over the same period of time. The controls also serve to establish the reproducibility of the plasma and erythrocyte free amino acid levels during the production of a blood loss anemia resulting from the withdrawal of 4–5 ml blood samples.

Nitrogen mustard was administered intravenously to 7 animals. 4 rabbits were given 2.5 mg/kg of mustard and 3 animals were given 1 mg/kg. All animals were

---

* Present address: Prof. OKINAKA's Clinic School of Medicine, University of Tokyo, (Japan).

References p. 447/448

studied with a control sample followed by samples at 5, 24, 48, 72, and 96 h. All of the animals in the 2.5 mg/kg group and 1 of the animals in the 1 mg/kg group were sacrificed at the end of 4 days since this was the time of maximum response of peripheral blood leukocytes to the mustard injection. The 2 remaining animals in the 1 mg/kg group were sacrificed at 9 and 13 days. The liver, kidneys, spleen, appendix, and bone marrow were removed from all animals and examined for free amino acids. The total leukocyte, granulocyte, and lymphocyte counts were determined for each blood sample and a micro-hematocrit value was obtained on each occasion.

The effects of phenylhydrazine on the free amino acid levels of plasma and cells in rabbits was determined on blood samples taken from 3 rabbits that had been injected extensively with phenylhydrazine. These animals were supplied by Dr. HENRY BORSOOK and Dr. GEOFFREY KEIGHLEY of the California Institute of Technology. These animals showed a marked reticulocytosis that represented 88, 91 and 94% of the circulating red cells, respectively, in the 3 animals. At autopsy the animals were seen to have marked changes in the liver and other organs. The animals were generally bled 1 or 2 days prior to the expected time of death where the maximum reticulocytosis was obtained.

RESULTS AND DISCUSSION

Effects of nitrogen mustard on free amino acids of blood in humans

The results of the study of the administration of 4 doses of 0.7 mg of nitrogen mustard intravenously on 4 consecutive days to a patient with chronic lymphatic leukemia are shown in Figs. 133–144. Figs. 133, 134 and 135 show plasma, erythrocytes and leukocytes examined just prior to injection of the first dose of nitrogen mustard. The plasma glutamine level was somewhat lower than normal and the plasma glutamic acid level was distinctly above the normal level in the control sample. The total erythrocyte free amino acid pool was above normal, while the leukocyte free amino acid pool was most notable for the very low level of glutamine (see discussion in part III).

Figs. 136, 137 and 138 show the plasma, erythrocytes and leukocytes, respectively on the third day of the study after a total dose of 1.4 mg of nitrogen mustard had been administered. The plasma amino acid levels were higher with the most noticeable increases seen for the leucines, valine, glycine, glutamic acid and aspartic acid. Taurine was decreased. The erythrocyte free amino acid pool showed a general decrease in free amino acids with the most marked change being a reduction in the glutathione level. The total free amino acid content of the leukocytes was increased with particularly large increases in taurine, valine and leucine plus isoleucine.

Figs. 139–141 show the plasma, erythrocytes and leukocytes obtained 1 day after the total dose of 2.8 mg of nitrogen mustard had been administered (the fifth day of the study). The plasma levels of the leucines and valine resembled the control levels and taurine was much nearer the control level. Most noticeable was the marked decrease in glutamine and the high level of glutamic acid (seen also in the previous plasma sample). The free amino acid pool of the erythrocytes shown in Fig. 140 represented a decrease of all free amino acids to a level even lower than seen in the previous blood sample. Glutathione and taurine were completely absent from the erythrocytes. As shown by Figs. 134, 137 and 140 there was a progressive drop in

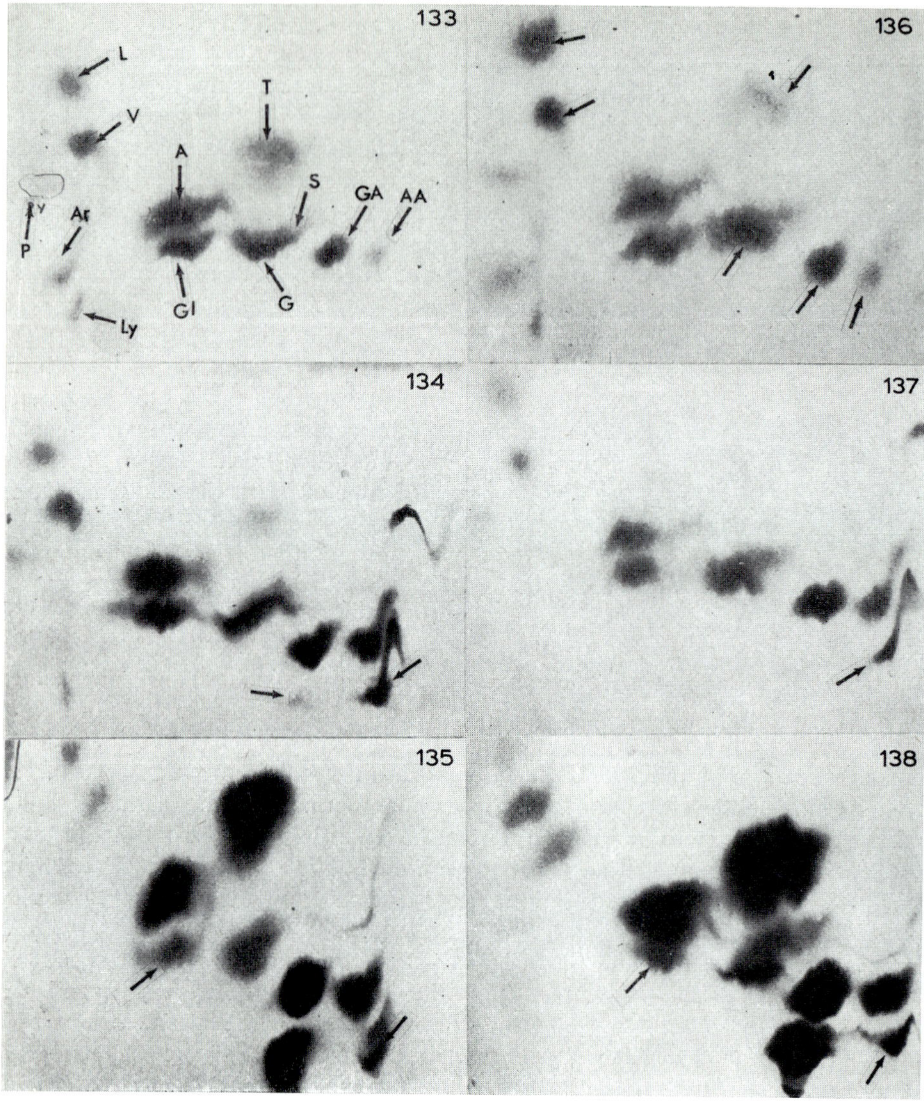

Figs. 133–138. From a patient with chronic lymphatic leukemia before and after treatment with nitrogen mustard. Aliquots equivalent to 0.25 ml of plasma and erythrocytes, and extracts of 100 mg of leukocytes. Figs 133–135 plasma, erythrocytes, and leukocytes before treatment; Figs. 136–138, 2 days after injection of nitrogen mustard.

taurine in erythrocytes after the drug was administered. Fig. 141 shows that the leukocyte free amino acid pool was similar to the previous sample except that glutathione was greatly reduced in leukocytes after the total dose had been administered.

Figs. 142–144 show plasma, erythrocytes and leukocytes 21 days after the administration of the total dose of nitrogen mustard. The plasma sample (Fig. 142) shows profound changes. The taurine level was low and the glutamic acid level was

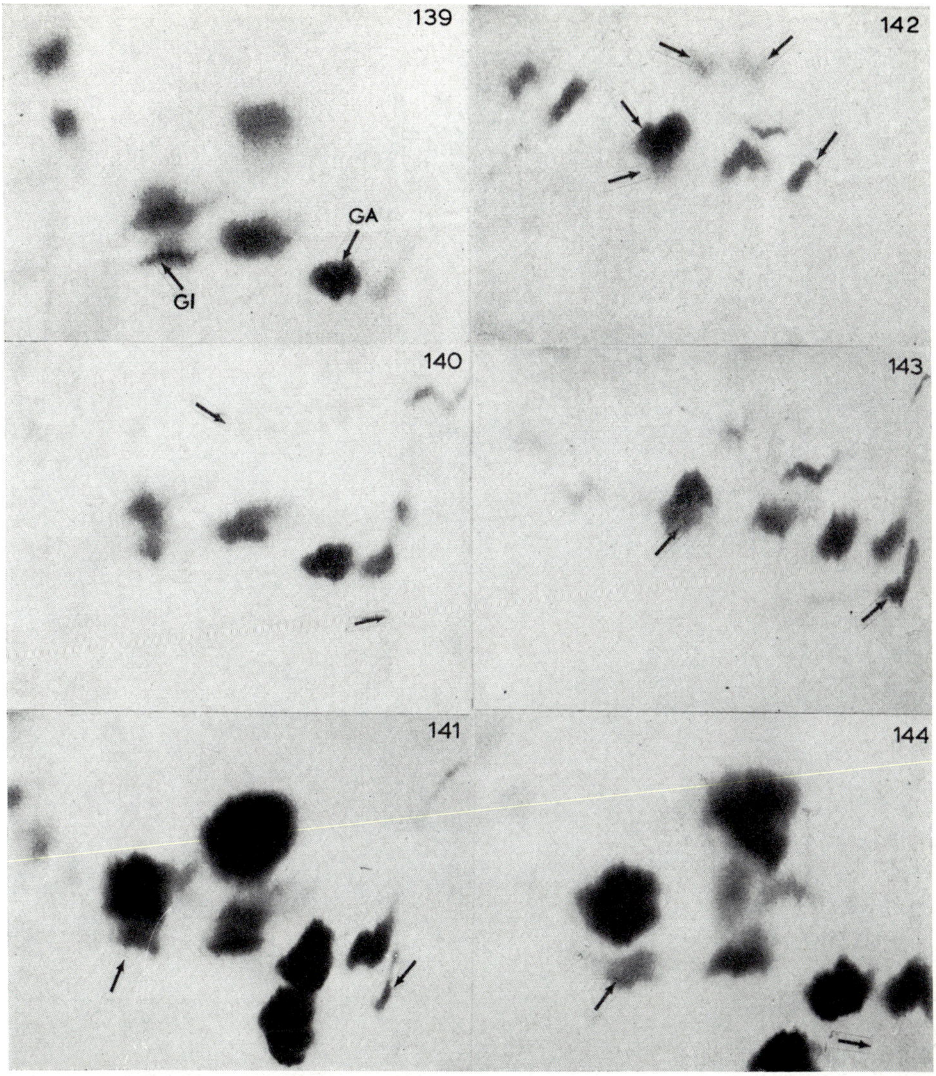

Figs. 139–141, 5 days from start of nitrogen mustard administration; Figs. 142–144, 21 days after nitrogen mustard. Note the changes in glutamine, glutamic acid, and glutathione (arrows). For abbreviations see p. 369.

below that seen in the control plasma sample. The most marked change was the almost complete absence of glutamine in plasma. The erythrocyte sample (Fig. 143) shows the return of glutathione to the red cell pool and a very low total free amino acid content. The most marked change in the leukocyte pattern was the complete disappearance of glutathione from these cells. Another blood sample (not illustrated) was examined 40 days later (61 days after the complete dose of nitrogen mustard had been given) and the free amino acid levels had returned to approximately the pretreatment state.

Before administration of nitrogen mustard, the patient's peripheral leukocyte

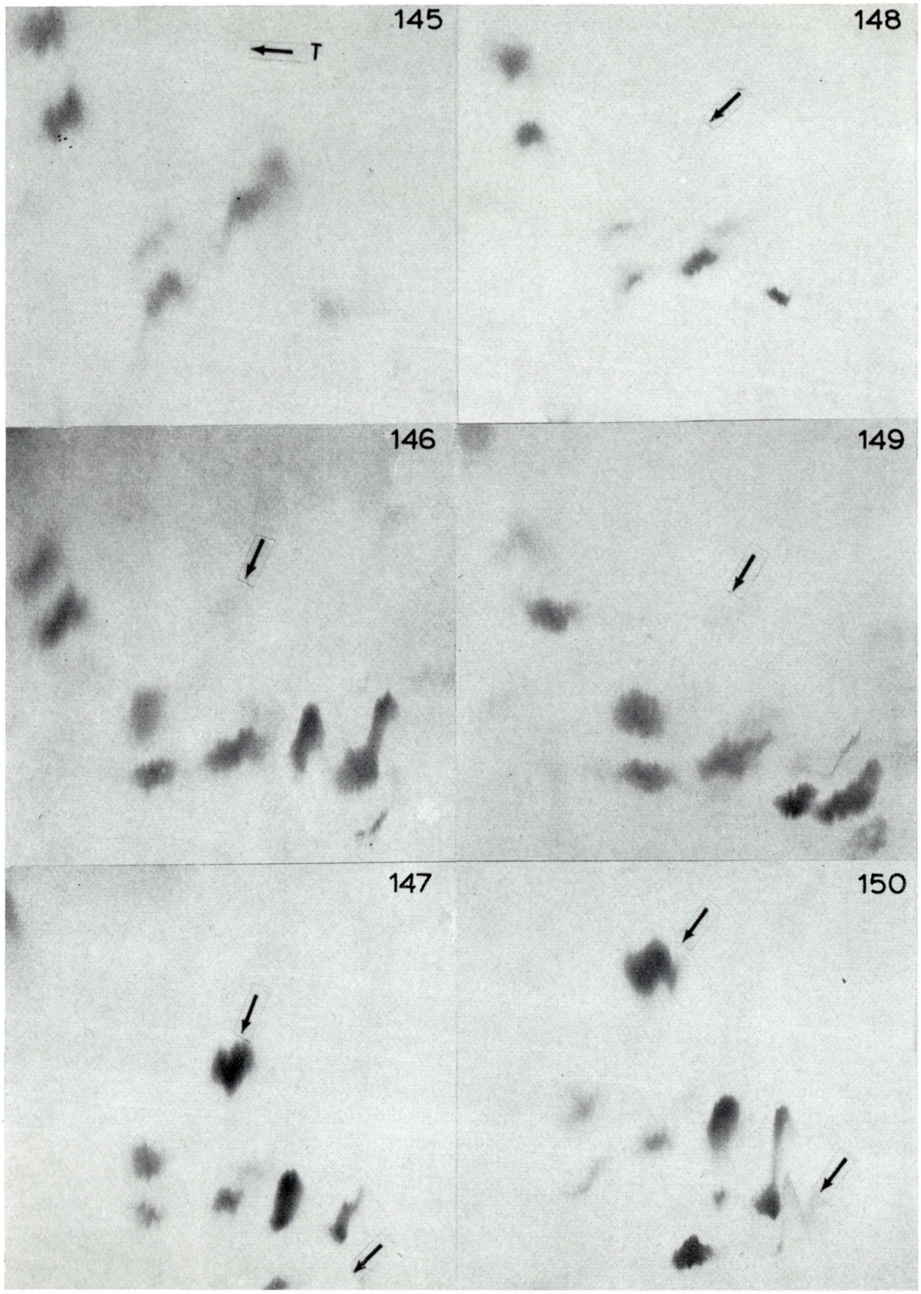

Figs. 145–156. Samples from a patient with chronic lymphatic leukemia afte· treatment with chlorambucil. Equivalent of 0.3 ml of plasma and erythrocytes, and extracts from 50 mg of leukocytes used. Figs. 145–147 prepared from plasma, erythrocytes, and leukocytes one day after completion of the total dose of chlorambucil; Figs. 148–150, 8 days after chlorambucil treatment; Figs. 151–153, 21 days after chlorambucil; and Figs. 154–156, 62 days after chlorambucil. See text for details. The upper arrows to taurine and lower arrows to glutathione (leukocytes only).

References p. 447/448

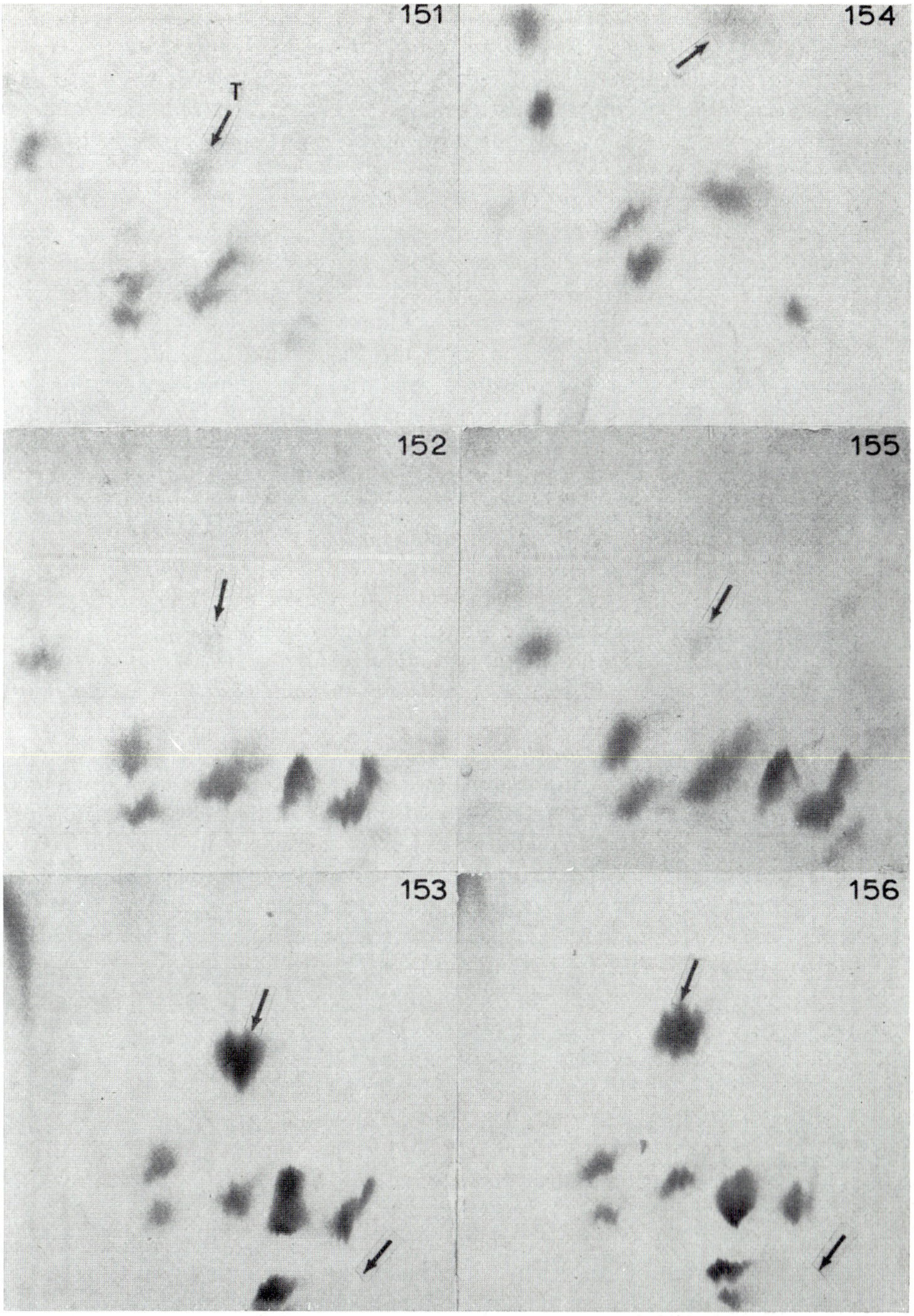

Figs. 151–156. For legend see p. 400.

count was 500 000/mm³ and fluctuated between 450 000 and 600 000 cells/mm³ during the course of the study. There was no definite change in the leukocyte count or any change in clinical status as a result of the administration of the drug. This study demonstrates that the free amino acids of plasma and cells may undergo marked changes as a result of drug administration without any apparent change in the hematological or clinical status of the patient.

In judging the marked changes produced by nitrogen mustard it is well to keep in mind the fact that the changes observed are distinctly different from those produced by *in vitro* incubation of blood of these patients as pointed out in part I. The *in vitro* incubation of blood from chronic lymphatic leukemia patients results in marked elevations of the levels of alanine, glutamic acid and aspartic acid in plasma and the complete disappearance of glutamine from the blood after several hours. These changes are distinctly different from those seen after drug administration. Patients without treatment did not show the marked changes observed after the administration of nitrogen mustard.

Effects of chlorambucil on the free amino acids of the blood of humans

The effects of chlorambucil on blood free amino acids of another patient (W. Ric. ♂) with chronic lymphatic leukemia are illustrated in Figs. 145–156. Figs. 145–147 show the plasma, erythrocytes, and leukocytes 1 day after completion of a dose of 14 mg given at the rate of 2 mg/day (for seven days). Figs. 148–150 show plasma and cells 8 days after the total dose was administered, while Figs. 151–153 and 154–156 show the plasma and cells on the twenty-first and sixty-second days after administration of the total dose. In the first post-treatment plasma sample free amino acid levels were much lower than for untreated patients. The plasma levels of free amino acids were still somewhat lower at the end of the study than were most commonly encountered in untreated patients. One of the most marked changes was for taurine which was virtually absent from the first post-treatment plasma sample and thereafter increased in amount throughout the period of study.

The erythrocyte free amino acid pool was lower at the first of the study than usually seen in untreated patients and remained at a somewhat reduced level throughout the course of the study. There were some dramatic effects on the leukocyte free amino acid pool. The most marked change was, as for plasma, that of the taurine concentration. Taurine was reduced to approximately 15% of the usual leukocyte level. Although other free amino acids were also reduced, these reductions were less pronounced. The most marked reductions aside from that of taurine were the reductions of ethanolamine-O-phosphate and glutathione. Throughout the course of the study the taurine level remained low, although it did increase in the second sample drawn 8 days after discontinuation of drug therapy.

The leukocyte count was 48 000/mm³ prior to taking chlorambucil and dropped to 30 000/mm³ at the end of the week of drug administration. The count then rose gradually to 77 000/mm³ 35 days after drug had been discontinued and remained at this level throughout the course of study. The depression of the leukocyte count was associated with a marked reduction in free amino acids of plasma, erythrocytes, and leukocytes. At all times the peripheral blood showed a preponderance of lymphocytes (84% of the cells were lymphocytes when the cell count was 30 000/mm³ and over 90% of the cells were lymphocytes on all other occasions).

References p. 447/448

The most dramatic changes aside from the total decrease of free amino acids were the low levels of taurine in plasma, erythrocytes and leukocytes and the marked decrease in the amount of glutathione in the leukocytes.

Chlorambucil thus produces changes similar to those produced by nitrogen mustard. Unlike nitrogen mustard, however, chlorambucil did not appear to affect the level of glutamine specifically. It is of considerable interest that the effects on the free amino acids of leukocytes noted above are exactly the opposite of those observed by McMenamy *et al.*[22] who reported that the free amino acids of leukocytes were 40% above the pretreatment level on the tenth day of a treatment series where the patient was given 6 mg/day of chlorambucil. A similar finding was reported by these investigators on the nineteenth day after completion of a three-week course of chlorambucil in another patient. The two samples from two different patients analyzed by these investigators were compared to other patients without treatment. Although a larger dose of drug was given in the studies of McMenamy *et al.*, the fact that only a single blood sample from each of two patients was examined probably accounts for the discrepancy in results. In our experience each patient must be studied both before and after treatment in order to obtain valid and reproducible results.

The effects of nitrogen mustard in rabbits

Several control studies were carried out as described under METHODS. The results with one control animal that was fasted, fed, and fasted again are shown in Figs. 157–162. The animal was fasted for 23 h (Fig. 157) and then allowed to eat over a 2 h period (57 g of vegetables and 15 g of Purina dog chow were consumed). Blood samples were drawn at the end of the feeding period (Fig. 158) and at 3, 8, 20, and 33 h (Figs. 159–162, respectively). The plasma samples showed minimal variations. A fall of total amino acids was evident 20 h after feeding, but the fall was not as great as that seen in animals given mustard (Figs. 163–168). The free amino acid levels of the erythrocyte samples (Figs. 157a–162a) were somewhat variable but no definite effect of fasting or feeding was apparent. The variations in the control study are quite different from the changes seen in mustard treated animals (Figs. 163a–168a).

Two control animals were sampled along with mustard treated animals. The only significant change in these animals was a small decrease in both alanine and glutamine in plasma during the studies.

The results with the control animal that was studied for the effects of multiple venipunctures spaced close together are presented in part I. Since the venipuncture response does occur, the mustard studies were designed to avoid the free amino acid lowering effect of repeated venipunctures by spacing the blood samples at intervals of several hours.

The general nature of the plasma free amino acid response after 1.0 or 2.5 mg/kg of nitrogen mustard was a steady decline in the levels of free amino acids that reached a low point on the second or third day at about the time the leukocyte count was minimal and was then followed by a rapid return to control levels on the fourth day. One animal (R-1) that was given 2.5 mg/kg of nitrogen mustard (Figs. 163–168) showed first a decrease of some plasma free amino acids 5 h after nitrogen mustard and then a rise in plasma amino acids at the end of 24 h. The rise was followed by a fall in total free amino acids to a rather low level at 72 h and then a return to slightly

References p. 447/448

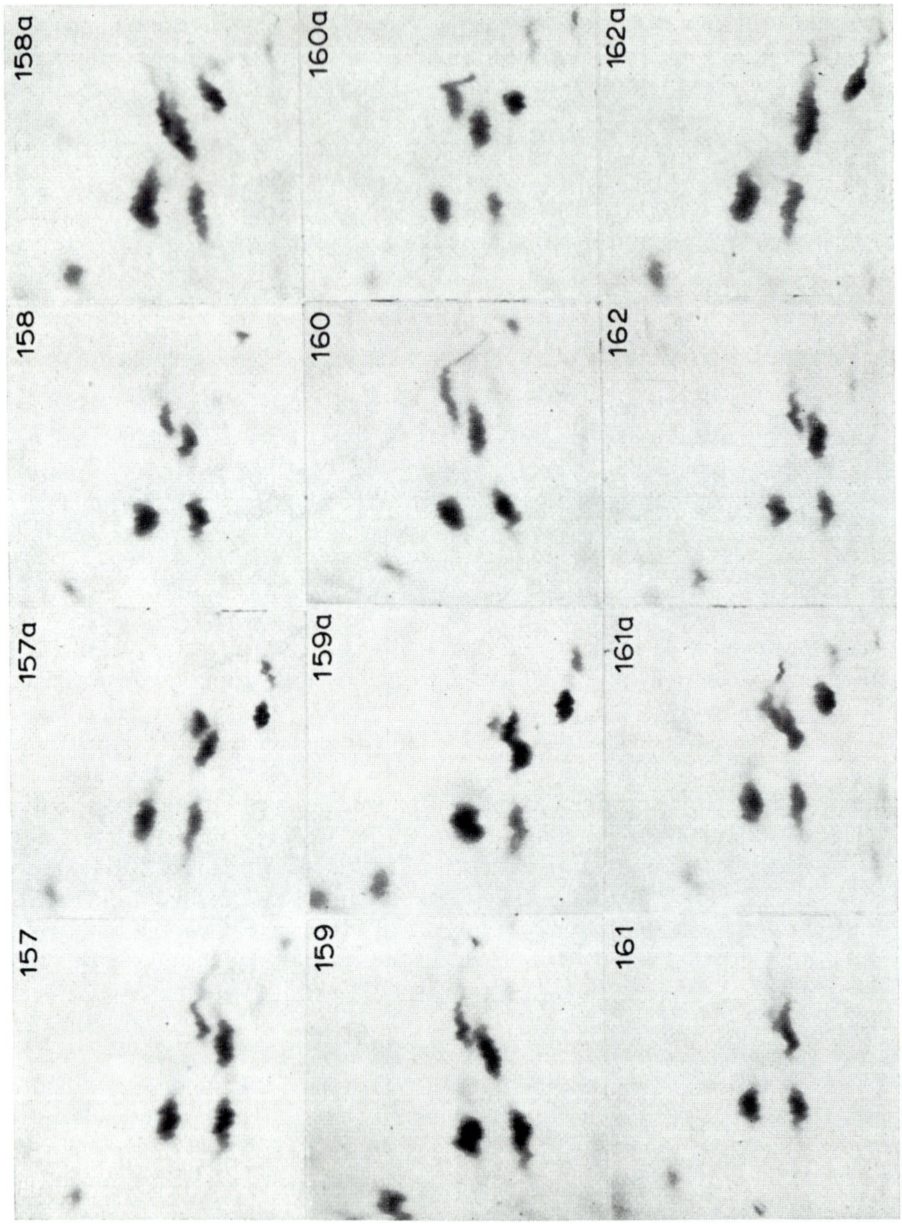

Figs. 157–162 from rabbit plasma, and Figs. 157a–162a from rabbit erythrocytes from a study of the effects of fasting and feeding. Extracts from 0.3 ml of plasma and erythrocytes. Figs. 157 and 157a, after a 20 h fast, Figs. 158 and 158a, immediately after feeding; and Figs. 159–162, 3, 8, 20, and 33 h following food (see text for details).

above the control level at 96 h. The response was different from that observed for other animals only in that there was a rise rather than a fall at 24 h (Fig. 165).

The erythrocytes of mustard treated animals showed a marked drop first in alanine

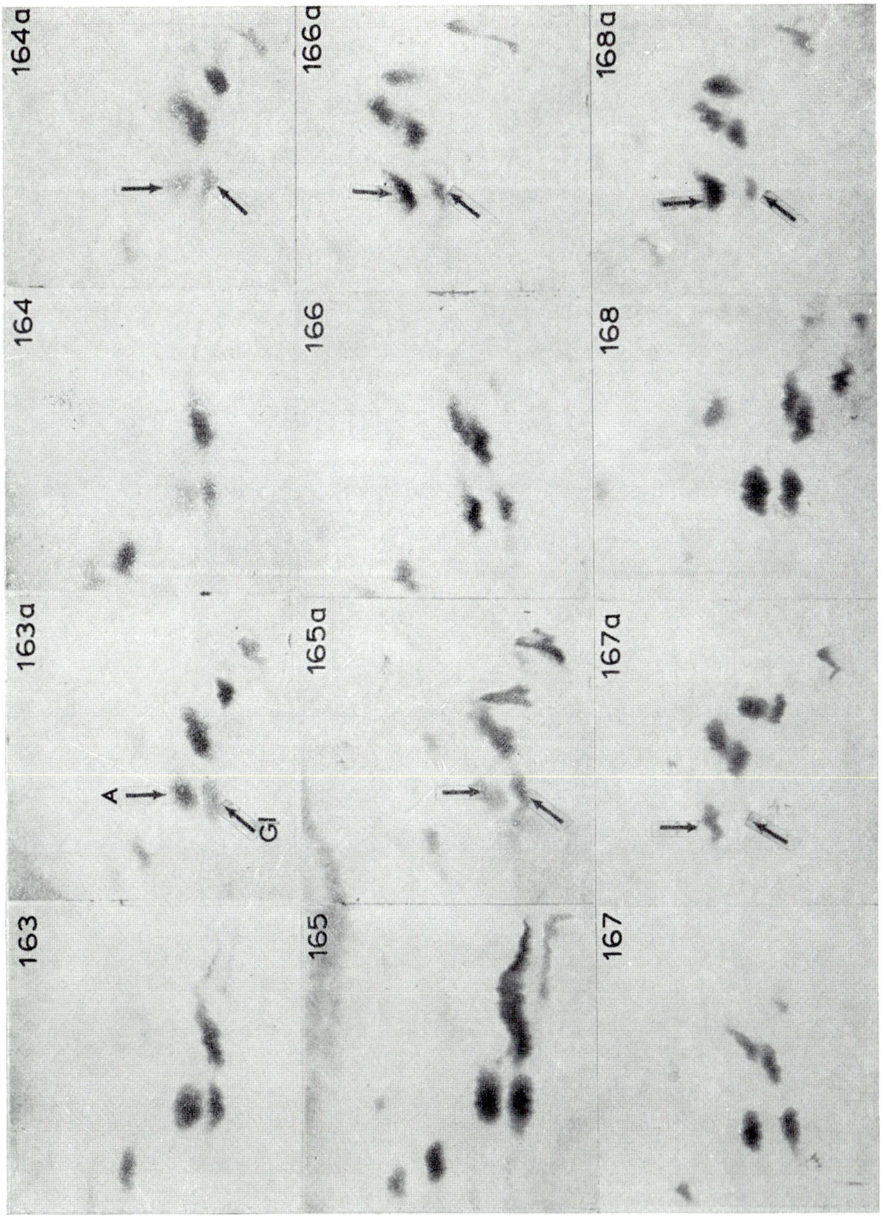

Figs. 163–168a Prepared from 0.3 ml of plasma and erythrocytes (a series) of a rabbit before and after administration of nitrogen mustard (see text). Samples taken before drug administration and 5, 24, 48, 72, and 96 h after drug. The upper arrows to alanine and the lower arrows to glutamine in erythrocytes.

and glutamine (Figs. 164a–168a) and then in valine and the leucines (Fig. 167a) with smaller reductions in glycine and serine. In general the total free amino acid pool of erythrocytes was decreased by mustard as observed in humans.

The largest fall in total plasma free amino acids of the mustard treated group was

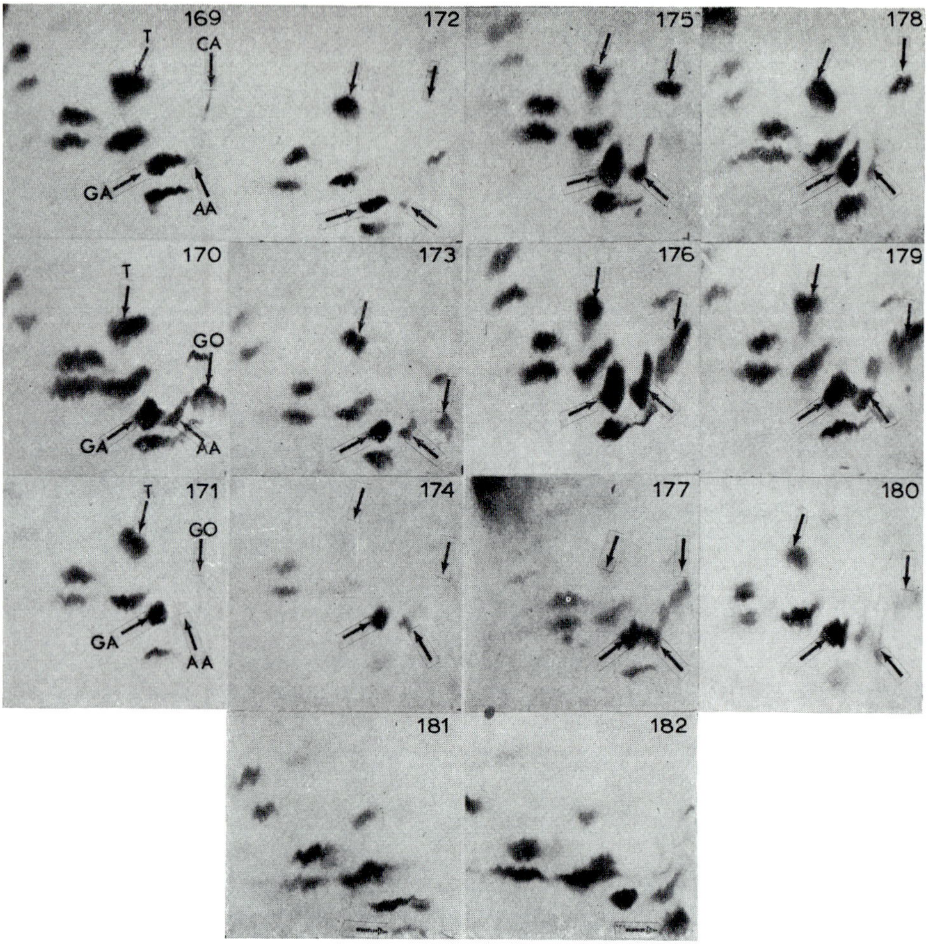

Figs. 169–180. Tissue samples (extracts of 50 mg) from control and mustard treated rabbits (4 days after drug administiation). Figs. 169–171, from kidney, spleen, and bone marrow of a control; Figs. 172–174, from kidney, spleen, and bone marrow after 2.5 mg/kg of mustard; Figs. 175–177, from kidney, spleen, and bone marrow from another animal after 2.5 mg/kg of mustard; Figs. 178–180, prepared from kidney, spleen, and bone marrow after 1.0 mg/kg of mustard.

Figs. 181 and 182. From 0.3 ml of erythrocytes of a rabbit prior to and 5 h after injection of nitrogen mustard. Note increase in glutathione (indicated by arrows) in the 5-h sample. For abbreviations see p. 369.

observed in animal R-11 that received 2.5 mg/kg of mustard. Fig. 192 shows the control plasma and Fig. 193 the 3 day plasma sample from this animal. All free amino acids except glutamic acid were greatly reduced. All of the animals in the 2.5 mg/kg group showed more pronounced changes than those in the 1.0 mg/kg group and no difference in the response of males or females was observed.

The hematological responses of the animals are shown in Fig. 183. The response is in full agreement with the findings of WEISBERGER AND HEINLE[31]. The means of the counts in each group are shown. There was a steady fall in the hematocrit value throughout the course of study. Part of this erythrocyte loss can be attributed to the withdrawal of the several 4.0-ml blood samples. There was an initial rise in the total leukocyte count followed by a marked drop. The differential counts showed clearly that the granulocytes were responsible for the increase. Subsequently the granulocytes dropped to very low levels. The lymphocyte counts showed a steady decline throughout the period of study.

Tissue samples removed from 4 animals sacrificed at the end of 4 days are shown in Figs. 169–180. Figs. 169–171 were prepared from samples of kidney, spleen, and bone marrow, respectively, of a typical control animal (subjected to the same conditions of study except not given mustard). Corresponding samples from 2 animals given 2.5 mg/kg of mustard and 1 animal given 1 mg/kg are shown in Figs. 172–180. Animal R-3 (Figs. 172–174) showed the maximum response with a marked fall in all free amino acids in all tissues. Animal R-11 (Figs. 175–177) showed the response typical of 3 of the 4 animals given 2.5 mg/kg of drug. Taurine was reduced in all tissues and cysteic acid (from cystine and cysteine) increased strikingly in kidney while

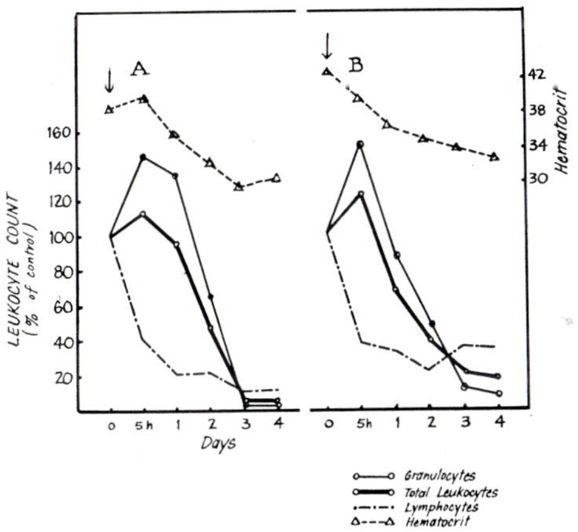

Figs. 183. Hematological response of rabbits after administration of nitrogen mustard. "A" the average of 4 animals after 2.5 mg/kg, "B" the average response of 3 animals after 1.0 mg/kg. The hematocrit value (△), granulocyte count (light line), total leukocyte count (heavy line), and the lymphocyte count (broken line) are plotted.

glutathione increased in spleen and bone marrow. Glutamate and aspartate increased in kidney, spleen, and bone marrow. The increase of glutamate in the marrow of animal R-11 (Fig. 177) was not as large as the increases observed for the other animals. Animal R-6 (Figs. 178–180) that received 1 mg/kg of mustard showed changes similar to those observed for R-11 (Figs. 175–177).

Liver and appendix were examined but are not illustrated. The changes in free amino acids seen in appendix were almost identical to those seen in spleen. The changes in the liver were similar to those in kidney except that the changes were not as great and taurine rose in liver.

Figs. 181 and 182 show erythrocyte samples removed from animal R-8 (1 mg/kg of mustard). The free amino acid level prior to drug administration is shown in Fig. 181 and the pattern 5 h after drug in Fig. 182. Glutathione and glutamic acid were increased in erythrocytes after 5 h. The rise in glutathione 5 h after nitrogen mustard administration was seen in erythrocyte samples from all 3 animals of this dosage group. This appears to be a part of the early response of the free amino acid pool to the injection of mustard.

The effects of phenylhydrazine on free amino acids of rabbit blood

One control and three phenylhydrazine treated animals were examined at the same time to minimize variations. Fig. 184 shows normal plasma and Figs. 185–187 show plasma samples from the 3 phenylhydrazine treated animals. A very marked elevation of free amino acids is seen in the plasma of treated animals. Plasma elevations are clearly apparent for the leucines, valine, tyrosine, proline, histidine, lysine, glutamine, alanine, glycine, taurine, glutamic acid, and aspartic acid. In the phenylhydrazine-treated animals the plasma showed a clear spot for α-amino n-butyric acid not seen in the normal plasma. Of particular interest in all of the plasma samples from the phenylhydrazine treated animals was the absence of arginine.

Erythrocyte free amino acids from the control animals are shown in Fig. 188 and reticulocyte samples from the animals treated with phenylhydrazine are shown in Figs. 189–191. Free amino acids of reticulocytes were increased markedly over the control level as observed for plasma. Valine, proline, α-amino n-butyric, tyrosine, alanine, glutamine, and threonine can be seen to be at very nearly the same levels in reticulocytes and plasma of treated animals (compare Figs. 185 and 189). The reticulocytes clearly contain more of the leucines, methionine, histidine, taurine, glycine, serine, and lysine. The very large amounts of α-amino n-butyric acid and ethanolamine-O-phosphate in reticulocytes from one animal (Fig. 191) are interesting.

It is evident that the reticulocytes from different animals varied, although there was a surprisingly uniform pattern considering the poor health of the animals after such large and repeated doses of phenylhydrazine. Evidently phenylhydrazine elevates both plasma and cellular levels of free amino acids.

Miscellaneous observations with regard to drug therapy

Three studies on the effects of triethylenemelamine (TEM) on the free amino acids of human plasma, erythrocytes, and leukocytes were undertaken in patients with chronic lymphatic leukemia. The results were obtained prior to the time that the venipuncture response was recognized clearly and effects due to this response were obtained by virtue of the fact that a routine blood sample was drawn from the patients just prior to sampling for the free amino acid study. This made it impossible to evaluate the effects of the drug properly. One effect was clear from the studies. A new compound appeared in leukocytes and occasionally in erythrocytes after administration of TEM. The migration of this substance is shown in Fig. 194. The new substance was readily hydrolyzed with 6 N hydrochloric acid and the migration was altered after treatment with hydrogen peroxide. This suggested that the compound was a sulfur-containing amino compound and the position on the chromatogram suggested the possibility that this substance might be γ-glutamylcysteine. γ-Glutamylcysteine was prepared from glutathione by the action of carboxypeptidase and found to cochromatograph with the substance seen on leukocyte chromatograms. The substance can be tentatively identified as γ-glutamylcysteine, although it has not been isolated and completely characterized.

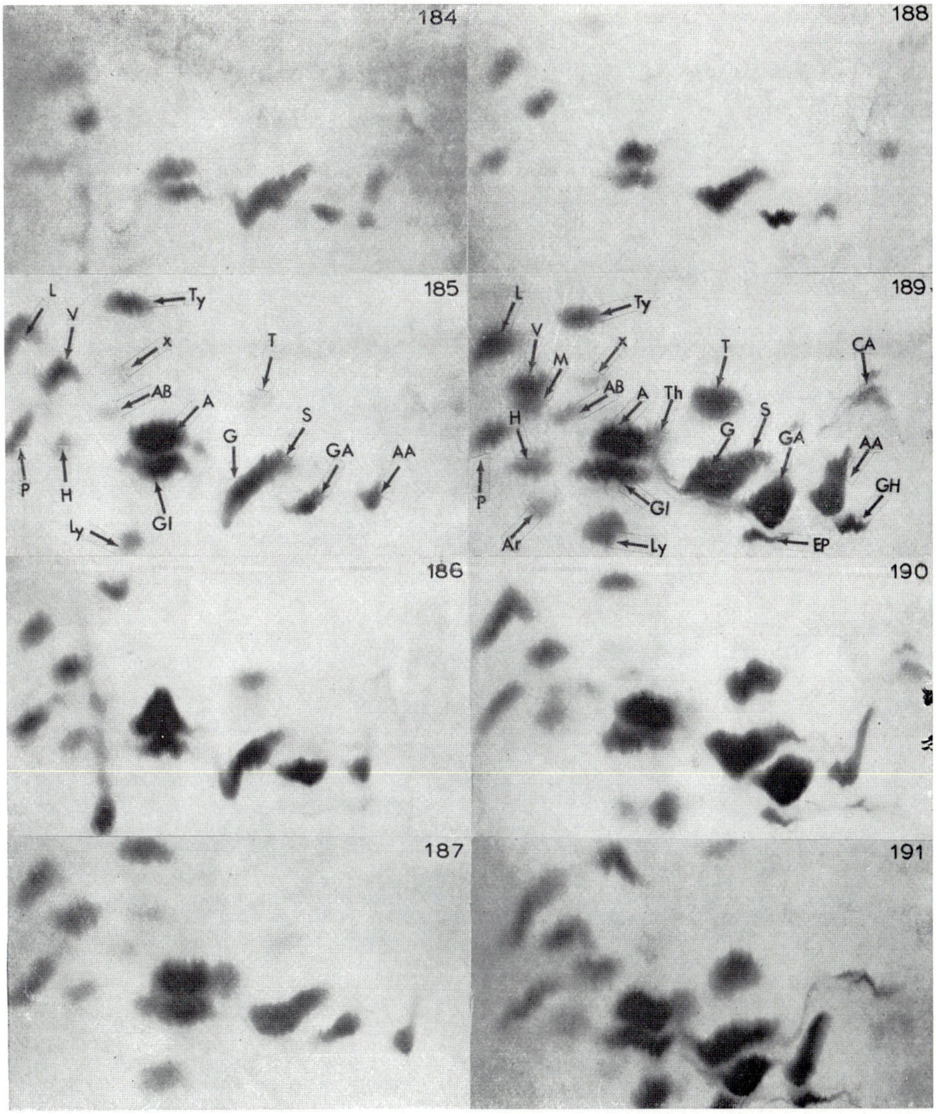

Fig. 184–191. From 0.25 ml of plasma, erythrocytes and reticulocytes to illustrate effects of phenylhydrazine. Figs. 184–187, plasma samples from a control and 3 different phenylhydrazine treated animals. Note the marked elevation of most of the amino acids in plasma of the phenylhydrazine treated animals (Figs. 185–189). Fig. 188 shows the free amino acid pool of the erythrocytes from the control blood sample, while Figs. 189–191 show the free amino acids of reticulocytes from the same blood samples. Note the marked increase in free amino acids of reticulocytes from the 3 phenylhydrazine treated animals. For abbreviations see p. 369.

DISCUSSION

The effects of nitrogen mustard on free amino acid levels in humans and rabbits are similar in most respects. Some humans are very sensitve to the drug and changes can be seen in blood free amino acids without observable effects on the hematopoietic

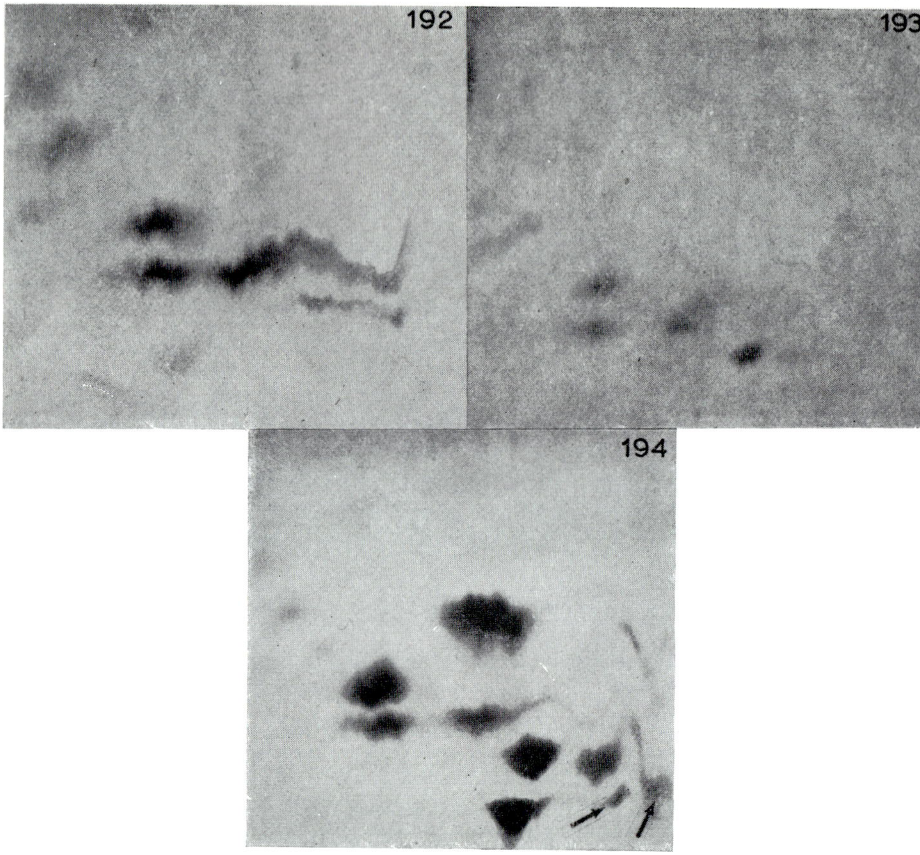

Figs. 192 and 193. From extracts of 0.3 ml of plasma obtained from a rabbit prior to and 3 days after injection with 2.5 mg/kg of nitrogen mustard to illustrate the marked fall in total free amino acids.

Fig. 194. Extract of 100 mg of lymphocytes from chronic lymphatic leukemia after triethylenemelamine (TEM) to show the presence of a compound tentatively identified as γ-glutamylcysteine, (arrow to right) and distinctly different from glutathione (arrow to left).

system. At low dosage levels and as an initial phase of the response to higher dosages, the blood and tissue free amino acids may increase. Glutamic acid in particular and to a lesser extent aspartic acid tend to be increased in blood and tissues and the levels tend to be maintained even when there is an overall decrease of free amino acids. Following a transient initial increase in free amino acids, the free amino acid levels of blood and tissues tend to fall. This decrease may be very great and can amount to an almost complete loss of the free amino acid pool. A marked fall in free amino acids has been observed also as part of the response to chlorambucil and dimethylmyleran (see part V).

The levels of sulfur-amino compounds are altered to a great extent by nitrogen mustard. In humans a decrease of glutathione in both erythrocytes and leukocytes is observed, while in rabbits at the peak of the peripheral blood leukocyte response cysteine, may be decreased in kidney and glutathione is decreased in spleen and bone marrow when there is a marked overall decrease in free amino acids (Figs. 172–174).

When the free amino acid pools are not reduced to such a marked extent, the cysteine–cystine level (as cysteic acid) may be increased in kidney, and glutathione may be increased in spleen and bone marrow at the peak of the peripheral blood cell response. Taurine levels of plasma, blood cells, and tissues always fell in both man and the rabbit after nitrogen mustard. The glutamine level is decreased in both species.

Nitrogen mustard is thought by some to act on nucleic acids as a cross-linking agent. This was postulated because mustards with 2 reactive groups are much more effective than those with one reactive group. The reaction with nucleic acids *in vivo* does not appear to have been demonstrated in animals. It is our feeling that a highly reactive substance like nitrogen mustard that decomposes in water within a few minutes, disappears from the blood within minutes after intravenous injection, and shows such strong reactivity toward many functional groups when tested *iu vitro* is unlikely to affect a single substance or enzyme system. On the other hand, it seems most probable that wide effects on enzymes and small molecules may be observed and that the particular sensitivity of certain functional groups related to their position or availability for interaction with nitrogen mustard may bring about or emphasize a reaction that might be thought to be less important on the basis of reactions observed in a test tube.

WEISBERGER AND HEINLE[31] have shown that cysteine administered to rabbits prior to intravenous injection of mustard exerts a marked protective effect. Leukocyte counts do not fall to the low levels observed with mustard alone. GOLDENTHAL *et al.*[32] have observed the formation of a mustard-cysteine derivative both *in vitro* and *in vivo* and they suggest that the sparing effect of cysteine is due to direct interaction of cysteine with nitrogen mustard. This is a very attractive hypothesis in view of the actual formation of a reaction product of the two compounds and its excretion in the urine of mice. A direct interaction is indicated also by the observation that cysteine protects against the effects of mustard if injected before, but not after the injection of the drug.

The marked changes in the levels of the sulfur compounds in blood and tissues in the present studies indicate marked effects on the metabolism of sulfur compounds. The very small amounts of mustard required to produce marked changes in humans make it apparent that the major effects of nitrogen mustard must be at the enzyme level. Many of the changes observed could be due to inhibition of activity of sulfhydryl dependent enzymes. The protective effect of cysteine is in keeping with this concept. Regardless of the mechanism, the marked fall in free amino acids of cells indicates marked changes in permeability with a decrease in the ability of cells to concentrate free amino acids. The increased levels of glutamic and aspartic acids may be explained by increased transamination of amino acids with α-ketoglutarate and oxalacetate from increased protein catabolism as a result of cell damage and death caused by nitrogen mustard.

Chlorambucil produced a marked fall in free amino acids of plasma and cells as noted for nitrogen mustard. This drug has not been studied as extensively as nitrogen mustard, but the free amino acid changes suggest that some of the *in vivo* interactions may be the same for both compounds .

The effects of phenylhydrazine 'are quite different from those produced by both chlorambucil and nitrogen mustard since blood free amino acids are increased greatly by phenylhydrazine. The increase is even greater than the increase in free amino acids seen in dog plasma and tissues by FLOCK *et al.*[33] after total hepatectomy. Since dogs survive only 1 day after total hepatectomy, the marked changes seen in rabbits injected with phenylhydrazine over a period of several days would not be observed. Phenylhydrazine undoubtedly exerts profound effects on extrahepatic tissues as well as the liver and hence overall amino acid metabolism is affected to a greater extent. Transamination and other metabolic reactions involving carbonyl compounds are undoubtedly greatly inhibited by phenylhydrazine and one of the major toxic effects of phenylhydrazine may be related to the inability to metabolize free amino acids with a resultant toxic effect due to the accumulation of amino acids and related metabolites. This may account for the fact that phenylhydrazine produces an elevation rather than a depression of blood amino acid levels. This flooding of the organism with free amino acids is observed for a short time after the administration of nitrogen mustard or dimethylmyleran and is followed by a phase where free amino acid levels in plasma and cells are greatly reduced.

References p. 447/448

FREE AMINO ACIDS IN THE BLOOD OF MAN AND ANIMALS

V. EFFECTS OF MYLERAN, DIMETHYLMYLERAN, AND RELATED COMPOUNDS IN CHRONIC GRANULOCYTIC LEUKEMIA

GEORGE ROUSER, KEITH KELLY, BOHDAN JELINEK AND DOROTHY HELLER

Departments of Biochemistry and Medicine, City of Hope Medical Center, Duarte, Calif (U.S.A.)

The previous parts of this series describe the methods for the chromatographic studies, the results with untreated patients, and the effects of nitrogen mustard and other drugs on free amino acid levels. The effects of the myleran series of compounds in patients with chronic granulocytic leukemia are presented here.

GENERAL METHODS OF STUDY

Five patients with chronic granulocytic leukemia were studied as indicated in Table II. Each patient was studied for several weeks prior to receiving any form of treatment. During the pretreatment period glutamine ingestion studies were carried out (see part VI). A sixth patient with chronic lymphatic leukemia was studied in a similar manner to provide information on the differences in response of the chronic leukemias to drugs. The body weights given in Table II are indicative only of the general size

TABLE II

Patient	*Sex*	*Age*	*Weight (lbs.)*	*Diagnosis*	*Pretreatment*	*Initial WBC/mm³*	*Months of study*	*DMM*	*M*	*CB 2432*	*Pentane Myleran*	*Total blood samples*
R. Tho.	♂	56	157	C. L.	none	147 000	2.5	1				14
M. Fal.	♀	14	100	C. G.	yes*	87 000	8	2				19
D. Fol.	♀	44	104	C. G.	none	135 000	2.5	1		1		17
H. Gol.	♂	42	185	C. G.	none	400 000	16	4				65
R. Tap.	♂	47	175	C. G.	none	137 000	11.5	2	1	1		28
Q. Tur.	♂	28	112	C. G.	none	98 000	11	—	1	1	1	10
						Total		10	2	3	1	153

Fairly typical differential counts were: R. Tho., polymorphonuclear neutrophils 6%, lymphocytes 93%, monocytes 1%; M. Fal., polymorphonuclear neutrophils 25%, metamyelocytes 17%, myelocytes 30%, blasts 7%, basophils 21%; D. Fol., polymorphonuclear neutrophils 23%, metamyelocytes 49%, myelocytes 6%, lymphocytes 4%, monocytes 3%, promyelocytes 10%, eosinophils 1%, blasts 4%; H. Gol., polymorphonuclear neutrophils 30%, metamyelocytes 36%, myelocytes 17%, lymphocytes 3%, promyelocytes 6%, eosinophils 3%, basophils 5%; R. Tap. polymorphonuclear neutrophils 48%, metamyelocytes 32%, myelocytes 12%, lymphocytes 5%; basophils 3%; Q. Tur., polymorphonuclear neutrophils 30%, metamyelocytes 37%, myelocytes 23%, lymphocytes 7%, eosinophils 3%.

* Myleran, 9 series of treatments up to 3 months before the amino acid studies. DMM = dimethylmyleran, M = myleran, CB 2432 = an isomer of dimethylmyleran.

of the individual since fluctuations were observed. Similarly, the differential counts and total white counts given in Table II were chosen as more or less representative for the patients before treatment. As shown in the table, a number of blood samples were examined from each patient. One patient was studied on four consecutive occasions after administration of dimethylmyleran, while two others were studied on two consecutive occasions, and the response of two other patients was studied for several months after a single administration of dimethylmyleran. One of the patients that was studied twice with dimethylmyleran was also given myleran and CB 2432, an isomer of dimethylmyleran, while one patient was given myleran, CB 2432, and a less active pentane myleran isomer for comparison. Each blood sample was divided into plasma, platelets, leukocytes, and erythrocytes when this was possible. Platelet samples were frequently too small and leukocyte samples were unobtainable when the counts were reduced after drug administration.

Dimethylmyleran was injected intravenously. It was dissolved by warming in aqueous ethanol and mixed with saline just prior to injection. Myleran was given orally.

RESULTS AND DISCUSSION

The general findings in untreated patients with chronic granulocytic leukemia are described in part II and further examples of the deviations from normal are apparent in the pretreatment control studies presented here. A summary of the effects of the drugs on free amino acids of plasma, erythrocytes, and leukocytes in the patients is presented, followed by detailed documentation of the responses of two patients to illustrate the extent of the changes.

Effects of dimethylmyleran and myleran on free amino acid levels of plasma

The individual response of the patient and the dosage level determined the magnitude of the changes in free amino acid levels and the total leukocyte count. It is to be emphasized that maximum changes in plasma were not always detected where maximum changes in leukocytes and erythrocytes were observed. Marked changes were observed in either plasma, erythrocytes, or leukocytes of some patients without similar changes being apparent in the other samples from the same patient.

The plasma changes varied from almost no definite deviation from the control level to the marked changes described below for patient R. Tap. (Figs. 250–267 p. 426). Four consecutive studies with patient H. Gol. with dimethylmyleran disclosed only minimal effects in plasma, consisting of a decrease of glutamic and aspartic acids. In only one patient (M. Fal.), at a low dosage level of dimethylmyleran (5 mg), was no effect observed in the plasma. Almost no effect was observed in the plasma of patient R. Tho. with chronic lymphatic leukemia (see Figs. 195–204). Intermediate effects on plasma free amino acids were related almost exclusively to marked drops in aspartic and glutamic acids and taurine. Taurine was not detected in patients M. Fal. and D. Fol. before drug administration. The most marked effects on plasma free amino acids are illustrated in Figs. 250–267. This patient showed not only the decreases of glutamic and aspartic acids and taurine observed in the other patients, but decreases in most of the other free amino acids as well. Alanine was decreased to a great extent and lesser decreases in the levels of valine, the leucines, glycine, and serine are apparent. From these findings it is evident that the major, consistent changes in the plasma free amino acid levels are changes in three compounds: aspartic acid, glutamic acid, and taurine.

Myleran produced changes in plasma free amino acid levels similar to those observed after administration of dimethylmyleran. After 10 mg of myleran per day for 10 days, patient Q. Tur. showed the typical drop in glutamic and aspartic acids and taurine. Some decrease of glycine and serine was evident also. After 10 mg of myleran per day for 10 days, the plasma of patient R. Tap. showed only the decreases of glutamic and aspartic acids and taurine.

The taurine level of plasma returned to or near the control level very slowly. After treatment the patient's free amino acid levels did not appear to return strictly to the original pretreatment levels. Prior to any form of treatment, the plasma free amino acid levels appeared to be higher. In particular, the glutamic acid levels were frequently above the normal levels. After treatment, there was a sharp decline and then a return toward the control levels for both aspartic and glutamic acids, but the levels seldom reached the levels observed before any form of treatment. This was observed in studies lasting for 100 days. It appears that there is a more or less permanent effect on the plasma levels of glutamic and aspartic acids for unknown reasons. This factor must be considered when plasma samples are examined for deviations from normal associated with the leukemic process. Evidently the characteristically high glutamic acid level of the untreated patient may not be observed if the patient receives a drug.

The free amino acid levels of blood were altered only when the drugs also produced a fall in the leukocyte count. Thus, an inactive pentane myleran (5 instead of 4 carbons in the chain) that did not change the leukocyte count, failed to change the free amino acid levels of plasma, leukocytes, or erythrocytes. Similarly, an ineffective dose of the isomeric dimethylmyleran (CB 2432) failed to produce significant changes in either the leukocyte count or the free amino acid levels of the blood. The patient with chronic lymphatic leukemia that did not respond to a relatively large dose of dimethylmyleran with a reduction of the white cell count also failed to show any but the most minor changes in free amino acids of plasma. The changes in leukocytes were actually in the reverse direction of those seen in chronic granulocytic leukemia, that is, there was an increase in free amino acids in lymphocytes after dimethylmyleran injection in contrast to decreases in the leukocytes of patients with chronic granulocytic leukemia.

Effects of dimethylmyleran and myleran on erythrocyte free amino acid levels

As with plasma, patients showed some variation in response. No effect was seen on the erythrocyte free amino acid levels of patient R. Tap. after 30 mg of dimethylmyleran. The same patient did show, however, minimal effects after 40 mg of dimethylmyleran. In this case there was a steady and marked fall in the taurine level and lesser decreases in alanine and glutamine. After 74 days, all but the taurine level had returned to pretreatment levels. Taurine was still very low 95 days after treatment. Another type of minimal effect was seen in patient H. Gol. the third time he received dimethylmyleran. The minimal changes in this case consisted of a decrease in glutamic acid, a lesser decrease of aspartic acid, and a considerable decrease of taurine.

Intermediate effects on erythrocyte free amino acid levels were observed in patient M. Fal. after administration of 5 mg of dimethylmyleran. These changes consisted

References p. 447/448

of an increase of most of the amino acids with a decrease in glutamic acid when the white cell count was reduced to a minimum. Taurine was not seen in the pretreatment samples. The most marked effects on the free amino acid levels were observed in patient M. Fal. after she received 20 mg of dimethylmyleran. Aspartic and glutamic acids, glycine, serine, alanine, and glutamine were decreased. Marked effects were also seen in patient H. Gol. in 3 of the 4 studies after administration of dimethylmyleran. The changes consisted of an overall decrease of free amino acids with the most marked decrease occurring at the time when the white cell count was reduced to a minimum. In one study the most marked decrease was for glutamic acid. Erythrocytes, like plasma, show the gradation from no effect to rather marked effects on most of the free amino acids.

The maximum effect on plasma free amino acids of patient R. Tap. was observed without any effects of the drug upon the erythrocyte free amino acid pool being apparent. The general tendency for the changes in the erythrocytes to correspond roughly to the changes seen in plasma is evident. The most consistent changes are in glutamic acid, aspartic acid, and taurine. Other amino acids may also be changed after dimethylmyleran, although these changes are less pronounced and less frequent. It is to be noted that patient H. Gol. was studied four times after administration of dimethylmyleran. Minimal effects on the erythrocyte free amino acid pool were observed in the third study, while marked effects were observed in the first, second, and fourth studies. Evidently variations in response exist for the same patient at different times that are similar to the variations observed for different individuals.

Effects of dimethylmyleran on leukocyte free amino acid levels

The most profound changes produced by dimethylmyleran were seen in the leukocytes. All patients showed some effect of the drug upon leukocyte free amino acid levels. Patient R. Tho. (chronic lymphatic leukemia) who received 40 mg of dimethylmyleran had a temporary increase of glutamic acid on the second, third, fourth, and fifth days following drug, and the glutamic acid level then returned to the control level. In contrast to this finding patient H. Gol. showed a decrease of glutamic acid (along with a decrease of white count) as the only definite change in leukocyte free amino acid levels. These changes represent the minimal effects of the drug upon the white cells. The effects of the drug were more pronounced in all other studies.

Patient H. Gol., after showing a minimal response in the first study, showed intermediate effects on the white cell free amino acid levels in three other studies. The changes consisted of a marked drop in the glutamic acid level with distinct decreases in the levels of valine and the leucines in the second study. During the third study the patient showed the same changes in free amino acids seen in the second study along with decreases in the levels of alanine and glycine. During the fourth study the patient showed decreases of all of the free amino acids of leukocytes. The most marked changes were for glutamic acid and taurine with moderate decreases in the levels of valine, the leucines, alanine, glutamine, glycine and serine.

Patient R. Tap. after 40 mg of dimethylmyleran showed a marked drop of glutamic acid only that persisted for 95 days. The same patient when studied on a second occasion after 30 mg of dimethylmyleran showed a reduction in total free amino acid that was maintained for the remainder of the period of study (95 days).

Marked effects were observed on leukocyte free amino acids of patient M. Fal. in two studies. The findings are described in detail below (see also Figs. 232–249). The total free amino acid pool decreased and new compounds appeared. The leukocyte count was reduced to a lower level than observed for other patients so that maximum free amino acid changes in leukocytes also appear to be associated with maximum changes in the leukocyte count.

The effects of CB 2432 (isomeric dimethylmyleran)

Patient R. Tap. showed minimal effect on free amino acids of plasma, erythrocytes, and leukocytes coincident with a minor effect of the drug on the leukocyte count. Patient D. Fol. showed more marked changes in both leukocyte count and blood free amino acids. A steady decline of plasma glutamic acid, aspartic acid, and taurine levels was observed from the sixth through the twentieth day post treatment. The changes were most marked when the white cell count was reduced to the greatest extent. A marked overall increase of erythrocyte free amino acids six days after drug administration was observed for this patient. The increase in free amino acids is in contrast to the effects of dimethylmyleran with which decreases of free amino acids were produced. Patient D. Fol. showed some white cell free amino acid changes after CB 2432. Six days after receiving drug there was an overall decrease of free amino acids. On the thirteenth day after drug administration control levels were observed for the first time and samples taken on the 28th, 32nd, 34th, 42nd, 50th, and 62nd days post treatment were similar to pretreatment samples.

Effects of dimethylmyleran on free amino acids in chronic lymphatic leukemia (patient R. Tho.)

The patient was studied in order to contrast the effects of dimethylmyleran in the two types of chronic leukemias. The patient received 40 mg of dimethylmyleran intravenously and the levels of free amino acids in blood were followed for 23 days. Dimethylmyleran did not change the leukocyte count.

Figs. 195–204 show the free amino acids of blood plasma of the patient after dimethylmyleran administration. Some of the pretreatment samples are presented in part VI (Figs. 324–336). Fig. 195 shows the plasma sample obtained 16 h after the administration of dimethylmyleran. No distinct change from the pretreatment level was observed. Fig. 197 shows that 2 days after administration of dimethylmyleran there was a fall in the levels of most of the free amino acids. The greatest reductions are seen in the aspartate and glutamate levels. A rapid return toward pretreatment levels was observed (Figs. 198, 199 on the 4th and 6th days) and was complete in 8 days (Fig. 200). Throughout the remainder of the study (Figs. 211–214) the plasma samples were relatively constant and no effects of the drug were apparent.

Figs. 205–214 show the chromatograms prepared for erythrocyte samples obtained at the same time intervals. A high degree of reproducibility of the erythrocyte pattern is clearly indicated. No drug induced changes are apparent. The reliability of the methods employed is demonstrated and the absence of uncontrolled effects that might be confused with drug effects is clear.

Figs. 215–224 show the findings with leukocyte samples from patient R. Tho. There was a distinct elevation of free amino acids in the leukocytes 2–4 days post treatment (Figs. 217, 218). The increase was particularly evident for glutamic acid

References p. 447/448

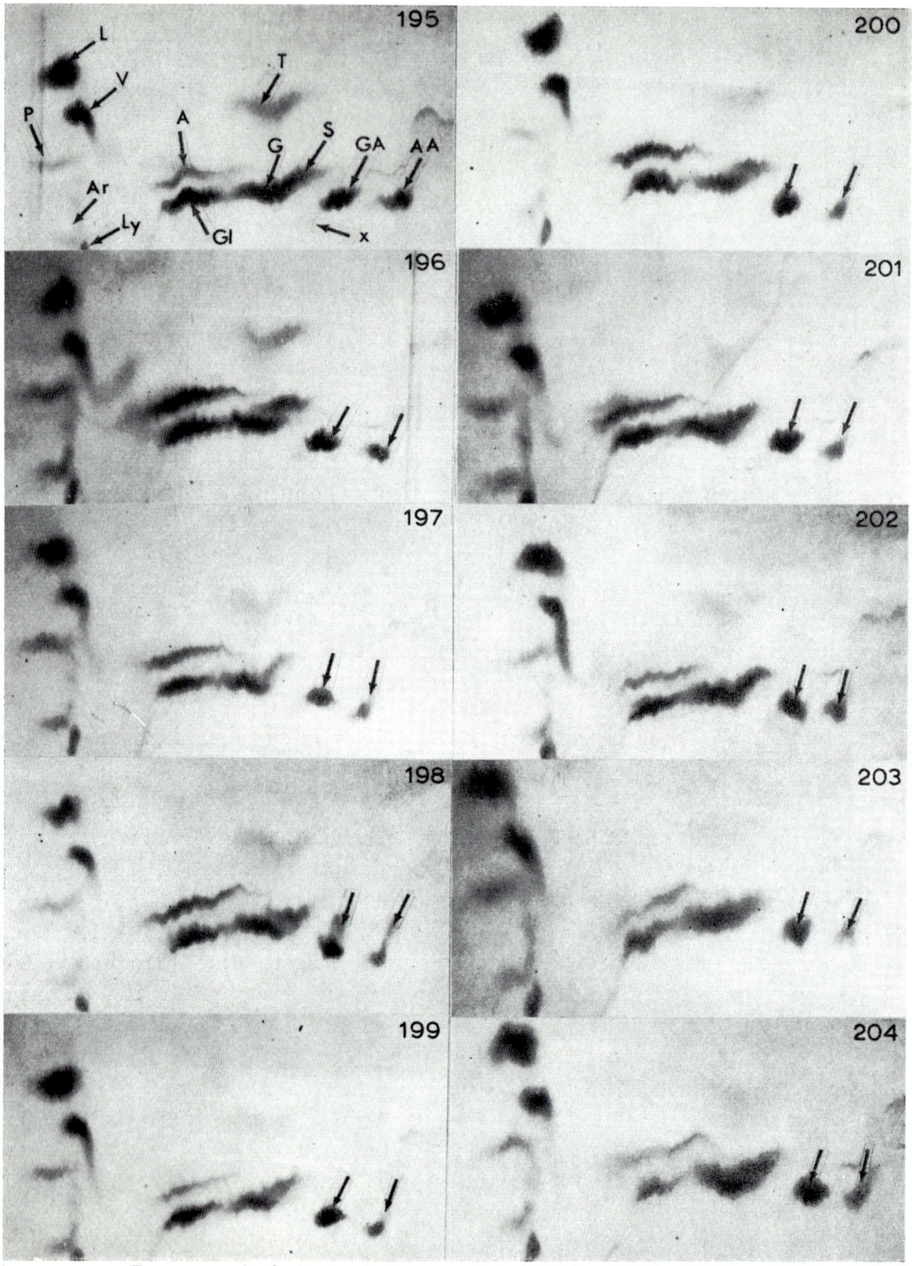

Figs. 195–204. From 0.5 ml of plasma from patient R. Tho. (chronic lymphatic leukemia) after dimethylmyleran. Samples obtained 16 h and 1, 2, 4, 6, 8, 12, 14, 16, and 23 days after drug. For abbreviations see p. 369.

and ethanolamine-O-phosphate. Subsequent leukocyte samples indicate the high degree of reproducibility of the methods and demonstrate the extent to which variations are observed when a patient is examined over a period of several weeks.

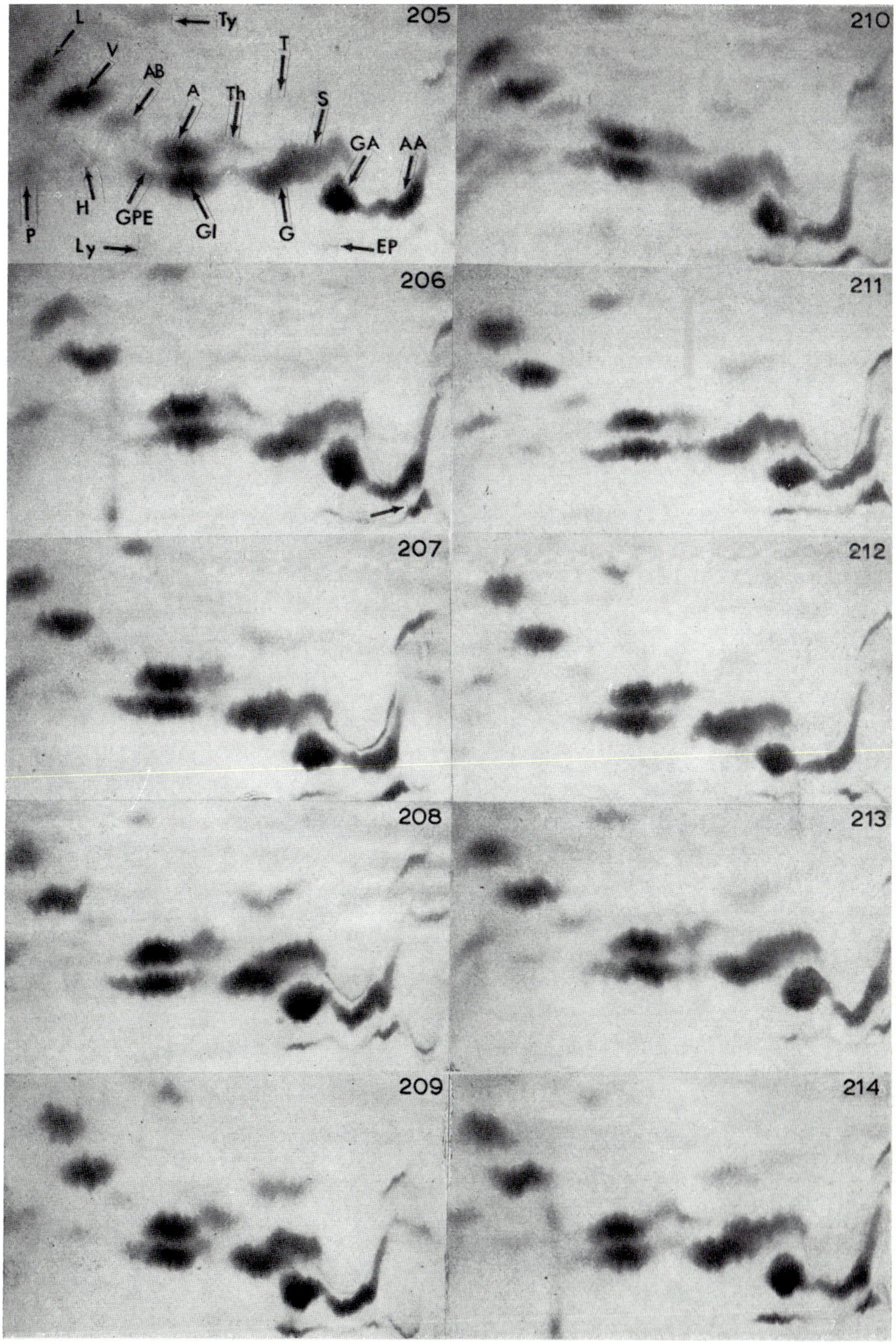

Figs. 205–214. From 0.5 ml of erythrocytes from R. Tho. (chronic lymphatic leukemia) after dimethylmyleran. From the same blood samples at the same time intervals listed in legend for Figs. 195–204. Note the consistent free amino acid pattern of the erythrocytes. The drug did not produce changes seen in granulocytic leukemia. For abbreviations see p. 369.

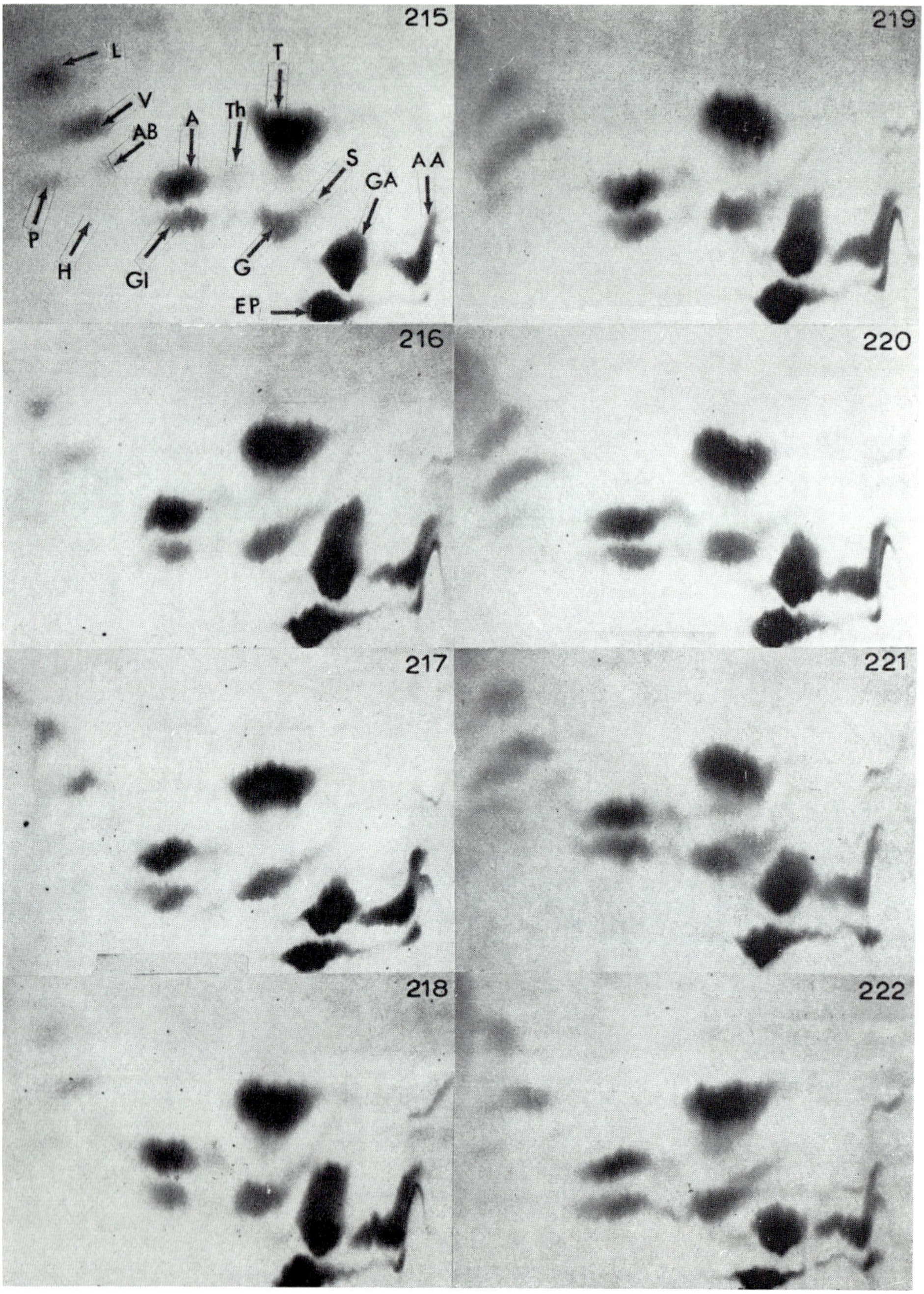

Figs. 215–224. From 75 mg of lymphocytes from R. Tho. after dimethylmyleran. Preparations from same blood samples at same time intervals indicated in legend for Figs. 195–204. Note the relatively high degree of uniformity of the free amino acid patterns. For abbreviations see p. 369.

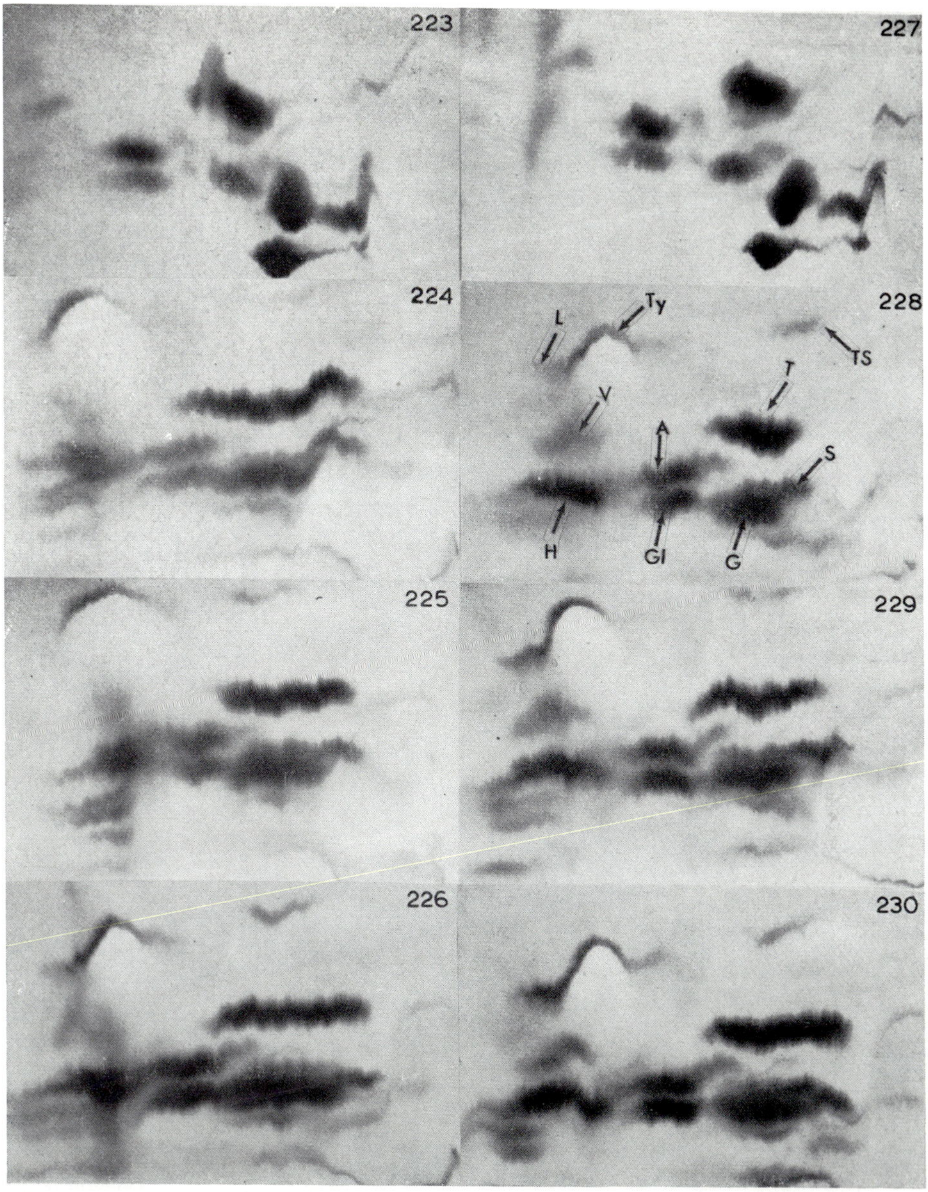

Figs. 225–230. From 0.5 ml of urine from R. Tho. after dimethylmyleran (see legend for Figs. 195–204). Samples obtained at 16 h and 1, 2, 4, 6, and 8 days after drug. For abbreviations see p. 369.

Evidently the fluctuations are so small that even relatively slight changes due to drug administration can be differentiated readily from the normal variations.

Figs. 225–230 show the findings with the urine samples after dimethylmyleran. Only the results from the first 6 urine samples are illustrated. The drug did not produce any definite changes in the urine free amino acid patterns. All of the urine samples showed spots from uncharacterized substances in the lower left portion of the chromatograms.

The response of patient R. Tho. to dimethylmyleran demonstrates the reproducibility of the methods and the relatively small effects produced by dimethylmyleran in lymphatic leukemia. This is in direct contrast to the marked changes in both the hematological status and free amino acid levels produced by the same amount of drug in patients with chronic granulocytic leukemia.

The effects of dimethylmyleran in chronic granulocytic leukemia (patient M. Fal.)

The two studies carried out with patient M. Fal. illustrate a number of the marked changes that can be produced by dimethylmyleran and the effects of different amounts of drug. Fig. 231 shows the effects on the leukocyte count of two different amounts (5 and 20 mg) of dimethylmyleran. On both occasions a marked fall in the leukocyte count was observed. Minimal changes were observed in the plasma free amino acids

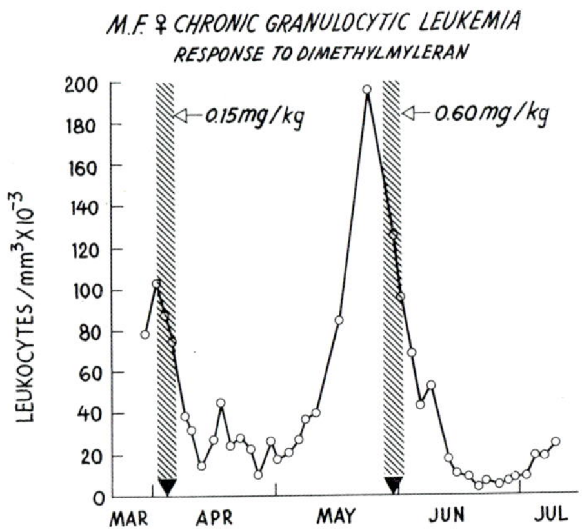

Fig. 231. Leukocyte response of M. Fal. (chronic granulocytic leukemia) after dimethylmyleran. See Figs. 232–249 for chromatographic results.

after 5 mg of drug. All free amino acids were increased only on the second day after drug administration. The erythrocytes showed a general increase of free amino acids with a decrease in glutamic acid at the peak of the white cell response. There was also a distinct reduction in the glutathione level on the first day post-treatment with a return to the control level in subsequent samples. The marked changes in the leukocyte free amino acid patterns at the lower dosage level are shown in Figs. 232–237. The most striking changes were observed on the second day post-treatment. The leukocyte free amino acid pool was reduced to extremely low levels (Fig. 235). The free amino acids were at the control level 5 days later.

Figs. 238–243 show the effects of 20 mg of dimethylmyleran on the leukocyte free amino acid pools. Marked shifts in the pool constituents are indicated and were most pronounced at 8, 11, and 13 days (Figs. 241–243). The changes consisted of marked decreases of most free amino acids with the appearance of two new compounds (Fig. 241). These changes persisted throughout the time that the leukocyte mass was large enough to sample. The glutathione level remained at or somewhat

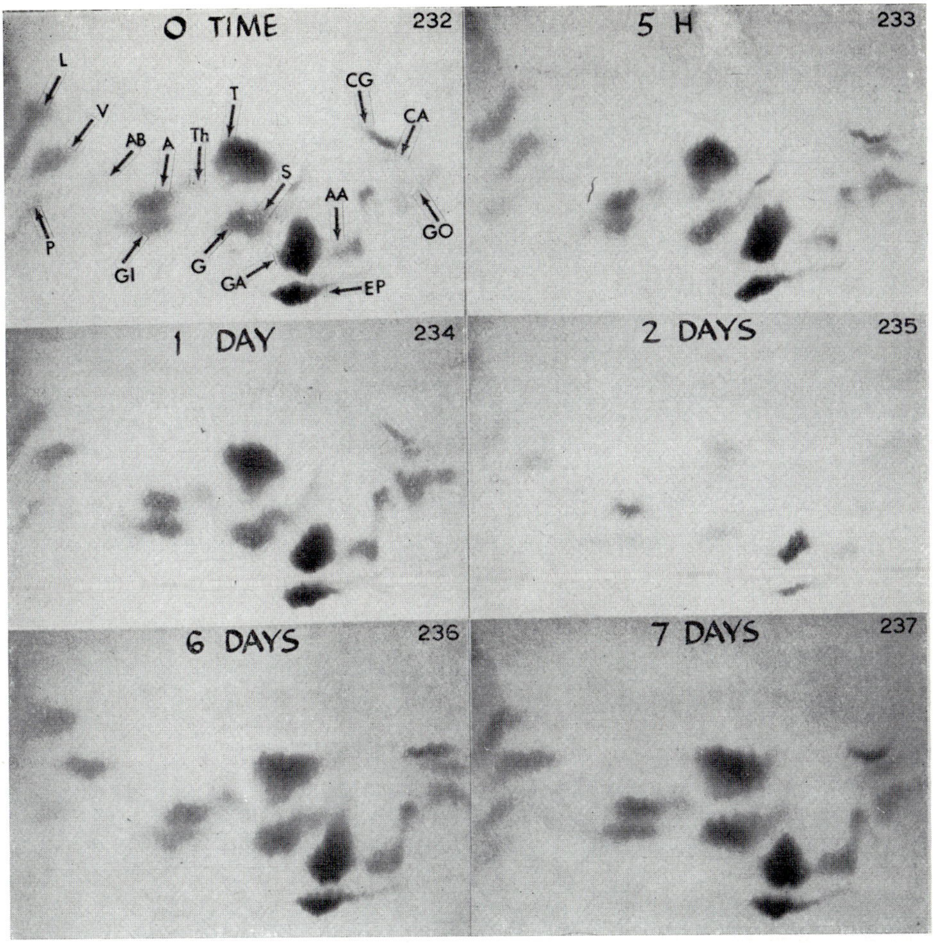

Figs. 232–237. Extracts of 50 mg of leukocytes from M. Fal. after 5 mg of dimethylmyleran. Note the marked fall of all free amino acids in the 2 day sample. For abbreviations see p. 369.

above the control level, while the cysteic acid and cysteinylglycine spots were markedly decreased.

The plasma glutamic and aspartic acid levels were greatly reduced after 20 mg of drug. It is of considerable interest that this change is in the reverse direction to the trend in the untreated patients (the level of glutamic acid is frequently increased above normal). Figs. 247–249 illustrate the findings with the erythrocytes. The glutamine level of the control erythrocyte sample was exceptionally low. The fall in aspartic and glutamic acids seen in the plasma was observed in erythrocytes. The fall in the erythrocyte glutamic acid level was the most pronounced change.

The findings with plasma at the high dosage level in this patient represent intermediate effects when compared to the findings in other patients. The leukocyte and erythrocyte responses, however, represent some of the most marked effects observed in any of the patients.

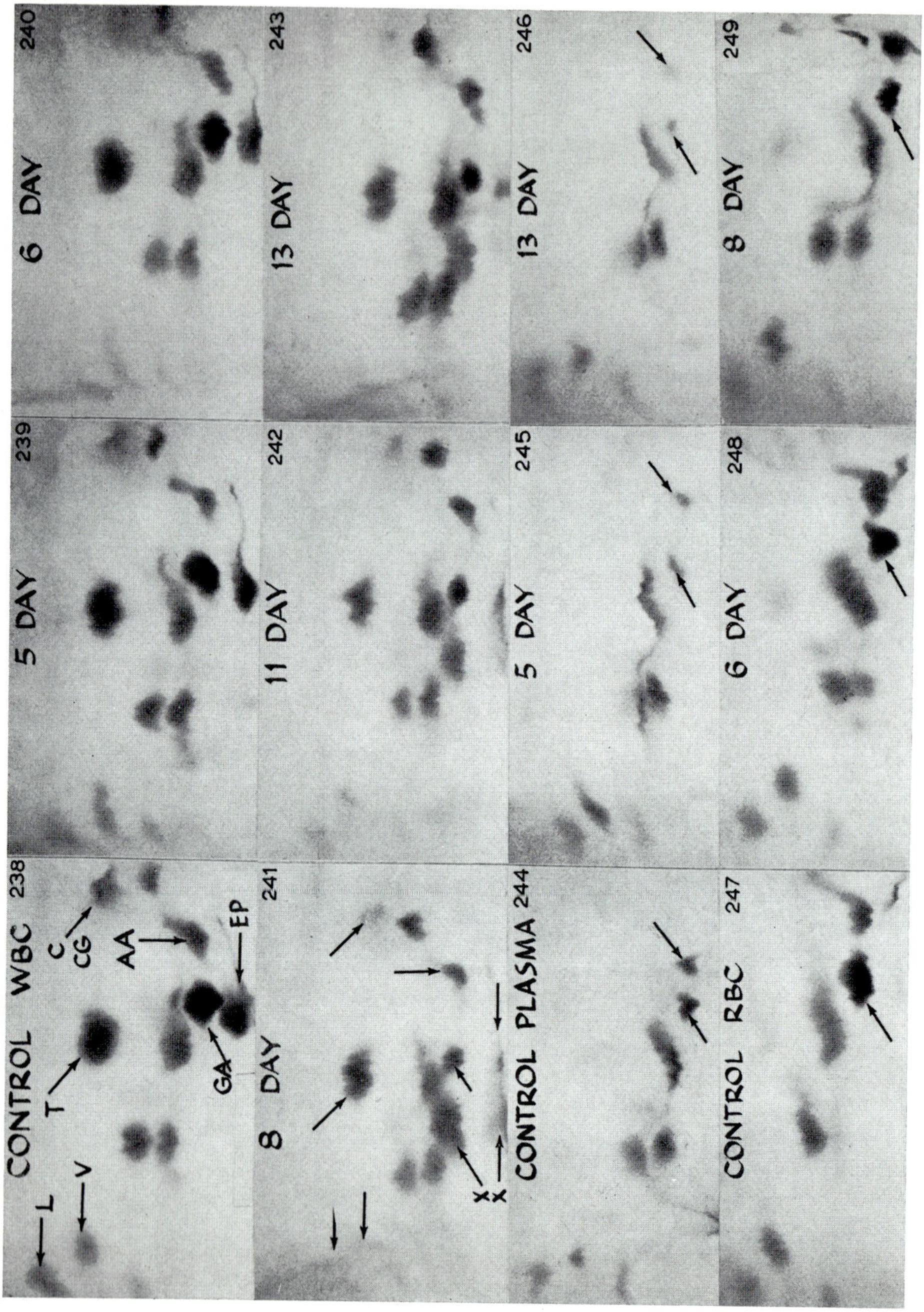

Figs. 238–249. Extracts of 50 mg of leukocytes and 0.3 ml of plasma and erythrocytes from M. Fal. after 20 mg of dimethylmyleran. Figs. 238–243 from leukocytes; Figs. 244–246, plasma; Figs. 247–249, erythrocytes. Note the marked changes in leukocytes including the appearance of two uncharacterized compounds (indicated by X) and the marked fall in plasma of glutamic acid and aspartic acid and the decrease of glutamic acid in erythrocytes (arrow).

Effects of myleran and dimethylmyleran (patient R. Tap.)

The effects of the drug on plasma free amino acids of patient R. Tap. are presented in Figs. 250–267. The chromatograms were selected for illustrations because the patient showed some of the most marked and consistent changes encountered. The figures illustrate the findings in two studies of dimethylmyleran, one study of the effects of myleran, and one study of the effects of CB 2432.

The patient first received 40 mg of dimethylmyleran (on 3-21-58) and Fig. 250 shows the free amino acid levels of plasma 4 days after drug. There were no changes from the control levels. A decline in the levels of some free amino acids is evident at 7 days (Fig. 251) and at 17 days (Fig. 252). Fig. 253 shows that after 74 days the plasma free amino acid levels were near the pretreatment levels. The patient received 30 mg of dimethylmyleran (on 7-24-58) and there was a prompt decline of the levels of most free amino acids one day later (Fig. 254). The general response was similar to that after the first injection of the drug, but the fall in free amino acids occurred much earlier and was more pronounced. The reduction of plasma free amino acids was still evident after 14 and 28 days (Figs. 255, 256). Free amino acid patterns similar to the pretreatment patterns were observed after 42 and 61 days (Figs. 257, 258). The plasma taurine level had returned to near pretreatment levels in contrast to the finding at 28 days. Dimethylmyleran produced a marked fall in taurine and the taurine level was slow to return to pretreatment levels.

The patient then received 10 mg of myleran per day for 10 days (beginning on 10-21-58). Figs. 259 and 260 show the plasma free amino acid levels 1 and 4 days after the complete dose of myleran was given. Note the lower taurine and glutamic acid levels in particular in the first sample. By the fourteenth day (Fig. 261) the free amino acids were near the control levels and high 28, 42, 49, 69, and 76 days after myleran administration. When the white cell count was reduced to the maximum extent the free amino acid levels were increased.

Essentially no changes were observed in the erythrocyte free amino acid levels following the course of myleran. There was a small increase in glutathione on the fourth day that persisted throughout the course of the study. As observed for dimethylmyleran, there was a total free amino acid decrease in the leukocytes following administration of myleran and the lowest point of free amino acids was coincident with the lowest leukocyte count. As the leukocyte count returned to pretreatment levels the free amino acids of the leukocytes returned toward the control levels. One interesting difference between the leukocyte count response and free amino acid response following myleran and dimethylmyleran is that both the white cell count and the free amino acid levels changed more gradually after myleran.

On 1-19-59 patient R. Tap. received 20 mg of CB 2432 (an isomer of dimethylmyleran) and Fig. 267 shows the findings in the plasma of the patient 14 days later. A marked reduction in the glutamic and aspartic acid levels and taurine can be observed. After 21 days (not illustrated) the levels of these three amino acids were still low. Taurine was not detectable in plasma.

CONCLUSIONS

The cytotoxic drugs of the myleran series depress the leukocyte count and alter the free amino acid levels of the blood of patients with chronic granulocytic leukemia.

References p. 447/448

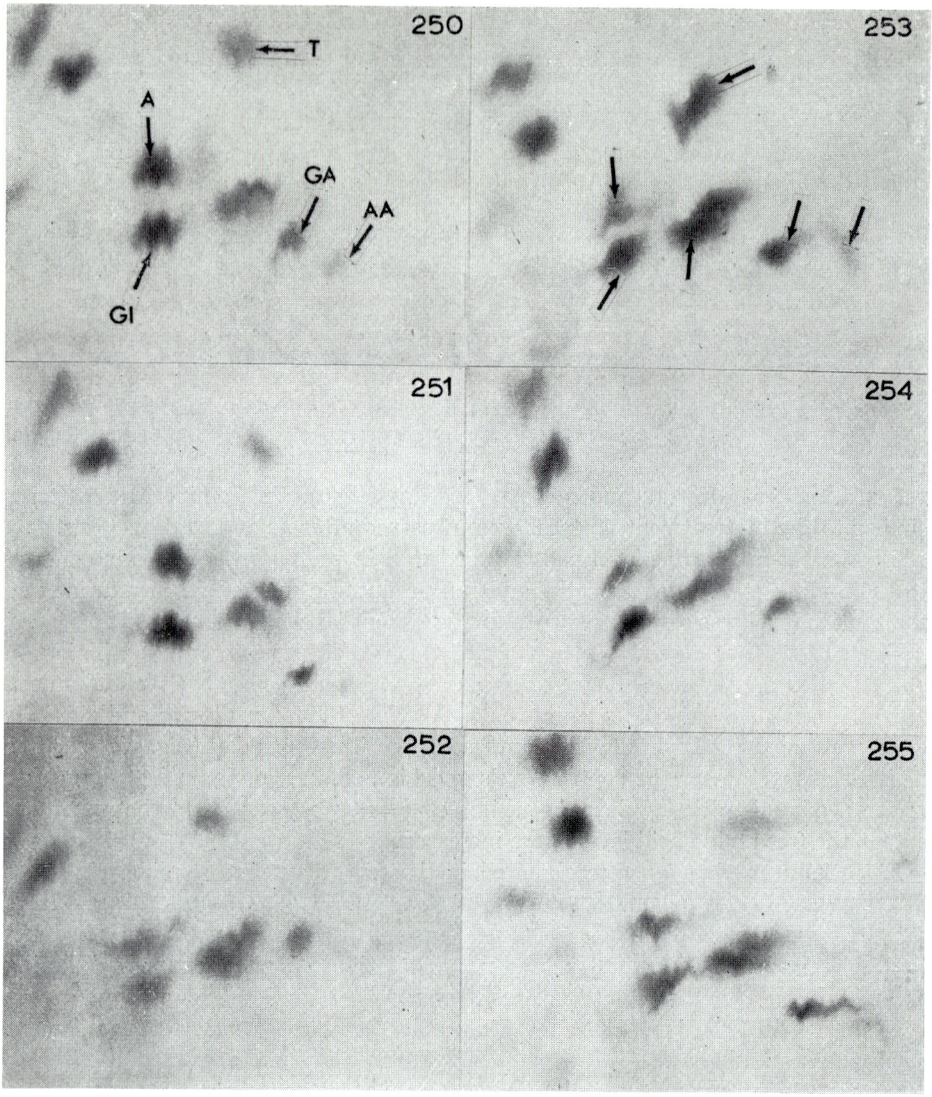

Figs. 250–267. Extracts of 0.3 ml of plasma from patient R. Tap. after drug administration. Figs. 250–253, plasma 4, 7, 17, and 74 days after 40 mg of dimethylmyleran. Figs. 254–259, plasma 1, 14, 28, 42, 61, and 85 days after 30 mg of dimethylmyleran. Figs. 260–266, plasma after myleran (10 mg/day for 10 days). Fig. 267, changes produced by injection of the isomeric dimethylmyleran (CB 2432). See text for discussion. For abbreviations see p. 369.

Little or no change is observed in free amino acid levels when the leukocyte count is not affected. The most marked changes in free amino acids are observed at the higher dosage levels when the white cell count is reduced to the lowest level. The relatively inactive pentane myleran that failed to produce a drop in leukocyte count failed to produce any changes in the free amino acid levels. The leukocyte count of a patient with chronic lymphatic leukemia did not decrease after the injection of dimethylmyleran and only very minor changes in blood free amino acids were observed.

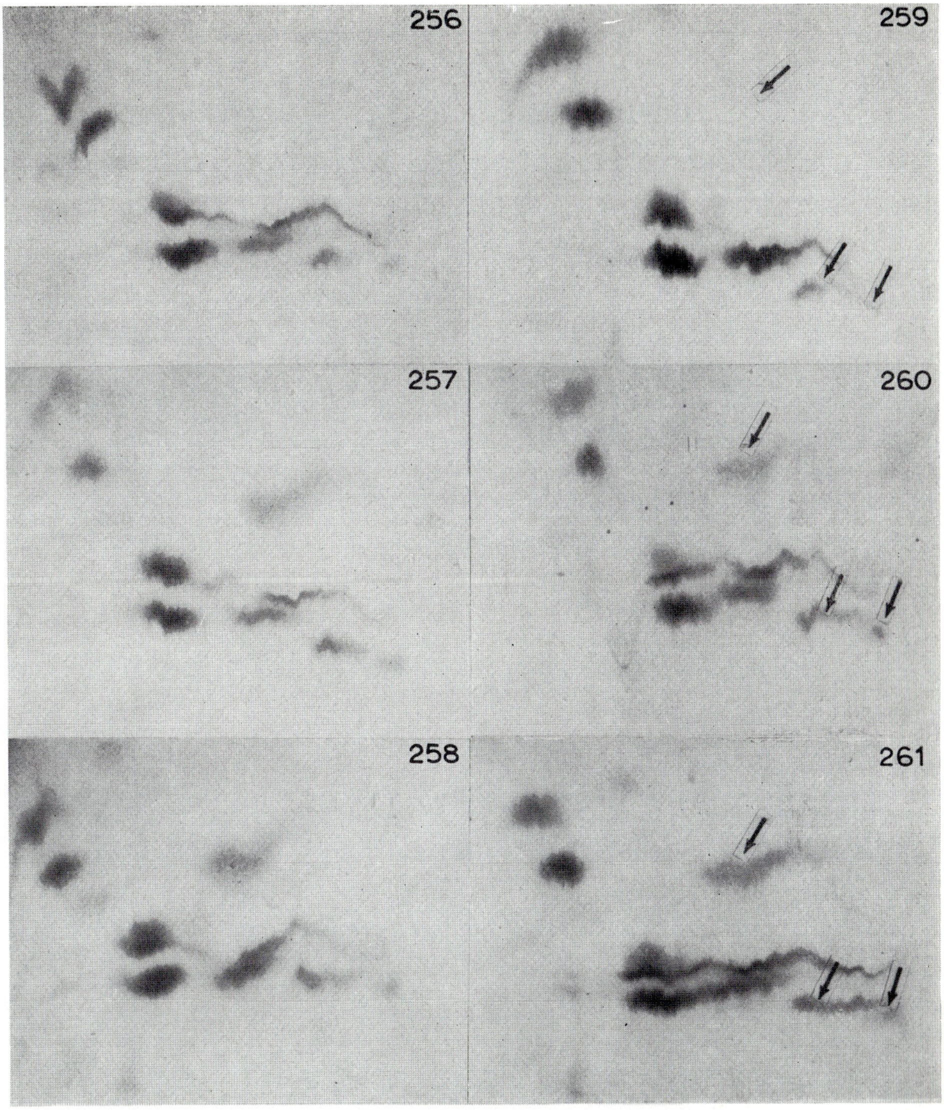

Figs. 256–261. For legend see p. 426.

These findings indicate that the changes in the levels of free amino acids are related to the overall reduction of the leukocyte count and indicate marked effects upon metabolism, particularly of leukocytes.

The compounds most affected are glutamic and aspartic acids and taurine. There are changes in the levels of other free amino acids in some cases. At times virtually all of the free amino acid levels of plasma and leukocytes are altered. The appearance of new compounds has been noted at higher dosage levels. The appearance of these compounds coincides with the marked fall of the leukocyte mass after drug administration and for this reason the isolation of the unknowns in adequate amounts for com-

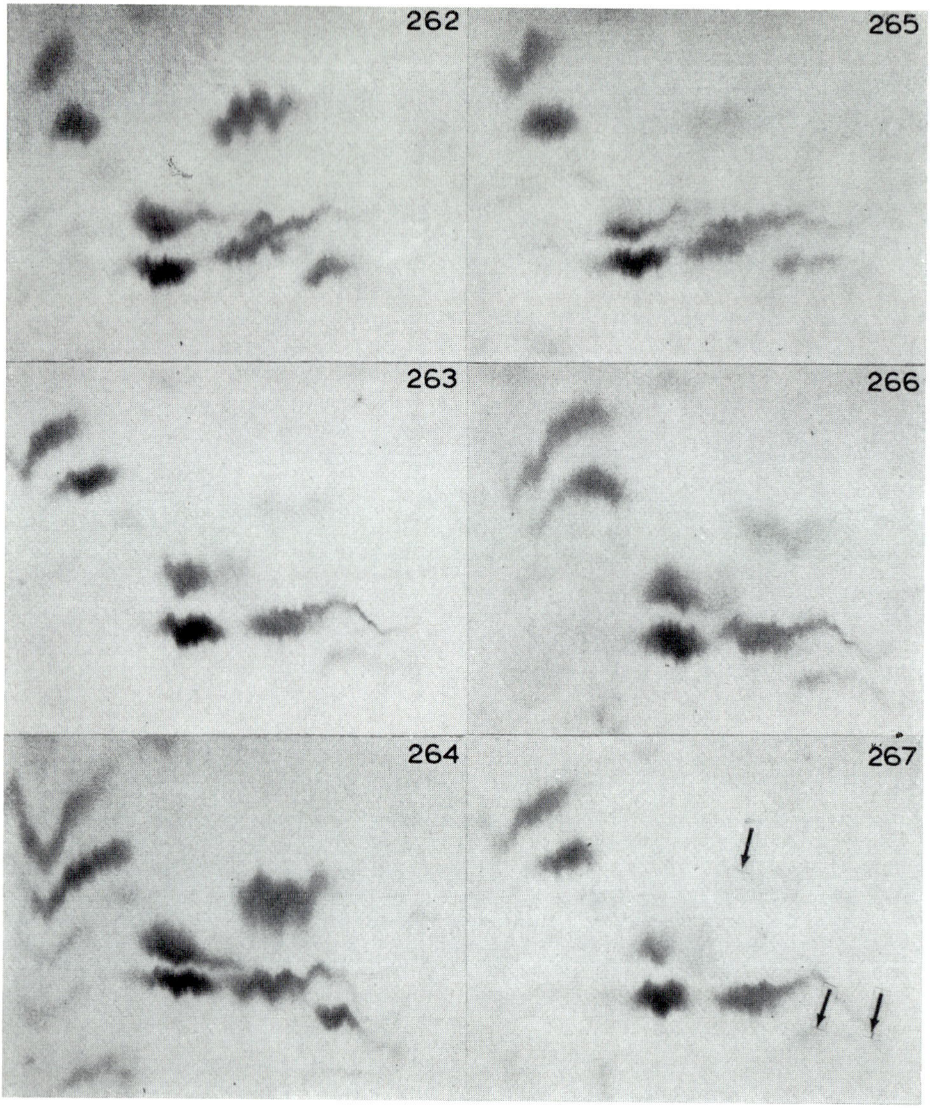

Figs. 262–267. For legend see p. 426.

plete characterization has not been possible. One of the unknowns, a substance moving to the left of glutamic acid on two-dimensional chromatograms (after oxidation with hydrogen peroxide), has the chromatographic characteristics of cystathionine. The unoxidized material moves in the ethanolamine-O-phosphate region.

One of the changes produced by dimethylmyleran and myleran is in the reverse direction to the change observed in the untreated patient. The higher than normal glutamic acid level seen in the untreated patient, is reversed by the drugs. Dimethylmyleran produces a drop in both glutamic and aspartic acids that can be seen in plasma, erythrocytes, and leukocytes.

All of the drugs that affect the hematopoietic system change the levels of sulfur amino acids. A marked drop in the taurine level is observed after myleran or dimethylmyleran administration. A similar finding is reported after nitrogen mustard injection (see part IV). In contrast to the effect of nitrogen mustard, glutathione has been observed to remain near the control level or to increase in cells after dimethylmyleran, while nitrogen mustard produces a drop in the glutathione level. One of the major differences between the effects of nitrogen mustard and the myleran series of compounds is the effect on the glutamine levels of cells and plasma. Nitrogen mustard produces a definite drop in the glutamine levels, while myleran and dimethylmyleran either do not change the level or increase it. Nitrogen mustard and the cytotoxic myleran series of drugs have one thing in common: the drugs tend to produce a marked fall in the free amino acids of leukocytes. This decrease may be seen in plasma and erythrocytes in some cases. This loss of the free amino acid pool is related to cell damage and indicates gross changes in permeability and concentrative uptake after drug administration.

The study of such labile constituents as free amino acids is complicated by the fact that a good deal of fluctuation of the levels occurs in plasma and cells in untreated patients and normal individuals. All or some of the free amino acids may first decrease (or increase) and then increase (or decrease) after drug administration. The appreciation of the nature of these changes is important and it must be understood that the study of one blood sample from a patient before (or after) treatment will not give reliable results. The variable nature of results reported in the literature is undoubtedly related in part to the inadequacies of studies based on small numbers of samples.

The results of several studies of the distribution and excretion of isotopically labeled myleran and related compounds and the *in vitro* reactions of these compounds have been published (refs. 34–39). These studies have demonstrated the interaction both *in vitro* and *in vivo* of myleran with cysteine and suggest that an important part of the mode of action of myleran may be related to reaction with sulfhydryl groups. The small amounts of drug necessary to produce an *in vivo* effect suggests to us that such reactions are most probably with protein sulfhydryl groups. The hypothesis that sulfhydryl compounds are involved in the action of myleran *in vivo* gains some support from our observations on free amino acids since the levels of sulfur-containing amino compounds have been found to change after administration of the drug. Additional effects, probably at the enzyme level, are indicated by changes in the levels of other amino acids.

FREE AMINO ACIDS IN THE BLOOD OF MAN AND ANIMALS

VI. CHANGES FOLLOWING GLUTAMINE INGESTION BY NORMAL INDIVIDUALS AND PATIENTS WITH CHRONIC LEUKEMIA

GEORGE ROUSER, KEITH KELLY, BOHDAN JELINEK, EUGENE ROBERTS AND FRANCIS W. SAYRE

Departments of Biochemistry and Medicine, City of Hope Medical Center, Duarte, Calif. (U.S.A.)

These studies arose from the observation that the patients with chronic leukemias tended to have a somewhat lower than normal plasma glutamine level and that some of the leukocytes appeared to metabolize glutamine relatively rapidly. It was then decided to investigate whether or not glutamine might be a limiting metabolite in these patients by observing the effects of ingestion of a large load of this amino acid.

GENERAL METHOD OF STUDY

The subjects were fasted overnight, 1 to 3 control blood samples and at least one urine sample were obtained; the subjects then ingested 50 g of L(+)-glutamine, and were followed at intervals up to 8 h. Following the ingestion of amino acid there was no further intake of food, but a limited intake of water was allowed in some cases.

Three adult males on the research staff were studied as controls (a total of 26 blood samples and 12 urine samples obtained). One patient with chronic lymphatic leukemia (R. Tho.) without any prior form of treatment was studied (13 blood samples and 5 urine samples obtained). The response of this same patient to dimethylmylcran is described in part V (see Figs. 195–230). Three patients with chronic granulocytic leukemia were studied before treatment and one of the patients (H. Gol.) was studied on two occasions after treatment with dimethylmyleran. The other two patients (R. Tap. and Q. Tur.) were studied by paper chromatography and glutamine levels of plasma and cells were determined with an enzymatic assay. The method of SAYRE AND ROBERTS[40] was used for the preparation of the purified enzyme. An 80% ethanol extract was prepared, evaporated to dryness, and an aliquot in buffer used for assay. A total of 40 blood samples and 12 urine samples were examined from the granulocytic leukemia patients. A total of 9 glutamine ingestion studies were carried out (a total of 79 blood samples and 29 urine samples examined).

The blood samples from the normal individuals were centrifuged at 1300 × *g* for 8 min, the plasma aliquot withdrawn, the white cell layer overlying the red cells was removed, and erythrocytes were then removed largely free of contamination with leukocytes and platelets. The samples were then extracted. The processing of the blood of the leukemic subjects was as described in part I. Blood samples from normal individuals required 15–20 min for processing (from the time of withdrawal of blood to extraction), while the blood samples from the patients required 30–35 min for workup.

References p. 447/448

RESULTS AND DISCUSSION

Normal individuals

Subject E. Rob. after an overnight fast ingested 50 g of glutamine with a total of 405 ml of water over a period of 10 min. Two pre-ingestion control blood samples were drawn and the subject was then sampled at 15, 30, 45, 60, 90, 120, 150, 180 and 240 min after ingestion of glutamine. Figs. 268–277 show the findings with plasma. The

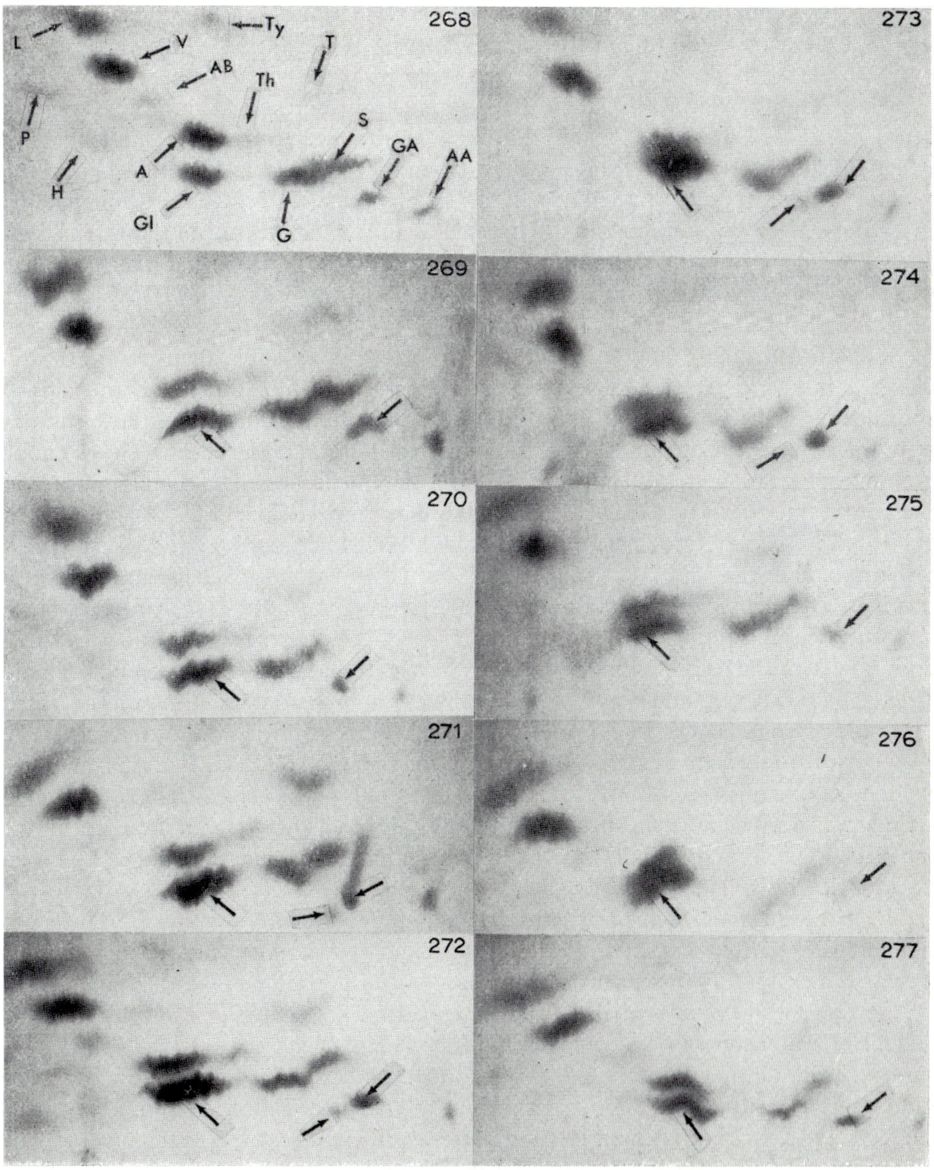

Figs. 268–277. Extracts of 0.3 ml of plasma from subject E. Rob. after ingestion of glutamine. Fig. 268 prior to glutamine ingestion, others 15, 30, 45, 60, 90, 120, 150, 180, and 240 min after ingestion of glutamine. For abbreviations see p. 369.

References p. 447/448

glutamine level gradually rose to a high point between 60–90 min (Figs. 272 and 273) and then fell gradually almost to the control level by 240 min. A rise in the glutamic acid level was observed along with the rise in glutamine. An uncharacterized substance migrating to the lower left of glutamic acid on chromatograms was also observed to rise and fall with the glutamine and glutamate levels. Some fluctuation of the taurine level is apparent in the plasma samples and this and fluctuations of the glutamic acid level are probably related to the venipuncture response described in part I. The venipuncture response may also account for a fall in the alanine level observed at 30, 60, and 90 min (Figs. 270–273).

Chromatograms prepared from the erythrocyte samples from subject E. Rob. are shown in Figs. 278–287. A distinct fall and then an increase of glutamine was observed in erythrocytes. The highest glutamine level was observed at 90 min (Fig. 283) when the plasma level was highest (Fig. 273). The glutamine level fell below the control level at 240 min. A general fall in free amino acid levels in erythrocytes was evident between 180 and 240 min. Of considerable interest are the variations that occurred during the study. There was a steady decrease of taurine in the first three samples (Figs. 278–280) with a rise, a fall, and another rise followed by a fall to below the control level at 240 min. These fluctuations are probably related to the venipuncture response and to the intake of water by the subject during the study. The alanine concentration of erythrocytes decreased then increased and finally returned to slightly below the control level at 240 min. The maximum level of alanine was reached at 90 min when glutamine was highest. Unlike plasma, the glutamic acid level of erythrocytes did not increase, and the uncharacterized compound seen in plasma was not detectable in red cells. A more marked venipuncture response in this subject may be related to the fact that he was the first to be studied and was apprehensive throughout the study. The subject complained of slight headache and mild nausea one hour after ingestion of glutamine but the symptoms passed quickly.

Typical urine findings are shown in Figs. 288–295. Figs. 288–292 show the free amino acids in the urine specimens from subject E. Rob. before ingestion of glutamine (Fig. 288) and 1, 3, 5, and 8 h after glutamine ingestion. A diluted urine was excreted for a time and then the free amino acid levels were again near the control levels. Figs. 293–295 show three samples obtained from subject H. Bie. after glutamine injection. A control specimen (Fig. 293) was collected 5 min before the control blood sample was drawn and then specimens were obtained 90 and 330 min after glutamine ingestion. The second sample was low in free amino acids and the final sample was back to nearly control levels.

The urine chromatograms show that the large amount of glutamine ingested does not increase the excretion of glutamine or other free amino acids to any extent. Urinary urea and uric acid were not increased after glutamine ingestion.

Another normal male subject (A. Knu.) ingested 50 g of glutamine in a total of 500 ml of water over a period of 4 min. One control blood sample was drawn and samples were obtained 30, 60, 90, 150, 210, 270 and 330 min following ingestion of glutamine. The subject was not as apprehensive as subject E. Rob., but like the latter subject experienced mild nausea that passed quickly about 1 h after ingestion of glutamine. The findings in blood plasma were similar to those with subject E. Rob. The highest levels of glutamine in plasma were noted between 60 and 90 min and the levels of both glutamic acid and the uncharacterized substance migrating just to

References p. 447/448

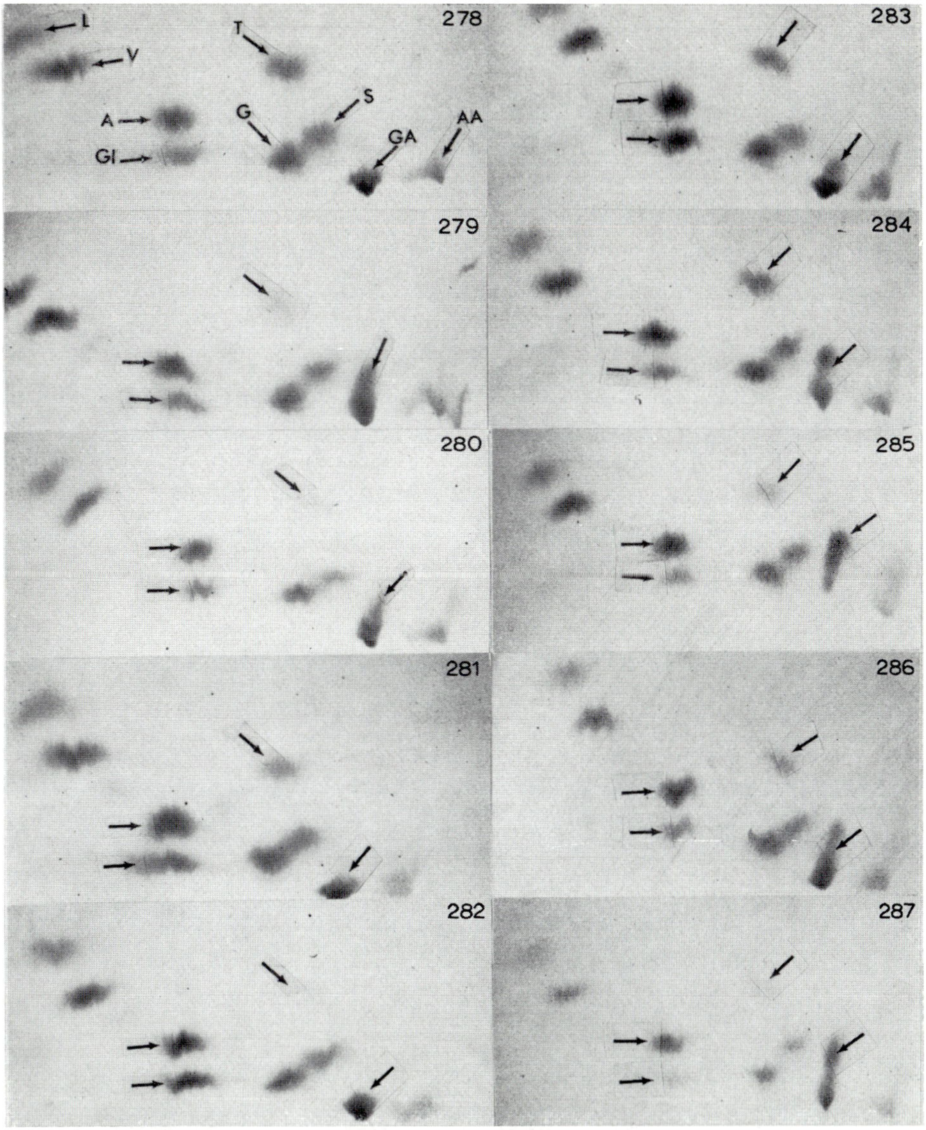

Figs. 278–287. Extracts of 0.3 ml of packed erythrocytes fıom subject E. Rob. following ingestion of glutamine. The samples were removed at the times indicated for corresponding plasma samples in the legend for Figs. 268–277. For abbreviations see p. 369.

the left of glutamic acid rose and fell along with glutamine. Figs. 296–303 show the findings with erythrocytes from A. Knu. The samples were extremely uniform. No changes of amino acid levels were produced from ingestion of glutamine or withdrawal of blood samples. The high degree of reproducibility of the methods employed is demonstrated. Note that subjects E. Rob. and A. Knu. showed grossly different responses; in the former, erythrocytes showed increases of both glutamine and alanine while in the latter no changes were observed. A marked difference in erythrocyte free amino acid

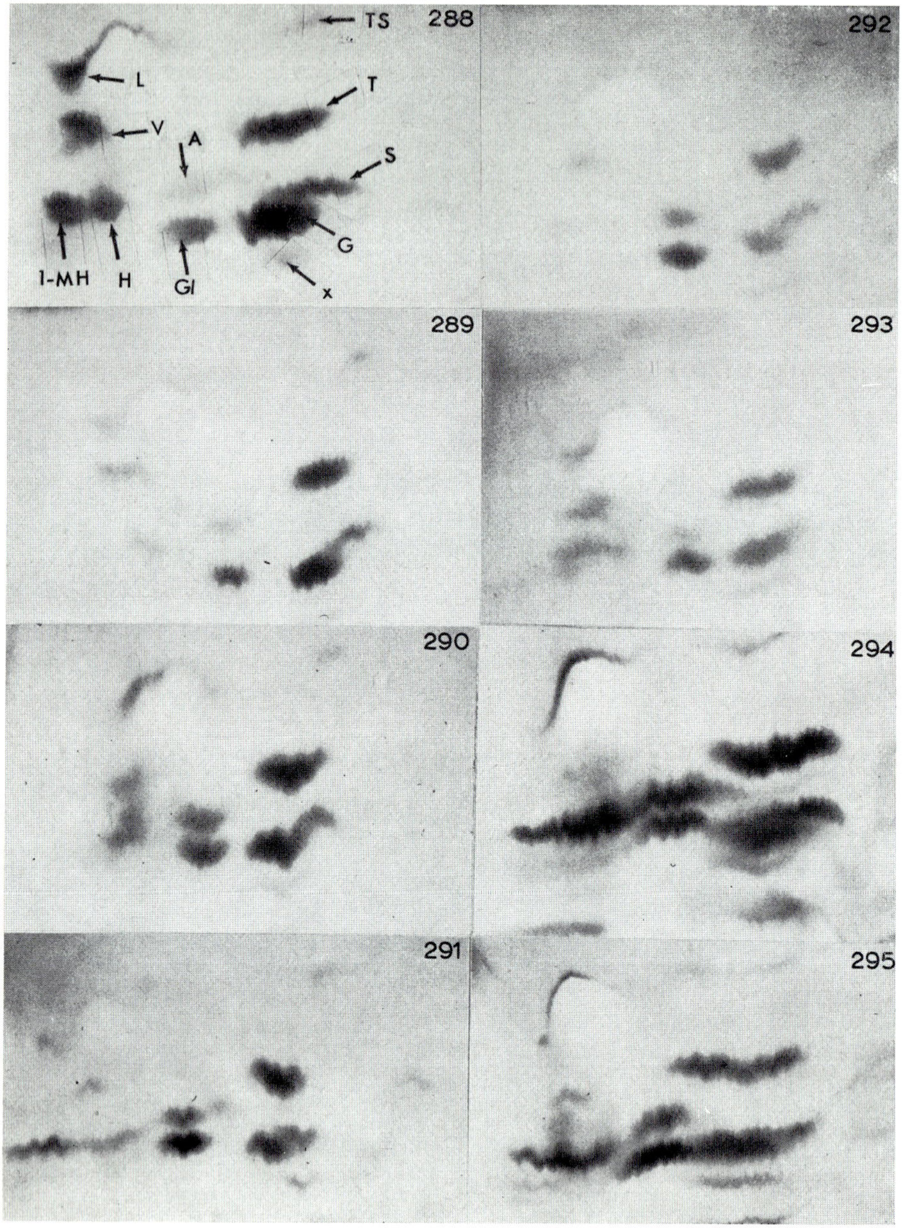

Figs. 288–292. From 0.5 ml of urine of subject E. Rob. and Figs. 293–295 from urine of subject H. Bie. after glutamine ingestion. Urine volumes (H. Bie) were 330, 600, and 340 ml. For abbreviations see p. 369.

pools from different individuals and a failure of erythrocyte free amino acids to reflect plasma changes has been pointed out in previous papers in this series.

A third normal male subject (H. Bie.) consumed 50 g of glutamine with a total of 480 ml of water over a period of 10 min. Two control blood samples were obtained

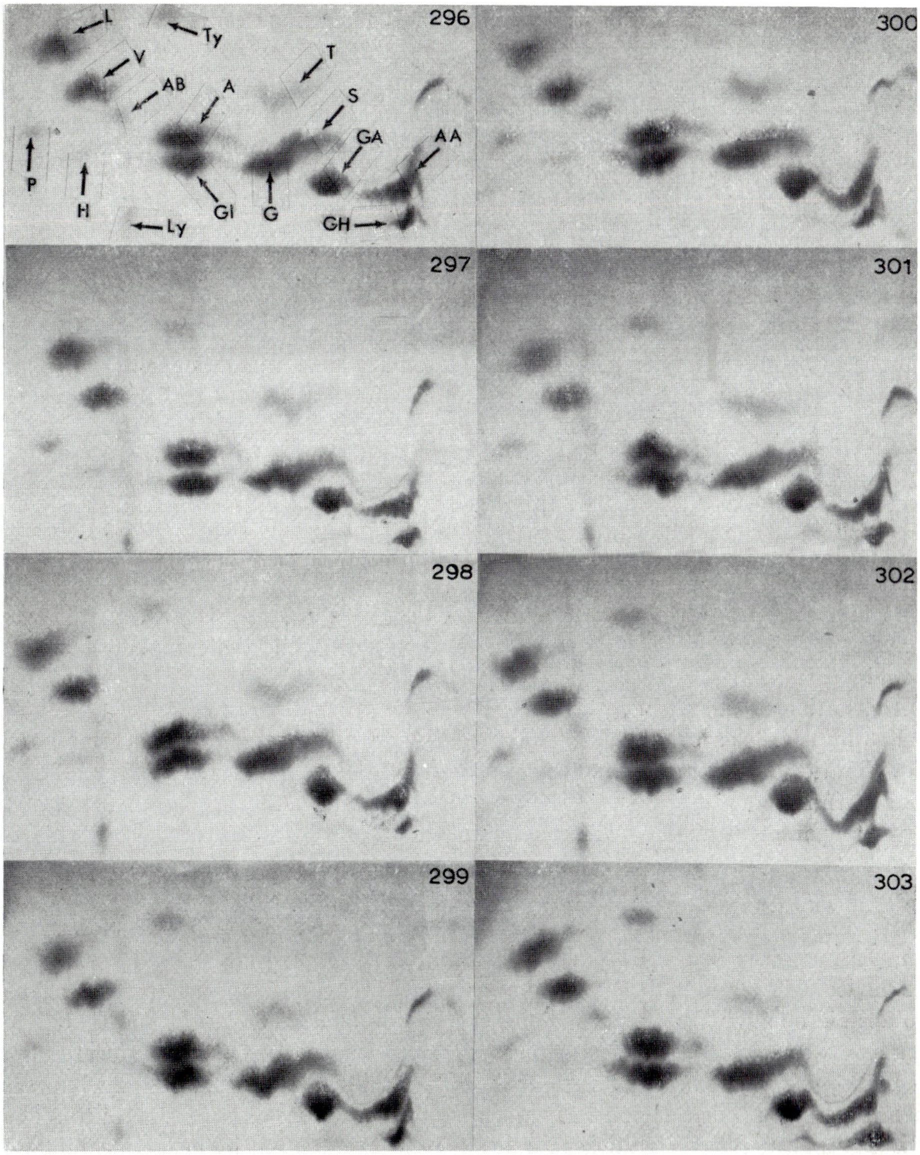

Figs. 296–303. Extracts from 0.5 ml of packed erythrocytes from subject A. Knu. following ingestion of glutamine. Samples prior to ingestion and 30, 60, 90, 210, 270, and 330 min after ingestion. Note the marked uniformity of the chromatograms and the failure of the glutamine level to rise appreciably. For abbreviations see p. 369.

and additional samples were drawn 15, 30, 50, 70, 145, 205, 265, and 340 min after ingestion of glutamine. One feature of the study of patient H. Bie. was unique. At the time glutamine was ingested, a saline infusion was begun in the antecubital vein and was continued for 1.5 h when a total of 200 ml of physiological saline had been infused . Thus, the two control samples were taken prior to the saline infusion, and the infusion was stopped 20 min after the 70 min post-ingestion sample was obtained.

This change in the method of study was made to determine whether a saline infusion would change the permeability of erythrocytes to glutamine. It was suspected that the red cells might show increased permeability to amino acid by virtue of some marked differences in results reported in the literature for the rate of equilibration of erythrocytes with plasma after ingestion as opposed to infusion of amino acids.

The plasma free amino acid levels of subject H. Bie. are shown in Figs. 304–308 and 314–318. The same rise and fall of glutamine, glutamic acid, and the unknown substance to the left of glutamic acid, as well as a gradual decrease of taurine followed by a return to somewhat above the control level, are in agreement with the findings with other normal subjects and patients with leukemia. Of considerable interest is the fact that the glutamic acid level did not return completely to the control level as observed for other subjects. The plasma glutamate level in this subject (H. Bie.) remained moderately high even after glutamine had returned to the control level.

A very clear venipuncture response is shown in Figs. 304–306. The fall in taurine, glutamic acid, and aspartic acid is very clear and was maximal 40 min after the first venipuncture.

The chromatographic results with erythrocytes from subject H. Bie. are shown in Figs. 309–313 and 319–323. The chromatograms are of considerable interest in that a rapid equilibration of plasma and erythrocyte glutamine levels is illustrated. The first sample drawn 15 min after ingestion of glutamine (Fig. 311) was very similar to the pre-ingestion controls (Figs. 309, 310), while the sample obtained at 30 min (Fig. 312) showed a greatly elevated glutamine level that corresponded to the elevation of the plasma glutamine level (Fig. 307). It is to be recalled that the saline infusion was begun just before the 15-min blood sample was drawn. Rapid equilibration of erythrocyte and plasma glutamine levels appears to have followed the beginning of the infusion. There was a decrease of taurine below the control level (apparent up to the 70 min sample) followed by a return to the control level.

All samples drawn after the saline infusion was begun showed close correspondence of plasma and erythrocyte glutamine levels. After the cessation of the saline infusion (between the 70 and 145 min samples) the erythrocyte levels were first higher than the plasma levels. The erythrocyte level then fell and again matched the plasma level as shown in Figs. 318 and 323. It can be concluded that the saline infusion was related to a rapid equilibration of plasma and erythrocyte glutamine levels, and that, after infusion was stopped, the rapid equilibration no longer existed. This interpretation is in keeping with the results from the other eight ingestion studies in which the absence of the saline infusion was associated with a failure of erythrocytes to reflect the changes in plasma glutamine levels to the same extent. The permeability of erythrocytes for amino acids appears to be altered *in vitro* when a solution of amino acid is added to plasma (see part I) in which the cells are incubated. This *in vitro* result is thus in keeping with the findings *in vivo* after saline infusion.

It is perhaps surprising that such a small quantity of saline given over a relatively long period of time (1.5 h) should have such an effect. It appears, however, that there are marked changes in erythrocyte permeability for some compounds *in vitro* and *in vivo* when relatively small amounts of salt solutions are added to blood. These permeability changes are not apparent for all of the amino acids in the erythrocyte free amino acid pool. The levels of some of the free amino acids remain relatively constant.

References p. 447/448

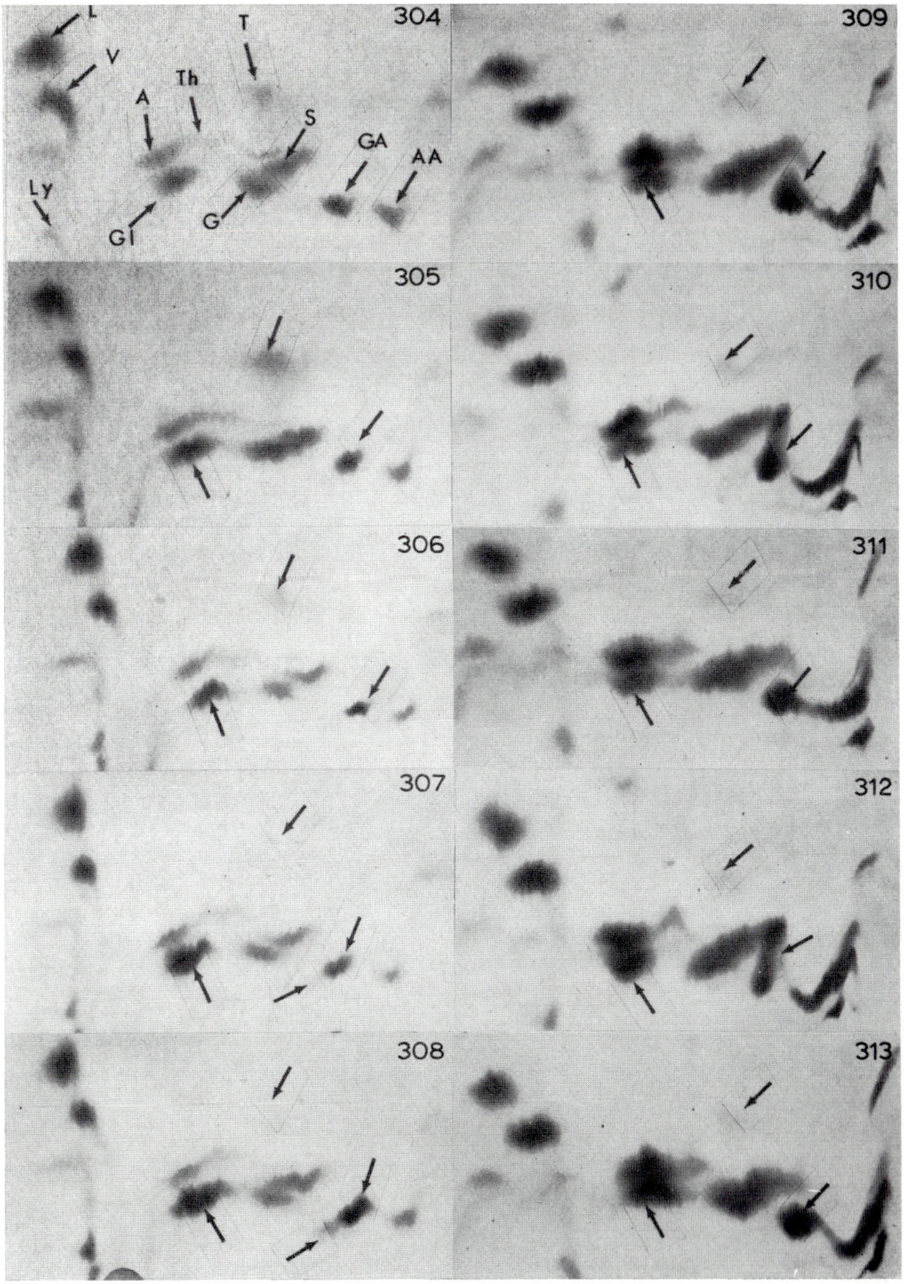

Figs. 304–323. Extracts of 0.5 ml of plasma and erythrocytes from subject H. Bie. following ingestion of glutamine. Plasma and corresponding erythrocyte samples shown side by side for ready comparison of the glutamine levels. Figs. 304 and 305 the control plasma samples and Figs. 309 and 310 the control erythrocyte samples. Figs. 306–308 and 314–318 show the plasma samples 15, 30, 50, 70, 145, 205, 265 and 340 min after ingestion of glutamine. Erythrocyte samples (Figs. 311–313 and 319–323) at the same time intervals. Note the close correspondence of the glutamine levels of plasma and cells. For abbreviations see p. 369.

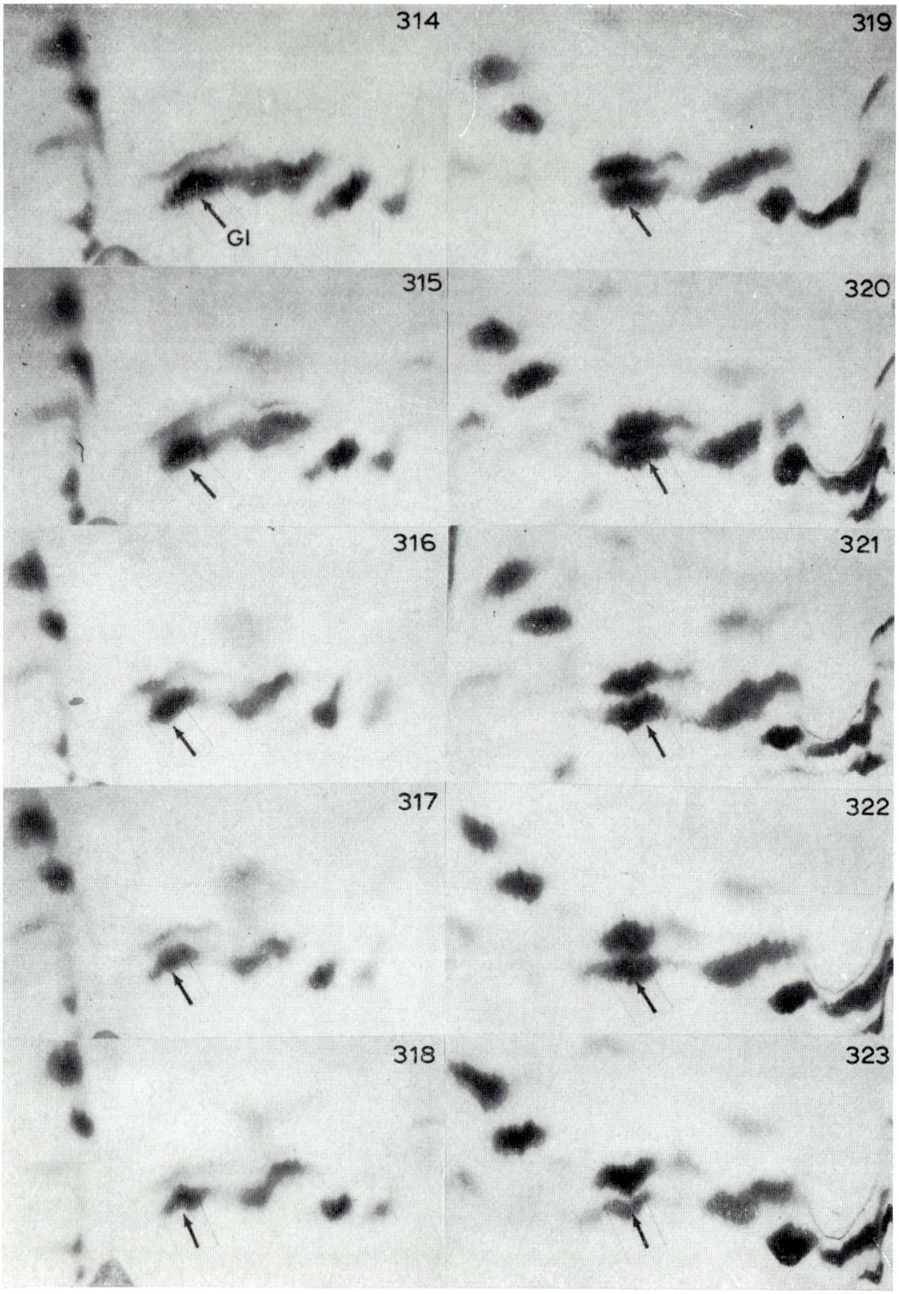

Figs. 314–323. For legend see p. 437.

Results with subject R. Tho. (chronic lymphatic leukemia)

Three control blood samples were removed after an overnight fast. The second control sample was drawn 20 min after the first and the third sample was drawn

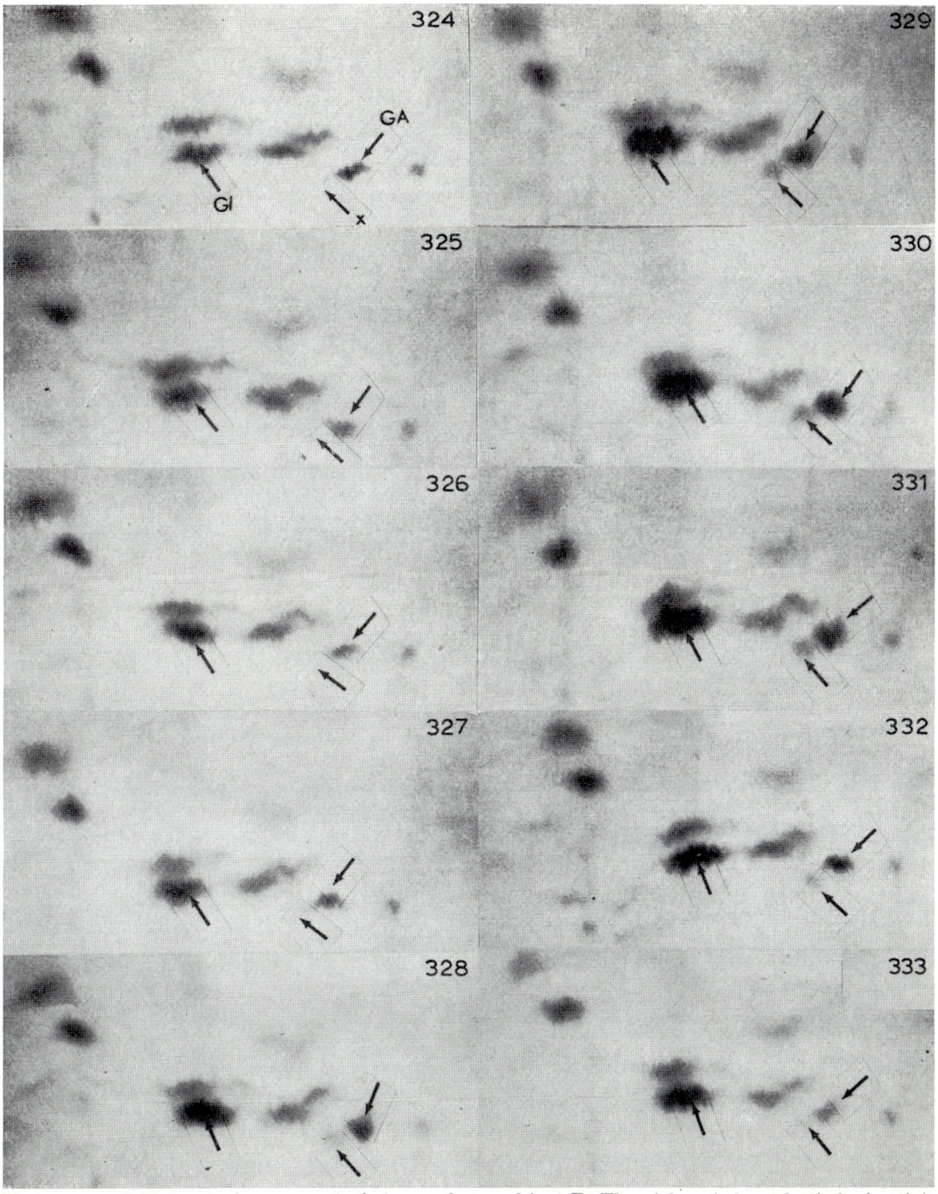

Figs. 324–336. Extracts from 0.3 ml of plasma from subject R. Tho. (chronic lymphatic leukemia) after ingestion of glutamine. Figs. 324–326 from 3 control samples (second sample 20 min after first, third sample 40 min after second. Figs. 327–336, 15, 45, 65, 85, 125, 175, 235, 295, 355 and 415 min after ingestion. For abbreviations see p. 369.

40 min after the second sample. The subject then ingested 50 g of glutamine with a total of 460 ml of water over a period of 10 min and additional blood samples were obtained 15, 45, 65, 85, 125, 175, 235, 295, 355 and 415 min after ingestion.

Figs. 324–336 show the chromatographic findings for the blood plasma samples from subject R. Tho. The three control samples (Figs. 324–326) show a relatively

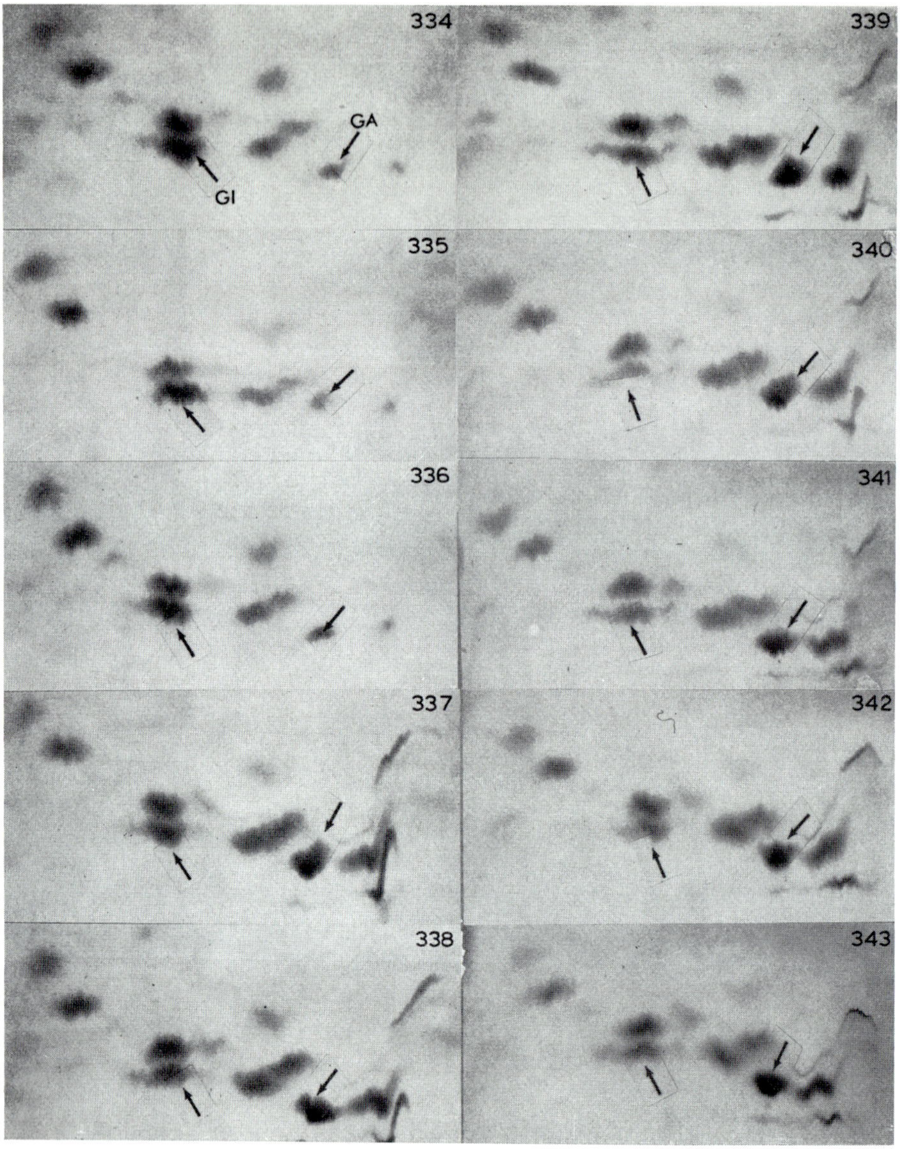

Figs. 337–349. Extracts from 0.3 ml of packed erythrocytes from subject R. Tho. after glutamine ingestion. The 3 control samples (Figs. 324–336) as indicated in legend for Figs. 324–336 and post-ingestion samples (Figs. 340–349) removed at times indicated in legend above.

constant free amino acid pattern prior to glutamine ingestion with only a small drop in taurine and glutamate (the venipuncture response). The venipuncture response in this case was relatively slight. The subject preferred to lie down during the study. He was not nervous, occasionally fell asleep, and showed no sign of disturbance at venipuncture. Following the ingestion of glutamine the plasma level rose to a maximum in the 65-min sample (Fig. 329). This high level was still observed in the 85-min sample (Fig. 330). The level then fell gradually to the control level at 355 min

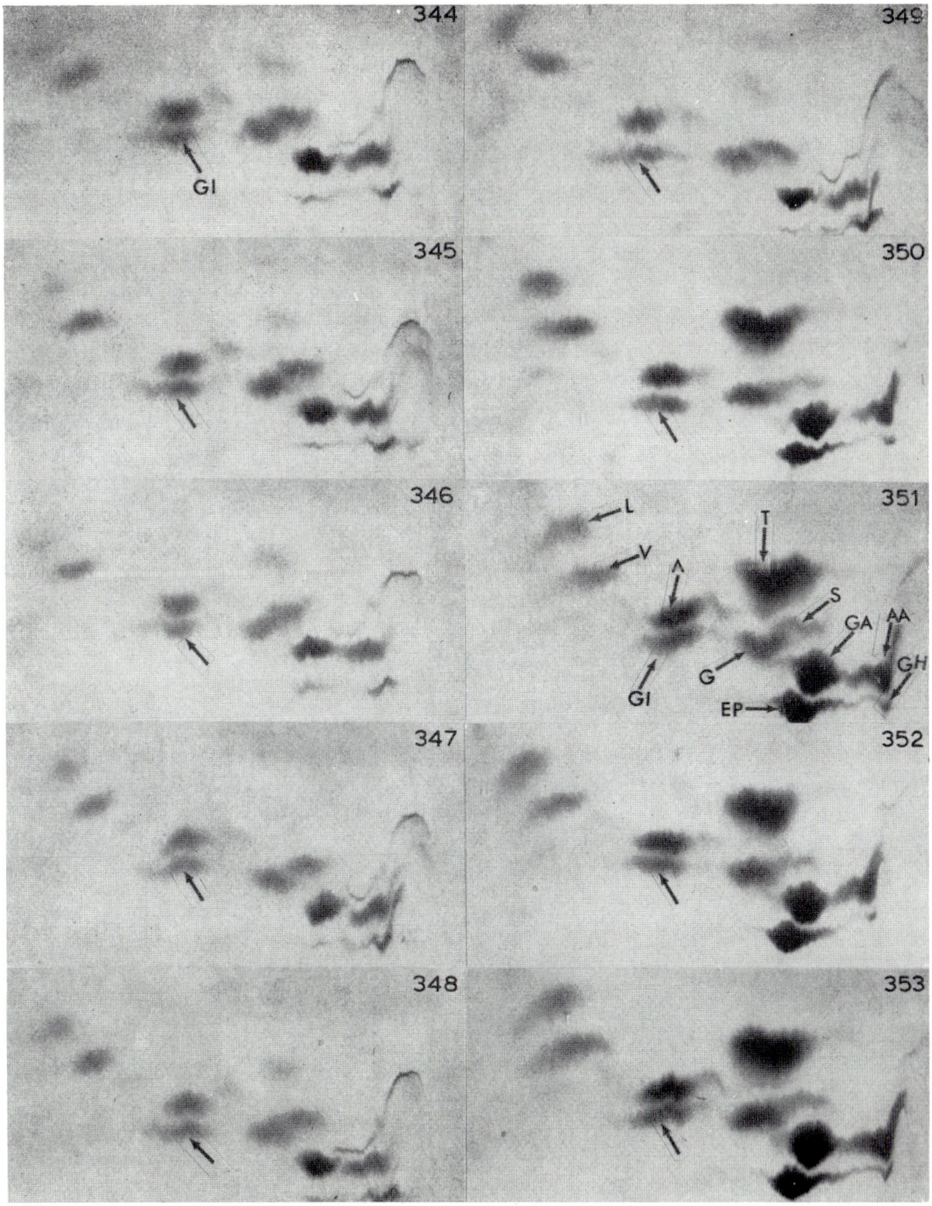

Figs. 344–349, see legend p. 440; Figs. 350–353, see legend p. 442.

(Fig. 335). As with normal individuals, plasma glutamic acid and the unknown substance to the left of glutamic acid increased and then decreased. The rise and fall of plasma glutamine in the patient could not be distinguished from the rise and fall observed in the three hematologically normal individuals. No difference from the normal manner of handling this load of glutamine was indicated. This is in spite of

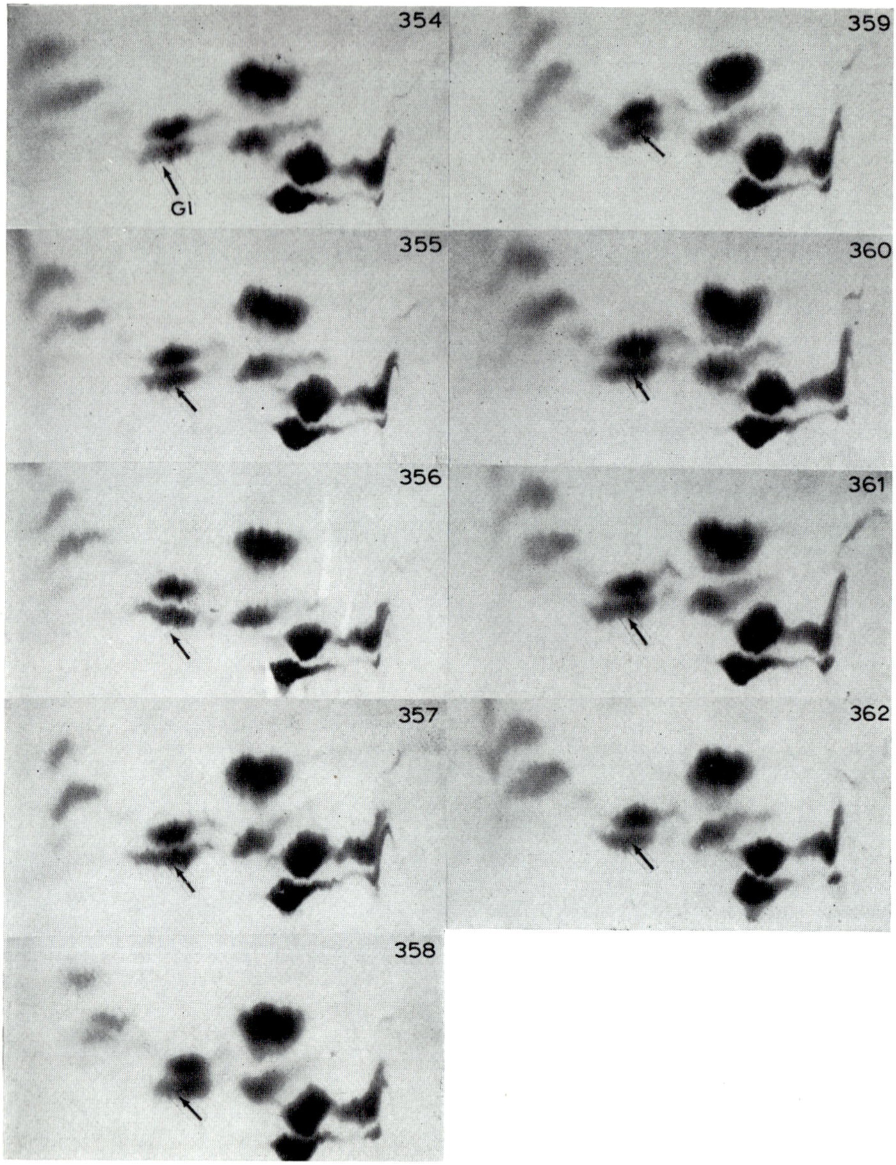

Figs. 350–362. Extracts from 75 mg of leukocytes from subject R. Tho. after glutamine ingestion. Samples at the same time intervals indicated for plasma samples (legend for Figs. 324–336).

the fact that the lymphocytes from this patient were found to degrade glutamine *in vitro* as described in part 1.

Figs. 337–349 show the chromatograms prepared from erythrocyte samples from subject R. Tho. The three control samples (Figs. 337–339) were similar. Following the ingestion of glutamine there was a general decrease of free amino acids in the erythrocytes that was not accompanied by any apparent packed cell volume changes (the hematocrit value remained constant). The level of glutamine was always lower

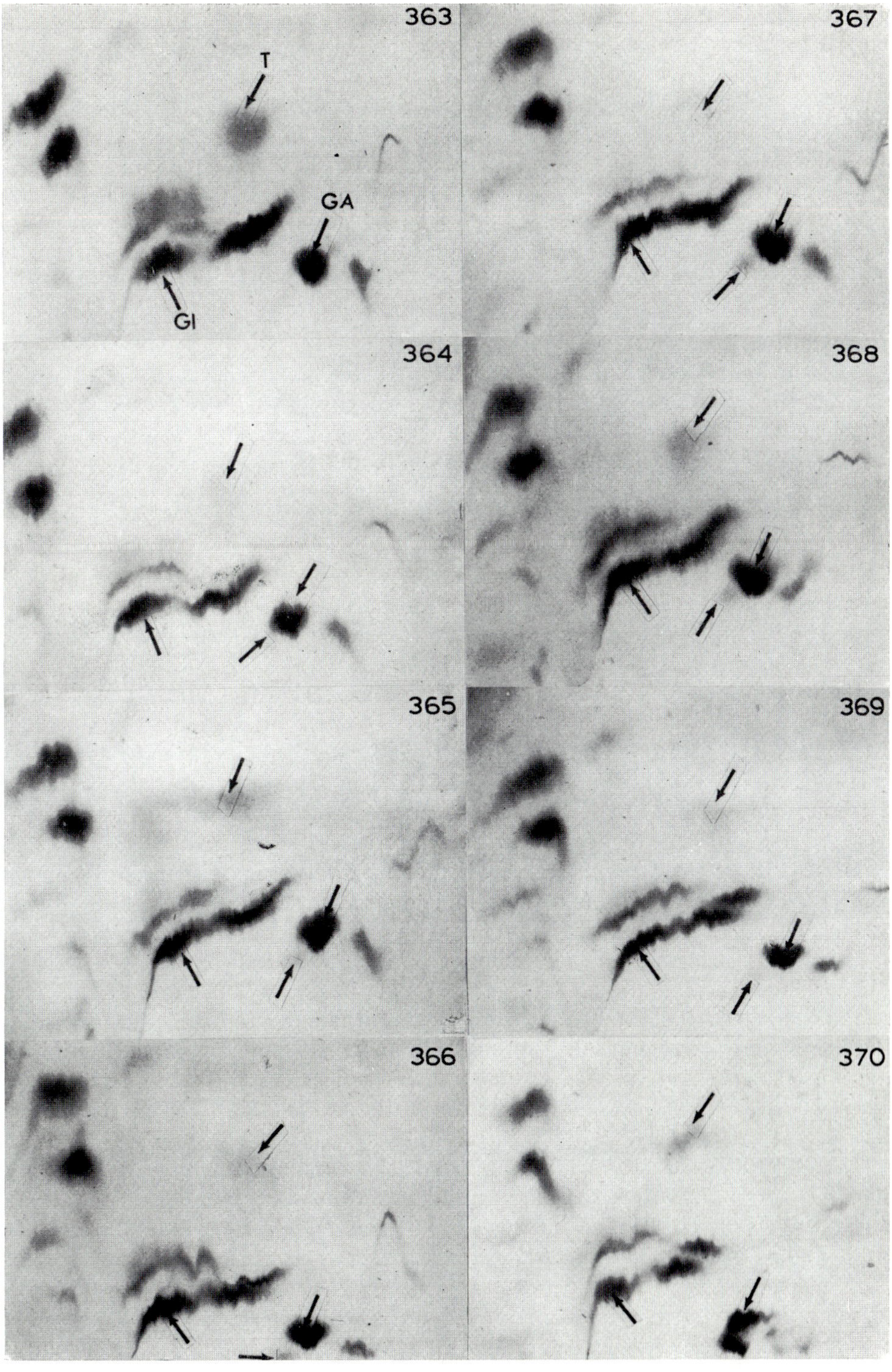

Figs. 363–372. Extracts from 0.3 ml of plasma from subject H. Gol. (chronic granulocytic leukemia) after glutamine ingestion. Fig. 363, control sample, and Figs. 364–372 obtained 15, 35, 60, 90, 120, 150, 180, 250 and 305 min after ingestion. For abbreviations see p. 369.

in the cells following ingestion of this amino acid than in the control samples. The high degree of reproducibility of the erythrocyte free amino acid pool after glutamine ingestion is illustrated in Figs. 340–349.

Figs. 350–362 show the free amino acid pools of the lymphocytes from subject R. Tho. The control samples in Figs. 350–352 are similar. The glutamine level in the leukocytes rose to reach a peak at 85 min (Fig. 356) that was maintained through the 175-min sample (Fig. 358). The level then fell to the level observed in control samples. The elevation of glutamine in the leukocytes of this subject was somewhat less than the elevation observed for granulocytes described below.

Each individual urine specimen from subject R. Tho. for 24 h before the glutamine ingestion study, all specimens obtained during the day of the study, and all specimens collected for 24 h after the study were kept separately and examined by paper chromatography (a total of 15 specimens). The free amino acid levels of the urine (not illustrated) were very uniform and the results were in keeping with those of the hematologically normal individuals in that no change in glutamine or other free amino acids was noted following the ingestion of glutamine.

Glutamine ingestion in chronic granulocytic leukemia

Patient H. Gol. (chronic granulocytic leukemia) without any previous form of treatment and a white cell count of 450 000/mm^3 was studied as follows. Three control blood samples were drawn. 50 g of glutamine was consumed in 400 ml of water over a period of 10 min, and additional blood samples were drawn 15, 35, 60, 90, 120, 150, 180, 250, and 305 min after ingestion of glutamine.

Figs. 363–372 show the plasma samples from patient H. Gol. Only the second of the three control samples is shown (Fig. 363). The three control samples are presented in part I of this series (Figs. 32–34, p. 362) to illustrate the marked venipuncture response. The plasma glutamic acid levels of the control samples were far above the levels encountered in normal individuals (compare Fig. 363 with Figs. 268, p. 431 and 304, p. 437) and most patients with chronic lymphatic leukemia (compare Fig. 363 with Figs. 324–326, p. 439). A definite elevation of the levels of most of the other amino acids can be observed also (see the complete discussion of the findings in chronic granulocytic leukemia in part II).

Following the ingestion of glutamine, the glutamic acid level rose to a peak at about 60 min and fell to below control levels at the end of the study. The level of the uncharacterized substance migrating just to the left of glutamic acid on chromatograms also rose and fell. The peak glutamine level was maintained through 120 min. The rise in plasma glutamine was not as marked as that observed in the three hematologically normal individuals or the patient with chronic lymphatic leukemia.

The glutamine levels of erythrocytes and leukocytes from subject H. Gol. (not illustrated) rose slightly and then returned to control levels.

Two other granulocytic leukemia patients without any previous form of treatment were studied in the same manner as for the other subjects. Each ingested 50 g of glutamine and control and experimental samples were drawn at intervals similar to the ones above. The samples were examined by paper chromatography and glutamine was assayed with glutaminase. Fig. 377 shows the glutamine levels determined by enzymatic assay in the plasma, erythrocytes, and leukocytes of subject Q. Tur., and

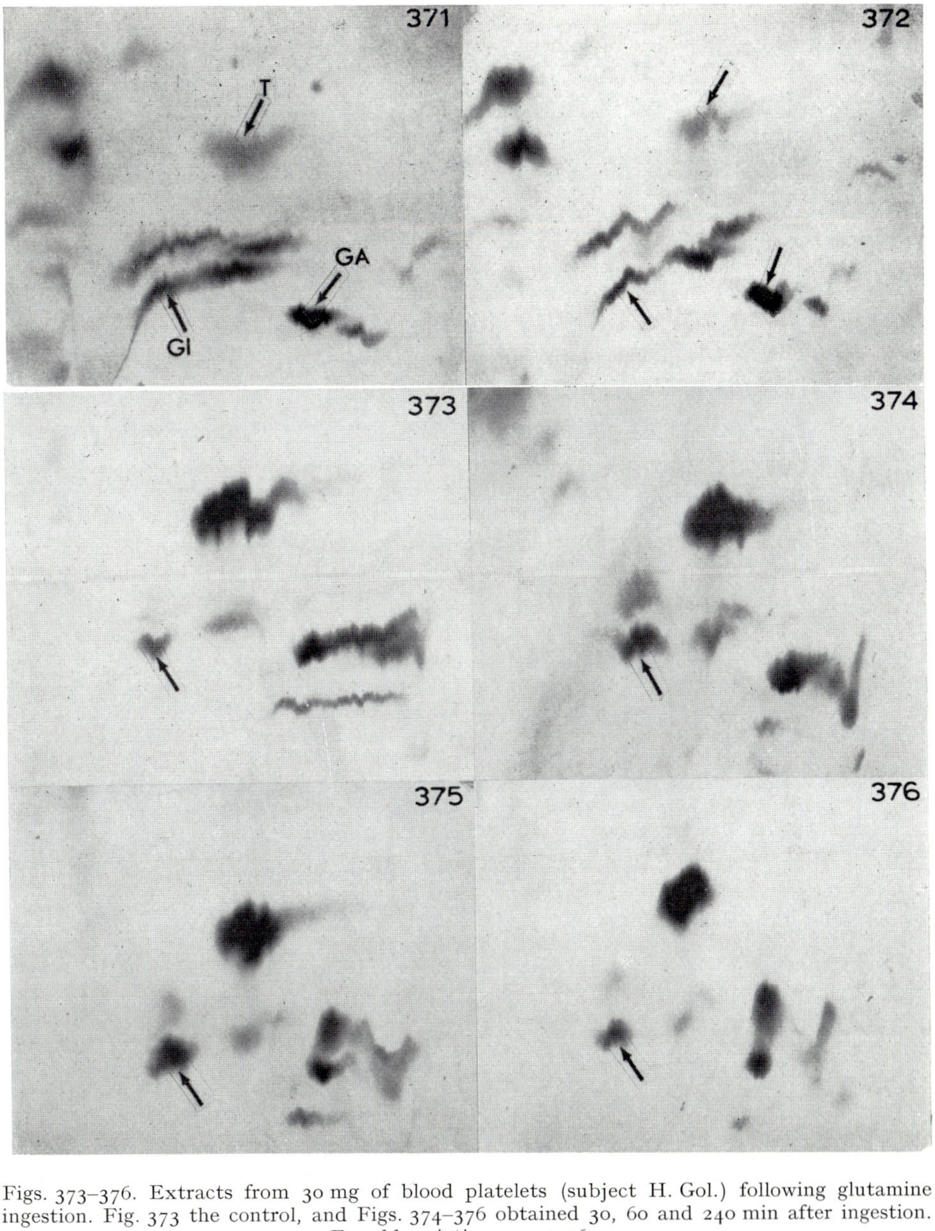

Figs. 373–376. Extracts from 30 mg of blood platelets (subject H. Gol.) following glutamine ingestion. Fig. 373 the control, and Figs. 374–376 obtained 30, 60 and 240 min after ingestion. For abbreviations see p. 369.

Fig. 378 shows the glutamine levels in plasma, erythrocytes, and leukocytes of subject R. Tap. The paper-chromatographic data obtained in these studies was similar to the data obtained in other studies in that the same compounds rose and fell.

The curve for subject R. Tap. was definitely abnormal. The results with subject Q.Tur. were not clearley different from those obtained from normal individuals. The glutaminase assays of erythrocyte glutamine levels show clearly the small increase of glutamine in red cells after amino acid ingestion.

Two additional studies were carried out with subject H. Gol. after treatment with dimethylmyleran when the leukocyte count had been reduced to about 40% of the pretreatment level. The findings in both studies (not illustrated) were much more similar to those obtained with normal individuals and the patient with chronic lymphatic leukemia.

The free amino acid levels of the blood platelets were examined by paper chromatography in one study of patient H. Gol. after treatment with dimethylmyleran. Figs. 373–376 show the free amino acid levels of platelets in the control sample (Fig. 373) and 30, 60, and 240 min after ingestion of glutamine (Figs. 374–376, respectively). The arrows in the figures point to the position of glutamine that can be seen to rise and fall. The study thus established that platelets can concentrate glutamine. The findings are similar to those obtained with leukocyte samples.

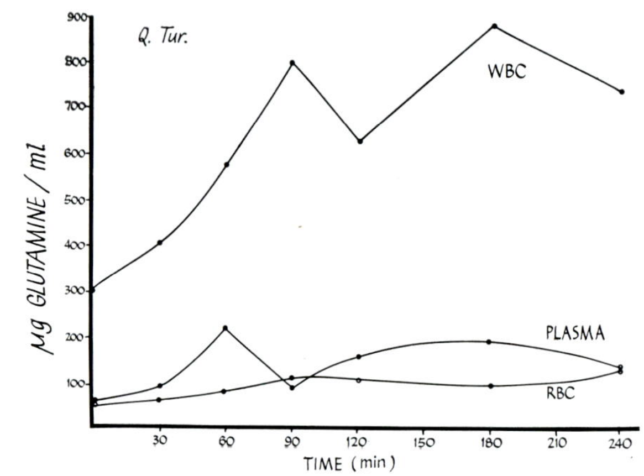

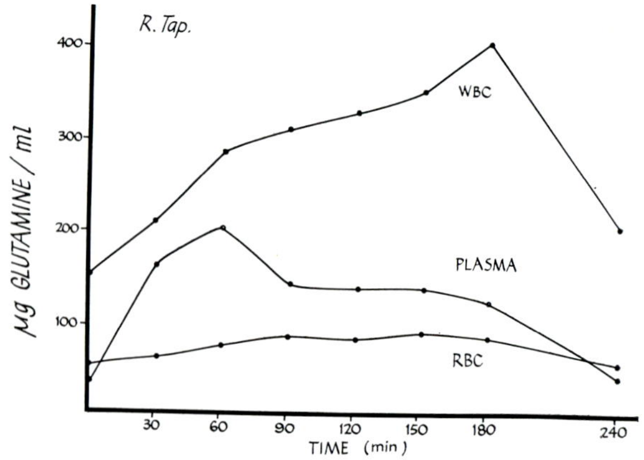

Figs. 377 and 378. Glutamine levels (glutaminase assay) of plasma and cells after glutamine ingestion. Subjects Q. Tur. (Fig. 377) and R. Tap. (Fig. 378).

CONCLUSIONS

After the ingestion of glutamine, the plasma level is elevated and an increase in the levels of glutamic acid and an uncharacterized compound migrating just to the left of glutamic acid on chromatograms is observed. The levels of all three compounds rise and fall together. The uncharacterized compound is probably a metabolite of glutamine. It is detectable at times in normal plasma and more frequently in plasma samples from patients with leukemia. No marked increase in urinary excretion of glutamine, other free amino acids, urea, or uric acid was observed after ingestion of 50 g of glutamine.

Leukocytes and blood platelets concentrate glutamine from plasma while erythrocytes do not. Erythrocytes are not normally freely permeable to glutamine. In some of the studies, the level of glutamine in erythrocytes was essentially unchanged when plasma glutamine was elevated. In other cases erythrocyte glutamine increased slightly. When the conditions in the blood are altered by infusion of saline, glutamine may enter red cells more rapidly and equilibrate completely with plasma.

The response of the leukemic patients was variable. The one patient with chronic lymphatic leukemia that was studied appeared to handle glutamine in a normal manner. One of the patients (H. Gol.) with chronic granulocytic leukemia had a very abnormal response and it appeared that glutamine was metabolized more rapidly in this patient. Two ingestion studies with the same patient after drug therapy indicated a nearly normal response to the ingestion of glutamine after reduction of the white cell mass. One patient (R. Tap.) with chronic granulocytic leukemia also showed an abnormal response to a glutamine load, but another patient responded in a manner not clearly distinguishable from normal. These findings indicate that glutamine metabolism is not significantly affected in all patients and further suggests that in the more extreme cases of untreated granulocytic leukemia glutamine metabolism may be distinctly abnormal. The conclusions from the glutamine ingestion studies agree with the overall impression gained from the study of blood free amino acids. Deviations from the normal levels are not observed at all times in the same patient, although on occasion the levels may be very abnormal.

The results of our studies of normal individuals are in agreement with previous studies. BESSMAN *et al.*[41] observed increases in the levels of glutamine and glutamic acid in plasma from cats after intra-intestinal administration of glutamine. MEISTER *et al.*[42] observed similar increases in glutamine and glutamate levels in phenylketonuric subjects after oral intake of glutamine. Previous investigatiors did not study the effects of glutamine ingestion on the free amino acid levels of erythrocytes, leukocytes, or blood platelets.

ACKNOWLEDGEMENTS

Mr. RICHARD RAY assisted in the preparation of the photographic illustrations. The experimental work was supported by Grant C-3134 from the U.S. Public Health Service.

REFERENCES

1 G. ROUSER, *Studies on the Biochemistry of White Blood Cells*, Ph. D. thesis, The University of Rochester, New York, 1952, 158 pp.
2 G. ROUSER, in J. W. REBUCK, F. H. BETHELL AND R. W. MONTO, (Eds.), *The Leukemias: Etiology, Pathophysiology, and Treatment*, Academic Press, New York, 1957, p. 361.

[3] G. Rouser, in D. Hammond, (Ed.), *Current Concepts in Leukemia*, Report of the 34th Ross Conference on Pediatric Research, Ross Laboratories, Columbus, 1960, p. 51.
[4] C. H. Greene and H. M. Conner, *A. M. A. Arch. Internal Med.*, 38 (1920) 167.
[5] J. M. Luck, *J. Biol. Chem.*, 77 (1928) 13.
[6] E. G. Schmidt, *A.M.A. Arch. Internal Med.*, 44 (1929) 351.
[7] S. Okada and T. Hayashi, *J. Biol. Chem.*, 51 (1922) 121.
[8] J. R. Beaton, W. J. McGanity, and E. W. McHenry, *Can. Med. Assoc. J.*, 65 (1951) 219.
[9] J. M. White, J. R. Beaton and E. W. McHenry, *J. Lab. Clin. Med.*, 40 (1952) 703.
[10] J. M. White, G. A. L. Ross, and E. W. McHenry, *Cancer Research*, 14 (1954) 173.
[11] J. M. White, G. Ozawa, G. A. L. Ross, and E. W. McHenry, *Cancer Research*, 14 (1954) 508.
[12] E. Roberts and S. Frankel, *Cancer Research*, 9 (1949) 645.
[13] E. Roberts and T. Tanaka, *Cancer Research*, 16 (1956) 204.
[14] H. A. Waisman, in J. W. Rebuck, F. H. Bethell, and R. W. Monto, (Eds.), *The Leukemias: Etiology, Pathophysiology, and Treatment*, Academic Press, New York, 1957, p. 339.
[15] J. J. Kelley and H. A. Waisman, *Blood*, 12 (1957) 635.
[16] F. Nour-Eldin and J. F. Wilkinson, *Brit. J. Haemat.*, 1 (1955) 367.
[17] G. Y. N. Iyer, *Indian J. Med. Research*, 46 (1958) 568.
[18] E. N. Sassenrath and D. M. Greenberg, *Cancer Research*, 14 (1954) 563.
[19] C. Wu and J. M. Bauer, *Cancer Research*, 20 (1960) 848.
[20] J. Bernard, R. Bourdon, J. Caen, R. Ardaillou, S. Ranisteano and M. Pays, *Le Sang*, 30 (1959) 32.
[21] P. E. Lutz, E. Larson, and W. F. Dunning, *J. Natl. Cancer Inst.*, 23 (1959) 1331.
[22] R. H. McMenamy, C. C. Lund, and D. F. H. Wallach, *J. Clin. Invest.*, 39 (1960) 1688.
[23] H. Meier, *Zentr. Veterinärmed.*, 4 (1957) 633.
[24] A. Brion and M. Fontaine, *Bull. acad. vét. France*, 30 (1957) 299.
[25] G. E. Gutman and B. Alexander, *J. Biol. Chem.*, 168 (1947) 527.
[26] H. N. Christensen, P. F. Cooper Jr., R. D. Johnson and E. L. Lynch, *J. Biol. Chem.*, 168 (1947) 191.
[27] P. Handler, H. Kamin and J. S. Harris, *J. Biol. Chem.*, 179 (1949) 283.
[28] H. N. Christensen and M. E. Henderson, *Cancer Research*, 12 (1952) 229.
[29] G. Wiseman and F. N. Ghadially, *Brit. Med. J.*, (1958, II) 18.
[30] G. Rouser and A. J. Samuels, *Proc.Am. Assoc. Cancer Research*, (Abstracts) 2 (1958) 340.
[31] A. S. Weisberger and R. W. Heinle, *J. Lab. Clin. Med.*, 36 (1950) 872.
[32] E. I. Goldenthal, M. V. Nadkarni and P. K. Smith, *Radiation Research*, 10 (1959) 571.
[33] E. V. Flock, F. L. Mann and J. L. Bollman, *J. Biol. Chem.*, 192 (1951) 293.
[34] J. J. Roberts and G. P. Warwick, *Nature*, 179 (1957) 1181.
[35] J. J. Roberts and G. P. Warwick, *Biochem. Pharmacol.*, 1 (1958) 60.
[36] J. J. Roberts and G. P. Warwick, *Nature*, 183 (1959) 1509.
[37] J. J. Roberts and G. P. Warwick, *Nature*, 184 (1959) 1288.
[38] E. G. Trams, M. V. Nadkarni, V. DeQuattro, G. D. Maengioyn-Davies and P. K. Smith, *Biochem. Pharmacol.*, 2 (1959) 7.
[39] W. E. Parkam and J. M. Wilbur Jr., *J. Am. Chem. Soc.*, 81 (1959) 607.
[40] F. W. Sayre and E. Roberts, *J. Biol. Chem.*, 233 (1958) 1128.
[41] S. P. Bessman, J. Magnes, P. Schwerin and H. Waelsch, *J. Biol. Chem.*, 175 (1948) 817.
[42] A. Meister, S. Udenfriend and S. P. Bessman, *J. Clin. Invest.*, 35 (1956) 619.

INVITED DISCUSSION

FREE AMINO ACIDS IN PLASMA, BRAIN AND MUSCLE FOLLOWING HEPATECTOMY

EUNICE V. FLOCK, AND JESSE L. BOLLMAN

Mayo Clinic and Mayo Foundation, Rochester, Minn. (U.S.A.)

In 1924 BOLLMAN, MANN AND MAGATH[1] demonstrated the importance of the liver in the maintenance of normal levels of free amino acids in blood and tissues. The formation of measurable amounts of urea ceased after removal of the liver from the dog. Increases in the concentration of free amino acids in blood, urine, and muscles[2] were found in the dehepatized dog, which were approximately equal to the amount required for urea that would have been formed by the normal animal.

When the free amino acids in the plasma of the dehepatized dog were examined by paper-chromatographic technics, increased levels of 15 amino acids or derivatives were found[3]. In muscle, in addition to small increases observed in the content of other amino acids, an increased concentration of glutamine was found associated with a decreased concentration of glutamic acid[4]. In brain tissue, however, a 5-fold increase in the amount of glutamine occurred without a corresponding decrease in glutamic acid, and smaller increases in the amounts of other amino acids such as phenylalanine and tyrosine were also found.

Intravenous administration of large amounts of glucose suppressed the accumulation of the free amino acids in plasma[5] and muscle[6] of the dehepatized dog so that normal levels could be maintained. Because administration of glucose did not prevent increases of the amino acids in plasma and muscles of the dehepatized-depancreatized dog, but glucose with insulin did, it was concluded that the effect of glucose on the level of free amino acids in the dehepatized dog was produced by stimulation of the secretion of insulin. It appeared that insulin would promote the synthesis of protein from free amino acids in plasma and muscles of the dehepatized dog. In the depancreatized dog with a liver, high levels of amino acids were not found in plasma or tissues due to formation of urea by the liver. Thus, this insulin effect on the levels of amino acids in plasma and tissues was most clearly seen in the dehepatized dog.

Curiously, the administration of glucose with or without insulin had no effect on the progressive increase in glutamine and other amino acids which occurred in the brain of the dog, after removal of the liver. The effect of the absence of the liver on the levels of free amino acids is, therefore, more clearly seen in the brain than in other tissues. These increases in content of free amino acids in the brain are of particular interest because liverless dogs under the most favorable conditions survive about 2 days and then die in hepatic coma.

Increases in the content of total amino acids in the brains of dogs with Eck's

References p. 460

fistula[7] are often as great as those in dehepatized dogs, and much of the increase is due to glutamine content. Glutamine is also increased many-fold in the cerebrospinal fluid where its concentration greatly exceeds that in blood plasma.

Increases in content of glutamine may be produced in the brain by other methods than removal of the liver or alteration of the circulation to the liver. Small increases have been produced by parenteral administration of glutamine into the brain of rats[8, 9] and dogs[10]. Elevated levels of glutamine in the cerebrospinal fluid have also been produced by oral administration of glutamine[11]. DE RUISSEAU *et al.*[12] have produced rapid increases in glutamine of the rat brain by administration of lethal doses of ammonium salts. Protective doses of arginine with or without glutamic acid and α-ketoglutaric acid, which increased the survival time of these rats, did not prevent the large increases in glutamine in the brain. EISEMAN *et al.*[13] administered 1% ammonium hydroxide in lactate–Ringer's solution to dogs by bilateral carotid infusion and found that the content of glutamine of the brain increased markedly with the time of infusion. No significant changes in the other amino acids of the brain were found after the administration of ammonia. In contrast, when the glutamine in the brain was elevated by removal of the liver or alteration of the circulation to the liver, smaller but definite increases in other amino acids, such as phenylalanine and tyrosine were found in the brain. The typical electro-encephalogram during coma induced by intoxication with ammonia differs from that seen in coma of liverless animals and from the essentially normal recordings obtained during the coma of animals with Eck's fistula.[7]

We have measured the total free amino acids in the brain, plasma, skeletal muscle and cardiac muscle of rats under a variety of conditions and have examined the relationship of the content of glutamine to the level of total free amino acids. The effect of administration of insulin on the amino acids and the effect of variations in body temperature have also been studied in normal rats, eviscerated rats, and eviscerated-adrenalectomized rats.

EXPERIMENTAL METHODS AND MATERIAL

Adult male rats of the Sprague-Dawley strain, weighing from 210 to 330 g and maintained on Fromm's dog food, were used in these studies. Hepatectomy was done by a two-stage evisceration. Ether anesthesia was used for both stages. In the first stage, the vena cava was ligated with surgical silk just above the kidneys and a cellophane band was placed around the portal vein so that adequate collateral circulation would be established. About 6 weeks later, the liver was removed with the stomach, intestines, spleen, and pancreas. Cystostomy with a small polyvinyl catheter in these and in control rats made continuous collection of urine possible. The adrenal glands were also removed from some of the rats at the time of evisceration, and in some others 1 or 2 days prior to evisceration.

A continuous intravenous infusion was administered at the rate of 1.25 ml/h to each rat. It contained physiologic saline solution, penicillin and streptomycin in sufficient quantities so that 1000 units of penicillin and 5 mg of streptomycin were given per 24 h. Insulin when used and glucose were also given in the continuous infusion. The dose of insulin was 4 units/24 h and it was given to some of the rats who were receiving glucose at the rate of 44–125 mg/100 g of rat per h. Other rats

References p. 460

maintained without insulin received glucose by continuous infusion at the rate of 10–100 mg/100 g of rat/h.

Rats after evisceration are unable to maintain normal body temperature and their temperature gradually approaches within a few degrees of that of the surrounding air[14]. In order to maintain more nearly normal body temperatures, some of the rats were placed in a warm cabinet. When the temperature of the cabinet was kept at 34°, the temperature of the eviscerated rats varied from 36.0 to 37.8°. INGLE AND NEZAMIS[15] have compared the times of survival of eviscerated rats at temperatures from 24° to 38°, and found the longest survival at 26°. When the rats at this temperature

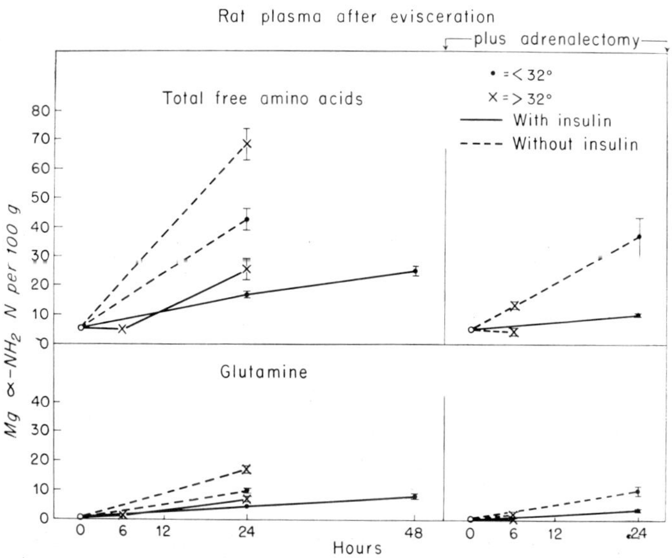

Fig. 1. Total free amino acids in plasma of ten normal rats (open circle)and in groups of 11 and 12 eviscerated rats maintained at less than 32° for 24 or 48 h with insulin or for 24 h without insulin. The average concentrations of amino acids are also shown for groups of 11 rats maintained at a higher temperature than 32° for 6 and 24 h without insulin and for 24 h with insulin. Standard errors are indicated by the vertical lines above and below the average values except when the standard error is less than the height of the symbol used. Values for glutamine content are those for the same groups of rats. From six to nine rats were included in each group that had both evisceration and adrenalectomy.

received penicillin and streptomycin, the average time of survival was extended to approximately 100 h. Our studies also show that rats with temperatures of less than 32° survived much longer than those of more than 32°.

Injections of ammonium citrate were given to one normal and one eviscerated rat at the rate of 1.0 mg of N/h and to another eviscerated rat at the rate of 0.25 mg of N/h. Glutamine was injected in normal and eviscerated rats for 3 and 6 h at the rate of 10 mg of α-amino N/h and to other rats for longer periods at the rate of 0.25 or 1 mg of α-amino N/h. A commercial preparation of amino acids for intravenous injection (elamine) was administered to a few rats.

Determination of total free amino acids and glutamine. Blood, brain (cerebrum), skeletal muscle and cardiac muscle (ventricle) were removed 6, 24 and 48 h after

References p. 460

evisceration. These tissues were also removed from normal rats. Plasma, brain, skeletal muscle, and the ventricular muscle of the heart were treated with picric acid, and the total α-amino nitrogen of the free amino acids in this filtrate was determined by the method of HAMILTON AND VAN SLYKE[17]. Glutamine was determined by the method of HAMILTON[18].

For separation of the free amino acids on paper chromatograms, protein-free extracts of tissue were made essentially by the method of SOLOMON *et al.*[19]. The extracts were desalted by the method of CONSDEN, GORDON AND MARTIN[20] and two-dimensional chromatograms were developed by the ascending technic with the frame

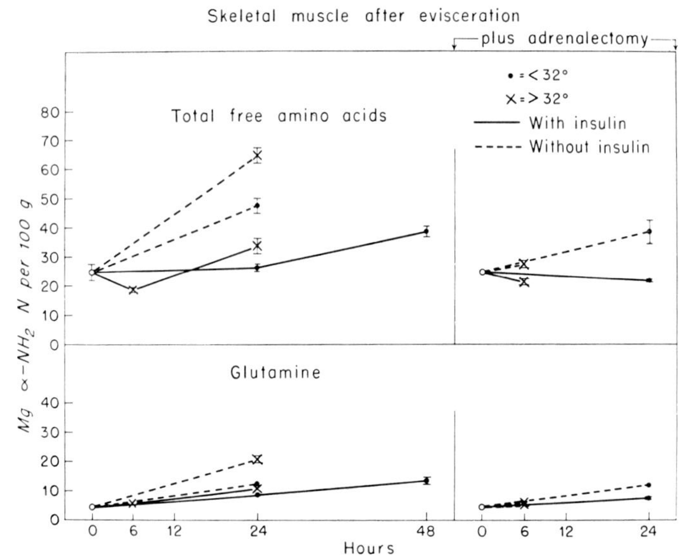

Fig. 2. Total free amino acids and glutamine in skeletal muscle of eviscerated rats and of eviscerated-adrenalectomized rats. The groups of rats are the same as in Fig. 1.

of DATTA, DENT AND HARRIS[21] as previously described[3] in a 75% solution of phenol containing sodium cyanide and in an ammoniacal atmosphere, and then in a 65% solution of lutidine containing diethylamine. The sheets were sprayed with 0.25% solution of ninhydrin in ethanol, and the color was allowed to develop at room temperature for 3 h. The chromatograms were photographed or used for quantitation of individual amino acids. The colored spots were outlined lightly with pencil before the color development was complete. At the end of the 3-h period, the spots were cut out and placed in test tubes containing 50% propanol, and allowed to stand except for gentle mixing for 30 min. The purple color was measured spectrophotometrically at 570 mμ. Two-dimensional chromatograms of extracts of plasma and tissues were made with chromatograms of a standard mixture of amino acids under identical conditions.

RESULTS

Effect of time after evisceration on levels of amino acids in the rat. The content of free amino acids increased slowly and progressively in the plasma and the skeletal and

References p. 460

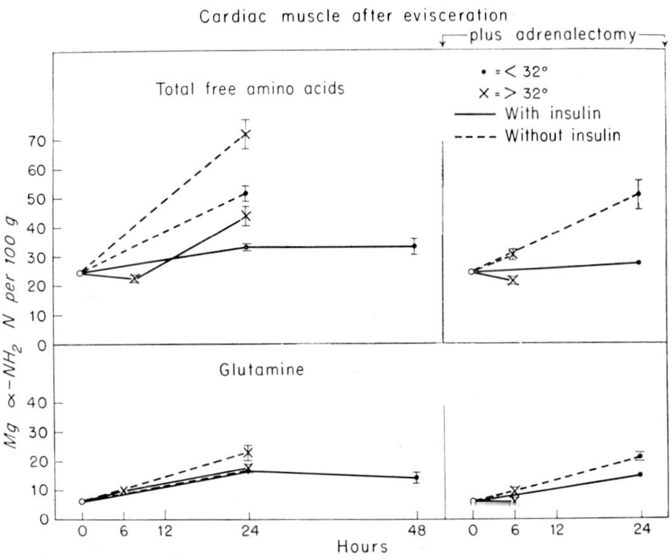

Fig. 3. Total free amino acids and glutamine in cardiac muscle (ventricle) of eviscerated rats and of eviscerated-adrenalectomized rats. The groups are the same as in Fig. 1.

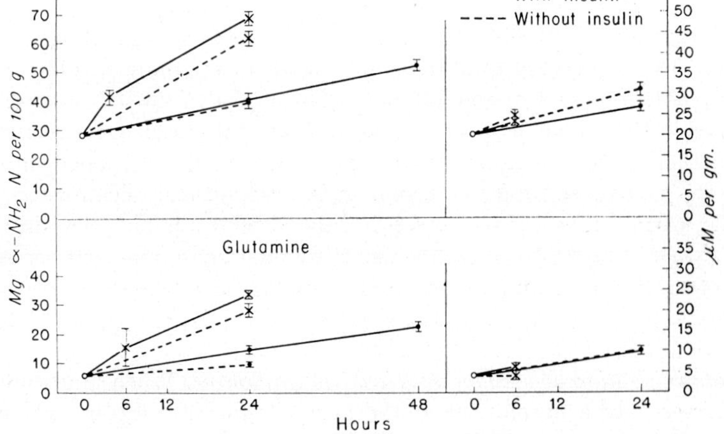

Fig. 4. Total free amino acids and glutamine in brain of eviscerated rats and of eviscerated-adrenalectomized rats. The groups of rats are the same as in Fig. 1.

cardiac muscles of the rat after removal of the liver when the rat was maintained with glucose and insulin (Figs. 1–3). There was a more rapid and greater increase in glutamine in the brain (Fig. 4).

Effect of insulin on amino acids. When the dehepatized rats were maintained with glucose but without insulin, the rate of increase of free amino acids in plasma and muscles was much greater than in the presence of insulin (Fig. 1–3). The amino acids in the brain, however, increased at the same rate whether or not insulin was administered. Glutamine usually increased more in the brain than in the plasma or muscles.

Effect of body temperature on the amino acids. When the eviscerated rats were separated into two groups depending on whether their body temperature was less or more than 32° at the end of the experiment, a clear-cut difference in the level of amino acids was found. The rate of increase of the amino acids in plasma and tissues was much slower in rats at the low temperature than at the more nearly normal temperatures. The levels of amino acids found in rats at temperatures exceeding 32° were higher than those found in dehepatized dogs[4]. This may be due to the higher rate of metabolism in rats. Approximately the same percentage of amino acids was found to be glutamine in plasma and skeletal muscle, whether the temperature was more or less than 32° and whether or not the rat was given insulin. The percentage of glutamine appeared to be characteristic of the tissue, not of the treatment. In the brain, however, glutamine accounted for approximately 20% of the free amino acids in the normal rat and for at least 50% of the extra free amino acids which appeared in the brain after evisceration of the rat.

Effect of adrenalectomy. When the adrenal glands were removed at the time of evisceration or 1–2 days earlier, little difference occurred in the rate of increase of the total free amino acids in plasma, brain, skeletal and cardiac tissue if the temperature of the rats did not exceed 32°. The survival time of rats with temperatures exceeding 32° was short; therefore, the rats at this temperature were studied only at 6 h after evisceration. The increase in total free amino acids in the brain was less at 6 h after evisceration with removal of the adrenals than in the rats with adrenals, whereas the increases in the amino acids of plasma and skeletal muscle were relatively small at 6 h whether or not the adrenals were present. Again, the changes in glutamine paralleled the changes in total free amino acids, and thus the accumulation of glutamine in the warmer rats was less at 6 h in rats without adrenals than in those with adrenals. INGLE, PRESTRUD AND NEZAMIS[22] found that adrenalectomy suppressed the level of amino acids of the eviscerated rat at each dose-level of insulin and glucose.

Effect of administration of ammonium citrate or glutamine. When ammonium citrate was administered to a normal rat at the rate of 1 mg of N/h for 24 h, no change was found in the glutamine content of brain or muscle (Table I). When this substance was administered to eviscerated rats for 17 to 19 h, however, the glutamine content of the brain increased greatly without any apparent increase in other amino acids. The levels of glutamine in the brain were greater than those found in dehepatized

References p. 460

rats which had not been given ammonium citrate. Increased levels of glutamine were found in the plasma with larger increases in the total amino acids.

When glutamine was administered at the rate of 0.25 mg of α-amino N/h for 18–22 h to eviscerated rats, maintained with insulin at temperatures of 32° or less, glutamine accumulated in the brain and accounted for approximately all the increase seen in the total free amino acids (Table I). The increase in glutamine in the brain

TABLE I

EFFECT OF INJECTION OF AMMONIA AND GLUTAMINE ON FREE AMINO ACIDS OF RAT BRAIN

Number of rats	*Injection of ammonium citrate*		Rectal temperature (°C)	*mg of α-amino N/g*					
				Plasma		*Brain*		*Skeletal muscle*	
	mg of N/h	*h*		*Total free amino acids*	*Glutamine*	*Total free amino acids*	*Glutamine*	*Total free amino acids*	*Glutamine*
10 N*				5.4 ± 0.1**	0.8 ± 0.1**	28.4 ± 0.7**	5.6 ± 0.6**	24.8 ± 2.6**	4.6 ± 1.6**
1 N	1.0	23.5	36.5			28.3	6.4		
1 E*	0.25	19	31.7	15.7	4.8	46.9	28.3	21.6	7.5
1 E	1.0	17	33.3	18.5	6.4	50.0	25.0	22.0	8.4
	Glutamine								
	mg of α-amino N/h	*h*							
1 E	0.25	18.8	31.6			65.3	35.1		
1 E	0.25	21.0	31.2	15.0	4.3	58.0	31.9	22.0	7.7
1 E	0.25	21.5	31.5	16.0	5.4	51.4	27.4	23.6	9.3
1 E	1.0	18.0	32.0	20.9	8.2	60.6	33.2	25.5	11.1
1 E	1.0	21.5	32.3	41.6	12.0	60.0	30.2	45.5	16.0
1 E	1.0	22.0	31.6	36.9	9.8	51.9	21.4	46.4	17.8

* N, normal; E, eviscerated.

** The number after the ± is the standard error of the mean for the ten normal rats. The eviscerated rats were given glucose and insulin, in addition to the ammonium citrate, or glutamine.

following the injection of ammonium citrate or glutamine to eviscerated rats with subnormal temperatures was no greater than that which accumulated spontaneously in other eviscerated rats at more nearly normal body temperatures. Increased levels of glutamine were found in skeletal muscle when glutamine was administered to eviscerated rats. In only two out of five of these eviscerated rats was a significant increase found in the other free amino acids. Glutamine levels were greatly increased in the plasma during the intravenous injection of glutamine. The great increase in the other free amino acids of plasma which also occurred helped to maintain a normal composition of the amino acid mixture in plasma. Intravenous administration of a protein hydrolysate (Table II) produced large increases in the content of free amino acids in the plasma and tissue of the eviscerated rat.

References p. 460

TABLE II

EFFECT OF INJECTION OF GLUTAMINE AND PROTEIN HYDROLYSATE ON FREE AMINO ACIDS

Number of rats	h after evisceration	Injection of glutamine: mg of α-amino N/h	Injection of glutamine: h	mg of α-amino N/100 g of rat — Plasma: Total free amino acids	Plasma: Glutamine	Brain: Total free amino acids	Brain: Glutamine	Skeletal muscle: Total free amino acids	Skeletal muscle: Glutamine	Cardiac muscle: Total free amino acids	Cardiac muscle: Glutamine
1 N		10	3	2.6	0.2	29.8	5.7	28.3	7.2	26.1*	7.9
1 N		10	3	3.0	0.2	29.2	2.9	24.4	7.2		
1 E	5	10	3**	17.4	10.1	35.7	10.3	53.0	21.3	43.5*	26.3
1 E	4	10	3	14.9	7.6	35.7	9.3	42.0	18.2		
1 E	15	16	3	33.3***	19.2	33.2	13.6	30.4	13.0	39.2*	23.6
1 E	26	16	3			38.0	17.8	37.9	16.9		
1 E	22	10	3	26.2	11.6	42.5	17.2	32.3	11.6	39.4*	21.4
1 E	21	10	3	24.1	9.6			37.2	14.7		
1 E	27	10	6	21.9	12.0	46.7	26.1	26.3	12.3	37.6*	23.4
1 E	29	10	6	23.1	13.3	39.3	19.3	30.5	16.8		
		Injection of protein hydrolysate: mg of N/h	h								
1 E	25	16	6			42.3	14.6	39.2	11.7	44.5**	19.0
1 E	24	16	6	27.5	7.6	38.5	13.0	30.6	9.4		
1 E	24	16	6	44.4	8.3	41.0	14.1	39.3	10.9	49.9	17.6
1 E	42	16	10	142.6	22.3	74.3	29.5	91.1	18.9	101.1	26.2

* Cardiac muscle from two rats was combined for analysis.
** Glutamine was administered toward the end of the experiment.
*** Plasma from two rats was combined.

CHROMATOGRAPHIC STUDIES

Brain. The large increase in concentration of glutamine described in the brain of the eviscerated rat was also conspicuous on the two-dimensional paper chromatograms of the brain (Figs. 5, 6). This increase in glutamine occurred without a decrease in the content of glutamic acid as in the dehepatized dog[4]. Phenylalanine and tyrosine, which were scarcely visible when a 100-mg sample of normal brain was chromatographed, were clearly visible on these chromatograms of brains of eviscerated rats. The γ-aminobutyric acid spot looked larger on many of the chromatograms of brains from eviscerated rats than on those of normal rats. It would be desirable to measure this amino acid by a more quantitative method.

Skeletal muscle. In addition to the marked increase in glutamine in the muscles of eviscerated rats which survived for 24 h without insulin, increases were noted in phenylalanine, arginine, lysine, tyrosine, alanine, threonine, glycine and serine (Fig. 7). The increases in these amino acids were less in eviscerated rats which received insulin during this period. Although in the dehepatized dog[4] the content of

References p. 460

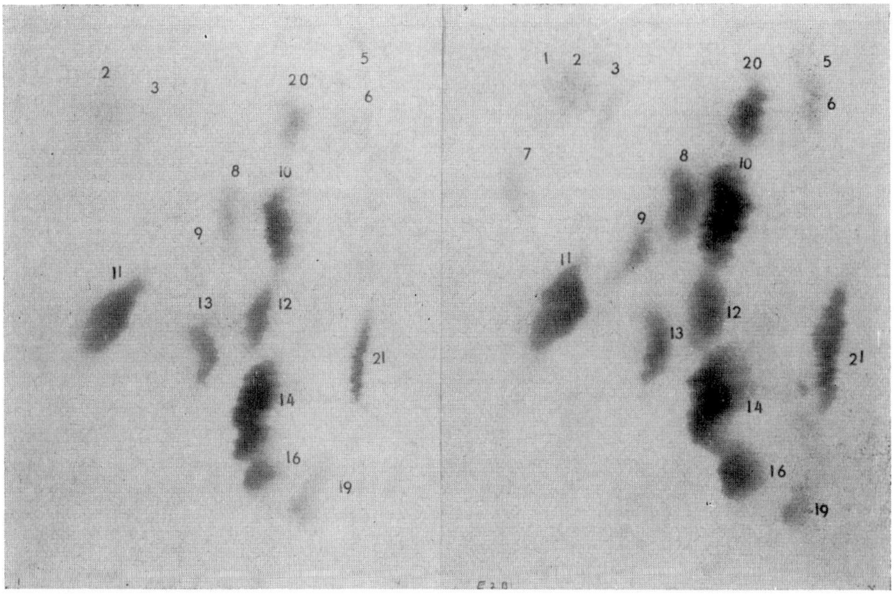

Fig. 5*. Chromatograms of the free amino acids of the cerebrum. Left, normal rat. Right, an eviscerated rat which had survived 1 day after operation and had been maintained with insulin and glucose; the body temperature of the eviscerated rat was less than 32°. Aliquots corresponding to 100 mg of tissue were applied to each sheet at spot X. The sheets were developed in the vertical direction with 75% solution of phenol in the presence of sodium cyanide and ammonia, and in the horizontal direction in 65% solution of lutidine containing diethylamine.

* Composite key to the numbers are: 1, phenylalanine; 2, leucine; 3, valine; 5, arginine; 6, lysine; 7, tyrosine; 8, alanine; 9, threonine; 10, glutamine; 11, taurine; 12, glycine; 13, serine; 14, glutamic acid; 15, carnosine; 16, aspartic acid; 18, histidine; 19, glutathione; 20, γ-aminobutyric acid; 21, unknown.

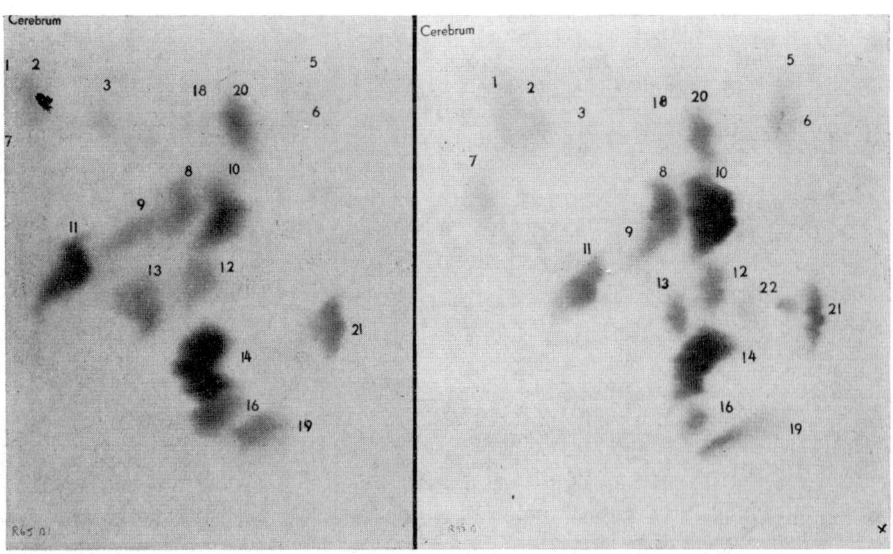

Fig. 6. Chromatograms of the free amino acids of the brain. Left, normal rat. Right, an eviscerated rat which had survived 1 day without insulin; at the end of the experiment in this rat, the rectal temperature was 32.9°. Samples of tissues used weighed 100 mg. (See footnote of Fig. 5 for key to the numbers).

glutamic acid in the muscle decreased as the content of glutamine increased, a definite decrease in the glutamic acid in the rat was not apparent.

Heart. Glutamine and many other amino acids increase after evisceration in the rat and particularly in rats without insulin (Fig. 8).

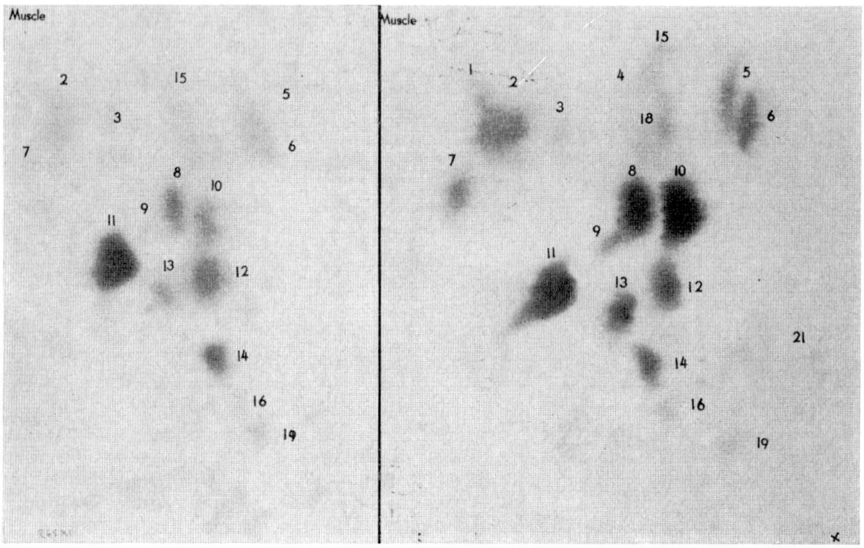

Fig. 7. Chromatograms of the free amino acids of skeletal muscle. Left, normal rat. Right, same eviscerated rat as in Fig. 6. (See footnote of fig. 5 for key to the numbers).

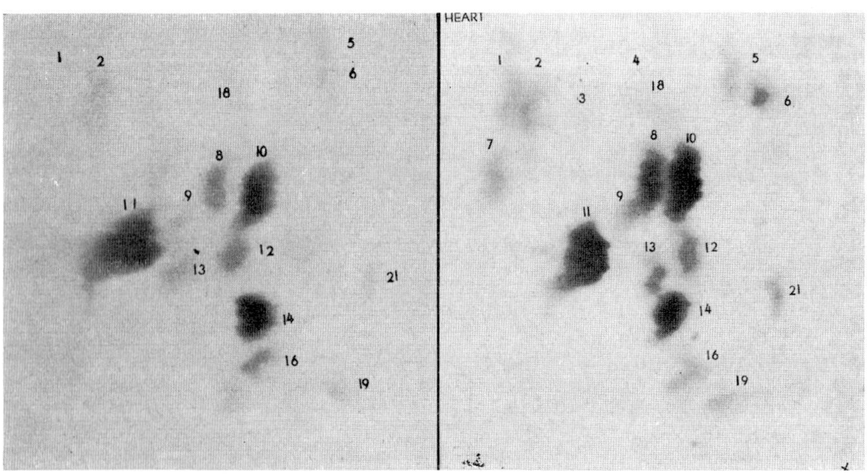

Fig. 8. Chromatograms of free amino acids from cardiac muscles of two rats (combined). Left, two normal rats. Right, two eviscerated rats maintained for 1 day after operation without insulin. Rectal temperatures of these two rats were 32.9° at end of study. (See footnote of fig. 5 for key to the numbers):

References p. 460

Plasma. Glutamine and all of the other amino acids in plasma increased markedly, particularly when insulin was not provided to the eviscerated rat (Fig. 9).

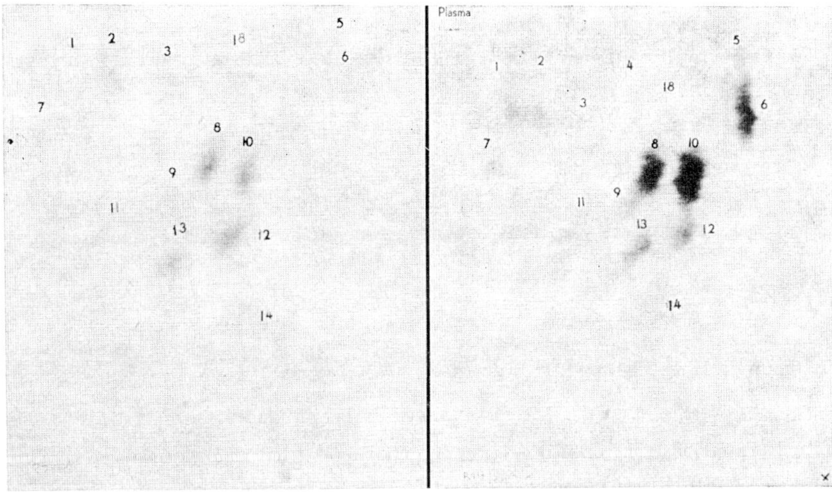

Fig. 9. Chromatograms of free amino acids in plasma. Left, normal rat; 200 μl of plasma was used. Right, same eviscerated rats as in Fig. 8. Sample consisted of 100 μl of plasma. (See footnote of fig. 5 for tey to the numbers.)

COMMENT

In each tissue there is a pool of free amino acids which provides building material for proteins and also is indicative of turnover and degradation of proteins as well as interconversions of amino acids. Likewise, the pool of amino acids in plasma reflects such changes as they occur in all the tissues. Mechanisms of stabilization of these pools of amino acids in the tissues are present, and the liver which is equipped to process the conversion of amino acids to urea can be removed without a great disturbance in the level of amino acids in blood plasma, skeletal muscle or cardiac muscle. Indeed, if adequate insulin was supplied with glucose to either the dehepatized dog or eviscerated rat, an increase in the level of amino acids could be prevented in these tissues. INGLE, PRESTRUD AND NEZAMIS[22] have made a similar observation on the plasma of the eviscerated rat.

Stabilization of the pool of free amino acids in the brain of the dog or the rat cannot be produced by insulin or glucose, and removal of the liver results in increased content of some free amino acids in the brain. Glutamine increases the most and actually spills over into the cerebrospinal fluid. It is not known to what extent the increase in brain glutamine is due to increased production of ammonia in the brain of the dehepatized animal or accumulation of ammonia which had been formed in other tissues. Nor is it known whether production of glutamic acid from glucose occurs more rapidly in the brain after removal of the liver and thus provides for the neutralization of ammonia. Possibly the glutamine accumulates because further metabolism of the substance is blocked by the removal of the liver.

From the data at hand concerning dehepatized dogs and rats no direct correlation

References p. 460

is apparent between the symptoms or survival time and the free amino acids of the brain. This suggests that these substances are not directly toxic to the brain. It is probable, however, that toxic substances do accumulate in the brain and account for the hepatic coma which develops within 1–2 days after removal of the liver and rapidly becomes fatal. It is also possible that a deficiency of some essential nutrient of the brain normally furnished by the liver develops and becomes a limiting factor in the survival of these animals.

REFERENCES

[1] J. L. Bollman, F. C. Mann and T. B. Magath, *Am. J. Physiol.*, 69 (1924) 371.
[2] J. L. Bollman, F. C. Mann and T. B. Magath, *Am. J. Physiol.*, 78 (1926) 258.
[3] E. V. Flock, F. C. Mann and J. L. Bollman, *J. Biol. Chem.*, 192 (1951) 293.
[4] E. V. Flock, M. A. Block, J. H. Grindlay, F. C. Mann and J. L. Bollman, *J. Biol. Chem.*, 200 (1953) 529.
[5] E. V. Flock, M. A. Block, F. C. Mann, J. H. Grindlay and J. L. Bollman, *J. Biol. Chem.*, 198 (1952) 427.
[6] J. L. Bollman, E. V. Flock, J. H. Grindlay, F. C. Mann and M. A. Block, *Am. J. Physiol.*, 174 (1953) 467.
[7] J. L. Bollman, E. V. Flock, J. H. Grindlay, R. G. Bickford and F. R. Lichtenheld, *A.M.A. Arch. Surg.*, 75 (1957) 405.
[8] P. Schwerin, S. P. Bessman and H. Waelsch, *J. Biol. Chem.*, 184 (1950) 37.
[9] H. Tigerman and R. MacVicar, *J. Biol. Chem.*, 189 (1951) 793.
[10] H. Kamin and P. Handler, *J. Biol. Chem.*, 188 (1951) 193.
[11] D. B. Tower, in R. O. Brady and D. B. Tower (Eds.) *The Neurochemistry of Nucleotides and Amino Acids*, John Wiley and Sons, Inc., New York, 1960, p. 173.
[12] J. P. De Ruisseau, J. P. Greenstein, M. Winitz and S. M. Birnbaum, *Arch. Biochem.*, 68 (1957) 161.
[13] B. Eiseman, H. Osofsky, E. Roberts and B. Jelinek, *J. Appl. Physiol.*, 14 (1959) 251.
[14] D. J. Ingle and J. E. Nezamis, *Am. J. Physiol.*, 159 (1949) 95.
[15] D. J. Ingle and J. E. Nezamis, *Am. J. Physiol.*, 160 (1950) 122.
[16] D. J. Ingle and J. E. Nezamis, *Am. J. Physiol.*, 166 (1951) 349.
[17] P. B. Hamilton and D. P. Van Slyke, *J. Biol. Chem.*, 150 (1943) 231.
[18] P. B. Hamilton, *J. Biol. Chem.*, 158 (1945) 375.
[19] J. D. Solomon, C. A. Johnson, A. L. Sheffner and O. Bergeim, *J. Biol. Chem.*, 189 (1951) 629.
[20] R. Consden, A. H. Gordon and A. J. P. Martin, *Biochem. J.*, 41 (1947) 590.
[21] S. P. Datta, C. E. Dent and H. Harris, *Science*, 112 (1950) 621.
[22] D. J. Ingle, M. C. Prestrud and J. E. Nezamis, *Proc. Soc. Exptl. Biol. Med.*, 67 (1948) 321.

Invited Discussion

INCREASE IN URINARY AMINO ACIDS ASSOCIATED WITH PANTOTHENIC ACID DEFICIENCY IN THE RAT

JOY D. MARKS AND HELEN K. BERRY

Department of Pediatrics, University of Cincinnati; The Children's Hospital Research Foundation, Cincinnati, Ohio (U.S.A.)

Chromatographic separation and identification of the amino acid constituents of urine indicate that a rather specific "pattern" of urinary amino acids may be associated with age, physiological condition and hereditary metabolic disease. Information about the metabolic state of the subject is provided by qualitative and quantitative data regarding the urinary amino acids.

The occasional finding of high levels of alanine in the blood and urine of children suggested the possibility of pantothenate deficiency impairing the coenzyme A metabolism of the child. The experiments described below were designed to test the hypothesis that alanine would be excreted by pantothenate-deficient rats. In the course of experiments aimed at producing congenital malformations in rats, pregnant rats fed pantothenic acid-deficient diet were also available. These rats were used as a means of investigating the amino acid composition of the urine during normal pregnancy and of studying the effect of pantothenate deficiency on the composition of the urine during the gestational period. The results of these investigations represent an example of a biochemical lesion that can be experimentally produced.

METHOD

Sprague-Dawley albino rats, of commercial stock, were placed in metabolic cages and maintained on the pantothenic acid-deficient diet (Nutritional Biochemicals, Cleveland). Urine was collected in graduated cylinders under toluene for 24-h periods or longer. Measured aliquots (usually 1/500 of the total daily volume) were chromatographed according to standard procedures[1] and total amino acid excretion (mg/day/rat) was determined. Pantothenic acid deficiency was produced by a method similar to that described by NELSON *et al.*[2]. Oral doses of ω-methylpantothenate (1–2 mg) were administered to assure the production of deficiency. With the pregnant rats several feeding schedules were tried to establish a state of pantothenic acid deficiency, but none of them was severe enough to cause resorption or premature termination of pregnancy.

RESULTS

Pantothenic acid deficiency in male weanling rats. Two male weanling rats were fed a complete diet and two were given a pantothenic acid-deficient diet plus ω-methylpantothenate for a 3-week period. The rats fed a complete diet excreted glutamic acid, glycine, and alanine (approx. 0.8–1.2 mg/day of each). Only traces of lysine, glutamine, threonine and serine were excreted (0.2 mg/day). The rats fed the deficient diet lost their hair and became weak and feeble by 3 weeks. Alanine excretion in-

References p. 464

creased progressively to a maximum of 10 mg/day. Excretion of glycine was not appreciably altered; there was a slight increase in glutamic acid. Addition of glutamine to the diet resulted in further increase in alanine excretion, but did not produce an elevation in glutamine excretion. During this period of acute deficiency β-alanine, aspartic acid, taurine and cysteic acid appeared in the urine. When the rats were placed on a complete diet, supplemented with oral doses of calcium pantothenate (1–2 mg/day) for approx. 2 weeks the urinary excretion pattern became normal. Average values of amino acid excretion by rats of various strains can be found in the publications of PENTZ[3] and others[4].

Amino acid excretion by pregnant albino rats fed a complete diet and pantothenic acid-deficient diet. Four rats were fed pantothenic acid-deficient diet beginning on the 1st day of pregnancy, which was established by vaginal smears. Oral doses of ω-methylpantothenate (1–2 mg/day) were fed on gestation days 10 to 13 inclusive while the deficient diet was continued. Three of the pregnant rats were given a complete diet from the 18th day of gestation until delivery to insure normal termination of pregnancy. As control, one rat was maintained on a complete diet throughout the pregnancy.

The rat fed a complete diet throughout the pregnancy excreted glutamic acid, glycine and an unidentified ninhydrin-positive substance in small amounts (0.8–1.5 mg/day), while alanine and taurine were present in relatively large amounts (2.6–3.0 mg/day) by the 13th day. The amounts of taurine and alanine excreted during the period are shown in Table I. The level of taurine increased during pregnancy to a peak in the middle of the gestational period. Post-partum urine samples showed little or no taurine.

Taurine concentration in urine is known to vary with the animal strain. In this strain little taurine was detected in urine specimens from males or non-pregnant females fed complete diets. As the pregnancy progressed taurine excretion of the rat fed a complete diet increased. These data suggest that in this strain of rats excretion of taurine is

TABLE I

AMINO ACID EXCRETION BY PREGNANT ALBINO RATS

Days gestation	*Alanine excretion*		*Taurine excretion*	
	Complete (mg/day)	*Deficient (mg/day)*	*Complete (mg/day)*	*Deficient (mg/day)*
0	0.20	0.10	0.93	—
2	0.89	0.10	0.90	—
5–6	0.44	5.2	1.7	1.1
7–9	1.6	11	2.7	3.2
10	—	39	—	2.3
13	3.2	50	2.5	3.3
15	—	—	2.8	—
16	—	—	2.1	—
18	—	75*	—	2.1*
19	—	180*	—	2.1*
20	2.0	7.5*	0.66	1.3*
Post-partum	—	5.5*		

* Deficient diet had been discontinued for this rat on the 18th day of gestation.
— Below minimum amount detectable.

References p. 464

associated with the metabolic changes which occur during pregnancy. Among the rats fed a pantothenic acid-deficient diet the onset of vitamin deficiency was first indicated by the composition of urine samples obtained after 7–9 days on the deficient diet. Initial excretion of amino acids was minimal (0.2–1.0 mg/day), but after 7 days on the diet, an elevation in the excretion of alanine was noted (11 mg/day). At this time, no physical symptoms of deficiency were apparent. Results of changes in excretion of taurine and alanine are presented in Table I. The increase in the excretion of taurine, which was observed in urine of pregnant rats on the complete diet, was also detected in specimens from pregnant rats fed the deficient diet. During pregnancy the values for taurine excretion observed in control and deficient rats were within a narrow range (2–4 mg/day). Alanine excretion, however, increased until the deficient diet was discontinued.

By the 13th day of gestation excretions of glycine, glutamic acid and taurine were slightly increased in urine specimens from pregnant rats fed pantothenic acid.

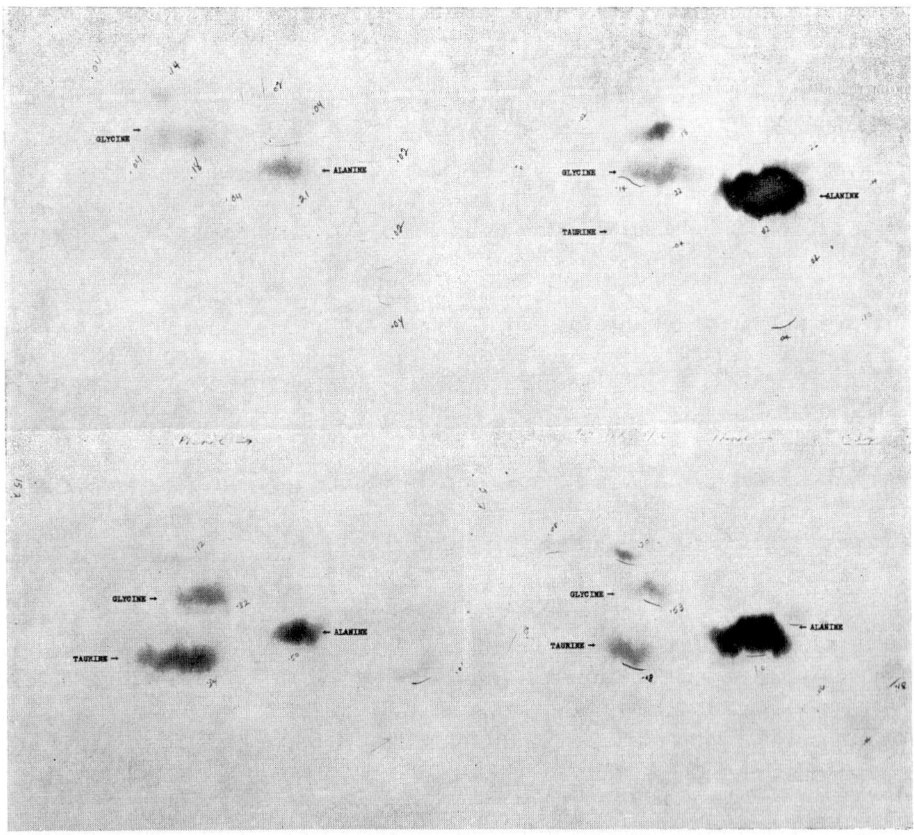

Fig. 1. Chromatograms of urinary amino acids in rats. Solvents: (1) buffered phenol; (2) 2,6-lutidine–water. Upper left: male weanling rat, pantothenate-deficient diet plus oral calcium pantothenate; upper right: male weanling rat pantothenate-deficient diet; lower left: pregnant female rat control, complete diet, 13 days gestation, urine volume 8.5 ml, weight 245 g; lower right: pregnant female rat, pantothenate-deficient diet, 13 days gestation, urine volume 3.3 ml, weight 235 g.

References p. 464

deficient diet compared to specimens from the pregnant female fed a complete diet- However, alanine excretion increased 16-fold in the non-deficient animal and 500-fold in the deficient animal during the same period. The extreme alteration in alanine excretion may be noted in Fig. 1. β-Alanine and aspartic acid were also detected during the period of acute deficiency.

DISCUSSION

The importance of pantothenic acid in the synthesis of Coenzyme A is well-established (SLATER, 1953). With-holding pantothenate from the diet of a rat would be expected to inhibit the tricarboxylic acid cycle in the tissues with resulting accumulation of pyruvate. In the presence of excess pyruvate, the glutamic–pyruvic transamination reaction should result in an excessive production and excretion of alanine. This possibility was demonstrated to occur. Reversal of the effect was produced by addition of pantothenate to the diet.

The excretion of taurine by pregnant rats, either in complete or pathothenatedeficient diet, and by pantothenate-deficient male rats is of interest. Taurine is known to be produced by way of several metabolic pathways[1, 3, 6–9]. The major source appears to be decarboxylation of cysteic acid. PENTZ *et al.*[9] reported that taurine excretion by patients with known or suspected adrenocortical disorders sharply increased after treatment with ACTH or 9-α-fluorohydrocortisone. These workers suggested that a relationship exists between adrenocortical function and taurine excretion.

In 1952 HURLEY *et al.*[10] reported that pantothenate deficiency in the rat caused release of ACTH, which stimulated the adrenals to secrete adrenocortical hormone until exhaustion occurred. Both pregnancy and pantothenate deficiency cause a metabolic stress. The results of this investigation support the suggestion of PENTZ *et al.* that taurine excretion is related to adrenocortical function. It is possible that taurine excretion is a biological index of adrenal activity and may be useful as a diagnostic criterion. Further experiments to test this hypothesis are planned.

Similar investigations may be useful for the interpretation of patterns of amino-aciduria observed in human beings. Cases of high alanine excretion in children, observed infrequently in urine samples in this laboratory, may be indicative of excessive requirements for pantothenate. Likewise, elevated taurine excretion may be indicative of a state of metabolic stress in the individual.

ACKNOWLEDGEMENTS

The authors are indebted to Mrs. E. Takacs for consultation and assistance with the management of pregnant rats in this study.

This investigation was supported in part by a research grant (MA-1175) from the National Institutes of Health, U.S. Public Health Service.

REFERENCES

1 H. K. BERRY, *Pediatrics*, 25 (1960) 983.
2 M. M. NELSON, H. WRIGHT, C. D. C. BAIRD AND H. M. EVANS, *J. Nutrition*, 62 (1957) 395.
3 E. I. PENTZ, *J. Biol. Chem.*, 231 (1958) 165.
4 R. MEFFERD, H. B. HALE AND H. H. MARTENS, *Am. J. Physiol.*, 192 (1958) 209.
5 E. C. SLATER, *Ann. Rev. Biochem.*, 22 (1953) 17.
6 J. AWAPARA, *J. Biol. Chem.*, 225 (1957) 877.
7 H. BLASCHKO, S. P. DATTA AND H. HARRIS, *Brit. J. Nutrition*, 7 (1953) 364.
8 J. B. GILBERT, Y. KU, L. L. ROGERS AND R. J. WILLIAMS, *J. Biol. Chem.*, 235 (1960) 1055.
9 E. I. PENTZ, W. T. MOSS AND C. W. DENKO, *J. Clin. Endocrinol. and Metabolism*, 19 (1959) 1126.
10 L. S. HURLEY AND A. F. MORGAN, *J. Biol. Chem.*, 195 (1952) 583.

INVITED DISCUSSION

FREE AMINO ACIDS IN BRAIN AFTER ADMINISTRATION OF IMIPRAMINE, CHLORPROMAZINE AND OTHER PSYCHOTROPIC DRUGS

H. H. TALLAN

Geigy Research Laboratories, Ardsley, N.Y. (U.S.A.)

The effects produced by psychotropic agents on the free amino acids of the brain have received much study in recent years, particularly in connection with changes brought about in the concentrations of γ-aminobutyric acid (GABA), glutamic acid, and glutamine. These studies derive much impetus from the fact that certain amino acids and their derivatives occur, to any large extent, only in the brain and, therefore, may be concerned in some way in the functioning of this tissue. Accordingly, the effect of a psychotropic drug on such an amino acid could provide a valuable clue to the mode of action of the drug. In the present work, the effects on the complete spectrum of amino acids of an antidepressant, imipramine, as well as of its monomethyl analog, desmethyl-imipramine, have been studied. These effects have been compared to those produced by the tranquilizers chlorpromazine and reserpine and to the action of hydroxylamine.

Female Wistar rats bred in our own colony were employed. The animals weighed between 130 and 230 g. Dosage schedules were as follows: imipramine, 25 mg/kg subcutaneously every day for 7 days, with sacrifice 4.5 h after the last injection; desmethyl-imipramine · HCl, the same, but at a dosage of 15 mg/kg; chlorpromazine · HCl, a single dose of 50 mg/kg intraperitoneally 5.5 h before sacrifice; reserpine, 5 mg/kg intraperitoneally, either 3 or 18 h before sacrifice; hydroxylamine · HCl, 60 mg/kg intraperitoneally 100 min before sacrifice. Except for a single rat given reserpine 18 h before death, pairs of animals of the same weight were treated together. The animals were decapitated, the brains were quickly removed, combined, and homogenized in 6% perchloric acid, and the precipitated proteins were separated by centrifugation. The cloudy supernatant solution was adjusted to pH 4.0 with 2 *N* KOH and, after being kept at 4° overnight, the precipitate of potassium perchlorate (and lipid-like material) was removed by centrifugation. Suitable portions of the clear supernatant fluid were analyzed on a Beckman Spinco Model 120 Amino Acid Analyzer, using the procedures for physiological fluids of SPACKMAN, STEIN AND MOORE[1]. Acetylaspartic acid was measured after a preliminary separation on Dowex-1 by hydrolysis and determination of the liberated aspartic acid on the Amino Acid Analyzer. A portion of the brain extract, usually 1.0 ml, was applied to a 0.55 × 5.5-cm column of Dowex-1-X-8 (screened wet through a 200-mesh sieve), which had been washed with 2 ml of 2 *N* NaOH and 5 ml of water. The sample was followed by 2 ml of water, 5 ml of 2 *N* acetic acid, and 5 ml of 2 *N* HCl. A drop of concentrated HCl was added to the HCl fraction, which contained the acetylaspartic acid, before the solution was made to exactly

5.0 ml. A 2-ml aliquot of the HCl fraction was heated for 1 h in a boiling-water bath to effect hydrolysis of the acetylaspartic acid. This solution and a second, unheated 2-ml aliquot of the HCl effluent were individually taken to dryness three times in a rotary evaporator; the residues were dissolved in citrate buffer of pH 2.2 and both were analyzed for aspartic acid. The amount of acetylaspartic acid was calculated from the increase in aspartic acid after hydrolysis.

The results of the analyses are presented in Table I. It is clear that, for the most part, profound changes do not occur in the concentrations of practically all of the amino acids. This is in accord with the studies of OKUMURA *et al.*[2], of WRIGHT *et al.*[3], and of ROBERTS and his group[4], who investigated the great majority of the amino acids for possible changes in concentration under the influence of various psychotropic drugs. Nevertheless, it was felt that in spite of the usual heavy emphasis in

TABLE I

AMINO ACIDS AND RELATED COMPOUNDS IN RAT BRAIN

Values are expressed in μmoles/g. N.D., non-detectable. For dosage schedules and treatment of animals, see the text. Reserpine Group 1 sacrificed 18 h after drug, Group 2 sacrificed at 3 h. The weights of the animals are given at the head of each column.

Constituent	*Normal*			*Imipramine*		*Des-methyl-imipra-mine*	*Chlorpromazine*		*Reserpine*		*Hydroxyl-amine*
	1	*2*	*3*	*1*	*2*		*1*	*2*	*1*	*2*	
	170 g	*230 g*	*150 g*	*153 g*	*153 g*	*143 g*	*155 g*	*180 g*	*130 g*	*150 g*	*170 g*
Glycerophosphoethanolamine	0.50	0.45	0.60	0.81	1.04	0.63	0.59	0.60	1.14	0.67	0.67
Phosphoethanolamine	1.90	1.72	2.03	1.46	2.09	1.91	2.02	1.84	2.16	2.18	1.76
Taurine	5.9	5.1	6.9	5.7	6.0	6.6	6.4	4.2	5.9	8.1	5.6
Urea	4.8	3.5	2.7	3.3	4.3	4.8	5.5	5.4	6.8	4.3	3.9
Aspartic acid	2.75	1.91	2.72	2.06	3.14	2.75	2.73	2.50	2.75	2.98	2.42
Threonine	0.69	0.42	0.57	0.45	0.55	0.61	0.80	0.57	0.78	0.64	0.81
Serine	0.91	0.74	0.90	0.68	1.04	0.97	0.97	0.83	1.07	0.89	0.73
Glutamine	5.4	4.9	6.0	3.9	5.5	5.3	6.2	5.4	5.6	6.1	5.3
Glutathione (total, as –SH)	2.40	1.38	1.14	0.93	2.07	1.19	2.26	1.11	2.51	1.07	1.64
Proline	0.04	0.06	Trace	0.06	Trace	Trace	0.02	0.04	N.D.	0.09	0.10
Glutamic acid	11.3	9.8	11.4	9.7	13.3	11.6	10.5	9.4	12.4	11.3	10.7
Glycine	1.20	1.04	1.20	1.04	1.36	1.38	1.25	1.08	1.27	1.18	1.25
Alanine	0.71	0.62	0.64	0.49	0.74	0.69	0.69	0.57	0.73	0.71	0.73
Valine	0.07	0.09	0.09	0.05	0.09	0.11	0.11	0.08	0.09	0.05	0.09
Cystathionine	0.06	0.03	0.05	0.05	0.05	0.06	0.04	0.05	0.06	0.06	0.06
Methionine	0.04	0.02	0.06	0.07	0.05	Trace	0.04	0.03	0.06	0.02	0.06
Isoleucine	0.03	0.02	0.03	0.02	0.03	0.04	0.04	0.03	0.03	0.03	0.04
Leucine	0.06	0.07	0.07	0.06	0.08	0.07	0.09	0.07	0.08	0.09	0.09
Tyrosine	0.04	0.03	0.04	0.05	0.05	0.05	0.04	0.04	0.06	0.06	0.09
Phenylalanine	0.02	0.02	0.03	0.04	0.05	0.04	0.06	0.05	0.04	0.07	0.04
γ-Aminobutyric acid	2.63	2.66	2.58	2.31	2.43	2.34	2.40	2.25	2.70	2.70	3.61
Ornithine	0.02	0.02	Trace	0.01	0.01	0.02	0.03	0.002	0.03	0.01	0.02
Ethanolamine	0.21	0.25	0.22	0.27	0.30	0.21	0.20	0.16	0.33	0.21	0.26
Ammonia	1.75	1.41	1.30	1.45	0.71	1.42	1.50	1.29	2.02	1.28	2.57
Lysine	0.41	0.28	0.38	0.33	0.22	0.36	0.41	0.32	0.42	0.42	0.52
Histidine	0.04	0.02	0.09	0.07	0.04	Trace	0.07	0.08	0.01	0.06	0.10
Arginine	0.10	0.08	0.12	0.10	0.14	0.12	0.08	0.03	0.11	0.08	0.14
Acetylaspartic acid	6.8	7.3	6.0	7.1	6.6	5.8	5.8	5.3	7.4	5.7	6.8

References p. 469/470

this type of study on GABA, glutamic acid and glutamine, the other amino acids should not be overlooked.

All of the values obtained in the present study, including the alterations to be noted in some of the substances, come within the "normal "range that may be constructed by collecting all the data for normal rat brain to be found in the literature*. Thus, actual concentrations of substances can be compared only within a given experimental series of the same strain. Furthermore, the response to a challenge also may vary from strain to strain. For example, with the Wistar rats reported here, hydroxylamine produced an increase in GABA concentration of 38%**. The same dose of hydroxylamine in Sprague-Dawley rats resulted in a rise of only 9%. In preliminary experiments with Royal Hart animals, it was found that the normal concentration of GABA and the response to hydroxylamine varied too much to show any consistent trend.

The changes that were observed in the concentration of some of the amino acids are as follows. In the hydroxylamine-treated animals, there is the expected increase in the concentration of GABA, though the effect is not as pronounced as has been found by BAXTER AND ROBERTS[5, 6] in Sprague-Dawley and Wistar rats. Large increases in GABA after hydroxylamine were also obtained by EIDELBERG *et al.* in cats[7] and monkeys[8]. Reserpine had no effect on GABA; BALZER, HOLTZ AND PALM[9, 10], on the other hand, have found a clear-cut decrease in mice.

A small, but definite decrease in GABA occurred after administration of imipramine, desmethyl-imipramine and chlorpromazine. (In the hands of OKUMURA *et al.*[2], chlorpromazine caused a slight increase.) The decrease in brain GABA concentrations associated with convulsions is a well-known phenomenon; it occurs after initiation of seizures by various agents: hydrazides[6, 9, 11, 12], KCN (ref. 13), and pyridoxine antagonists[9, 14–16]. Hydroxylamine, in complementary fashion, has anticonvulsant properties and raises the brain GABA concentration[7, 8]. However, it is by now quite certain that the decrease in GABA is itself not the "cause" of the seizures, but, rather, that the fall in GABA is, as PURPURA *et al.*[15] have stated, but one manifestation of the complex of metabolic changes initiated by the convulsive agent (*cf.* also refs. 9, 12, 16). There are many observations that support this view: some convulsive agents or procedures have no effect on GABA concentrations[3, 9, 11, 16, 17]; L-2,4-diaminobutyric acid causes seizures and raises the GABA concentration[12, 18]; pyridoxine prevents the convulsions produced by some hydrazides without preventing the decrease in GABA (ref. 9); thiosemicarbazide, given in conjunction with hydroxylamine, causes convulsions even though the GABA concentration is elevated[6]; applied topically, methoxypyridoxine causes a much smaller decrease in GABA than when administered intravenously, though the convulsive effect is as pronounced[15]; the seizure susceptibility of areas of the normal cortex is not related to the concentration of naturally occurring GABA (refs. 19, 20). The decreases in GABA noted in the present work provide some further evidence that a fall in GABA is not directly related to the convulsive state; the chlorpromazine-treated animals were sedated, and the imipramine- and desmethyl-imipramine-treated animals showed no marked changes in behavior.

* See the following paper.

** With $NH_2OH \cdot HCl$ at a dosage of 71 mg/kg, the increase in GABA concentration in the Wistar rats was 57%.

References p. 469/470

In assessing the effects of the drugs we have studied on glutamic acid, glutamine, and glutathione, the variation between animals in the same group prevents the assignment of any clear-cut action. OKUMURA *et al.*[2] do report increases in the three compounds after the administration of chlorpromazine. In mice, brain glutamic acid is unaffected by reserpine[9]. The reported actions of other psychotropic drugs on these substances may be noted here. The concentration of glutamic acid has been found to be unaffected by many convulsive agents[3, 9, 11, 14, 16, 21] and by methamphetamine[2]; to be decreased by other convulsants[9, 15] and in epileptogenic lesions[17], as well as by the anticonvulsant diphenylhydantoin[22], the tranquilizer meprobamate[23] and the anesthetic thiopentone[24]; and to be increased by ethanol[25] and KCN (ref. 13). The concentration of glutamine is reported to be unaffected by some convulsants[16, 26], by ethanol[25], and by nembutal[26]; to be decreased in cortical areas subjected to epileptogenic lesions[17]; and to be increased by an anticonvulsant[22] and a convulsant[15], in thiopentone anesthesia[24], and in ammonium poisoning[3] (*cf.* ref. 4). The concentration of glutathione has been found to be decreased by a pyridoxine antagonist[13] and in pyridoxine deficiency[27], and to be lowered in cortical areas subjected to epileptogenic lesions[17]. A common ground for these diverse actions is not readily apparent; the primary effect of these psychotropic agents is almost certainly not on the amino acids of the brain.

Ethanolamine and its derivatives are compounds that have received relatively little attention. Within the present series, definite elevations occur in the concentration of glycerophosphoethanolamine (GPE) after imipramine and reserpine. The possibility that an increase in GPE may be the result of *post mortem* breakdown of phosphatidylethanolamine (*cf.* ANSELL AND NORMAN[28]) must be considered, but is unlikely since all of the brain extracts were prepared in precisely the same way; whether the concentration of phosphatidylethanolamine itself had been increased was not investigated. OKUMURA *et al.*[2] found a decrease in GPE after chlorpromazine, not observed in the present work, as well as a decrease after fasting or the administration of insulin or methamphetamine. There is no clear-cut change in the present series in either phosphoethanolamine (PE) or ethanolamine. ANSELL AND SPANNER[29] found that insulin decreased the synthesis of brain PE without affecting the actual amount present, and OKUMURA *et al.*[2] found an increase in PE after fasting and after methamphetamine administration.

Acetylaspartic acid is of great interest in that it occurs normally only in brain[30], though DU RUISSEAU[31] has shown that large amounts are found in the liver of animals poisoned with ammonium acetate, which, however, has no effect on the brain acetylaspartic acid[31]. OKUMURA *et al.*[2] report an appreciable increase after administration of methamphetamine. MARGOLIS *et al.*[32]. found no change in nembutal narcosis; their data suggest a lowering after metrazole-induced convulsions. In the present work, there is a slight decrease after chlorpromazine (not seen, however, by OKUMURA *et al.*[2]), and perhaps after desmethyl-imipramine. Aspartic acid itself is unaffected by the drugs used in the present study. It is reported that no change in aspartic acid occurs following reserpine[9], various convulsants[9, 11, 16, 32], and imino-β,β'-dipropionitrile[33]; there is a decrease after nembutal[32] or ammonium acetate[3], an increase after ethanol[25] or insulin (*cf.* ref. 2).

Taurine is of interest because, as has been found by HOPE[34], the concentration in brain is not affected by extreme pyridoxine deficiency, although the urinary excretion, reflecting levels throughout the rest of the body, falls markedly. HOPE noted considerable variations between animals. There is a wide scatter of values in the present series and it is not possible to discern any changes in concentration after administration of the various drugs. Although OKUMURA *et al.*[2] noted increases under a variety of conditions, indicating a rather non-specific effect, others have reported no changes in taurine after convulsive hydrazides[11], ammonia poisoning[3] and ethanol[25].

In conclusion, it can be stated that the psychotropic drugs studied in the present work have but slight effect on the overall pattern of brain amino acids. The changes in concentration that do occur, in GABA, GPE, and possibly in acetylaspartic acid, are not correlated with the pharmacological actions of these drugs. A survey of previously reported experiments of this type also fails to reveal a consistent picture of drug effects on brain amino acids. The effects of the psychotropic drugs are to be sought in subtler aspects of brain metabolism; the alterations these drugs induce in the concentration of some amino acids are more likely by-products of a chain of metabolic reactions.

ACKNOWLEDGEMENTS

It is a pleasure to acknowledge the assistance, at various stages of this work, of Mrs. C. MCCANNON, Mrs. S. A. JOHNSON, and Mrs. YUN-OK SUH.

Compounds were generously supplied by Geigy Pharmaceuticals (Tofrānil[R], imipramine Geigy), Smith Kline & French Laboratories (Thorazine[R], chlorpromazine SKF), and Ciba Pharmaceutical Products, Inc. (Serpasil[R], reserpine Ciba).

REFERENCES

1 D. H. SPACKMAN, W. H. STEIN AND S. MOORE, *Anal. Chem.*, 30 (1958) 1190.
2 N. OKUMURA, S. OTSUKI AND H. NASU, *J. Biochem. (Tokyo)*, 46 (1959) 247.
3 G. J. WRIGHT, F. NAVAZIO AND M. K. HORWITT, *Federation Proc.*, 20 (1961) 323.
4 E. ROBERTS, in E. ROBERTS, *Inhibition in the Nervous System and γ-Aminobutyric Acid*, Pergamon, New York, 1960, p. 144.
5 C. F. BAXTER AND E. ROBERTS, *Proc. Soc. Exptl. Biol. Med.*, 101 (1959) 811.
6 C. F. BAXTER AND E. ROBERTS, *Proc. Soc. Exptl. Biol. Med.*, 104 (1960) 426.
7 E. EIDELBERG, C. F. BAXTER, E. ROBERTS, C. A. SALDIAS AND J. D. FRENCH, *Proc. Soc. Exptl. Biol. Med.*, 101 (1959) 815.
8 E. EIDELBERG, C. F. BAXTER, E. ROBERTS AND C. A. SALDIAS, in E. ROBERTS, *Inhibition in the Nervous System and γ-Aminobutyric Acid*, Pergamon, New York, 1960, p. 365.
9 H. BALZER, P. HOLTZ AND D. PALM, *Arch. exptl. Pathol. Pharmakol., Naunyn-Schmiedeberg's*, 239 (1960) 520.
10 H. BALZER, P. HOLTZ AND D. PALM, *Experientia*, 17 (1961) 38.
11 K. F. KILLAM AND J. A. BAIN, *J. Pharmacol. Exptl. Therap.*, 119 (1957) 255.
12 G. M. MCKHANN, R. W. ALBERS, L. SOKOLOFF, O. MICKELSEN AND D. B. TOWER, in E. ROBERTS, *Inhibition in the Nervous System and γ-Aminobutyric Acid*, Pergamon, New York, 1960, p. 169.
13 T. TURSKÝ, *Nature*, 187 (1960) 322.
14 G. RINDI AND G. FERRARI, *Nature*, 183 (1959) 608.
15 D. P. PURPURA, S. BERL, O. GONZALEZ-MONTEAGUDO AND A. WYATT, in E. ROBERTS, *Inhibition in the Nervous System and γ-Aminobutyric Acid*, Pergamon, New York, 1960, p. 331.
16 G. D. GAMMON, R. GUMNIT, R. P. KAMRIN AND A. KAMRIN, in E. ROBERTS, *Inhibition in the Nervous System and γ-Aminobutyric Acid*, Pergamon, New York, 1960, p. 328.
17 S. BERL, D. P. PURPURA, M. GIRADO AND H. WAELSCH, *J. Neurochem.*, 4 (1959) 311.
18 D. KESSEL, *Federation Proc.*, 18 (1959) 258.
19 C. F. BAXTER, E. ROBERTS AND E. EIDELBERG, *Federation Proc.*, 18 (1959) 187.
20 C. F. BAXTER, E. ROBERTS AND E. EIDELBERG, *J. Neurochem.*, 5 (1960) 203.
21 W. T. SULLIVAN AND L. M. STRONG, *Am. J. Physiol.*, 181 (1955) 43.

22 A. VERNADAKIS AND D. M. WOODBURY, in E. ROBERTS, *Inhibition in the Nervous System and γ-Aminobutyric Acid,* Pergamon, New York, 1960, p. 242.
23 V. FERRARI, *Minerva med.*, 2 (1956) 2192; Chem. Abstr., 51 (1957) 6845.
24 R. M. C. DAWSON, *Biochem. J.*, 49 (1951) 138.
25 H. M. HÄKKINEN AND E. KULONEN, *Biochem. J.*, 78 (1961) 588.
26 D. RICHTER AND R. M. C. DAWSON, *J. Biol. Chem.*, 176 (1948) 1199.
27 M. J. ASHWOOD-SMITH AND A. D. SMITH, *Nature*, 184 (1959) 2028.
28 G. B. ANSELL AND J. M. NORMAN, *Biochem. J.*, 55 (1953) 768.
29 G. B. ANSELL AND S. SPANNER, *J. Neurochem.*, 4 (1959) 325.
30 H. H. TALLAN, S. MOORE AND W. H. STEIN, *J. Biol. Chem.*, 219 (1956) 257.
31 J.-P. DU RUISSEAU, *Can. J. Biochem. and Physiol.*, 38 (1960) 763.
32 R. U. MARGOLIS, S. S. BARKULIS AND A. GEIGER, *J. Neurochem.*, 5 (1960) 379.
33 I. M. CHAK AND N. N. DE, *J. Sci. Ind. Research (India)*, 20C (1961) 98.
34 D. B. HOPE, *J. Neurochem.*, 1 (1957) 364.

INVITED DISCUSSION

A SURVEY OF THE AMINO ACIDS AND RELATED COMPOUNDS IN NERVOUS TISSUE

H. H. TALLAN*

The Rockefeller Institute, New York, N.Y. (U.S.A.)

The following tables present a compilation of the reported values for the concentration of amino acids and related substances in the normal nervous tissue of various species of animals. The accompanying text gives such discussion as is necessary to explain the tables and, also, assembles certain data that could not be included because of limitations of space. These data consist, for the most part, of the results of two types of studies: studies on the distribution of a compound in parts of the central nervous system and developmental studies on the variation in the concentration of a substance with the age of the animal.

The values presented appear as ranges, in those cases for which much information is available; as averages, in the few cases for which the literature values are almost the same; or as individual figures, when only one or two determinations have been made. In general, the values given as ranges will be found to be the most reliable, though perhaps somewhat all-encompassing. Values that are far outside the usual range of figures, or which are the result of techniques which seem of doubtful validity, have been omitted from the tables.

All values have been expressed as μmoles/g of fresh tissue. Since this has involved a great deal of recalculation and, occasionally, the making of what it is hoped are reasonable assumptions, it is also to be hoped that not too many errors have crept in. However, in dealing with such a wide variety of compounds, the usefulness of stating concentrations in moles becomes all the more apparent. Not only are comparisons among compounds facilitated, but any possible ambiguity is removed concerning the form in which a substance has been measured (as the · HCl or free base, for example). Furthermore, since modern analytical procedures most often give as the primary observed datum the molar quantities of some constituent part of the molecule, it seems far more logical to keep the results in moles, particularly, as is the case with the lipids, when a molecular weight must be assumed. The use of fresh weight of tissue as the base line is a personal preference of the author's, founded, however, on the fact that the wet weight is the original measurement.

Glutamic acid and glutamine (Tables IA and B). Glutamic acid and its derivatives are considered to be the characteristic amino acid constituents of nervous tissue. The central position of glutamic acid in the metabolism of nervous tissue has been discussed many times in recent years[93, 158, 165, 166, 171, 172, 174–176], and its function in

* Present address: Geigy Research Laboratories, Ardsley, N.Y. (U.S.A.).

TABLE IA

GLUTAMIC AND ASPARTIC ACIDS AND RELATED COMPOUNDS IN BRAIN

Concentrations in μmoles/g. Reference numbers indicated between ( ).

Species	*Glutamic acid*	*Glutamine*	*Glutathione*	*γ-Aminobutyric acid*	*γ-Guanidinobutyric acid*	*Aspartic acid*	*Asparagine*	*Acetylaspartic acid*
Man (frontal lobe)	10.5, 19.9* (80, 127)			2.0, 5.5, 4.6* (15, 127)	0.03–0.07 (88)			4.3*** 5.5** (89, 159)
Cattle	8.7* (171)	4.7* (171)	3.71* (112)	4.9 (15)	~0.03 (88)			5.7**, 7.1* (159)
Sheep	10.5* (100)	3.8* (100)	2.88* (112)			0.44 (99)	<0.05 (99)	
Dog	8.7** (67)	3.7**, 4.4 (67, 74)			~0.055 (88)			
Cat	8.7, 9.9 (100, 160)	5.3 (100)	0.88 (160)	2.2, 4.2* (144, 160)		2.2 (111, 160)	0.11 (160)	6.0 (111, 159)
Rabbit	8.8–10.3 (154, 164, 171)	3.1 (171)	3.07* (112)	2.7 (15)		2.1 (164)		5.4 (89)
Guinea-pig			2.77**, 2.92* (112)	2.4*, 7.1 (15, 20)				4.7 (159)
Rat	8.2, 9.1–12.8 (7, 13, 27, 39–42, 51, 66, 73, 76 79, 90, 94, 126, 136, 139, 151, 168, 180)	2.1–6.4 (27, 40, 41, 51, 73, 79, 90, 94, 126, 133, 136, 138, 151, 167, 180)	0.84–1.7, 3.44, 1.94–3.58* (7, 10, 27, 51, 112, 126, 148, 163)	2.0–6.1 (7, 15, 27, 39, 42, 51, 57, 73, 76, 90, 94, 126, 136, 139, 168)	0.05 (29)	1.7–2.8 (51, 73, 94, 126)		4.3–6.9 (56, 89, 126, 159)
Mouse	11.5, 17.5 (17, 151)	4.9 (151)		3.5–5.9 (17, 113, 142)				5.3 (89)
Pigeon	6.2–14.0 (136, 171)	4.3–7.0 (136, 171)		7.9 (15)				
Hen	9.9, 10.8 (126, 136)	1.9, 4.1 (126, 136)	0.86 (126)	2.7, 6.0 (126, 136)		2.5–3.8 (16, 126, 136)		2.6, 5.9 (126, 159)
Duck								6.3 (159)
Fish	5.0 (126)	6.2 (126)	0.46 (126)	1.8 (126)		0.29 (126)		0.8 (126)
Tortoise	4.6 (126)	2.1 (126)	0.17 (126)	1.7 (126)		0.47 (126)		1.3 (159)
Frog	4.4 (126)	5.5 (126)	0.44 (126)	2.7 (126)		0.66 (126)		<0.06 (159)
Lobster								<0.18 (159)
Horseshoe crab								<0.03 (159)
Locust (head)	15.8 (25)	1.5 (25)						

* Cortex.
** Cerebrum.
*** Parietal lobe.

References p. 482/485

the brain may, indeed, be accurately reflected by this emphasis on its metabolic role. In vertebrates, the concentration of free glutamic acid is higher in brain than in the other tissues, and in brain its concentration is greater than that of any other amino acid[160, 165]; in invertebrate nerve the most abundant amino acid is aspartic acid[46, 47, 96, 109] or taurine[46]. Glutamic acid and glutamine have been found in high concentrations in the brain of all species examined; in addition to the species listed in Table IA, the two compounds have also been found in alligator and opposum brains[145].

Coincident with the development of the brain from the fetal or neonatal state to that of the adult, there is an increase in the concentration of glutamic acid and glutamine. This has been demonstrated in the mouse[143], rat[79, 171], calf[100], rabbit[22], sheep[100], and in man[80, 127]. Marked variations (greater than 2 : 1) in the occurrence

TABLE IB

GLUTAMIC AND ASPARTIC ACIDS AND RELATED COMPOUNDS IN NERVE

Concentrations in μmoles/g. Reference numbers indicated between ( ).

Species	*Nerve*	*Glutamic acid*	*Glutamine*	*γ-Aminobutyric acid*	*Aspartic acid*
Rabbit	Spinal cord	2.9 (72)	5.8 (72)		
Rabbit	Sciatic nerve	2.2 (72)	1.8, 4.2 (72)		
Rat	Sciatic nerve	2.3 (72)	2.2 (72)		
Hen	Spinal cord	6.9 (136)	1.4 (136)	0.86 (136)	2.0, 3.2 (16, 136)
Hen	Sciatic nerve	1.7 (136)	0.50 (136)	1.5 (136)	1.1 (136)
Crab	Leg nerve	35 (109)			138 (109)
Lobster	Leg nerve	25 (109)			112 (109)
Cuttlefish	Axon	39 (109)			82 (109)
Squid *(Loligo pealii)*	Giant nerve axoplasm	13.5, 21.2 (46, 96)			62.5, 79.1 (46, 96)
Squid *(Dosidicus gigas)*	Giant nerve axoplasm	27.6 (46)			111.3 (46)

of glutamic acid and glutamine in the various parts of the central nervous system are not found[27, 72, 80, 100, 127, 171], though the concentration in nerve (hen sciatic) is much lower than in brain[136]. Large amounts of glutamic acid are found in retina[103].

The occurrence of glutamic acid in peptide form is discussed below. Lipid-bound glutamic acid has also been reported[77].

γ-Aminobutyric acid (Tables IA and B). This compound (reviewed in detail by ELLIOT AND JASPER[63] and the subject of a recent book[141]) has attracted considerable attention because of its unique occurrence in nervous tissue[11, 15, 142, 169] and because it is a component of extracts of mammalian brain which have inhibitory effects on nerve preparations, particularly the crayfish stretch receptor[23, 24, 62, 68, 110]. However, although γ-aminobutyric acid does exhibit marked pharmacological activity, there has grown up a considerable body of evidence demonstrating that the active principle of the extract (termed Factor I) is not solely γ-aminobutyric acid, and, indeed, preparations of Factor I may be obtained in which no γ-aminobutyric acid is detectable by paper chromatography[115]. The amount of γ-aminobutyric acid in a Factor I extract depends both on the time required for preparation and the length of storage[116].

McLENNAN[115] suggested earlier that the true active component of Factor I is a derivative of γ-aminobutyric acid, from which the amino acid may be split during the preparation of the extract. There is evidence for the existence of an "occult" form of Factor I (refs. 62, 64; see also ref. 63), and the presence in brain extracts of a combined form of γ-aminobutyric acid has been claimed[31], though the liberation of appreciable amounts of γ-aminobutyric acid by hydrolysis of brain extracts has not been noted by other authors[142, 160]. The presence of γ-aminobutyrylhistidine has been shown[134] (see below), and some evidence for the presence of γ-aminobutyrylcholine in dog brain has been presented[104]. McLENNAN[116] later used paper chromatography to separate Factor I into two fractions active on the crayfish stretch receptor. One fraction had chromatographic and pharmacological properties different from those of γ-aminobutyric acid; the other fraction contained γ-aminobutyric acid, but also a substance or substances with different pharmacological properties. More recently, McLENNAN[117] has presented evidence that neither active fraction is ninhydrin-positive; he suggests that the activity is due to the same compound, as yet unindentified, present as a free cation in one fraction and as an uncharged complex in the other.

Less γ-aminobutyric acid is found in the brain of immature animals than in the brain of adults, and, with growth, there is a gradual increase to the adult levels; this has been shown with the mouse[143–145], rabbit[22], chick embryo[140, 145, 153], frog tadpole[140, 145], and in man[127]. Differences have not been noted, however, between calf embryo and ox brains[11]. Besides its occurrence in the species listed in Table I, the compound is also present in alligator and opposum brain[145], and in crustacean nerve[58].

Distribution studies show that gray matter contains twice as much γ-aminobutyric acid as white matter[143]. In human brain, the highest values reported are in caudate nucleus[15, 127] and cerebellum[15], but this is not in complete accord with results in other species. The concentration of γ-aminobutyric acid in various parts of the rat[18, 19, 27], cat[21, 60], guinea-pig[20], and monkey[59] brain has been determined. γ-Aminobutyric acid occurs in spinal cord[136] (*cf.* ref. 68) at lower concentrations than in brain[136], but there seems to be little, if any, in peripheral nerves, though the compound does occur in retina[97, 103]. None was found in dog optic nerve and brachial plexus[144, 145], or in rabbit sciatic nerve[143]; it is found in hen sciatic nerve, but not consistently[136]. Small amounts of Factor I activity are found in bovine oculomotor and trochlear nerves[69]. A detailed study has been made of the distribution of Factor I activity in cattle brain[69]; however, the values obtained cannot be equated with γ-aminobutyric acid concentrations.

β-Hydroxy-γ-aminobutyric acid. This compound, which has much the same pharmacological actions as γ-aminobutyric acid[75, 114], has been proposed as the "true" chemical transmitter of inhibition in the mammalian brain[75]. Much indirect evidence points to its presence in dog brain[75, 76], and its occurrence in rat[76], human, cattle, mouse, and rabbit brain[125] has been reported. Estimates of the amount present range from 0.5 μmole/g in rat brain[76] to 4.1 μmoles/g in cattle temporal lobe[125]. However, the presence of the substance in mammalian brain has yet to be demonstrated unequivocally by isolation, and the use of isotopically labeled material has provided strong evidence against its presence[120].

References p. 482/485

Betaines. It is claimed that the Coenzyme A esters of γ-butyrobetaine, crotonbetaine and carnitine occur in mammalian brain, and that the free betaines accumulate in brain after chemically induced convulsions or electroshock[84, 85]. These observations have as yet not been verified.

Glycine betaine[47, 49] and homarine (*N*-methylpicolinic acid)[47, 49, 96] occur in the axoplasm of squid giant nerve. The concentrations found are, for betaine, 73.65 (*Loligo pealii*) and 119.11 (*Dosidicus gigas*) μmoles/g fresh axoplasm; for homarine, 20.36 (*Loligo*) and 21.40 (*Dosidicus*) μmoles/g (ref. 46).

γ-Guanidinobutyric acid (Table IA). This substance has been isolated from calf brain and identified unequivocally[87]. It occurs in brain in very small quantities[88]. McLennan[116] has presented evidence indicating that γ-guanidinobutyric acid occurs in Factor I (see under γ-aminobutyric acid) and accounts for some of the properties of the extract.

Aspartic acid and asparagine (Tables IA and B). Krebs[99] has reported that unless tissues to be analyzed are either frozen at once or extracted with two volumes of acid at least as strong as 0.5 *N* HCl, amounts of aspartic acid and asparagine 10–100 times the control values are found. Since values in the literature for the brain aspartic acid of different species of vertebrates are at least 50 times that reported by Krebs[99] for sheep brain, it is apparent that the question of "true" aspartic acid levels bears reinvestigation. Nevertheless, the figures for aspartic acid given in Tables IA and B are sufficiently consistent to warrant their inclusion as "normal" figures; however, most results obtained with vertebrate tissues not extracted with acid have been excluded from the table as being probably too high[7, 13, 39, 57, 127, 136, 154]. The figures for the hen[136] have been retained as illustrative of the variation between brain, spinal cord and nerve. The extremely high concentration of aspartic acid in invertebrate nerve[46, 96, 109] has already been noted.

The distribution of aspartic acid in parts of the human brain has been studied by Okumura *et al.*[127]. Studies on the variation with age indicate that there is less aspartic acid in immature brain than in adult brain[11, 22, 127, 143].

N-Acetylaspartic acid (Table IA). A major part of the non-protein aspartic acid in brain is present as the *N*-acetyl derivative. This compound is of particular interest in that it occurs only in nervous tissue[161], but whether it performs any special role related to the functioning of the brain is not yet known. In the newborn rat and rabbit, the acetylaspartic acid concentration in the brain is low (approx. 1.1 μmoles/g) and there is a rapid increase to adult levels by approx. 20 days of age[89, 159]. An increase with age occurs in human brain also[89, 127]. The concentration of the compound is highest in cerebral gray matter[159].

Glycine and alanine (Tables IIA and B). These amino acids are present in moderate amounts in vertebrate brain and nerves, and in large amounts in certain invertebrate nerves. The possible presence of a glycine-containing lipid in rat brain has been noted[6]. Fetal mouse brain contains more alanine than adult brain[143].

Serine, phosphoserine and threonine (Tables IIA and B). By virtue of its hydroxyl

TABLE II A

AMINO ACIDS IN BRAIN

Concentrations in μmoles/g. Reference numbers indicated between ( ).

Species	*Glycine*	*Alanine*	*Serine*	*Phosphoserine*	*Threonine*	*Taurine*	*Cystathionine*
Man (frontal lobe)			1.21 (127)		0.53 (127)	1.45 (127)	0.08, 1.1–2.5 (127, 162)
Monkey							0.45, 0.58 (127, 162)
Cattle							0.054 (162)
Cat	0.91, 1.3 (101, 160)	0.94 (160)	0.72 (160)		0.22 (160)	1.9 (160)	0.12 (162)
Guinea-pig	2.3 (101)					0.8 (12)	0.14 (162)
Rabbit	1.2, 2.2 (101, 154)	1.25 (154, 164)	5.8 (154)		0.58 (154)	1.4, 2.7 (12, 154)	
Rat	0.91–2.7 (7, 28, 57, 73, 94, 101, 126, 136)	0.25–0.96, 2.6 (7, 28, 51, 57, 73, 94, 126, 136)	0.66–1.8, 2.8 (7, 28, 57, 73, 94 126, 136)	0.21, 0.94 (118, 135)	0.30–1.6 (28, 57, 94, 121, 126, 136, 149, 150, 178)	2.3–5.4, 8.5 (7, 12, 13, 51, 57, 73, 81, 94, 126, 136)	0.18 (162)
Pigeon	1.8 (101)						
Duck							0.009 (162)
Hen	0.68, 1.8 (126, 136)	0.48, 1.0 (126, 136)	1.17 (126, 136)	0.07, 0.23 (16, 135)	0.35–0.66 (16, 126, 136)	0.97–3.2 (16, 126, 136)	0.027 (162)
Fish	0.8 (126)	0.25 (126)	0.28 (126)		0.66 (126)	3.4 (126)	
Tortoise	0.15 (126)	0.12 (126)	0.12 (126)		0.04 (126)	0.31 (126)	
Frog	1.6 (126)	0.24 (126)	0.20 (126)		0.15 (126)	0.048 (126)	
Horseshoe crab							0.67 (162)

group, serine plays a role in lipid as well as in protein chemistry. Phosphatidylserine itself is best discussed with the other "cephalins"; the presence of serine in incompletely characterized phospholipids not identical with phosphatidylserine has also been reported[34, 177]. Free phosphoserine was provisionally identified in brain by ROBERTS[140] and its presence, at concentrations of about one-fifth that of serine, was established by PORCELLATI AND THOMPSON[135, 136]. Since phosphoethanolamine and phosphocholine are both found in brain, the existence of phosphoserine was to be expected; glycerophosphoserine, on the other hand, has not yet been detected. The quantity of phosphoserine in nerve increases after section[135], as does the quantity of serine, threonine, and of the other amino acids[136]. The distribution of serine and threonine in parts of the human brain has been studied[127].

Taurine (Tables IIA and B). Extremely large variations in the taurine concentration of the nervous tissue of various species have been reported. In invertebrate nerve, the very high concentrations have been regarded as necessary to help balance the high cation concentration[96, 109]; it should be noted that in squid giant nerve axoplasm the related 2-hydroxyethanesulfonic acid (isethionic acid) is also present in large amounts for presumably the same purpose[46, 96]. Large amounts of taurine occur in the retina of many species[103]; an estimate of the concentration in cow retina is 12.2 μmoles/g (ref. 102). In the fetal mouse[143], calf[11] and human[127], the taurine concentration in the brain is higher than in adults. However, there is less taurine in tadpole brain than in frog brain[145]. The distribution of taurine in parts of the human brain has been determined[127].

Cystathionine (Table IIA). The occurrence of large amounts of cystathionine in human brain, but not in the brains of other animals, except the monkey[162], suggests that there may be a metabolic difference between primate brain and the brains of other species. The presence of the compound in cerebral cortex taken at biopsy[127], the relatively large amount found in monkey brain[127, 162], the virtual absence of cystathionine in fetal human brains[127], and the low values found in human cerebellum

TABLE IIB

AMINO ACIDS IN NERVE

Concentrations in μmoles/g. Reference numbers indicated between ( ).

Species	*Nerve*	*Glycine*	*Alanine*	*Serine*	*Phospho-serine*	*Threonine*	*Taurine*
Hen	Spinal cord	1.2 (136)	1.0 (136)	0.81 (136)	0.05, 0.29 (16, 135)	0.26, 0.50 (16, 136)	0.54, 0.84 (16, 136)
Hen	Sciatic nerve	0.56 (136)	0.44 (136)	0.34 (136)	0.21 (135)	0.20 (136)	0.52 (136)
Crab	Leg nerve	<5 (109)	33 (109)				65 (109)
Lobster	Leg nerve	35 (109)	33 (109)				~12 (109)
Cuttlefish	Axon	<5 (109)	21 (109)				103 (109)
Squid *(Loligo pealii)*	Giant nerve axoplasm	11.6, 14.0 (46, 96)	8.6, 9.0 (46, 96)	4.0, 5.7 (46, 96)		2.0 (46)	76.0, 106.7 (46, 96)
Squid *(Dosidicus gigas)*	Giant nerve axoplasm	11.0 (46)	9.8 (46)	1.0 (46)		0.32 (46)	33.1 (46)

and other tissues, provide evidence that the high cystathionine content of human cerebrum is not a *post mortem* artifact. (The appreciable amounts of cystathionine found in horseshoe-crab brain are a reflection of the fact that almost all of the amino acids occur in large amounts in this invertebrate nerve[162].) However, the cystathionine content of human brain is not consistently high[127], and the factors governing its concentration remain to be elucidated.

In the rat, a pyridoxine deficiency results in the accumulation of cystathionine in the brain and liver; very large amounts (approx. 3.0 μmoles/g) are found in the brain[82]. The highest concentrations in vitamin B_6-deficient rat brain are in the cerebellum (6.7 μmoles/g) and spinal cord (4.7 μmoles/g), decreasing to 1.5 μmoles/g in the cerebral hemispheres[83]. These findings may have a bearing on the accumulation of cystathionine in human brain, though it seems doubtful that the majority of the human brain samples could have come from pyridoxine-deficient individuals. Furthermore, preliminary experiments by the author have indicated that the cystathionine content of human cerebellum is low (0.1 μmole/g), in contradistinction to the findings in the deficient rat.

References p. 482/485

TABLE III

AMINO ACIDS AND RELATED COMPOUNDS IN NERVOUS TISSUE

Concentrations in μmoles/g. Reference numbers indicated between ( ). N.D., none-detectable.

Compound	*Man frontal lobe*	*Cat brain*	*Mouse brain*	*Rat brain*	*Hen brain*	*Hen spinal cord*	*Hen sciatic nerve*	*Tortoise brain*	*Frog brain*	*Fish brain*	*Squid giant nerve axoplasm Loligo*	*Squid giant nerve axoplasm Dosidicus*
β-Alanine		N.D. (160)										
β-Aminoisobutyric acid		0.0097 (160)										
α-Amino-*n*-butyric acid		0.019 (160)										0.05 (46)
Arginine		0.080 (160)		0.063–0.20 (29, 33, 126, 149, 150, 178)	N.D. (126)			0.10 (126)	0.069 (126)	0.075 (126)	3.46 (46)	4.32 (46)
Cystine				0.05 (94)								
Dihydroxyphenylalanine	0.001 (122)											
Glycocyamine				0.03, 0.14, 0.26–0.51 (29, 30, 33)								
Histidine		0.058 (160)		0.055–0.17 (33, 126, 149, 150, 178)	N.D. (126)			N.D. (126)	N.D. (126)	0.30 (126)		
Isoleucine	0.18 (127)	0.092 (160)		0.048–0.067, 0.12 (126, 149, 150, 178)	0.11 (126)			N.D. (126)	0.099 (126)	0.21 (126)	2.92 (46)	0.21 (46)
Leucine	0.56 (127)	0.14 (160)	0.073 (106)	0.067–0.26 (28, 126, 136, 149, 150, 178)	0.076, 0.19 (126, 136)	0.18 (136)	0.14 (136)	N.D. (126)	0.099 (126)	0.30 (126)		
Lysine		0.14 (160)	0.26 (105)	0.12–0.22 (126, 149, 150, 178)	0.12 (126)			N.D. (126)	0.082 (126)	0.21 (126)	2.6 (46)	0.22 (46)
Methionine		0.10 (160)		0.058–0.094 (149, 150, 178)							0.54 (46)	0.32 (46)
Ornithine		0.023 (160)									1.95 (46)	0.32 (46)
Phenylalanine		0.061 (160)		0.054–0.091 (149, 150, 178)							0.65 (46)	0.16 (46)
Proline		N.D. (160)		0.12–0.15 (149, 150, 178)							1.08 (46)	0.11 (46)
Sarcosine		0.30 (160)										
Taurocyamine				0.03 (29)								
Tryptophane		N.D. (160)		0.016–0.030 (149, 150, 178)								
Tyrosine		0.055 (160)		0.06–0.14 (28, 94, 136, 149, 150, 178)	0.14 (136)	0.14 (136)					0.76 (46)	0.32 (46)
Valine		0.15 (160)		0.09–0.21, 0.51 (28, 57, 94, 136, 149, 150, 178)	0.26 (136)	0.23 (136)	0.15 (136)				2.4 (46)	0.53 (46)

References p. 482/485

Other amino acids (Table III). The compounds grouped under this designation occur in amounts from the barely detectable to approx. 0.5 μmole/g in vertebrate nervous tissue (much larger amounts are present in invertebrate nerve) and include the essential amino acids and those that may be indispensable under certain conditions[165]. In addition to the substances listed in Table III, the following have been reported to occur in nervous tissue, though little information is available concerning the actual amounts present: hydroxyproline[165], citrulline[165] (not detected in the cat, however[160]), hydroxylysine phosphate[11], cysteinesulfinic acid[26], and hypotaurine[26]. The amide of cysteic acid occurs in squid giant nerve axoplasm[47, 48]; its concentration is 4.86 μmoles/g fresh axoplasm of *Loligo pealii*, 0.42 μmole/g in *Dosidicus gigas*[46]. Also present in small amounts in squid giant nerve axoplasm are methionine sulfoxide and citrulline[50]: 0.32 and 0.43 μmole/g respectively in *Loligo*, 0.21 and 0.16 μmole/g in *Dosidicus*[46].

The distribution of lysine in various parts of the monkey brain[71] and the distribution in human brain of dihydroxyphenylalanine[147], leucine and isoleucine[127] have been determined. In several species that have been studied, the concentration of tryptophane is much higher in the hypothalamus, pons and mid-brain than in other parts of brain[137]. In mice, the concentration of lysine in the brain is higher at birth (0.55 μmole/g) than in the young animal (0.35 μmole/g) or adult (0.26 μmole/g), (ref. 105). Leucine decreases from 0.15 μmole/g in the newborn to 0.07 in the adult[106]. There is more arginine, tyrosine, and γ-aminobutyric acid in fetal mouse brain than in the brains of the mothers[143]. In man, leucine and isoleucine tend to increase from the fetal to the adult brain[127].

Peptides. The following peptides are definitely known to occur in brain: glutathione (Table I), homocarnosine, carnosine and the so-called Substance P. Hydrolysis of protein-free extracts of cat brain results in unexplained increases of histidine, lysine, and of a substance in the chromatographic position of β-aminoisobutyric acid[160]; in hydrolyzed extracts of rat brain there are increases in the amounts of leucine, valine, histidine, isoleucine, and proline[149]. Boulanger and Biserte[31] have separated dog, pig and beef-brain extracts into a basic plus neutral fraction and three acidic fractions. After hydrolysis, the basic plus neutral fraction gave β-alanine, proline, aspartic acid and glutamic acid; one acidic fraction gave large amounts of β-alanine and γ-aminobutyric acid after hydrolysis; another, small amounts of several amino acids; and the third, large amounts of aspartic acid, glutamic acid and cysteic acid. None of the parent substances, which could be peptides, appear to have been studied further. It should be noted that in the work on the cat[160], there was an appearance after hydrolysis of glutamic acid, glycine, cysteic acid and aspartic acid that could be accounted for as arising from glutathione, glutamine and *N*-acetylaspartic acid.

The increase in histidine after hydrolysis suggests the presence of carnosine; carnosine was reported absent in dog brain[181] and only possibly present in ox brain[11], but there have been several recent reports of its occurrence[1, 86, 134]. Abraham *et al.*[1] state that carnosine was found in all brains examined, at values varying from almost none in human brain to 0.03 μmole/g in beef brain. It has recently been demonstrated that γ-aminobutyryl-L-histidine (homocarnosine) also occurs in brain[134], at values ranging up to 0.1–0.3 μmole/g in human, monkey, beef and rabbit brain[1]; 0.18 μmole/g was found in rabbit spinal cord, none was detectable in rabbit sciatic

References p. 482/485

nerve[1]. The presence of anserine in rat brain has been reported[86]; anserine is absent in rabbit and guinea-pig-brain extracts, though methylhistidine appears after hydrolysis[156].

Peptides are present in squid giant nerve axoplasm[47]. An unknown material, which on hydrolysis yielded threonine, other ninhydrin-positive substances, and phosphate, was found in the brains of a large number of species, though in especially high concentrations in fish, tortoise, and frog brain[126]. An unknown material yielding phosphoserine, phosphoethanolamine, glutamic acid and valine upon hydrolysis has been found in hen spinal cord[16]. Peptides occur, too, as part of some complex lipids[70, 107, 146, 177]; large numbers of lipophilic peptides have been detected in mouse brain[170].

Glutathione concentrations in brain have been measured recently by MARTIN AND MCILWAIN[112]; their figures, obtained by use of a yeast apoglyoxalase, are higher than those obtained by chemical reaction or by chromatography and undoubtedly represent more nearly the true values. The amount of oxidized glutathione present was found to be 2–4% of the total when the brain was frozen *in situ*; with other methods of preparation the proportion was much higher, though the total glutathione (–SH plus S–S) remained constant[112]. The distribution of glutathione in various parts of the rat brain has been studied[10, 27, 112]. Glutathione is present in bovine retina at a concentration of 2.5 μmoles/g; no oxidized form was detected[78].

Substance P, a peptide that stimulates smooth muscle, is found in intestinal muscle as well as in the central nervous system. The peptides from the two sources have identical properties with regard to counter-current distribution, paper electrophoresis, paper chromatography, enzymatic inactivation by chymotrypsin[61], and biological activity[61, 182]. Possibly there are two different fractions in brain, for after brain tissue has been thoroughly extracted with ethanol, boiling at pH 3 releases more Substance P activity; the two fractions behave differently upon autolysis and dialysis and are affected differently by pretreatment of the animals with various drugs[183]. The properties and distribution of Substance P are reviewed by PERNOW[131, 132] (for detailed studies, see ref. 2, 37, 38, 55, 65, 95, 98, 108, 129, 130, 152, 183, 184).

Another material from brain, which appears to be a peptide of large molecular weight, has the property of enhancing antithrombin activity; the best source of the material is reported to be infant human brain[179].

Ethanolamine (Table IV). The presence of ethanolamine in brain lipids was established at an early date (*cf.* ref. 52). More recently, the derivatives glycerophosphoethanolamine (GPE) and phosphoethanolamine (PE) have been shown to occur as such in brain tissue, and it is likely that there is also some free ethanolamine, though the amount of the latter is not certain[123, 160]. The exact amount of GPE normally present is also in doubt (for example refs. 160, 161), because of the lability of GPE and of phosphatidylethanolamine[5]. Other than in the brains of the species given in the table, GPE has been reported present in brains of the rabbit[156], guinea-pig[156], cow[173], and sheep[44], and in hen spinal cord[135]. GPE seems to be neither a metabolic breakdown product of phosphatidylethanolamine, nor to lie on the main synthetic pathway to it[5]. PE, which was first found in brain by AWAPARA, LANDUA AND FUERST[14], is not a breakdown product of phosphatidylethanolamine[4]; it is possibly on the synthetic pathway to the phosphatide, via cytidine-diphosphate-ethanol-

References p. 482/485

TABLE IV

ETHANOLAMINE, ITS DERIVATIVES, AND "PHOSPHATIDE SERINE" IN BRAIN AND NERVE

Concentrations in µmoles/g. Reference numbers indicated between ( ).

Species	*Tissue*	*Ethanol-amine*	*PE*	*GPE*	*"Phosphatide ethanolamine"*	*"Phosphatide serine"*
Man	Frontal lobe		1.17 (127)	1.05 (127)		
Cattle	Brain				14.2 (32)	6.9 (32)
Sheep	Brain				20.2, 24 (36, 45)	8.9, 11.7 (36, 45)
Dog	Brain		1.2 (157)			
Cat	Brain	3.4 (160)	3.0 (160)	3.9 (161)	18–23 (9, 36, 124)	6–12 (9, 36, 124)
Rabbit	Brain		2.2–3.0 (154, 155)			
Guinea-pig	Brain				23 (36)	3.5 (36)
Rat	Brain	0.2 (51)	0.95–2.9 (7, 8, 51, 94, 118, 126, 135, 155)	0.1–1.0 (5, 118, 126)		
Hen	Brain		1.0–3.25 (16, 126, 135)	0.80, 1.16 (16, 126)		
Fish	Brain		1.25 (126)	0.79 (126)		
Tortoise	Brain		0.72 (126)	0.69 (126)		
Frog	Brain		1.49 (126)	1.44 (126)		
Hen	Spinal cord		0.48 (16, 135)	0.61 (16)		
Hen	Sciatic nerve		0.19 (135)			

amine[3, 91, 92, 119]. Large amounts of PE in cow[173] and ox[11] brain have been reported though quantitative results are lacking. There is a decrease in PE in the developing rabbit[22] and human[127] brain. The distribution of GPE and PE in parts of the human brain has been studied; variations in PE content are very large[127].

It should be noted here that the phospholipids designated as "cephalins" by earlier workers (*cf.* ref. 53) are now known to consist of mixtures of phosphatidylethanolamine, phosphatidylserine, ethanolamine acetal phosphatide, serine acetal phosphatide, diphosphoinositide (*cf.* refs. 43, 53, 107), and other lipids as yet uncharacterized[34, 35, 77]. Each of the substances noted is, in fact, a family of related substances differing only in the composition of the fatty acids and aldehydes attached to the glycerol moiety (*cf.* refs. 43, 53, 107), so that it is not possible in most cases to speak of specific compounds. For the purpose of the present notes, it is of interest only to give some indication of the amounts of ethanolamine and serine present in brain in lipid form. Accordingly, values are given in Table IV for the concentrations of what are here termed "phosphatide-ethanolamine" and "phosphatide-serine". These figures are based on analyses of total lipid or total phospholipid extracts, and thus include ester and acetal phospholipid. The difficulties involved in the analysis of lipids for serine and ethanolamine content are discussed by OLLEY[128]; DITTMER *et al.*[54] have recently made a critical study of the appropriate conditions for such analysis, and DAWSON[45] has devised a procedure, using mild hydrolysis and chromatography, for the determination of all of the phospholipids.

References p. 482/485

REFERENCES

1 D. Abraham, J. Pisano and S. Udenfriend, *Federation Proc.*, 20 (1961) 344.
2 A. H. Amin, T. B. B. Crawford and J. H. Gaddum, *J. Physiol. (London)*, 126 (1954) 596.
3 G. B. Ansell and B. J. Bayliss, *Biochem. J.*, 78 (1961) 209.
4 G. B. Ansell and R. M. C. Dawson, *Biochem. J.*, 50 (1951) 241.
5 G. B. Ansell and J. M. Norman, *Biochem. J.*, 55 (1953) 768.
6 G. B. Ansell and J. M. Norman, *J. Neurochem.*, 1 (1956) 32.
7 G. B. Ansell and D. Richter, *Biochem. J.*, 57 (1954) 70.
8 G. B. Ansell and S. Spanner, *J. Neurochem.*, 4 (1959) 325.
9 C. Artom, *J. Biol. Chem.*, 157 (1945) 595.
10 M. J. Ashwood-Smith and A. D. Smith, *Nature*, 184 (1959) 2028.
11 T. Astrup, G. Carlström and A. Stage, *Acta Physiol. Scand.*, 24 (1951) 202.
12 J. Awapara, *J. Biol. Chem.*, 218 (1956) 571.
13 J. Awapara, A. J. Landua and R. Fuerst, *Biochim. Biophys. Acta*, 5 (1950) 457.
14 J. Awapara, A. J. Landua and R. Fuerst, *J. Biol. Chem.*, 183 (1950) 545.
15 J. Awapara, A. J. Landua, R. Fuerst and B. Seale, *J. Biol. Chem.*, 187 (1950) 35.
16 R. W. R. Baker and G. Porcellati, *Biochem. J.*, 73 (1959) 561.
17 H. Balzer, P. Holtz and D. Palm, *Arch. exptl. Pathol. Pharmakol., Naunyn-Schmiedeberg's*, 239 (1960) 520.
18 C. F. Baxter and E. Roberts, *Proc. Soc. Exptl. Biol. Med.*, 101 (1959) 811.
19 C. F. Baxter and E. Roberts, *Proc. Soc. Exptl. Biol. Med.*, 104 (1960) 426.
20 C. F. Baxter, E. Roberts and E. Eidelberg, *Federation Proc.*, 18 (1959) 187.
21 C. F. Baxter, E. Roberts and E. Eidelberg, *J. Neurochem.*, 5 (1960) 203.
22 C. F. Baxter, J. P. Schadé and E. Roberts, in E. Roberts, *Inhibition in the Nervous System and γ-Aminobutyric Acid*, Pergamon Press, New York, 1960, p. 214.
23 A. W. Bazemore, K. A. C. Elliott and E. Florey, *J. Neurochem.*, 1 (1957) 334.
24 A. W. Bazemore, K. A. C. Elliott and E. Florey, *Nature*, 178 (1956) 1052.
25 D. Bellamy, *Biochem. J.*, 70 (1958) 580.
26 B. Bergeret and F. Chatagner, *Biochim. Biophys. Acta*, 14 (1954) 297.
27 S. Berl and H. Waelsch, *J. Neurochem.*, 3 (1958) 161.
28 G. Biserte, P. Boulanger and P. Paysant, *Bull. soc. chim. biol.*, 40 (1958) 2067.
29 J. P. Blass, *Biochem. J.*, 77 (1960) 484.
30 H. Borsook and J. W. Dubnoff, *J. Biol. Chem.*, 138 (1941) 389.
31 P. Boulanger and G. Biserte, *Compt. rend.*, 233 (1951) 1498.
32 E. Chargaff, M. Ziff and D. Rittenberg, *J. Biol. Chem.*, 144 (1942) 343.
33 D. H. Clouet, M. K. Gaitonde and D. Richter, *J. Neurochem.*, 1 (1957) 228.
34 F. D. Collins, *Nature*, 182 (1958) 865.
35 F. D. Collins and L. W. Wheeldon, *Biochem. J.*, 66 (1957) 441.
36 F. D. Collins and L. W. Wheeldon, *Biochem. J.*, 70 (1958) 46.
37 P. Correale, *J. Neurochem.*, 2 (1958) 201.
38 P. Correale, *Arch. intern. pharmacodynamie*, 114 (1959) 435.
39 R. O. Cravioto, G. Massieu and J. J. Izquierdo, *Proc. Soc. Exptl. Biol. Med.*, 78 (1951) 856.
40 R. M. C. Dawson, *Biochem. J.*, 47 (1950) 386.
41 R. M. C. Dawson, *Biochem. J.*, 49 (1951) 138.
42 R. M. C. Dawson, *Biochim. Biophys. Acta*, 11 (1953) 548.
43 R. M. C. Dawson, *Biol. Revs. Cambridge Phil. Soc.*, 32 (1957) 188.
44 R. M. C. Dawson, *Biochem. J.*, 68 (1958) 357.
45 R. M. C. Dawson, *Biochem. J.*, 75 (1960) 45.
46 G. G. J. Deffner, *Biochim. Biophys. Acta*, 47 (1961) 378.
47 G. G. J. Deffner and R. E. Hafter, *Biochim. Biophys. Acta*, 32 (1959) 362.
48 G. G. J. Deffner and R. E. Hafter, *Biochim. Biophys. Acta*, 35 (1959) 334.
49 G. G. J. Deffner and R. E. Hafter, *Biochim. Biophys. Acta*, 42 (1960) 189.
50 G. G. J. Deffner and R. E. Hafter, *Biochim. Biophys. Acta*, 42 (1960) 200.
51 E. S. De Ropp and E. H. Snedeker, *Anal. Biochem.*, 1 (1960) 424.
52 H. J. Deuel, Jr., in *The Lipids*, Vol. 1, Interscience, New York, 1951.
53 H. J. Deuel, Jr., in *The Lipids*, Vol. 11, Interscience, New York, 1955, p. 741.
54 J. C. Dittmer, J. L. Feminella and D. J. Hanahan, *J. Biol. Chem.*, 233 (1958) 862.
55 H. Duner, U. S. Von Euler and B. Pernow, *Acta Physiol. Scand.*, 31 (1954) 113.
56 J. P. Du Ruisseau, *Can. J. Biochem. and Physiol.*, 38 (1960) 763.
57 J. P. Du Ruisseau, J. P. Greenstein, M. Winitz and S. M. Birnbaum, *Arch. Biochem. Biophys.*, 68 (1957) 161.
58 C. Edwards and S. W. Kuffler, *J. Neurochem.*, 4 (1959) 19.
59 E. Eidelberg, C. F. Baxter, E. Roberts and C. A. Saldias, in E. Roberts, *Inhibition in the Nervous System and γ-Aminobutyric Acid*, Pergamon Press, New York, 1960, p. 365.

[60] E. EIDELBERG, C. F. BAXTER, E. ROBERTS, C. A. SALDIAS AND J. D. FRENCH, *Proc. Soc. Exptl. Biol. Med.*, 101 (1959) 815.
[61] R. ELIASSON, L. LIE AND B. PERNOW, *Brit. J. Pharmacol.*, 11 (1956) 137.
[62] K. A. C. ELLIOTT AND E. FLOREY, *J. Neurochem.*, 1 (1956) 181.
[63] K. A. C. ELLIOTT AND H. H. JASPER, *Physiol. Revs.*, 39 (1959) 383.
[64] K. A. C. ELLIOTT AND N. M. VAN GELDER, *J. Neurochem.*, 3 (1958) 28.
[65] U. S. VON EULER AND E. ÖSTLUND, *Brit. J. Pharmacol.*, 11 (1956) 323.
[66] V. FERRARI, *Minerva med.*, 2 (1956) 2192; *Chem. Abstr.*, 51 (1957) 6845.
[67] E. V. FLOCK, M. A. BLOCK, J. H. GRINDLAY, F. C. MANN AND J. L. BOLLMAN, *J. Biol. Chem.*, 200 (1953) 529.
[68] E. FLOREY, *Arch. intern. physiol.*, 62 (1954) 33.
[69] E. FLOREY AND E. FLOREY, *J. Physiol. (London)*, 144 (1958) 220.
[70] J. FOLCH AND F. N. LEBARON, *Can. J. Biochem. and Physiol.*, 34 (1956) 305.
[71] S. FURST, A. LAJTHA AND H. WAELSCH, *J. Neurochem.*, 2 (1958) 216.
[72] R. W. GERARD, in K. A. C. ELLIOTT, I. H. PAGE AND J. H. QUASTEL, *Neurochemistry*, Thomas, Springfield, Ill., 1955, p. 458.
[73] H.-M. HÄKKINEN AND E. KULONEN, *Biochem. J.*, 78 (1961) 588.
[74] P. B. HAMILTON, *J. Biol. Chem.*, 158 (1945) 397.
[75] T. HAYASHI, *Nature*, 182 (1958) 1076.
[76] T. HAYASHI, *J. Physiol. (London)*, 145 (1959) 570.
[77] E. HECHT AND C. MINK, *Biochim. Biophys. Acta*, 8 (1952) 641.
[78] H. HERRMANN AND S. G. MOSES, *J. Biol. Chem.*, 158 (1945) 33.
[79] H. E. HIMWICH AND W. A. HIMWICH, *Research Publs., Assoc. Nervous Mental Disease*, 35 (1956) 19.
[80] W. A. HIMWICH, W. T. SULLIVAN, B. KELLEY, H. B. W. BENARON AND B. E. TUCKER, *J. Nervous Mental Disease*, 122 (1955) 441.
[81] D. B. HOPE, *J. Neurochem.*, 1 (1957) 364.
[82] D. B. HOPE, *J. Physiol. (London)*, 141 (1958) 31P.
[83] D. B. HOPE, *Federation Proc.*, 18 (1959) 249.
[84] E. A. HOSEIN AND H. MCLENNAN, *Nature*, 183 (1959) 328.
[85] E. A. HOSEIN AND P. PROULX, *Nature*, 187 (1960) 321.
[86] E. A. HOSEIN AND M. SMART, *Can. J. Biochem. and Physiol.*, 38 (1960) 569.
[87] F. IRREVERRE AND R. L. EVANS, *J. Biol. Chem.*, 234 (1959) 1438.
[88] F. IRREVERRE, R. L. EVANS, A. R. HAYDEN AND R. SILBER, *Nature*, 180 (1957) 704.
[89] K. B. JACOBSON, *J. Gen. Physiol.*, 43 (1959) 323.
[90] K. KAWAI, *Okayama-Igakkai-Zasshi*, 69 (1957) 2477; *Chem. Abstr.*, 52 (1958) 13832.
[91] E. P. KENNEDY, *Federation Proc.*, 16 (1957) 847.
[92] E. P. KENNEDY AND S. B. WEISS, *J. Biol. Chem.*, 222 (1956) 193.
[93] E. E. KLEIN, *Uspekhi Sovremennoĭ Biol.*, 41 (1956) 161; *Chem. Abstr.*, 51 (1957) 5961.
[94] H. G. KNAUFF AND F. BÖCK, *J. Neurochem.*, 6 (1961) 171.
[95] D. KOCIĆ-MITROVIĆ, *Arch. exptl. Pathol. Pharmakol., Naunyn-Schmiedeberg's*, 236 (1959) 37.
[96] B. A. KOECHLIN, *J. Biophys. Biochem. Cytol.*, 1 (1955) 511.
[97] K. KOJIMA, K. MIZUNO AND M. MIYAZAKI, *Nature*, 181 (1958) 1200.
[98] H. KOPERA AND W. LAZARINI, *Arch. exptl. Pathol. Pharmakol., Naunyn-Schmiedeberg's*, 219 (1953) 214.
[99] H. A. KREBS, *Biochem. J.*, 47 (1950) 605.
[100] H. A. KREBS, L. V. EGGLESTON AND R. HEMS, *Biochem. J.*, 44 (1949) 159.
[101] R. KRUEGER, *Helv. Chim. Acta*, 33 (1950) 429.
[102] R. KUBÍCEK AND A. DOLÉNEK, *Z. physiol. Chem., Hoppe-Seyler's*, 308 (1957) 249.
[103] R. KUBÍCEK AND A. DOLÉNEK, *J. Chromatog.*, 1 (1958) 266.
[104] K. KURIAKI, T. YAKUSHIJI, T. NORO, T. SHIMIZU AND S. SAJI, *Nature*, 181 (1958) 1336.
[105] A. LAJTHA, *J. Neurochem.*, 2 (1958) 209.
[106] A. LAJTHA, *J. Neurochem.*, 3 (1959) 358.
[107] F. N. LEBARON AND J. FOLCH, *Physiol. Revs.*, 37 (1957) 539.
[108] F. LEMBECK, *Arch. exptl. Pathol. Pharmakol., Naunyn-Schmiedeberg's*, 219 (1953) 197.
[109] P. R. LEWIS, *Biochem. J.*, 52 (1952) 330.
[110] K. LISSÁK AND E. ENDRÖCZI, *Naturwissenschaften*, 23 (1955) 630.
[111] R. U. MARGOLIS, S. S. BARKULIS AND A. GEIGER, *J. Neurochem.*, 5 (1960) 379.
[112] H. MARTIN AND H. MCILWAIN, *Biochem. J.*, 71 (1959) 275.
[113] G. M. MCKHANN, R. W. ALBERS, L. SOKOLOFF, O. MICKELSEN AND D. B. TOWER, in E. ROBERTS, *Inhibition in the Nervous System and γ-Aminobutyric Acid*, Pergamon Press, New York, 1960, p. 169.
[114] H. MCLENNAN, *J. Physiol. (London)*, 139 (1957) 79.
[115] H. MCLENNAN, *Nature*, 181 (1958) 1807.

[116] H. McLennan, *J. Physiol. (London)*, 146 (1959) 358.
[117] H. McLennan, *J. Physiol. (London)*, 151 (1960) 31.
[118] N. Miani, *Experientia*, 14 (1958) 34.
[119] N. Miani and G. Bucciante, *Experientia*, 14 (1958) 10.
[120] C. Mitoma, in E. Roberts, *Inhibition in the Nervous System and γ-Aminobutyric Acid*, Pergamon Press, New York, 1960, p. 236.
[121] P. Monaco, V. Castro and U. Quatrini, *Boll. soc. ital. biol. sper.*, 32 (1956) 1606.
[122] K. A. Montagu, *Nature*, 180 (1957) 244.
[123] E. Müller, *Z. Biol.*, 100 (1940) 249.
[124] S. Nojima and N. Utsugi, *J. Biochem. (Tokyo)*, 44 (1957) 565.
[125] K. Ohara, I. Sano, H. Koizumi and K. Nishinuma, *Science*, 129 (1959) 1225.
[126] N. Okumura, S. Otsuki and T. Aoyama, *J. Biochem. (Tokyo)*, 46 (1959) 207.
[127] N. Okumura, S. Otsuki and A. Kameyama, *J. Biochem. (Tokyo)*, 47 (1960) 315.
[128] J. Olley, *Federation Proc.*, 16 (1957) 845.
[129] M. K. Paasonen and M. Vogt, *J. Physiol. (London)*, 131 (1956) 617.
[130] B. Pernow, *Acta Physiol. Scandinav.*, 29, Suppl. 105 (1953) 1.
[131] B. Pernow, *Z. Vitamin-, Hormon- u. Fermentforsch.*, 7 (1955) 59.
[132] B. Pernow, in J. H. Gaddum, *Polypeptides which Stimulate Plain Muscle*, Livingstone, Edinburgh, 1955, p. 28.
[133] M. N. Pertseva, *Bull. Exptl. Biol. Med. (U.S.S.R.) (English Translation)*, 46 (1958) 827.
[134] J. J. Pisano, J. D. Wilson, L. Cohen, D. Abraham and S. Udenfriend, *J. Biol. Chem.*, 236 (1961) 499.
[135] G. Porcellati, *J. Neurochem.*, 2 (1958) 128.
[136] G. Porcellati and R. H. S. Thompson, *J. Neurochem.*, 1 (1957) 340.
[137] S. A. P. Price and G. B. West, *Nature*, 185 (1960) 470.
[138] D. Richter and R. M. C. Dawson, *J. Biol. Chem.*, 176 (1948) 1199.
[139] G. Rindi and G. Ferrari, *Nature*, 183 (1959) 608.
[140] E. Roberts, in S. R. Korey and J. I. Nurnberger, *Progress in Neurobiology: I. Neurochemistry*, Hoeber, New York, 1956, p. 11.
[141] E. Roberts, in *Inhibition in the Nervous System and γ-Aminobutyric Acid*, Pergamon Press, New York, 1960.
[142] E. Roberts and S. Frankel, *J. Biol. Chem.*, 187 (1950) 55.
[143] E. Roberts, S. Frankel and P. J. Harman, *Proc. Soc. Exptl. Biol. Med.*, 74 (1950) 383.
[144] E. Roberts, P. J. Harman and S. Frankel, *Proc. Soc. Exptl. Biol. Med.*, 78 (1951) 799.
[145] E. Roberts, I. P. Lowe, L. Guth and B. Jelinek, *J. Exptl. Zool.*, 138 (1958) 313.
[146] A. Rosenberg and E. Chargaff, *J. Biol. Chem.*, 232 (1958) 1031.
[147] I. Sano, T. Gamo, Y. Kakimoto, K. Taniguchi, M. Takesada and K. Nishinuma, *Biochim. Biophys. Acta*, 32 (1959) 586.
[148] E. F. Schroeder and G. E. Woodward, *J. Biol. Chem.*, 129 (1939) 283.
[149] P. E. Schurr, H. T. Thompson, L. M. Henderson and C. A. Elvehjem, *J. Biol. Chem.*, 182 (1950) 29.
[150] P. E. Schurr, H. T. Thompson, L. M. Henderson, J. N. Williams, Jr., and C. A. Elvehjem, *J. Biol. Chem.*, 182 (1950) 39.
[151] P. Schwerin, S. P. Bessman and H. Waelsch, *J. Biol. Chem.*, 184 (1950) 37.
[152] N. Serafimov and P. Stern, *Arch. intern. physiol.*, 66 (1958) 653.
[153] B. Sisken, K. Sano and E. Roberts, *J. Biol. Chem.*, 236 (1961) 503.
[154] L. C. Smith and S. R. Nelson, *Proc. Soc. Exptl. Biol. Med.*, 94 (1957) 644.
[155] L. C. Smith and F. M. Rossi, *Proc. Soc. Exptl. Biol. Med.*, 99 (1958) 754.
[156] S. S. Spicer and V. Weise, *Enzymologia*, 17 (1955) 263.
[157] W. E. Stone, *J. Biol. Chem.*, 149 (1943) 29.
[158] H. J. Strecker in D. Richter, *Metabolism of the Nervous System*, Pergamon Press, London, 1957, p. 459.
[159] H. H. Tallan, *J. Biol. Chem.*, 224 (1957) 41.
[160] H. H. Tallan, S. Moore and W. H. Stein, *J. Biol. Chem.*, 211 (1954) 927.
[161] H. H. Tallan, S. Moore and W. H. Stein, *J. Biol. Chem.*, 219 (1956) 257.
[162] H. H. Tallan, S. Moore and W. H. Stein, *J. Biol. Chem.*, 230 (1958) 707.
[163] O. Télupilová-Krestýnová and F. Šantavý, *Arch. ges. Physiol., Pflüger's*, 266 (1958) 473.
[164] W. Thorn, H. Scholl, G. Pfleiderer and B. Mueldener, *J. Neurochem.*, 2 (1958) 150.
[165] D. B. Tower, in *Annual Conference Protein Metabolism*, Rutgers University, Bureau Biological Research, 1957, p. 71.
[166] D. B. Tower, *Clin. Chim. Acta*, 2 (1957) 397.
[167] Y. Tsukada, G. Takagaki, S. Sugimoto and S. Hirano, *J. Neurochem.*, 2 (1958) 295.
[168] T. Turský, *Nature*, 187 (1960) 322.
[169] S. Udenfriend, *J. Biol. Chem.*, 187 (1950) 65.

[170] L. L. UZMAN, *Arch. Biochem. Biophys.*, 76 (1958) 474.
[171] H. WAELSCH, *Advances in Protein Chem.*, 6 (1951) 299.
[172] H. WAELSCH in K. A. C. ELLIOTT, I. H. PAGE AND J. H. QUASTEL, *Neurochemistry*, Thomas, Springfield, Ill., 1955 ,p. 173.
[173] D. M. WALKER, *Biochem. J.*, 52 (1952) 679.
[174] H. WEIL-MALHERBE, *Naturwissenschaften*, 40 (1953) 545.
[175] H. WEIL-MALHERBE, *Naturwissenschaften*, 41 (1954) 124.
[176] H. WEIL-MALHERBE, *Biochem. Soc. Symposia (Cambridge, Engl.)*, 8 (1952) 16.
[177] J. WESTLEY, J. J. WREN AND H. K. MITCHELL, *J. Biol. Chem.*, 229 (1957) 131.
[178] J. N. WILLIAMS, Jr., P. E. SCHURR AND C. A. ELVEHJEM, *J. Biol. Chem.*, 182 (1950) 55.
[179] W. C. WILSON, *Scot. Med. J.*, 2 (1957) 351.
[180] D. M. WOODBURY, A. KOCH AND A. VERNADAKIS, *Neurology*, 8, Suppl. 1 (1958) 113.
[181] N. A. YUDAEV, *Doklady Akad. Nauk S. S. S. R.*, 82 (1952) 447; Chem. Abstr., 49 (1955) 10474.
[182] G. ZETLER, *Arch. exptl. Pathol. Pharmakol., Naunyn-Schmiedeberg's*, 228 (1956) 513.
[183] G. ZETLER AND G. OHNESORGE, *Arch. exptl. Pathol. Pharmakol., Naunyn-Schmiedeberg's*, 231 (1957) 199.
[184] G. ZETLER AND L. SCHLOSSER, *Arch. exptl. Pathol. Pharmakol., Naunyn-Schmiedeberg's*, 224 (1955) 159.

INVITED DISCUSSION

FREE AMINO ACIDS IN BRAIN AFTER TREATMENT WITH PSYCHOTROPIC DRUGS

E. MUSSINI AND F. MARCUCCI

Department of Pharmacology, Medical School, University of Milano, Milano (Italy)

Only little is known about the distribution and function of free amino acids in brain, although the metabolism of a few has been extensively investigated in these last years[1, 2].
The lack of suitable analytical methods, at the same time reproducible, simple and sensitive, is one of the main handicaps in this field of research. This explains also why the data available on the relationship between central effects of drugs and changes in the composition of free amino acids in brain are so scarce.

TABLE I

RECOVERY OF STANDARDS

Amino acid	*Amount present* (μg)	*Amount found* (μg)	*Recovery* (%)
Cysteic acid	182	164	90
Taurine	180	161.8	90
Aspartic acid	190	203	106
Serine	150	130	87
Glutamic acid	186	163	88
α-Alanine	122	122	100
α-Amino-*n*-butyric acid	167	193	115
Valine	161	165	103
Cystine	150	142	95
Methionine	187	166	89
Isoleucine	162	149	92
Leucine	162	170	105
β-Aminoisobutyric acid	167	140	83
Tyrosine	212	199	94
Phenylalanine	206	192	93
γ-Aminobutyric acid	167	156	93
Ornithine	225	182	80
Lysine	212	173	81
Histidine	225	173.8	77

One of our first problems was the choice of a quantitative method. After a careful suvery it was decided to use the technique of column chromatography with ion-exchange resins. However, an additional problem was the fact that there are no methods able to extract free amino acids without also extracting peptides of small molecular weight. Since these peptides are eluted from the column in correspondence with some amino acids, and since their sensitivity to ninhydrin staining is not predictable[3], a change in peptide concentration might alter the values for the amino acids. In an attempt to overcome this hindrance two chromatographic separations of the amino acids were performed before and after a prolonged acid hydrolysis. On the other hand, hydrolysis may modify or destroy some amino acids. Therefore, a decrease of an amino acid after hydrolysis is considered as a chemical alteration,

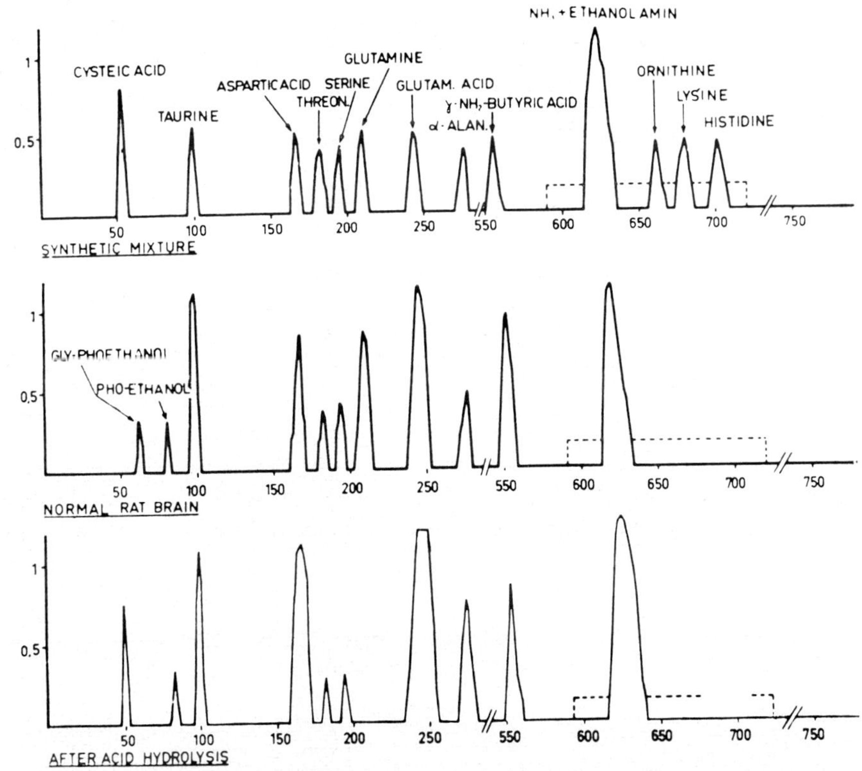

Fig. 1. Spectrum of the amino acids in rat brain.

while an increase is viewed as evidence of the presence of conjugated amino acids in the extract.

The method used is briefly described as follows: the brains of four Sprague-Dawley rats were pooled and homogenized with five times their weight of ice-cold distilled water in a Potter-Elvehjem homogenizer. The extraction of the amino acids was performed according to Hamilton and Van Slyke[3, 4] using picric acid. All details concerning the preparation of samples for chromatography, the preparation of the resins (Dowex-2-X-8 and Dowex-50-X-4) and the fractionation were similar to those described by Moore, Stein and Tallan *et al.*[3, 5, 6]. Total free amino acids before chromatography and the single fractions after chromatography were analyzed according to Moore and Stein[7] using the ninhydrin reaction. Hydrolysis was carried out with 6 N HCl at 100° for 24 h.

Table I shows that the recovery of a mixture of known amino acids was quanti-

References p. 492

tative with an average error of approx. 3%. The reproducibility was satisfactory in duplicate experiments. Table II shows that, after separation, the sum of the amino acids gave an average of 87% present in the extract added to the column. The spectrum of the amino acids present in rat brain is reported in Fig. 1 and includes cysteic acid, glycerophosphoethanolamine, taurine, aspartic acid, threonine, serine,

TABLE II

RECOVERY OF FREE AMINO ACIDS PRESENT IN POOLED RAT BRAINS

Sample No.	*Present before chromatography (mg/100 g)*	*Found after chromatography (mg/100 g)*	*Recovery (%)*
1	3.10	2.70	87.0
2	3.00	2.63	87.6
3	3.00	2.80	93.3
4	3.00	2.44	81.3
5	3.10	2.88	92.9
6	3.00	2.41	80.3
Average	3.03	2.64	87.1

asparagine, glutamine, glutamic acid, alanine and γ-aminobutyric acid (GABA).

Table III shows that the composition of the amino acids in rat brain changes according to the age of the rats. Therefore, when treatment was carried out, it was found necessary to use rats of the same age. The amino acids shown in Table IV will not be reported because they are present in concentrations lower than 6.0 mg/100 g of fresh tissue. At this concentration the method reaches the limit of sensitivity. Glutamic acid, alanine and aspartic acid were also present in a conjugated form.

TABLE III

FREE AMINO ACIDS OF RAT BRAIN ACCORDING TO THE AGE OF THE RATS

Values are expressed in mg/100 g.

Amino acid	*125 g*		*200 g*		*350 g*	
	Extract	*Hydrolyzed extract*	*Extract*	*Hydrolyzed extract*	*Extract*	*Hydrolyzed extract*
Cysteic acid	Trace	15.8	Trace	15.0	Trace	15.0
Taurine	29.8	32.2	52.9	37.2	39.2	25.7
Aspartic acid	28.9	68.4	40.1	67.7	38.0	63.6
Threonine	6.1	4.7	8.6	Trace	9.0	4.4
Serine	9.5	10.0	8.5	5.8	13.0	8.8
Asparagine, Glutamine	20.3	—	49.8	—	40.6	—
Glutamic acid	76.3	110.7	70.2	149.4	107.7	171.7
Alanine	11.3	16.3	8.0	30.7	16.3	34.6
GABA	26.0	32.0	33.6	33.2	30.0	28.3

The effect of a number of drugs on the free amino acids was evaluated. It was decided to use monoamineoxidase (MAO) inhibitors and reserpine for the following reasons:

TABLE IV

FREE AMINO ACIDS PRESENT IN RAT BRAIN AT CONCENTRATIONS OF <6.0 mg/100 g

Cysteic acid
Glycerophosphoethanolamine
Phosphoethanolamine
Methionine
Leucine
Isoleucine

(a) These drugs are known to be capable of increasing (MAO inhibitors)[8] or releasing (reserpine)[9] a number of amines in brain. This alteration of the amines may also induce a change in the amino acid composition.

(b) MAO inhibitors and reserpine respectively are active in preventing[10] or enhancing convulsions[11]. On the other hand, some types of convulsion seem to be related to changes of glutamic acid, glutamine and GABA[12, 13].

The effect of electroshock was also investigated in connection with these data.

I 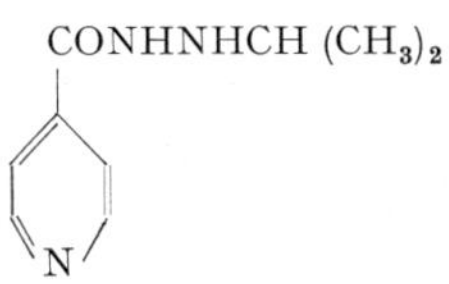

Iproniazid

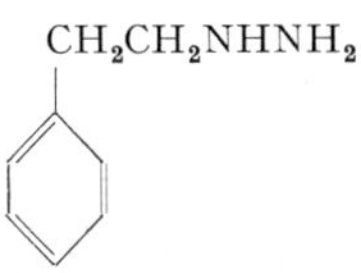 II

Phenelzine (W 1544)

III 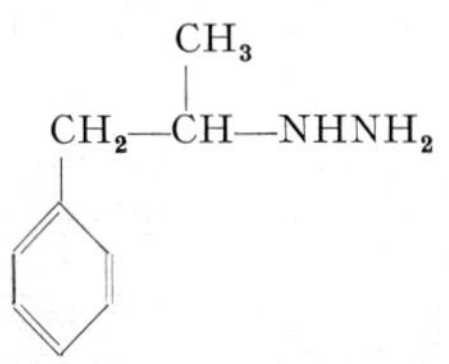

JB 516

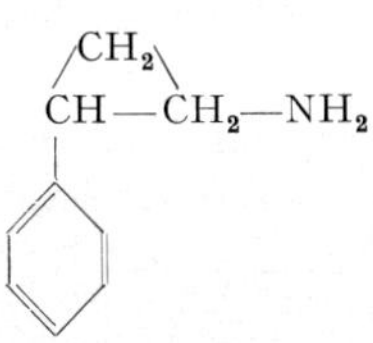 IV

Tranylcypromine SKF 385

MAO inhibitors were given intraperitoneally for 6 days at the doses reported in the tables. Table V shows that the conditions chosen were adequate for inducing a change of the brain amines in the expected way. MAO inhibitors such as reserpine, do not alter the concentration of the total amino acids before or after hydrolysis (see Table VI). The spectrum of amino acids in brain after MAO-inhibitor treatments is reported in Table VII. Although a certain MAO inhibitor may alter a given amino acid, a common pattern of effects was not observed. The hydrazine type of inhibitors, but not phenylcyclopropylamine, reduce the peak of free taurine. Conjugated aspartic acid was increased by all treatments although only phenelzine reduced the free aspartic acid. Threonine was usually reduced as was serine in the case of phenelzine treatment. Glutamic acid was not affected but the conjugated

References p. 492

TABLE V

EFFECT OF DRUGS ON AMINES IN BRAIN

Treatment	mg/kg × 6 days intraperitoneally	Brain	
		Serotonin (μg/g)	Noradrenaline (μg/g)
Controls		0.39	0.15
Phenelzine (W 1544) (II)	5	0.95	0.35
Tranylcypromine (SKF 385) (IV)	5	1.00	0.29
Iproniazid (I)	100	1.23	0.33
JB 516 (III)	5	1.24	0.29
Reserpine (single dose)	2.5	0.10	0.05

TABLE VI

EFFECT OF DRUGS ON FREE AMINO ACIDS BEFORE AND AFTER HYDROLYSIS

Treatment	mg/kg × 6 days intraperitoneally	Free amino acids in brain after picric acid extraction	
		Extract (mg/100 g)	Hydrolyzed acid extract (mg/100 g)
Controls		192	396
W 1544 (II)	5	189	421
SKF 385 (IV)	5	185	430
Iproniazid (I)	100	192	398
JB 516 (III)	5	200	400
Reserpine (single dose)	2.5	198	410

form was increased only by phenelzine. Alanine and GABA were practically unaffected. The results obtained with reserpine are summarized in Table VIII. There are practically no changes in the pattern of the amino acids.

Finally the effect of electroshock (electrodes applied to the ears — 110 V; 0.2 sec) is reported in Table IX. There are no changes in the concentration of the total amino acid before or after hydrolysis when the animals were killed 10 min, 6 h, 24 h or

TABLE VII

FREE AMINO ACIDS IN RAT BRAIN AFTER MAO-INHIBITOR TREATMENT

Values are expressed in mg/100 g. *E* means extract, *H* stands for hydrolyzed extract.

Amino acid	Controls		SKF 385		JB 516		W 1544		Iproniazid	
	E	H	E	H	E	H	E	H	E	H
Taurine	39.2	25.7	42.3	21.4	25.0	19.2	20.6	23.0	22.3	37.5
Aspartic acid	38.0	63.6	30.0	92.0	33.0	102.0	12.2	93.0	36.2	110.0
Threonine	9.0	4.4	5.4	4.5	6.5	6.2	2.0	6.7	5.7	7.3
Serine	13.0	8.8	11.1	8.3	12.5	13.7	4.0	8.8	9.8	13.2
Asparagine, Glutamine	40.6	—	39.3	—	53.0	—	23.0	—	30.0	—
Glutamic acid	107.7	171.7	113.0	192.0	109.0	194.0	105.9	241.0	109.4	143.0
Alanine	16.3	34.6	15.7	32.6	20.3	21.9	22.0	29.3	16.8	40.2
GABA	30.0	28.3	22.4	30.6	27.0	27.1	21.0	34.0	28.2	29.5

TABLE VIII

EFFECT OF RESERPINE ON BRAIN AMINO ACIDS

Reserpine: 2.5 mg/kg.

Amino acid	*Controls*		*Reserpine*	
	Extract	*Hydrolyzed extract*	*Extract*	*Hydrolyzed extract*
Taurine	39.2	25.7	42	35
Aspartic acid	38	63.6	26	68.5
Threonine	9	4.4	7.8	4.8
Serine	13	8.8	10	7.2
Asparagine, Glutamine	40.6	—	34.5	—
Glutamic acid	107.7	171.7	111.9	140.5
Alanine	16.3	34.6	15.1	32.2
GABA	30	28.3	26.3	29.5

48 h after electroshock. The pattern of amino acids is reported in Table X and indicates an increase in alanine and a decrease of threonine and serine.

In conclusion this investigation shows that it is possible to study quantitatively the distribution of amino acids in rat brain. Evidence for the presence of conjugated amino acids was also obtained. As an example the effect of MAO inhibitors, reserpine and electroshock were studied.

MAO inhibitors do not induce definite and constant changes in brain amino acids

TABLE IX

EFFECT OF ELECTROSHOCK ON TOTAL FREE AMINO ACIDS IN BRAIN

Treatment	*Time before killing*	*Free amino acids in rat brain after picric acid extraction*	
		Extract (mg/100 g)	*Hydrolyzed acid extract (mg/100 g)*
Controls	—	192	396
Electroshock	10 min	187	396
Electroshock	6 h	188	348
Electroshock	24 h	194	388
Electroshock	48 h	190	378

as they do on some brain amines. Conjugated aspartic acid was constantly increased, but other amino acids were affected only by some MAO inhibitors, and not by others. The observed differences could be related to some peculiar pharmacological effects not shared by all the drugs having in common the activity on MAO. Reserpine was without effect when given to rats in a single dose. During this investigation the claimed changes induced by MAO inhibitors (increase) and reserpine (decrease) on brain GABA[14] were not confirmed. Electroshock increased alanine concentration and decreased threonine and serine. The significance of this observation is still to be elucidated.

References p. 492

TABLE X

EFFECT OF ELECTROSHOCK ON FREE AMINO ACIDS IN RAT BRAIN

Values are expressed in mg/100 g. *E* means extract, *H* stands for hydrolyzed extract.

Amino acid	*Controls*		*Electroshock after*							
			10 min		*6 h*		*24 h*		*48 h*	
	E	*H*	*E*	*H*	*E*	*H*	*E*	*H*	*E*	*H*
Taurine	29.8	39.8	26.0	32.8	30.6	36.6	29.0	41.8	30.7	40.4
Aspartic acid	28.9	68.4	25.5	76.5	24.1	65.6	25.4	74.6	28.1	59.0
Threonine	6.1	4.7	Trace	Trace	Trace	Trace	Trace	Trace	Trace	Trace
Serine	9.5	10.0	Trace	Trace	Trace	Trace	Trace	Trace	Trace	Trace
Asparagine, Glutamine	20.3	9.6	25.1	5.2	24.2	4.5	22.2	6.4	20.2	6.3
Glutamic acid	76.3	110.7	71.4	102	68.6	128	72.3	95.8	82.2	98.5
Alanine	11.3	16.3	18.6	21.7	18.0	23.4	17.0	7.7	11.4	9.6
GABA	26.0	32.0	24.4	22.9	27.1	33.8	28.1	37.7	28.1	35.5

The results obtained may be considered only preliminary and further investigations are needed, particularly to study the effect of drugs on the free amino acid composition in selected parts of the brain.

REFERENCES

1 S. Berl and H. Waelsch, *J. Neurochem.*, 3 (1958) 161.
2 A. Lajtha, S. Berl and H. Waelsch, *J. Neurochem.*, 3 (1958) 322.
3 H. H. Tallan, S. Moore and W. H. Stein, *J. Biol. Chem.*, 211 (1954) 927.
4 P. B. Hamilton, *J. Biol. Chem.*, 158 (1945) 375.
5 S. Moore and W. H. Stein, *J. Biol. Chem.*, 211 (1954) 893.
6 R. S. de Ropp and E. H. Snedeker, *Anal. Biochem.*, 1 (1960) 424.
7 S. Moore and W. H. Stein, *J. Biol. Chem.*, 211 (1954) 907.
8 B. B. Brodie, S. Spector and P. A. Shore, *Ann. N.Y. Acad. Sci.*, 80 (1959) 609.
9 B. B. Brodie, A. Pletscher and P. A. Shore, *J. Pharmacol. Exptl. Therap.*, 116 (1956) 9.
10 D. J. Prockop, P. A. Shore and B. B. Brodie, *Ann. N.Y. Acad. Sci.*, 80 (1959) 643.
11 G. Chen, C. R. Ensor and B. Bochner, *Proc. Soc. Exptl. Biol. Med.*, 86 (1954) 507.
12 S. Berl, D. P. Purpura, M. Girardo and H. Waelsch, *J. Neurochem.* ,4 (1959) 311.
13 D. Purpura and H. Grundfest, *J. Neurophysiol.*, 20 (1957) 494.
14 H. Balzer, P. Holtz and D. Palm, *Experientia*, 17 (1961) 38.

INVITED DISCUSSION

THE EFFECT OF PSYCHOTROPIC DRUGS AND CHEMICALLY RELATED SUBSTANCES ON γ-AMINOBUTYRIC ACID AND GLUTAMIC ACID IN BRAIN TISSUE*

M. J. E. ERNSTING**, W. F. KAFOE**, W. TH. NAUTA***, H. K. OOSTERHUIS§ AND P. A. ROUKEMA**

Vrije Universiteit, Amsterdam and Research Department Brocades, Amsterdam (The Netherlands)

INTRODUCTION

Since the discovery of γ-aminobutyric acid (GABA) by ROBERTS AND FRANKEL[1] its function and metabolism with respect to the central nervous system has been studied at numerous laboratories. Several of these investigations have shown that GABA may be of value in controlling the neuron activity[2], that it can serve as a substrate in oxidative metabolism[3] and that dysfunction of the GABA system occurs in the case of pyridoxine deficiency[4, 5].

The GABA level in the brain is not affected by high doses administered parenterally; apparently the blood–brain barrier is relatively impermeable to this compound.

Under certain conditions, however, the GABA level may be changed *in vivo*. Thus a decrease was established, for instance, in the case of hypoglycaemia[6], after administration of convulsive hydrazides[4], and, in mice, under the influence of reserpine[5]. An increase was found by HÄKKINEN[7] with alcohol, by WOODBURY AND VERNADAKIS[8] with acetazolamide and by ROBERTS[9] with hydroxylamine.

In vitro investigations by TOWER[10] showed that both the GABA and the L-glutamic acid (Glu) level decreased if sliced rat brain cortex was incubated with glucose and 2-deoxyglucose as a substrate, and under 10% oxygen; the same effect was found with brain slices of epileptics. Malonate increased the amino acid content.

The influence of varying conditions on level and metabolism of GABA and Glu in brain tissue was investigated by ELLIOTT AND VAN GELDER[11], TAKAGAKI *et al.*[12], STERN *et al.*[13] and others.

Little is known, however, about the effect of psychotropic drugs on oxidative metabolism or on the level of the amino acids mentioned.

EXPERIMENTAL PART

We investigated the *in vitro* effect of a large number of psychotropic drugs, on oxygen consumption and on the GABA and Glu levels in the tissue and the incubation medium[14, 15, 16].

As the Parkinson drug orphenadrine (Disipal) finds extensive clinical application

* Presented orally by Dr. Nauta. This is the second part of Biochemical Studies on Psychotropic Drugs. For part I see *J. Neurochem.*, 5 (1960) 121.

** Research Department Brocades, Amsterdam.
*** Chemical Laboratory Vrije Universiteit, Amsterdam.
§ Laboratory for Chemical Physiology Vrije Universiteit, Amsterdam.

References p. 498

as an adjuvant in therapy with neuroleptics such as reserpine and chlorpromazine to counteract the extrapyramidal symptoms, and in the treatment of exogenous, (reactive) depressions, this substance and related compounds were included in the investigation.

The methods applied have already been described[17, 18]. The results have been summarised in Tables I, II and III, which also state the testing conditions.

TABLE I

THE INFLUENCE OF SOME PSYCHOTROPIC DRUGS ON THE METABOLISM OF RAT BRAIN CORTEX *in vitro*

*Conditions**	*Time (min)*	*% O_2 inhibition*	*GABA level (μg/100 mg)*			*Glu level (μg/100 mg)*		
			Medium	*Tissue*	*Total*	*Medium*	*Tissue*	*Total*
Control	0		11.4	14.7	26.1	83.2	70.6	153.8
	60	—	2.8	17.0	19.8	21.7	91.5	113.2
Phenobarbital	60	34	2.5	16.3	18.8	21.1	90.4	111.5
Azetazolamide	60	4	2.8	13.9	16.7	21.5	84.2	105.7
Orphenadrine	60	60	30.4	6.1	36.5	169.4	30.5	199.9
Chlorpromazine	60	90	34.7	6.1	40.8	197.9	24.6	222.5

* The concentration of the drugs was 2 m*M*. Glucose was used as a substrate (concentration: 0.02 *M*). Metabolism took place under 100% O_2, with a phosphate buffer containing 1 m*M* Ca. After dissection of the animal we found in the cortex 26.4 μg/100 mg for GABA and 135.0 μg/100 mg for Glu. (ERNSTING *et al.*[17] and ROUKEMA[18].)

Before metabolism is started part of the GABA and Glu leaks from the tissue to the medium. However, after 1 h metabolism with glucose as a substrate, the amino acids have been resorbed. Metabolism under air, without Ca in the buffer (Table III), produces an overall GABA and Glu content of medium plus tissue which approxi-

TABLE II

THE INFLUENCE OF PARKINSON DRUGS ON THE METABOLISM OF RAT BRAIN CORTEX *in vitro*

*Conditions**	*Time (min)*	*% O_2 inhibition*	*GABA level (μg/100 mg)*			*Glu level (μg/100 mg)*		
			Medium	*Tissue*	*Total*	*Medium*	*Tissue*	*Total*
Control	0		11.4	14.7	26.1	83.2	70.6	153.8
	60	—	2.8	17.0	19.8	21.7	91.5	113.2
Trihexyphenidyl (Artane R)	60	5	16.5	10.3	26.8	75.6	59.1	134.7
Procycline (Kemadrin R)	60	7	16.4	8.8	25.2	92.1	59.1	151.2
Cycrimine (Pagitane R)	60	6	15.2	13.8	29.0	59.0	79.7	138.7
Benzatropine methane sulfonate (Cogentin R)	60	88	26.3	3.0	29.3	149.4	20.9	170.3
Orphenadrine (Disipal R)	60	60	30.4	6.1	36.5	169.4	30.5	199.9

* For conditions see Table I.

References p. 498

mately equals the value found after dissection of the cortex. After metabolism under O_2, with 1 m*M* Ca in the buffer (Tables I and II), this value is lower.

After metabolism under 100% N_2 the amino acids are not resorbed while the GABA and Glu content of the medium far exceeds the value found after dissection of the control animals, the tissue being entirely depleted.

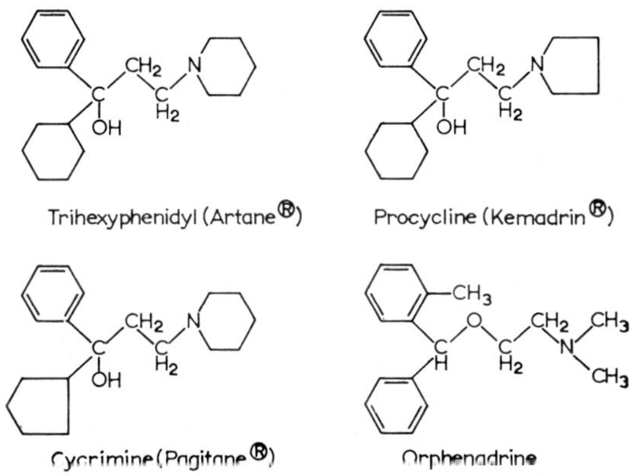

Fig. 1

The same picture is obtained if orphenadrine and chlorpromazin, are added at concentrations of 2 m*M*.

In both cases the depletion of the tissue is attended with inhibited oxygen consumption (60 and 94%, respectively).

However, after phenobarbital, at the same concentration, GABA and Glu are

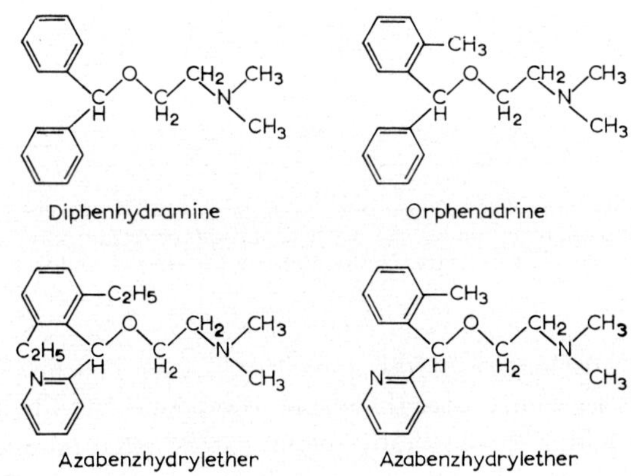

Fig. 2

resorbed while the oxygen consumption is inhibited by 34%. The same effect is observed after Diamox R, which does not affect the oxygen consumption. This led us to assume that a correlation existed between the highly inhibited oxygen consumption and the fact that the amino acids are not resorbed.

However, comparison of orphenadrine with a number of clinically useful Parkinson drugs refuted this assumption. The results have been summarised in Table II: Artane R, Kemadrin R and Pagitane R, which have closely related structures (see Fig. 1), hardly influence the oxygen consumption, while giving little, if any, resorption of GABA and Glu. Cogentin R and orphenadrine likewise prevent resorption of the amino acids, but in this case the oxygen consumption is inhibited by 60 and 88%, respectively.

It is further clear from Table III and Fig. 2 that small structural variations result in highly different effects.

Of the azabenzhydrylethers given in the table, the unsubstituted compound

TABLE III

THE INFLUENCE OF SOME AZABENZHYDRYLETHERS ON THE METABOLISM OF RAT BRAIN CORTEX *in vitro*

*Conditions**		*Time (min)*	*% O_2 inhibition*	*GABA level (μg/100 mg)*			*Glu level (μg/100 mg)*		
				Medium	*Tissue*	*Total*	*Medium*	*Tissue*	*Total*
		0		10.6	12.5	23.1	114.1	58.2	172.3
Control		60	—	1.3	20.6	21.9	23.7	137.2	160.9
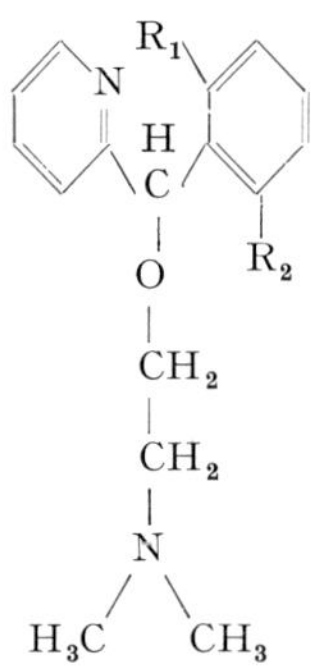	$R_1 = R_2 = H$	60	21	1.3	20.4	21.7	31.2	158.4	189.6
	$R_1 = H$, $R_2 = CH_3$	60	15	1.2	20.8	22.0	40.5	142.0	182.5
	$R_1 = R_2 = C_2H_5$	60	16	22.0	4.1	26.1	285.8	34.4	320.2

* The concentration of the drugs was 2 m*M*. Glucose was used as a substrate (concentration: 0.02 *M*). Metabolism took place under air, with a phosphate buffer which did not contain Ca. After dissection we found in the cortex 22.0 μg/100 mg for GABA and for Glu 177.4 μg/100 mg. (ROUKEMA[18].)

and the 2,6-diethyl derivative hardly affect the oxygen consumption, while GABA and Glu resorption are found after the former compound only. In the latter case the quantity of amino acid in the medium even far exceeds that found after dissection of control animals.

Like TOWER[19] we found that a highly reduced oxygen consumption is attended with GABA and Glu depletion of the tissue. However, in our case, the rise found for the incubation medium was much higher than was established by TOWER. This may be due to the fact that we used chopped brain instead of the slices used by this investigator. Undoubtedly, the disturbed energy supply plays an important part, as was found by STERN *et al.*[13] and others. However, this does not account for the rise in the GABA and Glu level.

References p. 498

An explanation might be that those enzymes are involved which determine the GABA and Glu levels. Orphenadrine was investigated for this purpose. No effect could be established on L-glutamic acid decarboxylase[18] and pyridoxal kinase; the effect on GABA–α-ketoglutaric acid transaminase is still being investigated. However, orphenadrine, at the test concentration (2 mM), inhibits the ATP-ase while it uncouples oxidative phosphorylation. This does not imply, however, that uncoupling oxidative phosphorylation invariably prevents the resorption of the amino acids. Thus, DNP (ref. 20), benactyzine, captodiamin and meprobamate[21], at concentrations which uncouple oxidative phosphorylation, do not interfere with GABA and Glu resorption. We do not know to what extent changes in permeability of the cell membrane affect the resorption. Orphenadrine is known to reduce the permeability of the blood–brain barrier for ^{24}Na in rats[22].

This was also established by LIEBALDT[23] with respect to the dye Evans Blue, during tests with dogs and rabbits. According to LIEBALDT this provides an explanation for the clinically established suppression of the reserpine-induced extrapyramidal effects. This substance appeared to increase the permeability for Evans Blue. A combination of reserpine and orphenadrine normalises the enhanced permeability.

During our investigations spectrophotometric determination of Evans Blue eluted from the brain tissue[24] showed that this dye is less well resorbed by the brain after administration of orphenadrine than in control animals; under the influence of orphenadrine the dye resorption decreased by approx. 20%.

In addition, preliminary results may be mentioned of some other recent investigations at our laboratories.

As reported by ROBERTS[25] administration of hydroxylamine to rats causes an increase in the metrazol seizure threshold while GABA levels in the brain rise concomitantly.

As we have been able to establish, rabbits also show such an increase in metrazol seizure threshold after administration of hydroxylamine (although in this case GABA levels remain unaffected).

Orphenadrine causes a transient lowering of the metrazol seizure threshold which returns to former levels in approx. 2 h time. However, when hydroxylamine is administered at that moment it fails to exert its expected effect on the seizure threshold. Although it would seem, therefore, that the gross GABA level is not determinative for the metrazol seizure threshold, orphenadrine evidently interacts with the system linking convulsion liability with GABA and hydroxylamine, perhaps by way of its influence on permeability, which may result in GABA being not available at its site of action.

Whether changes in permeability also play a part in the *in vitro* tests on resorption of amino acids must remain an open question for the time being.

We are trying to gain a better insight by measuring the resorption rate of the ^{24}Na added to the incubation medium. Further, it is uncertain whether the *in vitro* effect of psychotropic drugs on the amino acid levels is in any way related to the action mechanism of these drugs in the intact organism.

References p. 498

REFERENCES

[1] E. ROBERTS AND S. FRANKEL, *J. Biol. Chem.*, 187 (1950) 55.
[2] K. A. C. ELLIOTT, in *Biochemistry of the Central Nervous System*, 4th International Congress of Biochemistry, Vienna, Pergamon Press, New York, 1959, p. 251.
[3] G. M. MCKHANN AND D. B. TOWER, *Am. J. Physiol.*, 196 (1959) 36.
[4] K. F. KILLAM AND J. A. BAIN, *J. Pharmacol. Exptl. Therap.*, 119 (1957) 255.
[5] H. BALZER, P. HOLTZ AND D. PALM, *Arch. exptl. Pathol. Pharmakol. Naunyn-Schmiedeberg's*, 239 (1960) 520.
[6] H. G. KNAUFF AND F. BÖCK, *Klin. Wochschr.*, 38 (1960) 553.
[7] H. HÄKKINEN AND E. KULONEN, *Biochem. J.*, 78 (1961) 588.
[8] D. M. WOODBURY AND A. VERNADAKIS, *Federation Proc.*, 17 (1958) 420.
[9] E. ROBERTS, M. ROTHSTEIN AND C. F. BAXTER, *Proc. Soc. Exptl. Biol. Med.*, 97 (1958) 796.
[10] D. B. TOWER, *J. Neurochem.*, 3 (1958) 185.
[11] K. A. C. ELLIOTT AND N. M. VAN GELDER, *J. Neurochem.*, 3 (1958) 28.
[12] G. TAKAGAKI, S. HIRANO AND Y. NAGATA, *J. Neurochem.*, 4 (1959) 124.
[13] J. R. STERN, L. V. EGGLESTON, R. HEMS AND H. A. KREBS, *Biochem. J.*, 44 (1949) 410.
[14] W. TH. NAUTA, H. K. OOSTERHUIS, C. DE WAART AND M. J. E. ERNSTING, *Lancet*, ii (1958) 591.
[15] W. TH. NAUTA, in *Biochemistry of the Central Nervous System*, 4th International Congress of Biochemistry (discussion) ,Vienna, Pergamon Press, New York, 1959, p. 278.
[16] H. K. OOSTERHUIS, M. J. E. ERNSTING, W. F. KAFOE, W. TH. NAUTA AND C. DE WAART, *Acta Neurol. Psychiat. Belg.*, 61 (1961) 7.
[17] M. J. E. ERNSTING, W. F. KAFOE, W. TH. NAUTA, H. K. OOSTERHUIS AND C. DE WAART, *J. Neurochem.*, 5 (1960) 121.
[18] P. A. ROUKEMA, Thesis, Vrije Universiteit, Amsterdam, 1960.
[19] D. B. TOWER, in *Biochemistry of the Central Nervous System*, 4th International Congress of Biochemistry, Vienna, Pergamon Press, New York, 1959, p. 213.
[20] M. BERGER, *J. Neurochem.*, 2 (1957) 30.
[21] L. DECSI AND J. MÉHES, *Arch. intern. pharmacodynamie*, 119 (1959) 294.
[22] G. QUADBECK AND W. SCHMITT, *Arch. exptl. Pathol. Pharmakol., Naunyn-Schmiedeberg's*, 237 (1959) 94.
[23] G. LIEBALDT, *Med. Welt*, 36 (1960) 1876.
[24] D. F. CLAUSEN AND N. LIFSON, *Proc. Soc. Exptl. Biol. Med.*, 91 (1956) 11.
[25] E. ROBERTS, in S. S. KETY AND J. ELKES (Eds.), *Regional Neurochemistry, Proc. Intern. Neurochem. Symp.* 1960, Pergamon Press, New York, 1961.

INVITED DISCUSSION

EFFECT OF 4-METHOXYMETHYLPYRIDOXINE AND CARBONYL-TRAPPING AGENTS ON AMINO ACID CONTENT OF MAMMALIAN BRAIN AND OTHER TISSUES

CLAUDE F. BAXTER AND EUGENE ROBERTS

Department of Biochemistry, Medical Research Institute, City of Hope Medical Center, Duarte, Calif. (U.S.A.)

Evidence for the relative constancy of the levels of free or loosely bound amino acids in tissues of mature animals has been documented in a preceding paper at this Symposium[1]. Certain abnormal conditions, the administration of non-physiological doses of hormones and treatment with pharmacologically active agents, can elicit responses which may include a change in the level of one or more amino acids in specific tissues *in vivo*. These changes may be the result of metabolic changes affecting enzyme systems, permeability, transport mechanisms, etc. This paper is limited in

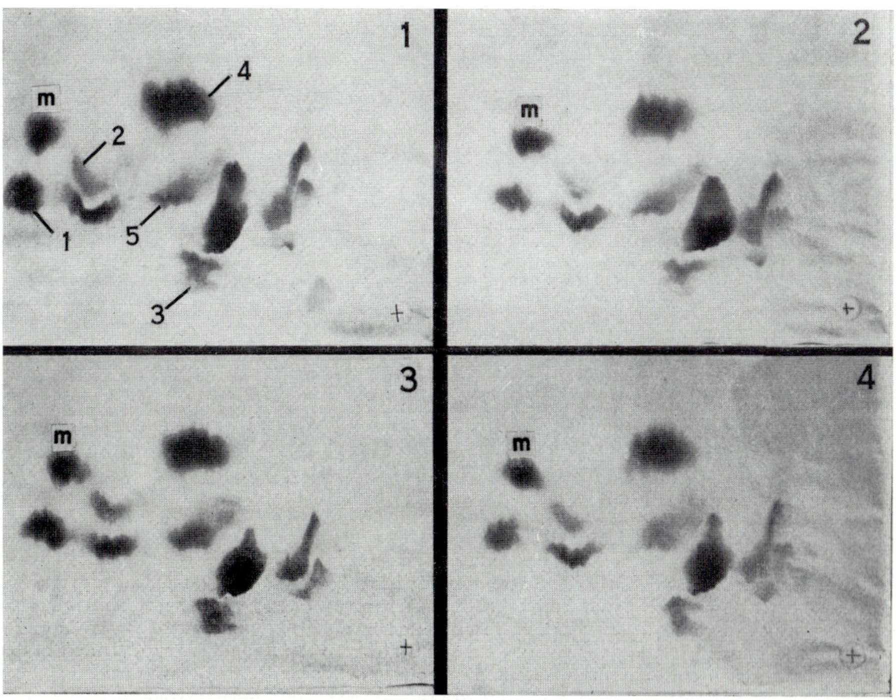

Fig. 1–4. Effect of 4-methoxymethylpyridoxine upon the concentration of amino acids of brain cortex in rats*. Injected dose: 250 mg/kg body wt. This was administered in two doses, 2.5 h apart. Animals were decapitated 3.5 h after the first injection. All odd numbered chromatograms represent tissues from normal control animals; all even numbered chromatograms illustrate changes in the tissues of treated animals.

* *For footnote see page 500.*

References p. 508

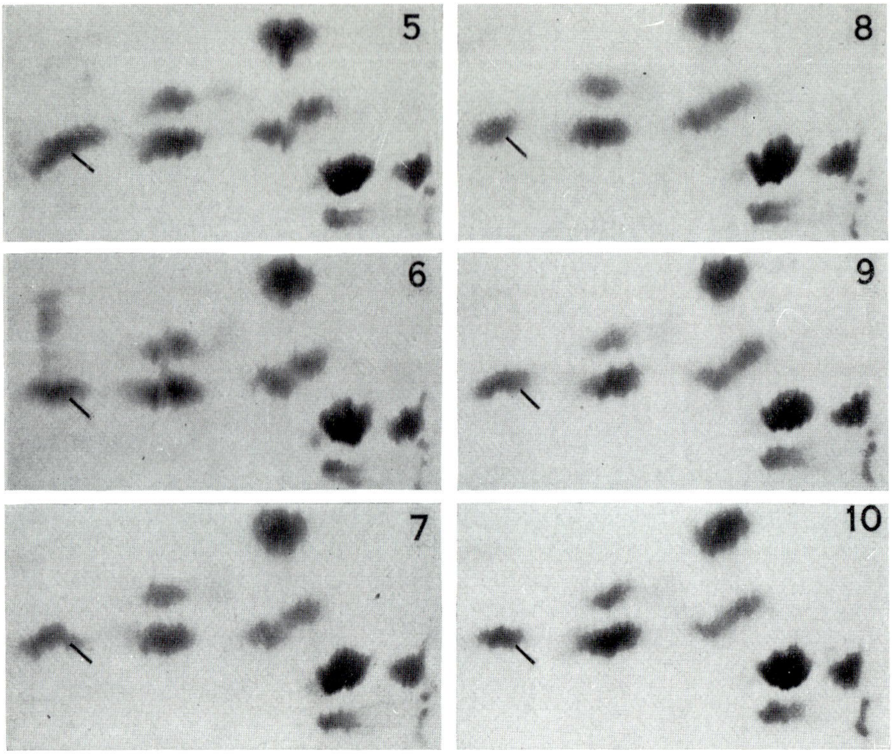

Figs. 5–10. Figs. 5,6: Effect of thiosemicarbazide upon the concentration of amino acids of brain cortex of rats. Injected dose: 20 mg/kg body wt. Figs. 7–10: Irreversibility of thiosemicarbazide effect with anticonvulsant doses of pyridoxine. Injected dose: 400 mg/kg body wt. Protocol of experiment: Fig. 5, control animals. Fig. 6; effect of thiosemicarbazide injected intraperitoneally. Some animals were decapitated 80 min after injection (Fig. 6); other animals were given anti-convulsant doses of pyridoxine at 80 min after thiosemicarbazide injection. These animals were decapitated 15, 45, 65, and 85 min after the vitamin B_6 administration (Fig. 7–10). This series shows that thiosemicarbazide depressed primarily levels of GABA (see arrow, Figs. 5, 6) and that subsequent administration of pyridoxine did not reverse this effect within 85 min. Each chromatogram represents the free or loosely bound amino acids of 25 mg of brain tissue (wet wt.)

Footnote to Figs 1–4.

* All chromatograms shown in this paper were developed at 26° by descending two-dimensional chromatography. Point of origin is at the lower right-hand corner of each chromatogram. The first solvent (horizontal, right to left) was water-saturated phenol for 18 h (a few drops ammonium were added to the chromatography boxes). The second solvent (vertical, bottom to top) was water-saturated lutidine for 24 h. These systems have been used and described previously[22]. Clear tissue extracts were prepared from ethanol (75%) homogenates by high speed centrifugation. Extracts were evaporated to dryness and resuspended in a suitable amount of water. No more than 25 μl was applied to the paper at one time and no more than 50 μl per spot. Unless stated to the contrary, each chromatogram represents the free or loosely bound amino acids of 30 mg of tissue (wet wt.). Amino acids were visualized by spraying with ninhydrin in methanol (3 g ninhydrin dissolved in 5 lb of methanol). Photographs were taken no earlier than 12 h and no later than 48 h after applying the spray reagent.

In Fig. 1 the numbered spots correspond to the following amino acids: (1) γ-aminobutyric acid; (2) alanine (and below it glutamine); (3) ethanolamine phosphate (and above it glutamic acid; the spot to the right of glutamic acid is aspartic acid); (4) taurine; (5) glycine (the upper right-hand portion of this spot is serine). The compound m, a marker is α-amino-n-butyric acid.

References p. 508

scope to a consideration of the effects of four potential vitamin B_6 antimetabolites upon amino acid levels and some related enzyme systems.

Almost every amino acid at some point of its metabolism undergoes transformations which are catalyzed by enzymes that require pyridoxal phosphate (PyP) as a cofactor. It is well known that *in vivo* some vitamin B_6-requiring enzymes are more sensitive to B_6 antimetabolites than others. The biochemical effects of 4-methoxymethylpyridoxine, administered to mice, have been studied[2]. The only significant change observed in brain tissue was a decrease in the concentration of γ-aminobutyric acid (GABA)[3]. In cats the injection of 4-methoxymethylpyridoxine affected amino acid levels in brain tissue by decreasing the level of GABA and increasing the level of glutamine[4]. 4-Methoxymethylpyridoxine affected the concentration of a number of amino acids in the brains of rats (Figs. 1–4). Levels of GABA, alanine and ethanolamine phosphate were decreased as the result of methoxypyridoxine administration. Glycine and taurine decreased to a lesser extent and glutamine, glutathione, serine, glutamic and aspartic acids appeared unchanged. It is possible that the levels of some other amino acids which were not detected in these chromatograms also were changed as a result of the vitamin B_6 antagonist[5].

Thiosemicarbazide, a carbonyl-trapping agent, is more limited in its effect. Early results with this compound were interpreted as showing that a specific inhibition of glutamic acid decarboxylase (GAD) in brain was produced[6]. Later results indicated that pyridoxal kinase, the enzyme which catalyzes the synthesis of PyP from pyri-

TABLE I

EFFECT OF PYRIDOXAL INJECTIONS UPON GABA LEVELS IN BRAINS OF THIOSEMICARBAZIDE-TREATED RATS*

*TSC Δ t*** (min)*	*Pyridoxal Δ t**** (h)*	*Cortex*	*Cerebellum*	*Colliculi*	*Diencephalon*
—	—	24§ (23–24)	27 (26–29)	50 (45–55)	61 (55–65)
95	—	17 (16–18)	20 (18–23)	31 (30–33)	43 (42–44)
85	1/4§§	20 (20–21)	25 (23–27)	39 (37–41)	50 (47–53)
85	1	16 (15–18)	23 (22–25)	40 (39–42)	40 (39–40)
85	2	13 (13–14)	19 (17–20)	33 (32–34)	35 (33–37)
85	3	17 (16–19)	22 (21–22)	37 (34–39)	38 (37–38)
85	5	18 (17–19)	23 (22–25)	43 (42–44)	38 (37–40)
85	7	20 (19–22)	25 (22–27)	47 (41–53)	50 (36–62)

* This experiment has been repeated twice (to 2 h after pyridoxal administration) with rats of different age and sex. In both trials each time period was represented by two animals. Results were in agreement with those shown for the trial recorded in Table I.

** Δ t, time period between the injection of TSC intraperitoneally (20 mg/kg, pH 8) and the time at which the animal was either decapitated or pyridoxal was injected.

*** Δ t, time period between the injection of pyridoxal HCl (50 mg/kg, pH 4.5) and the time at which the animal was decapitated.

§ The method for sampling and assay of brain areas for GABA has been described[12]. Above data are based upon one trial with two or three fed Sprague Dawley female rats (200 ± 15 g wt.) per point. Variability among animals in levels of GABA in corresponding brain areas is indicated.

§§ The administration of pyridoxal to thiosemicarbazide-treated rats appeared to produce a slight transient elevation of GABA levels at 15 min after intraperitoneal injection of the vitamin. The significance of this elevation, if any, remains to be evaluated.

References p. 508

doxal and adenosine triphosphate[7], was inhibited by thiosemicarbazide to a much greater extent than GAD[8, 9]. Nevertheless, when thiosemicarbazide was administered intraperitoneally to rats, the only significant effect on amino acids in brain was upon GABA levels (Figs. 5–10). The apparent specificity of thiosemicarbazide for GAD might be attributed to the loose attachment of PyP to the GAD apoenzyme. This would make GAD particularly susceptible to PyP depletion through

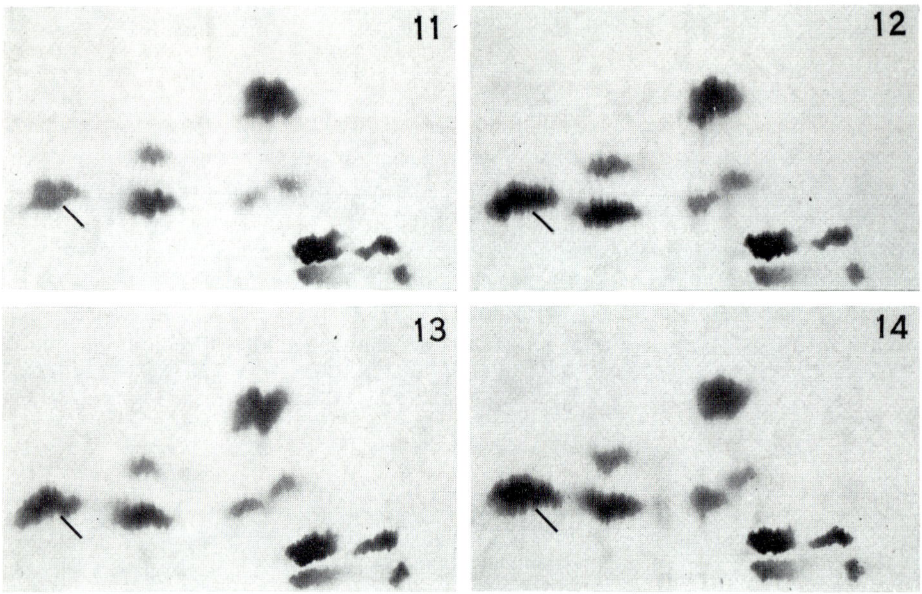

Figs. 11–14. Effect of hydroxylamine upon the concentration of amino acids in brain of rat. Injected dose $NH_2OH \cdot HCl$ (adjusted to pH 6): 75 mg/kg body wt. Animals were decapitated 90 min after injection. Chromatograms represent extracts of 25 mg equivalent wet weight of tissue. Figs. 11, 12: cortex. Figs. 13, 14: diencephalon. Arrows point to GABA. All odd numbered chromatograms represent tissues from normal control animals; all even numbered chromatograms illustrate changes in the tissues of treated animals.

inhibition of pyridoxal kinase. The depressed levels of cerebral GABA in thiosemicarbazide-treated rats could not be restored to normal levels by the injection of pyridoxal or pyridoxine (Figs. 5–10). Even 5 h after pyridoxal was administered, the levels of GABA in all areas of brain which were tested had not returned to normal (Table I). The results give added support to the belief that *in vivo* the primary inhibition by thiosemicarbazide was of pyridoxal kinase. Presumably the resultant vitamin B_6 depletion then decreased GAD activity and GABA levels. The effect of a number of hydrazides upon the levels of GABA, glutamic acid and aspartic acid in the brains of mice has been reported[10]. While thiosemicarbazide inhibited GAD in mouse brain, it did not affect the level of GABA. This contrasts with our results (Table I, Fig. 6) in rats where both GAD enzyme activity and GABA levels were affected by thiosemicarbazide. It was shown also in mice that the administration of vitamin B_6 could not rapidly restore levels of GABA which had been depressed by a prior injection of hydrazides other than thiosemicarbazide[10]. This finding is in

References p. 508

full agreement with our results using thiosemicarbazide in rats[11] (Table I, Figs. 6–10).

In the course of our studies with carbonyl-trapping agents we tested two which had remarkable specificity. Both hydroxylamine (NH_2OH)[12] and aminooxyacetic acid[13] appear to be specific inhibitors of γ-aminobutyric–α-ketoglutaric acid transaminase (GABA-T) in brain and β-alanine–α-ketoglutaric acid transaminase in liver. Present evidence favors the concept that these two enzymes in mammalian tissues may be identical[14].

Chromatographic evidence for the specificity of NH_2OH in brain and liver are shown in Figs. 11–16. Administration of NH_2OH resulted in increased levels of GABA in brain and β-alanine in liver. No significant changes were found in the level of other ninhydrin-reactive constituents in the tissues which were tested (Figs. 11–22).

Although GABA is synthesized and degraded by enzymes requiring PyP, the in-

TABLE II

EFFECT OF NH_2OH UPON GABA, GABA-T AND GAD *in vivo*

Trial No.	*No. of rats**	*Treatment***	*Time after injection (min)*	*Brain area*	*GABA***, § (mg%)*	*GABA-T activity, § (µmoles GABA/g tissue/h)*		*GAD activity, §§ µmoles GABA/g tissue/h*	
						No addn.	*+ 100 µg PyP/tube*	*No addn.*	*+ 50 µg PyP/tube*
1	2	Control	—	Whole brain§§§	22.1	29.5	—	—	27.1
	2	NH_2OH	75	Whole brain	35.7	17.2	—	—	25.5
		NH_2OH	210	Whole brain	47.5	15.1	—	—	24.4
2	3	Control	—	Cortex	21.0	43.9	31.0	12.2	27.5
	3	NH_2OH	90	Cortex	46.9	18.1	16.7	13.8	28.4
3	2	Control	—	Cortex		48.7	39.9		
	2	Control	—	Cerebellum		68.7	56.7		
	2	NH_2OH	90	Cortex		19.9	17.1		
	2	NH_2OH	90	Cerebellum		30.9	20.2		
4	2	Control	—	Cortex	20.8	38.7	28.7		
	2	Control	—	Cerebellum	18.5	53.1	37.1		
	2	NH_2OH	100	Cortex	43.1	19.5	16.4		
	2	NH_2OH	100	Cerebellum	28.5	27.6	23.6		
5	2	Control	90	Cortex				16.1	27.8
	2	NH_2OH	90	Cortex				13.4	27.3

* All rats were Sprague Dawley females 180–210 g in weight.

** $NH_2OH \cdot HCl$ was adjusted to pH 6 before injecting at dosage of 75 mg/kg body wt.

*** For additional confirmation see ref. 12.

§ GABA-T was determined by a method described previously[16] for acetone powder preparations. In these assays 60 mg equivalent of brain homogenate was incubated at 37° C for 1 h. GABA was determined enzymatically[12]. GAD was determined by incubating tissue homogenate and glutamic acid substrate for 15 or 20 min at 37°C with shaking. The incubation system consisted of 0.2 ml phosphate buffer (0.1 *M*, pH 6.3), 0.15 ml glutamic acid (0.5 *M* neutralized to pH 6.3), the equivalent of 30 mg of brain tissue homogenate in 0.2 ml buffer and 0.05 ml pyridoxal phosphate (1 mg/ml) or the equivalent amount of additional buffer. The reaction was terminated by the addition of 2.5 ml of 95% ethyl alcohol. The GABA formed, in excess of that present in the tissue at zero time, was determined enzymatically[12].

§§ These results have been confirmed measuring GAD activity by $^{14}CO_2$ evolution from uniformly labeled ^{14}C-glutamic acid substrate[15].

§§§ Whole brain less cerebellum, pons and medulla.

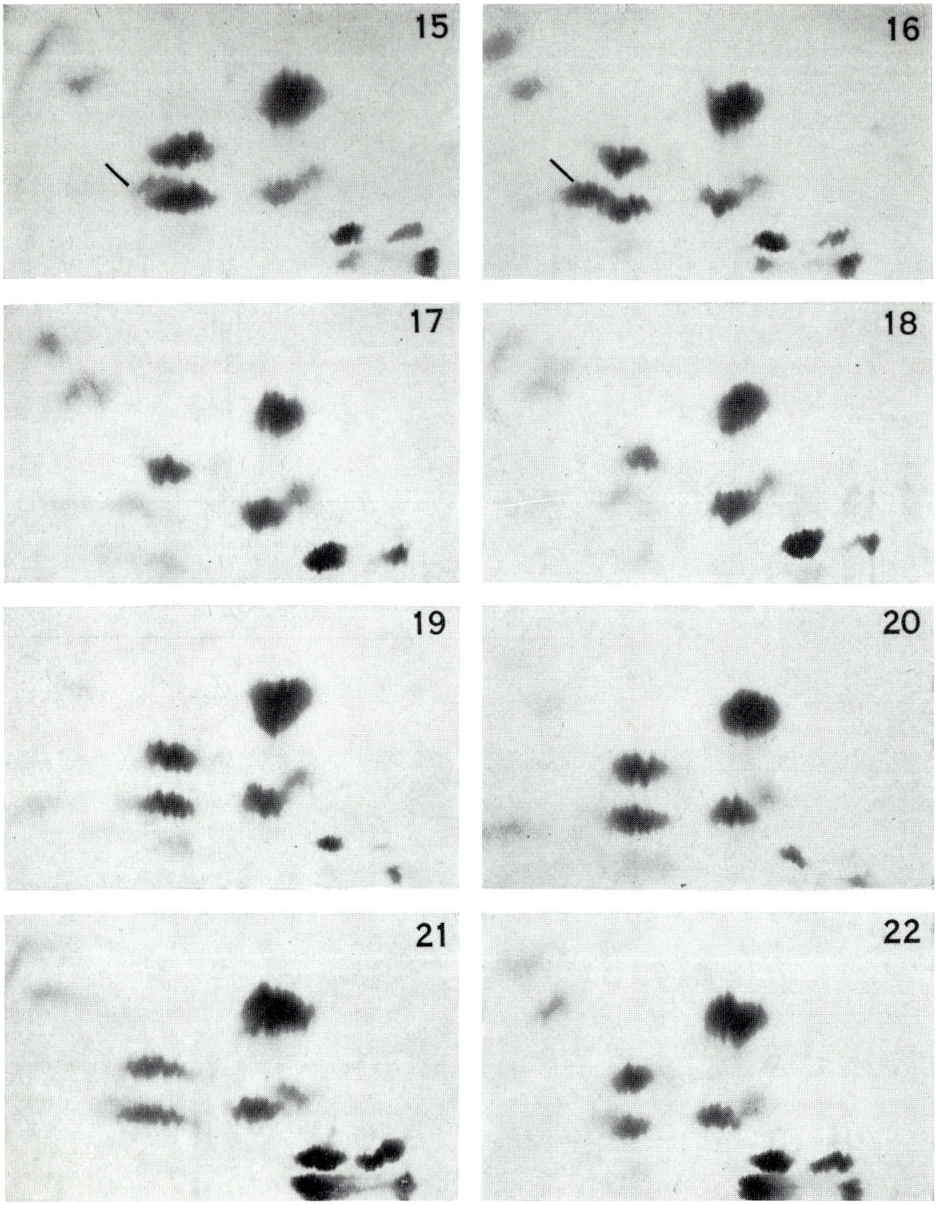

Figs. 15–22. Effect of hydroxylamine upon the concentration of amino acids in other body tissues of rat. Injected dose $NH_2OH \cdot HCl$, (adjusted to pH 6): 75 mg/kg body wt. Animals were decapitated 90 min after injection. Chromatograms represent 25 mg equivalent wet weight of tissue. Figs. 15, 16: represent the livers of control and treated rats. Arrow points to β-alanine. The identity of this amino acid was verified by the greenish color of the spot shortly after development with ninhydrin spray reagent. Since glyceryl phosphoryl ethanolamine has the same R_F as β-alanine in our solvent system, an aliquot of brain extract was hydrolyzed (6 N HCl at 100° in sealed tube for 24 h). The hydrolysate, after removal of most of the acid, was again examined chromatographically. The retention of the spot in question provided further evidence in favor of β-alanine. Figs. 17, 18: kidney. Figs. 19–20: muscle. Figs. 21, 22: spleen. All odd numbered chromatograms represent tissues from normal control animals; all even numbered chromatograms illustrate changes in the tissues of treated animals.

hibition with NH_2OH *in vivo* is primarily of the transaminase (Table II). A mechanism for this preferential inhibition has been discussed[17] and the formation of a non-dissociable complex with the tightly bound PyP coenzyme was suggested. Recent results in our laboratory have shown that pyridoxal oxime does not inhibit the GABA-T of rat brain *in vivo*[18]. This observation is in accord with the idea that the apoenzyme–coenzyme bond of GABA-T is very strong, permitting little if any substitution with the oxime directly.

Treatment of rats with aminooxyacetic acid caused the accumulation of β-alanine in liver, kidney, and spleen (Figs. 23–28). No changes were found in skeletal muscle, but in heart muscle aminooxyacetic acid produced a decrease in lysine content which might be related to possible shifts in electrolyte balance (Figs. 29–32). Levels

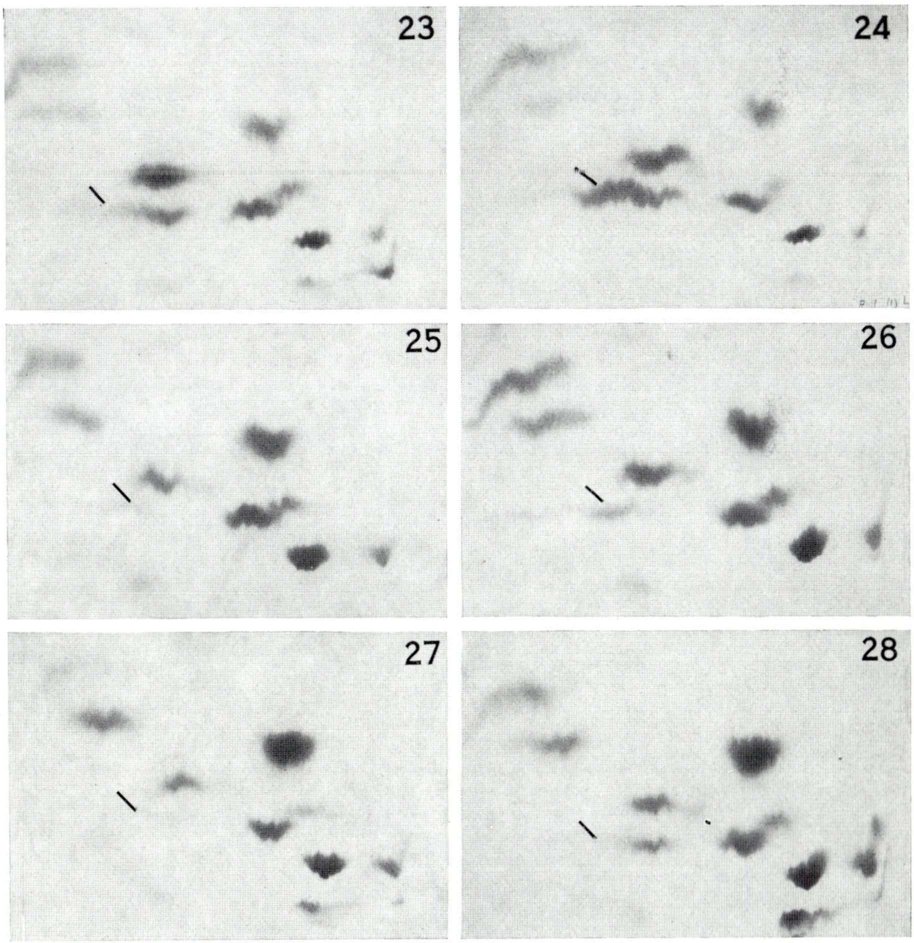

Figs. 23–28. Effect of aminooxyacetic acid upon the concentration of amino acids in tissues of rat. Injected dose NH_2OCH_2—$COOH \cdot {}^1/_2$ HCl (adjusted to pH 6): 100 mg/kg. Animals were decapitated 6 h after injection. Arrows point to β-alanine. Figs. 23–24: liver. Figs. 25–26, kidney. Figs. 27–28: spleen. All odd numbered chromatograms represent tissues from normal control animals; all even numbered chromatograms illustrate changes in the tissues of treated animals.

References p. 508

of lysine have been shown to increase in muscle of potassium-deficient rats[19]. The most remarkable change produced by aminooxyacetic acid was the spectacular increase in GABA concentration of brain (Figs. 33–34) at a time when no significant changes occurred in the level of any of the other amino acids which were detected by our chromatographic procedure. The highest levels of GABA in brain were observed at about 3 h after injection. This time differs from that reported by others[13, 20] who found that a maximal elevation of GABA was not attained until 6 h after the injection of aminooxyacetic acid.

The assay of brains from rats treated with aminooxyacetic acid showed that the inhibition of GABA-T and the elevation of levels of GABA correlated well in time for the first 7 h after intraperitoneal injection (Table III). During this time period

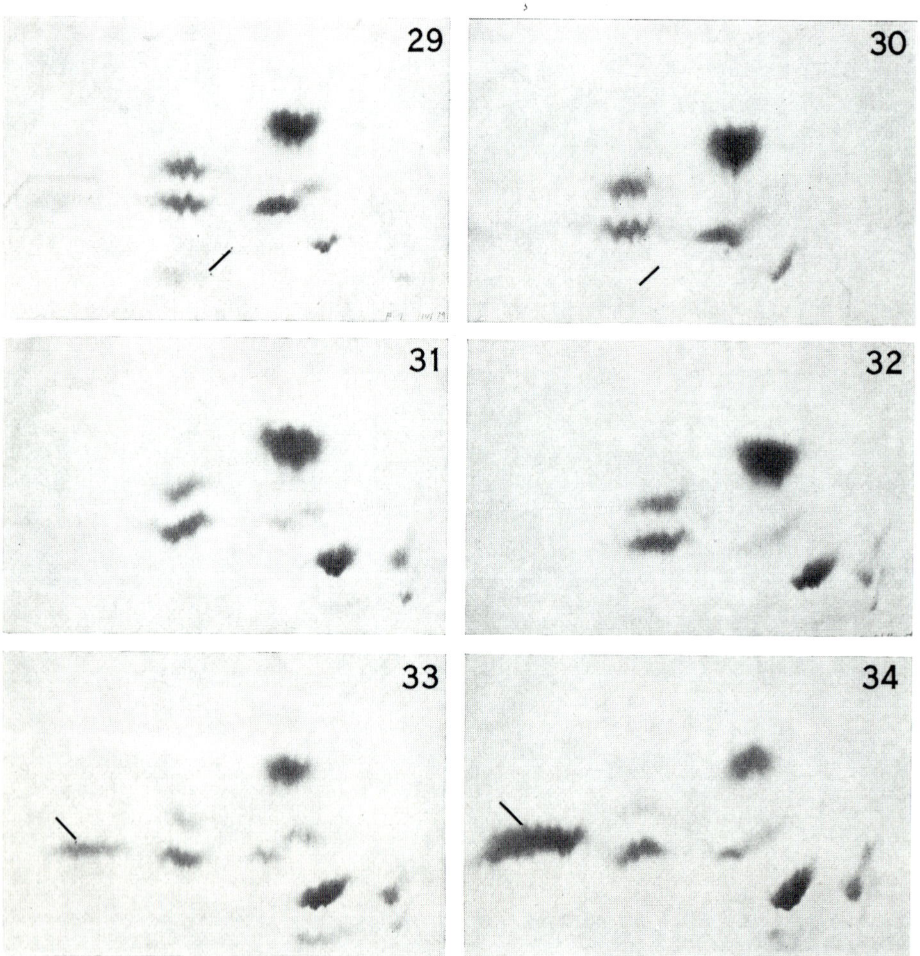

Figs. 29–34. Effect of aminooxyacetic acid upon the concentration of amino acids in tissues of rat. Injected dose NH_2OCH_2—$COOH \cdot {}^1/_2$ HCl, (adjusted to pH 6): 100 mg/kg. Figs. 29–30: heart; the arrows point to lysine. Figs. 31–32, muscle. Figs. 33–34: brain cortex; the arrows point to GABA. All odd numbered chromatograms represent tissues from normal control animals; all even numbered chromatograms illustrate changes in the tissues of treated animals.

References p. 508

TABLE III

ELEVATION OF GABA AND INHIBITION OF GABA-T BY AMINOOXYACETIC ACID*

Trial No.	*Time after injection (h)*	*GABA (mg %)*	*GABA-T μmoles GABA/g tissue/h*	
			No addn.	*+ 100 μg PyP/tube***
1	0	30	41	37
	1	42	20	17
	3	78	19	16
	7	59	21	19
2	0	29	53	52
	1	39	25	24
	3	93	14	13
	7	66	27	26
	22	53	25	24
	49	39	33	36
	69	34	31	30

* Methods used for the assay of GABA and GABA-T are indicated in Table II. Two animals were used for each value of this table. The dose of aminooxyacetic acid ($NH_2OCH_2COOH \cdot {}^1/_2$ HCl) used was 50 mg/kg body wt., injected intraperitoneally. Fed Sprague Dawley females weighing 200 ± 20 g were used exclusively.

** Pyridoxal phosphate was premixed with homogenates in Trial 2. The inhibition of GABA-T by aminooxyacetic acid was not reversed at any time period by *in vitro* addition of PyP.

no consistent inhibition of GAD was observed in the tissues of aminooxyacetic acid-treated rats. This is in sharp contrast to results obtained recently with GAD prepared from mouse brain acetone powders[21]. In these *in vitro* experiments, aminooxyacetic acid was shown to be one of the most potent inhibitors of GAD activity yet tested. Such results illustrate forcefully the subtle differences which may be observed in the effect of an agent *in vitro* and *in vivo*.

Although changes in the level of an enzyme system have been demonstrated *in vivo*, additional effects upon permeability and transport phenomena cannot be excluded. Conversion to a substance other than the compound originally injected also is possible. Thus other considerations may be involved in explaining the differences in specificity of *in vivo* and *in vitro* inhibition with aminooxyacetic acid.

The foregoing presentation has been used to illustrate the use of specific agents to change levels of free or loosely bound amino acids in animal tissues. This technique has and will continue to provide means to study the physiological significance of intrinsic amino acids in tissues and their biochemical and physiological interrelationship to each other and to other types of compounds.

ACKNOWLEDGEMENTS

This investigation was supported in part by Grants B-1615 and B-2655 from the National Institute of Neurological Diseases and Blindness, National Institutes of Health, and a grant from the National Association for Mental Health.

References p. 508

REFERENCES

1 E. ROBERTS AND D. G. SIMONSEN, in J. T. HOLDEN, *Amino Acids Pools,* Elsevier Publishing Company, Amsterdam, 1962, p. 284.
2 G. D. GAMMON, R. GUMNIT, R. P. KAMRIN AND A. A. KAMRIN, in E. ROBERTS, *Inhibition in the Nervous System and γ-Aminobutyric Acid,* Pergamon Press, 1960, p. 328.
3 R. P. KAMRIN AND A. A. KAMRIN, *J. Neurochem.,* 6 (1961) 219.
4 D. P. PURPURA, S. BERL, O. GONZALEZ-MONTEAGUDO AND A. WYATT, in E. ROBERTS, *Inhibition in the Nervous System and γ-Aminobutyric Acid,* Pergamon Press, 1960, pp. 331.
5 D. B. TOWER, *Am. J. Clin. Nutrition,* 4 (1956) 329.
6 K. F. KILLAM AND J. A. BAIN, *J. Pharmacol. Exptl. Therap.,* 119 (1957) 255.
7 E. ROBERTS AND S. FRANKEL, *J. Biol. Chem.,* 190 (1951) 505.
8 D. B. MCCORMICK AND E. E. SNELL, *Proc. Natl. Acad. Sci. U.S.,* 45 (1959) 1371.
9 D. B. MCCORMICK, B. M. GUIRARD AND E. E. SNELL, *Proc. Soc. Exptl. Biol. Med.,* 104 (1960) 554.
10 H. BALZER, P. HOLTZ AND D. PALM, *Arch. exptl. Pathol. Pharmakol., Naunyn-Schmiedeberg's,* 239 (1960) 520.
11 C. F. BAXTER AND E. ROBERTS, *The Pharmacologist,* 2 (1960) 78.
12 C. F. BAXTER AND E. ROBERTS, *Proc. Soc. Exptl. Biol. Med.,* 101 (1959) 811.
13 D. P. WALLACH, *Biochem. Pharmacol.,* 5 (1960) 166.
14 E. ROBERTS AND H. M. BREGOFF, *J. Biol. Chem.,* 201 (1953) 393.
15 B. SISKEN, K. SANO AND E. ROBERTS, *J. Biol. Chem.,* 236 (1961) 503.
16 C. F. BAXTER AND E. ROBERTS, *J. Biol. Chem.,* 233 (1958) 1135.
17 E. ROBERTS AND C. F. BAXTER, in F. BRUCKE, *Biochemistry of the Central Nervous System,* Pergamon Press, 1958, p. 268.
18 C. F. BAXTER AND E. ROBERTS, unpublished results.
19 M. IACOBELLIS, E. MURTWYLER AND C. L. DODGE, *Am. J. Physiol.,* 185 (1956) 275.
20 D. P. WALLACH, *Biochem. Pharmacol.,* 5 (1961) 323.
21 D. G. SIMONSEN AND E. ROBERTS, unpublished results.
22 R. CONSDEN, A. H. GORDON AND A. J. P. MARTIN, *Biochem. J.,* 38 (1944) 224.

DISCUSSION

Chairman: GEORGE ROUSER

MAYRON: Dr. SCHREIER, was free sialic acid found in either the blood or urine?

SCHREIER: One can find sialic acid in urine with special methods.

H. ROSENBERG: Dr. SOUPART, you mentioned the possibility that ultrafiltration might liberate amino acids that are adsorbed to proteins. Could you comment on this any further? I do not quite see how this could possibly happen, because you would not get a dilution or concentration of any filterable material during the entire process.

SOUPART: I was only suggesting that our present concept of renal handling of amino acids should be carefully reinvestigated in the light of the following facts. When considering the renal handling of plasma free (or so-called free) amino acids, it is generally assumed that they pass freely into the glomerular filtrate and are present in this filtrate at the same concentration as in the plasma, but this assumption has not been experimentally demonstrated up to now for obvious reasons. Moreover, what we call blood plasma free amino acid concentrations are values determined on plasma filtrates obtained by use of various deproteinization methods, such as a 1% picric acid solution. These procedures might well be able to break weak bonds linking amino acids to protein carriers. That such a possibility has to be kept in mind is suggested by recent papers of Dr. MCMENAMY[1,2] and his group in Boston, who claimed that there is evidence for protein binding, especially in the case of tryptophane, which they say is largely bound to albumin. As suggested by other speakers at this conference, "free" amino acid binding to other structures seems to happen quite often in tissues. If substantial evidence could be obtained that such a binding occurs in blood plasma, our present concept of the renal handling of "free" amino acids would have to be revised. I would also emphasize the fact that when preparing a sample for plasma free amino acid analysis one has to deproteinize the plasma immediately after the blood is drawn from the vein, because there is a possibility that free cysteine tends to combine with -SH groups of the plasma proteins.

ROUSER: I think it is interesting that tryptophane binding by plasma protein has been reported. Apparently it did not occur as a general phenomenon, and I would not expect this to be important with cystine except by disulfide exchange.

PERRY: I wanted to make one comment about Dr. WESTALL's paper, concerning amino acids in cerebrospinal fluid. We have had the opportunity in our laboratory to study pools of normal adult cerebrospinal fluid and cerebrospinal fluid from children with various genetically determined forms of mental defect using the MOORE AND STEIN technique. We were surprised to find no trace of γ-aminobutyric acid at all, so that even though the brain is rich in γ-aminobutyric acid and cerebrospinal fluid is formed by the brain and bathes it, none appears in the cerebrospinal fluid.

One other comment I might make is to emphasize the remarks of Dr. SOUPART about the importance of deproteinization. We found no trace of cystine in cerebrospinal fluid and finally concluded that it was because we failed to deproteinize immediately. Even though one deproteinizes cerebrospinal fluid 24 hours after it is withdrawn, and keeps it at 20° during that period, one still loses all of the cystine through sulfhydryl binding to protein.

PIEZ: I would like to clarify a question brought up by Dr. WESTALL concerning the free amino acids in saliva. A spot seen on chromatograms has been identified by some as γ-aminobutyric acid. Dr. FOSDICK and I showed some years ago that this material is not γ-aminobutyric but has a behavior closer to the next higher homologue, α-aminovaleric acid. More important is the fact that this is not a natural constituent of saliva, but a result of bacterial action in the mouth. In fact, almost all of the free amino acids in saliva arise from that source. The only exception, I think, is phosphoethanolamine, which is found in pure parotid saliva.

SHAW: I would like to say a few words concerning Hartnup disease, which was discussed earlier by Dr. WESTALL. We have been able to carry out some studies at Cal Tech on Hartnup disease through the courtesy of Professor DENT, and we find that the abnormal pattern of urinary indole compounds which has been described in this disorder results from the action of intestinal microorganisms on dietary tryptophane. When neomycin is administered in small doses the abnormal indoles disappear entirely. Other evidence points to a defective transport of amino acids as the key factor in this disorder.

References p. 511

CHRISTENSEN: I have one comment I would like to address to Dr. SOUPART. I notice that you correct your leucine level for that lost by washing to reach the conclusion that it may be three times as great in the red blood cell as in the plasma. When we add radioactive leucine to fresh heparinized blood we find that it enters extremely rapidly. In our hands, even with long delays, the ratio fails to surpass a value of 1.1, calculated on a water basis. Now, because of the extreme speed with which the amino acids with large hydrophobic side chains do move into the red cell, the correction for washing must be extremely difficult, and I wonder if, in this case, the correction may have been too large.

Also, I should like to comment to Dr. ROBERTS that we interpret quite differently the ease with which the levels of the free amino acids of the tissues can be changed. We find that it is desirable to consider the tissue levels always with reference to the extracellular level, since we have never found a situation where they are not highly responsive to their environmental levels. Suppose you had a section of heart muscle that is cut off by infarction. The lack of irrigation of that area could cause a backing up of amino acids being lost from the fiber into the interstitial fluid, which would perhaps prevent further loss from that area.

I think, however, a more important source of the disagreement in our philosophies about this arises from the low precision that one gets in comparing the amino acid patterns under a given change in condition by gazing at a pair of chromatograms. For example, partial hepatectomy causes an analytically measured increase of 70 per cent in 24 hours, but has apparently led to a rather insignificantly changed chromatogram. I wonder if changes highly significant to metabolic processes are not being overlooked in the inspection of chromatograms.

E. ROBERTS: Using paper-chromatographic methods we have examined the extracellular phase in some instances, for example, by studying the ascitic fluid in which ascites tumor cells exist. In this case, we found that both cells and fluid showed a remarkable resistance to change. Here there is no problem of circulation at all. In some experiments the cells looked like they were virtually falling apart, and yet the easily extractable amino acids were retained and did not leak into the fluid. There were some changes, but actually, if anything, the easily extractable intracellular pool was higher than in the untreated tumors.

In the case of the heart, we know that big molecules must be leaking out rapidly after infarction because there is a remarkable rise in lactic acid dehydrogenase activity and other enzymes in blood. I believe this has been traced to an actual leakage of the enzymes from the infarcted heart muscle. Thus, this is evidence that there is some interchange taking place between infarcted heart muscle and blood. If "backing up" occurred and the interchange were stopped we would get autolytic changes, which we know can take place under sterile conditions, as was shown in some of the experiments we reported here. We would have expected then an increase in many of the detectable constituents. Instead, there was a retention of the initial pattern for quite some time, 8 hours or so, and then a very slow loss. The kinetics of these changes, I am sure, are quite different from the kinetics of other processes, such as loss of large molecules and of the appearance of various pathological changes.

Probably, the most important metabolic aspects of the easily extractable ninhydrin-reactive constituents relate to the turnover rates of these constituents in the various tissue compartments. There is no question that adrenalectomy, hypophysectomy, and other experimental procedures employed produced changes in turnover and possibly in the compartmentalization within the tissues studied, but the total concentrations of the various constituents were observed to be remarkably constant. It is quite possible, for example, that the total amount of γ-aminobutyric acid found in a particular brain area is much less important than the quantity that is bound to some neuronal structure at which it affects the ionic movements.

I fully agree that what we have reported today is a very broad descriptive study, the purpose of which was to delineate areas for further investigation fully aware of all of the reservations. We were greatly surprised, indeed, by the remarkable stability of steady state concentrations of the easily extractable ninhydrin-reactive constituents that we detected in our chromatograms.

ROUSER: Actually, if we look at the studies that Dr. ROBERTS reported, some distinct differences on the chromatograms can be seen. I know he has been most impressed that an over-all uniformity exists, yet there are differences. I think Dr. CHRISTENSEN is pointing out that in contrast he can measure various changes. This sort of thing can also be seen on a chromatogram, but it does depend on what you are looking at. Dr. ROBERTS' over-all point is that such changes were not as impressive as one might have expected.

I would like to add some comment about the observations on the red cells. Red cell permeability is an intriguing subject, and we began to do some *in vitro* control studies in our work to determine the extent of change from *in vivo* conditions. The permeability for amino acids was always different *in vitro* judging from the responses seen when we fed glutamine to a person and followed the blood level afterwards. When we added so-called physiological saline or a buffer *in vitro*, the red cell became rather freely permeable. It did not concentrate amino acids particularly, but they would move in. On the other hand, in *in vivo* studies the striking thing was the slowness of penetra-

References p. 511

tion, particularly in some individuals. One person was quite different from another. Some went for 4 hours with a six-fold elevation of glutamine in plasma and very little change in the erythrocytes. At other times there would be a very slight and gradual elevation followed by a fall in erythrocytes.

The literature suggested some differences in perfusion as compared to ingestion studies. On one occasion, after feeding glutamine, we gave 250 ml of saline by slow intravenous drip over a period of about 2½ hours. This was the only time that we saw plasma and red cells go up and down together. We withdrew the needle to stop the saline infusion, and observed during the remainder of the study that plasma and red cell glutamine were not at the same level each time as before. This study illustrates the ease with which *in vivo* permeability can be changed to resemble the *in vitro* situation.

CHRISTENSEN: The entrance of glycine is very slow, of course, in the red cells. I will speak of that this afternoon. I do want to emphasize that the cellular levels of the free amino acids are in a dynamic equilibrium with the extracellular levels. For the liver or the Ehrlich cell and for most amino acids, the response is extremely fast; for muscle or the erythrocyte and for such an amino acid as glycine the response is considerably slower. But we know of no exception among the tissues of higher animals: the cellular level of an amino acid needs to be related to the extracellular level and should not be considered in isolation.

TALLAN: Dr. MUSSINI mentioned the presence of bound amino acids in brain. This ties in with what Dr. MITCHELL was discussing yesterday. I think that most of the "bound amino acids" can be accounted for by acetylaspartic acid, glutathione, and glutamine. In normal rat brain we find about 6.7 μmole/g of acetylaspartic acid, which, after hydrolysis, would give 89 mg of aspartic acid per 100 g of tissue. There are 1.6 μmole/g of glutathione, which would give rise to 55 mg/100 g of glycine, glutamic acid and cysteine, and 5.4 μmole/g of glutamine, which would yield 79 mg/100 g of glutamic acid. Any glutamine that had cyclized to pyrrolidone carboxylic acid during the preparation of the extract would not appear on the chromatogram, but would yield glutamic acid upon hydrolysis. If the whole extract is to be analyzed with ninhydrin before and after hydrolysis, the expected increase in ninhydrin-reactive compounds arising from the three substances I mentioned, calculated as "leucine", agrees very well with Dr. MUSSINI's figure of about 200 mg per cent of bound amino acids. I do not think therefore, that there can be very much peptidic material present.

L. MILLER: I just wanted to ask a few minor questions of technique of Dr. TALLAN. You mentioned the use of perchloric acid in preparing amino acids as a non-protein filtrate, and I wonder whether this affords a special advantage over the Hamilton-Van Slyke picric acid procedure that you did not make clear in your presentation. Then, again, Dr. TALLAN gave some quantitative results of glutamine content, and I thought, perhaps on the basis of inadequate information, that the conventional MOORE, SPACKMAN AND STEIN procedure does not give accurate glutamine levels. I wondered how these were obtained.

TALLAN: Perchloric acid is somewhat easier to use than picric acid; removal of the excess with KOH is simpler than the removal of excess picric acid with Dowex 2. With brain, perchloric acid gives a crystal clear extract, whereas a picric acid extract is usually cloudy and more difficult to work with. Our glutamine figures are really minimal values, uncorrected for any possible losses. However, the values we get are quite reproducible for a given extract.

H. ROSENBERG: Dr. ROBERTS mentioned serine ethanolamine phosphodiester in connection with the chick embryo. We have looked at this and found that eggs contain none, but as the embryo develops this compound appears on the fourteenth day, and can be detected from then on. We always examined extracts of whole embryos, without bothering about organ distribution.

REFERENCES

[1] R. H. MCMENAMY, C. C. LUND AND J. L. ONCLY, *J. Clin. Invest.*, 36 (1957) 1672.

[2] R. H. MCMENAMY, C. C. LUND, G. J. NEVILLE AND D. F. H. WALLACH, *J. Clin. Invest.*, 39 (1960) 1675.

Round Table Discussion

METHODOLOGY AND OCCURRENCE OF FREE AMINO ACIDS

Session Chairman: KURT SCHREIER
Transcript Editor: MILTON WINITZ

SCHREIER: I should like to ask Dr. SOUPART to tell us about his new analytical method.

SOUPART: In this presentation, I want to draw your attention to a new device for the counting of analytical samples and the continuous recording of chromatographic effluents. This method has been worked out by my colleagues, ERIC SCHRAM and ROBERT LOMBAERT, an electronic engineer, and permits the determination of tritium and ^{14}C in aqueous solution by means of scintillating anthracene powder.

The use of anthracene powder may be considered as an important improvement in the heterogenous counting of weak β emitters. In the present method, the aqueous solution to be counted is run through a cell of constant volume, containing anthracene particles at calibrated size, with the following advantages: *(a)* Counting of individual samples or continuous recording of chromatographic effluents is possible with a single cell; *(b)* reproducibility of the results and excellent stability of the efficiency of the scintillator are ensured due to the constant characteristic of the cell and to the absence of quenching; *(c)* no problems are encountered resulting from the slow settling or creeping along the walls of the powder, as in the case with counting vials; *(d)* sample preparation and counting technique are much simplified as compared with methods involving precipitates in planchettes or incorporation of difficultly soluble substances in liquid scintillators. Moreover, the method is non-destructive and the counted sample may be recovered for other uses.

H. ROSENBERG: Professor SOUPART, would the background you observe indicate any irreversible adsorption of radioactive material on the anthracene?

SOUPART: Not in this case, because the background remains constant at least for a few weeks.

SCHREIER: I will now ask WINITZ to tell us something about methods for the determination of D-amino acids in proteins.

WINITZ: What methods can be employed for the determination of D-amino acid residues in proteins is, of course, a problem that has plagued biochemists for a good many years and is one to which there is, as yet, no satisfactory answer. My own speculations concerning a possible approach to this problem—and I should emphasize that they are merely speculations—arise from a question that was asked of me by Dr. ROSENBERG with regard to which of the presently available methods would offer some hope for unequivocally demonstrating the existence of D-amino acid residues in proteins. His concern with this problem arose from his own very excellent work on the isolation of D-serine from the serine ethanolamine phosphodiester of the turtle and from the possibility that D-amino acids might likewise reside in the proteins of this same species.

In seeking a solution to this difficult problem, it would of course be most advantageous if one could completely hydrolyze a protein to its constituent amino acids, in the absence of accompanying racemization or decomposition of the amino acids, and then use some very sensitive assay method, such as oxidation with D-amino acid oxidase in a Warburg respirometer, in order to ascertain whether there are actually one or more D-amino acid residues present amongst the several hundreds or thousands of L-amino acid residues which constitute the protein molecule. However, if you attempt to completely hydrolyze the protein molecule with acid, racemization occurs in the case of certain residues, such as glutamic acid, and partial or complete destruction occurs in the case of others, such as serine, threonine and tryptophane. If you attempt to use alkaline hydrolysis, residue destruction and racemization pose even more severe problems. And if you attempt to use the less drastic enzymatic approach—the usual enzymatic hydrolysis employing trypsin, pepsin, pancreatin or some other mixture of proteolytic enzymes—it is reasonable to assume that any of the resulting smaller peptides that might contain D-residues would be highly resistant to complete hydrolysis by virtue of the L-directed stereospecificity of the proteases employed. The failure to detect D-amino acids in these hydrolyzates with D-amino acid oxidase would not eliminate the possibility of their presence in a peptide-bound form.

References p. 524

From what has already been said, it would appear that a reasonable step toward the solution of this problem would be to devise a means that would permit the protein molecule to be broken down by the usual enzymatic means and would, at the same time, permit the complete hydrolysis of those highly resistant peptides which incorporate D-amino acid residues. We became interested in this problem several years ago, not for the purpose of ascertaining whether or not there were D-amino acid residues in proteins, but because we sought a means that would allow us to determine the optical purity of small synthetic peptides. We were, at the time, engaged in the synthesis of a number of peptides of the D- and L-configuration for ultimate use as peptidase substrates, and we became increasingly concerned with the question as to whether the synthetic procedures we employed were accompanied by racemization. Thus, for example, after preparing the four stereoisomers of leucylleucine, how would it be possible to determine to what extent, if any, the L-L isomer was contaminated with its stereoisomeric D-L, L-D and D-D forms? Measurement of the optical rotation of these different isomers certainly couldn't provide the answer, because not only are such measurements of insufficient sensitivity to detect trace contamination, but they have little meaning in the case of molecules which contain two or more asymmetric centers. We therefore instituted the search for a method that would enable us to determine whether a given optically active peptide was contaminated with any of its stereoisomers to the extent of only a fraction of a per cent.

In this connection, it should be pointed out that some years ago GREENSTEIN isolated an aminopeptidase from hog kidney which exhibited an entirely different specificity than leucine aminopeptidase and which he unequivocally demonstrated was not identical with this latter enzyme. GREENSTEIN went on to demonstrate further that the D- and L-form of a large variety of dipeptides, all of which possessed an N-terminal glycine residue, were hydrolyzed at nearly equally rapid rates by this aminopeptidase, but the lack of suitable peptide substrates at that time did not permit him to pursue these specificity studies further. When we recently extended these specificity studies to each of the four stereoisomers of a large number of dipeptides, we found that not only were the L-L and L-D isomers rapidly hydrolyzed, as would be expected from the results observed earlier with the glycine-containing peptides, but the D-L and D-D isomers were also hydrolyzed at quite appreciable although somewhat slower rates. Even the stereoisomers of valylvaline, which are so slowly hydrolyzed by acid or alkali because of their high degree of steric hindrance, were readily cleaved by this enzyme. In any case, the observation of a peptidase activity toward which smaller peptides would be susceptible, irrespective of the configuration of the component amino acid residues, permitted us to develop a method for determining the optical purity of small peptides with a high degree of sensitivity.

The method briefly is as follows: 100 μmoles of an aqueous solution of the pertinent dipeptide, L-leucyl-L-leucine for example, in borate buffer at pH 8.5 is placed in the conventional Warburg vessel and treated with the renal aminopeptidase. An aqueous solution of hog renal D-amino acid oxidase is placed in the side arm of the vessel and the whole allowed to come to temperature equilibrium at 37°. Then the oxidase in the side arm is added to the peptide solution and the hydrolytic and oxidative reactions permitted to take place simultaneously. Oxygen consumption is recorded until it stops. Under these conditions, 1 μmole of a D-amino acid would consume 11.2 μl of oxygen so that the complete hydrolysis of each micromole of the contaminating optical antipode, D-leucyl-D-leucine in this instance, should yield 2 μmoles of the D-amino acid and consequently cause the consumption of 22.4 μl of oxygen. A control is always run which is identical to the above but to which, in addition, is added 1 μmole of the D-D isomer of the dipeptide. Of course, the optical purity of a dipeptide of the D-D configuration can be ascertained in the same manner, except that it is obviously necessary here to use L-amino acid oxidase to detect contaminating L-residues. In any event, the method is capable of permitting the detection of optical contamination with a sensitivity of less than 1 part in 100.

Until Dr. ROSENBERG raised the problem yesterday, I had not considered the application of this method to the detection of D-amino acid residues in proteins. In this case, it might prove of interest to hydrolyze a given protein with the usual proteolytic enzymes, treat the enzymic digest further with hog renal aminopeptidase in order to break down any small resistant peptides containing D-residues, and finally subject the whole mixture to the action of D-amino acid oxidase in a Warburg apparatus.

While we are in the realm of speculation, there is an alternative method that might be used here that does not involve the prior isolation of the aminopeptidase, since it is an *in vivo* method. I offer this method because of our experiences with chemically defined diets. Over the past 6 years, we have formulated and studied the effect of some 200 different chemically defined diets, composed of pure crystalline L-amino acids, vitamins, glucose, the pertinent salts and essential fatty acids, on growth, reproduction, lactation and pathologic conditions of various types. Now it has long been known that certain of the D-amino acids, such as D-methionine, can be used instead of the corresponding essential L-amino acid for growth and that, on the basis of *in vitro* studies, this is presumably due to the conversion of the D-amino acid to its L-antipode by the successive en-

References p. 524

zymatic processes of oxidation of the D-isomer to the α-keto acid and transamination of the latter to the L-isomer. Well, we became interested in this problem of the metabolic breakdown of D-amino acids, not in the test tube but the whole animal, and wondered whether we could, by adding sufficiently large quantities of non-essential D-amino acids to the diet, so overload the mechanism of the animal for converting these D-amino acids to their L-isomers that it would now be unable to utilize the D-isomer of an essential amino acid that was ordinarily utilizable for purposes of growth. This we were able to demonstrate. I won't go into this in detail, but we did show that we could actually pinpoint the enzymes responsible for the *in-vivo* conversion of a D-amino acid to an L-amino acid and that D-amino acid oxidase was the enzyme involved in the first step of this process. The important point here is that we did find that we could inhibit this D-amino acid oxidase activity in the whole animal to any desired extent merely by adding to the diet, sodium benzoate which was shown by earlier workers to be a very potent inhibitor of D-amino acid oxidase *in vitro*. Diets containing D-amino acids that were ordinarily utilizable became increasingly less so the greater the benzoate concentration. We also found that when benzoate was added to the diet, not only was the ability of the animal to convert a D-amino acid to the utilizable L-isomer impaired, but that most of the D-amino acid ultimately appeared in the urine in very large amounts. It is of course well known that D-amino acids are generally more rapidly excreted into the urine than the corresponding L-amino acids, which are reabsorbed by the kidney tubules, recycled and more efficiently utilized.

Returning now to the problem of the identification of D-amino acid residues in proteins—how can this system be employed here? Well, if the protein which is suspected of containing D-residues were fed to an animal—and I must again state that this is a speculative method that hasn't yet been tried experimentally—in a diet which contained a concentration of sodium benzoate sufficient to inhibit effectively the D-amino acid oxidase activity of that animal, then the protein would be hydrolyzed in the usual manner by the proteases of the gastrointestinal tract. The resulting enzymic digest, which might also include small resistant peptides containing D-residues, would then be absorbed through the gut wall into the circulatory system. These smaller peptides could ultimately come into contact with the aminopeptidase of the kidney where they would undergo hydrolysis to their component amino acids. As the D-amino acid oxidase is inhibited by the benzoate, any D-amino acids that might be present would ultimately appear in the urine and could subsequently be estimated *in vitro* using some suitable technique.

H. ROSENBERG: I am extremely grateful for this. This seems to be an excellent method, and we are certainly going to try it out. I may just add that perhaps a very good control to try the method out would probably be to put an animal on benzoate and then give it a diet of casein hydrolyzate plus one part in a thousand of, say, D-alanine and see whether you can pick it up from the urine.

The question I wanted to ask concerns the inhibitor. There have been a number of compounds similar to benzoate (salicylates, etc.) which are all known to be FAD antagonists. I suppose they will be active in this procedure. The second thing is, what levels of benzoate do you give to produce inhibition. Do you know offhand?

WINITZ: The levels of benzoate that are effective in our diets—and we feed our chemically defined diets as crystal clear solutions in water—are 2% of a diet solution. A 100-g animal will generally eat about 25 to 30 ml of 50% solution of diet daily, which is equivalent to about $12^1/_2$ to 15 g of a solid food diet.

Your first question, with regard to salicylate and various other things, touches on a very personal experience that I had some years ago when my older son, who was 3 years old at the time, somehow managed to get into the medicine cabinet, which my wife and I had thought was safely beyond his reach, appropriated the candy-flavored baby aspirin therefrom, and swallowed some 50 of those wonderfully tasting aspirin tablets. When we learned what had happened—which fortunately was very soon after the incident had occurred—we rushed the youngster to the hospital where he immediately had his stomach pumped in order to remove the unabsorbed aspirin. Well, the youngster came through the experience O.K., but it seemed incredible to me at the time that there really was no adequate means for counteracting salicylate toxicity; there still isn't. Yet the aspirin sales in this country are in the vicinity of $160,000,000 a year, and that's a lot of aspirin.

Well, what does this have to do with Dr. ROSENBERG's question? We looked into this benzoate phenomenon recently and found that when benzoate is added to a chemically defined diet which contains one of the essential amino acids as the D-isomer, and which will ordinarily support growth, then growth is markedly retarded due to the fact that the benzoate, by inhibiting the D-amino acid oxidase activity, effectively prevents adequate conversion of the D-isomer to its essential L-form. We further found that this inhibitory effect of benzoate on growth of the animal could be counteracted by adding sufficient glycine to the diet. Now it is well known that benzoate is detoxified in the whole animal by its conversion to hippuric acid, which is excreted in the urine as such, and that hippuric acid is a combination of benzoic acid and glycine in amide linkage.

References p. 524

It therefore clearly appeared that, in this case, the animal was not synthesizing glycine at a sufficiently rapid rate to overcome the toxic effects of benzoate, and that a supplementary source of glycine in the diet was necessary for this purpose.

Returning now to salicylate, one of the principal excretion products of salicylate is salicyluric acid, which is a combination of salicylic acid and glycine in amide linkage. Could it be possible that we will encounter a detoxification phenomenon here comparable to the benzoate situation? Such a study will be undertaken shortly.

ROBERTS: I would just like to say that both Dr. MILLER and I feel that the contributions that Dr. WINITZ has made, and is on the brink of making, really augur well for the future of nutritional science. It is again going to become an exciting frontier of biochemical endeavor. For many years the young people who were going into biochemistry were shying away from nutrition, and I think rightfully so, because much of the work had gotten into the hands of people who were just varying this or that constituent of a poorly defined diet and observing weight changes and other gross changes of that sort. But it is obvious that nutritional biochemistry is becoming an exciting field again.

WINITZ: Thank you very much for those very generous remarks. I think I should point out that when Dr. JESSE GREENSTEIN and I entered this area some six years ago, we entered it not as nutritionists but as chemists. I think it is important to remember, also, that the really wonderful and exciting early quantitative work with chemically defined diets in nutrition began with ROSE, and ROSE too viewed this area essentially as a chemist, not as a nutritionist. I myself have no doubt that through the use of chemically defined diets, the animal can be used essentially as an animated test tube which can be titrated with various chemical agents and made to respond in a quite predictable and reproducible manner. Such an approach will permit the study of a specific enzyme system or metabolic process in association with the tremendous amount of other metabolic activity that takes place in its natural environment—the whole animal. I really feel that this is much more meaningful than isolating a particular enzyme from a given tissue, putting it in a test tube, and then attempting to find out what it does.

E. ROBERTS: Without the *in vitro* work one wouldn't know that benzoate inhibits D-amino acid oxidase.

WINITZ: That is certainly very true. I don't mean to underestimate the importance of *in vitro* studies, but it should also be born in mind that when enzyme systems are studied in an artificial environment which bears no relation to their natural surroundings, then a distorted picture of the situation as it actually occurs is quite likely to emerge. For example, if the relative rates of oxidation of various D-amino acids by D-amino acid oxidase, as measured in the test tube, are compared with the relative rates of utilization of these same D-amino acids for purposes of growth, then an entirely different pattern becomes evident. The orders of oxidation *in vivo* and *in vitro* are very dissimilar, with some of the amino acids, that are very poorly utilized by the animals being very rapidly oxidized in the test tube, and *vice versa*.

In the case of isoleucine, for example, D-alloisoleucine which is quite effectively oxidized *in vitro* will not support growth if it replaces an equimolar amount of L-isoleucine in the diet, whereas D-tryptophane, which is more slowly oxidized *in vitro* is utilized *in vivo* to an appreciable extent. So it becomes a matter of studying renal reabsorption mechanisms and we find that when D-alloisoleucine is provided in the diet, it subsequently appears in appreciable quantities in the urine, whereas D-tryptophane does not. The way to make D-alloisoleucine utilizable for purposes of growth—and this we have done—is to provide it in the diet in very high excess, so that although the major portion of it is rapidly excreted into the urine unchanged, sufficient material is nevertheless available to permit it to be utilized in part. What we have done here is to overload the renal mechanism for disposing of this amino acid, so that a portion of it is recycled and eventually oxidized.

SCHREIER: May I now ask Dr. WITTER to present some of his studies on amino acids in the brain.

WITTER: I would like to take this opportunity to present the results of some studies that Dr. FARRIOR and I have carried out on the effect of DDT and dieldrin on the free amino acids of the rat brain.

Previous workers have investigated the effects of a variety of drugs on the free amino acids in the central nervous system. Most of these studies have been summarized in the first symposium on amino acids which was held here a few years ago[1]. However, no report has appeared on the effects of chlorinated hydrocarbon insecticides on the free amino acids of the mammalian brain. This is somewhat surprising, in view of the widespread use of these compounds and their known effects on the nervous system.

Accordingly, a study was made of the effects of two members of the group of chlorinated hydrocarbon insecticides on the free amino acids of rat brain. The compounds studied were DDT and dieldrin. Both of these compounds are stimulants of the central nervous system, but peripheral stimulation may play a role to an unknown degree in the pharmacological effects of these compounds.

References p. 524

The sequence of events after the administration of DDT to animals is quite different from those noted when dieldrin is given. With a single large dose of the former compound the animal first exhibits muscle twitching, which is followed by generalized body tremors gradually increasing in intensity that last for several hours. Finally the animal goes into a convulsion, collapses and dies[2]. On the other hand, with a lethal dose of dieldrin, the animal becomes hyperexcitable, and in a short time begins to have a series of epileptoid convulsions which terminate in collapse and death[3]. This marked difference in symptomatology suggests a difference in mode of action between the two insecticides. Under these circumstances, any change noted with one compound and not with the other might reveal some clue as to a specific mode of action.

In the present experiments, female albino rats were given approximately two times the LD_{50} dose of dieldrin or DDT by the oral route[4]. Then the animals were sacrificed at time intervals which previous experience had shown would produce severe poisoning but would not permit the rats to become moribund. The animals given DDT were killed after two hours of violent tremor; those given dieldrin were sacrificed during their fourth convulsions. The concentration of alanine, γ-aminobutyric acid, glutamine and glutamic acid in the brains of these animals and their controls was estimated by quantitative filter paper chromatography.

It was found that there were no significant changes in the levels of these amino acids in the brains of the animals given DDT. Where dieldrin was administered, a rise of 24% was noted in the level of γ-aminobutyric acid. However, the most striking change, easily visible on the chromatograms, was an increase of 80% in the concentration of alanine. These changes were shown to be significant by statistical analysis.

Dr. ROBERTS, in his review of metabolism of the free amino acids of the nervous system, points out that the most significant changes that have occurred in the levels of alanine, γ-aminobutyric acid, aspartic acid, glutamic acid or glutamine, have resulted from procedures which directly or indirectly affect the oxidation of intermediates of the KREBS' cycle. Therefore, it seems most probable that the above increases in the levels of alanine and γ-aminobutyric acid in the brains of rats poisoned with dieldrin are not the result of the direct effect of the insecticide on the metabolism of these amino acids, but rather manifestations of an inhibition of the oxidation of pyruvate through the KREBS' cycle. Changes in the levels of the keto acids in the KREBS' cycle could lead through transamination reactions to changes in the concentrations of these amino acids. As a working hypothesis, we suggest that dieldrin may inhibit some specific reaction of the cycle. The fact that DDT did not inhibit the respiration of the mammalian brain[5, 6] and does not change the level of brain free amino acids is consistent with this hypothesis. However, the effects of dieldrin on the oxidation of pyruvate and on the individual steps of the KREBS' cycle in the brain have not been investigated. Therefore, proof of this hypothesis must await further experimental work.

HOSEIN and co-workers[7, 8] have claimed that dieldrin and certain other convulsants bring about the release from the brain of a coenzyme A ester of γ-aminobutyrobetaine from a bound to a free form, and that this free ester is the cause of convulsion. The esters of γ-aminobutyrobetaine have pharmacological properties similar to those of acetylcholine[9]. However, the proof of the identity of this coenzyme A ester with the convulsant substance and evidence for its formation from γ-aminobutyric acid have not been published so far as we are aware. Oxidative reactions are known to be involved in a number of processes in which permeability of membranes are changed. Thus the present hypothesis that dieldrin may act to inhibit the oxidations of the KREBS' cycle is compatible with, but does not necessarily support, the theory proposed by HOSEIN.

At the previous symposium, MCKHANN and co-workers[10] reported confirmation of KESSEL's observation[11] that the level of γ-aminobutyric acid was increased in the brains of animals given the convulsant 2,4-diaminobutyric acid. BAXTER AND ROBERTS[12] showed that a rise in γ-aminobutyrate could occur if hydroxylamine and thiosemicarbazide were given together. Both groups concluded that seizures may occur whether the γ-aminobutyrate content of the whole brain is abnormally high or low. Our results, which show that the administration of dieldrin also increases the γ-aminobutyrate level of the brain, are in agreement with these conclusions.

E. ROBERTS: I think that the results first of all suggest a possibility that both the α-ketoglutarate oxidase and pyruvate oxidase might be inhibited by dieldrin.

With regard to the interest in raising or lowering levels of γ-aminobutyric acid (γABA) in the central nervous system and convulsive seizures, the relationships appear to be more complex than originally envisioned. Good correlations were found between the decreases in γABA levels and the occurrence of seizures when thiosemicarbazide was administered to animals and between the increases in seizure thresholds and elevations of brain γABA levels after injection of hydroxylamine. However, when both thiosemicarbazide and hydroxylamine were administered to rats, seizures occurred which were characteristic of those usually produced by thiosemicarbazide alone even though the γABA levels were elevated (see BAXTER AND ROBERTS[13], and ROBERTS[14]). These results indicated clearly that a general depression of γABA levels is not a necessary requirement

References p. 524

for the induction of seizures by thiosemicarbazide. When chemicals which are not ordinarily present are applied to living cells, one must keep in mind the myriad possible ways in which some of them may act and not merely focus on the one or two enzyme systems which are under investigation. Both hydroxylamine and thiosemicarbazide may form derivatives with a large variety of naturally-occurring carbonyl compounds, which derivatives may have various actions of their own. Thiosemicarbazide can chelate zinc. Zinc is required for the activity of various dehydrogenases and carbonic anhydrase. The hippocampus, an area of brain that is particularly sensitive to seizures, has a high zinc content. Could at least part of the effect of thiosemicarbazide be related to its zinc-chelating capacity?

There is now evidence that γABA may be bound physically to some constituent found in the tissue of the central nervous system[15]. It may well be that it is not the total quantity of γABA, but the amount of bound γABA which will determine the effects of this substance on the conductance of potassium and/or chloride ions, which is presently believed to be the way in which γABA influences neuronal activity. An agent which can either inhibit or enhance binding of γABA may have an effect on seizure threshold and other neuronal properties regardless of the effect on total content of γABA in a particular area of brain.

Thus, as in almost every other field of study related to the nervous system, the area related to γABA has turned out to be more complicated than one would have anticipated or wished.

TALLAN: Dr. ROBERTS has so well summarized the current status of studies on the amino acids in brain that there is very little one could add. I would like, though, to bring up one technical question for discussion. That is, what is the best way to extract the free amino acids of brain? Dr. MUSSINI uses picric acid, Dr. ROBERTS and his group use ethanol, and I have been using perchloric acid. There have been a number of side discussions on this point during the past few days, concerned with whether these are really comparable. For example, I found earlier that the use of TCA, followed by ether to remove the excess, could result in a loss of tyrosine. Dr. MAYRON has asked me whether perchloric acid might not leave glycoproteins in solution, as is the case with blood serum. Dr. BAXTER has brought up the question of whether perchloric acid extracts as much γABA from the brain as does ethanol. These are points we are going to check. Has anyone else had experiences along this line?

E. ROBERTS: We have not done the refined quantitative work, in general, outside of determinations of γ-aminobutyric acid and glutamate, which you have performed on your columns, Dr. TALLAN. Early in the game we tried various procedures, such as heat coagulation followed by dialysis, trichloroacetic acid precipitation and extraction of the trichloroacetic acid with ether, preparation of perchloric acid extracts and removal of excess perchlorate as the potassium salt. We ran chromatograms and got virtually identical chromatograms on extracts prepared by these various methods. When γABA is elevated after treatment with hydroxylamine, the effect can be seen by all the methods. Routinely we have been preparing alcoholic extracts, which are evaporated to dryness and taken up in water saturated with picric acid. I do not believe that it is a critical matter which of the above methods is used for study of free amino acids. When one studies substances which are liberated upon acid hydrolysis, the choice of method will be critical since what is precipitated and what is left in solution depends on conditions.

KIHARA: I would like to ask Dr. ROBERTS if he has considered the possibility that when hydroxylamine does raise γABA it might act by inhibiting pyridoxal phosphate formation.

ROBERTS: I think that thiosemicarbazide lowers γABA because it inhibits pyridoxal kinase. The results which Dr. BAXTER reported at this Symposium indicate very clearly that the γABA transaminase is inhibited very remarkably by hydroxylamine while the glutamic decarboxylase is not. Our idea at present is that hydroxylamine acts by inhibiting the γABA transaminase preferentially to the glutamic decarboxylase because it forms a stable oxime derivative of the pyridoxal phosphate which remains attached to the apoenzyme, and that the oxime may be hydrolyzed slowly or the inactive coenzyme may take a relatively long time to dissociate from the apoenzyme. Once free, the apoenzyme can reassociate with active coenzyme present in the tissues.

L. MILLER: In speaking about actions of drugs on the concentration of one or another substance in the brain, I believe one ought to try at least to designate what tissue is being studied; that is to say, whether it was the cerebral cortex or whether one has really looked carefully at the midbrain and medulla, where very powerful effects can be mediated. In just removing cerebral cortex, as one is more apt to do in an undiscriminating brain study, you may totally miss a point. So I wonder, for example, in your case, Dr. WITTER did brain analysis clearly include the medulla and midbrain?

WITTER: These were included in this study.

H. ROSENBERG: What were the concentrations of dieldrin which you found produced the near-lethal and lethal effects? Just for information purposes, if you happen to have it, how would this concentration compare with the amounts of dieldrin generally applied over areas of agricultural land and what would be a lethal dose for sheep?

WITTER: The first part of your question was answered in my talk. We gave the animals 160 mg/

References p. 524

kg orally or about four times the LD_{50} dose. Now in regard to the second part of your question about the lethal dose required with sheep, the toxicity of dieldrin to sheep is of the same order of magnitude as that to rats. Therefore the dosage level used in the present experiment would be more than sufficient to kill sheep. The amounts of dieldrin to which a grazing sheep might be exposed present no hazard to the sheep or the consumer.

WINITZ: With regard to the metabolic response to drug toxicity, several years ago we had occasion to carry out studies on ammonia toxicity in rats and found that arginine had a marked protective effect against ordinarily lethal doses of ammonia. Thus, when $LD_{99.9}$ doses of ammonium acetate were administered intraperitoneally to rats, the mortality was 100%. However, animals treated either concurrently or prior to such injection with protective doses of L-arginine invariably showed 100% survival. These studies were carried out by GULLINO, who is a very competent pathologist, and in these studies he used several thousand rats—so the results were pretty unequivocal. We subsequently pinned down the mechanism of action of arginine to its implication in the *in vivo* KREBS'–HENSELEIT cycle.

Later we extended these studies to hepatectomized rats. It was found here that ordinarily protective doses of L-arginine were completely ineffective in protecting partially hepatectomized rats—about 40–70% of the liver was removed—from $LD_{99.9}$ doses of ammonium acetate, even after the liver had been permitted to regenerate for several days following its removal. So despite the fact that the size of the liver had returned to normal or nearly normal, the liver function had not. How long it takes for liver function to get back to normal—this we don't know.

SCHREIER: I may add that in the terminal stages of various liver diseases, arginine, glutamine and similar compounds have been tried, but they did not show very much effect.

The last topic on the agenda is concerned with a discussion of amino acid balance. Although it has been very easy to get toxic reactions with different types of amino acids in animals, there are very few publications, to my knowledge at least, that have been concerned with such studies in human beings. In Marburg, Germany, STAVE recently carried out some studies with cystine and found that he could cause kidney damage with this amino acid. Of course, the relation to FANCONI's syndrome is clear.

WINITZ: Evidently, it has been pretty well established that this type of cystinosis is characterized by a marked deposition of cystine in the tissues. The question now arises as to whether the clinical manifestations of the syndrome can be alleviated if the deposition of cystine is prevented. Well, how can one prevent cystine deposition in the tissues? It seems that we have a situation here analogous to that found with phenylpyruvic oligophrenia, where diets are employed in which the level of phenylalanine is adjusted upwards to the level just required for growth, maintenance and tissue repair, and no higher, since providing higher levels of phenylalanine can lead to retardation of growth, mental defects and so on. Would it likewise be possible to control the effects of cystinosis by dietary means?

Recently, we had occasion to set up some studies in which nitrogen-free diets were provided to rats. For these studies, chemically defined diets were employed which contained the necessary glucose, vitamins, minerals and essential fats, but which contained no amino acids or protein. However, we neglected to note that in the process of leaving out the methionine and cystine components, we had also left out all traces of dietary sulfur. When we examined the amino acid plasma levels of animals that had been fed these diets for about 5 days, we found that, with the exception of methionine, the amounts and relative proportions of the free amino acids in the plasma were very nearly the same as the fasting amino acid plasma levels—amino acid levels of animals fed a complete diet but fasted for 24 hours prior to the removal of the blood sample. However, the methionine level of animals on the nitrogen-free and sulfur-free diet was only one-third that of the fasted animals. What presumably had happened was that the sulfur needed for certain of the essential metabolic processes was being supplied at the expense of the endogenous methionine.

If we now return to cystinosis, would it not be of interest here to employ nutritionally adequate chemically defined diets wherein the methionine component is provided as the sole source of dietary sulfur at a level just sufficient for growth, maintenance, tissue repair and conversion to the necessary cystine? Excess dietary sulfur for the formation and ultimate deposition of cystine in the tissues could thereby be avoided.

SCHREIER: Dr. WESTALL, would you like to comment on the suggestion of Dr. WINITZ concerning the use of cystine-free diets in FANCONI's cystinosis?

WESTALL: Well, we haven't done anything about dietary treatment of this type for cystinosis. Unfortunately, free amino acids in England are still very expensive, and we have to make a great deal of effort to get amino acids for any particular project of this type. This would seem to be a very long-term experiment, and extremely costly. I agree that this is a thing which will have to come, because there are a number of diseases now where, if the diagnosis is made early enough, you have an extraordinarily good chance of eliminating many of the effects of the metabolic block by a protective diet.

References p. 524

SCHREIER: A cystine-free diet will help the patient with FANCONI's syndrome only if cystine is the toxic substance responsible for the changes. If there is something else, an inborn error of a different type, and cystine deposition is only a secondary effect, one would not obtain very satisfactory results, don't you think?

WESTALL: I agree absolutely. The cystine effect in this cystinosis is so gross that I think you would have to eliminate, if possible, this effect before you can make further studies to see whether or not the cystine deposition was a secondary effect only.

May I change the subject slightly, to alter the drift of the discussion. There are just two things that I would like to put to the meeting here for some consideration. One is the general type of findings in phenylketonuria and in argininosuccinic aciduria, that it appears that brain damage in these cases occurs at a very early age. This, I think, pertains to the possibility that it might be due to amino acid imbalance, inasmuch as during that time, when there is a good deal of brain growth and myelinization takes place, it is more than likely that the nutrient medium from which these substances are synthesized may be considerably out of balance. A further note may be made, possibly, that particularly in these two diseases that I have just mentioned, it appears that they are not usually lethal, and that after the period of growth has finished, the disease doesn't seem to get any worse. Such patients learn to live with their defect and survive to a reasonable age.

The second point is this; the idea of amino acid imbalance has been much to the fore with nutritionists for a good number of years, particularly in view of the low protein diets which are used in underprivileged countries. Leucine and isoleucine are two amino acids that have been studied a good deal but these types of studies tend to be rather long term. We have had some sort of evidence recently, particularly in maple-syrup disease, where the effects of imbalance of leucine and isoleucine are more acute. Two years ago we picked up a patient in England, the first one that we had found, with maple-syrup urine disease. I might, perhaps, say to those who are not so familiar with the disease that the blood plasma amino acids in the condition are in a tremendous state of imbalance. The most obvious thing is that leucine, isoleucine and valine are up to 10 to 20 times their normal value. There seems to be an imbalance of the two sulfur amino acids, although the fact that the methionine is high and the cystine is low is not quite so unusual, now that we know a little more about the plasma levels of younger babies. The other thing is that the KREBS' cycle amino acids, the non-essential ones, are very low. The obvious course of treatment here is to try and correct this imbalance. We devised a diet containing only the minimum amounts of leucine, isoleucine and valine which were needed for growth. The amounts were calculated from information from ROSE's work and from some studies done by SNYDERMAN in New York. The response was quick. These particular amino acids in the blood fell to more reasonable levels, although not quite back to normal; the large urinary excretions of the keto acid analogs of these amino acids disappeared, as did the characteristic smell.

We then thought we would go a little further, because from the point of view of the genetics of this disease we like, if we can, to pinpoint the defect at one site, and we couldn't at that time live with the idea that three of the essential amino acids could be in a way affected by just one enzyme deficiency. So we decided that perhaps only one of the metabolic pathways was in error, and that the other two were involved by some sort of amino acid transport inhibition, so we started to put these amino acids back into the diet one at a time and to watch the effect. The first one we put back was valine because we felt at that time that it might possibly be least important. But as soon as the valine was put back into the diet the blood valine rose to a high level, and the blood α-ketoisovaleric acid came out in the urine, though not as high as it was before. We withdrew the valine, stabilized the child with the basal diet, and gave additional isoleucine. The same sort of thing happened. The isoleucine rose to an abnormally high level in the blood. Finally we put back the leucine to a normal dietary level and once more the leucine level in the blood rose, the other two amino acids tended to rise also, a high amount of the α-ketoisocaprylic acid came out in the urine, and the characteristic maple-syrup smell returned. Although we had the idea the leucine was probably doing more damage than the other two, we couldn't escape from the fact that all three were causing trouble.

There is just one other thing I would like to cite, and that is that we ran out of synthetic amino acids at that time, so I took a casein hydrolyzate, fractionated it on ion exchange columns, withdrew the fractions containing the leucine, isoleucine, and part of the valine, and fed the others plus a little bit of supplemental tryptophane and cystine. Once again the smell disappeared from the urine and the blood levels came down; in this case the blood level of leucine and isoleucine were just about normal. We tended to overdo the withdrawal. We were giving only about half the essential amount of these amino acids. But with regard to valine, which was only reduced by about a third of what it was in the original diet, the blood level rose to a phenomenal figure of 77 mg%. Our general conclusion here correlates with other ideas that have come about, that there is not only a sort of general balance of amino acids of this group, but that the withdrawal of one does have an effect on the others.

References p. 524

SELMA SNYDERMAN was doing some work on amino acids some years ago, and she was studying the requirement of leucine. She found that as leucine was withdrawn without any other change in the diet the level of isoleucine in the blood rose. This is a phenomenon that is now known, and I think it is a thing that might well be studied further. I would like to go just one step further, and make the point that there is a connection, possibly in an indirect way, with certain forms of hypoglycemia, whereby blood glucose has been altered in these cases by high protein, and especially by high leucine level in the diet, and this in turn, it is felt, is also connected with the alterations in the secretion of certain enzymes.

WINITZ: May I add a comment concerning the relation of leucine to hypoglycemia. Some years ago, we studied the effect of toxic doses of the essential amino acids on blood sugar, liver glycogen and muscle glycogen levels. Each of the essential amino acids, as a single $LD_{99.9}$ dose, was injected intraperitoneally into the rat and blood samples were withdrawn periodically by way of the caudal vein until the animal died. It was found that those animals treated with L-leucine, L-isoleucine, L-phenylalanine or L-tryptophane all died with a very marked hypoglycemia. A determination of the liver glycogen and muscle glycogen levels of the animals revealed a 25- to 100-fold decrease in these levels as compared to controls. That there was no apparent relationship between the toxicity of amino acids and the effect of these compounds on the blood sugar level was subsequently revealed by the fact that although the simultaneous administration of glucose did not alter the lethal effect of these amino acids, the animals now died with a marked hyperglycemia. It therefore appeared that the animals were not dying as a result of the hypoglycemia *per se*.

We have also carried out some studies with dietary leucine–isoleucine imbalances. The initial studies in this area were, of course, carried out by ELVEHJEM, HARPER and their associates, who found that a dietary balance between leucine and isoleucine is necessary for optimal growth. The ideal dietary ratio of these amino acids, in our own experience with chemically defined diets, appears to be in the vicinity of about 10 parts of leucine to 6 or 7 parts of isoleucine. If you go too far beyond these values in either direction, then a imbalance develops which is reflected in poorer dietary intake and consequently in poorer growth of the animals. Thus, for example, it is not necessary to withdraw a certain amount of either leucine or isoleucine from completely adequate diets in order to get poorer growth but merely to add one or the other of these components to the diet in an amount sufficiently large to disturb the ratio.

What is not generally realized, either, is that there is a delicate balance, not only between leucine and isoleucine, and perhaps valine enters the picture here too, but between lysine and tryptophane as well. It appears that the most efficient dietary ratio of lysine to tryptophane is about 6 or 7 to 1 on a molar ratio basis.

SCHREIER: I should like to make an addition to the remarks that Dr. WINITZ just made. In the recent medical literature one can find that leucine increases the excretion of insulin—PAYNE and others did a determination of insulin in blood and claimed to have found very high increases—let's say "eine Ausschwemmung."

The lysine–tryptophane imbalance has been stressed very much among the classical nutritionists. ALBANESE claimed that lysine and tryptophane should have, in human nutrition, a certain ratio, and he published several papers where he put in the value of the protein according to the tryptophane–lysine content. The nutritionists believe, and I believe it too, of course, that lysine is the amino acid which is needed most in growth because all the proteins of the different cereals, wheat and so on, are very poor in lysine; this amino acid is very important for the developing countries. Of course, some big firms wanted to sell their tons of lysine, and they proposed that lysine be added to flour and to different foods in the U.S.A. and in western Europe. We became interested to determine whether lysine had any toxic effect and whether it was meaningful to add lysine or whether it was much better to add some cheap protein to the diet. So we carried out some studies on infants and on rats. If we increased the lysine content of the food by adding 700 mg/kg body weight to the milk formula as ALBANESE has suggested, then we got not only a higher excretion of lysine but of many another amino acids too. We tried to show the same thing in rats with high doses of lysine, and we could produce in this way a permanent hyperaminoaciduria.

WINITZ: Some of our own experiences have led us to the belief that the degree of essentiality of lysine as a dietary component—for that matter of any of the so-called essential amino acids—is dependent to a large extent on the clinical state of the animal. In the case of rats bearing the WALKER tumor, lysine appears, ironically, to be one of the least essential of the essential amino acids. We made this somewhat bizarre observation some years ago when we force fed various chemically defined diets to rats bearing the WALKER tumor as well as to normal controls. These diets differed from one another in that, although they all incorporated nine of the ten essential amino acids, each lacked a different one; they were otherwise identical and contained all of the necessary dietary components. Diets which were devoid of one of the branched-chain amino acids, valine, isoleucine or leucine, led to the most toxic effects. The valine-free diets induced

References p. 524

especially severe reactions in that the animals became quite ill after only the third or fourth day and died as early as the seventh day of force feeding. In the case of the other essential amino acids, these toxic reactions were not so severe. After 11 days of force feeding, the animals were sacrificed, seven days in the case of valine, the tumors excised, and tumors and carcasses weighed separately. Not much variation with regard to tumor growth was shown between the different groups. In addition, the carcasses of the animals on each of the deficient diets showed significant weight losses with but two exceptions, the exceptions being the animals fed the lysine-free and tryptophane-free diets, where slight net gains in carcass weight were noted despite the usual rapid tumor growth. This finding was further substantiated by nitrogen balance studies, which revealed that the tumor-bearing animals on these two diets suffered virtually no net nitrogen loss over the entire eleven day period. Normal animals on these same diets showed a very marked negative nitrogen balance, however, as did the tumor-bearing animals on the other deficient diets. From these results, one is almost inclined to the impression that, in the case of the lysine-free diets, for example, the tumor-bearing animal has in some mysterious way developed the ability to synthesize a portion of its lysine requirements, although what actually occurred may well have been due to a very highly efficient mobilization of nitrogen between carcass and tumor.

L. MILLER: While Dr. WINITZ is talking in such a fascinating way, I wish he would go on and tell us a little more about the manner in which these animals die in only a few days from forced feeding of an amino acid mixture deficient in leucine, isoleucine or valine. This may not seem as striking to some of you as it does to me, because quite obviously an animal that is getting adequate calories and vitamins and minerals can carry on for a very long time with relatively little difficulty, and the fact that forcing this amino acid mixture on these animals, which they won't eat by themselves, incidentally—an expression of some sort of wisdom of nature—the fact that it leads to death in such a short time implies that there is a great deal more to amino acid balance, I think, than we really understand.

WINITZ: Yes, Dr. MILLER, this is undoubtedly very true. When you feed a diet that is completely devoid of only one of the essential amino acids, you are in effect imposing a more toxic condition than were you to supply no nitrogen source whatsoever. For example, were you to fast an animal overnight, don't feed it anything but just fast it, a 200-gram animal would lose 10% of its body weight, or about 20 grams. Now were you to feed this same animal a nitrogen-free diet—just leave all the nitrogen out of an otherwise adequate diet—the animal loses weight, of course, but it doesn't lose 20 grams per day; it uniformly loses only about 2 grams per day. So it takes an animal on a nitrogen-free diet a matter of ten days to lose the same amount of weight that a fasting animal loses overnight. Yet during all this time, the animal on the nitrogen-free diet appears healthy, happy and vigorous. If you were to add an amino acid mixture which lacks only one of the essential amino acids to this same nitrogen-free diet, this diet would now induce quite toxic effects. ROSE observed this quite dramatically in this studies on the amino acid requirements of man, especially in his study on the isoleucine requirement wherein he fed his students diets deficient in this amino acid. After a few days on isoleucine-deficient diets, pronounced psychological disturbances developed in these young men, who found it increasingly more difficult to tolerate the diet. In fact, their disturbed mental state became such that they actually pleaded with ROSE to place them on a more tolerable diet. And all of this developed from an isoleucine deficiency in the diet. It's really a remarkable phenomenon.

H. ROSENBERG: Rats will not take these mixed diets. How well do they take complete mixtures of amino acids?

WINITZ: Animals will, of course, accept these diets containing complete mixtures of amino acids—and very, very well—because when these diets are provided as the sole source of food, they will grow, undergo normal pregnancy, produce satisfactory litters and undergo normal lactation. This has been well established. It should also be pointed out that animals will accept, and again very well, nitrogen-free diets, although they will not grow on these diets. They will nevertheless ingest these diets at a remarkable rate. They will ingest them even more readily than they do a complete dietary mixture containing nitrogen, presumably because they are searching for a source of nitrogen. But should you give them a mixture from which only one essential amino acid has been left out, then they will soon reject it entirely. It now becomes necessary to force feed in order to study the effects induced by these diets.

E. ROBERTS: Do they eat even a little bit of this mixture?

WINITZ: Just a very little bit and then they reject it.

E. ROBERTS: That is sufficient signal so that they don't eat any more?

WINITZ: Apparently so. But, it's a very interesting thing, because if you now take this same mixture, in which you have only nine essential amino acids, and add the one that is missing, the animals will immediately accept the mixture and start eating again.

WESTALL: In view of the things that have just been said, it is extremely interesting that in the maple-syrup syndrome the first indication that these children might not be normal seems to occur at about the fourth day, when they have a disinclination to suck. The second point I would

References p. 524

like to say, too, that of all the metabolic diseases involving amino acids that we know so far, this one, which involves leucine, isoleucine and valine, is by far the most lethal, and many of the children with the disease die at about 11 to 12 days.

SCHREIER: I should like to generalize, if I may. Any shortage of an essential factor in food—and the same thing applies to any overdosage—will initially appear in the appetite of the human being. If you have too little vitamin D children won't eat and if you add too much they lose their appetites. Apparently it is due to something inborn. I further would like to add that if you give insufficient protein hydrolyzates to human premature infants, they don't gain weight. If these are added to human milk, they will not grow better but rather they will show less growth and they will show a high excretion of several nitrogen products because of an imbalance, and you can find it in the blood. The amino acid missing, which is usually tryptophane, drops to a very low value. It almost cannot be found, and that may be the reason why there is no growth.

E. ROBERTS: I am very much interested in this signal that goes to the central nervous system and is reflected in such a complex pattern as a refusal of the diet, on the basis of the absence of a single dietary constituent. I think that this certainly means that there must be centers in the brain which are constantly monitoring the total chemical composition in their environment and sending out signals which will then coordinate the activity of the organism with the environment on the basis of the findings. I think that is as clear as can be. And, therefore, I think there is a good reason for neurophysiologists to become interested in nutrition, which is a thing that is often furthest from their minds.

SCHREIER: There are studies which show that the hunger of the human being and of animals is very well localized in the hypothalamus, and you can increase and decrease the hunger by increasing blood sugar—that is well known and an old phenomenon. MELINKOFF found also that you can change the hunger feeling of the animal and the human being by increasing and decreasing amino acids in blood. And the same thing, I think—I am not sure about it—holds with vitamin B_1, which if given it in high doses, will immediately decrease the hunger feeling.

L. MILLER: Decrease it? This is putting it right in the diet?

SCHREIER: No, you inject it in high doses.

H. ROSENBERG: I would like to ask Dr. WINITZ, what is the time delay between tasting or consuming an incomplete diet and the rejection? Is this a matter of tasting or is it actually consumed and monitored?

WINITZ: I think you have several effects combined. Appetite is probably one of them. Again, several years ago, we studied the physiological response to high concentrations of glucosamine in chemically defined diets. We carried out these studies because the work of QUASTEL, some years earlier, had indicated that D-glucosamine might be a good tumor inhibitor. What we attempted to do initially was use the D-glucosamine both as a dietary source of non-essential nitrogen and as a source of carbohydrate. However, it was soon found that the animals completely rejected this diet, and as a result it became necessary to set up diets to which a source of non-essential nitrogen and glucose was added and which, in addition, contained varying amounts of glucosamine. It was now found that the animals would voluntarily ingest the diets, but that the amount of diet ingested varied inversely with the amount of glucosamine it contained. In fact, the variation of dietary intake with the glucosamine level was so quantitative and predictable that it actually became possible for us to control the amount of weight that an animal would ultimately gain over a 50- or 60-day period, to within only a few grams, merely by employing a pre-determined and fixed level of D-glucosamine in the diet. The animals on the lowest D-glucosamine diet—this was about 25 g of glucosamine per kg of dry weight of diet—ingested the diet in rather considerable amounts. But when the glucosamine level was raised to 150 g, the dietary intake fell off very markedly as, of course, did the growth.

We felt that these *ad libitum* feeding experiments were not an adequate test of QUASTEL's effect because at the high glucosamine levels, the dietary intake and hence the intake of glucosamine was insufficient. So we subsequently undertook forced feeding studies with tumor-bearing animals, and we forced these animals, now, to take in larger quantities of these higher glucosamine diets than they would voluntarily inject *per os*. We found, in these experiments, that a high glucosamine intake had no effect on tumor growth. But more to the point of the present discussion, we further found that animals, when force fed identical amounts of identical diets which differed only in the glucosamine level, showed exactly the same growth pattern, the same metabolic pattern, and the same degree of nitrogen retention. So it appeared that the relation between dietary intake and glucosamine concentration was not due to any toxicity of the glucosamine *per se*. It rather looked like an appetite factor here.

SCHREIER: May I try to answer, from the standpoint of human studies, the question raised by Dr. ROSENBERG? If you leave a vitamin out of the diet, of course the taste of the diet remains entirely the same and children don't notice it, so it would, depending on the missing vitamin, take several weeks or months before they stop eating.

H. ROSENBERG: My question was really directed toward these rat experiments. I was just

References p. 524

wondering, when you were feeding those rats on a complete synthetic amino acid mixture and then withdrew one essential amino acid, how long would it be before they stopped eating?

WINITZ: This would depend upon the particular essential amino acid you chose to withdraw.

H. ROSENBERG: Let's say the most severe condition.

WINITZ: Withdrawal of leucine, isoleucine or valine lead to the most severe conditions. But with regard to the animal not eating a diet, it is not necessary to leave an essential amino acid out of the diet. The same effect can be achieved by using a particular non-essential amino acid as the source of non-essential nitrogen. When we studied the ability of different non-essential amino acids to serve as the sole source of dietary non-essential nitrogen—L-alanine, D-alanine, glycine, serine, hydroxyproline and various others—we found that the growth response of the animals varied considerably with the amino acid employed. We further found that certain amino acids, as in the specific instance of hydroxyproline, were extremely toxic. The animals rejected the diets where this material was employed and consequently their weight fell off very markedly. The apparent toxicity of hydroxyproline in the diet was probably due to the inability of the animal to metabolize it effectively in large quantities. In any case, the animals refused the diet almost immediately. They would take only a few sips and that's all.

E. ROBERTS: In line with that, is feeding collagen toxic to animals?

WINITZ: No.

E ROBERTS: Well, how would we square that with hydroxyproline being toxic, and not collagen?

WINITZ: This I don't know, except that it may be a matter of hydrolytic rate—just how much the organism is actually exposed to at any given time. The collagen is very, very slowly hydrolyzed by the proteases of the gut as opposed to other proteins, probably due to the very fact that it contains so many hydroxyproline residues, and this may be part of the answer.

SCHREIER: In studies in human beings, methionine seems to be the most toxic amino acid of all. Dr. SOUPART, would you like to add something to this subject?

SOUPART: The only things I know about amino acid imbalance deal with protein in malnutrition, especially of the kwashiorkor type. When there is a slight protein deficiency, as in mild cases of this disease, the only abnormalities that are found in urine are high excretion levels of taurine and β-aminoisobutyric acid. Both of these are end-products of metabolism, and in order to find hyperexcretion of other amino acids, one must investigate the more severe cases of protein malnutrition. This probably results from the fact that the kidney itself is affected in severe cases; in mild cases, however, when high protein diets are given to the patients, the result of this therapy is to disclose the hyper-amino acid excretion. For example, when milk protein is given to children with kwashiorkor, a hyperexcretion of all of the other amino acids occurs for some time. This therapy is really effective, and taurine excretion disappears very quickly, in most instances in 24 hours. The other amino acids are elevated in the urine but return to a normal level within one or two weeks; the only one which remains elevated for a longer period in the urine is β-aminoisobutyric acid.

I would ask for suggestions as to the interpretation of these findings. Do the high taurine and β-aminoisobutyric acid levels result from a disturbance of the permeability of the cells? Do they result from the fact that there is a higher number of cells dying as a result of starvation? Or is there another interpretation of these facts?

WESTALL: With regard to β-aminoisobutyric acid, I would be inclined to think that it is probably due to an increased number of cells dying. In the examination of routine chromatograms over a number of years, we have been rather shaken by the alteration in taurine, and Dr. DENT has come to the general conclusion that if the patient has a fever it is usually reflected in an increasing taurine excretion in the urine.

To say something rather different, in general terms, I would like to make a plea here that studies in diseases and nutrition stress the condition of the amino acids in the blood. This, in my opinion, is far more important than what is shown in the urine. The intervention of the kidney tends, in a way, to give a false picture of what may be happening, because different amino acids have different rates of clearance. It is much easier, for example, to force histidine out into the urine than it is to force out isoleucine, so that the isoleucine level of the blood may be up three times and it would never be seen in the urine.

WINITZ: I most fully agree with Dr. WESTALL's statement concerning the importance of the blood amino acid picture. Recently, we have had occasion to use the fasting amino acid plasma levels of the rat to determine its essential amino acid requirements in only a few days with a greater degree of precision than can be achieved by the conventional long and tedious growth or nitrogen balance studies. The method we employed is extremely simple. The animals, 100g rats in this case, were fasted for 24 h, blood samples were taken, pooled and deproteinized, and the amino acid composition of the plasma determined with the aid of the amino acid analyzer. Chemically defined diets were then prepared which contained the essential amino acids in exactly the same molar proportion as they appeared in the fasting plasma. These diets, when compared with the best chemically defined diets we had been previously been able to devise over a period of six

References p. 524

years using the usual criterion of growth, showed a significantly greater growth response. All diets compared were isonitrogenous, but there were striking differences in the relative proportions of the essential amino acids as obtained by the *fasting plasma concept* and the earlier growth studies. A simple explanation as to why this concept is here applicable can possibly be attributed to the fact that since the vital body functions and metabolic processes must continue even in the absence of a dietary source of nitrogen, the condition found in the animal in the fasting state, the protein reserve of the body is broken down in the most efficient manner that will permit the organism to survive for the longest period of time. As the blood is the vehicle which distributes the amino acids liberated during this protein breakdown to the various tissues, its fasting amino acid composition reflects the nutritional needs of the organism. I should mention that this concept was not initially proposed by us but was developed by Dr. JAROWSKI of Chas. Pfizer and Co. Its experimental demonstration was carried out, in collaboration with members of his group, at our laboratory in Bethesda. This experience clearly indicated to us that the blood amino acid profile could tell us a good deal more about the nutritional needs and mechanisms of metabolic action of the organism than we had formerly thought possible.

ROCKLAND: It may be appropriate to recall that analogous amino acid imbalances have been observed with a variety of microorganisms including the lactic acid bacteria and ciliated protozoa, *Tetrahymena*. It is well known that minimal amounts of required amino acids limit the growth response of microorganisms in direct relation to their proportions in the basal medium and that under these conditions, high relative proportions of other amino acids will impose further limitations of their growth. It is perhaps more significant that even on an optimum basal medium, containing completely adequate amounts of required amino acids, the presence of high proportions of one or more amino acids may inhibit severely the growth response of microorganisms.

It is suggested that the mechanisms and physiological basis for inhibited growth or other effects of amino acid imbalance might be elucidated through investigations of this phenomenon using the convenient, germ-free, one-celled animal *Tetrahymena*.

SCHREIER: With this final note, we will close this session on free amino acids. Thank you very much, all of you.

REFERENCES

1 E. ROBERTS, *Intern. Symposium on Inhibition in the Nervous System and γ-aminobutyric Acid*, Pergamon Press, 1960, p. 144.
2 R. DOMENJOZ, *Arch. exptl. Pathol. Pharmakol., Naunyn-Schmiedeberg's*, 208 (1948) 144.
3 W. J. HAYES, F F. FERGUSON AND J. S. CASS, *Am. J. Trop. Med.*, 31 (1951) 519.
4 T. B. GAINES, *Toxicol. Appl. Pharmacol.*, 2 (1960) 88.
5 B. J. JANDORF, H. P. SARETT AND O. BODANSKY, *J. Pharmacol. Exptl. Therap.*, 88 (1947) 333.
6 J. D. JUDAH, *Brit. J. pharmacol.*, 4 (1949) 120.
7 E. A. HOSEIN AND H. MCLENNAN, *Nature*, 183 (1959) 328.
8 E. A. HOSEIN AND P. PROULX, *Nature*, 187 (1960) 321.
9 E. H. COLHOUN AND E. Y. SPENCER, *Science*, 130 (1959) 504.
10 G. M. MCKHANN, R. W. ALBERS, L. SOKOLOFF, O. MICKELSEN AND D. B. TOWER, *Intern. Symposium on Inhibition in the Nervous System and γ-Aminobutyric Acid*, Pergamon Press, 1960, p. 169.
11 D. KESSEL, *Federation Proc.*, 18 (1959) 258, A-1020.
12 C. F. BAXTER AND E. ROBERTS, *Proc Soc. Exptl. Biol. Med.*, 104 (1960) 426.
13 C. F. BAXTER AND E. ROBERTS, *Amino Acid Pools*, Elsevier Publishing Cy., Amsterdam, 1962, p. 499.
14 E. ROBERTS, in S. S. KETY AND J. ELKES, *Regional Neurochemistry*, Pergamon Press, 1961, p. 324.
15 K. SANO AND E. ROBERTS, *Biochem. Biophys. Research Comm.*, 4 (1961) 358.

PART TWO

DYNAMIC ASPECTS OF CELLULAR FREE AMINO ACID POOLS

I. PERMEABILITY AND AMINO ACID TRANSPORT

ON THE MECHANISM OF AMINO ACID TRANSPORT INTO CELLS

HALVOR N. CHRISTENSEN, HITOSHI AKEDO, DALE L. OXENDER AND CHARLES G. WINTER

Department of Biological Chemistry, The University of Michigan, Ann Arbor, Mich. (U.S.A.)

We intend in this presentation to consider some evidence bearing on the nature of the process by which amino acids are transported into cells. In selecting this emphasis we do not mean to disregard the unifying interest in the biological meaning of the free amino acids of the cell, around which this symposium is organized. Rather we hope whatever assistance we can give with that theme can be offered throughout the conference, so that we can reserve the present opportunity for directing your attention to the challenging problem of *mechanism*.

The picture of amino acid transport that has so far emerged for most cells divides the amino acids into at least three classes, the neutral, the anionic and the cationic. Competitive effects across these class lines are rarely seen and, as far as we are aware, not well documented. This division may well arise from the presence of three distinct types of transport sites, one for each of these classes. Or instead, the charged amino acids may not be able to approach the site for neutral amino acid transport because of repulsion by a nearby charged group. In passing we can comment that we have recently approached this question by synthesizing the new amino acid, DL-trifluoromethylalanine:

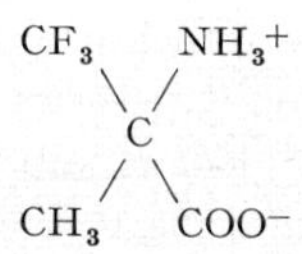

With remarkable values of 0.5 and 5.94, respectively, for the pK_a values of the carboxyl and amino groups, this amino acid is monocarboxylic and yet at pH 7.4 anionic. In various systems it inhibits the transport of neither neutral nor anionic amino acids; up to the present we do not know if it even has a mediated transport. We can not yet interpret this finding to exclude the possibility that dicarboxylic amino acids do use sites differing from those used by neutral amino acids only in the presence of a negative charge barrier; this new amino acid has such a strong carboxylic dissociation that conceivably the carboxyl group fails to hydrogen-bond to the usual bonding points, thereby explaining the absence of interaction with ordinary amino acid transport sites.

My main concern today, however, is with the neutral group of amino acids, and with what we may learn by considering interactions in their transport. Questions at issue are whether two- or three-point attachment occurs in their transport, and whether these amino acids can be arranged in a single consistent series of increasing affinities, whether estimated by relative rates or relative inhibitory actions, as has generally been assumed.

This assumption is not a necessary one. Fig. 1 shows schematically a widely known conception of active transport. Several other representations would serve as well. At the right-hand limit of the membrane, the transport site or carrier is assumed to be removed, in this case by conversion to a less suitable form, and then to be regenerated in the reactive form at the left-hand or outer limit. Energy is required for this portion of the cycle. The effect of this energy delivery process is to modify the transport site to make it more accessible or more attractive on side *o* of the energy-barrier and less so at side *i* so that the amino acid tends to be dissociated into the right-hand phase at higher concentrations. Investigators have generally

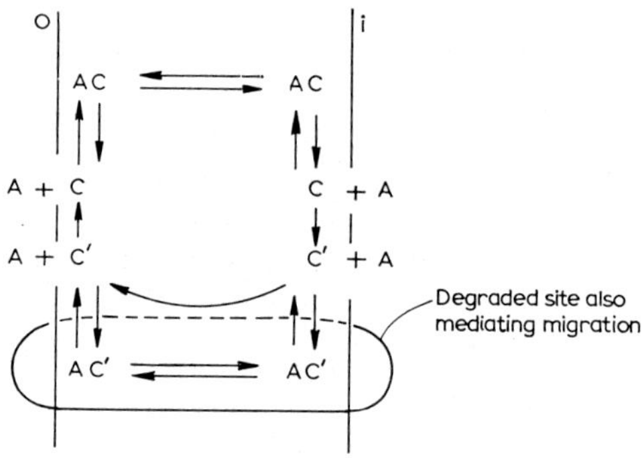

Fig. 1. Scheme for active transport. To a standard model we have added, at the bottom of the figure, the possibility that the transport carrier or site retains some affinity for amino acids after it has been modified or degraded, and can serve for the migration (primarily the exit) of some amino acids more than for others.

assumed that this modification of the transport site causes a complete loss of affinity. If so, a single affinity constant should describe all interactions within a class of amino acids that use the given site. But if substantial affinities are retained after the site modification, so that migration via this site may also occur, as illustrated in the lower encircled portion of the figure, then a second affinity, for this *unfavored* or *inferior* form of the site becomes relevant. In that event inconsistent sequences may be obtained, depending on whether the rate of transport or the steady-state asymmetry of distribution is being observed. An amino acid with a high affinity for the site in both its initial (preferred) and final (inferior) form may cross the membrane rapidly in either direction and act as a strong competitive inhibitor to the transport of other amino acids, and yet not be greatly concentrated. Another amino acid that is very easily pushed off the site by the site modification may eventually be strongly concentrated, although at a slow rate if the affinity even for the favored form of the site is low.

If this analysis is correct, the phenomenon under question may be signalled by inconsistencies between the order in which the *rates* and the *extents* of uptake take their place. That is, the extent to which an amino acid is concentrated at the steady state will measure a simple transport affinity only if the losses against which active uptake operates occur by diffusion or by other processes having much the same rate for every amino acid, as is usually assumed; but not if exodus occurs through mediation by a second site for which a second separate order of affinities applies. This second site could of course be an entirely distinct one rather than a modified form of the active transport site. In that event we would need to assume that its role in exodus is predominant only because our observations are usually made at high cellular gradients, causing an asymmetric use of this site for exodus, at the steady state just balancing the operation of the active transport site for entry.

Apparent inconsistencies between the order of affinity and the order of the extent of accumulation already can be found if we examine published work.

In 1952 we showed that the extents of uptake by the Ehrlich cell of the D- and L-forms of several amino acids (not including alanine), although generally preferring the L-form, are remarkably similar[1]. Subsequently PAINE AND HEINZ[2] found much larger differences between the D- and L-forms in their effectiveness in inhibiting the uptake of glycine. Phenylalanine and histidine were studied in both investigations and gave highly divergent results by the two methods. Presumably the comparisons by the two methods ought to be made simultaneously before the inconsistency is taken to support a double set of affinities.

In exploring for inconsistency in the affinity series, let us first look at how a group of amino acids arrange themselves as to affinity for transport into mature human erythrocytes. Although rather typical ability to accumulate amino acids is shown by nucleated erythrocytes[3] and at the reticulocyte stage[4], this ability for uphill transport has largely been lost on the maturation of these cells. Some amino acids, such as glycine, appear endogenously in erythrocytes at levels apparently somewhat higher than those in the plasma[3], but when added in the labeled form to whole heparinized blood, the isotope in the instances so far studied is only very slightly accumulated into cells, approaching distribution ratios of approx. 1.1 on a water basis[5]. These cells probably offer us the best chance to find a simple straight-

TABLE I

DISTRIBUTION RATIOS OF AMINO ACIDS, ON A WATER BASIS, BETWEEN RED BLOOD CELLS AND PLASMA REACHED IN VARIOUS INTERVALS OF TIME AT 37°

From WINTER AND CHRISTENSEN[5].

Amino acid	*30 min*	*90 min*	*120 min*	*150 min*	*240 min*
Glycine	0.26			0.51	0.72
Sarcosine	0.24		0.47		0.60
L-Alanine	0.42 (25 min)	0.54		0.74	0.89
D-Alanine	0.24 (25 min)			0.50	0.62
α-Methylalanine	0.20	0.35	0.38		0.60
α-CF_3-alanine	0.60	0.70			
L-Proline	0.78		0.95		0.99
L-Isovaline	0.39		0.77		0.99
D-Isovaline	0.27		0.56		0.86
L-Valine	1.01		1.07		1.04
L-Leucine	1.07		1.05		1.08
DL-Norleucine	1.05		1.07		1.05
Cycloleucine	0.96	1.10			

References p. 538

forward affinity sequence. Although the apparatus for uphill transport, functional at the reticulocyte stage, is probably still complete in the mature cell, the energy delivery may very well have become inadequate. In this event very little of the hypothetical site modification will occur, and transport in both directions may be mediated symmetrically by a single site. Some of you might tend to consider the uptake of amino acids by such cells as irrelevant to our problem since "pooling", if we understand the sense in which the term is now often used[1] hardly seems to occur.

Table I shows a comparison of the entry of a number of amino acids at 0.8–1.2-mM levels into human erythrocytes, as observed by ^{14}C-labeling. You will note in this table the sequence: glycine, alanine, valine, leucine, in which the rate of uptake increases as the length of the hydrocarbon chain increases. Either the N-methyl or the α-methyl group decreases the rate, as does the α-trifluoromethyl group. The latter derivative, an anionic amino acid, nevertheless enters the erythrocyte faster than it does other cells studied, probably because of the facilities for anion transport possessed by this cell. A transport advantage of the L- over the D-configuration is evident in two cases here, for alanine and isovaline.

The increase of rate with length of hydrocarbon chain might be interpreted in either of several ways. One could suppose that the increasing lipophilic nature might permit passage across a lipid barrier, perhaps in the form of a complex in which the amino and carboxyl groups are masked. This hypothesis fails to account for the advantage of having the longer side-chain in the position characteristic of the L-series. Furthermore one doubts that a liquid phase of randomly orientated lipid is encountered. More likely London forces, the so-called hydrophobic bond, between the hydrocarbon side-chain and a hydrocarbon structure in the membrane helps to orientate the amino acid molecule with reference to the two primary bonding points. One might propose that the side-chain forms a distinct and essential third point of attachment as suggested by PAINE AND HEINZ[2]. Against this idea stands the wide variety of side-chains that serve, as shown more extensively for the Ehrlich ascites tumor cell, and also the ability of a hydrogen atom to serve in the case of glycine.

Cycloleucine is the name used here for convenience for 1-aminocyclopentane-1-carboxylic acid*. The rapid uptake of this amino acid by all tissues studied refutes any idea that an α-hydrogen is necessary for typical amino acid transport. Its transport into red blood cells is inhibited by leucine, whereas it in turn inhibits the transport of α-aminoisobutyrate, as one would expect from the order of their rates of uptake.

Limited further exploration of the inhibitory effects in erythrocytes suggests that the other amino acids, except for trifluoromethylalanine, also belong in the same transport family. For example the uptake of valine can be inhibited by valine itself, or by alanine in a 20 : 1 excess. The uptake of proline is inhibited by leucine, norleucine and valine, as would be expected. Leucine and valine, as well as cycloleucine, inhibit the uptake of α-aminoisobutyrate (α-methylalanine), as one would expect,

* This amino acid has recently been shown to be toxic when fed to rats, and to have antitumor activity[6]. It does not specifically antagonize the utilization of any of 17 amino acids tested in several bacterial species[7]. Conceivably its transport behavior underlies its toxic action. We are investigating its metabolic fate. The labeled amino acid can be obtained from the California Corporation for Biochemical Research.

but proline and alanine failed to inhibit the uptake of this amino acid. The preliminary indications are that the uptake of α-aminoisobutyrate is probably not as readily inhibited as it should be if we are dealing with a simple, single-affinity sequence, but certainly no striking inconsistencies are yet evident for the red blood cell.

We turn then to the Ehrlich ascites cell, for which the uptakes of several amino acids have been rather extensively proved concentrative, and which accumulates a great variety of amino acids.

TABLE II

INITIAL UPTAKE RATES AND DISTRIBUTION RATIOS AT 30 MIN FOR A NUMBER OF AMINO ACIDS IN THE EHRLICH CELL

The steady-state external levels were in the range 0.5–0.9 m*M*.

Amino acid	*Uptake rate 1st min (μmoles/ml cell water/min)*	*Distribution ratio at 30 min*
L-Methionine	3.52	12.9
Cycloleucine	2.90	18.7
DL-Norleucine	2.87	10.2
L-Leucine	2.32	3.3
L-Histidine	2.08	12.7
L-Valine	1.55	5.0
α-Aminoisobutyrate	1.27	26.3
L-Proline	1.20	11.8
D-Alanine	1.0	4.66
L-Isovaline	0.85	10.1
D-Isovaline	0.45	8.2
Glycine	0.67	10.5

Table II compares the uptake behavior of a number of amino acids in molar terms. Here again we see the same sequence of increasing rates: glycine, alanine, valine and the leucines, using the 1-min values. The positions taken by L-isovaline and α-aminoisobutyrate are slightly different from those of the erythrocyte. But many of the very rapidly accumulated amino acids soon reach their steady-state levels, while the uptake of others continues much longer, until the inverted order: leucine, valine, alanine, glycine is assumed, the more slowly accumulated members of this aliphatic series being concentrated the most. This simple relationship does not apply outside the simple L-aliphatic series. For example methionine has a rapid uptake but also yields a steep gradient. Note that the initial rate for L-isovaline is nearly twice that for the D-isomer, whereas the distribution ratios at 30 min are quite similar. This is the result predicted above from comparing our results with those of PAINE AND HEINZ[2].

Fig. 2 illustrates the progressive uptake of three contrasting amino acids types: leucine, which is taken up quickly but briefly; glycine, which is taken up more slowly but continuing until a much higher distribution ratio is reached; and cycloleucine and methionine which are taken up quickly but to high gradients.

The aliphatic group in the L-position appears to assist in the initial combination with the transport site, but may be unfavorable to the loss of affinity on the hypothetical site modification. Either a polar atom in this aliphatic group, as in methionine, or the shift (as in D-alanine and D-isovaline) or extension of the aliphatic group

References p. 538

over into the D-area (as in α-aminoisobutyric acid and cycloleucine) appears to permit more effective depletion of affinity. Accordingly the D-amino acids show low affinities for influx but are eventually concentrated almost as well as the L-forms, apparently because they show proportionally as large an affinity difference between the sites predominating for influx and efflux, respectively. Perhaps it will be possible to reach by this pathway a complementary picture of the character of a site deformation that tends to dislodge amino acids.

If we suppose that an identical parcel of energy serves in the preparation of the transport site for amino acid binding, no matter which neutral amino acid is to be

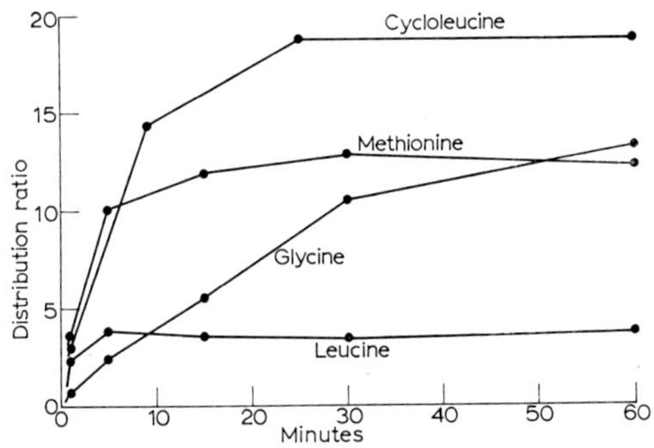

Fig. 2. Progressive uptake of several amino acids by Ehrlich ascites tumor cells at 37°. Initial level of the test amino acid in Krebs–Ringer–bicarbonate medium, 1 m*M*; 25 vol. of suspending medium used.

carried, energy wastage might be anticipated if considerable affinity is retained on site modification. Until we know better the character of the energy exchange, however, this concern is perhaps premature. It is of course possible that the apparent second affinity arises from efflux by an entirely distinct site; the results in our tests so far would be the same, although we would then have to explain why one site is available mainly for exodus, another for entry. In the present formulation we suppose that the preferred site is eliminated almost as fast as it appears at the inside limit of the membrane; whereas the degraded site is reactivated as quickly as it reaches the outer limit.

The speed of the uptake of the leucines and valine in comparison with methionine, histidine and other amino acids, out of all proportion to the extent to which they can be concentrated by the Ehrlich cell, does not arise merely from their susceptibility to uptake by exchange, since these rate superiorities are retained when cells have been depleted of amino acids, and are observed for erythrocytes, which have very small quantities of endogenous amino acids available for exchange. Note that the uptake of glycine and α-aminoisobutyrate reaches its relatively lowest rate in the erythrocyte, supporting the view that with this cell we are measuring mainly the affinities for a single unmodified site.

Other evidence also supports the existence of a mediated exodus process for solutes from cells, a process which can be modified without modifying the mediated entry. HORECKER *et al.*[8] detected an inducible exit process for galactose in *E. coli.*

References p. 538

Growth on mannose or succinate plus 10^{-5} *M* galactose increased the exit rate for galactose, thereby decreasing the equilibrium internal galactose level. 2,4-Dinitrophenol blocked the entry process so that the exit process could be studied separately. OSBORN AND McLELLAN[9] have reported that dinitrophenol also accelerates the exit process in the same strain, an effect which we suggest could arise from an interconvertibility of the transport mediator.

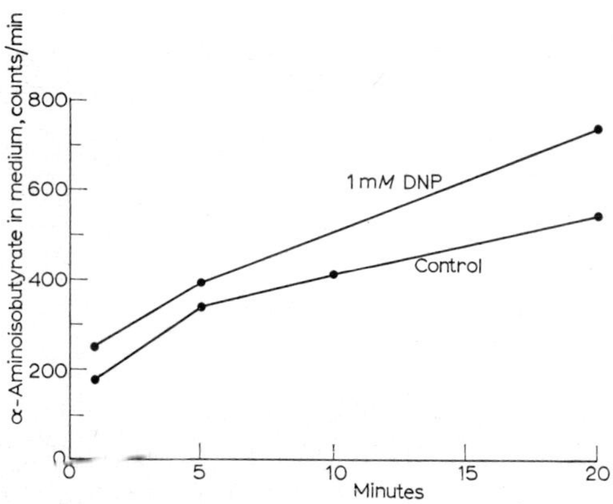

Fig. 3. Increase by 2,4-dinitrophenol (1 m*M*) of the exit rate for previously accumulated labeled α-aminoisobutyrate. The Ehrlich cells had been incubated 40 min in 20 m*M* α-aminoisobutyrate, washed in cold media, and then transferred to a large volume of Krebs–Ringer–bicarbonate buffer held at 37°.

OXENDER[10] and OXENDER AND ROYER[11] have shown that pyridoxal phosphate and related agents stimulate the extent of amino acid uptake by cells by slowing a mediated efflux, and not by accelerating the mediated influx. In the Ehrlich cell we find that dinitrophenol distinctly accelerates the efflux of either valine or α-aminoisobutyrate (Fig. 3) supporting the idea that the exit site may be an inactive or inferior form of the entry site, which is present in increased amounts when the energy supply of the cell is curtailed.

The intestinal transport of amino acids shows a remarkable resemblance to the behavior I have described so far. At low levels the leucines and valine are found to be most rapidly transported and most strongly concentrated into the everted gut sac of the rat, using the technique of CRANE AND WILSON[12]. With shorter side-chains slower rates are encountered. The amino acids with lipophilic side-chains are also the strongest inhibitors of transport. In fact they are such good inhibitors that at 20-m*M* levels their transport becomes the slowest, as WISEMAN showed in 1953, not the fastest. The intestinal transport of glycine and α-aminoisobutyrate can be extensively suppressed by valine or methionine, but unreasonably high levels are required, and the intestinal transport of these two symmetrical amino acids can scarcely be saturated by raising their levels (H. AKEDO, cited by CHRISTENSEN[13]). Hence, although glycine and aminoisobutyrate (AIB) apparently use the ordinary transport site, they are not displaced from it by other amino acids as readily as would be expected from their low transport rates. Again departure from a single affinity series is evident.

References p. 538

In Fig. 4 we have arranged results obtained by FINCH AND HIRD[14] on the uptake (not the absorption) of amino acids by isolated intestinal wall. You will note that the rates of uptake at 10 mM levels tend to be reversed from those seen at 1 mM. At high levels a high affinity solute may leave so few sites available that only rarely

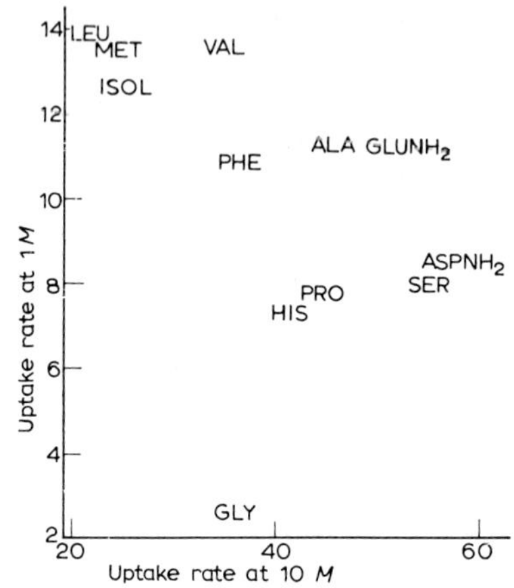

Fig. 4. Rates of uptake of amino acid by isolated small intestine of the rat at 1 mM plotted against the rates at 10-mM levels. Data of FINCH AND HIRD[14].

can a new molecule of solute combine with the transport site[15]. This plot also brings out the distinct position occupied by glycine. α-Aminoisobutyrate turns out to be a somewhat specific model for glycine; on a similar plot showing transport rates, it occupies a position even nearer the origin than glycine. Here, as for the ascites cell, glycine and α-aminoisobutyrate show distinctive behavior, including low affinity and access to a transport site that can scarcely be saturated. This behavior is not a consequence of the optical symmetry of these two amino acids, since cycloleucine

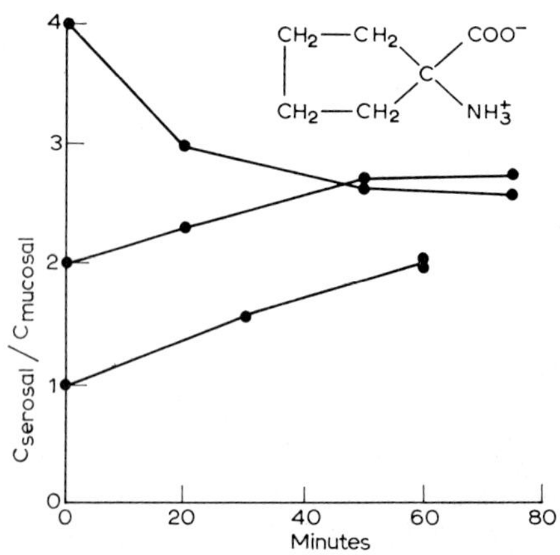

Fig. 5. Concentrative transfer of cycloleucine across the everted, isolated jejunum of the rat. The technique of CRANE AND WILSON was used[12]. The initial cycloleucine level on the mucosal side was 2 mM; on the serosal side it was 2, 4, or 8 mM as indicated above by the initial distribution ratio. Reproduced with permission from AKEDO AND CHRISTENSEN[20].

References p. 538

proves again for this system a model for methionine or valine, with a high rate of transport and a corresponding high inhibitory action. Fig. 5 illustrates that this interesting amino acid is concentrated approx. 2.7 times into the everted intestinal sac.

A distinct uphill transport of AIB can also be obtained (Table III) if one uses the strategem of JERVIS AND SMYTH[16] to minimize water migration. Ileal gut sacs are used, and phosphate buffer is substituted for the bicarbonate system in the medium.

TABLE III

CONCENTRATION OF α-AMINOISOBUTYRATE BY THE RAT ILEUM

Initial level in mucosal phase, 1 mM; in serosal phase, 1 mM in the first two experiments, 2 mM in the third. Incubation periods: 60, 110 and 100 min, respectively.

Experiment	*Final AIB levels*		
	Mucosal phase (mM/l)	*In tissue (mM/l cell water)*	*Serosal phase (mM/l)*
1	0.84	3.08	1.39
2	0.81	3.05	1.42
3	0.88	—	2.08

By this technique JERVIS AND SMYTH were able to show distinct concentration of D-methionine by the rat intestine.

We are currently exploring another line of evidence which at first also suggested to us a separate exit process. HEINZ AND WALSH[17] noted a paradoxic failure of accumulated L-methionine to drive glycine influx, in the way that accumulations of glycine, α-aminoisobutyrate and other amino acids do. This failure was attributed

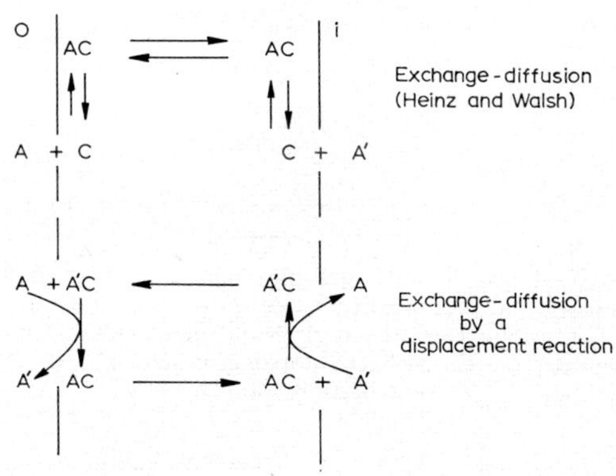

Fig. 6. Schemes for exchange diffusion. Above: as visualized by HEINZ AND WALSH[17]. Below: assuming a displacement reaction without the intermediate release of the transport site or carrier.

to a particularly strong inhibitory action of escaped methionine on glycine uptake at the external membrane surface. But why should the mass-action of methionine from outside predominate over its mass-action from inside, when for glycine the opposite relationship is found? A suggestion arises here of a different order of affinities

References p. 538

for one side of the membrane than for the other. JACQUEZ[18] has recently reported a similar failure of accumulated glycine to accelerate the subsequent uptake of L-tryptophane, although tryptophane and azaserine were mutually effective in driving exchange.

A priori, exchange rates could prove to be valueless for measuring affinities. HEINZ AND WALSH proposed that exchange occurs through a dissociation–association

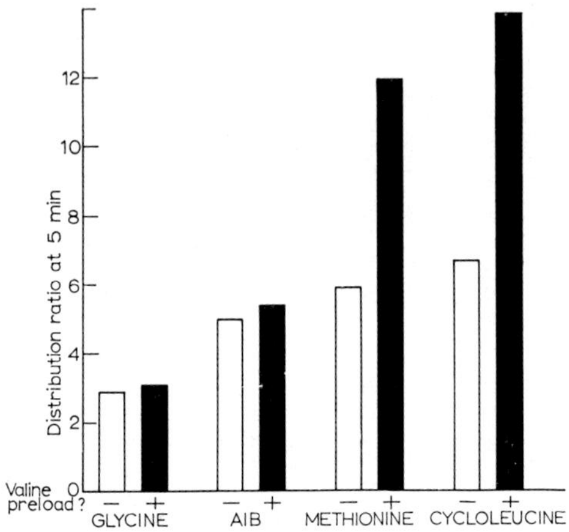

Fig. 7. Effect of valine previously accumulated from a 10-mM solution on the uptake of several amino acids at 1-mM levels by the Ehrlich ascites tumor cell.

sequence (Fig. 6, upper diagram). But if exchange with another amino acid occurs without the preliminary dissociation of the carrier–solute complex (lower diagram), factors quite distinct from relative affinities may determine the rate. We know for example that certain highly stable chelates will exchange their constituent parts more quickly than other less stable ones.

Nevertheless we have analyzed the exchange relationships between five amino

TABLE IV

STIMULATING ACTION OF METHIONINE PREVIOUSLY TAKEN UP ON THE UPTAKE OF AMINO ACIDS BY HUMAN ERYTHROCYTES

To heparinized whole blood was added L-methionine, 20 mM/l. After incubation at 37° for 40 min, the cells were separated and washed twice with ice-cold plasma and then resuspended in 10 parts of plasma containing the indicated ^{14}C-labeled amino acid at a 5-mM level. After incubation the cells were extracted with trichloroacetic acid and the ^{14}C measured by liquid-scintillation counting.

Amino acid	*Distribution ratio cells/plasma*			
	After 1 min		*After 5 min*	
	Control	*Preloaded*	*Control*	*Preloaded*
Glycine	0.21	0.18	0.18	0.19
Valine	0.38	0.53	0.81	1.06
Methionine	0.42	1.07	0.81	1.28

References p. 538

acids: valine, methionine, cycloleucine, AIB and glycine, in the manner illustrated in Fig. 7. Here valine preaccumulated by the Ehrlich cells from a 10-mM solution, is seen to be highly effective during 5 min in stimulating the exchange-uptake of methionine, cycloleucine and valine, but without significant influence on the uptake of glycine and AIB. In a similar way we find that the presence of 10 mM valine in the outside environment stimulates the exchange-loss of cycloleucine, but not of glycine or AIB. The relationships are the same from the two sides, indicating that exchange-diffusion probably operates on the same site or carrier at both faces of the membrane. In this way one of the four possible points for the introduction of

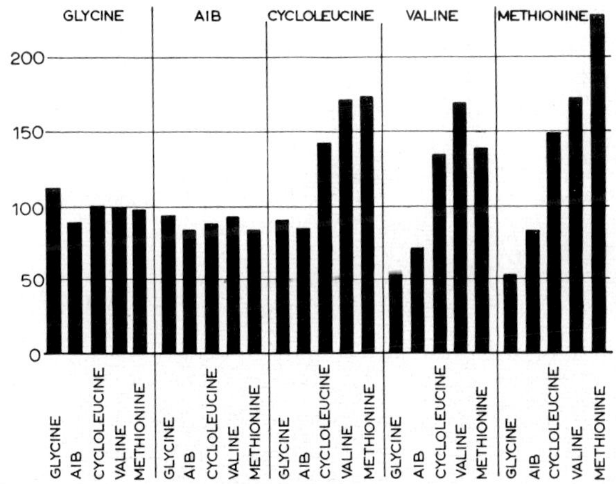

Fig. 8. Comparison of the ability of five previously accumulated amino acids to stimulate the uptake of each other. A single lot of Ehrlich cells was divided into six parts. Each of five of these portions was incubated 20 min in a 10-mM level in Krebs–Ringer–bicarbonate medium of the amino acid listed at the top. The cells were then washed once with ice-cold buffered medium, and then suspended for 1 min at 37° in fresh medium containing 1-mM levels of the amino acids shown at the bottom of the figure, in labeled form. The amount of radioactivity taken up is compared here with the amount taken up by the control cells previously incubated in the amino acid free medium, setting the latter at 100%. For example the first five bars show that previously accumulated glycine failed to stimulate the uptake of any of the five amino acids.

energy, as outlined by PATLAK[19] in his generalized model, namely to modify the solute–carrier combination, is made less probable.

Fig. 8 illustrates the approximate exchange relationships found among these five amino acids, allowing 1 min at 37°. The three amino acids, valine, cycloleucine and methionine, exert very potent drives on each other, whereas they have no noticeable tendency to drive counterflux of glycine or AIB. When we extended the period of observation, AIB was observed to drive the counterflow of glycine, in agreement with HEINZ AND WALSH[17]. Table IV shows that the same general situation applies for erythrocytes. Preaccumulated methionine is able to cause these cells to concentrate ^{14}C-methionine or valine, but not glycine.

This result again places glycine and AIB in a separate class. This separation no doubt arises from the marked difference in affinity between the two sets. An amino acid with a high affinity responds by counterflow to another also with high affinity; they tend to change places with high efficiency. In contrast the relatively slowly

References p. 538

moving glycine finds itself competing from the outside with escaped methionine before it has responded measurably to the rapid movement of methionine out of the cells. Suitable manipulation of the relative levels and intervals permits one to show the driving of counterflows in some of the less favorable circumstances.

Although this excursion into the area of exchange processes does not confirm the existence of different affinities for exchange from the two sides of the membrane, it does again show the dichotomy between the behavior of glycine and AIB, on the one hand, and on the other, of valine, cycloleucine and methionine.

In summary, although the neutral amino acids yield qualitative evidence of utilizing the same reactive site in their transport, when we attempt to place them in an affinity sequence on the basis of the rates of their transport, the extents of their accumulation, and the magnitudes of their competitive action, serious inconsistencies are uncovered, indicating that at least one additional affinity constant must be involved, probably for a mediated exodus, perhaps on the modified form of the original transport site.

ACKNOWLEDGEMENT

The experiments described here were supported in part by a grant (C-2645) from the National Cancer Institute, National Institutes of Health, U.S. Public Health Service.

REFERENCES

1 H. N. Christensen and T. R. Riggs, *J. Biol. Chem.*, 194 (1952) 57.
2 C. M. Paine and E. Heinz, *J. Biol. Chem.*, 235 (1960) 1080.
3 H. N. Christensen, T. R. Riggs and N. E. Ray, *J. Biol. Chem.*, 194 (1952) 41.
4 T. R. Riggs, H. N. Christensen and I. M. Palatine, *J. Biol. Chem.*, 194 (1952) 53.
5 C. G. Winter and H. N. Christensen, unpublished results, 1961.
6 T. A. Connors, L. A. Elson, A. Haddow and W. C. J. Ross, *Biochem. Pharmacol.*, 5 (1960) 108.
7 M. N. Mickelson, *Biochem. Pharmacol.*, 5 (1960) 165.
8 B. L. Horecker, J. Thomas and M. Monod, *J. Biol. Chem.*, 235 (1960) 1586.
9 M. J. Osborn and W. L. McLellan, Jr., *Federation Proc.*, 19 (1960) 131.
10 D. L. Oxender, unpublished results (1960).
11 D. L. Oxender and M. Royer, *Federation Proc.*, 20 (1961) 141.
12 R. K. Crane and T. H. Wilson, *J. Appl. Physiol.*, 12 (1958) 145.
13 H. Akedo and H. N. Christensen, *J. Biol. Chem.*, 237 (1962) 113.
14 L. R. Finch and F. J. R. Hird, *Biochim. Biophys. Acta*, 43 (1960) 278.
15 W. Wilbrandt, *J. Cellular Comp. Physiol.*, 47 (1956) 137.
16 E. L. Jervis and D. H. Smyth, *J. Physiol. (London)*, 149 (1959) 433.
17 E. Heinz and P. O. Walsh, *J. Biol. Chem.*, 233 (1958) 1488.
18 J. A. Jacquez, *Proc. Natl. Acad. Sci. U.S.*, 47 (1960) 153.
19 C. S. Patlak, *Bull. Math. Biophys.*, 19 (1957) 209.

INVITED DISCUSSION

SOME REMARKS ON ACTIVE TRANSPORT AND EXCHANGE DIFFUSION OF AMINO ACIDS IN EHRLICH CELLS

ERICH HEINZ

Johann Wolfgang Goethe-Universität, Frankfurt a. Main (Germany)

Transport of amino acids in Ehrlich mouse-ascites tumor cells is associated to a high extent with exchange diffusion. As shown in Fig. 1 preloading these cells with cold glycine or related amino acids strongly accelerates the influx of glycine[1, 2] We assume that this phenomenon, which represents a true one-to-one exchange, is related to similar ones later reported for other systems, *e.g.* by PARK *et al.*[3], by

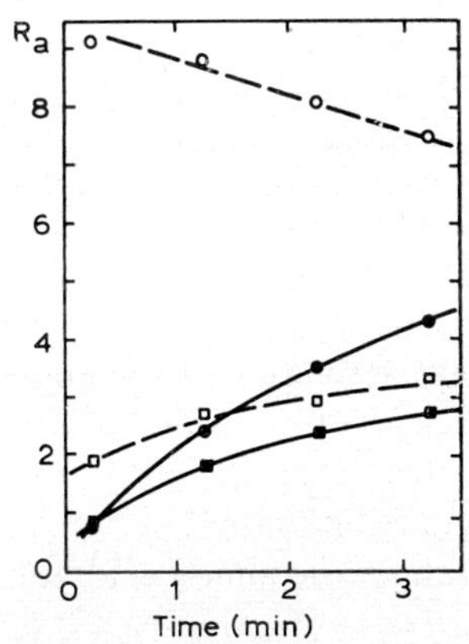

Fig. 1. Relative uptake of [1-^{14}C]glycine after preloading cells with unlabeled glycine. Solid lines, distribution ratios of radioactivity between cells and medium. Dotted lines, distribution ratios of total glycine. The circles and squares refer to preloading experiments and controls, respectively. Krebs'–Ringer–bicarbonate solution at 36° was used as medium. (From HEINZ AND WALSH[2], with permission).

ROSENBERG *et al.* under the term "uphill transport induced by counter flow"[4] and by HEINZ AND DURBIN[5]. With Ehrlich nouse-ascites tumor cells there is evidence, that this exchange diffusion refers to the same carrier mechanism as does the active transport. This follows from the wide agreement of both processes with respect to substrate specificity[2, 6] and to susceptibility to lipotropic inhibitors[7]. With KROMPHARDT we provided further evidence analyzing the glycine influx in terms of the Michaelis-Menten equation with and without preloading[8]. Preloading increases the maximum velocity whereas the Michaelis constant remains the same (Fig. 2).

Exchange diffusion indicates the degree of reversibility and hence the thermo-

References p. 544

dynamic efficiency of the transport system[9]. It must be accounted for in any model which is to represent the carrier mechanism. The following model (Fig. 3) which has been based on the kinetic observations, seems to meet this requirement[2]. The exchange diffusion requires that steps 1–4 are reversible. Since inhibitors of energy supply affect only the net transport and not the exchange diffusion[7], energetic coupling probably does not concern these steps, but rather step 5 in which the carrier is reactivated. JACQUEZ arrives at a similar conclusion through a mathematic analysis of

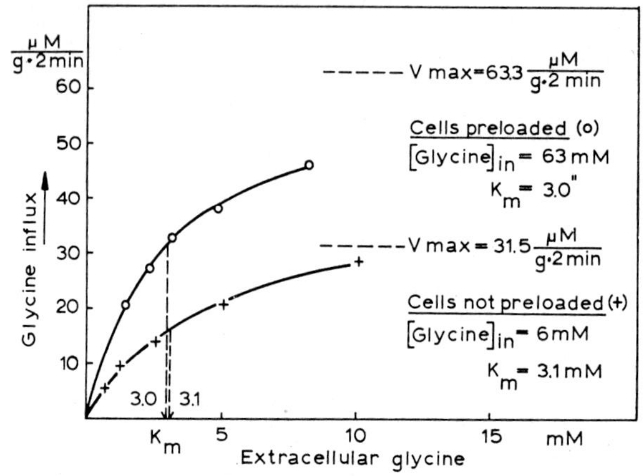

Fig. 2. Influx of [1-^{14}C]glycine at various extracellular glycine concentrations. Upper curve, cells preloaded by preincubation with unlabeled glycine. Lower curve, cells not preloaded[8].

the kinetic data available[10]. More recently we have tried to modify the above model along the lines of MITCHELL's hypothesis of group transfer[11]. Evidence, however, does not favor the intermediate activation of the amino acid by a covalent linkage, as seems to be implied by such a group transfer. Carboxyl-oxygen does not exchange during the transport[12], nor is the NH_2 group likely to be linked to a carbonyl group of the carrier by formation of a Schiff base[13].

A major contribution concerning the amino acid uptake by Ehrlich mouse-ascites

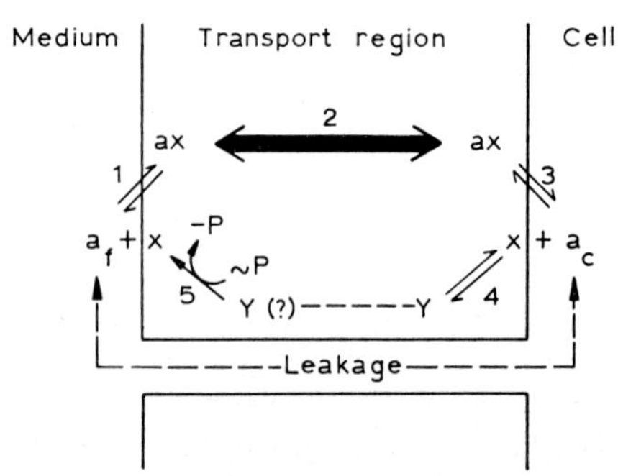

Fig. 3. Schematic model of active transport, associated with exchange diffusion; a_f and a_c refer to extracellular and intracellular substrates, respectively. x is the active; y the inactivated carrier; ax is the carrier–substrate complex; ~P and –P stand for reactant and product, respectively, of any exergonic reaction, energerically coupled to transport mechanism. (From HEINZ AND WALSH[2], with permission).

References p. 544

tumor cells was made by CHRISTENSEN *et al.* by the discovery, that pyridoxal and some other compounds markedly enhance glycine accumulation by these cells[14]. This observation has been confirmed by other laboratories including our own. Since however, we never succeeded in accelerating the glycine influx by pyridoxal under a variety of conditions we had to conclude on the basis of the above model, that the effect of pyridoxal increases accumulation by means other than stimulating the

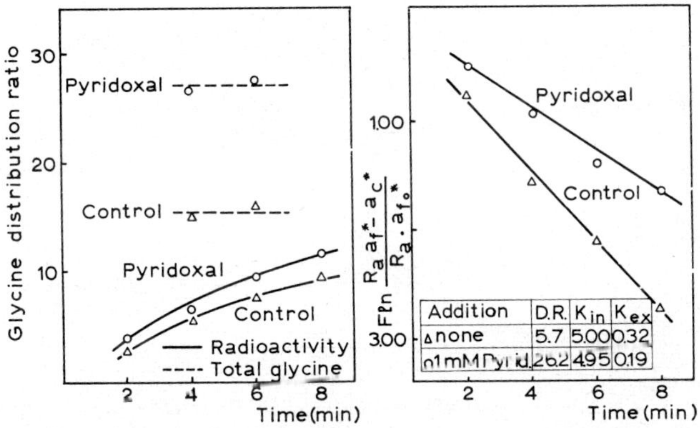

Fig. 4. Effect of 1 m*M* pyridoxal on glycine distribution and fluxes. Left side, distribution ratio (R_a) of total glycine (dotted line) and of radioactivity (solid lines) with and without pyridoxal. Right side, analysis of these data according to the equation of HEINZ[21], in order to determine the flux coefficients[15]. (Fig. reproduced by permission of Springer-Verlag).

transport mechanism*. In recent studies with BITTNER we confirmed this by showing that pyridoxal, while strongly enhancing the steady state level of glycine, leaves the influx unchanged but markedly reduces the efflux coefficient[15] (Fig. 4). This could mean, that pyridoxal reduces the exit of glycine without influencing the transport mechanism. The accumulation of α-aminoisobutyric acid, however, is hardly increased by pyridoxal (Table I). Since this amino acid probably enters and leaves this cell by the same pathways as does glycine, pyridoxal seems to affect neither the entry nor the exit mechanism of glycine. It rather should concern some

TABLE I

EFFECT OF PYRIDOXAL ON ACCUMULATION OF GLYCINE AND AMINOISOBUTYRIC ACID IN EHRLICH ASCITES TUMOR CELLS

After 1 h at 37°.

Addition to medium	*Distribution ratio of radioactivity*
2 m*M* [1-^{14}C]glycine	7.9
2 m*M* [1-^{14}C]glycine + 1 m*M* pyridoxal	18.1
2 m*M* [1-^{14}C]aminoisobutyric acid	10.2
2 m*M* [1-^{14}C]aminoisobutyric acid + 1 m*M* pyridoxal	11.6

* Unpublished observations.

References p. 544

step of glycine uptake which follows after, and is more specific than the transport process.

I would like to add a few comments on the action of various ions, in particular H^+ ions and alkali ions, on glycine accumulation, which had been previously observed by CHRISTENSEN *et al.*[16]. With KROMPHARDT we found H^+ ions to inhibit the influx instantaneously and reversibly[8]. The action is likely to concern the transport mechanism directly rather than via metabolic inhibition. Plotting glycine influx versus

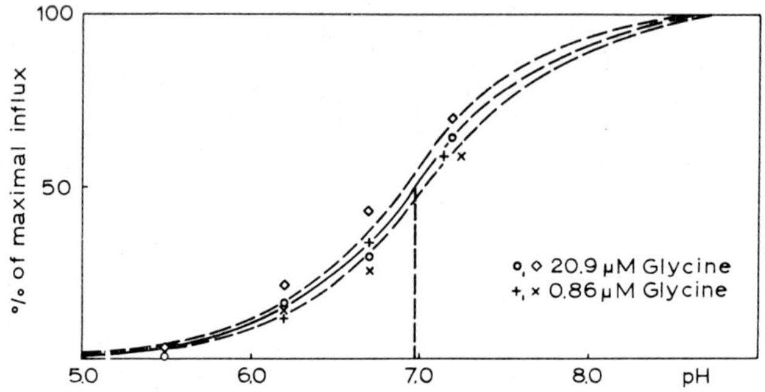

Fig. 5. Effect of H^+ ions on glycine influx at different concentrations of extracellular glycine. The flux values are corrected for a small pH-independent fraction of the influx. The maximum influx is obtained by extrapolation[8]. (Fig. reproduced by permission of Springer-Verlag).

pH and taking into account a constant pH-independent fraction of the influx we obtain a titration curve with an inflexion point around pH 6.9. This recalls a similar pH dependence of certain enzyme reactions. Since this inflexion point is not shifted by greatly increasing the glycine concentration there is no competition between substrate and H^+ ion for the same point of attachment (Fig. 5). We assume that

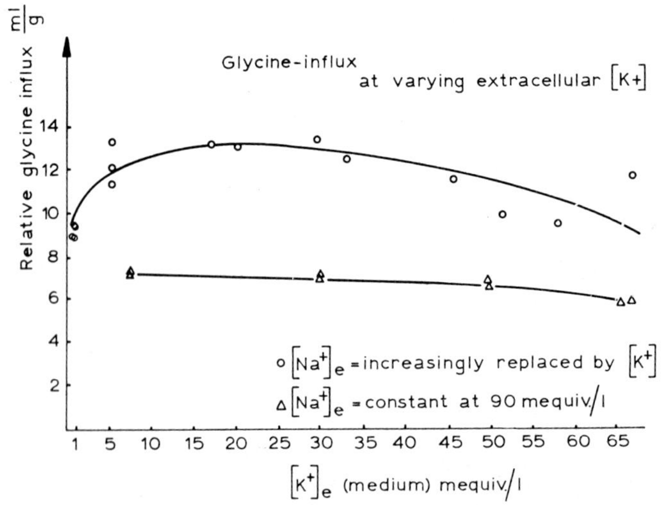

Fig. 6. Effect of extracellular K^+ concentration on glycine influx. Upper curve, K^+ is increased at the expense of Na^+. Lower curve, extracellular Na^+ remains constant at 60 mequiv./l. Choline ions are used to make up for deficient Na^+ ions[19].

References p. 544

there are certain basic groups at the cell surface, which are indispensable for the full function of the transport mechanism but which do not directly combine with the attaching group of the substrate amino acid.

A similar much weaker effect on glycine influx is exerted by short-term (2 min) increase of the K^+ ions (Fig. 6). This effect is much smaller than that observed previously by CHRISTENSEN in long-term experiments on the steady-state distribution of glycine[16]. The inhibitory effect of potassium on the flux, however, seems to be only apparent and rather due to the concomitant decrease in extracellular sodium. Accordingly an increase of potassium scarcely inhibits the glycine influx if by the use of choline the Na^+ concentration of the medium is kept constant (Fig. 6). On the other hand the glycine influx diminishes when the extracellular sodium is reduced below a value of approx. 90 mequiv./l, even at constant K^+ concentration (Fig. 7). Provided choline is as indifferent to transport systems and permeability as

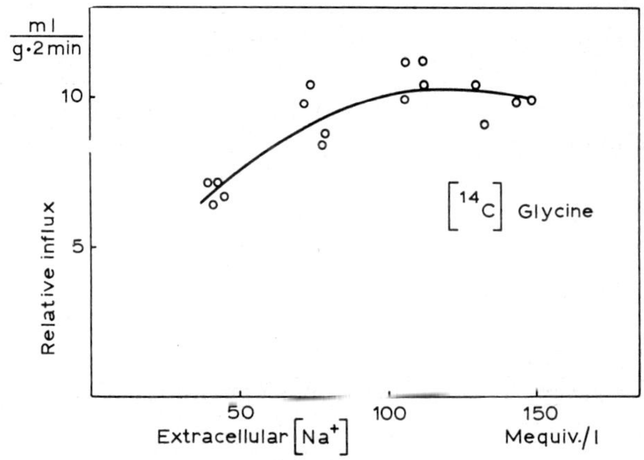

Fig. 7. Glycine influx and extracellular Na^+ concentration. K^+ concentration is kept constant at 6 mequiv./l. The decrease of Na^+ ions is compensated for by choline ions[19]. (Fig. reproduced by permission of Springer-Verlag).

TABLE II

INTRACELLULAR LEVELS OF Na^+ AND K^+ AND GLYCINE INFLUX

Extracellular concentrations are the same in all experiments, *i.e.* corresponding to Krebs-bicarbonate–Ringer solution.

Intracellular [*Na*+] (*mequiv./l*)	*Intracellular* [*K*+] (*mequiv./l*)	*Relative glycine influx (ml/g 2 min)*
61	168	9.7
167	74	9.8

References p. 544

is usually assumed; these results seem to indicate that Na^+ ions support the glycine transport. They recall similar findings by CSÁKY *et al.* with respect to the effect of Na^+ ions on the transport of sugars across the intestinal wall[17].

As to the energetic coupling between glycine transport and K^+ exit, as suggested by RIGGS *et al.*[18], some of our observations do not support this hypothesis. Short-term reduction of cellular K^+ by preincubation with K^+-free solutions, though greatly increasing K^+ influx, had no effect on glycine influx (Table II). The proposed mechanism of this energetic coupling would also imply a strong competition between K^+ and glycine inside the cell for the same carrier site. Hence increasing cellular glycine should depress K^+ efflux and favor K^+ accumulation. No such effect, however, has been found by KROMPHARDT AND GROBECKER in our laboratory[19]. Similar findings have been reported by HEMPLING[20]. We therefore assume that a relationship between the transport mechanisms for glycine and K^+, if it exists, is not simply through an energetic coupling between K^+ exit and glycine entry.

ACKNOWLEDGEMENT

This work has been supported by a Grant (NSF-10812) of the National Science Foundation.

REFERENCES

1 E. HEINZ, *J. Biol. Chem.*, 211 (1954) 781.
2 E. HEINZ AND P. M. WALSH, *J. Biol. Chem.*, 233 (1958) 1488.
3 C. R. PARK, R. L. POST, C. F. KALMAN, J. H. WRIGHT, JR., L. H. JOHNSON AND H. E. MORGAN, *Ciba Foundation Symposium Endocrinol.*, 9 (1956) 257.
4 T. ROSENBERG AND W. WILBRANDT, *J. Gen. Physiol.*, 41 (1957) 289.
5 E. HEINZ AND R. P. DURBIN, *J. Gen. Physiol.*, 41 (1957) 101.
6 C. M. PAINE AND E. HEINZ, *J. Biol. Chem.*, 235 (1960) 1080.
7 R. M. JOHNSTONE AND J. H. QUASTEL, *Biochim. Biophys. Acta*, 46 (1961) 527.
8 H. KROMPHARDT AND E. HEINZ, unpublished results.
9 E. HEINZ AND C. S. PATLAK, *Biochim. Biophys. Acta*, 44 (1960) 324.
10 J. A. JACQUEZ, *Proc. Natl. Acad. Sci. U.S.*, 47 (1961) 153.
11 P. MITCHELL AND J. MOYLE, *Proc. Roy. Soc. Edinburgh*, 27 (1958) 61.
12 H. N. CHRISTENSEN, H. M. PARKER AND T. R. RIGGS, *J. Biol. Chem.*, 233 (1958) 1458.
13 H. N. CHRISTENSEN, in *Membrane Transport and Metabolism, Symp. Czechoslovak Academy of Sciences*, Prague. Preliminary issue, p. 293.
14 H. N. CHRISTENSEN, T. R. RIGGS AND B. A. COYNE, *J. Biol. Chem.*, 209 (1954) 413.
15 J. BITTNER AND E. HEINZ, unpublished results.
16 H. N. CHRISTENSEN, T. R. RIGGS, H. FISCHER AND I. M. PALATINE, *J. Biol. Chem.*, 198 (1952) 1.
17 T. Z. CSÁKY AND M. THALE, *J. Physiol. (London)*, 151 (1960) 59.
18 T. R. RIGGS, L. M. WALKER AND H. N. CHRISTENSEN, *J. Biol. Chem.*, 233 (1958) 1479.
19 H. KROMPHARDT, H. GROBECKER AND E. HEINZ, unpublished results.
20 H. G. HEMPLING, *Physiologist*, 3 (1960) 78.
21 E. HEINZ, *J. Biol. Chem.*, 255 (1957) 305.

INVITED DISCUSSION

UPTAKE OF TYROSINE BY BRAIN *IN VIVO* AND *IN VITRO*

GORDON GUROFF AND SIDNEY UDENFRIEND

Laboratory of Clinical Biochemistry, National Heart Institute, National Institutes of Health, Bethesda, Md. (U.S.A.)

Our interest in the last few years has centered around the transport of tyrosine into the brain[1, 2]. The use of tyrosine was prompted by the availability of a rapid, specific, sensitive chemical method of assay[3] and by the realization that the insolubility of this amino acid would allow intraperitoneal depots to maintain high constant levels in blood plasma for an extended period. The selection of brain for these investigations involved, among other things, the long-term interest of our research group in the neuropharmacology of hydroxylated aromatic compounds.

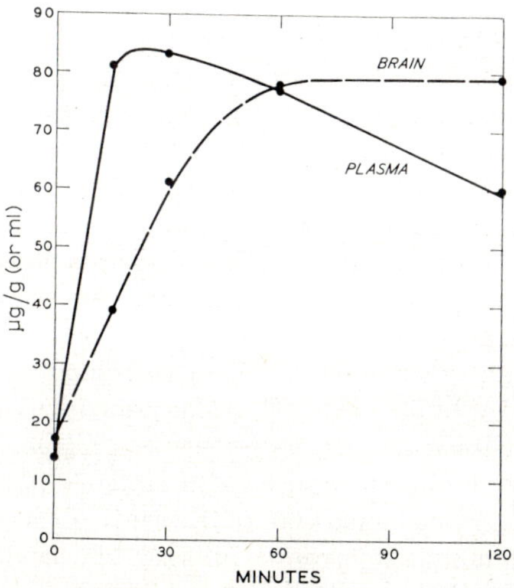

Fig. 1. Tyrosine distribution in brain and plasma after intraperitoneal injection of L-tyrosine.

The uptake of amino acids by brain cannot be reviewed here but it can be noted that certain specific amino acids have been extensively studied[4–8] and that the uptake of several different amino acids has been investigated under specified conditions[9, 10]. The prevailing conclusion drawn from these past studies is that excess amino acids from blood enter the brain to a very limited extent or not at all. On the other hand, administration of labeled amino acids leads to rapid exchange between blood and brain[6, 11] indicating that the brain substance is freely permeable to amino acids. These contradictory lines of evidence have not yet been resolved.

References p. 553

In previous work from this laboratory[1] L-tyrosine (400 mg/kg) was injected into fasted, 250 g rats and the plasma- and brain-tyrosine levels observed. It can be seen (Fig. 1) that the plasma-tyrosine level rose rapidly and remained relatively constant over the 2-h experimental period. The brain-tyrosine level rose more slowly but after 60 min equaled and exceeded the plasma level. The brain-to-plasma ratio during the experiment (Table I) clearly showed this relationship. In the fasting animal the endogenous tyrosine ratio was 1.3–1.4. When equilibrium was reestablished after approx. 90 min this ratio was again reached even though the absolute values were about 5-fold higher. To allow comparisons between the L- and D-isomers the endogenous tyrosine levels have been subtracted in both cases. The uptake of D-tyrosine (Table I) was much slower under similar conditions. It has been shown, however[19], that little or no racemization of the injected D-tyrosine occurred over the time period of these experiments and that, in fact, the increased tyrosine in the brain was the D-isomer.

TABLE I

BRAIN-TO-PLASMA RATIOS OF L- AND D-TYROSINE FOLLOWING INTRAPERITONEAL ADMINISTRATION

Time (min)	L-*Tyrosine** *(400 mg/kg)*	D-*Tyrosine** *(400 mg/kg)*
15	0.32	0.09
30	0.63	0.12
60	0.95	0.23
120	1.32	0.52

* Endogenous L-tyrosine subtracted from total tyrosine of brain and plasma.

The presence of certain other amino acids in the blood had a marked effect on the uptake of tyrosine by brain (Table II). The prior injection of aromatics such as tryptophane, or of aliphatics such as isoleucine or leucine, inhibited tyrosine uptake. On the other hand, molecules such as glutamate, arginine, or threonine had little or no effect. Also, certain unnatural amino acids, such as *p*-fluoro-DL-phenylalanine or norleucine[19], inhibited tyrosine uptake. The acid analogs, *p*-hydroxyphenylacetic acid and isovaleric acid[19] did not inhibit tyrosine uptake. In addition, the earlier suggestion that D-tyrosine was an inhibitor of L-tyrosine uptake has been examined and has not been substantiated. The effective inhibitors act only during the first approx. 30 min of these experiments, probably because they do not remain in the blood stream as long as does tyrosine.

Structural alterations changed the uptake of molecules of this type (Fig. 2). It can be seen that tyramine and α-methyl-DL-tyrosine also entered brain but that *p*-hydroxyphenylacetic acid did not. The figures in parentheses indicate the brain-to-plasma ratio of each compound 30 min after administration. Other tyrosine analogs with additional carboxyls added to the ring structure did not enter brain to any significant extent.

References p. 553

TABLE II

EFFECT OF OTHER AMINO ACIDS ON TYROSINE DISTRIBUTION

*Amino acid**	*Brain to plasma ratio (30 min)*
Control	0.72
L-Tryptophane	0.18
p-Fluoro-DL-phenylalanine	0.24
L-Leucine	0.24
L-Isoleucine	0.28
L-Valine	0.31
L-Cysteine	0.37
L-Histidine	0.38
L-Alanine	0.88
L-Serine	0.94
L-Threonine	0.77
L-Arginine	0.84
L-Lysine	0.83
L-Glutamate	0.81

* Amino acids injected 5 min before L-tyrosine (400 mg/kg). Between 0.8 and 1.7 mmoles of other amino acids used[1].

In a separate study[19] of the comparative uptakes of tyrosine by other tissues it has been concluded that stereoselectivity can also be observed in uptake by muscle (Fig. 3). It can be seen that the uptake of D-tyrosine by muscle was less than that of the L-isomer but that comparatively large amounts of the D-isomer entered muscle compared to the small amounts which entered brain under similar circumstances. On the other hand, the inhibition of tyrosine uptake by other amino acids observed in brain did not occur in muscle (Table III). Some non-specific reduction in tyrosine uptake did seem to occur regardless of the identity of the second compound. Muscle,

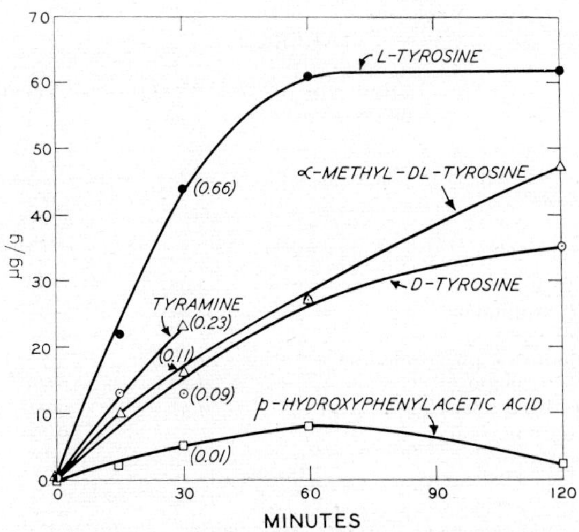

Fig. 2. Uptake of tyrosine and related compounds by rat brain *in vivo*.

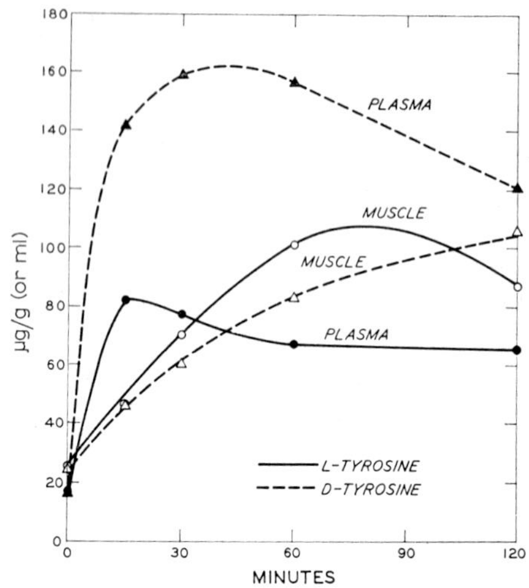

Fig. 3. Uptake of tyrosine by gastrochemius muscle of rat *in vivo*.

then, takes up L-tyrosine more rapidly than D-tyrosine but the marked stereospecificity and the inhibition of tyrosine uptake by other amino acids shown in brain is not evident in muscle. It can be noted that some stereoselectivity has also been observed in studies on tyrosine uptake by diaphragm *in vitro*[12].

Another phase of these studies has been the uptake of tyrosine by brain *in vitro*[2]. These experiments have been conducted in an attempt to localize the uptake system and obtain more information about its biochemical mechanism. Work on amino acid uptake by brain slices has appeared[13–18] but the mechanism of this uptake is obscure.

The uptake of L-tyrosine by brain slices exposed to 1 m*M* L-tyrosine dissolved in Krebs–Ringer–bicarbonate buffer (Fig. 4) was rapid and concentrative. Observation

TABLE III

EFFECT OF OTHER COMPOUNDS ON THE UPTAKE OF L-TYROSINE BY BRAIN AND MUSCLE *in vivo*

Compound*	Tissue-to-plasma ratio (30 min)	
	Brain	Muscle
Control	0.63	0.88
L-Tryptophane	0.20	0.70
L-Isoleucine	0.18	0.71
p-Fluoro-DL-phenylalanine	0.27	0.77
DL-Norleucine	0.29	0.75
p-Hydroxyphenylacetic acid	0.51	0.72
L-Glutamic acid	0.52	0.72
L-Glutamine	0.43	0.68

* Compounds were injected (1000 mg/kg) simultaneously with L-tyrosine (500 mg/kg).

References p. 553

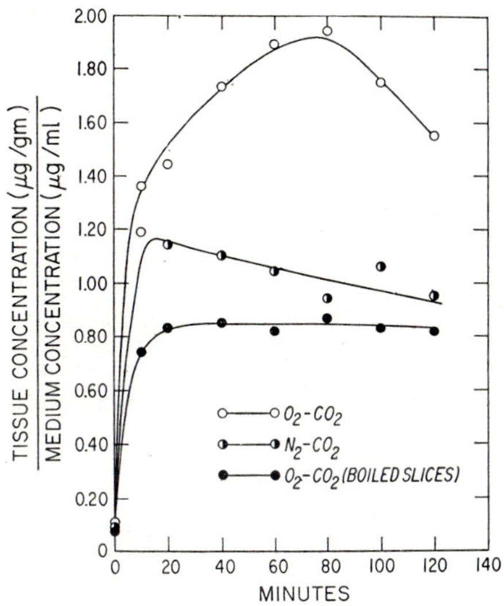

Fig. 4. Uptake of L-tyrosine by brain slices.

of the uptake of tyrosine under anaerobic conditions and the uptake by boiled slices made it apparent that tyrosine uptake can be divided into two components, a concentrative component, dependent on aerobic metabolism, and a non-concentrative component, probably involving diffusion.

When glucose or other hexoses (Table IV) were added to the buffer the uptake

TABLE IV

EFFECT OF VARIOUS CARBOHYDRATES ON L-TYROSINE UPTAKE BY BRAIN SLICES

Figures in parentheses indicate number of incubations. Slices incubated for 60 min in 10 ml of Krebs–Ringer–bicarbonate buffer containing L-tyrosine (1×10^{-3} M) and the indicated carbohydrate (1×10^{-2} M).

Compound	*Intracellular concn. (μg/ml)* / *Medium concn. (μg/ml)*
Control	2.73 ± 0.28* (36)
D-Glucose	4.63 ± 0.39 (12)
D-Galactose	3.59 ± 0.34 (8)
D-Fructose	3.99 ± 0.44 (8)
D-Mannose	4.71 ± 0.44 (8)
D-Ribose	2.87 (2)
2-Deoxyglucose	3.02 (4)
D-Glucose-6-phosphate	3.13 (2)
Mannitol	2.95 (4)
Sorbitol	2.93 (5)
Sucrose	2.91 (6)
Maltose	3.10 (4)
Lactose	2.84 (4)

* Mean ± standard deviation.

References p. 553

TABLE V

EFFECT OF AEROBIC SUBSTRATES ON L-TYROSINE UPTAKE BY BRAIN SLICES

Figures in parentheses indicate number of incubations. Slices incubated for 60 min in 10 ml of Krebs–Ringer–bicarbonate buffer containing L-tyrosine (1×10^{-3} *M*) and the indicated substrate (1×10^{-2} *M*).

Compound	*Intracellular concn. (μg/ml) / Medium concn. (μg/ml)*
Control	2.73 ± 0.28* (36)
Succinate	2.55 (4)
Citrate	2.13 ± 0.13 (4)
Acetate	2.53 (6)
α-Ketoglutarate	3.05 (6)
Pyruvate	3.51 ± 0.40 (14)
Fumarate	2.51 (2)
Isocitrate	2.38 (2)
Oxaloacetate	2.49 (2)
Malate	2.51 (2)

* Mean ± standard deviation.

was markedly stimulated. Pyruvate also enhanced uptake (Table V) but aerobic substrates were completely ineffective. In this regard it could be shown (Table VI) that dinitrophenol, iodoacetate, cyanide, and azide were effective inhibitors of the concentrative portion of the uptake. Raising the Ca^{2+} concentration of the buffer increased the tyrosine uptake whereas increases in the K^{+} concentration decreased it[2].

The inhibition of tyrosine uptake by other amino acids was investigated *in vitro* (Table VII) and it was observed that the same classes of amino acids were effective inhibitors of uptake by the slice as were inhibitors of uptake by the intact brain.

TABLE VI

EFFECT OF METABOLIC INHIBITORS ON THE UPTAKE OF L-TYROSINE BY BRAIN SLICES

Slices incubated for 60 min in 10 ml of Krebs–Ringer–bicarbonate buffer containing L-tyrosine (1×10^{-3} *M*).

Compounds	*Intracellular concn. (μg/ml) / Medium concn. (μg/ml)*
Control	2.63
+ 2,4-Dinitrophenol (10^{-3} *M*)	1.23
+ 2,4-Dinitrophenol (10^{-3} *M*) + glucose (10^{-2} *M*)	2.85
+ Iodoacetate (10^{-2} *M*)	1.69
+ Iodoacetate (10^{-2} *M*) + glucose (10^{-2} *M*)	2.57
+ Cyanide (10^{-3} *M*)	1.01
+ Azide (10^{-3} *M*)	1.48

TABLE VII

EFFECT OF OTHER AMINO ACIDS ON THE UPTAKE OF L-TYROSINE BY BRAIN SLICES

Slices incubated for 60 min in buffer containing L-tyrosine (1×10^{-3} M) and the indicated amino acids (1×10^{-2} M).

Compound	*Intracellular concn. (μg/ml)* / *Medium concn. (μg/ml)*
Control	2.60
L-Glutamate	2.51
γ-Aminobutyric acid	3.04
L-Arginine	2.55
L-Histidine	1.77
L-Tryptophane	1.28
L-Phenylalanine	1.15
L-Valine	1.15
p-Fluoro-DL-phenylalanine	0.90

Also those which were not effective *in vivo* were not effective *in vitro*. The inhibitors, however, acted only on the concentrative component of uptake by the slice and did not affect the diffusion uptake.

When the structural and steric specificity of the slice uptake were studied (Fig. 5)

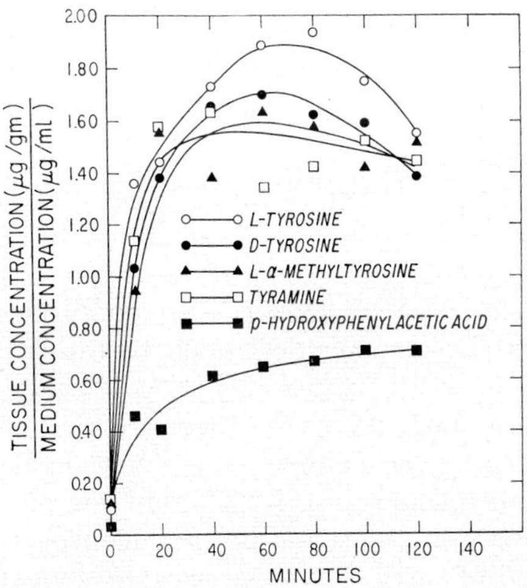

Fig. 5. Uptake of tyrosine and related compounds by brain slices.

it could be seen that the molecules taken up by brain *in vivo* were concentrated by brain slices (L-tyrosine, tyramine, α-methyl-L-tyrosine). On the other hand, D-tyrosine, which was taken up to a very small extent *in vivo*, was taken up as well as L-tyrosine *in vitro*. In addition, *p*-hydroxyphenylacetic acid, which did not enter the brain *in vivo*, did enter the slice, probably passively. The factors which modified the uptake of tyrosine by the slice had no effect on the uptake of *p*-hydroxyphenylacetic acid (Table VIII) indicating that they affect only the concentrating mechanism.

The metabolic component responsible for the concentrative uptake of tyrosine by the slice is not evident since hexoses stimulate uptake, anaerobic conditions eliminate it, and inhibitors of both anaerobic and aerobic metabolism depress it. The only conclusion that can be reached at this time is that the aerobic metabolism of hexose best supports the uptake mechanism.

Comparison between *in vivo* and *in vitro* data reveals that although the concentration by the brain is not radically different under the two conditions, the speed of uptake by the slice is much greater. The marked stereospecificity observed *in vivo* is not seen *in vitro*. Also, *p*-hydroxyphenylacetic acid, which does not enter the intact

TABLE VIII

EFFECT OF VARIOUS AGENTS ON UPTAKE OF L-TYROSINE OR *p*-HYDROXYPHENYLACETIC ACID BY BRAIN SLICES

Slices incubated for 60 min in 10 ml of Krebs–Ringer–bicarbonate buffer containing L-tyrosine (10^{-3} *M*) or *p*-hydroxyphenylacetic acid (10^{-3} *M*).

Compounds	*Intracellular concn. (μg/ml)* / *Medium concn. (μg/ml)*	
	L-*Tyrosine*	*p-Hydroxyphenylacetic acid*
Control	2.7	0.86
+ Glucose (10^{-2} *M*)	4.6	0.88
+ Citrate (10^{-2} *M*)	2.1	0.97
+ Iodoacetate (10^{-2} *M*)	1.7	0.84
+ Ca^{2+} (12.7×10^{-3} *M*)	3.8	0.95

brain, diffuses into the slice. The list of amino acid inhibitors is the same *in vivo* and *in vitro*, but in the intact animal some competitors eliminate all the tyrosine uptake, while in the slice they eliminate only a part, the concentrative part. In general, it seems that the concentrative component of uptake by the slice represents the uptake observed *in vivo*.

On the basis of these findings it may be suggested that two mechanisms are involved in tyrosine uptake by the brain *in vivo*. The cellular mechanism is concentrative, metabolically linked, and subjective to inhibition by other amino acids. The other mechanism, missing in the slice and possibly identified with the "blood–brain barrier", limits the speed of entry of tyrosine, possesses great stereospecificity, and prevents the entry of such molecules as *p*-hydroxyphenylacetic acid. The properties of this second mechanism cannot yet be investigated directly but it must be considered active in at least one sense: if a "blood–brain barrier" exists to the uptake of such molecules as *p*-hydroxyphenylacetic acid and D-tyrosine, some mechanism for facilitating the entry of L-tyrosine and other metabolites must exist at the same site.

Preliminary studies indicate that the uptake systems for tryptophane and phenylalanine are similar to that for tyrosine. It is of interest to determine how general these relationships are and which amino acids share common uptake mechanisms.

Our current effort is directed toward elucidating the mechanism of cellular concentration by studying its metabolic involvement and pursuing the properties of the second mechanism through *in vitro–in vivo* contrasts.

REFERENCES

1 M. A. Chirigos, P. Greengard and S. Udenfriend, *J. Biol. Chem.*, 235 (1960) 2075.
2 G. Guroff, W. King and S. Udenfriend, *J. Biol. Chem.*, 236 (1961) 1773.
3 T. P. Waalkes and S. Udenfriend, *J. Lab. Clin. Med.*, 50 (1957) 733.
4 P. Schwerin, S. P. Bessman and H. Waelsch, *J. Biol. Chem.*, 184 (1950) 37.
5 M. K. Gaitonde and D. Richter, *Biochem. J.*, 59 (1955) 690.
6 A. Lajtha, *J. Neurochem.*, 2 (1958) 209.
7 W. Dingman and M. B. Sporn, *J. Neurochem.*, 4 (1959) 148.
8 R. Kuttner, J. A. Sims and M. W. Gordon, *J. Neurochem.*, 6 (1961) 311.
9 F. Friedberg and D. M. Greenberg, *J. Biol. Chem.*, 168 (1947) 405.
10 H. Kamin and P. Handler, *J. Biol. Chem.*, 188 (1951) 193.
11 A. Lajtha, S. Berl and H. Waelsch, *J. Neurochem.*, 3 (1959) 322.
12 G. Guroff and S. Udenfriend, *J. Biol. Chem.*, 235 (1960) 3518.
13 J. R. Stern, L. V. Eggleston, R. Hems and H. A. Krebs, *Biochem. J.*, 44 (1949) 410.
14 G. Takagaki, S. Hirano and Y. Nagata, *J. Neurochem.*, 4 (1959) 124.
15 R. Rybová, *Nature*, 185 (1960) 542.
16 Y. Tsukada, Y. Nagata and S. Hirano, *Nature*, 186 (1960) 474.
17 S. Schanberg and N. J. Giarman, *Biochim. Biophys. Acta*, 41 (1960) 556.
18 K. D. Neame, *J. Neurochem.*, 6 (1961) 358.
19 G. Guroff and S. Udenfriend, *J. Biol. Chem.*, 237 (1962) 803.

CEREBRAL PASSAGE OF FREE AMINO ACIDS

ABEL LAJTHA

The New York State Psychiatric Institute and Department of Biochemistry, College of Physicians and Surgeons, Columbia University, New York, N.Y. (U.S.A.)

The question whether it is justified to treat cerebral transport mechanisms separately from those of other organs cannot be answered unequivocally at the present time. There is little doubt that because of its isolated circulation and anatomical localization the brain offers a number of convenient possibilities for the experimental approach to organ transport, although its structural complexity adds difficulties of interpretation. It is not certain whether transport processes operate in the brain that are unique for this organ; however, it has been shown that cerebral permeability is considerably lower than that of most organs.

Quantitative differences between the brain and other organs in penetration of substances have been amply documented for several classes of substances (for reviews discussing the subject see[1–6]). It is tempting to speculate that if net increase of some substances in organs is inhibited by means of active transport processes (partially transporting substances out of the organ), then at least quantitatively these processes have to be more active in brain than in most other organs. In *in vitro* experiments employing tissue slices, differences between the brain and other organs have been pointed out; for example, accumulation of α-aminobutyrate against a concentration gradient could be shown only in brain, among the tissues tried[7, 8]. On the other hand, glutamate uptake against a concentration gradient was found in several, though not in all, of the examined tissues[9]. It is likely that processes closely related to transport mechanisms studied in other tissues will be found in the central nervous system. Some aspects of this question have been reexamined lately by GUROFF AND UDENFRIEND[10, 11], and Dr GUROFF will discuss this point in more detail elsewhere in this volume.

EXCHANGE OF AMINO ACIDS BETWEEN PLASMA AND BRAIN

Our interest in cerebral amino acid transport was started with our observation, made in studying the amino acid and protein metabolism of the brain, that although net increase in brain levels is restricted if plasma concentrations of a number of amino acids are greatly increased, a rapid exchange between plasma and brain takes place. This finding, made with acidic, basic, and neutral amino acids such as glutamic acid[12, 13], lysine[14, 15], and leucine[16, 17], made it unlikely that amino acid passage in the brain is governed by passive diffusion only.

By studying this exchange in more detail[18, 19] it was found that rapid exchange occurred between plasma and brain even when after the elevation of brain amino acid levels there was a net efflux from the brain with time. The rate of exchange was to some extent proportional to the level of the amino acid in brain, in that increased cerebral levels resulted in a higher rate of exchange as measured by the appearance in the brain of label from the intravenously administered tracer doses (Table I)[18].

TABLE I

THE EFFECT OF INCREASED CEREBRAL LEVELS ON AMINO ACID EXCHANGE

Administered substance		*Counts/g fresh brain/min*
Intravenously	*Subarachnoidally*	
Lysine	—	3,300
,,	Lysine	4,800
,,	Leucine	4,300
,,	Saline	3,800
Leucine	—	5,500
,,	Lysine	5,500
,,	Leucine	7,100
,,	Saline	4,100

Intravenous injection: 0.2 μmole ^{14}C-lysine or ^{14}C-leucine (1.2×10^6 cts/min) in 0.02 ml.
Subarachnoid injection: 0.02 ml of 0.9% saline or 0.9% saline containing 1.9 μmole, ^{12}C-lysine or 2.1 μmole ^{12}C-leucine.

In the case of the two amino acids studied up to the present time this effect seems to be fairly specific, in that each amino acid increased its own exchange rate more than that of the other amino acid. The dependence of exchange rate on cerebral levels could be shown without disturbing physiological equilibria. Lysine levels varied from one brain part to another under normal conditions, without the administration of lysine; when exchange was estimated by measuring the appearance of the label following intravenous tracer doses, the rate of lysine exchange was higher in the parts of brain in which lysine level was higher. On the other hand, leucine distribution and consequently leucine exchange were more homogeneous (Table II)[18].

TABLE II

THE RELATIONSHIP BETWEEN DISTRIBUTION AND EXCHANGE OF CEREBRAL AMINO ACIDS

	L-Lysine		*D-Lysine*	*L-Leucine*	
	*Distribution**	*Exchange***	*Distribution**	*Distribution**	*Exchange***
Cerebellum-pons-medulla	150	140	140	100	100
Posterior half of cerebrum	98	99	120	98	95
Anterior half of cerebrum	73	79	60	100	110

* Distribution = μmoles/g fresh brain, whole brain = 100.
** Exchange = counts/min/g fresh brain, whole brain = 100.

Exchange was measured 5 min after the intravenous administration of tracer doses of the ^{14}C amino acid.

Exchange of an amino acid between plasma and brain may occur through any number of possible mechanisms and not necessarily through a single reversible process. Experiments such as those shown in Tables I and II may measure the final result or the average of a number of processes. Amino acid transport occurs not only at the cell membrane but also at other membranes, such as the nuclear membranes[20]. In the brain the complexity of the problem is further increased by the anatomical and histological complexity of the organ itself. Thus an investigation of cerebral transport processes may have to consider a bewildering number of membranes—organ membranes, membranes of the various cell types, and intracellular membranes—in addition to other factors peculiar to the brain, such as cell processes, myelin sheaths,

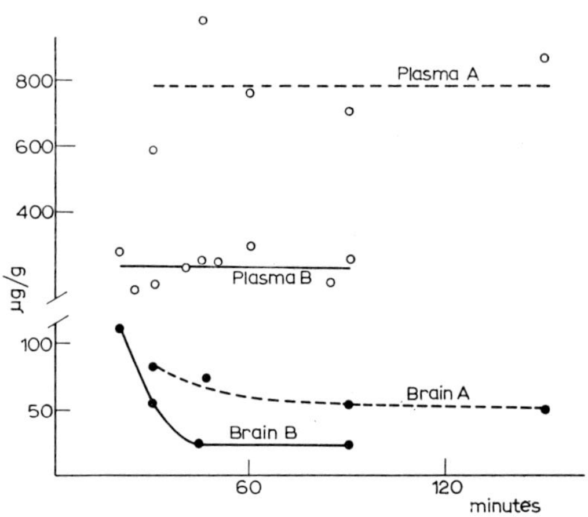

Fig. 1. Leucine transport from brain against a concentration gradient; experiment A against higher, experiment B against lower plasma levels. In both experiments, 2 μmoles of leucine were administered subarachnoidally to young rats in 0.02 ml 0.9% saline. Plasma levels were kept high by intraperitoneal injections.

and fluid spaces. It is likely, however, that the transport of any one amino acid in the brain will be governed or limited by one or only a few transport mechanisms and that valid comparisons and conclusions can be drawn from investigations concerning even such a complex organ.

Findings have been made in other systems comparable to the above findings of increased exchange caused by increased cerebral levels of an amino acid. Apparent uphill transport induced by counterflow has been shown in several systems (see[21]); in particular, amino acid transport in Ehrlich ascites tumor-cell has been studied in detail. Preloading the ascites cells with glycine[22, 23] increased glycine influx from the medium, the influx coefficient rising parallel with the internal concentration and the efflux coefficient remaining unchanged. It is possible that the model of exchange diffusion of amino acids (and it was suggested that exchange diffusion and active transport refer to the same mechanism) proposed for ascites cells[23] is applicable for amino acid exchange in the brain as well. I hope Dr. HEINZ will discuss his findings on exchange diffusion at greater detail in this volume. Similar to the finding of rapid exchange in spite of a barrier to net uptake, the effect of increased brain levels on the rate of exchange cannot be satisfactorily explained by passive diffusion.

References p. 563

TRANSPORT OF AMINO ACIDS FROM THE BRAIN AGAINST A CONCENTRATION GRADIENT

A more direct test for the existence of active transport processes in cerebral amino acid passage can be performed by studying transport against a concentration gradient. Plasma levels of an amino acid in rats can be kept elevated by constant infusion or repeated intraperitoneal injections, and subsequently cerebral levels can be increased

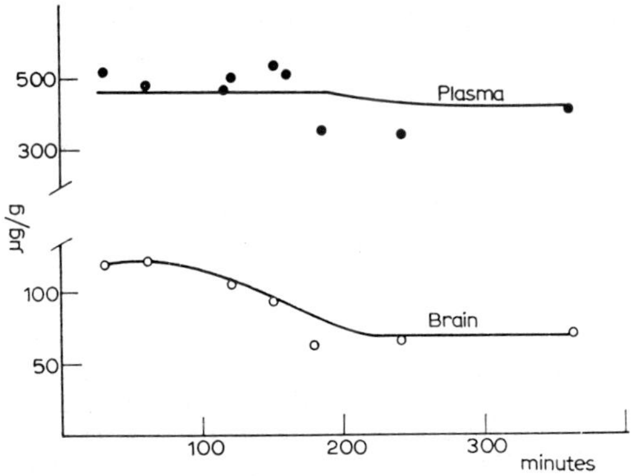

Fig. 2. Lysine transport from brain against a concentration gradient; 2 μmoles of lysine were administered subarachnoidally.

by intracerebral administration. The results of such experiments with leucine are shown in Fig. 1[17]. In the beginning of the experiment plasma and brain levels were elevated, plasma levels being considerably above that of brain. In time brain levels decreased against the concentration gradient of elevated plasma levels till a plasma

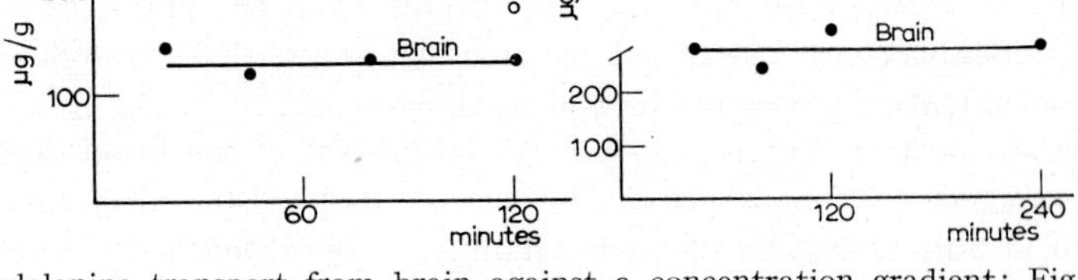

Fig. 3. Phenylalanine transport from brain against a concentration gradient; Fig. 3a against lower, Fig. 3b against higher plasma levels. In both experiments, 2 μmoles of phenylalanine were administered subarachnoidally.

to brain ratio of 11–15 was reached. Similar plasma to brain ratios were reached at higher (experiment A) or lower (experiment B) plasma levels, showing that the value of this ratio is to some degree independent of the absolute leucine levels in the plasma. Such experiments were performed with lysine (Fig. 2); here the decrease in cerebral levels took longer as compared to leucine efflux, and the plasma to brain ratio reached was about 5–6. Under such experimental conditions and at two different plasma levels no decrease in the cerebral concentration of phenylalanine could be shown (Fig. 3).

The above observations point out the two-directional nature of transport mechanisms and consequently of the control of cerebral metabolite levels. Substrate levels in the brain can be kept below that of the plasma by a barrier to entry, by an exchange in which upon entry an equivalent amount passes in the opposite direction, and by a *pumping mechanism* transporting substrates out of the organ. The differences

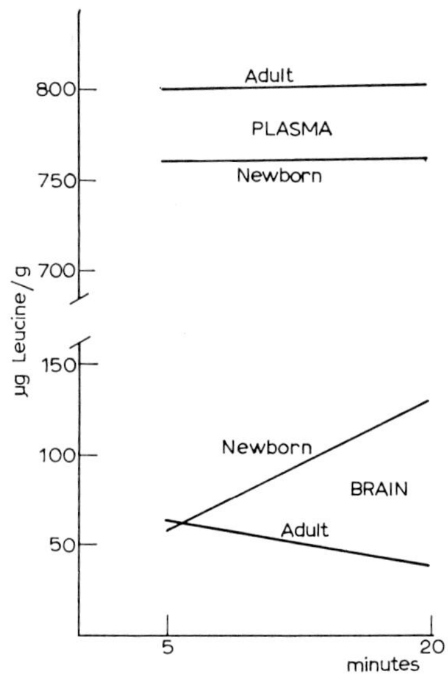

Fig. 4. Leucine transport from brain at different ages. Mice were injected with leucine at the beginning of the experiments. (Newborn with 0.8 mg intraperitoneally, adults with 16 mg intravenously and 16 mg intraperitoneally).

in plasma to brain ratios at equilibrium indicate the variations in the *efficiency* of this transport from one substrate to another—or conversely, the ability of the tissue to take up various amounts of different substrates. It is difficult to establish at present that no transport of phenylalanine occurs from the brain; perhaps experimental circumstances can be varied in such a fashion that phenylalanine efflux against a concentration gradient can also be shown.

The apparent lack of transport of phenylalanine out of the brain does not mean that no restriction to the uptake of this amino acid exists. After intravenous or intraperitoneal administration of phenylalanine, cerebral levels were below that of plasma, liver and muscle[17].

Leucine transport against a concentration gradient could be shown in adult mice but not in newborn under similar circumstances (Fig. 4). The great number of changes in the brain during development makes it impossible to single out permeability changes among the many possible reasons[1, 24]. One possible contributing factor to the greater uptake of substances in newborn also shown for glutamate[25] and lysine[15], could be a more fully developed transport out of the brain in the adult.

References p. 563

It has to be emphasized that the finding of a decrease in the cerebral level when the plasma level was above that of the brain can have, in addition to active transport, these alternate explanations: *(a)* incorporation into proteins; *(b)* metabolism instead of efflux; *(c)* efflux from a cerebral space where the level of the amino acid is above that of plasma.

(a) Control experiments with labeled amino acids[17] showed that the incorporation of the excess amino acid into protein—an unlikely event requiring net protein synthesis in the non-growing adult brain—did not occur. Of the label that penetrated maximally into the brain after one hour, 8% was incorporated into proteins, 10% was still left in the free amino acid fraction, and 82% left the brain.

(b) In the experiments on cerebral metabolism of the amino acids, the labeled amino acids were found to be present as such in the brain and no significant amount of other labeled metabolites could be found[26]. Decrease in cerebral levels may have occurred through slower metabolism and rapid exit of the formed metabolites from the brain, in which case no metabolic products would be found in the brain. Such an explanation does not seem very likely in these short-term experiments with amino acids that are not metabolized rapidly.

(c) The intracerebrally administered amino acid, instead of being restricted to the area near the site of administration, was found to be distributed throughout the brain, in one experiment evenly in the six gross parts, in another experiment evenly in the seven vertical sections into which the brain was divided[18, 26]. The restriction to the extracellular space is also not very likely since some of the label did get incorporated into cellular proteins. A restriction to a glial space is not likely, since to explain the different plasma to brain ratios at equilibrium (13 for leucine, 5.6 for lysine, 1.7 or 7.7 for phenylalanine) a *glial space* of different size would have to be postulated for each of the three amino acids studied.

The present experiments cannot establish the place where the transport against a concentration gradient occurs. Determination of a fraction of the cerebrospinal fluid (CSF) showed a rapid disappearance from the CSF as well[26], but the transport can occur between brain and plasma, brain and CSF, or CSF and plasma. Active transport of diodrast and phenolsulfonphthalein from CSF to blood was shown recently[27], with the site of active transport possibly at the choroid plexus of the 4th ventricle.

The existence of active transport mechanism in amino acid transport in brain slices has been known for some time. That slices can accumulate amino acids against a concentration gradient has been shown for L-glutamate[9], D-glutamate[28], γ-aminobutyrate[7, 8], aspartate[29], tyrosine[11], histidine, proline, lysine, ornithine, methionine, and arginine[30], and 5-hydroxytryptophan[31]. Interference with the source of energy decreased the rate of entrance and the net uptake of the amino acids in most of these experiments. The transport of amino acids from the brain against a concentration gradient *in vivo* (Figs. 1, 2), the active transport of substances from the CSF[27], the inhibition of tyrosine uptake *in vivo* by other amino acids[32], and the effects on exchange (Table I) show that the active and carrier mediated processes are operative in the living brain as well, perhaps in the passage of all amino acids. The active transport from the brain[17] shows that these processes operate in both directions. Both exchange and net unidirectional flow theoretically permit that influx or efflux should be governed by the same mechanism; flux rate in any one direction may be individually controlled. It is of interest to recall in this respect that under certain

References p. 563

conditions the glycine influx coefficient could be reduced without a change in the efflux coefficient in ascites[33] and that in galactose transport in *E. coli*[34] the exit process was found to be to some extent independent of the entry process in that efflux could be increased or decreased without corresponding changes in influx. Further studies are necessary to establish the relationship between cerebral influx and efflux of the amino acids.

The differences in behavior between the individual amino acids should caution us not to generalize from observations made on a single compound.

TABLE III

THE EFFLUX OF L-AMINO ACIDS FROM RAT BRAIN

Minutes after administration	*Increase over control µmoles/100 g fresh brain*		
	Leucine	*Phenylalanine*	*Lysine*
5	40	93	33
20	15	93	71
45	2.3	30	72
60	2.3	20	69
120	0	5.2	54
180		0	39
240			28

2 µmoles of the amino acid were injected in 0.02 ml 0.9% saline subarachnoidally into young male rats.

EFFLUX OF AMINO ACIDS FROM THE BRAIN

Efflux of the amino acids from the brain when plasma levels are not elevated was studied after the intracerebral administration of 2 µmoles of an amino acid[26]. With L-amino acids (Table III), maximal levels were reached at various times after administration, and the time to decrease to the physiological level was different for each of the amino acids studied.

TABLE IV

THE EFFLUX OF D-AMINO ACIDS FROM RAT BRAIN

Minutes after administration	*µmoles/100 g fresh brain*		
	Leucine	*Phenylalanine*	*Lysine*
5	59	71	66
20	47	59	84
45	18	38	78
60	12	26	74
120	1.6	4.9	63
180	0.7		49
240			43

2 µmoles of the amino acid in 0.02 ml 0.9% saline were injected subarachnoidally into young male rats.

References p. 563

In similar experiments with the D-isomers the results were parallel to those with the L-forms in that the rate of exit varied from one amino acid to another and the efflux rates of the three compounds were in decreasing order: leucine > phenylalanine > lysine (Table IV).

From the data presented in Tables III and IV the relative rates of efflux can be calculated. These are presented (Table V) as half times (time required for the brain level to decrease to half), and the two isomers are compared. The ratio of the relative exit rate of the two isomers (D over L) was highest in the case of leucine, lower in the case of lysine, and close to unity in the case of phenylalanine. It is of interest that these ratios—an indication in a sense of the stereospecificity of the efflux—showed the same relationship as did the relative transport against a concentration gradient

TABLE V

RELATIVE EFFLUX RATES OF L- AND D-AMINO ACIDS FROM RAT BRAIN

Time intervals minutes	Minutes for 50% decrease in level*					
	Leucine		Phenylalanine		Lysine	
	L	D	L	D	L	D
5–45	14	28				
20–120			27	32		
45–240					140	200
Ratio D/L	2.0		1.2		1.4	

$$* \frac{\text{time interval in minutes}}{\%\ \text{loss during interval}} \times 50.$$

from the brain, where the leucine *pump* was more *efficient* than the lysine *pump* and no phenylalanine *pump* could be shown. This parallelism suggest, that stereospecificity is a property of the active transport process.

The differences in efflux rates between the two isomers, however, do not exclude the possibility that the transport of the D-forms can also be mediated. This possibility has to be taken into consideration because of the finding of accumulation of D-glutamic acid against a concentration gradient in brain slices[28]. At present experiments are being conducted in our laboratory to test *in vivo* D-amino acid transport from the brain against a concentration gradient[35].

An indication that the passage of the two isomers of the amino acids is not completely independent is the finding of increased content of the L-amino acid in the brain following the administration of the D-form[26]. Since in a number of other experiments no racemization of the D-form was found, the most likely explanation for this finding is that some of the D-form in the brain was replaced by the L-form through exchange, that is, that the exchange mechanism of the L-form has some affinity for the D-form.

A net decrease in the content of the D-form in the brain thus occurs through net outflow and through exchange (replacement by the L-form), while a net decrease in the content of the L-form occurs through net outflow only. (It cannot be excluded

References p. 563

that one L-amino acid leaves the brain by being exchanged for another though no indication for this possibility was found in the present experiments by measuring total amino nitrogen in the extracts). It follows, then, that if the relative rates of the outflow of the two forms could be compared the differences would be greater than shown in Table V. The relationship of this unidirectional flow to exchange is not clear at the present time, but the possibility exists that they refer to the same mechanism[23, 36].

Although two isomers of any particular amino acid may be transported by the same mechanism, considerable experimental evidence makes it likely that all the amino acids are not transported through the same mechanism. Some specificity in the effects of increased cerebral levels on exchange rates (Tables I and II) have been discussed already. The fact that *in vivo* tyrosine transport was inhibited by some but not all amino acids was also interpreted as evidence of more than one amino acid transport mechanism[32]. Only a little competition could be found among amino acids studied in a comparison of individual with simultaneous cerebral entrance and exit[35]. A transport system specific for L- and D-tryptophan was found in *E. coli*[37].

These findings show the specificity of the transport processes with the possibility of individual control of the passage of a specific metabolite. It is likely that a separate carrier exists for each amino acid or group of amino acids in the brain.

In an investigation of stereospecificity it is important to distinguish between the rate of uptake and the levels reached at equilibrium. It was found in our laboratory[35] that D-leucine or D-phenylalanine penetrated the brain from the plasma or from the CSF at a lower rate than the corresponding L-form, but that at equilibrium the level of D-leucine was higher, and the level of D-phenylalanine lower, than that of the corresponding L-form, showing that the rate of passage is not the only factor determining the final equilibrium.

CONCLUSION

With our increasing knowledge of amino acid transport in simpler systems it becomes possible to investigate the systems operative in such a complex organ as the brain.

Although this organ has certain advantages for a study of transport processes, the picture is far from simple. Questions such as the mechanism of transport, the relationship of exchange, influx and efflux, the metabolic control of transport, and differences between cerebral transport mechanisms as opposed to other organs cannot be answered at the present time.

The evidence gained in the investigation of the passage of amino acids in the brain shows that at least in part this passage is carrier mediated and that it has an active, energy-requiring component. Quantitative and qualitative differences in passage of the various compounds were demonstrated; metabolite specificity as well as stereospecificity and homeostatic control reaching down to separate brain areas were found. The transport processes are two directional, are mediated in efflux as well as in influx, and show changes with development.

A closer understanding of cerebral transport mechanisms is essential, since it seems likely that they have significant influence on brain metabolism.

References p. 563

REFERENCES

1 H. Waelsch, in *Biochemistry of the Developing Nervous System*, Academic Press, New York, 1955, p. 187.
2 L. Bakay, *The Blood Brain Barrier*, C. Thomas, Springfield, 1956.
3 J. B. Brierley, in D. Richter, *Metabolism of the Nervous System*, Pergamon Press, Oxford, 1957, p. 121.
4 R. D. Tschirgi, in W. F. Windle, *Biology of Neuroglia*, C. Thomas, Springfield, 1958, p. 130.
5 D. Tower, in M. D. Yahr (Ed.), *Properties of Membranes and Diseases of the Nervous System*, Springer-Verlag, 1962, p. 1.
6 A. Lajtha, in K. A. C. Elliott, I. Page and J. H. Quastel (Eds.,) *Neurochemistry*, C. Thomas, Publ., Springfield, 1962, p. 399.
7 K. A. C. Elliott and N. M. van Gelder, *J. Neurochem.*, 3 (1958) 28.
8 Y. Tsukada, Y. Nagata and S. Hirano, *Nature*, 186 (1960) 474.
9 J. R. Stern, L. V. Eggleston, R. Hems and H. A. Krebs, *Biochem. J.*, 44, (1949) 410.
10 G. Guroff and S. Udenfriend, *J. Biol. Chem.*, 235 (1960) 3518.
11 G. Guroff, W. King and S. Udenfriend, *J. Biol. Chem.*, 236 (1961) 1773.
12 P. Schwerin, S. P. Bessman and H. Waelsch, *J. Biol. Chem.*, 184 (1950) 37.
13 A. Lajtha, S. Berl and H. Waelsch, *J. Neurochem.*, 3 (1959) 322.
14 A. Lajtha, S. Furst, A. Gerstein and H. Waelsch, *J. Neurochem.*, 1 (1957) 289.
15 A. Lajtha, *J. Neurochem.*, 2 (1958) 209.
16 A. Lajtha, *J. Neurochem.*, 3 (1959) 358.
17 A. Lajtha and J. Toth, *J. Neurochem.*, 8 (1961) 216.
18 A. Lajtha and P. Mela, *J. Neurochem.*, 7 (1961) 210.
19 A. Lajtha, in S. S. Kety and J. Elkes, *Regional Neurochemistry*, Pergamon Press, Oxford, 1961, p. 19.
20 V. G. Allfrey, R. Meudt, J. W. Hopkins and A. E. Mirsky, *Proc. Natl. Acad. Sci. U.S.*, 47 (1961) 907.
21 H. N. Christensen, *Advances in Protein Chem.*, 15 (1960) 239.
22 E. Heinz, *J. Biol. Chem.*, 211 (1954) 781.
23 E. Heinz and P. M. Walsh, *J. Biol. Chem.*, 233 (1958) 1488.
24 A. Lajtha, *J. Neurochem.*, 1 (1957) 216.
25 W. A. Himwich, J. C. Petersen and M. L. Allen, *Neurology*, 7 (1957) 705.
26 A. Lajtha and J. Toth, *J. Neurochem.*, 9 (1962) 199.
27 J. R. Pappenheimer, S. R. Heisey and E. F. Jordan, *Am. J. Physiol.*, 200 (1961) 1.
28 G. Takagaki, L. Hirano and Y. Nagata, *J. Neurochem.*, 4 (1959) 124.
29 S. R. Korey and R. Mitchell, *Biochim. Biophys. Acta*, 7 (1951) 507.
30 K. D. Neame, *J. Neurochem.*, 6 (1961) 358.
31 S. Schanberg and N. J. Giarman, *Biochim. Biophys. Acta*, 41 (1960) 556.
32 M. A. Chirigos, P. Greengard and S. Udenfriend, *J. Biol. Chem.*, 235 (1960) 2075.
33 E. Heinz, *J. Biol. Chem.*, 225 (1957) 305.
34 B. L. Horecker, J. Thomas and J. Monod, *J. Biol. Chem.*, 235 (1960) 1586.
35 A. Lajtha, J. Toth and J. Schwartz, unpublished.
36 R. M. Johnstone and J. H. Quastel, *Biochim. Biophys. Acta*, 46 (1961) 527.
37 J. A. Boezi and R. D. DeMoss, *Biochim. Biophys. Acta*, 49 (1961) 471.

DISCUSSION

Chairman: HALVOR CHRISTENSEN

L. ROSENBERG: I would appreciate a little further clarification of this concept of exchange diffusion from either Dr. CHRISTENSEN or Dr. HEINZ and perhaps a little further discussion of the mechanisms of the loading experiments that I think each of you reported.

HEINZ: If you preload these cells with unlabeled glycine, and then expose them to labeled glycine, the influx of the labeled glycine is greatly increased as compared to the control. I think this phenomenon is related to those, described later under other names, such as "induced uphill transport by counterflow", by Dr. ROSENBERG AND Dr. WILBRANDT, by PARK *et al.*, and by others.

CHRISTENSEN: Perhaps I might offer this word of explanation. If you let a given cell take up, let us say, leucine to a steady state, perhaps there will be three times as much inside as there is outside. But if you take cells that have already been loaded with leucine to quite a high level by exposing them to a high environmental leucine concentration, and then put on the outside ^{14}C-labeled leucine, you will have in the course of one minute a rise in the level of ^{14}C-label inside to approximately twelve times the outside level. Then leucine-^{14}C will eventually return to the 3 : 1 distribution, corresponding to a constant specific activity for an essentially unchanged load of leucine. This behavior means, while ^{12}C-leucine is moving out with the gradient, that ^{14}C-leucine is driven into the cell against the gradient. The transport site does not differentiate between these two forms. So far as site is concerned, nothing is happening, but so far as the observer is concerned, uphill transport is being produced by counterflow. In the case of two different amino acids the behavior may seem still more dramatic, but the site still considers them the same, except perhaps for slight affinity differences.

LAJTHA: I would like to add here that there may be some similarity between transport mechanisms in ascites cells and brain. I think exchange diffusion also occurs in brain. As with ascites in the experiments discussed just now, preloading the brain increases cerebral exchange. That is, the uptake of labeled amino acid from the plasma by the brain is greater in the brains in which amino acid levels were increased by intracerebral administration. However, there seems to be some specificity in that preloading the brain with leucine increases leucine exchange and not lysine exchange and, in turn, preloading with lysine increases lysine exchange more than leucine exchange. Dr. GUROFF discussed at this meeting, and we did several times at other places, that although net uptake in brain is restricted, a rapid exchange of most amino acids between brain and plasma can be shown. I find exchange diffusion again the most probable explanation for this phenomenon.

E. ROBERTS: I might just add a word of caution about such work with brain. When one studies exchanges in slices, it should be kept in mind that exchanges observed may be taking place in several different cell types. There seems to be an inference in the discussion that the exchange is occurring in a uniform preparation. But since there are several cell types, there may be uptake into one and exchange in another. I think that for the moment this presents such a complex problem that it is difficult to see how one can analogize those results with those obtained using single cell types such as bacteria or ascites tumor cells.

LAJTHA: Of course, Dr. ROBERTS is correct in reminding us of the great complexity of the brain and we should be aware of the many possible explanations and not oversimplify the picture. At present, however, at least to me, of all such possible explanations of the above discussed phenomena, including that uptake into one cell alters exchange in another, that of exchange diffusion is the most attractive and most plausible.

CHRISTENSEN: I wanted to take the opportunity to comment on the paper of Dr. HEINZ. At the last Federation meeting (April 1961), Dr. OXENDER AND Dr. ROYER reported further on the identification of the site of action of pyridoxal and pyridoxal phosphate, showing that it does work on a *mediated process outward*, a mediated efflux. This was established by showing that the process concerned is subject to saturation, to competition, and also serves for exchange. With these three properties, diffusion is ruled out as the efflux process being slowed.

We are coming to question, now, whether there really is very much migration by diffusion into or out of normal cells. Instead, mediations may be entirely responsible. I should like to call this possibility to Dr. GUROFF's attention, to question whether, in the absence of oxygen, he is limiting migration to diffusion. It may well be that mediation is still occurring but without a supply of energy to permit a concentrative mediation.

H. ROSENBERG: Mr. Chairman, I would like to ask you about the involvement of certain chelating mechanisms in the transport of amino acids, which you mentioned briefly. I was wondering whether chelators like EDTA have any pronounced effect on the process. One could, for instance, imagine a sexadentate chelate of cobalt, with two free positions which could involve a double attachment for an amino acid, and provide the correct built-in discriminator for an optical isomer.

CHRISTENSEN: Through the years we have very seriously considered this possibility. There is further evidence in that direction in the fact that the α-amino acids are concentrated, and there are mechanisms also for the β-amino acids, but at least for the tumor cells and the other cells we have studied, γ-aminobutyric and similar amino acids with the amino and carboxylate groups further removed, do not fit into these mechanisms at all. Similarly, with the diamino acids, the space between the two amino groups is critical so that 5- or 6-membered rings might be formed with a metallic ion, for example. Therefore, we have applied two kinds of tests. First we have determined the effect of EDTA and other binding agents on transport. These do tend to be inhibitory, but the inhibition does not appear at once. Their use is marked by a loss of potassium from the cells, and this kind of loss we have generally found to be associated with a gradual decline in the ability to transport amino acids, on what basis we are not sure at present. Assuming that they have access to the metal, chelators like EDTA and 8-hydroxyquinoline ought to begin much more quickly to inhibit uptake if the metal is really serving in the concentration mechanism.

The other approach has been to take the isotopic forms of these metal ions, put them into the environment of the cells in trace amounts, increasing as insignificantly as possible the concentration prevailing there, and then seeing if their flux is increased by a heavy load of amino acids presented for accumulation. One metal, manganese, has its influx into the Ehrlich cell increased by the uptake of α, γ-diaminobutyric acid, and in the presence of pyridoxal, by glycine and certain other amino acids. This is permissive kind of evidence, but it has no great force in indicating that manganese is operating as a carrier, and I have my doubts. The metals cobalt, zinc, and ferric iron were instead held out of the cell, as one might expect, by the presence of amino acid. Since the amino acid level was perhaps ten times as high inside as outside, obviously the metal was getting only to the extracellular compartment and failed to follow the amino acid into the cell, either passively or as part of the carrier.

One never is too sure about the form taken by metals in this part of the periodic table in solutions of pH 7.4, so far as ionic activity is concerned. The conditions are rather uncertain and no doubt natural chelators escaping from the cell define the final activity of the metal. True, the manganese was sometimes as low as 10^{-11} molar, but one still cannot be sure if this value represents a real activity. Hence, this experiment does not provide a final answer and it is only permissive.

TRANSPORT AND ACCUMULATION OF AMINO ACIDS BY MICROORGANISMS*

JOSEPH T. HOLDEN

Department of Biochemistry, Medical Research Institute, City of Hope Medical Center, Duarte, Calif. (U.S.A.)

INTRODUCTION

The discovery of freely extractable amino acids in microorganisms by GALE and his associates[26, 27] (*cf.* HOLDEN, p. 73) was followed by the pioneering studies of this group on the uptake and accumulation of amino acids by *Streptococcus faecalis* and *Staphylococcus aureus*. These studies, which have been summarized extensively by GALE[29, 31, 32], focused attention to this problem and demonstrated that microorganisms might be advantageously used in studies of transport. This review will summarize some of the commonly described properties of representative microbial amino acid accumulation systems subsequently studied and with this as a base discuss current views concerning the mechanism of this process.

At the outset it should be understood that there is considerable confusion and inaccuracy in terminology used by workers in this field. (*cf.* ROSENBERG[102] and CHRISTENSEN[15] for discussions of nomenclature). It is not always appreciated that the existence of a permeability barrier or the interaction of cell components with solute during uptake even in a catalytic manner are not necessarily indicative of active transport, nor is the additional demonstration of an energy requirement. The term "active transport" should be reserved to describe situations in which the solute studied occurs in the same state in the intra- and extracellular compartments and is present at a higher concentration in the compartment into which it has moved, *i.e.* movement from a lower to a higher chemical or electrochemical potential has occurred. While it has become customary, with considerable justification, to use the term transport in discussing many cases of amino acid accumulation, it should be understood that evidence sufficient to demonstrate *active* transport has not been presented yet for any specific microbial amino acid accumulation system. The inferential evidence in some cases is very strong, as will be seen for the systems studied in this laboratory. Nevertheless this evidence appears incomplete, largely because the intracellularly accumulated amino acid still has not been demonstrated beyond all doubt to be in the free state. Furthermore, some investigators after extensive studies retain the belief that the accumulations they have observed cannot be attributed entirely, if at all, to a classical active transport process. In view of this uncertainty,

* The following abbreviations will be used: DNP, 2,4-dinitrophenol; HB_6 cells, high vitamin B_6 cells, nutritionally normal cells grown in complete medium, containing an excess of this vitamin (*cf.* ref. 47); LB_6 cells, low vitamin B_6 cells grown in media containing none, or trace amounts, of vitamin B_6.

the more general terms, "accumulation" and "transport" will be retained in referring to this phenomenon.

Presently held views on the mechanism of amino acid accumulation have been influenced strongly by many exciting, provocative findings with other classes of solutes, in bacterial cytology, on the properties of protoplasts, etc. As a result, the specific status of the amino acid transport problem is not always clearly evident. This review, therefore, will discuss principally experimental observations obtained with amino acids, and their implications as they relate to amino acid transport. Findings from other systems, most of which have been elegantly reviewed by others (see below) will be introduced only when they are crucial to an understanding of this problem. Hopefully, this more exclusive treatment will permit a clearer understanding of the strengths and limitations of the evidence available as support for currently popular concepts depicting the mechanism of amino acid transport and accumulation in microorganisms.

Additional discussions of amino acid permeability and accumulation in microorganisms can be found in the reviews of MITCHELL[80, 84], COHEN AND MONOD[20], CHRISTENSEN[15], and BRITTEN AND MCCLURE[12]. Other aspects of microbial permeability and transport have been reviewed by MITCHELL[78, 81, 87, 90], ROTHSTEIN[103], MONOD[91], DAVIS[23, 24], CIRILLO[10A] and others[59] and will be discussed in a forthcoming review by COHEN AND KEPES[19].

PROPERTIES OF AMINO ACID ACCUMULATION SYSTEMS

Explorations of various properties of amino acid accumulation systems has permitted some limited deductions concerning the possible nature of this process. Thus, the operation of a simple passive diffusion process has generally been excluded by showing a non-linear relation between rate and external concentration, competition among structurally related substrates and a high temperature coefficient. A reliance on cell metabolism generally has been observed most often as a dependence on energy-yielding reactions. Finally high concentration gradients are usually reported although little evidence exists to show that these are true gradients involving solutes in the same state within and outside the cell. Examples of such studies will be considered below.

Methods

Experimental procedures sufficient to demonstrate accumulations are extremely simple, requiring only the exposure of cells to a solution of the test substance, separation of cells and medium after an appropriate time interval, and analysis of the cells (or extract thereof) and/or the supernatant fluid for the solute. Some investigators use exponentially growing cells in the steady state whereas others, following GALE's example, use relatively dilute suspensions of resting cells incubated in buffers or growth media rendered incapable of supporting cell division. Thick suspensions in which intra- and extracellular volumes are roughly equivalent are seldom used for accumulation studies, although they are frequently used in studies of permeability properties. By employing isotopically labeled amino acids and measuring isotope and the L-amino acid in both intra- and extracellular compartments it is possible to insure the absence of extensive metabolism of the accumulated amino acid which can seriously limit the utility of a system as a model for studying transport phenomena. Unfortunately, most studies have relied almost exclusively on measurements of isotope in cell extracts, occasionally supported by chromatographic evidence showing that most or all of the isotope is still present in the original amino acid. Seldom has the optical configuration of the amino acid been confirmed or the supernatant fluid examined for metabolites of the amino acid. Many of these difficulties can be avoided by using a non-metabolizable solute such as α-aminoisobutyric acid[16, 76] or by em-

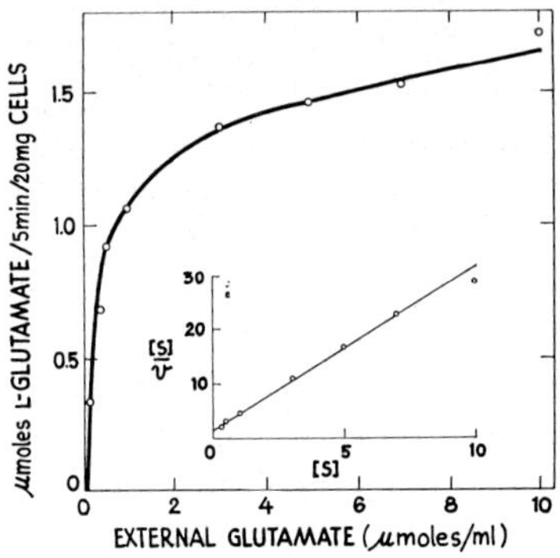

Fig. 1. The effect of external glutamic acid concentration on the rate of glutamate accumulation by *L. arabinosus*. Standard uptake conditions as described by HOLDEN AND HOLMAN[54]. The inset shows these data plotted according to the method of LINEWEAVER AND BURK[67].

ploying a mutant strain incapable of significantly metabolizing the test substance[56]. The associated movement or counterflow of other cell constituents or metabolites during amino acid accumulation, which could have an important bearing on an analysis of the reaction mechanism, has seldom been explored in microbial studies.

While the simple procedures described above generally suffice to demonstrate accumulation, decisions regarding the mechanism of accumulation pose vastly more difficult methodological problems. For example, direct evidence demonstrating the intracellular state of accumulated amino acids so far has proven difficult to obtain in most systems. It is anticipated that much more sophisticated techniques will be required to detect, measure and identify transport catalysts in cell fractions.

Effect of external concentration on accumulation rate and capacity

The effect of external concentration on the L-glutamic acid accumulation rate in *L. arabinosus* is illustrated in Fig. 1. The data fit a straight line when plotted according to the method of LINEWEAVER-BURK. Similar results have been observed for glutamate accumulation by *S. faecalis*[27], and for the uptake of various amino acids by *S. cereviseae*[41] and *E. coli*[12]. Such behavior indicates an interaction of exogenous amino acid with a cell component present in limiting quantity. It is clearly recognized that this does not permit a choice between the operation of rate-limiting catalysts or of internal binding sites. Few investigations include conclusive evidence for the participation of catalysts other than those concerned with the supply of energy.

Only a few examples have been reported in which a limiting rate was not observed

References p. 592/594

as the external amino acid concentration was varied within physiologically reasonable limits. A notable example is that of lysine accumulation by *S. faecalis* described in GALE's initial report[27] which was consequently thought for this and other reasons to occur by a passive diffusion process. Lysine accumulation by another strain of *S. faecalis* studied in this laboratory shows saturation kinetics and dependence on glucose metabolism[52]. The same is true for lysine accumulation by *S. cereviseae*[41].

A summary of accumulation rates observed in representative studies is presented in Table I. In Gram-negative bacteria accumulation is usually completed in approximately one minute at 37° so that accurate rate measurements are difficult to achieve. The values shown, therefore, should be regarded as minimal and, in any case, must be appraised cautiously since the process is undoubtedly complex and probably involves a number of components. There is no certain knowledge of the nature of the rate-limiting process. Accumulation rates in Gram-negative and Gram-positive bacteria as well as in yeasts appear to be comparable although in the two classes of bacteria they are achieved at substantially different external concentrations.

TABLE I

RATES OF AMINO ACID ACCUMULATION IN VARIOUS MICROORGANISMS

Organism	*Amino acid*	*Half maximum rate µmoles/min/100 mg*	*Concentration for half maximum rate µmoles/ml*	*Ref. Numbers*
L. arabinosus	Glutamic acid	0.80	0.50	54
	Alanine	1.1	2.0	52
S. faecalis R	Alanine	1.2	0.20	52
	Lysine	1.9	0.50	52
S. faecalis	Glutamic acid	0.45	0.50	27
E. coli	Proline	0.80	0.003	12
	Valine	1.0	0.0005	12
S. cereviseae	Glutamic acid	4.2	0.30	41
	Arginine	3.0	0.40	41
	Lysine	2.4	0.30	41

The capacities normally observed also differ markedly in these groups of organisms. Illustrative examples of the maximum capacity and the maximum apparent concentration ratios (apparent internal concentration to external concentration) for a variety of amino acids and organisms are presented in Table II. Gram-negative organisms generally accumulate amino acids to one-fifth or one-twentieth of the level observed in Gram-positive bacteria, but achieve gradients at least as high as those attained in Gram-positive organisms. However, BRITTEN, ROBERTS, *et al.* have shown that the size of the proline pool in *E. coli* is dependent on the osmotic strength of the extracellular medium[12, 100]. Using proline or a mixture of amino acids pool levels as high as 100 µmoles/100 mg have been obtained by incubating cells in sucrose-containing media. Exceptionally high apparent gradients appear to exist for endogenously synthesized amino acids in *E. coli* (valine 28 000, glutamic acid 7 300)[12]. Since this material is present in minute amounts, its state and, therefore, the reality of these gradients is open to question.

References p. 592/594

The values presented in Table II are cited only to show the general magnitude of pool sizes. Considerable variability originating from as yet undescribed causes has been encountered, for example, by separate groups working with *E. coli*. The culture of *L. arabinosus* used in this laboratory recently suffered an abrupt 35% loss in glutamate accumulation capacity which is as yet unexplained. GALE observed that over a period of a few years *S. aureus* increased its ability to accumulate glutamic acid. Such observations are experimental curiosities at present.

The possibility that pools are not homogeneous has been suggested by the studies of COWIE AND MCCLURE[22] who found that *Candida utilis* accumulates amino acids in

TABLE II

AMINO ACID ACCUMULATION CAPACITIES OF VARIOUS MICROORGANISMS

Organism	*Amino acid*	*Maximum capacity* μmoles/100 mg*	*Maximum concentration ratio***	*Ref. Numbers*
L. arabinosus	Glutamic acid	68	390	54
	Alanine	30	250	52
	Threonine	18		52
	Glycine	14		52
	Proline	14	58	52
	Valine	9	45	52
S. faecalis R	Glutamic acid	43	200	52
	Alanine	22	150	52
	Threonine	30		52
	Glycine	23		52
	Lysine	42	180	52
	Arginine	16		52
S. faecalis	Glutamic acid	22	300	31
S. aureus	Glutamic acid	47	600	31
E. coli	Proline	24***	1000	12
	Valine	6.0	2200§	12
E. coli	Valine	1.2	460	21
E. coli	Proline	3.4	50	58A
E. coli	Tryptophane	2.5		7
Dutch Top Yeast	Glutamic acid	61		111
S. cereviseae	Glutamic acid	45	211	41
	Arginine	29	985	41
	Lysine	22	815	41
C. utilis	Threonine	44		22
N. crassa	Proline	2.1	2	118

* Values are for 100 mg dry weight of cells. A majority of the values are taken from experiments in which the effect on pool size of varying the extracellular concentration was determined. Reasonable estimates are given for cases in which pool sizes increase slowly as external concentration is raised over a wide range. Where the effect of external concentration was not studied, the largest pool size reported is cited.

** In most cases, the values shown were cited in the original publication and refer to the ratio: internal concentration/external concentration, assuming the internal amino acid to be distributed throughout all the intracellular water. Generally maximal ratios are observed at relatively low extracellular concentrations, far below those which are required to saturate the pool.

*** Proline or a mix of amino acids can be accumulated to levels as high as 100 μmoles/100 mg by adding high concentrations of sucrose to the medium.

§ An extracellular proline concentration of 0.0029 μmoles/ml was used. A ratio of 18 000 was reported using an extracellular concentration of 0.016 mμmoles/ml.

an "expandable" osmotically sensitive pool from which exchange is possible and also in an "internal" pool which is not readily removed from the cells by osmotic shock, does not exchange with external amino acids and is composed of endogenously synthesized amino acids when growth occurs in the absence of exogenous amino acids. HALVORSON AND COHEN[42] also have encountered evidence for metabolically distinguishable amino acid pools in *S. cerevisease*. ZALOKAR[118] has shown with Neurospora that exogenous amino acids can by-pass a portion of the pool during incorporation into protein. BRITTEN *et al.*[12, 13] using *E. coli* have found that amino acids enter a very specific pool when extracellular levels are low, whereas at high concentrations a much larger and relatively non-specific pool is formed whose size, as indicated above, can be increased to very high levels by increasing the tonicity of the extracellular medium. The possible heterogeneity of pools in Gram-positive bacteria appears not to have been examined critically although HANCOCK's findings[43] suggest that proline pools having different exchange properties exist in *S. aureus*.

The relationship between the rate and final capacity of accumulation varies greatly. Among Gram-positive bacteria the initial rate at 37° can be maintained for thirty to sixty minutes before a limiting capacity is reached for most amino acids. Gram-negative bacteria however incubated at 37° achieve a saturating capacity usually within one minute. The basis of this difference has not been established but may be related to differences in cell wall rigidity.

Substrate specificity

Competitive effects by structurally related amino acids and their analogs provide additional evidence that amino acids interact with cell components during accumulation. In GALE's studies this aspect of the process was not extensively examined, although it was shown that asparate reduced glutamate accumulation in *S. aureus* and that it was accumulated under these conditions. A truly competitive interaction was not demonstrated. Glutamate accumulation also was inhibited by cysteine, alanine and glycine, but the accompanying appearance of extracellular peptide suggests that this was not a competitive interaction in the uptake system. COHEN AND RICKENBERG[21] showed with *E. coli* that isoleucine, leucine and valine interact in a common accumulation process which is relatively unreactive with other amino acids. Phenylalanine and methionine accumulation also were reduced only by structurally related substances. In all cases the process was specific for L-amino acids. A study of valine analogs showed that the amino and carboxyl groups must be unsubstituted and that modification of the side chain (substitution of dibutyl for dimethyl residues) resulted in an inactive molecule. An additional test of specificity was the ability of an amino acid or analogue to displace previously accumulated amino acids from the cell. Very good agreement was observed in the ability of a substance to reduce accumulation and to displace accumulated amino acid from the pool.

Glutamate accumulation in *L. arabinosus* likewise shows a high order of structural specificity[52, 54]. L-aspartic acid and glutamine are very effective competitors, the latter exceeding L-[^{12}C]glutamic acid in reducing the accumulation rate of L-[^{14}C]-glutamic acid. However, D-glutamic acid, asparagine, α-ketoglutaric acid and γ-aminobutyric acid are all essentially inactive in reducing the rate and the amount of glutamate accumulated.

References p. 592/594

The rather strict structural requirements cited above are not uniformly encountered. Thus, the D-isomers of methionine in *Alcaligenes faecalis*[61], phenylalanine in *S. cereviseae*[42] and tryptophane in *E. coli*[7] reduce accumulation of the corresponding L-amino acid. Broader specificity of the *S. cereviseae* systems is indicated also by the ability of valine to reduce phenylalanine accumulation and of phenylalanine and methionine to reduce valine accumulation. In such cases however, extracellular peptide formation does not seem to have been excluded as a cause of reduced accumulation. N-methylvaline and valine amide were inactive as inhibitors demonstrating again that both the amino and carboxyl groups are required for activity.

In *E. coli* accumulation of small amounts of proline from very dilute solutions is highly specific, being unaffected by a mix of fifteen amino acids each of them present at a concentration 100 times higher than proline[13]. When the concentration of extracellular proline is raised, much larger pools accumulate and competitive interactions are evident. On the other hand, these workers also encountered competitive interactions between isoleucine, leucine and valine at very low extracellular concentrations[12].

Confirmation of the structural specificity of some accumulation systems is found in the behavior of so-called transport mutants. The limited data available so far show that mutants incapable of accumulating D-serine also fail to accumulate L-alanine and glycine, while those incapable of accumulating canavanine accumulate alanine and glycine normally but accumulate subnormal amounts of L-arginine, L-lysine and L-ornithine[106]. LUBIN *et al.*[69] also have presented data suggesting specificity of amino acid accumulation in transport mutants.

Peptides are taken up rapidly by bacteria but apparently by routes independent of those responsible for amino acid uptake. In *E. coli* valine peptides do not interfere significantly with valine retention[21] suggesting that they do not interact with the valine accumulation system. Since these peptides support growth of valine-requiring strains, it is likely that a separate system exists for their uptake. LEACH AND SNELL[63] showed that glycine peptides are taken up more rapidly than glycine by *Lactobacillus casei*. The peptides do not accumulate as such but are hydrolyzed rapidly. The possibility exists that this rapid metabolism may account for the superior rate of uptake compared to the free amino acid. An *E. coli* mutant with a reduced ability to take up glycine grows well with glycine peptides again suggesting the existence of separate uptake systems for this amino acid and its peptides[66]. GALE and his co-workers also observed glutamate accumulation during exposure of *S. aureus* to a variety of glutamyl peptides, but the relation to the glutamic acid accumulation system was not established[104].

Requirement for energy

The dependence of glutamate accumulation in *L. arabinosus* on glucose metabolism is illustrated in Fig. 2. In this organism and *S. faecalis* all amino acids studied so far have been found to accumulate in large quantities only in the presence of a fermentable carbohydrate[52, 54]. Requirement for an energy source has been reported in many other investigations, *e.g.* in *S. aureus* (glutamic acid)[38], in yeast (glutamic acid)[111], (various amino acids)[41] and in *E. coli* (proline)[13], (various amino acids)[21]. In some cases, endogenous reserves appear to suffice as for α-aminoisobutyric acid accumulation in *B. megaterium*[76]. GALE has studied energy-yielding substrates other than glucose and observed that pyruvate and arginine have some activity in *S. aureus*[31]. This group also has studied the ability of various glutamate derivatives to permit the intracellular accumulation of this amino acid in the absence of fermentable carbohydrate[104]. N-phosphorylglutamate and diethylglutamate had some activity although the rates were inferior to those observed when the cell was exposed to glutamic acid and glucose. None of a variety of glutamyl peptides led to intracellular glutamic acid accumulation in the absence of an energy source. Since the accumulation of glutamate derived from N-phosphorylglutamate was sensitive to DNP and azide GALE suggested

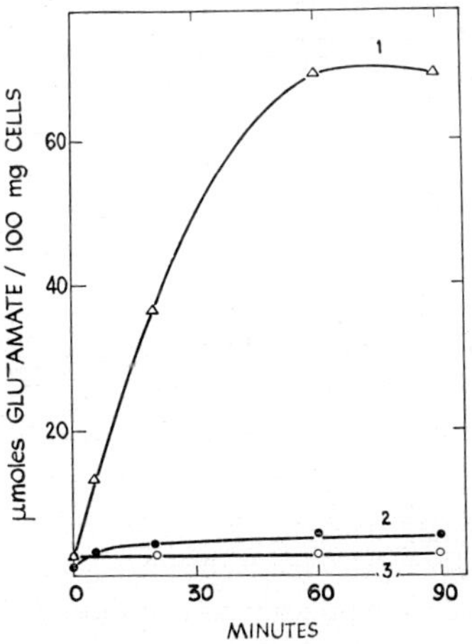

Fig. 2. Effect of glucose and reduced incubation temperature on glutamate accumulation by *L. arabinosus*. Glutamic acid of cell extracts was assayed by enzymatic decarboxylation and radioactivity. Standard uptake conditions[54] with, curve 1, △——△, glucose present; curve 2, ●——●, glucose omitted; curve 3, ○——○, glucose present, incubated at 2°.

that the breakdown of this compound might yield energy for its own transport or of the liberated glutamic acid. Apart from these few observations there is as yet no evidence indicating the manner in which energy-yielding reactions are coupled to the accumulation process.

Evidence for energy dependency in transport is also provided by studies showing the inhibitory effect of uncoupling agents such as DNP and azide as in the accumulation of tryptophane[7], and other amino acids[21] by *E. coli* and of various amino acids by *S. cereviseae*[41, 42]. It must be recognized that these substances may react in many ways within a cell and that specific interaction only in phosphorylation systems cannot be assumed. A number of investigators have in fact observed anomalous behavior with these substances. GALE[30] found that DNP and azide at some concentrations markedly *increased* glutamate accumulation by *S. faecalis*. As in previous studies with triphenylmethane dyes[35] this was attributed to a differential effect on competing reactions which normally consumed accumulated glutamate. Using isotopically labeled glutamic acid we have observed a marked resistance of glucose-dependent glutamate accumulation in *L. arabinosus* to DNP and azide inhibition except at high levels (DNP, 30% inhibition at 0.01 *M*; azide, 30% inhibition at 0.03 *M*). At lower concentrations there is a pronounced stimulation of accumulation by both substances. Unlike GALE's experience, there is little metabolic loss of glutamate in the absence or presence of these compounds. HORECKER *et al.*[57] and OSBORN *et al.*[94] have reported that the entry and exit reactions governing galactose accumulation in *E. coli* can be distinguished by different sensitivities to DNP inhibition. Some of our findings suggest that the exit of amino acid is more readily inhibited than entry by DNP[52].

Studies of temperature dependency generally support the conclusion that amino acid accumulation involves processes other than passive diffusion. In most cases a drastic reduction of the accumulation rate at 0° to 4° (Fig. 2), and temperature coefficients in excess of 1.8 in the range between 18° and 35° have been observed[27, 41]. Some investigators have studied slow accumulation at low temperatures[12, 21]. The possible dependence of such uptake on residual energy metabolism has not been excluded. Of course, in those cases where accumulation is markedly accelerated by providing cells with an oxidizable or fermentable substrate, the finding of anything but an elevated temperature coefficient would indeed be a surprise. Thus such evidence has limited value in deducing the nature of interactions which the accumulated amino acid undergoes during its passage into the cell.

Similar limitations apply in studies of the alteration in amount or rate of accumulation when the extracellular pH is varied. One would like to know, of course, which molecular species reacts most successfully in the uptake process. Unfortunately, the pH optimum is a summation of the optima of a number of reactions including those of the energy-producing reactions and, therefore, provides little information about the properties of the catalysts which react directly with the amino acid. In a number of cases[27, 54] the optimum pH for maximum accumulation corresponds to the pH at which the fermentation rate is most rapid. On the other hand, lysine accumulation in *S. faecalis*[27] increased steadily as the pH was raised to 9.5, which is very close to the isoelectric point of this amino acid. Tryptophane uptake (isoelectric point 5.9[40]) in *E. coli* showed a maximum at pH 8.0 to 8.5[7].

Retention and exchange of accumulated amino acids

One of the striking observations reported initially by GALE was that little or no accumulated glutamate was lost from *S. faecalis* during incubation in water or buffer at 37°. A slow leakage was observed with *S. aureus* which could be reduced markedly by providing the cells with glucose[31]. As shown in Fig. 3 glutamate accumulated by *L. arabinosus* similarly is retained with little loss for long periods in water or buffer. It is readily eluted (by exchange) from the cell when extracellular glutamate is present but only at an elevated incubation temperature. This shows that a source of energy is not required to maintain a pool even when gradients of several thousand-fold exist, and, for the case of suspension in water, that severe osmotic stress does not significantly diminish pool retention. In contrast, Gram-negative bacteria rapidly lose accumulated amino acids when suspended in water[11, 12] probably because of an inferior ability to oppose unfavorable osmotic forces. This is suggested by the finding that small pools of accumulated amino acid are lost very slowly from *E. coli* incubated in buffered media at 25° and that progressively larger fractions of the pool are eluted as the medium is diluted with water[12]. In *C. utilis* COWIE AND MCCLURE[22] observed that endogenously synthesized amino acids are not removed from the cell by water

References p. 592/594

washes whereas material accumulated from the medium and present in the so-called expandable pool is rapidly lost under these conditions. Such pools appear also to be distinguishable by their exchange properties; amino acids in the internal pool do not exchange with extracellular amino acids, whereas those in the expandable pool do. The tenacious retention of large pools during exposure to buffer at 37° in the absence of exogenous amino acids is not explicable by the so-called permease concept which predicts a sizable leak under these conditions. However, retention of the pool on adsorption sites is not a necessary alternate explanation. As will be seen below, studies with *L. arabinosus* suggest that this property is largely determined by the rigidity of the cell wall and that the phenomenon can be accommodated by either the adsorption site or membrane transport concepts of the pool.

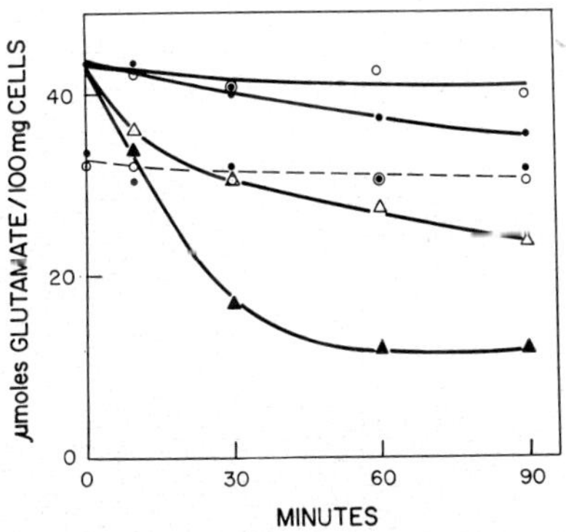

Fig. 3. Retention of accumulated glutamic acid by *L. arabinosus*. Cells were allowed to accumulate L-[^{14}C]glutamic acid for 75 min at 37°. In these and the reincubated cells intracellular glutamic acid was measured isotopically. After centrifugation the cells were resuspended in water or 0.12 *M* phosphate buffer (pH 6.5) containing the indicated supplements and incubated at 2° or 37° as follows: ●— — —●, water, 2°; ○— — —○, water + L-[^{12}C]glutamic acid (0.003 *M*), 2°; ●———●, water, 37°; ○———○, buffer, 37°; △———△, buffer + L-[^{12}C]glutamic acid (0.003 *M*), 37°; ▲———▲, buffer + L-[^{12}C]glutamic acid (0.003 *M*) + glucose (0.028 *M*), 37°. Separate batches of cells were used for experiments at 2° (– – – –) and 37° (———).

Accumulated amino acids generally can be eluted by exogenous amino acids and their analogs. As indicated previously, amino acids which compete with the test amino acid when uptake is measured will also elute previously accumulated amino acid from the cell[12, 21, 52]. In *E. coli* the studies of COHEN AND RICKENBERG do not show whether such displacements are dependent on cell metabolism. BRITTEN *et al.*[12] using growing cells have considered extensively the relation between the uptake and exchange reactions. They find that the rate of proline uptake into the pool is reduced by a factor of 270 when the incubation temperature is reduced from 25° to 0°, whereas the exchange rate is reduced only to one-fourth of the control value. Proline exchange at 0° does not follow a simple exponential curve and Britten concludes that there are at least two separate components of the pool which exchange at different rates. Both exchange rates appear to be independent of extracellular amino acid concentration and dependent on the intracellular pool size. HEINZ[45] has reported acceleration of glycine uptake into Ehrlich ascites carcinoma cells preloaded with this amino acid. A comparable finding using microorganisms has not been reported.

Action of inhibitors

A variety of substances other than those suspected to interfere with the supply of high energy compounds (*e.g.* azide and DNP) and structural analogues of the amino acid have been observed to inhibit accumulation. GALE[28] observed inhibition of glutamate accumulation by 8-hydroxyquinoline at levels below those which inhibit glycolysis. Reactivation was observed using Mg^{2+} and Mn^{2+} which were also effective in restoring activity of cells grown in media low in metallic ions. Stimulation of tryptophane accumulation in *E. coli* by Mg^{2+} has been observed[7] but an effect on the supply of energy was not excluded.

Although GALE observed penicillin inhibition of glutamate accumulation in *S.*

TABLE III

EFFECT OF PHYSIOLOGICAL AGE AND MEDIUM TONICITY ON SENSITIVITY OF GLUTAMATE ACCUMULATION TO PENICILLIN INHIBITION*

Cell type	*Penicillin (units/ml)*	*Sucrose (moles/l)*	*Glutamate accumulated (μmoles/100 mg)*
Late Exponential	0	0	76.3
	100	0	66.6
	100	0.6	75.5
Early Exponential	0	0	71.5
	100	0	28.9
	100	0.6	60.0

* *L. arabinosus* was incubated at 37° for 90 min with L-[^{14}C[glutamic acid (*cf.* ref. 54) and the indicated supplements. Intracellular glutamic acid was calculated from the isotope found in hot water extracts of centrifuged cells.

aureus this effect was dependent on prior growth in the presence of the antibiotic. In *L. arabinosus* more reliable interpretations of the effects of this antibiotic on the accumulation process have been achieved by taking advantage of physiological and nutritional variables[52]. Viewed in the light of current information regarding the mechanism of action of penicillin, an adequate explanation of GALE's findings is that growth with penicillin leads to the production of osmotically sensitive cells due to defective cell wall formation. The formation and retention of high intracellular glutamate pools would be impossible, if not lethal, due to the inability of such cells to tolerate the elevated intracellular osmotic pressures which might accompany accumulation. This view conflicts with MAAS' recent suggestion[70] that penicillin primarily affects the accumulation system and that the commonly recognized cell wall effects are secondary phenomena. Our findings using *L. arabinosus* at various physiological ages do not support this interpretation. As shown in Table III glutamate accumulation by cells from late exponential cultures is inhibited only slightly by penicillin whereas cells from early exponential cultures are much more sensitive. The penicillin inhibition in this case is almost completely reversed by high concentrations of sucrose or KCl. Early exponential phase cells contain less wall material. They increase markedly their capacity for glutamate accumulation during incubation in buffer, at the same time depositing additional cell wall substance in the absence of significant DNA synthesis and cell division. Clearly the accumulation system itself is insensitive to penicillin, since it functions normally when the cell is protected by sucrose, or in a cell containing sufficient quantities of wall substance. Reduced accumulation in the presence of penicillin appears, therefore, to be a secondary effect caused very likely by a primary interference in cell wall formation. HANCOCK[44] also has reported that the inhibitory effects of penicillin on amino acid accumulation in *S. aureus* can be prevented by adding sucrose to the incubation medium.

GALE observed little or no inhibition of glutamate accumulation by chloramphenicol and some inhibition by aureomycin at concentrations which did not significantly interfere with glycolysis[36]. In *L. arabinosus* chloramphenicol has consistently shown a small stimulatory effect on glutamate accumulation. Streptomycin causes leakage of intracellular constituents in sensitive strains of *E. coli*[4]. Damage to the permeability barrier of the cell is suspected and the effect is accompanied by a marked decline in ability to accumulate exogenous valine. Other antibiotic[14] and antifungal[75] compounds have been observed to cause leakage of intracellular constituents. It would be of interest to determine whether amino acid accumulation also is affected in these cases.

In *L. arabinosus* bacitracin, tyrocidin and gramicidin reduce glutamate accumulation markedly. Gramicidin, but not tyrocidin inhibition, is reversed by high levels of sucrose. Tyrocidin was shown by GALE AND TAYLOR[37] to release accumulated glutamic acid from *S. faecalis* cells presumably by disruption of the cellular permeability barrier. The steroid, deoxycorticosterone, reduces the ability of germinated Neurospora conidia to take up various low molecular weight constituents including inorganic ions and amino acids[65]. Subsequent studies have indicated that rubidium uptake occurs in two phases, an energy-dependent entry step followed by a binding step[64]. The latter process is believed to be inhibited by deoxycorticosterone. A comparable analysis of amino acid uptake, unfortunately, has not been reported. In our studies on glutamate accumulation deoxycorticosterone had no inhibitory effect in *L. arabinosus*.

As yet, inhibitor studies have not provided definitive clues to the nature of the accumulation system. A recurrent observation, that compounds which promote leakage of cell constituents also reduce accumulation, suggests that a functional permeability barrier is essential for the accumulation of large amino acid pools.

Miscellaneous observations

BRITT AND GERHARDT, during a study of lysine accumulation in cells and protoplasts of *Micrococcus lysodeikticus*[9], observed that a sizable fraction of the "pool" was adsorbed in the cell wall[10]. This underscores the importance of showing that any charged solute, such as a basic or acidic amino acid, does in fact occupy an intracellular locus when it is accumulated or thought to distribute in cellular water.

Changes in cell volume and methionine accumulation activity are not correlated in cells of *Alcaligenes faecalis* undergoing synchronous division[62]. Accumulation activity was found to remain constant per cell throughout the division cycle indicating that synthesis of the accumulation machinery was closely synchronized with the reproductive process. However a sizable increase in accumulation activity per cell was observed when stationary phase cultures were incubated in growth medium prior to the onset of exponential growth.

References p. 592/594

BOEZZI AND DE MOSS[7] have observed a ten-fold increase in tryptophane accumulation capacity when *E. coli* is grown with a complete amino acid mix including high levels of tryptophane. Glucose inhibits the activity of the accumulation system but not this apparently adaptive increase.

Studies on solute penetration monitored by its consumption intracellularly have been described occasionally (*e.g.* refs. 74, 116). Unfortunately, interpretation of such findings is greatly complicated by the necessity to dissociate the properties of the transport and the metabolizing systems. They appear more useful in studies on the functional aspects of accumulation systems, as, for example, in investigation of crypticity.

MECHANISM OF AMINO ACID ACCUMULATION

The mechanism by which exogenous amino acids are taken into the cell in opposition to apparent large concentration gradients still remains in doubt, although the active transport hypothesis is favored by most workers. The differences in findings with different organisms raises the possibility that more than a single mechanism may be found to operate. Speculations regarding the nature of this phenomenon have centered on two very different views of the cell and the state of its amino acid pool. In the first instance it has been proposed that intracellular pools are adsorbed or associated with internal polymers[12, 101] (*cf.* BRITTEN, p. 595 and COWIE, p. 633) in a cell which is readily penetrated by a wide range of nutrients and metabolites. The more frequently encountered proposal is that intracellular solutes are free within cells which are enclosed by a permeability barrier capable of retaining these solutes. In the latter instance, entry would be achieved by specific catalysts which permit net movement across the membrane into the cell even when the intracellular amino acid concentration exceeds that of the extracellular environment. In the most recent formulation of the adsorption site hypothesis BRITTEN AND MCCLURE have suggested that a mobile carrier catalyzes the reaction of exogenously derived amino acid with adsorption sites within the cell[12] (*cf.* BRITTEN, p. 595). Beyond establishing the metabolic dependency of accumulation phenomena and suggesting that specific catalysts function in this process, the properties considered in previous sections have had limited value in evaluating this question. A choice between these models could be made by demonstrating beyond doubt the state of the accumulated amino acid.

State of intracellular amino acid pools

If amino acids occur within the cell largely in a free form, unassociated with intracellular polymers, a permeability barrier must exist at the cell surface or at the surface of that portion of the cell which contains the amino acids, and the accumulation phenomena would then clearly involve active transport in view of the large concentration gradients which must consequently exist. If the pool is largely associated with adsorption sites a different type of catalytic reaction can be expected and the occurrence of a functional permeability barrier might not be required, although it could still exist and possibly even play a role in the accumulation process. Therefore, dependence of accumulation on an effective permeability barrier cannot by itself exclude the adsorption site mechanism, but is a necessary condition if the pool is in a free state. On the other hand, existence of a major portion of the pool in a bound state would not exclude active transport as the mode of entry, so long as the concentration of that portion of the pool in the free state still exceeded the extracellular concentration.

GALE adopted a cautious attitude when interpreting the results of his extensive studies recognizing that the data available did not exclude either of the abovementioned mechanisms[31]. Recently he has proposed that the process involves active transport[33, 34]. The suggestion that *E. coli* is freely permeable to most inorganic ions and the usual cell metabolites[101] led to considerable controversy which has greatly increased our understanding of the permeability properties of bacterial cell membranes and stimulated a critical examination of the intracellular state of accumulated solutes including amino acids. In the original hypothesis amino acids in *E. coli* were believed to be retained on specific adsorption sites located within the cell which appeared from space studies to be freely penetrated by a wide variety of low molecular weight solutes. Contradicting these findings, MITCHELL observed initially that most of the internal volume of *S. aureus* was not accessible to inorganic phosphate ion[77]. It should be noted that while high concentrations of phosphate were excluded from the cell in the absence of an energy source, a one for one exchange of intra- and extracellular phosphate did occur. This relative aspect of "impermeability" is not always appreciated. Inhibition studies showed that this exchange involved catalytic rather than stoichiometric adsorption sites. These studies were extended by MITCHELL AND MOYLE to show that most of the cell interior of this organism in the resting state also was not accessible to various amino acids and inorganic ions, thus indicating the occurrence of a permeability barrier near the cell surface[84, 89]. In their hands the cell water of *E. coli* likewise was inaccessible to NaCl and phosphate salts[84]. Subsequently, a large number of other investigators also have observed impermeability of intact bacterial cells, protoplasts and spheroplasts to various solutes thereby establishing the widespread existence of a functional permeability barrier near the bacterial cell surface[6, 39, 71, 81, 82, 84, 85, 114, 115].

As indicated above, the state of the intracellular solutes is not established by the demonstration of a permeability barrier. It is conceivable that an important fraction of the intracellular solutes might still be retained on adsorption sites. Therefore, as evidence grew for the existence of effective cytoplasmic membranes in bacteria, considerable effort was made to determine whether the intracellular constituents are osmotically active, since this would be one way of deciding whether or not they are free. The ingenious experiment of MITCHELL AND MOYLE[84, 89] in which cell pastes were exposed to atmospheres above a series of sucrose solutions of graded concentration and subsequently weighed to determine the variation in water content and thus the relative osmotic strength of the cell constituents and the sucrose solution indicated that the intracellular osmotic pressure in *S. aureus* approximately equalled that predicted if the total internal solutes extracted by cold TCA were in free solution. Unfortunately, this experiment was not sufficiently precise to show the extent to which the relatively small portion of the total solute pool made up by amino acids was contributing to the observed osmotic activity.

Another approach has involved measurement of changes in light scattering (indicative of changes in volume) when cells or protoplasts are exposed to penetrating solutes. MITCHELL AND MOYLE[82, 84, 85, 89], AVI-DOR *et al.*[6, 60], GILBY AND FEW[39] and others have observed swelling with some substances although generally not with

References p. 592/594

amino acids. ABRAMS, who initially observed an energy-dependent entry of sucrose into *S. faecalis* protoplasts accompanied by swelling[1, 2], subsequently found that a variety of amino acids used at high concentrations as osmotic stabilizers also cause swelling of protoplasts[3] (*cf.* p. 615). Since amino acids within the protoplasts cause an influx of water measured indirectly as swelling (*i.e.* change in optical density) it is inferred that they must be free and not bound to adsorption sites. In such experiments the amino acids were present extracellularly at extremely high concentrations and, therefore, moved down a concentration gradient into the protoplasts. A comparable phenomenon has been observed in this laboratory using protoplasts accumulating amino acids in opposition to concentration gradients (Fig. 4). ABRAMS also has

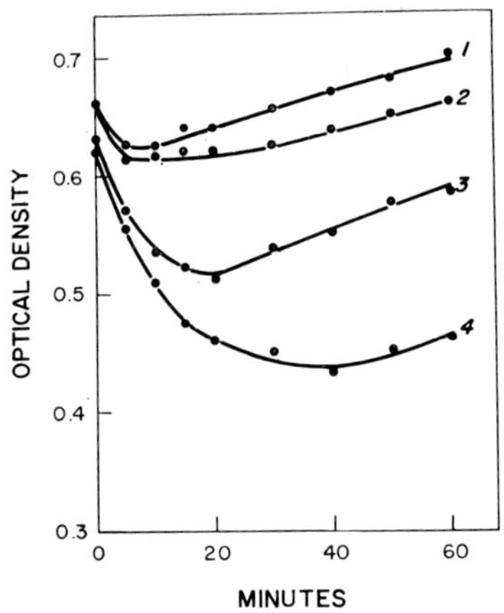

Fig. 4. Swelling of *S. faecalis* protoplasts associated with glutamic acid accumulation. Incubation at 37° in 0.18 *M* phosphate buffer (pH 6.5) containing 0.6 *M* sucrose. Optical density was measured at 640 mμ. Suspensions received supplements at the following concentrations: 1, no additions; 2, L-glutamic acid (0.003 *M*); 3, glucose (0.005 *M*); 4, glucose and glutamic acid.

observed this behavior. WACHSMAN AND STORCK[113] have described energy-dependent swelling of *B. megaterium* protoplasts induced by propionate, various amino acids and other organic acids. Recently MARQUIS[76] has shown that protoplasts of *B. megaterium* swell while accumulating α-aminoisobutyric acid. These experiments are comparable to those carried out earlier by SISTROM[107] with *E. coli* spheroplasts in which swelling was induced by accumulation of galactosides, and they suggest strongly that the intracellular amino acids are osmotically active. In none of these studies, however, has it been conclusively demonstrated that the osmotically active material is the accumulated solute and not some other displaced substance. However remote this possibility may seem, this concept will not be secure until the osmotically active material is conclusively identified. Furthermore, bacterial protoplast swelling induced by accumulated amino acids so far has not been studied quantitatively. In view of the evidence that amino acid pools in some organisms may be heterogeneous, the possibility must be considered that only a portion of the pool participates in the swelling phenomenon. Furthermore, explanations can be conceived for the phenomenon other than that the penetrating solute induces an osmotic influx of water. For example, there is the possibility that membrane stability and resistance to distention is diminished by adsorption of accumulated acidic substances to the inner surface of the membrane, or by neutralization of a protecting substance and that the apparent influx of water is not solely or even principally attributable to the osmotic activity of the accumulated solute. The enhanced stability of bacterial protoplasts in the presence of basic substances such a spermine[73] is consistent with this view. Thus, while the likelihood is great that at least a portion of the intracellular amino acid pool is osmotically active, the force of this evidence could hardly be considered as overwhelming.

Dependence of accumulation on osmotic factors as a function of physiological and nutritional status

Additional evidence suggesting that accumulated amino acids in *L. arabinosus* and *S. faecalis* may be osmotically active was encountered indirectly during our studies on the effects of various nutritional deficiences on the accumulation process. Stimu-

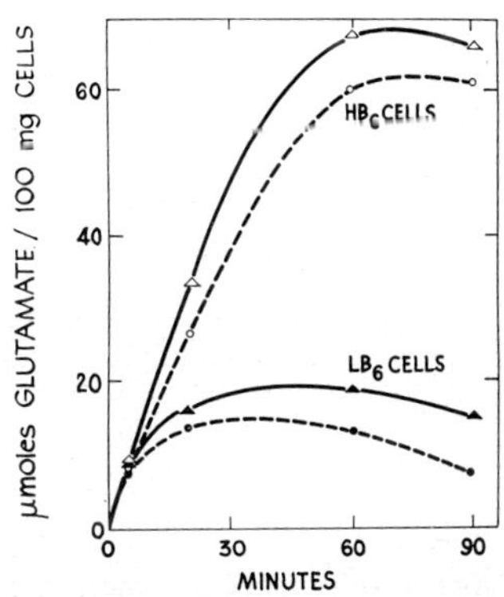

Fig. 5. Effect of vitamin B_6 deficiency on glutamic acid accumulation by *L. arabinosus*. See ref. 47 for experimental details. Glutamate content of cell extracts measured isotopically (———) and by enzymatic decarboxylation (– – – –).

lated by the provocative finding that pyridoxal increases the amount of amino acid accumulated by mouse Ehrlich ascites cells[18, 99], glutamate accumulation was compared in nutritionally normal and vitamin B_6-deficient cells of *L. arabinosus* [47, 54].

The results illustrated in Fig. 5 show that the initial rate of accumulation is not modified by the deficiency, but that there is a large decline in the amount of amino acid which can be retained. This and other evidence contra indicated a catalytic function for vitamin B_6 in amino acid accumulation (this is true also for *S. faecalis*). Furthermore, the large change in capacity appeared to be an indirect effect, not attributable to the direct involvement of vitamin B_6 in the uptake process[47].

Subsequent studies showed that vitamin B_6-deficient cells were morphologically abnormal[53] and that they leaked unusually large amounts of intracellular nucleotides during incubation in phosphate buffers[46]. Suspecting a cell wall defect with secondary alterations in permeability properties, osmotic aspects of the accumulation process were investigated. It was found[48] that the addition of high concentrations of sucrose and other substances to the incubation buffer permitted essentially normal glutamate accumulation by severely vitamin B_6-deficient cells. Fig. 6 shows that precisely

References p. 592/594

equal stimulatory effects are obtained with isomolal concentrations of KCl and sucrose[50]. Additionally, substances such as lysine, glucose, NH_4Cl and KNO_3 at high concentrations permit large increases in accumulation capacity, whereas substances such as glycine or glycerol, which is known to penetrate the cell rapidly, are inactive. In contrast to the behavior of large glutamate pools accumulated by normal B_6-adequate cells, these pools are instantaneously lost when the cells are exposed

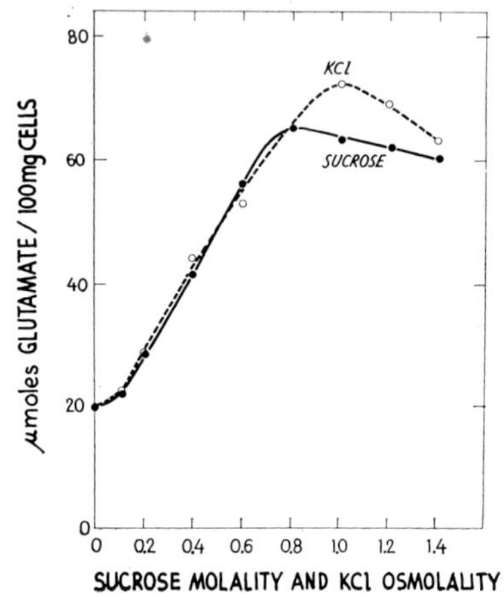

Fig. 6. Effect of sucrose and KCl on glutamic acid accumulation by vitamin B_6-deficient *L. arabinosus*. Cells were incubated for 100 min under standard uptake conditions modified to provide the indicated concentrations of sucrose or KCl.

to dilute buffer (Fig. 7). However, the accumulation system appears to suffer no significant damage under these conditions since resuspension in sucrose buffer containing a fresh supply of amino acid leads to accumulation of a second large pool. The time of addition of glutamate to glycolizing cells has little effect on the initial rate of uptake suggesting that there is not a progressive degeneration of the uptake system in the absence of glutamate accumulation.

It is likely that the beneficial effect of sucrose on accumulation depends on its ability to reduce water content in some part of the cell. The region of the cell so

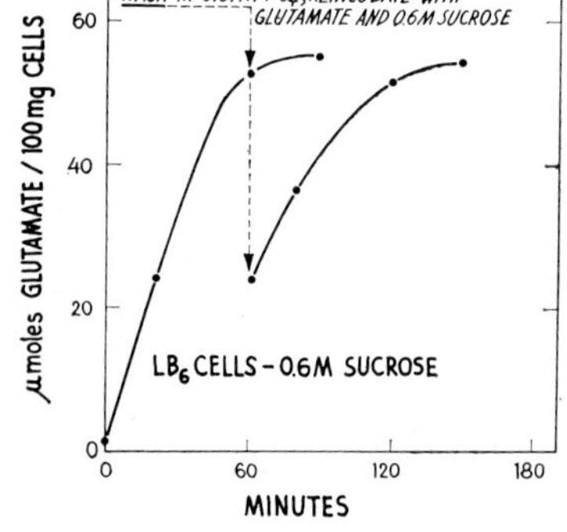

Fig. 7. Effect of rapid resuspension in dilute buffer on retention of previously accumulated glutamic acid and reaccumulation of this amino acid by vitamin B_6-deficient *L. arabinosus*.

References p. 592/594

affected should not be penetrated by sucrose. To determine the magnitude of this space the distribution of sucrose in pellets of normal and vitamin B_6-deficient cells was measured. The results expressed as impermeable volumes are summarized in Table IV. The sucrose-impermeable volume is only slightly lower than the total cell volume (the difference between total pellet volume and extracellular space measured by dextran and inulin). In pellets of B_6-deficient cells the slightly larger difference between these values may be due partly to the presence of a small population of dead, completely permeable cells and possibly to a more porous cell wall which

TABLE IV

PERMEABILITY OF *L. arabinosus* TO SUCROSE*

Space measured	*Test substance*	*HB_6 cells*	*LB_6 cells*
		Volume (ml/g dry wt.)	
Total pellet	—	4.16	4.43
Total cell	Dextran	3.15	3.13
Total cell	Inulin	3.04	3.19
Sucrose impermeable	Sucrose	2.87	2.58
Sucrose impermeable	[^{14}C]Sucrose	2.43	2.23

* Approximately 100 mg of centrifuged cells were resuspended and incubated at 37° for 45 min in an equal volume of the test substance dissolved in 0.12 *M* phosphate buffer. Dextran (avg. M. W. 60 000–90 000) was used at 100 mg/ml, inulin at 50 mg/ml, and sucrose at 0.2 or 0.3 *M*. The cells were centrifuged and the supernatant analyzed as follows: dextran by the anthrone method[25], inulin by the anthrone and ROE methods[5], and sucrose by radioactivity or the ROE method. The values shown are averages of 4 experiments with each cell type.

admits a greater amount of sucrose relative to inulin or dextran than occurs in control cells. It is clear that at least 75%, and possibly much more, of the cell interior is not penetrated by sucrose, suggesting strongly that sucrose is excluded from that part of the cell lying within the surface membrane, and, therefore, that it must increase accumulation by preventing water influx into this region of the cell. These data, however, give no certain information whether water influx is deleterious to accumulation because the membrane is distended or because intracellular structures are disrupted.

These findings could be interpreted with greater confidence if a procedure was available which could reduce membrane distention without influencing movement of water into the protoplast. The finding that the cell wall of vitamin B_6-deficient cells is defective and that conditions which permit deposition of additional cell wall material also restore normal accumulation activity appeared to satisfy these requirements. Electron microscopic examination has shown that B_6-deficient cells of *L. arabinosus* contain a much thinner than normal cell wall layer external to the membrane (Fig. 8)*[55]. Deficient cells also yield only one-half the amount of cell wall

* The occurrence of complex intracellular membranous structures appearing as extensions of the cytoplasmic membrane (as shown in Fig. 8A) indicates the shortcomings of the simple view that the bacterial cell consists of a largely disorganized cytoplasm contained within a sack (the cytoplasmic membrane). Earlier speculative suggestions that accumulation phenomena in microorganisms might involve interactions with intracellular membranes must now be given more serious attention especially in attempting explanations for pool heterogeneity.

material recoverable from normal cells by standard isolation procedures. Essentially normal glutamate accumulation occurs within vitamin B_6-deficient cells if the incubation buffer is supplemented with acetate, ammonium ion and vitamin B_6 (Fig. 9)[48]. Furthermore, pretreatment of deficient cells with these substances and [^{12}C]-glutamic acid results in an increased ability to accumulate [^{14}C]glutamic acid or alanine in the *absence* of the stimulatory substances or sucrose. A large increase in isolatable cell wall occurs during the pretreatment period. Although there was a

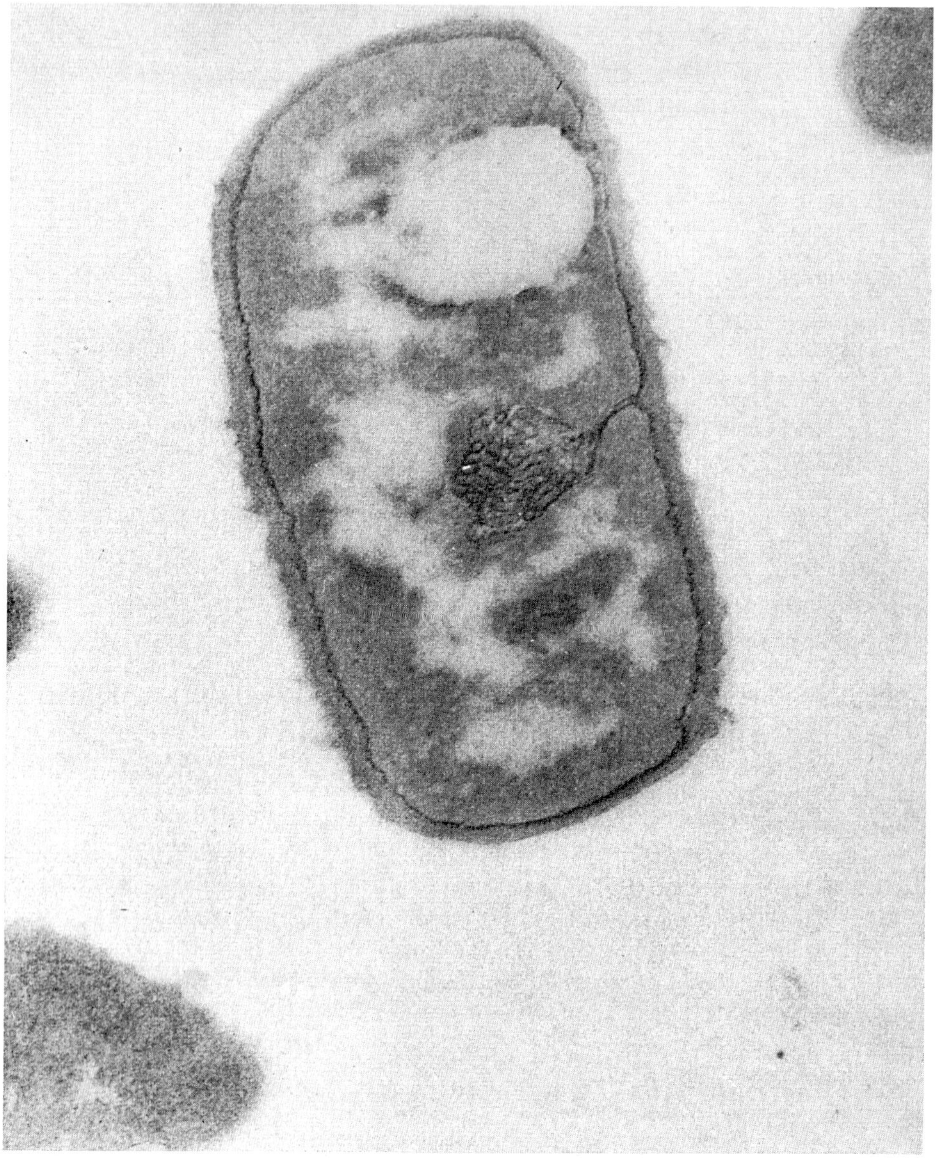

Fig. 8(a). Electron micrographs of (a) nutritionally normal, and (b) vitamin B_6-deficient cells of *L. arabinosus*. Total magnification 160 000 ×

References p. 592/594

clear association of improvement in accumulation capacity and increase in cell wall mass, this might be a coincidental relationship and not of a causal nature. Therefore, the fate of [^{14}C]acetate was investigated under pretreatment conditions[51]. As shown in Table V, this substance is incorporated principally in two cell fractions, the ethanol-soluble and the cell wall fraction. In vitamin B_6-deficient cells the incorporation of label in the cell wall is specifically reduced. In the presence of ammonium ion, vitamin B_6 and glutamate this incorporation is increased to levels observed in control B_6-adequate cells (Table VI), without modifying the incorporation of acetate into other cell fractions. In other words, those conditions required for a substantial improvement in accumulation activity are identical with those required for normal entry of acetate into cell wall material, and of all the cell fractions the wall specifically shows this effect. It can be concluded with greater confidence, therefore, that an increase in cell wall substance in vitamin B_6-deficient cells leads to a restoration of nearly normal accumulation activity.

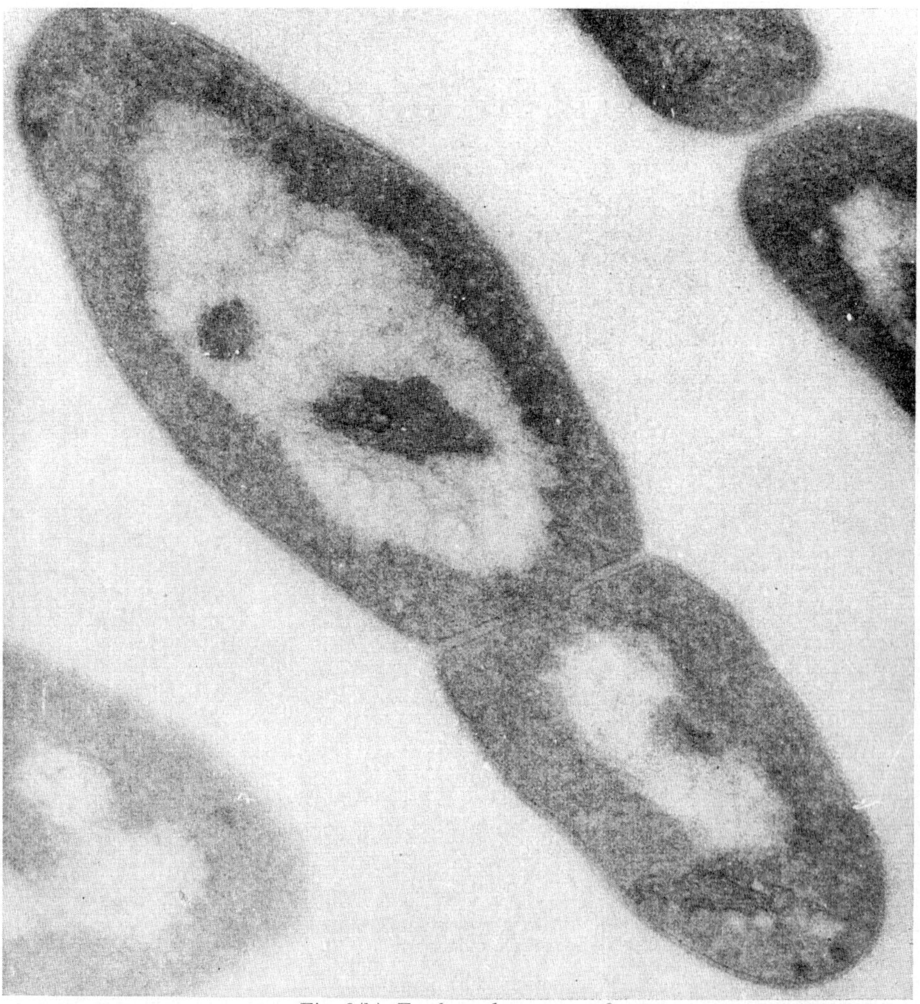

Fig. 8(b). For legend see page 584.

We can now reconsider the mechanism by which sucrose, although excluded from most of the internal volume of the cell, promotes a large increase in the amount of glutamate accumulated by B_6-deficient cells. Cells which have been pretreated with acetate, ammonium ion, glutamate and vitamin B_6, and allowed to synthe-

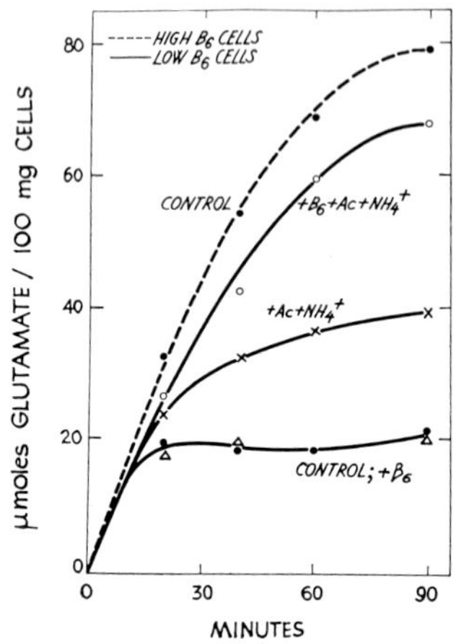

Fig. 9. Effect of acetate, NH_4^+ and vitamin B_6 on glutamic acid accumulation by vitamin B_6-deficient cells of *L. arabinosus*. See ref. 48 for complete experimental details. The indicated supplements were added to standard uptake buffer as follows: ●——●, no additions; △——△, pyridoxamine (1 µg/ml); X——X, CH_3COOK (0.0057 *M*) and NH_4Cl (0.003 *M*); ○——○, CH_3COOK, NH_4Cl and pyridoxamine; ●- - - -●, no additions. (——) B_6-deficient cells; (- - - -), B_6-adequate cells.

size more cell wall material, are still morphologically abnormal (swollen shape) It is unlikely that water influx, especially into intracellular structures, would be completely restrained in such reactivated cells, nor, therefore, that internal accumulation sites would be significantly protected by this procedure. However is likely that an unfavorably large swelling of the internal protoplast with a conse-

TABLE V

DISTRIBUTION OF [2-^{14}C]ACETATE IN *L. arabinosus* CELL FRACTIONS

Fraction	*Cell type*	
	HB_6	*LB_6*
	*(counts/min/10 mg cells)**	
Cold TCA	3 000	6 000
Hot TCA	100	300
Ethanol	114 000	128 000
Trypsin	1 600	500
Residue (Wall)	38 000	19 000

* Washed cells were incubated at 1.6 mg/ml in 0.12 *M* phosphate buffer[54] containing glucose (0.028 *M*), sucrose (0.5 *M*), K[2-^{14}C]acetate (0.0059 *M*). After 60 min at 37° cells were centrifuged, the pellets frozen and extracted as described by PARK AND HANCOCK[95]. Appropriate aliquots were plated and radioactivity measured in a gas-flow counter.

References p. 592/594

quent distention of its membrane would be inhibited by the deposition of additional wall substance. Since only one procedure (sucrose supplementation) would be expected to reduce significantly the influx of water to intracellular sites, whereas both stimulatory conditions have in common the prospect that membrane distention would be inhibited, it can be concluded that the latter effect is decisive and that the membrane plays an active role in the accumulation phenomenon. Cell membrane distention could reduce accumulation capacity by allowing an equilibration of influx and efflux rates at a lower than normal intracellular concentration. Since the rate of accumulation is the same no matter when the labeled amino acid is added to glycolizing cells, it is likely that any distention which does occur in unprotected

TABLE VI

ACETATE INCORPORATION INTO CELL WALL OF *L. arabinosus*

Addition to buffer	*LB_6 cells*	*HB_6 cells*
	*(counts/min/10 mg cells)**	
[2-^{14}C]Acetate	19 700	38 000
[2-^{14}C]Acetate + NH_4^+ + B_6 + L-glutamic acid	44 600	45 000

* Incubation was for 90 min at 37° using conditions and extraction procedures described in Table V. Supplements to the incubation buffer were provided at the following concentrations: NH_4Cl (0.0033 *M*), pyridoxamine, 2HCl (0.084 μg/ml), and L-glutamic acid (0.003 *M*).

vitamin B_6-deficient cells is associated with the accumulation of amino acid and, therefore, that the latter is osmotically active. With the exception of the protoplast swelling studies cited previously, efforts to obtain more direct evidence for the existence of accumulated amino acids in an osmotically active form have been inconclusive.

Additional support for this interpretation of our findings is that these effects are not confined to vitamin B_6-deficient cells. Biotin- and pantothenate-deficient cells as well as those harvested in the early exponential phase of growth from nutritionally-adequate media all accumulate amino acids in an abnormal manner[47]. Normal accumulation is restored in biotin-deficient cells either by high concentrations of sucrose (Fig. 10) or by osmotically inconsequential levels of acetate (Fig. 11)[49]. Comparable responses are observed using acetate and pantothenate or sucrose with pantothenate acid-deficient cells and acetate alone or sucrose alone with nutritrionally normal early exponential phase cells. All such cell types contain smaller amounts of cell wall material than do the late exponential phase cells normally used as controls in such studies. It appears then that *any* procedure which reduces the quantity of cell wall also reduces the accumulation capacity (without reducing the initial accumulation rate). Conversely when conditions are provided which permit increases in wall mass there is a concomitant expansion in the amino acid accumulation capacity. This is true, of course, only in cells in which accumulation capacity is submaximal to begin with.

From these and the previously cited studies it seems highly likely that a number of bacterial pools are accumulated in a free, osmotically active form. Definitive and incontrovertible proof, however, has not yet been produced for any microbial system and a number of inconsistent observations must be reckoned with. For example, the simple interpretation that accumulated amino acids control their own uptake in vitamin B_6-deficient *L. arabinosus* by causing an influx of water which distends the membrane and causes a premature equilibration of the entry and exit rates is not supported by the finding that alanine accumulation, which in normal cells is no

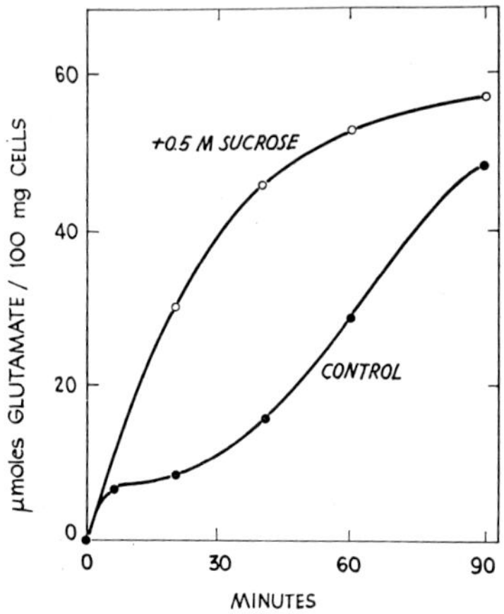

Fig. 10. Effect of sucrose on glutamic acid accumulation by biotin-deficient cells of *L. arabinosus*. Incubation at 37° under standard uptake conditions in the absence and presence of 0.5 *M* sucrose.

higher than the amount of glutamate accumulated by deficient cells, also is only 30–50% of normal in deficient cells[52]. It is not readily apparent why normal alanine accumulation of 20 µmoles/100 mg cells is not possible in a B_6-deficient cell which can accumulate this much glutamate, nor why such cells loaded with glutamate will take up a sizable amount of alanine while losing much less than an osmotically equivalent amount of glutamic acid. The occurrence of a non-exchangeable pool in

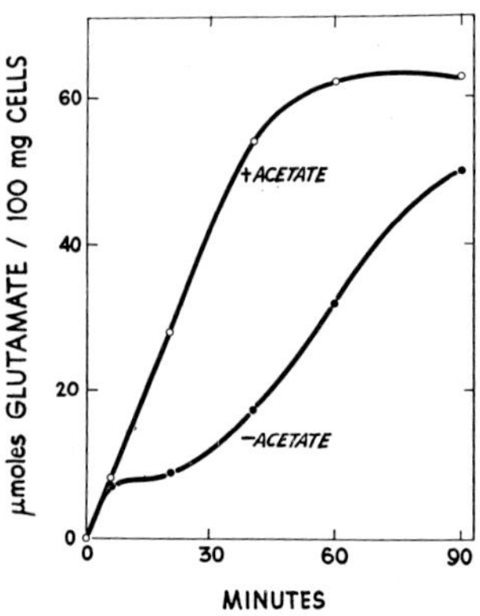

Fig. 11. Effect of acetate on glutamic acid accumulation by biotin-deficient cells of *L. arabinosus*. Incubation at 37° under standard uptake conditions in the absence and presence of 0.0057 *M* CH_3COOK.

References p. 592/594

Candida utilis also is not readily understood in terms of a single free pool. Indeed, this applies to all instances of pool heterogeneity. It should be noted however, that much of the evidence for pool heterogeneity comes from experiments with exponentially growing cells and that the properties of these accumulation systems closely resemble those we have found using early exponential phase cells of *L. arabinosus*. It is clear from our studies that cells in different phases of growth contain pools with different elution properties. This raises the possibility that the heterogeneity of pools cited by COWIE, for example, may be a heterogeneity in the cell population and indicative of a difference in the type of pool a given cell accumulates rather than a heterogeneity of pools within each of the cells in the population.

It may also be appropriate to note that considerable evidence exists in addition to the osmotic observations which is apparently inconsistent with the binding site theory. Thus, the evidence for a bound amino acid pool in *E. coli* is strongest when properties of small pools are considered. Large pools, on the other hand, are osmotically sensitive and show a remarkable lack of specificity, properties which are not easily accommodated by the binding site theory. In *L. arabinosus* loss of as much as 25% of the internal RNA and a substantial amount of protein can occur with little or no diminution of glutamate accumulation capacity[53]. Even prior to this loss the number of amino acid molecules accumulated *exceeds* the total number of nucleotide residues contained in the cell. Except for the adsorption of amino acids to cell wall[9, 10, 76] there is no direct evidence in broken cell studies for an association of sizable amounts of pool amino acids and cell fractions. It is clear that in their purest form both theories can be challenged, that much more evidence is required and that each case must be examined individually. In view of these inconsistencies, it would come as no surprise if a composite mechanism operated in most organisms, *i.e.*, that amino acids are actively transported against real gradients, but that a sizable portion of the accumulated "pool" exists in association with labile internal binding sites, or as some other sequestered form.

Nature of the accumulation catalysts

Most workers, regardless of experimental material utilized observe that accumulation involves interaction of the amino acid with some cell component probably having catalytic properties. Our extensive studies on the effects of nutritional deficiencies, some of which have been described briefly above, indicate that vitamin B_6, biotin and pantothenic acid do not fulfill a direct catalytic function in amino acid accumulation. This is based largely on the observation that in no case is there a significant change in the initial rate of accumulation when cells are depleted of these vitamins. While all these deficiencies markedly influence the amount of amino acid which can be accumulated, these appear to be secondary effects arising from unfavorable structural changes in the cell. While one may argue, as we have elsewhere[47], that such negative evidence has limited value in excluding these substances from consideration as integral parts of a catalytic system, the exclusion of vitamin B_6 as a carrier seems most secure now that we have conclusively demonstrated the completely indirect nature of its effect on uptake capacity and shown that normal accumulation can be achieved by cells possessing as few as 40 molecules of the vitamin. Although evidence continues to appear which suggests some relation between vitamin B_6 and

References p. 592/594

amino acid transport in mammalian tissue, CHRISTENSEN now also has concluded, that in the Ehrlich ascites tumor cell the vitamin does not participate in the entry reaction[15].

On the basis of evidence indicating that bacteria are bounded by a membrane which is virtually impermeable to many solutes in the resting state and the finding that bacterial protoplast membranes contain a considerable number of enzymes[83, 88, 110]. MITCHELL has proposed that solute penetration is achieved by a translocation of chemical groups catalyzed by enzymes in the membrane for which the compound in question is a normal substrate[79, 80, 84, 86, 87, 90]. He has proposed the name "translocases" for such systems. The most convincing evidence for the involvement of enzymatic or, at least, protein catalysts in solute accumulation has come from MONOD, COHEN and their associates who proposed the permease concept (*cf.* 20, 91, 98.) It was demonstrated that galactoside accumulation in *E. coli* occurs with the intervention of an inducible, specific catalyst whose formation can be prevented by interfering with protein synthesis. Kinetic evidence was relied on originally to exclude stoichiometric binding as the mechanism of accumulation. Subsequently, SISTROM[107] demonstrated that protoplasts accumulating galactoside undergo swelling, suggesting that this substance is osmotically active and not bound internally. Thus, it was proposed that the cell contains protein catalysts presumably within the membrane serving to promote the entry of specific organic solutes. Although originally envisaged as being distinct from metabolic catalysts, ZABIN *et al.*[117] presented preliminary evidence suggesting the possible identity of the so-called galactoside permease with an enzyme which acetylates thiomethylgalactosides. Subsequent experiments summarized elsewhere in this volume (ZABIN, p. 613) now make this appear unlikely.

A completely analogous set of observations for amino acid accumulation has not yet been reported. These systems appear to be constitutive in the organisms investigated so far, although BOEZZI AND DE MOSS[7] recently have reported that the tryptophane accumulation system in *E. coli* might be inducible. SCHWARTZ *et al.*[106] and LUBIN *et al.*[69] have described the isolation of viable mutants of *E. coli* lacking specific amino acid accumulation systems. It is not yet clear whether these mutants lack a protein which is the accumulation catalyst or a protein which synthesizes some essential component of a catalyst. Further study of such mutants offers one of the best available tools for identifying the components of the accumulation system.

Additional comments on the possible identity of such systems must be based largely on speculations with little or no experimental support. The activating enzymes presumed to be involved in protein synthesis have been repeatedly proposed as possible accumulation catalysts (*e.g.* ref. 81). This is supported by reports that they may occur in the protoplast membrane[58, 92, 93], (*cf.* ref. 81) although a conflicting recent report has appeared[112]. One would expect that their loss by mutation would be lethal in view of their apparent involvement in protein synthesis. However MITCHELL's cautioning suggestion[80] might be appropriately mentioned here, namely that transport mutants could result simply by a change in location of the catalyst in the cell. In this specific case the activating enzymes might be synthesized but not incorporated into the membrane. Their participation in accumulation might thereby be prevented without significantly reducing their effectiveness for protein synthesis. Perhaps the most impressive argument against the possibility that the well-known activating enzymes function as transport catalysts is the large difference in substrate specificity observed for the two processes.

A number of restrictive findings similar to the results of our nutritional studies may be cited. For example, BOYER AND STULBERG[8] studied the transfer of ^{18}O-labeled amino acids from growth medium to cell protein in *Leuconostoc mesenteroides* and concluded, on the basis of a large retention of isotope in the protein amino acids, that transfer across the cell membrane probably did not require reactions in which oxygen loss from the carboxyl group was possible. Since the exchange observed could be accounted for by the activation step presumably preceding incorporation into protein, one is tempted to conclude that transport involving separate catalysts would not lead to formation of bonds which permit exchange in the carboxyl group. Alternately, one could suggest that activating enzymes are involved in transport but that the activated amino acids proceed directly through the protein-synthetic reactions without equilibrating with the free amino acid pool in exponentially dividing cells. This concept of transport predicts a by-pass of the pool during protein synthesis which in fact has been observed in *S. cereviseae*[42] but not in *E. coli*[13]. In the latter case, this conclusion rests on the behavior of very small pools having an uncertain relation to the large pools found in Gram-positive organisms. In view of this uncertainty it would be of interest to determine directly whether accumulation of amino acids in non-growing cells leads to exchange. CHRISTENSEN has concluded that in Ehrlich ascites carcinoma and in the rat there is no significant loss of carboxyl-^{18}O from labeled α-aminoisobutyric acid despite entry, presumed to be active, into many tissues of the animal[17].

BOEZZI AND DE MOSS observed that there was not an increase in tryptophane activating enzyme, tryptophane synthetase or tryptophane-α-ketoglutarate transaminase in cells of *E. coli* which had undergone a large adaptive increase in tryptophane accumulation activity, therefore, apparently excluding these enzymes as components of the accumulation system[7]. Glutamine has repeatedly been proposed as a possible intermediate in glutamate uptake [31, 80].Recently we have isolated a mutant strain of *L. arabinosus* with an absolute growth requirement for glutamine. The organism nevertheless accumulates glutamic acid at a normal rate compared with its parent. The mandatory intervention of the glutamine synthetase, therefore, would seem to be excluded in this system.

In other transport systems more definitive progress has been achieved in relating cellular enzymatic activities and solute accumulation or permeation. Potassium and sodium movement in nerve[108] and erythrocyte[96, 97] may be associated with the operation of a potassium and sodium stimulated ATPase located in the surface structures of these cells. Bacterial mutants have been isolated which are defective in the ability to accumulate potassium ion[68, 105]. SOLOMON[109] has now shown that the activity of a potassium-dependent ATPase normally present in the membrane is greatly reduced in this mutant compared to wild-type strains. Such observations support the recurrent models suggesting that accumulation systems may consist of, or at least include, enzymatic catalysts located in the cell membrane. It should be noted that, in contrast, BRITTEN AND MCCLURE's analysis of experimental findings on amino acid accumulation in *E. coli* has led to the formulation of a model containing a *mobile* relatively small molecular weight intracellular catalyst[12].

There is extensive evidence showing that net accumulation (although not neces-

References p. 592/594

sarily retention) of amino acids proceeds effectively only when the energy delivery systems of the cell are functioning. The manner in which these activities are coupled is completely unknown, although it is easily seen how the ATPase or activation systems cited above could consume high energy phosphate compounds during transport. How the activities of these enzymes could lead to movement of molecules through a membrane is somewhat less certain. The reader is referred to previously cited reviews for numerous intriguing speculations on the possible mode of operation of membrane catalysts. My reluctance to repeat once again or add to these well publicized hypothetical proposals reflects a conviction that the most urgent need at present is for more data, some of it of a rather basic nature, as, for example, on the composition of the protoplast membrane. It can be hoped that rapid progress of studies with mutants defective in amino acid accumulation will delineate the possibilities for coupling of transport and energy-yielding systems.

CONCLUSIONS

Many microorganisms have been observed to accumulate amino acids in opposition to apparent high concentration gradients using an energy-dependent process which generally requires that the amino acids react with at least one cell component having varying degrees of structural specificity. Extensive evidence for the occurrence of a permeability barrier in bacteria and the demonstration of apparent osmotic activity of intracellular solutes suggests that amino acids within many microorganisms may be largely unassociated with binding sites and thus that accumulation in such cells occurs by active transport. Direct, incontrovertible evidence that a major portion of the amino acid pool accumulated within microorganisms is in a free state has not yet been obtained and, consequently, a final decision that this mechanism alone accounts for accumulation in these organisms must still be regarded as being premature. Recent studies on mutants deficient in accumulation activity for various groups of amino acids offer great promise of resolving this difficult problem.

ACKNOWLEDGEMENT

Investigations of the author described here were supported by Grant E-1487 from the U.S. Public Health Service and by Contract No. NONR-2702 (00) between this institution and the Office of Naval Research.

REFERENCES

[1] A. ABRAMS, *J. Biol. Chem.*, 234 (1959) 383.
[2] A. ABRAMS, *J. Biol. Chem.*, 235 (1960) 1281.
[3] A. ABRAMS, G. YAMASAKI, J. REDMAN AND C. G. MACKENZIE, *Federation Proc.*, 19 (1960) 129.
[4] N. ANAND AND B. D. DAVIS, *Nature*, 185 (1960) 22.
[5] G. ASHWELL, in S. P. COLOWICK AND N. O. KAPLAN, *Methods in Enzymology*, Academic Press, New York, 1957, Vol. 3, p. 75.
[6] Y. AVI-DOR, M. KUCZYNSKI, G. SCHATZBERG AND J. MAGER, *J. Gen. Microbiol.*, 14 (1956) 76.
[7] J. A. BOEZZI AND R. D. DE MOSS, *Biochim. Biophys. Acta*, 49 (1961) 471.
[8] P. D. BOYER AND M. P. STULBERG, *Proc. Natl. Acad. Sci.*, 44 (1958) 92.
[9] E. M. BRITT AND P. GERHARDT, *J. Bacteriol.*, 76 (1958) 281.
[10] E. M. BRITT AND P. GERHARDT, *J. Bacteriol.*, 76 (1958) 288.
[11] R. J. BRITTEN, *Science*, 124 (1956) 935.
[12] R. J. BRITTEN AND F. T. MCCLURE, *Bacteriol. Revs.*, (1962) In press.

[13] R. J. Britten, R. B. Roberts and E. F. French, *Proc. Natl. Acad. Sci. U.S.*, 41 (1955) 863.
[14] T. D. Brock and M. L. Brock, *Arch. Biochem. Biophys.*, 85 (1959) 176.
[15] H. N. Christensen, *Adv. Prot. Chem.*, 15 (1960) 239.
[16] H. N. Christensen, A. J. Aspen and E. G. Rice, *J. Biol. Chem.*, 220 (1956) 287.
[17] H. N. Christensen, H. M. Parker and T. R. Riggs, *J. Biol. Chem.*, 233 (1958) 1485.
[18] H. N. Christensen, T. R. Riggs and B. A. Coyne, *J. Biol. Chem.*, 209 (1954) 413.
[18a] V. P. Cirillo, *Ann. Rev. Microbiol.*, 15 (1961) 197.
[19] G. N. Cohen and A. Kepes, in I. C. Gunsalus and R. Y. Stanier, *The Bacteria*, Academic Press, New York, Vol. 4, In press.
[20] G. N. Cohen and J. Monod, *Bacteriol. Revs.*, 21 (1957) 169.
[21] G. N. Cohen and H. V. Rickenberg, *Ann. Inst. Pasteur*, 91 (1956) 693.
[22] D. B. Cowie and F. T. McClure, *Biochim. Biophys. Acta*, 31 (1959) 236.
[23] B. D. Davis, in O. H. Gaebler, *Enzymes: Units of Biological Structure and Function*, Academic Press, New Yoık, 1956, p. 509.
[24] B. D. Davis, *Arch. Biochem. Biophys.*, 78 (1958) 497.
[25] N. J. Fairbairn, *Chem. and Ind.*, (1953) 86.
[26] J. C. Freeland and E. F. Gale, *Biochem. J.*, 41 (1947) 135.
[27] E. F. Gale, *J. Gen. Microbiol.*, 1 (1947) 53.
[28] E. F. Gale, *J. Gen. Microbiol.*, 3 (1949) 369.
[29] E. F. Gale, *Bull. Johns Hopkins Hosp.*, 83 (1949) 119.
[30] E. F. Gale, *Biochem. J.*, 48 (1951) 286.
[31] E. F. Gale, *Adv. Prot. Chem.*, 8 (1953) 287.
[32] E. F. Gale, in *Active Transport and Secretion, Symp. Soc. Exptl. Biol.*, 8 (1954) 242.
[33] E. F. Gale, in W. D. McElroy and B. Glass, *Amino Acid Metabolism*, The Johns Hopkins Press, Baltimore, Maryland, 1955, p. 119.
[34] E. F. Gale, *Synthesis and Organization in the Bacterial Cell*, John Wiley and Sons, New York, 1959, p. 20.
[35] E. F. Gale and P. D. Mitchell, *J. Gen. Microbiol.*, 1 (1947) 299.
[36] E. F. Gale and T. F. Paine, *Biochem. J.*, 48 (1951) 298.
[37] E. F. Gale and E. S. Taylor, *J. Gen. Microbiol.*, 1 (1947) 77.
[38] E. F. Gale and E. S. Taylor, *J. Gen. Microbiol.*, 1 (1947) 314.
[39] A. R. Gilby and A. V. Few, *J. Gen. Microbiol.*, 20 (1959) 321.
[40] J. P. Greenstein and M. Winitz, *Chemistry of the Amino Acids*, John Wiley and Sons, New York, 1961, Vol. 1, p. 486.
[41] H. Halvorson, W. Fry and D. Schwemmin, *J. Gen. Physiol.*, 38 (1955) 549.
[42] H. O. Halvorson and G. N. Cohen, *Ann. Inst. Pasteur*, 95 (1958) 73.
[43] R. Hancock, *Biochim. Biophys. Acta*, 26 (1957) 211.
[44] R. Hancock, *Biochem. J.*, 70 (1958) 15P.
[45] E. Heinz, *J. Biol. Chem.*, 211 (1954) 781.
[46] J. T. Holden, *Biochim. Biophys. Acta*, 29 (1958) 667.
[47] J. T. Holden, *J. Biol. Chem.*, 234 (1959) 872.
[48] J. T. Holden, *Biochim. Biophys. Acta*, 33 (1959) 581.
[49] J. T. Holden, *Federation Proc.*, 18 (1959) 248.
[50] J. T. Holden, *Federation Proc.*, 19 (1960) 130.
[51] J. T. Holden, *Proceedings Fifth International Congress of Biochemistry, Moscow, 1961*, p. 304.
[52] J. T. Holden, In preparation.
[53] J. T. Holden and J. Holman, *J. Bacteriol.*, 73 (1957) 592.
[54] J. T. Holden and J. Holman, *J. Biol. Chem.*, 234 (1959) 865.
[55] J. T. Holden, T. Kakefuda, and N. Utech, in preparation.
[56] B. L. Horecker, J. Thomas and J. Monod, *J. Biol. Chem.*, 235 (1960) 1580.
[57] B. L. Horecker, J. Thomas and J. Monod, *J. Biol. Chem.*, 235 (1960) 1586.
[58] G. D. Hunter, P. Brookes, A. R. Crathorn and J. A. V. Butler, *Biochem. J.*, 73 (1959) 369.
[58a] D. Kessel and M. Lubin, *Biochim. Biophys. Acta*, 57 (1962) 32.
[59] A. Kleinzeller and A. Kotyk, *Membrane Transport and Metabolism*, Czechoslovak Academy of Sciences, Prague, 1961.
[60] M. Kuczynski-Halmann, Y. Avi-Dor and J. Mager, *J. Gen. Microbiol.*, 18 (1958) 364.
[61] K. G. Lark, *Can. J. Microbiol.*, 5 (1959) 381.
[62] K. G. Lark and C. Lark, *Biochim. Biophys. Acta*, 43 (1960) 520.
[63] F. R. Leach and E. E. Snell, *J. Biol. Chem.*, 235 (1960) 3523.
[64] G. Lester and O. Hechter, *Proc. Natl. Acad. Sci.*, 44 (1958) 1141.
[65] G. Lester, D. Stone and O. Hechter, *Arch. Biochem. Biophys.*, 75 (1958) 196.
[66] E. M. Levine and S. Simmonds, *J. Biol. Chem.*, 235 (1960) 2902.
[67] H. Lineweaver and D. Burk, *J. Am. Chem. Soc.*, 56 (1934) 658.
[68] M. Lubin and D. Kessel, *Biochem. Biophys. Res. Communs.*, 2 (1960) 249.

[69] M. LUBIN, D. H. KESSEL, A. BUDREAU AND J. D. GROSS, *Biochim. Biophys. Acta*, 42 (1960) 535.
[70] W. K. MAAS, *Biochem. Biophys. Res. Communs.*, 1 (1959) 13.
[71] R. E. MACDONALD AND P. GERHARDT, *Can. J. Microbiol.*, 4 (1958) 124.
[72] K. MCQUILLEN, in I. C. GUNSALUS AND R. Y. STANIER, *The Bacteria*, Academic Press, New York, 1960, Vol. 1, p. 249.
[73] J. MAGER, *Biochim. Biophys. Acta*, 36 (1959) 529.
[74] J. MANDELSTAM, *Biochim. Biophys. Acta*, 22 (1956) 313.
[75] F. MARINI, P. ARNOW AND J. O. LAMPEN, *Bacteriol. Proc.*, (1960) 160.
[76] R. E. MARQUIS, Doctoral Thesis, University of Michigan, 1961.
[77] P. MITCHELL, *J. Gen. Microbiol.*, 9 (1953) 273.
[78] P. MITCHELL, in *Active Transport and Secretion, Symp. Soc. Exptl. Biol.*, 8 (1954) 254.
[79] P. MITCHELL, *Nature*, 180 (1957) 134.
[80] P. MITCHELL, in E. M. CROOK, *The Structure and Function of Subcellular Components*, Cambridge Univ. Press, 1959, p. 73.
[81] P. MITCHELL, *Ann. Rev. Microbiol.*, 13 (1959) 407.
[82] P. MITCHELL AND J. MOYLE, *J. Gen. Microbiol.*, 15 (1956) 512.
[83] P. MITCHELL AND J. MOYLE, *Biochem. J.*, 64 (1956) 19P.
[84] P. MITCHELL AND J. MOYLE, in E. T. C. SPOONER AND B. A. D. STOCKER, *Bacterial Anatomy*, Cambridge Univ. Press, 1956, p. 150.
[85] P. MITCHELL AND J. MOYLE, *J. Gen. Microbiol.*, 16 (1957) 184.
[86] P. MITCHELL AND J. MOYLE, *Nature*, 182 (1958) 372.
[87] P. MITCHELL AND J. MOYLE, *Proc. Royal Phys. Soc. Edinburgh*, 27 (1958) 61.
[88] P. MITCHELL AND J. MOYLE, *Biochem. J.*, 72 (1959) 21P.
[89] P. MITCHELL AND J. MOYLE, *J. Gen. Microbiol.*, 20 (1959) 434.
[90] P. MITCHELL AND J. MOYLE, *Proc. Royal Phys. Soc. Edinburgh*, 28 (1960) 19.
[91] J. MONOD, in O. H. GAEBLER, *Enzymes: Units of Biological Structure and Function*, Academic Press, New York, 1956, p. 7.
[92] B. NISMAN, *Biochim. Biophys. Acta*, 32 (1959) 18.
[93] B. NISMAN AND M. L. HIRSCH, *Ann. Inst. Pasteur*, 95 (1958) 615.
[94] M. J. OSBORN, W. L. MCLELLAN, Jr. AND B. L. HORECKER, *J. Biol. Chem.*, 236 (1961) 2585.
[95] J. T. PARK AND R. HANCOCK, *J. Gen. Microbiol.*, 22 (1960) 249.
[96] R. L. POST, C. R. MERRITT, C. R. KINSOLVING AND C. D. ALBRIGHT, *J. Biol. Chem.*, 235 (1960) 1796.
[97] R. L. POST AND A. K. SEN, *Proceedings, Fifth International Congress of Biochemistry, Moscow*, 1961, p. 122.
[98] H. V. RICKENBERG, G. N. COHEN, G. BUTTIN AND J. MONOD, *Ann. Inst. Pasteur*, 91 (1956) 829.
[99] T. R. RIGGS, B. COYNE AND H. N. CHRISTENSEN, *Biochim. Biophys. Acta*, 11 (1953) 303.
[100] R. B. ROBERTS, in R. E. ZIRCKLE, *A Symposium on Molecular Biology*, Chicago Univ. Press, 1959, p. 201.
[101] R. B. ROBERTS, P. H. ABELSON, D. B. COWIE, E. T. BOLTON AND R. J. BRITTEN, *Studies of Biosynthesis in Escherichia coli*, Carnegie Inst. Wash. Publ. No. 607, 1955.
[102] T. ROSENBERG, in *Active Transport and Secretion, Symp. Soc. Exptl. Biol.*, 8 (1954) 27.
[103] A. ROTHSTEIN, *Bacteriol. Revs.*, 23 (1959) 175.
[104] D. A. ROWLANDS, E. F. GALE, J. P. FOLKES AND D. H. MARRIAN, *Biochem. J.*, 65(1957) 519.
[105] S. G. SCHULTZ AND A. K. SOLOMON, *Nature*, 187 (1960) 802.
[106] J. H. SCHWARTZ, W. K. MAAS AND E. J. SIMON, *Biochim. Biophys. Acta*, 32 (1959) 582.
[107] W. R. SISTROM, *Biochim. Biophys. Acta*, 29 (1958) 579.
[108] J. C. SKOU, *Biochim. Biophys. Acta*, 42 (1960) 6.
[109] A. K. SOLOMON, *Oral Presentation, International Biophysics Congress, Stockholm*, 1961.
[110] R. STORCK AND J. T. WACHSMAN, *J. Bacteriol.*, 73 (1957) 784.
[111] E. S. TAYLOR, *J. Gen. Microbiol.*, 3 (1949) 211.
[112] J. T. WACHSMAN, H. FUKUHARA AND B. NISMAN, *Biochim. Biophys. Acta*, 42 (1960) 388.
[113] J. T. WACHSMAN AND R. STORCK, *J. Bacteriol.*, 80 (1960) 600.
[114] C. WEIBULL, *Exptl. Cell. Research*, 9 (1955) 139.
[115] C. WEIBULL, *Ann. Rev. Microbiol.*, 12 (1958) 1.
[116] V. R. WILLIAMS, *J. Bacteriol.*, 79 (1960) 125.
[117] I. ZABIN, A. KEPES AND J. MONOD, *Biochem. Biophys. Res. Communs.*, 1 (1959) 289.
[118] M. ZALOKAR, *Biochim. Biophys. Acta*, 46 (1961) 423.

THE MECHANISM OF AMINO ACID POOL FORMATION IN *E SCHERICHIA COLI*

R. J. BRITTEN AND F. T. McCLURE*

Carnegie Institution of Washington, Department of Terrestrial Magnetism, Washington, D.C. and Johns Hopkins Applied Physics Laboratory, Silver Spring, Md. (U.S.A.)

SUMMARY OF EXPERIMENTAL RESULTS

Bacteria maintain internally synthesized small molecules at high internal concentrations and in addition have the capacity to concentrate many compounds from the environment. Since the majority of these compounds are intermediates in synthesis they are collectively termed the pool of metabolic intermediates, or, simply,

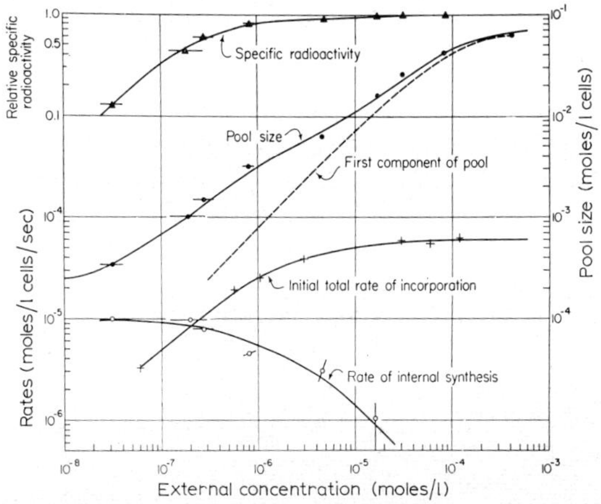

Fig. 1. Proline pool formation in *E. coli* at 25° (log log plot). The experimental points are derived from a set of eight simultaneous measurements of the time course of incorporation of radioactive proline, except the point at 4.2×10^{-4} *M* which is the average of several measurements of the saturation pool size. ▲, ratio of specific radioactivity of the pool (at the time it reached its maximum value) to that of the added radioactive proline. ●, maximum pool size reached *versus* concentration present at that time. – – –, prediction of the carrier model for the size of the major component of the pool alone. +, initial total rate of incorporation of external proline. O, calculated average rate of internal synthesis of proline during the time required for the pool to reach its maximum value. (By courtesy of the editor of *Biophys. J.*)

the "pool". However, the state of organization and ultimate chemical fate of exogenous compounds concentrated by the cell may be different from that of identical compounds synthesized by the cell.

Since the mechanism by which high internal concentrations are maintained is not understood and the processes are obviously complex, it appears fruitless to enter

* Research Associate of the Carnegie Institution of Washington.

References p. 609

into an extended discussion of the meaning of the term "pool". Therefore, we will simply define the "pool" as the total quantity of low molecular weight compounds which may be extracted from the cell under conditions such that the macromolecules are not degraded into low molecular weight subunits, for example, brief exposure to 5% trichloroacetic acid at room temperature.

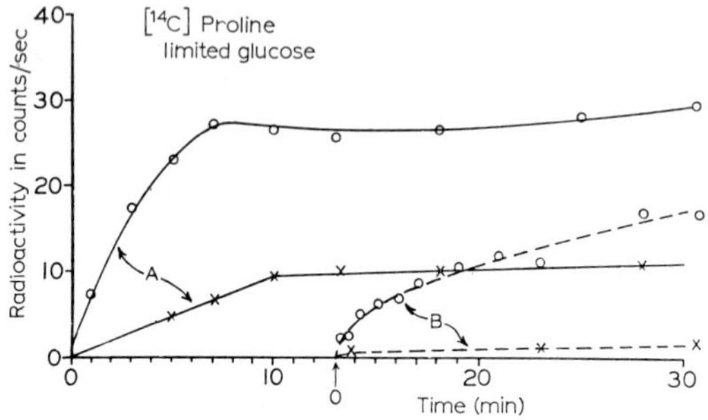

Fig. 2. Maintenance and exchange of pool proline in the absence of glucose. In both experiments, growing cells were suspended at time zero in medium containing 10 μg/ml glucose and 0.87 μg/ml [^{12}C]proline. For curve A a small quantity of [^{14}C]proline was added at time zero. For curve B an equal quantity of [^{14}C]proline was added at 13 min. In each case the upper curve (○) represents the total [^{14}C]proline taken up, and the lower curve (×) the ^{14}C incorporated into protein. The difference is the [^{14}C]proline in the pool.

Experiments have, in general, been designed to answer the following three questions: What are the mechanisms by which exogenous compounds are concentrated? What states of organization exist for compounds in the pool? What is the relationship of the pool to the mechanism of macromolecular synthesis?

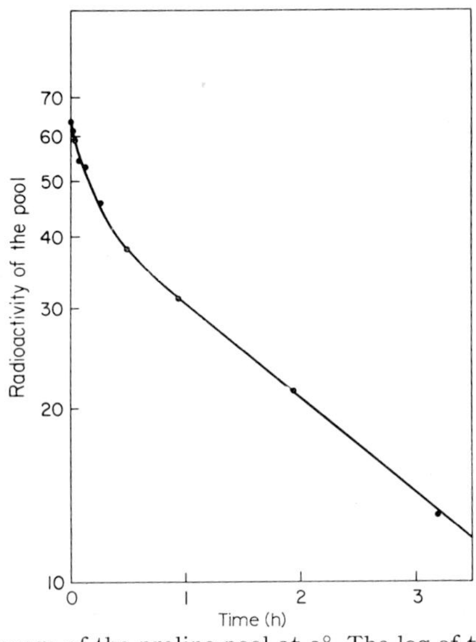

Fig. 3. Time course of exchange of the proline pool at 0°. The log of the radioactivity of the pool is shown as a function of time after [^{12}C]proline was added (10^{-4} M) to a suspension containing a [^{14}C]proline pool of 2.9×10^{-6} moles/g wet wt., in equilibrium with external [^{14}C]proline (1.9×10^{-6} M) at 0°.

At the present time unequivocal answers do not exist for any of these three questions. However, the large body of experimental evidence does provide a restrictive set of conditions which theoretical models must satisfy, and supplies a background for the formulation of more refined questions.

The following discussion of the mechanism of pool formation has been abstracted from a monograph[12] and corresponds to the central part of the paper presented at the conference. Unfortunately, due to limitations of space, it is impossible to present here the experimental observations supporting the 19 items listed in Table I. As a compromise five figures are included, having self-explanatory captions.

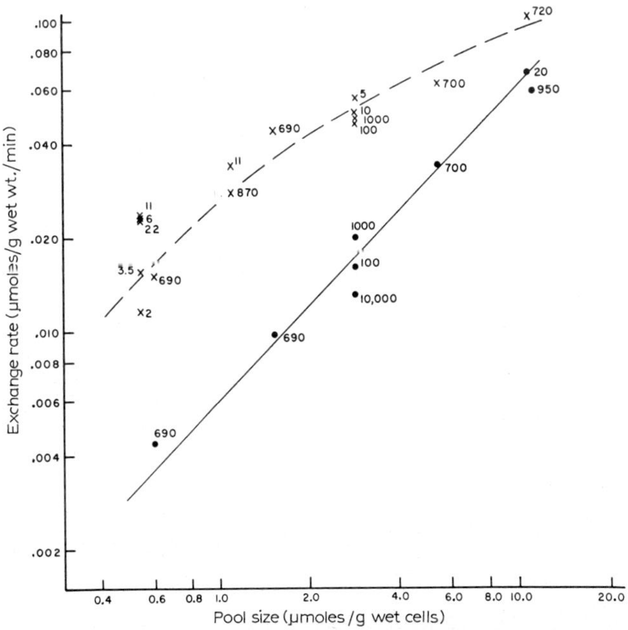

Fig. 4. Rate of exchange as a function of pool size (log log plot). The points were obtained from experiments such as that shown in Fig. 3 by fitting the time course of exchange to curves derived from the sum of two exponential decays. The numbers beside each point are the external concentrations during exchange in µmoles/l. The straight line shown would result if the exchange rate were proportional to pool size. (By courtesy of the editor of *Biophys. J.*)

The experiments summarized in the figures were performed with *Escherichia coli* strain B (ATCC 11303) (except that of Fig. 5) using the membrane-filter technique previously described[1].

It must be pointed out that all of the curves in Fig. 1 except the uppermost one are theoretical predictions from the carrier model. These calculations are described in the mathematical appendix of ref. 12.

A partial description of the experimental results has also appeared[1, 2, 3, 4, 9]. The following summary of the principal features of the pool may also be useful:

1) Passage through the pool appears to be an obligate step for incorporation of an exogenous amino acid into protein.

2) Amino acids present in the pool are incorporated into protein at random, regardless of the length of time they have been in the pool.

3) Peptides do not appear to be intermediates in protein synthesis.

4) An energy source (such as glucose) is required for pool formation to occur at normal rates but is not required for maintenance of the pool for relatively long periods.

5) Specific pool formation mechanisms exist for each amino acid or group of structurally similar amino acids.

6) For any given amino acid there appears to be maximum pool size (or saturation value) at large external concentrations.

7) Any damage to the cells' integrity, or even mild treatments (for a bacterial cell), such as osmotic shock, leads to loss of the pool.

8) Exchange between pool and external amino acids occurs at a high rate, not only when there is steady flow through the pool but also in absence of glucose or at 0° when the flow through the pool is strongly suppressed. (Conditions which also suppress pool formation.)

DISCUSSION OF THE MECHANISM OF POOL FORMATION

Introduction

The purpose of this discussion is to examine the implications of the large number of experimental observations[12]. It appears necessary, in order to bring some clarity to a problem of this complexity, to start out by postulating models of the process. The discussion of the experimental observations in relation to the models allows the implications to be brought out more clearly.

The qualitative predictions of the relatively simple models considered here can be deduced easily. However, our knowledge of their quantitative predictions must depend on a fairly crude analysis. What are undoubtedly complex reaction sequences are taken to be single steps subject to the simpler equations of chemical kinetics. The coupling of the concentration process to the cell's energy supply is, for example, included in an almost purely symbolic way.

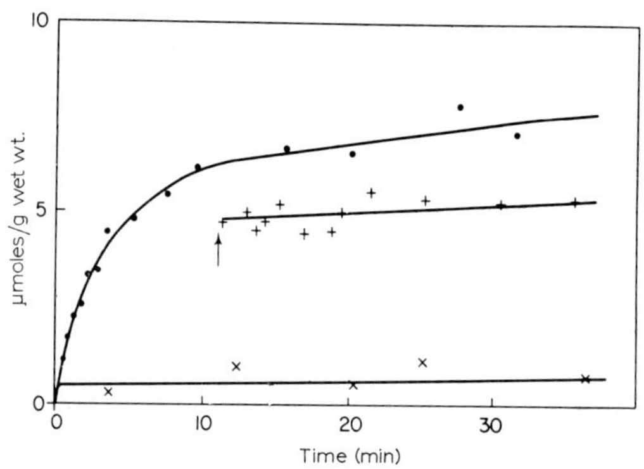

Fig. 5. Stability of the proline pool after reduction of the external concentration. *E. coli* strain 15 T⁻ A⁻ U⁻ suspended at 0.5 mg wet wt./ml at 25° in the absence of thymine, arginine and uracil. Glucose present at 2 mg/ml. At time zero [^{14}C]proline was added (●) to a concentration of 10^{-5} *M* (20 μmoles/g wet wt.). At 11 min (shown by arrow), part of the culture (+) was diluted by a factor of 15 and correspondingly large samples were taken. The concentration of [^{14}C]proline after dilution was 5×10^{-7} *M*. The incorporation into protein is shown by the lower set of points (×). The initial rate of uptake was 1.5 μmoles/g wet wt./min. The rate of loss from the pool is not measurable, but certainly less than 1/100 of the rate of uptake.

The state of the subject at this writing is that the two simple models that have been widely discussed, termed by MONOD *permease* and *stoichiometric site*, both fail in a qualitative way to explain the presently known facts. Following a discussion of the reasons for these failures, a new model is proposed which we have called the *carrier model*. The deductions from this model are consistent with practically all of the observations.

The permease model

The permease model is shown schematically in Fig. 6. COHEN AND MONOD state[8]: "Obviously the actual mechanism of specific permeation must be more complex than the deliberately bare and abstract model we have set up". Further, they use the term "permease" nearly synonymously with "specific permeation mechanism". However, for reasons of clarity, in this discussion, the term "permease model" will be restricted to a model with the essential features listed in Fig. 6. The list has been abstracted from ref. 8 and is written for the case of an energy-dependent amino acid-concentrating system.

The permease model accounts for a great many of the observations regarding the concentration of substances by bacteria. There is no need here to repeat the extensive review of the evidence given in ref. 8. It is sufficient, for the moment, to say that the evidence clearly shows that there are stereospecific sites which act as catalysts for the concentration of compounds by the cell. In other words, there are sites which participate in the early steps of the process but do not, by themselves, hold the compounds in the cell. Also some evidence indicates that the catalytic sites are protein in nature. Such sites will certainly have to be retained as a part of any adequate model for the pool-forming process.

The permease model predicts a number of features of the concentration process which are contradicted by the evidence. In order to visualize these contradictions, we will first consider the way in which a pool is formed according to the permease model, and then discuss the experimental observations listed in Table I.

An external amino acid first forms a complex with the permease (y) in the osmotic barrier. In the presence of an energy supply, the complex dissociates on the inside of the barrier. The free amino acid within the cell may then pass out of the cell by means of a non-specific leak (z). Thus there will be a continuous circulation of amino acid through the pool at equilibrium when the two rates balance. For a given external concentration, the pool size will be determined by the rate at which the permease can pump amino acid into the cell. At the external concentration at which this rate saturates, the pool size will saturate.

The observations (Table I, items 1 and 2) that glucose is not required to maintain a large internal concentration but is required to form a pool at normal rates are clearly not consistent with the above description of the pool-forming process. Similarly, the permease model fails to explain the observation (items 3 and 4) that, while pools are formed very slowly at 0°, pools formed before chilling are maintained indefinitely (if not too large a size). From the point of view of the permease model these results imply that when glucose is exhausted or the temperature reduced to 0°, not only is the active permeation process suppressed but also the "passive" leak is suppressed to an *identical degree*. Further, the pool is maintained at 25° in the

References p. 609

absence of the amino acid, whether or not glucose is present. Thus, item d in the list of properties of the permease model is contradicted.

These observations clearly establish that a passive leak, as measured by pool maintenance experiments, is very much too small to control the steady state pool size by balancing against a simple active input process. They imply either that the leak must be made a rather strange function of the environmental conditions or that a more sophisticated active process must be considered. As will be illustrated in the discussion of the carrier model, a self-suppressing input mechanism seems to provide a more direct and satisfactory explanation of the empirical data.

The evidence on the exchange between external and pool amino acids summarized in items 12, 13 and 14 of Table I raises further difficulties for the model. Exchange under normal conditions, for example at 25°, is expected, due to the circulation through the pump and leak. The measured rate of exchange is comparable to the initial formation rate, as the model predicts. However, the high rates of exchange in the absence of glucose or at 0° are not predicted. It might be suggested that the complex $\overline{Ay}$ can exchange with both A and P, at a rate higher than the actual association or dissociation in the absence of energy. The fact that the rate of exchange (at 0°) saturates at a low concentration would suggest that the permease itself is

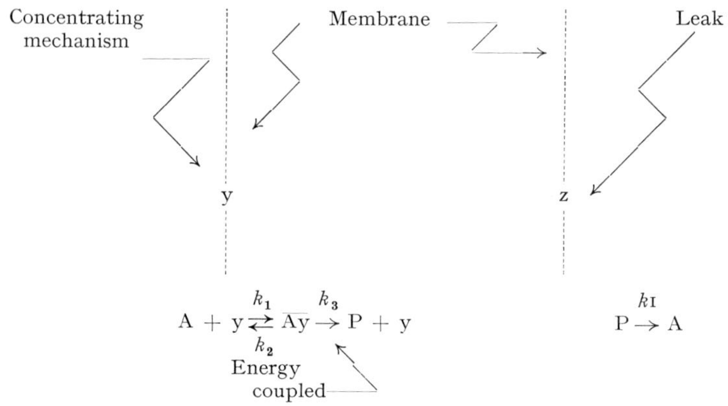

Fig. 6. Permease model. Properties of the permease model: (a) The bacterial cell is enclosed with an osmotic barrier which is highly impermeable to amino acids. (b) The impermeability is not absolute and leakage may occur, tending to slowly equilibrate the inside and outside concentrations. (c) Within the barrier exist proteins (the permease) capable of forming specific complexes with the amino acid. (d) The complex associates and dissociates reversibly with amino acid either inside or outside of the barrier, and catalytically activates the equilibration of internal and external concentrations. (e) When coupled to an energy donor, the internal association reaction is, in effect, inhibited, and amino acid accumulates inside the cell. (f) As the interna- concentration rises, the non-specific leakage increases until its rate balances the rate of accumulation at equilibrium. (g) The amino acid is presumed to be in a free state within the cell.

saturated. In addition, some property of the permease must limit the rate of exchange. This is perhaps reasonable since the complex cannot be exposed on the inside and outside of the membrane at the same moment. The permease model has been given a minus score in Table I for items 12, 13, 14 and 15 since an exchange mechanism must be added. In fact this mechanism would have to be specified in detail in order that a comparison with the observations be made adequatelv.

References p. 609

Item 12 shows clearly that the pool has more than one component. This argument has been given in detail in the experimental section on exchange. The evidence summarized in items 6, 8 and 18 also strongly supports this conclusion. From the point of view of the permease model, one might say that the cell has several pool-containing compartments separated by osmotic barriers, each containing an appropriate permease, but this is not a pleasant prospect. It seems much more likely that at least part of the pool is not in free solution within the cell, as will be discussed below.

TABLE I

RELATIONSHIP OF PROPERTIES OF THE POOL TO THE MODELS

Observation	*Score*	
	Permease model	*Carrier model*
Formation and Maintenance		
1. Glucose required for formation	+*	+
2. Glucose not required for maintenance	—	+
3. Pools formed slowly at 0°	—	+
4. Pools maintained at 0°	—	+
5. Pool maintained at 25° in absence of either the amino acid or glucose	—	+
6. Pool size *versus* concentration not Michaelis	—	+
7. Initial rate of formation not proportional to pool size	—	+
8. Small pools not generally influenced by other AA but large pools are suppressed	—	+
9. Evidence for catalytic site in general	+	+
10. Pools may be very large	+	—
Exchange		
11. Exchange occurs in addition to steady flow through the pool	+	+
12. Rapid exchange occurs in absence of glucose or at 0°	—	+
13. Fast and slow components in exchange at 0°	—	+
14. The 0° exchange rate saturates at low external concentration	—	+
15. The 0° exchange rate increases with pool size	—	+
Osmotic Behavior		
16. Pools removed by sudden reduction in osmotic strength	+	(+)**
17. Pools immediately re-formed after removal by shock	+	+
18. Different pools removed to different extent by shock	—	(+)**
19. Maximum pool size increases with osmotic strength of medium	—	—

* +, indicates that the model satisfactorily explains the observation. —, indicates that there is a contradiction or that a modification may be required by the experimental evidence.
** Assuming that the sites may be osmotically sensitive.

The observation (item 6) that the pool size, far below saturation, does not rise in proportion to the concentration could be taken to indicate that there is more than one component in the pool or that there are permeases with a variety of affinities for a given compound. For a number of reasons this item cannot be considered a crucial argument against the permease concept. However, since the permease model we are discussing does not predict such a result, it is given a minus score for item 6.

The initial rate of pool formation (item 7) at the time the amino acid is supplied shows a much smaller variation with concentration than does the pool size. In fact it saturates at a relatively low concentration. A similar conclusion can be drawn from the observation that a low concentration of isoleucine will block valine incorporation. Both these observations show that the catalytic site for pool formation saturates at a concentration far below that at which the pool itself saturates. For this reason, these observations supply good evidence for the existence of a catalytic site. A characteristic of the permease model is that the circulating flow is proportional to the pool size. This circulating flow is, in turn, identical to the initial rate of pool formation. The failure of this proportionality further supports the conclusion that the pool is maintained by mechanisms other than the balance between a rapid active process and corresponding rapid leak.

It is clear from the above discussion that the simple permease model is inadequate because it fails to agree with experimental data in a number of ways. In addition there is a philosophical objection to the model as written. This may be seen by noting that P is the same material as A except on the other side of the barrier. Why then don't P and y interact with the same rate constants (k_1 and k_2) as A and y? Of course, the energy coupled reaction (k_4) may occur only inside the cell because the energy carrier may be so localized. Now it might be suggested that if the energy coupled reaction (k_4) is fast enough it will be so dominant that the ordinary reactions (k_1 and k_2) could be neglected inside the cell. However, note that the input rate is not greater than k_1yA. Therefore $k_3\overline{\text{Ay}}$ is also not greater than k_1yA. But k_1yP is much greater than k_1yA simply because P is ordinarily much greater than A. Thus it is seen that this philosophical objection is hardly pedantic and one in fact has assumed the existence of some rather tricky means of distinguishing the performance inside and outside the cell. This mechanism, in fact, is really the key to producing the desired behavior and the failure to display it explicitly simply avoids the whole question. In this sense the model really isn't a scientific model at all. If we are to say that the materials inside the cell are somehow physically or chemically different from those outside, doesn't the real elucidation of the problem lie in explaining the nature of this difference?

From the experimental side it is again worth noting that to explain the relationship between pool sizes, loss rates and initial-formation rates it would appear necessary to postulate a mechanism whose details provide that the size of the pool is not determined solely by the loss rate increasing until it equals the input but rather by a heavy contribution from the input rate being lowered as the pool increases in size. In other words, the pool should inhibit its own formation.

One can postulate detailed mechanisms which meet both the philosophical objection to the simple model and the inhibiting requirement mentioned above. With such improvements the permease model becomes more satisfactory, but in order to make it fit the full variety of the experimental facts, even greater complexity and sophistication is apparently required. It seems out of place in this review to illustrate the possibilities of a number of models of increasing complexity. Suffice it to say that it may be possible to add enough special features to make a reasonably satisfactory model in which the elements of the simple permease model may perhaps still be recognized. In particular, the crucial part played by the osmotic barrier would presumably still be dominant in such an extended model.

References p. 609

The stoichiometric site model

The central feature of any "site" model is that the amino acid is held by association with the macromolecules of the cell. For the moment the nature of this association is unspecified and it is presumed that the osmotic barrier which may exist near the surface of the cell is of minor importance in the maintenance of the pool. The simplest possible model of this type is shown schematically in Fig. 7.

It is immediately obvious that this model is unsatisfactory since it does not explicitly contain a step equivalent to the catalytic site. Further, the rate of pool formation should be proportional to the amino acid concentration and the number of unoccupied sites.

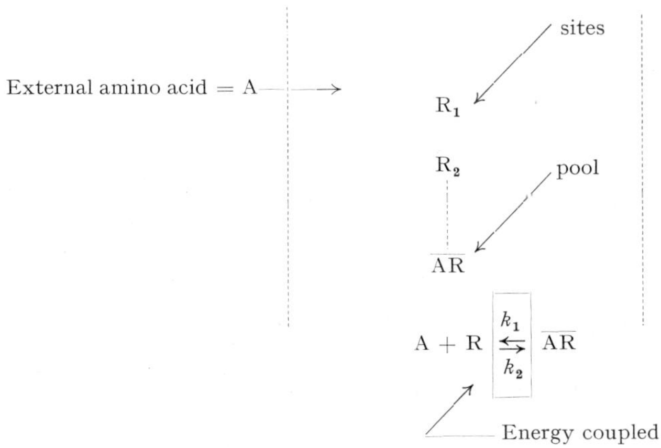

Fig. 7. Stoichiometric site model.

In order that the pool size increases with the external concentration (at low concentrations), there must be a process by which the site-amino acid complexes dissociate. The fraction occupied would then be determined by a balance between the rate of formation and dissociation. The rate of dissociation, however, would have to be energy-dependent, in order that the pool be maintained as implied by items 2, 3, 4 and 5. Since all of these difficulties can be resolved together, we will immediately pass to the discussion of the "carrier" model.

The carrier model

The carrier model is consistent with practically all of the known facts concerning amino acid pools. We have been led to postulate its central features by the failure of the previously discussed models.

The fact that pools are maintained under adverse conditions where they might be expected to leak out, combined with other evidence, has led us to include sites as the major mechanism for maintaining the pool. The strong evidence that a catalytic site participates in pool formation has led us to include such an intermediate step in the formation of the site-amino acid complex (pool). We have, therefore, postulated that the catalytic site is part of a molecule of moderate molecular weight

References p. 609

termed the "carrier". The carrier molecule is assumed to be large enough to form a stereospecific complex with the amino acid, but still small enough to diffuse within the cell with some freedom. The mobility of the carrier is necessary since there are few carriers to transfer amino acids to the many pool-holding sites.

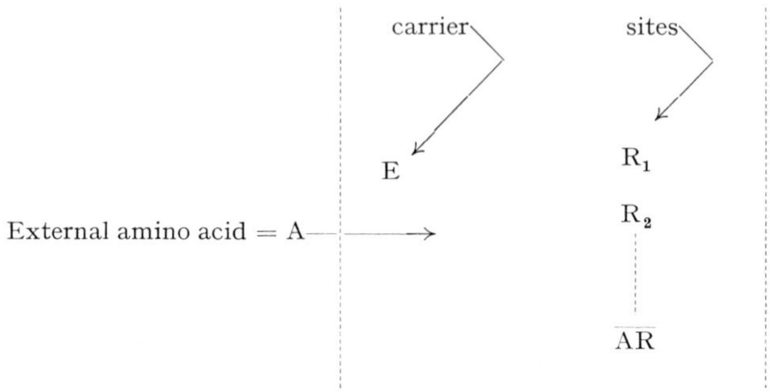

$$A + E \underset{k_2}{\overset{k_1}{\rightleftarrows}} \overline{AE}$$

Energy coupled

$$\overline{AE} + R \underset{k_4}{\overset{k_3}{\rightleftarrows}} \overline{AR} + E$$

Fig. 8. Carrier model. Properties of the carrier model: (a) The cell contains a small quantity of mobile stereospecific carriers which freely form complexes ($\overline{AE}$) with amino acids, without participation of energy donors. (b) The cell also contains a relatively large quantity of non-mobile groups (the sites) which form complexes ($\overline{AR}$) with the amino acids. (c) The site complex $\overline{AR}$ can only be formed by a reaction with the carrier-complex $\overline{AE}$ and this reaction is coupled to an energy donor. (d) Exchange may occur between the site-associated amino acids and those associated with carriers, without coupling to energy donors. (e) Exchange also occurs between free amino acids and carrier associated amino acids, but *not* between free amino acids and site-associated amino acids. (f) There are several classes of sites, some stereospecific and some non-specific. (g) There may be an osmotic barrier near the surface of the cell and "free" amino acid may not diffuse through the protoplasm at the same rate as in water, but the formation of the carrier complex nevertheless occurs at a sufficient rate, without the participation of an energy donor.

A schematic diagram of the carrier model is shown in Fig. 8, along with equations indicating the reactions that are proposed. The listed properties of the model have been specified quite sharply so that the deductions may be analyzed quantitatively. It cannot be ruled out that certain "forbidden" processes, such as exchange between free amino acid and site-associated amino acids, proceed at slow rates. Further, the evidence is insufficient to specify which of the two processes is actually coupled to an energy donor.

According to this model, the pool is formed in the following way. An external amino acid diffuses into the cell and collides with an unoccupied carrier. A complex with the stereospecific carrier is formed ($\overline{AE}$); the complex diffuses through the cell and collides with an unoccupied site. In a reaction coupled to an energy donor, the amino acid is transferred from the carrier to the site. In turn an unoccupied carrier may collide with an occupied site and remove the amino acid. The evidence at the moment does not specify whether or not the reverse reaction is also energy-requiring.

References p. 609

As the pool rises, the reverse reaction ($\overline{AR} + E \rightarrow \overline{AE} + R$) reduces the quantity of free carrier (E). Thus, the rate of formation of carrier complex with free amino acid falls until it equals the rate required for protein synthesis. This is then the steady state.

Referring again to Table I, the deductions from this model will now be compared with the observations. Item 1 is obvious, since the energy requirement has been built in. Two distinct properties of the model lead to maintenance of the pool when formation is suppressed (items 2, 3, 4 and 5). We could assume the reverse reaction ($\overline{AR} + E \rightarrow \overline{AE} + R$) to be energy-dependent, and then clearly the pool would be mai tained in the absence of glucose. Alternatively, we are at liberty to choose a small value for the constant k_2 without influencing any other properties of the model, and therefore the loss rate under all conditions (short of damaging the cell) can be set as low as necessary. The choice of a small k_2 simply means that the carrier (or catalytic site) has a high affinity for the amino acid, and thus the amount of carrier complex will saturate at low concentrations. This is consistent with the saturation of the exchange rate (item 14) and of the formation rate (item 7) at low external concentrations.

The evidence on exchange between pool and external amino acids led to the postulation of the exchange processes described in items d and e in Fig. 8. These processes are sufficient to explain all the observations on exchange (items 11, 12, 13, 14 and 15 of Table I). Saturation of the exchange rate at low external concentrations is predicted if the natural assumption is made that the exchange rate between free amino acid and carrier is more rapid than the exchange rate between the amino acids of the carrier complex and the site complex. Thus, when exchange is studied with labeled amino acid, the specific radioactivity of the amino acid associated with the carrier complex would always be close to that of the external amino acid. The amount of the carrier complex is saturated under the conditions of the experiment at 0°. The collisions between the carrier complexes (constant specific activity and quantity) and the site complexes would control the exchange. The rate of exchange would therefore be independent of the external concentration and proportional to the pool size.

In the mathematical appendix of ref. 12, the carrier model is examined in some detail. Allowance has been made for two different kinds of sites. In this model it is important to account for the utilization of amino acid for protein synthesis, the native pool and for the competition between internally synthesized and exogenous amino acid if quantitative evaluation of the model is to be made. The constants of the model (for proline in *E. coli*) which give a relatively good quantitative correlation with the data are presented in the appendix[12]. These constants were obtained from experiments on pool sizes and rates of formation. It is very pleasing that so much information on the details of the concentration relationships and the competition between synthetic processes and concentrating processes can be encompassed in one conceptually simple model. One interesting consequence of the treatment is that the two sets of sites correspond to two components of the pool of quite different binding, one of which has a saturation value 20 times the other. The larger component has a half-saturation value of external concentration which is 100 times that of the smaller and, as a consequence is only dominant at the higher end of the concentration range. It is very satisfying that the existence of two components of the pool of such

References p. 609

different characteristics provides qualitative agreement with conclusions deriving from experiments on exchange and osmotic shock.

The removal of the pool by osmotic shock (items 16, 17 and 18) is not an obvious prediction of the carrier model. The simplest explanation, from the point of view of this model, is that the sites themselves are temporarily affected during a transient period of distension of the cell. The implication is that the macromolecules on which the sites are located are temporarily distorted in such a way that the affinity of the site for amino acid is drastically reduced. This could result from a direct change in the hydration of the macromolecule itself or could result from a mechanical coupling of the macromolecule to major cell structures. It must be pointed out that a semi-permeable membrane is not necessary in order that osmotic phenomena occur. An ion-exchange column (Dowex-50, 2% cross-linked) will undergo striking volume changes when sucrose solutions of different concentrations are passed over it.

From this point of view, some interesting speculation about item(s) 19 (and 10) may be indulged in. The size of a very large pool, which is non-specific, is roughly proportional to the osmotic strength of the medium. Thus, the maximum pool is somehow related to the osmotic balance of the cell although even the largest amino acid pools account for only a small fraction of the total osmotically-active material that may be released from the cell. It may be suggested that there exist non-specific associations between the amino acids and the dense protoplasm (25% dry material) of the bacterial cell. The maximum quantity of amino acid in such an association might decrease with increasing hydration of the protoplasm. No such associations are observed in relatively dilute protein solutions or in disrupted suspensions of cells. However, a carrier or energy donor present in the living cell might be a necessary condition in the formation of such an association.

The fact that different pool compounds are removed in varying degrees by a given osmotic shock (item 18) probably reflects the differences in sensitivity of site complexes.

Finally, it is difficult to leave the discussion of this model without some speculation on the nature of the carrier. The properties required of the carrier, for its function in this model, are that it be a large enough molecule to form a stereospecific association with an amino acid and, on the other hand, that it be small enough to diffuse with some freedom within the cell. Further, it must have a high affinity for free amino acid and be able to give up amino acid freely to form a site complex. In turn, unoccupied carriers must be able to accept amino acids from the site complexes.

At least two possible candidates for the carrier are known at present. The lipid–amino acid complex observed by HENDLER[10] in the hen oviduct has been observed in *E. coli*. The quantity and rapidity of labeling of such complexes in tracer experiments are consistent with the possibility of their function as carriers in the sense of this model. However, the molecular weight is unknown. The S-RNA–amino acid complex discovered by HOAGLAND also occurs in *E. coli* in very small quantities. However, there is no indication[11] that the rate of turnover of the amino acid in this complex is fast enough to carry out the function of the carrier.

A crude lower limit on the amount of carrier present in the cell can be set from its turnover number and rate of diffusion. At the maximum rate of proline pool formation, there are 40 000 molecules entering the pool per second per cell. The time required for a small molecule such as proline to diffuse 1 μ is approx. 1 msec. The

References p. 609

mean distance the carrier must diffuse between taking up an amino acid and delivering it to a site is much smaller than this. However, the molecular weight is probably much larger and the diffusion constant is probably much smaller in the protoplasm than in water. Thus the turnover number probably would be considerably less than 1000 per carrier per second, and therefore the number of carriers greater than 40 per cell. If the number of carriers were as small as this, it would certainly be difficult to observe the carrier complex directly.

The properties of "cryptic" mutants have had an important place in discussion of bacterial concentrating mechanisms[8]. They have not, so far, been mentioned since we have limited ourselves to amino acid concentrating systems. For our purposes, the observations may be summarized as follows: There are strains of *E. coli* (y^-, z^+, i^-) which contain large quantities of the enzyme β-galactosidase, but will not utilize lactose or concentrate galactosides. The enzyme is fully active in preparations of these cells treated with toluene. Undamaged cells will only split the test-substrate (ONPG) at low rates when it is supplied at high concentrations. Other strains which utilize lactose contain both the enzyme and an operative concentrating system. The enzyme in these cells will split ONPG at high rates whether or not treated with toluene and when the concentration of galactosides has been blocked by metabolic inhibitors.

The fact that the mechanism for the concentration of galactosides and the enzyme β-galactosidase are controlled by distinct genetic loci is consistent with the carrier model. Clearly the carrier, the sites and the enzyme would be three distinct elements in the cell. However, in the carrier model no osmotic barrier limits the rate of entry of substrate. What then limits the rate of splitting of the substrate (ONPG) by the enzyme present in the cryptic mutant (y^-, z^+, i^-)? An additional hypothesis is necessary: In the organized cell, the free substrate does not have full access to the enzyme, but when associated with the carrier, it can reach the active site and be attacked at maximal rate. In defense of this hypothesis, it may be pointed out that there are a large number of examples of enzymatic reactions which are suppressed or absent in whole cells but occur at high rates in disrupted cell preparations. In some cases, the suppression has been attributed to an impermeable barrier but there are a number of examples for which the substrate is known to be present in the cell and such an explanation is clearly invalid. The fact that (in y^+ strains) metabolic inhibitors block the concentration process but do not reduce the rate of splitting of ONPG further implies that the carrier complex is formed rapidly without the participation of energy donors.

Conclusion

In this paper we have outlined the rather diverse set of experimental observations of the amino acid concentrating processes in *E. coli*. Any satisfactory model of these processes must be formulated with all of this information in mind.

We have reviewed the simple "permease" and simple "stoichiometric site" models and pointed out their failures. Then we have developed the "carrier" model in some detail and demonstrated that it can be made to correlate almost all of the data in a highly satisfying manner.

While it is tempting to do so, it should not be asserted that the function of sites

References p. 609

and carriers has been rigorously proved or that the function of an osmotic barrier has been demonstrated to be unimportant. All that can be said is that the equations derived from the carrier model accurately describe the experimental data. Possibly these equations are not unique to that model. Any model which gave essentially the same equations would also be satisfactory (and one which gave essentially different equations clearly unsatisfactory).

The permease model might be elaborated to remove some of its difficulties. Part of such an elaboration would have to consist of specifying the mechanism by which asymetry in reaction constants on the two sides of the barrier is obtained. This mechanism might introduce additional features which would modify greatly the properties of the permease model. We must ask whether any mechanism can be proposed which has the necessary properties, if the pool is assumed to be free in solution. The minimal requirements are as follows: rapid pool formation; slow loss in absence of glucose and/or amino acid; increase of pool with external concentration until saturation is reached; lack of proportionality between formation rate and pool size. If an exhaustive search does not yield a model with these properties, then there would be no reasonable alternative but to assume that the amino acid pool is not in free solution within the cell. In this connection it should be remembered that experimental data conclusively demonstrate that the pool has at least two components, so that if we are to assume the pool is unbound amino acid we must introduce at least one more osmotic barrier. Possibly it would be necessary to resort to a combination of a site and barrier.

Of course, there are many problems which deserve further investigation. The meaning of the term "free in solution" needs to be examined both experimentally and theoretically. A 25% solution of protein, RNA, etc., organized in subtle ways, is certainly an unusual solvent from a chemical point of view. The activity of the amino acids might be strikingly depressed. The results of such a study would have very broad implications for other processes in living cells.

Finally, a more detailed experimental study of the rates of loss and exchange is needed. The study of these phenomena with pools of several amino acids and other compounds, such as galactosides, would supply quantitative information that might be helpful in deciding among the alternative mechanisms.

With present knowledge, alternative interpretations of the concentration process in bacteria still remain possible, in spite of the large amount of experimental evidence which has given insight into many aspects of the process. The simplicity and the degree to which individual steps may be understood, from a chemical point of view, differ among the various models. It is for the future to decide which of the alternative approaches will be most useful.

ACKNOWLEDGEMENTS

In writing this article we have used freely the experiments and concepts of our colleagues at the Department of Terrestrial Magnetism, especially R. B. ROBERTS, D. B. COWIE and E. T. BOLTON. They have also liberally given guidance and advice. We are much indebted to the wife (B. H. B.) of one of us for assistance with the typing and editing.

References p. 609

REFERENCES

[1] R. J. BRITTEN, R. B. ROBERTS AND E. F. FRENCH, *Proc. Natl. Acad. Sci. U. S.*, 41 (1955) 863.

[2] E. T. BOLTON, R. J. BRITTEN, D. B. COWIE, E. H. CREASER AND R. B. ROBERTS, *Carnegie Inst. Wash. Year Book*, 55 (1956) 110.

[3] E. T. BOLTON, R. J. BRITTEN, D. B. COWIE, J. J. LEAHY, F. T. MCCLURE AND R. B. ROBERTS, *Carnegie Inst. Wash. Year Book*, 56 (1957) 118.

[4] E. T. BOLTON, R. J. BRITTEN, D. B. COWIE AND R. B. ROBERTS, *Carnegie Inst. Wash. Year Book*, 57 (1958) 127.

[5] R. B. ROBERTS, P. H. ABELSON, D. B. COWIE, E. T. BOLTON AND R. J. BRITTEN, *Carnegie Inst. Wash. Publ. No. 607*, 1955.

[6] K. MCQUILLEN, R. B. ROBERTS AND R. J. BRITTEN, *Proc. Natl. Acad. Sci. U.S.*, 45 (1959) 1437.

[7] G. N. COHEN AND H. B. RICKENBERG, *Ann. inst. Pasteur*, 91 (1956) 693.

[8] G. N. COHEN AND J. MONOD, *Bacteriol. Rev.*, 21 (1957) 169.

[9] R. J. BRITTEN AND F. T. MCCLURE, *Proceedings of the 1st National Biophysics Conference*, Yale University Press, 1959, p. 414.

[10] R. W. HENDLER, *Science*, 128 (1958) 143.

[11] A. I. ARONSON, E. T. BOLTON, R. J. BRITTEN, D. B. COWIE, J. D. DUERKSEN, B. J. MCCARTHY, K. MCQUILLEN AND R. B. ROBERTS, *Carnegie Inst. Wash. Year Book*, 59, (1960) 229.

[12] R. J. BRITTEN AND F. T. MCCLURE, *Bacteriol. Rev.*, in the press.

INVITED DISCUSSION

SOME PROPERTIES AND POTENTIAL USES OF BACTERIAL MUTANTS DEFECTIVE IN AMINO ACID TRANSPORT

MARTIN LUBIN

Department of Pharmacology, Harvard Medical School, Cambridge, Mass. (U.S.A.)

The study of active transport is vexing. In spite of considerable work on transport of sugars and amino acids in a variety of cell types, there remains in doubt no less than the entire sequence of biochemical reactions. The availability of suitable bacterial mutants has enlarged our information on β-galactoside transport, and the recent isolation of a series of mutants defective in amino acid transport provides hope for a fresh approach to the problem of amino acid accumulation.

Of the *(E. coli)* mutants we have isolated[1], defective in the transport of either histidine, glycine, or proline, only the last has as yet received extensive study in our laboratory. The transport-defective mutant (Tr^{-}_{pro}) was derived from a parent strain blocked in the biosynthesis of proline, and has the property of requiring a high proline supplement in the medium (250–500 μg/ml) for rapid growth. Using [^{14}C]proline, KESSEL[2] has found that in the Tr^{-}_{pro} strain, the free pool of proline equilibrates with the medium without evidence of a concentrating mechanism. The relative rates of leak of proline from parent strain (Tr^{+}_{pro}) and mutant strain (Tr^{-}_{pro}) have been measured both at 37° and 0°, and found to be similar, at a given temperature, for both strains. This leads to the unequivocal conclusion that the defect in the mutant lies in the system responsible for active uptake.

A surprising observation was made when exchange of proline at 0° was studied. It should be recalled that BOLTON *et al.*[3] reported very rapid exchange of labeled for unlabeled proline at 0° in the absence of net transport. KESSEL has confirmed this for the parent (Tr^{+}_{pro}) strain, but finds this exchange is nearly absent in the Tr^{-}_{pro} mutant, although exchange of histidine or glycine is unimpaired. This result is consistent with the notion that the mutant has lost a catalytic site or a shuttle-system involved both in exchange and active transport, the latter process requiring coupling to an energy source. It will be of interest to determine, by examining other transport-defective mutants, including those of the β-galactoside system, how often loss of transport by mutation is accompanied by loss of exchange.

The second feature of interest which has emerged from our study concerns the physiological localization of the transport process. The result was not surprising, but the experiment had not previously been done. Back-mutants with the ability to synthesize proline were selected on minimal plates. The Tr^{+}_{pro} and Tr^{-}_{pro} properties were identified by the extreme difference in the ability to concentrate [^{14}C]proline. Both back-mutants, however, grew at nearly identical rates in minimal medium. The conclusion we have reached is that the defect in this Tr^{-}_{pro} mutant is not accom-

References p. 611

panied by any defect in the processes which utilize internally synthesized proline for protein biosynthesis.

A systematic attempt has recently been made to find a difference in the amount of [^{14}C]proline attached to the "cytoplasmic membrane" fraction of Tr^{+}_{pro} and Tr^{-}_{pro} strains, but the result has been negative. Although the clarification of the biochemical mechanism involved in transport is of greatest interest to us, some indirect and partial information may soon be obtained by the methods of microbial genetics and physiology. Among these, we look forward to: (1) chromosomal mapping of the transport genes; (2) the determination of the number of enzyme determining units (cistrons) of the chromosome involved in transport; and (3) measurement of the rate of phenotypic expression of the transport character in a zygote formed from a Tr^{-} recipient and Tr^{+} donor.

As a by-product of the isolation of these mutants, control may be exerted over the internal level of the amino acid pool, which nearly equilibrates, at least for proline, with the concentration in the medium. At least one application is of current interest: the determination of the minimum concentration of internal amino acid needed for the net synthesis of RNA which occurs when bacterial cells are treated with chloramphenicol.

In summary, our studies on a bacterial mutant show that the same mutation which causes a defect in proline *transport* is responsible for loss of the capacity for rapid *exchange* of proline at zero degrees. The defect in the mutant strain is not located in any biochemical step involved in the biosynthesis of protein from internally synthesized amino acids.

REFERENCES

1 M. LUBIN, D. KESSEL, A. BUDREAU AND J. D. GROSS, *Biochim. Biophys. Acta*, 42 (1960) 535.

2 D. KESSEL AND M. LUBIN, *Biochim. Biophys. Acta*, 51 (1962) 32.

3 E. T. BOLTON, R. J. BRITTEN, D. B. COWIE AND R. B. ROBERTS, *Carnegie Inst. Wash. Yearbook*, 57 (1957) 127.

DISCUSSION

Chairman: HALVOR CHRISTENSEN

CHRISTENSEN: It should be noted that Dr. HOLDEN's findings and those with Ehrlich cells are in agreement; in neither case is the initial rate accelerated by pyridoxal. At one time we published a curve purporting to show initial stimulation of α-aminoisobutyric acid uptake, but we do not get this effect with other amino acids nor is it very convincing evidence.

I understand from Dr. SNELL that pyridoxal added to protoplasts of *Streptococcus faecalis* causes stimulation of amino acid uptake. Perhaps we can hear from Dr. MORA about that.

MORA: During studies of the role of the cell wall in the transport of amino acids, we have found that pyridoxal stimulates the rate of uptake of glycine only in protoplasts and not in intact cells of *Streptococcus faecalis*. Pyridoxal was used in 1×10^{-3} M concentration and a 30 per cent increase in the rate of uptake was observed.

CHRISTENSEN: The work of Dr. GERHARDT has been mentioned, I wonder if I could ask him to comment?

GERHARDT: A study of bacterial permeability to a metabolically inert amino acid recently has been completed in my laboratory by Dr. R. MARQUIS. I appreciate the opportunity afforded me by our convenor to present a generalized summation of this work. Several of its basic premises and results may be pertinent to some of the papers presented at this conference.

First, we sought to meet the need for dissociating uptake of a compound from its subsequent metabolism, which enables one to determine just what transport alone involves. This concept has been a notable contribution resulting from the studies of Dr. COHEN AND Dr. MONOD, and later others, on the "permease" system in bacteria. Analogs of galactosides provided suitable experimental material, unfortunately not equally met in studies of amino acid permeases. A model of amino acid uptake meeting these requirements, however, had been established by the studies of Dr. CHRISTENSEN and his associates on the accumulation of methyl analogs in ascites carcinoma cells. Following their lead, we investigated α-aminoisobutyric acid (AIB) as a substrate which test bacteria might accumulate but not metabolize.

A second premise of our studies was to provide for direct comparison between bacterial and mammalian cells. Employment of a common substrate and a relatable experimental approach in two different laboratories we hoped would allow reliable interpretations of similarities and differences in the two types of cells. In all too many instances, such comparisons become almost impossible.

Third, our studies sought to interpret the complex of uptake reactions occurring simultaneously in a single bacterium and for a single substrate in terms of known structural features. Bacteria have a cell wall in addition to the usual plasma membrane. A basic aim was to identify the permeability associated with each of these structures rather than the collective surface. In turn, the accumulation of amino acid into an internal pool could be distinguished from that into cell wall, the latter sometimes operationally termed the "expandable pool".

Although *Staphylococcus aureus* was employed in our intial expeiiments, a more suitable test system was provided by use of *Bacillus megaterium* strain KM. Using the so-called space technique to measure the uptake of ^{14}C-labeled AIB, it was found that: (1) AIB is accumulated to significant extents by both these organisms; (2) in contrast to most other analogs of natural amino acids, AIB is not toxic; and (3) once accumulated, the AIB appears to be inert to further metabolism, as shown by chromatographic and other evidence.

The bacillus accumulated AIB by constitutive mechanisms to a considerably greater extent than the coccus and so came to be used exclusively. Under anaerobic conditions, uptake occurred by characteristically passive processes of readily-reversible absorption and less-reversible chemadsorption, the cell wall *in situ* or isolated having a particular affinity.

Under aerobic conditions, AIB was in addition accumulated by a metabolism-dependent mechanism capable of building up an intracellular AIB concentration in excess of one hundred times the extracellular concentration. This active transport was stimulated by increasing aeration, inhibited by cyanide or fluoride, and uniquely driven from endogenous energy reserves, glucose exerting little effect. Active AIB uptake was temperature-dependent (21° optimum, 1.8 Q_{10}) but relatively insensitive to external osmotic pressure or to external pH over the range from 6 to 8. That AIB entered the cell by normal pathways for amino acids was shown by the specific competition of alanine and glycine, the relative affinities being AIB > alanine > glycine.

The extent of leakage of internally accumulated AIB into distilled water was not great. However, if cells were lysed there was an immediate release of most of the AIB, a small fraction remaining bound with the cell debris. Accumulated AIB was also released when aerobic metabolism was inhibited. AIB was taken up rapidly by protoplasts with resultant swelling. This result, together with the lytic release and the exchange (below), suggested the existence of the amino acid free in an internal pool.

Exchange reactions between intracellular and extracellular AIB were discovered. These reactions depended on both internal and external concentrations of AIB, even at low levels of accumulation, and complete exchangeability was demonstrated. It was postulated that mobile carriers mediated the reaction.

The accompanying figure summarizes diagrammatically the postulated mechanisms of AIB entry and exit and their relationship to surface structures of a bacillus.

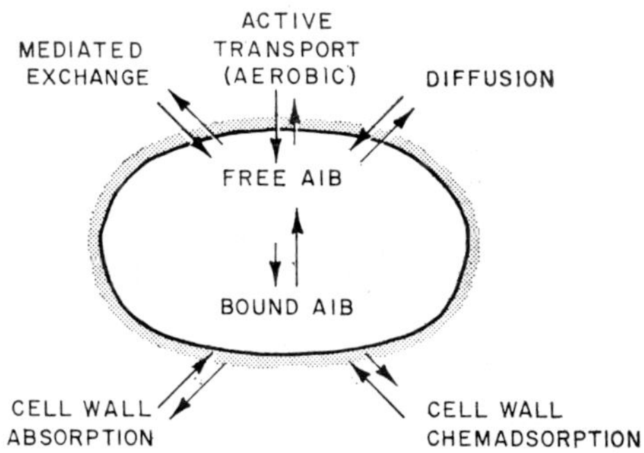

Fig. 1. Postulated mechanisms of the permeability of *Bacillus megaterium* to α-aminoisobutyric acid and the relationship to the cell wall and cytoplasmic membrane.

CHRISTENSEN: Dr. ZABIN's studies also have been mentioned. I wonder if we can ask him to comment on his work in Paris.

ZABIN: Permease has been mentioned here. It has been defined by Dr. COHEN AND Dr. MONOD in the case of galactoside permease as a specific inducible substance which is responsible for the accumulation of galactosides by certain strains of *E. coli*. This name implies, of course, the existence of an enzyme or enzyme system which carries out this process.

The original studies of RICKENBERG, COHEN, BUTTIN AND MONOD which gave the observations that led to this concept, used thiogalactosides as substrates because these materials were not metabolized by the enzyme β-galactosidase. Actually, it turned out that thiogalactosides are not completely inert. Under some conditions one can show with whole cells the formation of a derivative which accumulates to the extent of 5 per cent or less of the total amount of thiogalactoside concentrated by the bacterial cell. This derivative was isolated by Dr. HERZENBERG and was shown to be a 6-*O*-acetyl thiogalactoside. One very interesting finding which was obtained by Dr. HERZENBERG and others was that only those mutant strains which could carry out the accumulation reaction could form the derivatives. At that time it was not possible to decide whether or not this was simply a coincidence.

KEPES, MONOD and I were able to demonstrate in cell-free extracts of various mutant strains the formation of this acetyl thiogalactoside. This was a simple reaction which required an enzyme, acetyl coenzyme A, and an acceptor. With this system we were then able to test a variety of mutants for the presence or absence of the enzymatic activity forming the acetyl galactoside in extracts and to compare this with the presence or absence of the permease reaction in the whole cells from which these extracts were derived.

We found that there was a 1 : 1 correlation. Every time permease was induced the acetylation activity was present. If the strain was not induced, and, therefore, did not have permease, there was no acetylation activity. Since certain of these mutants were single step mutations, this suggests very strongly that the acetylation activity is very closely associated with the permease activity. Yet the acetyl compound cannot be an intermediate in the accumulation process, because it is quite inert. It is not active in a number of important ways; it is not concentrated, it does not act as an inducer, it does not inhibit induction, and it does not inhibit concentration. Further, it is not hydrolyzed either by whole cells or by extracts. Therefore, its formation can be considered only to be in the nature of a side reaction.

This, therefore, suggests that the enzyme responsible for the acetylation reaction may be doing something else which takes a more direct part in the permeation reaction. In order to try to get evidence on this point, we have more recently purified the enzyme and looked at certain of its properties. On a biochemical level, these properties do not lend support to the idea that this enzyme is a part of the permease reaction mechanism.

I would like to mention three observations. The first is that the pattern of specificity for the acetylation reaction, while similar to that for the accumulation or permease reaction, is in some respects vastly different. For instance, phenyl S- and O-glucoside are acetylated, but neither of these substrates can be accumulated in the permease reaction.

Secondly, the apparent Michaelis constant for the permease reaction has been measured for a variety of substrates and has been found to be generally in the neighborhood of 10^{-4} or 10^{-5} moles per liter. In the case of isopropyl thiogalactoside, the Michaelis constant for this substrate in the acetylation reaction is extremely high, 3×10^{-1} moles per liter.

Thirdly, one would expect that this enzyme, if it is involved in the permease reaction, might be associated with or be found in the membrane fraction. When protoplasts of *E. coli* are prepared and then shocked with water, the particles have little or no enzymatic activity. The enzyme is found in the supernatant solution. Therefore, while evidence based on a correlation in a variety of mutants strongly suggests an identity of this enzyme with a component of the permease reaction, biochemical evidence so far does not support this view.

CHRISTENSEN: An important principle in transport studies has been carefully recognized here. If a solute is modified to a new form during transport, is this event an essential part of transport, or is it merely an incidental event? In informal conversation elsewhere I have heard the present work misinterpreted to suggest that an essential transport intermediate has been identified. Dr. ZABIN has dealt with this distinction very carefully.

HENDLER: We have been concerned with the possible participation of the lipid–amino acids in transport and in collaboration with Dr. BRITTEN and Dr. R. ROBERTS at the Carnegie Institute we ere in the process now of determining whether we can observe such a role in *E. coli*. We are able to observe the formation of lipid-soluble forms of amino acid in this organism and to isolate them from the cell wall and ribosomes. These complexes turn over rapidly with respect to the amino acid. One experiment in particular was rather suggestive. When cells loaded with cold proline were exposed to a small amount of highly radioactive extracellular proline, the specific activity of the internal pool rose while, the specific and total activity of the external pool was falling. The fact that the radioactivity of the lipid-amino acids also decreased during this period indicates that they are closer to the external pool in their formation as opposed to being formed from the internal pool.

Dr. DUGGAN also has been working on this problem in my laboratory, and we have recently obtained the mutant mentioned by Dr. HOLDEN. This mutant isolated by Dr. LUBIN at Harvard is not able to concentrate proline. The experiments are quite preliminary with this system, but using high specific activity, ^{14}C-proline, we are studying the metabolism of the lipid-bound forms of the amino acid in an attempt to pick up significant differences in the mutant and wild type.

HALVORSON: I want to confirm one point made by Dr. BRITTEN; that is, when we have looked at the soluble RNA fraction in yeast, we found that it picks up radioactive amino acids at only about a fourth or fifth of the rate that we would expect to explain the rate of their incorporation into proteins.

I would like to address a question to Dr. BRITTEN. I am intrigued by the fact that he has a by-pass for uracil. Together with Dr. COHEN we have been concerned with a phenomenon in yeast where, under some conditions, a similar by-pass appears to exist. I wonder if he would care to comment on by-passes for amino acids.

BRITTEN: The evidence for by-pass on kinetic data essentially alone is very clear. Even though you pretreat cells with unlabeled uracil, at the moment you add labeled uracil there is incorporation into RNA with a delay of less than five seconds. The initial rate corresponds to about 40 per cent of the ultimate rate after the main storage pool has come to equilibrium.

I would like to make two brief philosophical comments. The first has to do with the role of a model such as the one that I described. The type of reactions that are postulated which lead to a certain set of equations, will not ultimately be lost sight of. They, in fact, in some modified form, will enter into the ultimate explanation of proline pool formation. However, the relationship between the equations and the object with which they have been associated is quite open. We have been unable to find suitable models in which sites are not the principal mechanism for holding the pool. But I do not think that we have demonstrated, after considerable effort, that in fact the sites are necessary. There are alternate models which may ultimately prove suitable.

The last point I would like to make is that there has been considerable discussion of free amino acids, and I do not really believe that this term has any meaning in the present state of knowledge. The amino acid present in a bacterial cell, for example, is dissolved in material which is not only 25 percent protein, RNA, and lipid but is also highly organized. Under these circumstances the amino acid is not in solution in the classical sense.

CHRISTENSEN: I should like to comment on that point. It seems to me that by working at levels of the order of 10^{-8}, 10^{-5}, or 10^{-4} molar, one provides the maximum opportunity for binding to obscure the potentiality for uphill transport. These low solute levels must have the best chance to be of the same order as the levels of the coenzymes, and it seems to me not so strange that one would find intracellular amino acids at these levels bound to coenzymes or holoenzymes. Multiple pools are most to be expected under these conditions. This may be an unavoidable difficulty for a given organism. But when we come to cases where amino acid levels of several hundred millimolar are obtained, or where gradients of that magnitude are obtained, then the opportunities are greatly reduced for sufficiently significant binding or changes of activity coefficients to occur.

Also, in the experiment Dr. OXENDER has done in our laboratory in which cells were arranged into the form of a membrane, so that two extracellular phases could be studied, application of conditions stimulatory to transport from one side or the other produced concentration gradients whose reality are very hard to question. Although normal secretion perhaps already does this, the experiment seems to me to show that the production of real concentration gradients has to be reckoned with.

BRITTEN: It seems quite obvious that at low concentrations there are indeed opportunities for association of a variety of small pools with sites. The question that I was directly raising was with regard to very large pools. In this case, it is not easy for me to visualize the type of affinity with the protoplasm which might take place. Part of the difficulty stems from the fact that we do not know enough about solution chemistry in dense protoplasm. However, it was precisely with regard to very high concentrations that I made my comment.

ABRAMS: I would like to make a few remarks with regard to downhill penetration of solutes into bacterial protoplasts of *Streptococcus faecalis* which may bear on this question. These protoplasts have no endogenous energy metabolism. If they are suspended in high concentrations of compounds which do not penetrate the membrane, the protoplasts are stable; they do not swell. On the other hand, if protoplasts are suspended in solutes which penetrate, they will swell and eventually lyse. The rate of swelling in a particular solute is an indication of the rate of penetration.

We have examined the swelling or lack of swelling of *S. faecalis* protoplasts suspended in a great variety of amino acids and, as one might expect, there has been a wide variation in the rate of penetration. L-Alanine penetrates, as indicated by the rate of swelling, at a very rapid rate. On the other hand, D-alanine penetrates only at a very slow rate. L-Serine penetrates at a rather rapid rate, somewhat less than L-alanine, but its optical isomer penetrates only very slowly. L-Alanine penetrates far faster than β-alanine; α-aminobutyric acid penetrates much more rapidly than γ-aminobutyric acid. We have studied the behavior of approximately twenty amino acids in this way. All I want to point out is that with regard to spontaneous downhill penetration, the membrane appears to be able to recognize the difference between closely related amino acids. Furthermore, with L-alanine and L-serine, we see a lag period in the rate of penetration, which would indicate that some preliminary reaction must precede penetration.

One further remark with regard to the composition of isolated cell membranes of *S. faecalis* may be of interest. This membrane contains a large amount of lipids, as has been found with other membranes from bacteria. In addition, it contains a small amount of RNA, and since the proposition has been made that RNA might be involved in some manner in the transport process, I suggest that the presence of RNA in the membrane would be eminently suited for such a role.

GUROFF: Could you comment, Dr. CHRISTENSEN, on the maintenance of pools in ascites cells with respect to absence or presence of energy sources.

CHRISTENSEN: The Ehrlich cell has very great endogenous reserves of energy, apparently in the form of lipid, so that this cell is able to continue to produce energy for a great many hours. The cell also has a considerable glycolytic capacity, so one cannot, by inhibiting either respiration or glycolysis alone, prevent amino acid accumulation. Apparently, one has to strike at both of these things at once. Under anaerobic conditions adding glucose or fructose is necessary if the accumulation is to continue. Eventually the pH falls enough through the formation of lactic acid to produce a handicap. Accumulation continues experimentally as long as we have cared to follow amino acid uptake. I think, Dr. HEINZ has attempted to exhaust the energy supplies of the cells by various means. Would you care to comment on this same question?

HEINZ: Unfortunately, we did no experiments to exhaust the energy supply. We did, however, try to wash all the glycine out of the cell but without success. A certain amount, somewhat less than 10 millimoles per liter, always stays inside the cell. Also, the kinetic data indicate that 5 to 10 per cent of the cellular glycine may not be directly exchangeable. This could represent the "internal pool", but I do not think one can say more about it.

GUROFF: What about cooling to 0°? Has that been done with ascites cells?

CHRISTENSEN: Yes. It is possible to wash the cells very quickly at ice temperatures without loss. Letting them stand in such a solution leads to the loss of the gradients.

E. ROBERTS: The total amino acid pool of ascites cells incubated in ascitic fluid *in vitro* for rather prolonged periods of time without any added glucose, is maintained very tenaciously. However, when we tried to incubate these cells in a variety of buffers, even for brief periods, we found considerable leakage into the fluid of the easily extractable ninhydrin-reactive constituents. We have not yet succeeded in finding *in vitro* conditions other than incubation in the protein-rich ascitic fluid, under which we could maintain the pool as we see them on chromatograms. This finding has puzzled me in terms of what it might mean in relation to the various kinetic experiments that have been done in buffers rather than in the ascitic fluid itself.

II. AMINO ACID POOL TURNOVER

LIMITING FACTORS IN BIOSYNTHESIS OF MACROMOLECULES

JOHN M. REINER

Department of Microbiology, Emory University, Atlanta, Georgia (U.S.A.)

It is a source of some satisfaction to me to be invited here, not in my capacity as an experimenter, but in my alternative role of mathematical theoretician. The satisfaction stems not from vanity satiated—for my preceptors in applied mathematics, CARL ECKART AND NICOLAS RASHEVSKY, set a standard of skill and ingenuity that would deflate the most swollen head—but from the evidence that quantitative theoretical analysis is today an accepted partner in my chosen field, instead of the half-ignored, half-resented stranger that it was when I began.

This happy situation demands a commensurate candor from the theorist; and so I must warn you not to expect to hear earth-shaking revelations today. Every problem has a history, during which the apt strategies shift as understanding develops. There are, and always will be, areas in which it is easier to do the experiments, however complex and demanding these may be, than to try to predict the answers theoretically. This holds, in particular and as of today, for the problem of the mechanisms of macromolecular biosynthesis: it would be presumptuous now (though in five years it may turn out to be reasonable) to try to derive the mechanisms of protein synthesis from physicomathematical first principles. For this problem, at this time, the callouses must be put on the feet rather than the seat; and we have a model, in the work of ARTHUR KORNBERG and his collaborators, of a beautifully simple solution to what seemed a few years ago to be the terrifying complex problem of the synthesis of DNA*, achieved by the classical biochemical stratagems of clean and assiduous labor.

A more modest role must be assigned, for the present, to the mathematical theorist. What he can and should do is to formulate in quantitative terms the mechanisms we know, or think we know, or hope to substantiate the day after tomorrow, and to inquire concerning the more interesting consequences of these mechanisms. In short, here as elsewhere, the function of the applied mathematician is to take what we know (or what we are willing to assume) and examine it from every angle—to turn it over and sideways and inside out by means of the devices of mathematics, in order to ascertain its full content.

When mechanisms interact in complex networks, a mathematical formulation becomes particularly helpful. As patterns become increasingly complex, their *modus operandi* ever more successfully eludes our unaided intuitions. But mathematics is *par excellence* the science of pattern; and it should help us to see how the intricate patterns will function. An obvious example is the field of kinetics, where we ask

* Abbreviations to be used are: DNA for deoxyribonucleic acid; RNA for ribonucleic acid; ATP for adenosine triphosphate.

References p. 632

how systems comprising various interacting mechanisms will evolve with time, and how rates of reaction will depend on the various factors involved. The material I shall present here is a modest attempt to illustrate this point, and to present frankly the difficulties and limitations as well as the valuable features of the mathematical treatment.

PROTEIN SYNTHESIS: A SIMPLE MODEL

Let us consider what we know or have fairly good reason to suppose true about the mechanisms of protein synthesis. Proteins are formed from amino acids; and these may enter the cell from without, or be formed inside the cell from simpler precursors, which eventually enter from without. Both free amino acids and precursors enter by some device more complex than simple diffusion; one commonly speaks of active transport. Numerous schemes of active transport have been proposed, some concrete and some vague. The one that seems to me to be most satisfying will be presented at this Conference by Dr. ROY BRITTEN, to whom I am indebted for the privilege of reading his manuscript in advance of presentation. However, for the sake of mathematical simplicity, I shall consider here chiefly a somewhat simpler scheme, the so-called permease of COHEN AND MONOD[1]. The essence of this mechanism is the assumption that the rate of transport of a compound from outside to inside of the cell is formally the same as the rate of an enzymatic mechanism; the intermediate complex discharges the substance in question into the cell interior in its free form*,**. Inside the cell, amino acid precursor is converted to free amino acid; we simplify the complexities of this sort of process by writing it as a single reaction.

Amino acid, whether of endogenous or exogenous origin, is activated, according to the scheme resulting from the work of HOAGLAND *et al.*[2], with the help of ATP; each type of amino acid is acted upon by a specific enzyme, and its activated form is a compound with a specific RNA, usually present in solution in the cell sap, of relatively low molecular weight[3], and commonly referred to as S-RNA (S for *soluble*).

We must now speculate a little. The work of HOAGLAND cited above[2] suggests that protein synthesis from activated amino acids commonly occurs in submicroscopic particles (microsomes, ribosomes) rich in RNA of substantial molecular weight (about 0.6×10^6 and the dimer of this type[4])***. The current notion, unsupported by any real evidence but in accord with generally held views as to how the genetic control of protein specificity is mediated, is that the ribosomal RNA (or some specific fraction of it[5,6]) acts as a template to determine the specific polypeptide sequences that are synthesized.

The term *template* is somewhat vague. What is essential to the idea is that a large

---

* The permease mechanism treats the rate of *exit* as a diffusion *leak*. The possibility of a mechanism of active outward transport (*exitase*) might be considered. In this paper we neglect outward movement for the sake of simplicity, and because this seems unlikely to be significant in the presence of active utilization for synthesis.

** The relation also resembles a Langmuir adsorption isotherm. The formal properties do not permit the inference that an enzymatic or an adsorption mechanism is at work. For further discussion of this problem, see ref. 15.

*** The case for synthesis in ribosomes rests substantially on evidence for energy-requiring incorporation of labelled amino acids. No really good evidence for net protein synthesis in purified ribosomal systems is yet available.

RNA molecule has specific segments or areas, that will attract and hold specific activated amino acids in appropriate spatial relations to form a chain with a given sequential order. The events that precede completion of a polypeptide chain might be of various kinds. For example, it is possible that activated substrate molecules find their places in random temporal order; when the template is full, the formation of all peptide bonds is somehow completed. The most plausible alternative is the deposition of the amino acids in sequential order, peptide bonds being formed one at a time.

We have shown[7] that turnover rate studies alone are unlikely to discriminate between these two major alternatives. However, there is some experimental evidence favoring the second choice[8,9]. We shall therefore adopt it as a working hypothesis. To this we shall add the assumption that a single enzyme, specific merely for the formation of the peptide bond, is involved at every step. This is modelled on the findings of KORNBERG *et al.*[10] on DNA synthesis, where a single polymerase forms all the phosphodiester linkages, the sequential specificity being determined by a DNA template, usually termed *primer*. The final step of protein synthesis is conceived as the release of the completed polypeptide chain from the template; this step might very well be non-enzymatic.

THE BIOSYNTHETIC EQUATIONS

We now propose to write down the reaction equations that represent our model, and the differential equations that are their consequences. But first a little about notation. We shall adorn the various symbols with subscripts, primes, or other affixes as may be suitable to make distinctions and denote relationships. We shall denote the amino acid precursors by x with suitable subscripts; thus x_1 shall stand for the precursor of the first amino acid of some polypeptide chain, x_2 for the precursor of the second, and so forth. The corresponding amino acids themselves will be y_1, y_2, and so on, while the activated forms will be distinguished by primes ('), y_1', y_2', etc. Concentrations outside the cell will be marked by the added subscript 0: x_{01} for the external concentration of x_1, and so on. The S-RNA classes will have the subscripts of their amino acids, so that r_1 will stand for the S-RNA that couples to amino acid #1. All enzymes will be denoted by the letter E with suitable affixes; in particular the amino acid-activating enzymes will be E_1, E_2, etc., numbered according to their specific substrates, while the *polymerase* that forms peptide bonds will be E'. Permeases will be marked by the letter P instead of E. As for the ribosomal RNA template, the empty template will be T_0, the template with the first amino acid added T_1, and so forth to the template with completed chain, T_n if there are n amino acids in the chain.

The same symbol will be used to refer to the *substances* in the stoichiometric reaction equations, and to their molar *concentrations* in the kinetic differential equations, a usage which has been adopted[11] to save superfluous brackets and parentheses, inasmuch as occasions for confusing the senses are extremely rare.

Classes of substances distinguished by subscripts may be referred to in general by literal subscripts: *e.g.*, r_i is any of the S-RNA series, E_j might be any of the activating enzymes, and the like. The usual conventions will apply to the use of the summation sign Σ, as applied to expressions with such variable subscripts.

References p. 632

We now write the stoichiometric equations for the reactions involved in synthesizing one particular protein—which one, for the moment, need not be specified:

$$\text{(A)} \qquad x_{0i} \xrightarrow{P_i} x_i \quad ; \quad y_i \xrightarrow{P_i'} y_{0i} \quad ; \quad x_i \xrightarrow{E_{1i}} y_i$$

$$y_i + r_i \xrightarrow{E_i} y_i'$$

$$y_i' + T_{i-1} \xrightarrow{E'} T_i + r_i$$

$$T_n \rightarrow \text{Protein} + T_0$$

It should be noted that we are treating all the reactions as irreversible. This is best viewed as an approximation adopted for the sake of mathematical simplicity; there may be occasions when we would like to investigate the consequences of the fact that some of the early reactions, at least, are in principle reversible.

We may now write down the differential equations corresponding to the reactions of Scheme A. For abbreviation we will adopt one more notational convention with respect to enzymatic or permease reactions. We know[11] that enzymatic reactions involving one step may often have their rates expressed as k(Substrate) (Enzyme)/ $[K +$ (Substrate)$]$, while bimolecular reactions will take the form

$$k(\text{Substrate 1})\ (\text{Substrate 2})\ (\text{Enzyme})/[K + aS_1 + bS_2 + cS_1S_2].$$

We will abbreviate such expressions, in a great deal of what follows, by such forms as A(Enzyme); and we will term the abbreviating symbol A the *efficiency* of the enzyme*. We shall in particular use the letter H, with suitable affixes, when referring to permeases. When it is desirable to express the efficiency of a step explicitly, we will frequently expand in a linear form, which would be a good approximation at low values of the concentrations—*e.g.*, k(Substrate) or k(Substrate 1) (Substrate 2). However, the complete forms indicated above guarantee that every efficiency has a finite limiting value when the substrate concentration is sufficiently great.

With these conventions, we may write:

$$\text{(1)} \qquad dx_i/dt = H_iP_i - A_iE_{1i} \quad ; \quad dy_i/dt = A_iE_{1i} + H_i'P_i' - B_iE_i;$$

$$dy_i'/dt = B_iE_i - M_iE' \quad ; \quad dT_i/dt = M_iE' - M_{i+1}E'$$

$$(\text{when } i = 1, 2, \ldots, n-1)$$

$$dT_0/dt = aT_n - M_1E' \quad ; \quad dT_n/dt = M_nE' - aT_n$$

$$d(\text{Protein})/dt = aT_n.$$

---

* The efficiency is very nearly the same as what is usually termed the specific activity of an enzyme. However, we are talking about the activity per mole of protein, not the activity per milligram or per gram. Moreover, we want the term to cover permeases, for which we have as yet no clearcut evidence for an enzymatic function. Finally, the term *specific activity* already has a well-established usage in tracer isotope work. It seems preferable not to use the term in two ways.

In the above, the efficiency abbreviations are:

(2) $H_i = H_i x_{oi}/(L_i + x_{oi})$; $A_i = k_{1i} x_i/(K_{1i} + x_i)$;
$H_i' = h_i' y_{oi}/(L_i' + y_{oi})$; $B_i = k_i r_i y_i/(K_i + a_i r_i + b_i y_i + c_i r_i y_i)$;
$M_i = m_i y_i' T_{i-1}/(D_i + Q_i T_{i-1} + R_i y_i' + S_i T_{i-1} y_i')$.

(Symbols like h_i, L_i, etc., are constants).

It should be noted that we are treating the T_i equations in such a way as to imply (from equation 2) that the T_i and y_i' both enter as substrates in a simple bimolecular enzyme-catalyzed reaction[11]. Since we know nothing of the actual kinetics of such template reactions, this is speculative. We might well argue instead that the (empty or partly filled) template should enter linearly in the same way as the enzyme, in which case we would have $M_i = m_i T_{i-1} y_i'/(D_i + y_i')$. The linear approximation to the efficiencies M_i of equation 2, with respect to the partial templates T_{i-1}, will probably be good in any case, since the total level of template may be very small. The precise form of M_i is not too important for most of our discussion, so we may regard the symbol M_i as non-committal between the two possibilities (and others) if we like.

THE STEADY-STATE APPROXIMATION

To expect to solve the complete system (1) rigorously is quixotic. One might solve it, for a given set of assumed values of the constants, with the help of a computer; but such a solution would not be very instructive. What we want is to know how the parameters affect the general properties of the solution, in particular for those parameters that represent measurable or controllable factors. We can get at this problem by dumping overboard most of the mathematical complexities of the complete system. It seems reasonable to assume that, once we wind up a system like this and start it going, the *intermediates* of synthesis may rapidly reach steady values, where they are formed and removed at equal rates, while the end-products (*e.g.*, the proteins) progressively change.

This means, mathematically speaking, that (if we are willing to get along without the solution for the rapidly changing transient state while the intermediates approach their steady state) we can say that the time derivatives on the left sides of the equations of (1) are equal to zero, except for the time derivative of protein concentration. This means that the right sides are equal to zero. And this means that the differential equations become simply algebraic equations, which of course are a good deal less troublesome to solve.

The problem is still simpler when we decide what we want to solve for, which is the value of $d(\text{Protein})/dt$, the rate of protein synthesis. We note that this equals aT_n. Looking at the last of the T_i equations, we see that $aT_n = M_n E'$. Tracing our way back through the system of T_i equations, we wind up with the result $aT_n = M_1 E'$, since it appears that $M_n = M_{n-1} = \ldots = M_2 = M_1$. This brings us to the equation for y_1', from which $M_1 E' = B_1 E_1$; this in turn leads to the y_1 equation from which $B_1 E_1 = A_1 E_{11} + H_1' P_1'$. The x_1 equation gives $A_1 E_{11} = H_1 P_1$, hence, $B_1 E_1 = H_1 P_1 + H_1' P_1'$. Thus we have finally:

(3) $$d(\text{Protein})/dt = H_1 P_1 + H_1' P_1'.$$

References p. 632

The subscript 1, let us remember, refers to the first (*e.g.*, the N-terminal) amino acid in the polypeptide sequence of the *protein*. The efficiencies H_1 and H_1' are functions of x_{01} and y_{01}, the extracellular concentrations of the amino acid precursor and amino acid itself. The P_1 and P_1' are the permeases for these compounds.

Thus (not altogether surprisingly) it appears that, if a protein is synthesized in sequential order, the rate of synthesis in the steady state will vary directly with the rates of permeation of the first amino acid in the chain and of its precursor.

Can we achieve somewhat less banal results? The permeases are presumed to be proteins; hence equation (3), which applies to any protein, should apply to each of them as well. This, however, does not immediately get us out of trouble. For if we put, say, P_1 for protein in (3), then on the right side there will in general appear a new set of Ps—namely, the permeases for the first amino acid of P_1 and its precursor; and this amino acid will not in general be the same as the first amino acid of another protein, for which P_1 is the permease. We appear to be going in circles, or in a messy system of loops and regresses.

An interesting game could be played by talking about a conceivable case—that in which all the permeases can be arranged in a cyclical order, such that each permease works on the terminal amino acid of the preceding permease. This system can be solved fairly easily. Nevertheless, the cyclical order assumed seems too implausible to justify the mad pursuit of mathematical curiosity.

Let us try another approach. Suppose permeases are like group-specific rather than compound-specific enzymes (*i.e.*, like phosphatases or esterases). Then we may assume that there is a single permease for all of the x_i, and another for all of the y_i. The rates of transport of different "substrates" by these permeases will differ only in the efficiency factors, which involve the substrate concentrations and rate constants. This leaves us, as far as amino acids and their precursors are concerned, with just two permeases, which we may denote by P and P'. A given one of the x's may be transported with efficiency E_1, rate E_1P, another of the x's with efficiency E_2, and so on. The equations for the two permeases may then be written, modelled on (3):

$$dP/dt = E_1P + F_1P' \quad ; \quad dP'/dt = E_2P + F_2P' \tag{4}$$

where E_1 is the efficiency factor for the precursor of the terminal amino acid of P, F_1 the factor for the amino acid itself, and E_2 and F_2 the corresponding factors for P'.

The pair (4) has the solution:

$$P = a_1e^{m_1t} + a_2e^{m_2t} \quad ; \quad P' = -a_1f_1e^{m_1t} - a_2f_2e^{m_2t}; \tag{5}$$

with

$$\begin{aligned} a_1 &= (P_0f_2 + P_0')/(f_2 - f_1) \quad ; \quad a_2 = -(P_0' + P_0f_1)/(f_2 - f_1); \\ m_1 &= \tfrac{1}{2}[S + (S^2 - 4G)^{\frac{1}{2}}] \quad ; \quad m_2 = \tfrac{1}{2}[S - (S^2 - 4G)^{\frac{1}{2}}]; \end{aligned} \tag{6}$$

$$S = E_1 + F_2;\ G = E_1F_2 - E_2F_1;\ f_1 = (E_1 - m_1)/F_1;\ f_2 = (E_1 - m_2)/F_1,$$

where P_0 and P_0' are the initial values of P and P'. From (6) it is easy to show that $m_1 > m_2$, that m_2 has the same sign as G, that $m_1 > 0$, that $f_1 < 0$, and that $f_2 > 0$. Thus, both P and P' increase exponentially, but one of the two terms in each may decrease exponentially to zero in case $m_2 < 0$.

References p. 632

If we introduce the result (5) into equation (3), and solve:

$$(7) \qquad Pr = Pr_0 + (P_0' + P_0 f_2)\,(H_1 - H_1' f_1)\,(e^{m_1 t} - 1)/m_1 - (P_0' + P_0 f_1)\,(H_1 - H_1' f_2)\,(e^{m_2 t} - 1)/m_2.$$

Now $m_1 > m_2$. Hence, as t increases, the first exponential term will more and more outweigh the second one; and the time course of the protein concentration will tend to be governed by a simple exponential increase:

$$(7') \qquad Pr \rightarrow (P_0' + P_0 f_2)\,(H_1 - H_1' f_1) e^{m_1 t}/m_1.$$

Cell composition

One consequence of these results is easily inferred. To make it easier to see, we may abbreviate the expression (7) as $Pr_0 + Q_1(Pr)R_1(t) + Q_2(Pr)R_2(t)$, where the notation $Q_1(Pr)$, for instance, means that the factor depends on constants characteristic of the specific protein, while $R_1(t)$ is a factor common to all proteins, dependent on the time t and on constants of the twin permease system. We now consider what fraction Pr is out of the total cell protein. By definition, this is $Pr/\Sigma Pr$. This is, in general:

$$[Pr_0 + Q_1(Pr)R_1(t) + Q_2(Pr)R_2(t)]/\Sigma[Pr_0 + Q_1(Pr)R_1(t) + Q_2(Pr)R_2(t)].$$

Without further assumptions, this expression cannot be appreciably simplified. It means that the given Pr is a fraction of total protein that depends on its individual characteristics and on the time t, so that the cell composition, in general, is variable.

However, consider times long enough for (7') to be a good approximation. The fraction of Pr will be

$$Q_1(Pr)R_1(t)/\Sigma Q_1(Pr)R_1(t).$$

Here the numerator and denominator have a common factor $R_1(t)$. When this is removed, the fraction of Pr is simply $Q_1(Pr)/\Sigma Q_1(Pr)$. This is independent of the time t. We conclude that, as a system such as we are considering begins the biosynthetic process, its protein composition will shift with time. However, it will approach a constant composition, characteristic for each protein, as time passes. This result is reminiscent of the changes that take place in a bacterial culture during the early stages of growth, and of their tendency toward a constancy of cell composition during the logarithmic phase of growth.

It should be noted that the characteristic composition depends on the composition of the medium, since the H_1 and H_1' of each protein depends on the external concentration of amino acids and their precursors. Thus, if one amino acid is omitted from the medium, the H_1' for all proteins in which it is terminal will vanish, and those proteins will have a lower representation in ΣPr.

Competition among proteins

So far, we have been treating a somewhat idealized situation, in which each amino acid occurs but once in a protein, and in which we ignore the fact that it is also used in the synthesis of other proteins. The key equation is in (1), in the expression for dy_i'/dt, which is written as $B_iE_i - M_iE'$. Since the i-th amino acid in the protein we are considering, in its activated form y_i', also enters, in general, other places in

References p. 632

the same protein and also other proteins, the term M_iE' should actually be $(\Sigma M_i)E'$, where the summation is over all occurrences. Consequently, at the step (in the steady-state treatment) where we solve for M_1E', we must write the equation as $B_1E_1 = M_1E'(\Sigma M_1)/M_1$. The final result will then be, instead of (3):

$$d(\text{Protein})/dt = w(H_1P_1 + H_1'P_1') \tag{3'}$$

where w is an abbreviation for $M_1/\Sigma M_1$. Formally, the effect is to give all the results as before, but with a factor w incorporated into each permease efficiency factor — not only the E and F factors, but the H_1 and H_1', the factor w being characteristic for each amino acid, provided that it appears terminally in some protein.

Inasmuch as amounts of protein and rates of synthesis are, in a general way, determined by efficiency factors, the result of these considerations is that amounts and rates will be lowered according to the intensities of competition denoted by the various w's.

To see a little of what is involved, let us take a closer look at w. Since, in the steady state, $M_1 = M_2 = \ldots$ for a given protein, the summation of the M terms for each protein may be written as M_1 multiplied by the number of times the terminal amino acid occurs in that protein. If we denote this number, for protein a, by n_a, and so forth, we may write the w_a of protein a (and of its terminal amino acid) as

$$M_1(a)/[n_aM_1(a) + n_bM_1(b) + \ldots].$$

If we were to assume that the M_1 factors for all proteins are nearly the same (which implies that the rates of synthesis are the same, and is certainly not strictly true) as an approximation, this expression would be simply $1/(n_a + n_b + \ldots)$. That is, the competitive factor for any terminal amino acid will be inversely proportional to its total number of occurrences in the proteins of the system.

More generally, recalling that $dPr/dt = M_1E'$, and multiplying numerator and denominator of w_a by E', it appears as

$$(dPr_a/dt)/[n_a(dPr_a/dt) + n_b(dPr_b/dt) + \ldots].$$

That is, the competition factor is the rate of terminal incorporation in the given protein as a fraction of the total rate of incorporation into all its occurrences in all proteins.

We get a particularly interesting view of the situation if we take account of the fact that, for a given terminal amino acid, the sum $\Sigma M_1E' = \Sigma n_idPr_i/dt$ is equal to the permeation rate for that amino acid and its precursor, an expression of the form $HP + H'P'$. Abbreviating this expression by $\emptyset$, we have a linear equation $\Sigma n_idPr_i/dt = \emptyset$ for the rates of synthesis of the various proteins. For each different kind of terminal amino acid, we can write a similar linear equation. If exactly x amino acids are terminal in all proteins, there will be x equations in the unknown rates of synthesis.

There are more than x proteins, even if $x = 20$ (the number of different kinds of amino acids); and this introduces an indeterminacy into the system of equations. However, this may be removed by an approximate procedure as follows.

For two proteins, a and b, with the same terminal amino acid, $(dPr_b/dt)/(dPr_a/dt) = M_1(b)/M_1(a) \simeq T_o(b)/T_o(a)$. We can show that this ratio is proportional to the ratio of total template material for the respective proteins. For if we use the linear approxi-

References p. 632

mation for the M_i ($M_i = my_i'T_{i-1}$, treating the rate constant m of peptide bond synthesis as being the same for all bonds), we can solve the recursive set of equations $M_1 = M_2 = \ldots$, getting $T_i = T_o y_1'/y_i'+_1$, $T_n = my_1'T_0E'/a$. From the conservation equation

$$T_{total} = \sum_{i=o}^{n} T_i,$$

it then follows that $T_o = cT_{tot.}$, where

$$c = \left(\sum_{i=1}^{n-1} y'_i/y'_{i+1} + my_1'E'/a \right)^{-1}.$$

Since the y_i' is constant in the steady state, and the rate of unpeeling from the template, a, is either a universal constant or characteristic of the given protein, and constant with respect to time, we may consider the c for each protein as a characteristic constant. For our two proteins a and b, therefore, we may restore the affixes and write $T_o(b)/T_o(a) = [T_{tot.}(b)/T_{tot.}(a)](c_b/c_a)$. Abbreviating this ratio by the symbol c_{ab}, we have $(dPr_b/dt)/(dPr_a/dt) = c_{ab}$. Thus, in an equation containing the two terms $n_a dPr_a/dt$ and $n_b dPr_b/dt$, we may merge the two terms as $(n_a + c_{ab}n_b)dPr_a/dt$. This effectively reduces the number of unknown rates to the number of terminal amino acids x. It has the disadvantage that the c_{ab} presently cannot be computed independently of rate measurements. They must be treated as empirical constants to be determined from rate data. There are compensations for the somewhat inelegant character of this procedure, however; inferences about the relative availability of templates for various proteins would not be without interest.

The problem of indeterminacy arises from our assumption that a common pool of activated amino acids is formed, and fed to all proteins. If activation occurred in small units (*e.g.*, ribosomes) where only a few proteins are made, this indeterminacy would be removed. But we would have a separate set of biosynthetic equations for each kind of particle, not one set for the cell as a whole.

The system of linear equations could in principle be solved straightforwardly by determinants; and from our knowledge of the expansion of determinants we can say that the solutions will be linear functions of the $\emptyset$'s. Since these are linear functions of P and P', we are led to predict that each dPr/dt will be a linear function of P and P', just as in (3) and (3'), but with coefficients that are linear functions of the various H and H' efficiency factors, multiplied by functions of the n_i only.

In principle, therefore, we could find exact equivalents of (3) and (3'), which would still lend themselves to the same elementary methods of integration as before. In practice, however, the evaluation of determinants of high order whose elements are algebraic, not numerical, is not an attractive procedure. It is satisfying, however, to see that competitive interaction in protein synthesis would not seem to change the basic pattern of the equations of synthesis, when compared with a simpler treatment where this interaction is neglected. It is also interesting to see that the coefficients of P and P' in the rigorous rate expressions will depend on the permeation rates and external supplies of all the amino acids, and on the composition of all the proteins, which seems like an eminently reasonable result.

References p. 632

LIMITATIONS OF SUPPLY

Up to the present, we have treated the factors H and H′ as constant, implying a constant extracellular supply of amino acids and of their precursors. It may be of some interest to consider what might happen where this does not hold, as in the ordinary course of events during the growth of a bacterial culture.

To keep the mathematical developments from becoming unduly complex, let us consider the special case where we have only one of the two key permeases, say P'. Let the supply of its terminal amino acid be limited so that the equations (1) would include:

$$(1') \qquad dy_o/dt = -H'P'.$$

There is no possible steady-state solution for this equation, so we must solve it as it stands. We take a linear approximation to H', and write it as $h'y_o$. To eliminate P' from (1′), we note that, by (3), it is governed by the equation $dP'/dt = H'P'$, or, using (1′), $dP'/dt = -dy_o/dt$, hence $P' = k - y_o$, where $k = P'_o + y_{oo}$, the sum of the initial values. Eliminating P' from (1′) and solving, we have:

$$(8) \qquad y_o = kCe^{-kh't}/(1 + Ce^{-kh't}),$$

where $C = y_{oo}/P'_o$. Since $P' = k - y_o$, we have at once:

$$(9) \qquad P' = k/(1 + Ce^{-kh't}),$$

the familiar *logistic* relation.

Now consider one of the other enzymes, say E. If E has the same terminal amino acid as P', then $c_{ep}dE/dt = dP/dt$, or:

$$(10) \qquad E = E_o + c_{ep}(P' - P_o).$$

If E has a terminal member distinct from that of P', then its rate of synthesis, using the same linear approximation as we did for H' above, is given by:

$$(11) \qquad dE/dt = -dy_a/dt = h_a'y_aP',$$

where we use y_a to represent the external concentration of the terminal amino acid of E. Solving (11) for y_a, with P' given by (9), we get:

$$(12) \qquad y_a = y_{ao}[(1 + C)e^{-kh't}/(1 + Ce^{-kh't})]^{h_a'/h'}.$$

Inasmuch as $dE/dt = -dy_a/dt$, we have (with $k' = E_o + y_{ao}$):

$$(13) \qquad E = k' - y_a.$$

(y_{ao} is of course the initial value of y_a). This will have a course similar to the logistic of (9). The latter will arise initially at the rate $h'y_{oo}P_o'$, the former at the rate $H_a'y_{ao}P'$. With sufficient time P' will approach the value $y_{oo} + P_o'$, while E approaches $y_{ao} + E_o$. That is to say, the proteins P and E rise to values limited, as should be the case, by their supplies of terminal amino acid. The only other noteworthy feature of the results is that, while P rises to its maximum at steadily decreasing rate of increase, E may have a period of positive acceleration; the condition for this to occur is that $h_a' > h'$ and that $2y_{oo}/P_o' > h_a'/h' - 1$.

Thus, when the supply of amino acid is limited (the same argument could have been applied to amino acid precursor), the permeases and enzymes approach maximum concentrations in a manner similar to the so-called logistic curve of growth. Since protoplasmic mass is proportional *inter alia* to cell protein, this agreement is not only reasonable in itself, but is also in concordance with the idea that the logistic course of growth depends on synthesis of cell material by systems similar to the one we have been considering[12]. Equations (5) and (7) can be viewed as approximately the early phases of (9), (12), and (13) in the same sense as the simple exponential growth equation approximates the logistic.

From the foregoing, it would appear that one could influence the protein composition of a cell rather readily, by supplying some amino acids in definitely limiting amount, others at much higher levels, since each protein tends toward a limit set by the total supply of its terminal amino acid. However, let us recall the rigorous analysis of competition between proteins, which predicts that each rate of synthesis will in general be a function of all the efficiencies for all the amino acids that occur terminally in any protein. In view of this, it will not be possible to treat individual proteins separately in terms of the supply of their respective terminal amino acids alone. Accordingly, while some cases may occur in which a specific juggling of the amino acid supply will affect the protein composition of the cell, we may expect in most instances to find cells relatively insensitive to such differential procedures. The exceptions would arise with respect to protein classes of exceptional composition (like the histones, for example), where the buffering effect of competitive interaction might fail.

PROTEOLYSIS

We now consider how proteolytic enzymes might influence biosynthetic processes. As a very crude first approximation, we may consider that the proteolytic enzyme E_p (which we will take as representative of proteolytic effects in general) does not vary with time. We ask about the fate of a permease and another enzyme (say P and E) as in the preceding section. The equations for these compounds are:

(14) $$dP/dt = HP - mPE_p \quad ; \quad dE/dt = H_1P - nEE_p,$$

approximating the rate of proteolytic activity, as we have done previously, by expressions that are linear in the concentration of substrate. The solutions are:

(15) $$P = P_0e^{(H - mE_p)t} \quad ; \quad E = Ce^{-nE_pt} + H_1P_0e^{(H - mE_p)t}/(H - mE_p + nE_p),$$

where C is a constant of integration, obtained as usual by setting $E = E_0$ and $t = 0$ in the equation.

Evidently, there are just two possible outcomes for constant E_p. If $H - mE_p > 0$, both P and E will rise without limit, but at a slower rate than in the absence of E_p. If $H - mE_p < 0$, both P and E will fall steadily to zero levels.

Suppose E_p is not constant, but increasing. Then it may reach levels at which dP/dt and/or dE/dt may become zero or even negative. Under these circumstances, however, the solution (15) is no longer valid, and we must begin from the beginning once more. Suppose first of all that E_p is given by:

(16) $$dE_p/dt = H_2P.$$

Then, eliminating P between (16) and (14), we can get the relation:

$$P = (HE_p - \tfrac{1}{2}mE_p^2 + A_0)/H_2, \tag{17}$$

where $A_o = H_2P_o - (HE_{op} - \frac{1}{2}\, mE_{op}^2)$. Putting (17) into (16), we have :

$$E_p = (E_{1p} - E_{2p}e^{-mdt})/(1 - Be^{-\frac{1}{2}mdt}), \tag{18}$$

where:

$$\begin{aligned} B &= (E_{op} - E_{1p})/(E_{op} - E_{2p}) \quad ; \quad d = E_{1p} - E_{2p}; \\ E_{1p} &= [H + (H^2 + 2A_om)^{\frac{1}{2}}]/m \quad ; \quad E_{2p} = [H - (H^2 + 2A_om)^{\frac{1}{2}}]/m. \end{aligned} \tag{19}$$

In this case E_p will rise steadily to the limiting value E_{1p}, which is, by (19), greater than $2H/m$ if $A_o > 0$; if $A_o < 0$, E_{1p} lies between H/m and $2H/m$.

The consequences of this setup are obvious. From (16), it is clear that E_p will stop rising when $P = 0$; and from (14) it is evident that P will reach its maximum and begin declining already when E_p reaches the value H/m. From (14) for dE/dt, one can see that dE/dt will be negative when $P = 0$ and E_p is at its maximum. Thus E will eventually decline and reach a zero value, after rising to some maximum previously. It would be farfetched to regard this as a model of life and death; and as a model of normal biosynthesis and growth it could serve only if we added some assumption that would stop E_p early enough to prevent the total disappearance of P and E.

This could be done in a natural way by assuming that E_p itself is subject to proteolysis. To avoid introducing another enzyme to devour E_p, let us approximate this by having E_p act on itself. We can rewrite (16) as:

$$dE_p/dt = H_2P - gE_p^2. \tag{16'}$$

This modification prevents the simple mathematical device that permitted us to derive (17). However, we may assume that g is small enough so that the term gE_p^2 will be negligible while E_p is small. Thus, we assume that (17) still holds approximately. Putting (17) into the complete equation (16′), as the next better approximation, we solve for E_p and have the same form as (18), with the modification that $\frac{1}{2}m$ is replaced by $q = \frac{1}{2}m + g$. The maximum value of E_p, namely E_{1p}, is now approximately H/q instead of $2H/m$*. At this point, when $dE_p/dt = 0$, instead of P being also zero, we have from (16′) $P = gH^2/H_2q^2$. It is no longer necessary that dE/dt be negative; substituting $E_{1p} = H/q$ in (14), $dE/dt = (H/q)\ (H_1gH/qH_2 - nE)$. If this is to be zero, and E at its maximum, at this time, we must have $E = H_1gH/qH_2n$.

That this result for E, rather informally obtained, is reasonable can be seen by taking the value (17) for P, substituting into (16′) and (14), treating A_0 as negligible to minimize mathematical complications. This permits us to combine the equations (14) and (16′) and get an equation for dE/dE_p. Solving for E, we get:

$$E = C(H - qE_p)^{n/q} + \tfrac{1}{2}mH_1(H - qE_p)/H_2q(n - q) + H_1gH/qH_2n, \tag{20}$$

where C is a constant of integration. When E_p reaches its maximum value H/q, the first two terms of (20) vanish leaving $E = H_1gH/qH_2n$ as the limiting value of E.

* Whether E_{1p} is equal to H/q exactly, or a little less or a little more, depending on the sign and magnitude of A_o, is immaterial for the argument. The important point, evident from (16′), is that $dE_p/dt = 0$ does not imply $P = 0$ as was true in (16). We use the approximate limiting value H/q merely for the sake of some illustrative formulas.

References p. 632

The interpretation of this result is unremarkable but makes sense. The maximum value of E increases with its rate of synthesis (H_1) and that of its permease (H). It decreases with the rate of synthesis of the protease $E_p(H_2)$, and with the proteolytic rate constant (n). It decreases with the proteolysis constant of P (m, via q), and increases with the autoproteolysis constant of the protease itself (g). All the factors in the system enter into its capacity to make protein in about the way one would intuitively expect. But intuition might not have predicted the remarkably simple way most of the factors enter the result. And one might have guessed wrong about the role of g: as g increases without bound, the ratio $g/q = g/(\frac{1}{2}m + g)$ approaches the upper limit 1, so that the rise of P and E is not unbounded even if the protease destroys itself at an infinite rate.

COMPETITION AS A LIMITING FACTOR

The competitive system of equations discussed previously has one interesting feature which we have not mentioned so far: The coefficients of P and P' are not necessarily intrinsically positive, as in the simple non-interacting system. This has two consequences: (a) it is no longer necessarily true that at least one of the exponential terms of P and P' has a positive exponent; and (b) the relations among the constants may be such as to permit maxima or minima in the time curves of various proteins while this could be excluded by elementary calculations for the non-competitive system.

We shall not enter into these points extensively. It will suffice to give a couple of elementary examples that will point up the possibilities.

Consider for simplicity, as we have at times earlier, the case of just one permease, say P'. Suppose this is governed by $dP'/dt = -aP'$. The solution is $P' = P_0'e^{-at}$. Now take a protein governed by $dPr/dt = bP'$. Using the value just found for P' in this equation, we can solve for Pr, finding $Pr = Pr_0 + (b/a)P_0'(1 - e^{-at})$. This rises from Pr_0 to $Pr_0 + (b/a)P_0'$. At the same time, P' decreases to zero. This might serve as a simple-minded model of *irreversible maturation*: the various proteins increase, but always to limiting values, while the permease vanishes, so that it cannot be regenerated (since its development is autocatalytic). The model does not admit of turnover in the mature organism; but such turnover, independent of cell death and concomitant multiplication, is questionable in any case.

Consider a second example. Suppose both permeases are present. Let the values of the various constants of the system be such that the equation of a certain protein comes out as $dPr/dt = AP_0e^{at} - BP_0'e^{-bt}$, where the constants as written are assumed positive. We ask under what conditions this expression may equal zero, denoting a maximum or minimum of Pr. The condition is: $e^{(a+bt)} = BP_0'/AP_0$. Since the left side of the condition is a positive quantity, and greater than one, for real and positive values of t, it must be true that $BP_0' > AP_0$. (The same holds for $b < 0$, if $a + b > 0$; if $a + b < 0$, we demand $BP_0' < AP_0$.) If this condition is fulfilled, there will be *a minimum* of Pr, as is evident from setting $t = 0$ in dPr/dt, and seeing that it will be negative. Eventually, of course, Pr will increase exponentially; but there may be a long deep trough if $BP_0' \gg AP_0$ and a and b are not too large. If Pr were a protease, we would have a mechanism of *growth triggering and control* in which the protein-destroying activity at first decreases (while presum-

References p. 632

ably all the other proteins increase), and then begins to increase and exert the kind of control over further synthesis that we have studied in the preceding section. Crude as the model may be, it is suggestive.

The essence of these cases is that, under conditions of competition or interaction, it is no longer necessary for all proteins to rise or fall together, nor is it necessary that they rise uniformly or fall uniformly. And (despite the practical difficulties, at the moment, of working out the whole system in all its glory) it is worth stressing again that the composition of all the interacting proteins, as well as the relative availability of their templates, will largely determine who does what and with which and to whom.

DISCUSSION

Perhaps more than any of you, I am conscious of the shortcomings of what has been presented above, of the issues omitted, the mathematical difficulties glossed over or only partly solved. I do not feel that the development is at a stage where it can be deemed to lead to critical tests of the mechanisms assumed. I do believe that it illustrates the nature of the problems to be solved in the kinetic analysis of such systems, and the kind of algebraic debris that besets the path to a solution. I hope that some of the results presented will also have some intrinsic interest or at the very least some heuristic value.

The problems of amino acid uptake and pool sizes are too important to be considered as casual incidents of a kinetic analysis. I have therefore taken a relatively simple view of the matter, for the sake of getting something done, since others are treating these problems in greater depth. One thing might well be pointed out: the considerations that led Dr. BRITTEN to his more complex mechanism of transport and pool size determination are very relevant to the kinds of data he is talking about; but they do not altogether vitiate our simpler approach to another set of problems. The reversible internal binding of amino acids in the cell serves, from our point of view, as a sort of buffer mechanism or reservoir for available amino acids to be used in synthesis. The simple permease mechanism will give the wrong answers for pool sizes and the details of exchange reactions; but I do not believe that it seriously falsifies the issues of synthesis. It is not difficult to show, for the BRITTEN model, that the rate of the first step of protein synthesis (*e.g.*, amino acid activation) in the steady state is equal to the rate of the transport step (carrier or permease) and independent of the properties of the internal binding system.

We have given only a cursory treatment of the problems of growth. Among items deserving analysis in the future are three obvious ones: (a) the fact that quantities like the r_i and T_i will not be constant in a growing system, but will increase in a way that will depend on another set of permeases (*e.g.*, for the purine and pyrimidine bases and their precursors); (b) the changes in cell volumes and surfaces that occur in growing systems, their mechanisms (*e.g.*, osmotic water uptake balancing internal synthesis of macromolecules), and their effect on transport (the surface–volume ratio is involved in the equations for transport from extracellular to intracellular phase, though this point was ignored in our treatment); (c) the relations between mass increase and division in growing systems. The third problem has been considered extensively by RASHEVSKY[13], whose ingenious analysis should be brought up to date biochemically and adequately tested. A possible approach to some aspects of

References p. 632

the second problem is implicit in an earlier study of ours on non-equilibrium thermodynamics[14]; but the mathematical difficulties seem likely to be formidable.

The assumption of the steady state for intermediate steps of synthesis has implications that deserve to be considered more fully. In relations like $H_iP_i = A_iE_{1i}$, the factors P_i and E_{1i} are, we now know, exponential functions of the time. The coefficients of these exponentials will have to have certain relations if these *steady-state* equations are indeed to hold independently of time. In the non-competitive model, the relations that will arise are obvious, and can always be fulfilled; they will, among other things, define the internal pools of amino acids and their precursors in terms of the extracellular concentrations and the various constants of the system. For the competitive case, however, the greater freedom of the constants with respect to sign means that we will have to ask seriously whether the steady state conditions can always be fulfilled, or whether instances may arise when this will not be possible.

Of all our results, the emergence of the competitive system poses the greatest challenges to patience and ingenuity, and hence may perhaps be deemed the most interesting feature. The (admittedly oversimplified) illustrative cases, implying mechanisms of differentiation and control, may hint at the possible rewards awaiting further work on this feature.

In connection with the competitive system, it may be well to examine more closely the apparent paradox of 20 equations for vastly more than 20 unknown rates of protein synthesis—a paradox which we side-tracked by an approximate procedure. Let us start counting equations and unknowns more carefully. If there are 20 amino acids, and thus also 20 more activated amino acids, the first two lines of equation (1) exhibit equations of the type that will govern them; the same holds for precursors. The solutions for the activated amino acids will, however, involve all the T_i for all the proteins. Let us ask how many there are. If protein a contains N_a residues, its template has $N_a + 1$ states $T_i(a)$; there is a conservation equation that says their sum equals the total amount of that kind of template, reducing the number of independent $T_i(a)$ to N_a. The recursive equations $M_1(a) = M_2(a) = \ldots = aT_{N_a}$ are N_a in number. Hence there are in our system, as written for the steady state, enough equations to determine all the unknowns. However, if we proceed to eliminate the y_i', after solving the recursive equations, we get finally a set of non-linear equations for the T_i, each equation involving T_is from all the proteins. It is clear that a general solution is out of the question, and that results will have to come from considering special cases that are more tractable.

The general equations suggest one interesting conclusion, however. If we try to write them down carefully rather than sketchily, it becomes evident that our notation for enumerating amino acids according to their sequential positions in one arbitrarily chosen protein is ambiguous and confusing when we try to apply it to all the proteins simultaneously. It becomes necessary to have a notation that separates the enumeration of the *kinds* of amino acids from the specification of their *positions* in any one protein. Such a notation can be devised, and will be discussed in a forthcoming publication. The new notation does not in any way reduce the mathematical difficulties of a general solution. What it does, however, is to make it clear that the complete system is not only defined by the *compositions* of the proteins (in addition to the other factors), but also by their *sequential orders*—a result which I find intriguing, and which was certainly not intuitively obvious to me.

References p. 632

Clarification of the precise nature of this dependence is one of the results that one would hope to get from working out the details of special cases.

REFERENCES

1 G. N. Cohen and J. Monod, *Bacteriol. Rev.*, 21 (1957) 169.
2 M. Hoagland, E. B. Keller and P. C. Zamecnik, *J. Biol. Chem.*, 218 (1956) 345.
3 A. Tissières, *J. Molec. Biol.*, 1 (1959) 365.
4 C. G. Kurland, *J. Molec. Biol.*, 2 (1960) 83.
5 A. I. Aronson and B. J. McCarthy, *Biophys. J.*, 1 (1961) 215.
6 B. J. McCarthy and A. I. Aronson, *Biophys. J.*, 1 (1961) 227.
7 J. M. Reiner, *Atomlight*, No. 17, 1961.
8 R. Loftfield, *Proc. 4th Intern. Congr. Biochem. Brussels*, 8 (1960) 222.
9 H. M. Dintzis, *Proc. Natl. Acad. Sci. U.S.*, 47 (1961) 247.
10 M. J. Bessman, I. R. Lehman, E. S. Simms and A. Kornberg, *J. Biol. Chem.*, 233 (1958) 171.
11 J. M. Reiner, *Behavior of Enzyme Systems*, Burgess Publishing Co., Minneapolis, 1959.
12 F. Morin and J. Monod, *Rev. sci.*, 80 (1942) 227.
13 N. Rashevsky, *Mathematical Biophysics: Physico-mathematical Foundations of Biology*, Dover Publications, New York, 1960.
14 J. M. Reiner and S. Speigelman, *J. Phys. Chem.*, 49 (1945) 81.
15 J. M. Reiner, *Experientia*, 17 (1961) 457.

METABOLIC POOLS AND THE BIOSYNTHESIS OF PROTEIN

DEAN B. COWIE

Carnegie Institution of Washington, Department of Terrestrial Magnetism, Washington, D.C. (U.S.A.)

The study of the endogenous flow of carbon in microorganisms has been extremely valuable in elucidating many of the essential steps in the biosynthesis of protein. Since this carbon flow is dependent upon many environmental factors, often affecting cellular biochemistry and final organic composition, it is important to ascertain what

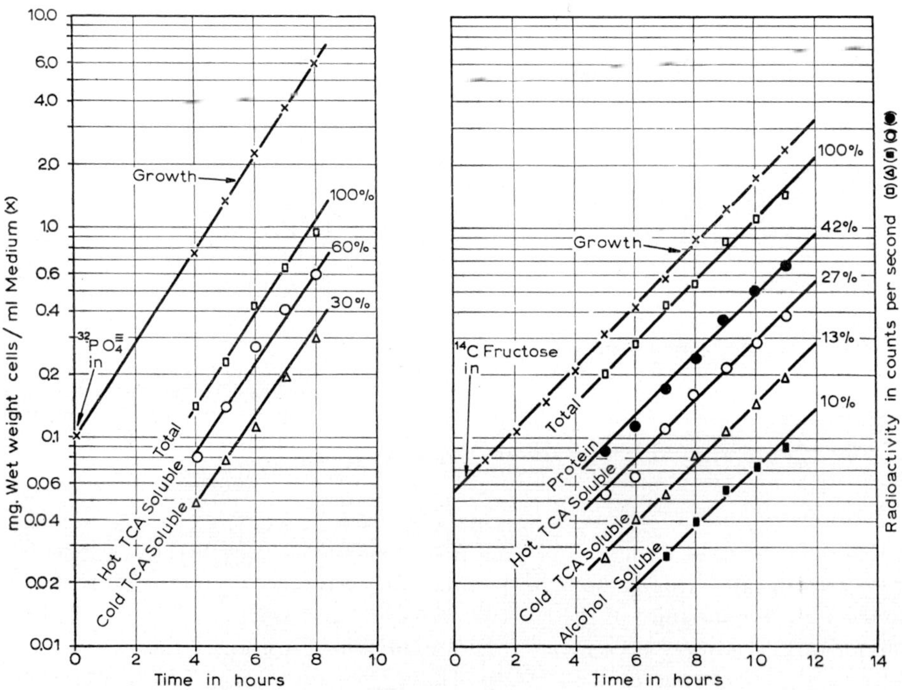

Fig. 1. Steady-state distribution of carbon and phosphorus among chemical fractions of the cell.

processes are directly involved in protein synthesis and to describe how these processes are influenced by environmental alterations.

Microorganisms, capable of exponential growth in a well-defined synthetic medium, provide a most valuable system for the investigation of such problems. With such a system a "standard" carbon flow may be defined and essential metabolic steps recognized. As an example of such a system the discussion here will center

References p. 645

on the yeast *Candida utilis* growing exponentially in C medium* containing fructose as the carbon source.

Distribution of cellular carbon

It is extremely important to determine the steady-state distribution of the cellular carbon among the chemical fractions of the cell.

For these studies the yeast cells were grown for many generations in C medium containing [^{14}C]fructose as the carbon and energy source. Fig. 1 shows the distribution of carbon (and phosphorus) among the chemical fractions of the cell during exponential growth. Samples of the culture suspension were removed at intervals, the cells washed, and chemically fractionated by a modification of the SCHNEIDER

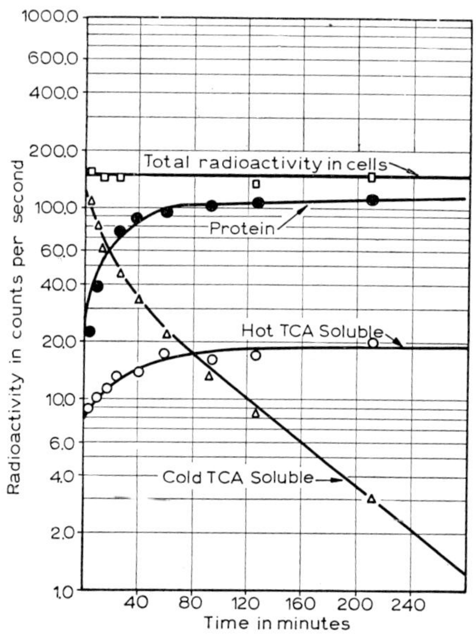

Fig. 2. Transfer of ^{14}C from the metabolic pool (TCA-soluble fraction) to the protein fraction of the cell. Exponentially growing cells were briefly immersed in medium containing [^{14}C]fructose and transferred after washing to non-radioactive medium.

method[2]. The rate of formation of each fraction is shown to be directly proportional to the growth rate. Approx. 50% of the total carbon is found in the proteins, 8–10% in nucleic acids (one-third of the hot trichloroacetic acid fraction), 10% in the lipids and 12–14% is always present as "free amino acids", easily extractable with *hot* water or 5% trichloroacetic acid (TCA). These "free amino acids" have been shown to be contained in a necessary and essential metabolic pool, on the main line of synthetic events leading to the formation of protein[3,4].

Specific labeling of the amino acid pool

Several minutes' immersion of rapidly growing cells in C medium containing [^{14}C]-fructose results in the specific labeling of these pool amino acids. Under these con-

* C medium: 2 g NH_4Cl, 6 g Na_2HPO_4, 3 g NaCl, 0.01 g Mg as $MgCl_2$, 3 g KH_2PO_4, 0.026 g S as Na_2SO_4, 100 ml 10% maltose and 900 ml distilled H_2O.

ditions as much as 90% of the incorporated radiocarbon may be found in the TCA-soluble fraction. If these specifically labeled cells are washed and transferred to non-radioactive medium, there is a rapid loss of pool radioactivity, the major portion of the carbon being transferred to the protein of the cell (TCA-precipitable fraction). These results indicate that there is a precursor–end-product relationship between the pool amino acids and the proteins of the cell. As shown in Fig. 2 very little if any of the labeled carbon is lost to the medium.

The transfer of carbon from individual amino acids of the pool to protein has also been investigated. Fig. 3 shows the rate of loss of radiocarbon from several

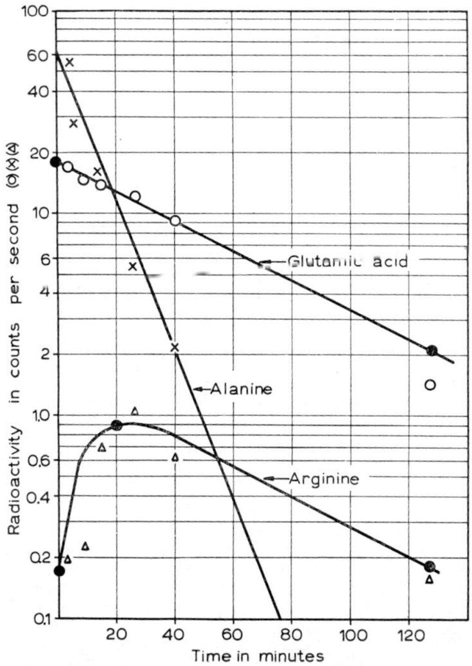

Fig. 3. Time course of ^{14}C concentration in amino acids of the cold TCA-soluble fraction. Data represents the measured radioactivity of individual amino acids appearing on chromatograms.

pool amino acids. The data were obtained from cells briefly labeled as above with [^{14}C]fructose and reinoculated into non-radioactive medium. Samples were removed at the indicated times, the cells chemically fractionated, and two-dimensional chromatograms made of aliquots of hydrolyzates of the cold TCA-soluble and -precipitable fractions. Chromatographic examination of the TCA-soluble material immediately after the labeling period showed that "parental" amino acids were highly radioactive while related "member" amino acids were found to be relatively non-radioactive.

The exponential rate of loss of radiocarbon from pool alanine and glutamic acid indicate that this loss is a simple process. Each alanine has an equal probability of being incorporated into protein, whether radioactive or not.

At the start of the experiment arginine was found to be only slightly radioactive (Fig. 3) but increased in radioactivity for 25 min. This increase was then followed by an exponential loss of ^{14}C equal to the rate observed for glutamic acid. These kinetics are characteristic of a more complex process in which *pool* glutamic acid is converted to *pool* arginine before protein incorporation occurs. In *Candida utilis* arginine carbon is supplied from glutamic acid[1]. Similar pool amino acid interconversions have been observed among other "parental" amino acids and their family "members" and this pool has been designated as the "internal pool" of amino acids[4, 5].

The appearance of ^{14}C in the individual amino acids of the proteins was also shown to be directly correlated with the loss of radiocarbon from the corresponding amino acids of the internal pool[3]. The rate of transfer of each pool amino acid to protein is therefore dependent upon the protein requirements.

Isotopic competition studies

a. Competition between exogenous amino acids and fructose carbon.

Exogenous amino acids reduce the flow of fructose carbon into protein (ref. 1, chapt. 13). Thus when a non-radioactive amino acid is supplied to medium containing [^{14}C]fructose, a reduction occurs in the radiocarbon incorporation in the same amino acid in the proteins. When a family "head" amino acid is used to supplement the radioactive medium, that amino acid and all of its family "members"

TABLE I

EFFECT OF EXOGENOUS [^{12}C]AMINO ACIDS ON THE LOSS OF RADIOCARBON FROM THE TCA-SOLUBLE FRACTION IN EXPONENTIALLY GROWING *Candida utilis*

Time (min)	*Mg wet wt./ml medium*	*Total radioactivity in TCA-soluble fraction (radioactivity in counts/sec)*	
		*[^{12}C]Amino acids present**	*No amino acids present*
5	0.59	20.8	20.8
17	0.67	12.2	12.4
35	0.82	8.2	8.3
62	1.00	4.7	4.9
92	1.35	2.9	3.0

* [^{12}C]Amino acids present as competitors (mg/ml medium): aspartic acid 0.75; glutamic acid 0.75; alanine 0.50; valine 0.25; glycine 0.25; lysine 0.25; serine 0.25; arginine 0.19; leucine 0.19; proline 0.12; threonine 0.12; isoleucine 0.12; methionine 0.10; tyrosine 0.10.

will contain reduced quantities of fructose carbon. The preformed amino acid thus effectively competes with the fructose carbon for protein formation.

b. Competition between exogenous amino acids and internal pool material.

Since the flow of carbon from fructose has been demonstrated to pass through the amino acid internal pool the effects of exogenous amino acids upon pool formation and utilization have been investigated.

An exponentially growing culture of *Candida utilis* immersed for 10 min in medium containing [^{14}C]fructose, was washed, divided into two equal fractions, and quickly transferred to fresh C medium containing [^{12}C]fructose. More than 70% of

References p. 645

the total ^{14}C in the cells was found in the TCA-soluble fraction. To one culture was added a mixture of [^{12}C]amino acids at concentrations as shown in Table I. Samples were removed from both cultures at intervals following this transfer.

Table I shows the loss of ^{14}C from the TCA-soluble fraction for both the control culture and the culture containing the exogenous [^{12}C]amino acids. No difference could be detected in the loss rates in the two cultures. Both cultures continued to grow exponentially at the same rate as observed prior to and during the labeling period.

Measurements were made of the radioactivity contained in individual amino

TABLE II

EFFECT OF EXOGENOUS [^{12}C]AMINO ACIDS ON THE APPEARANCE OF RADIOCARBON IN PROTEIN AMINO ACIDS

Protein amino acid	*Radioactivity in protein amino acids (radioactivity in counts/sec)*	
	*[^{12}C]Amino acids present**	*No amino acids added*
Glutamic acid	7.1	7.0
Threonine	6.5	6.8
Aspartic acid	13.0	11.5
Arginine	4.5	4.2
Alanine	7.0	6.5
Glycine	8.2	8.3

* [^{12}C]Amino acids present at concentrations as shown in Table I.

acids of the proteins obtained from cells at the end of the experiment. Table II shows that no significant differences exist in the distribution of ^{14}C in these amino acids of the two cultures.

This experiment demonstrates that the exogenous [^{12}C]amino acids have no effect upon the transfer of pool radiocarbon to the protein amino acids. The competitive effects of exogenous amino acids observed when both [^{14}C]fructose and [^{12}C]-amino acids are present during the labeling period therefore occur either before, or at the time the amino acids enter the internal pool. Those amino acids already associated in the pool are not affected by the presence of the exogenous competitors.

More direct evidence confirming this conclusion may be obtained by following the fate of the carbon of a single pool amino acid. When exponentially growing cells briefly immersed in medium containing carrier-free high-specific radioactive glutamic acid were washed and transferred to fresh C medium containing no radioactive supplement, there was a rapid loss of the labeled glutamic acid of the pool as shown in Fig. 4. This radiocarbon eventually appeared as protein [^{14}C]glutamic acid, [^{14}C]-arginine and [^{14}C]proline.

The same rate of loss of pool [^{14}C]glutamic acid occurs whether or not [^{12}C]arginine is present. Similar evidence that the exogenous competitor does not influence the flow of the pool material was obtained using cells labeled with carrier-free [^{14}C]aspartic and exogenous [^{12}C]threonine, also shown in Fig. 4.

It may be argued that the exogenous amino acids do not enter the cell rapidly

References p. 645

enough to compete with the pool amino acids. The kinetics of incorporation of [^{14}C]-glutamic acid, shown in Fig. 5 indicate that such is not the case. The exogenous amino acid is rapidly incorporated by the cell with significant quantities appearing in the TCA-soluble and protein fractions within a few minutes incubation.

When higher concentrations of exogenous amino acid are available, the quantity of amino acids contained in the cold TCA-soluble fraction increases. Table III

TABLE III

DISTRIBUTION OF RADIOCARBON AMONG POOL AND PROTEIN AMINO ACIDS

Cells grown from light inoculum to approx. 2.3 mg wet weight of cells per ml medium.

Concentration of exogenous [^{14}C]threonine (µmoles/ml medium)	µmoles [^{14}C]threonine/g dry wt. cells*	
	Cold TCA-soluble fraction	Cold TCA-precipitable fraction
50.0	442	520
8.4	354	446
6.7	268	328
5.0	190	272
3.4	104	149
1.7	32	82
0.8	9	41

* Calculated on the basis that all the radiocarbon incorporated remained [^{14}C]threonine.

shows the ^{14}C content of the cold TCA-soluble fraction for various concentrations of exogenous [^{14}C]threonine. Other amino acids may be also concentrated in the cold TCA-soluble fraction as shown in Table IV. When these accumulations are compared with the steady-state amino acid concentrations in the internal pool shown in Table V, it is apparent that not only are exogenous amino acids rapidly incorporated (Fig. 5) but that accumulation to levels greatly in excess of the internal-pool concentrations are possible.

TABLE IV

AMINO ACID ACCUMULATION OF THE COLD TCA-SOLUBLE FRACTION

In each experiment, cells grown exponentially for 150 min in C medium containing the labeled amino acid.

Experiment	Exogenous [^{14}C]amino acid	Concentration (µmoles/ml medium)	Cold TCA-soluble fraction (µmoles/g dry wt. cells)
1	Arginine	1.1	158
2	Proline	1.7	170
3	Valine	1.7	225

References p. 645

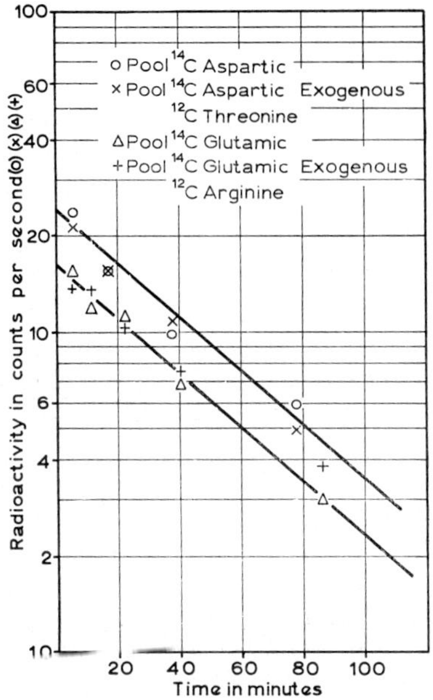

Fig. 4. Loss of [^{14}C]glutamic and [^{14}C]aspartic acid carbon from the cold TCA-soluble fraction. Data obtained from cells (approx. 1.0 mg wet wt./ml medium at $t = 0$) briefly immersed in medium containing either [^{14}C]glutamic or aspartic acid and transferred after washing to non-radioactive medium containing amino acid supplements (1.0 mg/ml medium) or to non-radioactive medium without amino acid supplements. ○, pool [^{14}C]aspartic; ×, pool [^{14}C]aspartic plus exogenous [^{12}C]threonine; △, pool [^{14}C]glutamic; +, pool [^{14}C]glutamic plus exogenous [^{12}C]arginine.

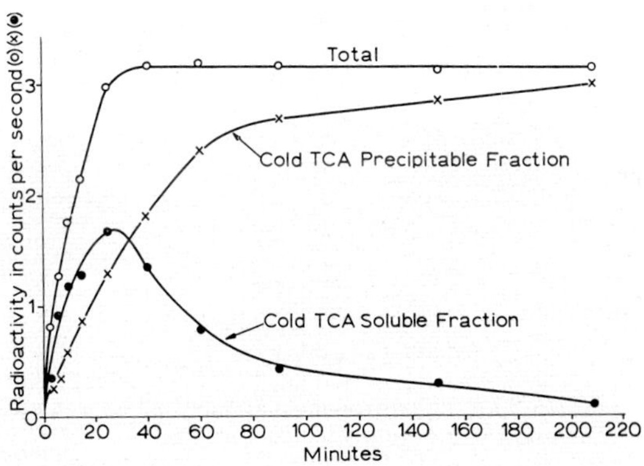

Fig. 5. Time course of incorporation of [^{14}C]glutamic acid. Initial exogenous concentration was 0.002 mg glutamic acid/ml medium.

DISCUSSION

Endogenously formed amino acids

At some point in the synthesis from fructose the family heads (glutamic, aspartic and pyruvic acids) are formed and become available to the internal pool where

References p. 645

conversion to their respective family members occurs. During exponential growth the production of the "parental" amino acids is sufficient to supply all the material required for the internal pool. This pool remains fixed in size and composition. Attempts to expand the pool by reducing the rate of protein synthesis by lowering the temperature quickly to 14° resulted in an equal reduction in the flow of internal-pool carbon. When the temperature was raised concurrent resumption of the synthesis of protein and pool occurred[6].

This close correlation between the internal pool and protein synthesis becomes even more apparent when exogenous supplements are employed to alter the fructose carbon flow.

Expandable pool material

When exogenous amino acids are present the cell accumulates these amino acids to levels exceeding their external concentrations. This system resembles the amino acid permeases found in *Escherichia coli*[7]. The amino acids accumulated are contained in a concentrating or "expandable" pool[4, 5] and have significant differences in

TABLE V

STEADY-STATE DISTRIBUTION OF RADIOCARBON AMONG POOL AND PROTEIN AMINO ACIDS*

Component	*Pool quantity of compound (μmoles/g dry wt.)*	*Protein quantity of compound (μmoles/g dry wt.)*
Isoleucine–leucine	117	785
Lysine	24	625
Glutamic acid	290	640
Aspartic acid	9	762
Valine	65	512
Alanine	240	695
Threonine	8	455
Serine	8	600
Proline	7	287
Arginine	63	210
Glycine	108	488
% accounted for	87	85

* Data obtained from cells growing exponentially for many generations in C medium containing [^{14}C]fructose.

their characteristics which permit their easy distinction from the internal pool amino acids.

1. The degree of accumulation is dependent upon the kind of exogenous amino acid supplied, their external concentration and the presence of other exogenous material. Its size and composition are therefore highly variable.

2. No amino acid interconversions take place in this pool. When radioactive threonine was made available (50 μmoles/ml medium; Table III, line 1) 90% of

References p. 645

the radioactivity of the TCA-soluble fraction was shown to be contained in threonine. The remainder was found in isoleucine, an amino acid that derives four of its carbons from the four carbons of aspartic acid. This isoleucine and a *small* fraction of the total pool threonine are believed to be components of the *internal* pool since the threonine and the isoleucine in the proteins of the cell both became equally radioactive[4].

3. Unlike the internal pool the material in the expandable pool is sensitive to osmotic shock, extractable with cold water, and readily exchanges with exogenous amino acids[4].

Diffusion

In addition to the material accumulated by the concentrating system amino acids enter the cell by diffusion. This process is most easily observed when the concentrating system becomes saturated (or is absent) and amino acids (or their analogs) contained in the cell increase directly in proportion to an increase in exogenous levels[5]. Indeed, as shown by HALVORSON AND COHEN[8], exogenous amino acids could be used preferentially for protein synthesis without equilibrating with preformed expandable pool material.

Thus at least three processes describe the means by which amino acids become available for incorporation in the internal pool: (a) endogenous formation of family head amino acids; (b) concentrating or permease system; (c) diffusion.

At high external concentrations of amino acids the probability of an endogenously synthesized family head amino acid entering the internal pool is very small. The accumulated amino acids in the expandable pool, and those which have diffused into the cell, are at such high levels that dilution of the endogenous material is great. In fact with both *E. coli*[1] and *C. utilis*[6], the addition of exogenous [^{12}C]amino acids (approx. 1 mg each amino acid/ml medium) in medium containing randomly ^{14}C-labeled sugar results in the appearance in the medium of large quantities of [^{14}C]glutamic acid, [^{14}C]aspartic acid and [^{14}C]alanine. Significantly only traces of ^{14}C-labeled end-product amino acids such as threonine, isoleucine, methionine, proline and arginine appear in the medium. Apparently the amino acid production system continues to synthesize ^{14}C parental amino acids despite the large excesses of [^{12}C]-amino acids present.

While a simple mass-law effect on equilibria is satisfactory as an explanation for the competitive reactions which occur between exogenous amino acid and fructose carbon, other explanations are necessary to account for the fact that accumulated amino acids do not rapidly equilibrate with internal pool material. It has been suggested that the amino acids contained in the *internal* pool form amino acid–protein or amino acid–nucleic acid–protein complexes which provide for this distinction[3, 4]. The large quantity of pool material precludes any small molecule–amino acid association. Only the macromolecules of the cell are present in sufficiently large amounts to provide for a single substance with which the bulk of the internal pool may be associated.

Furthermore, metabolic transformations have been observed among the individual amino acids of the pool. During incorporation of ^{14}C from the fructose of the medium the "heads" of each "family of amino acids" become radioactive before the other family members; later when the source of ^{14}C is removed from the medium these

References p. 645

members gain in radioactivity at the expense of the family "head". These transformations are not affected by the presence of external amino acids. It can be concluded therefore that the amino acid–macromolecule complex once formed is not broken during the series of reactions required for the conversion of one pool amino acid to another.

The schematic diagram shown in Fig. 6 is useful as an aid in presenting one interpretation of the results cited above.

An exponentially growing cell rapidily forms from fructose, family head amino acids which are adsorbed on specific sites on protein surfaces. Aspartic acid, for example will be adsorbed on several types of sites indicated in the figure as ASP,

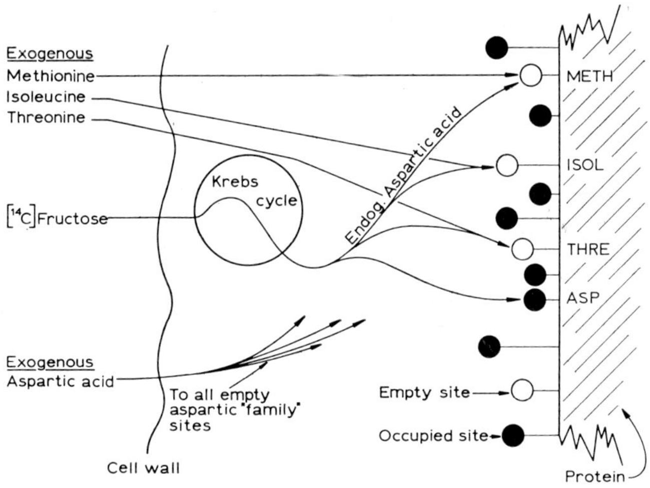

Fig. 6. Site model to explain behavior of aspartic family of amino acids.

THRE, METH and ISOL. When adsorbed on ASP it ultimately is incorporated into protein as aspartic acid. Adsorbed on a METH site, however, it must be transformed to methionine before incorporation into protein occurs. Conversion of other aspartic acid molecules on sites THRE and ISOL to the amino acids threonine and isoleucine respectively, are also necessary before final protein utilization.

If exogenous amino acids are also present in the medium they will compete with the endogenously formed amino acids (from fructose) for *empty* amino acid sites but will not displace amino acids already adsorbed. For example exogenous [^{12}C]-methionine could fill any unoccupied METH site and thus exclude ^{14}C derived from [^{14}C]fructose from reaching protein methionine. However exogenous [^{12}C]methionine would not displace either aspartic acid or methionine already adsorbed on the METH sites and hence could not compete with radioactive pool materials. Only when the pool methionine is withdrawn for protein incorporation (and a site left unoccupied) may another aspartic acid (or methionine) be adsorbed.

The "site" described in the above model obviously represents a very complex system. For example the aspartic-family pathway shown briefly below involves a number of enzymatic reactions:

References p. 645

ASP → ASP-P* → HS → HSP → threonine → α-keto butyric → isoleucine
↘
HC
↘
methionine

One interpretation of the system pictures the amino acid site as a separate, linked enzymatic system schematically indicated in Fig. 7. To account for the experimental observations no branched reactions as shown in the aspartic pathway above may exist. Once aspartic acid has entered the ISOL system its final destination must be isoleucine, and *not* methionine (or even threonine). An independent linked enzymatic system is required for the conversion of aspartic acid to each of the above amino acids. If this were not the case it would be possible to saturate the ASP sites shown in Fig. 6 with high concentrations of exogenous methionine and exclude any endo-

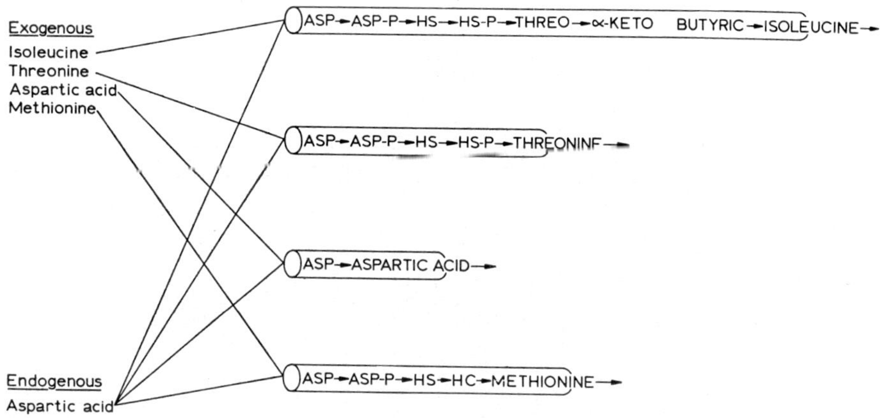

Fig. 7. Linked-enzyme system model.

genously formed aspartic acid from entering the system. Further evidence supporting the concept of separate linked biosynthetic enzymatic systems has been recently shown by STADTMAN *et al.* with their observation that in *E. coli* a different aspartokinase molecule is involved in the conversion of aspartic acid to threonine from that found in the biosynthesis of lysine from aspartic acid[9].

As in the site model, entry into the linked-enzyme system must occur at the first step of the synthetic events. The ISOL system will accept isoleucine and any precursor or intermediate of isoleucine but the ultimate product will be only isoleucine. Entry along any part of the conversion sequence would result in competitive effects, effects not detected in the conversion of internal-pool [^{14}C]aspartic acid to protein isoleucine.

Control of biosynthetic reactions

At least three kinds of mechanisms have been postulated which account for the preferential utilization of exogenous metabolites by microorganisms[10]: (1) simple *mass-law equilibria* along the biosynthetic chain; (2) cessation of *synthesis* of at

* Abbreviations: ASP-P: aspartyl-phosphate; HS: homoserine; HSP: homoserinephosphate; HC: homocysteine.

References p. 645

least one enzyme in the biosynthetic pathway; (3) inhibition of the *activity* of at least one enzyme in the biosynthetic pathway.

The mass-law equilibrium type of reaction certainly applies to the competitive effects observed among the endogenous and exogenous amino acids which occur prior to or at the time of incorporation of these amino acids into the internal pool.

The inhibition of biosynthetic enzyme *activity* through a negative feedback mechanism has been shown in a number of *in vitro* examples[10–14]. It is significant that in each of these investigations the end-product has been shown to inhibit the earliest step leading specifically to its own synthesis[10].

The data obtained from exponentially growing *Candida utilis* indicates that in these cells the earliest steps are naturally *maximally inhibited*. The steady-state

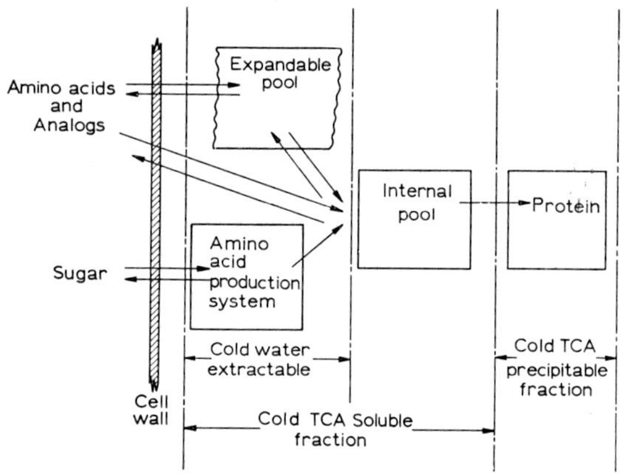

Fig. 8. Carbon flow in *C. utilis*.

production of family head amino acids is sufficient to saturate the internal (or conversion) pool of amino acids thereby preventing immediate entry of exogenous material along the chain of reactions of the biosynthetic pathway. *Entry at the first step of the linked enzyme system is possible, however, as long as the correct end-product (or any of its intermediates) is available and the site (or system) is empty.*

It is significant that at the 1961 Federation Meetings F. LYNEN reporting on the biosynthesis of fatty acids showed that an enzyme–substrate complex (molecular weight $\sim 2 \times 10^6$) was capable of at least five sequential biosynthetic reactions which could not be interrupted by the addition of intermediary compounds.

Carbon flow in Candida utilis

Fig. 8 summarizes what is believed to be the carbon flow in *C. utilis*. While both the expandable and internal pool appear to have functional differences, no direct chemical distinctions have ever been obtained even though a variety of extraction methods have been employed. Despite this the amino acids contained in the internal pool appear to be closely coupled to final protein formation. Attempts to prevent or interfere with their incorporation have always been negative. In fact it has been demonstrated that while these cells maintain two mechanisms selecting an amino

References p. 645

acid in preference to its analog[5], no further selection occurs after entry of the analogue into the internal pool. One wonders whether the amino acids and analogs contained in the internal pool have not already been selected by the protein-forming templates, but have yet to be linked together in polypeptide strands.

REFERENCES

1 R. B. Roberts, P. H. Abelson, D. B. Cowie, E. T. Bolton and R. J. Britten, *Carnegie Inst. Wash. Publ. No. 607*, 1955.
2 W. C. Schneider, *J. Biol. Chem.*, 161 (1954) 295.
3 D. B. Cowie and B. Walton, *Biochim. Biophys. Acta*, 21 (1956) 211.
4 D. B. Cowie and F. T. McClure, *Biochim. Biophys. Acta*, 31 (1959) 236.
5 E. S. Kempner and D. B. Cowie, *Biochim. Biophys. Acta*, 42 (1960) 401.
6 D. B. Cowie, unpublished results.
7 G. N. Cohen and J. Monod, *Bacteriol. Rev.*, 21 (1957) 169.
8 H. O. Halvorson and G. N. Cohen, *Ann. inst. Pasteur*, 95 (1958) 73.
9 E. R. Stadtman, G. N. Cohen, G. LeBras and H. de Robichon-Szulmajster, *J. Biol. Chem.*, 236, (1961).
10 H. E. Umbarger and B. Brown, *J. Biol. Chem.*, 233 (1958) 415.
11 H. E. Umbarger, *Science*, 123 (1956) 848.
12 R. E. Yates and A. B. Pardee, *J. Biol. Chem.*, 221 (1956) 757.
13 H. J. Strecker, *J. Biol. Chem.*, 225 (1957) 825.
14 H. S. Moyed and B. Magasanik, *J. Am. Chem. Soc.*, 79 (1957) 1812.
15 F. Lynen, *Federation Proc.*, 20 (1961) 544.

THE FUNCTION AND CONTROL OF INTRACELLULAR PROTEIN TURNOVER IN MICROORGANISMS

HARLYN O. HALVORSON

Department of Bacteriology, University of Wisconsin, Madison, Wisc. (U.S.A.)

Microorganisms contain metabolic pools of nucleotides and amino acids which are on the main line of synthesis of macromolecules. As had just been described in this Symposium by Dr. COWIE, these endogenous metabolic pools are derived either from exogenously provided components or synthesized *in vivo* from NH_3 and glucose. The recent demonstrations in bacteria, yeast and mammalian cells that under certain conditions nucleic acids and proteins are labile and are degraded to their constituent precursors (see review by MANDELSTAM[1]) provides yet another mechanism for internally replenishing these metabolic pools. Under conditions permitting protein and nucleic acid synthesis, a dynamic turnover exists due to the simultaneous breakdown and reutilization of the degradation products for synthesis.

Two of the central questions raised by the demonstration of protein and nucleic acid turnover are:

1. What are the control mechanisms governing its function?
2. Does this turnover have any significance to the physiology of the cell?

Demonstration of intracellular protein turnover

For the present discussion we will define protein turnover as the coupled degradation of protein to amino acids and their re-incorporation into newly synthesized protein. Given a population of microbial cells, at least three mechanisms can be envisaged for such turnover:

1. *Cell turnover:* $\text{Protein} \xrightarrow[\text{proteolysis}]{\text{cell lysis or secretion}} \text{AA} \xrightarrow[\text{intact cells}]{\text{remaining}} \text{protein}$

2. *Exchange incorporation:* $\text{Protein} \rightleftharpoons \text{AA}$

3. *Intracellular degradation and resynthesis:* k_2 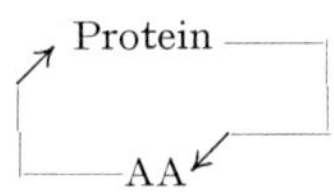 k_1

Exchange incorporation[2] was proposed to account for a chloramphenicol-resistant incorporation of glutamic acid and glycine into the hot trichloroacetic acid-soluble fraction of staphylococci. Presumably amino acids in the protein were replaced by identical free amino acids. Failure to demonstrate such chloramphenicol-resistant

References p. 653/654

incorporation in *Escherichia coli*[3] and *Bacillus cereus*[4] led to the demonstration in staphylococci that in the presence of chloramphenicol, glutamic acid and glycine were involved in *de novo* cell wall synthesis rather than protein synthesis[3, 5].

Examples of turnover due either to cell turnover or to intracellular degradation and resynthesis exist in the microbial literature. In order to demonstrate the latter, it is necessary to show that cell turnover does not occur or is insignificant. An example of this has been provided in yeast[6] as shown in Fig. 1. When phenylalanine- or leucine-labeled yeast are incubated under conditions of carbon and nitrogen starvation, protein breakdown leads to a linear replenishment of the cold trichloroacetic acid-

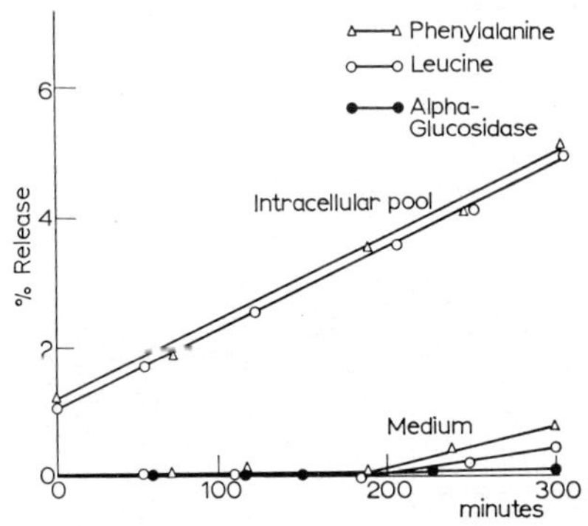

Fig. 1. Kinetics of replenishment of the free amino acid pool and the release of components into the medium. Data from HALVORSON[6].

soluble pool at the rate of 0.76%/h. Similar results were observed with a number of amino acids examined suggesting that overall protein degradation is taking place. The initial replenishment may involve the internal pool since exchange with the medium is minimal and addition of glucose leads to a rapid reincorporation. During the extended protein degradation, the level of the intracellular amino acids rises and eventually amino acids appear in the medium. It is not clear whether the amino acids flow through the expandable pool during their excretion.

Cell lysis could be ignored in the experiment in Fig. 1 since: (a) only traces of amino acids are released to the medium; (b) an intracellular inducible enzyme, α-glucosidase, appeared in the medium only at the rate of 0.03%/h; and (c) single cell isolates from 5-h starved cells showed only 2 out of 124 cells to be non-viable. MANDELSTAM[3] arrived at similar conclusions from the failure to detect β-galactosidase release during protein turnover in *E. coli*. From these experiments it seems clear that the replenishment of the metabolic pool in non-dividing cells arises primarily from intracellular protein degradation.

Synthetic capacities under conditions of limited growth

The rate of protein turnover can be measured either by measurements of protein degradation (k_1) in the presence of a sufficient trap for amino acids or by direct measurement of reincorporation (k_2). This is illustrated in Fig. 2. When yeast cells are briefly exposed to labeled glycine and then incubated in N-free medium, there is a rapid transfer of label from the free amino acid pool to the protein fraction, a synthesis of the inducible enzyme α-glucosidase, and a dilution in the specific activity of the glycine free amino acid pool. In *E. coli* both protein degradation and reincorporation occur at the rate of 5%/h (ref. 3). Chloramphenicol inhibits reincorporation but does not abolish protein degradation.

Is protein synthesis preferential in non-dividing cells? Several lines of experimentation provide a negative answer to this question. First, the differential rate of synthesis ($\Delta E/\Delta$ newly formed protein) of β-galactosidase in *E. coli* and of α-glucosidase in yeast (HALVORSON, unpublished results) is the same for growing and non-

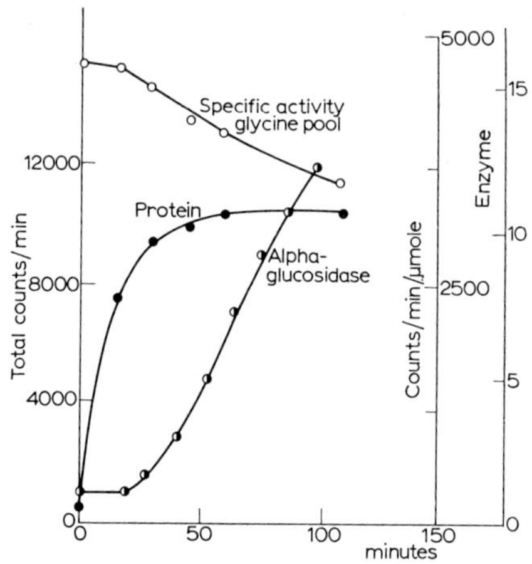

Fig. 2. Utilization of [2-^{14}C]glycine-labeled pools during nitrogen starvation. Data from HALVORSON[6].

growing cells. Secondly, JANECEK[8] has shown in *E. coli* that the electrophoretic pattern of [^{14}C]phenylalanine incorporation into proteins under limited nitrogen supply is identical to that observed in exponentially growing cells.

Effect of growth rate on protein turnover

It was initially observed by PODOLSKY[9] that, when arginine-labeled *E. coli* was transferred to unlabeled medium, approximately 8% of the isotope was lost from the cells over a long semi-stationary phase. From these data MANDELSTAM[1] calculated a turnover rate of 5%/h. Protein turnover rates in non-dividing cells of between 2–5%/h have been reported for *E. coli*[3, 10, 11], 7%/h for *B. cereus*[4] and slime molds[12] and 0.7%/h for yeast[6].

In rapidly growing cultures of bacteria and yeast, the proteins are essentially stable and accurate measurements of turnover are difficult. The assumptions and limitations of the methods employed have been discussed elsewhere[1, 13, 14]. Employing analysis of the kinetics of dilution of prelabeled pools and also an internal trapping method, the rate of protein turnover in rapidly growing cells has been estimated to be less than 0.1%/h (refs. 13–16). FOX AND BROWN[17], employing a less sensitive method, estimated the rate of protein turnover in *E. coli* to be approx. 2.7%/h. One clear exception has appeared. URBÁ[4] has shown that in growing *B. cereus* the turnover rate is 1.4%/h.

Although the turnover rates in growing cells are uncertain, one can conclude from the present data that the absolute rate of protein turnover is appreciably greater in non-dividing cells than in growing cells.

Degradation of cellular components

Do all of the cellular proteins participate in intracellular turnover? That heterogeneity towards protein turnover exists is suggested by the findings that under conditions in which protein turnover is taking place, certain inducible enzymes are stable in the absence of their inducer[1, 18, 19] whereas others[20, 21], as well as the alcohol-soluble proteins of *E. coli*[22], are labile.

TABLE I

TURNOVER OF PROTEIN AND NUCLEIC ACID FRACTIONS IN NITROGEN-STARVED BACTERIA

Fractions	*% breakdown**		*% synthesis**	
	RNA	*Protein*	*RNA*	*Protein*
Soluble	5	25	14.5	16.5
Ribosomal	24	21.5	6.0	21.5

* 4 h starved. Data from MANDELSTAM AND HALVORSON[23].

Turnover can involve a significant proportion of cellular protein. Approximately half of the total protein of *B. cereus*[4] and one quarter of the protein of *E. coli*[3] are subject to turnover. Such turnover involves both structural (ribosomal particles) and soluble proteins[23, 24]. As shown in Table I, in non-dividing *E. coli* the proteins of the ribosomes and of the soluble fraction are equally labile and both are degraded at approx. 5%/h. This process is approximately balanced by the rate of resynthesis. Ribosomal nucleic acid is degraded at about the same rate as protein, however, the rate of synthesis of RNA occurs at a rate of only 1.5%/h. Intracellular stability may depend upon starvation conditions. In *E. coli* the alcohol-soluble protein is labile during sulfur starvation[22] but stable in nitrogen starvation.[1]

Initiation of turnover

At least two mechanisms can be involved to explain the increased turnover rate in non-dividing cells:

1. Synthesis of intracellular degradation system(s) by induction or depression.
2. Activation of the pre-existing degradative system(s) because of the removal of an endogenous inhibitor or unmasking from a latent form.

Some insight into this problem can be achieved by examining the conditions which lead to intracellular breakdown and turnover. In general, specific starvations, imposed either through exhaustion of nutrients, by resuspending cells in deficient medium, or by the use of selective inhibitors lead to both RNA and protein degradation. Table II summarizes some of the methods employed. In many cases both RNA and protein breakdown have not been followed in the same system. It is clear that imposing conditions which lead to unbalanced synthesis favor intracellular breakdown. For example, BARNER AND COHEN[26] showed that when *E. coli* strain $15z_{T-U-}$ was incubated in the absence of pyrimidines, protein synthesis continued coupled with an increased turnover of RNA.

One would expect that if intracellular turnover was dependent upon the *de novo* synthesis of an essential degradative enzyme(s) that its production should be dependent upon protein synthesis. The data summarized in Table II argue against this

TABLE II

METHODS FOR INITIATING BREAKDOWN AND TURNOVER

Conditions employed	*Degradation*		*Reference*
	Protein	*RNA*	
Starvation			
Carbon	+	+	6
Nitrogen	+	+	6, 23
Specific amino acids	+	+	3, 11, 24
Sporulation	+	+	25
Pyrimidines	0	+	26
Mg^{2+}		+	27
Phosphate		+	28
Inhibitors			
Penicillin	+		16
Amino acid analogs	+		29
Phage infection	±	+	30

hypothesis. When rapidly growing cells are incubated under conditions restricting protein synthesis (nitrogen and amino acid starvation; in the presence of amino acid analogs), protein breakdown ensues. Since a limited synthesis of protein might take place during the starvation of the intracellular amino acid pool, such observations do not directly rule out the protein synthesis hypothesis. Chloramphenicol, which immediately blocks protein synthesis was shown by MANDELSTAM[3] to have no effect on the rate of protein breakdown in *E. coli* during leucine starvation. After 90 min a progressive inhibition of protein breakdown was observed in the presence of chloramphenicol which is probably due to an indirect effect. HANCOCK[31] observed a chloramphenicol-induced accumulation of pool amino acids in *Staphylococcus aureus*, however, it is not clear in this case to which extent the free amino acids are derived from cellular proteins. Finally, dosages of ultraviolet light, which inhibit protein synthesis, lead to an increase in the free amino acid pool of yeast via protein breakdown[32].

References p. 653/654

Metabolic regulation of turnover

The synthesis of protein and nucleic acids are coupled to energy-yielding reactions. Thus restricting the energy supply reduces the rate of reincorporation of pool components into macromolecules. Of particular interest was the finding that the breakdown of protein and nucleic acid in non-dividing cells is suppressed by agents which inhibit energy-yielding reactions[3, 6, 33, 34]. An example of this is shown in Table III. When prelabeled yeast is incubated with 2,4-dinitrophenol (DNP), azide or arsenate, protein and nucleic acid degradation is almost completely abolished. MANDELSTAM[3] has observed a similar effect with DNP in *E. coli*. Azide inhibited protein breakdown only after a 45-min delay suggesting that it causes a gradual accumulation within the cell of some inhibitory metabolite.

TABLE III

EFFECT OF ENERGY INHIBITORS ON PROTEIN AND NUCLEIC ACID BREAKDOWN IN YEAST

Data from HALVORSON[6]

Incubation conditions	*Starvation period (h)*	*% inhibition of release*	
		Protein	*Nucleic acid*
$4 \cdot 10^{-4}$ *M* dinitrophenol	1	89	
	2	91	
	4	92	98
$2 \cdot 10^{-2}$ *M* azide	5		100
$2 \cdot 10^{-2}$ *M* arsenate	5		100

The above inhibition of the degradative system(s) can be understood either if its activation or function required energy or if its activity were regulated by metabolites whose levels are controlled by inhibitors. An energy dependence of the degradative system(s) would seem unlikely on several grounds. First, as will be discussed later, RNA and protein degradation appear to be mediated by strictly hydrolytic reactions. Secondly, at least for RNA degradation, conditions can be established in which these inhibitors have no effect. HORIUCHI[28] observed a 5%/h breakdown of RNA in *E. coli* suspended in phosphate-free medium which was unaffected by DNP and azide and stimulated by arsenate.

If the degradative system(s) is regulated by the intracellular concentration of low molecular-weight metabolites then conditions which alter the steady-state level of metabolic pools should influence degradation. Examples of this occurs during glucose metabolism under reduced growth rate or during selective starvation when anabolic reactions are unlinked from catabolic reactions. Phosphorylated metabolites may play an essential role in this regulation since inhibitors of oxidative phosphorylation inhibit breakdown and phosphate deficiency both stimulates this reaction and renders the system insensitive to such inhibitors. These observations suggest that the activation of the degradative reactions may be linked to reduced levels of some phosphorylated metabolites.

References p. 653/654

Mechanism of protein and RNA degradation

From the previous discussion it is clear that the degradative enzymes pre-exist to the initiation of RNA and protein turnover. This was confirmed by the observations that the same content of proteolytic activity exists in extracts of growing and non-dividing cells of *E. coli*[35]. The *in vitro* rate of degradation was about equal to that observed *in vivo* with non-dividing cells. Proteolytic activity in extracts was increased by conditions leading to a breakdown in cellular structures. An explanation for this phenomenon has been recently provided by the demonstration of a latent aminopeptidase associated with *E. coli* ribosomes[36], which are themselves labile. In yeast ribosomal disruption activates the proteolytic actively (HALVORSON, unpublished results).

A similar situation has been observed for ribonuclease. This enzyme is almost exclusively present in bacteria[37–39] and yeast[40] as a latent, ribosomal-bound enzyme. These same particles contain approx. 90% of the RNA and 30% of the protein of microbial cells[41]. Ribosomal stability is dependant upon the concentration of Mg^{2+} (refs. 27, 41). When the Mg^{2+} is removed the larger particles dissociate into smaller particles, the latent enzymes are released and the fragments eventually undergo autodegradation[42]. When intact cells are starved for carbon and nitrogen or Mg^{2+} (refs. 27, 43), an analogous ribosomal breakdown is observed. When starved cells are transferred to complete medium, the larger ribosomal particles are stabilized.

These findings provide a model for turnover based on the stability of the ribosomal particles[1]. In actively growing cells ribosomes are intact and the hydrolytic enzymes inactive. When cells are starved, a fall in the level of phosphorylated intermediates follows which is linked either to Mg^{2+} removal or some other process which renders the ribosomal particles labile. During the breakdown of these particles, RNA, protein and hydrolytic enzymes are liberated.

Support for this scheme is provided by the observation *in vivo* that in non-growing *E. coli* there is a balanced degradation of both RNA and protein[23, 24]. Degradation of the ribosomal components proceeded at the same rate as the soluble components. The observation that under some conditions differential breakdown occurs (Table II) suggests that the hydrolases can either be selectively inhibited or that alternate degradation mechanisms exist.

Significance of intracellular turnover

Intracellular turnover provides a mechanism whereby the phenotype of a cell can be changed in non-dividing cells. Under such conditions growth is dependent upon the turnover rate; the amino acids and nucleotides required for synthesis and growth are supplied by protein and RNA breakdown. In several cases, such as endotropic sporulation[25] and differentiation in slime molds[12], cellular morphogenesis is observed in non-dividing cells which involves changes in cell morphology, chemical composition, enzyme patterns, metabolism and in the types and distribution of macromolecules. It is of interest that these occur in cells with high turnover rates. To a lesser extent, some of the precursors required for the synthesis of phage in infected cells are also derived from endogenous macromolecules[30].

Turnover can also lead to selective enrichment of an enzyme in non-dividing cells

References p. 653/654

in several ways. If, under conditions of turnover, an inducer is added[7] or depression established, such as phosphate starvation for alkaline phosphatase synthesis[44], reincorporation is accompanied by an increased differential rate of synthesis of a particular enzyme. A second type of enrichment can occur following extensive turnover if the newly synthesized enzyme is selectively stable to degradation[18, 19]. An example of this may occur in the case of β-galactosidase in *E. coli*.

The interesting question remains as to whether turnover exists in rapidly growing cultures and if so, whether it has any physiological significance. This problem is technically difficult and any turnover if it occurs is at a low rate. In recent years several types of RNA have been shown to participate in protein synthesis and in turnover, at least in part, *in vivo*[45]. Of particular interest are the observations of SYLVEN *et al.*[46] who observed a cyclic variation in the activities of peptidase and proteinase during the synchronous growth of yeast. The highest cellular levels of enzyme occurred immediately prior to cell division, declining to minimum levels during subsequent budding. These enzymes could provide alternating conditions for intracellular protein degradation associated with a particular stage in cellular division. Similar variations have been observed in the phosphatase activity of yeast[47] and in the metaphosphate content of *Corynebacterium diphtheriae*[48] during synchronous cell division.

ACKNOWLEDGEMENTS

This work was supported in part by research grants from the National Institute of Health, National Science Foundation and the U.S. Air Force Office of Scientific Research of the Air Research and Development Command.

REFERENCES

1 J. MANDELSTAM, *Bacteriol. Rev.*, 24 (1960) 289.
2 E. F. GALE AND J. P. FOLKES, *Biochem. J.*, 55 (1953) 721.
3 J. MANDELSTAM, *Biochem. J.*, 69 (1958) 110.
4 R. C. URBÁ, *Biochem. J.*, 71 (1959) 513.
5 R. HANCOCK AND J. T. PARK, *Nature*, 181 (1958) 1050.
6 H. HALVORSON, *Biochim. Biophys. Acta*, 27 (1958) 255.
7 J. MANDELSTAM, *Nature*, 179 (1957) 1179.
8 J. JANEČEK, *Biochem. Biophys. Research Communs.*, 2 (1960) 349.
9 R. J. PODOLSKY, *Arch. Biochem. Biophys.*, 45 (1953) 327.
10 E. BOREK, L. PONTICORVO AND D. RITTENBERG, *Proc. Natl. Acad. Sci. U.S.*, 44 (1958) 369.
11 J. CHALOUPKA, *Folia microbiol. (Prague)*, 5 (1960) 287.
12 B. E. WRIGHT AND M. L. ANDERSON, *Biochim. Biophys. Acta*, 43 (1960) 67.
13 A. L. KOCH AND H. R. LEVY, *J. Biol. Chem.*, 217 (1955) 947.
14 H. HALVORSON, *Biochim. Biophys. Acta*, 27 (1958) 267.
15 D. B. COWIE AND B. P. WALTON, *Biochim. Biophys. Acta*, 21 (1956) 211.
16 R. CHAYEN, J. CHAYEN AND E. R. ROBERTS, *Biochim. Biophys. Acta*, 31 (1959) 186.
17 G. FOX AND J. W. BROWN, *Biochim. Biophys. Acta*, 46 (1961) 387.
18 D. S. HOGNESS, M. COHN AND J. MONOD, *Biochim. Biophys. Acta*, 16 (1955) 99.
19 B. ROTMAN AND S. SPIEGELMAN, *J. Bacteriol.*, 68 (1954) 419.
20 A. I. VIRTANEN AND J. DELEY, *Arch. Biochem.*, 16 (1948) 169.
21 J. J. ROBERTSON AND H. O. HALVORSON, *J. Bacteriol.*, 73 (1957) 186.
22 R. B. ROBERTS, P. H. ABELSON, D. B. COWIE, E. T. BOLTON AND R. J. BRITTEN, *Carnegie Inst. Wash. Publ.*, No. 607 (1955).
23 J. MANDELSTAM AND H. O. HALVORSON, *Biochim. Biophys. Acta*, 40 (1960) 43.
24 A. GOLDSTEIN AND B. J. BROWN, *Biochim. Biophys. Acta*, 44 (1960) 491.
25 J. W. FOSTER AND J. J. PERRY, *J. Bacteriol.*, 67 (1954) 295.

[26] H. D. Barner and S. S. Cohen, *Biochim. Biophys. Acta,* 30 (1958) 12.
[27] S. Dagley and J. Skyes, *Nature,* 179 (1957) 1249.
[28] T. Horiuchi, *J. Biochem.,* 11 (1959) 1467.
[29] G. N. Cohen, H. O. Halvorson and S. Spiegelman, in R. B. Roberts, *Microsomal Particles and Protein Synthesis,* Pergamon Press, New York-London, 1958, p. 100.
[30] A. D. Hershey, *J. Gen. Physiol.,* 37 (1953) 1.
[31] R. Hancock, *Biochim. Biophys. Acta,* 37 (1960) 47.
[32] H. Halvorson, W. Fry and D. Schwemmin, *J. Gen. Physiol.,* 38 (1955) 549.
[33] M. V. Simpson, *J. Biol. Chem.,* 201 (1953) 143.
[34] D. Steinberg and M. Vaughan, *Biochim. Biophys. Acta,* 19 (1956) 584.
[35] J. Chaloupka and J. Liebster, *Folia microbiol. (Prague),* 4 (1959) 167.
[36] E. T. Bolton, R. J. Britten, D. B. Cowie, B. J. McCarthy, R. McQuillen and R. B. Roberts, *Carnegie Inst. Wash. Yearbook,* 58 (1959) 275.
[37] Y. Tashiro, *J. Biochem.,* 43 (1958) 937.
[38] D. Elson, *Biochim. Biophys. Acta,* 27 (1959) 216.
[39] P. F. Spahr and B. R. Hollingworth, *J. Biol. Chem.,* 236 (1961) 823.
[40] J. J. Stenesh, *Thesis,* University of California, Berkeley, 1958.
[41] A. Tissières and J. D. Watson, *Nature,* 182 (1958) 778.
[42] H. E. Wade, *Biochem. J.,* 78 (1961) 457.
[43] J. K. Ashikawa, in R. B. Roberts, *Microsomal Particles and Protein Synthesis,* Pergamon Press, New York-London, 1958, p. 76.
[44] A. Torriani, *Biochim. Biophys. Acta,* 38 (1960) 460.
[45] G. N. Cohen and F. Gros, *Ann. Rev. Biochem.,* 29 (1960) 525.
[46] B. Sylven, C. A. Tobias, H. Malmgren, R. Ottoson and B. Thorell, *Exptl. Cell Research,* 16 (1959) 75.
[47] H. Suomalaenen, M. Linko and E. Oura, *Biochim. Biophys. Acta,* 47 (1960) 482.
[48] S. Sall, S. Mudd and A. Takagi, *J. Bacteriol.,* 76 (1958) 640.

DISCUSSION

Chairman: JOHN REINER

REINER: I would like to exercise the privilege of the Chair and ask Dr. COWIE about this suggested set of tunnels here. You are assuming, as I understand it, that up to a point, namely the point at which the final amino acid is synthesized, all the reactions are really the same in all the tunnels. Are you taking care of the fact that things go just the right way by implying that some mechanism picks off the final product just as soon as it is formed, or are you proposing that a varying number of terminal enzymes are missing in the various tunnels?

COWIE: Every attempt to determine the state of the amino acids in the internal pool during their transfer to protein has been negative. We have, as I have indicated, specifically labeled the internal pool and tried, without success, to interfere with its transfer to protein by adding a large quantity of exogenous material. Furthermore, amino acid analogs, once in the internal pool show no final selection steps for protein incorporation even though discrete selection steps could be observed prior to internal pool incorporation of these analogues. One interpretation of the model presented is that the amino acids in the internal pool (or in the "tunnels") are already on the final sites awaiting appropriate conversion and final peptide linkage.

HENDLER: Dr. COWIE, could you give me some idea of your thoughts on the type of bonds which may be involved in holding your internal pool in terms, for example, of the type of washing procedures which leave it intact or remove it?

COWIE: One of the distinctions between the two kinds of amino acid pools is the following: hot water or 5 per cent cold TCA will extract both pools together; cold water will extract only the expandable pool, leaving intact the internal pool. We have no knowledge of what kind of affinity or association these amino acids form to account for this distinction or how the internal pool material is adsorbed. Perhaps VAN DER WAAL forces or hydrogen binding might be the answer.

A. MILLER: I would like to ask Dr. COWIE if he has considered models that go beyond the macromolecular stage. There is a great deal of fine structure in an organism such as *Candida*.

COWIE: I think I am in enough trouble already, but since you ask, I would like to suggest that the pool might be associated with the cellular ribosomes; that the internal pool is actually complexed with the protein-forming system, and that the ribosomes are indeed this system.

A. MILLER: I was wondering if you had considered models in which the external amino acids enter vacuoles rather than mix with the cell contents.

COWIE: We have looked into all kinds of partitioning systems in which the amino acids of the internal pool are in bags or sacks and even in vacuoles. When you do this, however, these sacks or vacuoles must be able to account for all the experimental data available. The concentrating mechanism, the conversion system, and the final transfer from the bag or sack to protein (without competition) must be accounted for. It would be a pleasure for me if someone would use the bag or vacuole system and construct a satisfactory model taking into account all the information on the pools that has accumulated. So far, a bag system has not been postulated which reasonably accounts for the kinetic and other experimental evidence observed.

A. MILLER: I would like to ask Dr. HALVORSON if he has any data on recovery from the increased turnover during starvation. How does the cell get rid of all these lytic enzymes?

HALVORSON: I really have no information as to the level of the proteolytic activity on reinoculation into fresh media. In our yeast experiments we have tried to make measurements of the proteolytic activity as a function of growth; that is, the $\triangle$proteolytic activity/$\triangle$ mass. We found that this ratio dramatically increases as the cell begins to decrease its growth rate, so that you get what looks like a preferential synthesis but which we expect is a release of enzyme from the particulate fraction. When the cells are put into fresh growth media, this ratio again reverts to normal. In the experiments of SYLVEN *et al.* one wonders whether the proteolytic systems themselves are subject to autodegradation. Another possibility is that the proteolytic enzymes are masked in some way when functional particles are reformed.

We have had a similar problem in examining de-adaptation in yeast where, upon removal of the inducer, a net loss of the enzyme occurred. The question of whether this enzyme is being degraded or inactivated is difficult to answer. We have studied this protein using specific antisera against the enzyme but this has not as yet convincingly established whether the loss of activity is accompanied by the degradation of the apoenzyme. The same question would remain for the proteolytic enzymes.

HOROWITZ: Dr. COWIE, did I understand you to say that *Candida* is maximally repressed? The reason I ask is that I have almost arrived at the conclusion that feedback repression is shown only in bacteria, and if there is evidence that it occurs also in yeast, this would be very interesting.

COWIE: If I did, I made a mistake. I am only talking about inhibition. There are two processes one has to consider; one is the *synthesis* of an enzyme, the other the *activity* of the enzyme *per se*. My remarks are restricted to the *activity* of the enzymic system. It is postulated that there is sufficient material provided by the endogenous flow of carbon to keep the internal pool, which is fixed in size, almost completely saturated during exponential growth. This material as well as exogenous amino acids may compete to form the internal pool by complexing with protein (and nucleic acid). This complex prevents any further competition until protein is formed and these sites again become available for competitive selection.

The kind of negative feedback reactions observed *in vitro* takes place in systems which have lost the internal organization of the living cell. In such cellular extracts one can introduce materials which inhibit specific enzyme reactions. In the living cell the system appears to remain saturated and complexed until proteins are made, and inhibition by end-product amino acids cannot occur as long as the complex sites are filled.

HOROWITZ: Do you know of any evidence in yeast that indicates derepression of the kind that has been found in bacterial mutants?

COWIE: Not to my knowledge.

HALVORSON: We have performed experiments designed to test over-all repression in yeast. Two cultures of yeast cells were grown, the first on synthetic media and the second on a rich mixture of amino acids, vitamins and nucleic acid precursors. The proteins of both were extracted and fractionated on DEAE cellulose columns. We expected to find a significant difference between the two extracts as a result of repression of the biosynthesis enzymes. To our surprise the protein fractions were completely identical.

EAGLE: In animal cells also we have seen no example of negative feedback inhibition in the biosynthesis of serine, glycine, homocystine or proline. Large amounts of exogenous amino acid have no effect on neosynthesis from the appropriate precursor.

DYNAMICS OF AMINO ACIDS IN PLANTS

S. ARONOFF

Department of Biochemistry and Biophysics and the Institute for Atomic Research, Iowa State University, Ames, Iowa (U.S.A.)

Amino acid metabolism is among the very oldest of studies in plant biochemistry and physiology, going back to the discovery of asparagine in 1806[1]. The relation of the appearance of asparagine to protein catabolism was the object of controversy between SCHULZE and PFEFFER as early as the 1870's. The tremendous amount of subsequent analytical work required for even meager additional insight continued to almost 1940, and is summarized in CHIBNALL's classic *Silliman Lectures at Yale*[1]. Progress since then has been sporadic. Almost explosive efforts have been followed by periods of relative moderation, as is common in most areas of human endeavor. Thus, there was the pioneering work of STEWARD[2] in the paper-chromatographic elucidation of rare, and even novel amino acids in plants. Similarly, the VIRTANEN school has been prominent in the identification of heretofore unknown amino and imino acids[3].

However, it was only with the advent of isotopic techniques that a study of the kinetics of amino acid transformations could be made. Thus, it is only 11 years ago since CALVIN's group first demonstrated[4] the formation of some of the more common amino acids as being among the early products of photosynthesis by algae. And it is only this year that a kinetic study has appeared of the rate of formation of these amino acids under steady-state conditions[5]. Similarly, our initial studies on amino acid formation in leaves appeared a decade ago[6] and kinetic experiments are still in progress.

Changes in free amino acids during germination

The germination of dicotyledonous plants is accompanied by hydrolysis of protein in the cotyledon, and the amino acids thereby formed are assumed to be transported to regions of growth. This proteolysis was, in fact, much of the basis of the historical work noted above. However, more recent chromatographic evidence provides direct knowledge of the changes of the individual amino acids of germinating bean seeds[7] and lupines[8]. As shown in Table I, initially the seedlings are virtually devoid of free amino acids. Indeed, glutathione, notably scarce in mature tissue, is conspicuous by its relative prominence. The increase in asparagine was known historically, but the prominence of arginine was less understood as was that of the less obvious histidine or lysine. Furthermore, as might be expected now, quantitative differences occurred even as to direction. Thus, in the case of the bean, germination resulted in a decrease of free glycine, whereas in lupines, the same stage produced a decided increase. The comparisons are not completely valid, however, as the lupine was germinated in the dark to avoid the complications resulting from amino acid formation during photosynthesis. Other complications also occur. For example, in the light, arginine gra-

References p. 666

dually disappears. Its incorporation into protein formed during photosynthesis is real, though minor. In contrast arginine may be the *major* amino acid of seed protein. Nevertheless, to the extent to which seedlings are unable to photosynthesize until the maturation of chloroplasts, the type of analysis described above may have meaning.

Rough qualitative variation of the reciprocal relation (the variations in free amino acids in pea seeds during their development) has been given by ROBERTSON *et al.*[9]. The virtual disappearance of numerous amino acids coincides (Fig. 1) with the onset of rapid protein synthesis, so that what may be construed as a general lowering of the amino acid steady-state level is, in fact, a virtual extinction of some, followed by the eventual increase of others, as protein synthesis is concluded.

TABLE I

FREE AMINO ACIDS AND PROTEIN AMINO ACIDS OF GERMINATING ETIOLATED SEEDLINGS OF *Lupinus*

The proteins represent approx. 71 % of the seed N. From VENEKAMP[8].

Amino acid	*Free amino acids*				*Protein amino acids*			
Days after germination:	*0*	*7*	*14*	*21*	*0*	*7*	*14*	*21*
Leu	0	0	1	1	36	27	23	5
I-leu	0	1	1	1	12	11	14	5
Phe	0	2	4	6	15	13	12	5
Val	0	2	3	4	16	13	13	7
Met	0	0	0	0	2	1	1	1
Tyr	0	2	3	1	8	4	4	0
Pro	—	—	—	—	6	9	9	2
Glu	1	3	4	1	93	59	40	7
Ala	1	2	—	—	28	10	9	7
Thr	0	1	2	2	12	8	7	6
Asp	1	2	3	5	64	58	45	6
Ser	1	4	5	6	24	15	11	8
Gly	0	1	3	4	55	25	20	13
Asp-NH_2	0	98	169	351	—	—	—	—
Glu-NH_2	0	2	3	5	—	—	—	—
Arg	3	37	65	102	85	64	26	29
Lys	1	5	9	10	51	28	10	5
His	1	9	14	17	28	19	15	3
Cys	1	2	9	8	18	12	—	3
?	—	—	—	—	19	5	4	1

Leaf dynamics

Despite the most sensitive and precise chromatographic procedures, these methods alone provide only *net* changes, and it was only the use of isotopic tracers which allowed for considerations of metabolic rates. Thus it was shown in soybean[16] that in brief (15 sec) photosynthesis in $^{14}CO_2$, the ^{14}C content of the free amino acids exceeded that of the sugars, though the latter predominate within 5 min time. The major amino acids thus produced were alanine, serine and glycine, in decreasing

References p. 666

order of total and specific activity. Furthermore, the equivalent 15-sec activities of the α- and β-carbons of serine and of alanine, as well as the two of glycine, showed that considerable photosynthetic cycling had occurred even in that brief period. These rapid transformations were often lost in the changes of pool size with time. Thus, there was very little aspartic acid in the initial photosynthesis of soybean

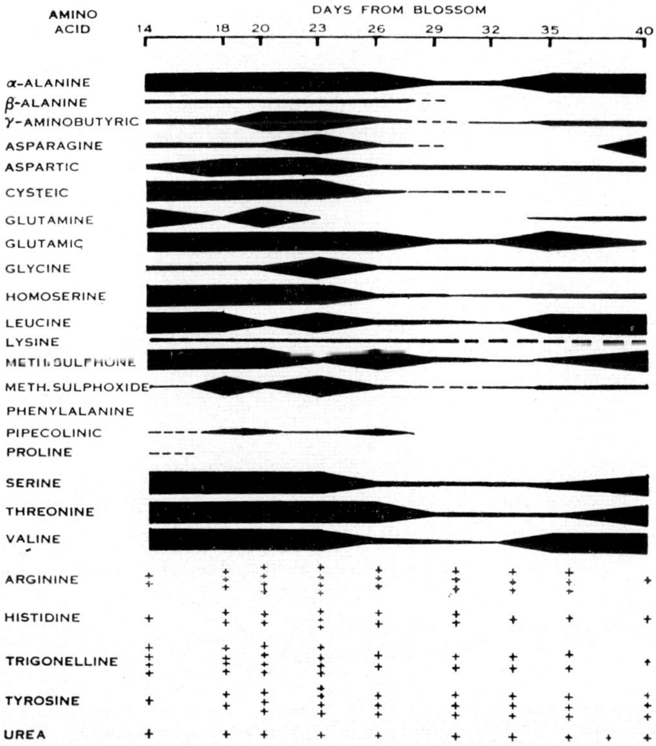

Fig. 1. Changes in amino acids of pea seeds with time after blossoming. The width of the line represents the amount of amino acid as estimated from the size of the chromatographic spot. From McKee, Nestel and Robertson[9].

leaves, and, on the contrary, there was considerable in tobacco, but the similarity of the distribution of radioactivity in the leaves of the two plants following photosynthesis was remarkable (Table II).

Furthermore, the change, with time, of the specific activities, showed that the size of the pools bore no simple relation to the rate of formation of the amino acids (Table III).

Indeed, a variety of models must be considered in interpreting such data. For example, pools may be either fixed or variable in size. If fixed, there may be only a single pool, or they may be multiple. If variable, the variability may be functions either of concentration, or time, or both; and without knowledge of the specific functions, discrete models cannot be constructed. The kinetics of fixed pools, while relatively straightforward in themselves, may be complicated by dilutions, *i.e.* synthetic pathways other than the one under consideration. In any event practical

References p. 666

considerations of the ratio of matter input to pool size must also be considered, especially when their values approach within a single magnitude of each other. Thus, it might be expected that alanine, which is known from short-time experiments to be in virtual equilibrium with the photosynthetic cycle, would have isotopic kinetics of the simplest type: a single input and output, where the log specific activity *versus* t should be linear. As Fig. 2 shows, this is certainly not the case. Furthermore,

TABLE II

DISTRIBUTION OF RADIOACTIVITY IN FREE AMINO ACIDS OF TOBACCO AND SOYBEAN LEAVES AFTER 1 H PHOTOSYNTHESIS IN $^{14}CO_2$

From RACUSEN AND ARONOFF, unpublished results.

Amino acid	*Soybean*	*Tobacco*
Asp	17.4	14.2
Asp-NH_2	11.1	14.7
Glu	14.3 }	31.3
Glu-NH_2	1.0 }	
Gly	3.5	6.4
Ser	44.5	25.5
Ala	8.0	7.9

how does one interpret the case of glycine, or aspartic acid, which virtually does not change over the entire period, or of serine, which appears to have an initial metabolism, followed by a fall to a relatively constant value? The simplest interpretation of the latter data is to presume the existence of isolated pools, *e.g.* the vacuoles,

TABLE III

SPECIFIC ACTIVITY OF FREE AMINO ACIDS OF TOBACCO LEAF PUNCHES AS A FUNCTION OF TIME FOLLOWING 2 MIN PHOTOSYNTHESIS IN $^{14}CO_2$

From RACUSEN AND ARONOFF, unpublished results.

Amino acid	*5 min*	*15 min*	*30 min*	*60 min*
Asp	72.0	85.4	75.6	69.2
Asp-NH_2	0.0	22.6	23.6	71.5
Glu	9.2	71.0	100	152
Gly	37.4	31.2	22.4	31.2
Ser	336	219	120	124
Ala	700	214	59.5	38.2

which do not metabolize and turn over very slowly; but this is not the sole interpretation.

It is interesting to compare these data with those obtained by SMITH *et al.*[5] where $^{14}CO_2$ was fed continuously to a steady-state system of photosynthesizing *Chlorella*. Here, too, saturation was shown to be quite early for alanine and aspartic acid; but saturation was never attained for glutamic acid and glutamine in the half-generation

References p. 666

time of the experiment. Glycine presented an anomaly, saturating only slowly, although present in large quantities in protein. In this case, as in the glutamic acid, it was recognized that dilutions from other reactions and/or other pools, may complicate the kinetics. Indeed, one of the major conclusions by these investigators is that the isotopically active pools constitute less than half the total pools, and that there may be two or more pools per amino acid.

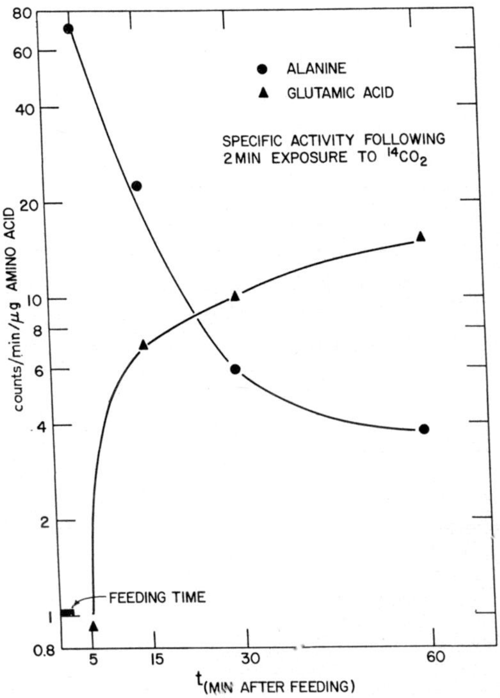

Fig. 2. Time changes of alanine and glutamic acid activity in tobacco leaf punches exposed to $^{14}CO_2$ for 2 min.

An experiment indicative of the complexity of interpretation involved the feeding of 1 mg of [2-^{14}C]glycine to an excised tobacco leaf. Analysis of the major fraction of the leaf showed the distribution summarized in Table IV. The extensive conversion of glycine to starch in this excised leaf may reflect the general pathway of protein degradation in excised leaves. The surprising variability of this metabolism is illustrated in the wide distribution of the initial glycine activity into the other amino acids of protein simultaneously synthesized in the excised leaf (Table V).

Effects of light and age

Certain other observations made at about this time appear to bear repeating[10], for example, the effect of light and of age. The feeding of $^{14}CO_2$ for a 20-h period to a soybean leaflet in the dark, resulted in the distribution shown in Table VI in which it is seen that there is a remarkable incorporation of ^{14}C into polymeric substances. The ethanol-soluble fraction, however, was by far the most prominent, and of these compounds, the most prominent is arginine, which accounts for 35%

References p. 666

TABLE IV

DISTRIBUTION OF ACTIVITY IN EXCISED TOBACCO LEAF FED [2-^{14}C]GLYCINE

From RACUSEN AND ARONOFF, unpublished results.

Leaf fraction	*Activity (counts/min)*
80% aq. ethanol extract	248 000
80% aq. ethanol residue	400 000
a) crude fiber	19 800
b) starch	280 000
c) protein	120 000

of the free amino acid fraction, although *none* is found in the protein (Table VII). The major portion of the radioactivity (75%) is in the guanido-carbon, and consequently parallels, in part, the distribution found in the Cyanophyceae (*e.g. Synechococcus* and *Nostoc*) where citrulline is generally a major product of $^{14}CO_2$ fixation[11] both in the light and in the dark. The small amount of photosynthetic arginine found as free

TABLE V

DISTRIBUTION OF ACTIVITY IN AMINO ACIDS OF TOBACCO LEAF PROTEIN FOLLOWING UPTAKE OF [2-^{14}C]GLYCINE IN THE LIGHT

From RACUSEN AND ARONOFF, unpublished results.

Amino acid	*% activity in protein*	*Amino acid*	*% activity in protein*
Asp	9.4	Ala	11.7
Glu	10.3	Phe	4.7
Ser	26.2	Leu	5.6
Gly	22.5	Others*	9.4

* Roughly equally distributed between tyr, val, pro, arg and cys (cystine not identified positively).

amino acid in soybean leaves is not far from that resulting in the dark with an equivalent amount of $^{14}CO_2$ and equal time. The relatively small amount is therefore interpreted not as an inhibition, but simply as not being accelerated by photosynthesis, as are others. On the other hand, the extremely low values (~ 1%) of free glycine and alanine, compared to corresponding values in the light (12.7 and 5.4%, respectively) suggest that in the light these amino acids are being made more rapidly than they can be incorporated into protein.

The experiments described above were all conducted with mature, that is, non-expanding leaves, in an effort to separate problems of turnover from those of accretion. The amino acids from $^{14}CO_2$ fed to young leaves differed from the mature in two major respects, both related to the so-called "indispensible" amino acids. First,

TABLE VI

DISTRIBUTION OF RADIOACTIVITY IN SOYBEAN LEAF FED $^{14}CO_2$ IN THE DARK

From RACUSEN AND ARONOFF, unpublished results.

Fraction	*% Activity*	
80% aq. ethanol solution	70.8	
a) cation		54.0
b) neutral + anion		16.8
80% aq. ethanol residue	28.3	
a) cation		25.7
b) neutral + anion		2.6
Crude fiber (5.5 mg)	0.9	

the amounts of these amino acids in the protein of young leaves is of the order of 10-fold more than in mature leaves (37.5% compared to $<$ 5%); and, secondly, the free amino acids of young leaves are present only as traces, virtually non-existent, whereas in mature leaves they constitute an analytic amount, though totaling less than 5%. Thus, young leaves have not only a greater capacity for producing the essential amino acids, but for incorporating them more completely into protein.

TABLE VII

DISTRIBUTION OF RADIOACTIVITY IN CATION FRACTION OF SOYBEAN LEAF FED $^{14}CO_2$ 20 H IN THE DARK

From RACUSEN AND ARONOFF[10].

Amino acid	*% of cation activity*	
	Free amino acids	*Protein amino acids*
Asp	6.8	23.8 (Asp + Asp-NH$_2$)
Asp-NH$_2$	34.2	
Glu	13.5	19.0 (Glu + Glu-NH$_2$)
Glu-NH$_2$	0.0	
Ser	6.8	14.8
Gly	±1	31.0
Ala	±1	4.9
Arg	35.2	0.0
Others	2.3*	7.1**

* Unknown; not urea.

** Phe, tyr, pro, leu, arg; altogether less than 5% of total cationic.

Amino acids and boron deficiency

Experiments on the biochemical function of boron in plants demonstrated that even severe deficiency symptoms did not affect appreciably the rate (per unit area of leaf) or qualitative aspects of photosynthesis. Quantitatively, however, the anionic

References p. 666

fraction of the deficient leaves was considerably richer, as shown in Table VIII, and this was reflected in the accumulation, in deficient plants, of phenolic compounds, notably chlorogenic and caffeic acids. In deficient plants (— B) the levels of these phenolic acids ranged up to 10 times that of the + B leaves. The sunflower contains

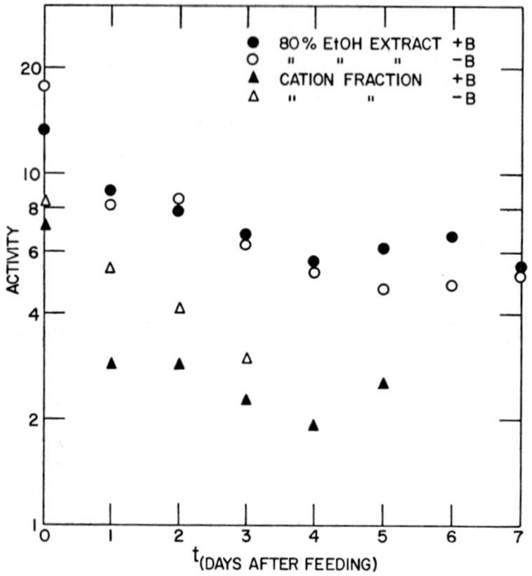

Fig. 3. Time changes of activity in 80% aq. ethanol extract and free amino acids of — B and + B plants.

relatively low levels of free amino acids; it is the classical "carbohydrate plant". Nevertheless, — B plants consistently demonstrated a higher free amino acid content, a difference which persisted despite variations of overall level with time (Fig. 3). The relative excess of free amino acids was found to be characteristic not only of sunflower, but of — B lettuce, tomato and radish, as well as peach, apricot, prune and cherry. The differences in the leaves, while real, appeared to be much larger

TABLE VIII

DISTRIBUTION OF RADIOACTIVITY IN BORON-DEFICIENT AND NORMAL PLANTS AFTER 3 H PHOTOSYNTHESIS IN $^{14}CO_2$

From PERKINS AND ARONOFF, unpublished results.

State	*Petr. ether*	*Anionic*	*Cationic*	*Neutral*
		(counts/min)		
+ B	63 200	37 400	53 100	11 000 000
— B	39 400	180 000	74 400	4 400 000

in the stems. More intriguing, the most readily identifiable amino acids were the aromatics, tyrosine, phenylalanine and tryptophane, which were apparent only as traces in the corresponding + B stems. We do not know, as yet, the immediate basis for the excessive amounts of the aromatic acids in the — B stems. They may arise, of course, from proteolysis, or an increased rate of synthesis, or a decreased rate of utilization.

References p. 666

(The increased rate of synthesis would have to occur in the stem, as leaf differences are relatively minor.) The correlation with shikimic acid metabolism is an obvious one, as both the phenolic acids and the benzenoid aromatic amino acids arise from shikimate. A hypothesis we are pursuing at present is that of boron control of membrane permeability, possibly the chloroplast membranes, allowing the diffusion of sedoheptulose, or even of its phosphate. If shikimate formation was controlled normally by one of the metabolic steps subsequent to sedoheptulose formation, its excess in the cytoplasm by virtue of — B membrane transmission would result in excess phenols and amino acids. However, this would also require translocation of these amino acids down the stem, and, at present, there is no satisfactory evidence of appreciable movement of amino acids out of leaves.

In fact, the question of tryptophane synthesis in leaves of higher plants is itself a troubled one. Thus, in our early days, we were never able to demonstrate the formation of radiotryptophane by the feeding of $^{14}CO_2$ to excised soybean leaves, although phenylalanine was identified readily and shown to be labeled uniformly[10]. It was possible, of course, that tryptophane was formed in the roots and translocated, as needed, to leaves. Furthermore, it had been reported previously[12] that the level of

TABLE IX

RADIOACTIVITY OF TRYPTOPHANE FROM SOYBEAN LEAVES FED $^{14}CO_2$
From PERKINS AND ARONOFF, unpublished results.

Sample	*Free try, relative specific activity*	*Protein try, relative specific activity*
	(counts/min/mg)	
Leaf, young, excised	9.2	93.1
Leaf, mature, excised	7.9	16.9
Leaf, young, intact	2.8	38.2
Leaf, mature, intact	16.8	96.1
Root, mature, intact	1.1	14.9

tryptophane in legumes was a function of their boron content. We therefore repeated our earlier studies with relatively massive $^{14}CO_2$ feedings to both mature and young leaves, excised and intact, and examined these, as well as the mature root, with special regard to tryptophane, both as a free amino acid and as part of protein. The results (Table IX) demonstrate that excised leaves do, indeed, contain radiotryptophane, though in rather meager amounts. Of course, the radioactivity could well lie in the alkyl portion of the molecule, and thus bear no relation to the shikimate problem. The answer lay in the degradation of the molecule to some fragment of the indole nucleus. Because of the paucity of activity, we limited ourselves to a reaction known to yield demonstrable product, namely the degradation of tryptophane to anthranilic acid by liver homogenates[13, 14]. Unfortunately, the requirements for purification of the anthranilic acid involved continuing losses, and only the preparations from the two tryptophanes of highest specific activity gave anthranilic acid with reasonable activity. Even these were so low that all one can say was that highly purified anthranilate was active to a level roughly corresponding to a uniform distribu-

References p. 666

tion* and, by inference, the indole was also. On this rather weak basis it thus appears that the young leaf is able to synthesize the indole moiety of tryptophane, but the problem obviously needs further study.

This matter has been pursued somewhat further by the feeding of [U-^{14}C]shikimic acid to − B and + B sunflower leaves, but as Table X shows, there was no appreciable differential metabolism of the − B plants. Our efforts to resolve this problem, especially with studies of stem-section metabolism, are continuing.

TABLE X

DISTRIBUTION OF RADIOACTIVITY IN NEUTRAL, ANIONIC AND CATIONIC FRACTIONS OF TOBACCO LEAVES FED [U-^{14}C]SHIKIMIC ACID

From PERKINS AND ARONOFF, unpublished results.

State	*Neutral*	*Cationic*	*Anionic*
		(counts/min)	
+ B	16 500	2 760	174 000
− B	11 800	4 000	205 000

ACKNOWLEDGEMENTS

This work was supported by grants from the National Science Foundation and the U.S. Atomic Energy Commission.

REFERENCES

1 A. C. CHIBNALL, *Protein Metabolism in the Plant*, Yale University Press, New Haven, 1939.
2 F. C. STEWARD, in H. NEURATH AND K. BAILEY, *The Proteins, IIA.*, Academic Press, New York, 1954.
3 A. I. VIRTANEN, in *Festschrift* ARTHUR STOLL, 1957, B. Schwabe, Basel, 1957, p. 565.
4 J. A. BASSHAM, A. A. BENSON AND M. CALVIN, *U.C.R.L.*, 584 (1950).
5 D. C. SMITH, J. A. BASSHAM AND M. KIRK, *Biochim. Biophys. Acta*, 43 (1961) 447.
6 L. P. VERNON AND S. ARONOFF, *Arch. Biochem. Biophys.*, 29 (1950) 179.
7 N. C. GANGULI, *Naturwissenschaften*, (1954) 140.
8 J. H. VENEKAMP, *Acta Botan. Neerl.*, 4 (1955) 487.
9 H. S. MCKEE, L. NESTEL AND R. N. ROBERTSON, *Australian J. Biol. Sci.*, 8 (1955) 467.
10 D. W. RACUSEN AND S. ARONOFF, *Arch. Biochem. Biophys.*, 51 (1954) 68.
11 P. LINKO, O. HOLM-HANSEN, J. A. BASSHAM AND M. CALVIN, *J. Exptl. Botany*, 8 (1957) 147.
12 V. I. SHELDON, W. G. BLUE AND W. A. ALBRECHT, *Plant and Soil*, 3 (1951) 33.
13 M. MASON AND C. P. BERG, *J. Biol. Chem.*, 195 (1952) 515.
14 C. YANOFSKY, *J. Biol. Chem.*, 217 (1955) 345.

* For example, 22.8 mg of tryptophane of a specific activity of 93.1 counts/min/mg (total activity, 2120 counts/min) would have given us (7/11) (2120) = 1350 counts/min of anthranilic acid on a 100% yield basis. However, the eventual yield was only 10%. Furthermore, the relatively large amounts of acid required chromatography onto three sheets. Each sheet thus had 135/3 = 45 counts/min. The observed count per eluted chromatogram was 40 counts/min, thus suggesting that the anthranilate was uniformly labeled.

THE FREE NITROGEN COMPOUNDS IN PLANTS CONSIDERED IN RELATION TO METABOLISM, GROWTH AND DEVELOPMENT

F. C. STEWARD AND R. G. S. BIDWELL*

Department of Botany, Cornell University, Ithaca, N.Y. (U.S.A.); Department of Botany, University of Toronto (Canada)

SECTION I. DISTINCTIVE FEATURES OF THE FREE OR SOLUBLE NITROGEN COMPOUNDS OF PLANTS

Certain major differences exist between the economy of plants and animals toward nitrogenous compounds. This different economy has consequences which are not always appreciated; it leads to a very great variety of metabolic reactions and to cells and organs with very different nitrogenous metabolism and characteristics, as well as to the need to re-cycle and re-use nitrogen compounds in ways that higher animals commonly do not do.

Whereas plants are capable of what might be called "primary" protein synthesis from inorganic sources, like nitrate and ammonia, and they derive the necessary carbon ultimately from photosynthesis, it is a salient characteristic of animals that they are dependent on essential amino acids. This dependence may vary somewhat from organism to organism and even within the organism. For example, the mature human is said to require eight essential amino acids, but certain cultured tissues required 12 (ref. 11). Although whole plants may utilize nitrate or ammonia, their individual growing cells may receive the nitrogen for synthesis in already elaborated forms, for either roots or leaves may be the prime organs of nitrate reduction. This situation creates a consequential demand for movement and subsequent storage of what have been called nitrogen-rich forms. Of such nitrogen-rich forms, the amides asparagine and glutamine as well as arginine have been historically prominent, though such compounds as allantoin and allantoic acid, as stressed by MOTHES[24] and by BOLLARD[5] have also to be considered. Of these "nitrogen-rich storage and mobile reserves", glutamine is now recognized as an equally important metabolite in the economy of animals as of plants, but asparagine is still conspicuous as peculiarly a plant product and seems only to be used in protein formation by ruminants amongst higher animals[22]. In fact, much of the asparagine which is ingested by higher animals is said to be secreted in the urine. Whereas the body of higher animals is organized about the massive ingestion of preformed protein, or of elaborated nitrogen compounds, and permits the elimination of nitrogenous waste in quantity, this does not occur in plants. On the contrary, nitrogenous end products of plant metabolism are rarely lost as volatile compounds, nor are they excreted, and certainly not so in quantity; hence, higher plants store their nitrogenous end-products of metabolism for later use, or they translocate them to other regions in which they are required.

* Participation of R.G.S.B. in the Symposium was made possible by assistance from the National Research Council of Canada.

References p. 692/693

Thus, even in a single cell the large aqueous plant vacuole may function as a reservoir for stored, soluble, free amino acids and other nitrogenous compounds; older and mature leaves may function somewhat in the same way; and fleshy storage organs such as tubers, bulbs, corms, rhizomes, etc., are proverbially rich in soluble nitrogen compounds. Seeds and fruits also deposit rich stores of nitrogen, often in the form of proteins, especially in fleshy cotyledons. All this makes for an overall pattern of re-cycling nitrogen and re-using it with appropriate mechanisms of movement, storage, and re-entry of the stored forms of nitrogen into the main stream of metabolism.

Thus, once nitrogen is reduced from nitrate or is incorporated into organic compounds from ammonia, the organic nitrogen is largely retained by plants. Furthermore, the actual formative or growing regions in the plant body are strictly confined to very limited areas. To nourish and furnish these regions with their essential requirements, specific regions of the shoot and root may be physiologically active in the *de novo* synthesis of the nitrogenous and carbohydrate compounds which are required, but their period of active function in this way may be relatively brief; senescence of cells and organs may set in earlier than is often supposed. Such mature or senescent organs (*e.g.* older leaves), though often available for storage, may themselves make demands on more active centers of metabolism for their maintenance.

In rapidly growing plant cells, as for example in tissue cultures, the ratio of their alcohol-soluble to alcohol-insoluble nitrogen (largely protein) tends to be small, whereas their quiescent counterparts in storage organs may deposit a large part of their nitrogen in the form of the free amino acids which have been described in an earlier paper. Whereas in the potato tuber some two thirds of the total nitrogen and in the carrot root approximately half is in the free or soluble form, the proportion may be very much less in the corresponding cultured or growing tissue[53,58]. Indeed, one often finds that the unusual nitrogen compounds that have recently been discovered in plants have been first recognized in such storage organs as seeds, fruits, tubers, rhizomes, etc. On examination, the actual growing region of the shoot of an angiosperm, *e.g. Lupinus*[60] did yield a reasonably complete range of amino acids which are needed to make protein, although these were in somewhat low concentration. By contrast, examination of the growing apex of a fern, *Adiantum*[60], disclosed a surprising amount of soluble nitrogen which was in the form of an unexpected metabolite which proved to be γ-hydroxy-γ-methylglutamic acid[13], and this represented approx. 90% of the total soluble nitrogen of this growing region!

Therefore, before reaching the later theme of this discussion, which is to be the interactions of nitrogen metabolism with other physiological functions, mainly respiration, it is well to summarize some of the factors which are known to determine the composition of the soluble nitrogen pool in plants.

Even though a given organ may be in approximate nitrogen balance with respect to its total content of protein, the composition of the soluble nitrogen fraction may, nevertheless, be subject to change. The source of these changes is often obscure, except that they may be events in a sequential process of development. Such an example appeared in the fruit of the edible banana[47]. Early in its development the very young fruit contains soluble nitrogen compounds. In the phase of most rapid growth of the banana fruit, its soluble nitrogen content falls to a very low level and, as the organ subsequently grows by cell enlargement, it again deposits storage material and its content of amides increases, although whether this amide is glutamine or asparagine seems also to be a function of the time of year in which the fruit develops. In this later phase of growth of the fruit, an unusual feature is the accumulation of relatively large amounts of free histidine and, after passing the crisis known as the climacteric, so that ripening may ensue, there is a relatively massive conversion of what was previously amide nitrogen into free histidine. This example is but one of what may be many kinds of conversion which have hitherto passed unnoticed for lack of the ready means of their detection which chromatography has provided. When such massive accumulations of soluble compounds occur, this seems to be associated either with the slowing down of growth and of the consequential demand for nitrogenous compounds to build protein, or it may be due to a block in metabolism for genetic or environmental reasons.

It is now apparent that the course of nitrogen metabolism and the storage of the soluble nitrogen compounds in plants is dramatically influenced by the inorganic nutrients. A typical example is the mint plant, where effects due to the so-called macro nutrient elements have been described[46]. Potassium tends to promote one type of change in leaves and calcium another, whereas deficiencies of an element such as sulphur may result in massive accumulations of amide, mainly glutamine, and of the amino acid arginine.

However, not only nutrition but also the environment in which the plant grows has a determining effect on the course of its nitrogen metabolism and on the soluble compounds which are formed and stored. A diurnal cycle in the mint plant tends to promote glutamine in leaves in the light and asparagine in the dark, and this in turn is associated with a general trend toward protein synthesis in the light and toward protein breakdown in the dark[46]. A similar conversion of [^{14}C]proline, when added to leaf discs of tobacco, to glutamine in the light and to asparagine in the dark (Fig. 1) has also been observed in this laboratory by POLLARD AND ROCHOW[28], working with one of us (F.C.S.). MOTHES also reported to the International Botanical Congress at Montreal in 1959 the formation from [^{14}C]urea of asparagine in the dark and glutamine in the light. A very important feature of nitrogen metabolism is the way in which it may be modified by such features of the environment as the length of day and night and the fluctuating diurnal cycle of temperature. For the mint plant, where these relationships have been worked out in some detail, it is of interest to note that the environmental variables also interact with the nutritional ones, so that effects which are due to length of the day and to potassium supply tend to work together; whereas effects due to calcium nutrition and to short days and long nights also tend to work together. But, surprisingly, the overriding variable is that of night temperature. Low night temperature causes asparagine to appear in otherwise long-day-high-potassium mint plants, in the leaves of which it would not normally occur but where it replaces the amide glutamine. Such effects as these clearly present a complex picture of the soluble nitrogen pool which, for its interpretations, requires both physiological and biochemical investigation.

Effects upon the course of nitrogen metabolism are not confined to the so-called macro-nutrients, for the trace elements, which are required only in very minute amounts, may also have a very prominent effect as shown by ZACHARIUS (see ref. 46), and they also interact both with each other and with other physiological variables. A prominent example of this concerns molybdenum and manganese, the effects of which on the nutrition and nitrogenous composition of the tomato plant require the

References p. 692/693

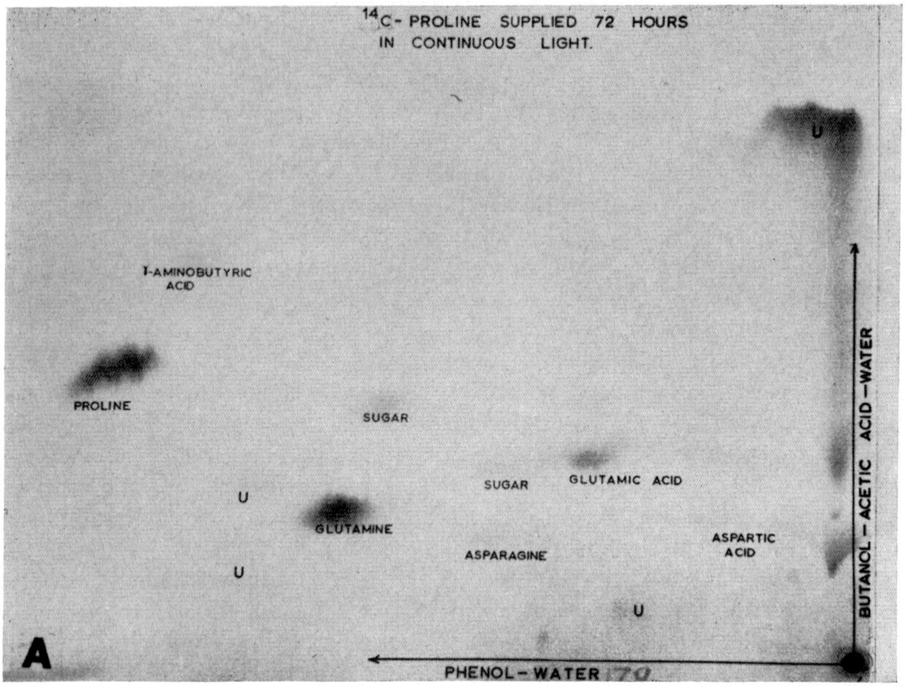

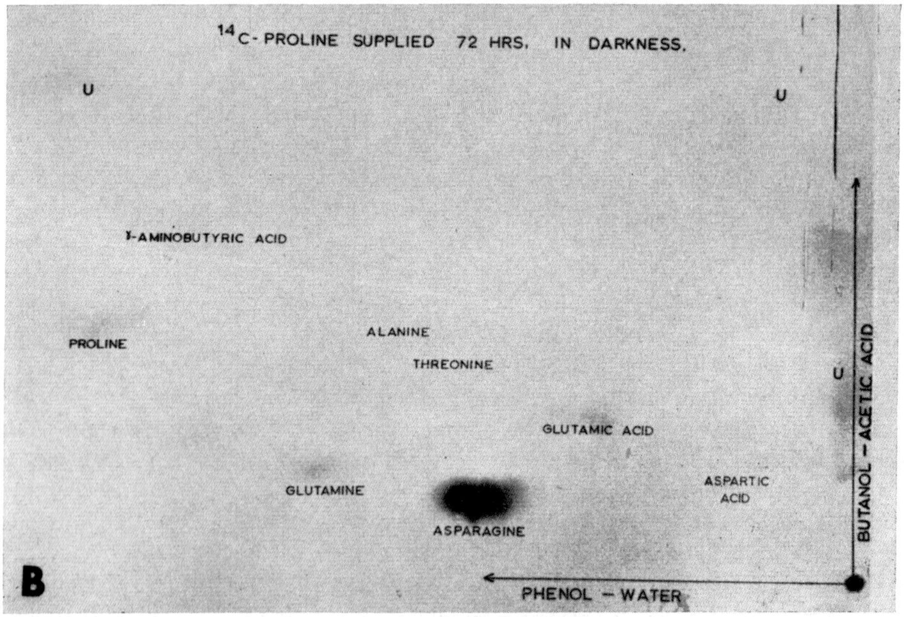

Fig. 1. Radioautographs of the chromatograms of the alcohol-soluble nitrogen fraction of tobacco leaf discs. [^{14}C]proline supplied 72 h (A) in continuous light and (B) in darkness. (U designates a radioactive but unidentified spot.)

References p. 692/693

consideration of their interactions with each other and with the source of nitrgen supply, *i.e.* whether this is nitrate or ammonia. A brief summary of work on this interesting problem by MARGOLIS[21] has already been made[46].

Another interesting example of the complex factors that affect the relative composition of the soluble nitrogen compounds of plants is that provided by the observations of VOSKRESENSKAYA[67, 68], which was also referred to by CAYLE AND EMERSON[8]. The first observations suggested that the balance between the amount of the $^{14}CO_2$, fixed in the light, which enters the protein or the carbohydrate is a function of wavelength. On re-examination of this question, CAYLE AND EMERSON[8] did see some effects which were due to wavelength on the specific activity of the carbon in certain amino acids. Recent work by HAUSCHILD, NELSON AND KROTKOV[16a] shows that *Chlorella* cells produced 25–30% more aspartic acid and 20% less serine and glycine in red light plus 4% of blue light than in red light alone, even when the comparisons were made under conditions that relate to the same amount of carbon being fixed.

Sufficient has, therefore, been said in this introduction to show that the nitrogenous metabolism in plants is far more intricate than would be supposed from a mere preoccupation with the idea that amino acids are elaborated as the principal soluble products and that the chief destiny of these is to combine directly in the form of protein. The array of nutritional, environmental, and developmental factors that impinge upon nitrogen metabolism to determine the compounds which are stored free is not comprehensible on this idea alone.

In short, the problem of nitrogen metabolism of higher plants should be seen in its entirety, and the role of the free or soluble compounds is not merely, or immediately, in relation to protein synthesis but should be seen in many other contexts. When pushed to the point of ultimate explanation, most physiological functions in plants make contact with nitrogen metabolism, but this is especially true where the courses of nitrogen metabolism and carbohydrate metabolism interlock, prominently at the level of the keto acids. The need of plants to synthesize, break down and re-use their nitrogen compounds, whether in different parts of the plant body or within a single cell, and the interlocking and mutually dependent cyclical processes of nitrogen metabolism and carbohydrate metabolism, constitute the main theme of the remainder of this discussion which will emphasize the distinctive part which these events play in plants.

SECTION II. DIFFERENT BIOSYNTHETIC PATHWAYS OF NITROGEN COMPOUNDS WHICH OCCUR IN DIFFERENT METABOLIC SITUATIONS

In Section I reference has been made to various considerations and circumstances which determine the overall composition of the pool of soluble nitrogen compounds which may occur in plants. In this section it will be emphasized that the common constituents of the soluble nitrogen moiety may arise by entirely different biosynthetic pathways under different situations. An outstanding example of this concerns the principal amides glutamine and asparagine. In many situations the occurrence of glutamine is associated with conditions which are favorable to growth and protein synthesis, and indeed glutamine is often added to tissue-culture media as a preferential source of nitrogen. By contrast, the amide asparagine often seems to be

References p. 692/693

associated with relatively adverse conditions for growth of the organ in which it is formed or with conditions where protein synthesis has ceased, where protein breakdown occurs and where the storage of soluble nitrogen is in excess. Similarly, as mentioned for the mint plant[46], the occurrence of arginine in quantity in leaves is often associated with quite marked mineral deficiency conditions as observed by MILLAR (*cf.* ref. 50). When recovery from sulphur deficiency occurs in the light the amide present is glutamine; when recovery takes place in the dark asparagine is formed from arginine (*cf.* also MILLAR in ref. 46).

These considerations became apparent when plants which normally contain asparagine as a principal storage product, as in the case of both potato tuber tissue and carrot root tissue, were brought into a state of active growth; the asparagine disappeared almost entirely, whereas some glutamine tended to persist[58].

But even though glutamine may be present in both a resting organ and in the corresponding growing cells, the method by which it originates may be quite different in the two types of system. In the storage organ, as for example in the beet root, the general idea is that the glutamine originates from glutamic acid by amidation after the manner worked out originally by SPECK[39]. In fact, this was virtually proven by HOOD, LYNAM AND TATUM[18], who showed that in the formation of glutamine in the beet root the amide nitrogen derived from exogenous ammonia, but the amino group came from endogenous sources, *i.e.* glutamic acid. However, in tissue cultures derived from storage organs, for example the carrot root, it has been shown that glutamine may arise readily from exogenous γ-aminobutyric acid (γ-AB)[45]. In other words, the carbon framework which appeared in the glutamine of the storage organ derives prominently from α-ketoglutaric acid originating in the Krebs' cycle, whereas the carbon framework in the glutamine of the actively growing cell may originate in a somewhat different way via γ-AB, perhaps from succinic semialdehyde.

A similar contrast applies to glutamine as it occurs in leaves. In normally photosynthesizing and full nutrient leaves it has been shown that glutamine may be derived directly from $^{14}CO_2$ in photosynthesis[4]. In nutritionally deficient leaves, notably mint leaves suffering from sulfur deficiency, glutamine is known to accumulate in great quantity. However, in the attempt to use the latter system to label the glutamine so formed directly with carbon (by supplying the sulfur-deficient leaves with $^{14}CO_2$) it was found that the glutamine simply did not become labeled (R.G.S.B., unpublished result). Hence the carbon which entered the glutamine in the sulfur-deficient mint plants must have come from some endogenous source, presumably via protein breakdown. Similar attempts to label the glutamine of beet roots by supplying the $^{14}CO_2$ to their leaves also failed, because the immediate fate of the carbon was to enter other compounds. Hence the glutamine which was formed in the roots under these circumstances was able to get its carbon much more readily from some other source than immediately from the metabolism of the sugar being produced in photosynthesis[4].

It is, therefore, quite apparent that glutamine may originate in different ways in different metabolic situations, and different compounds may be the source of its carbon framework. It has, of course, been well known for a long time that excised and senescent leaves, notably of barley, form glutamine readily when their protein is undergoing breakdown[74].

Although the carbon of glutamine may be labeled in different ways, albeit in

References p. 692/693

different situations, as for example by the use of both exogenous urea and carbon dioxide[27, 50], the carbon of asparagine is almost impossible to label intensely. For example, in the white Lupin *(Lupinus)*, which is a classical asparagine plant, the asparagine seems not to be readily labeled either from applied $^{14}CO_2$ or even from radioactive sugar applied exogenously[72]. The interpretation here is that the asparagine arises not directly from sugar but by the reworking of the products of protein breakdown.

In recent experiments with wheat leaves it has been possible to obtain a *small* amount of radioactively labeled asparagine which, however, contrasts very markedly with the massive synthesis of glutamine which also takes place. The labeling

TABLE I

DISTRIBUTION OF RADIOACTIVITY IN SOLUBLE AMINO ACIDS

Wheat leaves supplied with [1,4-^{14}C]succinic acid (Data of BIDWELL, unpublished) and (2) bean leaves supplied with $^{14}CO_2$ (Data of NELSON AND KROTKOV[25]).

Amino acid	*Carbon atom(s)*	*% of radioactivity found in*	
		Wheat leaf (1,4-^{14}C)Succinic	*Bean leaf $^{14}CO_2$*
Aspartic acid	$C_1 + C_4$	—	32
Asparagine	C_1	50	26
Glutamic acid	C_1	75	4
Glutamine	C_1	75	10
Alanine	C_1	1.2	34

of the asparagine was accomplished by the use of succinate radioactively labeled in the two carboxyl groups. The resultant labeling of the carbon in the asparagine corresponded precisely to the labeling in the succinic acid supplied, showing that there must have been a direct conversion via fumaric acid or oxaloacetate to aspartic acid and then to asparagine (BIDWELL, unpublished results). By contrast, however, in bean leaves this seems not to be so (see Table I). NELSON AND KROTKOV[25] allowed bean leaves to photosynthesize in $^{14}CO_2$ for 15 min. The small amount of radioactive asparagine when isolated proved to have 75% of its ^{14}C in the two middle carbons. This result contrasts sharply with that obtained with the wheat leaves and does not lead to any readily understood line of synthesis of asparagine via a Krebs' cycle intermediate. Perhaps an even more striking example is the labeling to be found in alanine when [1,4-^{14}C]succinic acid is supplied to wheat leaves. In the absence of ammonia a small amount of radioactive alanine was formed which had 96% of the label in the carboxyl group. When ammonia was added to the leaves together with the labeled succinate, somewhat more radioactive alanine was formed, but the carboxyl carbon now only contained 2% of the label.

Therefore, with alanine as with asparagine and glutamine, one is forced to the same conclusion; namely that the same compound may arise in different situations by quite different routes, and one should not, therefore, be obsessed with the more obvious reactions but try to find out what actually does occur in these different systems.

References p. 692/693

SECTION III. SOLUBLE POOLS IN RELATION TO PROTEIN SYNTHESIS AND BREAKDOWN

The classical view is that all carbon for protein enters via the soluble amino acids which occur free in the cell. Much evidence indicates that this is probably true for animals; but in plants, which possess a much greater reservoir of soluble nitrogen compounds, this is by no means as obvious, for the constituents of this pool may need to be extensively re-worked prior to protein synthesis. Also, the early products of photosynthesis seem to be especially accessible to protein formation, in fact much more so than the stored amino acids. $^{14}CO_2$ is stored during photosynthesis to a very surprising degree in the protein, or at least in the alcohol-insoluble fraction and, therefore, presumably in protein; and, concurrently, very few soluble amino acids are formed in the light in such a way that they receive the label. A suggestive quotation is the following taken from BENSON AND CALVIN (*cf.* ref. 2, p. 30): "The major portion of the insoluble products formed in the first few minutes by algae ... was protein. ... Protein obtained from longer experiments (5–10 min) contained more activity than that found in several amino acids present in the cell extracts. Glutamic acid, by far the largest amino acid reservoir, was not converted into protein in amounts commensurate with its concentration". By contrast, however, the conditions of dark fixation produce a great variety of amino acids in such a way that they become labeled, and it was this contrast that led CALVIN *et al.* for a while to assume, erroneously, that respiration in the light virtually ceased[3]. This indicates at the outset, therefore, that the carbon which enters protein in the light must do so by a more direct route than via the soluble amino acids which are stored in bulk in the cell.

Recent experiments by this group[38] have led to the conclusion that protein amino acids are synthesized rather directly from the intermediates of the carbon reduction cycle. It has been shown that protein synthesis occurs, not at the expense of the total pool of free amino acids, but from small "active" pools which are separated in compartments from the larger "inactive" pools.

An earlier Russian experiment is interesting: ANDREEVA[1] added ^{14}C-labeled sugars, ^{15}N-labeled ammonium sulphate and non-isotopic carbon dioxide to tobacco leaves in light. He found that whereas ^{15}N rapidly entered proteins, ^{14}C did so much more slowly. This indicated that the carbon precursors of protein synthesis are derived more or less directly from photosynthetic carbon, and are not readily mixed with those derived from soluble compounds in the cell.

Similar conclusions emerged from experiments on carrot tissue cultures in which the availability of the carbon in various exogenously applied and labeled substrates for protein synthesis was investigated. The compounds used consisted of labeled sugar and two labeled nitrogen compounds, namely glutamine and γ-AB. Prior to these experiments one might have anticipated that the labeled sugar on absorption would have intervened directly in the oxidative reactions of respiration and have produced carbon dioxide which would tend to approach the specific activity of the source. Similarly, the nitrogen compounds might have been held to enter into the nitrogen metabolism in such a way that their carbon would have entered promptly into the protein which was being synthesized in the growing cells. In point of fact, the exact opposite occurred, for it was shown that the carbon of the labeled sugar entered the protein much more readily than that of the labeled amino acids. Furthermore, the specific activity of certain of the protein amino acids (*e.g.* glutamic acid)

References p. 692/693

was much higher when sugar was the source of labeled carbon than when labeled amino acids were supplied. This is shown in Table II. Furthermore, the labeling of the carbon dioxide was more closely parallel to the labeling which occurred in the protein than to that in the free compounds; this suggests that carbon dioxide emerged from the products of protein breakdown (Table III). Not only this, but as may be

TABLE II

SPECIFIC ACTIVITIES OF FREE AND PROTEIN-BOUND AMINO ACIDS IN CARROT EXPLANTS AFTER SUPPLY OF C^{14}-LABELED GLUTAMINE, γ-AMINOBUTYRIC ACID OR GLUCOSE

Data of STEWARD, BIDWELL AND YEMM. Specific activities based on the specific activity of the compound supplied = 100. N.D., not determined. —, no detectable activity.

Amino acid	*^{14}C-Substrate supplied*					
	Glutamine		*γ-Aminobutyric acid*		*Glucose*	
	Soluble	*Bound*	*Soluble*	*Bound*	*Soluble*	*Bound*
Glutamic acid	81	28	26	8.5	100	46
Aspartic acid	—	22	—	10.2	N.D.	43
Threonine	—	6.4	12	4.9	N.D.	26
Alanine	N.D.	2.5	N.D.	—	100	50
Proline	8	14	—	3.7	—	—

seen from Table II the specific activity of protein-bound amino acids did not relate in any way to the specific activities of their counterparts in the soluble pool.

This type of result led to a scheme[44] in which the absorbed and labeled sugar enters into those phases in the cell which are intimately associated with the anabolic reactions, in which the Krebs' cycle operates as a line of synthesis and, therefore, the carbon of sugar enters promptly into the protein. By contrast, the stored amino

TABLE III

SPECIFIC ACTIVITIES OF CARBON IN SOLUBLE AND PROTEIN-BOUND COMPOUNDS AND IN RESPIRED CO_2 FROM CARROT EXPLANTS SUPPLIED WITH ^{14}C-LABELED COMPOUNDS

Data from STEWARD, BIDWELL AND YEMM. N.D., not determined. —, no detectable activity.

(^{14}C)Compound supplied	*Rate of growth of tissue*	*Time (h)*	*Specific activity of C (Supplied compounds = 100)*			
			Sugar	*Glutamic acid*		*CO_2*
				Soluble	*Bound*	
Glucose	Rapid	22	17–38	—	N.D.	41
Glucose	Rapid	109	71	100	46	74
Glucose	Slow	40	26–29	—	N.D.	34
Glucose	Slow	61	29–33	59	24	35
Glutamine	Rapid	40	0	84	22	18.4
Glutamine	Rapid	109	1.4	81	28	24.6
γ-AB	Rapid	40	0	33	2.6	7.7
γ-AB	Rapid	109	1.1	26	8.5	14.1

acids in the cell were regarded as having originated largely from the protein breakdown in phases in the cell which are concerned largely with catabolism. The nitrogen from protein breakdown had to be donated in acceptable form to the site of protein-re-synthesis, whereas the carbon skeletons of these protein breakdown products were respired away as carbon dioxide. Therefore, the point was made that any amino acids that occurred as intermediates of protein synthesis at the site of protein synthesis were not free to mingle with the large pool of soluble nitrogen compounds in the cells. In this sense the free amino acids which exist in quantity in plant cells are not to be regarded as the immediate precursors of protein, for the carbon of the protein can be derived more readily from the sugar entering the cell than from the free amino acids.

The main idea that emerged, therefore, from these experiments on carrot tissue explants is as follows. Since the same substances may exist in different regions of the cell as free compounds, the availability of their carbon for incorporation into protein will depend greatly on the location in the cell and upon the ease of access to the site of synthesis.

However, even in the carrot tissue, evidence was obtained that some amino acids, *e.g.* proline, were directly incorporated into a protein fraction, albeit into a protein fraction which was not subsequently metabolized[44, 45]. This occurred so readily, however, that when ^{14}C-proline was externally applied it did not accumulate extensively in the free soluble fraction, nor did it enter that fraction even when the tissue was allowed to metabolize for a further 72 h in the absence of exogenous labeled proline[26]. Thus, in the growing carrot stimulated by coconut milk, the soluble pool of amino acids is reduced to a low level, and there is obvious evidence of ready withdrawal of the metabolites from the soluble pool into the metabolic system.

The situation, however, in cells of the Jerusalem artichoke tuber is now known to be somewhat different. This was first shown by the analysis of the soluble nitrogen as it exists in artichoke tissue explants which were prompted to grow in the presence of coconut milk and/or naphthaleneacetic acid. In this work, carried out recently by Dr. ROBINSON and Dr. DURANTON with one of us (F.C.S.), it is evident that the cultured artichoke cells retain in the soluble pool a much higher concentration of a variety of amino acids and of proline in particular. This strong retention is so apparent that the tissue utilizes exogenous supplies of amino acid to form protein even much more readily than it does the endogenous supplies, a result which contrasts sharply with that obtained with carrot.

Other areas of metabolism not specifically involving amino acids and protein synthesis nevertheless lead to somewhat similar considerations; namely that there are different compartments, or pools, in the cells in which the fate of a given metabolite may be quite different. VITTORIO, KROTKOV AND REED[66] supplied ^{14}C-labeled glucose to tobacco leaves. The specific activity of the sucrose which was formed in the leaves and was isolated from them proved to be actually higher than the specific activity of the free glucose or fructose of the leaves. This obviously means that there was a pool or compartment which contained non-radioactive glucose and which was not available to take part in sucrose synthesis. Much more recent work of VICKERY AND ZELITCH[64] leads to somewhat similar ideas. These investigators supplied tobacco leaves with labeled pyruvic acid, and they investigated the specific activity of the citric acid in the leaves. They were led to conclude that there must be two pools of citric acid in the cells, one which is concerned with the simple storage of citric acid and the other associated with the rapid metabolism of citric acid as in respiration. There is also to be found in work from the laboratory of Dr. H. PORTER a similar indication in the carbohydrate metabolism of tobacco leaves that there are geographical areas or pools of sucrose which are distinct[29].

Therefore, in both the relation of nitrogen compounds to protein synthesis and of carbohydrates and organic acids to respiration, we are led to the view that different phases or compartments in the cells exist, and in these the same substances may be associated with different degrees and kinds of metabolic activity.

A possible method whereby amino acids could be formed in the soluble phase of a cell from newly incorporated carbon and then passed directly to the protein synthesizing system without mixing with the soluble amino acids has been indicated by HANFORD AND DAVIES[16]. An enzyme from pea hypocotyls forms phosphoserine via phosphohydroxypyruvate from phosphoglyceric acid, a compound which plays a direct part in both photosynthesis and respiration. The phosphoserine so formed could easily be incorporated into protein without ever mixing its carbon with the free serine which occurs in the cell.

In the work of ROBERTS *et al.*[35] on micro-organisms there is also evidence that all the free amino acids in the cell are not equally accessible. For example, certain acids (notably threonine and lysine) when supplied exogenously to *Escherichia coli* enter completely different sequences of reactions than do the threonine and the lysine which are already within the cells.

In later work by COWIE AND WALTON[10] on *Torulopsis*, reference is made to distinctive pools of amino acids which are here regarded as being segregated by binding on more complex compounds. In the view of COWIE *et al.*, it is these latter amino acids that are more immediately accessible to protein synthesis than those which exist freely. (This idea has been amplified in another contribution to this Symposium, see p. 633.)

Thus, in several examples which have been cited, one has to consider the location of a given metabolite in the cell in order to understand its metabolic role. However, these considerations obviously do not apply to experiments which are carried out with cell free preparations which lack the organization of the intact cell. For example, RABSON AND NOVELLI[31] used particulate cell free preparations from *Zea mays* to show that these tend to behave like animal systems, for they incorporate amino acids into protein in a manner essentially comparable to the mammalian cell free systems which have been studied. This sort of work, including the results of WEBSTER[70] with cell free preparations, contrasts with the work done on intact cells of higher plants, for the latter have conspicuous vacuoles and extensive soluble fractions, and to these systems additional and somewhat different considerations must apply.

Stress has been laid upon the ability of exogenously supplied sugar to furnish carbon directly to protein in the synthesis of protein in growing carrot tissue cultures. Some recent and hitherto unpublished work by HELLEBUST AND BIDWELL with corn seedlings and corn root tips also bears on this question. The soluble compounds in the leaves of corn seedlings became strongly radioactive during 2 h of photosynthesis in $^{14}CO_2$, and the protein amino acids also acquired ^{14}C. This gave a high level of activity in the sugars and in the soluble nitrogen compounds. Subsequently the plants were allowed to continue photosynthesis for 64 h on a 16-h day and to metabolize, but only in the presence of non-radioactive carbon dioxide. During this period there was an increase in the total quantity of the protein but only a relatively slight increase in its radioactivity, although the soluble components were still heavily labeled. This result shows, therefore, that the newly synthesized protein must have derived its carbon from an unlabeled source immediately accessible in the products of photosynthesis and not from either the preformed and stored sugar or from the amino acids in the cell, which were heavily labeled. The data which support these conclusions are given in Table IV.

TABLE IV

SOURCE OF CARBON FOR PROTEIN SYNTHESIS IN CORN

Data of HELLEBUST AND BIDWELL, unpublished. Corn leaves were supplied with $^{14}CO_2$ for 5 h, then $^{12}CO_2$ for 64 h. Corn root tips were supplied with [^{14}C]glucose for 2 h, then [^{12}C]glucose for 7 h.

	Corn leaves	*Corn root tips*
Time after supply of ^{14}C (h)	24–64	2–7
Average specific activity of soluble sucrose (counts/min/μg carbon*)	75	240
Average specific activity of soluble hexose (counts/min/μg carbon*)	75	48
Average specific activity of soluble amino acids (counts/min/μg carbon*)	144	213
Average specific activity of protein amino acids (counts/min/μg carbon*)	84	23
Increase in carbon of protein amino acids (μg carbon)	35	3.0
Increase in activity of protein amino acids (counts/min)	280	243
Carbon entering protein: if from soluble sugars (μg carbon)	3.8	1.0
if from soluble amino acids (μg carbon)	1.9	1.1
Protein synthesized from non-active source (%)	90–95	66

* These values did not change much during the experimental period; hence the average values are sufficiently accurate for the calculations.

Similar conclusions follow from experiments on excised roots of corn which received exogenous supplies of [^{14}C]sugar followed, in an ensuing period, by non-radioactive sugar. These data are also cited in Table IV.

Thus the overall conclusion is that the ability of a given compound to furnish carbon for protein is not only determined by its nature but by its location within the cell.

Another example flows from the work of RACUSSEN AND HOBSON[33]. This work shows that carbon supplied exogenously as $^{14}CO_2$, like the endogenous sugar in the carrot explants, is much more readily accessible to the site of protein synthesis. Thus the long-known efficiency of the green leaf in the synthesis of protein is seen to be due to the ready availability at the site of synthesis of newly made photosynthetic products. RACUSSEN AND HOBSON showed the prompt conversion of the ^{14}C-containing sugar to [^{14}C]aspartic and [^{14}C]glutamic acid in the protein and the even more prompt use of the $^{14}CO_2$ for the formation of arginine and lysine in the protein. In other words, the accessibility of carbon for protein synthesis is much greater in this system at the site where CO_2 is being reduced.

It has already been mentioned that BASSHAM *et al.* have arrived at similar conclusions. These workers now believe that the ^{14}C which enters protein of chlorella cells from $^{14}CO_2$ in the light does so from a small but metabolically active pool, and in so doing it bypasses the larger and more sluggishly metabolized pool of the conventional soluble compounds which there exist in bulk.

SECTION IV. PROTEIN TURNOVER AND THE SOLUBLE POOLS

The classical picture of nitrogen metabolism in plants, dating back to the work of PFEFFER, SCHULZE AND PRIANISHNIKOV, involved the idea that protein synthesized in one organ may be broken down, translocated and re-synthesized in another. Although this is protein "turnover" within the organism, this is a quite different phenomenon from what is commonly meant by protein turnover within a cell. Even the breakdown of one protein and the re-synthesis of another protein within a cell, although this may commonly occur during development, is still not what is commonly implied by cyclical turnover. It is true that the loss of one protein and the gain of another has been prominently invoked to explain events that occur during development. However, what is here in question is the cyclical breakdown and re-synthesis of metabolically reactive protein, this being regarded as a normal concomitant of metabolism.

This idea is by no means new, because it was first given prominence for plants by GREGORY AND SEN[12] as early as 1937 to explain certain data which arose from the study of respiration of barley plants as this was affected by nutrition. Briefly, their idea was as follows.

A certain amount of the protein in the organism was regarded as cyclically broken down to give amino acids as the immediate products of protein breakdown, the nitrogen was re-cycled and synthesized back into protein while the carbon skeletons were respired away. The idea of GREGORY AND SEN was that the pace of the protein cycle determined in part the pace of the respiration, although the net amount of protein present might remain steady.

With the subsequent availability of ^{15}N, two now classical papers by VICKERY *et al.*[65] on the one hand and by CHIBNALL AND WILTSHIRE[99] on the other furnished the evidence upon which the ideas of cyclical protein breakdown and re-synthesis in higher plants became familiar. These papers drew attention to the fact that ^{15}N, supplied originally as ammonia, entered the bulk protein even though the total amount of protein present either failed to increase or even decreased. It came, therefore, as a surprise that enough re-synthesis of protein occurred to incorporate so much of the ^{15}N, even in isolated leaves of tobacco and of beans where there was little need for bulk synthesis of protein and even where actual breakdown should occur.

Almost concurrently, work with slices of potato tuber (STEWARD *et al.*[54], p. 418) led to the view that a part of their respiration, when in dilute salt solutions, proceeded over pathways in which the respiration was closely linked to protein synthesis and the metabolism of soluble nitrogen compounds, whereas another part of this respiration seemed to be independent of the nitrogenous metabolism. In fact, there was a linear regression of CO_2 produced on protein synthesized, and this relation held despite a wide range of salt and oxygen conditions to which the tissue was subjected.

The work on potato discs required a similar interpretation to that of GREGORY AND SEN, namely that the very actively metabolizing cells synthesize protein from endogenous carbohydrate and nitrogen drawn from the nitrogen-rich reserves in the the cells. The synthesis of protein, the use of sugar and the disappearance of soluble nitrogen were all affected *pari passu* by several variables. The conclusion was that the carbon skeleton from the erstwhile storage nitrogen compounds contributed via the Krebs' cycle to the extra respiration which was so maintained. Such now nitrogen-free residues, which enter the Krebs' cycle as keto acids, spare the use of sugars via glycolysis, and enter the cycle under such low concentrations of carbon dioxide and such efficient conditions of carbon dioxide removal that the carboxylation of pyruvate could well be limiting (STEWARD AND STREET[55], *fc* p. 482–486).

RACUSSEN AND ARONOFF[32], working with isolated soya bean leaves, produced somewhat similar evidence, which in this case pertained to the source of the carbon which entered into the protein. They showed that $^{14}CO_2$ entered into the protein in

TABLE V

RELATIVE INCREASES IN FRESH WEIGHT, TOTAL PROTEIN, TOTAL RESPIRATION, AND TOTAL RADIOACTIVITY OF THE PROTEIN DUE TO GROWTH; *i.e.* RATIO OF THE QUANTITY IN THE FAST- TO THAT IN THE SLOW-GROWING CULTURES

From STEWARD AND BIDWELL[43].

	Ratio	
Fresh weight	3·7	
Total protein	3·7	
Respiration	5·1	
^{14}C-Activity in amino acids of the protein		
Glutamic acid	5·6	5·1 average
Aspartic acid	4·6	
Threonine	5·2	
Proline	3·9	3·0 average
Hydroxyproline	2·0	

the light by the synthesis of new protein, even though the total protein in the leaf was on the decline. From data of this kind one may conclude that there is a normal process of breakdown and re-synthesis of protein in the leaf, and that the immediate products of photosynthesis supply in part the carbon used in that re-synthesis. (This idea has also been mentioned in the earlier consideration of the data of Table IV on corn seedlings.)

But the idea of a cyclical system of protein synthesis and breakdown as a normal concomitant of metabolism and respiration comes most readily from the consideration of cells which are actively growing, or which are about to embark on active growth and development.

These concepts were necessary to explain the data which were obtained by the use of carrot tissue explants induced to grow under the stimulus of coconut milk under tissue culture conditions[44, 45]. It should be made clear, however, that in the first instance these experiments were not designed especially to this end, but their purpose was to study the fate of labeled sugars, glutamine and γ-AB in the actively growing cells, with a view to comparing the metabolism of amides in growing and non-growing cells, and so to determine the extent to which glutamine could furnish nitrogen and/or carbon for protein synthesis.

The crucial evidence for protein turnover, accentuated in the growing carrot cells, was as follows. Under the stimulus of coconut milk a net increase of protein occurred but the incorporation of ^{14}C into certain of the protein amino acids was increased by coconut milk several times more than the increase of the total protein, as shown in Table V. This led to the idea that far more of these amino acids were entering the protein fraction than could be accounted for by the increase in the total bulk of protein in the cells.

This general picture of protein turnover was substantiated by the consideration of the specific activity of glutamine in the cells. The essential arguments are re-

TABLE VI

TOTAL AMOUNT AND SPECIFIC ACTIVITY OF GLUTAMINE IN CARROT EXPLANTS SUPPLIED WITH ^{14}C-LABELED COMPOUNDS

Data from STEWARD, BIDWELL AND YEMM.

(^{14}C)Compound supplied	Time (h)	Slow-growing cultures (no C.M.)		Fast growing cultures (+ C.M.)	
		Amount (μg C)	Specific activity (counts/min/μg C)	Amount (μg C)	Specific activity (counts/min/μg C)
γ-AB	22	31	285	95	120
γ-AB	61	58	182	287	101
γ-AB	109	237	118	371	103
Glutamine	22	47	750	266	583
Glutamine	61	166	615	563	408
Glutamine	109	311	386	916	348

capitulated below, and the main data on which they are based are shown in Table VI.

(1) The fact of protein synthesis is established by the entry of ^{14}C from the exogenous substrate into the protein.

(2) Free glutamine in the cell had a continuously declining specific activity, and especially so when [^{14}C]glutamine was exogenously supplied. This unexpected fact indicated substantial entry of non radioactive glutamine into the soluble pool (see Table VI). Thus, the only evident source for the non-radioactive carbon of the glutamine which was being formed in the cell must have been protein by breakdown.

(3) The non-radioactive glutamine being synthesized in the cell did not come readily from sugar, because if [^{14}C]sugar was furnished the glutamine did not become radioactive.

(4) The [^{14}C]glutamine which was supplied exogenously to the cell could readily appear as CO_2.

(5) Furthermore, the formation of non-radioactive glutamine with time was much more rapid in the rapidly growing than in the slower growing cells. This was shown by the decline in its specific activity concurrently with the rapid increase in total quantity.

References p. 692/693

All these facts show that the rate of synthesis of non-radioactive glutamine from protein was even greater in the cells which were more rapidly growing. Thus, protein breakdown was also greater when synthesis was also greater. In other words, the pace of the protein turnover cycle had been increased in the more rapidly growing cells.

It was in fact shown that the chief effect of the coconut milk, which induced growth in the otherwise resting cells, was to cause first a net synthesis of protein, partly at the expense of the stored reserves in the cell and partly by the use of exogenous nitrogen, and also to accentuate the pace of protein turnover. The coconut milk thus acts to stimulate vital activity inasmuch as it accelerates the operation of the protein cycle of synthesis and breakdown, which in turn furnishes carbon substrates for respiration and causes a demand for new carbon for the re-formation of the protein so broken down. In this way the pace of the "protein cycle" draws carbon from sugar into the metabolism of nitrogen compounds.

The study showed that the amino acids which were present in the protein hydrolyzate fell into two groups. In one group, of which proline and hydroxyproline were the prominent examples, the increment of bound amino acids caused by the coconut milk was simply in proportion to the increase in the bulk of total protein. In the other group of amino acids, of which glutamic and aspartic acid, and threonine were prominent examples, the increase as indicated above far exceeded the increase of total. These data, briefly summarized in Table V, suggested that there were two distinct ways in which amino acids were entering protein in these cells.

In the case of proline, it seemed that the free proline which existed in the soluble phases of the cell could be directly incorporated into the synthesized protein; whereas other amino acids, which existed in bulk in the soluble constituents of the cells, did not appear to be the immediate precursors of the corresponding amino acid in the synthesized protein, for, in these cases, the carbon of these amino acids in the protein seemed to come more directly from sugar.

Thus the investigation of carrot explants led to the idea that there were two functional types of protein in the cell. Some protein, which incorporated certain amino acids directly, seemed to be structural and non-metabolized; whereas other protein seemed to be in a sufficiently active state that it was metabolized by breakdown and re-synthesized at rates which were linked to respiration. It should be recognized, and indeed it was implicit in the original thought, that one cannot at this point distinguish between metabolically active protein as an entire molecule, completely broken down or completely re-synthesized, and an alternative view that there is a metabolically active part of the protein which is merely a fraction of a larger molecule, so that the amino acids which are not susceptible to re-metabolism (*e.g.* proline) would be in the metabolically inaccessible part. Until work is done with isolated and completely purified proteins, these possibilities cannot be distinguished. However, the present working hypothesis is that there is some structural moiety of the protein of growing carrot cells which does not actively participate in metabolism and which incorporates proline directly.

Some indirect evidence which relates to protein turnover also accrues from the following. SUTCLIFFE[61] has returned to the idea (*cf.* summary in STEWARD AND SUTCLIFFE[56]; also see p. 424) that the intake and accumulation of ions by discs of tissue (carrot and beet) is linked to protein synthesis and turnover. It has been shown that chloramphenicol inhibits protein synthesis in bacteria, and SUTCLIFFE utilized this

References p. 692/693

reagent to study its effect on ion uptake and on respiration. Cells which have previously developed a salt-induced respiration by contact with KCl have this component unaffected by the chloramphenicol but concurrently their ability to accumu late ions is decreased. Therefore, SUTCLIFFE believes that protein synthesis and turnover in these cells is directly associated with ion uptake, though not with respiration in the general manner which has been postulated (*cf.* STEWARD AND SUTCLIFFE[56]).

WRIGHT AND ANDERSON[73] have studied the incorporation of [^{35}S]methionine into the free amino acid pool and into the protein of a slime mold. This is a very illuminating study, and the results are very relevant to this discussion. Briefly, these investigators find that the [^{35}S]labeled methionine readily enters into the protein of the slime mold, and this protein is regarded as in active turnover, even though the total protein in the organism is declining during its development. It is estimated that the rate of turnover of the protein in the slime mold is about 7% of the total protein per hour, as compared to such other rates of turnover as 5% for resting bacterial cultures, 1–2% for mammalian cells in culture. These figures may be compared with the rate of turnover of tobacco or of corn leaf protein which is estimated at $^1/_2$–1% from unpublished data in the laboratory of one of us (R.G.S.B.).

Furthermore, WRIGHT AND ANDERSON now distinguish between two broad classes of protein in the slime mold. These classes are distinguished as the ethanol soluble and trichloroacetic acid precipitable protein on the one hand the ethanol-insoluble trichloroacetic acid-precipitable protein on the other. ^{35}S-labeled methionine enters the protein of both these classes, but the interesting feature is that the ethanol insoluble moiety "turns over" much more rapidly than the ethanol soluble fraction. In both of these classes of protein the incorporation of the [^{35}S]methionine occurs even though the total protein is on the decline. This is evidence for turnover. But it is also important that the incorporation of the [^{35}S]methionine into the insoluble fraction is very much greater than that in the soluble fraction; and this ratio, which achieves a maximum value of 4, is a function of the development of the slime mold. Furthermore, WRIGHT AND ANDERSON point out that the incorporation of the [^{35}S]methionine into the more actively metabolizing protein moiety, that is the alcohol-insoluble fraction, which is a function of its turnover, seems also to be related to such other vital activities of the slime mold as oxygen consumption and certain enzymatic activities.

One can, therefore, conclude that a variety of evidence from a variety of plants and botanical systems all points in the general direction that cyclical turnover of plant protein certainly occurs, although it must be recognized that the absolutely final and critical proof may still remain to be produced. This can only emerge from the study of the specific activities of amino acids which have been derived from single and purified proteins, isolated from the cell, and from the changes which these incur during metabolism. Experiments to provide this type of evidence are now being planned.

In animal systems the synthesis of single proteins has been approached by selecting those proteins that have either enzymatic or hormonal properties and which can, therefore, be readily identified and purified. SCHAPIRA *et al.*[36] refer to rat muscle aldolase and can be quoted as follows: "It seems that a direct demonstration of the turnover of a well-defined protein has never been given. In our study we tried to make such a demonstration from muscle aldolase. The finding of an exponential decrease

References p. 692/693

of the radioactivity of rat muscle aldolase, labeled by injection of [^{14}C]glycine, provides the direct demonstration of an intra-cellular dynamic state of this protein."

The evidence seems to be as follows. [^{14}C]glycine was allowed to be incorporated into muscle, and the content of ^{14}C in the isolated aldolase was determined. Thereafter the animal was maintained in fresh weight and protein balance so that, if the protein was stable, its ^{14}C content should have remained steady. Either secretion of the enzyme into the plasma or the death and renewal of some cells would give an appearance otherwise consistent with turnover. However, the authors demonstrate that neither of these factors significantly affected their results, which nevertheless showed an exponential decrease with time of the [^{14}C]glycine content of the isolated aldolase. This decrease is explained by the breakdown of the first formed ^{14}C containing protein and its re-synthesis from unlabeled precursors.

SECTION V. PROTEIN SYNTHESIS AND TURNOVER IN RELATION TO DEVELOPMENT

In much current thinking on development it is implicit that stages of differentiation in the organism may be marked by the ascendancy of given protein moieties which are characteristic of different stages of development. This may occur by breakdown of one kind of protein formed at one stage of development followed by synthesis of another type of protein at a later stage of development (though how these modifications may be accomplished by changes at the presumed RNA template of synthesis is another matter).

Using roots of higher plants, BROWN *et al.*[6, 7] have traced the changing patterns of protein and of enzyme content in cells with their ontogenetic development as this can be seen along the axis of roots. There is clearly a stage of growth, predominantly by cell multiplication, in which the self-duplicating structures of the cell are reproduced, and one would expect the protein synthesis at this stage to reflect the formation of all the essential constituents of the organelles. Later, however, growth may be largely by cell enlargement. In the coleoptile of grasses and in the cells of roots it is now accepted that a measure of protein synthesis occurs throughout both these broad phases of the growth and development of the cell. However, BROWN has placed the data of protein or enzyme content upon a cellular basis, as distinct from a fresh weight basis and has noted that the content of protein in the cell increases up to a maximum and it then falls. It is interesting also that BROWN finds the point of maximum respiratory activity per cell coincides with the point of maximum protein content[7]. Moreover, BROWN sees the possibility that, as cells differentiate along the axis of a root, a different complement of enzymatic proteins comes into play.

Populations of bacterial cells furnish statistically convenient systems on which to seek evidence of synthesis and turnover. Although it had become rather widely accepted that protein turnover could occur in the more quiescent or resting cells, it has seemed that there was little or no conclusive evidence from cells in the logarithmic phase of growth.

Such conclusions had been reached for cultured mammal cells by KING *et al.*[20] using exogenous [^{14}C]leucine, who believed that there was positively no turnover in the actively growing cells. However, JORDAN AND SCHMIDT[19] have now repeated hese experiments, using much lower cell densities and much higher concentrations of the [^{14}C]substrate in the ambient medium. Under these latter circumstances a constant turnover of protein does take place, even in the exponential phase of growth, and these authors estimate that the bulk protein had a half-life of 3–4 days.

Therefore, protein turnover does appear to take place even in actively growing cells, though the degree of importance may well be different at different stages along the growth curve and in the development of the cell population.

The ability of cells in plant storage organs to re-synthesize protein may have some significance here. In the resting cells of a potato tuber the balance between the soluble nitrogen and the protein nitrogen remains quite stable during the life of the storage organ. It has, of course, been shown that this is drastically disturbed when

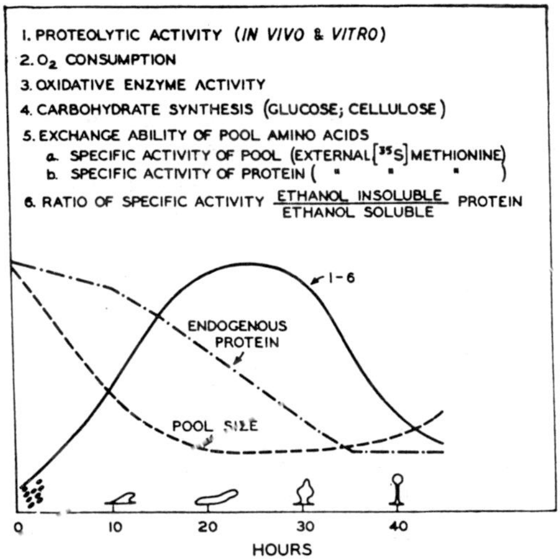

Fig. 2. A schematic representation of various biochemical changes occurring during differentiation in the slime mold (From WRIGHT AND ANDERSON[73]).

the cells commence to grow again, as in the behavior of cells at the surface of a cut slice. However, this condition of balance in the cells of the tuber, which may well be maintained by a constant but slow process of breakdown and re-synthesis, is drastically disturbed if the tubers are stored at a low temperature, particularly at a temperature of the order of +1°. It has been shown (*cf.* ref. 42, p. 244) that under these circumstances the ability of the cell to re-synthesize protein entirely disappears, so that the suggested normal cycle of breakdown and re-synthesis would be disturbed. Under these circumstances the tissue is unable to maintain many of its vital activities, notably the ability to retain solutes against distilled water and to absorb ions against an adverse diffusion gradient.

Another critical point in the life of resting or storage cells is the climacteric, as this is shown in various fruits; notably the apple and the banana. Again it appears that there is a relatively stable content of protein in the mature organ, but the climacteric, which is associated with a temporary increase of respiration and also of protein, is

References p. 692/693

also marked by irreversible changes which eventually lead to death (for references see STEWARD AND THOMPSON[57], p. 519). Here again there is a critical event which disrupts what would otherwise be the normal and stable level of turn-over.

At this point it is profitable to return to the work of WRIGHT AND ANDERSON[73] on slime molds. In addition to their finding that there are different protein moieties in the slime molds which have different rates of turnover, these authors also reach the interesting conclusion that the pace of that turnover is itself a function of development. In response to starvation, the slime mold enters upon a reproductive phase and its total protein content declines, but the rate of turnover of that protein which remains is actually much increased. This maximum rate of turnover is accompanied by a corresponding increase in the activity of enzymes in respiration, as measured by oxygen consumption, in carbohydrate synthesis and in the exchange of amino acids in the soluble pool with labeled amino acids that are supplied exogenously (Fig. 2). The interesting and applicable idea is that the rate of turnover in slime molds is associated with their morphogenetic development. In the carrot tissue cultures also the rate of protein turnover is associated with a degree of growth induction which, in free cells, leads on to such morphogenetic developments that whole plants may finally emerge[48].

The work on growth induction in plant cells has, however, provided even more direct evidence. As already stated, the cells which are treated with coconut milk synthesize protein in bulk, and they display accentuated turnover of their metabolically active protein. In addition, however, they synthesize a metabolically inactive moiety which has the distinctive property that it acquires an unusual hydroxyproline content by oxidation of proline after the latter has been built into the protein molecule[51]. This result is of special significance in two ways. First, there is no direct incorporation into this protein of any free hydroxyproline which they may absorb from the ambient medium[26]; second, this hydroxyproline-containing protein is quite essential for the growth induction and the morphogenesis which follows. By the use of either L-hydroxyproline or some of its derivatives, the incorporation of L-proline into the protein of the growing cells may be reversibly inhibited and, when this occurs, growth also stops[52].

SECTION VI. SOME IMPLICATIONS OF THE MECHANISM OF PROTEIN SYNTHESIS

Any ideas upon protein synthesis now tend to be dominated by the general working hypothesis that the DNA of the nucleus controls the synthesis of the RNA of the cytoplasm, and this in turn acts as the template which governs the sequence of amino acids in the protein that is synthesized. However, not all the protein synthesis may proceed over the same pathways. It would indeed be surprising if, in plants, the resting proteins of seeds which are elaborated from preformed nitrogenous compounds; the large amount of chloroplast protein in a leaf; the storage protein in the vacuoles of fleshy storage organs, and the metabolically active proteins formed in growing, dividing cells, as in a growing point, were all made by an identical type of chemical machinery.

Indeed, even in the animal kingdom there is some evidence to show that more than one method exists for the incorporation of amino acids into protein. Here one has to

References p. 692/693

admit to the difficulty of distinguishing what is commonly called "incorporation" from what would be more strictly termed "net synthesis". PROSSER *et al.*[30] experimented upon a pellet of material which was centrifuged after the separation of microsomes from rat-liver preparations. They were able to show that [^{14}C]leucine could be incorporated in the pellets, by an ATP-dependent mechanism, even though ribonuclease was present. In this way they demonstrated what they regard as ribonuclease-insensitive amino acid incorporation. By contrast, the microsomes and the supernatant sap lost the power of incorporating the leucine into protein when in the presence of ribonuclease, and this demonstrates what is regarded as ribonuclease-sensitive amino acid incorporation. There is, therefore, the suggestion that certain animal cells may not incorporate all the amino acid into their protein by the now conventional RNA-activation route. (It is interesting to note in this connection that HENDLER[17] (see also this Symposium) has described amino acid–lipid complexes which may also have some role in protein synthesis.)

Whereas recent well-known biochemical work (for summary see ref. 76) regards amino acid–RNA complexes as the immediate precursor of protein, there is other evidence that bears a somewhat different interpretation. RIETH[34] supplied cells of ascites tumors with ^{14}C-labeled aspartic acid and was then able to isolate from the cells a radioactive complex which hydrolyzed to ^{14}C-labeled aspartic acid and unlabeled adenylic acid or uridylic acid, as the case may be. This work, therefore, supports the idea that such complexes could also function as donors of amino acids at the site of synthesis. Furthermore, it is not even universally accepted that individual amino acids are invariably added singly into the orderly sequence in the chain which composes the protein molecule. ANFINSEN[40, 63] has supplied radioactively labeled amino acids to systems (hen oviduct or calf pancreas) which synthesize ovalbumin, ribonuclease and insulin, and has then hydrolyzed the resultant protein to recover the labeled amino acids from the different parts of the protein chain. ANFINSEN observed that a given amino acid may be differently labeled if it is isolated from different parts of the protein chain. For example, differences of up to 400% in the specific activity of alanine were observed. It is therefore inferred that single amino acids are not necessarily incorporated into the chain but that larger preformed units may be incorporated. However, using the rabbit, SIMPSON[37] concluded that the individual amino acids were incorporated singly into the enzymatically active protein that he studied (aldolase and glyceraldehyde-3-phosphodehydrogenase). Using various labeled amino acids, he found that the specific activities of the different amino acids and their relations to each other remained constant throughout the whole of the molecule. This evidence is held to be consistent with the incorporation of individual amino acids rather than of preformed peptides. Different types of amino acid incorporating systems also seem to be required in higher plants, as indicated below.

WEBSTER has followed a general pattern of investigation which was suggested by the work on animal systems and on microorganisms and has sought from higher plants evidence comparable to that which is produced by HOAGLAND and others (WEBSTER[71]). One may now summarize some of WEBSTER's latest findings.

Using preparations from pea seedlings (ribonucleoprotein particles), WEBSTER has followed the synthesis of a "soluble" protein which has ATPase activity. This is followed by measuring the release of phosphate from ATP. The evidence of synthesis

References p. 692/693

of the protein rests on measurements of ATPase activity or upon the incorporation of ^{14}C from amino acids into non-particulate protein. When the preparations are supplied with ^{14}C-labeled amino acids, complexed with what is described as "polynucleotide (sRNA)", both the radioactivity of the total protein and the enzymatic activity toward ATP were increased by a factor of 2.7; this increase was compared with the effect observed when the free amino acids were supplied alone. From this it is concluded that the nucleotide–amino acid complexes are more effective donors into this specifically identifiable protein than the free amino acids alone. In other words, this system which requires ATP, GTP and RNA constitutes the amino acid-activating system of protein synthesis. But some alternative suggestions for plants have also been made, or can be adduced, and these will now be mentioned.

In *Torulopsis* COWIE AND WALTON[10] now visualize an amino acid pool from which protein is constituted. However, the pool amino acids are not regarded as those which are free in the cell, but they are associated with macromolecules. Thus the

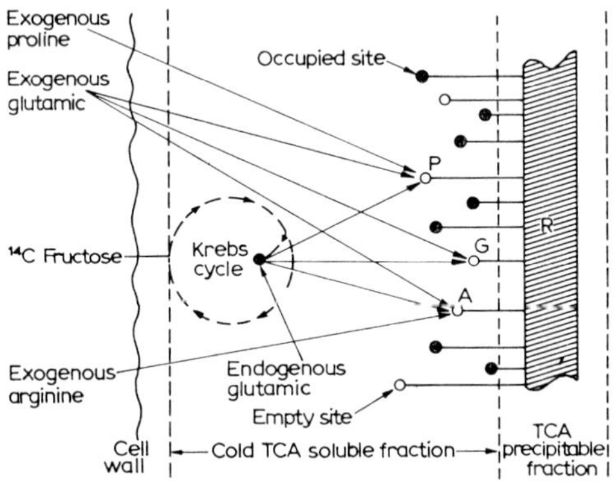

Fig. 3. Model to explain behavior of glutamic family of amino acids (From COWIE AND WALTON[10]).

binding of these amino acids is really equivalent to segregating them in a separate compartment. The normal source of the pool amino acids is by synthesis from sugar via the Krebs' cycle, and once an amino acid becomes part of the pool its conversion to protein is not interfered with by the competitive supply of exogenous amino acids. However, in occupying the binding sites, exogenous and endogenous amino acids may compete. This scheme (illustrated in Fig. 3) differs from that which was elaborated for carrot cells, principally because it does not include any concept of protein turnover, or of the independent metabolism of protein breakdown products in the phase or compartment which is associated with the storage of the soluble nitrogen compounds. This difference, however, may well be intelligible in terms of the different morphology and organization of the two systems. In yeast, more of the reactions seem to occur on the periphery of the cell, metabolic products are more readily excreted to the medium, and there is no analogue for the large aqueous vacuole of the parenchyma cell of plants in which so much organic material is stored internally.

MEDVEDEV[23] labeled the cytoplasmic proteins with ^{14}C, extracted them and then subjected them to partial or complete hydrolysis with pepsin or HCl respectively.

The labeled products of hydrolysis were then supplied to the petiole of similar bean leaves, and their reincorporation in leaf protein studied. More radioactivity was found in the synthesized leaf protein when the partial hydrolyzate was supplied than when the free amino acids of complete hydrolysis were furnished. From this it could be inferred that peptides were preferred intermediates in the synthesis of this protein. However, there is the possibility that the partially hydrolyzed proteins may have entered the plant cells more easily than the completely hydrolyzed, and would thus be more readily available for protein synthesis even if they had first to be converted to their constituent amino acids before being incorporated into protein.

It is still a very puzzling feature that one of the best documented cases of peptide bond formation in plants is that of HANES *et al.*[14, 15], which, however, leads to γ-glutamyl bonds which are not the kind required to build protein directly. It is suggestive that these γ-glutamyl bonds involve glutamine which so often seems to play a key role in protein synthesis in plants, and that there is also a mechanism for incorporating energy from ATP into this γ-linkage. It was suggested earlier (*cf.* ref. 49, p. 381) that the protein-synthesizing surface might well act first as a γ-glutamyl peptidase, and in this way the γ-glutamyl peptide could be regarded as a form in which the nitrogen for synthesis and the energy in the γ-linkage could be presented acceptably to the synthesizing surface. On this view, glutamine would in effect become an amino acid carrier; this would be consistent with its ubiquitous occurrence in active centers of protein synthesis in plants. It may also be mentioned that in the work of WAELSCH[69] glutamine often acts as a "starter" of protein synthesis in bacterial cultures. An interesting development along these lines is that, as communicated to this conference, THOMPSON (*cf.* this Symposium, p. 54) now finds it possible to detect in plants a variety of γ-glutamyl peptides, some of which correspond to the protein amino acids.

All the foregoing discussion indicates, therefore, that the immediate precursors of protein in plants are amino acids in some form or another, but it is equally evident that the amino acids which are immediately involved in the synthesis are certainly not those which constitute the often large stores of free soluble compounds. These amino acids which occur in bulk are in fact often in different compartments in the cell. Moreover, the composition of these soluble stores never reflects the amino acid composition of the resultant protein[58]. The problem, therefore, is to identify correctly the sites of both amino acid synthesis and of their incorporation into protein.

The interpretation of the carrot tissue culture sytem was that this site of protein synthesis was remote from the stored amino acids, was more accessible to sugar as a source of carbon than to the stored amino acids and, if amino acids were separately formed prior to their entry into the protein as they must surely be, then these individual molecules were not free to mingle with the amino acids of the soluble pool[43, 44]. YEMM AND FOLKES[75] have erroneously interpreted this scheme to mean that amino acids are not intermediates of protein synthesis. This was neither implied nor implicit in the scheme, nor is it necessary to make such a restriction until both the site and the mechanism of the primary synthesis of protein in the plant cell are precisely known. What is clear is that the amino acids that participate intimately in the actual formation of protein at the site of synthesis may be, and apparently are, formed there, and they do not mingle freely with those that are stored in bulk in the soluble phases of the cell.

References p. 692/693

SECTION VII. THE IMPORTANCE OF ORGANIZATION IN RELATION TO PROTEIN IN AMINO ACID METABOLISM

The purpose of this concluding section is to engender a respect for the organization of cells and of the plant body in the interpretation of the amino acid and nitrogenous metabolism of plants.

It has been shown that amino acid and protein metabolism are not fully understood without appreciating their interactions with carbohydrate and organic acid metabolism as well as with respiration; these interactions go even further because they include a variety of other physiological functions that are linked to protein synthesis and to growth, as for example water and ion intake in the growing cells.

Thus, important as individual enzyme reactions are, as in the present trend toward the understanding of activating enzymes which are specific for each amino acid, these considerations alone cannot properly yield a full understanding of the protein metabolism of even a single cell, which is a far too heterogenous system to be understood solely in this way. Although it is of great interest to see how much, or how little, synthesis may be achieved by particles in cell free preparations, this also is too attenuated a system to represent a whole cell, because it does not include the variety of organelles and loses everything gained by their interactions. Moreover, in the subdivision of the cells the essential features of the growing system are invariably lost. (An example is that isolated mitochondrial particles from carrot will incorporate [^{14}C]proline readily, but they do not form hydroxyproline from it[62].

Even isolated organs of the plant body also give but a partial view. Rarely can isolated leaves be induced to grow, and when they are excised and fully mature their nitrogen metabolism leans heavily toward protein breakdown. The attached leaf on the contrary, controlled by the factors which maintain integration in the plant body, preserves its greater ability to maintain itself in nitrogen balance. Throughout development each organ, and in particular each leaf primordium, traces out a sequential series of nitrogenous changes, the full significance of which in terms of both total protein synthesis and the kinds of proteins synthesized are still largely to be understood. One may, however, anticipate from observations already made that the kinds of protein metabolized in the phase of active cell division may be different from those that are largely involved during growth which is predominantly by cell enlargement.

Difficult as it is to discern these sequential changes by reference only to the relative amino acid composition of the total soluble pool and of the total protein hydrolyzate, nevertheless certain observations have been made on the shoot growing points of a flowering plant, *Lupinus*, and of a fern, *Adiantum*[60]. Briefly, and not unexpectedly, the metabolism of the dividing cells seems to stress the basic amino acids more than the metabolism of the cells that only enlarge. It is interesting to note that changes which are characteristic of the transition from the dividing to the vacuolated state in cells were detected, and it was the growing point of *Adiantum* that gave the first evidence of the presence in large amount of a then undiscovered amino acid, which later turned out to be γ-methyl-γ-hydroxyglutamic acid[59].

During development, however, there is an apparently brief phase in which the young leaf is most active in exporting its synthesized materials to other parts of the plant body and, by analogy with the distribution of other solutes, such as salts, and

References p. 692/693

by the strictly polarized movement of stimuli, like auxin, one may expect that this export is directed within the plant body to specific regions. The green leaf is, of course, accepted as a prime organ of protein synthesis, though much of that protein may be broken down for export to the growing regions. The same, of course, applies to starch. Using $^{14}CO_2$, HAUSCHILD *et al.*[16a] have re-examined the stage at which a tobacco leaf is most effective as an "exporter" of carbon to the growing regions and the stages when it actually makes demands by importing compounds which are synthesized elsewhere. They have found that it was only the very young leaves which exported any significant amount of newly fixed carbon. Therefore, only the young leaves really participate very actively in the metabolism of the whole plant; the rest are rather passive appendages. Thus it is mainly during the period of its active growth that each leaf is most significant as a nutritional unit.

Of particular interest are the interactions which have been observed between such environmental factors as length of day and night temperature, which interact with inorganic nutrition as in the case of potassium and calcium, and determine the balance between the soluble or free amino acids and the combined or protein amino acids in the plant. But it goes much further than this because these environmental factors in some way must make their impact on the enzymatic machinery of the cells to cause the great differences in the relative composition of the stored soluble nitrogen compounds which have been encountered[46]. In fact, it seems as though the point of contact between these environmental factors and the metabolism lies in that region of metabolism where, via keto acids, carbohydrate metabolism impinges upon nitrogen metabolism (*loc. cit.* see Figs. A–D, p. 157).

The nitrogen metabolism of shoots, however, lacks something essentially attributable to roots. This was early recognized by CHIBNALL and described as "hormonal control", but the difficulty of culturing minute portions of the shoot apex is also significant here. Minute apical segments of the shoot large enough to form angiosperm root primordia invariably grow well and will synthesize protein, but the central apex of the growing point and the smallest primordia will apparently not do so.

The facts of rest and dormancy also pose their problem because the quiescent cells, despite their frequent richness of soluble nitrogen compounds in both quantity and variety, fail to utilize this in protein synthesis and in growth. Perhaps many of the recently recognized and unusual soluble nitrogen compounds which have been found free in such organs may operate as metabolic inhibitors to suppress some line of synthesis. This may well be true of azetidinecarboxylic acid, which, like hydroxyproline, can act as a competitive proline inhibitor[52] and will, therefore, suppress protein synthesis in otherwise growing carrot cells. The marked intervention of factors that stimluate cells to grow by reactivating synthesis is significant here. The characteristic formation in all the rapidly proliferating cells so stimulated which have been examined (carrot, potato, crown gall tumor cells, etc.) of a protein moiety unusually rich in hydroxyproline is again suggestive that protein metabolism and morphogenetic activity may well be correlated. In the climacteric of fruits, after which cells seem irreversibly to have lost their capacity to grow again, some irreversible changes involving protein synthesis and turnover undoubtedly occur.

From all these points of view, therefore, the problems of amino acid metabolism, protein synthesis and turnover are not merely to be regarded as problems of biochemistry because their full understanding will only come when the biochemistry is linked to an understanding of the organization of the cell, the organ and the organism. This requires investigation by methods that take account of these complexities of the problem. While for many features of the individual reactions one must seek the simpler and more homogeneous systems, the work should not rest there, for this will not solve the problem of amino acid and protein metabolism of higher plants; it will merely evade many of them that are dependent upon the organization of cells and of the entire plant body. In this context the changes in nitrogenous compounds of higher plants that may be induced by such variables as photo- and thermo-period, as illustrated by the mint plant[46], are very much concerned with the effect of the variables in question, and their interaction, on the organized plant body. Similarly, the dramatic effects of growth induction upon the synthesis and turnover of protein in cells in culture should be thought of in relation to the equally dramatic and visible changes in the cells as they pass from the resting to the actively growing state (*cf.* STEWARD[41], Fig. 3, p. 452) for the comparison of carrot, potato and banana cells at rest in the intact organ and as they appear when in active growth in the free state). It is to explain such situations that one needs to invoke the concepts that have been developed in this paper.

REFERENCES

1 T. F. ANDREEVA, *Fiziol. Rastenii Akad. Nauk S.S.S.R.*, 3 (1956) 157.
2 A. A. BENSON AND M. CALVIN, *Ann. Rev. Plant Physiol.*, 1 (1950) 25.
3 A. A. BENSON AND M. CALVIN, *J. Exptl. Botany*, 1 (1950) 63.
4 R. G. S. BIDWELL, G. KROTKOV AND G. B. REED, *Arch. Biochem. Biophys.*, 48 (1954) 72.
5 E. G. BOLLARD, *Symposia Soc. Exptl. Biol.*, 13 (1959) 304.
6 R. BROWN AND P. BROADBENT, *J. Exptl. Botany*, 1 (1950) 249.
7 R. BROWN AND E. ROBINSON, in *Biological Specificity and Growth*, Princeton University Press, Princeton, N.J., 1955, p. 93.
8 T. CAYLE AND R. EMERSON, *Nature*, 179 (1957) 88.
9 A. C. CHIBNALL AND G. H. WILTSHIRE, *New Phytologist*, 53 (1954) 38.
10 D. B. COWIE AND B. P. WALTON, *Biochim. Biophys. Acta*, 21 (1956) 211.
11 H. EAGLE, *Federation Proc.*, 17 (1958) 985.
12 F. G. GREGORY AND P. K. SEN, *Ann. Botany (London)*, 1 (1937) 521.
13 N. GROBBELAAR, J. K. POLLARD AND F. C. STEWARD, *Nature*, 175 (1955) 703.
14 C. S. HANES, G. E. CONNELL AND G. H. DIXON, in W. D. MCELROY AND B. GLASS (Eds), *Symposium on Phosphorous Metabolism*, The Johns Hopkins Press, Baltimore Md., 1952.
15 C. S. HANES, F. J. HIRD AND F. A. ISHERWOOD, *Nature*, 166 (1950) 288.
16 J. HANFORD AND D. D. DAVIES, *Nature*, 182 (1958) 532.
16a A. H. W. HAUSCHILD, C. D. NELSON AND G. KROTKOV, *Can. J. Botany*, 40 (1962) 179.
17 R. W. HENDLER, *J. Biol. Chem.*, 234 (1959) 1466.
18 D. W. HOOD, C. M. LYMAN AND F. A. TATUM, *Arch. Biochem. Biophys.*, 30 (1951) 351.
19 H. C. JORDAN AND P. A. SCHMIDT, *Biochem. Biophys. Research Communs.*, 4 (1961) 313.
20 D. W. KING, K. G. BENSCH AND R. B. HILL, *Science*, 131 (1959) 106.
21 D. MARGOLIS, *Ph. D. Thesis*, Cornell University, Ithaca, N.Y., 1958.
22 C. M. MCCAY, in *Nutrition of the Dog*, Comstock Publishing Co., Inc., Ithaca, N.Y., 1949.
23 Z. A. MEDVEDEV, *Biokhimiya* (translated), 24 (1959) 84.
24 K. MOTHES, in W. RUHLAND, *Encyclopedia of Plant Physiology*, Vol. 8, Springer-Verlag, Berlin, 1958.
25 C. D. NELSON AND G. KROTKOV, *Can. J. Botany*, 34 (1956) 423.
26 J. K. POLLARD AND F. C. STEWARD, *J. Exptl. Botany*, 10 (1959) 17.
27 J. K. POLLARD, E. G. BOLLARD AND F. C. STEWARD, *Plant Physiol.*, 32, Suppl. XIV (1957).
28 J. K. POLLARD, W. F. ROCHOW AND F. C. STEWARD, *Plant Physiol.*, 33, Suppl. XII (1958).
29 H. PORTER, R. V. MARTIN AND I. F. BIRD, *J. Exptl. Botany*, 10 (1959) 264.
30 E. J. T. PROSSER, H. J. HIRD AND H. N. MUNRO, *Biochem. Biophys. Research Communs.*, 4 (1961) 243.
31 R. RABSON AND G. D. NOVELLI, *Proc. Natl. Acad. Sci. U.S.*, 46 (1960) 484.
32 D. RACUSEN AND S. ARONOFF, *Arch. Biochem. Biophys.*, 51 (1954) 68.
33 D. RACUSEN AND E. L. HOBSON, *Arch. Biochem. Biophys.*, 82 (1959) 234.

[34] W. S. RIETH, *Nature*, 178 (1956) 1393.
[35] R. B. ROBERTS, P. H. ABELSON, D. B. COWIE, E. T. BOLTON AND R. J. BRITTEN, *Carnegie Inst. of Wash. Publ. No. 607*, 1955.
[36] G. SCHAPIRA, J. KRUH, J. C. DREYFUS AND F. SCHAPIRA, *J. Biol. Chem.*, 235 (1960) 1738.
[37] M. V. SIMPSON, *J. Biol. Chem.*, 216 (1955) 179.
[38] D. C. SMITH, J. A. BASSHAM AND M. KIRK, *Biochim. Biophys. Acta*, 48 (1961) 299.
[39] J. F. SPECK, *J. Biol. Chem.*, 179 (1949) 1405.
[40] D. STEINBERG AND C. B. ANFINSEN, *J. Biol. Chem.*, 199 (1952) 25.
[41] F. C. STEWARD, *Can. J. Botany*, 39 (1960) 441.
[42] F. C. STEWARD, W. E. BERRY, C. PRESTON AND T. K. RAMAMURTI, *Ann. Botany (London)*, 7 (1943) 221.
[43] F. C. STEWARD AND R. G. S. BIDWELL, *J. Exptl. Botany*, 9 (1958) 285.
[44] F. C. STEWARD, R. G. S. BIDWELL AND E. W. YEMM, *Nature*, 178 (1956) 734, 789.
[45] F. C. STEWARD, R. G. S. BIDWELL AND E. W. YEMM, *J. Exptl. Botany*, 9 (1958) 11, 37.
[46] F. C. STEWARD, F. CRANE, K. MILLAR, R. M. ZACHARIUS, R. RABSON AND P. MARGOLIS, *Symposia Soc. Exptl. Biol.*, 13 (1959) 148.
[47] F. C. STEWARD, S. R. FREIBERG, A. C. HULME, M. P. HEGARTY, R. BARR AND R. RABSON, *Ann. Botany (London)*, 24 (1960) 117.
[48] F. C. STEWARD, M. O. MAPES AND K. MEARS, *Am. J. Botany*, 45 (1958) 693.
[49] F. C. STEWARD AND F. K. MILLAR, *Symposia Soc. Exptl. Biol.*, 8 (1954) 367.
[50] F. C. STEWARD AND J. K. POLLARD, in W. D. MCELROY AND B. GLASS (Eds.), *Inorganic Nitrogen Metobolism*, The Johns Hopkins Press, Baltimore Md., 1956, p. 377.
[51] F. C. STEWARD AND J. K. POLLARD, *Nature*, 182 (1958) 828.
[52] F. C. STEWARD, J. K. POLLARD, B. WITKOP AND A. A. PATCHETT, *Biochim. Biophys. Acta*, 28 (1958) 309.
[53] F. C. STEWARD AND C. PRESTON, *Plant Physiol.*, 15 (1940) 23.
[54] F. C. STEWARD, P. R. STOUT AND C. PRESTON, *Plant Physiol.*, 15 (1940) 409.
[55] F. C. STEWARD AND H. E. STREET, *Ann. Rev. Biochem.*, 16 (1947) 471.
[56] F. C. STEWARD AND J. C. SUTCLIFFE, in F. C. STEWARD (Ed.), *Plant Physiology; a Treatise*, Vol. 2, Academic Press, New York, 1959, p. 253.
[57] F. C. STEWARD AND J. F. THOMPSON, in H. NEURATH AND K. C. BAILEY (Ed.), *The Proteins*, Vol. 2, Part A, Academic Press, New York, 1954, p. 513.
[58] F. C. STEWARD, J. F. THOMPSON AND J. K. POLLARD, *J. Exptl. Botany*, 9 (1957) 1.
[59] F. C. STEWARD, R. H. WETMORE AND J. K. POLLARD, *Am. J. Botany*, 42 (1955) 946.
[60] F. C. STEWARD, R. H. WETMORE, J. F. THOMPSON AND J. P. NITSCH, *Am. J. Botany*, 41 (1954) 123.
[61] J. F. SUTCLIFFE, *Nature*, 188 (1960) 294.
[62] J. F. SUTCLIFFE, E. G. BOLLARD AND F. C. STEWARD, *J. Exptl. Botany*, 11 (1960) 151.
[63] M. VAUGHEN AND C. B. ANFINSEN, *J. Biol. Chem.*, 211 (1954) 367.
[64] H. B. VICKERY AND I. ZELITCH, *J. Biol. Chem.*, 235 (1960) 1871.
[65] H. B. VICKERY, G. W. PUCHER, R. SHOENHEIMER AND D. RITTENBERG, *J. Biol. Chem.* 135 (1940) 531.
[66] P. V. VITTORIO, G. KROTKOV AND G. B. REED, *Can. J. Botany*, 32 (1954) 369.
[67] N. P. VOSKRESENSKAYA, *Trudy Inst. Fiziol. Rasteniĭ im. K. A. Timiryazeva, Akad. Nauk S.S.S.R.*, 8 (1953) 42.
[68] N. P. VOSKRESENSKAYA, *Fiziol. Rasteniĭ Akad. Nauk S.S.S.R.*, 3 (1956) 49.
[69] H. WAELSCH, *Advances in Enzymol.*, 13 (1952) 237.
[70] G. C. WEBSTER, *Plant Physiol.*, 30 (1955) 351.
[71] G. C. WEBSTER, *Arch. Biochem. Biophys.*, 89 (1960) 53.
[72] D. G. WILSON, G. KROTKOV AND G. B. REED, *Science*, 113 (1951) 695.
[73] B. E. WRIGHT AND M. L. ANDERSON, *Biochim. Biophys. Acta*, 43(1960) 62.
[74] E. W. YEMM, *Proc. Roy. Soc. (London)*, 123 (1937) 243.
[75] E. W. YEMM AND B. F. FOLKES, *Ann. Rev. Plant Physiol.*, 9 (1958) 245.
[76] The Chemistry of Life: How Cells Synthesize Protein in Chemical and Engineering News, 39 (1961) Nr. 19, p. 80.

AMINO ACID POOLS, PROTEIN SYNTHESIS AND PROTEIN TURNOVER IN HUMAN CELL CULTURES*

HARRY EAGLE** AND KARL A. PIEZ

U.S. Department of Health, Education and Welfare, Public Health Service, National Institutes of Health, Bethesda, Md. (U.S.A.)

Although the amino acid metabolism of cultured human cells presents many points of similarity to the systems which have already been discussed at this Symposium, there are also important points of difference which deserve further exploration.

These cells may be grown either as a "monolayer" adherent to a glass surface, or in suspension cultures operationally similarly to cultures of bacteria in liquid media. We have found no difference in the amino acid metabolism of cells grown

TABLE I

A BASAL MEDIUM INCORPORATING THE MINIMUM GROWTH REQUIREMENTS OF CULTURED MAMMALIAN CELLS (FROM EAGLE[4])

Optional supplementation is as follows: (i) "non-essential" amino acids (alanine, asparagine, aspartic acid, glycine, glutamic acid, proline, serine), each at 0.1 m*M*; (ii) sodium pyruvate (1 m*M*). Of these, asparagine, serine, glycine, and pyruvate have proved necessary for the growth of certain cell lines, in a dialyzed serum medium, and serine is similarly required for the growth of single cells.

Compound	*Concentration*		*Compound*	*Concentration*	
	(mM)	*(mg/l)*		*(mM)*	*(mg/l)*
L-*Amino Acids*			*Salts*		
Arginine	0.6	105	NaCl	116	6800
Cystine	0.1	24	KCl	5.4	400
Glutamine	2.0	292	$CaCl_2$	1.8 (0) †	200 (0) †
Histidine	0.2	31	$MgCl_2 \cdot 6H_2O$	1.0	200
Isoleucine	0.4	52	$NaH_2PO_4 \cdot 2H_2O$	1.1 (11) †	150 (1500) †
Leucine	0.4	52	$NaHCO_3$	23.8	2000
Lysine	0.4	58	*Vitamins*		
Methionine	0.1	15	Choline		1
Phenylalanine	0.2	32	Folic acid		1
Threonine	0.4	48	Inositol		2
Tryptophane	0.05	10	Nicotinamide		1
Tyrosine	0.2	36	Pantothenate		1
Valine	0.4	46	Pyridoxal		1
Carbohydrate			Riboflavin		0.1
Glucose	5.5	1000	Thiamine		1
			Serum Protein		
			Whole or dialyzed serum, 5–10%		

† In suspension culture.

* The present paper describes the results from a single laboratory. No attempt has been made to review the literature in detail, or to document the results obtained in other systems.

** Present address: Department of Cell Biology, Albert Einstein College of Medicine, New York, N.Y. (U.S.A.).

References p. 705

by these two methods, and they are not distinguished in the following presentation. In contrast to the results reported yesterday by Dr. ROBERTS, who found significant and important differences in the amino acid pools of normal and tumor tissues, we have found no demonstrable differences between cultured cells deriving from such tissues[1].

Again in contrast to some of the systems which have here been described, the amino acid pool of these cells is not an intrinsic part of the cell, the composition of which is largely independent of the environment, but is instead in immediate dynamic equilibrium with the medium. The absolute amounts of amino acids within the cells must therefore be considered in relation to the composition of the medium at the time the cells were harvested.

Amino acid metabolism

Table I summarizes the growth medium used in the present experiments, and describes the minimum requirements for the serial propagation of mammalian cells in culture. Every factor there indicated is essential, and if any one is removed the cells die. Although eight amino acids suffice for nitrogen balance in man, these cultured

TABLE II

BIOSYNTHESIS OF AMINO ACIDS IN MAMMALIAN CELL CULTURES

Amino acid synthesized	*% of carbon skeleton deriving from uniformly ^{14}C-labeled*								*% of α-Amino N deriving from (^{15}N)-glutamic acid*
	Glucose	*Serine**	*Glycine**	*Ribose***	*Pyruvate***	*Glutamine**	*Glutamic****	*Arginine**	
Alanine	71–101	0.4–1.4	0.2	0.8–4	85	7	7	—	80
Serine	70–111	22–102	10	63–66	1.7	2	9.9	—	75
Glycine	69–106	25–99	50	58–60	1.9		7.0	—	76
Asparagine Aspartic acid	8–23	0.1–1.3	<0.1	0.48–12	4.7	71–87	100	—	78
Glutamine Glutamic acid	4–11	0.02–0.2	<0.1	0.31–0.63	4.1	101	84	—	83
Proline	3–18	0.09–0.44	0.15	0.2–0.3	3.1	59	27	26	58

* In basal growth medium, containing glucose.
** In medium lacking glucose.
*** In medium lacking glutamine.

cells require a minimum of 13: the eight "essential" amino acids plus arginine, cystine, glutamine, histidine and tyrosine. The remaining amino acids are synthesized from glucose and glutamine.

Glutamine occupies a central role in mammalian cell metabolism, and is the immediate precursor for the synthesis of glutamic acid, proline, aspartic acid and asparagine[2, 3]. To a minor extent, the proline derives also from arginine, by way of ornithine[4, 5]. Glucose provides the carbon skeleton for serine, glycine and alanine. If ribose and pyruvate are used in lieu of glucose, the alanine derives from the pyruvate, and to only a minor extent from ribose; conversely, serine and glycine derive from ribose, but not from pyruvate[4]. In these cell cultures there is no significant

References p. 705

metabolic interconversion between serine and glycine on the one hand, and alanine on the other (*cf.* Table II).

The α-amino group of the nutritionally non-essential amino acids derives solely from glutamic acid, this despite the fact that, as BARBAN has shown[6], cultured human cells contain a broad spectrum of transaminases. This further illustrates the point that the presence of a given enzymatic activity in a cell extract is no necessary indication of its physiological significance.

In these cultures we have not yet encountered end-product inhibition ("negative feedback") in amino acid biosynthesis. In all examples so far examined, the provision of pre-formed nutritionally non-essential amino acid has not inhibited its biosynthesis[5].

Intracellular amino acids and amino acid transport

The nutritionally essential amino acids are concentrated by the cell to varying degrees[7] (*cf.* Table III). The nutritionally non-essential amino acids, which are

TABLE III

THE AMINO ACID POOL OF THE HELA CELL

From PIEZ AND EAGLE[7].

		(A) Concentration in cell water (mM)	*(B) Concentration in medium (mM)*	*(A) (B) Distribution ratio*
Nutritionally essential amino acids, present in medium	Arginine	(0.03)	0.017	2
	Glutamine	8.1	0.72	11
	Histidine	0.26	0.032	8
	Isoleucine	1.00	0.125	8
	Leucine	0.73	0.097	8
	Lysine	0.29	0.119	3
	Methionine	0.19	0.035	5
	Phenylalanine	0.52	0.069	8
	Threonine	0.96	0.178	5
	Tryptophane	<0.1	(0.012)	<8
	Tyrosine	0.81	0.087	9
	Valine	0.79	0.111	7
Nutritionally non-essential amino acids, synthesized by cell	Alanine	1.43	0.081	18
	Asparagine	0.15	<0.005	>30
	Aspartic acid	1.27	0.012	106
	Cystine	0.05	0.023	2
	Glutamic acid	10.80	0.794	14
	Glutathione	4.42	<0.1	>44
	Glycine	0.79	0.005	158
	Proline	0.80	<0.005	>160
	Serine	0.03	0.003	10
Related compounds produced by cell	β-Alanine	0.50	<0.002	>250
	γ-Aminobutyric acid	0.17	<0.002	>85
	Ammonia	4.96	1.646	3
	Glycerophosphoethanolamine	1.36	<0.002	>680
	Ornithine	0.10	0.046	2
	Phosphoethanolamine	0.02	<0.002	>10
	Taurine	14.30	(0.002)	7150
	Urea	(0.55)	(0.43)	1

References p. 705

synthesized by the cell from glutamine and glucose, are concentratively retained by the cell to a generally higher degree; and a wide variety of compounds deriving in part from amino acids are also present in the pool, as indicated in the bottom section of Table III.

The absolute amounts of the intracellular amino acids there shown depend on the composition of the medium at the time that the cells were harvested. The important aspect of those data is the fact that there is a concentrative retention of newly synthesized amino acids, and a concentrative transport of amino acids from

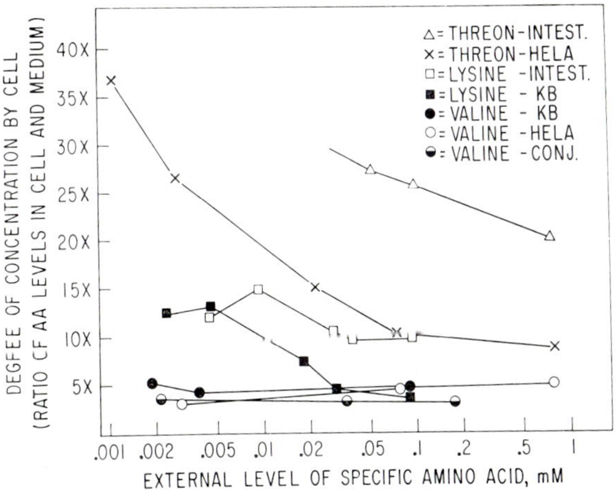

Fig. 1. The concentration of threonine, lysine, and valine by cultured human cells (from EAGLE AND PIEZ[18]).

the medium. As shown in Fig. 1, the degree to which the amino acids are concentrated by these cells is in some cases (*e.g.* valine) essentially independent of the absolute concentration of that amino acid in the medium. With other amino acids, as with lysine and threonine, the degree of concentration is concentration-dependent, as if there was a transport mechanism which was being saturated at the higher levels of amino acid.

In bacteria, the addition of large concentrations of one amino acid may have an inhibitory effect on the concentrative uptake of a second structurally or metabolically related amino acid. In these animal cells, although a given amino acid may similarly inhibit the transport of another amino acid, there is often no clear biochemical relationship between the compounds involved (Table IV).

As shown in Figs. 2 and 3, the rate of equilibration between the amino acid pool and the external environment is extremely rapid. Whether a high concentration of amino acid is added to the medium, or whether the external concentration is suddenly reduced, essential equilibration is attained within 15–30 minutes.

Amino acid depletion

1. When a cell is depleted of any one amino acid, there are profound structural changes, visible under the light microscope within 12–24 h. Those structural changes are evident in thin-section electron microscopy. Fig. 4(a) shows a normal cell, with a highly developed endoplasmic reticulum. Within 24–48 h after the cell is placed in

TABLE IV

THE EFFECT OF "PRE-LOADING" CELLS WITH AMINO ACIDS ON THE TRANSPORT OF [^{14}C]VALINE OR [^{14}C]THREONINE

Suspension cultures were depleted of valine (or threonine) by resuspension for 1 h at 37° in a medium from which that amino acid had been deleted. A single amino acid was then added in large excess (10 m*M*). After 1 h labeled valine or threonine was added for 5 min at 37°, and the culture rapidly chilled to 0°. The cells were collected by centrifugation in the cold, and the radioactivity of the cell pool determined.

L-*Amino acid used for pre-loading*	*Relative amount of labeled amino acid in cell pools after 5 min*	
	Cell	
	Conjunctiva	*Intestine*
	Label	
	Valine (0.1 mM)	*Threonine (0.1 mM)*
—	100	100
Arginine	100	98
Histidine	36	56
Lysine	73	112
Tryptophane	19	90
Phenylalanine	18	125
Methionine	54	36
Leucine	23	62
Isoleucine	55	84
Tyrosine	33	76
Cells exposed to label at 0°:		
a) with no pre-loading	—	26
b) pre-treated with 10 m*M* phenylalanine	14	27

a medium lacking a single essential amino acid, one may observe the profound structural alterations illustrated in Fig. 4(b). The endoplasmic reticulum largely disappears; the cell becomes vacuolated, and huge amounts of electron-dense material may accumulate. At this point, if one adds back the missing amino acid, there is a rapid recovery process. The normal appearance of the cell may be re-established within 24 h, and the cell resumes growth at a normal rate[8].

A surprising aspect of these single amino acid deficiencies is the fact that although the essential amino acids are not metabolized by the cell to an important degree, and are used primarily for protein synthesis, the microscopic appearance of these amino acid-deficient cells varies according to the specific amino acid which is deleted[9, 10].

2. We have now encountered a number of situations in which cells have a rigorous requirement for a metabolite which they can synthesize in amounts which should suffice for sustained growth, but in the absence of which the cells nevertheless die. In all cases so far examined, this paradoxical requirement is population-dependent, disappearing at a sufficiently high population density.

The first example encountered involved the growth of small numbers of cells. In the case of bacteria, small inocula often have specific nutritional requirements over and above those necessary for the growth of mass populations. Similarly, although

References p. 705

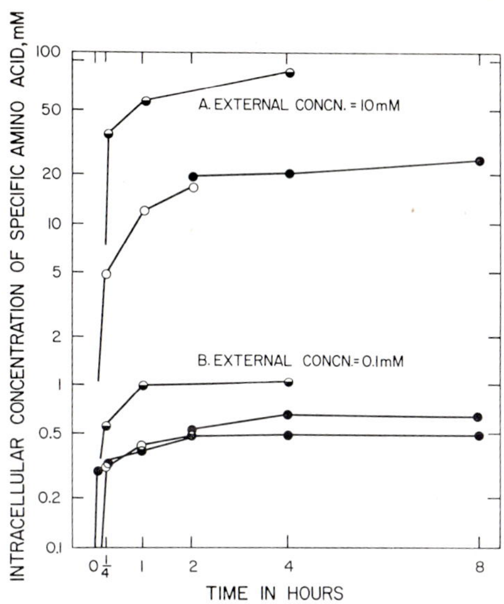

Fig. 2. The concentrative uptake of phenylalanine (○, ●) and threonine (◒) by HeLa cells. Suspension cultures of HeLa cells were resuspended in a phenylalanine-(or threonine-)free medium for 1 h at 37°. The amino acid was then added at a concentration of 0.1 or 10 mM, and the intracellular concentration of free amino acid determined on aliquot portions of the suspension after varying intervals at 37°. The rapid uptake, reaching essential equilibrium in less than 1 h, is shown in the figure.

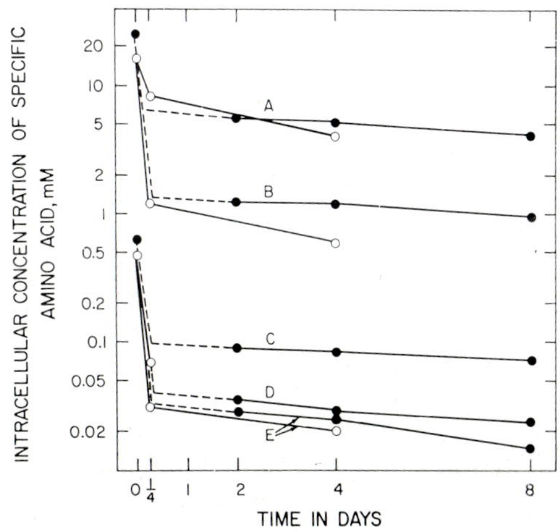

Fig. 3. The rapid loss of free amino acids from the cell to the medium. HeLa cells were preloaded with uniformly labeled [^{14}C]phenylalanine at 10 or 0.1 mM, in a complete growth medium. Aliquot portions were then diluted 1 : 10 (curves A, C), or 1 : 100 (curves B, D), in a phenylalanine-deficient medium, and the cell free amino acid content determined at the intervals indicated in the figure. In curve E, the cells were resuspended in a phenylalanine-deficient medium. The rapid loss to the medium, reaching essential equilibrium in considerably less than 1 h, is apparent in the curves.

References p. 705

single human cells will grow in the minimal medium of Table I if that minimal medium is supplemented with whole serum, they do not survive if one uses dialyzed serum instead, while large cell populations grow in the dialyzed serum medium as rapidly as they do with whole serum. The growth factor supplied by the whole serum, and necessary for the growth of small inocula, proved to be explicitly serine[11]. With its provision, a single cell could be grown in 10^{10} vols. of minimal medium. The effective serine concentration was on the order of 0.01 mM, and the population density at which the serine became unnecessary for growth was on the order of 100–500 cells/ml (ref. 5). A serine-requiring strain of rabbit fibroblast[12] (RT6) differed from the generality of mammalian cells only in the population density at which the serine requirement disappeared. The inoculum permitting survival growth and in an initially serine-free medium was approx. 10 000–50 000/ml, instead of 100/ml.

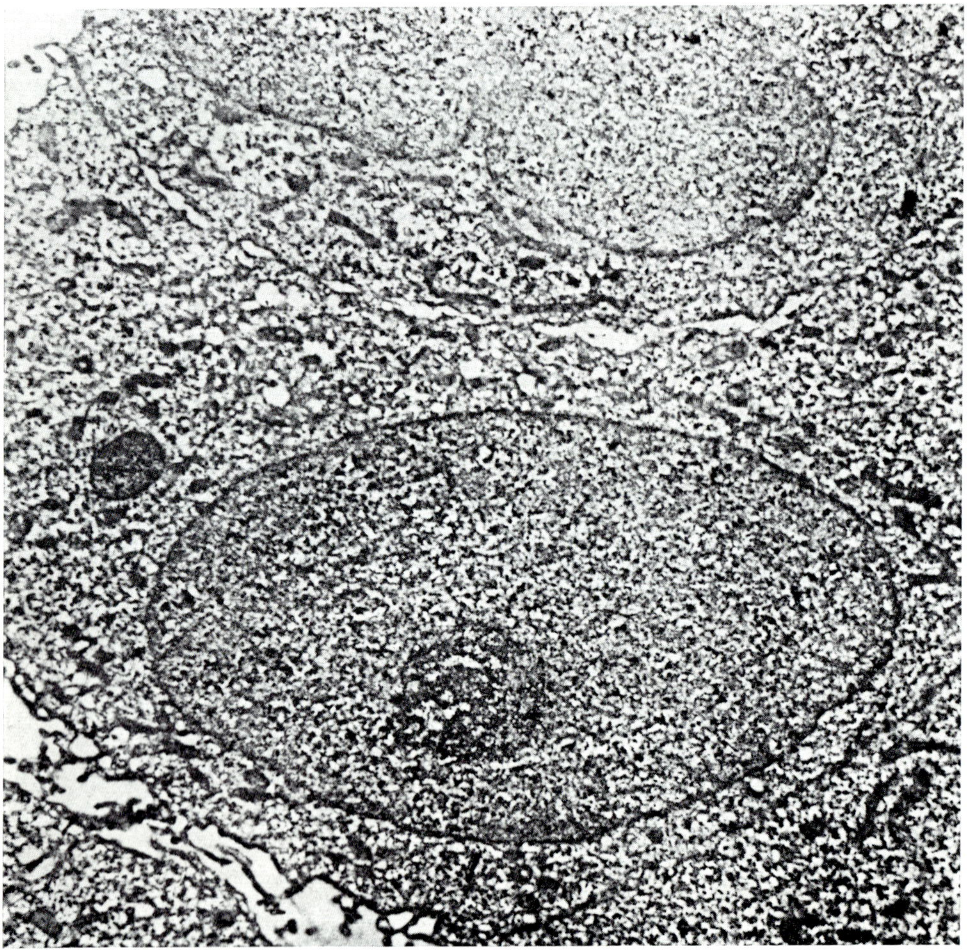

Fig. 4 (a). An electron-microscope photograph of a normal HeLa cell in thin section (from COHEN, NYLEN AND SCOTT[8]).

References p. 705

A number of similar examples of a population-dependent requirement have now been observed. Thus, all but one of the mammalian cell cultures so far examined had a rigorous requirement for inositol, despite the fact that when they were given pre-formed inositol, approx. 25% of the total inositol residues of the cell were being synthesized from the glucose of the medium[13]. At a population density of approx. 200 000 HeLa cells/ml, the inositol became unnecessary. The cells grew in the absence of added inositol, provided only that the population density was maintained in excess of 200 000 cells/ml.

Another example of this phenomenon was a mouse leukemia cell cultivated by HERZENBERG AND ROOSA[14], which required either serine or pyruvate for growth. This requirement disappeared if the population density of this cell line was maintained in excess of approx. 50 000–150 000 cells/ml.

Glutamine provides yet another example of population-dependent requirements

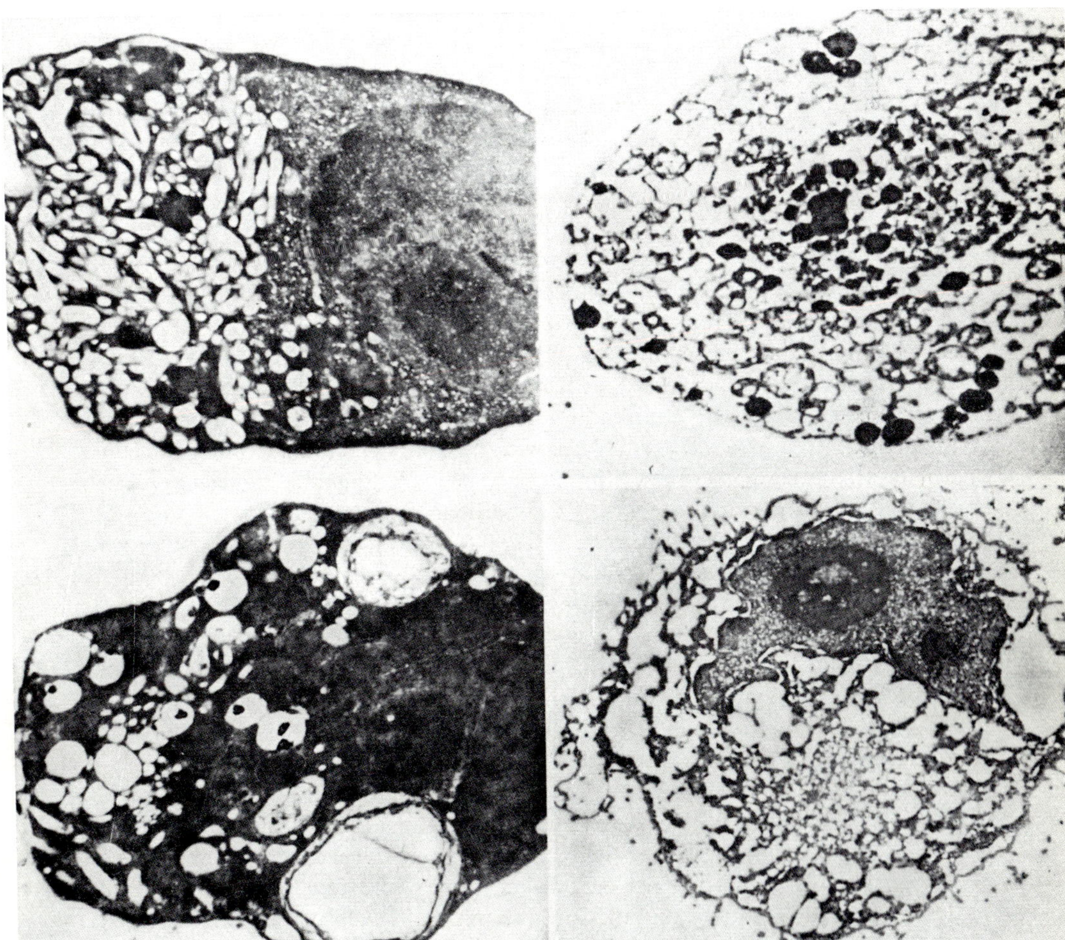

Fig. 4 (b). The profound structural alterations produced in the HeLa cell by valine deficiency (from COHEN, NYLEN AND SCOTT[8]), reversible on the restoration of valine to the medium.

References p. 705

for metabolites the cells can synthesize. Cultured human cells can be adapted to grow in the absence of added glutamine if they are exposed for varying periods of time to extremely high and non-physiological levels of glutamic acid[15]. As DEMARS showed[16] the cells then formed increased amounts of glutamine synthetase; having "adapted", they would then grow at relatively low levels of glutamic acid. However, it has recently been found that such adapted cells can grow in the absence of exogenous glutamine only at high population densities. If the population density is reduced below a critical level, even the glutamic acid-adapted cells have a glutamine requirement.

Perhaps the most striking example we have encountered of a population-dependent requirement is the case of cystine. Cystine is one of the five amino acids required for the growth of mammalian cells, even though it is not required by the whole organism. It has developed that every serially propagated human cell so far examined can indeed synthesize cystine from methionine and glucose by the classic pathway which involves the biosynthesis of homocysteine and serine as intermediates, their condensation to cystathionine, and the cleavage of the latter to homoserine and cystine[17]. However, if the cells are grown in a medium containing only methionine and glucose, in which the cell is under the necessity of making and retaining metabolically effective levels of homocystine, serine, cystathionine and cystine, it will not survive unless the population density is maintained in excess of about 500 000 cells/ml. If the cells are given pre-formed homocystine, so that they must now make and retain only serine and cystathionine en route to cysteine, they will grow with inocula of approx. 50 000–100 000/ml. If both homocystine and serine are provided, the critical population density becomes 50–500/ml; and if given both cystine and serine, one cell will grow in 10^{10} vols. of fluid.

The general explanation for these population-dependent requirements is probably the fact that at low population densities, the cells are unable to make enough of the specific compound to "condition" the medium, *i.e.* to bring its concentration in the medium up to a level in equilibrium with a metabolically effective intracellular pool, before the cells die of what is in effect a specific amino acid deficiency. At the critical population density, the task of conditioning the medium with respect to the product is shared by enough cells so that the necessary levels can be reached before the cells die; and in all the situations so far examined, at that critical population (but not at lower levels) one does indeed find that the concentration in the medium attains the minimum level necessary for the growth of smaller inocula[5].

3. Closely related to the foregoing is the size of the intracellular amino acid pool required for protein synthesis. To determine this point, the cells had to be grown at extremely low external concentrations under steady state conditions, such that the concentration of amino acids inside and outside the cell was not changing. This proved possible under cloning conditions, with an extremely large volume of fluid relative to the volume of cells, and a large amount of amino acid relative to that used for protein synthesis[18]. Under these circumstances, the rate of protein synthesis, measured by the reciprocal of the generation time, was a function of the external concentration of amino acids, and increased sharply with relatively small increments in the concentration of the external amino acid. From the data of Fig. 1 showing the degree to which the individual amino acids were concentrated by specific cell lines, one could obtain the internal concentration of amino acid corresponding to

References p. 705

each of these external concentrations. Fig. 5 shows the rate of growth, and thus of protein synthesis, as a function of these intracellular concentrations. There was a critical threshold level of amino acid (0.01–0.04 mM) below which protein synthesis proceeded at an insignificant rate. In excess of this critical level, the growth rate increased sharply to reach maximal levels at internal pool concentrations only two or three times the minimal effective level. Further increase in the intracellular amino acid concentration had no effect on the rate of protein synthesis.

Amino acid pools and virus synthesis

In the synthesis of poliovirus by HeLa cells, the only metabolites required by the cell for maximal production of poliovirus were found to be glucose and glutamine[19]. Clearly, the cell can provide from its own substance everything else needed for the

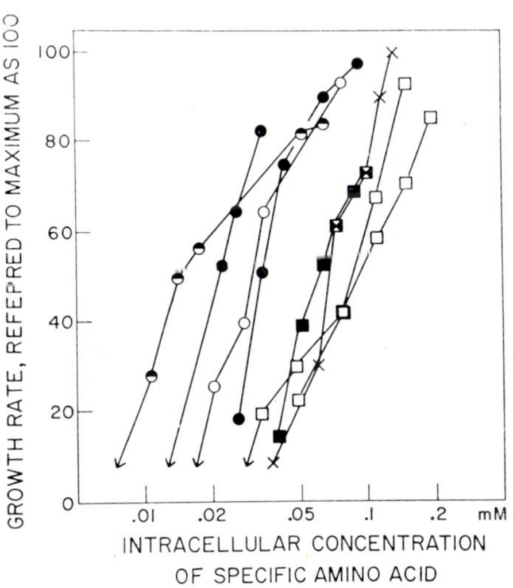

Fig. 5. The rate of growth of four human cell lines as a function of the intracellular concentration of valine, lysine and threonine (from EAGLE AND PIEZ[18]).

Symbol	*Cell strain*	*Amino acid*
◓	Conjunctiva	Valine
●	KB	Valine
○	HeLa	Valine
■	KB	Lysine
□	Intestine	Lysine
×	HeLa	Threonine

synthesis of viral protein and nucleic acid. What that early study did not make clear was whether the pool amino acids and nucleotides were being used, or whether instead the cells' own protein and nucleic acid were being broken down to provide the materials for viral synthesis. It was subsequently found that if the cell population density were reduced in an amino acid-free medium, and the cellular amino acid pool thereby drastically depleted, the cells did not elaborate poliovirus. The competence of the cell to make virus was totally restored if amino acids were added back to the medium[20].

References p. 705

This strongly suggested that the cell was making poliovirus from the amino acid pool. This was conclusively shown by DARNELL AND LEVINTOW in labeling experiments with purified virus[21]. With "cold" cell protein and a "hot" amino acid pool, the specific activity of the virus protein was essentially as high as that of the pool; conversely, with highly labeled protein and relatively unlabeled pool, the viral protein was unlabeled. The viral nucleic acid was similarly found to derive from the nucleotides of the pool[22].

Protein turnover

In cultured animal cells, as in bacteria and yeast, there is active protein turnover. There are, however, important qualitative and quantitative differences in this respect between animal cells and microorganisms. The turnover rate is approx.

TABLE V

THE CONTRIBUTION OF CELL PROTEIN TO THE AMINO ACID POOL

Cells were pre-labeled for 7 days in suspension culture in a growth medium containing [^{14}C]valine, washed and re-suspended for varying periods in unlabeled medium. The specific activities of the free valine in the medium and in the cell pool were then determined. In the table, these specific activities have been referred to that of the valine residues in the cell protein as 100.

Cell Strain	*Time for which labeled cells were grown in cold medium (h)*	*Specific Activities of valine residues in*			*% of free valine in cell pool deriving from protein*	*% of protein valine present in pool**
		Cell protein	*Cell pool*	*Medium*		
KB	5	100	4.0	2.2	1.86	0.070
HeLa	24	100	14.2	10.4	4.2	0.115

* Protein-derived free valine in pool, referred to total valine in protein.

1%/h, of the same order of magnitude as that observed in yeast, and less than the 5%/h rate in bacteria. This is, however, an extraordinarily high turnover rate considered in relation to the growth rate. Animal cells grow at a rate of approx. 3%/h. A 1% turnover rate per h implies that for every four molecules being synthesized by the cell, one is being broken down. Teleologically, protein turnover could be considered a mechanism for the replacement of inactive enzymes. On this basis one may ask why the mammalian cell needs this mechanism to so much larger an extent quantitatively than do either bacteria or yeast. An interesting qualitative difference between animal cells and microorganisms is the further fact that in the former, protein turnover proceeds at essentially the same rate in growing as in resting cells. One may estimate from the data of Table V that in cells growing on the medium of Table I, approx. 2–4% of the free amino acid of the pool derives directly from the cell protein by some turnover mechanism, rather than from the amino acids of the medium. The amount of that protein-derived amino acid corresponds to approx. 0.1% of the cell protein.

Dr. HALVORSON has pointed out the possible mechanisms which may be operative in this turnover process. The current orthodoxy is that protein turnover represents

References p. 705

the degradation and resynthesis of protein molecules; and this may very well be correct. Recently, however, we have obtained some experimental data which we find difficult to reconcile with this view.

In animal cell cultures, protein turnover can be stopped by a wide variety of metabolic inhibitors, but only at relatively high and toxic concentrations, at which the cells are rapidly killed. There is, however, a range of inhibitor concentration at which net synthesis of protein is stopped, while protein "turnover," *i.e.* the incorporation of amino acid residues into protein in peptide linkage, continues at essentially an unchanged rate of 1%/h.

It is difficult to see why an inhibitor which stops net synthesis should have no effect on a process which is supposed to involve degradation and resynthesis, unless one makes one of three unlikely assumptions. Thus, it is possible that protein degradation is a constant process, which in the growing cell is merely exceeded by the rate of synthesis. On this basis, however, one must assume that over a wide range of inhibitor concentrations, and for a number of quite different inhibitors, the rate of degradation and resynthesis are fortuitously equivalent, and equal to that observed in a medium in which net synthesis is blocked by the omission of essential amino acids. A second possible explanation is that there are two qualitatively distinct kinds of protein synthesis, and that the resynthesis associated with turnover is not stopped by inhibitors which block one or more metabolic steps essential for the net synthesis associated with growth. The latter is, however, merely a restatement of the problem. A third possibility is that amino acid residues in a polypeptide chain can be replaced without its total degradation to the amino acid level. This is consistent with the experimental observations; however, it involves a mechanism for which no convincing experimental evidence has yet been adduced.

REFERENCES

1. H. Eagle, V. I. Oyama and M. Levy, *Arch. Biochem. Biophys.*, 67 (1957) 432.
2. L. Levintow, H. Eagle and K. A. Piez, *J. Biol. Chem.*, 227 (1957) 929.
3. L. Levintow, *Science*, 126 (1957) 611.
4. H. Eagle, *Science*, 130 (1959) 432.
5. H. Eagle and K. A. Piez, unpublished results.
6. S. Barban and H. O. Schulze, *J. Biol. Chem.*, 234 (1959) 829.
7. K. A. Piez and H. Eagle, *J. Biol. Chem.*, 231 (1958) 533.
8. E. P. Cohen, M. U. Nylen and D. B. Scott, *Exptl. Cell Research*, 23 (1961) 443.
9. H. Eagle, *J. Biol. Chem.*, 214 (1955) 839.
10. H. Eagle, *J. Exptl. Med.*, 102 (1955) 37.
11. R. Z. Lockart, Jr. and H. Eagle, *Science*, 129 (1959) 252.
12. R. F. Haff and H. E. Swim, *J. Gen. Physiol.*, 41 (1957) 91.
13. H. Eagle, B. W. Agranoff and E. E. Snell, *J. Biol. Chem.*, 235 (1960) 1891.
14. A. Herzenberg and R. A. Roosa, *Exptl. Cell Research*, 21 (1960) 430.
15. H. Eagle, V. I. Oyama, M. Levy, C. L. Horton and R. Fleischman, *J. Biol. Chem.*, 218 (1956) 607.
16. R. DeMars, *Biochim. Biophys. Acta*, 27 (1958) 435.
17. H. Eagle, K. A. Piez and V. I. Oyama, *J. Biol. Chem.*, 236 (1961) 1425.
18. H. Eagle and K. A. Piez, *J. Biol. Chem.*, 236 (1961) 2039.
19. H. Eagle and K. Habel, *J. Exptl. Med.*, 104 (1956) 271.
20. J. E. Darnell, Jr., and H. Eagle, *Virology*, 6 (1958) 556.
21. J. E. Darnell, Jr., and L. Levintow, *J. Biol. Chem.*, 235 (1960) 74.
22. N. P. Salzman and E. D. Sebring, *Virology*, 13 (1961) 258.
23. H. Eagle, K. A. Piez and R. Fleischman, *J. Biol. Chem.*, 228 (1957) 847.
24. H. Eagle, K. A. Piez, R. Fleischman and V. I. Oyama, *J. Biol. Chem.*, 234 (1959) 592.

DISCUSSION

Chairman: JOHN REINER

ROBERTS: I wonder if Dr. EAGLE has compared the pools in cells and medium when more complex materials were used in the environment, with those found in these pools when cells are suspended in the relatively purified culture media which contain only a small amount of protein. I ask this question because we found no similarity whatsoever, between the pools observed in ascites tumor cells grown in animals and the pools found in the ascitic fluid in which the tumor cells were suspended. However, if we separated these cells from the fluid and put them into various buffers the pools began to look quite similar. I was just wondering whether we might be altering relationships between what Dr. COWIE has described as "internal" and "external" pools. Perhaps some factors in the ascitic fluid might shift more of the amino acids to the so called "internal" pool in mammalian cells while removing them into buffer might cause a shift to the "external" pool.

EAGLE: In the range of 0 to 20% serum, which means, therefore, in the range of 0 protein to approximately 1.5% protein in the environment, there was no effect whatever on the steady state relationship between the cell and the medium. We have studiously avoided using salt solution as the medium first, because it is lethal to these cells, and rapidly so, and second, because the equilibration between cell pool and medium is so rapid. The only way we can get meaningful pools in these cells under these conditions is to chill monolayer cultures and wash just once, very quickly with cold salt solution. If one washes three or four times, one begins to lose amino acids from the pool. Parenthetically, we have found that amino acid pools obtained from suspension cultures by centrifugating the cells preliminary to treatment with TCA result in artifact pools, due to cell proteolysis in the course of the centrifugation and packing.

HALVORSON: All of the claims in microorganisms for exchange incorporation, have essentially been explained by the finding that one was dealing with cell wall synthesis. At the moment there is no convincing example of such exchange incorporation. Dr. EAGLE's rather provocative observations lead me to ask this question. We have always been impressed with the extent to which turnover takes place when you carry out long-term experiments, suggesting that the bulk of the protein is involved. Now, in the experiments you just mentioned at the end of your talk, were these carried out long enough to implicate a large fraction of the protein? Also, could you comment on the characterization of the newly incorporated material in the presence of the inhibitor?

EAGLE: If turnover is studied in a non-growth medium, in which protein synthesis is prevented quite simply by leaving out one or more essential amino acids from the environment, then, as Dr. HALVORSON indicated, in a good experiment the cells continue to incorporate for as long as 72 hours, and at the same rate of 1 per cent per hour. One must conclude, therefore, that most of the cell proteins are involved in the turnover process. Since the turnover is observed with all the amino acids studied, essentially all of the amino acid residues of the cell protein are similarly involved.

If metabolic inhibitors are used at effective concentrations, the cells die and slough off in less than 72 hours. In our best experiments, however, we have gone 24 to 48 hours, at a turnover rate of 0.5–1 per cent per hour. The amino acids incorporated are in the hot TCA residue, and presumably in peptide linkage. We can only conclude that, at least to this extent, we are getting the continuing incorporation of labeled amino acids in peptide linkage in protein. I hold no brief for the thesis of exchange incorporation. I only say that it is consistent with our data, and I find it difficult to reconcile these data with the classic view of protein turnover as degradation and resynthesis unless one assumes two quantitatively different kinds of protein synthesis.

GUROFF: The isolated intact diaphragm and also the brain slice will maintain its normal endogenous concentration of amino acid when incubated for long periods of time in the presence or absence of substrate or in the presence of metabolic inhibitors. However, it will rapidly lose additional amino acid which has been accumulated *in vitro*. I would like to know, Dr. EAGLE, if the tissue culture maintains a constant low level of amino acid while losing the newly accumulated pool?

EAGLE: If you leave out of the medium any one amino acid, that amino acid essentially disappears from the cell pool and falls to levels of less than 0.001 μmoles per ml cells.

I should have mentioned earlier (and actually the observation is consistent with either view of protein turnover) that with appropriate isotope labeling experiments you can show that at any one moment in growing cells 2–3 per cent of the cell amino acid pool has derived from the cell protein, either by a process of degradation and resynthesis, or by a process of amino acid exchange.

LUCY: In this discussion of the control of protein turnover, Dr. HALVORSON suggested that increased protein breakdown may result from repression or induction of a latent degradative system that is normally present in the bacterial cell. As an example of the activation of a latent degradative system, he mentioned the liberation of ribonuclease that occurs on breakdown of the ribosomes of *E. coli*.

I would like to extend this suggestion to the control of protein breakdown and turnover in animal cells. The lysosome particles of animal cells contain a number of hydrolytic enzymes, including the cathepsin proteases. These particles provide a preexisting system that can, upon suitable activation, yield an active proteolytic enzyme.

At the Strangeways Laboratory in Cambridge, England, we have been studying a system that may possibly be important in the control of proteolysis in animal cells. It has been found that the addition of vitamin A to a suspension of lysosomes from rat liver results in the release of a protease from these particles. As an example of the possible function of vitamin A in controlling proteolytic activity, we have considerable evidence which indicates that the loss of polysaccharide from embryonic chick cartilage that occurs on treatment with excess vitamin A in tissue culture is the result of increased protease activity. The observations indicate that, in this instance, excess vitamin A alters the permeability of the lysosomes thereby liberating a proteolytic enzyme that attacks the protein–polysaccharide complex of the cartilage matrix.

ARONOFF: I would like to ask Dr. STEWARD two questions. The idea has occasionally been suggested here today that parts of proteins may metabolize, and others not. I would like to ask him initially, does he have evidence for this for a specific protein? Secondly, as a corollary to this, does the glutamine of TMV protein metabolize while the proline does not?

STEWARD: In answer to the first question, I was rather careful to point out that we could not distinguish between these two possibilities; that is, whether the whole protein is turning over or whether parts of it were being lopped off leaving an inaccessible part which would not be so removed. At first, when we obtained these data, we thought in terms of a turnover of a distinct protein, and I still think this may be a possibility. What we are actually doing, however, is to try to purify these substances, and by critical electrophoresis on the new acrylamide gels we are getting very characteristic patterns of bands. What we now hope to be able to do is to localize the labeling in a homogenous component, of which there are very many. Then we may be able to distinguish between these two possibilities. However, I think if you will ask Dr. POLLARD this question, he may say that he thinks in terms of a central part of a protein which receives the proline but, for some reason, is not accessible to turnover; for my part, I just don't know which possibility will on the evidence prove to be correct.

As for the second question, I was using the turnover of TMV as an example of a protein which did not turn over, so far as we know and which had different degrees of accessibility to different soluble components in the cell as potential sources of carbon.

THE ROLE OF THE LIVER AND THE NON-HEPATIC TISSUES IN THE REGULATION OF FREE AMINO ACID LEVELS IN THE BLOOD

LEON L. MILLER

Departments of Radiation Biology and Biochemistry, School of Medicine and Dentistry, University of Rochester, Rochester, N.Y. (U.S.A.)

It is the purpose of this report to describe observations made with the isolated rat-liver perfusion technique and with the eviscerated surviving rat in connection with the metabolism of the amino acids.

In the first part of this paper, studies are described in which ^{14}C-labeled amino acids were used from which it will become apparent that the liver plays an important, but not exclusive, role in the oxidation of all of the amino acids, and that the essential amino acids histidine, lysine, arginine, methionine, threonine, phenylalanine and tryptophane are almost exclusively oxidized by the liver. Further, it will be seen that the non-hepatic tissues are fully as capable as the liver of oxidizing a large group of amino acids, including leucine, isoleucine and valine.

The second portion of this report will deal with the changes produced in the plasma amino acids in the course of isolated rat-liver perfusion, and these changes will be compared briefly with the amino acid picture seen in the plasma of the eviscerated surviving rat. Here it will be emphasized that not only is the liver concerned with the oxidative destruction of excessive quantities of a variety of amino acids but it is also concerned with the active production or release of amino acids serving to maintain blood levels above minimum values. These latter observations, based originally upon paper-chromatographic procedures, have been extended and refined more recently by the application of the automatic amino acid analyzer technique of MOORE, SPACKMAN AND STEIN[1].

METHODS

Before discussing these experiments, some reference should be made to the isolated rat-liver perfusion technique as it has been developed in our laboratories. The operative procedure and the properties of the isolated rat liver in the perfusion have been described in some detail[2, 3]. The apparatus used is diagrammed in Fig. 1. The glass components readily made from commonly available standard laboratory glassware are now commercially available* as is also the stripping action pump originally described by CRISP AND DE BROSKE[4]. The system developed in our laboratories allows sampling of blood, expired CO_2, and the liver without interrupting the course of an experiment.

The criteria of the functional vitality of the isolated perfused liver have been discussed elsewhere[3].

* Blaessig Glass Specialties, 645 Atlantic Avenue, Rochester 9, N.Y. (U.S.A.).

References p. 721

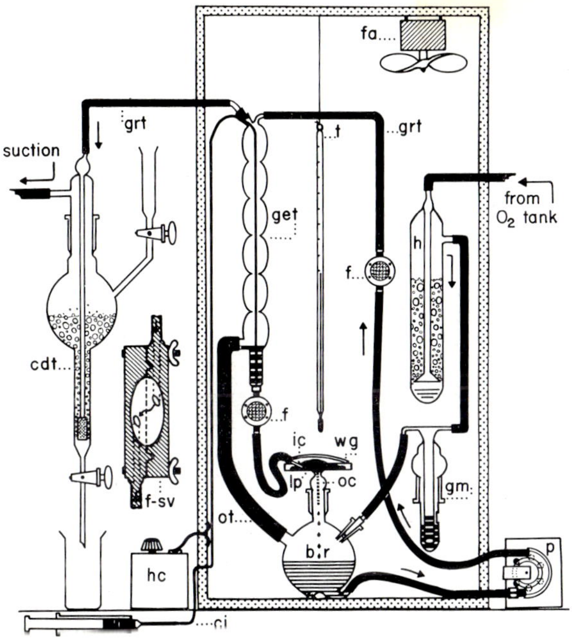

Fig. 1. Diagram of the apparatus used in the perfusion of the isolated rat liver. h, humidifier; gm, gaseous manometer; p, pump; f, filter; t, thermometer; fa, fan; get, gaseous exchange tube; ic, inflow cannula; wg, watch glass; oc, outflow cannula; lp, liver platform; br, blood reservoir; ot, overflow tube; f-sv, filter side view; grt, gaseous return tube; cdt, CO_2 trap; hc, heater control; ci, constant infusion. From GREEN AND MILLER[7].

TABLE I

OXIDATION OF ^{14}C-LABELED AMINO ACIDS* BY THE LIVER AND BY NON-HEPATIC TISSUES IN 5-H EXPERIMENTS

Amino acid	*Dose (mg)*	*$^{14}CO_2$ as per cent dose*		
		Liver perfusion	*Eviscerated rat*	
			Liver out	*Liver in***
L-Phenylalanine	14	25.8	0.13	27.7
DL-[2-^{14}C]Tryptophane	4.4	16.8	1.6	21.2
L-Arginine · HCl	156	16.1	3.0***	
L-Threonine	8	66	1.2	
DL-[2-^{14}C]Methionine	1.2 + 9.6 L-	47.7	2.0	22.0
L-[2-^{14}C]Histidine · HCl	10	36.7	0.35	
DL-[6-^{14}C]Lysine · HCl	0.6 + 21 L-	25	1.1	
L-Valine	11		19.3	31
L-Leucine	34	7.4	13.3	14.3
L-Isoleucine	15	4.3	27.5§§	17.7
L-Glutamic acid	43	47.8	40.2	62.5§§§
L-Glutamine	1.0	43.0	20.3	
[1-^{14}C]Glycine			8.4	19.0
L-Aspartic acid	2		53†	63
L-Proline	2		17.2†	

* All amino acids uniformly labeled unless otherwise indicated.
** Liver left in supplied with hepatic artery.
*** Dose, 2 mg.
§ Dose, 17.4 mg, duration, 4 h.
§§ Duration, 4 h.
§§§ Dose, 5.5 mg.
† Kidneys also removed.

References p. 721

General procedures and details of the chemical methods used in the studies described here have been given in previous publications[5].

Paper-chromatographic studies have been carried out using essentially the technique of DENT, phenol saturated with water, and butanol–acetic acid–water (4 : 1 : 5), being the two solvent systems used in two-dimensional chromatographic studies.

The technique of evisceration used is that of INGLE[6] and entails removal of all abdominal viscera excepting kidneys. In a number of instances the liver was left functioning with the intact hepatic artery supplying blood.

RESULTS

Part 1. Oxidation of labeled amino acids to $^{14}CO_2$ by the isolated perfused rat liver compared with the non-hepatic tissues as exemplified by the eviscerated surviving rat

Table I lists first a group of amino acids which are oxidized extensively by the isolated perfused liver and to a small or insignificant extent by the non-hepatic

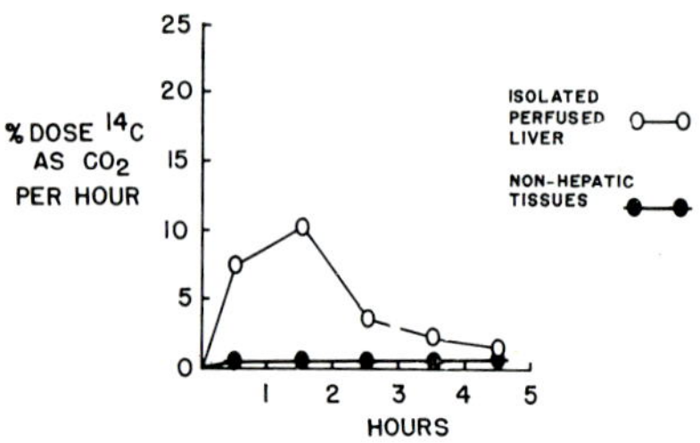

Fig. 2

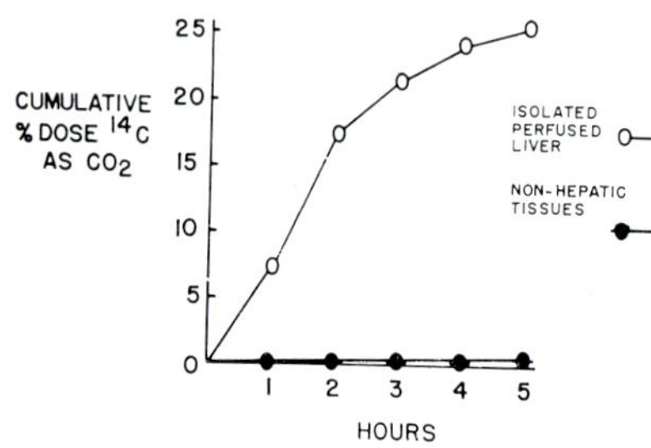

Fig. 3

Figs. 2–3. Oxidation of L-[^{14}C]phenylalanine by the liver compared with non-hepatic tissues. 18-h fasted Sprague-Dawley male (weight, 261 g), used as liver donor for perfusion (RLP 371); blood volume, 202 ml, to which was added at the outset 14 mg of L-[U-^{14}C]phenylalanine (5 μC) and 500 mg of glucose. Duration of perfusion, 6 h. Eviscerated rat (EVR-PA-6), Sprague-Dawley male (weight, 212 g), eviscerated under ether anesthesia. 24 mg L-[U-^{14}C]phenylalanine in 3 ml of Ringer's solution injected intravenously at zero time. Duration of experiment, 5 h.

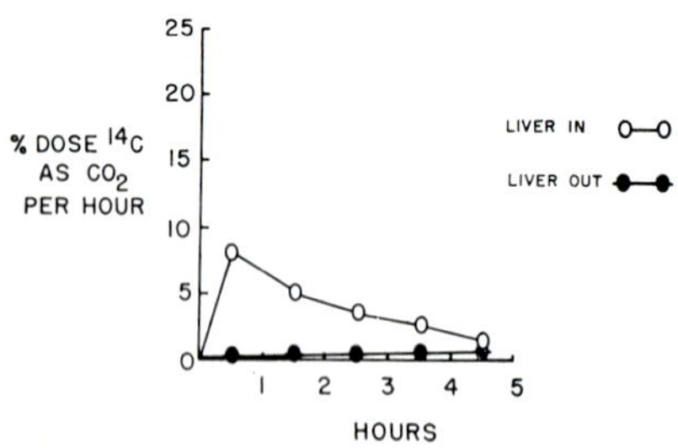

Fig. 4

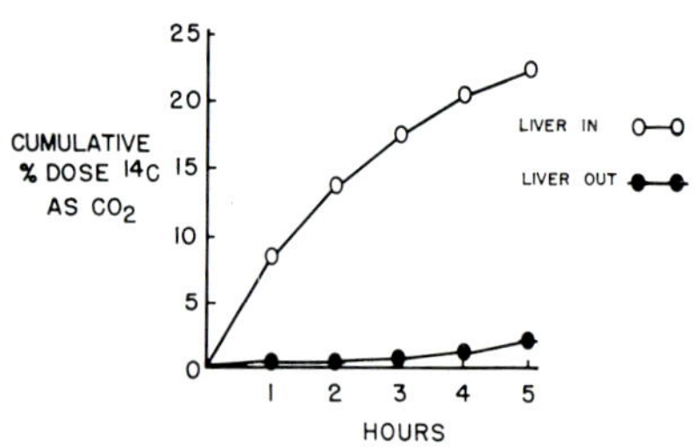

Fig. 5

Figs. 4–5. Oxidation of DL-[2-^{14}]Cmethionine in the eviscerated rat.
Eviscerated surviving rat with liver left functioning supplied by the hepatic artery (EVR-ME-4), normal Sprague-Dawley male (weight, 259 g), received a total of 1.3 mg DL-[2-^{14}C]methionine plus 9.6 mg of L-methionine in approx. 2 ml of Ringer intravenously at the outset of the experiment. Duration of experiment, 5 h. Eviscerated surviving rat, with liver out, (EVR-ME-2) received the same dose of methionine intravenously as above. Duration of experiment, 5 h.

References p. 721

tissues. Here, we have used the eviscerated surviving rat with the liver left functionally *in situ* supplied by the hepatic artery as an alternative to an isolated liver perfusion study. Figs. 2–5 give a more complete hour-to-hour picture of the course of oxidation of L-[U-^{14}C]phenylalanine and of DL-[2-^{14}C]methionine. These results are typical of those also obtained with the essential amino acids histidine, lysine, arginine, threonine and tryptophane, in which oxidation by the liver is most extensive within the first few hours. Figs. 6 and 7 compare the oxidation of DL-[1-^{14}C]leucine by the liver with that by the non-hepatic tissues. It is at once apparent that in contrast to the first group of essential amino acids, leucine is as extensively oxidized to $^{14}CO_2$ by the non-hepatic tissues as by the liver. As Table I reveals, isoleucine and valine are similarly as extensively metabolized by the non-hepatic tissues as by the liver.

The relatively low oxidation of L-[U-^{14}C]leucine by both the liver and non-hepatic tissues is reflected in the unusually high incorporation of L-leucine into both liver and plasma proteins which we have described recently with GREEN[7].

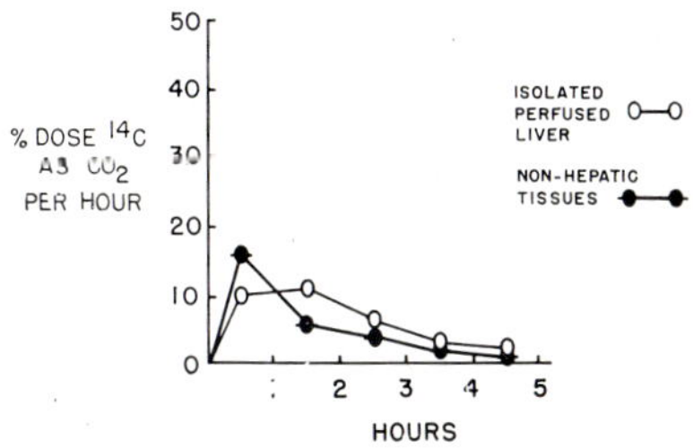

Fig. 6

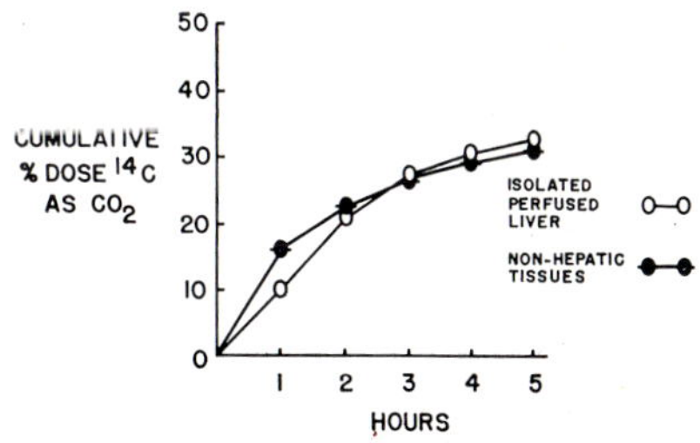

Fig. 7

Fig. 6–7. Oxidation of DL-[^{14}C]leucine by the liver compared with non-hepatic tissues. Rat liver perfusion (RLP 547). Fed Sprague-Dawley male (weight, 350 g) as liver donor. Liver weight, 9.3 g; 17.5 mg DL-[1-^{14}C]leucine (3.3 μC) plus 500 mg glucose added to the perfusion blood (83 ml) at the outset of the experiment. Duration of experiment, 5 h. Eviscerated surviving rat (EVR-LE-16), male, Sprague-Dawley (weight, 245 g), eviscerated under ether anesthesia. 3.5 mg (3.3 μC) DL-[1-^{14}C]leucine plus 31.6 mg L-leucine injected intravenously in 2.2 ml of saline at the outset of the experiment. Duration of experiment, 5 h.

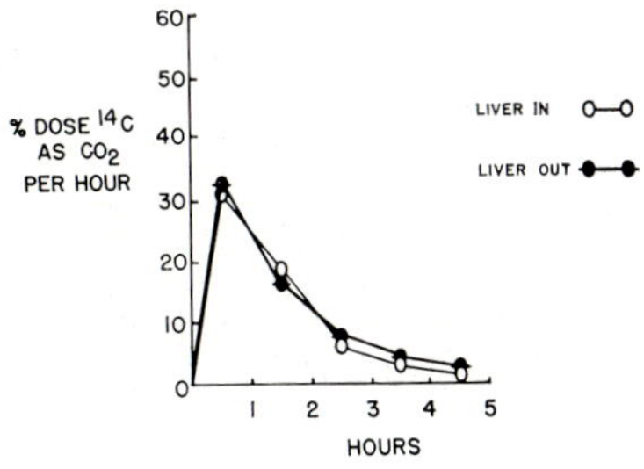

Fig. 8

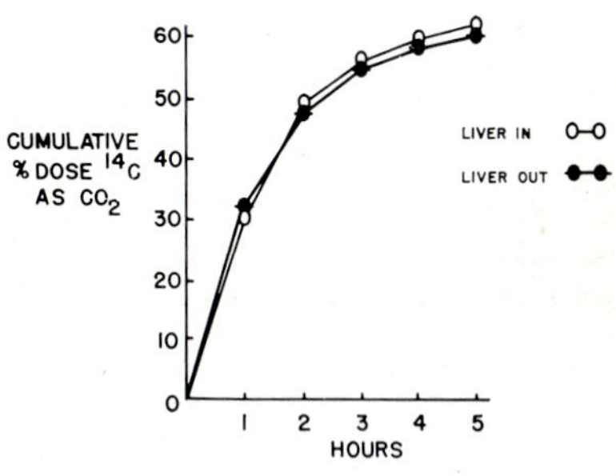

Fig. 9

Fig. 8–9. Oxidation of L-[^{14}C]alanine in the eviscerated rat. Eviscerated surviving rat (EVR-AL-3), male, Wistar (weight, 287 g); dose of 38.25 mg L-[U-^{14}C]alanine, 0.05 mg L-[U-^{14}C]alanine (5μC) plus 22 mg DL-alanine in a volume of 2 ml of saline given intravenously at the outset of experiment. Duration of experiment, 5 h. Eviscerated surviving rat (EVR-5), normal male, Sprague-Dawley (weight, 351 g), liver left functioning supplied with hepatic artery; dose identical with EVR-AL-3, given at outset of experiment. Duration of experiment, 5 h. Liver weight, 8.6 g.

References p. 721

Table I also summarizes comparative observations on the oxidation of a group of non-essential amino acids which are as extensively oxidized to $^{14}CO_2$ by non-hepatic tissues as by the liver. Figs. 8–11 present some more detailed data on the oxidation of L-[U-^{14}C]glutamic acid and L-[U-^{14}C]alanine. In summary, the experimental results here clearly separate the amino acids into two large groups, one group of seven essential amino acids which are almost exclusively oxidized by the normal liver and another substantial and larger group which contains the amino acids that are as avidly oxidized by non-hepatic tissues as by the liver.

DISCUSSION

The observations of BOLLMAN, MANN AND MAGATH[8] on gross changes in the blood amino acid level in hepatectomized dogs led some to conclude that the liver was the sole site of deamination and presumably of oxidation of amino acids in the mammalian organism. Our observations lead to significant modification of this point of view. The results we have obtained with [^{14}C]amino acids more sharply focus attention on the group of essential amino acids, excluding leucine, isoleucine and valine, in connection with experimental and clinical hepatic insufficiency. They make more understandable how hepatic insufficiency may produce the qualitative and quantitative changes in the blood amino acid picture[9–11] described in the literature. Thus, in terms of our results, severe hepatic insufficiency connotes inability to handle lysine, arginine, histidine, phenylalanine, methionine, tryptophane and threonine. Of this group, methionine and phenylalanine and their related derivatives, cysteine and tyrosine, have received attention from a variety of observers[12, 13]; and it may not be entirely fortuitous that the non- hepatic tissues of the rat oxidize phenylalanine and methionine to the lowest extent of any of the amino acids.

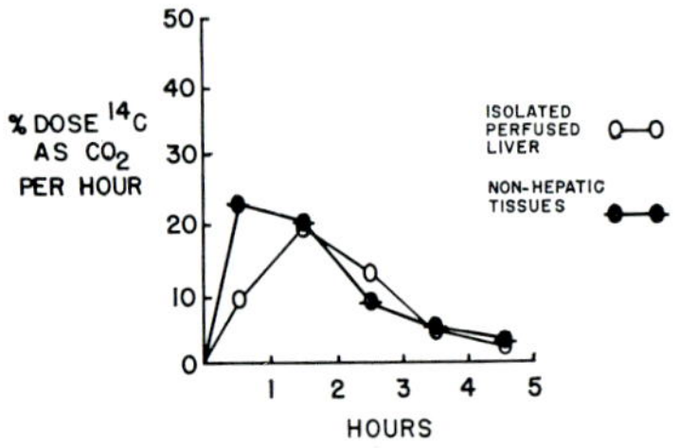

Fig. 10.

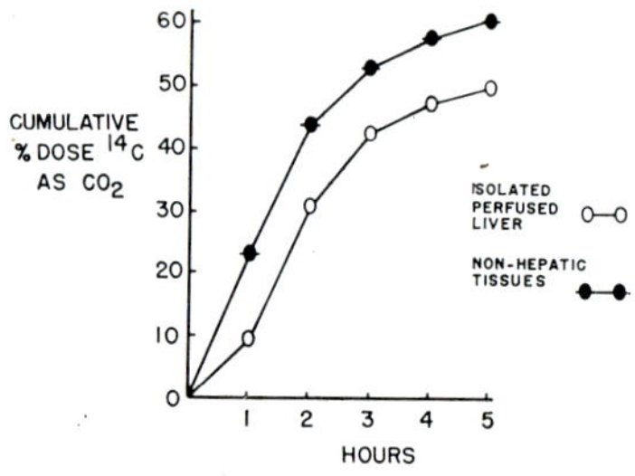

Fig. 11.

Fig. 10–11. Oxidation of L-[^{14}C]glutamic acid by the liver compared with non-hepatic tissues. Rat liver perfusion (RLP 515). Normal Wistar male (weight, 328 g) as liver donor, liver weight, 7.2 g. Dose of 0.5 mg of [U-^{14}C]glutamic acid (4 μC) mixed with 51 mg L-glutamic acid and 500 mg of glucose added to perfusion blood, 191 ml. Eviscerated surviving rat (EVR-Gl-AC-16), Sprague-Dawley male (weight, 183 g); dose of L-glutamic acid same as for the perfusion above.

In general, it becomes clear that the non-hepatic tissues are capable of oxidizing amino acids corresponding to approx. 50–65% of the nitrogen in the average protein. In terms of our current knowledge of intermediary amino acid metabolism, the production of glutamine in the peripheral tissues serves as a relatively non-toxic means of transport of nitrogen to the liver. It is of some interest, we believe, that although severe changes in hepatic function incidental to the precancerous state or

References p. 721

to experimental cirrhosis are associated with impairment of capacity to oxidize amino acids and to produce urea from amino acids[14] that the capacity to produce urea at a normal rate from glutamine is retained[15]. This is all the more remarkable since this occurs even at a time in the development of experimental hepatic insufficiency when the production of urea from arginine is also obviously impaired. We have pointed to this observation in the past[16] as one imposing bit of evidence in favor of a pathway of urea synthesis other than that conventionally associated with the Krebs-Henseleit urea cycle.

Part 2. Changes in blood amino acid levels seen during isolated liver perfusion studies

We have previously[5] summarized our observations pointing to the fact that when a complete amino acid mixture is added to the blood in the isolated perfused rat-liver system, the normal liver rapidly clears such amino acids from the circulating blood. Thus, a test dose consisting of a mixture of all the essential and non-essential

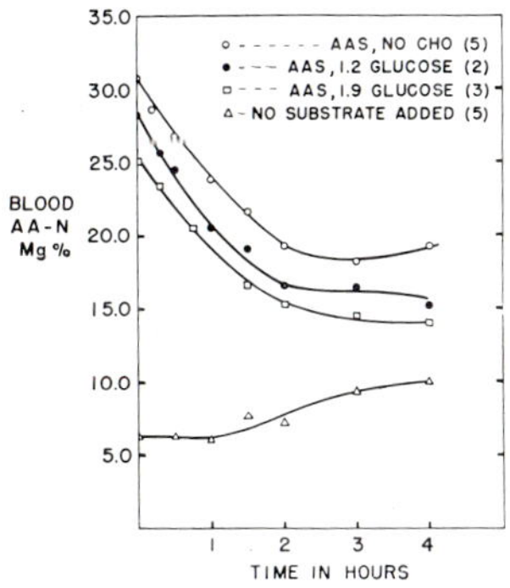

Fig. 12. Changes in blood amino acid concentration during isolated rat-liver perfusion. For detailed explanation, see MILLER, BURKE AND HAFT[5].

amino acids whose composition is described in detail elsewhere[2] is rapidly cleared from the blood with an equilibrium value of 14–18 mg% of α-amino acid nitrogen being reached at the end of 4–6 h (Fig. 12). In order to obtain a more detailed picture of the specific amino acid changes occurring during the course of perfusion of a complete mixture of amino acids[5], we have separated the amino acid from equal volumes of plasma taken at zero time and at the end of 5 or 6 h of liver perfusion. Fig. 13 compares the initial and final chromatograms and reveals that during the course of the perfusion, there was a general decrease in the apparent concentration of most amino acids. However, there is a striking persistence, if not an actual increase, in the leucine–isoleucine and valine spots. This latter observation may, perhaps, be thought to represent the result of the quantitatively smaller degree of oxidation of leucine, isoleucine, and valine as noted previously for the isolated perfused liver. However, it will become apparent from subsequent observations that the persistence of, or a real increase in leucine, isoleucine and valine probably results from protein breakdown by the liver. In fact, in a liver perfusion (exemplified by Fig. 14), where no extra amino acids have been added to the perfusing blood, enhanced leucine–isoleucine spots stand out prominently in chromatograms of the

References p. 721

plasma amino acids at the end of the experiment. In such experiments we have previously alluded to the fact that the total free α-amino acid nitrogen of the perfusion blood is maintained relatively constant for the first 2 or 3 h of the perfusion and then increases somewhat in the 5th and 6th h (see Fig. 12)[5, 18].

In such experiments, although no amino acids are added to the perfusion mixture, there is a continuous, essentially linear production of urea[5, 18] and maintenance or small increase in the total α-amino acid nitrogen level of circulating perfusing blood. Fig. 14 reveals that in such liver perfusions there has probably been a quantitative redistribution with an enhancement of the leucine–isoleucine and valine content. It

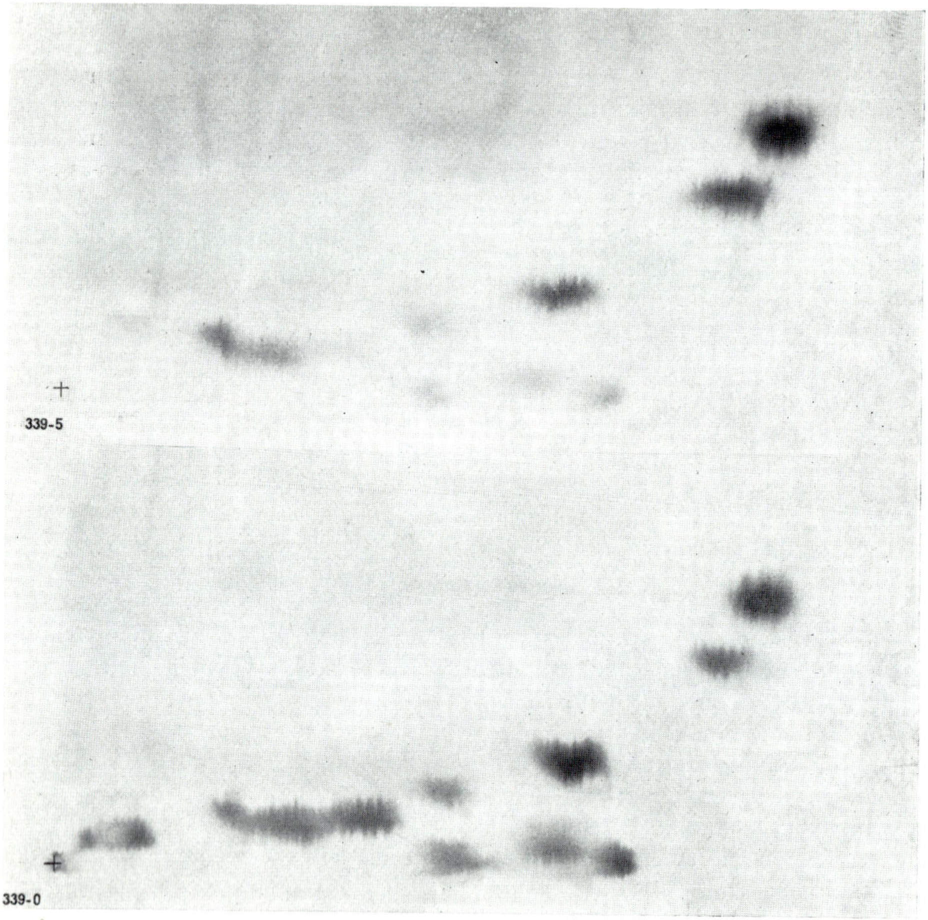

Fig. 13. Comparison of paper chromatograms prepared from free amino acid concentrates of plasma specimens taken at 0 and 5 h from isolated rat-liver perfusion. Liver donor, normal male Wistar rat (weight, 378 g); liver weight, 10.1 g; blood volume, 199 ml; standard complete amino acid mixture with glutamic acid replaced by an equal weight of glutamine[16] and 500 mg glucose added at the outset of experiment. Duration of experiment, 5 h. For this and subsequent chromatograms, equal volumes of plasma were precipitated with 10 vols. of 5% trichloroacetic acid. The clear filtrate was extracted with ether to remove excess trichloroacetic acid and the resulting aqueous solution concentrated *in vacuo* almost to dryness at 40°. The residues were taken up in a total volume of 1 or 2 ml of water and after thorough mixing, equal aliquots of the initial and final plasma free amino acids were placed at the origin of the chromatograms.

References p. 721

is of some interest that in the perfusion of livers taken from alloxan-diabetic rats the increases in leucine–isoleucine and valine are exaggerated and reach an extreme in the perfusion of the liver taken from a ketotic diabetic rat as seen in Fig. 15. In this regard, it is of considerable interest that IVY, SVEC AND FREEMAN, studying changes in the blood free amino acids in the experimental alloxan diabetes of the dog, noted that only levels of leucine, isoleucine and valine were significantly above their normal range[19]. Such evidence as we have presented strongly supports the view

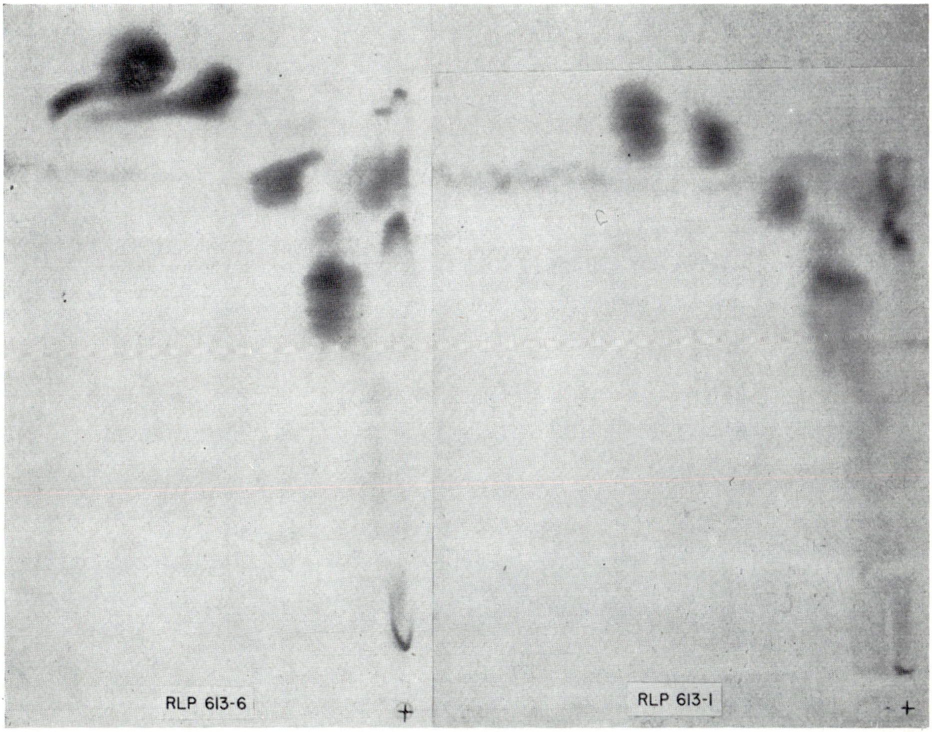

Fig. 14. Isolated rat liver perfusion (RLP 613). Liver donor, normal male, 18-h fasted Wistar (weight, 356 g); liver weight, 7.1 g. Dose of 3.5 mg DL-[1-^{14}C]leucine plus 500 mg glucose added to perfusion blood, volume 100 ml, at outset. Total duration of experiment, 6 h.

that the liver contributes, in a substantial way, to the maintenance of the blood free amino acid level under circumstances where amino acids are not forthcoming from the diet.

The introduction of a single dose of 43 mg of L-glutamic acid at the onset of the perfusion of a normal liver is associated ultimately not only with the above-mentioned changes in leucine–isoleucine and valine but also with the persistence, if not actual increase, of some of the non-essential amino acids, particularly alanine (Fig. 16).

Fig. 17 again shows changes in the free amino acids of the blood incidental to perfusion of an isolated liver taken from an 18-h fasted rat with 1.5 mmoles (219 mg) of L-glutamine added to the blood at the outset of the perfusion. As was established by independent measurements of the disappearance of glutamine from the blood and the appearance of stoichiometric amounts of urea[3, 16], the glutamine has disappeared well before the end of the perfusion. The increases in leucine–isoleucine, valine and alanine, seen in Fig. 17, are striking, and the maintenance or actual increase of glycine and serine is obvious.

Fig. 18 reveals changes incidental to the addition of a single large dose (1.5 mmoles) of L-arginine to the isolated perfusion of an 18-h fasted rat liver. Here again, on the basis of independent chemical determinations, the arginine was completely cleaved to measured quantities of urea[15]. At the end of the perfusion, not only are

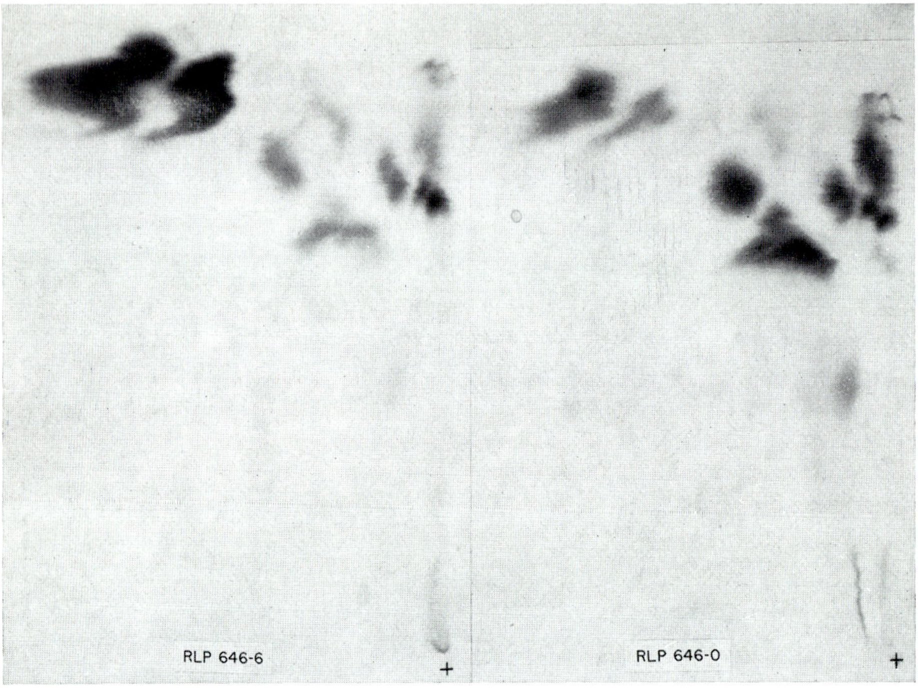

Fig. 15. Rat liver perfusion (RLP 646). 18-h fasted male Wistar alloxan-diabetic rat (weight, 218 g). Urine, 4+ for glucose and 4+ acetone before operation. Dose of DL-[1-^{14}C]leucine and glucose the same as in Fig. 14. Liver weight, 9.6 g.

leucine, isoleucine and valine prominent but alanine and glutamic acid spots are also increased in intensity. These results call to mind the probable role of transamination from glutamic acid, glutamine and glutamic acid derived from arginine via ornithine in the elaboration of the non-essential amino acids whose concentrations were either enhanced or at least maintained during the course of liver perfusions as depicted in Figs. 15–17.

Fig. 19 affords, for the sake of comparison, changes in plasma free amino acids seen after a 5-h survival of an eviscerated rat. In this particular experiment, a dose of glutamine had been given intravenously at the outset of the experiment. In contrast to the isolated liver perfusions, it is at once clear that many of the amino acids, *with the striking exceptions of leucine, isoleucine and valine,* have increased in

References p. 721

TABLE II

CHANGES IN PLASMA FREE AMINO ACIDS PRODUCED BY ISOLATED PERFUSED RAT LIVER

Amino acid	*Plasma level in µmoles/100 ml*		*Ratio (b)/(a)*
	0 h (a)	*6 h (b)*	
Alanine	39.6	66.3	1.67
α-Amino-*n*-butyric acid	Slight trace	Trace	
Valine	21.4	101.5	4.74
Methionine	3.14	1.02	0.324
Isoleucine	14.8	75.3	5.10
Leucine	15.5	74.0	4.78
Tyrosine	7.62	0.80	0.105
Phenylalanine	6.73	4.31	0.64
β-Alanine	—	Trace	
Phosphoserine	Trace	0.71	
Phosphoethanolamine	—	6.80	
Taurine	22.9	29.4	1.28
Urea	417.0	2090.0	5.02
Methionine sulfoxides	1.04	8.83	8.50
Aspartic acid	1.65	3.03	1.84
Threonine	20.2	11.8	0.585
Serine	24.6	10.0	0.406
Asparagine	38.2	48.2	1.26
Proline	12.6	7.6	0.604
Glutamic acid	25.1	107.8	4.30
Citrulline	5.85	—	
Glycine	36.0	8.23	0.228
Galactosamine	18.2	8.64	0.474
γ-Amino-*n*-butyric acid	2.53	0.87	0.344
Ornithine	6.05	15.3	2.53
Ethanolamine	3.24	7.77	2.40
Ammonia	296.0	160.0	0.54
Lysine	35.2	24.2	0.68
1-Methylhistidine	0.57	2.90	5.10
Histidine	5.27	7.71	1.46
3-Methylhistidine	Very slight trace	Very slight trace	
Arginine	13.8	—	
Cysteic acid	0.93	9.20	9.90

References p. 721

concentration. These observations are grossly in agreement with changes noted by FLOCK *et al.*[20] and FREEMAN AND SVEC[21] incidental to their studies in hepatectomized dogs.

Some of the changes noted in the comparison of initial and final paper chromatograms from isolated liver perfusions have been more firmly documented by quantitative amino acid estimation made by the procedure of MOORE, SPACKMAN AND STEIN*. The following significant observations have been made on initial and final plasma specimens from a perfusion in which only glutamine was added at the outset of the experiment: (a) As should be apparent from Table II, the gross observations made in paper chromatograms concerning the increase in leucine–isoleucine and valine are quantitatively most prominent in the column-chromatographic analysis. In each case, the relative increase in concentration at the end of 6 h perfusion was approx. 5-fold. (b) The major decreases in free amino acid concentration were

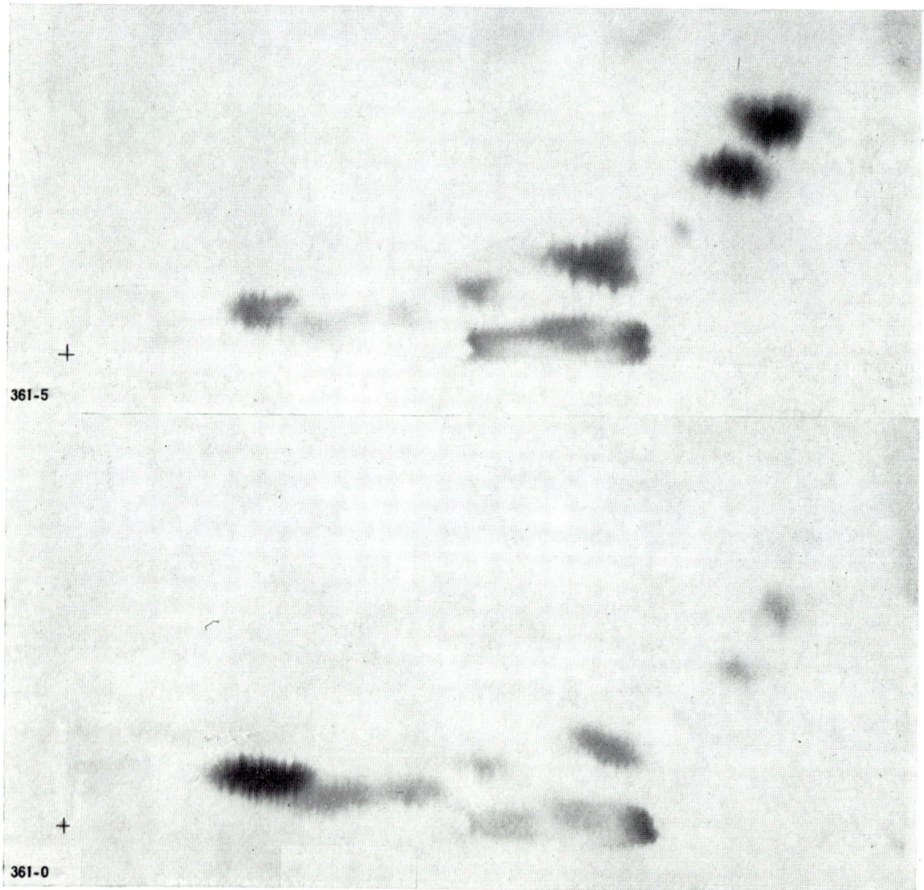

Fig. 16. Rat liver perfusion (RLP 361). Liver donor, 18-h fasted male Sprague-Dawley (weight, 257 g); liver weight, 8.0 g. Dose of 43.5 mg L-[U-^{14}]glutamic acid (5 μC). L-glutamic acid and 500 mg glucose added to 200 ml of blood at start of perfusion.

* We are indebted to Dr. E. S. NASSET for carrying out these analyses.

References p. 721

associated with the essential amino acids previously noted to be oxidized almost exclusively by the liver. (c) Remarkable increases are noted not only in ethanolamine but also in phosphoethanolamine. (d) The increase noted in 1-methylhistidine and histidine are small and may be referable to the breakdown of hemoglobin from lysis of red cells. (e) The striking *total* conversion of arginine, presumably chiefly to ornithine and urea, points to the fact that the isolated liver contributes no net

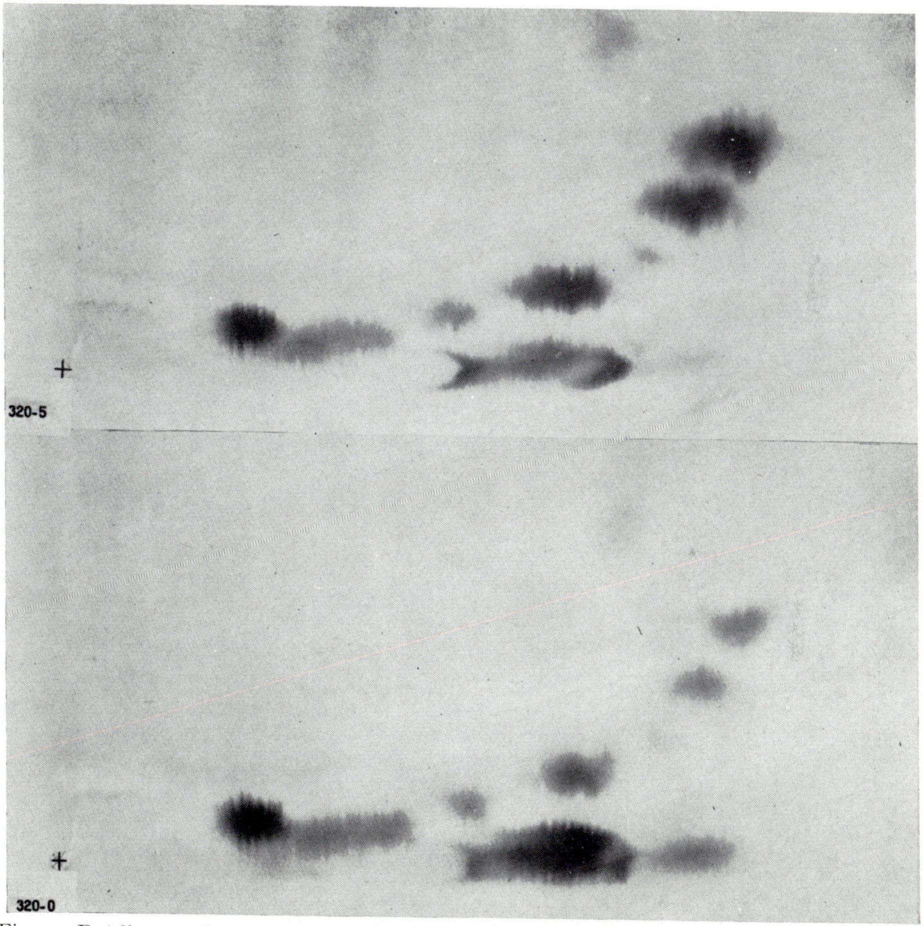

Fig. 17. Rat liver perfusion (RLP 320), normal, 18-h fasted Sprague-Dawley, male; liver weight, 9.5 g; 219 mg L-glutamine (1.5 mmoles) plus 500 mg glucose added to perfusion blood, 177 ml at outset.

arginine to the circulating blood and that free arginine, present in the blood of an intact animal, is probably derived from the diet or from non-hepatic tissues.

ACKNOWLEDGEMENTS

In the course of the development and use of the isolated liver perfusion technique,

References p. 721

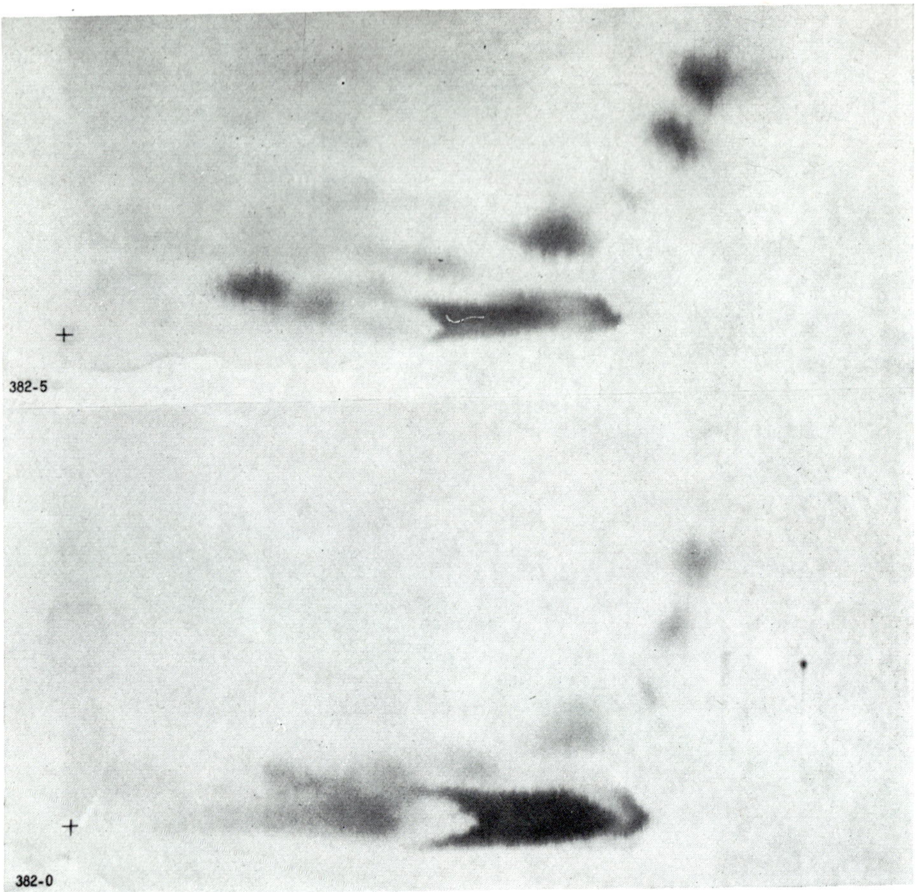

Fig. 18. Rat liver perfusion (RLP 382), 18-h fasted normal male, Sprague-Dawley, liver donor (weight, 269 g); liver weight, 8.7 g. Dose, 2 mg L-[U-^{14}C]arginine · HCl (5 μC) plus 314 mg (1.5 mmoles) L-arginine · HCl and 500 mg glucose added to 203 ml perfusion blood at outset of experiment.

we have been particularly fortunate in our association with W. F. BALE, C. G. BLY, W. T. BURKE, M. GREEN, D. E. HAFT, M. TOPOREK and M. L. WATSON. We wish also to acknowledge the able technical assistance of Miss H. R. HANAVAN, Miss C. SANTOMIERI, Mrs. D. WEMETT and Miss M. HAMSHIRE.

These studies were performed under contract with the United States Atomic Energy Commission at the University of Rochester Atomic Energy Project, Rochester, N.Y., and were aided in part by a research grant from the Medical Research and Development Board, Office of the Surgeon General, Department of the Army, under Contract No. DA-49-007-MD-451, and by a grant from the Jane Coffin Childs Memorial Fund for Medical Research.

References p. 721

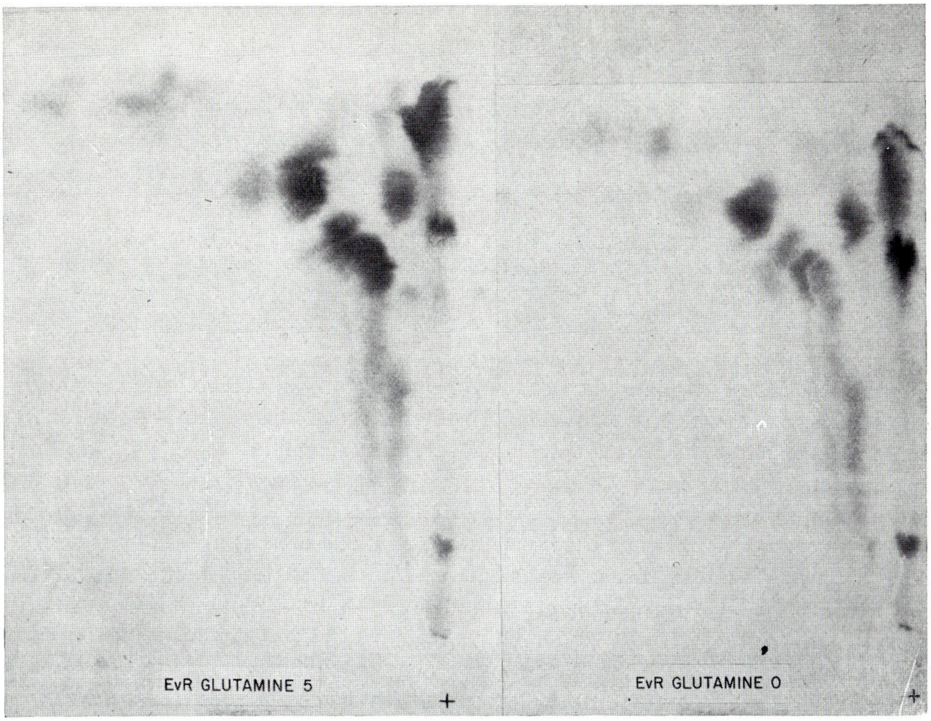

Fig. 19. Eviscerated surviving rat, normal adult male. L-glutamine (50 mg in 2 ml of saline) given intravenously at outset, approx. 2 ml of blood withdrawn at 10 min (sample 0); experiment terminated in 5 h to give final plasma sample.

REFERENCES

1 D. H. Spackman, W. H. Stein and S. Moore, *Anal. Chem.*, 30 (1958) 1185, 1190.
2 L. L. Miller, C. G. Bly, M. L. Watson and W. F. Bale, *J. Exptl. Med.*, 94 (1951) 431.
3 L. L. Miller, W. T. Burke and D. E. Haft, *Some Aspects of Amino Acid Supplementation*, Rutgers University Press, 1956, p. 44.
4 L. R. Crisp and J. M. F. de Broske, *Rev. Sci. Instr.*, 23 (1951) 281.
5 L. L. Miller, W. T. Burke and D. E. Haft, *Federation Proc.*, 14 (1955) 707.
6 D. J. Ingle, *Exptl. Med. Surg.*, 7 (1949) 34.
7 M. Green and L. L. Miller, *J. Biol. Chem.*, 235 (1960) 3202.
8 J. L. Bollman, F. C. Mann and T. B. Magath, *Am. J. Physiol.*, 78 (1926) 258.
9 C. E. Dent and J. M. Walshe, *Brit. Med. Bull.*, 10 (1954) 247.
10 C. Wu, J. L. Bollman and H. R. Butt, *J. Clin. Invest.*, 34 (1955) 845.
11 W. Frankl, H. Martin and M. S. Dunn, *Arch. Biochem.*, 13 (1947) 103.
12 D. Muting, *Z. fur d. ges. inn. Med. u. Grenzgebiete*, 24 (1957) 1107.
13 J. Blass, M. Cachin and J. Durlach, *Bull. soc. med. hop. Paris*, 66 (1950) 1253.
14 W. T. Burke and L. L. Miller, *Cancer Research*, 16 (1956) 330.
15 W. T. Burke and L. L. Miller, *Cancer Research*, 20 (1960) 658.
16 W. T. Burke and L. L. Miller, *Cancer Research*, 19 (1959) 148.
17 W. T. Burke and L. L. Miller, *Cancer Research*, 19 (1959) 622.
18 L. L. Miller, *Recent Progress in Hormone Research*, Vol. 17, Academic Press, New York, 1961, p. 539.
19 J. H. Ivy, M. Svec and S. Freeman, *Am. J. Physiol.*, 167 (1951) 182.
20 E. V. Flock, F. C. Mann and J. L. Bollman, *J. Biol. Chem.*, 192 (1951) 293.
21 S. Freeman and M. Svec, *Am. J. Physiol.*, 167 (1951) 201.

IN VIVO COMPARTMENTS OF GLUTAMIC ACID METABOLISM IN BRAIN AND LIVER*

HEINRICH WAELSCH

New York Psychiatric Institute and Columbia University, College of Physicians and Surgeons, New York, N.Y. (U.S.A.)

For the biochemist primarily engaged in the study of brain metabolism the relationship between structure and intermediary metabolism of this organ is a problem of immediate urgency. The brain's structure, the most complex of any organ of the mammalian body, must be correlated with its metabolism in order to understand and evaluate the metabolism, as such and its implications for cerebral function.

It has been suggested that the concept of metabolic compartments or metabolic pools satisfies a unifying concept of structure and metabolism[1–3]. In living matter there exist metabolic compartments on all levels of organization: the organ, the tissue, the cells and the subcellular units. On each level of organization there are compartments of different order of homogeneity characterized by their concentration of enzymes and metabolites and separated by structural features such as membranes, fibers and interstitial components. The compartments are in communication with each other: they are not closed to one another, but are in a dynamic relationship. Some metabolites are produced in one and used in another, some stored here and adsorbed on active sites there, some transported by the blood or interstitial fluid to receptor sites, some being part of one transport system and some diffusing from one compartment to another. Macromolecules such as enzyme proteins might stay in one place, but are dependent in their activity on the concentration of coenzymes and on their own rate of rejuvenation.

Neurophysiologists in particular have used the concept of compartments and their interrelationships consciously or unconsciously for many years in their studies of the forces responsible for the separation and movements of ions across membranes. In an organ such as the nervous system function is essentially a membrane phenomenon; the properties of the membranes are the results of the dynamic interaction of the metabolic compartments on either side and the specific structure.

In examining the central nervous system we become aware of the number of metabolic compartments in different levels of organization. A rough survey from the microscopic to the subcellular is given in Table I. The significance of metabolic compartments for the metabolism of the central nervous system immediately becomes apparent if we consider the cerebral glucose and energy metabolism. Although at present the mechanism of transport of glucose into the brain from the blood is unknown, its utilization appears to be controlled by the rate of its metabolism rather

* This presentation is based in part on a lecture given on the occasion of the Sesquicentennial Anniversary Symposium at McLean Hospital, Belmont, Massachusetts. Since the author was unable to attend, the paper was not presented at this Conference but is included in the Proceedings.

References p. 730

than by a regulation of its transport. Glucose is the respiratory substrate of the brain, and its oxidation provides all the utilizable energy, the major portion by direct oxidation and the rest through alternate routes such as the metabolism of amino acids or lipid constituents into which the intermediates of glucose metabolism have been incorporated. In the oxidation of glucose and in the utilization of the generated adenosine triphosphate (ATP), we are faced with the major problem of metabolic pools on all levels of cell organization, since ATP is generated in the mitochondrial compartment but is utilized for rejuvenation of structure and support of function in various compartments to differing degrees. For the purpose of this discussion the yield of ATP derived from glycolysis may be neglected. The resynthesis of ATP will also

TABLE I

NERVOUS-SYSTEM COMPARTMENTS ON DIFFERENT LEVELS OF ORGANIZATION

Arrow indicates increasing degree of attempted biochemical interpretation.

Level of organization	*Compartments*
Organ	CNS, Blood
Tissue	Extracellular space, neuron, glia
Cell	Cell body, dendrites, axon, synapse
↓ Subcellular	Nucleus, mitochondria, endoplasmic reticulum

depend on the flow of the necessary substrates and coenzymes into the compartment of oxidative phosphorylation. The most significant consequence of the compartmentation of ATP formation and utilization is the control and regulation of these processes.

The concept of metabolic compartments as a tool for the understanding of intermediary metabolism was forced upon us when we were unable to interpret experimental results obtained *in vivo* by the conventional assumptions[4–7]. It also pointed to a way to study metabolic compartments in the intact mammalian organism, and in particular, in the brain *in vivo*. Practically all experimental results obtained in other laboratories concerned with regulation and controls of metabolism, and thereby indirectly with metabolic compartments, were obtained either on tissue preparations, yeast, or ascites cells. All our experiments dealt with the metabolism of glutamic acid in the central nervous system, and in particular, with the conversion of glutamic acid to glutamine. They serve as a model of how the compartmentation of metabolic events may be demonstrated and studied *in vivo*. Glutamic acid occupies a central position in the intermediary metabolism of all tissues but particularly in the nervous system. Indicative of this fact may be the exceptionally high concentration of 10–12 μmoles/g of brain. Besides the participation in the metabolism of the citric acid cycle, by virtue of ketoglutarate, and in the metabolism of amino acids, glutamic acid is the precursor of γ-aminobutyric acid (GABA). The conversion of glutamic acid to glutamine represents, as will become apparent, the major removal mechanism for ammonia in nervous tissue. This conversion is catalyzed by glutamine synthetase and is an ATP-dependent synthesis of the amide linkage of the γ-carboxyl of glutamic acid. This pathway is, as far as we know (and this will be of significance later on), the only way by which glu-

References p. 730

tamine may be synthesized in the mammalian body. If [^{14}C]glutamic acid is administered to animals in order to study glutamine synthesis, the administered glutamic acid will first, if we deal with a homogeneous system, be diluted by the tissue glutamic acid, before the conversion to glutamine, the labeled glutamine again being diluted by the unlabeled tissue glutamine. Therefore, in a homogeneous system the glutamine will have not more than the specific activity of the glutamic acid but probably less, since not all of the glutamine will, in the experimental period, be derived from the labeled glutamic acid.

TABLE II

CEREBRAL METABOLISM OF INTRACISTERNALLY ADMINISTERED [^{14}C]GLUTAMIC ACID (RATS)

Time (min)	*Substrate*	*Specific activity counts/μM/min*			*Total counts in 1 g fresh brain (Glutamic + glutamine)*
		Plasma	*Brain*	*Liver*	
1/4	Glutamic		7 700		92 000
	Glutamine		3 300		
1	Glutamic	180 000	5 500	200	92 000
	Glutamine	1 200	7 300	66	
2	Glutamic	130 000	3 200	800	92 000
	Glutamine	2 200	12 000	600	
5	Glutamic	34 000	2 700	770	97 000
	Glutamine	6 100	14 000	760	
15	Glutamic	8 900	2 200	190	77 000
	Glutamine	6 500	11 000	280	
30	Glutamic	3 300	2 100	180	58 000
	Glutamine	870	7 300	240	

Let us compare this prediction with the actual data. The experiments were carried out in collaboration with BERL and LAJTHA[3, 4, 6]. [^{14}C]glutamic acid was injected intracisternally into rats, and the animals were decapitated 15 sec–60 min after the administration of the labeled amino acid (Table II). Only the experiment of 15 sec duration shows the expected relationship, glutamic acid having a higher specific activity than the amide. In all other experiments, glutamine has a higher specific activity than its precursor amino acid, glutamic acid, up to five times in the 15-min experiment. That this is not due to dilution by cold glutamic acid fed into the system or other metabolic reactions may be seen from the fact that during the first 15 min the sum of counts of glutamic acid and glutamine stays constant per g of brain. Neither GABA nor glutathione ever reaches the specific activity of glutamic acid when this acid is administered as the labeled compound. On the basis of the glutamic acid–glutamine data it may be concluded that the administered glutamic acid is converted into glutamine before it is mixed with the whole tissue glutamic acid, *i.e.* only a small fraction of the glutamic acid takes part in this conversion while most of the tissue glutamic acid is inert (Fig. 1).* Where is this compartment

* In this and also in Figs. 2 and 3 "inert" glutamic acid, glutamine or aspartic acid refers to the large tissue compartment of the respective compound with a rate of turnover which is lower than, or relatively inert, when compared with the small compartment of high metabolic activity.

Abbreviations: GA: Glutamic acid; GLINE: glutamine; ASP: aspartic acid; GSH: glutathione; MT: mitochondria; EPR: endoplasmic reticulum.

of the metabolically active glutamic acid? Since we have administered the labeled glutamic acid intracisternally we cannot exclude with certainty that this conversion proceeds in the cerebral surface membranes or the ependymal linings. We have no decisive evidence against such an assumption, but it is significant in this connection that only 2 min after the administration of labeled aspartic acid the isolated brain glutamine has five times the specific activity of brain glutamic acid[6]. One would have to assume that the whole chain of events leading from aspartic to ketoglutaric acid, glutamic acid and glutamine would essentially occur outside the brain tissue. Similarly, some time ago ROBERTS[8] mentioned the results of experiments in which, after the intracisternal administration of [^{14}C]GABA, glutamine showed a higher activity than glutamic acid. Since glutamine synthetase occurs in high concentrations in the microsomes[1] it was suggested that the administered glutamic acid is

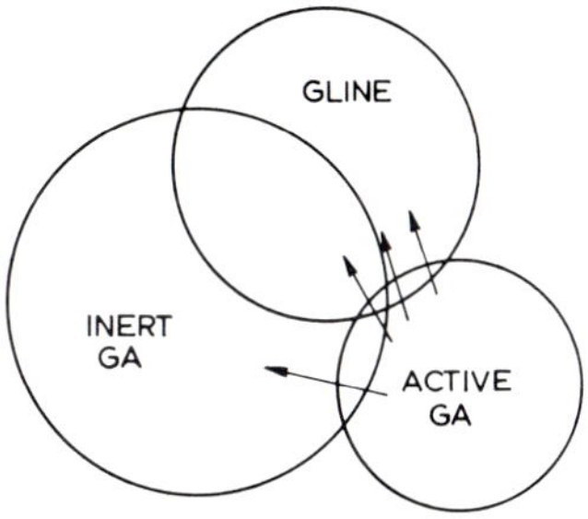

Fig. 1. Pools of administered glutamic acid (measured by glutamine synthesis). For abbreviations see p. 724.

converted into glutamine in the endoplasmic reticulum before it mixes with the total tissue content. A very approximate estimate of the size of the active glutamic acid compartment may be of interest. With a ratio of 5 : 1 of the specific activity of glutamine to that of glutamic acid, the glutamic acid compartment cannot be larger than 20% of the tissue glutamic acid.

Because of the uncertainty of the site of conversion of administered glutamic acid to glutamine the metabolism of glutamic acid synthesized in the brain and liver was studied. In collaboration with BERL, TAKAGAKI, CLARKE AND PURPURA[7], $^{15}NH_3$ was administered to cats by intracarotid infusion and the various members of the glutamic acid family isolated. 1 mmole of ammonium acetate was infused per minute and the animal exsanguinated at the end of the period of infusion. There are several observations which should be stressed (Table III). Upon infusion of ammonia all constituents analyzed stayed constant with the exception of cerebral glutamine, the concentration of which rose by at least 50%. This finding, in accord with results from other laboratories[9], shows that in the brain glutamine formation is the major removal mechanism for ammonia. Pertinent to our discussion is the observation that the ^{15}N concentration of the α-amino group of glutamine was about ten times higher than that of the α-amino group of glutamic acid. If we do not assume a still undiscovered pathway for the synthesis of glutamine, we have to conclude that this glutamine is synthesized from newly made glutamic acid before the acid has time to mix with the total tissue glutamic acid. What has already been demonstrated for glutamine formation from administered glutamic acid is also true for glutamine formation from glutamic acid synthesized in the tissue, namely, that this biosynthetic event occurs in a metabolic compartment. Since, according to fractionation studies, glutamic dehydrogenase is localized in the mitochondria[10], one might suggest that the glutamic acid newly formed from ketoglutarate and labeled NH_3

TABLE III

LABELING OF GLUTAMIC ACID, GLUTAMINE AND ASPARTIC ACID DURING [^{15}N]AMMONIA INFUSION (CAT)

All values represent ^{15}N atom % excess

Compound	*Blood*		*Brain cortex*		*Liver*	
	15 min	*20 min*	*15 min*	*20 min*	*15 min*	*20 min*
Glutamine amide	12	14	35	45	38	26
α-Amino	2	3	9	13	9	6
Glutamic acid						
α-Amino	5	10	0.7	1.6	28	35
Aspartic acid						
α-Amino	—	—	0.3	1.0	48	46

is converted in those particles into glutamine before it mixes with the tissue glutamic acid. It is of interest to compare these data with those obtained from liver in the same experiments. In liver, the α-amino group of glutamic acid always contains a higher isotope concentration than the α-amino group of glutamine, contrary to the situation in brain. This finding suggests that glutamine in the liver is not formed

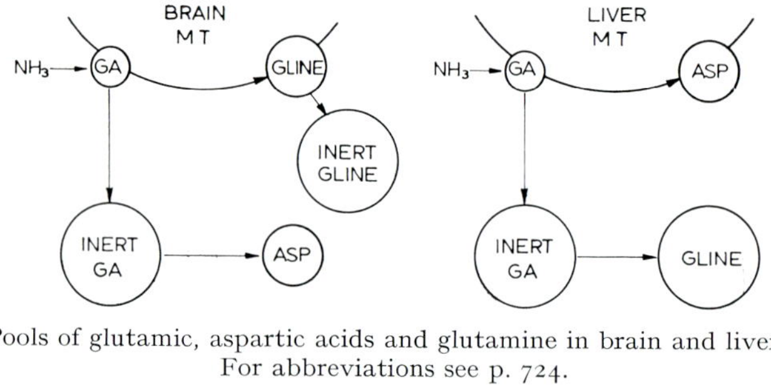

Fig. 2. Pools of glutamic, aspartic acids and glutamine in brain and liver ($^{15}NH_3$). For abbreviations see p. 724.

from a small compartment of rapidly metabolized glutamic acid but rather from the total tissue glutamic acid or a large portion of it. On the other hand, the isotope concentration of the α-amino group of aspartic acid soon overtakes that of the α-amino group of glutamic acid, while in brain the latter contains an isotope concentration considerably higher than that of aspartic acid. These findings suggest that aspartic acid in liver is formed by transamination of a glutamic acid newly synthesized, while in brain it is formed by transamination with the total tissue glutamic acid or a large part thereof. A tentative picture of this relationship is given in Fig. 2. It is interesting in this context to consider the possibility of enzyme mapping by a study of metabolic rates. In experiments of longer duration the isotope concentration of liver aspartic acid equals, or is lower than that of glutamic acid.

Without going into details of the calculation it is of some interest to obtain approximate values for the half-lives of the different glutamic acids. This estimate can only be very approximate and give an indication of the range of values.

On the basis of the isotope concentration in the free ammonia of the brain and in the α-amino group of glutamic acid, the bulk of glutamic acid has a half-life time of approx. 6 h; on the other hand, the glutamic acid used for glutamine synthesis has a half-life time of less than 1 h. The pool size of this labeled glutamic acid would be approx. 15% of the total. On the other hand, in liver the half-life time of

TABLE IV

LABELING OF GLUTAMIC ACID, GLUTAMINE AND ASPARTIC ACID AFTER INFUSION OF [^{14}C]$NaHCO_3$

Infusion: 0.11 mmole [^{14}C]$NaHCO_3$, 0.65 mC in 25 min

Compound	*Counts/min/µmole*		
	Blood	*Brain cortex*	*Liver*
Glutamine	390	500	1800
Glutamic acid	260	360	2000
Aspartic acid	420	1800	5300

the total glutamic acid is close to that of the small pool of rapidly metabolized cerebral glutamic acid, around 30 min, the compartment of glutamic acid used for transamination in the synthesis of aspartic acid having a faster turnover.

As pointed out before, the α-amino group of brain glutamine showed an ^{15}N concentration on the average ten times higher than that of glutamic acid, and in brain the amide group had the highest isotope concentration of all compounds analyzed. Glutamine was also the only compound which increased in brain. The question, therefore, arose as to the origin of the glutamic acid moiety of cerebral glutamine. A comparison of the isotope concentration in blood and brain glutamic acid and glutamine makes it unlikely that these blood constituents are the sources of the glutamic acid moiety of cerebral glutamine. Direct measurement of the uptake of labeled glutamic acid from the blood demonstrates that during the experimental period only a small fraction of glutamic acid could have come from the circulating blood. In order to maintain the tricarboxylic acid cycle if it should be depleted by a removal of ketoglutaric acid as glutamic acid for glutamine formation it would be necessary to replenish its intermediates by CO_2 fixation. Therefore, a study was made of the ability of brain tissue to fix CO_2, (ref. 11) a faculty which has been denied in varying degree by several authors. It can be seen that upon intracarotid administration of [^{14}C]bicarbonate to cats, considerable radioactivity appears in cerebral aspartic and glutamic acids (Table IV). Aspartic acid has the highest specific activity, a finding which is in accord with the assumption that the fixation occurs on the oxaloacetate level. From a comparison of the specific activities of liver and brain aspartic acid, it

References p. 730

appears that brain cortex has a potent CO_2 fixing mechanism. It is interesting to note that when $^{15}NH_3$ is administered together with the [^{14}C]bicarbonate, there is a shift in counts to glutamine, as expected from the fact that glutamic acid synthesis is increased in order to remove ammonia by glutamine formation (Table V). Although the final demonstration of CO_2 fixation in the brain is, of course, not dependent on the compartments of metabolism, the proof that cerebral glutamine cannot be derived from the blood was facilitated by the demonstration of a compartment of glutamine synthesis. It may be suggested that an understanding of metabolic pools and compartments is sometimes essential for a critical interpretation of metabolic events.

What are the particular characteristics of the experiments which provided the

TABLE V

LABELING OF GLUTAMIC ACID, GLUTAMINE AND ASPARTIC ACID AFTER INFUSION OF AMMONIUM ACETATE AND [^{14}C]$NaHCO_3$

Infusion: 25 mmoles ammonium acetate; 0.2 mmole [^{14}C]$NaHCO_3$ 0.65 mC in 25 min.

Compound	*Counts/min/µmole*		
	Blood	*Brain cortex*	*Liver*
Glutamine	180	830	300
Glutamic acid	210	290	600
Aspartic acid	150	370	1340

results discussed in this presentation? What are the implications of these results for an understanding of intermediary metabolism as far as turnover rates, control of metabolism and cerebral functions are concerned? The experiments presented are characterized by their short duration, none lasting longer than 60 min. Therefore, the initial fate of the labeled intermediate is studied before the isotope has penetrated all metabolic pools. In such experiments fast reactions will be measured; in long-time experiments some of the rapidly metabolized compounds will have lost their label, and mainly the metabolism of compounds with long half-life times will be estimated. Since in experiments of long duration some of the pools will have been penetrated by the isotope precursor and its metabolic derivatives, usually the precursor product relationship of the classical isotope experiments will be found.

Turnover values of metabolites are the sum of the turnover values of the metabolites in different metabolic compartments. Therefore, any overall turnover value of a tissue constituent has only limited biological significance. This conclusion, a clear consequence of the existence of metabolic compartments, becomes of particular importance if intermediary metabolism is to be related to the function of an organ. It is apparent that the glutamic acid with a half-life time of several hours will have a different functional involvement than that with a half-life of less than 1 h. The former may serve as a relatively stable cellular anion. In the experiment presented, the rate of conversion of the metabolite was the indicator of the existence of a pool. From the comparative rates conclusions were drawn as to the enzymatic make-up

References p. 730

of the pool, and its size. Tentative conclusions may also be drawn as to the localization of the cerebral metabolic compartments. The occurrence of glutamine synthetase in the microsomal fraction[1] led to the suggestion that the conversion of administered glutamic acid to glutamine occurs in the endoplasmic reticulum. On the other hand, the occurrence of this enzyme[1], and in particular, of glutamic dehydrogenase in mitochondria[10] suggests that the conversion to the amide of glutamic acid synthesized in the cell occurs at mitochondrial sites (Fig. 3). In these instances we have been able to connect tentatively the compartmentalized metabolic event with definite cell structure. This may not always be possible or justified. In addition to these "visible" metabolic compartments, there may be compartments which are

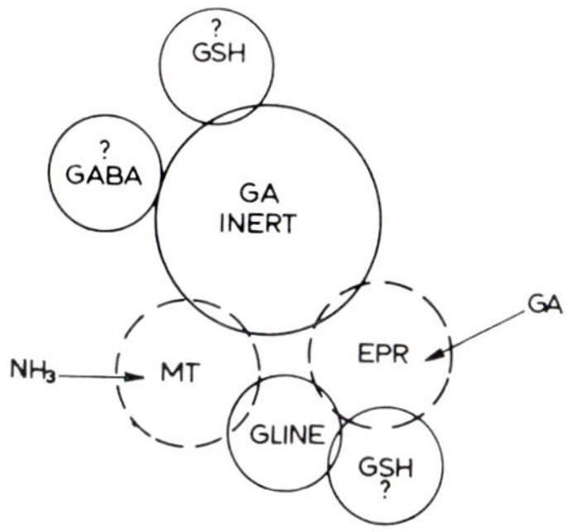

Fig. 3. Compartments of metabolism of administered glutamic acid and of glutamic acid synthesized in the tissue. The question marks indicate that the compartments of origin of GABA and glutathione are at present not determined with certainty. For abbreviations see p. 724.

formed concomitant with metabolism and maintained by it. They may be called "invisible" compartments.

It is clear that the control of intermediary metabolism and also the relation between function and metabolism cannot be interpreted rationally without a knowledge of the metabolic compartments. In the control and regulation of metabolism compartmentation of metabolic events and also the exchange between, and interdependence of, metabolic compartments will play a decisive role. This is apparent not only from the data presented here but also from the work of many other laboratories.

It does not seem necessary to elaborate on the significance of metabolic compartments for an understanding of the relation between cerebral metabolism and function. Some time ago we studied the metabolic changes in experimental epileptogenic lesions which exhibit paroxysmal electric activity[12]; the administration of small amounts of GABA eliminated the low frequency discharge. When the lesions are analyzed it was found that the glutamic acid, glutamine and glutathione concentrations have dropped to less than 50%, while the concentration of GABA had not changed. Upon systemic administration of GABA the low-frequency discharge was eliminated long before the GABA concentration rose[13]. These results show clearly that only a small fraction of GABA is affecting the properties of the lesion while the bulk of this compound is inert.

The experiments with glutamic acid discussed here only serve as a model as to how the problem of the compartmentation of metabolic events may be attacked

References p. 730

in vivo. Future studies will provide many more examples of the compartmentation of metabolism and of the modification of metabolic compartments under various physiological and pathological conditions. The concept of metabolic compartments removes the dichotomy between structure and metabolism and, we hope, also between metabolism and function. The isotope technique may be used in order to study metabolic compartments *in vivo* and characterize the intermediary metabolism of various organs in a novel way. It is surely not a matter of chance that the question of integration of structure and metabolism became most urgent in the central nervous system and stimulated our laboratory to embark on these investigations.

ACKNOWLEDGEMENTS

The work from the author's laboratory described in this paper has been supported in part by grants (B-226 and B-557) from the National Institute of Neurological Disease and Blindness, U.S. Public Health Service, by contracts between the Office of Naval Research and the Air Force Office of Scientific Research and the Psychiatric Institute, and a grant from the Supreme Council, 33rd Scottish Rite Masons of the Northern Jurisdiction, U.S.A.

REFERENCES

1 H. WAELSCH, in F. BRÜCKE, *Biochemistry of the Nervous System*, Proc. 4th Internatl. Congress Biochemistry, Pergamon Press, London, 1959, p. 36.

2 H. WAELSCH, in D. B. TOWER AND J. P. SCHADÉ, *Structure and Function of the Cerebral Cortex*, Elsevier Publishing Co., Amsterdam, 1960, p. 313.

3 H. WAELSCH, in S. S. KETY AND J. ELKES, *Regional Neurochemistry; Regional Chemistry Physiology and Pharmacology of the Nervous System*, Pergamon Press, London, 1961.

4 A. LAJTHA, S. BERL AND H. WAELSCH, in E. ROBERTS *et al.*, *Inhibition of the Nervous System and γ-Aminobutyric Acid*, Pergamon Press, London, 1960, p. 460.

5 A. LAJTHA, S. BERL AND H. WAELSCH, *J. Neurochem.*, 3 (1959) 322.

6 S. BERL, A. LAJTHA AND H. WAELSCH, *J. Neurochem.*, 1(1961)186.

7 G. TAKAGAKI, S. BERL, D. D. CLARK, D. P. PURPURA AND H. WAELSCH, *Nature*, 189 (1961) 326.

8 E. ROBERTS, in S. R. KOREY AND J. I. NURNBERGER, *Neurochemistry*, Hoeber-Harper, New York, 1956, p. 11.

9 J. P. DU RUISSEAU, J. P. GREENSTEIN, M. WINITZ AND S. M. BIRNBAUM, *Arch. Biochem. Biophys.*, 68 (1957) 161.

10 W. C. SCHNEIDER, *Advances in Enzymol.*, 21 (1959) 1.

11 S. BERL, D. D. CLARKE, G. TAKAGAKI, D. P. PURPURA AND H. WAELSCH, *Federation Proc.*, 20 (1961) 342.

12 S. BERL, D. P. PURPURA, M. GIRADO AND H. WAELSCH, *J. Neurochem.*, 4 (1959) 311.

13 S. BERL, G. TAKAGAKI AND D. P. PURPURA, *J. Neurochem.*, 1(1961)198.

DISCUSSION

Chairman: HARRY EAGLE

EAGLE: Dr. WESTALL, would you care to open the discussion?

WESTALL: There is one thing that is rather interesting, not with respect to the leucine, isoleucine and valine, where their high plasma levels build up so dramatically in maple-syrup disease, but in the case of argininosuccinic aciduria. When we first started to study the disease we were still of the opinion that urea production was a hepatic function only and we were very interested to hear of some results of SPORN, I think it was, and his collaborators, who seemed to be able to show that many of the enzymes necessary for urea production were present in brain tissue.

L. MILLER: With respect to SPORN AND DINGMAN's observations, one has only to look at their data in terms of quantitation to appreciate the fact that although the brain may produce urea, it could produce urea only from labeled arginine. I was not convinced that they showed any kind of net urea synthesis, labeled or otherwise, from carbon dioxide and ammonia. In terms of the total mass of urea produced in any one day to maintain normal nitrogen economy, the brain plays a very insignificant role.

EAGLE: Dr. MILLER, your experiments offer a very good explanation for the fact that in cell cultures most of the essential amino acids, in general, are not metabolized to any significant degree. I am puzzled, however, as to why isoleucine, leucine, and valine are not metabolized if their metabolism is not limited to the liver but seems to be as general as your results would indicate.

L. MILLER: Do I understand you correctly to state that you have recorded observations on the failure of cells in tissue culture to oxidize leucine, isoleucine, and valine? I can only tell you that of these three we have studied only the metabolism of randomly labeled leucine, but here again, not with any of the cells that you have used. We have used a strain of a Walker tumor grown in cell culture under sterile conditions and we find that one gets very definite oxidation to CO_2. It is hard to compare it with what goes on in the intact animal, but in terms of a relatively small mass of tissue this was significant. We have also observed that the same Walker tumor cell preparation would not oxidize ^{14}C-lysine in cell culture.

EAGLE: We had better go back and take another look at it.

FREE AMINO ACIDS AS OBLIGATORY INTERMEDIATES IN PROTEIN SYNTHESIS*

ROBERT B. LOFTFIELD

John Collins Warren Laboratories Huntington Memorial Hospital of Harvard University, Massachusetts General Hospital, Boston, Mass. (U.S.A.)

After two days of discussing amino acids, it is appropriate to consider that function of amino acids which makes them uniquely important in metabolism. The particular aspect of protein synthesis I would like to comment on is the question of whether all protein synthesis proceeds from free amino acids or whether there are mechanisms that involve intermediates no smaller than peptides.

During the last decade there had been a series of experiments on a number of biological systems which persuasively indicated that no substantial fraction of protein synthesis involves the incorporation into protein of fractions larger than monomeric amino acids. MONOD[1, 2] and SPIEGELMAN[3, 4] had demonstrated that no part of newly synthesized β-galactosidase is derived from the pre-existing protein of *Escherichia coli*. ASKONAS *et al.*[5] isolated some 30 peptides from partial hydrolysates of casein and β-lactoglobulin after *in vivo* labeling with [^{14}C]valine and [^{14}C]lysine in a goat. The two amino acids possessed constant specific radioactivity in every peptide; an impossibly unlikely coincidence unless all residues were derived from a single pool, namely, the free amino acid pool. Entirely different experimental designs by SIMPSON AND VELICK[6] and HEIMBERG AND VELICK[7] showed that the incorporation of several amino acids, *in vivo*, into purified rabbit-muscle proteins must proceed through a single pool rather than the multiple pools which must be postulated for peptide intermediates. In my laboratory we[8] were able to show that the valine and leucine residues of pure newly synthesized rat-liver ferritin possessed exactly the same specific activity as the free valine and leucine inside the liver cells. Thus, when I accepted the invitation to speak here I expected to be able to discuss critically some of the consequences of these conclusions. As I reacquainted myself with the recent literature I realized that many workers do not consider the problem to be settled. In fact, since there have been so many suggestions of "proof" that peptides participate in protein synthesis I cannot even discuss the individual experiments. Nevertheless I feel it may be worthwhile to discuss the various lines of experimentation and argument and to consider whether these are unambiguous.

One argument for peptide intermediates involves the many cases in which it has been shown that the amino acid residues of a labeled protein or peptide appear to be incorporated into new protein more rapidly than the free amino acid. MEDVEDEV AND CHIANG-HSIA[9, 10], BABSON AND WINNICK[11] and others have described such experiments. In general, a comparison is made between incorporation of a labeled amino acid and a labeled peptide into the proteins of an intact animal or an isolated tissue.

* This is publication No. 1066 of the Cancer Commission of Harvard University.

References p. 737

It is found that the two substrates are metabolized differently and that added cold amino acid has little effect on the incorporation of labeled amino acid from polypeptides. From this it has been easy to conclude that amino acids from one peptide can be incorporated into another protein without passing through the free amino acid pool. Frequently there has been a failure to prove that, in fact, a new protein has been formed; the "isolated" protein may include the intact unchanged starting material. Where this is not a proper criticism, there has been no demonstration that both the peptide and the amino acid have both reached the intracellular sites of synthesis. Thus it is to be noted that peptides may be taken into a cell many times faster than amino acids[12]. Finally, before one can claim that the new protein was derived from peptides and not amino acids, it would be necessary to demonstrate that the peptide is not hydrolyzed. In every case where hydrolysis has been looked for it has been found. BURNETT AND HAUROWITZ[13, 14] found that intracellular valylleucine was 90% hydrolyzed in reticulocytes and LEACH AND SNELL[12] found that alanylglycine was completely hydrolyzed in *Lactobacillus casei*.

A second line of argument involves the apparent exchange of one amino acid residue for another in experiments using doubly labeled proteins. In a sense this can be regarded as the use of a very large intact peptide and only one or two new amino acids.

Thus WIGGANS *et al.*[15] have reported that when rat albumin containing both labeled valine and labeled leucine is injected into a recipient rat, the circulating albumin loses the two amino acids at different rates. Similar observations have been made by others (FRIEDBERG AND WALTER[16], FRANCIS AND WINNICK[17], among others). These workers interpreted their data as proving that individual amino acids could exchange (perhaps on the biosynthetic template) with free amino acids and that the rates of such exchange varied with the nature of the amino acid and its position in the protein molecule. The flaw common to all these arguments is that the administered doubly tagged protein may be hydrolyzed to its constituent amino acids and that these labeled amino acids will be diluted to a varying extent by endogenous free amino acids. Newly formed protein would then contain labeled amino acids possessing the relative specific activities of the pool amino acids and hence there would appear to be a change in the relative abundance of the two. There is certainly an extensive breakdown and resynthesis of protein through the free amino acid intermediate. Thus BUCHER AND FRANTZ[18] found that massive amounts of non-radioactive alanine or glycine decreased the apparent half-life of glycine or alanine labeled serum protein in a rat and, very much to the point, that glycine and alanine were not at all equally effective in interfering with the reincorporation of the respective labeled residues. PENN, MANDELES AND ANKER[19] elegantly showed that the half-life of labeled serum protein depends on whether the other proteins of the animal are also labeled. Thus the radioactivity was lost much more slowly in the donor rabbit than in the recipient of a plasma transfusion. The only interpretation is that a rapid breakdown of many labeled tissue proteins results in a relatively highly radioactive amino acid pool in the donor and hence the continued synthesis of labeled serum proteins, whereas in the recipient, amino acids derived from the breakdown of labeled plasma proteins are thoroughly diluted by the vast continuing supply of non-radioactive amino acids released from the non-labeled body protein. We ourselves[8] have established that, following a single injection of [^{14}C]leucine, the

References p. 737

specific activity of the free leucine in a rat's liver remains high and nearly constant for one day and is reduced only by 50% after 3 days of continuous infusion of large amounts of non-radioactive leucine. Protein breakdown to amino acids and resynthesis are such active processes in many tissues that experimental determinations are necessary to establish the specific activities of amino acid pools. No amount of exogenous amino acid can reduce the specific activity of the intracellular amino acid to zero.

A third type of observation which has been interpreted as suggesting that peptides or larger protein fragments may participate in protein synthesis is the reversible incorporation of certain amino acids such as glutamic acid or alanine into "protein" under conditions where net protein synthesis was not taking place[20]. As Dr. HALVORSEN pointed out this morning many of these experiments are now invalidated because the incorporation of radioactive amino acid was into cell wall material rather than protein[21].

A fourth source of support for peptide intermediates comes from teleologic reasoning. RAACKE[22] found in developing pea seeds an increase in peptide material followed by a drop when protein synthesis became rapid. TUBOI AND HUZINO[23] have found that several peptides are activated by enzymes similar to those that activate amino acids. KONINGSBERGER[24] has isolated peptide-nucleotide compounds from yeast. HANES, HIRD AND ISHERWOOD[25] observed that a variety of peptides can be formed by the enzymatic reaction of various amino acids with glutathione and FRUTON[26] has discovered many systems in which transamidation reactions can yield peptides of considerable length. It is appropriate to inquire: "What significance do these observations have?" But it is a *non sequitur* to conclude, if you can think of no other meaning, that they have a bearing on protein synthesis.

A final type of argument in favor of peptides as intermediates in protein synthesis involves those experiments in which a given labeled amino acid is found to be incorporated into different parts of a protein molecule at different rates. If every amino acid residue in a newly formed protein is derived from the same amino acid pool and if the time required for the synthesis of a protein molecule is short relative to the rate of change of specific activity of the amino acid pool, all residues must have the same specific activity. There are several reports to the contrary[27-31]. In most of these it is possible to suggest sources of artifact. Suffice it to say that in the most carefully conducted experiment of this type, ASKONAS *et al.*[5] found, as mentioned above, constant activities of some 30 valine and lysine containing peptides isolated from goat milk. There are, of course, circumstances in which non-uniform labeling of a protein is to be expected because the duration of the labeling experiment was of the same order as the time required for the synthesis of a single molecule. Thus in the schematic synthesis shown here (Fig. 1) valine residues close to one end of the chain will be more radioactive than residues isolated from the other end until the time when there are many more fully labeled protein molecules than partly labeled molecules. The kinetics of labeled amino acid incorporation do indeed correspond to this scheme[32] and not to a scheme that involves peptide intermediates. Recently DINTZIS[33] has shown that, as predicted by this scheme, hemoglobin peptides isolated from reticulocytes after brief exposure to labeled leucine are not equally labeled but that after longer exposure all peptides become equally radioactive. These experiments might be construed to suggest that peptides are intermediates in

References p. 737

protein synthesis and indeed peptides are intermediates in the sense that the entire protein molecule does not spring into existence in an instant. However, the unit being added to the growing chain is a monomeric amino acid, not a peptide. Furthermore, as BURNETT[14] has shown for hemoglobin synthesis, the growing peptide chain does not exist as a free peptide but it is distinguished by some chemical or physical conditions which make it non-interchangeable with a corresponding free peptide.

In a general way I have tried to show the flaws in most of the arguments and experiments offered in support of the proposition that peptides are intermediates in protein synthesis. I have quite deliberately avoided invoking the current theories of protein synthesis, for my argument would have seemed cyclic. That is to say, most

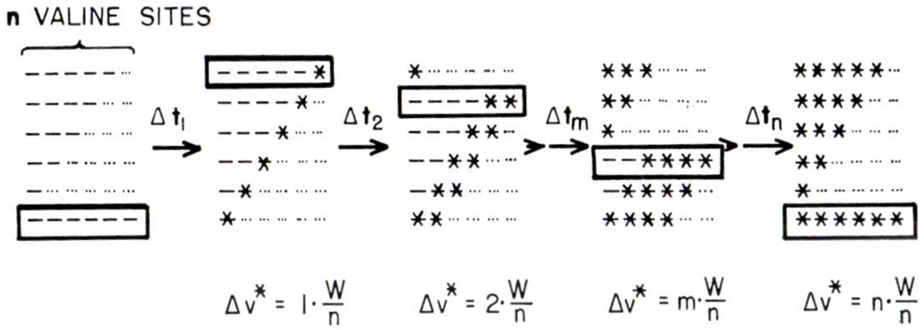

Fig. 1. Scheme for the labeling of protein molecules formed by the stepwise assembly of activated amino acids. Each horizontal line represents one ferritin template with valine sites occupied by [^{12}C]valine (—), by [^{14}C]valine (*), or unoccupied (...). An array is shown such that one ferritin chain is to be found in each stage of assembly. In each time interval, $\triangle t$, one valine residue is added to each growing chain and one ferritin molecule (indicated by the box) is completed. The increment in radioactivity in ferritin in any interval, $\triangle t$, is seen to be proportional to the number of such intervals and hence proportional to the time since the labeled amino acid was first introduced (from ref. 32).

of the contemporary theories begin with the assumption that the immediate precursors of protein are monomeric activated amino acids and hence neither the theories nor the experiments are likely to provide evidence for peptide intermediates. However, it might be worth considering what has been established experimentally and asking whether it would be difficult to accommodate the experimental facts to a theory in which free peptides were intermediates in protein synthesis.

The first and most solid fact is that proteins are chemical entities each with a unique arrangement of amino acid residues. Not only have such unique structures been established for several proteins but the fact that specific genetic changes result in the substitution of a single residue at a single locus[34] convinces one that the biosynthetic mechanism is capable of extremely high precision, making errors only when these are, so to speak, directed by the genetic blueprint.

Beyond this there is strong evidence that there exists in every organism for every natural amino acid one and only one activating enzyme whose function is to convert the free amino acid first to an aminoacyladenylate and then to an aminoacyl S-RNA in which the natural amino acid is bound to a terminal ribose of a transfer ribonucleic acid molecule that is specific for the particular amino acid. Then the aminoacyl-S-RNA diffuses to the ribosome where, as far as can be told, the aminoacyl-S-RNA is oriented and the amino acid residue is incorporated into a polypeptide. The mechanism for orientation is uncertain but presumably it involves some sort of base-pairing

References p. 737

by which the genetic information is used in the synthesis. The mechanism for the polymerization reaction is also largely unknown, although GTP is required and certain drugs can interfere with the process.

How well does a peptide mechanism fit these observations? It is extremely difficult to understand how even 20 different amino acids can be distinguished from each other by the activating enzymes which must also select among twenty different S-RNA's; it is nearly inconceivable that there should be a sufficient number of enzymes to distinguish among the thousands of potential peptide intermediates and the similar number of S-RNA's that would be postulated for the peptides. In truth, since there is a limited amount of S-RNA present in the cytoplasm and since at least 90% of it can be shown to react with one or another of the twenty natural amino acids there simply is not any significant amount of peptide specific S-RNA. It might be suggested, of course, that peptides are incorporated by a mechanism that does not involve S-RNA, for TUBOI AND HUZINO[23] have shown that peptides can be activated. However, both BERG[35] and we[36] have found that the specificity of amino acid-activating enzymes is not absolute, that other natural and unnatural amino acids may be activated but that the Michaelis constants and the transfer of the activated amino acids to S-RNA are such as to preclude the operation of an activating enzyme on any but its natural substrate. Although we have not examined TUBOI's peptides, I am certain that we will find that they are not activated to a significant extent under conditions of concentration and competition such as would exist in a cell.

One might then say that perhaps the peptides are incorporated into protein by a route that doesn't involve even the activating enzymes. Such systems have in fact been described[37]. In such cases it may be that protein is actually not being synthesized, that the apparent synthesis is a manifestation of an artifact. However, at this point I think it would be wisest to reserve judgment. There are, after all, many syntheses of polypeptide or protein substances which clearly do not involve the reaction scheme presented above. Glutathione synthesis proceeds through a free peptide and the free amino acids are activated by a different mechanism[38]. The polypeptide material of cell walls involves not only free peptide intermediates but the coupling of one peptide to another[39]. The "synthesis" of trypsin from trypsinogen is clearly not mediated by microsomes and activating enzymes. It might be surprising but surely not impossible to find a protein whose final completion requires transpeptidation rather than hydrolysis.

I would end on this note. I feel that there has been no persuasive evidence that free peptides are, in fact, intermediates in protein synthesis. There have been many conscientious efforts to discover evidence for such intermediates and in each case the evidence is negative. The question remains unsettled only because an argument cannot be proved conclusively by an accumulation of negative data. Further negative data will add little to the argument and the unlikelihood of getting positive evidence of peptide participation seems so small that I can think of several more promising areas of research.

ACKNOWLEDGEMENTS

The preparation of this material was supported by AEC Contract AT(30-1)2643 and U.S. Public Health Service Grant C-2387.

References p. 737

REFERENCES

1 J. MONOD AND M. COHN, *Intern. Congr. Microbiol. ,6th Congr., Rome*, 4 (1953).
2 J. MONOD, A. M. PAPPENHEIMER JR. AND G. COHEN-BAZIRE, *Biochem. Biophys. Acta*, 9 (1952) 648.
3 B. ROTMAN AND S. SPIEGELMAN, *J. Bacteriol.*, 68 (1954) 419.
4 H. O. HALVORSON AND S. SPIEGELMAN, *J. Bacteriol.*, 64 (1952) 207.
5 B. A. ASKONAS, P. N. CAMPBELL, C. GODIN AND T. S. WORK, *Biochem. J.*, 61 (1955) 105.
6 M. V. SIMPSON AND S. F. VELICK, *J. Biol. Chem.*, 208 (1954) 61.
7 M. HEIMBERG AND S. F. VELICK, *Federation Proc.*, 13 (1954) 227.
8 R. B. LOFTFIELD AND A. HARRIS, *J. Biol. Chem.*, 219 (1956) 151.
9 A. Z. MEDVEDEV, *Biokhimiya* (English translation), 24 (1959) 84.
10 A. Z. MEDVEDEV AND S. CHIANG-HSIA, *Biokhimiya* (English translation) 24 (1959) 650.
11 A. L. BABSON AND T. WINNICK, *Cancer Research*, 14 (1954) 606.
12 F. LEACH AND E. SNELL, *Biochim. Biophys. Acta*, 34 (1959) 292.
13 J. P. BURNETT AND F. HAUROWITZ, *Federation Proc.*, 20 (1961) 389.
14 J. P. BURNETT, *Ph. D. Thesis*, University of Indiana, 1960.
15 H. W. RUMSFELD Jr., W. W. BURR Jr., AND D. S. WIGGANS, *Federation Proc.*, 16 (1957) 240.
16 W. FRIEDBERG AND H. WALTER, *Federation Proc.*, 14 (1955) 214.
17 M. D. FRANCIS AND T. WINNICK, *J. Biol. Chem.*, 202 (1953) 273.
18 N. L. R. BUCHER AND I. D. FRANTZ Jr., *Cancer Research*, 10 (1950) 207.
19 N. PENN, S. MANDELES AND H. S. ANKER, *Federation Proc.*, 14 (1955) 262.
20 E. F. GALE AND J. P. FOLKES, *Biochem. J.*, 55 (1953) 721.
21 J. MANDELSTAM AND H. J. ROGERS, *Nature*, 181 (1958) 956.
22 I. D. RAACKE, *Biochem. J.*, 66 (1957) 101, 110, 113.
23 S. TUBOI AND A. HUZINO, *Arch. Biochem. Biophys.*, 86 (1960) 309.
24 V. V. KONINGSBERGER, C. VAN DER GRINTEN AND J. T. G. OVERBEEK, *Proc. Koninkl. Ned. Akad. Wetenschap.*, B60 (1957) 144.
25 C. S. HANES, F. J. R. HIRD AND F. A. ISHERWOOD, *Nature*, 166 (1950) 288.
26 J. S. FRUTON, *Congr. intern. biochim., 2e Congr., Paris*, 5 (1952).
27 J. KRUH, J. C. DREYFUS, G. SCHAPIRA AND P. PADIEU, *J. Biol. Chem.*, 228 (1957) 113.
28 M. VAUGHN AND C. B. ANFINSEN, *J. Biol. Chem.*, 211 (1954) 367.
29 D. STEINBERG AND C. B. ANFINSEN, *J. Biol. Chem.*, 199 (1952) 25.
30 M. FLAVIN AND C. B. ANFINSEN, *J. Biol. Chem.*, 211 (1954) 375.
31 K. SHIMURA, H. FUKAI, J. SATO AND R. SACKI, *J. Biochem., (Tokyo)*, 43 (1956) 101.
32 R. B. LOFTFIELD AND E. A. EIGNER, *J. Biol. Chem.*, 231 (1958) 925.
33 H. M. DINTZIS, *Proc. Natl. Acad. Sci. U.S.*, 47 (1961) 247.
34 V. M. INGRAM, *Nature*, 180 (1957) 326.
35 F. H. BERGMAN P. BERG, J. PREISS, E. J. OFFENGAND AND M. DIEKMANN, *Federation Proc.*, 18 (1959) 191.
36 R. B. LOFTFIELD AND E. A. EIGNER, *Federation Proc.*, 20 (1961) 1.
37 S. OCHOA AND M. BELJANSKI, *Proc. Natl. Acad. Sci. U.S.*, 44 (1958) 1157.
38 J. E. SNOKE AND K. BLOCH, *J. Biol. Chem.*, 199 (1952) 407.
39 E. ITO AND J. L. STROMINGER, *J. Biol. Chem.*, 235 (1960) PC 5.

DISCUSSION

Chairman: HARRY EAGLE

HENDLER: Dr. LOFTFIELD has very clearly presented the generally accepted scheme for incorporation of amino acids into proteins. I think that it would be appropriate at this time to indicate several of the points on which I disagree with this scheme and the reasons for disagreement.

Peptide intermediates were ruled unlikely by showing the inadequacies of experimental schemes which purported to show their existence. I agree with his conclusion that these schemes do not show that there are peptide intermediates. But I think we can use the very same approach and analyze the validity of the experimental evidence which was cited to show that there are *no* peptides. For instance, there is the experiment in which the protein of certain cells is made radioactive, the free amino acids are depleted and then cold amino acid is introduced. The protein subsequently synthesized is unlabeled. This could say that if peptide intermediates exist, they are present only in very low quantities and are turning over rapidly. But it cannot say that they do not exist. In addition, if the peptides interchange rapidly with free amino acid while the depletion of radioactive free amino acid is being carried out, their activity might also be depleted.

It is not entirely a question of semantics, to say that the amount of peptide would have to be rather small and rapidly turning over. The question is, can we rule out the existence of such peptides which are difficult to demonstrate? The incomplete peptides growing on the template surface can be considered as peptides, but I am really talking about peptides that have not yet reached the template, and I think that the possibility of their occurrence deserves further consideration.

The second argument is that peptides having a sequence known to exist in a protein are not incorporated. Well, even if peptide pools were involved in the formation of that protein, there would be an astronomical number of possible peptides which could exist in the pool. You may have picked a particular dipeptide; perhaps you should have taken the tripeptide. Another problem is permeability barriers. How do you know that the peptide is reaching the site of synthesis? Another is that even though these peptides may be intermediates before they reach the template they may be intermediates in another bound sense, and that you are not able to approach this binding by introducing the peptides.

Another series of experiments mentioned to show that there are no peptide intermediates was based on the observation that one could add "specific inhibitors" for a given amino acid and find that the incorporation of all the other amino acids into the protein was inhibited. If there were a pool of peptide intermediates, the argument is that these other amino acids should at least be able to supply the requirements for the protein until the peptides have been depleted. But it seems very possible that if you throw in a specific analog blocking the incorporation of a given amino acid that all these interrelated actions may also come to a halt. The fact that interference with a single amino acid has blocked incorporation of all the others does not rule against peptides.

The argument was also raised that there is enough of a problem for enzymes to determine the specificity of a single amino acid, let alone having enzymes handle the recognition of all the peptides. But peptide intermediates do not have to work without a template. You could still have the template, and the job of recognizing the final sequence of the protein could still be handled by whatever sequence mechanism handles it for the template.

The argument also was advanced that since we have a well-established scheme for explaining protein synthesis, it is unlikely that peptides are involved because there is no place for them in this scheme, as we understand it. It would be superfluous at this point, apparently, to introduce a stage with peptides going through the same sequence. I think that if you critically examine the currently accepted scheme for protein synthesis as I have done, there is no doubt that all of these reactions occur, and the specificity as indicated occurs. But the only "evidence" that I have been able to find for the involvement of this scheme in protein synthesis is the query "Why else are these things around?"

I want to make it clear that I am not questioning the very fine experiments showing that compounds in the form of the soluble RNA and the AMP-amino acid can transfer their amino acid into TCA precipitates of a non-described character, and that the amino acid in that precipitate is held by bonds having the strength of peptide bonds. But this does not show that protein is being made; it does not show that the amino acid is found in any normal protein that this tissue can make. There are other arguments, for example, that equally uncharacterized systems for incorporation of amino acid into protein have been described which incorporate without the presence of amino acid activating enzymes or the presence of soluble RNA. Also in studies using intact cells that are making their normal proteins, the kinetics of passage of amino acid through the nucleic acid fraction is much too slow, or appears much too slow, to account for the total rate of incorporation into protein. On the other hand, this argument also is not conclusive, because the possibility exists that the RNA examined, which shows very little activity, may contain a small component which is turning over very rapidly.

While I agree that there is no good evidence for peptide intermediates, I want to point out that there is also no good evidence saying that they do not exist. I think it is still valid for us to ask, are there stages less complex than the completed protein, perhaps less complex than partial protein on a template? I think that experimentally we are still not able to give an answer to this question, and we have to keep our minds open to this possibility.

LOFTFIELD: There are a couple of things that I had not actually said. The last point is the only one I can remember and that is the question of the kinetics. Since it is not my work at all, I can defend it quite objectively. François GROS has studied the kinetics of incorporation of valine and tyrosine, I think, in a bacterial preparation and has shown that some 90 per cent of the recorded rate of protein synthesis can be accounted for by the rise and subsequent fall in the amount of tyrosine and valine attached to the soluble RNA. Similar work has been done in ascites cells by LITTLEFIELD and by HOAGLAND, so that I do not think that this is a good argument against the system.

But my point here was not to defend the generally accepted scheme for protein biosynthesis. I very pointedly said that I did not want to use that scheme to defend my proposition that free peptides are probably not intermediates in protein biosynthesis. I wanted to argue as far as possible on the basis that the experiments are faulty. I also said that there was not any incontrovertible proof that peptides are not intermediates in protein synthesis.

LEAHY: I would like to add a few remarks in regard to Dr. HENDLER's comment concerning the isolation of an identifiable protein using the soluble RNA ribosome system. In the hemoglobin system studied by SCHWEET, reticulocytes produce 85 per cent or more of just one protein, hemoglobin. A protein labeled with radioactive amino acids and released from microsomes *in vitro* has the same behavior as hemoglobin on columns and in its ammonium sulfate fractionation. The ratios of different amino acids in the isolated product are the same as the known ratios of those amino acids in hemoglobin, rather than the ratios of the amino acid concentrations in the incubation mixture. The per cent of N-terminal ^{14}C valine is the same as that found in hemoglobin from intact cells.

ANDERSON: I would like to speak for just a moment to this point of what peptides might be doing. It is rather easy to isolate them from microsomes, for example, and ALBRIGHT and I showed some time ago that they turn over very rapidly. In a system where the machinery is breaking down at a very substantial rate, as is the case with RNA, for example, I do not believe that the cell is such a perfect system that it never makes any mistakes. It is not improbable that in a high percentage of instances when synthesis of a protein is started some mistake is made, and the pieces are subsequently broken down and reused, so that it may well be that the peptides which we isolate are merely mistakes.

GRIFFITH: It has been adequately demonstrated by Professor W. C. ROSE and by others that an amount of total nitrogen which will maintain nitrogen balance if fed as protein may fail to do so if fed as an equivalent mixture of free amino acids unless additional calories are supplied. This observation implies that a mixture of free amino acids is not absorbed and used as efficiently in the body as is the mixture of the same amino acids entering the gastrointestinal tract in the form of protein. Some offer the explanation that small amounts of certain peptides are normal residual products of digestion, are absorbed as such and serve as particularly useful building stones in protein synthesis. Will the speaker comment on this possibility, please?

LOFTFIELD: I think that there is one very likely answer which is that amino acids do not get through cell walls or through cell membranes very easily. We did an experiment in which we used radioactive albumin labeled *in vivo*, and compared the rate at which this and intravenously injected leucine entered rat liver cells. The albumin was taken into the cells about seven times more rapidly than the free leucine. Given the same amounts of radioactivity, the same specific activities of leucine in albumin or in free leucine, the specific acitivity of free intracellular leucine was seven times higher with albumin than with free leucine. I think that this probably corresponds to the observations that Dr. EAGLE has on glutamic acid. Does it get in very well?

EAGLE: This may be the explanation, but on the basis of our experience with cultured cells it would not be, because in cultured cells amino acids get into cells much more readily than does intact protein. As a matter of fact, we have had great dfficulty in demonstrating, at least to our own satisfaction, that small peptides get in as such. Our results are consistent with the idea that peptides are either hydrolyzed extracellularly by extracellular peptidase, or on the surface of the cell, with only occasionally an indication that the small peptides are taken in as such. The situation in an intact animal may be very different.

SCHREIER: There may be another answer to the question of Dr. GRIFFITH. We know that if you feed monosaccharides to an animal, part of it is excreted in the urine, and much more is metabolized than if you feed polysaccharides. The same thing, of course, is true for proteins. The liberation time during digestion apparently has a great influence on the incorporation of the substances into the cells. The human body, or any animals' body, is not used to having high peaks. It wants to have very slow increases of any substances in blood because it is much easier for it to synthesize all the structural substances. In other words, much of the substances are lost by oxidation and other side reactions if they are given in a free uncombined from and not as polypeptides, polysaccharides, and so on.

BORSOOK: I would like to ask the speakers what they think now as to whether or not changes in the specific activity in the pool might or might not account for the claims of varying specific activity in proteins into which amino acids have been incorporated. This, I am aware, is a separate question from whether peptides or amino acids are incorporated, but it is conceivable that at some intermediate state, regardless of how it was arrived at, there might be exchange, and I would welcome any comments the speakers would care to make to this question.

LOFTFIELD: You have specifically in mind an experiment like KRUH's with labeled hemoglobins? I think that most of these experiments suffer from their experimental conditions, the rather low incorporation, the rather extended times, and the question of whether there was something going on besides protein synthesis. KRUH, for instance, gets a hemoglobin isolated after some hours of *in vivo* synthesis and finds that on partial hydrolysis the radioactive amino acid is not released at the same rate as the ^{12}C-amino acid. Presumably this must involve, the formation of something besides a peptide bond or something besides the accepted scheme of synthesis. I know of no simple interpretation of the results.

L. MILLER: Why couldn't you interpret the result of this experiment in terms of the difference in specific activity of the precursor in the pool at the time when the hemoglobin was being produced? Was it clearly demonstrated that the precursor's specific activity at the site of the hemoglobin synthesis was the same or different?

LOFTFIELD: No. The actual experiment, as I recall, was between just glycine or phenylalanine injected into a rabbit.

L. MILLER: You know that the red cells are formed in the bone marrow, and that the circulation of different parts of the bone marrow is not the same at different times, so that the pool, in one part of the bone marrow is not going to be the same at any particular time as it is at another. You are going to get all kinds of things there, and I do not see how it could be used to support one argument or the other.

BORSOOK: I think I may add a couple of facts to this discussion. I know how KRUH's experiments were done, because they were begun in my laboratory. Some of them were done with reticulocytes which were suspended in a large volume of reaction medium. We know now that equilibration in terms of specific activity is obtained between the reticulocytes and the external medium in a very short time, a matter of a minute or two, so that in this case, I do not believe variations in specific activity can account for the result. My own question is on the analysis of the subsequently synthesized protein. I have always been puzzled that there should be a progressive change of specific activity according to the time at which the amino acid was released from the protein. This seems too orderly, somehow, for this kind of thing, and I had hoped that you might have some information to add to this.

LOFTFIELD: No, I haven't, although I would certainly think it appropriate for KRUH now to repeat this experiment and to isolate the peptides, a procedure which has become fairly routine. It remains, I think, the clearest example, or the most unchallenged example, of unequal labeling.

ARONOFF: Professor BORSOOK, is there any possibility that there may be an isotopic fractionation? If this were to occur in a series of consecutive steps it could result in extremely large differences inspecific activity.

BORSOOK: I do not know.

HALVORSON: I would like to ask Dr. LOFTFIELD a question on his system involving ferritin formation. As I understand it, you have a 5-minute half life in this experiment. How does that agree with the over-all protein synthesis of these cells? Is this a delay in the release, or are you looking at the heterogeneity in the completion of polypeptides?

LOFTFIELD: In these particular experiments we had induced the synthesis of ferritin, so that by prior experiment we knew exactly the rate at which ferritin was being made, that is, the absolute rate of synthesis. Then, at zero time we injected ^{14}C-valine or ^{14}C-leucine. During the next 5 minutes there was a gradually upswinging curve of incorporation into released ferritin. At the end of 5, or actually at the end of 6 minutes the rate of release of radioactive ferritin corresponded exactly with the known rate of ferritin synthesis. In other words, at the end of 6 minutes every valine in the newly released ferritin had been derived from the pool which was established at zero time, and we had, in fact, established the speed with which the pool reached its steady state activity.

L. MILLER: I would like to have a few more facts in connection with this matter of differential labeling and its interpretation. I think one consideration almost conspicuous by its absence, although some workers in the field have briefly alluded to it, has to do with the relationship between what one might call the molecular generation time of a protein and the rate of change of specific activity with respect to that molecular generation time.

If the molecular generation time of the protein is exceedingly short, then obviously it seems almost impossible to design an experiment to pick up changes in specific activity of different parts of a protein molecule even though they may exist in terms of changes per micro-second, let us say. On the other hand, if you have a protein molecule with a sufficiently long molecular generation time, then it becomes altogether practical to examine this question critically. That is, can one change the specific activity fast enough between the time when the first amino acid residue is laid down and the time when the last one is laid down, to see changes that would make or break the argument. I do not know whether this has really been examined critically by anyone.

LOFTFIELD: Well again, in the case of our ferritin experiment we did establish that we changed the specific activity of the intracellular leucine or valine from zero to a relatively steady position within 15 seconds after the intravenous injection, and that the molecular generation time, to use your word, was about 6 minutes. DINTZIS found that he had what amounts to a molecular generation time of about a minute and a half, and that the amino acid got in extraordinarily fast, within 10 seconds or something on that order, very much as Prof. BORSOOK has mentioned.

E. ROBERTS: Shouldn't one be concerned in all such studies with whether there is a good net synthesis of protein or whether such a small net synthesis of protein is taking place that the protein isolated after the experiment will include an appreciable amount of protein made from partially filled templates. In the latter case one would end up with different specific activities in different positions. If there is a good net synthesis, so that protein made from originally partially filled templates represents only a very small proportion of the total, uniform labeling would be observed.

LOFTFIELD: I think that this is the case in both the experiments I have cited. In our case we found that after 10 minutes or 80 minutes the ferritin was substantially homogeneously labeled, and DINTZIS did exactly the same thing in his most recent and more elegant experiment. During the first minute and a half he was getting unequally labeled protein, and by the time he had gone 10 minutes, and this is substantially an *in vivo* situation since these were intact reticulocytes, there had been the complete synthesis of three, four, or five molecules of hemoglobin. Accordingly, the specific activity was leveling off for all positions.

EAGLE: I think it is pertinent to point out that this does not explain KRUH's experiment.

LOFTFIELD: That's right.

MILLER: Are you referring to the *in vivo* ones or the *in vitro* ones?

EAGLE: In this instance to the *in vitro* ones.

H. ROSENBERG: We have always been worried about experiments in which one has to base one's deductions on time intervals of 5 or 10 seconds after intravenous injections. For instance, if you inject isotope into a rabbit's ear vein and put a counter on the second ear, it takes a considerable time for the count to reach maximum. I wonder, therefore, if such short experiments can be controlled properly. I think you said you have used intravenous injections, and I wonder how long it takes to equilibrate the blood stream completely with respect to the isotope?

LOFTFIELD: We did this with the rat, and the earliest sample of tissue that we could take was about 5 seconds, more or less, after we had done the intravenous injection. At this point the total amount of radioactivity in the liver had already reached its maximum. The distance between a rat's tail and its liver is somewhat less than between the rabbit's ears, in addition to which it has a much more vigorous circulation. Experiments, of the type performed by DINTZIS and by SCHWEET, of course, get away from this problem because a suspension of reticulocytes is used. I might say, actually, that HAUROWITZ and BURNETT, in Indiana, even worrying about this, found that they could ease the substrate into the cells by cooling them to zero and making the medium slightly hypotonic. They would get their substrate in without any other reactions and then quickly warm up the cells. This was an effort to achieve this same security against the question of how long it takes for things to get in and out.

BIOCHEMICAL PRODUCTION OF LIPO-AMINO ACID COMPOUNDS

BERNARD AXELROD, JOSEPH L. HAINING AND TOSHIO FUKUI

Department of Biochemistry, Purdue University, Lafayette, Ind. (U.S.A.)

To begin with we must say that our interest in the problem of lipo–amino acid compounds was stimulated by a provocative suggestion by HENDLER that lipo–amino complexes might be concerned with the biological synthesis of proteins. His suggestion was based mainly on his observations that an incorporation of labeled amino acid into lipoidal material occurred in incubating minced hen oviduct and

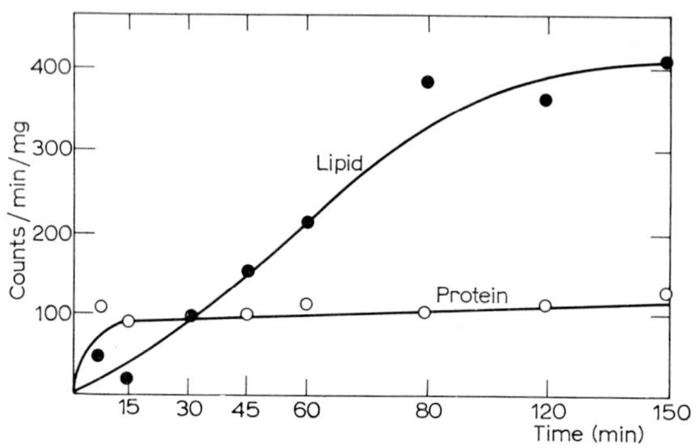

Fig. 1. Time course of the incorporation of [^{14}C]phenylalanine into protein and lipid fractions of a rat-liver microsome preparation. Each vessel contained 0.7 ml of whole supernatant solution, 1 μmole of ATP, 10 μmoles of 3-phosphoglyceric acid, 0.23 μmole of DL-[3-^{14}C]phenylalanine (500 000 counts/min), and water to a final volume of 1 ml. After incubation at 37° under N_2–CO_2 (95 : 5) the reaction was stopped by adding trichloroacetic acid to the vessels and the protein and lipid fractions prepared from the precipitated material. (By courtesy of the *J. Biol. Chem.*)

that the relative rates of labeling of protein and of lipoidal substance was not inconsistent with this view[1]. At that time, meeting with frustrations on another aspect of protein synthesis, and having a goodly supply of rats on hand we attempted to see if this phenomenon occurred in rat–liver preparation, and if it did, how it compared with "conventional" incorporation into protein.

AMINO ACID INCORPORATION INTO PROTEINS AND LIPIDS BY CELL-FREE PREPARATIONS OF RAT LIVER

The supernatant obtained by centrifuging a homogenate of rat liver at 10 000 × *g*, is so easily prepared and so frequently shown to be suitable for showing amino acid incorporation into protein, that we confined our attention to this system. Using the

References p. 749

time-honored conditions of ZAMECNIK AND KELLER[2] we had no difficulty in detecting incorporation of radioactive phenylalanine into protein, and in finding, at the same time that lipid material was also labeled[3, 4] (Fig. 1). The behavior of whole supernatants of rat spleen and pancreas resembled that of liver. Here it must be emphasized that the lipoidal material in question was that which was extracted from the precipitate obtained with trichloroacetic acid. Such a precipitate would consist not only of protein and nucleic acid, but also would contain the microsomes as well. It should be noted that incorporation into lipid continued long after the level of activity in the protein became constant. In another experiment, when the microsomes and

TABLE I*

EFFECT OF VARIOUS TREATMENTS ON [^{14}C]PHENYLALANINE INCORPORATION INTO PROTEIN AND LIPID

Whole supernatant solutions (and the controls) were incubated for 5 min at 37° with the enzymes specified after which the other components of the incubation mixture were added. Total incubation time: 2 h.

Treatment	Relative specific activity in	
	Protein	*Lipid*
None	100**	100**
ATP and 3-phosphoglyceric acid omitted	21	115
NaF (10^{-2} *M*; ATP and 3-phosphoglyceric acid omitted)	33	99
p-Chloromercuribenzoate (10^{-3} *M*)	15	86
Chloramphenicol ($6.2 \cdot 10^{-3}$ *M*)	70	84
RNAase (0.1 mg/ml)	42	107
Crotoxin (20 μg/ml)	13	393

* By courtesy of the *J. Biol. Chem.*
** Actual specific activity (counts/min/mg) of protein and of lipid separately set equal to 100. The values in one column cannot be compared directly to those in the other.

the microsome-free supernatant were examined separately, radioactivity was found in the protein and lipid material in both fractions. Incubation of the supernatant fraction alone gave even better incorporation into lipid than when microsomes were present.

In the course of these incubations 3-D-phosphoglycerate-ATP was used in conjunction with the endogenous glycolytic system as a source of energy. Omission of these chemicals resulted in a substantial reduction of incorporation into protein, but incorporation into lipid was not effected (Table I). Similarly, as expected, the presence of ribonuclease reduced incorporation into protein but had no effect on the radioactivity found in the lipid. Crotoxin, which HENDLER[1] found depressed the overall incorporation of labeled amino acid into hen-oviduct mince, severely depressed the incorporation into protein but caused a marked increase in the amount of label found in the lipid (Table I).

In our earlier work we found a preferential diminution of labeling in protein to be caused by lipoxidase but we have not been able to repeat this.

The relative degree of incorporation of several other amino acids into the protein and lipid fractions from microsomes is shown in Table II. Phenylalanine and tryptophane surpassed all of the other radioactive amino acids in their capacity to label the lipid as compared to the protein. A similar pattern of incorporation was noted in the lipid obtained from the particle-free supernatant. Leucine was relatively the most effective in labeling the protein.

Finding activity in a lipid fraction however, does not on the face of it prove that a radioactive amino acid, as such, has actually become joined into a lipo–amino acid compound. The existence of fat-soluble substances containing amino acids derived by extraction from biological systems has often been the subject of controversy

TABLE II

RELATIVE INCORPORATION OF SEVERAL AMINO ACIDS INTO MICROSOMAL PROTEIN AND LIPID

Incubation conditions are similar to those described in Fig. 1. Each vessel contained 1.15 µmoles of the amino acid specified. The measured specific activities in the protein and lipid fractions were adjusted for the differences in the specific activities of the various amino acids (shown in the table).

Amino acid	*Specific activity (µC/µmole)*	*Radioactive specific incorporation**		*Specific activity ratio** (Lipid/protein)*
		Lipid	*Protein*	
DL-[3-^{14}C]Phenylalanine	4.34	100	100	0.76
DL-[3-^{14}C]Tryptophane	1.86	105	94	0.84
DL-[1-^{14}C]Leucine	5.33	19	140	0.13
DL-[1-^{14}C]Lysine	3.19	9.5	67	0.10
[2-^{14}C]Glycine	4.87	9	22	0.32
DL-[1-^{14}C]Valine	2.92	9	96	0.07
DL-[1-^{14}C]Alanine	4.05	3.2	29	0.07

* Adjusted specific activity (counts/min/mg) in lipid and in protein separately set equal to 100 for phenylalanine incorporation in order to facilitate comparison of the incorporation of each amino acid into each fraction.

** Ratio of adjusted counts/min/mg of lipid to that of protein to illustrate the absence of any simple relationship of amino acid incorporation into the two fractions.

because of the ease with which natural mixtures of lipids may become contaminated with amino acids. The presence of phospholipids in solutions of lipids often permits the solubilization of polar substances in apparent violation of the rules which are based on studies made with pure solvents. Even if covalent combinations are not formed, separation of amino acids from crude natural mixtures of lipids can be difficult. WREN AND MITCHELL[5] as well as GABY AND SILBERMAN[6] are among those who have used paper electrophoresis to establish freedom from contaminating amino acids in lipo–amino compounds. The possibility has been shown that cyclized amino acids, which have lost much of their polar nature can act as particularly tenacious contaminants. WREN[7] has considered at some length the formation of lipo–amino artifacts by the chemical combination of auto-oxidized lipids with amino acids. Such combinations could well be favored with the highly active fatty oxidation products such as free radical compounds and carbonyl compounds. Finally,

References p. 749

one is always faced with the possibility that a radioactive amino acid can serve as a metabolic precursor of a radioactive lipid.

The lipid which was obtained in incubation of the rat liver whole supernatant with [^{14}C]phenylalanine was separated by silicic acid column chromatography[8] into the phosphatidic and non-phosphatide fractions. The radioactivity appeared in the latter fraction. When this fraction was subsequently refractionated the radioactive material was eluted after the sterols, triglycerides and sterol esters.

Chromatography of the radioactive fraction on paper gave but one radioactive area and this was quite distinct in position from that expected for phenylalanine. Control experiments with phenylalanine showed that there was no tendency for the free amino acid to change its mobility in the presence of the lipid material. Hydrolysis of the radioactive lipid (6 N HCl, 110°, 18 h) released a water-soluble substance which was radioactive and migrated coincidentally with free phenylalanine. Milder conditions (2 N HCl, 100°, 2 h) liberated only negligible amounts of phenylalanine.

The radioactive lipoidal fraction also released other ninhydrin-positive substances, of varying mobilities presumably amino acids which were not radioactive. Although these experiments were carried out with DL-phenylalanine similar results were obtained with the natural isomer.

FORMATION OF LIPO–AMINO ACID COMPOUNDS BY "PURIFIED" RAT-LIVER ENZYME, SPECIFICITY OF THE REACTION AND NATURE OF THE PRODUCT

In order to learn more of the nature of the amino-acid incorporating system as well as of the product it was necessary to separate the enzyme system from the endogenous lipid acceptor. To this end[9], rat liver was extracted in a Waring Blendor with acetone at −20° and the resulting powder extracted with diethyl ether. Despite the removal of considerable lipoidal substance suspensions of the extracted material were still effective in catalyzing the incorporation of added [^{14}C]phenylalanine into lipid-soluble material without the necessity of additional lipid. Furthermore the enzymatic activity remained substantially particle-bound. In order to remove the residual lipid, a suspension of the extracted material was allowed to autolyze for 3 h at 37° and then extracted again with cold acetone. The residue thus obtained exhibited little activity until some of the extracted lipid was returned. Also, an appreciable portion of the activity could now be extracted with buffer. The activity could be concentrated by ammonium sulfate precipitation; dialysis yielded a product which was not only free of endogenous lipid, but was also low in endogenous free amino acids.

The lipid obtained in the second acetone extraction was separated into phospholipid and non-phospholipid fractions. Both were effective as "acceptors" when tested with phenylalanine. Results obtained with some specific lipids appear in Table III. Monoolein, as the α-isomer, was the most effective of the glycerides tested. Triolein and tripalmitin were almost without effect. Rather unexpectedly free oleic and palmitic acids functioned better than any of the glycerides. Lecithin showed good activity.

In addition to phenylalanine, a number of other amino acids were tested using monoolein as the lipid substrate, under comparable conditions. [2-^{14}C]Glycine, DL-[1-^{14}C]leucine, DL-[1-^{14}C]lysine, DL-[1-^{14}C]valine and DL-[1-^{14}C]alanine were incorporated in the amounts of 2240, 1685, 825, 770 and 425 counts/min respectively, as compared to DL-[3-^{14}C]phenylalanine, 2270 counts/min. In all cases 1.15 μC of the amino acids were employed, and their respective amounts were: 0.24, 0.22, 0.36, 0.39, 0.28 and 0.27 μmole.

There is little doubt about the enzymatic nature of the reactions. The quantity of the amino acid–lipid compound formed depends on the quantity of enzyme solution used. No product is formed in the absence of enzyme. The activity of the enzyme preparation is lost on heating.

The possibility that the observed reaction might be due to a reversal of the hydrolytic action of a glyceride esterase was tested by adding a 1090-fold molar excess of glycerol in an effort to depress the formation of the radioactive phenylalanine–lipid by the soluble enzyme in the presence of palmitic acid, but the reaction was not greatly affected.

TABLE III*

INCORPORATION OF [^{14}C]PHENYLALANINE USING PURE LIPIDS

Each vessel contained 1 ml of the soluble enzyme, 5 mg of lipid, and 0.4 μmole of DL-[3-^{14}C]phenylalanine (550 000 counts/min) in a final volume of 1.6 ml. Incubation was carried out for 2 h.

Lipid	*(^{14}C) Phenylalanine incorporation (Counts/min)*
None	90
Crude lipid extract	950
α-Monoolein	1670
α-Monopalmitin	535
Triolein	25
Tripalmitin	25
Oleic acid	2050
Palmitic acid	2315
Lecithin	1725

* By courtesy of the *J. Biol. Chem.*

Adding a 50-fold molar excess of unlabeled L-phenylalanine produced virtually no depression in incorporation. In order to examine the possible relationship between this enzyme and previously described esterases and carboxypeptidase, several specific inhibitors were tested. The formation of *N*-palmitoylphenylalanine from palmitic acid and phenylalanine was inhibited 68% by 6.8 · 10^{-3} *M* diisopropylfluorophosphate and 48% by 1.0 · 10^{-3}*M* physostigmine, but not by 2.0 × 10^{-3} *M* hydrocinnamic acid. The latter is a specific inhibitor of carboxypeptidase. The enzyme solution was incubated with the inhibitors for 40 min before the addition of the substrates.

Crystalline carboxypeptidase was entirely without effect in catalyzing the combination of DL-[3-^{14}C]phenylalanine and palmitic acid, ruling out the possibility that an enzyme of this type was responsible for the reaction.

The product formed from DL-phenylalanine and monoolein was highly purified by column chromatography and isolated in weighable amounts. Its specific radioactivity on a weight basis agreed with a 1 : 1 molar ratio of phenylalanine to oleic

References p. 749

acid. The phenylalanine isolated on hydrolysis was tested with L-amino acid oxidase and proved to be the L-isomer, showing that the incorporation reaction was specific for the natural form.

When the lipo–amino acid was reduced with lithium aluminum hydride before the hydrolysis at 110°, no radioactive phenylalanine appeared on paper chromatography, but a new radioactive component which coincided in position with α-amino-β-phenylpropanol was present. It thus appears that phenylalanine is bound through its amino group, with its carboxyl group free. This mode of combination was further confirmed by the fact that the phenylalanine–lipid compound was hydrolyzed almost completely by the prolonged action of carboxypeptidase. It is therefore probable that the compound formed from phenylalanine and monoolein is *N*-oleoylphenylalanine.

The nature of the lipo–amino acid was further established by chromatography of synthetic *N*-palmitoyl-DL-phenylalanine with the product derived enzymatically from phenylalanine and palmitic acid.

When the isolated lipo–amino acid was incubated with the soluble enzyme, it was almost completely hydrolyzed to a radioactive substance which was no longer ether-soluble.

SIGNIFICANCE OF LIPO–AMINO ACID COMPLEXES IN NATURE

Among the roles suggested for lipo–amino acids (or lipo–amino acid complexes) is that of a precursor to proteins.

However attractive it might be to ascribe a topically important role to the compounds on which we have lavished so much effort we do not feel that our own studies provide evidence for such a view, insofar as the limited system which we have studied is concerned. We have been inclined to interpret our kinetic evidence in support of the contrary view, but as long as the possibility of heterogenous pools of lipo–amino acid complexes exists, we must admit that an absolutely unequivocal interpretation is not possible. Our own studies do not provide evidence for this view. This is not to deny to possibility that amino acids present in combination with the lipids may be released and subsequently transferred via the commonly accepted pathway involving nucleotide activation. Nor do these studies have anything to say about the lipid portion of the microsome in protein synthesis.

What is the significance of the acylamino acids? We are not prepared to say. We can only say (with respect to the system we have studied) that they are formed by an enzymatic system and that they probably exist in the cell. Their formation does not require a source of triphosphonucleoside, and if they are in the pathway of protein synthesis their formation does not occur subsequent to the well-known ATP–amino acid activation. Since the amino acid activation is known to occur directly with free amino acids in relatively purified systems, it is difficult to see, how, in our system, the acylamino acid can be obligatorily involved. However, we have not worked with intact cells and cannot on the ground of this work provide a direct evaluation of the claims based on work with more complete systems.

The best reasons for implicating lipo–amino acids (or complexes) are based on work with more elaborate systems such as protoplasts, minced tissue or entire *Drosophila*. It is emphasized that in all of these cases elaborate proof has been provided that the lipo–amino acid complex is not an artifact. Larvae of *Drosophila* have been shown in MITCHELL's laboratory[10] to contain lipo–amino acid complexes, probably

References p. 749

phospholipids, the quantity of which decreases sharply when protein synthesis begins. The investigations of HENDLER on oviduct mince have provided stronger insinuations. These investigations have been thoroughly pursued and extended to other organisms[11, 12]. Fortunately HENDLER is present so we can look forward to an authoritative presentation and a critical evaluation of the issue.

Strong claims for the participation of lipo–amino acid complexes in protein synthesis have been put forth by HUNTER AND GOODALL[13]. These workers following up an earlier lead which HUNTER *et al.*[14] obtained in studying the incorporation of amino acids into cytoplasmic membranes of *Bacillus megaterium*, fed ^{14}C-labeled amino acids to protoplasts. They isolated therefrom lipo–amino acid complexes which were composed of both phospholipid and non-phospholipid fractions. Incubating in the presence of chloramphenicol or crotoxin prevented their formation. A number of amino acids were suitable for the purpose of preparing radioactive lipo–amino acid complexes. Strangely enough, phenylalanine behaved in an aberrant manner. A complex prepared with radioactive threonine was isolated and incubated with protoplasts. In 30 min over 70% of the activity appeared in the protein. If a great excess of unlabeled threonine was added there was no diminution in the amount of activity taken up. It was therefore claimed that the amino acid of the complex can be taken directly into protein without prior conversion to free amino acid. However, radioactive threonine itself was readily incorporated into protein and lipid. If the quantity of unlabeled threonine used in the above experiment was of an order of magnitude required to saturate the system (and this would have to be the case for the conclusion to be valid) then it would be necessary to argue that the lipo–threonine complex or the olive oil used as its vehicle, suppressed the normal incorporation, or that there are a tremendous number of sites for deposition of the amino acid.

In our studies we have found that dilution of radioactive phenylalanine with unlabeled amino acid reduced incorporation into protein, but had no effect on the lipo–amino acid. The amino acid-activating enzymes are characterized by low K_m values, while the lipo–amino acid forming system we found, is not saturated by high concentrations of phenylalanine.

GABY *et al.*[15] have obtained phospholipid–amino acid complexes from *Penicillium chrysogenum*. Using radioactive amino acids they have concluded that amino acids are taken up at a faster rate by the phospholipids than by the water-soluble amino acid pool. While they have expressed no conclusions concerning the role of the lipo–amino acid complex in protein synthesis they indicate that they have an active role in the utilization of amino acids.

We are indebted to LEDERER for drawing our attention to the existence of a naturally occurring fatty acid–amino acid combination, found in some species of *Mycobacterium*[16]. Thus one obtained from *M. marinum*[17] and another from *M. avium*[18] contain together with three unusual deoxy sugars, a pentapeptide, whose N-terminal amino acid is attached to a long chain carboxylic acid.

Bacteria are fruitful sources of strange compounds. Thus *B. circulans* provides in the antibiotic circulin another example of a peptido–acyl combination. Here L-α, γ-diaminobutyric acid is attached via its γ-amino group to 6-methyloctanoic acid[19].

MAALØE has pointed out that if lipo–amino acids can serve as a sort of transient storage site for amino acids they cannot be ignored in estimating the amino acid pool size when studying enzyme induction and repression in the cell.

References p. 749

REFERENCES

1 R. W. HENDLER, *Science*, 128 (1958) 143.
2 P. C. ZAMECNIK AND E. B. KELLER, *J. Biol. Chem.*, 209 (1954) 337.
3 J. L. HAINING, T. FUKUI AND B. AXELROD, *J. Am. Chem. Soc.*, 81 (1959) 1259.
4 J. L. HAINING, T. FUKUI AND B. AXELROD, *J. Biol. Chem.*, 235 (1960) 160.
5 J. J. WREN AND H. K. MITCHELL, *J. Biol. Chem.*, 234 (1959) 2823.
6 W. L. GABY AND R. SILBERMAN, *Arch. Biochem. Biophys.*, 87 (1960) 188.
7 J. J. WREN, *Nature*, 185 (1960) 295.
8 J. HIRSCH AND E. H. AHRENS, Jr., *J. Biol. Chem.*, 233 (1958) 311.
9 T. FUKUI AND B. AXELROD, *J. Biol. Chem.*, 236 (1961) 816.
10 J. WESTLEY, J. J. WREN AND H. K. MITCHELL, *J. Biol. Chem.*, 229 (1958) 131.
11 R. W. HENDLER, *J. Biol. Chem.*, 234 (1959) 1466.
12 R. W. HENDLER AND E. LOVE, *Federation Proc.*, 20 (1960) 390.
13 G. D. HUNTER AND R. A. GOODALL, *Biochem. J.*, 78 (1961) 564.
14 G. D. HUNTER, P. BROOKES, A. R. CRATHORN AND J. A. V. BUTLER, *Biochem. J.*, 73 (1959) 369.
15 W. L. GABY, R. N. NAUGHTEN AND C. LOGAN, *Arch. Biochem. Biophys.*, 82 (1959) 34.
16 E. LEDERER, *Compt. rend.* 246 (1958) 858.
17 M. CHAPUT, G. MICHEL AND E. LEDERER, *Experientia*, 17 (1961) 107.
18 P. JOLLES, T. GENDRE, E. BIGLER AND E. LEDERER, *Bull. soc. chim. biol.*, 43 (1961) 177.
19 H. KOFFLER, *Science*, 130 (1959) 1419.

ON THE METABOLIC IMPORTANCE OF AMINO ACID–LIPID COMPLEXES

RICHARD W. HENDLER

National Heart Institute, National Institutes of Health, Bethesda, Md. (U.S.A.)

Although amino acids have been observed in lipid fractions for many years[1], they have been regarded generally as contaminants which the investigator should try to remove by further purification. During a study of the formation of the proteins of the hen oviduct from radioactive amino acids, it was learned that an extremely non-polar form of the amino acid was rapidly formed[2]. It was found that protein synthesis was markedly inhibited by low concentrations of lecithinase A (crotoxyn) and other lipolytic agents, and that CoA and cytidine triphosphate exerted a stimulatory effect. Further studies with the hen oviduct have shown that amino acids can enter and leave these lipid complexes at a high rate when compared to their entry and exit from associations with nucleic acids and their entry into protein[3]. It was found* that each

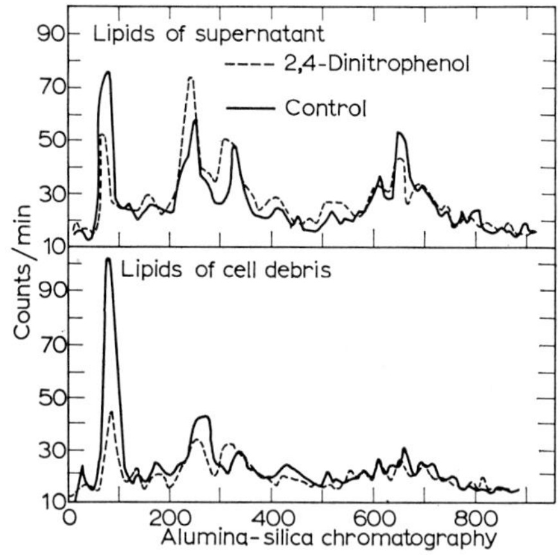

Fig. 1. The abscissas designate volume of effluent in ml. Lipid fraction was obtained from 4 g hen oviduct after incubation for 15 min with ^{14}C-labeled algal protein hydrolyzate prior to homogenization and separation into cell debris and supernatant fractions. 5 g of adsorbent in a column of 1 cm dia. was used. The 2,4-dinitrophenol (10^{-3} M) was present during a 15-min preincubation prior to addition of radioactive amino acids and subsequent 15-min incubation. Solvent changes were as follows: hexane–chloroform (1 : 1) to 60 ml effluent, hexane–chloroform (1 : 2) to 100 ml, hexane–chloroform (1 : 4) to 150 ml, chloroform to 210 ml. Methanol–chloroform (1 : 99) to 490 ml, methanol–chloroform (4 : 96) to 590 ml, methanol–chloroform (8 : 92) to 625 ml, methanol–chloroform (15 : 85) to 685 ml, methanol–chloroform (25 : 75) 730 ml methanol–chloroform (36 : 65) to 785 ml, methanol–chloroform (50 : 50) to 840 ml and methanol to end.

* This material was presented at a meeting of the Faraday Society, March 1960, University of Reading, Reading (Great Britain).

References p. 758

of 11 different amino acids tested were capable of forming such complexes and that phenylalanine was unusual in that it was able to form certain types of complexes to about ten times greater extent than the other amino acids[4]. That the amino acids were not present in the free form was shown by the fact that as an alcoholic solution of these complexes was diluted with water, they were more easily extracted with hexane; radioactive amino acid could be added to the biologically labeled complexes and then be separated away by simple extraction or chromatography; free amino acids were not observed before hydrolysis either directly or as DNP derivatives (but could be observed after hydrolysis), these complexes could be chromatographed on silicic acid under conditions where free amino acid could not[3,4].

Successful fractionation of amino acid–lipid complexes from hen oviduct was achieved by graded counter-current distribution and chromatography on columns of alumina–silica and silicic acid. It was found that the structural elements of hen oviduct contained a greater proportion of the most non-polar of the mixture of amino acid–lipid complexes obtained from this tissue (Fig. 1).

Incubations of the tissue with radioactive amino acid under conditions in which protein synthesis was inhibited yielded mixtures of amino acid–lipid complexes with an altered distribution of radioactivity. That is, there was a decrease in the proportion of radioactivity found in the least polar components. One rather interesting observation was that after treatment with 2,4-dinitrophenol, it was found that there was an increase in the amount of radioactivity found in the lipid fraction. Upon fractionation of the tissue and chromatography of the complexes, it was found that there was marked decrease in the radioactivity of the least polar complex obtained from the particulate fraction and a more than equivalent increase in the radioactivity of the more polar complexes of the supernatant fraction[4] (Fig. 1).

HAINING, FUKUI AND AXELROD have studied the formation of lipid–amino acid complexes in rat-liver microsome and supernatant preparations[5, 7]. Although their experiments increase our understanding of the type of compound with which we are concerned, I must take some time to examine critically the interpretation of these studies as regards protein synthesis. It has never been established that the system under study is capable of normal protein synthesis. Aside from the fact that the incorporated radioactivity cannot be demonstrated to be present in the proper internal position for normal liver proteins, it is known that at least one type of artifactitious incorporation, that of disulfide bonding, takes place in cell-free liver preparations that does not take place in intact cells[8]. In their studies on relative uptake of amino acid into protein and lipid as well as the experiments on effects of metabolic poisons, the specific activity of the whole lipid fraction was determined.

In the hen oviduct, it has been observed that many different amino acid–lipid complexes are formed for a given amino acid and that the specific activities for each vary. Furthermore, when the hen oviduct system was poisoned with 2,4-dinitrophenol, there was a marked depression in the labeling of a complex associated with the structural elements and an increase in the labeling of a different amino acid–lipid complex obtained from the supernatant fraction. The overall labeling of the lipids was enhanced by this poison. This observation illustrates the danger of drawing conclusions from a study of the total lipid fraction. Many of the data presented by HAINING, FUKUI AND AXELROD were obtained with phenylalanine. It has been observed that phenylalanine is an unusual amino acid in that it can label certain

References p. 758

lipid complexes to an extent approx. 10-fold greater than the other amino acids[4, 9]. HAINING *et al.* also found that phenylalanine and tryptophane were similarly of a different order of magnitude in their ability to label lipid complexes when compared to other amino acids. In their metabolic studies, they isolated the total lipid fraction after hot trichloroacetic acid treatment. By so doing, they may have destroyed certain amino acid complexes not able to stand these conditions. In one of their studies, AXELROD *et al.* found that by adding unlabeled phenylalanine to a small amount of labeled phenylalanine the total radioactivity of the lipids did not decrease, whereas the radioactivity of the protein was decreased. Since we have shown that the lipid sites are far from saturation while the protein-forming system is near saturation, such experiments are not inconsistent with the amino acid passing through the lipid en route to the protein.

The existence of an enzyme for catalyzing the formation of *N*-fatty acylphenylalanine was well established, as well as various properties of the enzyme, and it will be worthwhile to investigate the possible metabolic significance of this new finding. However, I think it important to emphasize that these studies cannot be interpreted at present in terms of the relation of amino acid–lipid complexes to protein synthesis.

HUNTER *et al.* have recently become very much interested in the possible role of lipids in protein synthesis in *B. megaterium*[9]. They have been able to repeat in their system many of the characteristics reported for the hen oviduct system such as: the ability of many amino acids to form such complexes; the relatively rapid kinetics for amino acid labeling of the lipid complexes compared to the labeling of the ribonucleic acid and protein; the unusual nature of lipid complexes formed from phenylalanine; the fact that lipid–amino acid complexes containing little or no phosphorus could be obtained; and that the amino acid incorporation process is quite sensitive to lecithinase A. In addition, they have reported that radioactivity could be transferred from an isolated lipid–amino acid complex to the protein of their membrane fraction and that the incorporation of amino acid into the lipid complexes is inhibited by chloramphenicol.

GABY *et al.* have been concerned with the possible involvement of phospholipid in the transport of amino acid across cell walls[10, 12]. They have observed that amino acid becomes firmly bound to phospholipid upon incubation with the metabolizing cells. Furthermore, the amino acid is held in the phospholipid in a metabolically exchangeable form as opposed to being held in the form of stable end-product. Similar evidence for the involvement of phospholipid in the metabolism of amino acid was obtained with cells of *Penicillium chrysogenum*, rabbit liver, rabbit lymph node and Ehrlich ascites tumor. Similar findings were made for several of the amino acids. These studies are clear in their implication and are being pursued further. It is to be hoped that the examination will be extended to the less polar lipids.

WESTLEY, WREN AND MITCHELL have found several chromatographically different amino acid–lipid complexes in *Drosophila melanogaster*, whole human blood and other sources[1, 13, 14]. During maturation of the *Drosophila* larvae, a major quantity of amino acid appears to move from a lipid-soluble form eventually into newly formed protein. These authors provide an excellent review of papers since the early 50's reporting the presence of amino acid in lipid fractions. The question of artefacts has been seriously considered in these works and various procedural operations are

References p. 758

suggested to minimize the possible occurrence of amino acid derivatives as contaminants. In the latest paper[15], WREN has decided that although some amino acid–lipid complexes are artefacts, others are undoubtedly real. WREN has defined artefact to include those complexes that form when amino acid or amino acid analogs are added to a metabolically inactive form of the tissue being worked up. Although it is wise to exercise caution in the study of these complexes, I believe that WREN is not justified in applying the label of artefact to these complexes. His criterion is the fact that the complexes are formed too efficiently, that is by simply extracting the tissue in the presence of the amino acid. I can see no reason why this high affinity of the non-polar lipid for highly polar amino acids cannot be used in the normal metabolic sequences of tissues. If we are considering the possibility of certain stages of metabolism occurring in membranes or non-polar cellular media, then why should we decide *a priori* that complexes which are formed so easily are artefacts as opposed to real metabolic intermediates? Other criteria must be employed before making such a categorization.

TRIA, BARNABEI AND FERRARI have described a lipid-soluble "peptidylphosphatide" isolated from rat liver. Upon perfusing the isolated rat liver with heparinized rat blood containing [^{14}C]amino acids, it was found that at early time points, the specific activity of the peptide isolated from the lipid fraction was considerably higher than the specific activity of the average protein[16].

In recent experiments with the hen oviduct, we have been concerned with further criteria for the metabolic importance of these complexes and have studied the structural and optical specificity requirements for the amino acid. We have tried to dilute a small quantity of radioactive L-valine with unlabeled L-valine, D-valine and L-serine respectively. If the formation of the amino acid–lipid complex required L-valine then dilution could be accomplished with L-valine, but not with D-valine or L-serine. If the specificity was directed only towards the gross aspects of the valine

TABLE I

A TEST OF SPECIFICITY OF LIPID–AMINO ACID SITES

	Relative dilution of valine specific activity by			*Relative internal concentration of valine*	*Relative radioactivity in lipids*	
	L-valine	*L+D-valine*	*L+D-valine + serine*		*Total silicic acid peaks*	*$CHCl_3$ peak from silicic acid*
Expt. 1						
L-[^{14}C]Valine (83 μg)	1.0	1.0	1.0	1.0	1.0	1.0
+L-[^{12}C]Valine (6 mg)	10	10	10	8.9	0.93	0.70
+D-[^{12}C]Valine (6 mg)	0.97	5.4	5.4	6.6	1.0	1.48
				Serine		
+L-[^{12}C]Serine (6 mg)	0.75	0.75	13.6	5.5	1.0	0.92
Expt. 2						
				Valine		
L+D-[^{14}C]Valine (4 mg)	1.0	1.0	1.0	1.0	1.0	1.0
+L-[^{12}C]Valine (20 mg)	8.4	6.2	6.2	4.7	0.43	0.57
+D-[^{12}C]Valine (20 mg)	1.0	4.0	4.0	4.7	1.33	1.29
				Serine		
+L-[^{12}C]Serine (20 mg)	0.84	0.88	5.4	11.4	0.91	0.97

References p. 758

structure, then perhaps D- or L-valine could dilute but not serine. For very low specificity requirements, all three of the amino acids should be able to dilute the labeled valine equally well. In Table I, two such experiments are shown. Expt. 1, Case 1 shows the situation with high-specific activity radioactive valine (83 μg) added to the incubation flask. Case 2 shows the result of adding 6 mg of unlabelled L-valine. In Case 3, 6 mg of unlabeled D-valine was added and in Case 4, 6 mg of unlabeled L-serine. The change in specific activity of the intracellular valine pool was verified by isolating the amino acid and determining its specific activity. With this information and a knowledge of the specific activity of the added valine and total uptake of radioactivity into the cells, the dilution of the intracellular pool could be calculated for the three situations of decreasing specificity as indicated in Columns 1, 2 and 3. The change in relative size of the valine pool compared to the other amino acids was verified by chromatography of the intracellular supernatant fluid on Dowex-50, and these data are shown in Column 4, except in the fourth line where the increase is shown for serine. Column 5 shows the relative amount of radioactivity recovered in the lipid complexes which were chromatographed on silicic acid, and the last column shows the recovery of radioactivity in the most non-polar amino acid–lipid complex. Expt. 2 is basically the same. The major differences were that both labeled L- and D-valine were used in the initial case. A larger initial amount of labeled valine was used (2 mg each of L- and D- and larger amounts of unlabeled valine and L-serine were used for dilution (20 mg of each). Both experiments show that the most non-polar material showed effects of dilution by L-valine, but not by D-valine or L-serine. The other lipid components also show distinct signs of specificity at the higher concentrations. The increase in radioactivity obtained from the lipid complexes upon addition of the D-amino acid may be related to the D-amino acid's ability to raise the pool of L-amino acid slightly by competing with L-valine in other reactions of low specificity or by virtue of its having a low competitive affinity with respect to the lipids.

Since total radioactivity is experimentally measured, it is important to take into account the increased amount of amino acid in the form of the complex as a consequence of the higher concentration of amino acid inside the cell as a result of dilution. For a monomolecular reaction, where the amount of lipid acceptor is not limiting, the total radioactivity held in the form of a complex will be constant even after dilution with unlabeled amino acid. If the reaction is multimolecular with respect to the amino acid, the total lipid-bound radioactivity will start to increase upon addition of unlabeled carrier until the number of multisite lipid acceptors becomes limiting. The total radioactivity held will then level off and decrease. If a given lipid fraction can accommodate amino acid by two independent reactions, one multimolecular and the other monomolecular, and if the number of multimolecular sites become saturated early, then the total radioactivity held in the form of a lipid complex will first rise (as unlabeled amino acid is added), reach a maximum, start to decrease and level off at a value determined by the affinity of the monomolecular site for the amino acid. Actually the saturation curves for valine in the lipid complexes in the hen oviduct resemble this behavior. The initial phase for the chloroform-eluted material from silicic acid behaves like a mixed mono- and multimolecular reaction. The multimolecular site becomes limiting when the external valine concentration is approx. 7 mM (5 mg added) and the internal free valine pool is approx. 2 mM after a 15-min

References p. 758

incubation. The total radioactivity in the lipid complex rises from zero addition of carrier to a maximum at approx. 5 mg added. The total radioactivity then decreases to a value which is approx. 70% of the zero carrier amount and 40% of the maximum radioactivity value. This value is maintained even after extensive dilution.

We also have been interested in the nature of the bonds which hold amino acid in such non-polar forms. In order to see if the carboxyl group of the amino acid was

TABLE II

FE^{3+} IRC-50 FRACTIONATION

Mixture			*% of radioactivity in*		
Amino acid	*Proline hydroxamate*	*Palmitic hydroxamate*	*H_2O wash*	*0.3 N HCl wash*	*CH_3OH : 1 N HCl 5 : 1*
—	^{14}C	—	3	97	—
$[^{14}C]$Proline	^{14}C	^{12}C	45	55	—
	Leucine hydroxamate				
$[^{14}C]$Glycine	^{12}C	^{12}C	93	5	2
$[^{14}C]$Alanine	^{12}C	^{12}C	86	4	10
$[^{14}C]$Alanine + Alanine hydroxamate					
$[^{14}C]$proline + Proline hydroxamate		—	96	2	2

covalently linked to a lipid by an ester bond, we studied the reactions of our various amino acid–lipid complexes with hydroxylamine. It was first necessary to have an effective means of distinguishing between free amino acid, amino acid hydroxamate, and fatty acid hydroxamates. This was done by forming a column of weakly acidic ion-exchange material which was previously equilibrated with Fe^{3+} under controlled

TABLE III

FE^{3+} IRC-50 FRACTIONATION

Silicic acid fractions	*% of radioactivity in*		
	H_2O wash	*0.3 N HCl wash*	*CH_3OH : 1 N HCl 5 : 1*
$CHCl_3$ peak	11	65	24
$CHCl_3$ peak	12	70	19
$CHCl_3$: CH_3OH (9 : 1): 1st part	41	39	20
2nd part	53	20	27
$CHCl_3$: CH_3OH (4 : 1): 1st part	25	15	60
2nd part	9	71	20
$CHCl_3$: CH_3OH (1 : 1)	34	10	56

References p. 758

conditions. The column was light tan in color and iron did not wash off. Free amino acids however, could be removed with water. Amino acid hydroxamates formed a red band at the top of the column and were not eluted with water but were eluted with 0.3 N HCl. Palmitic hydroxamate, due to its insolubility, stayed at the top of the column, produced no color and was not eluted. An alcoholic HCl solution later revealed the fatty acid hydroxamate as a purple band which could easily be eluted from the column. Table II shows how some control mixtures behaved in this system. When various fractions of amino acid–lipid complexes (obtained from chromatography on silicic acid) were subjected to this treatment, it was found that a high percentage of the radioactivity was not in the free amino acid form, (Table III). The amount of radioactivity in the water wash is an upper limit since a certain amount of lipoidal material could be washed through in this fraction. Furthermore, there was a considerable amount of red material which could be eluted with 0.3 N HCl from the most non-polar peak from silicic acid. Upon paper-chromatographic analysis with authentic samples of the amino acid hydroxamates, it was found that the red Fe hydroxamate obtained from this tissue fraction was not in the form of free amino acid hydroxamates. A peptide hydroxamate was considered as a possibility, especially since the dilution experiments previously described indicated multisite positions for the amino acid valine. The presence of peptide material in this fraction was further indicated by the LOWRY modified biuret reaction and the liberation of a complex spectrum of amino acids by 6 N HCl hydrolysis overnight. It was found that after a 15-min incubation with radioactive alanine and proline, the specific activity of the average alanine and proline liberated from this total fraction by hydrolysis was 15 times higher than the average specific activity for each of these amino acids liberated from the total proteins. One other finding of interest was that a radioactive sugar moiety was also isolated from the hydrolyzate of this fraction.

Further evidence was sought to determine if it is the carboxyl group of the amino acid derivative which is linked to the lipid by an ester bond. To study this question, we quantitatively determined the amount of hydroxamate material and amount of LOWRY modified biuret material present in two successive pooled cuts from chromatography on silicic acid. It was found that the ratio of hydroxamate to biuret – FOLIN material was the same for both of these chromatographic fractions and we take this as further evidence that it is the amino acid derivative which is linked by its carboxyl group to the lipids. Since this material in the past has been obtained by many transfers in a counter-current distribution system between an aqueous acid phase and hexane, the amino group must also be bound. Reaction with dinitrofluorobenzene followed by hydrolysis and chromatography have revealed the absence of free amino groups for the amino acids in general. DNP-serine was observed as would be expected since this lipid fraction contained cephalin. Very faint DNP-aspartic and glutamic acid were also observed. Free amino acids were revealed by ninhydrin after electrophoresis. It thus appears that one way in which the highly polar character of amino acid compounds is suppressed involves binding both polar ends of the molecule.

Certain studies are in progress using *Escherichia coli*, and they are still in an incomplete form. Since these studies represent a collaborative effort and will be reported separately when ready, I will not go into detail at this time. However, in rounding out the current picture with respect to the metabolic role of

References p. 758

amino acid–lipid complexes, some features of these studies should be mentioned.

First, it was found in collaboration with Dr. R. B. ROBERTS and K. MCQUILLAN at the Carnegie Institution of Washington, that amino acid–lipid complexes do form. Amino acid in these complexes was in a dynamic state as opposed to a stable end product type of labeling. When in subsequent experiments with Dr. R. B. ROBERTS, Dr. R. BRITTEN and Dr. F. DUGGAN, we took special pains to obtain a clean ribosome preparation from *E. coli* we found that more radioactive amino acid–lipid complex was obtained from the ribosome preparation than from the cell wall fraction. Other experiments showed that the amino acid lipid complexes appear to form at the cell wall site before the internal pool became labeled. We currently are studying mutants of *E. coli* obtained from Dr. M. LUBIN of the Harvard University. These mutants can no longer concentrate proline and we are trying to see if this defect is in any way related to the manner in which these mutants handle proline–lipid complexes.

At the present time, I think we can safely say that associations between amino acids and peptides with lipids do occur in a wide variety of living tissues and that they are not artefacts of preparation. In those cases where the metabolic state of the amino acid has been examined using radioactive tracers, it has been found that the amino acid association is dynamic and not static. Many of the properties of these complexes, as already described in this paper, indicate that they are undergoing active metabolism. Briefly restated, these are: generality with respect to different tissues and amino acids; sensitivity of formation of certain complexes to chloramphenicol, dinitrophenol and conditions which inhibit protein synthesis; sensitivity of protein synthesis to conditions with interfere with lipids; structural specificity regarding the amino acid; demonstration of the participation of chemical bonds requiring active metabolism such as peptide bonds, ester bonds and acylamides and the rapid labeling of these complexes with amino acid as compared to the kinetics for nucleic acids and proteins.

The idea that non-polar forms of amino acid may be of metabolic importance is not unrealistic. The first stage in the metabolism of amino acid, that is, getting it into the cell, is a problem of bringing the amino acid through a barrier of a non-polar medium. One way of doing this would be to confer upon the amino acid, lipid solubility for its passage through the cell membrane. This possibility is currently being actively pursued in the laboratories of GABY and TRIA as well as ROBERTS, BRITTEN and our own. The other major possibility for the metabolic significance of these complexes is the process of protein synthesis[4]. Considerations which tend to support this picture are as follows*: the microsomal system has been revealed as the structural element most directly concerned with large-scale protein synthesis in the cytoplasm; in systems where protein synthesis and secretion are particularly active processes, the microsomal system consists of a lipid membrane containing in it, or on it, ribonucleoprotein-rich areas. In other active systems lacking a demonstrable endoplasmic reticulum, other lipid surfaces such as the cellular membrane may be similarly involved or the ribosome themselves may possess a membrane-like component (in this connection, there is a lipid-rich high-specific activity component isolated from labeled pea seedling ribosomes by T'so). Recent electron-microscope studies and refined fractional centrifugation indicate an intimate association of lipid and nucleo-

* References for statements in the following discussion can be found in ref. 4.

References p. 758

protein. Studies with secretory tissues suggest that newly synthesized proteins are associated with lipid. In a wide variety of tissues, amino acid has been observed to enter quickly into a lipid-soluble form. We thus have a situation in which the machinery for protein synthesis as well as its initial reactants and final products have been found in lipid soluble forms. The working hypothesis that amino acid derivatives may be condensed in the lipid medium to form protein, seems at the present time to be extremely practical. Indeed, the possibility should be seriously considered that transport and synthesis may be linked metabolic events.

One other aspect of the problem seems worthy of mention. That is the method of energizing peptide bond formation. Contrary to popular belief, this question at the time of writing is not settled. The wealth of literature concerning amino acid activation and S-RNA becomes very sparse as we approach biological systems making discrete proteins. The passage of electrons through components of lipid membranes during which process energy is released and stored is currently being elucidated. In the most familiar energy-requiring process, this energy is stored in water-soluble "high-energy compounds" which circulate in the aqueous part of the cell where they are utilized.

If there are complex processes in a lipid phase which require energy, it seems possible that energy could be directly utilized in the lipid phase without the participation of the common aqueous "high-energy" carriers. If this is so, one must be very careful in studying the energy requirements of such a system on the basis of techniques which in the past have been used in connection with energy-requiring steps in the aqueous phase.

The considerations developed here have been presented in general terms. At the present state of our knowledge or more accurately, lack of knowledge, more detailed speculations and presentation of models is premature. We must realize that at the present time, the metabolic importance of amino acid–lipid complexes, AMP-amino acids, and S-RNA-amino acids has not been firmly established and therefore the further investigation of these materials must be pursued in various laboratories.

REFERENCES

1 J. Westley, J. J. Wren and H. K. Mitchell, *J. Biol. Chem.*, 229 (1957) 131.
2 R. W. Hendler, *Science*, 128 (1958) 143.
3 R. W. Hendler, *J. Biol. Chem.*, 234 (1959) 1466.
4 R. W. Hendler, *Biochim. Biophys. Acta*, 49 (1961) 297.
5 J. L. Haining, T. Fukui and B. Axelrod, *J. Am. Chem. Soc.*, 81 (1959) 1259.
6 J. L. Haining, T. Fukui and B. Axelrod, *J. Biol. Chem.*, 235 (1960) 760.
7 T. Fukui and B. Axelrod, *J. Biol. Chem.*, 236 (1961) 811.
8 E. A. Peterson and D. M. Greenberg, *J. Biol. Chem.*, 194 (1952) 359.
9 G. D. Hunter and R. A. Goodsall, *Biochem. J.*, 78 (1961) 564.
10 W. L. Gaby, R. N. Naughten and C. Logan, *Arch. Biochem. Biophys.*, 82 (1959) 34.
11 W. L. Gaby and R. Silberman, *Arch. Biochem. Biophys.*, 87 (1960) 188.
12 W. L. Gaby, H. L. Wolin and I. Zajac, *Cancer Research*, 20 (1960) 1508.
13 J. J. Wren and H. K. Mitchell, *Proc. Soc. Exptl. Biol. Med.*, 99 (1958) 431.
14 J. J. Wren and H. K. Mitchell, *J. Biol. Chem.*, 234 (1959) 2823.
15 J. J. Wren, *Nature*, 185 (1960) 295.
16 O. Barnabei and R. Ferrari, *Arch. Biochem. Biophys.*, 94 (1961) 79.

DISCUSSION

Chairman: HARRY EAGLE

ROUSER: I would like to know how Dr. HENDLER and Dr. MITCHELL define a lipid. This is not always easy, but one thing that generally is excluded in the definition of a lipid is a complex of lipid with some other substance held together by ionic or secondary forces. Such complexes can be dissociated without breaking covalent bonds. The data presented thus far indicate to me that the combinations of amino acids and lipids under discussion can be referred to most appropriately as lipo–amino acid complexes, but not as lipids containing amino acids or abbreviated and called lipids. If amino acids are covalently bound into a molecule that contains a long hydrocarbon chain (of a long chain fatty acid or fatty aldehyde) or some cyclic hydrocarbon, then the substance can be called a lipid (containing amino acids). If the amino acid or peptide is bound ionically to a true lipid, the combination should be designated as a lipid–amino acid or lipid–peptide complex.

The significance of this differentiation is important for metabolic considerations. If amino acid is bound by covalent bonds into a lipid, the metabolism of lipids and amino acids is interrelated. If lipid and amino acids or peptides are merely associated by ionic bonds, then the complex may only be related to the method of isolation and not necessarily to metabolic interrelationships.

Solubility criteria are by no means adequate to establish covalent linkage because of the well-known ability of lipids to solubilize water-soluble compounds. Clear solutions of lipids and amino acids can be prepared with solvents in which the amino acids are insoluble. Also, some peptides may be appreciably soluble in organic solvents and thus appear with lipids. Amino acids present in lipid extracts can be eluted from silicic acid columns along with lipids, and thus, silicic acid column chromatography alone is not a reliable indicator that amino acids occur in covalent linkage in a lipid molecule. Is there any evidence then that the amino acids in the lipo–amino acid complexes are linked by covalent bonds to hydrocarbon material? Stated another way, are there data to show that a compound has been isolated that maintains a constant composition when chromatographed in several different systems, when dialyzed, and when subjected to electrophoresis?

HENDLER: My interest in these compounds has been in their possible metabolic significance. If we are considering that reactions of significance may occur in a non-polar area within the cell, then so far as I can see the most important criterion is its solubility, and if this is not satisfactory in terms of defining lipids in general, I think that is another question. In my analysis of these systems attention has been placed on solubility because we are considering their reacting in non-polar areas of the cell.

As for the relation to Dr. AXELROD's system, there is no reason why this compound should necessarily be characteristic of the major compounds I have been studying. Now that I know the properties of this material perhaps I will be able to isolate it. However, for any given amino acid, I find a number of different complexes. We are studying not one but several substances.

L. MILLER: Just by way of trying to get some notion of the possible biological implication of these lipid–amino acid complexes, to what extent have these materials been isolated from living animals? To what extent do they turn over rapidly or not so rapidly? What kind of evidence do we have that they have any role in biology as we understand it?

AXELROD: That is easy for me to answer, because I have made no claim for these things. We have not attempted to isolate a known compound from natural material. We have only worked with the labeled material which we obtained incidental to the protein in our first attempt to incorporate phenylalanine into lipid material, and I have really no idea what the physiological importance of this material is. I think I did say that we wished that we could have attributed something of great value to it, but as things stand, we will just have to let someone else see if they can find a place for it.

ROUSER: Dr. MITCHELL, would you agree with what Dr. HENDLER said? I want to get back to this definition. Are you willing to state that these bound amino acids have lipid residues?

MITCHELL: Well, I can say I agree with everybody! First, I should say that I have been slightly misquoted here twice. Our interest in this field has been chemical, and our work has been primarily on the chemistry of these substances. Anything biological about it has been purely accidental. We have encountered compounds of the nature described by Dr. AXELROD and we have encountered other classes of compounds, too, such as those described by Dr. HENDLER. I think there are a very large number of substances in the amino acid containing lipid preparations, but as to their biological functions, I have no idea.

As to your question about the lipids; perhaps we have been a little misleading in nomenclature. We have always found fatty acids in the same fractions as the amino acids, on hydrolysis. We have not proven the structure of any one compound, nor have we ever obtained any one compound in sufficient yield to do so. The substances have been called lipids on the basis of solubility and fractionation characteristics.

EAGLE: If I understood Dr. MITCHELL correctly, although he has isolated such compounds from nature, it is a whole complex of substances rather than a single entity.

ROUSER: On the basis of what I have heard up to now, I want to insist that we consider one point. None of these so-called lipid–amino acid complexes really seems to be at this time characterized as lipid-containing amino acids, and I think that this is important. Now, as for lipids and bound amino acids associated with each other, this is something different. On the basis of this latter definition, I can say that I have isolated such material which I feel is not an artifact. For example, we did find amino acids with lipids in rabbit appendix, which is a rapidly growing tissue. On the other hand, we found essentially none in brain. However, I am not at all certain that these are lipids in the classical sense of the word.

EAGLE: Your scepticism can hardly apply to the *in vitro* experiments with a defined lipid.

ROUSER: No. I am not talking about Dr. AXELROD's experiments. He has clearly demonstrated the presence of a fatty acid in his compound. I am talking about the variety of other substances that have not so far been characterized as lipid material, but are associated with lipids.

EAGLE: You are not prepared to extrapolate from these *in-vitro* phenomena to what is isolated from tissue? You do not want to extrapolate from Dr. AXELROD's compound to the compounds which are extracted from tissues?

ROUSER: Oh, no; by no means.

HENDLER: Dr. ROUSER, I must be missing the point you are making. You say that you have isolated these complexes and that you are convinced that they are not artifacts of preparation, that they actually occur in your tissue, and that they dissolve in organic solvents. Then what I do not understand is why you feel that if they do not also contain a long carbon chain that they could not be metabolized in non-polar regions of the cell into which they might penetrate.

ROUSER: That is quite another point. I am not questioning that at all; just the decision that these substances are lipids. This is a very distinct and different situation, and I am simply suggesting that it is causing a great deal of confusion to call them lipids. They might be very important metabolic substances and still not be lipids in the practical sense. I think that this is most likely the case.

LOFTFIELD: I do not want to meddle in this little thing, but it seems to me that polyglutamine, polyphenylalanine, polylysine, all of the polyamino acids with the exception of polyglycine and polyproline, are lipid-soluble as opposed to water-soluble, so that these things which have the properties that Dr. ROUSER is talking about could indeed not contain any lipid at all.

GRIFFITH: There is one fatty acid–amino acid complex of venerable age that ought to be mentioned in this connection, *viz.*, benzoic acid, or phenylformic acid, and its combination with glycine to form hippuric acid, also phenylacetic with glutamine to form a similar compound. I couldn't help but think of this in connection with the solubility matter, because sodium benzoate in the blood stream is eliminated very slowly by the kidney whereas hippurate is eliminated very rapidly. As far as I can see, the detoxication of benzoic acid that we have known about so long is primarily a matter of changing a less to a more diffusible molecule in order to hasten its elimination from the body.

ARONOFF: Dr. HENDLER, did you not use any fatty acids instead of amino acids?

HENDLER: You mean labeled fatty acids? No. I will have to look into this point.

L. MILLER: Dr. HENDLER, did you find any correlation whatsoever between the known protein synthetic activity of various tissues, let's say in the sense in which CASPERSSON looked at them, and the occurrence of these amino acid–lipid complexes? This is an impertinent question on several grounds.

HENDLER: I have not made the studies of CASPERSSON. There are numerous reports now of various types of tissue that have been found to contain such compounds, but no systematic effort has been made to correlate the existence of these things and the rate of protein synthesis in these tissues.

SEGAL: I would like to ask Dr. AXELROD how his system differs from that described by TAGGART in a series of papers in the *J. Biol. Chem.* in 1953 and 1954, using the kidney to study the formation of acylamino acids.

AXELROD: There is no acyl activation involved here at all. There is no requirement for coenzymes; you do not need anything but the free fatty acid and the free amino acid.

LOFTFIELD: I was interested in comparing Dr. AXELROD's experiences with the work that OREKOVICH has done in Russia on GREENSTEIN's acylase which he has purified extensively. He finds that it does do a very nice job of reversing the hydrolysis of acylamino acid. The equilibrium constants have been determined, and at reasonable physiological concentrations there is a quite substantial (1 or 2 per cent) synthesis of the acyl compound, and of course, the very compounds that are synthesized are those that would be hydrolyzed by carboxypeptidase again. I just wonder Dr. AXELROD, if you have found a GREENSTEIN enzyme in the liver, so to speak? Do you know if it has other comparable properties and have you done kinetics to measure the relative rates of reaction?

AXELROD: We have not done any kinetics at all. We have only shown that we could split the product that was being formed, and we have not determined the equilibrium constant for the reaction. We have tried to block the reaction by using inhibitors for esterases. DFP inhibits it to a rather good degree, as does physostigmine. We also tried hydrocinnamic acid, which is a specific inhibitor for carboxypeptidase, but found that it had no influence on the formation of the compound.

WINITZ: If I might add to Dr. LOFTFIELD's statement with regard to OREKOVICH's data, it is indeed perfectly valid data. However, OREKOVICH uses extremely high concentrations of substrate, in the vicinity of about 2 or 3 M, whereas in the GREENSTEIN resolution procedure, concentrations of about 0.1 to 0.2 M are employed in order to ensure complete hydrolysis. With concentrations of one molar or higher, the hydrolysis of an acylamino acid is in the vicinity of about 95 per cent or less. With regard to the distribution of acylase I, it is present in various tissues such as liver, brain and heart. However, the highest concentrations are found in kidney, and that is the tissue we generally use for its isolation.

Round Table Discussion

STATE OF THE INTRACELLULAR AMINO ACIDS

Session Chairman: JOSEPH. T. HOLDEN
Transcript Editor: ERICH HEINZ

HOLDEN: It was apparent during the discussion last night that one of the most urgent problems in this field concerns the state of the intracellular amino acids. That is: are they in fact free or associated in some way with intracellular structures? I believe, this to be, therefore, the best place to begin this discussion. In view of his extensive experience with various aspects of this problem, I would like to ask Dr. CHRISTENSEN to express his views on the subject.

CHRISTENSEN: I think that life tends to find a single answer only, or very few answers, to each of its problems. Accordingly I think that if we can prove that amino acids are concentrated by some organisms we may assume that the same thing occurs in many others, and probably by a similar process. No one can question that the animal organism is able to form an extracellular fluid and various secretions with compositions totally different from that of its aqueous environment, and from each other. Further it is able to maintain these differences in spite of the more or less constant movement of water and solutes across the separating barriers of cells. The barriers that accomplish this effect may be illustrated by epithelial cell layers of the alimentary tract and of the renal tubule. This situation shows that at least specialized cells are able to transport solutes against concentration gradients, in some cases producing concentration ratios of ten million to one, as in the case of hydrogen ion secretion by the gastric glands.

We might evade the apparently endless argument about the nature of solutions within an unharmed cell, by seeing whether cells not specialized for secretion can also generate gradients between two isolated segments of extracellular fluid. In this case two phases should be fully accessible to study, without such serious uncertainties as to activity coefficients.

As I pointed out earlier in this meeting remarkably similar properties and similar affinity series are seen between, on the one hand, the intestinal mucosa in concentrating amino acids from the mucosal to the serosal phases, and, on the other hand, the Ehrlich cell in taking up amino acids. With the intestinal system we have two abundant extracellular phases that are highly susceptible to examination. Obviously one is dealing here with activities of the free amino acids and not with bound forms. Activities have indeed been measured for calcium ion transport across the intestinal mucosa, and for the hydrogen ion across the stomach wall.

But perhaps more to the point is an experiment in which OXENDER and I arranged the Ehrlich cell as a barrier some three or four cells thick on a *millipore* filter, to form an artificial barrier or membrane[1]. We then put saline solution on the two sides of this membrane and placed a test amino acid, let's say glycine, at equal levels in the two solutions. If we now added alanine to one side at a substantially higher level, we could observe the phenomenon of exchange. That is, while the alanine was moving down gradient, glycine was driven in the opposite direction so that the initially alanine-rich phase became enriched in glycine at the expense of the opposite phase. This effect persisted for more than an hour, until the two amino acids eventually came to be distributed uniformly between the two saline solutions.

A glycine gradient could also be produced by placing pyridoxal on one side of the membrane. This agent appears to stimulate amino acid transport, for reasons I won't discuss here; it caused glycine to be accumulated somewhat on the opposite side of the barrier.

These results lead us to the hypothesis that secretion across membranes composed of cells probably occurs by the existence of an unidentified asymmetry between the two phases presented by the cells on the respective sides of the membrane. KOEFOED-JOHNSON AND USSING[2] earlier proposed the parallel idea that a cell layer of the frogskin transports Na^+ by extruding it more effectively to the inside than to the outside of the skin. The natural basis for such an asymmetry is not clear. In the intestinal mucosal cell the transport from the luminal side may well be faster because of the much greater surface area of the brush border, and the consequent presence of more transport sites at that surface of the cell. A type of profile of solute levels has repeatedly been

References p. 777

shown across such cells, with the level within the cells higher than on either the mucosal or serosal side. This situation supports the idea that cells secrete solutes from one phase to another by pumping the solutes into themselves more strongly from one side than from the other.

The force of the above experiment was increased by showing point-by-point an identical specificity for secretion across our barrier of Ehrlich cells and for the accumulation of amino acids by the same cells. We know, for example, that an α-methyl group on a neutral amino acid does not interfere substantially either with transport into the cells, or with the development of an asymmetric distribution across the cells. But with the dicarboxylic amino acids, the α-methyl group completely prevents accumulation, and it also stops the formation of a gradient across the intestinal wall.

Similarly the same aldehydes, pyridoxal, 5-deoxypyridoxal, and 4-nitrosalicylaldehyde stimulate the two types of transport. The requirement of the cells for oxygen, unless glucose or fructose were supplied, was identical for the two processes. The temperature optima were the same; both required the presence of potassium ion. Every criterion that we investigated, indicated that we were dealing with the same phenomenon of cellular uptake in both cases, which had, however, been made asymmetric in the one case so that a fraction of the total concentrative activity was actually realized between two separate extracellular phases.

To me this experiment has seemed rather decisive about the matter. There are, however, other strong lines of evidence that one solute or another in one cell or another is actually present at a substantially higher free concentration inside than outside. One such line of evidence arises from the very high levels that can sometimes be achieved inside the cells. We have heard here about enormous taurine gradients in marine invertebrates in which taurine may represent 3 or 4% of the fresh weight of the cell. What binding structure can be present at an equivalent level? Apparently the only possible candidate is the peptide bond. Almost every peptide bond would need to bind one taurine to account for such a degree of accumulation. If we take into account the other amino acids contained in such organisms we reach even less probable requirements.

I can also cite the transport of water that has been shown to accompany the uptake of solutes. The swelling of the Ehrlich cell is very nearly proportional to the magnitude of the glycine gradient reached by these cells. The same type of phenomenon has been shown by SISTROM for lactose accumulation by spheroplasts of *E. coli*. In both cases the cells appear to respond almost linearly by water uptake to changes in osmotic gradients between their interiors and their environment.

These are perhaps the strongest lines of evidence, but there are other indications of a somewhat more equivocal nature, having to do with the ease with which the accumulated solute may be obtained from the cell in the free state, or with the necessity of metabolic energy for the uptake.

I do not mean by these comments to urge that uphill transport be taken for granted in any new case. Certainly we ought to be cautious about this. Nowadays transport has become a popular word, and quite commonly active transport is invoked in situations where no real criterion even for transport has been demonstrated. Uphill transport can be produced by a wide variety of cells, and in certain instances uphill transport seems to be established. But we certainly have greater cause to deplore an excessive enthusiasm for taking active transport for granted every time a solute is found to be accumulated by cells, than we have for excessive doubt that the phenomenon can occur at all. On the other hand a uniform tendency to require quotation marks around the word *free* in free cellular amino acids seems to me unnecessarily cautious.

REINER: It seems to me that first of all we ought to have some comparative biochemistry or some comparative transport chemistry, because obviously different people have been working with different materials. Therefore, we should try to find out how many of the differences are due to the different organisms that have been worked on. Differences in organisms may also in part explain why osmotic effects as mentioned by Dr. CHRISTENSEN occur in some but not in all systems.

I would also like to say something about the basic general principles, because I don't think it is quite clear as to what *free* or *bound* means. The general problem is transport between two phases. We may at first forget about the membranes and just consider the distribution of a solute between two phases. This is most easily done by a diagram in which the potential energy of the solute is plotted against position relative to these two phases. In each phase a given solute has a certain energy level due to the attractions and repulsions caused by the solvent and other components of the solution. At the interface perhaps there is a transition which may or may not involve a hump of energy.

This gets slightly more complicated if you put in a third phase, which you call a membrane. It is my personal prejudice that before talking about a membrane in connection with any cell, I would like to see a lot of evidence that there is a discrete or nearly discrete structure at the interface having finite thickness. But this isn't really important, because a membrane is merely an extension of these notions to three phases, of which the middle one is thinner than the others.

In view of all these differences of potential energy and potential barriers the problem of transport is almost identical in some respects with the problem of a chemical reaction. You have somewhat similar types of exponential terms involving activation energy.

As to the problem of *free* versus *bound*, no solute is wholly free in anything except in the state of a dilute gas. It will have, in general, an activity coefficient other than unity, which is determined by these energy relations. It will also be hydrated to an extent determined by the activity of water. The activity of water depends on all the other components that are around.

The cases that have been talked about here are extreme cases: On the one hand, the *binding* of the solute you are interested in to some structure, by which you will mean a relatively fixed relationship in which the solute and the binding agent come within, less than one Ångström of each other and stay that way; and the free state, which I think is wholly fictitious in solution chemistry but which we could represent by considering the solute as being in pure water with nothing else around.

There is evidence that no solute in the liquid system is ever completely bound, and osmotic effects will depend on the activity of water and not merely on the presence of discrete membrane

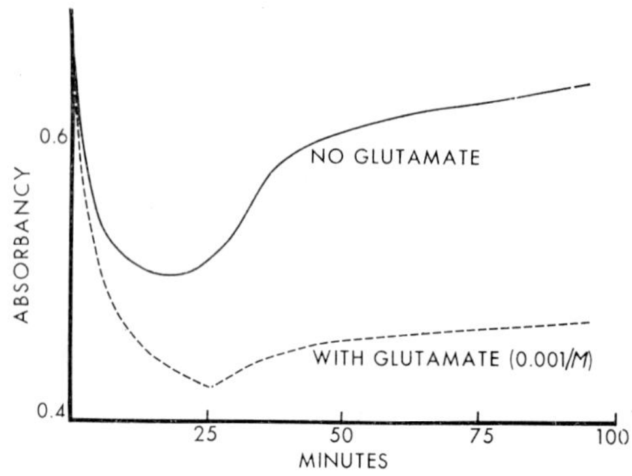

Fig. 1. The osmotic effect of metabolic glutamate uptake in protoplasts. Protoplasts of *S. faecalis* were stabilized in solutions of sucrose (0.4 *M*) –K phosphate (0.075 *M*, pH 7.2), with and without glutamate (0.001 *M*). Glucose (0.01 *M*) was added at zero time and swelling was observed photometrically[3]. The enhanced swelling in the presence of glutamate may be ascribed to the osmotic effect of glutamate uptake.

phases. It is, therefore, useless to discuss this problem as if we were dealing with an osmometer. In most of the cases it is physically and cytologically true that you are not dealing with an osmometer in the classical sense; that is, a bag with a solution inside and another solution outside, in which all effects are due to holes through which some molecules physically can not pass.

Whatever the relation of molecules to the interface of the cell is, there is no question of there being holes that are too small for something to squeeze through. All the effects, in other words, have to result from physicochemical forces acting between molecules. Hence there is a gradation of possibilities from the idealized completely fiee to the chemically completely bound. Besides the question of comparative biochemistry it would therefore be important to try to answer the question of *free* versus *bound*: *(a)* in terms of how much free and how much bound, rather than just *yes* or *no*, and *(b)* in terms of a specific mechanism. In other words, we have to ask what in a given case may affect the activity of water in such a way as to produce certain osmotic effects in some cases but not in others, and whether this is chemically a reasonable hypothesis.

ABRAMS: I would like to describe some experiments with *Streptococcus faecalis* protoplasts which have a bearing on the present discussion. These protoplasts have no endogenous energy metabolism and such metabolism, which is entirely glycolytic, can be introduced by adding glucose. The resting protoplasts (no glucose) are osmotically stable in sucrose (0.4 *M*), thus sucrose is essentially impermeable under these conditions. However, on addition of glucose, a swelling takes place (Fig. 1) which can be shown to be correlated with the penetration of sucrose as indicated by labeling experiments with sucrose-^{14}C. The reversal of swelling takes place subsequent to the completion of glycolysis (ABRAMS[3]). If a small amount of glutamate (0.001 *M*) is added to this system, an additional swelling takes place, as can be seen in Fig. 1, and it can be shown that during this process glutamate is taken up. The additional swelling may be ascribed to the osmotic effect of free glutamate concentrated in the protoplasts during glycolysis (ABRAMS AND MACKENZIE[4]).

References p. 777

This observation is similar to that described by SISTROM, which Dr. CHRISTENSEN mentioned, and also resembles the observation of Dr. HOLDEN on the osmotic effects of glutamate uptake into bacteria with weakened cell walls (Fig. 1).

By assuming that swelling of protoplasts suspended in a solute at high osmotic pressure represents the osmotic effects of a penetration of extracellular solute, the permeability of the membrane to a variety of compounds, including amino acids, can be tested (ABRAMS, YAMASAKI, REDMAN AND MACKENZIE[5]). The rate of swelling becomes a measure of rate of penetration. An illustration of this type of investigation with amino acids is shown in Fig. 2. The experiment described in Fig. 2 is designed to test the permeability of L- and D-alanine and L- and D-serine in the absence of glycolysis. The curves in Fig. 2 indicate that L-alanine and L-serine penetrate rapidly and spontaneously, the latter with a long lag period. On the other hand, the D-isomers penetrate the non-glycolyzing protoplasts very slowly if at all. However, on the addition of glucose, the D-isomers also penetrate very rapidly. It appears that the membrane barrier distinguishes sharply between the optical isomers of alanine and serine with respect to their spontaneous down-

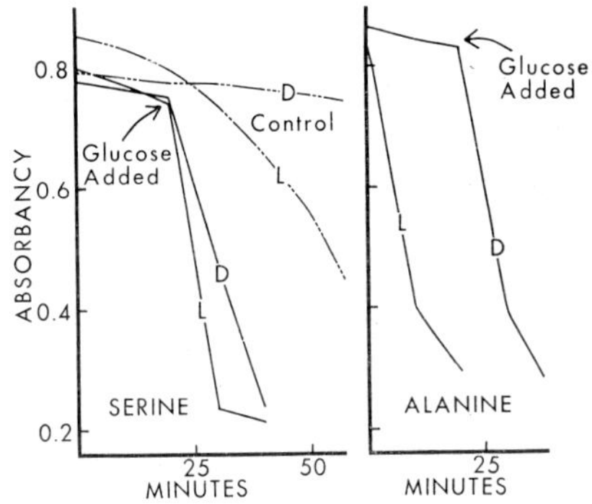

Fig. 2. The effect of glycolysis on protoplasts stabilized with the optical isomers of serine and alanine. Protoplasts from *S. faecalis* were suspended in solutions of L- and D-alanine (0.4 *M*) and L- and D-serine (0.4 *M*) containing K phosphate (0.075 *M*, pH 7.2). Glucose (0.01 *M*) was added as indicated. The rate of swelling, photometrically observed, indicates the rate of penetration of the amino acids.

hill penetration, and during glycolysis the membrane seems to open up to allow a far more rapid penetration of both L- and D-isomers.

Similar experiments have been carried out with many of the common amino acids with results which indicate a more or less unique behavior for each amino acid. Thus, both L- and D-valine penetrate spontaneously very rapidly, whole both L- and D-threonine are practically impermeable. However, the addition of glucose causes a very rapid penetration of L- and D-threonine. In the light of these findings it appears that the structure of the membrane is constituted in such a way as to discriminate between the different amino acids with regard to their spontaneous penetration into the cell. Moreover, the membrane seems to change its permeability under the influence of glycolysis allowing, in most cases, more rapid penetration.

CHEN: In this connection I should like to mention the recent work of J. E. TREHERNE[6] on the absorption of amino acids in the gut of the locust *Schistocera gregaria*. He injected into the alimentary canal an experimental solution which had about the same concentration and osmotic pressure as the hemolymph. Using ^{14}C-labeled glycine and serine he found that these amino acids are rapidly absorbed in the mid-gut region, especially from the caeca. An important finding was the significant increase in concentrations of glycine and serine above those in the body fluid, obviously due to removal of water, after injection of the experimental solution into the gut. He concluded that the transfer of these amino acids across the gut wall is mainly due to a diffusion gradient thus established.

I think his experiment demonstrates that here the movement of water plays an important role, and there is, at least in insects, no evidence of active transport of amino acids against concentration gradients.

BRITTEN: I don't think anybody has denied that an amino acid can be in solution, particularly in water, and that indeed the presence of different concentrations in different cellular fluids is a classical observation. The issue is, I think, to decide to what extent within a given cell the amino acid is actually free in solution, and to what extent it is bound, and how this binding is connected with special metabolic processes.

I would like to cause a little confusion about the nature of an osmotic membrane. Anyone who has used Dowex-50 ion exchange resin, 2% cross-linked, knows that he takes the risk of blowing up his chromatograph tubes because of the large volume changes that occur when concentrated aqueous solutions of sucrose or glycerol are passed over the resin. The resin has no membrane in the biological sense, but a molecular structure which corresponds to the resin itself. I think it would be possible to make models with this resin that behaved osmotically like certain classes of cells, with the clear knowledge that no membrane in the traditional sense exists.

The experiments of Dr. ABRAMS and Dr. SISTROM are quite suggestive that the amino acid is in solution if we ignore for the moment the confusing issue of the nature of solution and of the osmotic situation just mentioned. However, I think it is difficult to transfer this evidence to the undamaged cell.

When spheroplasts are made from *E. coli* by treatment with lysozyme followed by a change in sucrose concentration from 10%–5%, they commonly appear in the phase microscope as large transparent spheres with a dark saddle of about the volume of the original cell. The relationship between the spheroplast and the original cell is uncertain, in my mind, particularly with regard to the location of the original cellular fluids. In any case the clear area is a large trapped volume with osmotic properties. According to E. BOLTON's experiments the capability of the spheroplast to form amino acid pools is identical to that of the original cell. The rate of pool formation, final pool size and rate of incorporation into protein at a moderate ^{14}C-proline concentration are unchanged. The spheroplasts, however, can not be treated roughly and their pool is sensitive to washing with buffer. In any case the time course and the size of the proline pools are identical for these two structures in spite of the fantastic change in the organization of the free water inside the cell, or internal osmotic solution, or whatever it is.

COWIE: If one measures in yeast or in *E. coli* the accumulation of *free amino acids* as a function of external concentration, a complex incorporation curve is obtained. At low external concentrations the accumulation may greatly excede the external concentration. At higher external concentrations the incorporation appears to be proportional only to the external level.

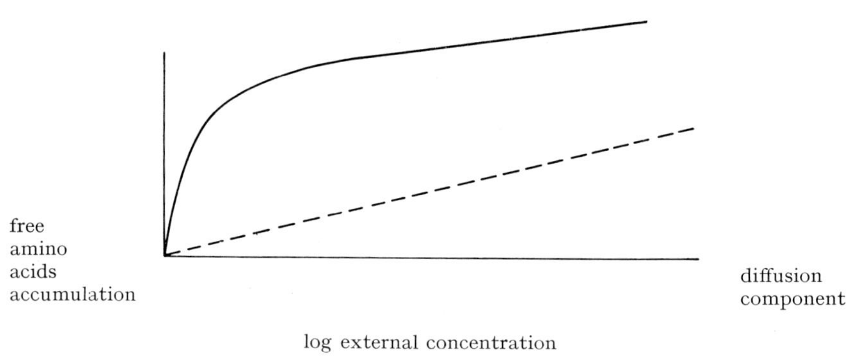

Obviously there is an accumulation mechanism which saturates and which is independent of a second process (presumably diffusion) dependent directly upon external concentrations. The total quantity accumulated is, in my estimation, contained in two states: a free, and a bound form. In yeast the quantity of tyrosine or guanine which can be concentrated exceedes the solubility of these compounds in the culture medium. One interpretation of such data is that some of the accumulated material complexes with internal substances thereby changing their solubility characteristics. There would not be *free amino acids or bases*. On the other hand some of the accumulated material certainly would be in the free state.

CHRISTENSEN: I am not sure that I can at this moment help to explain the form of Dr. COWIE's curves. It seems to me that one can accept an asymptotic approach to a horizontal line, representing a state of saturation of a transport process, only when one measures rates, preferably initial rates, rather than the steady state distributions. The steady-state distribution represents the algebraic sum of all fluxes. One might have a second transport site mediating a transport, perhaps

References p. 777

passively; diffusion will make a contribution, and indeed a stoichiometric binding of a portion of a solute may well occur in addition to an uphill transport.

Chemical mediation of the entrance of solute molecules into cells seems to be extremely variegated. It is not limited to metabolically useful solutes. For instance, one cannot imagine that the human red blood cell needs its enormous capacity to admit glycerol. The cell boundary may present such a wide spectrum of mediating sites that the chemist can hardly synthesize a hydrophilic molecule that will not somehow interact with any of these sites. Why some of these mediating groups can release the molecule at a higher chemical potential than they acquired it, and into a different phase, is the interesting question here.

To an earlier comment I want to reply that none of us here, I think, looks on the membrane in a completely morphologic sense. We are thinking in the physicochemical sense of a boundary between two phases, and no assumption is made about the thickness of the membrane phase; nor do we assume that the dissociation from a transport site (or series of such sites) occurs at a given distance from the association reaction. How deep the penetration goes before another event intervenes may be merely a statistical matter. In short we look on the membrane in a functional sense, and let the morphologist try to find a structural basis for the function that we observe.

We are also trying to take into account possible general changes in activity coefficients arising from the unusual character of the cytoplasmic environment. In this connection the largest distribution ratios are interesting, because they demand the most improbable diminutions of the activity coefficient. When conscientious efforts are made to measure the activity coefficients of ordinary solutes in broken cell preparations, no great changes are seen from those applying outside the cell. One may, of course, say that as soon as we touch the cell the vital state within has been entirely changed, so that no such observations are valid. But consider that for many solutes the activity coefficients would need to fall from values or nearly unity to perhaps 0.01 or 0.001 within the cell. Without ignoring this uncertainty, we need to consider how large it can plausibly be. By selecting our solute and the level used we can get extremely high distribution ratios—and I see no reason why in exploring this matter—we should limit ourselves to amino acids of biological occurrence and to physiological levels. By selecting favorable models we can require that a general fall in activity coefficients must be enormous to account for the extent of accumulation by cells, just as we can also set the requirement for binding sites improbably high by achieving high gradients. Once model solutes have been shown to set such extreme demands on the binding and inactivation hypotheses, we need to show that these models are appropriate models and that the conclusions may be applied to the ordinary amino acids at ordinary levels.

LAJTHA: There are some experiments of ours mentioned in more detail elsewhere in this volume, which I interprete as transport into what I call a free solution of amino acids. I find active transport a more reasonable explanation of these experiments than the binding of amino acids. In the higher organisms (rats *in vivo*) one can produce and maintain high plasma levels by continuous infusion. Upon subsequent intraspinal or intracerebral injection, brain levels also increase but still stay below the elevated plasma levels. During the experiment the brain level decreases in time against a concentration gradient of the high plasma levels. In this case amino acids are actively transported from an organ into the plasma, unless we suppose that most of the amino acid in the plasma is bound. In some of these experiments plasma levels were increased more than 10–15 fold.

The transport process is fairly specific in that under the same experimental circumstances leucine is transported out of the brain until a plasma to brain ratio of 11–15 fold is reached, whereas lysine transport stops at a plasma to brain ratio of 5–6. No transport of phenylalanine is found. Leucine transport, present in adults, could not be shown in newborns in similar experiments; we interpreted this as showing that the transport system for this amino acid is not fully developed in newborns.

These experiments show transport from the brain into a pool which most likely consists of free amino acids.

HOLDEN: I wonder whether Dr. REINER's formulation can be made consistent with the apparent dependence of some swelling phenomena on metabolism.

REINER: It seems to me that you could make an argument for the potential difference between the two phases being the barrier in the extreme case. If you calculate conventionally a distribution ratio, for example, between ether and water as two adjacent phases, there will be certain substances that will distribute themselves mostly in the water and very little in the ether. The barrier is their solubility in the ether relative to their solubility in water. Although a cell is a much more complicated system than this, I think that the same fundamental principles hold. Obviously there is some evidence for special molecular layers at the surface of some cells. So you probably have not a simple jump of potential energy but whole series of gradations of energy. There may also be a potential barrier, as I drew it originally, extending not over a couple of Ångströms but over several molecular layers. The principle, however, would be the same. You can keep a solute out, without the intervention of various mechanisms such as have been discussed this past week, merely by a difference in solubility, which could mean a difference in the forces that attract or

References p. 777

repel the solute itself, and a difference in the activity of the available water. This reminds me that Dr. CHRISTENSEN's remarks about the possible values of activity coefficients may give a key to some of the species differences that seem to be observed between animal and bacterial cells. For example, there is no question that *E. coli* is not isoosmotic with frog Ringer solution, as anybody knows who has worked on the stabilization of *E. coli* protoplasts. In other words, in some bacterial cells the osmolarity is clearly very much higher than it is in animal cells, and it may be, although I can't produce figures for it at the moment, that the activity coefficient of water is a great deal lower and may be down to the point that Dr. CHRISTENSEN was implying would never be reached. It would have to be painstakingly calculated in terms of analyses that I don't think have been made yet.

I would like to add, since Dr. BRITTEN referred to it one of the experiments with protoplast. If you make penicillin protoplasts of *E. coli* and lyse them thoroughly by homogenization you get something that looks like a badly rumpled protoplast, but not like a membrane. Some people lyse protoplasts and publish nice photographs and say: "These are photographs of cell membranes." I think they are photographs of cell ghosts, at least in the cases I have worked with. But at all event, if you lyse protoplasts of *E. coli* in this fashion, after having induced the formation of β-galactosidase to the maximum capacity of the protoplast system to form it, and if you then assay it by means of ONPG, you find that unless you treat the protoplast with toluene, you get only about 25% of the β-galactosidase activity that has been developed. In other words, you could speak here of some kind of osmotic barrier, that survives the rather thorough osmotic and mechanical disruption of the intact protoplast. I feel that this is partial evidence, that the osmotic effects are not all at the surface of the protoplast.

CHRISTENSEN: I wonder if I may shift your attention elsewhere for a moment, before I have to leave this discussion. I refer you to a line of evidence that Dr. HEINZ has pioneered, namely the driving of exchange processes (HEINZ AND WALSH[7,8]). Experiments we are now making have emphasized the value of this approach, namely by comparing the effect of an amino acid *inside* the cell with its effectiveness *outside* the cell, in driving the counterflow of another amino acid. We find that valine, for example, is almost exactly as effective in driving the counter movement of other amino acids, whether the valine is inside and the other amino acid outside or whether they are in the reversed positions. If you make a profile of the response of five selected amino acids this profile of migration in response to valine is essentially the same whatever the relative initial position of the two amino acids under study, inside or outside. Essentially the same result is obtained down the line, using each of the amino acids to provide the driving force for exchange. The behavior certainly leads one to feel that he is dealing with the same mass-action situation at the two faces of the membrane, that the amino acid reactant is the same in the two cases, and that the measured levels are approximately correct.

I hope that Dr. HEINZ will continue the discussion of this behavior.

HEINZ: I think Dr. CHRISTENSEN is referring to what we call the *preloading effect*, or the *counter-flow-effect* as it is called by others. This effect, which we interpreted as due to a partial exchange diffusion has already been discussed earlier in this meeting. To review it briefly, we found many years ago that the uptake of some labeled amino acids by Ehrlich cells is greatly enhanced after preincubation of these cells with the same—or a related—amino acid in the unlabeled form[7,8]. In other words, the influx of the labeled amino acid, at constant extracellular level, rises almost proportionally with the intracellular concentration (*trans*concentration) of the amino acid with which the cell has been preloaded. Normally after adding the label the cellular radioactivity rises asymptotically towards the steady state value. After preloading, however, the cellular radioactivity, apart from rising more rapidly, may temporarily exceed the final steady state value. Furthermore we found evidence that the increment in influx, as it is caused by the preloading, is almost stoichiometrically equivalent to the carrier bound exit (counterflow) of the cellular amino acid. Such an effect obviously corresponds to the definition of exchange diffusion, as originally given by USSING. As already mentioned there is also evidence that this exchange diffusion refers to the same mechanism as the active transport of the amino acid, and for this reason we call it *partial* exchange diffusion.

The preloading effect appeared to us as strong evidence of an active transport process mediated by a mobile carrier. The details of this argumentation have been reported elsewhere[7,8]. The phenomenon seemed also to suggest that the major part of the accumulated amino acids is in a free state, or at least that binding to an intracellular, non-mobile site, if it occurs, is not responsible for the accumulation process. I have to admit, however, that this evidence, though very suggestive, is not entirely conclusive. It is known that *e.g.* activated complexes may exchange their ligands much faster than they are formed *de novo*. So HOLZER recently reported, that labeled acetaldehyde is bound to a thiaminpyrophosphate much faster if this coenzyme had been preincubated with unlabeled acetaldehyde. Such kind of mechanism should, of course, be considered in explaining the accumulation of amino acids and in particular the preloading effect. The fact that the formation of an activated complex requires the expenditure of energy would agree with the high sensitivity of the accumulation process towards metabolic inhibitors and anoxia. It should be kept in

References p. 777

mind, however, that the formation of an activated complex such as that between acetaldehyde and thiaminpyrophosphate, involves a highly specific and energy-rich covalent bond. There may be plenty of sites available inside the cell, which could be imagined to function as more less unspecific adsorption sites. I find it difficult, however, to visualize enough specific sites able to form activated complexes with the above-mentioned characteristics, in order to accomodate the tremendous amounts of amino acids accumulated.

Furthermore, a similar dependence of a unidirectional flux on its transconcentration has later been observed with other systems where such an activated binding was strictly excluded. So we found with DURBIN[9] that the chloride flux across the gastric mucosa towards the lumen is also increased by the chloride *trans*concentration. A counter-flow effect was also reported by PARK *et al.*[10], and by ROSENBERG AND WILBRANDT[11] with the flux of certain sugars out of the red cell. Here the *trans*concentration, contrary to our system, referred to the extracellular concentration of the substrate which was undoubtedly in a free dissolved state. Of special importance are probably the more recent experiments of Dr. CHRISTENSEN *et al.* which reproduced the preloading or counter-flow effect between two extracellular compartments, which were separated by a layer of ascites cells arranged to a solid membrane. So I think that by analogy, the preloading effect as originally observed is strongly suggestive that the crucial amount of accumulated amino acids in our cells is in a free dissolved state. By this, of course, I do not mean a state equal to that of an ideal gas but a state comparable to that of the extracellular amino acid. I have to concede that a certain extent of intracellular adsorption more or less loose and unspecific cannot be rigidly excluded. Such adsorption would, in our view, not be metabolically linked and hence rather incidental with respect to the active accumulation mechanism.

In this connection I would like to draw your attention to another phenomenon which we reported several years ago (HEINZ[12]). If a cell has accumulated a certain amino acid, for example glycine, then—according to the CARNEGIE foundation—I would assume that there would be a concentration or activity of the free amino acid similar to that in the medium, and in addition, there would be a certain amount of glycine which is what you call *bound*. Is that right?

COWIE: Yes.

HEINZ: This *bound* amount is with normal cells. If you now do the same experiment, but add an efficient metabolic inhibitor as DNP, then you get, of course, the same amount of free amino acid as in the uninhibited control. Owing to this inhibitor much less binding could be expected.

If we now measure the initial rate of efflux in this case, it should be proportional to the concentration (or activity) of the free glycine, and not of the total glycine. Does this agree with your views?

BRITTEN: I think I indicated that the efflux in such a case would be proportional to the number of loaded carriers and their dissociation constant. This is a free parameter, so that I can adjust it to meet the results of your experiment. That is, there are circumstances in which you may get very slow losses, and others in which you can get relatively rapid ones, depending on the nature of the carrier. This carrier has taken a great burden of properties.

HEINZ: We actually found that the efflux, with and without inhibition, was almost proportional to the total cellular glycine. In other words, metabolic inhibition or anoxia reduced the steady state efflux by a factor of up to six or seven. Accoiding to Dr. BRITTEN's statement, that the efflux is "proportional to the number of loaded carriers", our finding would indicate that in metabolic inhibition this number of loaded carriers is drastically reduced, *e.g.* 6–7 times. In Dr. BRITTEN's carrier model, however, it is not the loading of the mobile carrier which is coupled to an energy donor, but the transfer of the amino acid from this carrier complex to the non-mobile site—*i.e.* a reaction which clearly tends to decrease the number of loaded carriers, whether they are saturated or not. This model would, therefore, predict that metabolic inhibition, if anything, increases rather than decreases the efflux. As already mentioned we found precisely the opposite. Provided that our results give the true steady state efflux values, and provided that the metabolic inhibition is specific enough, these results support the view that the accumulation of amino acids is brought about by an active transport process and not by an endergonic binding to non-mobile sites. I do not see how one could adjust the parameters of Dr. BRITTEN's model to meet these results except by the assumption that either these parameters or the passive permeability of the cell membrane are profoundly altered by metabolic inhibition or anoxia, an assumption, which to my knowledge is not made in your model. I wonder whether DR. REINER would like to comment on this point.

REINER: I think that theoretical models may appear to be fundamentally different—judging by the elements that are involved in them. That is, a pump or a membrane model can be made to give varied predictions. By changing some relationships in this—and I think this could quite possibly be done—you could perhaps make its predictions come out to be very much like the predictions of the carrier model. In fact, this philosophical problem of distinguishing the object that you are talking about from the mathematical interpretation of the quantitative aspects of an experiment is extremely difficult. To be resolved it needs a crucial experiment, and crucial experiments very often get denuded when you analyze them.

References p. 777

In this particular case, the first interpretation that I would give to the fact that the efflux is proportional to the total glycine concentration is that if a carrier did mediate the outflow, it would not have been saturated in this circumstance. So that in fact the concentration, the number of carriers filled with amino acid as amino acid complexes, would be simply larger in one case than in the other.

LOFTFIELD: Then shouldn't they be proportional to the amount of the bound material?

REINER: Broadly speaking that would be true, and in this case they are proportional, as I judge the statement, to the actual amount of concentrate. Is that correct? This is to say that under the circumstances the carrier is not saturated, so that it reflects the amount that is bound.

HEINZ: With our system we never found any saturation phenomena of the glycine efflux, even at intracellular glycine concentrations of more than 60 mM.

GUROFF: Some observations that we made a couple of years ago with the isolated rat diaphragm may be of interest here. In the *intact diaphragm*, prepared by the technique of KIPNIS AND CORI, you observe an endogenous level of 20 μg of tyrosine per gram of tissue. You can incubate this diaphragm under various rather unfavorable conditions without removing significant amounts of tyrosine even in the presence of metabolic inhibitors. If you expose a diaphragm to large concentrations of tyrosine, tyrosine enters the diaphragm. If you resuspend the diaphragm in tyrosine-free buffer, most of the tyrosine comes out except for an amount comparable to the original endogenous concentration. We further observed that the initial rate of tyrosine entrance into the diaphragm was proportional to external concentration and not to the difference between this concentration and the endogenous 20 μg per g. Since the entrance of tyrosine into diaphragm cells conforms largely to the criteria of simple diffusion and does not depend on metabolism the endogenous tyrosine or an equivalent amount does not seem to equilibrate with the entering tyrosine.

ANDERSON: I wanted to mention some experiments on what is in some respects a model system. The isolated rat liver nucleus, as shown by the interference microscope or the phase contrast microscope, is very permeable to proteins. If you add to the solution a very small amount of calcium chloride or magnesium chloride, "*blebs*" begin to appear on the surface of the nucleus, which ultimately rise completely off the nucleus, and it is not difficult to show that these "*blebs*" are impermeable to protein. It can also be shown that protein put into the nucleus prior to the addition of calcium chloride is now inside these "*blebs*" and does not come out.

A small amount of sucrose added to this solution prevents these "*blebs*" from forming, so that the membranes seem to have large pores with some mobile lipid component which can flow together over these pores. My point is that the membrane is a function of the experimental conditions, so that we may be dealing with different cells in different solutions. In other words, the structure of the membrane is not a constant.

I would like to ask whether there are any experimental data on binding of amino acids to isolated cell components? Has anyone done simple experiments such as centrifuging these components down to see whether or not the amino acids fall *e.g.* with soluble proteins, or with microsomes?

HEINZ: Several such experiments have been carried out with Ehrlich cells by Dr. CHRISTENSEN and his collaborators, as far as I remember, and by ourselves, but the amino acids appeared always in the supernatant, whether the cells had been broken by pressure release or by grinding in the frozen-dried state. Dr. CHRISTENSEN also tried then to equilibrate a suspension of broken cells within a cellophane bag with normal extracellular medium containing amino acids. Also here the final distribution ratio of the amino acids was only slightly higher than unity so that hardly more than a negligeable amount of amino acids could have been bound to cellular components.

BAXTER: We have done some simple experiments with γ-aminobutyric acid which seem to indicate that this amino acid is bound to specific cell constituents in brain. We were initially interested in localizing, within the cellular components of brain tissue, those enzyme systems which were responsible for the synthesis and degradation of γ-aminobutyric acid. As a by-product of this study, which involved the differential centrifugation of guinea-pig brain homogenates, we found that a surprisingly large portion of the intrinsic γ-aminobutyric acid of these tissues was contained in the slow-speed sediment. We have not degraded this fraction any further.

ANDERSON: How dilute is this preparation?

BAXTER: The medium we used is identical to that recommended by BASFORD for the separation of brain mitochondria. It consists of 0.4 M sucrose supplemented with some heparine and EDTA. The homogenate is made up of one part of brain tissue to six parts of medium (w:v).

ANDERSON: This would have a bearing on it, and also the solution in which the homogenate was made would have quite an effect.

BAXTER: The experimental conditions will determine to a large extent whether γ-aminobutyric acid is bound or unbound in a brain homogenate. ELLIOTT AND VAN GELDER[13] were the first to demonstrate this. They showed that in a brain homogenate suspended in 0.88 M sucrose solution, only 10% of the total γ-aminobutyric acid of the tissue was bound to the 18,000 $\times$ g sediment. A similar brain homogenate suspended in a Ringer-type salt solution retained 60–70% of the total

References p. 777

γ-aminobutyric acid in the 16,000 $\times$ g sediment. Results of experiments by SANO AND ROBERTS[14] suggest that sodium ion may be required to bind optimally exogenously supplied γ-aminobutyric acid. In these latter studies it was shown that isotopically labeled γ-aminobutyric acid was bound to brain tissue but not to several other tissues which were tested. I think that all available data show that the binding of γ-aminobutyric acid is fairly specific, but the degree of binding in any particular brain homogenate depends on many factors including the molarity, temperature, ionic composition and concentration of the suspending medium. This dependence upon the suspending medium has made it difficult to determine which cellular fraction has the greatest capacity to bind γ-aminobutyric acid. In the Ringer solution which SANO AND ROBERTS used originally, the ^{14}C-labeled γ-aminobutyric acid was found in the slow speed sediment. On the other hand, in their more recent studies in which fractionation of brain homogenate was performed by differential centrifugation in a sucrose medium, it was found that most of the isotope was bound by the mitochondrial fraction and less by the slow speed sediment. In these experiments the precipitates were resuspended in Tris buffer containing sodium chloride and the binding capacity for ^{14}C-labeled γ-aminobutyric acid was tested by equilibrium dialysis. All of these observations tend to emphasize the need for more experimental data with regard to the variables which influence the binding of this amino acid to brain tissues.

GUROFF: Some investigators in our own laboratory have been interested in amines which under almost any fractionation conditions are extensively bound to subcellular fractions. But attempts to isolate these sedimentable fractions have led to the conclusion that this binding is very non-specific. We have done the same type of experiment with tyrosine, using sucrose fractionation of spleen. Parts of the free tyrosine did come down in each fraction, but we attached no particular significance to this.

COWIE: I like Dr. ANDERSON's remarks about the choice of organism and variability of conditions which accounts in part for experimental difference. I would like to point out, however, that in the yeast, we have at least three states of so-called free amino acids. Referring to Dr. HEINZ's description, a certain amount of amino acids may enter the cell through diffusion. In addition material may be accumulated and concentrated in excess of the external concentration. In the yeast we have still a third form of amino acid, the internal pool which is always present and does not exchange with external material and shows functional differences from the other forms of endogenous amino acids. Biosynthetic conversions take place in this third pool of amino acids. This pool is insensitive to osmotic shock. Thus there are three states of pool amino acids. One of these states, presumably that of the internal pool amino acids certainly must represent some form of binding inside the cell.

HOLDEN: In addition to the possibility that differences in results may derive from the use of different organisms, there is a strong likelihood that different results also may be obtained in a given organism depending on whether growing or non-growing cells are studied. You will recall that in our experiments with *L. arabinosus* we found a marked difference in the size of the glutamate pool and its response to extracellular osmotic strength when cells from early and late exponential phase cultures were compared. The proportion of actively dividing cells in these two populations would be expected to differ greatly. Amino acid pooling by early exponential phase cells had many characteristics in common with the process in growing. *E. coli* whereas the reverse was true of late exponential stationary phase cell. In this connection, I would like to ask Drs. BRITTEN and COWIE if they have used non-growing cultures in any of their experiments.

COWIE: In yeast these different amino acid systems have always been defined during exponential growth. Here the formation of the internal pool will be directly proportional to the quantity of growth. However just before the cells go into the resting phase the internal pool anticipates this condition and you can actually see a different rate of formation of the internal pool before the optical density changes. Since, however, I have carefully avoided this condition I can't say anything more than this.

BRITTEN: Our work in general has been with the growing cell, because we have found the incorporation into protein a very useful clue as to what is going on. We have also done experiments in the absence of energy sources and at zero degrees, and these do certainly change the pool situation.

In our system, the exhaustion of glucose leads to a behavior of the amino acid pool which, to use a far fetched analogy, suggests the presence of a rachet. While glucose is present, external amino acids enter the expandable pool. At the moment glucose is exhausted the amount of amino acid in the pool stops at the level it has reached and remains stable at this level for long periods. The amino acid is not lost from the cells even when you remove the external amino acid in the absence (or presence) of glucose. Nevertheless a very rapid exchange occurs in the presence of external amino acid. The rate of exchange depends on the amount of amino acid in the pool much more than it does on the external amino acid concentration. A generally similar behavior is observed when pool formation is suppressed by placing the cells at 0°.

Beyond these studies we have explored one other example of non-growing cells. Chloramphenicol

References p. 777

strongly inhibits protein synthesis in *E. coli*. It apparently directly affects the pool forming mechanism apart from its interference with synthesis.

ANDERSON: I would like to ask what the requirements are for your binding sites, because it seems to me that if the binding is rather loose no one will isolate this material in a bound form. If these sites are more or less ion exchanging sites then they may already have been demonstrated sufficiently in experiments that—as far as I know—haven't been described in detail, but have been inferred here.

The inside of a cell is very hard to reproduce in a laboratory, for this reason. We have a large number of poly-anions and low molecular weight cations, so that, for instance, the concentration of chloride in liver cells is very low. We would expect under these conditions that cationic substances would be bound to the macromolecules present, and this is probably what occurs. When we try to make a homogenate generally we have to put in a little bit of salt. But we don't usually make our homogenate under conditions such that the cation is small, and the anion that we add is an extremely large one. We would have to add such substances if we wanted to maintain the original condition which exists inside the cell.

So far as the polyamines are concerned, I would expect putrescine, cadaverine, and spermine could be bound tightly to RNA and loosely to almost any proteins that are present. Again it would be very difficult to decide whether or not there are any specific binding sites for these.

COWIE: I would like to comment on Dr. ANDERSON's remarks. Actually we have tried to find some binding sites in the cell to account for our data and were forced to some rather unusual conclusions. There is so much material in the internal pool of non-exchangeable amino acids that the only single component in the cell to complex with this material is the protein. We would like to include the binding sites of nucleic acid for this particular pool. Taking the DNA and the RNA of the cell alone you would have 3.3 times too many amino acids in the pool per nucleoside residue, so this material level obviously cannot be the source of sites. If you add the proteins there are plenty of available sites for the pool material.

BRITTEN: Not until this moment, did I realize that we had experiments demonstrating a very large amount of binding in cell preparations. This particular experiment deals with a system which has a negative heat capacity. It demonstrates that there is a fantastic interaction between water and protoplasm. We know, however, very little beyond the crude experimental observations. The freezing point of a very thick suspension of *E. coli* cells suspended in water was measured. The thick *goop* contained approximately 60% wet cells by volume, and the cells did not need to be intact for the phenomenon to occur. An ordinary freezing point system was used with an effective motor-driven stirrer, and the temperature was continuously recorded. When the cell was filled with water a typical cooling curve was observed. The temperature fell well below 0° and remained there while ice steadily froze out. When the cell was filled with the suspension, supercooling was observed. However, when an ice crystal was added the temperature suddenly rose to 0° and then slowly rose perhaps to as high as +0.7°.

HOLDEN: Have you performed freezing point determinations using cells loaded with large quantities of proline?

BRITTEN: No. We haven't explored this system, which is obviously a trap of some sort. We have done freezing point determinations on released pools in an attempt to estimate the total osmotic constituents, and in the case of *E. coli* it looks as if, without extensive measurements, the osmotic strength within the cell is comparable to that of the medium, and depends on it. This is quite distinct from what one would observe in the Gram positive cell.

HENDLER: Would you go just far enough to explain the relation of the negative heat coefficient to the problem?

BRITTEN: Well, the existence of the negative heat capacity shows an interaction between water and the protoplasm, because the protoplasm strangely elevates the freezing point of water. It is presumably some gel phenomenon that would be very hard to understand. The change in freezing point is, for our present argument, the only element of significance, and I hadn't realized until this moment how clearly it demonstrates a gross association of a small molecule with protoplasm in a cell preparation.

ANDERSON: I want to mention some work of Dr. MAZUR on freezing of yeast in our laboratory. An ice crystal in order to freeze through the pore of a membrane must deform. The freezing point of the ice that seeds through therefore, is a measure of the size of the pore. It is a very neat system, and yeast cells in distilled water show a survival curve something like this (at blackboard). At −5° all the water outside the yeast cells is frozen, and the survival rate is very high. When you get down to −30°, the survival curve drops very suddenly which means that seeding through the pores had begun. The functions of the radius of curvature are not ours, but they seem to be pretty well worked out.

ABRAMS: I would like to mention some experiments which indicate that K^+ and Na^+ ions have profound effects on the permeability of bacterial protoplast membranes. Earlier in the discussion I presented evidence that the penetration of oligosaccharides and certain amino acids is dependent

References p. 777

on glycolysis. In the case of oligosaccharides, evidence that K^+ ions are also required and that Na^+ ions are inhibitory was set forth in published reports.[3] A K^+ requirement for metabolically dependent penetration of short chain monocarboxylic aliphatic anions has also been observed. This is illustrated with propionate, butyrate, and valerate in Fig. 3. The same results were obtained with formate and acetate. The failure of metabolic penetration in the presence of Na^+ (0.55 *M*) can be overcome by addition of small quantities of K^+ (*e.g.* 0.01 *M*). With appropriate levels of Na^+ and K^+ ions, it turns out that the rate of metabolic penetration is a function of the carbon chain length, the rate decreasing with increasing chain length.

Of interest also is the marked difference between the behaviour of monocarboxylate anions and chloride as shown in Fig. 4. Glycolysis seems to have no effect on the penetration of KCl while K formate penetrates rapidly. In a like manner, glycolysis appears to have no effect on the penetration of NaCl.

HENDLER: This has bearing on a point raised by Dr. ANDERSON, as to what could bind the amino

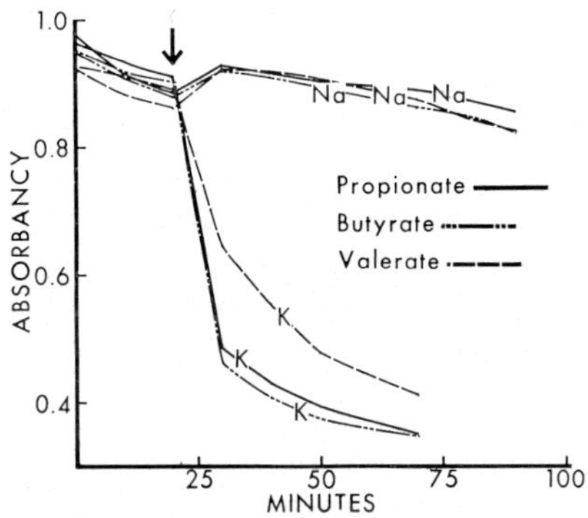

Fig. 3. The effect of glycolysis, K^+ and Na^+ on protoplasts stabilized with proprionate, butyrate and valerate. Protoplasts of *S. faecalis* previously depleted of K^+ were osmotically stabilized in solutions of salts of monocarboxylic acids (0.4 *M*) containing phosphate (0.075 *M*, pH 7.2). The cations were either all K^+ or all Na^+ as indicated on the curves. Glucose (0.01 *M*) was added at the time shown by the arrow. The rate of swelling during glycolysis, observed photometrically, indicates that the rate of penetration of the K salts during glycolysis becomes very rapid while the penetration rate of the Na salt remains unchanged. The substitution of Na^+ for K^+ does not alter the initial rate of glycolysis as measured by continuous titration of acid production at constant pH (pH stat).

acid in sufficient quantity. Dr. CHRISTENSEN raised the question whether in view of the tremendous amounts of intracellular amino acid there are sufficient sites for them. Dr. COWIE was saying that in his system perhaps protein or nucleic acid might have sufficient surface. But in those experiments I reported yesterday, where attempts were made to dilute the specific activity of a small quantity of radioactive valine with a rather large quantity of unlabeled valine, it was found that the monomolecular site I talked about with respect to amino acid did not give any indication of approaching saturation after the tremendous quantity of 200 mg unlabeled valine were added. The total volume was about 9 ml, including external medium and tissue, and there was a linear amount of total radioactive valine which was held in the lipid-soluble form as opposed to the water-soluble form (as a function of added valine), so that there was a capacity for holding valine in a form which altered its solubility and was able to handle tremendous quantities of valine and gave no indication of approaching saturation even out of this high level.

REINER: Before we get too far from Dr. ABRAMS' remarks I would like to point out their relevance to what we were discussing before; that is, the experimental evidence for binding. This seems to be one of a number of pieces of evidence that indicate that for studying the binding of amino acids to cell components, the composition of the medium is extremely critical, and it is impossible to decide whether your experiments are successful until a lot of work has been done just along these lines.

What is the optimum medium for studying binding, if there is any binding? There is a lot of other evidence available, but I don't think it is necessary because Dr. ABRAMS' experiments are about as dramatic evidence as you could ask for.

HEINZ: I am not quite satisfied with the explanation given, for instance for the swelling and the

References p. 777

osmotic effects, nor do I see dramatic evidence for binding in Dr. ABRAMS' experiments. If the osmotic effect is equivalent to the amount of molecules accumulated, how could you explain this otherwise than by assuming that these molecules are free, because the activity of water outside the cell can be considered as rather constant. If water goes into the cell there must be something which decreases the activity of the water inside, and how could we explain this by simple binding in an undissociated form?

BRITTEN: I don't think that from our point of view it is necessary to say that, for instance, the sucrose in Dr. ABRAMS' system is bound. It might very well be in free solution. We, in fact, have never taken the position that all of these phenomena can be explained on the basis of binding, and I think it is quite clear that there are osmotic constituents in free solution in the cell.

I would like to discuss for a moment some observations on pools of uracil compounds formed with labeled ^{14}C-uracil by *E. coli*. These observations have made clear to me some novel properties of the pool in *E. coli*.

The first observations were preparatoıy to studies of RNA synthesis and we got into the pool

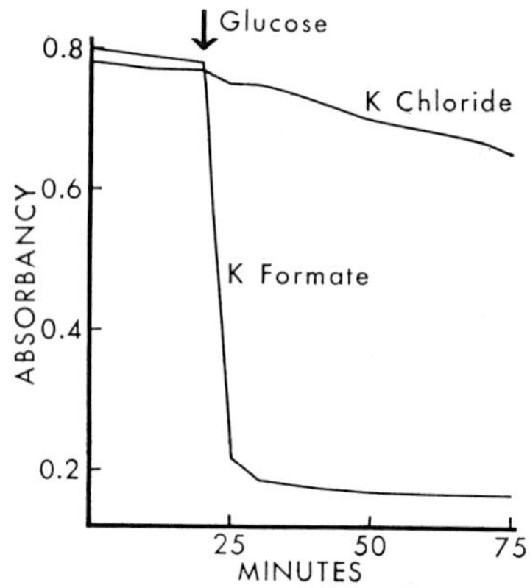

Fig. 4. The effect of glycolysis on protoplasts stabilized with 0.2 *M* K formate and 0.2 *M* K chloride. Protoplasts of *S. faecalis* were stabilized in solutions of K-formate (0.2 *M*) and KCl (0.2 *M*), each containing 0.075 *M* K-phosphate, pH 7.2. Glucose (0.01 *M*) was added to both suspensions at the time shown by the arrow. The rate of swelling observed photometrically indicates that the rate of penetration of K formate during glycolysis becomes very rapid while the rate of penetration of KCl remains unchanged. The initial rates of glycolysis in both cases are the same.

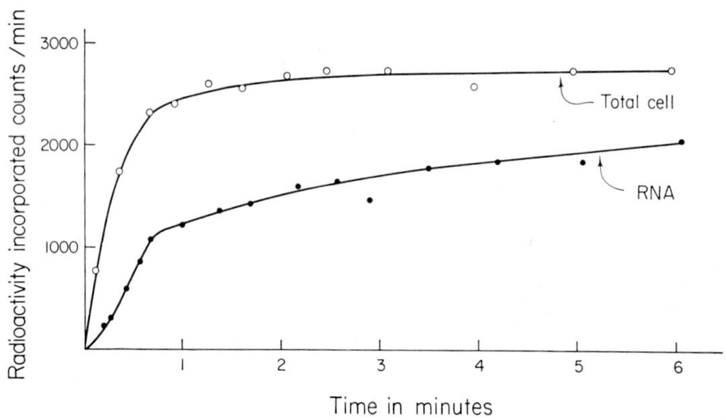

Fig. 5. Incorporation of [^{14}C]-uracil into the metabolic pool and the RNA of *E. coli* ML-30, per mg (wet)/ml. The difference between the two curves is the radioactivity in the metabolic pool. Initial uracil concentration was 10^{-7} *M*. Temperature: 37°.

References p. 777

question, because the complicated results influence the interpretation of our experiments on the labeling of RNA. When a small quantity of ^{14}C-uracil (10^{-37} M) is added to a culture of growing cells, it is completely taken up within 40 sec (Fig. 5). The radioactivity rapidly enters RNA with a delay of less than 5 sec and continues nearly linearly for 40 sec. At this time about 40% of the added uracil has been incorporated into RNA. Just when the external uracil is exhausted the rate of incorporation into RNA drops abruptly. The curve for the incorporation into RNA after this time corresponds to an exponential decay of the radioactivity in a 10-minute pool of uracil compounds.

In a comparable experiment done at a very high concentration of uracil (10^{-3} M) the pool was relatively so small that it could not be measured accurately. The incorporation of the radioactivity into RNA starts again without delay, and slowly accelerates to about twice the initial rate with a time constant of about 10 min. These observations as well as those made at a variety of intermediate concentration may be interpreted by means of the following diagram.

Internal synthesis
↓
External uracil ——→ P ——→ Incorporation into RNA
⇅
S

P represents a small pool or sequence of reaction steps leading from uracil to a chemical form suitable for incorporation into RNA. The quantity in P corresponds to the requirement for about 5 sec of RNA synthesis. S represents a large pool of compounds which exchange with some uracil compound in P. These compounds are principally UDP and UTP, and there is conversion to CDP and CTP. The quantity in S corresponds to the requirement for 5 to 10 min of RNA synthesis.

The rate of exchange between S and P is not fast and equilibrium between the specific radioactivity of P and S requires several minutes at least. P then effectively forms a by-pass around the large pool, S, for the entry of external uracil into RNA.

The rate of uptake of uracil into the cell has a maximum value corresponding roughly to the rate of utilization of uracil compounds for RNA, and it turns out that these pools are not expandable. The cell very avidly concentrates uracil, with a K_s of 10^{-7} molar, but in fact does not increase its internal concentration of uracil compounds. Here you have a concentrating process without increase in internal concentration.

Fig. 6 shows the results of an experiment in which the cells were pretreated with 4×10^{-5} M ^{12}C-uracil for 10 min before the ^{14}C-uracil was added. An identical curve within small limits of error was obtained in a simultaneous control experiment in which the ^{14}C-uracil and the ^{12}C-uracil were both added at zero time. The fact that there is still a direct entry of ^{14}C-uracil into RNA is shown in the upper set of curves which are simply a magnification of the early time region. If the pool P were expanded there should be an initial delay, which is not observed. The expansion of S, on the other hand, would cause a major difference between the control and the experiment in which ^{12}C-uracil was added beforehand. The increasing rate of entry of radioactivity into RNA shown in Fig. 6 is due to the increasing specific radioactivity of S. If the pool S were expanded, its specific radioactivity should rise more rapidly in the control since the new material flowing in would be radioactive. On the other hand in the experiment in which ^{12}C-uracil was added beforehand, the specific radioactivity should rise more slowly since a large amount of unlabeled compounds would have been present at the time the ^{14}C-uracil was added.

I think the argument is fairly rigid that compounds in P must be on special sites, whose properties are comparable to my carrier. In the first place, something limits the maximum rate of entry of uracil from outside, which cannot go in at a rate faster than it is used, so there is a very adequate feedback control, on the internal concentration. At the same time there is a feedback control on the internal synthesis of uracil. When the concentration is approached where the entry rate meets the requirement for RNA synthesis, then the internal synthesis is shut down.

The mechanisms involved in the binding of the compounds in P are also capable of carrying out the conversion of the uracil through all of the stages to whatever compound is required for incorporation into RNA, and including immediate delivery to the sites of RNA synthesis. Thus P represents four processes, which are intimately related with each other, and which can by no stretch of the imagination be considered to occur at the same location in the cell. They are internal synthesis, concentration of external uracil, chemical conversion and transfer to sites for incorporation into RNA.

I think we can presume rather definitely, that the very small quantity of compounds in P is associated with some sort of site which I would like to call a carrier. It is rigidly controlled in amount as is the large pool with which it is in equilibrium.

The stability of the large pool, S, leads me to believe that S also is bound to sites. This is supported by other experimental evidence, but I don't think that kinetic arguments could establish

References p. 777

it. In fact a rigid control of the quantity associated with the carrier, if it is in equilibrium with S, might also rigidly control the concentration of compounds in S.

The equilibria we are discussing here are complicated by many chemical transformations also going on. The evidence that the compounds in S are in fact organized in some special way in this cell is provided by experiments with osmotic shock. Every pool we have measured in *E. coli*—as distinct from the pools in yeast—is osmotically sensitive. If cells are washed with water at room temperature the large pool (S) of uracil compounds as well as the amino acid pools are totally extracted. However, if the cells are washed with water at 0° the amino acids are still totally extracted while the uracil compounds are only partly extracted. This shows quite clearly a difference in organization in the cell between the two different kinds of pools.

LAJTHA: I would like to use the structures of Dr. BRITTEN to illustrate alternative interpretations

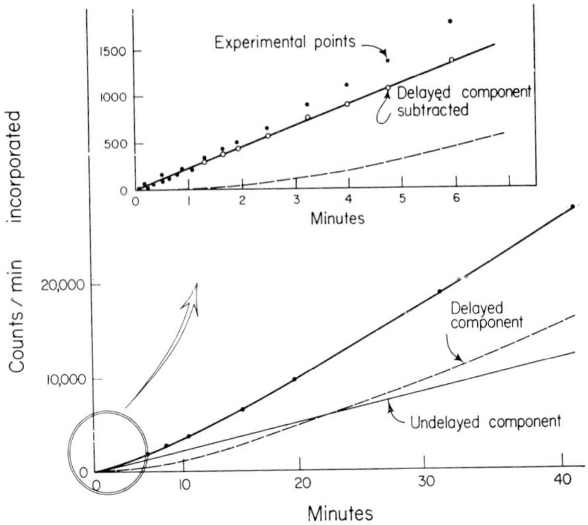

Fig. 6. Incorporation of [2-^{14}C]-uracil into the RNA of *E. coli* ML-30 after 10 min pretreatment with [^{12}C]-uracil at $4 \cdot 10^{-5}$ M. The upper figure presents the data at early times with expanded scales. The light solid and dashed curves represent the best separation of the curve into a linear component and a component delayed by the pool (S). Solid circles are experimental values. For the open circles the component delayed by S was subtracted. The abscissa scale is proportional to increase in cell mass, with sample times indicated.

of some of his experimental data. Such data, though they may be consistent with the existence of heterogenous substrate pools do not necessarily indicate such pools, if the enzymes metabolizing the substrate are spatially oriented in a special way. Though I am thinking of amino acid metabolism, the compounds on the blackboard can be used as an example. If the enzymes synthesizing RNA were at the surface or on a structure which is preferentially accessible from the surface, then labeled RNA precursors added from outside would be incorporated into RNA at a higher rate than endogenous precursors. In spite of this preferential incorporation, however, there could still be a homogenous precursor pool in the cell. Whenever in short term experiments the labeled substrate in the cell has lower specific activity than its metabolic product, a compartmentation of the substrate can be deduced.

BRITTEN: But would you go so far as suggesting that the concentrating process, incorporation into RNA, internal synthesis, and the chemistry for conversion from uracil to nucleotide phosphate, that all of these can actually occur in the membrane region? Although I did say, that this was beyond the stretch of imagination, I think now it could be visualized.

LAJTHA: I agree with Dr. BRITTEN that in his system this is unlikely. We have, however, some evidence that for hippurate synthesis, which involves mitochondrial enzymes, glycine from plasma is used in preference to endogenous organ glycine. This might be explained equally well by compartmentation, in a sense, of the enzymes as by heterogeneous glycine compartments.

BAXTER: I would like to ask Dr. GUROFF a question relating to his work with tyrosine in brain. Much emphasis at this symposium has been placed upon processes involving incorporation of amino

References p. 777

acids into cells and tissues. Very little has been said about excorporation of amino acids. In your presentation yesterday you showed rather impressively the uptake of tyrosine by brain tissues, but all of your slides seemed to end at the time of maximum uptake. I am interested what happens beyond this point. Is there any active excorporation of tyrosine out of brain tissue similar to that found by Dr. LAJTHA for lysine and leucine?

GUROFF: I think Dr. LAJTHA is probably more entitled to talk about efflux from the brain than I am. However, I can tell you that *in vivo* the endogenous tyrosine ratio between brain and plasma is about 1.3–1.4. This ratio is re-established in about 60–90 min when you have tyrosine in the blood at a high concentration. With the decrease of tyrosine concentration of the blood the tyrosine concentration of the brain also falls, but the endogenous ratio is maintained. That is, the brain remains higher than the plasma. After 4 h, the endogenous concentration had not been restored, but the pattern was clear.

HOLDEN: Do you find this for any other amino acid?

GUROFF: We have looked only very cursorily at the other aromatic acids, phenylalanine and tryptophane, and not in the efflux sense at all.

HOLDEN: There is a report by LINDENBERG AND MASSIN that different amounts of tyrosine can be extracted from yeast cells using hot water and cold TCA suggesting that the tyrosine pool is heterogenous and differs in this regard from that of the other amino acids.

LAJTHA: I would like to comment briefly on the important point brought up by Dr. BAXTER, concerning the direction of active transport. I think that at least in the brain levels of substrates are controlled by active processes working in two directions. Probably there is a number of systems where active transport out of the cell takes an important or even dominating role in homeostatic control. It would be important to learn more about influx and efflux, *e.g.* whether they refer to the same mechanism or to separate ones.

HOLDEN: In closing I would like to remind you and those who read these proceedings that this session was held primarily to expose all of us as intimately as possible to the existing, and in some cases divergent, views and data regarding transport and accumulation phenomena. There was no expectation that major issues would be resolved. However, we did hope that such discussion and contact might reduce the time required to attain a clearer understanding of these processes.

COWIE: I would like to say that it isn't so much a matter of divergency. I think in the CARNEGIE group we do recognize that there are free amino acids and as well as bound amino acids; in other words, all we ask is that the so-called *free amino acid* advocates recognize that there might be some *bound* ones also.

HOLDEN: I think that would be an appropriate note on which to end. Thank you very much.

REFERENCES

1 D. L. OXENDER AND H. N. CHRISTENSEN, *J. Biol. Chem.*, 234 (1959) 2321.

2 V. KOEFOED-JOHNSON AND H. USSING, *Acta Physiol. Scand.*, 42 (1958) 298.

3 A. ABRAMS, *J. Biol. Chem.*, 234 (1959) 383; 235 (1960) 1281.

4 A. ABRAMS AND C. G. MACKENZIE, *Federation Proc.*, 18 (1959) 178.

5 A. ABRAMS, G. YAMASAKI, J. REDMAN AND C. G. MACKENZIE, *Federation Proc.*, 19 (1960) 129.

6 J. E. TREHERNE, *J. Exptl. Biol.*, 36 (1959) 533.

7 E. HEINZ, *J. Biol. Chem.*, 211 (1954) 781.

8 E. HEINZ AND P. M. WALSH, *J. Biol. Chem.*, 233 (1958) 1488.

9 E. HEINZ AND R. P. DURBIN, *J. Gen. Physiol.*, 41 (1957) 101.

10 C. R. PARK, R. L. POST, C. F. KALMAN, J. H. WRIGHT JR, L. H. JOHNSON AND H. E. MORGAN, *Ciba Foundation Symposium: Endocrinol.*, 9 (1956) 257.

11 T. ROSENBERG AND W. WILBRANDT, *J. Gen. Physiol.*, 42 (1958) 289.

12 E. HEINZ, *J. Biol. Chem.*, 225 (1957) 305.

13 K. A. C. ELLIOTT AND N. M. VAN GELDER, *J. Physiol. (London)*, 153 (1960) 423.

14 K. SANO AND E. ROBERTS, *Biochem. Biophys. Research Communs.*, 4 (1961) 358; unpublished results.

AUTHOR INDEX

SUBJECT INDEX

In preparing this index it was decided that the citation of every occurrence of an amino acid in the text was an impractical objective which, in any case, would have had dubious value, since many amino acids are found in almost every tissue examined. Therefore, amino acids are cited only when they have unusual structures or distributions, or are singled out for special attention in the text. On the other hand, an effort was made to record all textual citations of organisms and tissues in which amino acid composition or metabolism are discussed.

LIST OF PAPER CHROMATOGRAPHY AND PAPER CHROMATOGRAMS